WITHDRAWN
NDSU

CORN AND CORN IMPROVEMENT

AGRONOMY

A Series of Monographs Published by the
AMERICAN SOCIETY OF AGRONOMY

General Editor Monographs 1 to 6, A.G. NORMAN

1. C. EDMUND MARSHALL: The Colloid Chemistry of the Silicate Minerals, 1949
2. BYRON T. SHAW, *Editor:* Soil Physical Conditions and Plant Growth, 1952
3. K.D. JACOB: Fertilizer Technology and Resources in the United States, 1953
4. W.H. PIERRE and A.G. NORMAN, *Editors:* Soil and Fertilizer Phosphate in Crop Nutrition, 1953
5. GEORGE F. SPRAGUE, *Editor:* Corn and Corn Improvement, 1955
6. J. LEVITT: The Hardiness of Plants, 1956

---

7. JAMES N. LUTHIN, *Editor:* Drainage of Agricultural Lands, 1957
General Editor, D.E. GREGG

8. FRANKLIN A. COFFMAN, *Editor:* Oats and Oat Improvement, 1961
Managing Editor, H.L. Hamilton

9. C.A. BLACK, *Editor-in-Chief,* and D.D. EVANS, J.L. WHITE, L.E. ENSMINGER, and F.E. CLARK, *Associate Editors:* Methods of Soil Analysis, 1965.
Part 1 — Physical and Mineralogical Properties, Including Statistics of Measurement and Sampling
Part 2 — Chemical and Microbiological Properties
Managing Editor, R.C. DINAUER

10. W.V. BARTHOLOMEW and F.E. CLARK, *Editors:* Soil Nitrogen, 1965
Managing Editor, H.L. HAMILTON

11. R.M. HAGAN, H.R. HAISE, and T.W. EDMINSTER, *Editors:* Irrigation of Agricultural Lands, 1967
Managing Editor, R.C. DINAUER

12. R.W. PEARSON and FRED ADAMS, *Editors:* Soil Acidity and Liming, 1967
Managing Editor, R.C. DINAUER

13. K.S. QUISENBERRY and L.P. REITZ, *Editors:* Wheat and Wheat Improvement, 1967
Managing Editor, H.L. HAMILTON

14. A.A. HANSON and F.V. JUSKA, *Editors:* Turfgrass Science, 1969
Managing Editor, H.L. HAMILTON

15. CLARENCE H. HANSON, *Editor:* Alfalfa Science and Technology, 1972
Managing Editor, H.L. HAMILTON

16. B.E. CALDWELL, *Editor:* Soybeans: Improvement, Production, and Use, 1973
Managing Editor, H.L. HAMILTON

17. JAN VAN SCHILFGAARDE, *Editor:* Drainage for Agriculture, 1974
Managing Editor, R.C. DINAUER

18. GEORGE F. SPRAGUE, *Editor:* Corn and Corn Improvement, 1977
Managing Editor, D.A. FUCCILLO

Monographs 1-6, published by Academic Press, Inc., should be ordered from: Academic Press, Inc., 111 Fifth Avenue, New York, New York 10003

Monographs 7-18 should be ordered from: American Society of Agronomy, 677 South Segoe Road, Madison, Wisconsin, USA 53711

CORN AND CORN IMPROVEMENT

EDITOR:
G.F. SPRAGUE

Professor of Plant Breeding and Genetics,
University of Illinois,
Urbana, Illinois

Managing Editor:
D.A. FUCCILLO

Assistant Editor:
L.S. PERELMAN

Editor-in-Chief, ASA Publications:
MATTHIAS STELLY

Number 18 in the series
AGRONOMY

American Society of Agronomy, Inc., Publisher
Madison, Wisconsin, USA
1977

Copyright © 1976 by the American Society of Agronomy, Inc.

ALL RIGHTS RESERVED

No part of this book may be reproduced or utilized in any form or by any means, electronic or mechanical, including xerography, photocopying, microfilm, and recording, or by any information storage and retrieval system, without permission in writing from the publisher

Library of Congress Cataloging in Publication Data

Sprague, George Frederick, 1902 -
   Corn and Corn improvement.

   (Agronomy; no. 18)
   Includes bibliographies and index.
   1. Corn.   2. Corn — Breeding.   I. Title.
II. Series: Agronomy, a series of monographs; 18. SB191.M2S6254 1976 633'.15'3 76-29528 ISBN 0-89118-043-5

The American Society of Agronomy, Inc.
677 S. Segoe Rd., Madison, Wisconsin, USA 53711

Printed in the United States of America

FOREWORD

Could any American alive in the year our nation was founded have possibly imagined that 200 years later corn would achieve its present importance to the nation and the world? Could the editor and authors of the first edition of this monograph have envisioned at that time the improvement in yields, in nutritional value and overall efficiency in production of calories, oil, and protein which would take place between then and now? Perhaps the best answer is given in the Editor's preface to the First Edition: "The limits of further improvement are not yet definable by corn breeders."

Certainly we are now a long way from seeing the end to the improvement and to the increasing importance of this marvelous American crop. Continuing improvement in yield, quality, and value of this crop will be limited only by the talent and effort devoted to it. It is, indeed, most appropriate that the second edition of this Monograph entitled *Corn and Corn Improvement* is published by the American Society of Agronomy as our nation finishes its bicentennial celebration.

Dr. George F. Sprague, Editor of both the First and the Second Editions, has been a distinguished leader of corn research for decades. His leadership is recognized wherever corn research is in progress. After serving with distinction as a scientist and program leader for the Agricultural Research Service of the U.S. Department of Agriculture, he has "retired" to a University faculty position where he can be found in office, field, or laboratory almost any working day.

Dr. Sprague, the first recipient of the Crop Science Award of the American Society of Agronomy, has been a Fellow of the Society for 30 years and has also served as President. He has had numerous honors and awards. He has been elected a member of the National Academy of Sciences, America's most distinguished scientific body.

Urbana, Illinois
February, 1977

MARLOWE D. THORNE, *President*
American Society of Agronomy

GENERAL FOREWORD

Interest in corn as food, feedstuff, and various products has continued at an unabated pace over the two decades since the first edition of the Corn Monograph. Recognizing this fact, the American Society of Agronomy Executive Committee in 1972 asked the Monographs Committee to proceed with the preparation of another Corn Monograph. Like the first Corn Monograph, this publication is largely a new contribution. The society was fortunate in obtaining the services of the previous Editor, George F. Sprague, who remains an active worker in corn research now with many more years of accumulated experience. Dr. Sprague and members of his committee circulated a Table of Contents of the new Corn Monograph widely among researchers and other scientists knowledgeable on the subject. The authors contributed several years of effort, and the result of their cooperation is this volume.

Corn and Corn Improvement is the 18th monograph in the series prepared by the American Society of Agronomy since 1949. Publication of the first six volumes was by the Academic Press, Inc., of New York but since 1957 the society has become the publisher. A complete list of the titles in the series may be found among these first pages. These publications represent a significant and continuing activity of the American Society of Agronomy, its officers, and its more than 9,000 members located in more than 100 countries around the world.

The American Society of Agronomy is associated with the Crop Science Society of America and the Soil Science Society of America. The three societies share many objectives and activities in promoting these branches of agriculture and scientific disciplines. Members of these three educational and scientific societies contribute generously of their time and talents in producing various publications, including monographs, and in pursuing other activities of an educational nature for public use.

This book should be of great interest and benefit to corn scientists, producers, and users. The American Society of Agronomy considers it as one of its major contributions to human welfare because of the worldwide importance of corn as a field crop. Through the presentation of this up-to-date scientific and practical material on corn, the society believes that this effort will help to make corn an even more useful crop to mankind.

In behalf of society members and myself in particular, I sincerely thank Dr. Sprague for his repeated performance as editor, and many authors, the managing editor, and all others who have contributed directly or indirectly to the accomplishment of this worthy project.

Madison, Wisconsin
June, 1977

MATTHIAS STELLY
Editor-in Chief
ASA Publications

PREFACE TO FIRST EDITION

Corn was indigenous in the Americas at the time of their discovery. Its production and utilization played an important role in successful colonization. In fact one historian has stated "The maize plant was the bridge over which English civilization crept, tremblingly and uncertainly, at first, then boldly and surely to a foothold and permanent occupation of America."

Tremendous developments have been made in agricultural methods since the first colonization, and throughout this whole period corn has remained one of the major crops grown in the New World. In the beginning the only corn available was the type or types being grown by the natives. Although the early history of corn domestication is still largely shrouded in mystery, by 1800 a large number of varieties were in existence. Through the efforts of private and public agencies intensive work was done on mass selection followed in order by varietal hybridization and ear-to-row breeding. Finally the present method of inbreeding and hybridization emerged from the early studies of Shull, East, Jones, and others. This present method is continuing to show progress as evidenced by the rapid replacement of the first commercial hybrids by newer and better combinations. The limits of further improvement are not yet definable by corn breeders.

Concomitant with the development of improvements in breeding methods has been the acquisition of increased knowledge of plant morphology, of mode of pollination, and most recently of the genetics and cytology of corn. At the present time more is known concerning the genetics and cytology of the corn plant than of any other plant species.

Although the most spectacular improvement in corn yields have come from plant breeding, marked progress also has been made in production practices. Fertilization of the crop has changed from the use of one or two fish per hill in certain restricted areas to the extensive use of balanced starter and supplemental commercial fertilizers. Production practices also have changed from a laborious hand culture to a high degree of mechanization.

Utilization has evolved from the various types of hand grinding used by the Indians through the water powered grist mill to present day types of milling which produce a bewildering array of products and by-products. Even the use of corn as a feed has undergone change; from reliance on exclusive use of corn in a fattening ration, practices have become diverse through adoption of extensive supplementation and have resulted in decreasing by at least one half the pounds of feed required per 100 pounds of gain.

Obviously the present Monograph cannot hope to cover all of these developments in any detail. In fact many of them will be largely ignored, and emphasis will be placed primarily upon the present status of our knowledge. Even with this limitation the preparation of this Monograph has posed many problems. It has been assumed that the particular agronomic audience to be served is the research worker and advanced student interested in corn. Even within this limited group interests are so varied that no choice of topics could be universally satisfactory.

The subject matter covered in the Monograph may be divided in three broad divisions; breeding, production, and utilization. Under the broad classification of breeding are included such topics as history and origin, morphology, cytogenetics, and breeding. The chapters included under the general topic of production treat with climatic requirements, mineral nutrition, culture, production of hybrid seed, the special crops, sweet corn and popcorn, and diseases and insect pests. Utilization is treated under the headings of industrial utilization and corn as a livestock feed.

Any bias which is apparent in the choice of the various sub-topics and in the relative space assigned is to be charged primarily to the editor although some contributors did not use all of the space and others exceeded the space allocations originally made. In any volume involving multiple authorship it is extremely difficult to obtain uniformity of treatment among chapters. This is particularly true where all coordination must be done by correspondence. Certainly others serving in an editorial capacity might well have chosen a somewhat different array of topics and made quite different space allocations than were used here.

The volume includes a rather large number of illustrations. The general policy was followed of using adequate illustrations where this was felt to be an essential part of the presentation, and limiting illustration in other chapters where they did not serve an equally important function.

The editor wishes to express his sincere appreciation to the authors of the several chapters who were persuaded to take the necessary time, often against their better judgment, from their already full research and teaching schedules to prepare their separate contributions.

Ames, Iowa
May, 1955

G.F. SPRAGUE

PREFACE TO SECOND EDITION

The first edition of *Corn and Corn Improvement* was published in 1955 and has been out of print for many years. Possibly the acceptance and usefulness of the first edition can best be judged by the fact that it was translated into several languages. The prime justification for the preparation of a new edition of any book is to provide an updating of information and thereby increase its overall usefulness.

Since 1955 major advances have been achieved in breeding and genetics, in production techniques, and in utilization. The new edition has been rewritten to cover these developments.

The origin of corn has received renewed attention and concepts have been materially modified in the last 20 years. The race concept has been developed and much of the world germplasm has been collected and classified. New chapters on each of these topics should provide information which will facilitate a more intelligent use of the great wealth of genetic diversity available within the species.

The new chapter on genetics provides the first comprehensive review in many years. The chapter on cytogenetics has been extensively revised.

Since publication of the first edition, average yields of corn have more than doubled. This increase has been achieved through a combination of improved cultivars and management practices. New breeding procedures have been devised, evaluated, and are being incorporated into standard breeding practice. The changes in techniques of hybrid seed production are also detailed.

A new chapter has been added covering the development and use of special corn types arising from genetic modification of the carbohydrate, oil, and protein reserves of the corn kernel.

With increased intensity of production, diseases and insects and their control have assumed an increased importance. The more common pests are described as are the most effective means of control.

Improved management practices involving simplified systems of planting, weed control, and harvesting have come into general use. There has been a continuing trend toward heavier plant populations and increased fertilizer applications.

Improvements have also occurred in the utilization of corn in livestock feeding and in industrial processing.

The current revision attempts to provide an interpretive overview of the current situation in each of the major areas mentioned.

The current volume was prepared for the research worker, graduate and advanced undergraduate students and others interested in any or all aspects of genetics, breeding, production, or utilization. Adequate illustrations and extensive literature citations should enhance usefulness.

The editor expresses his sincere appreciation to the several contributors for their respective chapters and to the many others who have provided assistance in this effort.

Urbana, Illinois
November 25, 1976

G.F. SPRAGUE

CONTRIBUTORS

D.E. ALEXANDER, *Department of Agronomy, University of Illinois, Urbana, IL 61801.*

WILLIAM L. BROWN, *Pioneer H-Bred International, Inc., 1206 Mulberry Street, Des Moines, IA 50308.*

WAYNE CARLSON, *Department of Botany, University of Iowa, Iowa City, IA 52240.*

E.H. COE, JR., *Agricultural Research Service, U.S. Department of Agriculture, University of Missouri, Columbia, MO 65201.*

WILLIAM F. CRAIG, *Funk Seeds International, Inc., 1300 W. Washington Street, Bloomington, IL 61701.*

ROY G. CREECH, *Department of Agronomy, Mississippi State University, State College, MS 39762.*

F.F. DICKE, *Pioneer Hi-Bred International, Inc., Johnston, IA 50131.*

S.A. EBERHART, *Agricultural Research Service, U.S. Department of Agriculture, Iowa State University, Ames, IA 50010; now with Funk Seeds International, Inc., 1300 W. Washington Street, Bloomington, IL 61701.*

WALTON C. GALINAT, *Suburban Experiment Station, University of Massachusetts, Waltham, MA 02164.*

MAJOR GOODMAN, *Department of Statistics, North Carolina State University, Raleigh, NC 27607*

J.J. HANWAY, *Department of Agronomy, Iowa State University, Ames, IA 50010.*

W.E. LARSON, *Agricultural Research Service, U.S. Department of Agriculture, University of Minnesota, St. Paul, MN 55101.*

M.G. NEUFFER, *Department of Genetics Research, University of Missouri, Columbia, MO 65201.*

JOHN E. SASS *(deceased), Iowa State University, Ames, IA 50010.*

ROBERT H. SHAW, *Department of Climatology and Meteorology, Iowa State University, Ames, IA 50010.*

G.F. SPRAGUE, *Department of Agronomy, University of Illinois, Urbana, IL 61801.*

ARNOLD J. ULLSTRUP, *Department of Botany and Plant Pathology, Purdue University, W. Lafayette, IN 47907.*

STANLEY A. WATSON, *CPC International, Inc., Moffett Laboratories, Argo, IL 60501.*

CONTENTS

	Page
FOREWORD	v
GENERAL FOREWORD	vii
PREFACE TO THE FIRST EDITION	ix
PREFACE TO THE SECOND EDITION	xi
CONTRIBUTORS	xii

Chapter 1   The Origin of Corn 1

WALTON C. GALINAT

    Corn and Its Relatives 1
    Corn-*Tripsacum* Hybrids 27
    The Contribution of Corn's Relatives to Corn Improvement 35
    Summary 39
    Literature Cited 43

Chapter 2   Races of Corn 49

WILLIAM L. BROWN AND MAJOR M. GOODMAN

    Races of Mexico, Central America, and the West Indies 53
    Races of South America 62
    Races of the USA 72
    Effective Sources of Germplasm Among Lowland Tropical Races 79
    Literature Cited 84

Chapter 3   Morphology 89

JOHN E. SASS

    Development of the Caryopsis 89
    Histology of the Vegetative Plant Body 99
    Literature Cited 109

Chapter 4   The Genetics of Corn 111

E. H. COE, JR., AND M.G. NEUFFER

    Genetic Factors of the Species 113
        Nomenclatural Conventions 113
        Genetic Loci and Their Map Locations 114
        Dosage Effects, Pleiotropisms, Cell Autonomy,
          and Duplicate Factors 123
    Genetic Systems 129
        Form and Texture of the Kernel 129
        Anthocyanins and Related Pigments 135
        Chlorophyll and Carotenoid Pigmentation 148
        Plant Form 156

        Leaf Cuticle Waxes 163
        Kernel Lipids... 164
        Fluorescent Seedlings and Anthers 164
        Enzyme Variations in Electrophoresis 165
        Reaction to Pests and to Adverse Conditions. 168
        Gametophyte Formation and Development 170
        Regulatory Systems Controlling Gene Expression and Mutation 173
        Paramutation... 181
        Aberrant Ratio 182
        Extrachromosomal Inheritance 183
    Methods of Genetic Analysis 186
        Analysis of Genetic Constitution and Behavior 186
        Locating Factors to Chromosome and to Map Position. 189
        Developmental Studies................................. 193
        Mutation Studies 196
        Analysis of Genetic Fine Structure 204
    Literature Cited 210

Chapter 5   The Cytogenetics of Corn 225

WAYNE R. CARLSON

    I. Chromosome Morphology 225
    II. Nucleolus Organizer Region 229
    III. The Euploid Series..................................... 232
    IV. Aneuploidy... 236
    V. Meiosis in Corn 239
    VI. Mutations of Meiosis, Mitosis, and Chromosome Structure........ 241
    VII. Synapsis 243
    VIII. Crossing Over in Corn................................. 250
    IX. Chromosomal Rearrangements........................... 260
    X. Supernumerary Elements............................... 282
    Literature Cited 293

Chapter 6   Corn Breeding 305

G.F. SPRAGUE AND S.A. EBERHART

    I. Early History.. 305
    II. Development of Inbred Lines 313
    III. Evaluation of Inbred Lines............................. 318
    IV. The Prediction of Double-Cross and Three-Way Cross Performance. 323
    V. Type of Gene Action 324
    VI. Genotype by Environmental Interactions 327
    VII. Recurrent Selection................................... 330
    VIII. A Comprehensive Breeding System 339
    Literature Cited 354

Chapter 7   Breeding Special Industrial and Nutritional Types 363

D.E. ALEXANDER AND ROY G. CREECH

    High Oil Corn... 363
    Protein: Quantity and Quality.............................. 370

Carbohydrates. 373
Popcorn 385
Literature Cited. 386

Chapter 8   Diseases of Corn. 391
A.J. ULLSTRUP

I. General Considerations 392
II. Seed Rot and Seedling Blights. 395
III. Stalk and Root Rots 398
IV. Ear Rots 411
V. Storage Rots. 420
VI. Leaf Diseases. 421
VII. Downy Mildew. 448
VIII. Virus and Mycoplasma Diseases. 453
IX. Corn Smuts. 462
X. Corn Rusts 467
XI. Nematode Diseases 472
XII. Witchweed. 474
XIII. Non-Infectious Diseases 476
Literature Cited. 483

Chapter 9   The Most Important Corn Insects. 501
F.F. DICKE

Research Trends. 502
Soil Insects 503
Insects Attacking the Leaf, Stalk, and Ear. 518
Lesser Recognized Groups 543
Insects in Relationship to Corn Diseases 546
Stored-Grain Insects. 552
Insect Resistance in Corn. 562
Chemical Control Status. 574
Appendix: Common and Scientific Names of Corn Insects 576
Literature Cited 578

Chapter 10   Climatic Requirement 591
ROBERT H. SHAW

Worldwide Corn Production and Climate. 591
Stages of Growth and Development 593
Moisture Stress Concept. 595
Growing-Degree Unit Concept 598
Effect of Weather on Certain Periods of Plant Growth 600
Literature Cited 617

Chapter 11   Corn Production. 625
W.E. LARSON AND J.J. HANWAY

Plant Growth and Development. 625

Requirements for Plant Growth............................. 630
Cultural Practices.. 636
    Preplant Tillage Systems 636
    Planting for Grain 643
    Plant Spacing.. 647
    Planting for Silage 648
    Fertilizer 649
    Previous Crop and Erosion Control..................... 655
    Weed Control... 657
    Corn Harvesting and Storage 660
Literature Cited.. 663

Chapter 12   Production of Hybrid Corn Seed.......................... 671
WILLIAM F. CRAIG

Parent Seed Stocks 673
Producing the Seed Crop................................... 675
Pollen Control.. 681
Harvesting the Seed Crop 692
Processing the Seed Crop.................................. 697
Quality Control... 703
Storage and Distribution.................................. 709
Literature Cited.. 713

Chapter 13   Industrial Utilization of Corn.......................... 721
STANLEY A. WATSON

Formulated Animal Feeds................................... 723
Wet Milling... 725
Dry Milling... 745
Fermentation Industries 751
Corncob Utilization....................................... 755
Literature Cited.. 756

Index **765**

Chapter 1   The Origin of Corn

WALTON C. GALINAT
University of Massachusetts
Waltham, Massachusetts

The origin of corn (maize) continues to be intriguing to botanists, archaeologists and agronomists, partly because of its apparent cataclysmic transformation from the wild and partly because of its great economic importance as the staff of life for all of the New World civilizations, including our own. With present concern about the genetic vulnerability of uniform corn and the effort to adapt corn to a technical Green Revolution to cope with the human population explosion, there is interest in tapping the wild relatives and primitive varieties of corn for genes controlling special attributes no longer present in the eroded variability of our advanced varieties. An efficient exploitation of the variation that lingers from the past requires an understanding of the evolutionary pathways and, in some cases, special techniques to circumvent natural crossability barriers. Insights of what came "in the beginning" provide an understanding of how corn evolved to its present state and, thereby, indicate possible extrapolations for a future evolution in symbiotic harmony with man.

CORN AND ITS RELATIVES

Corn is known only as a cultivated species *(Zea mays);* the wild ancestor is either its closest relative, teosinte *(Zea mexicana)* or, according to some, a wild corn known only through a few 7,000-year-old archaelogical corn cobs. Whereas the corn tribe, Maydeae, comprises seven genera, only two, *Zea,* and *Tripsacum,* are New World. The remaining genera are Oriental *(Coix, Chionachne, Schlerachne, Trilobachne,* and *Polytoca.)* The Oriental Maydeae are usually considered to have no immediate bearing on the origin of corn although interest continues in a possible remote involvement of the genus *Coix* with the origin of the American Maydeae.[1] The genus *Manisuris* of the tribe Andropogoneae has also been considered and it will be included in the descriptions here (Fig. 1).

All genera of the Maydeae usually have unisexual flowers borne on the same plant (monoecious); all except corn have forms with clustered spikes; all except *Tripsacum* tend to have their female spikes enclosed in at least one leaf sheath; all except corn and the male spikelets of all Maydeae usually

---

[1]Galinat, W.C. 1974. Inquiries into the origin of the American Maydeae. Proc. symp. on the origin of *Zea mays* and its relatives. Bot. Mus. Harv. Univ. (unpubl.)

have single female spikelets although the sessile female spikelet may be paired with a sterile pedicel in the Oriental Maydeae; all usually have a suppressed lower floret within the functional female spikelet and all except corn usually have an indurated organ that forms a fruit case.

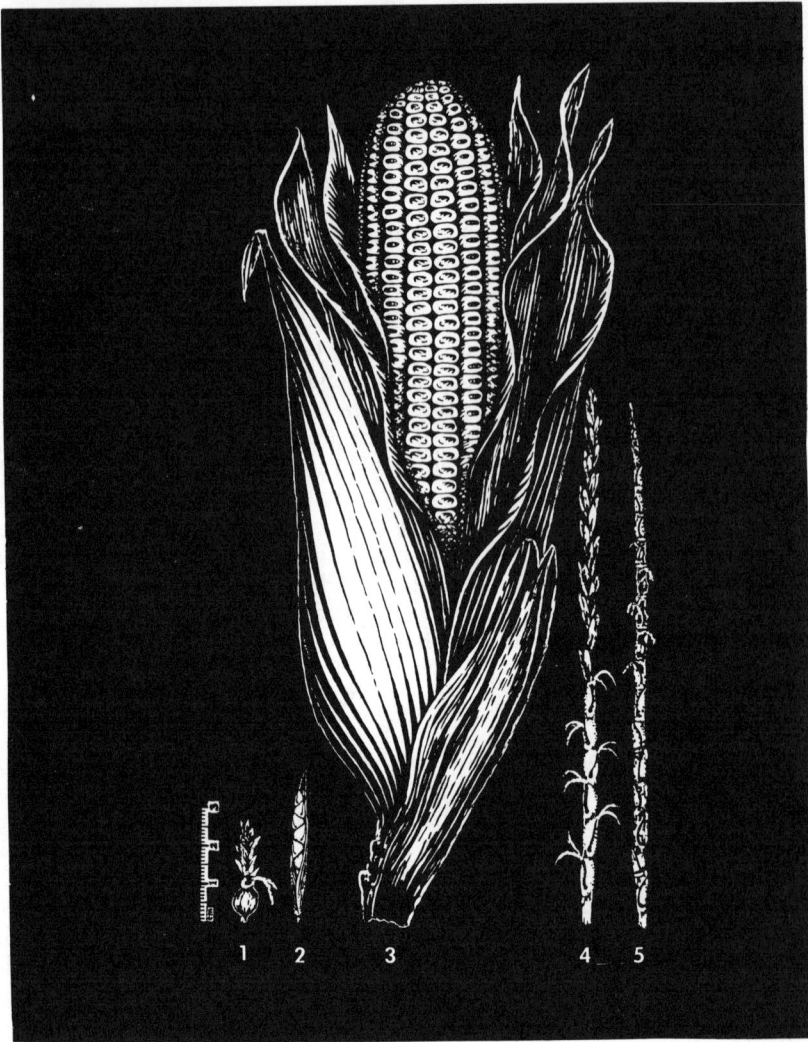

Fig. 1. Corn and its relatives: 1. *Coix* sp.: — female spikelet in fruit case formed by indurated leaf sheath. Its long styles dangle to one side. The terminal rachis bears staminate spikelets. 2. Teosinte: Pistillate spike showing the combined protection of a husk enclosure and individual cupulate fruit cases. The styles, which are extremely long as in corn, have been removed. 3. Corn: Pistillate spike with many rows of paired spikelets with the ear as a whole protected by a multiple husk system. 4. *Tripsacum:* As in teosinte, the grains are individually protected by cupulate fruit cases. There is little or no protection from leaf sheaths and the terminal staminate region is extended. 5. *Manisuris:* With an incipient cupulate fruit case approaching that of *Tripsacum, Manisuris* differs primarily in being perfect flowered.

Various organs have become specialized to form the protective fruit case during radiation from a stem stock common to all of the Maydeae (Galinat, 1956; Stebbins, 1956). In the non-*Coix* members from the Orient, it is chiefly an indurated outer (first) glume of the spikelet; in *Coix* it is primarily an indurated, ovoid, uppermost leaf sheath; in the wild members of the American Maydeae, namely in the genus *Tripsacum* and in corn's closest relative, teosinte, it is chiefly a thick, indurated, rachis segment and/or its cupule within which solitary spikelets are embedded. The lateral wings of the cupule clasp the outer glume of the enclosed spikelet, forming with it, at maturity, a smooth corneous fruit case (Galinat, 1963).

Because the genus *Manisuris* of the subtribe Rottboelliinae (tribe Andropogoneae) more closely than any other genus morphologically approaches the cupulate fruit case of the American Maydeae and because its basic chromosome number is nine pairs, or half of that of the genus *Tripsacum*, it has been considered the logical connecting link to the Andropogoneae, and more closely related to *Tripsacum* and *Zea* than is *Coix* of the Oriental Maydeae (Celarier, 1957a; Weatherwax, 1935). The fruit case of *Manisuris* is formed by a protective cavity resulting from a groove in the rachis, together with a sterile pedicel appressed along one of its edges; the opening is sealed by a specialized outer glume from the enclosed spikelet. Differences in the depth of the groove in the rachis segment, and in the degree of participation of the sterile pedicel in forming a cavity, represent variations leading toward the type of cupulate fruit case found in *Tripsacum* and teosinte. When the sterile pedicel was completely reduced, when the remaining sessile spikelet that is remnant from a perfect flowered pair was reduced to a solely female condition, when the cavity in the rachis segment was compacted into a cupule, and when both cupule and outer glume of the enclosed spikelet were indurated, the fruit cases of *Tripsacum* and teosinte were evolved out of variations that began in *Manisuris*.

The species of *Manisuris* are characterized by distinctive types of sculpturing in the outer glume. The genus, like most other grasses, and in contrast to the Maydeae, is perfect flowered. It includes at least five Oriental species described by Bor (1960) and five New World species described by Hitchcock (1951) and others. They are distributed throughout the tropics and subtropics around the world.

In *Tripsacum*, as in *Manisuris*, the mature spike is elevated above the sheath of the uppermost leaf. Disarticulation is accomplished by a ball-and-socket connection between successive rachis segments. The rind at the upper end of each fruit case remains erect to form a socket which contains a hemispherical mass of nodal parenchyma attached to the underside of the overlying fruit case. The cupule is not as well defined as in teosinte but greater seed protection is usually derived from the outer glume of the enclosed spikelet. The cupule of *Tripsacum* is intermediate between that of teosinte and that of *Manisuris*.

Tripsacum is a more distant and diverse American relative of corn than teosinte. It comprises at least five diploid ($2n=36$), and four tetraploid ($2n=72$), species all of which are perennial, but of diverse morphology, and adapted to a wide range of habitats throughout much of the New World (Cutler and Anderson, 1941; Randolph, 1970; Tantravahi, 1968). It has survived for long periods in diverse habitats ranging from dry rocky canyon walls in Mexico, brackish banks in New England, and tropical lowlands of Colombia representing elevations from sea level to nearly 2.50 km. While *Tripsacum* has been hybridized with corn (Mangelsdorf and Reeves, 1939) and some *Tripsacum* germplasm has been transferred to corn (Galinat, 1974b; Maguire, 1962), it seems doubtful that *Tripsacum* had a role in the origin of either corn or teosinte (de Wet et al., 1971; Galinat, 1971).

Teosinte *(Zea mexicana)* is the closest relative to corn and probably part of the original stem stock of corn. It now comprises six races of annual teosinte, including the distinct Guatemalan race (Wilkes, 1967), that are adapted to habitats from the Nobogame Valley of North Mexico south to Honduras, from low western elevations in Guerrero to 2,133 km in the Valley of Mexico. Teosinte grows as a wild plant in successful competition with other wild plants, as well as in competition with corn in cultivated fields despite man's attempts to exterminate it. Because teosinte crosses readily with corn and is cytogenetically similar to corn (Beadle, 1932; Emerson and Beadle, 1932; and others), it serves to inject heterotic vigor into the domestic species in their sympatric areas and, thereby, to increase its productivity (Beadle, 1972; and others).

There is a perennial tetraploid of teosinte *(Zea perennis* R. & M.) in which the tetraploidy appears to be of recent origin (Randolph, 1955; Shaver, 1962), although its perennialism may stem from an extinct perennial diploid. Even the tetraploid is now extinct in the wild, although it survives in experimental cultures derived from the collection of Collins made in 1921 at the site of Hitchcock's 1910 discovery in Jalisco, Mexico (Hitchcock, 1922).

In teosinte, there is a double system to protect the grain, combining a cupulate fruit case similar to, but more compressed than, that of *Tripsacum,* with an enclosing uppermost leaf sheath that is homologous, to, but less specialized than, the protective spathe of *Coix.* Additional protection for the teosinte spike may come from the physical overlapping of many such spikes, each enclosed in its own protective leaf sheath. This clustering of teosinte spikes appears to result from a ramosa type of spike proliferation terminal to the branch, and the condensation of nodes bearing leaves and axillary spikes into a massive spike cluster, borne in a position close to the main culm. Superficially the entire mass of teosinte spikes resembles a husk-enclosed ear of corn and they are probably homologous. Although this massive condensation of teosinte spikes may interfere and/or delay natural seed dispersal, it greatly increases the efficiency of harvestability when done on an individual spike or spike cluster basis. The harvesting of entire plants would be difficult on the steep hillsides and/or in the brush often associated with teosinte. When the spikes are borne singly or in widely separated small clusters, the mature fruit cases are lost through their open husk tops as the plant bends in the wind and/or the birds tear open the husks.

Corn

In the domestic species of *Zea (Z. mays l.)* corn, the original protective function of the cupules became obsolete and protection for an elaborated spike (ear) as a whole is dependent upon numerous overlapping leaf sheaths (husks) borne at the nodes of the condensed shank. These numerous husk leaves, enclosing a massive uppermost spike, were liberated from their previous function of protecting the many, now suppressed, lateral spikes from a condensed cluster of teosinte spikes. Concurrently all nourishment became concentrated in the development of a single uppermost spike, i.e., the emergent ear of corn. Iltis[2] has compared this process to the origin of the monocephalic head and stout single stem of the cultivated sunflower *(Helianthus annuus L.)*, through a suppression of the many small lateral heads of the wild sunflower. The potential for development of ears in the axils of the husks has long been recognized. Anderson (1949) observes that,

> If one will examine a number of ear shoots in any corn field, he will find at least a few in which there are little undeveloped ears hidden by the husks at the lower nodes. They usually do not develop but if we carefully break out the (uppermost) ear without damaging the shoot on which it was borne, one of two of the axillary ears may be stimulated into development. There are a few varieties of popcorn in which these axillary ears develop under normal conditions.

The resulting cluster of ears is comparable to a cluster of teosinte spikes. Through the suppression of such axillary spikes, the nourishment is channeled into a single depository, the giant husk-enclosed ear of modern corn.

The styles (silks) of both teosinte and corn have become adapted to function in the presence of surrounding leaf sheaths, by acquiring an ability to continue to elongate from their base until they have protruded from the apex of the enclosing husks or until they are pollinated.

The important floral characteristics that distinguish the corn cob from the female spike of teosinte are: 1) Polystichous vs. distichous spikes; 2) Nonseparating vs. disarticulating rachis segments; 3) Paired vs. single spikelets; 4) Coriaceous vs. corneous outer glumes; 5) Reflexed vs. erect spikelets; 6) Elongate vs. sessile rachillae (spikelet axes); 7) Condensed vs. elongate internodes and cupules; and 8) Spike enclosed in many husks rather than one. Other characteristics in which corn and teosinte usually differ such as leaf width, tillering, and response to photoperiod are not considered as definite in distinguishing the species. If one were to judge solely on the basis of this list without recognition of the great power of human selection (well known since Darwin's time), then a taxonomist might place these relatives in separate genera, as they were once placed: teosinte in the genus *Euchlaena* and corn, the sole species, in the genus *Zea.* Now, however, many authorities agree that corn and teosinte belong at least in the same genus, if not consolidated further as subspecies of *Z. mays* (Iltis, 1972).

Beadle (1972), Iltis (1972), and others emphasized that the above-listed morphological differences are precisely those which might be expected if corn were a domesticated form of teosinte. Randolph[3] and Mandelsdorf (1974) have countered that these same corn traits occur in most of the 7,000-year-old cobs from Tehuacan, Mexico, which they assume to be wild corn, and that there is no known archaeological record that clearly demonstrates a

---

[2]Iltis, H.H. 1974. The maize mystique: a taxonomic-geographic interpretation of the origin of corn. Proc. symp. on the origin of *Zea mays* and its relatives. Bot. Mus. Harv. Univ. (unpubl.)

[3]Randolph, L.F. 1974. Contributions of maize relatives to the origin of domesticated maize: a synthesis of conflicting views. Proc. symp. on the origin of *Zea mays* and its relatives. Bot. Mus. Harv. Univ. (now: Econ. Bot. 30:321-345. 1976).

transition from teosinte to corn. However at least one of the oldest Tehuacán cobs is similar to a corn-teosinte hybrid (described later) despite the absence of any record of teosinte (past or present) in the valley. While many of these tiny slender cobs from Tehuacán are broken (fragmented) in at least one position, they are remarkably intact for such fragile specimens that have been shelled by man and trampled under foot. They possess no functional natural system of disarticulation which characterizes wild grasses such as teosinte. Archaeological teosinte is known from Tamaulipas (Mangelsdorf et al., 1967a) and from the Chalco area (Lorenzo and Gonzalez, 1970), but in neither case is it old enough to be significant to the problem.

Long before the remarkable archaeological record of 7,000 years of corn evolution was revealed in the remains from Tehuacán (Mangelsdorf et al., 1967b), the present wide differences in points of view about the origin of corn had already been firmly cast. Those who still reject the idea that corn came from teosinte hold to their earlier statements. Mangelsdorf (1947, 191) emphasizes both the magnitude of the change and a presumed lack of time for such a change during domestication: "If maize has originated from teosinte, it represents the widest departure of a cultivated plant from its wild ancestor that comes within man's purview," and suggests that "one must indeed allow a considerable period of time for its accomplishment, or one must assume cataclysmic changes, of a nature unknown, have been involved." Similar ideas were expressed by Weatherwax (1955, 10):

> To have changed into something like corn, it (teosinte) would have had to do the unlikely thing of undergoing a despecialization in two or three ways, and these changes would have had to occur so closely together that they immediately gave the plant an economic value which it did not previously have.

Weatherwax then concludes that "some future discovery may also compel us to reconsider the whole question, but, at present (1955), those who have seen much of both plants are inclined to look somewhere else for the direct ancestor of corn." Elsewhere Weatherwax (1954) was more open in saying,

> We are still far from any positive evidence that maize came directly from teosinte, and there is much ground for theoretical objection to the idea, but it would be well not to close the door too soon on such a possibility.

At both San Marcos and Coxcatlan Caves, the two most important of the five corn-yielding caves excavated by MacNeish in the Tehuacán Valley, there were both eight-rowed and four-rowed (two ranks of paired spikelets) cobs in the oldest remains. This two-ranked (distichous) feature is important as one of the key traits distinguishing corn from teosinte. Although the two-ranked ear sometimes occurs in secondary and stunted ears of present-day, eight-rowed corn, there is an important difference between the modern and some of the ancient two-ranked specimens. One of the latter (Tc50 Zone IX, 4000 B.C.) has teosinte-like abscission layers extending about halfway across the rachis, with complete disarticulation being prevented by condensation. There is also a partial reduction of the second or pedicellate spikelet with the overall effect suggestive of an F1 hybrid of Confite Morocho corn by Guerrero teosinte.

In the specimens from Coxcatlan Cave, one of the six whole cobs from Zone XIII and three of the six whole cobs from Zone XI were admitted to the column of "Early Cultivated" because "they were somewhat larger in size, and these are presumably the first products of cultivation" (Mangelsdorf et al., 1967b, 194, and Table 22). This interpretation of size difference as the discriminating factor between the wild and domestic state of corn is the prime tenet of those who hold that corn's evolution has been chiefly just an increase in size and that the divergence of corn and teosinte occurred long before the advent of man in the New World.

The long sequence of early archaeological corn cobs that do not show the effects of increased induration apparent in derivatives from modern corn-teosinte hybrids (Galinat et al., 1956; Galinat and Ruppe, 1961;) (Mangelsdorf et al., 1967b; Mangelsdorf and Smith 1949) may indicate only that teosinte introgression does not produce significant induration in a background of primitive corn. Such a situation might be expected if primitive corn contained an intermediate tunicate allele, such as those studied by Mangelsdorf and Galinat (1964), and occurs in the primitive race Chapalote and probably others. Beadle's[4] suggestion that a tunicate mutation in teosinte, which transformed its hard fruit case to a soft one, was selected early during a domestication of teosinte and later discarded as the modern cob evolved. Such a mutation to a higher tunicate allele has been described by Randolph (1974) for a chimeric sector of a teosinte plant collected on a field trip in Guerrero, Mexico, organized by G.W. Beadle in November 1971.

A counterpart to such a soft-shelled teosinte, as conceived by R.A. Emerson (Beadle, 1972), has been developed by transferring a tunicate allele from corn into teosinte. A parallel domestic change in a different type of fruit case has taken place in *Coix lachryma-Jobi.* In the *Coix* cereal variety, mayuen, man has selected for a threshable soft-shelled spathe to replace the highly indurated wild type spathe (Jain and Banerjee, 1974). Although high induration of the fruit case as in teosinte and *Coix* may be indicative of evolutionary advancement, man may undo this achievement to serve his own purposes. Thus the fact that the pistillate spike of teosinte is more indurated than that of corn does not in itself prove which of the two is the older species.

The delayed appearance of teosinte introgression in archaeological corn, as estimated by the degree of induration in the cob, might be a product of secondary intergradation when wild and cultivated forms came together after a previous separation. Thus as man carried corn from Mexico into the U.S. Southwest, there would be a period of escape from teosinte until teosinte also extended its geographical range. The reintroduction of teosinte germplasm into the highly evolved types of Chapalote frequently replaced its tunicate allele with the lower one of teosinte and generally increased cob induration in this area by A.D. 900. Then by at least A.D. 1200, the introduction of Maiz de Ocho into the complex, thought by some to carry exotic germplasm from the Colombian corn, Cabuya (Roberts et al., 1957) further increased the level of variability, and its associated heterosis, at a critical time when the corn of the Southwest had to adapt to a drier climate (Galinat et al., 1956). Thus, at present, the discussion about the origin of corn centers primarily on differences of opinion concerning the adequacy of the archaeological record to confirm or deny any possibility of the role of ancient human selection in a

---

[4] Beadle, G.W. 1974. Teosinte and maize. Proc. symp. on the origin of *Zea mays* and its relatives. Bot. Mus. Harv. Univ. (unpubl.)

divergence of corn and teosinte. The alternative is that these species diverged by natural processes of variation, selection, and adaption to environmental factors independent of man's activities, such as his relentless search for food.

Mangelsdorf (1974) objects to viewing the cupule as being derived from the ancestral wild type because this development is "equivocal," a term used previously by Weatherwax (1954) in objecting to the thesis, supported by Mangelsdorf, that pod corn was derived from the ancestral wild type. Certainly the expression of cupule development is no more equivocal than that of the pod corn (tunicate) alleles. The weak expression of cupule development in the elongate and partially staminate tip of a pod corn ear, as illustrated by Mangelsdorf (1974, 80) is only to be expected in the transition zone between the noncupulate staminate tip and cupulate base of the rachis. Obviously a slender rachis cannot bear a wide cupule. The presence or absence of the cupule is an inherited trait (Galinat, 1969), and it cannot be denied that the oldest archaeological cobs from Tehuacán had cupules.

If the assumption is made that corn is older than teosinte and that man had no role in their divergence, it is difficult to account for the presence of cupules in all of the oldest known archaeological cobs and a level of compaction in the rachis that is so high as to interfere with any natural dissemination of seed. The cupule is an integral part of the fruit case that protects the grain in teosinte and the extension of the trend of increasing degrees of compaction to the point of interfering with disarticulation and dissemination, as in the Tehuacán cobs, would best be explained as, at least, an indirect product of human selection or harvesting activity.

Was the Oldest Tehuacán Corn a Wild Plant?

How good is the evidence that the oldest corn from Tehuacán was an indigenous plant collected in the wild, rather than an introduced plant already along the way to domestication? Other plants which were part of the gathering phase were either under domestication or approaching domestication at Techuacán before the appearance of corn, namely avocado *(Persea americana),* maguey *(Agave spp.),* prickly pear *(Opuntia spp.),* and *Setaria* (Smith, 1967).

An analysis of the rise and decline of *Setaria* in relation to the appearance of corn, as shown in the data of Smith (1967, Table 26) is significant because it contrasts what is obviously the usage curve for a plant approaching domestication to that for corn. According to Smith (1967, 232), during the El Riego Phase in zones 23 to 15, representing 1,500 years prior to corn at Coxcatlan Cave,

> Increasing amount of *Setaria* seed may be interpreted as an indication that *Setaria* may have been sown and harvested near the site, which would require occupation at the beginning and end of the rainy season.

MacNeish (1967, 291) is not convinced by the evidence for *Setaria* being a pre-corn cultivar because "the seeds from later levels show no increase in size as do seeds of many cultivated plants." But Yarnell (1965) has shown that in some plants, such as *Ambrosia trifida,* there is no relationship between plant size and seed size such as might result from cultivation. There is no evidence of genetic variability for seed size in modern *Setaria* and so the objection of MacNeish does not seem to apply to this case.

During the El Riego Phase, the increased dependence upon *Setaria* is reflected by an increase in weight of specimens from 0.28 g to a peak of 538.6 g at which time it may have been cultivated. At this high point in reliance upon *Setaria,* there was a sudden appearance of corn at a substantial, but increasing, level of usage. This intrusion came despite a search of the Tehuacán countryside for edible plants resulting in known usage of 23 other species during the previous 1,500 years, all without a trace of corn. Is it not more probable that corn was introduced at this point rather than overlooked? The decline and displacement of *Setaria* and the other plants appears to result from the adaptation, improvement, and later by the reintroduction of a more indurated form of corn. Because the original semidomestic types of Tehuacán corn were isolated from the swamping effects of large populations of teosinte, such as might occur in Guatemala, Oaxaca, or the Valley of Mexico, they were free to stabilize into a tiny soft-glumed version of the present-day corn cob. But later, with a secondary introduction of new variation from a more indurated, so-called teosintoid form of corn from the outside, there were pronounced heterotic effects that increased the size and yield of corn to a level that corn agriculture effectively replaced that of *Setaria* by the Abejas Phase (3400 to 2300 B.C.)

The absence of corn during the collecting period known for *Setaria* and other plants, followed by the sudden appearance of corn long after these other plants had become cultivated, makes appear untenable the postulate (Mangelsdorf et al., 1967b), that the oldest corn at this site was an indigenous plant collected from the wild. Moreover the morphological features of these oldest corn cobs and plant fragments suggest those of a semidomesticated rather than a wild plant (Beadle, 1972; Galinat, 1971a).

The position now held by the author is that all of the oldest cobs from Tehuacán belong in the early cultivated, rather than the wild, category. As an item of trade and cultural diffusion, an early domestic type of corn would be readily accepted by a people who already had an agricultural tradition involving another annual grass *(Setaria).* The introduction of other cultivars from outside the valley in the same phase as that of corn "at periods when agriculture was still subordinated to plant-collecting" is known from a pod of common bean *(Phaseolus vulgaris)* from Zone XI of Coxcatlan Cave (Kaplan, 1967, 205). Kaplan suggests that the region of Tehuacán was not the site of an early domestication of beans and "evidently they were an item of highland-lowland transport in the Tehuacán Valley." If domestic beans were traded in, why not also corn?

The archaeological evidence for a shift from gathering to cultivation of *Setaria* and the known collection of 23 food plants at a high level of usage, prior to the introduction of corn is prima facie evidence that: 1) Corn was not an indigenous wild plant in the valley and, therefore, was not collected in the wild; and 2) The oldest corn was a semidomesticated type that was introduced from outside the valley. The further adaptation of corn as a food plant is indicated by its change to more productive types during its displacement of *Setaria* cultivation.

Pollen Evidence

Even if the oldest corn cobs from Tehuacán were those of cultivated, rather than wild, corn, it is claimed that the much older fossil pollen grains extracted from deep drill cores prove that the ancestor of cultivated corn was wild corn and not teosinte. Because these ancient pollen grains have greater lengths measured on collapsed spheres than that of present day Mexican teosinte and are, therefore, more similar to that of modern corn, this has been interpreted as "definitive proof" that corn did not stem from teosinte (Barghoorn et al., 1954; Bartlett et al., 1969).

Pollen size alone, however, as a key taxonomic character, does not always distinguish corn and teosinte. Guatemalan teosinte has much larger pollen than Mexican teosinte and it overlaps significantly with that of the oldest Bat Cave pollen (Galinat, 1973b). It is also possible that the large size fossil pollen is from an extinct species of *Zea*. Environmental factors, as well as degree of inbreeding, also dramatically influence the size of pollen (Kurtz et al., 1960[5]). Electron microscopic studies of the oldest known fossil pollen from the Valley of Mexico have not revealed any differences in spinule pattern from that of either present-day corn or teosinte (Banerjee and Barghoorn, 1972). While these studies did show spinule clumping in the pollen from one plant of the highly evolved race, Cuzco giganti from Peru, that was similar to that of *Tripsacum* derivatives of modern corn, no light was shed on the origin of corn. Unlike the cupulate fruit case which appears to have its origins in the Andropogoneae (genera *Manisuris* and *Elyonurus*), the large pollen diameters of *Zea* appear to have affinities to the Oriental Maydeae, especially *Coix lachryma-Jobi* (Galinat, unpubl.) In summary, the data on pollen size are confusing and ambiguous and I believe that they do not solve the problem of the origin of corn.

On the Time Available for *Zea* Domestication

One of Mangelsdorf's (op. cit.) objections to teosinte being in the direct line of corn's ancestry is that, since the 7,000-year-old archaeological corn from Tehuacán is clearly corn and not teosinte, there would have been insufficient time for such a cataclysmic transformation of the female spike without any intermediate adaptive stages. But if the process of deliberate selection by man were preceded by "a considerable period of time," accumulating the products of unconscious human selection during preagricultural gathering, a gradual transformation from a teosinte-type rachis to the corn cob could indeed be accomplished without assuming "that cataclysmic changes of a nature unknown" were involved (Galinat, 1974 c,d). Long before the oldest known corn cobs were discarded at Tehuacán, man had been collecting and possibly planting various food plants in the area (MacNeish, 1964; Smith, 1967).

---

[5] Johnson, C.M., D. Mulcahy, and W.C. Galinat. 1973. Effect of inbreeding in *Zea mays* on pollen size. Contributed paper. Soc. Study Evol. 21 June. Univ. of Mass., Amherst.

THE ORIGIN OF CORN

At Tehuacán, the gathering of *Setaria* grain was underway by 7000 B.C. and it culminated in a form of agriculture based on indigenous plants (*Setaria*, avocado, maguey, and prickly pear) prior to 5000 B.C., according to the data of Smith (op. cit.). Subsequently the first corn cobs appeared and these were followed by other introduced, semidomestic food plants. During the time before corn's appearance at Tehuacán (6000 to 8000 B.C.), pollen from the Guila Naquitiz Cave in the adjacent state of Oaxaca (to the south) indicates that incipient agriculture involving a *Zea*-like plant was underway (Schoenwetter, 1974). The Oaxaca *Zea*-like pollen grains were found in areas where the plant was (p. 302) "unlikely to have grown unaided in the local environment." The late appearance of corn at Tehuacán may have been the result of introductions by man of semidomestic forms of *Zea* from Oaxaca or elsewhere.

Evidence for agriculture earlier than that known in Mesoamerica is claimed from the San Pablo site of coastal Ecuador where a large kernel of corn was embedded in a Valdivia V-VI sherd (Zevallos, 1971). This evidence of early domestic corn is substantiated at the Real Alto site nearby, where kernel impressions in rim sherds from the Valdivia III period which, after the Suess correction, date at about 3000 B.C. (Zevallos et al., unpubl.). The earlier use of corn-type monos and metates suggests that the use of evolved types of corn in the area extended back to earlier than 4000 B.C. or about 1,000 years before such advanced races of corn are presently known to have appeared in either Mesoamerica or Peru. Lathrap (1973, 1974) concludes that proto-agricultural systems developed as a supplement to marine sources of food in the tropical riverine prior to the Pleistocene-Holocene boundary. Thus Lathrap has presumptive evidence in ceramic remains to support Sauer's (1952) hypothesis that there was an earlier agriculture in the tropical lowlands. In any case, this evidence seems to support the theory of multiple centers of corn domestication but in itself does not resolve the problem of corn's origin.

It appears that at a very early date in the New World, man had to depend more upon the flora than the fauna, and this suggests that plant domestication started at least as early in the Americas as in the Near and Far East, and long before the oldest remains of corn were deposited at Tehuacán. Martin (1973) suggests that the larger animals were decimated by man in Mesoamerica about 11,000 years ago, requiring a gradual shift from a dependence upon hunting to a greater exploitation of the flora. There were only traces of the horse and the antelope left in the Tehuacán Valley by 8000 B.C., although there was a lingering use of white-tailed deer, rabbits, and other small animals increasingly supplemented by plant foods (Flannery, 1967, Fig. 95). The spread of man into these areas also may have occurred earlier than that suggested by Martin (op. cit.) because there is early evidence of man at sites in the New World, such as at Ayacucho, Peru, dating from 22,000 years ago (MacNeish, 1971). Dating deduced from aspartic acid racemization on several Indian skeletons from California suggest that man was present in North America at least 50,000 years ago (Bada et al., 1974), presumably as a result of migration from Asia during an earlier ice age than that represented by the 12,000-year-old migrations plotted by Martin (op. cit.).

One factor which led Mangelsdorf to believe that the oldest Tehuacán cobs were those of a now extinct wild corn was his observation that any domestic transformation from a teosinte spike would require a considerable period prior to 7,000 years ago, while man was still only in the plant gathering phase of cultural development. In order to explain the cytogenetic similarity of corn and teosinte, Mangelsdorf (1974, 52) suggests "teosinte is essentially a mutant form of maize" and that this postulated evolution "may have occurred over a period of millions of years." He explains,

> Grasses and grazing animals began to become abundant in the Miocene epoch some 20 or 30 million years ago. If Galinat's most recent theory (1970) that *Tripsacum* is an allopolyploid hybrid of a species similar to teosinte in its characteristics and one similar to *Manisuris* is valid, . . . time must be allowed for the still earlier divergence of teosinte from maize.

The timing of such a natural divergence is placed by Randolph[3] in the late Pleistocene when disturbances in the environment were conducive to rapid speciation.

In contrast to the above, the more primitive genera, *Tripsacum* and *Manisuris*, have cupulate fruit cases more similar to that of teosinte than to the cob of corn. Also corn, more highly compacted and many ranked, is more specialized than its relatives; therefore teosinte must be older. The gathering phase age of the oldest Tehuacán cobs may be explained as a result of changes accidentally introduced into campsite cradles of *Zea* during harvesting.

Campsite Colonies as Cradles Leading to *Zea* Domestication

Whenever primitive man accidentally became involved in the dispersal and planting of seeds during the hunting and gathering phase, plants with better harvesting qualities, especially a retention of mature seed, became concentrated near the campsites as a secondary by-product, as discussed for Old World cereals by Helbaek (1959) and New World corn and beans by Lynch (1973). These campsite colonies became divergent evolutionary cradles. They started from seed gathered by man but accidentally lost, perhaps during threshing, thousands of years before deliberate plant breeding was practiced. The plants with less efficient methods of natural seed dispersal would be automatically, unconsciously and more efficiently gathered and would form the campsite cradles. The gathering of cereals, especially teosinte, is not a random process but, quite unintentionally on man's part, a highly selective one. Not until man made a deliberate effort to select seed with better utilitarian qualities from within the cradle population for active sowing, did the intentional process of domestication actually begin (Hawkes, 1970).

The present-day variation in teosinte includes both the elongate habit, from which "seeds" (fruit cases) cannot be efficiently gathered at maturity, and a more condensed type, which would become concentrated in the cradles because of its preadaptability for abundant gathering. Because the spikes of the former are widely separated on elongate branches, the wind shatters and disperses the seeds readily at maturity, as indicated by field observations (Galinat and Galinat, 1972). Most of the gathered seeds come from those plants that have a higher-than-average degree of compaction in the branches homologous to the ear shank. The gathering into the campsite areas of such telescoped types, unconsciously selected from distant populations in the wild, would gradually establish a polygenic capacity for a high level of compaction to be expressed at some phase during development. A reprogramming of the pattern of condensation for expression later during the floral phase of development would cause *Zea* to become permanently fixed for tardy disarticulation. Once man discovered the advantages of sowing, cultivating, irrigating, and harvesting these semidomestic types with a more compressed rachis, he would deliberately select out additional changes in rachis morphology. Increasing levels of compression would exclude spikelets within the cupule and reduce the internode space between cupules within a row. Under such extreme compression, selection for a functional integration of parts would favor an elongation of the rachilla (kernel stalk) and its relaxation to a position at right angles to the rachis as well as an elevation of the kernels above the glumes. (Figure 2 illustrates this interpretation of how the harvesting action of man may have inadvertently transformed teosinte into corn, probably from 10,000 to 7,000 years ago.)

The shift from condensation in the shank to that in the ear appears to involve regulatory type genes or substances, e.g. auxin peroxidase (J.L. Brewbaker, unpubl.), rather than structural genes, operating during phase change. Gross changes in plant and ear morphology result from minor changes in regulatory genes, which shift the phase specificity and degree of

Fig. 2. An interpretation, based on present knowledge, of how man's harvesting action may have inadvertently transformed teosinte into corn, probably between 10,000 to 7,000 years ago. Extreme clustering of the spikes (condensation within the branch or shank) was followed by a triangularization of their fruit cases (condensation within the rachis). When successive fruit cases overlapped, disarticulation was inhibited by a fusion of the apex of the cupule onto the internode above and a domestication of these most condensed forms followed. By 5,000 to 7,000 years ago, the primitive popcorn races, such as represented by the Chapalote-Nal Tel complex, were well established. By the time of the Pilgrims, shown here with a northern flint type, almost all of the 300 races of corn known today had differentiated. The final step shown here, with a modern farmer, was the Corn Belt dent, the world's most productive race.

expression of structural genes. For example, the corn grass and teopod genes of corn reduce the rate of phase change giving various intermediates between vegetative and floral development. Other mutants may transpose part of the usual floral program, such as high condensation and decussate phylotaxy, upon the vegetative axis. Teosinte types recovered out of segregations from hybrids with corn tend to have their spikes in clusters because of condensation derived from corn. These various bits of evidence are all consistent with the idea that the expression of high condensation accumulated during harvesting first in the vegetative phase of the branch axis could be delayed and, thereby, manifested during the floral phase. The result would be high condensation such as usually occurs in the female rachis of the Tehuacán cobs in contrast to the low condensation in the teosinte spike.

Mangelsdorf's assumption that the oldest Tehuacán corn was wild is based partly on a lack of intermediate steps between corn and teosinte. This lack of evidence, however, may well be misleading because the open campsites of a mobile people in transition from a hunting and gathering existence would not preserve the transformation series (Schoenwetter, 1974).

Origin of Polystichy as an Effect of Condensation

In both their polystichous (many-ranked) condition and their high degree of compaction, the oldest ears of archaeological corn and, in more extreme degree, modern corn, are more specialized or evolved than teosinte or than any other of corn's known relatives. The importance of condensation in increasing kernel row number in corn was demonstrated by Anderson and Brown (1948), following a more general principle on the effects of compaction described by Arber (1934) and Mangelsdorf (1945). Anderson and Brown pointed out, among other things, that the degree of condensation affects row number in both tassel and ear; at extremely high levels of condensation, the tassel branches of corn may also become polystichous although ordinarily they are distichous. Because the tassel branches of typical present-day corn are more like the spikes of teosinte than the central spike of a corn tassel, both in phyllotaxy and degree of condensation, some have supported the early suggestion of Weatherwax (1918) that teosinte evolved from a corn-like ancestor by a suppression of the central spike (as well as the pedicellate member of paired female spikelets). Because the polystichous trait is a key one separating corn and teosinte, the nature of its origin is paramount to the question of whether man was the selective agent.

As the differentiating spikelet primordia attempt to expand under the extreme compression, or lack of internode elongation, at the base of a distichous-type of corn ear, they may be physically pinched off to one side or crowded off into the open space available at one side, creating a simple three or four-ranked phyllotaxy in this region. Even today in many races of modern corn, the higher kernel row numbers at the butt of the ear are followed by drop-rowing and reduced row numbers toward the tip of the ear, as a consequence of such a gradient in compression. The hypothesis of the origin of polystichy by yoking and twisting advanced by Collins (1919) many years ago is similar in some respects to the observed effects of compression. Collins, however, carried his concept too far by stating erroneously that drop-rowing occurred simultaneously on opposite sides of the ear and not from adjacent paired rows, an observation correctly pointed out by Weatherwax (1935) and Mangelsdorf (1945). There is also an independent system for the direct induction of polystichy which may be combined with that of compression effects (Galinat, 1973c). The gradual assemblage of a polygenic complex for increasing levels of compaction during the development of corn may be reflected today in the polygenic nature (about six genes) of the high kernel row number of modern corn, as described by Emerson and Smith (1950). Once a polystichous spike with softer glumes became available to man, its domestic utility would be increased further by man's selection for paired female spikelets, by an enlargement of kernels, and by a further proliferation in the number of kernel rows. As the cupule collapsed and fused upon itself, its original function as part of a protective fruit case would shift to that of structural support for a modern-type cob. During this sequence of events, the original protective function of the cupule is presumed to have been gradually transferred to a condensed rosette of husk leaves enclosing the entire ear. The effects of condensation in transforming the teosinte-type of relationship between spikelet and cupule to that of Corn Belt corn is illustrated in Fig. 3.

Little more than the cupule in the modern corn cob remains as evidence of the cob's apparent origin from the teosinte spike in "the widest departure of a cultivated plant from its wild ancestor that comes within man's purview." The smooth, continuous nature of the transformation from the deep, vertical cupule of teosinte to the compressed, horizontal cupule of corn has been described (Galinat, 1970). At a point early in the series where the female spikelets change from single to paired, there is an abrupt shift from the narrow, vertically shaped cupule to a triangular shape. This is not a valid taxonomic discontinuity, however, because the same change in cupule shape in association with spikelet pairing occurs on a single cob showing unstable phenotypic expression of the *pd* (single female spikelets) gene.

On the basis of comparative morphology of the oldest cobs from Tehuacán to the female spike of teosinte, it has been shown 1) that man could have had a role in the elaboration of corn from teosinte by increasing the level of condensation during harvesting and 2) that the cupule of the Tehuacán cobs, like that of modern cobs, is a remnant from the teosinte fruit case (Galinat, 1970).

Possible Areas of Predomestication

If corn was introduced into the Tehuacán Valley as a domesticated derivative of teosinte, it seems probable that the initial selections leading toward domestication occurred outside the valley but within the known geographic ranges of teosinte. It has been suggested that the original stem stock from which the Mexican corn and Mexican teosinte originated was located in Guatemala (Kempton and Popenoe, 1937; Longley, 1941). This postulated stem stock may have been a restricted ecotype, such as that adapted to marshland (e.g., Lake Retaná area) that excluded it from the Tehuacán Valley. If the initial semidomestic types diffused as highly variable forms, they could have diverged into the present-day sympatric races of Mexican corn and teosinte. Further domestication would cannalize cob development on the one hand, while volunteers or escapes would revert to the wild and reestablish the shattering rachis and other essential features of teosinte by the process of disruptive selection (Galinat, 1973a). To some extent, this would be consistent with Mangelsdorf's present contention (1974, 52) that at least some types of teosinte (present-day Mexican) represent a "mutant" form of corn and yet be consistent with the condensation and cupule sequences leading through *Manisuris, Tripsacum,* teosinte, and corn (Galinat, 1956; Weatherwax, 1935). But Mangelsdorf's new speculation is that, rather than represent a reversion from a domesticated form of teosinte sometime after about 10,000 years ago, teosinte differentiated from corn by

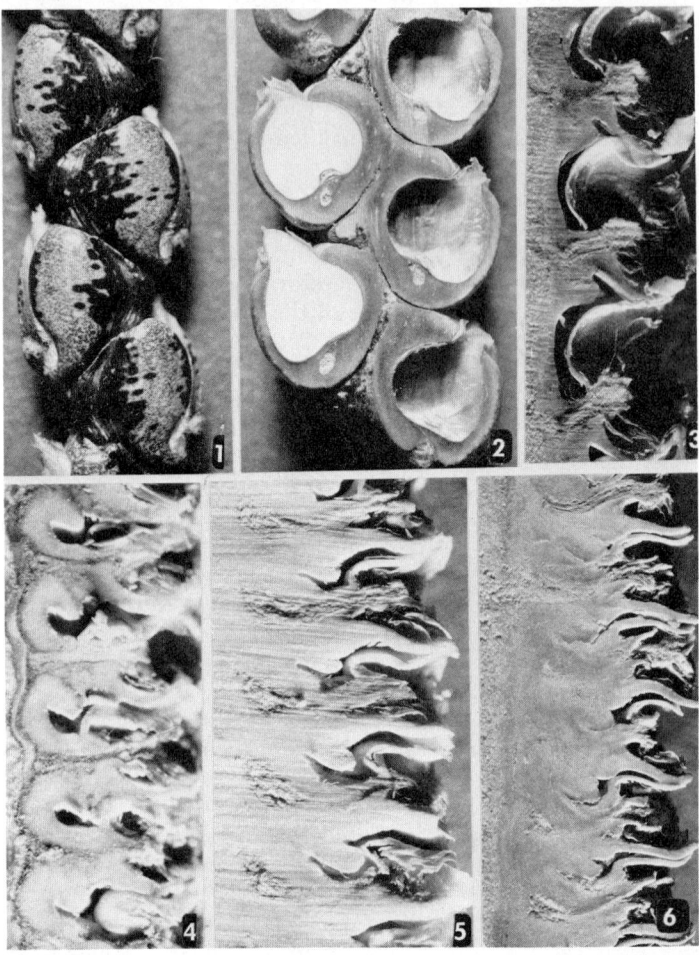

Fig. 3. Increasing levels of condensation within the rachis will transform the type of relationship between spikelet and cupule found in teosinte to that of corn: (1) Condensed form of teosinte; (2 to 6) Longitudinal sections; (2) Corn-teosinte hybrid show grain emerging between glume and cupule; (3) Confite Morocho corn with long shallow cupules; (4) Inbred A158 corn with cupules folded over; (5) Inbred B2 corn with cupules closing; (6) Condensed B2 corn with cupules fused floor to roof.

THE ORIGIN OF CORN

natural processes in the late Pleistocene quite independent of man. Because the presently available, oldest archaeological remains of teosinte (Lorenzo and Gonzales, op. cit.) are of about the same age as those of corn, and the older fossil *Zea* pollen does not settle the problem, it may well be that the older intermediate steps between teosinte and corn might be discovered in Guatemala where the more primitive forms of teosinte now occur. The lifestyle of the people living in temporary, open-campsites during the predomestication period may preclude the preservation of the essential intermediate steps in dry caves as at Tehuacán. In any case, the isolation of early domestic types of *Zea* at Tehuacán would be important in stabilizing a genetic complex for cob development. When man carried these semidomestic types into isolation from teosinte, they were protected from genetic swamping by dense stands of teosinte such as once grew in Guatemala (Kempton and Popenoe, 1937). Thus the early Tehuacán corn was free to diverge rapidly into utilitarian perfection as a botanically correct corn cob. The importance of isolating a sub-population by genetic drift away from the swamping effect of a large population for rapid evolutionary response was first recognized by Wright (1931).

Multiple Domestications

Several researchers have suggested that there may have been multiple domestications of corn (or teosinte), which is partly responsible for its present diversity. Randolph (1959, 10) states,

> The diversity of existing types of maize with respect to cytological, morphological and physiological characteristics, and their widespread distribution in very early times suggests that more than one species of wild corn was involved in the origin of cultivated maize.

McClintock (1959, 465) concludes "that cultivated maize may have had several independent origins, from plants whose knob-forming regions had distinctly different capacities for producing knob substance." Mangelsdorf and Sanoja (1965, 111) suggest that,

> Pollo represents the domesticated form of a wild maize which once grew in Colombia and Venezuela. As a result of their genetic studies of the tunicate locus in maize, Mangelsdorf and Galinat (1964) have concluded that there were once at least two races of wild maize in Mexico. There may well have been additional wild races elsewhere in America, of which one was the ancestor of Pollo.

More recently Mangelsdorf (1974, 113) has expanded on the independent suggestions of Randolph and McClintock with "the idea that there might have been as many as six different geographical races of wild corn and that primitive races stemming from these still exist." One of these six postulated lineages combines Pollo into a Chapalote-Nal Tel complex.

Accidental selection of a small sample of variability in his food plants for transport during seasonal migrations in search of food would have a drift or founder's principle effect on establishing new colonies that may diverge into distinct semidomestic and wild races. Once the nonshattering rachis was established near a dwelling site, locally adapted, semidomestic races would emerge. These semidomestic forms, fringing on the wild type, would often revert back to the wild shattering forms from volunteer colonies, only perhaps to undergo a redomestication at a later time, as the selective pressure for nonshattering or condensed spikes caught up with them once again during a later harvesting. The outcome would certainly be multiple races of teosinte and of primitive corn.

If there were multiple domestications of corn or teosinte, then there were probably several different cradles leading to these domestications that correspond to some extent to the presumed races of wild corn. Takahashi (1955) demonstrated that barley was domesticated twice in western and central Asia, the antidispersal mechanism having been developed by different genetic systems. Pickersgill (1972) has noted that *Phaseolus vulgaris* and *P. lunatus* have undergone multiple domestications and that four species of chili peppers *(Capsicum annuum)* were domesticated independently from separate wild ancestors. Thus even though the domestication of a wild plant may be a slow process, the occurrence of multiple cradles leading toward domestication would enhance the potential magnitude of any transformation into a radically different domestic cultigen.

Because the archaeological remains of corn now available do not provide an indisputable link with teosinte, except for their common cupules and at least one of the oldest Tehuacán rachises (which is intermediate between corn and teosinte), it is concluded that these remains are either inadequate in quantity at the oldest levels and/or of insufficient age to reveal the origin from the wild type. In the absence of such indisputable evidence, we may seriously consider studies of the surviving taxa of this evolution.

The reliability of such experimental evidence may be judged in part by the degree of its conservativeness. The attributes which become involved in the speciation process in one group may be independent of such a change in another group. In certain genera such as *Clarkia* that are in a process of rapid speciation, there are cataclysmic differences in both number and morphology of chromosomes with few, if any, associated changes in floral morphology (Lewis, 1966; Stebbins, 1971).

Floral Diversity in Corn

In corn and its relatives, however, it is clear that in the present stage of evolution, greater reliability may be assigned to comparisons of genetic architecture than to those of floral structure alone. The well-known extreme floral diversity in corn is a product of genetic changes during domestication. As Darwin (1858) observed, the plant parts that are most important to man, in this case the female spike, receive his greatest attention and, thereby, diversify at the most rapid rate. In the genus *Zea*, a floral transformation from a teosinte spike to a primitive ear of corn is in many respects more than matched by the present-day incredible floral divergence between some of the races of corn (e.g., Argentine popcorn vs. Cuzco flour corn) that are not associated with gross cytological differences.

Until the advent, starting in the 1920's, of hybrid corn in the USA and much more recently elsewhere, especially in the now technologically emerging countries, the farmers saved and selected their own corn seed from crops grown in many diverse environments and selected for different ethnic culinary traditions (Hernandez, 1973). Much of the resulting variation, originally described under 219 races of Latin American corn including many synonyms, has been consolidated to 169 relatively distinct races in a statistical analysis of their taxonomy by Goodman and Bird[6]. When these races were grouped on the basis of ear morphology, kernel size was found to be the principal definitive component (Bird and Goodman)[7]. The natural variation in these numerous open-pollinated varieties, often enhanced in Mexico by gene flow from teosinte, provided the genetic options and buffering necessary to cope with fluctuations in climate and disease in such a way that a crop of some sort was insured.

Although some of the different uses of corn were originally based on single gene differences effecting the kernel (e.g., flinty, floury, waxy, sugary, dent), the deliberate isolation of these kernel types tended to allow some definitive value for classifying corn (Sturtevant, 1899), at least at a varietal level.

Within the race Máiz de Ocho, we may have floury endosperm (Northern Flour), flinty endosperm (Northern Flint), or sugary endosperm (Golden Bantam sweet corn). The repatterning of chromosomes would be at the generic or specific level while knob differences and allelic changes would more likely be at the racial or varietal level, although either the presence of, or polymorphism for, knobs in contrast to their complete absence in a species may have special significance. As a racial trait, knobs are usually conservative, fixed in patterns that characterize certain races of teosinte and corn, as discussed below.

Complexity of Inheritance of Floral Traits Distinguishing Corn and Teosinte

The various traits distinguishing corn and teosinte have been listed in an earlier section. The problem of priority of genetic state is obviously paramount to the problem of which is older, teosinte or corn. Because teosinte and corn are almost identical, genetically and cytologically, the change from one species to the other must have happened relatively recently, geologically. There are modifying factors that have accumulated to stabilize the mutant forms and in doing so they give some resemblance to polygenic inheritance and certain cytological factors that reduce crossing-over within certain essential segments that give block inheritance, as discussed later.

There are conflicting results for the linkage and inheritance of the key floral traits that distinguish corn and teosinte when different stocks of corn and teosinte are parental to the segregations (Galinat, 1971). These varying results appear to result from the segregation of genes for different degrees of compaction and vascularization in different corn-teosinte segregations. In developing low-condensation strains of "string cob" corn, the author (Galinat, 1974a) has found that the polystichous condition is often unstable, reverting to the distichous phenotype in the upper portion of the ear. On the other hand, when a gene for a distichous spike *(tr)* that is stable in a string

---

[6] Goodman, M.M., and R. McK. Bird. 1974. The races of maize. IV. Tentative grouping of 219 Latin American races. Proc. symp. on the origin of *Zea mays* and its relatives. Bot. Mus. Harv. Univ. (unpubl.)

[7] Bird, R. McK., and M.M. Goodman. 1976. The races of maize: V. Grouping maize races on the basis of ear morphology. Econ. Bot. (in press).

cob background is introduced to thick cob corn, the penetrance of the gene is weak or unstable. Essentially the same results are obtained with the *pd* (single female spikelets) gene, in that a stable phenotype is promoted by the slender rachis of string cob. While the double mutant, *trtr, pdpd,* is stable in the string cob background, it is female sterile through a lack of style development, a result reported also for a normal cob type background (Daniel, 1973). But when teosinte chromosome 4 is introduced, a harmonious balance is reached which allows a fertile expression of the double mutant combination.

The necessity of teosinte chromosome 4 to produce a congruous background for a fertile expression of the combination *trtr, pdpd* is important to the origin problem because it would appear to rule out the suggestion of Mangelsdorf (1974, 52) that teosinte is a mutant derived from wild corn, such as that which he assumes to be represented by the oldest remains at Tehuacán. The chromosome 4 complex of teosinte is described as producing indurated glumes and rachises with erectoid spikelets, while the oldest Tehuacán remains are described as having soft glumes and rachises with reflexed spikelets. Thus a combination of the *tr* and *pd* mutants would probably be infertile in a background of the so-called "wild corn" of Tehuacán and, therefore, could not produce teosinte by assembling these mutations in corn. Would it not be more probable that the reverse mutations had taken place because the combination of paired female spikelets together with the polystichous condition is fully fertile in the presence of teosinte chromosome 4? A breakdown of the chromosome 4 complex might be expected in semidomestic corn older than the oldest Tehuacán material because of the loss of its co-adaptive value when it functioned with the *tr tr, pd pd* combination of alleles; this chromosome 4 complex also creates domestic disadvantages by interfering with the ease of shellability of the grain.

While the recovery rate of parental types in F_2 segregations from corn-teosinte hybrids indicates only four or five inherited units are involved (Beadle, 1972; Collins and Kempton, 1920; Mangelsdorf and Reeves, 1939), studies on the inheritance of the individual key taxonomic traits by which these taxons differ indicate a much more complex type of inheritance based on the total number of genes involved. The chromosome 4 complex (Mangelsdorf and Reeves, 1939), as well as single vs. paired female spikelets, two vs. four-ranked, and sessile vs. pedicellate female spikelets all involve at least two genes each although the number and location of the genes involved is variable in different experiments with various races of corn and teosinte (Galinat, 1971).

Tha anomalous situation of a simple inheritance based on high parental recovery rates and a complex inheritance based on total number of genes involved with definitive traits has been reconciled by Mangelsdorf (1947) in his discovery that various teosinte chromosome segments tend to have duplicating effects, based on multigenic systems involving the female spike and that each segment consists of a number of closely linked genes. Until recently, these segments were often interpreted as being introgression segments derived from *Tripsacum* chromosomes involved in a hybrid origin of teosinte, as postulated in the tripartite hypothesis of Mangelsdorf and Reeves (1939). Objecting to this hypothesis, Beadle (1939) suggested "that these major differences arose by mutation, either genetic or chromosomal in nature, and were preserved by man through deliberate selection." De Wet and Harlan (1972) reasoned that since experimental introgression of *Tripsacum* into corn has not produced teosinte, the natural origin of teosinte from such a hybrid is unlikely. Galinat (1971a, 1974b) reported that the essential chromosome 4 complex of teosinte could not be derived from *Tripsacum* introgression because it does not occur in *Tripsacum*. Also he (1973a) supported the concept that the essential blocks of genes were assembled by disruptive selection between man and nature. The recent concession by Mangelsdorf (1974) that teosinte is not a corn-*Tripsacum* hybrid, together with various types of evidence to be discussed later, all tend to support the early conclusion drawn by Beadle (1939).

Cytology of Corn and Teosinte in Relation to Their Evolution

Chromosome Uniformity in Corn

The karyotype of the many races of corn is remarkably uniform and stable for a species which is so variable in chromosome knobs and also so genetically dynamic. If there were cytological races of corn separated by inversions and translocations as might originate spontaneously or from segmental interchanges with the chromosomes of *Tripsacum*, then this might now appear in some wide racial hybrids within corn. Any heterozygous translocations and inversions could result in incomplete chromosome pairing, associations of three or four chromosomes, or else by inversion loops leading to chromosome bridges and fragments. In tests for such gross variability in the chromosomes of corn, as expressed by pollen abortion, Cooper and Brink (1937) had negative results with 68 varieties from Canada, Mexico, Central America, Peru, and Bolivia, as did Rhoades and Dempsey (1953) with 85 strains from Latin America. This gross chromosome uniformity among races of corn is partly an evolutionary artifact, resulting from man's selection for only well-filled ears for seed stocks. Despite this selection, the relative chromosome uniformity between isolated races of corn makes it unlikely that there were extensive changes in architecture and number of chromosomes such as would be necessary if corn came from the very different chromosomes of *Tripsacum*, as suggested by Cutler (1947), or extensive introgression of alien *Tripsacum* germplasm into corn during the origin of certain South American races of corn, as suggested by Roberts et al. (1957).

The "abnormal" chromosome 10 (Rhoades, 1952; 1955) and the various spontaneous aberrations, such as the inversion on the short arm of chromosome 8 (McClintock, 1933), are random exceptions to gross chromosome uniformity in the races of corn. While some of these chromosomal variations of corn may also occur in teosinte, according to Randolph (1955) there is no evidence that they were derived from *Tripsacum*. In the case of one form of abnormal chromosome 10, Ting (1958a) has evidence that it originated as a simple translocation involving the normal chromosome 10 and a B-chromosome. In any case, the chromosomal aberrations of corn occur at random and do not characterize any natural race. It is generally agreed that the genomes of the various races of corn are more or less completely homologous despite their differences in chromosome knobs and allelic frequencies.

Chromosome Variability in Teosinte

While corn and teosinte have the same chromosome number, n=10, and are cytogenetically similar (Longley, 1937) with essentially the same frequencies of crossing-over in the regions tested (Emerson and Beadle, 1932), except for the short arm of chromosome 9 (Beadle, 1932), there is increasing evidence of diverse cytological polymorphism between them for small inversions, knob sizes, and knob positions.

Inversions

Except for the chromosome 9 inversion (Durango, Florida, Xochimilco, Nobogame), the several other large inversions now known in various teosintes were not detected genetically in the early studies mentioned above. Later the teosinte chromosome 9 inversion was confirmed cytologically by O'Mara (1942) in studies that also revealed an inversion in the long arm of chromosome 8 of Durango teosinte and small rearrangements in the long arm of chromosome 4 of Florida teosinte. The chromosome 9 inversion was also detected genetically in tetraploid perennial teosinte by Shaver (1960) and then confirmed cytologically by Ting (1964). A structural difference on chromosome 5 of Florida teosinte, observed first by Arnason (1936), was later identified cytologically as an inversion on the long arm of this chromosome in Nobogame teosinte by Ting (1958b, 1964) who also discovered an inversion in the long arm of chromosome 3 (Xochimilco) and an inversion in the short arm of chromosome 8 (Xochimilco, Chalco, Nobogame). The latter was similar to a chromosome 8 inversion in corn described by McClintock (1933) and, if the two inversions are identical, they probably had a common origin. The chromosome 8 (Nojoya) rearrangement previously described by O'Mara (op. cit.) was, in contrast, in the long arm. More recently Kato (1976) has found inversions in the long arms of chromosomes 1 and 7 in Nobogame teosinte hybrids.

According to Ting, some of the large inversions are homozygous in various wild populations of teosinte. This is to be expected because large heterozygous aberrations induce a high rate of sterility in plants carrying them but fertility may be restored in the structural homozygotes. Except for the chromosome 9 inversion, appropriate genetic markers for the other inversions have not been used to pinpoint their location. Neither have the inversions been tested for linkage with the essential genes controlling developmental differences in the two distinct types of female spikes separating corn and teosinte. The possibility that such linkages occur is suggested by the observation of Wilkes (1967) that the large inversions of teosinte are most frequent where hybridization with corn is most extensive. These points certainly need to be tested.

The small structural differences observed by O'Mara (op. cit.) in the long arm of chromosome 4 has particular significance because this chromosome carries many of the genes that distinguish the teosinte spike from the corn cob (Mangelsdorf and Reeves, 1939; Rogers, 1950). According to O'Mara (p. 10, 11), the long arm of chromosome 4 in a Florida-Nojoya teosinte hybrid,

> . . . was paired regularly to a point half-way towards the end from the centromere. Beyond this point various different configurations were observed. In some instances the ends were not paired. In others, various kinds of loops occurred which suggested inversion configurations. The Florida homologue, which could be identified by its large knob, was slightly longer than the Nojoya homologue. While no exact determination of the nature of the differences between the chromosomes was possible, the existence of one extensive or of several small rearrangements seemed most possible.

O'Mara's cytological observations of apparent small rearrangements in the long arm of chromosome 4 of a northern Guatemalan (Nojoya) teosinte in comparison to the structure of a southern Guatemalan (Florida) teosinte have some genetic confirmation (Galinat, 1973a). In the latter studies, still in progress, there is reduced crossing over to a level of about 20% in the sugary glossy-3 *(Su Gl3)* region in segregations from a tester gene corn hybrid with a Mexican (Nobogame) teosinte chromosome 4 derivative of corn. Isogenic segregations within corn itself as well as with a chromosome 4 extracted from Guatemalan (Florida) teosinte have about 36% recombination in this sugary glossy-3 region. Because the short arm of chromosome 4 is involved in fruit case development and the long arm of this chromosome is involved in rind abscission, the reduction in crossing over in teosinte that is sympatric with corn is indicative of gene cluster (super gene) formation or block inheritance. Chromosomal races of sympatric populations often maintain their specializations by a suppression of crossing over.

Small paracentric inversions such as that suggested in the long arm of teosinte chromosome 4 mentioned above, reduce crossing over without the sterility that is associated with large inversions. Thus small inversions would survive as heterozygotes and be able to spread in a population, especially if they became co-adapted through linkage with the definitive fitness traits of teosinte. The experimental demonstration of any such small inversions is difficult. Although small inversions are usually manifest as loose pairing in the heterozygote rather than in loops usually characteristic of heterozygous large inversions, a search for small inversion loops at pachytene may reveal them in somewhat less than 1% of the cells. There is also difficulty in obtaining genetic evidence of small inversions because it requires appropriate marker genes in the small region of the inversion and segregation data from large progenies.

The Significance of Inversions

Inversions can have a significant role in the genetic isolation of sympatric species when they are tightly linked to the definitive alleles that separate these species. Whenever the harmonious balance within combinations of alleles that are essential to the survival of one species is upset by hybridization with the other species, any associated inversions would become heterozygous. As a result, the ill-adapted hybrid types would be partially sterile and/or recombination within the inverted segment restricted. Thus inversions may provide one of the mechanisms to maintain intact a cluster of alleles cooperating in the development of a particular complex trait, despite repeated outcrossing to a sympatric race that has a different adaptive trait controlled by another set of alleles. The introgression of such inversions from teosinte to corn would become so difficult that it would maintain genetic isolation between these sympatric species. Thus because of the four known large inversions and the indications of various small inversions in certain teosintes, especially those with fewer chromosome knobs, one may speculate that these small rearrangements may be an alternate system for the genetic separation of some pairs of the species during the evolution of corn. A somewhat similar role for inversions occurs in certain sympatric species of *Drosophila* (Carlson, 1965; Epling et al., 1953) as pointed out by Ting (1964). While inversions may reduce crossing over within the rearrangement, they increase the rate of recombination elsewhere, as is known from experiments in both *Drosophila* and corn (Bellini and Bianchi, 1963). Such an effect would not only protect certain blocks of inverted germplasm, but would also promote the exchange of variability in other regions such as that involved in the mutually instilled heterosis between sympatric corn and teosinte (Wilkes, 1972).

Chromosome Knobs

Chromosome knobs are deeply staining (heterochromatic) regions that are relatively stable markers for fixed positions on the chromosomes. Thus they may be useful criteria to reveal structural changes, as well as relationships between certain races of corn and between corn and its sympatric partner, teosinte, even though all of the American Maydeae, as well as *Coix lachryma-Jobi* of the Oriental Maydeae are polymorphic for size, position, and frequency of knobs. The number of chromosome knobs has been used as evidence for the hybrid origin of the Corn Belt dents (2 to 8 knobs) from a cross between the Southern Dents (5 to 12 knobs) and the northern flints with few if any knobs (Anderson and Brown, 1952; Brown, 1949).

Differences in the number of active knob positions are often assumed to be a result of natural selection. Corn that is adapted to either high latitudes or altitudes tends to have a low number of knobs while that found in either low latitudes or altitudes has a high number of knobs (Brown, 1949; Longley, 1938; Mangelsdorf and Cameron, 1942). The knob constitution or pattern of knob distribution in corn may change during the adaptation of a race to a new locality. This is apparent in the development of a knobby form of Maíz de Ocho (Harinoso de Ocho) in northwestern Mexico in contrast to its nearly knobless putative ancestor (Cabuya of highland Colombia) and its essentially knobless descendants (northern flour and northern flint) as discussed by Galinat and Gunnerson (1963.

While little is known about the origin of knob positions or the process by which they become manifest in the form of a visible knob or the degree of knob stability in various genotypes and cytoplasms, there is evidence that knobs may be reduced in size on the one hand or increased in size through duplication on the other (Longley and Kato, 1965). The origin of polymorphism for knobs in the American Maydeae and in *Coix* of the Oriental Maydeae, however, is still a mystery. Several genera in the Andropogoneae, namely *Manisuris, Coelorhachis, Rottboellia,* and one species of *Elyonurus (E. tripsacoides,* n=10), that have been suggested as possible ancestors of the American Maydeae, appear, on the basis of presently available evidence, to be knobless; the exception is *Elyonurus argenteus,* a South African species with five pairs of long chromosomes, two of which have terminal knobs (Celarier, 1957b).

If we assume that *Manisuris* and *Elyonurus* are essentially knobless, as the presently available evidence indicates, then based on knobs and/or heterochromatic regions, we may give more serious consideration to *Coix,* especially *C. lachryma-Jobi,* as a possible source of knobs leading to those of the American Maydeae. Longley (1941) was the first to point out a morphological similarity between the chromosomes of *Zea* and this species of *Coix.* Both *Coix* and *Zea* vary considerably with respect to the presence of chromosome knobs. Nirodi (1955) observed five terminal knobs and one internal knob in the pachytene stage of *C. lachryma-Jobi,* while P.N. Rao (1973) observed only two terminal knobs and no internal knobs in the same species. In *C. aquatica* and *C. gigantea* P.N. Rao (1973)[8] also observed many internal and terminal heterochromatic regions that, through differential condensation and/or reduction during crossing over, could be precursors of knobs or knob positions. All of this suggests a primitive relatedness through a common stem stock that could serve as a source for both the terminal knobs of *Tripsacum* and Guatemalan teosinte, as well as the internal knobs of Mexican corn and Mexican teosinte with *Coix,* in its many diverse forms, as an ancient by-product.

Role of Chromosome Knobs

While both corn and teosinte are polymorphic for number and size of chromosome knobs, the variation in number of knob forming positions and size of knobs is greater in teosinte. The Guatemalan teosintes are uniform for many terminal knob positions that are unknown in both corn and Mexican teosinte (Longley, 1937). Kato (1976) has located at least one knob position. unknown in corn on each chromosome in the various Mexican teosintes. Kato has also found differences in size of knobs between corn and teosinte. For example, the 8L2 knob of teosinte is sometimes larger than its counterpart in corn. Kato found a type of abnormal chromosome 10 in teosinte that is unreported in corn.

Heterozygous chromosome knobs are known to reduce crossing over in their region, as shown by a subterminal knob in chromosome 3 (Rhoades and Dempsey, 1957) and a terminal knob on 9 in which the size of the knob is related to the degree of reduction in recombination values (Kikudome, 1959). Thus when the chromosome knobs of one member of a synaptic pair are closely linked to allelic combinations, essential to its specific characteristics, the knobs may prevent these combinations from breaking up when their counterpart from the other member is knobless in the same positions. This is exemplified by the presence of numerous knobs in Chalco teosinte in relation to the near knobless condition of its Chalco corn partners, such as

---

[8] Rao, P.N. 1973. Cytogenetic studies in *Coix* with some observations on other oriental Maydaea. Ph.D. Thesis. Andhra Univ., Waltair, India.

Cacahuacintle (avg. ca. 1.0 knobs), Conico (avg. 1.0 knobs), and recently Chalqueno (avg. 6.8 knobs). Although there is some confusion about the correct knob number for the ancient race of Chalco corn, Palomero Toluqueno, Wellhausen el al. (1952) report an average of only 1.2 knobs for the type collection. Kato (1976) reports that many of the knobs of Chalco teosinte are unknown in any of the various races of Chalco corn.

Because the above cytological mechanisms may reduce crossing over within the essential blocks of teosinte germplasm, they make a complete evolutionary divergence of the species unnecessary and allow them to coexist and exchange germplasm to the extent that they develop similar plant types. Wilkes (1972) observed that the plant type of Chalco teosinte is more similar to that of its ancient sympatric partner, Palomero Toluqueño corn, in color (red) and pubescence than it is to its more recent corn partners, namely Conico and Chalqueño that are greenish and less pubescent. Although an introgression of plant type from these recent corn partners may have some adaptive value in corn mimicry, the time lag and gene flow have not yet been sufficient to allow teosinte to catch up with the change in corn. Wilkes counted the present field ratio as 25 of Chalco teosinte — 1 hybrid: 500 Chalqueño corn. In the inbred and/or isolated strains of either teosinte or corn having few or no chromosome knobs, the genetic function of these knobs in protecting the integrity of essential blocks of alleles would be redundant and they tend to become lost (Longley and Kato, 1965). Teosinte requires protection from genetic swamping by its sympatric partner since man controls the size and propagation of the corn population. This may account for the well-known fact that the teosintes from Central Mexico (which excludes Nobogame) have, on the average, more knobs than Mexican corn. It is not yet clear if Jutiapa teosinte from southern Guatemala has more or less than its share of large inversions; however, during its isolation and divergence, it did retain many terminal knobs, not only unknown in Mexican corn and teosinte, but also unknown in Guatemalan corn as well (Kato, 1976).

Although the genetic amalgamation of the northeastern flints and the southern dents has proceeded for nearly a century, some of the characteristics of the northern flints are still somewhat linked in blocks of germplasm in the modern Corn Belt corn (Brown and Anderson, 1947). The known differences in chromosome knobs between these parents of the Corn Belt dent may well be a factor in maintaining block inheritance of many genes that distinguish these races; this factor plays a role similar to that of knob differences in preserving teosinte from genetic swamping by its sympatric corn.

If, as the evidence suggests, an ancient teosinte-like plant represented a species prior to corn, then there would have been more time for it to accumulate teosinte's present-day, greater cytological variability, more fully revealed by recent intensive cytological studies of teosinte. During the divergence through disruptive selection of present-day corn and teosinte from this ancient common ancestor, any knob positions and small inversions in teosinte that happened by chance to be tightly linked with traits essential to its survival as a wild plant, especially those involving seed dispersal, would become co-adaptive with these wild traits and spread in the populations. Through natural selection for less vulnerability to genetic swamping from the large sympatric populations of corn grown and selected by man, those cytological features peculiar to teosinte would cause their associated essential genes to become inherited in blocks suggestive of gene complexes. Such inheritance of teosinte's essential traits in four or five major blocks of linked genes is well known from the high parental recovery rates from segregations of corn-teosinte hybrids (Beadle, 1972; Collins and Kempton, 1920; Mangelsdorf and Reeves, 1939).

Longley (1941) concluded that, from a cytological point of view, a shift in the activation patterns of knob positions could have converted the southern Guatemalan teosintes into the northern types; and subsequently these northern teosintes might have been changed into corn. The debate as to whether there was an ancestral teosinte with both terminal knobs, as in the southern Guatemalan teosintes, and internal knobs, as in Mexican corn and Mexican teosinte, or if, on the other hand, the internal knobs of teosinte resulted from hybridization and recombination with a "wild corn," is not at this time reconciled. An attempt to determine whether the combination of both terminal and internal knobs in the teosintes from northern Guatemala is a result of introgression and recombination or if it is a relic stem stock condition is crucial.

There is an urgent need for intensification of research on hybrids between sympatric partners of corn and teosinte in order to discover the cytogenetic basis of the difference between the parents in ability to self-perpetuate and, thereby, to gain an understanding of the basis of *Zea* domestication.

CORN—*TRIPSACUM* HYBRIDS

Our present understanding of the origin of corn has been influenced by an intensification of studies on the cytogenetics of corn-*Tripsacum* hybrids. These studies have theoretical significance for our understanding of the processes of chromosome repatterning observed in the American Maydeae. Because this is a wider cross to a broader gene pool than that of teosinte, it is of applied significance, providing the essential groundwork necessary for the agronomic use of this new source of genetic variation, especially in locating valuable alleles which might contribute to parasite resistance and heterosis.

Production and Maintenance of the Hybrid

Although no spontaneous *Zea-Tripsacum* hybrids have been confirmed, experimental hybrids between these genera first were developed by Mangelsdorf and Reeves (1931), and have subsequently been repeated by others. The first crosses involved a pruning back of the styles of corn to about the same lengths as those of *Tripsacum*. A mixture of corn and *Tripsacum* pollen promoted the development of hybrid kernels when adjacent to normally developing pure corn kernels. The hybrids were grown with surface sterilization and with help by peeling of the pericarp.

These results were confirmed by Randolph (1955), who employed a somewhat different technique. Rather than prune back the husks and styles, the husks were pulled apart sufficiently to allow one to sift a mixture of corn and *Tripsacum* pollen down to near the style attachment to the ovary and then securing the husks closed with a glassine bag fastened with rubber bands. After 2 weeks, the developing hybrids were reared by embryo culture techniques and eventually transferred to soil.

Similar techniques were used by Galinat et al. (1964) who found considerable differences in crossabilities when using different strains of corn and *Tripsacum;* the *T. floridanum* tested crossed easily on corn giving hybrid kernels requiring no special germination techniques (Galinat, 1961). Stimulated by this work, agronomists at the University of Illinois have produced many corn-*Tripsacum* hybrids (Anand and Leng, 1963), especially when tetraploid *T. dactyloides* is used as the pollen parent (Lambert, 1965). The resulting allotriploid (TTZ) hybrids tend to produce eggs with just the diploid (TT) *Tripsacum* genome (de Wet et al. 1970). *Tripsacum*-corn hybrids may also be produced on *Tripsacum* (Farquharson, 1957). This hybrid has the additional value of providing a new source of cytoplasmic variation to the corn genome at a time when studies of the cyto-male sterile-restorer gene systems of hybrid corn seed production are being redeveloped (Duvick, 1972) because of the susceptibility of the formerly used "T" cytoplasm to the "T" race of southern leaf blight *(Helminthosporium maydis)* (Smith et al., 1971; Tatum 1971). Crosses of corn made onto *Tripsacum* as well as the introgression of *Trip-*

Fig. 4. The plant habits of a tester gene stock of corn and *Tripsacum dactyloides* together with their F1 hybrid at their first cycle of flowering. Left: *Tripsacum dactyloides* 2n (Bussey Clone) of Manhattan, Kan. Age ca. 365 days. Center: F1 hybrid of MMT corn × *T. dactyloides*. The parents are shown on either side. Age 284 days. Right: MMT corn bearing the recessive marker genes $bm2$, lg, a, su, pr, y, gl, j, wx, r, with one marker located on each of its ten chromosomes. Age 80 days.

sacum germplasm into corn may, in some cases, require that the *Ga* gene which produces cross sterility be present in the corn parent (de Wet et al., 1972).

The plant habits of a tester gene stock of corn and *Tripsacum dactyloides* together with the F1 hybrid at their first cycle of flowering are illustrated in Fig. 4. The young tillers or shoots shown at the base of the hybrid stalk are continuously produced at higher nodes so that, although the plant is not a true perennial, it may be maintained indefinitely by cuttings and/or raising the soil-level in relationship to the tillering zone. Such a hybrid has now been maintained in this manner for 15 years in the author's cultures.

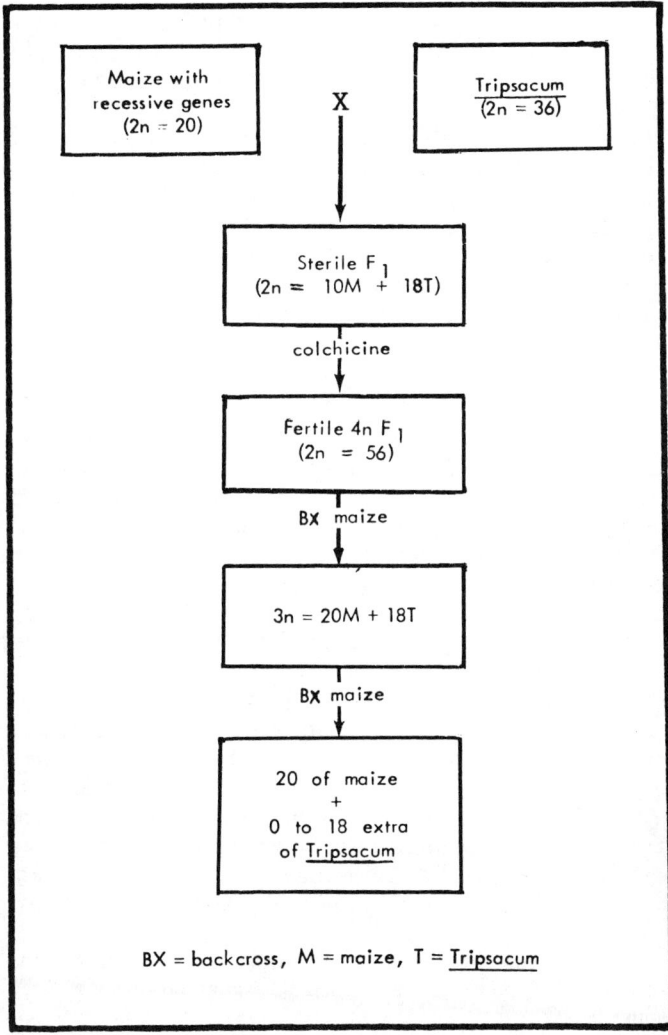

Fig. 5. Breeding chart illustrating the crosses involved in developing the various addition monosomic stocks of corn that carry extra cytogenetically identified chromosomes from *Tripsacum*. (adapted from Galinat, 1974).

Isolation and Identification of
Allotrisomics from Corn-*Tripsacum* Hybrids

The development of alien addition monosomics (allotrisomics) and their addition disomic condition requires several years to accomplish and the results are fortuitous. The procedure is outlined in the breeding chart of Fig. 5.

The Mangelsdorf multiple tester (MMT) stock of corn is a key to the isolation and identification of the 20M+1 Tr condition. In hybrids with this stock, one can simultaneously test for loci marking 9 out of the 10 corn chromosomes. Although the F1 hybrids are male sterile, they do produce a low frequency of viable megaspores from unreduced gametes. In our case, it was more efficient to develop the amphidiploid with the use of colchicine which produces almost 100% viable megaspores, although the plants remain male sterile for some unexplained physiological reason. The first backcross of the amphidiploid to the marker gene corn produces an allotriploid that has two genomes of corn and one from *Tripsacum* and this once again is usually about 80 to 90% female sterile and completely male sterile. The progeny resulting from the second backcross to tester stock corn has a variation in the number of extra chromosomes derived from *Tripsacum* ranging from 0 to 18. Certain of these *Tripsacum* chromosomes that have become isolated in a given corn plant can then be identified genetically in terms of the ability of their dominant alleles to suppress one or more of the known corn recessives.

Maintenance and Loss of an Extra *Tripsacum* Chromosome

The segregation of the 18 *Tripsacum* chromosomes in the progeny of the allotriploid hybrid (ZZT) is skewed to the low end of the scale far in excess of that expected on a basis of a random distribution. The actual frequencies for the 20, 20+1, and 20+2 classes in a population of 150 plants from the segregation of a genome of *T. floridanum* chromosomes on tester stock corn was 19.3%, 30.6%, and 14.0%, respectively, with the theoretical mode of 20+9 being only 1.3% instead of 50.0% (Chaganti, 1965). This distribution pattern favoring the low chromosome numbers is surprisingly similar to that estimated by Mangelsdorf and Reeves (1939) on the basis of pollen abortion frequencies; however, the latter system is now regarded as erroneous, as pointed out by Maguire (1957) and then conceded by Reeves and Mangelsdorf (1959). Because the individual transmission frequency of six genetically marked *Tripsacum* chromosomes (*Bm2, Lg, A, Su, Gl, Wx*) is about 32% (Galinat et al., 1964), there should be a massive elimination of classes with more than one extra *Tripsacum* chromosome, although sometimes there are exceptions (Maguire, 1963). The following reasons for the nonrandomness of the distribution for *Tripsacum* chromosomes have been given by Chaganti (1965, 42).

> Added to the inherent variation in the transmissibility of the individual chromosomes of *Tripsacum* (when added to a genome of corn), factors like chromosome length, interchromosomal interactions between maize and *Tripsacum* chromosomes and interactions arising out of the fact that the *Tripsacum* chromosomes are now situated in alien cytoplasm could further complicate the survival and compatibility patterns of gametes of the triploid hybrid.

THE ORIGIN OF CORN

Rapid elimination of *Tripsacum* chromosomes at frequencies higher than the 20+1 class makes any attempt to incorporate a multiple tester for a particular corn chromosome, before the population has been screened through the general MMT tester, futile in the first or second backcross to corn. It is essential to incorporate the appropriate multiple tester from corn to a given homeolog from *Tripsacum* as soon as possible because once the *Tripsacum* chromosome becomes altered by interchanges or fragmentation, its alteration makes correct cross-mapping with corn impossible. Practically, however, the best route of extraction is screening through the first and second backcrosses to Mangelsdorf's multi-chromosome tester before concentrating on the appropriate multiple markers for an individual chromosome. When a given *Tripsacum* chromosome is extracted more than once, and its cytogenetic characteristics are repeated, the data on comparative mapping with corn are regarded as valid.

In the reciprocal allotriploid hybrid (TTZ) composed of two genomes of *Tripsacum* and one genome of corn, there is a similar but even more extreme exclusion of the alien (corn) chromosomes from the functional megaspores. De Wet et al. (1972) found that triploid hybrids from crossing corn by the tetraploid (2n=72) of *T. dactyloides* from Illinois had, in their first backcross to corn, exclusively 36 chromosomed megaspores, except for a few percent of unreduced gametes. By the fifth backcross to corn, they found that the megaspores carried a low percentage of extra chromosomes presumed to represent either those of corn or products of corn-*Tripsacum* interchanges.

Identification of the Addition Disomic

Once a *Tripsacum* chromosome has become isolated on a corn background and genetically identified, the addition bivalent (20+2) may sometimes be established, tending to remain unaltered by avoiding interchanges with corn. In the bivalent condition, its morphological features are suitable for cytological identification by comparison with the idiogram of the original *Tripsacum* genome prepared by Chandravadana et al. (1971).

The addition bivalent may be established in either of two ways depending on the behavioral characteristics of the particular chromosome. The most stable bivalent condition is generally derived through self-pollination, in which a 10+1 microspore fertilizes a 10+1 megaspore producing the 20+2 condition in the sporophyte. Although the same result may be achieved through nondisjunction of the extra chromosome during either mega or microsporogenesis, such bivalent products are generally unstable since they are as readily lost by the process of nondisjunction as they were produced by it.

Interchanges and Transfer of *Tripsacum* Segments

The cross-mapping between corn and *Tripsacum* chromosomes (Galinat, 1974b) and studies of meiosis of corn hyperploid for various known *Tripsacum* chromosomes indicate which of them may be utilized in corn breeding as permanent alien addition disomics and which of them are so unstable and transitory as intact additions, that they can only be transferred to corn as segmental substitutions introgressed within the corn chromosomes.

The meiotic behavior of Tr7 from *Tripsacum* illustrates the type that functions best in corn as an alien addition disomic (Rao and Galinat, 1974). As an addition monosomic, it has a low transmission only about 10%, with irregular meiotic behavior, and some associated sterility. Of the six corn chromosome 4 loci tested, it carries only the *Su* locus in common and pairs at random with individual arms of various other corn chromosomes. But once Tr7 attains the disomic condition, usually on selfing, its behavior becomes regulated and coordinated with that of corn and thereby adds a new eleventh bivalent to the corn complement. In the progeny resulting from selfing corn plants with Tr7 as an eleventh bivalent, there were almost no individuals with more or less than 20+2 chromosomes, such as might result from non-disjunction of the Tr7 chromosome pair during either microsporogenesis or megasporogenesis. Thus in these families, the Tr7 has established itself as a stabilized viable intergenomic addition disomic with 2n=22 chromosomes like that synthesized by Sears (1953) in *Triticum* with extra pairs of *Haynaldia* chromosomes.

The cytogenetic behavior of Tr9 that has a series of linked loci in common with the short arm of corn chromosome 2, and of Tr5 with an assemblage of loci in common with both arms of corn chromosome 9, are better predisposed for use in segmental substitutions with their corn homeologs. They undergo viable interchanges with their corn counterparts and the derived recombinant chromosomes may be readily propagated with higher fertility than their original unaltered hyperploid state with the corn genome. Tr9-M2 interchanges between the *Lg* and *Gl-2* loci have been observed by Maguire (1960) at a greatly reduced frequency of less than 1% and at a somewhat higher rate of about 5% by Rao and Galinat (1970), as compared to 19% for the same region in pure corn. In both of these independent studies, the *lg Gl2* corn-*Tripsacum* interchange lost its terminal knob from *Tripsacum* while the *Lg-gl2* recombinant chromosome gained the terminal knob showing that the sequence of terminal knob, *Lg, gl2,* in Tr9 places the *Lg* to *gl2* loci in the same order as in corn M2, although the shorter length of Tr9 (about 70% of the short arm of M2) and reduced crossing over in this region in *Tripsacum* suggest that these loci may be located closer together in *Tripsacum* and apparently the pairing between them is loose.

The assemblage of loci in *Tripsacum* chromosome Tr5 is sufficiently similar to that of corn M9, as reflected by their known possession of at least eight loci in common *(Yg2 C Sh Bz Wx Gl-15 Bk2 Bm4),* as to be able to function as a homozygous substitution (18M + 2Tr) in the corn complement. The plants are largely both male and female sterile, however, due to a high frequency of asynapsis. Interchange chromosomes involving the transfer of the terminal knob of Tr5 to M9 are known in our lines and these have a much higher degree of fertility than the intact Tr5 so that the agronomic significance of at least portions of the *Tripsacum* Tr5 chromosome may be evaluated. This may also prove to be the case with the *Tripsacum* chromosomes such as Trl3 which carries the *Gl3* locus in common with corn chromosome 4, but like Tr7 it appears to have little else in common with this linkage group as indicated by the 5 loci tested on this chromosome.

In general, crossing over between homeologous chromosomes of corn and *Tripsacum* that have been tested so far is about one-tenth the frequency of that in corn alone. Such crossovers have been genetically identified with corn chromosomes 2, 7, and 9.

Some Evolutionary Considerations from the Cytogenetic Mapping of Corn and *Tripsacum*

As mentioned previously, the chromosome 4 complex of genes which is essential to the definitive separation of corn and teosinte could not be derived from *Tripsacum* introgression because a corresponding segment does not occur in *Tripsacum*. This has been demonstrated conclusively by the known repatterning of the fourth linkage group loci in corn on several different chromosomes in *Tripsacum*. The *Su* locus is on Tr7 and the *Gl3* locus is on Tr13 while four other M4 loci are located on other as yet unidentified *Tripsacum* chromosomes (Galinat, 1971a; Rao and Galinat, 1974). Thus it appears unlikely that the germplasm of teosinte chromosome 4 came from *Tripsacum* introgression into corn as according to an earlier hypothesis on the hybrid origin of teosinte proposed by Mangelsdorf and Reeves (1939). This point has recently been conceded in a footnote by Mangelsdorf (1974, 49) primarily on a basis of electron microscope studies by Banerjee and Barghoorn of the pollen grain exine of corn, teosinte, and *Tripsacum*, as well as that of *Tripsacum* derivatives of corn, experimentally developed by Galinat.

The observation of Maguire (1961) that Tr9 germplasm has a dominant effect in reducing the number of kernel rows has special significance. When we altered the background of our chromosome 2 tester gene stock to the lower condensation level of eight-rowed (four-ranked) corn, the reduction in row number was to a two-ranked condition (sometimes three or four-ranked at the base of the ear where condensation is slightly higher). Through a study of the effects of various types of Tr9-M2 interchange chromosomes, it should be possible to pinpoint the location of the gene that produces two-ranks at low condensation levels. Dr. Chandravadana (unpub.) has observed, in the heterozygous interchange substitution for knob-*Lg*, *gl2*, that the knobbed member of the pair extended some distance beyond the end of its knobless corn partner at pacytene in all cells examined from one plant and the two-rank phenotype was still manifest. But this heteromorphy does not appear to be the case in the heterozygous corn-Nobogame teosinte chromosome-2 stocks studied by Kato (unpub.) Furthermore the Tr9-M2 interchange chromosomes studied by Maguire (1957) do not appear to have a longer member associated with the terminal knob from *Tripsacum* and we have been unable to repeat our results. If it should be demonstrated that the two-rank gene is subterminal on the short arm of teosinte chromosome 2, then a deficiency for this terminal segment would allow expression of a recessive factor for a many-ranked spike as in corn and, thereby, indicate that the direction of the evolution was teosinte to corn, rather than corn to teosinte. Investigation of the problem is complicated by two or more other two-ranked types including its phenocopy resulting from a depauperate condition and by a low condensation and/or low vascularization of the rachis. These equivocal types of expression for two-ranked corn are usually recessive and difficult to study with repeatable results although they appear to have an inherited basis of variable low penetrance. They may well act as stabilizing factors for the dominant Tr9-M2 factor but be inherited, independently, possibilities which can and will be tested.

Ancient Amphidiploid Origin of Diploid *Tripsacum*

As a result of their data on the frequency of genetically markable vs. unmarkable extra chromosomes from *Tripsacum* segregating in the second backcross to MMT corn of an MMT corn × *Tripsacum* hybrid, a new hypothesis as to the origin of *Tripsacum* was suggested (Galinat et al., 1964). In these studies, Chaganti discovered by pachytene analysis that *Tripsacum* had only one homeologue for each of the seven genetically marked corn chromosomes. Some of the plants had up to nine extra chromosomes from *Tripsacum* that were unmarkable in terms of the seven recessively marked corn chromosomes of the MMT parent. Of the nine unmarkable chromosomes in one plant, at least six and perhaps all nine were alien to the MMT recessives. If we speculate that *Tripsacum* has two distinct genomes consisting of nine markable chromosomes in terms of the corn loci and nine unmarkable ones, then we may compare the theoretical and observed results on this basis. Because the classification for *j* was found unreliable in this segregation, only six out of the possible nine markable chromosomes were considered. On this basis in the 163 plants studied, the theoretical ratio of marked and unmarked *Tripsacum* chromosomes becomes 54:109. The observed ratio of 63:100 was considered as a good fit.

Based on this information, together with the fact that Maguire (1961) had found several linked loci on the short arm of corn chromosome 2 in common with a segment of a *Tripsacum* chromosome, as well as the fact that *Tripsacum* is intermediate between corn and *Manisuris* in a number of characteristics, we suggested that the apparent unmarkable genome in *Tripsacum* was from *Manisuris* and that which was markable represented a genome from corn. Furthermore this hypothesis had some precursors that came close to the same conclusion. Anderson (1944) suggested that the basic 18 pairs of chromosomes of the functionally diploid species of *Tripsacum* may have arisen through allopolyploidy in which distantly related genomes designated as XX and YY came from the genus *Manisuris* with nine pairs of chromosomes. He suggested further that the 36 pairs of chromosomes in the so-called tetraploid species of *Tripsacum* originated as allotetraploids between XXYY and XXZZ genomes of functionally diploid species. There is no indication in Anderson's paper that the "Z" symbol was intended to represent a modified genome of *Zea,* as it was in the hypothesis of Galinat et al. (1964). Maguire (1961) also approached the suggestion that *Tripsacum* might contain a genome of corn chromosomes as follows:

> . . . if similarities of the sort demonstrated here are widespread, then the chromosomes of *Tripsacum* must be basically more like those of corn than a comparison of their idiograms would suggest. There is evidence to support the view that the 36-chromosome diploid *Tripsacum* actually arose as an allotetraploid (Randolph, 1955). In this case the basic genome of *Tripsacum* would be 9 chromosomes compared to 10 of corn.

Tantravahi (1968, 1971) has demonstrated that the tetraploid (2n=72) species of *Tripsacum* regularly form at least some quadrivalents and show some segregation for the morphological characteristics which distinguish their putative parents and may thus be considered as segmental allotetraploids. This concept of segmental divergence within the tetraploid

Tripsacum implies some exchanges occurred between their homeologous genomes before the event of doubling.

More recently Galinat (1970, 1974b) has modified his previous hypothesis on the origin of *Tripsacum*. The *Zea* parent would be more teosinte-like than corn-like according to the assumption that teosinte was the ancestor and not the progeny of corn. As should have been anticipated in our earlier hypothesis of *Zea* and *Manisuris* genomes, the obvious differences in the Idiograms of these genera in comparison to that of *Tripsacum* are a result of extensive repatterning both within and between the apparent genomes involved in the origin of *Tripsacum*. This repatterning now tends to obscure the ancient linkage groups except for certain common gene combinations such as the long arm of *Tripsacum* chromosome 9 compared to the short arm of corn chromosome 2. While the comparative mapping of corn and *Tripsacum* is little more than started (Galinat, 1974b), some important evolutionary facts have come to light such as the absence of a *Zea* chromosome 4 complex in *Tripsacum,* as previously discussed in terms of the historic but now defunct hypothesis that teosinte originated as a corn-*Tripsacum* hybrid. It is also clear that an ancient allopolyploid origin of the genus *Tripsacum* is inconsistent with the conclusion of Weatherwax (1935) that corn, teosinte, and *Tripsacum* have diverged along parallel pathways from a remote common ancestor.

THE CONTRIBUTION OF CORN'S RELATIVES TO CORN IMPROVEMENT

The richest source of variability for corn improvement lies potentially in some of the recombinants from hybridization with its wild relatives. The gene combinations in corn's wild relatives are highly selected for survival under the adversities of a natural environment and their genetic variation is not restricted by the artificial inbreeding of corn that is selected and nurtured by man. The freedom in both cross pollination and seed dissemination in a natural environment, allows the wild relatives to accumulate a wider and different pool of genetic variability than that of their cultigens.

Teosinte has a high potential usefulness for corn improvement because of its relative freedom of gene flow into corn. Once a desirable agronomic trait, such as parasite resistance, is screened from teosinte, it may be transferred to corn with comparative ease. In contrast, the more alien cytogenetic nature of *Tripsacum* germplasm makes its incorporation into corn extremely complex. There is, however, a varying level of difficulty depending upon the *Tripsacum* chromosome, some of which are virtually impossible to transfer to corn while others are more readily incorporated, at least as interchange chromosomes (Galinat, 1974b).

Teosinte Introgression

By increasing the level of genetic variability and its associated heterosis in corn, introgression from teosinte has increased corn's productivity and adaptability to cope with new environmental stresses (Lambert and Leng, 1965; Sehgal, 1963).

The hybridizations responsible for teosinte introgression into corn have probably occurred for several thousand years in regions of Mexico and Guatemala where these wild and cultivated species are sympatric. Despite these repeated hybridizations which allow some introgression between them, the process of disruptive selection maintains the two parental types. This was clearly described by Weatherwax (1954, 181) as follows.

> With the one species preserved by its biological fitness and the other by man — and hybrids between them finding no friend in either nature or man — the two species continue to live in the same area and remain more or less distinct. Some permanent interchange of germplasm occurred between them, of course, but teosinte continued to be predominantly teosinte and maize to be maize.

Introgressed individuals of either corn or teosinte that contain some of the essential traits of the female spike of the other species, described earlier, are normally eliminated by disruptive selection between man and nature and this process is accelerated by their linkage in four or five blocks (Beadle, 1972; Collins and Kempton, 1920; Mangelsdorf and Reeves, 1939). Such an introgression process is known to occur between at least seven other sympatric pairs of wild and cultivated plant species, as reviewed by Heiser (1973). In the case of corn and teosinte, there are also polygenic differences between the species involving high levels of induration in teosinte and high levels of condensation in corn, both of which tend to flow more readily within certain limits into the other species than do their more essential taxonomic traits.

The induration effects of teosinte introgression are apparent in 22 of the 25 described races of corn in Mexico (Wellhausen et al., 1952). Corn carrying such effects has been transported by man to distant areas, such as the U.S. Corn Belt, more than 1,000 miles away from teosinte and yet this dispersed introgression continues to contribute significantly to the heterosis of U.S. hybrid corn (Johnston, 1966; Sehgal and Brown 1965). The induration of the lower glume and rachis may carry with it effects on the inclination and length of the rachilla as well as elongation of the cupule all tending in the direction of teosinte and duplicating the effects of known teosinte chromosomes (Sehgal, 1963). In primitive corn, such induration effects that might be descendant from teosinte are presumed by Beadle (1972 to have been removed by man's selection of a weak tunicate allele that was later discarded as the larger types of modern cobs evolved. In such cobs, indurated and condensed cupules provide the structural support necessary to prevent a precocious shrinkage of the cob that would result in a loss of grain by shattering prior to a deliberate shelling (Galinat, 1970).

Although the utilization of teosinte introgression in corn improvement has been underway on a modest scale by a number of seed companies and research institutions for some time (i.e., Pioneer Hi-bred Int.; Funk Seeds Int.; Crow's Hybrid Corn Co.; Cornell Univ.; Neb. Agric. Exp. Stn.; Mass. Agric. Exp. Stn.; Texas Agric. Exp. Stn.; Univ. of Hawaii at Manoa) a direct screening of teosinte for nonheterotic traits has been limited.

There is some suggestive evidence that Jutiapa teosinte is a source of broad spectrum resistance. Resistance to the Hawaiian strains of both leaf rust *(Puccinia sorghi)* and Maize Mosaic Virus I has been discovered recently in Jutiapa teosinte from Guatemala while the Balsas and Chalco teosintes from Mexico tested were very susceptible (J.L. Brewbaker, personal communication). If the resistance happens to be dominant, then its transfer to a corn inbred by repeated backcrossing is direct and it would only need to be incorporated into one parent of an F1 hybrid. But if, on the other hand, the desired trait is recessive, the breeding process becomes complex. It is then necessary to pair the backcrossing with selfed progenies in order to identify the segregant plants carrying the recessive trait and finally it must be fixed in both divergent parents of a good hybrid.

A factor on teosinte chromosome 9 for elongation of the uppermost internodes in the ear shows promise for use in thick-cob types of Corn Belt corn. This factor stimulates internode elongation late in cob development so that instead of becoming flat or fasciated, there is a vertical accommodation for the developing distal cupules, spikelets, and kernels. The final result of combining a large vascular system, initiated early in ear development, with increased internode elongation, coming in later, has produced giant ears of corn when grown under either low plant densities or at high altitudes where there is high light intensity. For intensive agriculture, as in underdeveloped countries, double crop plantings are economical when the late giant-eared corn is spaced out by alternating pairs of rows with a short term crop (Galinat, 1972a, 1973d) (field corn with early soybeans; sweet corn with early lettuce or snapbeans, etc.) The late corn would take over the empty space after the early crop is harvested and neither crop would compete with the other for solar radiation. The advantages include less water stress in late summer, such as recently suffered under our current system of high density plantings, and less nitrogen demands early in the season.

Tripsacum Introgression

While introgression from *Tripsacum* was suggested in the origin of two South American races of corn, Chococeno (Roberts et al., 1957) and Enano (Grobman et al., 1961), natural hybrids of corn and *Tripsacum* have never been discovered and until such hybrids are found, the evidence for such introgression will remain circumstantial. The basis for *Tripsacum* introgression in Chococeno is its primitive plant characters that adapt this race to slash agriculture without cultivation. The basis in Enano is a tendency for induration of the lower glume as occurs in *Tripsacum*. Evidence of *Tripsacum* introgression in corn based on pollen dated at 2000 to 1600 B.C. from Peru has been claimed by Banerjee and Barghoorn (1973).

If there are cytological differences because of *Tripsacum* introgression in the South American races of corn, it could only involve small segments of germplasm which would be observed as loose pairing at pachytene in hybrids with non-Tripsacoid corn. Apparently gross rearrangements do not characterize any particular race of corn.

It has also been implied by Brown and Anderson (1947) that the northern flints of United States may carry South American *Tripsacum* germ plasm. Like *T. australe* of South America, the northern flints have few or no chromosome knobs. Like the "Tripsacoid" but high knob number corn of Guatemala (Mangelsdorf and Cameron, 1942), they have wide kernels borne on cylindrical, few-rowed, low condensation, strong cobs that terminate a stout shank. In addition to these traits, the northern flints tend to have some induration in the lower glume and cupule as occurs in more extreme form in *Tripsacum*. Their ancestry including the postulated *Tripsacum* introgression could trace back through Cabuya and Clavo of Colombia (Roberts et al., 1957) to Confite Morocho of Peru (and indirectly to the various "interlocked" races from South America). Thus the low condensation of the northern flints, which is the lowest of all United States corn (Anderson and Brown, 1948) may stem from even lower condensation in these South American races of corn.

The extremely low condensation in Confite Morocho could have been derived from *Tripsacum* introgression into a Pollo-like corn. This possibility is consistent with the fact that the oldest cobs from Ayacucho, Peru, the site location of Confite Morocho, in the collections of MacNeish et al. (unpub.) were condensed and Pollo-like while the more elongate Confite Morocho types came in later on (Galinat, 1972b). This observation is in contradiction to the suggestions of Grobman et al. (1967) and Mangelsdorf (1974) that Confite Morocho may be derived from a separate wild corn. Mangelsdorf (op. cit. 118) now considers Pollo to be "part of the Chapalote - Nal Tel complex" that has been ancestral to many Mexican races and now through Pollo, to many South American races as well. The idea presented here that the elongate internodes of Confite Morocho may come from *Tripsacum* introgression is supported by our observation that experimental introgression from *Tripsacum* into corn will elongate the internodes of the rachis and reduce kernel row number independently of any effects upon induration (Galinat, unpub.).

The long internodes in the rachis of Cacahuacintle in Mexico are thought to trace to Salpor of Guatamela (Wellhausen et al., 1952) and that, in turn to Sabanero and its relatives, Pollo of Colombia and Confite Morocho of Peru, as described by Roberts et al., (1957).

At least *Tripsacum* does appear to have a larger pool of genetic variability in its nine or more species than does the genus *Zea* and some of it can be experimentally tapped for corn improvement. It seems that the potential usefulness of introgression from *Tripsacum* is severely limited however, by the low degree of homology which certain alien chromosomes have with their counterparts in corn (Galinat, 1974b). There are other opinions; Engel et al. (1973) concluded that the degree of intergeneric pairing of chromosomes in backcross offspring derived from a corn-*Tripsacum* hybrid suggest that an extensive amount of gene transfer is feasible from *Tripsacum* into corn.

The first success in improving a corn inbred with experimental introgression from *Tripsacum* was reported by Reeves and Bockholt (1967). Since then a strongly fasciated inbred out of Havel's Dent has been improved in terms of producing a longer, more slender ear of better heterotic combining ability by work in progress at the Massachussetts Agricultural Experiment Station. This H.D. inbred is one parent of the giant-eared Waltham Dent that carries teosinte chromosome 9 in the other parent (Galinat, unpub.).

Resistance to corn rootworm *(Diabrotica virgifera)* larvae has been discovered in *Tripsacum dactyloides* (Branson, 1971); resistance to corn leaf aphis *(Rhopalosiphum maidis)* in *T. floridanum* (Branson, 1972) and resistance to the northern leaf blight *(Helminthosporium turcicum)* in derivatives from a corn × *T. floridanum* hybrid supplied by Galinat to A.L. Hooker has, after many years of breeding, finally been transferred to dent corn (Hooker, unpub.) The problem with screening for resistance directly on *Tripsacum,* unlike that of teosinte, is that once the desired trait is discovered, the chances of transferring it to agronomically useful corn are remote.

SUMMARY

Of the various hypotheses on the origin of corn, essentially only two alternatives now remain as viable options: 1a) Present-day teosinte is the wild ancestor of corn; 1b) A primitive teosinte is the common wild ancestor of both corn and Mexican teosinte; 2) An extinct form of pod corn was the ancestor of corn with teosinte being a mutant form of this pod corn.

The second proposal represents a major revision in the tripartite hypothesis advanced by Mangelsdorf and Reeves in 1939. The other two parts of this classic hypothesis are as follows: 1) Teosinte descended from a corn-*Tripsacum* hybrid; 2) The evolution of modern corn was advanced by introgression from its relatives.

For the past three or four decades, the parts of this hypothesis have served to stimulate studies of the cytological, morphological, and archaeological evidence on corn's origin and evolution. But the old problem of which came first, corn or teosinte, has been clearly reopened by Mangelsdorf's (1974) recent abandonment of that part of the hypothesis suggesting a hybrid origin for teosinte. The evidence that finally persuaded Mangelsdorf, according to his book, is the electron-microscopic studies of pollen from corn, teosinte, *Tripsacum,* and corn-*Tripsacum* hybrid derivatives by Barghoorn, Banerjee, and Galinat (unpubl.) Other earlier evidence on the significance of cupules in the oldest corn cobs and the lack of the essential teosinte-like chromosome 4 complex in *Tripsacum* must have also had their effect (Galinat, 1970, 1971a). Mangelsdorf (op cit.) still continues to maintain that the ancestor of cultivated corn is an extinct wild pod corn. Teosinte, instead of being a product of *Tripsacum* introgression into corn, is now considered by Mangelsdorf (p. 52) as "essentially a mutant form of corn." This new hypothesis on the mutant origin of teosinte seems to ignore two important points: 1) Teosinte is intermediate between corn and its more primitive ancestors, such as *Manisuris,* in terms of condensation within the rachis (Galinat, 1974d); and 2) the oldest Tehuacán cobs have nonfunctional cupules remnant from the cupulate fruit case of teosinte (Galinat, 1970).

The origin problem is not resolved by arguing about the probable direction of simple reversible types of mutations such as single vs. paired female spikelets. Rather greater significance lies with polygenic traits in which the genes have been assembled by slow and constant selective pressures. An identification of the selective forces that assembled the polygenic complexes is indicative of whether they were natural or man-directed.

In the case of modern corn vs. teosinte, Mangelsdorf and his associates have shown polygenic differences in induration of the glumes and rachis. But the induration effects of this polygenic system may be reversed by a single gene mutation to a higher tunicate allele such as that purported to be in a chimeric section of a teosinte plant collected in Guerrero, Mexico by Randolph[3]. Such a tunicate mutation, perhaps originating through gene duplication in teosinte, favors a teosinte origin for corn because in one step it would elongate the rachilla and soften the glumes as in the oldest Tehuacán cobs (Beadle[4]). The present-day lower tunicate allele of most corn may have been returned from teosinte or resulted from a reverse mutation. The tunicate locus is known to be complex and it may be either dissected or elaborated by crossing over between its heterozygous components (Mangelsdorf and Galinat, 1964).

It seems reasonable that the high level of compaction in the oldest Tehuacán cobs, relative to that of teosinte, was polygenic in origin just as the still higher level of compaction in certain modern races of corn is polygenic in relation to that of the more primitive races. The polygenes would have been assembled during a transition from the gathering phase, starting outside the Tehuacán Valley, as indicated by Smith's data (1967), in open-site, predomestic cradles (Galinat, 1974c, d). Such cradles would stem from seed gathered in the wild and accidentally lost at the campsites. The gathering process would not only increase the level of compaction in the rachis but other domestically useful changes would tend to accumulate in these cradles. High levels of heterozygosity within the various cradles would allow the stabilization of domestic forms once man began to select and deliberately plant the more useful types. An alternative to the concept that the six basic lineages of corn stem from six races of wild corn, as postulated by Mangelsdorf (1974), is that they reflect independent domestications from geographically isolated cradles.

The difference between natural and domestic specializations has been brought into the controversy. Since the many-ranked spike is unknown in any of the wild relatives of corn and since it is also difficult to imagine what natural adaptive value a many-ranked spike would have over its two-ranked ancestor, it is logical to consider the many-ranked trait as a specialization selected by man, in contradiction to the opinions of Weatherwax (1954) and Mangelsdorf (1974). It is also argued by these researchers that the teosinte traits of solitary female spikelets and high induration are evolutionary specializations and, therefore, that corn must be ancestral to teosinte. It is equally as logical to consider that the pairing of female spikelets, like the proliferation to many rows and many husk leaves as well as the softer glumes and emergent kernels (elongate rachillas) are all domestic specializations because they would all make a teosinte type spike more useful as a food plant for man. Furthermore an evolutionary sequence for increasing degrees of condensation within the rachis is not only culminated by the ear of corn, but increasing levels of condensation may be manifest within a given corn race as more evolved forms of it develop, such as early forms of Nal Tel-Chapalote compared to evolved Chapalote.

The well known segregation of the essential traits that demark corn and teosinte into four or five blocks of linked genes facilitates a trickle of gene flow between these sympatric species without compromising their taxonomic integrity. If this block inheritance evolved under disruptive selection between teosinte and corn, then the cytogenetic structure of the more variable wild species would have a greater potential for various mechanisms such as linkage to knobs and/or small inversions that tend to restrict crossing over in the heterozygote. Small inversions that are fully viable and that happen to be linked to or include genes for essential taxonomic traits would serve as possible locations for gene cluster formation of the type giving this block inheritance. On this basis, teosinte with its greater cytological variability including features unknown in corn would probably be ancestral to corn. The necessary experimental work of attempting to correlate the cytological features peculiar to teosinte with its definitive traits has yet to be undertaken. Sympatric pairs of corn and teosinte from different geographical areas would probably acquire slightly different cytogenetic systems to cope with possible genetic swamping of their essential gene combinations.

That the evolution of modern corn was advanced by introgression from its relatives, the third part of the tripartite hypothesis, is accepted for teosinte and is in question or at least based on inference in the case of *Tripsacum*. Apparently the introgression of germplasm from corn's relatives as well as gene transfer between races of corn has always been important in corn's evolution. This flow of genetic variability involves hybrids between parents at three levels of relatedness: interracial, interspecific, and intergeneric.

Interracial hybridization as a source of corn improvement has been established by evidence from archaeology, history, cytology, and genetics. The origin of the Corn Belt dent, the most productive race, as a hybrid of the northern flints and the southern dents has been well documented (Anderson and Brown, 1952), while its complete genealogy extending back thousands of years directly involved at least twelve races (Grobman et al., 1961). The ancestries of Jala, the largest-eared race (Wellhausen et al, 1952) and Cuzo Gigante, the largest-kerneled race, are equally as complex.

Part of the success of interracial hybridizations is attributed to the introgression of genetic variability from previous interspecific hybridizations with teosinte. This repeated contact with a storehouse of variation from teosinte left corn in a genetically flexible condition. As a result, the isolations and selections made by man often became genetically distinctive and adaptive to his selective whims. The exchange of germplasm between sympatric corn and teosinte has been repeatedly observed as evidenced by hybrids and similar plant types except for the female spikes. Field observations of such corn-teosinte crossings are known from some of the earliest literature and archaeological evidence of teosinte introgression in corn cobs extends back thousands of years (Mangelsdorf, 1974).

The southern dent component of the Corn Belt dent carries with it various amounts of teosinte germplasm that can be matched with the experimental introgression of teosinte into corn (Johnston, 1966; Sehgal and Brown, 1965). The influence of teosinte germplasm on the corn cob causes the structures that are homologous to the teosinte fruit case to become modified in that direction, that is the rachilla of the spikelet is short and upward inclined, internode elongated, and the lower glume and cupule indurated.

The potential of teosinte as a source of variation for corn breeding purposes is great. If a certain trait such as disease resistance is recognized in teosinte, it may be screened out during backcrossing to corn. Variability for increasing the combining ability of a given inbred of corn may be identified in the second backcross to the inbred. In contrast, only the first backcross to an inbred may be necessary to start the screening of exotic corn races for their potential contribution to the heterosis of Corn Belt corn (Mangelsdorf, 1974). Even just the individual four or five blocks of linked genes that are essential to demark corn from teosinte (Mangelsdorf and Reeves, 1939; and others) can significantly increase heterosis and yield by several percent in isogenic crosses (Sehgal, 1963). Therefore selection for the cupulate fruit case tending traits in the derivatives of corn-teosinte hybrids can contribute to combining ability for overdominance when hybridized with inbreds lacking in these traits (Johnson, 1966; and others). The homozygous effects of certain essential segments of teosinte may be deleterious to yield when in a corn background. This type of segment would not respond favorably to the convergent improvement system of developing solemates for hybrid corn (Mangelsdorf, 1974).

The northern flint parent of the Corn Belt dent may carry with it, South American *Tripsacum* germplasm, as we inferred here from an analysis of the relevant literature and from the writer's own work with derivatives from experimental hybrids between corn and *Tripsacum*. The total heterosis from interracial heterozygosity, interspecific heterozygosity (corn and teosinte), and intergeneric heterozygosity (corn and *Tripsacum*) may tend to be accumulated in the best Corn Belt dent hybrids. By the expeditious addition of teosinte germplasm to southern dent type inbreds and *Tripsacum* germplasm in northern flint type inbreds, it may be possible to further increase the level of this heterosis and its associated yield in Corn Belt corn. The variability to insure continued progress in breeding including the resistance to prevent epidemics is available in a vast storehouse held in corn's wild relatives and in the primitive varieties some of which are preserved in cold storage banks.

The process of transferring *Tripsacum* germplasm into corn is so technically difficult, the transmission rate of single extra *Tripsacum* chromosomes added to the corn genome is so low, the rate of corn-*Tripsacum* crossing over so reduced, as to practically exclude the general use of experimentally introduced *Tripsacum* germplasm in corn improvement. By close cooperation with cytologists, a few plant breeders may be able to manipulate this foreign germplasm, possibly with spectacular results for corn improvement.

The question remains as to where lie the more distant roots of the American Maydeae. The morphological evidence based on formation of the cupulate fruit case would seem to point in the direction of Africa. In contrast the more conservative evidence from chromosome structure including the formation of chromosome knobs would seem to indicate *Coix* in the direction of the islands in the Pacific Ocean. The monoecious trait in common with the Oriental Maydeae and the several reports of *Coix*-corn hybrids made but unsuccessfully reared add increments of evidence indicating an Oriental connection.

THE ORIGIN OF CORN

ACKNOWLEDGMENTS

Paper from the Mass. Agric. Esp. Stn., University of Massachusetts at Amherst. This research supported in part from Experiment Station Hatch Projects Nos. 258R and 352-NE66 as well as by a grant (GB-15767 #1) from the National Science Foundation and by the Bussey Institution of Harvard University. Grateful acknowledgment is made for frequent discussions with Dr. R.V. Tantravahi, Dr. W.D. McEnroe, Dr. P. Chandravadana, and Dr. T.A. Kato.

LITERATURE CITED

Anand, S.C., and E.R. Leng. 1963. Corn × *Tripsacum* hybrids. Maize Genet. Coop. Newsl. 37:41.

Anderson, E. 1944. Cytological observations on *Tripsacum dactyloides*. Ann. Mo. Bot. Gard. 31:317-323.

----. 1949. The corn plant of today. Pioneer Hi-bred Corn Co. Des Moines, Iowa.

----, and W.L. Brown. 1948. A morphological analysis of row number in maize. Ann. Mo. Bot. Gard. 35:323-336.

----, and ----. 1952. Origin of Corn Belt maize and its genetic significance. *In* J.W. Gowen (ed.) Heterosis. Iowa State College Press, Ames, Iowa.

Arber, A. 1934. The Gramineae. Cambridge Univ. Press., Cambridge, England.

Arnason, T.J. 1936. Cytogenetics of hybrids between *Zea mays* and *Euchlaena mexicana*. Genetics 21:40-60.

Bada, J.L., R.A. Schroeder, and G.F. Carter. 1974. New evidence for the antiquity of man in North America deduced from aspartic acid racemization. Science 184:791-793.

Banerjee, U.C., and E.S. Barghoorn. 1972. Fine structure of pollen grain ektexine of maize, teosinte and *Tripsacum*. *In* C.J. Arceneaux (ed.) 30th Annu. Proc. Electron Microscopy Soc. Am. Los Angeles, Calif.

----, and ----. 1973. Palynological evidence for natural introgression between *Tripsacum* and *Zea* (abstr.). Am. J. Bot 60:(4)(Suppl.):34.

Barghoorn, E.S., M.K. Wolfe, and K.H. Clisby. 1954. Fossil maize pollen from the Valley of Mexico. Bot. Mus. Leafl. Harv. Univ. 16:229-240.

Bartlett, A.S., E.S. Barghoorn, and R. Berger. 1969. Fossil maize from Panama. Science 165:389-390.

Beadle, G.W. 1932. The relation of crossing over to chromosome association in *Zea-Euchlaena* hybrids. Genetics 17:481-501.

----. 1939. Teosinte and the origin of maize. J. Hered. 30:245-257.

----. 1972. The mystery of maize. Field Mus. Nat. Hist. Bull. 43:1-11.

Bellini, G., and A. Bianchi. 1963. Interchromosomal effects of inversions on crossover rate in maize. Z. Verebungs. 94:126-132.

Bor, N.L. 1960. Grasses of Burma, Ceylon, India and Pakistan. Pergamon Press, New York, N.Y.

Branson, T.F. 1971. Resistance in the grass tribe Maydeae to larvae of the Western Corn Rootworm. Ann. Entomol. Soc. Am. 64:861-863.

----. 1972. Resistance to the corn leaf aphid in the grass tribe Maydeae. J. Econ. Entomol. 65:195-196.

Brown, W.L. 1949. Numbers and distribution of chromosome knobs in U.S. maize. Genetics 34:524-536.

----, and E. Anderson. 1947. The northern flint corns. Ann. Mo. Bot. Gard. 34:1-28.

Carlson, H.L. 1965. Chromosomal morphism in geographically widespread species of *Drosophila*. p. 503-531. *In* H.G. Baker and G.L. Stebbins Jr. (eds.) Genetics of colonizing species. Academic Press, New York, N.Y.

Celarier, R.P. 1957a. Cytotaxonomy of the Andropogoneae II. Subtribe *Ischaeminae, Rottboelliinae,* and the *Maydeae.* Cytologia 22:160-183.

----. 1957b. *Elyonuras argenteus,* a South African grass with five chromosome pairs. Bull. Torrey Bot. Club 84:157-162.

Chaganti, R.S.K. 1965. Cytogenetic studies of maize-*Tripsacum* hybrids and their derivatives. Bussey Inst. Harv. Univ. 93 p.

Chandravadana, P., W.C. Galinat, and B.G.S. Rao. 1971. A cytological study of *Tripsacum dactyloides.* J. Hered. 62:280-284.

Collins, G.N. 1919. Structure of the maize ear as indicated in *Zea-Euchlaena* hybrids. J. Agric. Res. 17:127-135.

----, and J.H. Kempton. 1920. A teosinte-maize hybrid. J. Agric. Res. 19:1-37.

Cooper, D.C., and R.A. Brink. 1937. Chromosome homology in races of maize from different geographical regions. Am. Nat. 71:582-587.

Cutler, H.C. 1947. A comparative study of *Tripsacum australe* and its relatives. Lloydia 10:229-234.

----, and E. Anderson. 1941. A preliminary survey of the genus *Tripsacum.* Ann. Mo. Bot. Gard. 28:249-269.

Daniel, L. 1973. The synthesis of two-rowed maize ears. Acta Agron. Acad. Sci. Hung. 22:13-18.

Darwin, C. 1958. (orig. 1859) The origin of species. Mentor Books, New York.

de Wet, J.M.J., L.M. Engle, C.A. Grant, and S.T. Tanaka. 1972. Cytology of maize-*Tripsacum* introgression. Am. J. Bot. 59:1026-1029.

----, and R.J. Lambert. 1972. Origin of maize: The tripartite hypothesis. Euphytica 20:255-265.

----, ----, J.R. Harlan, and S.M. Naik. 1970. Stable triploid hybrids among *Zea-Tripsacum-Zea* backcross populations. Caryologia 23:183-187.

Duvick, D.N. 1972. Potential usefulness of new cytoplasmic male steriles and sterility systems. Proc. 27th Annu. Corn Sorghum Conf. A.S.T.A. Pub. 27:192-201.

Emerson, R.A., and G.W. Beadle. 1932. Studies of *Euchlaena* and its hybrids *Zea.* II. Crossing over between the chromosomes of *Euchlaena* and those of *Zea.* Z. Abstram. Verebungsl. 62:305-315.

----, and H.H. Smith. 1950. Inheritance of number of kernel rows in maize. Cornell Univ. Agric. Exp. Stn. Mem. 296:1-30.

Engle, L.M., J.M.J. de Wet, and J.R. Harlan. 1973. Cytology of backcross offspring derived from a maize-*Tripsacum* hybrid. Crop Sci. 13:690-694.

Epling, C., D.F. Mitchell, and R.H.T. Mattoni. 1953. On the role of inversions in wild populations of *Drosophila pseudoobscura.* Evolution 7:342-365.

Farquharson, L.I. 1957. Hybridization of *Tripsacum* and *Zea.* J. Hered. 48:295-299.

Flannery, K.V. 1967. Vertebrate fauna and hunting patterns. p. 132-177. *In* D.S. Byers (ed.) The prehistory of the Tehuacan Valley (Vol. 1) Univ. of Texas Press, Austin, Texas.

Galinat, David W., and W.C. Galinat. 1972. A possible role of condensation in a domestication of teosinte. Maize Genet. Coop. Newsl. 46:109-110.

Galinat, W.C. 1956. Evolution leading to the formation of the cupulate fruit case in the American Maydeae. Bot. Mus. Leafl. Harv. Univ. 17:217-239.

----. 1961. *Tripsacum floridanum* crosses readily with corn. Maize Genet. Coop. Newsl. 35:38-39.

----. 1963. Form and function of plant structures in the American Maydeae and their significance for breeding. Econ. Bot. 17:51-59.

----. 1969. The evolution under domestication of the maize ear: String cob maize. Mass. Agric. Exp. Stn. Bull. 577:1-19.

----. 1970. The cupule and its role in the origin and evolution of maize. Mass. Agric. Exp. Stn. Bull. 585:1-20.

----. 1971a. The origin of maize. *In* H. Roman (ed.) Annu. Rev. Genet. 5:447-478, Annual Reviews, Inc., Palo Alto, Calif.

----. 1971b. The evolution of sweet corn. Mass. Agric. Exp. Stn. Bull. 591:1-20.

----. 1972a. Some contributions of corn's relatives to the development of its modern varieties. Proc. 27th Annu. Corn Sorghum Conf. A.S.T.A., Washington, D.C.

----. 1972b. Common ancestry of the primitive races of maize indigenous to the Ayacucho area in Peru. Maise Genet. Coop Newsl. 46:107-108.

----. 1973a. Preserve Guatemalan teosinte, a relict link in corn's evolution. Science 180:323.

----. 1973b. Pollen size and the origin of maize. Maize Genet. Coop. Newsl. 47:105-108.

----. 1973c. Two systems that transform a two-ranked spike into a four-ranked spike. Maize Genet. Coop. Newsl. 47:102-103.

----. 1973d. A formula for giant ears in maize. Maize Genet. Coop. Newsl. 47:104-105.

----. 1974a. A congruous background for the *tr* and *pd* genes. Maize Genet. Coop. Newsl. 48:93.

----. 1974b. Intergenomic mapping of maize, teosinte, and *Tripsacum*. Evolution 27:644-655.

----. 1974c. The domestication and genetic erosion of maize. Econ. Bot. 28:31-37.

----, and R.S.K. Chaganti, and F.D. Hager. 1964. *Tripsacum* as a possible amphidiploid of wild maize and *Manisuris*. Bot. Mus. Leafl. Harv. Univ. 20:289-316.

----, and J.H. Gunnerson. 1963. Spread of eight-rowed maize from the prehistoric Southwest. Bot. Mus. Leafl. Harv. Univ. 20:117-160.

----. P.C. Mangelsdorf, and L. Pierson. 1956. Estimates of teosinte introgression in archaeological maize. Bot. Mus. Leafl. Harv. Univ. 17:101-124.

----, and R.J. Ruppé. 1961. Further archaeological evidence on the effects of teosinte introgression in the evolution of modern maize. Bot. Mus. Leafl. Harv. Univ. 19:163-181.

Grobman, A., W.W. Salhuana, and R. Sevilla with P.C. Mangelsdorf. 1961. Races of maize in Peru. N.A.S.N.R.C. Publ. 915.

Hawkes, J.G. 1970. The origins of agriculture. Econ. Bot. 24:131-133.

Heiser, C.B., Jr. 1973. Introgression re-examined. Bot. Rev. 39:347-366.

Helbaek, H. 1959. Domestication of food plants in the Old World. Science 130:365-372.

Hernandez-X.,E. 1973. Genetic resources of primitive varieties of Mesoamerica p. 76-115. *In* O.H. Frankel (ed.) Survey of crop genetic resources in their centers of diversity. AGP:CGR/73/7 FAO-IBP. United Nations, New York.

Hitchcock, A.S. 1922. A perennial species of teosinte. J. Wash. Acad. Sci. 12:205-208.

----. 1951. Manual of the grasses of the United States. USDA Misc. Pub. No. 200. Washington, D.C.

Iltis, H.H. 1972. The taxonomy of *Zea mays* (Graminae). Phytologia 23:248-249.

Jain, S.K., and D.K. Banerjee. 1974. Preliminary observations on the ethnobotany of the genus *Coix*. Econ. Bot. 28:38-42.

Johnston, G.S. 1966. Manifestations of teosinte and *"Tripsacum"* introgression in Corn Belt maize. Bussey Inst., Harv. Univ. 73 pp.

Kaplan, L. 1967. Archaeological *Phaseolus* from Tehuacan. p. 201-211. *In* D.S. Byers (ed.) The prenistory of the Tehuacan Valley. Univ. of Texas Press, Austin, Texas.

Kato, Y.T.A. 1976. Cytological studies of maize and teosinte in relation to their origin and evolution. Mass. Agric. Exp. Stn. Bull. 635. 172 p.

Kempton, J.H., and W. Popenoe. 1937. Teosinte in Guatemala. Contr. Am. Arch. 23:200-217.

Kikudome, G.Y. 1959. Studies on the phenomenon of preferential segregation in maize. Genetics 44:815-831.

Kurtz, E.B., Jr., J.L. Liverman, and H. Tucker. 1960. Some problems concerning fossil and modern corn pollen. Bull. Torrey Bot. Club 87:85-94.

Lathrap, D.W. 1973. Gifts of the Cayman: Some thoughts on the subsistence basis of Chavin. Essay VIII. p. 91-105. *In* D.W. Lathrap and J. Douglas (eds.) Variation in anthropology. Illinois Archaeological Survey. Urbana, Ill.

----. 1974. The moist tropics, the arid lands, and the appearance of great art styles in the New World. Spec. Publ. Mus. Texas Tech. Univ. 7:115-158.

Lambert, R.J., and E.R. Leng. 1965. Backcross response of two mature plant traits for certain corn-teosinte hybrids. Crop Sci. 5:239-241.

Lewis, H. 1966. Speciation in flowering plants. Science 152:167-172.

Longley, A.E. 1937. Morphological characters of teosinte chromosomes. J. Agric. Res. 54:835-862.

----. 1938. Chromosomes of maize from North American Indians. J. Agric. Res. 56:177-195.

----. 1941. Chromosome morphology in maize and its relatives. Bot. Rev. 7:263-289.

----, and T.A. Kato. 1965. Chromosome morphology of certain races of maize in Latin America. Int. Cent. Impr. Maize and Wheat (CIMMYT), Chapingo, Mexico. 112 p.

Lorenzo, J.L., and Q.L. Gonzales. 1970. The oldest teosinte. Boletin INAH. Museum Nacional, Dep. of prehistory, Moneda 16, Mexico D.F.

Lynch, T.F. 1973. Harvest timing, transhumance and the process of domestication. Am. Anthropol. 75:1254-1259.

MacNeish, R.S. 1964. The origins of New World civilization. Sci. Am. 211:29-37.

----. 1967. A summary of the subsistence. p. 290-309. *In* D.S. Byers (ed.) The prehistory of the Tehuacán Valley. Univ. of Texas Press, Austin, Texas.

----. 1971. Early man in the Andes. Sci. Am. 224(4):36-46.

Maguire, M.P. 1957. Cytogenetic studies of *Zea* hyperploid for a chromosome derived from *Tripsacum*. Genetics 42:473-486.

----. 1961. Divergence in *Tripsacum* and *Zea* chromosomes. Evolution 15:394-400.

----. 1962. Common loci in corn and *Tripsacum*. J. Hered. 53:87-88.

Mangelsdorf, P.C. 1945. The origin and nature of the ear of maize. Bot. Mus. Leafl. Harv. Univ. 12:33-75.

----. 1947. The origin and evolution of maize. p. 161-207. *In* M. Demerec (ed.) Advances in genetics. Academic Press, New York, N.Y.

----. 1974. Corn: its origin, evolution and improvement. Harv. Univ. Press, Cambridge, Mass.

----, and J.W. Cameron. 1942. Western Guatemala; a secondary center of origin of cultivated maize varieties. Bot. Mus. Leafl. Harv. Univ. 10:217-252.

----, and R.G. Reeves. 1939. The origin of Indian corn and its relatives. Texas Agric. Exp. Stn. Bull. 574:1-315.

----, and W.C. Galinat. 1964. The tunicate locus in maize dissected and reconstituted. Proc. Natl. Acad. Sci. USA. 51:147-150.

----, and M. Sanoja O. 1965. Early archaeological maize from Venezuela. Bot. Mus. Leafl. Harv. Univ. 21:105-112.

----, and C.E. Smith, Jr. 1949. New archaeological evidence on evolution in maize. Bot. Mus. Leafl. Harv. Univ. 13:213-247.

----, R.S. MacNeish, and W.C. Galinat. 1967a. Prehistoric maize, teosinte, and *Tripsacum* from Tamaulipas, Mexico. Bot. Mus. Leafl. Harv. Univ. 22:33-63.

----, ----, and ----. 1967b. Prehistoric wild and cultivated maize. p. 178-200. *In* D.S. Byers (ed.) The prehistory of the Tehuacán Valley. Univ. of Texas Press, Austin, Texas.

Martin, P.S. 1973. The discovery of America. Science 179:969-974.

McClintock, B. 1933. The association of non-homologous parts of chromosomes in the midprophase of meiosis in *Zea mays*. Z. Zellforsch. Mikrosk Anat. 19:191-237.

----. 1959. Chromosome constitutions of Mexican and Guatemalan races of maize. Annu. Rep. Dep. Gen. Carnegie Inst. Washington 59:461-472.

Nirodi, N. 1955. Studies on Asiatic relatives of maize. Ann. Mo. Bot. Gard. 42:103-130.

O'Mara, J.G. 1942. A cytogenetic study of *Zea* and *Euchlaena*. Mo. Agric. Stn. Bull. 341:3-16.

Pickersgill, B. 1972. Cultivated plants as evidence for cultural contacts. Am. Antiq. 37:97-104.

Randolph, L.F. 1955. Cytogenetic aspects of the origin and evolutionary history of corn. p. 16-57. *In* G.F. Sprague (ed.) Corn and corn improvement. Academic Press, New York, N.Y.

----. 1959. The origin of maize. Indian J. Genet. Plant Breed. 19:1-12.

----. 1970. Variation among *Tripsacum* populations in Mexico and Guatemala. Brittonia, 22:305-337.

Rao, B.G.S., and W.C. Galinat. 1974. The evolution of the American Maydeae: 1. The characteristics of two *Tripsacum* chromosomes (Tr7 and Tr13) that are partial homeologs to maize chromosome 4. J. Hered. 65:335-340.

Reeves, R.G., and A.J. Bockholt. 1964. Modification and improvement of a maize inbred by crossing it with *Tripsacum*. Crop. Sci. 4:7-10.

Rhoades, M.M. 1952. Preferential segregation in maize. p. 66-80. *In* J.W. Gowen (ed.) Heterosis. Iowa State College Press, Ames, Iowa.

----. 1955. The cytogenetics of maize. p. 123-219. *In* G.F. Sprague (ed.) Corn and corn improvement. Academic Press, New York, N.Y.

----, and E. Dempsey. 1953. Cytogenetic studies of deficient-duplicate chromosomes derived from inversion heterozygotes in maize. Am. J. Bot. 40:405-424.

----, and ----. 1957. Further notes on preferential segregation. Maize Genet. Coop. Newsl. 31:77-80.

Roberts, L.M., U.J. Grant, R. Ramirez E., W.H. Hatheway, D.L. Smith, with P.C. Mangelsdorf. 1957. Races of maize in Colombia. N.A.S. N.R.C. Publ. 510:1-153.

Rogers, J.S. 1950. The inheritance of inflorescence characters in maize-Teosinte hybrids. Genetics 35:541-558.

Sauer, C.O. 1952. Agricultural origins and dispersals. Am. Geographical Society Press, New York, N.Y.

Schoenwetter, J. 1974. Pollen records of Guila Naquitz Cave. Am. Antiq. 39:292-303.

Sears, E.R. 1953. Addition of the genome of *Haymaldia villosa* to *Triticum aestivum*. Am. J. Bot. 40:168-174.

Sehgal, S.M. 1963. Effects of teosinte and *"Tripsacum"* introgression in maize. Bussey Inst., Harvard Univ., Cambridge, Mass. 63 p.

----, and W.L. Brown. 1965. Introgression in Corn Belt maize. Econ. Bot. 19:83-88.

Shaver, D.L. 1960. Linkage in tetraploid hybrids of maize and perennial teosinte. Maize Genet. Coop. Newsl. 34:59.

----. 1962. A study of meiosis in perennial teosinte, in tetraploid maize and in their tetraploid hybrid. Caryologia 15:43-57.

----. 1964. Perennialism in *Zea*. Genetics 50:393-406.

Smith, C.E. Jr., 1967. Plant remains. p. 220-225. *In* D.S. Byers (ed.) The prehistory of the Tehuacán Valley. Univ. of Texas Press, Austin, Texas.

Smith, D.R., A.L. Hooker, S.M. Lim, and J.B. Beckett. 1971. Disease reaction of thirty sources of cytoplasmic male-sterile corn to *Helminthosporium maydis* race T. Crop Sci. 11:772-773.

Stebbins, G.L., Jr. 1956. Cytogenetics and evolution of the grass family. Am. J. Bot. 43:890-905.

----. 1971. Chromosome evolution in higher plants. Addison-Wesley Co. Menlo Park, Calif.

Sturtevant, E.L. 1899. Varieties of corn. USDA Off. Exp. Stn. Bull. 57:1-108.

Takahashi, R. 1955. The origin and evolution of cultivated barley. Adv. Genet. 7:227-266.

Tantravahi, R.V., 1968. Cytology and crossability relationships of *Tripsacum*. Bussey Inst. Harvard Univ., Cambridge, Mass.

----. 1971. Multiple character analysis and chromosome studies in the *Tripsacum lanceolatum* complex. Evolution 25:38-50.

Tatum, L.A. 1971. The southern corn leaf blight epidemic. Science 171:1113-1116.

Ting, Y.C. 1958a. On the origin of abnormal chromosome 10 in maize. Chromosoma 9:286-291.

----. 1958b. Inversions and other characteristics of teosinte chromosomes. Cytologia 23:239-250.

----. 1964. Chromosomes of maize-teosinte hybrids. Bussey Inst., Harvard Univ., Cambridge, Mass.

Weatherwax, P. 1918. The evolution of Maize. Bull. Torrey Bot. Club 45:309-342.

----. 1935. The phylogeny of *Zea mays*. Am. Midl. Nat. 16:1-71.

----. 1954. Indian corn in old America. The Macmillan Co., New York, N.Y.

----. 1955. Early history of corn and theories as to its origin. p. 1-16. *In* G.F. Sprague (ed.) Corn and corn improvement. Academic Press, New York, N.Y.

Wellhausen, E.J., L.M. Roberts, and E. Hernandez X. with P.C. Mangelsdorf. 1952. Races of maize in Mexico. Bussey Inst., Harvard Univ., Cambridge, Mass.

Wilkes, H.G. 1967. Teosinte: The closest relative of maize. Bussey Inst., Harvard Univ., Cambridge, Mass.

----. 1972. Maize and its wild relatives. Science 177:1071-1077.

Wright, S. 1931. Evolution in Mendelian populations. Genetics 16:97-159.

Yarnell, R.A. 1965. Early woodland plant remains and the question of cultivation. Fla. Anthropol. 18:78-81.

Zevallos, M.C. 1971. La agricultura en el formativo temprano del Ecuador. Centro de Investigaciones Historicas, Guayaquil, Ecuador (Oct. 1966).

Chapter 2 Races of Corn

WILLIAM L. BROWN
Pioneer Hi-Bred International, Inc.
Des Moines, Iowa

MAJOR M. GOODMAN
North Carolina State University
Raleigh, North Carolina

It was about 30 years ago that Anderson and Cutler (1942) pointed up the need for a more natural classification of the variability known then to exist among the varieties of corn, *Zea mays* L. In the late 1800's, Sturtevant (1899) separated corn into six main groups, five of which were based upon the composition of the endosperm. Anderson and Cutler recognized the artificial nature of Sturtevant's system and suggested that knowledge from archaeology and genetics, accumulated since Sturtevant's work, should make possible a more natural classification. They indicated that variability in maize was comparable to that found in mankind and proposed the use of a racial classification based to the extent then possible on natural relationships.

The first thorough assessment of the variation present in corn had been attempted more than a decade prior to Anderson and Cutler's proposal by Vavilov's associates (Kuleshov, 1929, 1930). Although those studies were partly completed before the downfall of Vavilov, the published results are not easily accessible. More limited studies by Girola (1919), Parodi (1932, 1935), and Marino (1934) also received limited distribution and little recognition.

Anderson and Cutler defined race "as a group of related individuals with enough characteristics in common to permit their recognition as a group." In genetic terms, a race according to Anderson and Cutler, "is a group of individuals with a significant number of genes in common, major races having a smaller number in common than do sub-races."

In a series of publications, Anderson (1943, 1944a, 1945, 1946, 1947), Anderson and Cutler (1942), Carter and Anderson (1945), Cutler (1946), Brown and Anderson (1947, 1948), and Anderson and Brown (1952) defined further the racial concept, described the morphological characters thought to be most useful in delimiting races, and classified in a preliminary way the corn of Mexico, Central and South America, and parts of the United States. At about the same time the cytological variation of corn and its relationships to varietal and regional diversity were being described by Longley (1938, 1941), Mangelsdorf and Cameron (1942), and Brown (1949).

While much progress has been achieved toward a natural system of classification of maize since that time, relatively little progress has been made in determining which characters are the most definitive for the study of racial differences in corn. Anderson and Cutler (1942) suggested a number of useful tassel, ear, and kernel characters, many of which were used by Wellhausen et al. (1951); more recently Goodman and Paterniani (1969) presented data suggesting that several ear and kernel characters were superior. However, the latter study was based on a narrow sample of germplasm and may not be representative of the species as a whole. Essentially none of the archaeologically important cob characters (Nickerson, 1953, 1954) have been tested to determine the relative importance of the genotype, the environment, their interaction, and sampling error on the expression of the characters.

Wellhausen et al. (1951) published the results of a comprehensive study of an extensive collection of Mexican corn made in order to organize the variation for use in a corn breeding program initiated in Mexico in 1943 and sponsored jointly by the Rockefeller Foundation and the Mexican government. This report, *Razas de Maiz en Mexico,* enlarged upon previous studies and set the stage for a series of race studies which followed over the next 12 years. The procedures followed in each of these studies were similar, although they varied considerably in emphasis and detail. First the corn grown in the region under study was sampled by collectors who usually obtained from 5 to 15 ears from each farm visited. Usually several thousand collections were made per study. These ear collections were then assembled, spread out on laboratory benches, and similar collections were tentatively assigned to the same race. Emphasis was placed upon such characters as ear shape, row number, kernel denting, etc., which were thought to be polygenic, rather than simply inherited ones such as pericarp and aleurone colors or endosperm differences such as flint vs. floury. The collections tentatively assigned to a given race were then grown together and those collections which differed in plant and tassel characters were removed from the set. Usually three to five collections were chosen on the basis of ear and plant characteristics (and sometimes as a result of breeding true to type) as being most typical of each race. Several ears from each of the most typical collections were saved as museum specimens, while the others were shelled and later increased. Descriptions of the races, including photographs and racial averages for many plant, tassel, ear, kernel, and sometimes physiological characters, as well as chromosome morphology, were published for Mexico and Central America (Wellhausen et al., 1952, 1957); Cuba (Hatheway, 1957); the West Indies (Brown, 1960); Venezuela (Grant et al., 1963); Colombia (Roberts et al., 1957); Ecuador (Timothy et al., 1963); Peru (Grobman et al., 1961); Bolivia (Ramírez et al., 1960); Chile (Timothy et al., 1961); and eastern South America (Brieger et al., 1958). This series of "race bulletins" was followed or accompanied by similar studies on European corn races (Sanchez-Monge, 1962; Leng et al., 1962; Edwards and Leng, 1965; Paviličić and Trifunovic, 1966; Brandolini and Mariani, 1968; Brandolini, 1969, 1970b, 1971; Costa-Rodrígues, 1969; Paviličić, 1971; Covor, 1972) and on Asian corn (Stonor and Anderson, 1949; Anderson and Brown, 1953; Suto and Yoshida, 1956; Mochizuki, 1968), and further studies on the races of Chile (Parker and Paratori, 1965); Bolivia (Rodriguez et al., 1968; Brandolini and Avila, 1971); Mexico (Hernandez and Alanis, 1970); and Brazil and adjacent areas (Blumenschein and Deuber, 1968; Paterniani and Goodman, in press).

The series of "race bulletins," which followed *Razas de Maiz en Mexico,* stemmed from the interest and concern of the late Ralph Cleland of Indiana University. Cleland was, at that time, Director of the Division of Biology and Agriculture of the National Academy of Sciences-National Research Council (NAS-NRC). Under his leadership, the Committee on the Preservation of Indigenous Strains of Maize (1954, 1955) was established within NAS-NRC. The committee was charged with the responsibility for collecting, classifying, and preserving the corn germplasm of the Western Hemisphere, a formidable task when viewed in retrospect. The committee, nonetheless, was unusually successful. Over a period of years, it provided stimulus, guidance, and direction, and obtained sufficient financial support to collect, classify, and publish the results of studies of most of the corn of the Western Hemisphere. In addition, and in cooperation with many USA and Latin American agencies, it helped to organize seed storage centers in Mexico, Colombia, and Brazil.

The classification of Western Hemisphere corn as reported in the NAS-NRC bulletins was considered by many of the authors of these reports to be of a preliminary nature. It was looked upon as a starting point and a basis for more definitive studies which it was hoped would follow. So far, this has occurred to a limited extent only. Goodman (1967, 1968, 1972) and Goodman and Bird (in press) have restudied many of the Latin American collections and have attempted, through the use of numerical taxonomy, to determine more precisely racial relationships between the previously described races.

A study of the chromosome knobs of Latin American races was initiated in the late 1950's by McClintock, Kato, and Blumenschein. This work is now nearing completion and should add much to our knowledge of the origin and relationships of the major races of maize of the Western Hemisphere. Preliminary reports of the results of these studies, (McClintock, 1959, 1960; Longley and Kato, 1965; Kato and Blumenschein, 1967; Blumenschein, 1973; Kato, 1975), indicate the presence of well-defined knob patterns associated with specific geographical areas and maize races. More importantly, these studies reveal the migration patterns of certain definitive knobs and knob complexes and apparently provide the best information to date on the movement of races from one geographic area to another.

Mangelsdorf (1974) has attempted to survey the variability found among the various Latin American corn races. Rather than delineating racial groups, complexes, or "super races," he assigned races to lineages, lineage being used in the sense of "descent in a line from a common progenitor." He described six such lineages, each of which he postulates was derived from a wild race of corn. "From north to south, the still-living ancestral races of these lineages are:
1. Palomero Toluqueño, the Mexican pointed-seeded popcorn.
2. The Chapalote-Nal-Tel complex of Mexico.
3. Pira Naranja of Colombia, the progenitor of the tropical flint corns with orange-endosperm color.
4. Confite Morocho of Peru, the progenitor of eight-rowed corns.
5. Chullpi of Peru, the progenitor of all sweet corns and related starchy-seeded forms with globular ears.

6. Kculli, the Peruvian dye corn, the progenitor of all races with complexes of pericarp and aleurone colors."

The only other recent attempt at comprehensive description of the maize races is that of Brandolini (1970a), who lists a number of primary and secondary centers of differentiation and races associated with such centers.

An understanding of the variability of corn is important for several reasons. It should shed light on the history and relationships of the peoples whose lives are closely associated with corn. Corn breeding should benefit from a better understanding of the evolutionary history and genetic variability within the genus. Finally, and perhaps more importantly, increased knowledge of the racial composition of corn should point the direction to the most efficient and effective ways of minimizing the genetic vulnerability of the commercial corn of the Americas. As indicated in a recent study by the Committee on Genetic Vulnerability of Major Crops (1972) of the National Academy of Sciences, corn has undergone a gradual but continual decrease in genetic diversity over the past 50 years. The decrease in genetic diversity has been accompanied by an increase in genetic vulnerability. As the genetic base of corn germplasm used in commerce is diminished, the risk of economic loss of the crop due to diseases, insects, or unusual stress conditions increases correspondingly. The most recent example of the hazards associated with the widespread use of uniform genetic material is that of the southern corn leaf blight epidemic of 1970 (Sprague, 1971). That experience has brought to the fore again the realization of the hazards associated with the erosion of genetic diversity in any widely grown crop species.

Most of the corn germplasm in use in the USA today is derived from mixtures of only two major races (Wallace and Brown, 1956). The simplest means of correcting this situation and of increasing the genetic diversity in this important crop is to introduce unrelated sources of germplasm, most of which are found in the tropics and subtropics. To do this intelligently and efficiently is a formidable task. There is a vast store of corn germplasm outside the USA which differs greatly in its potential usefulness within the USA. Though our knowledge of the races of corn found in the tropics and subtropics is still incomplete, the knowledge which is available should, if used, simplify the task of reducing the genetic vulnerability of our most important feed grain.

In the discussion which follows, no attempt is made to provide detailed descriptions of the races. This information is already available in the various publications cited earlier. Our objective, rather, is to recognize those earlier described races which seem still to be valid and to shed some light on apparent relationships between them. Many of these relationships are most simply indicated in the text figures. Readers interested in detailed descriptions are referred to earlier cited papers.

This report contains no reference to corn found outside the Western Hemisphere. There are many reasons for this, the foremost of which is the limited information available on the variability of corn of Europe, Africa, and Asia. In our opinion, much more information is needed before a complete and orderly classification of the corn of the Old World is possible. Supposedly, corn had its origin in the Western Hemisphere, yet it is apparent that distinct races have evolved in many areas excluded from this report.

Table 1—Races for which relationships are uncertain

Race	Country	Reference
Conejo	Mexico	
Mushito	Mexico	
Complejo Serrano de Jalisco	Mexico	
Zamorano Amarillo	Mexico	Wellhausen et al. (1952)
Maíz Blando de Sonora	Mexico	
Onaveño	Mexico	
Dulcillo del Noroeste	Mexico	
Apachito	Mexico	
Rosita	Mexico	
Gordo	Mexico	Hernandez and Alanis (1970)
Azul	Mexico	
Bofo	Mexico	
Tablilla de Ocho	Mexico	
White Pop	Cuba	
Yellow Pop	Cuba	Hatheway (1957)
White Dent	Cuba	
Nal-Tel Ocho	Guatemala	Wellhausen et al. (1957)
Huesillo	Costa Rica	
Maíz Dulce	Colombia	Roberts et al. (1957)
Maíz Harinoso Dentado	Colombia	
Mishca	Ecuador	
Chillo	Ecuador	
Cónico Dentado	Ecuador	
Uchima	Ecuador	
Gallina	Ecuador	Timothy et al. (1963)
Cholito	Ecuador	
Yunga	Ecuador	
Enano Gigante	Ecuador	
Yungeño	Ecuador	
Rabo de Zorro	Peru	
Chancayano	Peru	
Marañón	Peru	
Chuncho	Peru	
Jora	Peru	
Coruca	Peru	
Sarco	Peru	
Ajaleado	Peru	Grobman et al. (1961)
San Gerónimo	Peru	
Perlilla	Peru	
Tumbesino	Peru	
Colorado	Peru	
Chancayano Amarillo	Peru	
Polulo	Chile	
Chutucuno Grande	Chile	Timothy et al. (1961)
Choclero	Chile	
Morotí Guapí	Paraguay	Paterniani and Goodman (in press)
Cristal Semi-Dentado	Paraguay	
Achilli	Argentina	
Culli	Argentina	
Oke	Argentina	
Bola Blanca	Argentina	Brieger et al. (1958)
Brachytic Popcorn	Argentina	
Avati Moroti Mitá	Paraguay	
Avati Djakaira	Paraguay	
Chavantes White Soft	Brazil	

RACES OF MEXICO, CENTRAL AMERICA, AND THE WEST INDIES

With the exception of the races described by Rodriguez et al. (1968) for which no representative collections were cited, all the named Latin American corn races are included in this survey. For a number of races, the authors lack sufficient knowledge to accurately circumscribe the racial boundaries or to correctly describe their relationships to other races. Such races are listed in Table 1, although such a listing does not preclude mentioning such a race elsewhere.

Mexican Races

Most of the 25 recognized races of corn in Mexico (Wellhausen et al., 1952) can be reasonably assigned to what appear to be well-defined racial groups (Goodman, 1972). Figure 1 gives a general view of the apparent racial relationships. Only a single race, Maíz Dulce, appears not to have any close relatives among the other Mexican races.

The long, narrow-eared corns typical of northwestern Mexico, whether popcorns, flints, or flours, are all similar to the Indian corns of the southwestern USA (Anderson, 1944b, 1946; Brown et al., 1952; Carter and Anderson, 1945). These races include Chapalote, Reventador, Harinoso de

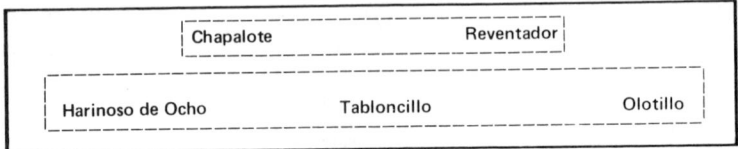

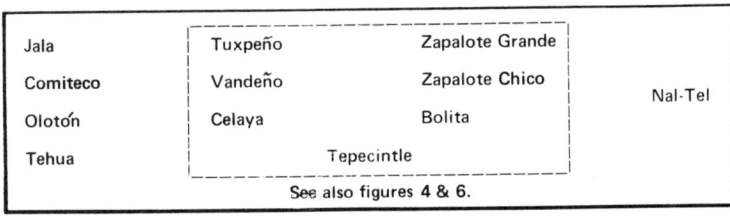

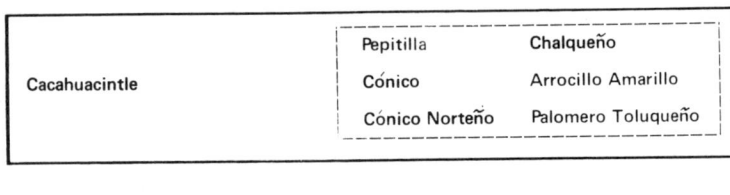

Fig. 1. Racial relationships of the corn of Mexico. Those races within cells outlined by solid lines are thought to be more closely related than races found in different cells. Similarly, races outlined by broken lines are assumed to be more closely related to each other than to races found outside the sub-cells. This method of depicting racial relationships is used in all text figures (Fig. 1 through Fig. 10).

Ocho, Tabloncillo, and Olotillo (the latter from southern rather than northwestern Mexico) as well as several more recently described races (Hernandez and Alanis, 1970). It appears (Nelson, 1952, 1960) that Harinoso de Ocho is one of only two Mexican races which shared the ga_1/ga_1 genotype with the most prevalent USA corn types (Northern Flints, Southern Dents, and Corn Belt Dents).

A second distinctive racial group consists of the conically-eared corns with pointed kernels mostly found at relatively high elevations in central Mexico. Endosperm varies from dented floury to pop, but the plant type — pubescent leaves; sparsely branched, thick tassels; weak roots — is consistent. Cónico, Cónico Norteño, Chalqueño, Arrocillo Amarillo, and Palomero Toluqueño each fall into this category. Pepitilla seems to be related to these races, but it may include sources of germplasm from lower altitude races. Cacahuacintle seems to be related more closely to the Guatemalan race Salpor than to any Mexican race, yet it shares both ear shape and the general aspect of the plant with the Cónico group. It differs most by its very large floury kernels. Kato and Blumenschein (1967) suggested that Cónico, Chalqueño, Palomero Toluqueño, Cacahuacintle, and Arrocillo Amarillo share an essentially knobless karyotype whose influence they have traced to the southern USA. More recently Blumenschein (1973) has suggested that Palomero Toluqueño is distinct from the other members in this group in its chromosome knob morphology. He gave no details, but this view is supported by the report by Nelson (1960) that Palomero Toluqueño, unlike most Mexican corn, possesses the ga_1/ga_1 genotype at the gametophyte locus on chromosome four. Palomero Toluqueño is the ancestral race for one of the six lineages described by Mangelsdorf (1974). He traces its influence as far as the Andean pointed popcorns and their relatives.

The remaining well-defined Mexican races fall into a set of clines or gradients between large-eared races (Jala, Olotón) and small-eared races (Zapalote Chico, Bolita, Nal-Tel). These appear to include the widest assortment of dent races found in a single area. Included are the widely used Tuxpeño, Vandeño, and Celaya dents, as well as races such as Comiteco, Tehua, Tepecintle, and Zapalote Grande, which appear to be closely related to them. Nal-Tel appears to be related to both the small-eared dents and to Chapalote, a popcorn from northwestern Mexico. Within this group, Kato and Blumenschein (1967) found two distinct chromosome knob complexes. The Tuxpeño knob complex of many medium-sized knobs was found along the east coast of Mexico from Yucatán to the southern USA, in central and northeastern Mexico, and along the southwest coast of Mexico. Its influence reached as far as Panama and Cuba and is found in scattered parts of South America. The Zapalote Chico complex of many knobs of various sizes encompasses the races Zapalote Chico, Zapalote Grande, and Bolita, and is found from the southwest coast of Mexico to western Costa Rica. It is suspected (Blumen schein, 1973) that this complex also reached Venezuela.

Chapalote and Nal-Tel, two races which today appear to be distinct in both ear shape and plant type, are listed as ancestral to one of Mangelsdorf's (1974) lineages. That lineage includes, among others, Reventador, the Mexican dents described above, and the Central American Nal-Tels.

Central American Races

The races from Central America have neither been as carefully collected or described as the races of Mexico. Only for Guatemala (see Fig. 2) are the collections or descriptions reasonably thorough. The Guatemalan races, especially those from higher altitudes, resemble South American types of corn as closely as they resemble Mexican types. Very limited tests of R alleles, which condition aleurone color, by Van der Walt and Brink (1969), however, suggest that, at least for those corn races possessing aleurone color, the corn found north of the Andes is distinct from that of the Andes. They found that all tested Andean corn possessed nonparamutable R alleles, while all Central and North American corn tested, as well as most non-Andean South American corn, has paramutable R alleles. A number of lowland Guatemalan collections resemble Mexican Nal-Tel from Yucatán, although they are phenotypically and apparently genotypically (Nelson, 1960; McClintock, 1960) rather variable. Wellhausen et al. (1957) report that the influence of Nal-Tel extends as far south as Panama.

The Guatemalan Imbricado complex shares many features of both the ear and plant with the pointed Mexican popcorn, Palomero Toluqueño, although having larger kernels, less kernel pointing, and much less uniformity than Palomero Toluqueño. In these respects it more closely resembles the Colombian race Imbricado. As mentioned above, the Guatemalan Salpors seem to be closely related to the Mexican race Cacahuacintle. Both show great similarity in ear and plant characteristics to the Mexican Cónicos and their relatives but have distinctive large, floury, white kernels.

Comiteco, a medium to high altitude, long-eared semident and several low altitude dent races are common to Mexico and Guatemala. These include Tepecintle and Tuxpeño, as well as Dzit-Bacal, an eight-rowed, white dent described as a subrace of Olotillo in Mexico. Guatemalan Dzit-Bacal is much faster maturing and shorter than Olotillo, and appears more closely related to the low altitude collections of Nal-Tel than does Olotillo. Kato and Blumenschein (1967) reported that the southern Guatemala chromosome knob complex with numerous large knobs has influenced lowland corn from Panama to northern Mexico along the west coast, in Yucatán, and in Chiapas and Oaxaca. This complex, found especially in the race Nal-Tel, is reported to have reached Cuba and the Greater Antilles from Costa Rica.

Although the long-eared Guatemalan race Olotón is also found to a minor extent in Mexico, it resembles closely a South American racial complex of medium to high altitude, large-kernelled, long-eared flints (the Montañas, Amagaceño), Two other Guatemalan races, Serrano and Quicheño, also resemble South American races. There seems to be essentially continuous variation between Serrano (a small-eared flint, often with an enlarged base) and Olotón, with Quicheño being intermediate in altitudinal distribution. Olotón, and to a lesser extent, Quicheño, have very long ears borne on tall, late maturing plants. This series of clinal variants corresponds to the Pollo-Sabanero-Montáña complex found in northwestern South America. San Marceño seems to be basically a Serrano with eight rows of partially dented kernels, while Nal-Tel de Tierra Alta appears to have some Quicheño ancestry. A complex of small chromosome knobs is reported to be typical of high altitude collections from Guatemala (Kato and Blumenschein, 1967).

RACES OF CORN

The influence of this complex extends to high elevations from southwestern Mexico to Panama. There is also an indication in Nelson's (1960) data that highland Guatemalan corn carries the ga_1 allele rather than the alleles common to the tropical races. Finally, Negro de Chimaltenango, a fairly long, slender, tapering-eared, flinty flour corn with rounded kernels having deep blue aleurone, appears to be similar to Güirua and Negrito from northwestern South America. Wellhausen et al. (1957) suggest that this medium altitude race has introgressed into both high altitude Serrano and several lowland races to form the subraces Negro de Tierra Fría and Negro de Tierra

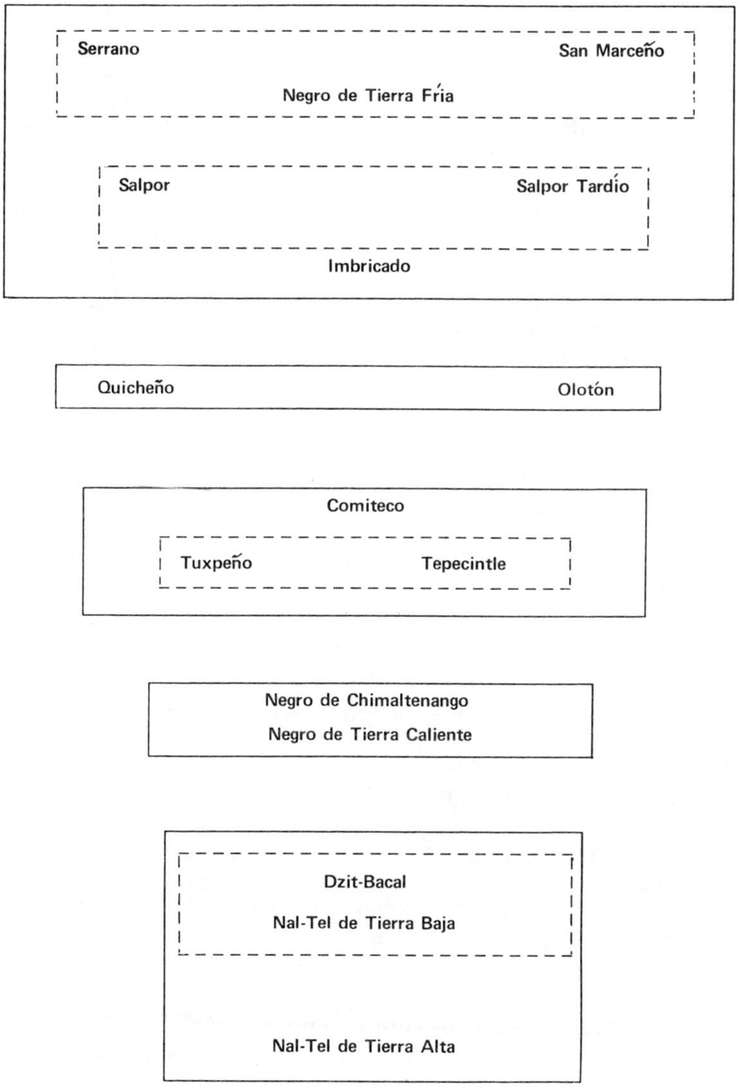

Fig. 2. Racial relationships of the corn of Guatemala.

Caliente, respectively.

The principal commercial corn from El Salvador, Honduras, Nicaragua, and Costa Rica is a low altitude, early maturing, flinty dent with tapering ears called Salvadoreño. In Costa Rica and Panama, a slender-eared white flint, called Clavillo by Wellhausen et al. (1957), resembles the Colombian race Clavo. Huesillo, an 8 to 10-rowed, white-capped flint resembling the Tabloncillos of Mexico, is also found in Costa Rica. Two Colombian races, Chococeño (Nickerson and Covich, 1966) and Cariaco, are found in southern Panama.

West Indian Races

The classification of West Indian corn into clearly defined races is difficult for several reasons. The area is lacking in natural barriers; because of the close proximity of the islands, one to the other, there has long been free migration of peoples between the islands. Likewise, the geographic position of certain of the islands relative to the South American mainland has encouraged the movement of peoples from the mainland to the islands since the days of the Arawak. The free movement of peoples in the area has apparently been accompanied by transport of corn varieties. As a result, there has been much mongrelization of varieties and a masking of racial differences. Despite these difficulties, it has been possible to identify several reasonably well-defined races in the West Indies. The validity of certain of these races has been confirmed through the subsequent discovery of similar counterparts in other areas of Latin America.

The first study of West Indian corn was that of Hernandez (1949), who described six provisional races from collections made primarily in Cuba. This was followed by a preliminary study by Brown (1953) of collections from 11 Caribbean islands. Brown recognized eight provisional races. Hatheway (1957) reported on a detailed analysis of Cuban corn and described seven races existing in Cuba at the time of his study. Brown (1960) did a more detailed study of West Indian maize in which he attempted to correlate his earlier findings with those of other authors. This work included the progenies of 135 collections which were grown and studied in Trinidad. He concluded that West Indian corn consisted essentially of seven distinct races plus a number of intermediate types which presumably were the products of interracial hybridization. The races recognized by Brown were Cuban Flint, Haitian Yellow, Coastal Tropical Flint, Chandelle, Early Caribbean, Tusón, and St. Croix.

This classification seems as valid today as when it was made in 1960. To our knowledge, no new, indigenous races have been reported from the West Indies since that date. On the contrary, many of the varieties formerly used as commercial corn have recently been replaced by modern hybrids. In Jamaica, for example, we estimate that 90% or more of the corn now grown consists of hybrid varieties.

The two most prominent races occupying the West Indies are Coastal Tropical Flint and Tusón. Both are widely distributed throughout the islands from Cuba to Trinidad. The two are also probably closely related. Brown (1953, 1960) postulated that Tusón arose as a hybrid between Coastal Tropical Flint and some unidentified dent maize. Even though related, the two races are distinct. Coastal Tropical Flint is more flint-like than Tusón, has a lower number of kernel rows, and has shorter and somewhat narrower kernels. The ears of Coastal Tropical Flint are distinctly tapered, whereas the ears of Tusón are cylindrical. The name Tusón, meaning "large cob," seems particularly appropriate for this race, since it has distinctly larger cobs than found in any other West Indian race. Tusón is slightly earlier in maturity than Coastal Tropical Flint and has somewhat fewer primary and secondary tassel branches. Both races are tall, without tillers, and have ears placed high on the culm.

Haitian Yellow seems to be more closely related to Coastal Tropical Flint than to other West Indian races. Its ears are characteristically pyramidal with conspicuously enlarged bases and with 8 to 14 rows of kernels. The pointed kernels vary from flint to slightly dented. Plants are tall and late maturing; they usually exhibit purple anthocyanin coloration. The plants possess more extensively developed husk leaf blades, "flag leaves," than do other West Indian races. The distribution of Haitian Yellow, apparently limited to Haiti, suggests relatively recent origin. Its similarity to Coastal Tropical Flint indicates it may have originated from crosses of that race with other Haitian corn.

The four remaining races found in the West Indies, Cuban Flint, Chandelle, Early Caribbean, and St. Croix, are well defined and appear not to be closely related to each other or to the three races referred to previously.

Cuban Flint is the only true flint found in the West Indies. Its distribution is limited to Cuba and the evidence is fairly clear that it was introduced from Argentina in the early 1900's (Hatheway, 1957). The ears of Cuban Flint are relatively short (17 cm), are slightly compressed at the base, and are gently tapered from the base to the tip. Row numbers vary from 12 to 16, with 14 being the most frequently found number. Kernels are short and deep orange yellow in color. Tassel branches are also short in comparison with those of other West Indian races. The plants are usually without tillers, are frequently two-eared, and the leaf sheaths are usually strongly pubescent in contrast with the glabrous leaves of other races.

Although the similarity between Cuban Flint and the Cateto and Cateto Sulino flints of eastern South America [described by Brieger et al. (1958) and Paterniani and Goodman (in press)] is obvious, Hatheway (1957) has suggested that Cuban Flint may be the result of accidental crossing of Argentine flint with local Cuban varieties. He feels that Argentine flints are poorly adapted to Cuba and that Cuban Flint is more similar to other Cuban varieties in plant type than to Argentine flints. Some introgression has undoubtedly occurred between introduced Argentine flint and Cuban varieties, yet the relationship between Cuban Flint and Cateto (and Camelia of Chile as well) appears to be still quite strong. Blumenschein (1973) suggested, on the basis of chromosome knob mapping, that the Cateto flints of eastern South America originated in the Antilles as a result of crosses between representatives of the Tuxpeño knob complex and the two Guatemalan knob complexes.

Chandelle is found in Cuba, Haiti, the Dominican Republic, Trinidad, and the Caicos Islands. It is a dominant race in the Dominican Republic. Two forms of Chandelle occur in Cuba and the Dominican Republic — a flint and a dent. In Haiti, Trinidad, and the Caicos Islands, it occurs only as a flint. Because of the nature of the ear, Chandelle is never confused with other races. The ears are long, slender, strongly compressed at the base, and gently tapered to the tip. Kernel rows range from 10 to 16, and the kernels are long in relation to the diameter of the cob resulting in a high shelling percentage. The cob is slender and flexible even when mature and dry, a trait which occurs infrequently in corn and in none of the other West Indian races. Chandelle is of late maturity, and the plants are taller than those of other West Indian races; however, some early maturing tropical lines have been extracted from the race. Considerable red anthocyanin color is usually present in the leaf sheaths.

Because of its early maturity, short stature, and distinctive internode pattern, Early Caribbean is unique among West Indian corns. When grown in the lowland tropics, it flowers 46 to 48 days after planting, which is 10 days to 2 weeks earlier than other West Indian races. It is sensitive to changes in photoperiod, however, requiring about 95 days to flower in Raleigh, North Carolina. Early Caribbean has few condensed internodes above the ear, resulting in an internode pattern more typical of early USA varieties than of those from the tropics. The ears are relatively short (18 cm) and slightly tapered with 10 to 12 rows of relatively wide, semi-flint kernels. Pericarp color is frequently red or reddish, although colorless pericarp is not unusual. The husks are quite loose compared with other races of the area. Insofar as is known, the distribution of Early Caribbean is limited to Martinique and St. Kitts. Its characteristics are such as to suggest relationship to some of the early flint corns of North America or Europe. Introgression with introduced early flints from North America or Europe could have occurred, yet in the absence of any supporting evidence for such we are inclined instead to consider Early Caribbean a relic of an old, indigenous race.

St. Croix apparently arose from introgression between local varieties and introductions from southeastern USA. Since many of the characteristics of St. Croix are similar to those of the old white endosperm variety, Hickory King, and because of the political connection between the USA and certain of the Virgin Islands, it seems likely that Hickory King is involved in the ancestry of St. Croix. Ears of St. Croix are the longest found in the West Indies; they are cylindrical to slightly tapered and consist of 10 to 14 rows of wide, well dented, flat-topped kernels. Endosperm is light yellow in color, suggesting again the possibility of some white endosperm parentage. The plants of St. Croix also suggest relationships to temperate zone corn. Ear placement is low compared to other races, and the husks are relatively short and loose. The tassels have relatively few branches which are arranged more or less at right angles to the primary axis. Except for Early Caribbean, St. Croix is of earlier maturity than other West Indian races. To our knowledge, the race is limited in its distribution to the island of St. Croix. Brown (1960) suggested that St. Croix might have arisen from the introgression of local varieties with Hickory King of the USA or Olotillo of Mexico. While one cannot completely rule out the possibility of relationship to Olotillo, the phenotype of St. Croix is that which one would more reasonably expect to result from introgression with temperate varieties.

RACES OF CORN

Four of the seven races found in the West Indies have similar counterparts in South America (Fig. 3). These are Cuban Flint, Coastal Tropical Flint, Chandelle, and Tusón. Their relationships are discussed further in the next section. The three remaining races, Haitian Yellow, Early Caribbean, and St. Croix, have not been collected from other areas and are apparently limited in their distribution to the Caribbean islands.

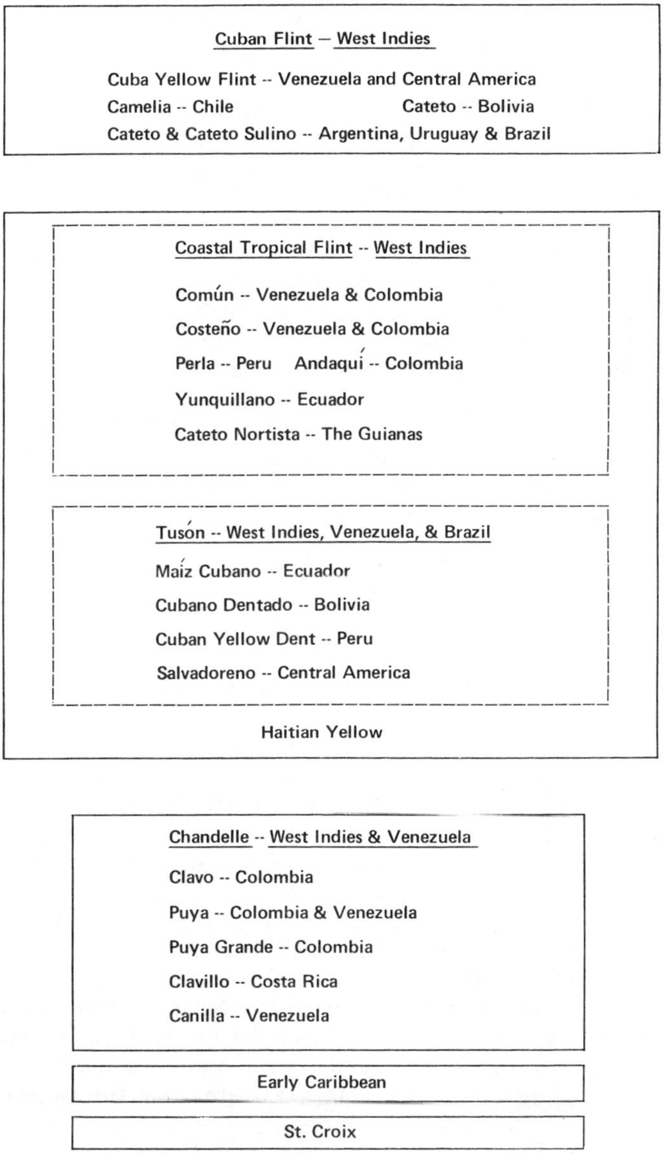

Cuban Flint — West Indies
Cuba Yellow Flint -- Venezuela and Central America
Camelia -- Chile   Cateto -- Bolivia
Cateto & Cateto Sulino -- Argentina, Uruguay & Brazil

Coastal Tropical Flint -- West Indies
Común -- Venezuela & Colombia
Costeño -- Venezuela & Colombia
Perla -- Peru   Andaquí -- Colombia
Yunquillano -- Ecuador
Cateto Nortista -- The Guianas

Tusón -- West Indies, Venezuela, & Brazil
Maíz Cubano -- Ecuador
Cubano Dentado -- Bolivia
Cuban Yellow Dent -- Peru
Salvadoreno -- Central America

Haitian Yellow

Chandelle -- West Indies & Venezuela
Clavo -- Colombia
Puya -- Colombia & Venezuela
Puya Grande -- Colombia
Clavillo -- Costa Rica
Canilla -- Venezuela

Early Caribbean

St. Croix

Fig. 3. Relationships of the races of corn of the West Indies and their Central and South American counterparts. (See also Fig. 4).

Kato and Blumenschein (1967) report that the corn of the West Indies appears to be influenced by three sources of chromosome knob variation. They suggest that the Tuxpeño complex of medium-sized knobs from the eastern coast of Mexico and the southern Guatemalan, mostly large-knobbed complex were introduced to the Greater Antilles. These introductions combined to form a secondary knob complex which they labelled Northwest Caribbean. This new, secondary complex spread southward through the Lesser Antilles to South America, then spread along the coast to Uruguay and northeastern Argentina, and inland to Paraguay and southwestern Brazil. The third chromosome knob complex found in the West Indies is Kato and Blumenschein's Venezuelan complex. This complex of large knobs apparently spread through the Lesser Antilles and along the coast of South America to Panama, Ecuador, Brazil, and Uruguay.

RACES OF SOUTH AMERICA

Much of the corn variation in South America can also be discussed on a regional basis. There are, of course, races which do not fit regional patterns closely, but several racial complexes, each largely confined to one of the four regions below, appear to encompass the realm of corn variation in South America. Detailed results of one set of numerical analyses of most of these races have been presented elsewhere (Goodman and Bird, in press). Other analyses are currently in preparation (Bird and Goodman, unpublished data). The summary presented here is largely based upon the results of these analyses tempered somewhat by field experience with the materials. We have also attempted to consider the sometimes extensive reviews of the histories and geographical distributions of the races which are included in the various "race bulletins." Nevertheless, we would like to emphasize that the racial groupings presented here are, at best, working hypotheses which need much further study.

Lowland Northern South America

Many of the flint and semi-flint races of northern South America were discussed in the section on West Indian corn, and their relationships are presented diagrammatically in Fig. 3. Cuban Yellow Flint of Venezuela (Grant et al., 1963), the Catetos and Cateto Sulinos of eastern South America (Brieger et al., 1958; Paterniani and Goodman, in press), and Camelia of Chile (Timothy et al., 1961) appear to be mainland counterparts of Cuban Flint. The currently available typical collections of Cuban Flint from Venezuela are phenotypically rather variable and appear to include some Tusón and Chandelle influence.

Coastal Tropical Flint (Fig. 3) occurs under the names of Perla in Peru (Grobman et al., 1961), Cateto Nortista in the Guianas (Paterniani and Goodman, in press), and Costeño and Común, both in Venezuela and Colombia (Grant et al.,1963; Roberts et al., 1957). However, the range of variation described for the latter two races seems to encompass both Coastal Tropical Flint and Tusón. The Costeño collections from Colombia are quite variable in ear shape, while the Común collections from that country have a much larger kernel size than those from Venezuela. This would appear to be the result of a higher percentage of high altitude germplasm in Colombian Común. The Colombian race Andaqui' and the very similar Ecuadorian race Yunquillano, (Roberts et al., 1957; Timothy et al., 1963), also appear to belong to this racial group. Andaqui' is a small-eared, white, semi-flint with many chromosome knobs. It is found at low elevations in the southern interior of Colombia.

Tusón is widespread in South America (see Fig. 3). It has been described from Ecuador as Maíz Cubano (Timothy et al., 1963), from Venezuela and Brazil as Tusón (Grant et al., 1963; Paterniani and Goodman, in press), from Bolivia as Cubano Dentado (Ramirez et al., 1960), and from Peru as Cuban Yellow Dent (Grobman et al., 1961). Salvadoreño from Central America (Wellhausen et al., 1957) and many of the collections of Tuxpeño from Venezuela and Puya Grande from Venezuela could also be classified as Tusóns. Because of its productivity, Tusón has undoubtedly been widely dispersed as a commercial corn in recent times. Whether the initial migration of Tusón was from the mainland to the islands or vice versa is still unknown. Yet, the antiquity of corn in South America suggests that Tusón probably reached the West Indies from the mainland.

Races very similar to Chandelle are found in Venezuela (Grant et al., 1963), Colombia (Roberts et al., 1957), and Central America (Wellhausen et al., 1957) — see Fig. 3. The collections from Venezuela have been given the names Canilla, Puya, and Chandelle, whereas in Colombia they are known as Clavo and Puya. Clavo seems to be very close to Chandelle, whereas Puya is probably less closely related. Venezuelan Canilla has somewhat smaller, more flinty kernels than Chandelle. Clavillo from Costa Rica (Wellhausen et al., 1957) is apparently also a member of the same racial complex. Puya Grande from Colombia appears also to be fairly closely related to Chandelle, whereas the Puya Grande collections from Venezuela seem more closely related to the Tusóns, described above.

There is also a group of dents and semidents (Tuxpeño of Ecuador and Venezuela; Yucatán from Colombia; Alazán and Arizona from Peru; Puya Grande from Venezuela) which presumably trace back to Mexican dents of the Tuxpeño-Vandeño type (see Fig. 4). Several Peruvian (Alemán, Chuncho, Jora) and Ecuadorian (Uchima, Gallina) races have apparently arisen as a result of admixture between Caribbean flint or dent races and lower altitude Andean races. The source of denting in all of these races presumably traces to Mexico.

The narrow-eared characteristics of Puya (see discussion above) and Chandelle appear to trace to the Canilla-Clavo-Tusilla-Clavito group of narrow-eared flints (Fig. 4). The latter appear to be related to Rienda, Chimlos, and Pagaladroga (and perhaps Rabo de Zorro) of Peru and Güirua of Colombia. Güirua has elongate ears similar to the Chandelles but with blue aleurone; it is an Indian corn collected from non-Spanish speaking Indians.

The relationships, if any, between the narrow-eared flints and flours discussed above and the phenotypically similar popcorns, such as Pira (Fig. 4) from Venezuela and Colombia (a white, cigar-shaped popcorn with a very thin, often almost disarticulating, cob) and Pira Naranja (a yellow, cigar-shaped popcorn from Colombia, which is quite late maturing and very tall), are unclear. Mangelsdorf (1974) suggests that Pira Naranja was ancestral to a lineage leading to the Cateto flints of eastern South America. Since Pira Naranja is not only extremely sensitive to photoperiod (Stevenson and Goodman, 1972) but also is quite late maturing even under short day conditions, it, unlike Mangelsdorf's other postulated ancestral races, would appear to be quite resistant to successful adaptation to new environments, particularly those with longer days and shorter growing seasons. Both of the latter are characteristic of the environments wherein the Catetos are found.

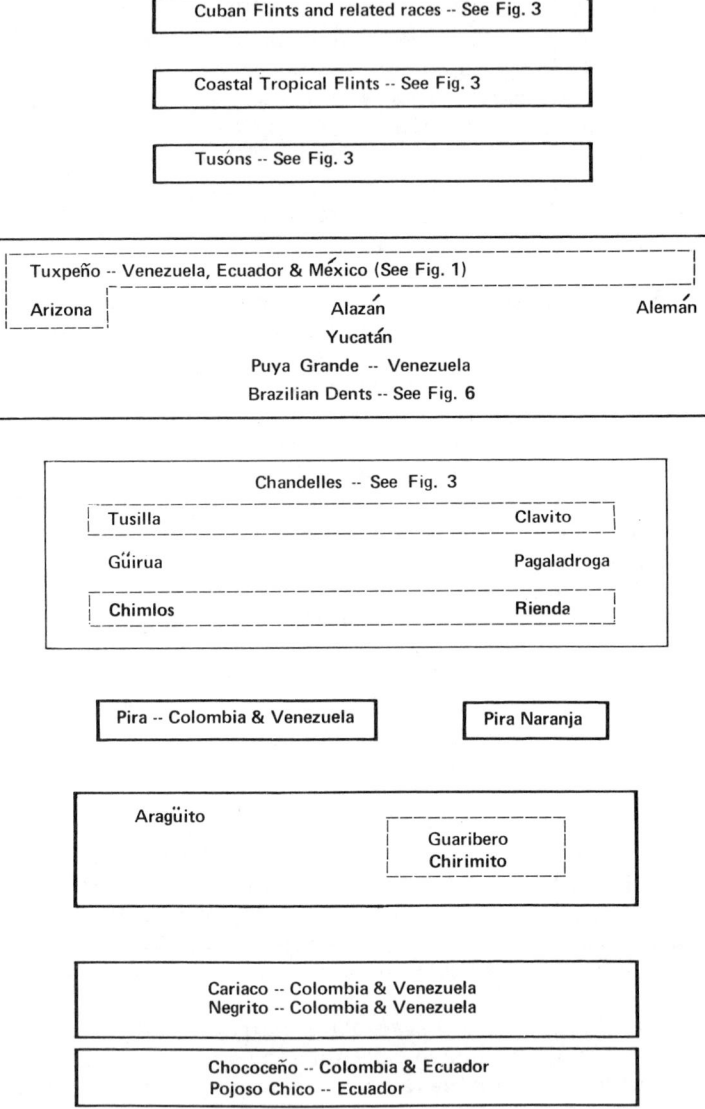

Fig. 4. Racial relationships of the corn of lowland northern South America.

Whether the Chirimito, Aragüito, and Guaribero small flints or popcorns (Fig. 4) from Venezuela are related to larger-eared races such as Pira (see above) is not yet known. The former (especially Araguito) are quite short and very early, but lack the brittleness and upright leaf and tassel branch habit of the somewhat similar collections of Enano (see next section and Fig. 5) from Bolivia and Peru. In plant type they more closely resemble two early maturing floury races with short, broad ears, Cariaco and Negrito of Venezuela and Colombia (Fig. 4). The latter especially appears to share its ear shape with the shorter-eared Caribbean tropical flints (discussed above) and both appear to share their kernel coloration with Amazonian races described in the next section. Negrito is often quite flinty.

The Chococeños (Fig. 4) of western Colombia and Ecuador (Patino, 1956) share the same general ear shape with Enano, Chirimito, Guaribero, and Aragüito (described just above), but the plants are tall, heavily tillered, and very late. While there are floury collections of Chococeño, it typically is a popcorn reportedly grown by sowing rather than row planting (Roberts et al., 1957).

The Amazon Basin and Surrounding Lowlands

Throughout the interior lowland area east of the Central Andes, a single racial complex predominates. First described by Cutler (1946) under the name Coroico (Fig. 5), it has also been called Piricinco and Pojoso (Grobman et al., 1961) as well as Entrelaçado (Brieger et al., 1958). The characteristic features of this racial complex are its long, narrow ears, often with bulging butts, interlocked rows of kernels instead of the customary paired-kernel rowing, and strongly attached shanks. The very low condensation, especially near the ear tip, which is a distinctive feature of these races, has been described by Galinat (1970). The kernels are usually floury with bronze or orange-colored aleurone, although lemon yellow aleurone is often found; there are collections with white, flinty kernels. Recently Wolf et al. (1972) reported that many collections of this racial complex have multiple aleurone layers.

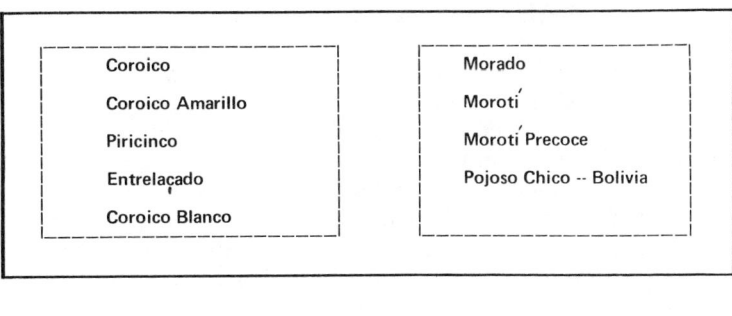

Fig. 5. Racial relationships of the corn of the Amazon basin and surrounding lowlands.

Around the periphery of this region, a number of races share similar kernel coloring and/or ear shape with the Coroico types described just above. Whether these features indicate genetic relationships remains to be ascertained, but the combination of geographic proximity and phenotypic similarity is suggestive. Several of these peripheral races are grouped together in the upper right of Fig. 5. The following races all appear to share their kernel coloring and texture with the Coroicos: Cacao and Cariaco from Colombia and Venezuela, Candela from Ecuador, Cabuya from Colombia, Morotı and Pojoso Chico from Brazil, Bolivia, and Paraguay, and perhaps Marañon from Peru. The ear shape of the Coroico types described above seems to be shared with the Montañas from Colombia and Ecuador, Amagaceño from Colombia, Morado from Bolivia, and Rienda, Chimlos, Pagaladroga, and Rabo de Zorro from Peru, as well as with most of the former (Cariaco and Marañon are exceptions).

Sympatric with the Coroicos, but apparently much more limited in distribution is Enano (Fig. 5), a dull, ivory-kernelled popcorn with short, tapering ears, very strongly attached shanks, and very stiff plants with erect leaves and tassel branches. The relationship between the two racial types is uncertain, but some collections of Enano approach the smaller collections of Coroico Blanco in ear size and shape.

The two types of genetic markers studied in these materials (chromosome knobs and presence of paramutable R alleles) present somewhat conflicting evidence for these races. With a few exceptions (Cacao, Rienda, possibly Cariaco, Montaña, and Amagaceño), these races basically have an "Andean" knob pattern with a medium to small knob in the long arm of chromosome 7 and a small or no knob at the lower position on the long arm of chromosome 6 (Roberts et al., 1957; Grobman et al., 1961). Other knobs are rare (McClintock, 1959). On the other hand, the limited evidence accumulated by Van der Walt and Brink (1969) suggests that at least some of these materials (Entrelaçado, possibly Cariaco) have paramutable R alleles rather than the nonparamutable R alleles typical of Andean maize.

Lowland Southern South America

Eastern and southern South America is characterized by a group of white (Cristal and Cristal Sulino from Brazil, Paraguay, Uruguay, and Argentina; Perola from Bolivia; Curagua Grande from Chile) and yellow or orange (Cateto and Cateto Sulino from Brazil, Bolivia, Uruguay, and Argentina; Cristalino and Camelia from Chile) flints with cylindrical to tapering, medium-sized ears (see Fig. 6). In addition, floury variants of the Coroico complex (Morotí from Brazil and Paraguay; Pojoso Chico from Bolivia) discussed above were spread as far south as northern Argentina by the Guaraní Indians.

In recent years, introduction of USA, Mexican, and Caribbean dents has resulted in a number of dent and semident races (Fig. 6) of hybrid origin (the dents of southern Brazil; Cubano Dentado and Argentino from Bolivia; Dentado Comercial from Chile). It also appears that flints from the northern part of the USA were introduced into both Chile (Cristalino Norteño, Araucano) and Argentina (Canario de Ocho). Similarly, USA sweet corns were collected and described in Chile (Timothy et al., 1961).

Several distinct floury races (Camba, a large, tapering white dent from Bolivia; Caingang, a long-eared, cylindrical, white dent from Brazil; Cholito, a short-eared, cylindrical to tapering, predominantly white dent segregating for purple or dotted aleurone, from Bolivia; Lenha, a stubby-eared, white-kernelled race with soft cobs, from Brazil; Harinoso Tarapaqueño, a wide-kernelled Chilean race) are found along the northern edge of the region (Fig. 6). Cateto Sulino Grosso from Uruguay is quite unlike most of the Catetos. It has thick ears with high row numbers, sharing its ear shape with Lenha and Cravo, both found in southern Brazil.

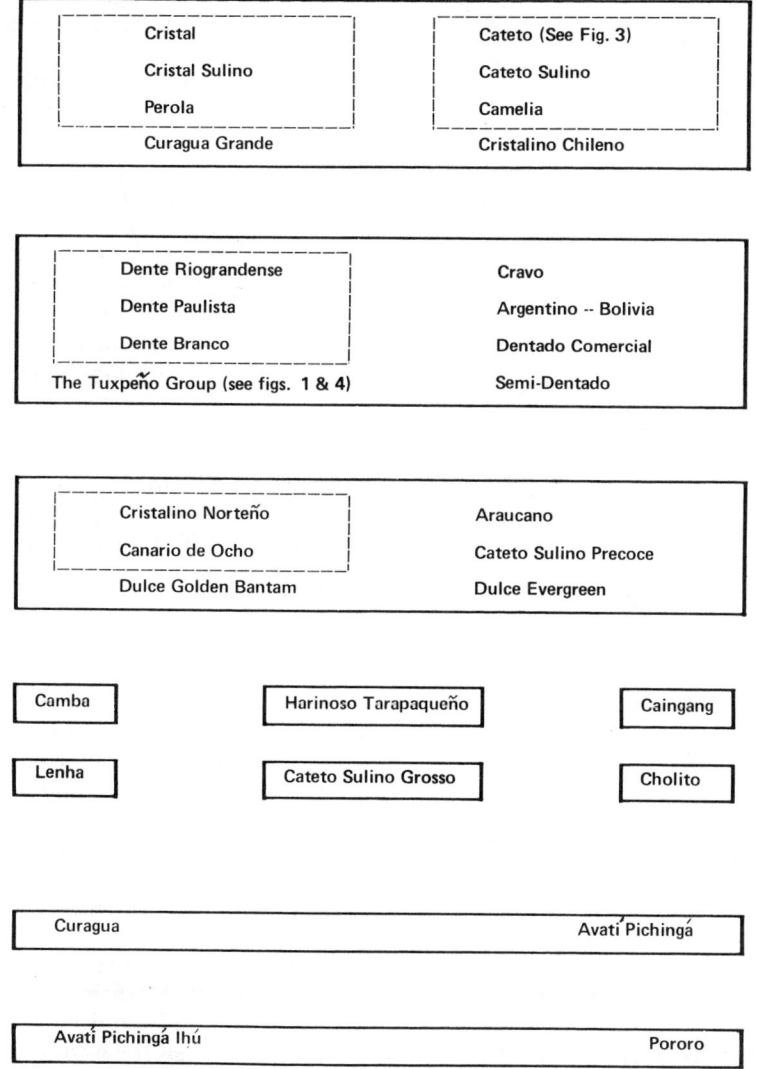

Fig. 6. Racial relationships of the corn of lowland southern South America.

Several cigar-shaped popcorn (Fig. 6) races (Avatí Pichingá Ihú from Brazil and Paraguay; Pororo from Bolivia) with predominantly white, rounded kernels (sometimes with red pericarp) are scattered throughout the region. Neighboring races with predominantly white, pointed kernels and more conical ear shape (Avatí Pichingá from Brazil and Paraguay; Curagua from Chile) may be related to higher altitude Andean pointed popcorns (see next section). These races tend to have multiple ears, flag leaves, and several tillers.

The Andean Region

The geographic diversity found in the region and the isolation imposed by the terrain has, in conjunction with trading and migration, resulted in a diverse set of mostly interrelated maize races. Figure 7 illustrates the sort of

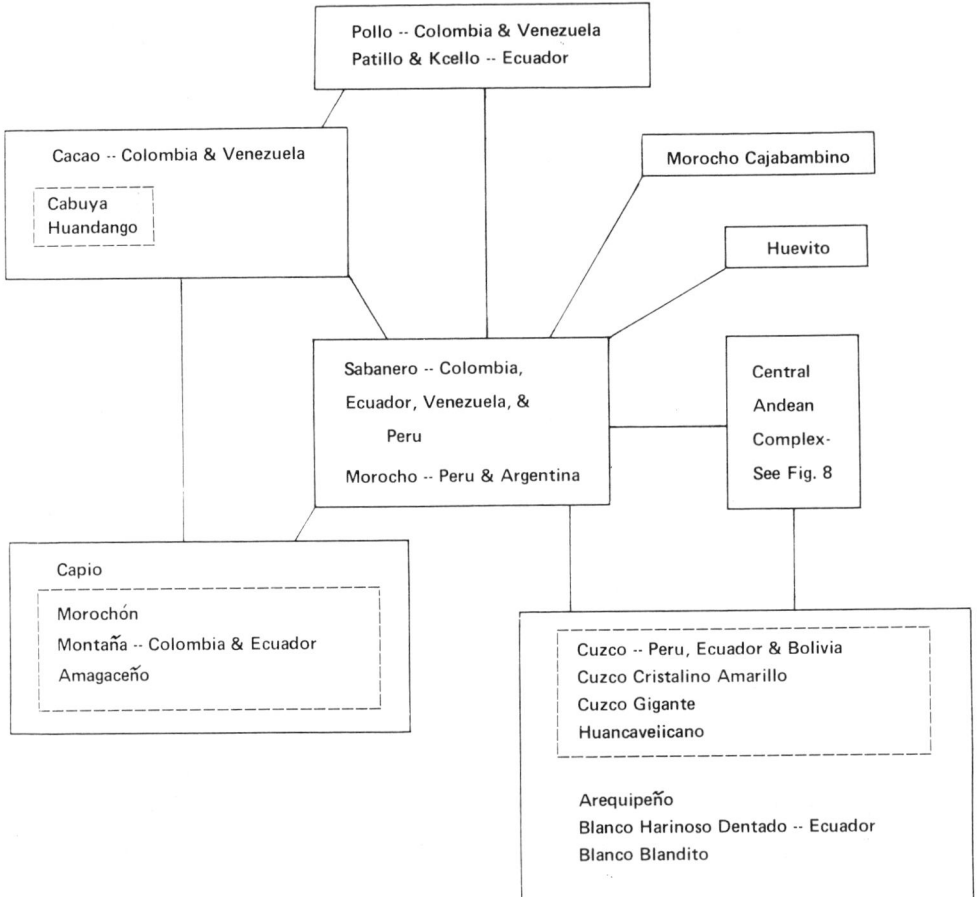

Fig. 7. Racial relationships of the major portion of the corn of Andean South America. Lines joining cells indicate closer relationships (often essentially continuous gradients) between racial groups.

variation that is most commonly encountered in mid to high altitude Andean corn. Sabanero from Venezuela, Colombia, Ecuador, and Peru and, to a lesser extent, Morocho from Peru are races which are central to a number of clines of variation throughout the north Andean region. Sabanero is a tapering, conical-eared flint or flour race with an enlarged base. It typically has very large yellow kernels, but both white and colored forms are common. Huevito seems to have somewhat smaller ears with narrower butts than has Sabanero, while Morocho Cajabamino has smaller kernels and higher row numbers.

Pollo from Venezuela and Colombia, and Patillo and Kcello from Ecuador (Fig. 7) have smaller ears than does Sabanero (see above). These small-eared flints have relatively large, rounded kernels.

Cacao from Colombia and Venezuela appears to be a brown or bronze variant of floury Sabanero (Fig. 7). Cabuya from Colombia and Huandango from Ecuador tend to have lower row numbers than Sabanero; Huandango especially seems to lack the enlarged ear base and conical shape common to Sabanero. Both Cabuya and Huandango appear to share common aleurone coloration with Cacao [and with the lowland Colombian and Venezuelan race Cariaco (discussed in the section on northern South America) as well (Mangelsdorf, 1974)].

Montaña from Colombia and Ecuador, Morochón from Ecuador, and Amagaceño from Colombia have larger (especially longer) ears than Sabanero. Amagaceño appears to include smaller-kernelled variants resembling Pira Naranja or Clavo (see section on northern South America) from Colombia. Capio from Colombia seems to be essentially a floury form of the Montañas.

In the Central Andes, the Cuzcos from Peru, Ecuador, and Bolivia; Huancavelicano and Arequipeño from Peru; and Blanco Harinoso Dentado and Blanco Blandito from Ecuador (Fig. 7) generally have wider, more floury kernels and lower row numbers than Sabanero. These races appear to have been spread northward as a consequence of the agriculture of the Incas and/or their predecessors. The Huancavelicano kernels are pointed as are the kernels of Chillo and Mishca from Ecuador, which may be yellow, floury variants of the same group.

Most of the races of the Central Andean complex in Fig. 8 basically have grenade-shaped ears, the size of which varies inversely with the altitude to which they are adapted. Plant height follows a similar pattern. Additional distinctions among these races are based upon kernel coloring (Kulli from Bolivia, Racimo de Uva from Ecuador, and Morado Canteño and Kculli from Peru are essentially black; Piscorrunto from Peru and Checchi from Bolivia are speckled), row numbering (the Chulpis, Capia from Argentina, Paro from Peru, etc., tend to have very high row numbers), or combinations of these two features with geographic distribution and kernel shape and texture.

In Fig. 8 the races are divided, perhaps somewhat artificially, into three groups. The upper group consists of races having relatively low row numbers (and often highly colored kernels). The middle group consists largely of those races having very high row numbers. This group includes essentially all of the indigenous South American sweet corns (the various Chulpi, Chullpi, and

Chuspillu races from Ecuador, Chile, and Argentina; Peru; and Bolivia, respectively). The third group consists of the very high altitude races Patillo from Bolivia and Confite Puneño from Bolivia and Peru. The latter are adapted to altitudes of about 3.50 km above sea level. They are probably ecologically specialized relatives of the second group in Fig. 8, having reduced ear and plant size as a result of their adaptation to extremely high elevations. Some of the collections having low row numbers, however, may be more closely related to members of either the first group in Fig. 8 or to the Uchuquilla group of 8 to 10-rowed flints to be described next.

Ancestral races for two of Mangelsdorf's (1974) postulated lineage relationships are included in Fig. 8. Mangelsdorf suggests (based mostly on ear shape, row number, and the deleterious nature of the su locus) that most, if not all, sweet corns trace their source of the su gene to the Andean Chulpi. While the argument weakens as the distance from the Andes increases, it appears to hold for most of the indigenous South American sweet corns. He also suggests that many of the races possessing dark red pericarp and/or bronze, orange, or brown aleurone in addition to floury endosperm are descended from Kculli of Peru.

A group of 8 to 10-rowed, knobby-kernelled flints (Fig. 9) is found in southern Peru and Bolivia in the east central Andes. Representative races include Uchuquilla from Peru and Bolivia, and Karapampa, Chake-Sara, Patillo Grande, and Kcello from Bolivia. Aysuma, Niñuelo, and some of the Patillo-Confite Puneño types from Bolivia also seem to belong to or intergrade into this group. All tend to have broad, flat flinty kernels which easily shell off the rather narrow cobs.

Two Andean popcorns with white, pointed kernels, Canguil from Ecuador and Confite Puntiagudo from Peru, have ears which are very similar to those of Pisankalla and Pisincho of Bolivia and Argentina (see Fig. 9). The Colombian pointed "popcorn" Imbricado also appears to be related to these

Kulli	Granada	Huilcaparu
Kculli	Checchi	Piscorrunto
Racimo de Uva	San Geronimo -- Huancavelicano	
Morado Canteno	Altiplano -- Bolivia and Argentina	

Huayleño	Paro	Capia
Shajatu	Marcame	Capio Chileno
Ancashino	Chullpi -- Peru	Chulpi -- Argentina, Ecuador & Chile
Paru	Negrito Chileno	Chuspillu -- Bolivia

| Confite Puneño -- Bolivia & Peru |
| Patillo -- Bolivia |

Fig. 8. Racial relationships of the corn of the Central Andean Complex.

races (as does the Chilean popcorn, Curagua), but Imbricado often has large kernels similar in size to the flinty Sabaneros (Fig. 7 and adjacent text). Both McClintock (1960) and Mangelsdorf (1974) present evidence that at least some of these Andean pointed popcorns are related to similar Mexican materials.

A third Andean popcorn, Confite Morocho of central Peru (Fig. 9), has yellow kernels, often arranged in eight rows on a very thin cob. It has been hypothesized to be a primitive ancestor to a number of more productive races (Grobman et al., 1961). In what appears to be Mangelsdorf's (1974) most speculative lineage, Confite Morocho is regarded as the ancestral race of a lineage leading to Harinoso de Ocho of Mexico and the Northern Flints of the United States, the so-called "Maíz de Ocho" of Galinat and Gunnerson (1963). Chutucuno Chico, a small, yellow, pearl popcorn from Chile appears to be distinct.

At lower elevations along the west coast, a series of small-eared, early, mostly white, floury corns (Fig. 9) is often found. These include races such as Mochero, Chaparreño, and Huachano from Peru. Large-eared Chancayano from Peru may also belong with this group, but several collections of this race appear to also have Caribbean flint germplasm. At low to intermediate locations in the eastern Andes are found two morphologically similar races,

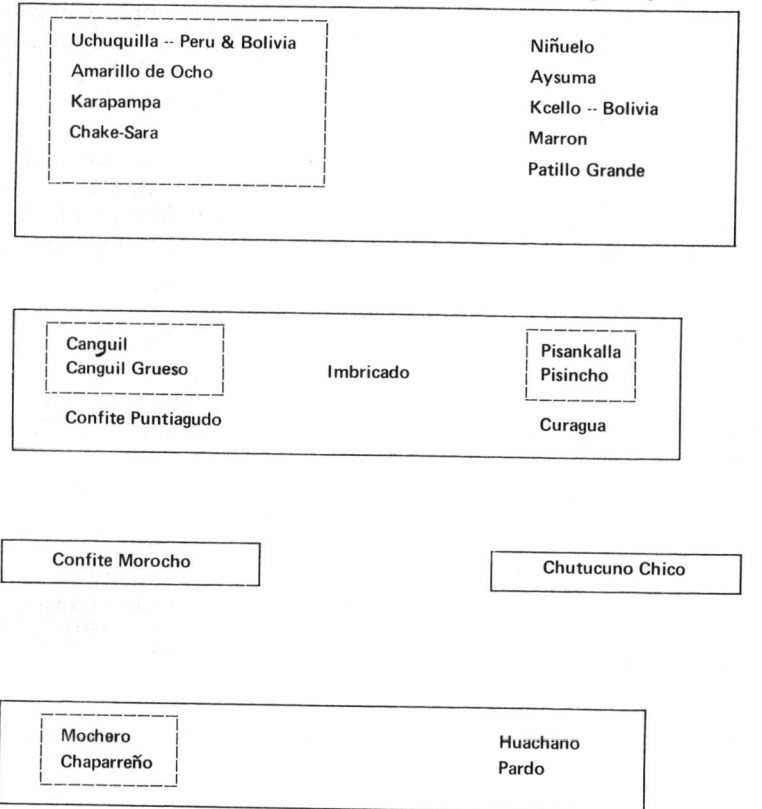

Fig. 9. Racial relationships among some additional Andean races.

Marañon and Chuncho, with dissimilar chromosome knob patterns (Grobman et al., 1961). Both have very tall, late maturing plants with ear types varying from conical, with many rows, usually with rounded to dented kernels, to elongate, often with pointed kernels. Despite their morphological similarity, Chuncho has many chromosome knobs while Marañon has few — it is essentially "Andean" in knob constitution.

With only a few exceptions (the Montañas, Amagaceño, Capio, possibly Pollo, Pisankalla, and Canguil, and especially Cacao) the races included in Figs 7, 8, and 9 basically appear to have the "Andean" chromosome knob pattern described on the basis of McClintock's work (Ramirez et al., 1960). In addition, the results obtained by Van der Walt and Brink (1969) suggest that Andean corn is also unique in possessing nonparamutable R alleles.

RACES OF THE USA

Unlike that of most countries of the world, much of the indigenous corn of the USA was replaced by modern hybrids prior to the implementation of an organized program of germplasm preservation.

A significant factor in the success of hybrid corn was the availability of highly selected adapted varieties as sources of the first inbred lines. Many of these varieties had undergone selection for almost a century in the hands of intelligent and ingenious farmers. Progress in population improvement by Robert and James Reid, Isaac Hershey, George Krug, Chester Leaming, and others gave the early USA developers of hybrid corn a source of elite germplasm not duplicated elsewhere in the world.

Unfortunately, after usable inbred lines were developed, these original sources of germplasm were largely ignored. Even less attention was given to the numerous Indian varieties and minor land races found outside the central Corn Belt. Consequently, when, at a later date, interest developed in preserving the corn germplasm of the USA, much of it had long since disappeared.

The literature, beginning with *The History and Present State of Virginia* (Beverly, 1705), contains descriptions of numerous varieties of corn found in specified geographic areas of the USA. Among the more comprehensive treatments are those of Atkinson and Wilson (1914), Will and Hyde (1917), Carter and Anderson (1945), and Brown and Anderson (1947, 1948). Yet, to date, no comprehensive attempts have been made to identify all of the major races of corn known to have existed in this country prior to the advent of hybrid corn. While much of this germplasm has disappeared from cultivation, some of it is still available, albeit frequently in a modified form, in germplasm banks, breeder's collections, etc.

The bulk of USA corn, exclusive of the popcorns and sweet corns, can be assigned to one of the nine broad racial complexes (Fig. 10) as follows:
1. Northern Flints
2. Great Plains Flints and Flours
3. Pima-Papago
4. Southwestern Semidents
5. Southwestern 12 Row
6. Southern Dents
7. Derived Southern Dents

RACES OF CORN

    8. Southeastern Flints
    9. Corn Belt Dents

There is considerable variation within each of these racial groups, yet to split them into additional entities seems unjustified on the basis of present knowledge.

Northern Flints

The Northern Flints, which also include floury endosperm types, are characterized by ears possessing 8 to 10 rows of crescent-shaped kernels, relatively short, highly tillered, frequently two-eared plants with narrow leaves, well developed husk-leaf blades, and slender culms. The shanks tend to be both long and thick, and the ears are frequently enlarged at the base.

The Northern Flints were, until the early 1800's, the dominant type of corn of eastern North America. Detailed descriptions, illustrations, and the geographic distribution of this racial complex have been given by Brown and Anderson (1947).

The origin of the Northern Flints is still unclear. Galinat and Gunnerson (1963) trace the eight-rowed flint and flour corns of North America to the Mexican race Harinoso de Ocho. Yet, the only important traits the two races seem to have in common are eight rows of kernels and early maturity (the latter is evident only when Harinoso de Ocho is grown under short day conditions). Mangelsdorf and Reeves (1939) suggested the Northern Flints reached eastern North America from the southwestern USA where corns with similar ear types are found. The latter, derived primarily from western

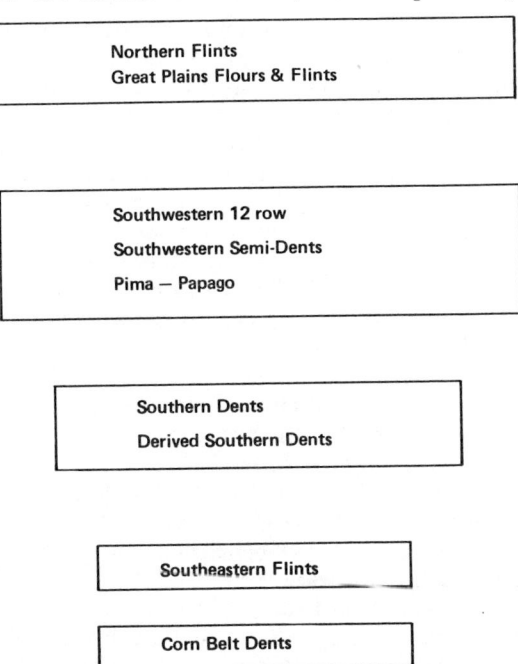

Fig. 10. Racial relationships of USA corn.

Mexico and from the Central Mesa of Mexico, differ from the Northern Flints in several traits including internode and chromosome knob patterns. Most of the southwestern USA corns possess many knobs in contrast to the knobless or nearly knobless Northern Flints (Longley, 1938).

Races which are quite similar in ear morphology to the Northern Flints are found in the highlands of Guatemala (Brown and Anderson, 1947). Within the races San Marceño and Serrano, from highland Guatemala, ear types are found which are quite similar to the Northern Flints. These are eight-rowed ears with crescent-shaped, flint kernels and distinctly enlarged butts. In contrast, Harinoso de Ocho typically possesses flat-topped kernels, distinct husk striations, and compressed butts. However, the route or routes by which these Guatemalan materials may have reached eastern North America is not clear, if indeed they do represent the origin of this race.

The role of the Northern Flints in the evolution of Corn Belt Dents has been described in detail by Anderson and Brown (1952) and Wallace and Brown (1956). The mixing of Northern Flints with the later, higher yielding dents of the South by Colonial farmers resulted in a new race which later completely dominated the corn-producing areas of the USA.

Some of the common varieties of Northern Flints and Flours are Longfellow, Dutton, Smut Nose, Canada Flint, Mammouth Yellow, Wilbur's Flint, Parker's Flint, Thompson Flint, New England Flint, and New York Council Flour Corn.

Great Plains Flints and Flours

Whereas practically all the corn found east of the Mississippi and north of northern Georgia in pre-Colonial times was typical Northern Flint, that found in the Northern Great Plains was much more variable. The flints and flours of this region are obviously related to the Northern Flints. Many of the varieties suggest mixtures between Northern Flints and varieties from the American Southwest. Apparently this mixing is not recent, since specimens from three village sites in South Dakota thought to have been occupied in the mid-1400's also show evidence of similar mixing (Cutler and Agogino, 1960).

Although much of this corn has 8 to 10-rowed ears and wide, crescent-shaped kernels, the frequency of 12 to 14-rowed ears with small square kernels is much higher than that found east of the Mississippi. Also, the Great Plains Flints and Flours have shorter and more highly tillered plants and shorter tassel branches than do the Northern Flints. They also have a larger number of chromosome knobs than the Northern Flints, although the range of chromosome knob numbers in the two races overlap.

One of the best descriptions of the Great Plains Flints and Flours is that of Will and Hyde (1917). These authors describe a large number of Indian varieties in considerable detail and also give a wealth of information on corn culture by the various tribes, the uses of corn as food, and the place of corn in the Indian ceremonial organization.

Among the more important varieties included in this racial complex are Pawnee White, Mandan Yellow Flour, Arikara Flints, Rhee Flint, Omaha Flour, Otoe White Flour, Ponca Red Flour, Winnebago White, Winnebago Blue, Gehu, Assiniboine, Dakota Squaw, and Santee.

Pima-Papago

Because of the archaeological and ethnological interest in the southwestern USA., corn of the area has been extensively collected. Both prehistoric and modern varieties are reasonably well represented in museum collections, yet only a small part of the historic corn of the area is still available in viable form.

Corn of the southwestern USA has apparently undergone a complicated history and includes germplasm and mixtures of germplasm from at least three geographic areas.

Apparently, one of the earliest types of corn in the area was that of the prehistoric "Basketmakers" (Carter and Anderson, 1945). Some of the modern varieties showing strong Basketmaker influence have been designated Pima-Papago by Anderson and Cutler (1942). This race is characterized by ears which are strongly compressed at the base and which taper gently from a subbasal position to the tip. Row numbers range from 10 to 16, with 10 rows being most common (Carter, 1945), and the tessellate kernels frequently exhibit strong husk striation. The ears are small (about 12 cm in length and 2.5 cm in diameter). The endosperm can be either flinty or floury, with the latter being more frequent. Plants usually possess two or more well developed tillers. Leaves are narrow and the leaf sheaths frequently exhibit purple anthocyanin coloration. The tassels are not particularly distinctive compared with other southwestern USA races. The race is most frequently found among the corn of the Pima, Papago, and Yuma tribes (Carter and Anderson, 1945) and is also represented by the variety Kokoma of the Hopi (Brown et al., 1952).

Pima-Papago or its progenitor probably reached the American Southwest from western Mexico where similar types (Chapalote, Reventador) are still grown. When or how it migrated to southwestern USA is still not clear. Mangelsdorf (1954, 1974) has suggested that Chapalote reached New Mexico (Bat Cave) as early as 1500 B.C. Pima-Papago, while undoubtedly related to prehistoric Chapalote, has apparently undergone considerable introgression with other races (Galinat and Gunnerson, 1963). Cutler and Blake (in press) suggest that a form of Pima-Papago spread eastward through Oklahoma and Arkansas and up the Mississippi River to northern Illinois by 100 A.D., and to the Gulf Coast of Georgia even earlier.

Southwestern Semidents

This race is apparently the product of the introgression of Tuxpeño type dent germplasm from Mexico into the typical 12 to 14-row flint and flour corns of the area. Except for a slight degree of denting, higher number of kernel rows, and more pronounced tassel condensation, the race is quite similar to the Southwestern 12 Row described later. While we have indicated the genes for denting found in Southwestern Semidents probably came from Mexico, the source could have been the Northern Periphery (central and northern Utah, western Colorado) where prehistoric dents have been found (Carter, 1945; Anderson, 1959, n.d.).

Although we feel Southwestern Semidents merit racial recognition, our experience suggests that such material is not recognized as a distinct variety by the tribes who grow it. As an example, typical ears of this race are usually present in collections of Hopi White Flour corn. Yet, to our knowledge, the Hopi, who place considerable emphasis on maintaining varietal purity, make no effort to separate the semident and flour forms of their white corn.

For an excellent illustration of Southwestern Semident, the reader is referred to Plate 3 of Carter and Anderson (1945). Those authors suggest that this race is found at highest frequency near the four corners region of the states of Arizona, New Mexico, Colorado, and Utah.

Southwestern 12 Row

This race of southwestern corn is synonymous with the Eastern Complex of Carter and Anderson (1945). Although these corns are similar morphologically to the Northern Flints of eastern North America, they differ in several important traits. Furthermore, the name Eastern Complex implies an eastern origin, which is questionable.

The plants are short (105 to 120 cm), highly tillered, and the tassels are characterized by long central spikes which average 30+ cm in length. The ears are usually enlarged at the base and are attached to large, indurated shanks. The kernels are wide (though not as strongly crescent-shaped as the Northern Flints) and are arranged in long straight rows. Endosperm is either flinty or floury. Row numbers are usually 12 to 14.

As indicated by Carter and Anderson (1945), Southwestern 12 Row corn is more prevalent among the easternmost Pueblos (mostly east of the Rio Grande River of central New Mexico) than among those in the westernmost part of the area. This lends some credence to the notion that this kind of corn germplasm reached the Southwest from the East, rather than having migrated from the Southwest to the East.

The morphological similarities between Southwestern 12 row and Northern Flints are great, suggesting a common origin or at least a close phylogenetic relationship. Yet, the two races differ markedly in two important traits, that of number of kernel rows and chromosome knob numbers and pattern. The Northern Flints possess few knobs. Many varieties are knobless and only a few varieties are known to possess as many as four to five knobs. In contrast, corn of the southwestern USA, including the Southwestern 12 Row, has many knobs (Longley, 1938; Brown, 1949). Furthermore, the chromosome knob patterns characterizing the Northern Flints and the Southwestern 12 Row are distinctly different (Kato, personal communication).

Southern Dents

This racial complex, prior to its replacement by hybrids, was concentrated in the southeastern USA. In Colonial times, it reached as far north as northern Virginia and Maryland. The western limits of its distribution are not clear, yet it is known to have been present in central Texas.

The Southern Dent complex was described and illustrated by Brown and Anderson (1948). The plants are consistently taller than other USA races. Internodes are condensed above the ear in contrast to the long upper internodes of the Northern and Great Plains Flints. Ears are enveloped in many tight, thick husks, which are usually without husk leaf blades. The plants are usually without tillers. Tassels are highly branched and condensed in contrast to the sparsely branched, non-condensed tassels of the Northern Flints. Numbers of rows of kernels are variable, ranging from 8 to 10 rows in the variety Hickory King to 24 to 26 rows in some selections of Gourdseed. Kernels are usually well dented and the endosperm color is usually white.

Many varieties of Southern Dents appear to be closely related to the dent corns of central Mexico. For example, Tuxpan of the southeastern USA is similar to Tuxpeño; except for cob color, Shoepeg is much like Pepitilla, and Hickory King is quite similar to Tabloncillo and Olotillo. Consequently, certain of the Southern Dents seem simply to be northern counterparts of Mexican races still prevalent in Mexico. The time at which these corns reached the southern USA is still unknown and so are the route or routes by which they migrated northward and eastward from Mexico. We know they were widely grown in the southern and southeastern USA in Colonial times, but little is known of their pre-Colombian history. That they provided at least one-half of the parentage of the more recently evolved Corn Belt Dents is well documented (Brown and Anderson, 1948; Anderson and Brown, 1952).

Some of the better known varieties of Southern Dents are Gourdseed, Shoepeg, Tuxpan, Hickory King, Jellicorse, and Mexican June.

Southeastern Flints

This race seems to represent simply a northern extension of the Caribbean flints from the West Indies and eastern South America. According to the historical record, the race was never an important one in the USA. Yet, it was grown to a limited extent in the southeastern states in Colonial and post-Colonial times. It apparently has also played some role in the evolution of certain of the Derived Southern Dents described below. The name usually applied to this race in the Southeast is Creole Flint. Morphologically, the race is very similar to Cuban Flint or to the Catetos of Argentina, Uruguay, and Brazil. The ears, consisting of 12 to 14 rows of kernels, are slightly compressed at the base and gently tapered toward the tip. Cobs are usually white; the kernels are flinty, and endosperm color varies from deep yellow to orange.

The Southeastern Flints referred to here are limited to historic corns. Prehistoric materials from this area seem to be more closely related to the Northern Flints (Cutler and Blake, in press).

Derived Southern Dents

This complex, although closely related to the Southern Dents, has apparently arisen through the hybridization of the latter with other races. The gross morphology of the ears and plants of the Derived Southern Dents suggests that, in addition to Southern Dents, at least three other races may be involved in the ancestry of this complex. These are the Southeastern Flints, Northern Flints, and Corn Belt Dents.

The Derived Southern Dents tend, on the average, to have somewhat less dented kernels, to have fewer chromosome knobs, and to be more prolific than the Southern Dents. Some varieties have many traits in common with the Southeastern Flints and others suggest admixture with Corn Belt Dents. As a group, the Derived Southern Dents represent the primary source of the genes for prolificacy present in southern corns which has found its way into many modern hybrids grown in that area.

A number of the more prominent varieties of Derived Southern Dents are Caraway's Prolific, Giant Yellow Dent, Horsetooth, Jarvis Golden Prolific, Latham's Double, Mosby's Prolific, Neal's Paymaster, Southern Snowflake, and Whatley's Prolific.

Corn Belt Dents

This racial complex dominated the major areas of corn production in the USA for more than half a century preceding the advent of hybrid corn. Although variable, the race is a well defined entity and easily distinguished from other types of North American corn. The ears are slightly tapered or cylindrical and consist of 14 to 22 straight rows of kernels which are distinctly dented at the tip. Cobs are usually red in color, although white cob varieties are known. Yellow endosperm is predominant, but included in the race are a number of white endosperm varieties. The plant typically has a single stem with strong arching leaves and a heavy, many-branched tassel.

The Corn Belt Dents are of recent origin, having arisen in the 19th century in the eastern part of the USA Corn Belt. They are largely the product of crosses of white Southern Dents with long-eared Northern Flints of the eastern USA. The historical literature is quite clear as to how Corn Belt Dents were developed. Some of the mixing of Southern Dents with Northern Flints was, undoubtedly, accidental and occurred when fields with poor stands of Southern Dents were partially replanted with early flints. However, much of it was the result of deliberate, controlled hybridizing on the part of Colonial farmers who recognized and documented the superior yield of the hybrid populations (Anderson and Brown, 1952; Wallace and Brown, 1956).

Even though Corn Belt Dents did not exist prior to the 19th century, they have provided the basic germplasm for virtually all of the corn now produced in the USA as well as in most other temperate regions of the world. Their contribution, therefore, to total corn production in the world is unparalleled.

Hundreds of varieties of Corn Belt Dents have been described, most of which, unfortunately, have long since disappeared. A few of the more prominent varieties were Reid's Yellow Dent, Lancaster Sure Crop, Krug Yellow Dent, Boone County White, Leaming, Osterland Yellow Dent, and Midland Yellow Dent.

Excluded Races

As indicated earlier, no attempt has been made in this brief treatment of USA corn to include the various sweet corns and popcorns. To date, we have not had an opportunity to study these two types of corn sufficiently to feel confident in assigning them to meaningful racial groups. Many of the sweet corns of the USA are simply sugary counterparts of Northern Flints or higher-row-number dents (Galinat, 1971). These two types, which encompass most of the Bantam and Evergreen varieties, and a third type, Country Gentleman, comprise the bulk of USA sweet corn. Yet, there are other older sorts which do not fit these categories, among which are the sweet corns of the Southwestern Indians.

The popcorns are far more complex and will require much more study before their relationships are understood.

In addition to the popcorns and sweet corns, we have excluded two other corn groups — the existence of which we should at least recognize. These are the corns of the Seminole of Florida and the "tamale" corns of southern California. Some of the Seminole corns are quite distinct from the flints and flours of the northeastern USA. But here again, our experience with these corns does not permit us to attempt to classify them. Neither have the "tamale" corns of the southern West Coast of the USA been studied in detail, yet these appear to represent northern counterparts of western Mexican varieties, which probably reached southern California by way of Arizona.

EFFECTIVE SOURCES OF GERMPLASM AMONG LOWLAND TROPICAL RACES

Among the collections of corn germplasm of the Western Hemisphere now existing in the various seed banks, there is undoubtedly much material which should be useful to the breeder. However, to date, little of this material has been systematically evaluated with respect to its usefulness. It is primarily for this reason that breeders have not made full use of the germplasm that is available. Most breeding programs are operated in such a way as to make impractical for the individual breeder the screening of vast numbers of accessions in order to identify the few genotypes of interest or value at the moment.

A major problem encountered in evaluating tropical materials for possible use in temperate regions is the very late maturity which these materials often exhibit. One reason for this is that almost all corn flowers more rapidly under short day growing conditions given equivalent temperatures. Tropical varieties adapted to 12 to 14-hour daylengths flower much later under the longer days typical of temperate regions. It is for this reason that use of these materials in the USA has generally either been confined to the South or limited to very early (hence inherently low yielding) types such as Zapalote Chico of Mexico.

While extreme response to daylength is probably fairly simply inherited, it effectively blocks most evaluation of tropical materials in temperate regions. For example, most collections of the closely related races Tuxpeño and Vandeño have about equally late maturity under temperate, long day conditions. However, most typical Tuxpeño collections from Mexico are later than Vandeño collections under short day conditions; and thus Tuxpeno is probably less adaptable for use in temperate regions despite its favorable yield potential. The daylength response may totally mask the true maturity of the material. One collection of Pepitilla flowers in less than 65 short days (south Florida, winter) while requiring about 150 long days (North Carolina, summer). (A widely used, adapted commercial hybrid flowers in about 75 and 77 days each, respectively, under the same conditions).

Such responses to daylength suggest that the degrees of *apparent* adaption of tropical materials to a temperate environment are probably negatively correlated (if correlated at all) with their breeding potentials for that region. This conflicts with the conclusions reached by Kramer and Ullstrup (1959), although their data can be interpreted to support this viewpoint. They showed that yields of topcrosses of semiadapted exotics onto adapted materials were closely correlated with maturity. In fact, almost all exotics offering potential yield improvement are very late maturing under USA conditions (regardless of their maturities in the tropics), while a number of low yielding exotics (often popcorns) are much earlier. Thus, such a correlation of exotic topcross yields with maturity would be expected for both developmental and genetic reasons.

Limited published work with synthetics in the USA adds support for the view that "preadapted" tropical materials (i.e., those that are very early in the tropics) are not good candidates for yield improvement in temperate regions. Troyer and Brown (1972) have shown that a synthetic involving the very early maturing Mexican races Zapalote Chico and Zapalote Grande initially outyielded two other synthetics involving other, later Mexican materials and West Indies materials, respectively. (These synthetics also included germplasm from elite USA inbreds). Selection for earliness in all three composites appears to have lowered the yield of the Zapalote Chico synthetic to the point where the other synthetics (involving late Mexican and West Indian materials) have higher yields. In contrast, Hallauer and Sears (1972) have shown that late maturing but variable tropical materials — in this case ETO, a synthetic described later — can effectively be selected for much earlier maturity and lower ear height, using relatively few cycles of mass selection or by a single generation of crossing with early maturing elite lines. In addition, Eberhart (1971) and Hallauer (1972) have shown that several synthetics containing large proportions of tropical germplasm (which in itself would largely be very poorly adapted) do show promise as source materials for breeding programs in the USA.

Much work with exotics has been conducted with composites or synthetics, whose ancestry often cannot be clearly traced. While this work has apparently been successful, we have learned less from it than from earlier corn breeding efforts with open-pollinated varieties. In the earlier work, specific sources (such as the varieties Lancaster and Reid's) of successful inbred lines were fairly quickly identified. Such identification has occurred to a lesser extent with the exotic races, in part due to the emphasis on synthetics.

Despite the lack of a systematic evaluation of the so-called "exotic" corn germplasm, some information has been obtained on the relative usefulness in breeding of a number of races, varieties, composites, etc. We have attempted to summarize the accessible published information below; much, of course, remains unpublished.

Wellhausen (1965) has reported on the yield potential of a number of Mexican and Caribbean races and populations which he feels merit additional use in breeding. Considerable information on yields of Mexican races comes from the results of a yield test of all possible crosses of the 25 recognized races of Mexico. This test was conducted near Celaya in Central Mexico. The inter-racial crosses were compared with an adapted hybrid, H352. Sixteen of the inter-racial crosses yielded more than the check hybrid. More importantly, 15 different races were represented among the 16 highest yielding inter-racial crosses. Among the highest yielding crosses were Pepitilla × Chalqueño, Pepitilla × Tuxpeño, Maíz Dulce × Comiteco, Comiteco × Celaya, and Pepitilla × Maíz Dulce.

The data reported by Wellhausen (1965) did not include many important agronomic traits, and, for this reason, it is not possible to assess from these data the practical usefulness of this obviously high yielding germplasm. (Chalqueño and Pepitilla have very weak roots, while Tuxpeño and Comiteco tend to have very late maturity even when grown under short day conditions). However, such work is fairly typical of what needs doing in many tropical regions: widescale testing of crosses among a diverse set of materials. For example, Darrah et al. (1972) found that crosses between a local Kenyan variety and a collection of the Ecuadorian race Montaña were suitable for production as one of Kenya's first hybrids. Similarly, in Brazil, Paterniani (1970, 1972) has reported that intervarietal and topcross hybrids involving Caribbean germplasm (mostly Cuban and Caribbean Tropical Flints) with Tuxpeño derivatives outyield the local H6999B double cross check by as much as 20% with improved agronomic traits as a bonus.

Among the Caribbean germplasm evaluated by Wellhausen (1965), a collection belonging to the race Tusón and designated Antigua 2D and a composite of collections called Antigua Group 2 have been shown to combine well with Tuxpeño and with Corn Belt Dents. This material is relatively early and short in stature as compared to most other tropical varieties.

The Cuban Flints are reported by Wellhausen (1965) to combine well with the Mexican dents, especially Tuxpeño. Timothy (1963) reported that the Cuban Flints have been widely used in Colombian corn breeding programs along with Tuxpeño, Costeño, Puya Grande, and the synthetics ETO and Venezuela I, discussed later.

Our experience in tropical corn breeding supports, in general, Wellhausen and Timothy's conclusions relative to the usefulness of the germplasm discussed above. [We might note that the evaluation of germplasm utility would be greatly facilitated if the International Maize Adaptation Nurseries (IMAN) program (CIMMYT 1972a, 1972b, 1974) were expanded and publicized. These trials included materials from 13 countries tested in 47 countries in 1971. Outstanding materials in the 1970 trials were from Australia, Jamaica, Kenya, Brazil, Mexico, and Peru.] The sections which follow include additional observations relative to the performance of the races mentioned earlier and refer also to other sources of germplasm which we believe to be of interest.

Mexican Dents

The most outstanding single source of Mexican Dents is the race Tuxpeño. This race, widely distributed on the east coast of Mexico, possesses excellent combining ability, good stalk quality, and good resistance to *Helminthosporium spp.* Its principal disadvantages are late maturity, tall plants with high ear placement, poor roots, and susceptibility to sugarcane mosaic.

There are several derivatives of Tuxpeño, both white and yellow. Some of these have given excellent performance in crosses with lines derived from Coastal Tropical Flint and Tusón. Some Tuxpeño lines have also exhibited good combining ability with material derived from Cuban Flint and Coastal Tropical Flint (especially the varieties Tiquisate, Mayorbela, and Venezuela I, described later).

Tuxpeño germplasm is now widespread throughout the tropical world and is an important constituent of many improved varieties and hybrids.

Tusón

Tusón is likely one of the best sources of tropical corn germplasm known today. Lines derived from this race combine well with lines from Tuxpeño, Cuban Flint, Chandelle, and southern USA dents. The race possesses excellent ear type, good grain quality, and a high degree of tolerance to sugarcane mosaic.

Coastal Tropical Flint

This race is widely spread throughout the lowland tropics and its influence is apparent whenever tropical corn is grown on a commercial scale. Among the many outstanding tropical varieties which have Coastal Tropical Flint in their background are ETO, Tiquisate Golden Yellow, Metro, Mayorbela, and Venezuela I. Tiquisate was developed in the 1940's by Dr. I.E. Melhus and his coworkers at the Iowa State University Tropical Research Station in Guatemala. It was derived from two Cuban varieties (probably Cuban Flint and Coastal Tropical Flint) to which were added some local varieties from coastal Guatemala. The plants and ears of Tiquisate resemble Cuban Flint. A selection of Tiquisate, developed at Bogor, Indonesia, and known as Metro, is used extensively as a commercial variety in that country. Mayorbela was developed in Puerto Rico in the 1930's. It traces back to the Mayor farm and the Isabela substation of the Mayaguez Experiment Station. In the late 1930's, this source of material was outcrossed to an inbred line (probably from the USA) followed by selection within the resulting population. Mayorbela is a reasonably good source of earliness, stalk strength, low ear placement, and grain quality. Venezuela I was developed by Langham (1942) in Venezuela primarily from Caribbean races and varieties.

Coastal Tropical Flint has been a productive source of elite tropical lines. In addition to combining ability, these lines contribute to excellent grain quality, good husk cover, stalk quality, good roots, and resistance to *Helminthosporium* spp.

Cuban Flint-Cateto

This racial complex, known as Cuban Flint in the Caribbean Islands, as Cateto in Brazil, and Argentine Flint in Argentina, is particularly useful in hybrid combinations with dents. The race and lines derived from it combine well with Tusón, Tuxpeño, and Coastal Tropical Flint. Cuban Flint is susceptible to virus diseases and must therefore be combined with virus-resistant lines, if it is to be used in areas where corn virus diseases are prevalent.

Chandelle

Among tropical germplasms, Chandelle is an excellent source of prolificacy, low ear placement, and virus resistance. It combines well with Coastal Tropical Flint and Haitian Yellow. The race and lines derived from it tend to have poor roots.

Haitian Yellow

Collections of this race from the vicinity of Jeremie, Haiti, have proven to be an excellent source of root and stalk strength, traits which are not easy to come by in tropical corn. Because of its outstanding stalk and root qualities, Haitian Yellow should play an increasingly important role in tropical corn improvement.

ETO

This synthetic variety developed in Colombia in the 1940's by Chavarriaga (1966) has contributed significantly to corn improvement in the tropics and subtropics. The variety has a broad genetic base, which includes the Colombian races Común and Chacoceño and the synthetic Venezuela I. The latter consists primarily of Caribbean races and varieties. To this broad mixture, known as Colombia I, were later added numerous lines and varieties from Mexico, Puerto Rico, Cuba, Venczuela, Brazil, Argentina, and the USA. Selection over a period of years resulted in the new variety ETO. The abbreviation represents "Estacion Tulio Ospina," the station at Medellín where the selection work was done.

ETO is available in both white and yellow endosperm forms. It is high yielding and combines well with the Tuxpeños. The variety includes a wide range of maturities, the earliest of which are no later than some lines in use in the Central USA Corn Belt.

Despite the excellent performance of ETO in many tropical areas, it appears to have generally poor adaptation to conditions in the Caribbean (Sehgal, 1966). It is also quite susceptible to sugarcane mosaic.

ACKNOWLEDGMENTS

Paper No. 4444 of the Journal Series of the North Carolina Agric. Exp. Stn., Raleigh, NC 27607. This investigation was supported in part by National Institutes of Health Research Grant GM 11546 from the National Institute of General Medical Sciences.

LITERATURE CITED

Anderson, Edgar. 1943. Races of *Zea mays*. II. A general survey of the problem. Acta Americana 1:58-68.

----. 1944a. Homologies of the ear and tassel in *Zea mays*. Ann., Mo. Bot. Gard. 31:325-344.

----. 1944b. Maiz Reventador. Ann. Mo. Bot. Gard. 31:301-316.

----. 1945. Maize in the New World. p. 27-42. *In* C.M. Wilson (ed.), New crops for the New World. MacMillan, New York. 295 p.

----. 1946. Maize in Mexico — A preliminary survey. Ann. Mo. Bot. Gard. 33:147-247.

----. 1947. Field studies of Guatemalan maize. Ann. Mo. Bot. Gard. 34:433-467.

----. 1959. Zapalote Chico: An important chapter in the history of maize and man. Actas del Congr. Int. de Americanistas (San José) 33:230-237.

----. n.d. Corn before Columbus. Pioneer Hi-Bred Corn Company, Des Moines, Iowa. 24 p.

----, and W.L. Brown. 1952. Origin of Corn Belt maize and its genetic significance. p. 124-148 *In* J.W. Gowen (ed.) Heterosis. Iowa State Coll. Press, Ames.

----, and ----. 1953. The popcorns of Turkey. Ann. Mo. Bot. Gard. 40:33-48.

----, and H.C. Cutler. 1942. Races of *Zea mays:* I. Their recognition and classification. Ann. Mo. Bot. Gard. 29:69-89.

Atkinson, Alfred, and M.L. Wilson. 1914. Corn in Montana — history, characteristics, and adaptation. Montana Agric. Coll. Exp. Stn. Bull. 107.

Beverly, R. 1705. The history and present state of Virginia. London. (Republished 1947. Univ. of North Carolina Press, Chapel Hill. 366 p).

Blumenschein, Almiro. 1973. Chromosome knob patterns in Latin American maize. p. 271-277 *In* Adrian M. Srb (ed.), Genes, enzymes and populations. Plenum Publ. Corp., New York.

----, and R. Deuber. 1968. Milhos cultivados no Nordeste Brasileiro. O Solo 60 (2):15-27.

Brandolini, A. 1969. European races of maize. Annu. Corn Sorghum Res. Conf. Proc. 24:36-49.

----. 1970a. Maize. p. 273-309 *In* O.H. Frankel and E. Bennett (eds.), Genetic resources in plants — their exploration and conservation. F.A. Davis Co., Philadelphia.

----. 1970b. Razze Europee di maiz. Maydica 15:5-27.

----. 1971. Preliminary report on South European and Mediterranean maize germ plasm. p. 108-116. *In* I. Kovacs (ed.) Proc. of the Fifth Meeting of the Maize and Sorghum Sect. of EUCARPIA. Akadémiai Kiadó, Budapest. 290 p.

----, and G. Avila. 1971. Effects of Bolivian maize germ plasm in South European maize breeding. p. 117-135 *In* I. Kovács (ed.), Proc. of the Fifth Meeting of the Maize and Sorghum Sect of EUCARPIA. Akademiei Kiadó, Budapest. 290 p.

----, and G. Mariani. 1968. Il germplasma italiano nela fase attuale del miglioramento genetico del mais. Genet. Agric. 22:189-206.

Brieger, F.G., J.T.A. Gurgel, E. Paterniani, A. Blumenschein, and M.R. Alleoni. 1958. Races of maize in Brazil and other eastern South American countries. Natl. Acad. Sci. — Natl. Res. Council Publ. 593. Washington, D.C. 283 p.

Brown, W.L. 1949. Numbers and distribution of chromosome knobs in United States maize. Genetics 34:524-536.

----. 1953. Maize of the West Indies. Trop. Agric. 30:141-170.

----. 1960. Races of maize in the West Indies. Natl. Acad. Sci. — Natl. Res. Council Publ. 792. Washington, D.C. 60 p.

----, and E. Anderson. 1947. The northern flint corns. Ann. Mo. Bot. Gard. 34:1-29.

----, ----. 1948. The southern dent corns. Ann. Mo. Bot. Gard. 35:255-268.

----, E.G. Anderson and Roy Tuchawena, Jr. 1952. Observations on three varieties of Hopi maize. Am. J. Bot. 39:597-609.

Carter, G.F. 1945. Plant geography and culture history in the American Southwest. Viking Fund Publ. Anthropol. No. 5. New York. 140 p.

----, and Edgar Anderson. 1945. A preliminary survey of maize in the southwestern United States. Ann. Mo. Bot. Gard. 32:297-322.
Chavarriaga, E. 1966. Maíz ETO, una variedad producida en Colombia. Inst. Colombiano Agropecuario. Separata de la Revista I.C.A. 1(1):5-30.
CIMMYT. 1972. International maize adaptation nurseries (IMAN). Annu. Rep. 1970-91:96-98. Mexico City. 115 p.
Committee on Genetic Vulnerability of Major Crops. 1972. Genetic vulnerability of major crops. Natl. Acad. Sci. Washington, D.C. 307 p.
CIMMYT, 1972a. International maize adaptation nurseries (IMAN). Annu. Rep. 1970-91:96-98. Mexico City. 115 p.
----. 1972b. Results of the first international maize adaptation nursery (IMAN) 1970-71. CIMMYT Information Bull. No. 7. Mexico City. 51 p.
CIMMYT. 1974. Results of the second and third international maize adaptation nursery (IMAN) (1971-72, 1972-73). CIMMYT Information Bull. No. 12. Mexico City. 118 p.
Committee on Genetic Vulnerability of Major Crops. 1972. Genetic vulnerability of major crops Natl. Acad. Sci. Washington, D.C. 307 p.
Committee on Preservation of Indigenous Strains of Maize. 1954. Collections of original strains of corn, I. Natl. Acad. Sci. — Natl. Res. Council. Washington, D.C. 300 p. (mimeo).
----. 1955. Collections of original strains of corn, II. Natl. Acad. Sci. — Natl. Res. Council. Washington, D.C. 298 p. (mimeo).
Costa-Rodrigues, L. 1969. Races of maize in Portugal. Agron. Lusitana 31:239-284.
Covor, Alexandru. 1972. Rasele De Porumb Din Romania. Acad. De Ştünte Agricole Şi Şilvice. Bucuresti 186 p.
Cutler, H.C. 1946. Races of maize in South America. Bot. Mus. Leafl., Harvard Univ. 12:257-299.
----, and G.A. Agogino. 1960. Analysis of maize from the Four Bear Site and two other Arikara locations in South Dakota. Southwestern J. of Anthropol. 16:312-316.
----, and L.W. Blake. In press. North American Indian Corn. Handbook of North American Indians. Vol. 3. Smithsonian Inst. Washington, D.C.
Darrah, L.L., S.A. Eberhart, and L.H. Penny. 1972. A maize breeding methods study in Kenya. Crop Sci. 12: 605-608.
Eberhart, S.A. 1971. Regional maize diallels with U.S. and semi-exotic varieties. Crop Sci. 11:911-914.
Edwards, R.J., and E.R. Leng. 1965. Classification of some indigenous maize collections from southern and southeastern Europe. Euphytica 14:161-169.
Galinat, Walton C. 1970. The cupule and its role in the origin and evolution of maize. Massachusetts Agric. Exp. Stn. Bull. 585. Amherst. 24 p.
----. 1971. The evolution of sweet corn. Massachusetts Agric. Exp. Stn. Res. Bull. 591. Amherst. 20 p.
----, and J.H. Gunnerson. 1963. Spread of eight-rowed maize from the prehistoric Southwest. Bot. Mus. Leafl., Harvard Univ. 20:117-160.
Girola, C.D. 1919. Variedades de maiz cultivadas en Argentina: maices Argentinos y aclimatados. Talleres Graficos J. Weiss y Preusche, Buenos Aires. 169 p.
Goodman, M.M. 1967. The races of maize: I. The use of Mahalanobis' generalized distances to measure morphological similarity. Fitotec. Latinoamer. 4(1):1-22.
----. 1968. The races of maize: II. Use of multivariate analysis of variance to measure morphological similarity. Crop Sci. 8:693-698.
----. 1972. Distance analysis in biology. Syst. Zool. 21:174-186.
----, and R. McK. Bird. In press. The races of maize. IV. Economic Botany.
----, and E. Paterniani. 1969. The races of maize. III. Choices of appropriate characters for racial classification. Econ. Bot. 23:265-273.
Grant, U.J., W.H. Hatheway, D.H. Timothy, C. Cassalett D., and L.M. Roberts. 1963. Races of maize in Venezuela. Natl. Acad. Sci. — Natl. Res. Council Publ. 1136. Washington, D.C. 92 p.

Grobman, Alexander, Wilfredo Salhuana, and Ricardo Sevilla, with P.C. Mangelsdorf. 1961. Races of maize in Peru. Natl. Acad. Sci. — Natl. Res. Council Publ. 915. Washington, D.C. 374 p.

Hallauer, A.R. 1972. Third phase in the yield evaluation of synthetic varieties of maize. Crop Sci. 12:16-18.

----, and J.H. Sears. 1972. Integrating exotic germ plasm into Corn Belt maize breeding programs. Crop Sci. 12:203-206.

Hatheway, W.H. 1957. Races of maize in Cuba. Natl. Acad. Sci. — Natl. Res. Council Publ. 453. Washington, D.C. 75 p.

Hernandez X., E. 1949. Plant exploration in Cuba. Report to Dr. J.G. Harrar, Director of the Rockefeller Agricultural Program in Mexico. April 1, 1949. Mexico City. (mimeo).

----, and G. Alanis F. 1970. Estudio morfologico de cinco nuevas razas de maiz de la Sierra Madre Occidental de Mexico: Implicaciones filogeneticas y fitogeograficas. Agrociencia 5:3-30.

----, and G. Alanis F. 1970. Estudio morfologico de cinco nuevas razas de maiz de la Sierra Madre Occidental de Mexico: Implicaciones filogeneticas y fitogeograficas. Agrociencia 5:3-30.

Kato Y., T.A. 1975. Cytological studies of maize *(Zea mays* L.) and teosinte *(Zea mexicana* Schrader Kuntze) in relation to their origin and evolution. Mass. Agric. Exp. Stn. Bull. 635. Amherst, Mass.

----, and A. Blumenschein. 1967. Complejos de nudos cromosómicos en los maices de América. Fitotec. Latinoamer. 4(2):13-24.

Kramer, H.H., and A.J. Ullstrup. 1959. Preliminary evaluation of exotic maize germ plasm. Agron. J. 51:687-689.

Kuleshov, N.N. 1929. The geographical distribution of the varietal diversity of maize in the world. Trudy po Prikladnoi Botanike, Genetike, i Seleksii (Bull. Appl. Bot., Genet. and Plant Breeding, Lenin Acad. Agric. Sci., U.S.S.R.) 20:506-510.

----. 1930. The maize of Mexico, Guatemala, Cuba, Panama, and Colombia. p. 492-502 *In* S.M. Bukasov (ed.) The cultivated plants of Mexico, Guatemala, and Colombia. Trudy po Prikladnoi Botanike, Genetike, i Seleskii (Bull. Appl. Bot., Genet. and Plant Breeding, Lenin Acad. Agric. Sci., U.S.S.R.) Prilozhenie (Suppl.) 47.

Langham, D.C. 1942. Venezuela I, una seleccion de maiz recomendable. Ministerio de Agricultura y Cria. Cir. No. 2. Caracas, Venezuela.

Leng, E.R., A. Tavčar, and V. Trifunović. 1962. Maize of southeastern Europe and its potential value in breeding programs elsewhere. Euphytica 11:263-272.

Longley, A.E. 1938. Chromosomes of maize from North American Indians. J. Agric. Res. 56:177-195.

----. 1941. Chromosome morphology in maize and its relatives. Bot. Rev. 7:263-289.

----, and T.A. Kato Y. 1965. Chromosome morphology of certain races of maize in Latin America. Int. Center for the Improvement of Maize and Wheat Res. Bull. No. 1. Chapingo, Mexico. 112 p.

Mangelsdorf, P.C. 1954. New evidence on the origin and ancestry of maize. Am. Antiquity 19:409-410.

----. 1974. Corn, its origin, evolution, and improvement. Harvard Univ. Press. Cambridge, Mass. 262 p.

----, and J.W. Cameron. 1942. Western Guatemala — A secondary center of origin of cultivated maize varieties. Bot. Mus. Leafl., Harvard Univ. 10:217-252.

----, and R.G. Reeves. 1939. The origin of Indian corn and its relatives. Texas Agric. Exp. Stn. Bull. 574. College Station, Texas. 315 p.

Marino, A.E. 1934. La agricultura en la Quebrada de Humahuaca (Jujuy). Lab. de Bot. de la Fac. de Agron. y Vet. Buenos Aires. 65 p. (mimeo).

McClintock, Barbara. 1959. Genetic and cytological studies of maize. Carnegie Inst. Washington Yearb. 58:452-456.

----. 1960. Chromosome constitutions of Mexican and Guatemalan races of maize. Carnegie Inst. Washington Yearb. 59:461-472.

Mochizuki, N. 1968. Classification of local strains of maize in Japan and selection of breeding materials by application of principal component analysis. p. 173-178 *In* Symposium on Maize Production in Southeast Asia. Agriculture, Forestry, and Fisheries Research Council, Ministry of Agriculture and Forestry, Kasumigaseki, Chiyoda-ku, Tokyo, Japan.

Nelson, Jr., O.E. 1952. Non-reciprocal cross sterility in maize. Genetics 37:101-124.

----. 1960. The fourth chromosome gametophyte factor in some Central and South American races. Maize Genet. Coop. News Letter 34:114-116.

Nickerson, N.H. 1953. Variation in cob morphology among certain archaeological and ethnological races in maize. Ann. Mo. Bot. Gard. 40:79-111.

----. 1954. Morphological analysis of the maize ear. Am. J. Bot. 41:87-92.

----, and A.P. Covich. 1966. A collection of maize from Darién, Panama. Econ. Bot. 20:434-440.

Parker, V., I., and O. Paratori B. 1965. Distribución geográfica, clasificación, y estudio del maiz *(Zea mays)* en Chile. Agric. Tecnica 25(2):70-86.

Parodi, L.R. 1932. Notas preliminares sobre plantas subamericanas cultivadas en la provincia de Jujuy.Gaea 4(1):19-28.

----. 1935. Relaciones de la agricultura prehispánica con la agricultura Argentina actual. Anales Acad. Nac. Agron. y Vet. de Buenos Aires 1:115-167.

Paterniani, E. 1970. Heterose em cruzamentos intervarietais de milho. Universidade de São Paulo, Escola Superior de Agricultura "Luiz de Queiroz," Departmento e Instituto de Genética Relatório Cientifico 1970:95-100. Piracicaba, S.P., Brazil. 158 p.

----. 1972. Capacidade de combinaçao de linhagens exóticas em cruzamento intervarietal. Universidade de São Paulo, Excola Superior de Agricultura "Luiz de Queirox," Departmento e Instituto de Genética Relatório Científico 1972:83-85. Piracicaba, S.P., Brazil. 128 p.

----, and M.M. Goodman. In press. Races of maize in Brazil and adjacent areas. Int. Center Improvement of Corn and Wheat Publ. Mexico City.

Patiño, V.M. 1956. El maiz Chococito, noticia sobre su cultivo en America ecuatorial. Am. Indigina 16:309-346.

Pavilićić, J. 1971. Contribution to a preliminary classification of European open-pollinated maize varieties. p. 93-107 *In* I. Kovács (ed), Proc. Fifth Meeting of the Maize and Sorghum Sect. of EUCARPIA. Akadémiai Kiadó, Budapest. 290 p.

----, and V. Trifunovic. 1966. A study of some important ecologic corn types grown in Yugoslavia and their classification. J. Sci. Agric. Res. 19(66):45-63 [also in Arhiv za Poljoprivredne Nauke 19(66):44-62].

Ramirez E., Ricardo, D.H. Timothy, Efrain Diaz B., and U.J. Grant, with G.E. Nicholson Calle, Edgar Anderson, and W.L. Brown. 1960. Races of maize in Bolivia. Natl. Acad. Sci. — Natl. Res. Council Publ. 747. Washington, D.C. 159 p.

Roberts, L.M., U.J. Grant, Ricardo Ramirez E., W.H. Hatheway, and D.L. Smith, with P.C. Mangelsdorf. 1957. Races of maize in Colombia. Natl. Acad. Sci. — Nat. Res. Council Publ. 510. Washington, D.C. 153 p.

Rodríguez, A., M. Romero, J. Quiroga, and G. Avila, with A. Brandolini. 1968. Maices Bolivianos, F.A.O. Rome, Italy. 243 p.

Sanchez-Monge, E. 1962. Razas de maíz en España. Ministerio de Agricultura Monografias No. 13. Madrid. 179 p.

Sehgal, S.M. 1966. Inbred-hybrid method of maize improvement. Proc. Caribbean Foods Crops Soc. 4:45-51.

Sprague, G.F. 1971. Genetic vulnerability in corn and sorghum. Proc. Annu. Corn and Sorghum Res. Conf. 26:96-104.

Stevenson, J.C., and M.M. Goodman. 1972. Ecology of exotic races of maize. I. Leaf number and tillering of 16 races under four temperatures and two photoperiods. Crop Sci. 12:864-868.

Stonor, C.R., and E. Anderson. 1949. Maize among the hill peoples of Assam. Ann. Mo. Bot. Gard. 36:355-404.

Sturtevant, E.L. 1899. Varieties of corn. USDA Office of Exp. Stn. Bull. 57. Washington, D.C. 108 p.

Suto, Tihara, and Yoshio Yoshida. 1956. Characteristics of the Oriental maize. p. 375-530. *In* H. Kihara (ed.), Land and crops of Nepal Himalaya. Vol. II. Fauna and Flora Research Society, Kyoto University. Kyoto, Japan. 563 p.

Timothy, D.H. 1963. Genetic diversity, heterosis, and the use of exotic stocks in maize in Colombia. p. 581-593 *In* Statistical genetics and plant breeding. Natl. Acad. Sci. — Natl. Res. Council Publ. 982. Washington, D.C. 623 p.

----, W.H. Hatheway, U.J. Grant, Manuel Torregroza C., Daniel Sarria V., and Daniel Varela A. 1963. Races of maize in Ecuador. Natl. Acad. Sci. — Natl. Res. Council Publ. 975. Washington, D.C. 147 p.

----, Bertulfo Peña V., and Ricardo Ramirez E., with W.L. Brown and Edgar Anderson. 1961. Races of maize in Chile. Natl. Acad. Sci. — Natl. Res. Council Publ. 847. Washington, D.C. 84 p.

Troyer, A.F., and W.L. Brown. 1972. Selection for early flowering in corn. Crop Sci. 12:301-304.

Van der Walt, W.J., and R.A. Brink. 1969. Geographic distribution of paramutable and paramutagenic R alleles in maize. Genetics 61:677-695.

----, Alejandro Fuentes O., and Antonio Hernández Corzo, with P.C. Mangelsdorf. 1957. Races of maize in Central America. Natl. Acad. Sci. — Natl. Res. Council Publ. 511. Washington, D.C. 128 p.

---. L.M. Roberts, and E. Hernández X., with Paul C. Mangelsdorf. 1951. Razas de maíz en México. Su origen, características, y distribución. Folleto Técnico Secretaria de Agricultura y Ganaderia No. 5. Mexico, D.F. 237 p.

----, L.M. Roberts, and E. Hernández X., with Paul C. Mangelsdorf. 1952. Races of maize in Mexico. The Bussey Inst., Harvard Univ. Cambridge, Mass. 223 p.

Wallace, H.A., and W.L. Brown. 1956. Corn and its early fathers. Michigan State Univ. Press. East Lansing. 134 p.

Wellhausen, E.J. 1965. Exotic germ plasm for improvement of corn belt maize. Proc. Annu. Hybrid Corn Ind.-Res. Conf. 20:31-50.

Will, G.F. and G.E. Hyde. 1917. Corn among the Indians of the Upper Missouri. St. Louis. (Republished 1964, Univ. of Nebraska Press, Lincoln. 323 p.)

Wolf, M.J., H.C. Cutler, M.S. Zuber, and U. Khoo. 1972. Maize with multilayer aleurone of high protein content. Crop Sci. 12:440-442.

Chapter 3 Morphology

JOHN E. SASS
Iowa State University
Ames, Iowa

DEVELOPMENT OF THE CARYOPSIS

Development of the Embryo

The Proembryo. The vegetative phase of the life of a corn plant begins with the zygote, which is derived by fusion of the egg and sperm, approximately 24 hours after pollination (Randolph, 1936). Division of the zygote occurs approximately 12 hours after fertilization (Randolph, 1936) and produces a two-celled proembryo with a transverse cell wall. The planes of subsequent cell divisions conform to the "asterad" type of proembryo development (Johansen, 1950). The term "proembryo" is used for stages prior to detectable organ initiation, and the term "embryo" for subsequent stages.

The clavate proembryo is at first externally symmetrical in longitudinal and transverse aspects; however, the internal cellular organization exhibits asymmetry by the fifth day. Localized acceleration of cell division produces an oblique zone of highly meristematic, deeply stainable cells, located on the distal-anterior end of the proembryo. This zone of activity foreshadows organ formation (Fig. 2).

Organogeny. External expression of organ formation consists of three lobes that are usually evident 10 days after pollination. The prominent "posterior lobe" gives rise to the mid-portion and proximal portion of the scutellum. The oblique "distal lobe" produces the distal extension of the scutellum. The "anterior lobe" gives rise to the plumular axis (Fig. 3). The orientation of the embryo in the caryopsis is shown in Fig. 1.

The coleoptile primordium arises immediately above the axial lobe, approximately 10 days after pollination (Fig. 4). In transverse section, the coleoptile primordium is at first a short, oblique crescent, which becomes a complete ring by progressive activation of cells under the ends of the crescent. This mode of initiation and growth is typical of the grass leaf.

The radicle primordium is usually well defined in the central region of the embryo by the 10th day (Fig. 5); however, specialized activity in this zone can be detected as early as the 5th day. The planes of successive cell divisions comprise an arc, which eventually becomes the dome-shaped tip of the radicle.

On the basis of the foregoing criteria of organ initiation, it may be concluded that the embryonic organs, root, stem, coleoptile, and scutellum become clearly defined 10 days after pollination.

The primordium of the first foliage leaf is detectable by the 12th day, as a protuberance on the anterior side of the plumular apex, opposite the coleoptile primordium (Fig. 8, 9). The first detectable activity that initiates leaf formation occurs in the surface layer, the tunica, followed immediately by cell division in the adjacent subsurface zone, the corpus (Sharman, 1942; Ledin, 1954).

In transverse section, the leaf primordium is a short crescent. Cell division in the apical dome under the ends of the crescent progresses laterally until an annular meristematic ridge is formed. This ring produces the encircling, adnate base of the leaf (Fig. 9). The distal edge of the primordium

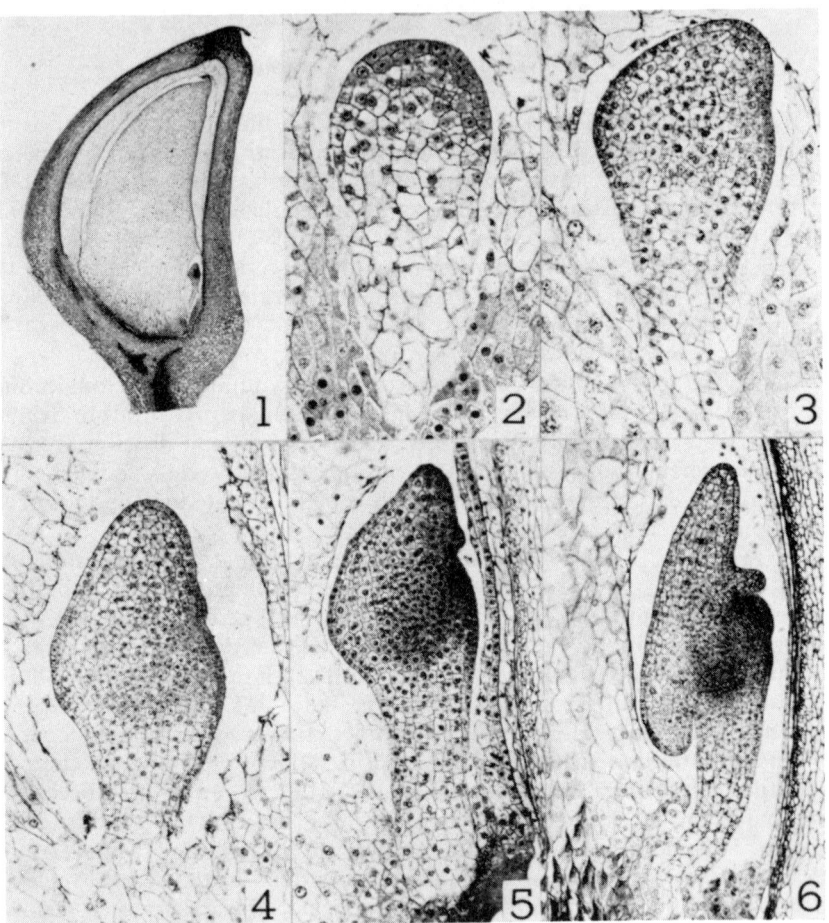

Fig. 1. Longitudinal section of kernel of popcorn, 10 days after pollination. 12×. (Posterior lobe at left, anterior at right, distal toward top of page in Figs. 1-6.)
Fig. 2. Embryo of yellow dent corn, 5 days. 130×.
Fig. 3. Embryo of yellow dent corn, 10 days. 130×.
Figs. 4, 5. Embryo of yellow dent corn, 10 days. 98×.
Fig. 6. Embryo of "smoky flint" corn, 10 days. 92×.

MORPHOLOGY

produces the free, overlapping edges of the sheath, and also the blade of the leaf.

The second foliage leaf is initiated before the 15th day, on the side of the apical dome opposite the first leaf (Fig. 10, 11). By this time the blade of the first leaf usually encircles the stem, and the leaf edges meet or overlap (Fig. 14). The coleoptile becomes an oblique cone that encloses the plumular axis by the 15th day, except for a narrow vertical slit (Fig. 11).

The time of initiation of the third foliage leaf lags appreciably in inbred lines, many of which have only two foliage leaves at 15 days, or at most a zone of activity at the locus of initiation of the third leaf (Fig. 14), whereas many hybrids have three distinct leaves. The fourth leaf is usually present by the 20th day in some hybrids (Fig. 16), whereas many inbreds and some hybrids

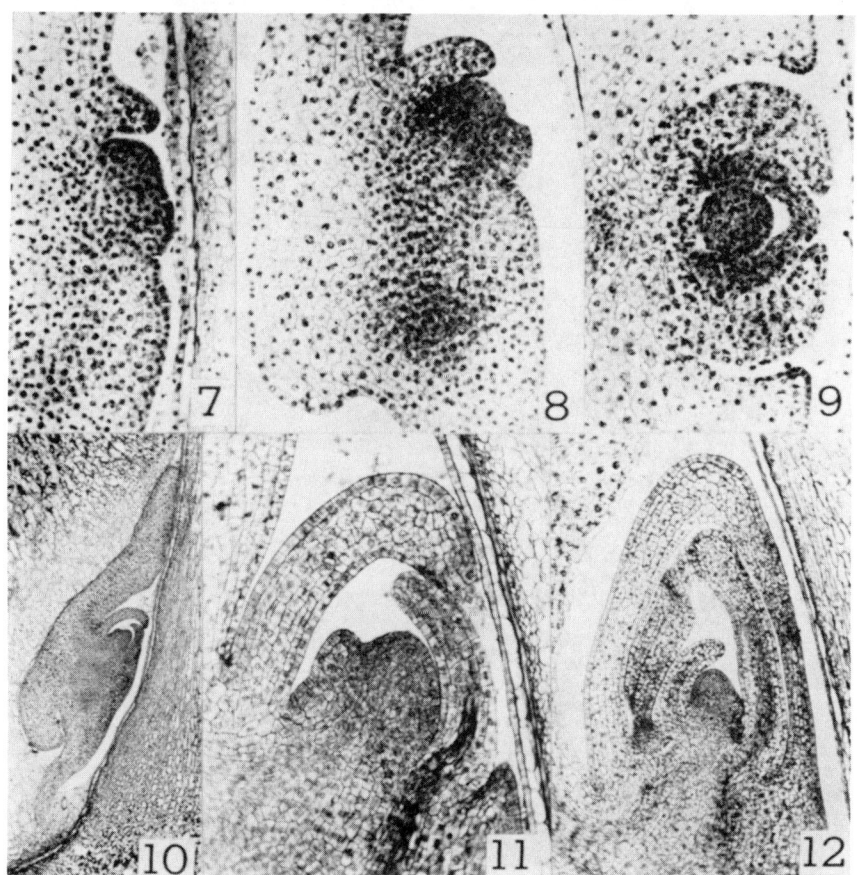

Fig. 7. Plumular primordia of "brittle" yellow dent corn, 12 days. 130×.
Fig. 8. Plumule-radicle axis of sweet corn, 12 days. 130×.
Fig. 9 Transverse section of plumular organs of waxy yellow dent corn, 12 days. 30×.
Fig. 10. Structurally normal, but retarded embryo of "germless" inbred line, 15 days. 31×
Fig. 11. Detail of plumule of inbred sweet corn, 15 days. 130×.
Fig. 12. Plumule of yellow dent hybrid, 15 days. 66×.

have only three leaves. The fifth and last embryonic leaf is commonly present 30 days after pollination (Fig. 17), and is almost invariably present before the 40th day in the numerous lines, hybrids, and varieties that have been examined (Fig. 18). Some weak inbreds have only a zone of activation at the locus of the fifth leaf.

The leaf blade expands by the activity of a continuous marginal meristem (Fig. 14, 15), and the edges of the older leaves overlap. Successive leaves attain progressively less overlapping, and the fifth leaf in the mature embryo is usually a short crescent.

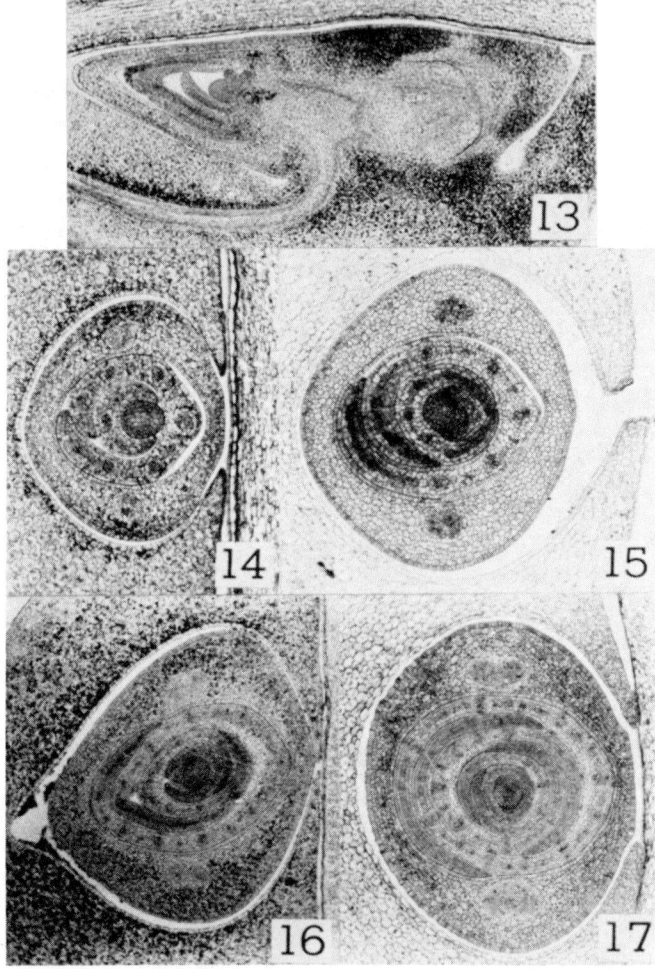

Fig. 13. Plumule-radicle axis of Iowax hybrid, 15 days. 32×.
Fig. 14. Transverse section of plumule of Iowax hybrid, 15 days. 66×.
Fig. 15. Plumule of yellow dent inbred, 20 days. 66×.
Fig. 16. Plumule of Iowax hybrid, 20 days. 40×.
Fig. 17. Plumule of yellow dent, 30 days. 40×.

In the inbred lines and hybrids that have been grown in central Iowa and studied in the writer's laboratory, the first foliage leaves are initiated earlier than in the corn described by Randolph (1936) and Kiesselbach (1949). Regional surveys of the developmental rates of adapted stocks, as well as the behavior of unadapted and exotic varieties, would provide useful information for genetic, physiological, and other studies. There is limited evidence that some very early, short varieties that have few foliage leaves at maturity, may prove to have fewer than five embryonic leaves. The exceedingly tall Central and South American types deserve further study to determine whether five is their maximum embryonic leaf number (Sass, 1951a).

Abbe and Stein (1954) described the dimensional changes of the shoot apex during embryonic plastochrones, the intervals between the initiation of successive leaves. The length and width of the apex increase during successive plastochrones, cell size remains constant, and cell number increases. The rate of leaf initiation is deceleratory during embryogeny, and acceleratory during seedling growth.

The rate of initiation and development of embryonic leaves is sufficiently consistent under given conditions to be used in studies of heterosis (Groszmann and Sprague, 1948). Fairchild (1953) has shown that leaf development is faster in the embryos of certain crosses than in the inbred parents; however, at maturity, these inbreds as well as the hybrids have five embryonic leaves.

Leaf initiation and development in the numerous kernels on an ear were found by Bell (1954) to be remarkably uniform. This makes possible the use of a relatively small sample, from any part of the ear, for diagnostic or comparative purposes.

Vascularization in the plumule is well under way 15 days after pollination, when the two strands of the coleoptile and at least the median strand of the first foliage leaf are evident (Fig. 11, 14). As a leaf continues its lateral growth, procambium strands are initiated near the active edge. (Fig. 15 to 18). The median strand is the first to exhibit differentiation of vascular elements. One protoxylem and one protophloem element can be unmistakably identified 20 days after pollination in vigorous hybrids. A comparable condition is not evident in some inbreds until 30 days. Very little further vascular differentiation occurs in the leaves. The oldest leaf bundle of the mature embryo has three or four protophloem sieve tubes, one thick-walled protoxylem element, and one or two enlarged, but thin-walled, nucleate protoxylem initials.

The two strands of the coleoptile enlarge by divisions of peripheral cells, and the compact arrangements and stainability of the cells defines the kidney-shaped strands sharply. Most of the numerous varieties that have been examined have no fully differentiated xylem cells in the coleoptile at maturity; however, a very early popcorn consistently has one or two fully differentiated, lignified protoxylem elements in each coleoptile bundle in the mature, dried grain.

The internode below the plumule, between the coleoptile node and the scutellar node, is a characteristic and unique feature of the Gramineae. The term "mesocotyl" has been used extensively in the literature to designate the first internode. This internode is not evident 15 days after pollination, when the coleoptile node is virtually in contact with the scutellar node (Fig. 13). Subsequent intercalary growth produces the long internode (Fig. 20).

Anatomically, most of the first internode in the embryonic axis resembles the root. The broad parenchymatous cortex and the stele are delimited by a poorly defined endodermis. The enlarged metaxylem initials of the exarch xylem can be identified with certainty in the stele (Fig. 22), but the other vascular elements do not become so obvious until after germination. Toward the distal end of the first internode, the vascular system undergoes transition to discrete collateral bundles scattered throughout the ground parenchyma. The first internode is inserted at the top of a long disc, the scutellar node. If the scutellum is regarded as the homologue of a cotyledon, all structures

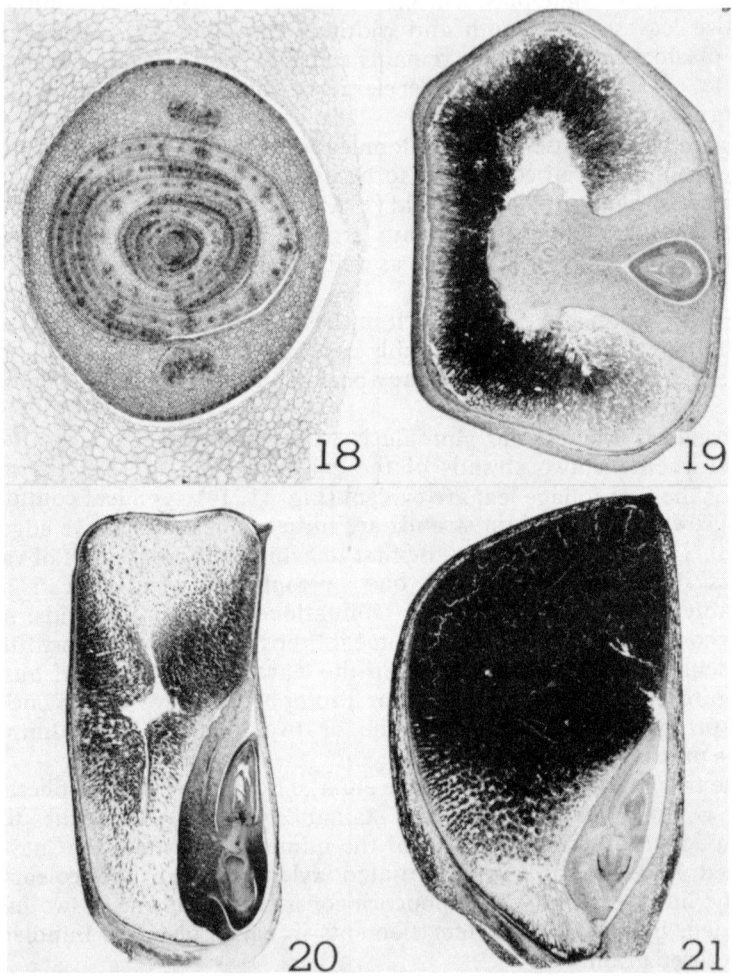

Fig. 18. Plumule of yellow dent single cross, 40 days. 41×. (From Fairchild, 1953.)
Fig. 19. Transverse section of kernel at level of stem tip, hybrid 939, 30 days. 8×.
Fig. 20. Longitudinal section of kernel of Iowax hybrid, 20 days. 5×.
Fig. 21. Longitudinal section of kernel of inbred white popcorn, 20 days. 9×.

MORPHOLOGY

above the scutellar node comprise the epicotyl, and all structures below the node are hypocotyl. The radicle and the associated coleorhiza would then comprise the hypocotyl.

The early initiation of the embryonic radicle has been mentioned briefly in a previous section. Cell division continues rapidly in the zone of radicle initiation, and a zone of cells in the form of a dome becomes well defined between the 10th and 15th day. Near the periphery of this dome, a substance of

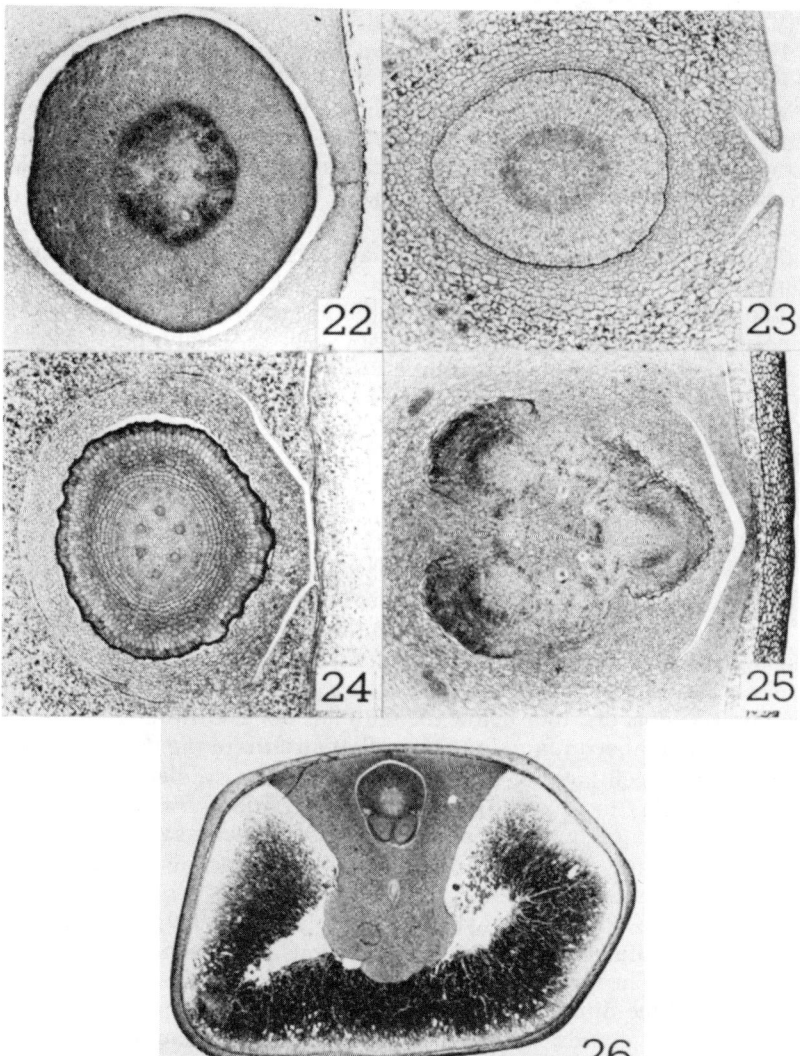

Fig. 22. Transverse section of first internode (mesocotyl) of dry mature kernel, yellow dent. 31×.
Fig. 23. Transverse section of radicle in embryo of yellow dent inbred Os420, 20 days. 65×.
Fig. 24. Transverse section of radicle and coleorhiza of yellow dent, 30 days. 39×.
Fig. 25. Section through seminal roots of yellow dent hybrid 939, 30 days. 31×.
Fig. 26. Transverse section of kernel of hybrid 939, at base of first internode, 35 days. 8×.

unknown composition is deposited between the cell walls. This deposition, the dark-stained layer in Fig. 13, 21, and 23, is assumed to interrupt organic continuity between the adjacent cells, and a cleft appears before the 20th day. The cleft elongates during subsequent elongation of the radicle, and separates the radicle from the cylindrical coleorhiza (Fig. 20, 24). The cleavage does not extend across the tip of the radicle (Fig. 13); therefore, continuity of tissues between the root tip and the coleorhiza is maintained until germination.

A transverse, stratified zone of cells develops across the end of the radicle between the 15th and 20th day, and the calyptrogen is evident 20 days after pollination (Fig. 20).

Vascular development in the radicle begins 12 to 15 days after pollination. A central core, the future stele, becomes evident by virtue of the difference in cell size, cell orientation, and staining density in the stele, in contrast with the encasing cortex. Linear columns of metaxylem vessel initials become appreciably enlarged by 15 days (Fig. 13), and protoxylem initials are considerably enlarged by 30 days (Fig. 24). Protophloem does not become fully differentiated until germination (Picklum, 1953).

The scutellum is a distinctive organ that has neither the mode of origin nor mature structure of any other embryonic organ. A major part of the bulk of a 10-day embryo is tissue that will be incorporated into the scutellum (Fig. 4). The most rapid growth occurs at first in the distal lobe, but in 15 to 20 days the proximal lobe curves under the coleorhiza and ruptures the suspensor (Fig. 10, 13). Lateral growth of the edges of the scutellum is also very rapid. An active ridge arises on the anterior face by 12 days (Fig. 9), extends part way around the axis by 15 days (Fig. 14), and completely encloses the axis by 20 days (Fig. 16).

A median procambium strand is barely detectable in the scutellum of weaker inbreds by 15 days (Fig. 10), whereas vigorous hybrids have a large, well-developed strand extending far into the distal lobe at 15 days (Fig. 13). Minor lateral procambium strands ramify irregularly into the distal portion of the scutellum. The proximal lobe of the scutellum has a few strands that extend only a short distance below the scutellar node. Differentiated vascular elements have not been found in the scutellar strands in the mature kernel.

The surface cell-layer of the scutellum has been studied and discussed by various authors in relation to the enzyme-secreting and food-absorbing function of the scutellum. The "epidermal" cells are elongated anticlinally, and they have dense, deeply stainable protoplasm. In some lines of corn the surface of the scutellum is smooth, with only minor creases and lobes, whereas in some lines and hybrids, notably Iowa 939, the surface is remarkably irregular, with infoldings, crenations, and even haustorium-like lobes (Fig. 19). The presence of folds, which may not have been studied adequately in three-dimensional aspect, has led to the interpretation that the invaginations are glands. This question should be re-examined by precise determinations of relative enzyme activity along the smooth, as well as the infolded, locations on the scutellum.

The scutellum is attached over a large area of a long, cylindrical portion of the root-stem axis. This scutellar node is rootlike in having a broad cortex, a distinct stele, and a radial, exarch vascular pattern. In the mature kernel protophloem and protoxylem initials can be identified, and the metaxylem elements are greatly enlarged, but not structurally mature. Vascular traces pass from the scutellum to the scutellar node near the top of the node, slant abruptly downward, and extend across the mid-region of the node, where they comprise the vascular plate (Fig. 13, 20).

The mature kernel of corn almost invariably contains four roots, the radicle and three "seminal" roots. The latter are initiated before the 30th day after pollination in many hybrids and some inbred lines, but initiation lags in some inbreds. The two posterior seminal roots arise at the outer limits of the stele, near the top of the scutellar node. They grow upward and emerge into the notch between the scutellum and the first internode (Fig. 26). The anterior seminal root arises somewhat lower and grows through the cortex nearly horizontally, as in Fig. 25. Few cross sections show the three seminal roots in one plane. The genesis and development of tissue systems is apparently identical in seminal roots and in the radicle. The relative time of initiation and degree of development at maturity in inbred lines and hybrids needs further study in relation to expressions of heterosis (Fairchild, 1953).

In the foregoing review of the embryology of corn, some salient features are the unique character of the scutellum and coleoptile, and the early initiation of the organ systems. The scutellum is entirely unlike any other organ, with respect to place and mode of initiation, vasculation, and function. The coleoptile arises from the scutellum, not from an axis, its mode of initiation is leaflike, but its subsequent development, tissue structure, and vasculation are unique.

The inception of the root-stem axis is detectable at 5 days, and it is possible that further critical study may reveal that the pattern of cell division in the proembryo foreshadows axial initiation less than 5 days after pollination. The time intervals in the formation of organs will certainly receive further attention, in relation to heterosis, as well as in regional, seasonal, and varietal studies. It is particularly desirable to study Central and South American varieties, grown in their habitats as well as in new environments.

The principal goal of this chapter is the presentation of the histological development of the vegetative plant body. However, the intimate developmental relationship between the embryo and the rest of the kernel made desirable a brief review of the extraembryonic features of the kernel.

The Integuments and Nucellus

The development of the integuments and the nucellus prior to fertilization has been described in another chapter. Disintegration of the integuments begins 3 days after fertilization and progresses so rapidly that remnants are difficult to recognize by 20 days. Varieties of maize certainly differ in the persistence of the integuments, but in no case does integumentary tissue remain as a significant "seed coat" in the mature kernel.

The enlarging endosperm invades the nucellus rapidly after fertilization. In 10 to 12 days the nucellus is almost obliterated (Fig. 1), and at maturity only an obscure "nucellar membrane" remains (Randolph, 1936). Whether the remnants of the integuments and nucellus have any functional significance in the dormant or germinating kernel has not been determined.

The Endosperm and Pericarp

The derivation of the endosperm as one of the products of double fertilization has been known for many years. The fusion of one of the sperms with the two polar nuclei produces the triploid primary endosperm nucleus. This nucleus divides within 4 hours after fusion, and subsequent rapid division of the derivative nuclei produces a free-nucleate endosperm. Between the 3rd and 4th day after fertilization, cell wall formation is initiated, and by the 5th day the endosperm is entirely cellular.

The endosperm enlarges very rapidly in linear and transverse dimensions, and apparently compresses and ruptures the thin-walled cells of the nucellus. By the 12th day, the endosperm occupies nearly the entire nucellar cavity. Mitotic figures are infrequent in preparations of 12-day endosperm, and very rare at 20 days. Cell division ceases first in the central region of the endosperm and persists longest in the peripheral cells, where the predominantly periclinal divisions produce a stratified zone. The outermost cell layer of the endosperm differentiates into the aleurone (Fig. 27).

Considerable starch is evident in the endosperm approximately 12 days after pollination, a little earlier in popcorn than in dent corn. Younger kernels have large plastids in the endosperm, but the starch does not give a convincing color test with iodine. Starch does not accumulate in sufficient quantities for the iodine test until the endosperm has filled the nucellar space.

The polarizing microscope reveals starch grains that are too small to detect with iodine. The characteristic appearance of starch by polarized light is shown best by grains in regions of the kernel in which the grains are not tightly packed in the cells (Fig. 28). The narrow, dark central cross rotates as the analyzer is rotated. A strikingly different pattern is evident in grains that are tightly packed, as in the distal region of the popcorn kernel, or at the level of the scutellar tip in dent corn (Fig. 27, 29). This change in the relationship

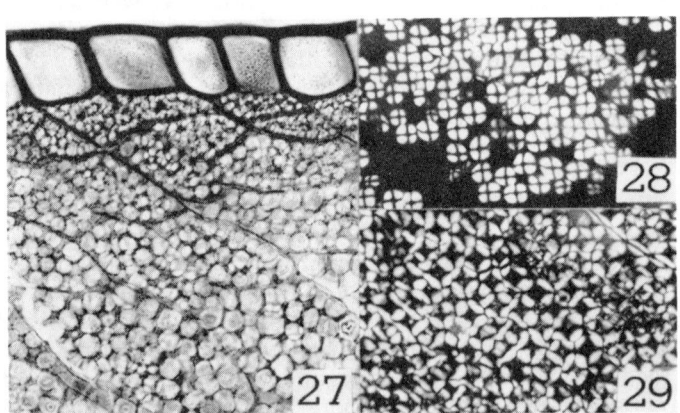

Fig. 27. Detail of endosperm starch and aleurone of yellow dent, at level of scutellar tip, 35 days. 280×.
Fig. 28. Starch by polarized light, near base of kernel of popcorn, 35 days. 280×.
Fig. 29. Starch by polarized light, at stylar end of popcorn, 35 days. 280×.

between the light and dark zones suggests a change in the structure of the "compressed" granules.

The pattern of starch distribution in the kernel becomes well established in 15 to 20 days. The numerous kinds of corn fall into two types with respect to this pattern. In flint and popcorn, the starch is most densely packed in the stylar region, and the density tapers off toward the base (Fig. 21). In dent corn, the zone of greatest density is in the form of an open cylinder that partly encircles the embryo (Fig. 19, 20).

HISTOLOGY OF THE VEGETATIVE PLANT BODY

Cytohistology of Germination

Germination of the kernel follows an orderly pattern of biochemical and morphological processes. Bernstein (1943) and Toole (1924) have described the mobilization and movement of food reserves during germination. Picklum (1953) studied imbibition of water by the kernel and the cytohistology of germination. He demonstrated that water enters primarily through the pericarp, and to a lesser degree through the fractured pedicel. The cells of the coleorhiza and radicle tip are the first to become turgid. Linear growth of cells of the coleorhiza pulls the coleorhiza away from the root cap (Fig. 30), and the elongating coleorhiza emerges. Subsequently, mitosis and cell elongation begin in the radicle, approximately 24 hours after the beginning of imbibition. Within 48 hours after the emergence of the primary root, the primordia of the lateral roots may become evident.

Mitosis begins in regions of the embryo in the following sequence: Midway between the root tip and the scutellar plate, progressing toward the tip; the cortex at the base of the first internode (mesocotyl); the base of the coleoptile, the leaf margins; the stem tip. Emergence of the plumule and its elevation to the surface of the soil occur mainly by the activity of the meristematic zone below the coleoptile node. Picklum (1953) found that under the specified conditions postembryonic leaves were initiated at 60-hour intervals.

During germination, the chromatin stains more deeply in the epithelial cells of the scutellum than in the internal cells, but no other cytological changes occur in the epithelial cells. The parenchymatous cells within the scutellum are uninucleate prior to germination. During germination, the nuclei become deeply lobed, and the lobes may become abstricted, forming multinucleate cells.

The pericarp over the scutellar node is ruptured by the expansion of a cushion-like mass of tissue on the anterior side (Fig. 31). Pressure exerted by the elongating first plumular internode may be a minor factor in rupturing the pericarp. The histological development of the plumule and radicle are described in subsequent sections.

The Stem

The Apical Meristem. The stem apex, which is initiated on the anterior face of the proembryo approximately 10 days after pollination, maintains structural and functional continuity as a vegetative apex until the tassel primordium is initiated. The shoot apex of the seedling, like that of the embryo, has a single surface layer of cells, the tunica, in which the planes of cell division are almost exclusively anticlinal. The second layer structurally resembles the tunica in some apices, but in other apices, especially very active ones, the second layer has many periclinal as well as anticlinal divisions. Such activity is intermediate between that of the tunica and the more or less random planes of division in the central corpus. The danger of rigid designations of meristematic zonation is evident.

Organogeny. The formation of foliage leaves on the stem apex is resumed soon after the onset of germination (Fig. 32). The factors that incite this reactivation and control the rate of leaf initiation will require further study. The mechanism of leaf formation on the seedling apex appears to be the same as on the embryonic apex. The process is terminated by a transition phase, which leads to the initiation of the inflorescence (Bonnet, 1953, Leng, 1951).

Axillary bud primordia arise very early in stem ontogeny, near the base of the third or fourth leaf below the tip. The primordium actually arises in the internode above the subtending leaf. These axillary primordia become vegetative tillers near the base of the plant, and potential ear shoots at the upper levels.

Histogenesis. The structure of the histogens of the stem apex has been reviewed by Bonnet (1953). The derivation of tissue systems will be outlined here. Procambium can be detected in the stem axis of corn at the approximate level of the second leaf primordium below the tip. Strands arise from ground meristem cells, which in turn are derived from corpus cells of the apex. It is possible that a strand arises from a single vertical column of cells, but such derivation is difficult to demonstrate because a strand can not be identified with certainty until it consists of 4 to 6 cells in transverse aspect, and after these cells have undergone some vertical elongation (Fig. 33). The individual procambial cell is an elongated prismatic cell with relatively dense cytoplasm and a large, elongated nucleus. The strand increases in diameter by longitudinal cell division. The tissue in which the strands are encased is best designated residual meristem (Esau, 1943), in which cell division occurs with decreasing frequency as the stem matures. Most of this tissue differentiates into the interfascicular "pith" parenchyma of the mature stem.

The differentiation of the procambium strand into a vascular bundle has been described in detail by Sharman (1942) and Esau (1943). The unique stainability of the first protophloem element makes this cell readily distinguishable very early in the ontogeny of the bundle (Fig. 34). The first protoxylem element becomes evident soon thereafter (Fig. 35). It is possible that further development of staining and optical techniques may show that protophloem and protoxylem arise essentially simultaneously in the strand.

The first xylem element is an annular trachea, also known as a tracheal tube, or vessel. Before this xylem element has become mature, a centrifugally adjacent column of procambium cells begins to enlarge. These cells also develop into a protoxylem trachea. Similarly, two or more tracheae develop progressively in centrifugal sequence (Fig. 36). The first and second tracheae invariably have annular secondary wall thickenings. Subsequent tracheae may be annular or spiral protoxylem tracheae. A vessel segment can have both annular and spiral thickenings. The bundles toward the periphery of the

MORPHOLOGY

stem, which become mature after a given level of the stem has ceased elongation, do not have spiral tracheae, and may have little or no protoxylem (Sass, 1951b).

A zone of cells external to the outermost protoxylem goes through a short period of cambiform activity and produces a zone of stratified derivatives (Fig. 37). Cell division is not limited to a single layer, and the useful but figurative term "cambiform" does not imply homology with the vascular cambium of dicotyledonous plants. On the lateral edges of this zone, two procambial cells undergo rapid enlargement (Fig. 37) and develop into pitted metaxylem tracheae, the most prominent elements in the vascular bundle in grasses. Dissolution of the end walls of the original cells produces a continuous tube with "porous" ends between the segments. The side walls of

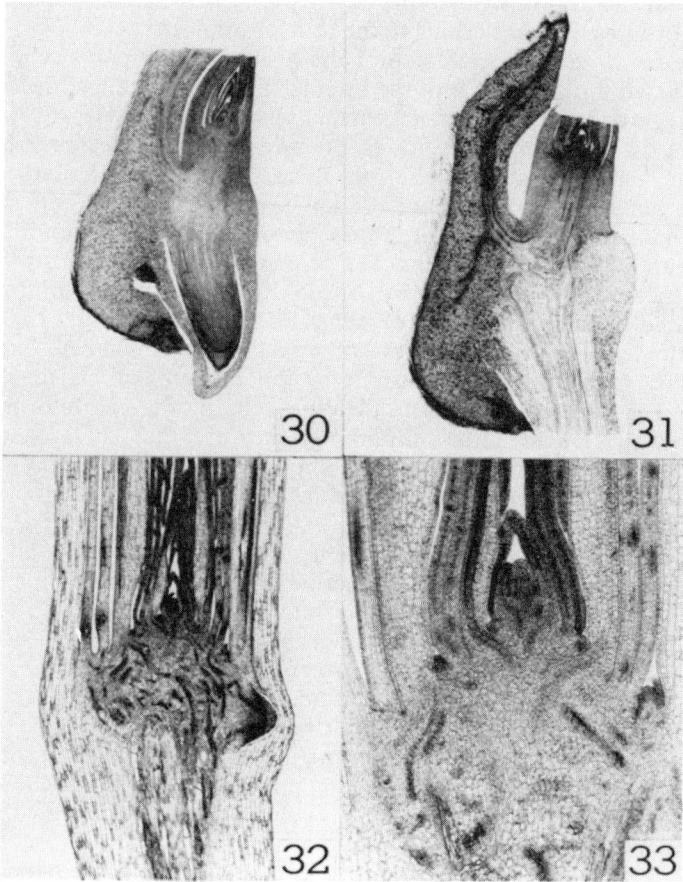

Fig. 30. Embryo of yellow dent corn, 24 hours after beginning of imbibition. 7×. (From Picklum, 1953).
Fig. 31. Portion of embryo after 36 hours. 7×. (From Picklum, 1953.)
Fig. 32. Plumule at coleoptile node, 96 hours. 13×. (From Picklum, 1953.)
Fig. 33. Longitudinal section of seedling stem, 1 week after emergence above soil level. 40×.

these metaxylem tubes are profusely pitted. The spirally arranged pits may be so close and numerous that the wall is literally a reticulum. The inner derivatives of the aforementioned cambiform zone, between the large metaxylem vessels, differentiate into metaxylem elements. These cells are prismatic in three-dimensional aspect, and their transverse or slightly oblique end walls have several large pits, forming a reticulate end. Esau (1943) uses the term "tracheary elements" for these cells. The term "trachea" would not be inappropriate, but the term "tracheid" is not valid.

During the period of rapid elongation and increase in diameter of the young stem, the first-formed protoxylem elements, the annular and spiral tracheae, become stretched beyond the limit of elasticity of the primary wall. The cells become ruptured and the annular and spiral thickenings become dislodged. Several ruptured vessels merge into an irregular vertical cavity, the protoxylem lacuna (Fig. 38). This cavity probably contains standing water that contributes little or nothing to the transpiration stream.

A zone of cells on each side of the protoxylem remains relatively undifferentiated during the life of the bundle. These thin-walled, nucleate cells may be regarded as xylem parenchyma, of obscure function (Fig. 38). The ontogeny of the diverse components of the xylem has been reviewed here with unbroken continuity, and the phloem will be discussed in a similar manner.

The remarkably early differentiation of the first protophloem sieve tube is now widely recognized. Several more sieve tubes develop rapidly, without the formation of companion cells. The protophloem sieve tubes are enucleate when mature. The outer derivatives of cell division in the cambiform zone in the bundle develop into metaphloem. Each phloem initial cell divides longitudinally and produces a sieve tube and a companion cell. Esau (1943) has shown that the orderly pattern of the metaphloem is the result of derivation from the tangentially dividing cells of the cambiform zone of procambium. During the differentiation of the metaphloem, the protophloem becomes progressively compressed and almost obliterated. In the mature bundle the crushed protophloem persists as an irregular layer of stainable material (Fig. 38).

The peripheral cells of a developing vascular bundle continue to divide and elongate during the period of internodal elongation. After the cessation of elongation, these cells differentiate into a sclerenchymatous sheath that encases the vascular bundle (Fig. 39). A typical mature sheath cell is hexagonal in section and the small lumen is circular in section. The thick, laminated, lignified wall is sparsely pitted. The cell length is at least 50 times the diameter, and the cell ends are slightly oblique.

The peripheral bundles of the stem have wider sheaths than the bundles toward the center (Fig. 39). Bundle shape is fairly constant in the lines and hybrids that have been studied, and some lines have strikingly distinctive bundle shape.

The tissue between bundles in the center of the stem is typical "pith" parenchyma. The cells are large, the thin cellulose walls give a cellulose reaction, and large intercellular spaces are present (Fig. 38). The cytoplasm is a thin film around the large central vacuole.

MORPHOLOGY

Outward from the center of the stem, the parenchyma cells are progressively smaller and the intercellular spaces are also smaller. In this transitional zone the cells resemble collenchyma, as pointed out by Esau (1943). The cell walls become thicker as the periphery of the stem is approched, until a zone is reached in which the interfascicular cells are typical sclerenchyma (Fig. 40). The hypodermal zone is especially strikingly sclerified. The epidermal cells are small in transverse section, and the walls are greatly thickened. Sunken stomates occur in the linear pattern that is typical in grasses (Fig. 41). Chlorenchyma is present in the hypodermis, in the region of the stomates.

The zone of cells that includes the epidermis and extends inward through the schlerified parenchyma has been designated the "rind," for lack of a better term. This zone is of particular interest in relation to stalk stiffness, to lodging resistance, and possibly to resistance to insect and fungal damage. Magee (1948) has shown that a high percentage of sheath per bundle and a wide schlerified peripheral zone are correlated with stalk stiffness.

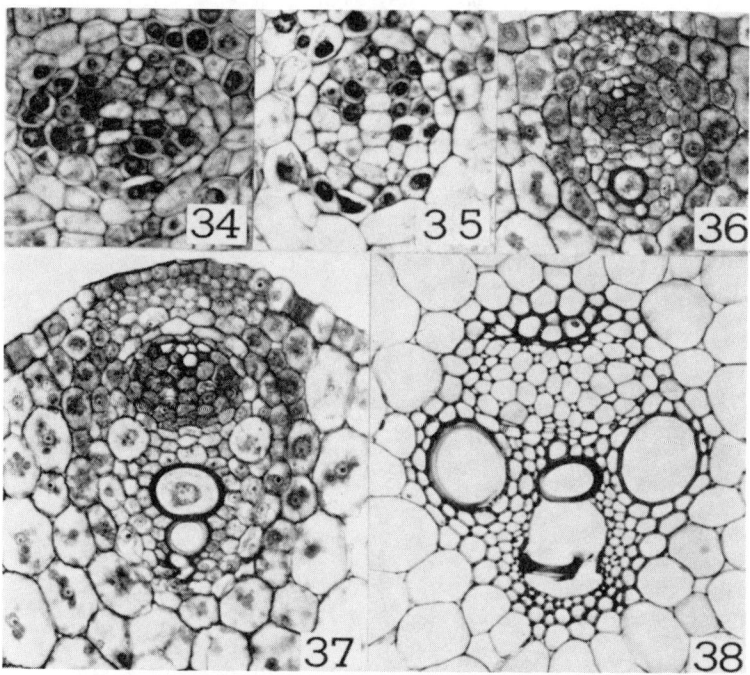

Fig. 34. Procambium strand with 1 protophloem sieve tube. 490×. (From Esau, 1943.)
Fig. 35. Procambium strand with two protophloem sieve tubes and one mature protoxylem element. 490×. (From Esau, 1943.)
Fig. 36. Developing vascular bundle with group of sieve tubes, cambiform layer and differentiating xylem elements. 260×.
Fig. 37. Vascular bundle with enlarged metaxylem vessels and cambiform zone. 260×.
Fig. 38. Bundle of central region of stem of stiff-stalked synthetics. 198×.

The Root

Histogenesis in the Apical Meristem. The early differentiation of the apical meristem of the radicle has been described in the section on embryology. The apical histogens become well defined long before the kernel is mature, and remain structurally constant during the subsequent emergence and growth of the root. The outermost histogen or tissue-generating zone is the calyptrogen, a single, transverse, saucer-shaped layer of cells from which the root cap is derived. Some cell division occurs in the stratified derivatives of the calyptrogen. The outermost region of the root cap becomes spongy, and the cells eventually slough away (Fig. 42). Continuous replacement occurs by activity of the calyptrogen.

The second generating layer, located immediately behind the calyptrogen, consists of the dermatogen-periblem initials. This transverse layer,

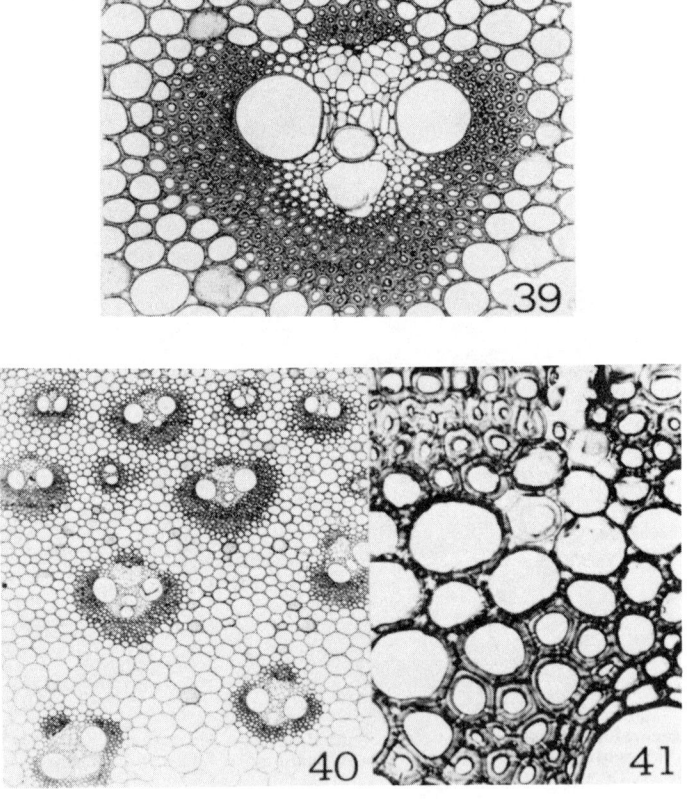

Fig. 39. Peripheral vascular bundle of stiff-stalked synthetic. 134×.
Fig. 40. Pattern of bundle distribution near periphery of stem of stiff-stalked synthetic. 40×.
Fig. 41. Detail on periphery of stem of stiff-stalked corn. 350×.

MORPHOLOGY

one cell thick and three to five cells wide, can be identified by activity rather than by structural differentiation. The cells in the central region of the layer divide in anticlinal planes, parallel to the axis of the root. The marginal cells divide periclinally, tangential to the domed apex. The outer derivatives of this division divide almost exclusively anticlinally and thereby maintain the continuity and identity of a surface layer. The term "dermatogen" may be applied to this surface layer, which ultimately becomes the hair-producing epidermis.

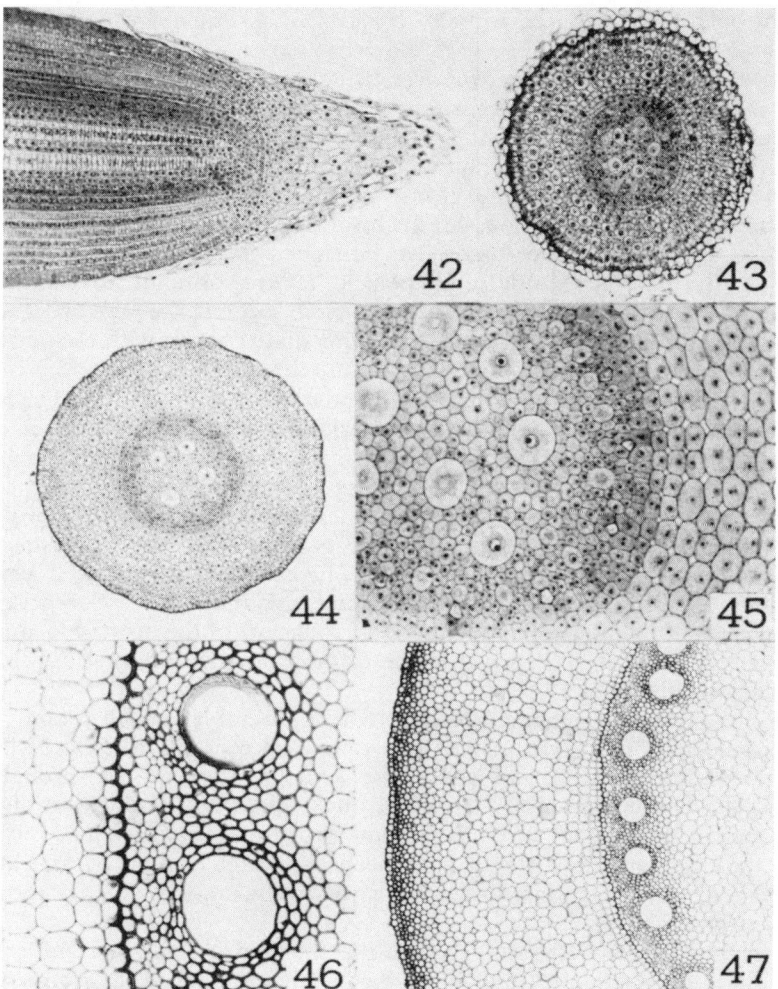

Fig. 42. Longitudinal section of root tip. 63×.
Fig. 43. Transverse section of root near tip, in zone of active cell division. 78×.
Fig. 44. Transverse section in zone where cell division has ceased and tissue differentiation has begun. 78×.
Fig. 45. Detail at stele-cortex interface of immature root. 130×.
Fig. 46. Detail at stele-cortex interface of root with fully differentiated tissues. 120×.
Fig. 47. Mature root showing all tissues from epidermis to pith. 40×.

The inner derivatives of the dermatogen-periblem initials give rise to the cortex. These derivatives undergo numerous periclinal divisions and thereby broaden the periblem, a zone that represents a transitional phase in the development of the cortex. The inner zone of the periblem is particularly active, producing a stratified zone by tangential divisions that may well be designated "cambiform" activity (Fig. 42, 43). With the cessation of this activity, the innermost cell layer becomes the immature endodermis, the inner limiting layer of the cortex (Fig. 45).

The third generating layer of the root tip, located immediately behind the dermatogen-periblem initials, consists of a single layer of cells, the plerome initials, which give rise to the entire plerome, a transitional phase in the development of the stele. However, evidence is accumulating that the concept of the plerome as a transitional phase is losing its validity, because vascular elements are known to arise almost immediately from the generating initials. As shown in connection with the development of the radicle in the young embryo, linear columns of metaxylem initials can be traced almost to the tip of the radicle (Fig. 42, 43). In favorable preparations of the emerged radicle, the metaxylem columns can be traced to the second or third cell behind the plerome initials. Phloem cells are difficult to identify in longitudinal sections of root tips. In transverse sections near the tip, it is easy to identify protophloem sieve tubes, in immediate contact with the pericycle. The latter is a single layer of cells adjacent to the endodermis (Fig. 45).

Vascular elements are differentiated centripetally. The first protophloem cell, in contact with the pericycle, is a sieve tube, which arises directly from a histogen cell without the formation of a companion cell. Two or more protophloem tubes arise successively in a similar manner. This is followed by the differentiation of metaphloem, which consists of sieve tubes and companion cells. A sieve tube and its companion cell are sister cells, derived from a histogen cell. The protophloem is not appreciably crushed during the development of the metaphloem, but the latter comprises the major part of the phloem. The functional longevity of the protophloem is not known. The mature phloem is in the form of strands, alternating with the xylem arcs (Fig. 46).

Protoxylem elements become distinguishable much later than protophloem or metaxylem. The first protoxylem cells are in contact with the pericycle. Successive zylem cells arise centripetally and go through the enlargement phase progressively. The thickening and lignification of walls then begins in the outermost element and progresses inward. Thus, it is clear that the large innermost elements of the xylem, which begin to enlarge almost at the root tip, are the last to become structurally mature, and are therefore metaxylem (Fig. 47).

A zone of cells between the xylem and the phloem has poorly understood affinities. The cells are small in diameter, closely packed, and have thick, lignified, pitted walls. The cells are structurally identical to the cells on the radial and inner sides of the metaxylem. A critical study of the derivation and pitting of the nontracheal lignified elements may clarify the affinities of such cells (Fig. 46).

The central tissue of the root is largely parenchymatous. The peripheral zone merges progressively into the previously mentioned lignified cells, whereas the nucleate central cells are loosely packed and have thin, cellulose walls. The term "pith" is a convenient designation for the parenchymatous central tissue.

Returning to the development of the cortex, it will be recalled that the maximum number of layers of cells in the cortex is established within a few millimeters of the root tip. The cell walls of the endodermis become thickened and lignified, and intercellular spaces develop between the enlarging, radially stratified cortical parenchyma (Fig. 46). At the upper limits of the root hair zone, the epidermal hairs and one or more layers of hypodermal parenchyma begin to collapse, and several layers of closely fitted subepidermal cells develop thick, lignified walls. This tough hypodermal sclerenchyma becomes the permanent protective layer of the root (Fig. 47).

Organogeny. The place, time and manner of initiation of lateral organs of the root differs strikingly from the course of events in the stem. The meristematic potential of the pericycle of the root is well known. Lateral roots emerge a considerable distance from the tip, at a level where the differentiation of vascular tissues is well advanced; however, the place of initiation is much nearer the tip. The locus of initiation is the pericycle, opposite the protoxylem "points." The histogens of the lateral root become organized before the root has grown halfway through the cortex. The structure of the histogens and of the derivative tissues is identical in a lateral root and in the root from which it arose. Vascular differentiation occurs first in the lateral root at its base, establishes vascular continuity with the original root, and progresses distally.

The Leaf

The intimate relationship between the stem apex and leaf initiation made it desirable to describe leaf initiation in a previous section. After the encircling base of the leaf becomes established, the highly meristematic distal edge of the leaf primordium produces the free edges of the blade. Continued activity of the marginal meristem of the leaf broadens the leaf, and the edges soon overlap (Fig. 14). The edge consists of a single row of cells. The number of cell layers increases rapidly, and the maximum number is attained a short distance from the edges (Fig. 15). In the surface layers, cell division parallel to the surface ceases approximately three cells from the edge, thereby establishing the identity and continuity of the "dermatogen," which becomes the epidermis. Anticlinal divisions continue in the surface cells, associated with the increase of leaf area. Increase in leaf thickness occurs thereafter by increase in the depth of epidermal cells, as well as by enlargement of the mesophyll cells, which become highly spongy (Fig. 49).

Procambium strands can be identified approximately 10 cells from the leaf margin. Strands arise from the meristematic mesophyll near the leaf edge. The differentiation of the major bundles of the blade is identical with the process in stem bundles, and the same pattern of vascular tissues occurs at maturity; however, the sheath of these major leaf bundles is distinctive. On the upper and lower sides of the bundle, a strand of sclerenchyma extends lengthwise of the leaf, in contact with the epidermis (Fig. 49). The sheath cells along the sides of the bundle have thickened and lignified inner walls. This is suggestive of endodermis, but the cells contain large chloroplasts, most of which are against the outer wall.

A second type of leaf bundle lacks the annular and spiral protoxylem elements and the two large metaxylem vessels. These bundles have only a few metaxylem and metaphloem elements. The prominent bundle sheath consists of large, thin-walled cells that contain large chloroplasts, mostly against the outer wall. These bundles lack the strands of sclerenchyma, and thus are completely enveloped in the spongy parenchyma of the mesophyll. Relatively few blade bundles are structurally intermediate between the major and minor bundle types.

In the midrib and in the adjacent blade the bundles are close to the lower surface. Most of the bundles are of the major type, anchored to the lower epidermis by sclerenchyma (Fig. 48). The mesophyll of the rib consists of thin-walled, polygonal, closely fitted parenchyma with few chloroplasts.

During the differentiation of the lower epidermis, the cells enlarge to approximately uniform size, except in the vicinity of the stomates. The upper epidermis develops a pattern that is characteristic of the grasses. The cells adjacent to the stomates are small in transverse section. Cell size increases progressively away from a stomate, reaches a maximum, and then becomes smaller to the next stomate (Fig. 49). The large epidermal cells are the

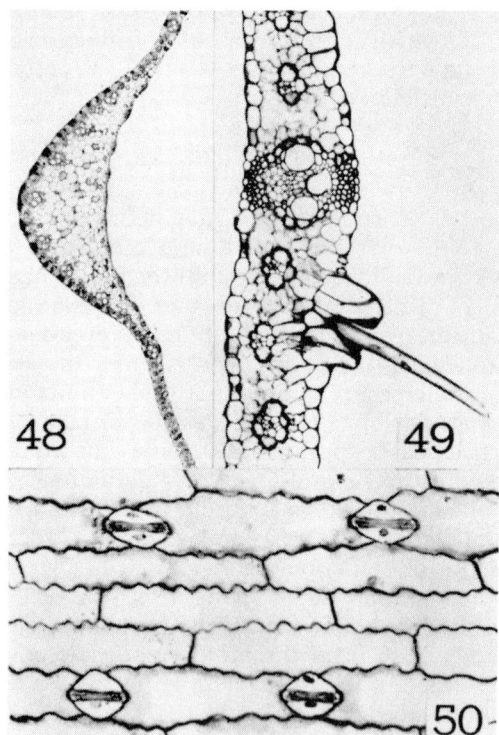

Fig. 48. Transverse section of leaf of sweet corn. 9×.
Fig. 49. Detail of leaf blade of dent corn. 64×.
Fig. 50. Surface view of epidermis. 256×.

bulliform or motor cells. Their change of turgor in relation to water supply and transpiration is associated with the curling and uncurling of the leaf blade.

The surface view of the epidermis conforms to the well-known pattern of the grass leaf (Fig. 50). The stomates occur in rows extending lengthwise of the leaf. The guard cells are surrounded by the accessory cells. The ordinary epidermal cells are long and narrow, with undulating side walls.

Inbred lines, hybrids, and varieties of corn exhibit striking differences in the morphology of the epidermis, as well as other leaf characters, such as spacing of vascular bundles, the frequency of cross veins, the amount of bundle sclerenchyma and many other features. Such morphological information has not been systematically and extensively accumulated, and only limited correlation with agronomic properties is possible at present.

EDITOR'S NOTE

This chapter is reprinted, with minor alterations, from the chapter, "Vegetative Morphology," in the original Corn Monograph, Agronomy 5. The author is deceased. Headings and text have been changed to conform to the rest of this book.

LITERATURE CITED

Abbe, E.C., and Stein, O.L. 1954. The growth of the shoot apex in maize embryogeny. Am. J. Bot. 41-285-293.

Bell, M.E. 1954. The development of the embryo of *Zea* in relation to position on the ear. Iowa State Coll. J. Sci. 29:133-139.

Bernstein, L. 1943. Hybrid vigour in corn and the mobilization of endosperm reserves. Am. J. Bot. 30:801-809.

Bonnet, O.T. 1953. Developmental morphology of the vegetative and floral shoots of maize. Univ. Ill. Agric. Exp. Stn. Bull. 568:5-47.

Esau, K. 1943. Ontogeny of the vascular bundle in *Zea Mays*. Hilgardia. 15:327-356.

Fairchild, R.S. 1953. Comparative development of the embryos of inbred and hybrid maize. Iowa State Coll. J. Sci. 27:381-405.

Groszmann, A., and Sprague, G.F. 1948. Comparative growth rates in a reciprocal maize cross. I. The kernel and its component parts. J. Am. Soc. Agron. 40:88-98.

Johansen, D.A. 1950. Plant embryology; embryogeny of the spermatophyta, Chronica Botanica Co., Waltham, Mass. p. 305.

Kiesselbach, T.A. 1949. The structure and reproduction of corn. Nebraska Agric. Axp. Stn. Res. Bull. 161:3-96.

Ledin, R.B. 1954. The vegetative shoot apex in *Zea Mays*. Am. J. Bot. 41:-11-17.

Leng, E.R. 1951. Time-relationships in tassel development of inbred and hybrid corn. Agron. J. 43:445-449.

Magee, J.A. 1948. Histological structure of the stem of *Zea mays* in relation to stiffness of stalk. Iowa State Coll. J. Sci. 22:257-268.

Martin, J.N., and Hershey, A.L. 1935. Ontogeny of the maize plant. The early differentiation of stem and root structures and their morphological relationships. Iowa State Coll. J. Sci. 9:489-503.

Picklum, W.E. 1953. Histological and cytological changes in the maize embryo during germination. Ph. D. thesis. Iowa State College, Ames, Iowa.

Randolph, L.F. 1936. Developmental morphology of the caryopsis in maize. J. Agric. Res. 53:881-916.

Sass, J.E. 1951a. Comparative leaf number in the embryos of some types of maize. Iowa State Coll. J. Sci. 25:509-512.

----. 1951b. "Reduced" vascular bundles in maize. Iowa State Coll. J. Sci. 26:95-98.

Sharman, B.C. 1942. Developmental anatomy of the shoot of *Zea mays* L. Ann. Bot. (London) 6:245-282.

Toole, E.H. 1924. The transformations and course of development of germinating maize. Am. J. Bot. 11:325-250.

Chapter 4 The Genetics of Corn

E.H. COE, JR.
ARS-USDA, and Department of Agronomy
University of Missouri
Columbia, Missouri

M.G. NEUFFER
Department of Agronomy
University of Missouri
Columbia, Missouri

It is the fortune of corn to be a crop with high potential in productivity and utility, yet suited at the same time to be a powerful tool in basic biological research. Presumably the earliest cultivators selected for plant habit that would be convenient for artificial culture and would yield storable ears, but it may be less reasonable to suppose that conscious selection was used toward convenience for artificial hybridization. In any case the species is easy to culture systematically on any scale from single plants or small nurseries to hundreds of hectares; the pollen-bearing inflorescence (the tassel) atop the plant is handily separated from the female inflorescence (the ear) along the culm, such that both can be easily manipulated, removed, or bagged; and the harvested ear, though bulky, is readily labelled, scanned, and stored as a unit. Pollen production is so prodigious (as many as 10^7 grains/day for a plant in the peak of a 7-day flowering period) and pollination is so convenient that it is often possible to pollinate 50 or more ears with a single collection of pollen. It is not unusual for one experienced person to complete 300 to 500 controlled pollinations in a single day under excellent conditions, each ear yielding several hundred kernel progeny. The very bulk and stature of the corn plant contributes greatly to the ease with which these manipulations can be done and to the ease of observation of traits, especially in the kernel.

Uncommonly great genetic variability is present in and among the diverse lines, varieties, and races of corn in the world. Although only part of this variability has been examined systematically, unit factors have been identified determining colors, forms, structures, constituents, or processes in every part and tissue. While the authors are aware that the cataloguing of these factors and their relationships is of interest in itself, we aspire also to assist in the analysis of the exciting practical and theoretical problems of today by providing an organized introduction to the detailed genetics of corn for persons unfamiliar with it, along with organized aids for the practicing research worker. We will be pleased if knowledge acquired through use of this compendium facilitates solutions to crop production problems in corn or other plant species, deeper penetration of the bases for physiological systems such as the photosynthetic process, clarification of the mechanisms of growth and differentiation, or comprehension of the processes of gene and chromosome function in eukaryotes. The journal papers, however, are the core of any research field and should be preferred to this summary in citations; we have attempted to select those references that are the most informative regarding the systems under discussion.

The debt the authors of this summary owe to those who have contributed to it, intentionally or unintentionally, is enormous. We could not have proceeded without a tacit assumption that whatever special knowledge we may have absorbed unlabelled from colleagues was acceptable for us to subsume into the compilation; that our understandings, when flawed, might be tolerated by the cognizant worker; and that corrections would be communicated to us in due course. This assumption is founded in the happy tradition, among corn research workers, of free and generous exchange of information, ideas, techniques, materials, and, above all, critiques. The vehicle by which this tradition is furthered (and the example for parallel informal exchange now in many organisms) is the *Maize Genetics Cooperation Newsletter*. (At the time of this writing, Marcus M. Rhoades, who maintained the *Newsletter* and its traditions from its inception, and Ellen Dempsey, the exacting keeper of its excellence, are relinquishing their graciously kept responsibility, so warmly appreciated over the years by their colleagues.)

We are aware that items of linkage data, genetic expressions, and other detailed information drawn from the *Newsletter* are properly cited and attributed only by permission of the writers. In the face of the enormous 75-year volume of formal literature to be selected and attributed, that formidable propriety must be foregone in this summary. We have, in fact, attempted to avoid citations to the *Newsletter* and to prefer formal publications, inasmuch as the *Newsletter* is not widely available in library collections. We can only hope, as Emerson et al. (1935) expressed it 40 years ago in introducing their classic compilation, "that this summary presentation will prove to be sufficiently helpful to the contributors to compensate them in some measure for their aid in its preparation."

No comprehensive summary of corn genetics has been made since the summary of linkage data and interactions compiled by Emerson et al. (1935) and the present summary can only update, not replace, that aid. Among several other aids to research are two indexes, the first (1962) to symbols and the second (1970) to authors and names in the *Newsletter* (both available from Coe). The 170-odd photographs, mostly in color, in the book *The Mutants of Maize* (Neuffer et al., 1968) complement the present paper and convey information that is not possible to express in brief phrases. Recent special-purpose listings of factors, with reference leads, have been presented in the *Handbook of Biochemistry and Molecular Biology* (Neuffer and Coe, 1976) and in the *Handbook of Genetics* (Neuffer and Coe, 1974). A concise survey of the cytogenetics and genetics of corn, prepared by K. R. Sarkar, is to be published shortly (In Sharma, D. *Maize*. Indian Counc. Agric. Res., New Delhi).

GENETIC FACTORS OF THE SPECIES

Studies in corn have identified genetic variation attributable to unitary factors at perhaps 600 to 1,000 or more loci, loosely defined; if rigorous criteria were applied with regard to supporting data, tests for allelism, and pleiotropisms, perhaps 350 loci could be documented unequivocally. Because of losses of strains, pending work, and preliminary reports in the *Newsletter* that have not been carried further, it has not been possible to include some factors. Systematic description and cataloguing of genetic variations is dependent upon the nomenclatural conventions, listings of the known factors, and linkage maps; in addition to these tools, brief consideration is given below to examples of dosage effects, pleiotropisms, cell autonomy, and duplicate factors.

Nomenclatural Conventions

As symbolization of genetic factors has developed, systematic conventions have formed, based on recommendations accepted by consensus. These recommendations were substantially revised in 1974 and the present summary follows the new conventions. Briefly, the significant conventions used in literature published before the revision were as follows: A one or two-letter symbol designated a locus; for different members of an allelic series, superscripts composed of numerals or letters or both were used (e.g., R^r, R^g, r^r, r^g; A, A^b, A^d, a^{m-1}, a^{m-2}, etc.); for different loci having the same class symbol, numerical subscripts were used (e.g., a_1, a_2; v_1, v_2, v_3, etc.); normal alleles of a recessive offtype were represented with a $+$ sign or capitalized symbol (e.g., An), while normal alleles of a dominant offtype were represented with $+$ or a lower case symbol (e.g., tu). Nonconforming symbols have occasionally been used in the literature. The revised nomenclatural conventions adopted in 1974 by the annual Maize Genetics Conference are reproduced here from the 1974 *Maize Genetics Cooperation Newsletter (MNL)*, vol. 48, p. 201-202.

Revised Genetic Nomenclature for Maize

During the 1974 Allerton Park meetings, there was consideration of the proposed nomenclature changes (1973 MNL 47:229-230). Following discussion of possible difficulties, the group voted to accept the recommended changes which are outlined below. It is hoped that these changes will be implemented in all journal papers written after this date.

Recommendation 1: Each locus be designated by a lower case italicized symbol. Traditionally, this has been a one or two letter symbol, but some three letter symbols have been used. We recommend that all newly assigned symbols have three letters in the future.

Recommendation 2: As previously, different loci at which mutations produce the same general phenotype are distinguished by italicized numbers following the gene symbol, but the number one will be omitted in the designation of the first locus identified, i.e., the first locus identified would be *sh* and the second *sh2*. The number will appear on the line both when the gene name is written out and when the symbol is used: e.g., *brittle-2* and *bt2*.

Recommendation 3: A mutational site or event is designated by an isolation number, laboratory number, or previous designation following the gene symbol and set off by a dash: e.g., *sh2-6801.*

The dominant allele at a locus should be designated by the gene symbol with a capital letter, *Sh2.* Where it is desirable to designate a particular dominant, this can be done as *Sh2-W22.*

The mutation by which a locus was first detected should be designated by a capital R or Ref. as *sh2-R* to indicate the reference allele.

The superscripts that currently indicate different alleles at a locus will be written after the dash following the locus designation. As examples, R^r would become *R-r* and P^{RR} would become *P-RR.*

Recommendation 4: A mutation at an unknown locus conditioning a phenotype similar to that conditioned by mutations at one or more known loci can be designated by an appropriate gene symbol, an * to indicate that the locus is unknown and a laboratory number as *bt*-7011.* After tests establish allelism with mutations at a given locus, the number of that locus can be substituted for the * but the laboratory isolation number retained, as *bt2-7011.* It would be preferable if the mutations within the locus that appear to represent independent mutational events were designated only by isolation numbers that do not purport to furnish any information about the characteristics of the allele.

Since these recommendations provide only a framework for changes, uncertainties in application are certain to arise. It is suggested that all queries be referred to Dr. R.J. Lambert, Maize Genetics Cooperation, University of Illinois, Urbana, IL 61801. Dr. Lambert has agreed to act as a clearinghouse for all questions relating to gene symbols.

C.R. Burnham	E.B. Patterson
E.H. Coe	M.M. Rhoades
O.E. Nelson	

Genetic Loci and Their Map Locations

The accompanying list of documented loci (Table 1) includes, in addition to map locations, descriptions of one or more characteristic alleles at each locus. Usually the "type" allele — i.e., the unusual, non-normal, or mutant variant by which the locus was first identified — is given for the symbol and for the description. Gaps in the list, for loci that are listed in Emerson et al. (1935) but not listed here or for members in sequence in a symbol class, are either the result of corrections following findings of allelism (a few of which are specifically entered for current clarification) or were excluded because authentic stocks with which to test allelism are no longer in existence. Listings of genetic and cytogenetic strains available from the Stock Center are given regularly in the *Newsletter*; copies of the listings or small quantities (20 to 50 kernels) of specific strains may be obtained from Dr. R.J. Lambert, Maize Genetics Coop., Univ. of Illinois, Urbana, IL 61801. Not all of the listed factors, however, are available in the stocks maintained by the Stock Center or in current collections of individual investigators. Literature sources cited in Table 1 have been selected to include those published reports that are cited in the text of this summary and some others chosen to lead to or give documentation; the documentary references cited in Emerson et al. (1935) are not systematically repeat here. A few symbols in Table 1 are identified as those used or designated for certain unusual phenomena, chromosomal aberrations, cytoplasmic traits, or locus subunits.

THE GENETICS OF CORN

Table 1—Genetic loci, map locations, phenotypes, and literature sources. Location (†) is indicated by chromosome and map position (3-111), long or short arm (3L, 3S), or other available information; parentheses indicate uncertainty. Photographs (‡) in Neuffer et al.(1968)

Symbol	Location†	Name, phenotype, literature
a	3-111	*anthocyaninless‡:* colorless aleurone, green or brown plant, brown pericarp with *P-RR*; 57 81 82 92 138 145-149 151-153 176-180 210 212 218 225-227 233-238 258 259 263-265 288 343 348 362.
(∝)	3-111	*alpha‡:* component at *A* (see *β*); pale aleurone, red-brown plant, dark brown pericarp with *P-RR*; 146 149 151-153 210 218 226.
a2	5-15	*anthocyaninless‡:* like *a*, but red pericarp with *P-RR*; 57 92 138 148 177 179 233 236 238 257-259 362.
a3	3L	*anthocyanin‡:* red pigment in sheath, culm, and husks; resembles *B* but is recessive; 84.
Ac	various	*activator‡:* transposable factor, regulates *Ds* activity; 16 34 92 176-179 194 203.
ad	1-(108)	*adherent‡:* seedling leaves, tassel branches, and occasionally top leaves adhere; 84.
Adh	1-near *lw*	*alcohol dehydrogenase (Adh2* of Scandalios): electrophoretic mobility; hybrid bands occur; null allele is known; 36 89 90 93 96 171 204 295 306-310 312.
Adh2	—	*alcohol dehydrogenase:* electrophoretic mobility; 96 204.
ae	5-(37)	*amylose extender‡:* glassy, tarnished endosperm; high amylose content; 39 68 94 101 121 191 202 204 229 316 318.
ag	1-14	*grasshopper resistant:* leaf feeding reduced; from Amargo cultivar; 123.
al	2-4	*albescent‡:* erratic development of chlorophyll; pale yellow endosperm; 279-282 287.
alh	—	*histone Ia:* electrophoretic mobility; 358.
am	5-0	*ameiotic‡:* meiosis fails, sporogenous tissue degenerates; 230.
Amy	—	*alpha amylase:* electrophoretic mobility; no hybrid bands; 48 49 204.
Amy2	near *Cat*	*beta amylase:* electrophoretic mobility; no hybrid bands; 47 49.
an	1-104	*anther ear‡:* dwarf with anthers in ear florets; few tassel branches; responds to gibberellins; 202 247 248.
Ap	—	*acid phosphatase:* electrophoretic mobility; 79 204.
Ap2	—	*acid phosphatase:* electrophoretic mobility; 79.
Ap3	—	*acid phosphatase:* electrophoretic mobility; 79.
ar	9-65	*argentia‡:* virescent seedling, greens rapidly; 84 86.
(AR)	various	*aberrant ratio:* distorted ratios following virus infection; 335 336.
as	1-56	*asynaptic‡:* synaptic failure of meiotic prophase chromosomes; 15 188.
B	2-49	*colored plant‡:* anthocyanin in major plant tissues; some alleles (see *R2*) affect seed color; 58 62 82 362 363.
(β)	3-111	*beta:* component at *A* (see ∝); purple or red aleurone and plant color, red pericarp; 146 149 151-153 210 218 226.
(B chr)	—	*B chromosome‡:* supernumerary chromosome; 11 46 51 273 274 369.
ba	3-72	*barren stalk‡:* ear shoot and most tassel florets missing; 84.
ba2	2-	*barren stalk:* like *ba*, but tassel more normal; 84.
bd	7-109	*branched silkless‡:* branched ear and tassel; silks absent; 223.
Bf	9-134	*blue fluorescent‡:* homozygous seedlings (homozygous or heterozygous anthers) fluoresce blue under ultraviolet; anthranilic acid present; 364.
bf2	10-see map	*blue fluorescent:* similar to *Bf* in expression; shows earlier, stronger seedling fluorescence than *Bf* (Anderson, E.G., MNL 27:5).
Bh	6-49	*blotched‡:* colored patches on colorless (*c*) aleurone; 84.
bk2	9-79	*brittle stalk‡:* brittle plant parts after 4-leaf stage (Langham, D.G., MNL 14:21).
bl	2-	*blotched:* yellow necrotic lesions in leaf; 84.
bm	5-21	*brown midrib‡:* brown pigment over vascular bundles of leaf sheath, midrib and blade; 102 142 143 173 174 196 202.
bm2	1-161	*brown midrib;* like *bm*; 143.
bm3	4-near *su*	*brown midrib:* like *bm*; 143 196.
bm4	9-138	*brown midrib:* like *bm*; 143.
Bn	7-71	*brown aleurone:* yellowish brown aleurone color; 84.
bp	9-44	*brown pericarp‡:* changes red pericarp of *P-RR* to brown; 84.
br	1-81	*brachytic‡:* short internodes, short plant; no response to gibberellins; 84 130.
br2	1L	*brachytic:* like *br*; 207 313 353.
br3	5-	*brachytic:* like *br* (Singleton, W.R., MNL 33:3).

Table 1—continued.

Symbol	Location	Name, phenotype, literature
bt	5-22	brittle‡: mature kernel collapsed, angular, often translucent and brittle; 39 68 92 94 101 190 202.
bt2	4-near su	brittle: like bt; ADP glucose pyrophosphorylase electrophoretic mobility; 39 68 94 101 190 202 204 318 371.
bt4	(allele bt2)	
bv	5-27	brevis: short internodes, short plant; 84.
Bx	—	benzoxazin: blue color reaction of crushed root tip with $FeCl_3$, indicating cyclic hydroxamates present; 65.
bz	9-31	bronze‡: modifies purple aleurone and plant color to pale or reddish brown; anthers yellow-fluorescent; 57 92 144 148 177-179 192 193 258 259 270 362.
bz2	1-106	bronze‡: like bz; anthers not fluorescent; 57 179 225 258 259.
C	9-26	colored aleurone‡: c = colorless, C-I = dominant colorless; 50 57 59 60 78 81 92 126 127 138 176-179 233 258 259 270 302.
c2	4-123	colorless‡: colorless aleurone, reduced plant color; 32 57 92 112 179 258 259.
Cat	near Amy2	catalase (was Ct): electrophoretic mobility; hybrid bands occur; 20 202 294 296.
Cat2	—	catalase (was Ct2): electrophoretic mobility; hybrid bands occur; 296.
Ce	—	curled entangled: rolled leaves tend to be entangled (Pawar, S. E., and C. Mouli, MNL 47:17).
Cg	3-31	corngrass‡: narrow leaves, extreme tillering; 99 221 329 351.
Ch	2-155	chocolate‡: dark brown pericarp; 84.
cl	3-38	chlorophyll: white to green seedlings, depending upon Clm; pale yellow endosperm; 14 67 280-282 284.
clh	—	histone Ic: electrophoretic mobility; 358.
Clm	8-	modifier of cl (was Cl_M): modifies seedling, not endosperm; 14 281 284.
Clt	8-	clumped tassel: variable dwarfing, developmental anomalies; 104.
cm	10-near R	chloroplast mutator: like ij; 359.
(cms-C)	—	cytoplasmic male sterility: female-transmitted male sterility, C type; 19 77 154 331.
(cms-S)	—	cytoplasmic male sterility: female transmitted male sterility, S-type; restored by Rf3; 19 40 76 77 154 326 331.
(cms-T)	—	cytoplasmic male sterility: female-transmitted male sterility, Texas-type; restored by Rf Rf2; 18 19 76 77 106 154 155 189 317 323 331.
cp	7-near vp9	collapsed: endosperm collapsed and partially defective; 219.
cp2	7-near in	collapsed‡: endosperm rough, collapsed, partially defective; seedling very light green with darker streaks; 219.
cr	3-0	crinkly‡: plant short; leaves broad, crinkled; 84.
ct	8-	compact: semi-dwarf plant; 207 208.
ct2	1S	compact (was codw): semi-dwarf plant with club tassel; (Glover, D.V., MNL 42:151).
Ct		catalase (see Cat).
Cx	10-near du	catechol oxidase: electrophoretic mobility; no hybrid bands; null allele is known; 254.
d	3-18	dwarf‡: plant andromonoecious, short, compact; responds to gibberellins; 27 202 247 248 353.
d2	3-	dwarf: like d; 27 202 248.
d3	9-62	dwarf: like d; 27 202 247 248.
d5	2-34	dwarf: like d; 27 202 248.
D8	1-see map	dwarf‡: dominant, resembles d; not responsive to gibberellins; 247 248.
de	4-0	defective endosperm‡: kernels small, distorted; 84.
de16	4-74	defective endosperm‡: like de; semi-dwarf plant; 84.
de17	—	defective endosperm: like de; semi-dwarf plant; 31.
Df	various	deficiency: general symbol for loss of segments of chromosome.
dp	4-143	distal pale (was di): seedling leaf tip virescent (E.G. Anderson).
Ds	various	dissociation‡: transposable factor, associated with chromosome breakage and/or control of expression of adjacent genes; regulated by Ac; 92 176-179 194 203.
Dt	9-0	dotted‡: regulates controlling element at A; responding a-m alleles express colored dots on colorless kernels and purple sectors on brown plants; 92 176-179 212 225 227 263-265.
Dt2	6-43	dotted: like Dt; 225.
Dt3	7L	dotted: like Dt, but expression variable; 225.
Dt4	—	dotted: like Dt, but dots chiefly on crown of kernel; 73.

THE GENETICS OF CORN

Table 1—continued.

Symbol	Location	Name, phenotype, literature
Dt5	9-near yg2	*dotted:* like *Dt*; 73.
du	10-see map	*dull:* glassy, tarnished endosperm; 39 68 94 101 202 204 318.
dv	—	*divergent:* spindle nonconverging in meiosis in microsporocytes; 56 356.
dy	—	*desynaptic:* chromosomes unpaired in microsporocytes; 205.
E	7-	*esterase:* electrophoretic mobility; hybrid bands occur; 36 202 300 303 311.
E2	—	*esterase:* presence-absence; 305.
E3	3-near E4	*esterase:* electrophoretic mobility; hybrid bands occur; 36 38 202 304.
E4	3S	*esterase:* electrophoretic mobility; no hybrid bands; null allele is known; 36 37 118 140.
E5-I	—	*esterase:* electrophoretic mobility; duplicate factor with *E5-II*; 164.
E5-II	—	*esterase:* see *E5-I*; 164.
E6	—	*esterase:* presence-absence; 164.
E7	—	*esterase:* presence-absence; 164.
E8	—	*esterase:* presence-absence; 164.
E9	—	*esterase:* electrophoretic mobility; null allele is known; 164.
E10	—	*esterase:* electrophoretic mobility; 164.
E12	—	*esterase:* electrophoretic mobility; no hybrid bands; 36 38.
E16	7-	*esterase:* electrophoretic mobility; no hybrid bands; 38.
eg	5L	*expanded:* glumes open at right angle (Burnham, C.R., MNL 32:93).
Ej	10-57	*extension of japonica‡:* extreme expression of *j*, *sr2*, etc.; closely associated with certain *r* alleles; 84 219.
el	—	*elongate‡:* chromosomes uncoiled during meiotic metaphase and anaphase; frequent unreduced gametes; 272 358.
En	various	*enhancer‡:* transposable factor, regulates *pg14-m* mutation; equivalent to *Spm*; 92 179 231-238.
Ep	6-near Y	*endopeptidase:* electrophoretic mobility; no hybrid bands; 183.
et	3-122	*etched‡:* pitted, scarred endosperm; virescent seedling; 66 (Stadler, L.J., MNL 14:26).
f	1-86	*fine stripe‡:* virescent seedling, fine white stripes on base and margin of older leaves; 71 84 110.
fl	2-68	*floury‡:* endosperm opaque, soft; dosage effect; 39 68 94 101 206 305.
fl2	4-63	*floury:* like *fl*, but phenotypically dominant; 110 190 202 204 206.
fv	—	*flavones:* polyphenols in silks absent; 157.
g	10-43	*golden‡:* seedling and plant with distinct yellow cast; 84 283.
g2	7-	*golden:* like *g*, but more extreme; sheaths whitish yellow-green; 84.
g5	3S	*golden:* like *g*, but more extreme and variable, leaves yellow-green, sheaths whitish yellow-green; vigor reduced (Beckett, J.B., et al., MNL 47:147)
Ga	4-35	*gametophyte factor: Ga* pollen grains competitively superior to *ga* on *Ga* silks; 83 125 129 170 297.
ga2	5-35	*gametophyte factor: Ga2* pollen grains competitively superior to *ga2*; 41 160.
ga6	1-15	*gametophyte factor: ga6* pollen grains nonfunctional on *Ga6* silks (see *Ga4* of Emerson, R.A., MNL 20:4).
ga7	3-128	*gametophyte factor: ga7* pollen from heterozygotes 10-15% functional regardless of silk genotype (Rhoades, *M.M., MNL 22:9).*
ga8	9-44 to 59	*gametophyte factor: Ga8* pollen grains competitively superior to *ga8* on *Ga8* silks; 24.
ga9	4-	*gametophyte factor: Ga9* pollen grains competitively superior to *ga9* on *Ga* silks (indifferent re *Ga9* or *ga9* silk constitution); 129.
gl	7-36	*glossy‡:* cuticle wax altered; leaf surface bright, water adheres; 23 120 286.
gl2	2-30	*glossy‡:* like *gl*; 23 120.
gl3	4-118	*glossy:* like *gl*; 23 120.
gl4	4-86	*glossy:* like *gl*; expression poor; 84.
gl5	5-	*glossy:* like *gl*; 84.
gl6	3-50	*glossy:* like *gl*; 84.
gl7	—	*glossy:* like *gl*; 84.
gl8	5L	*glossy:* like *gl*; 84.
gl9	(allele *gl*)	
gl10	(allele *gl8*)	
gl11	2-near B	*glossy:* like *gl*; abnormal seedling morphology (Sprague, G.F., MNL 12:2).
gl12	—	*glossy:* like *gl* (Sprague, G.F., MNL 29:6).

Table 1—continued.

Symbol	Location	Name, phenotype, literature
gl14	—	*glossy:* like *gl* (Sprague, G.F., MNL 29:6).
gl15	9-69	*glossy‡:* like *gl*; expressed after 3rd leaf; 23.
gl16	(allele *gl4*)	
gl17	5-14	*glossy:* like *gl* but semi-dwarf with necrotic crossbands on leaves (Rhoades, M.M., and E. Dempsey, MNL 28:58).
gl18	8-	*glossy* (was *glg*): like *gl*; expression poor (Anderson, E.G., MNL 29:6).
gm	—	*germless:* embryo fails; 84.
gs	1-135	*green stripe‡:* grayish green stripes between vascular bundles on leaves; tissue wilts; 84.
gs2	2-54	*green stripe‡:* like *gs*, but pale green stripes; no wilting; 84.
gt	—	*grassy tillers:* numerous basal branches; vegetatively totipotent in combination with *id* and *pe*; 319.
h	—	*soft starch:* endosperm soft, opaque; 39 68 94 101 206.
hcf	—	*high chlorophyll fluorescence:* strong red fluorescence under long-wave ultraviolet irradiation; very low CO_2 fixation; 187.
hcf2	—	*high chlorophyll fluorescence:* like *hcf*, but lethal; 187.
hcf3	—	*high chlorophyll fluorescence:* like *hcf*, but lethal; 187.
hm	1-64	*susceptibility to Helminthosporium carbonum‡:* disease lesions on leaves, black masses of fruiting bodies on ears with race I; 209.
hm2	9-	*susceptibility to Helminthosporium carbonum:* like *hm*; masked by *Hm*; 209.
Hs	7-0	*hairy sheath‡:* abundant hairs on leaf sheath; 84.
Ht	2-121	*resistance to Helminthosporium turcicum:* 44 65 122.
I	—	(= *C-I*, inhibitor allele at C locus; also commonly used as a general symbol for inhibition and for the controlling elements responding to *En*.)
id	1-near *an*	*indeterminate growth:* requires extended growth and short days for flowering; vegetatively totipotent with *gt* and *pe*; 100 319 328.
Idf	—	*diffuse* (allele *c2*).
ig	—	*indeterminate gametophyte:* polyembryony, heterofertilization, polyploidy, androgenesis; 132 135.
ij	7-52	*iojap‡:* many variable white stripes on leaves; conditions chloroplast defects that are cytoplasmically inherited; 76 266-268 325.
in	7-20	*intensifier‡:* intensifies anthocyanin pigments; 57 258 259.
In	various	*inversion‡:* general symbol for inversion of a segment of chromosome.
is	—	*cupulate interspace:* 97.
j	8-28	*japonica‡:* white stripes on leaf and sheath (see *Ej*); not expressed in seedling; 84.
j2	4-112	*japonica‡:* extreme white striping of leaves, etc.; 84.
(K-10)	10-99	*abnormal-10‡:* heterochromatic appendage on long arm of chromosome 10; neocentric activity distorts segregation of linked genes; 271.
Kn	1-127	*knotted‡:* scattered proliferation of tissue at vascular bundles on leaf; 103 105 224.
l	10-	*luteus‡:* yellow pigment in white tissue of chlorophyll mutants *w*, *j*, *ij*, etc.; 84.
l2	10-99	*luteus‡:* lethal yellow seedling; 84.
l3	—	*luteus:* like *l2*; 14 67.
l4	—	*luteus:* like *l2*; 14 67.
l6	9S	*luteus:* like *l2*; 84.
l7	9S	*luteus:* yellow seedling and plant; 67 84.
l8	10-near *du*	*luteus:* like *l2*; 276.
l10	6-near *Y*	*luteus:* like *l2* (Robertson, D.S., MNL 47:82).
l11	6-	*luteus* (was *l*-4120*): yellow seedling with green leaf tips; 14.
l12	6-near *Y*	*luteus* (was *l*-4920*): like *l2*; 67.
la	4-60	*lazy‡:* prostrate growth habit; 222 367.
Lc	10-61	*red leaf stripe:* red color in leaf surface; 133.
lg	2-11	*liguleless‡:* ligule and auricle missing; leaves upright, enveloping; 84.
lg2	3-83	*liguleless‡:* like *lg*, less extreme; 84.
Lg3	3-46	*liguleless‡:* dominant, no ligule; leaves upright, broad, often concave and pleated (Perry, H.S., MNL 13:7).
li	10-see map	*lineate‡:* fine, white striations on basal half of mature leaves; 84.
ln	—	*linoleic acid:* lower ratio of oleate to linoleate in kernel; 70 249-251.
lo	4-73	*lethal ovule:* ovules containing *lo* gametophyte abort; 84.
lo2	9-53	*lethal ovule:* like *lo*; 205.
Lp	—	*leucine aminopeptidase:* electrophoretic mobility; no hybrid bands; 21 202.

Table 1—continued.

Symbol	Location	Name, phenotype, literature
Lp2	—	*leucine aminopeptidase:* electrophoretic mobility; no hybrid bands; 21 202.
lu	5-9	*lutescent:* pale yellow-green leaves; duplicate factor with *lu2*; 322.
lu2	—	*lutescent:* see *lu*; 322.
lw	1-128	*lemon white‡:* white seedling, pale yellow endosperm; 14 67 280-282.
lw2	5-46+	*lemon white:* like *lw*; 67 280-282.
lw3	5-	*lemon white:* like *lw*; duplicate factor with *lw4*; 67 280 282.
lw4	4-71 to 84	*lemon white:* see *lw3*; 67 280 282.
ly	5-	*lycopenic:* similar to *ps* (possible allelism untested), but not viviparous; accumulates lycopene; 88 124 282 368.
M	—	(commonly used as a general symbol for mutator or modifier.)
mi	1-	*midget:* small plant; 84.
mn	2-near *fl*	*miniature seed‡:* small, somewhat defective kernel; fully viable; 162.
mn2	7-	*miniature seed:* small kernel, loose pericarp; extremely defective but will germinate (R.J. Lambert).
Mp	1-26, various	*modulator of pericarp:* transposable factor affecting *P* locus; parallel to *Ac - Ds*; 16 34 92 111 114 179 228.
Mr	9-40	*mutator of R-m‡:* transposable factor, regulates *R-m* mutation; 219 (Neuffer, M.G., MNL 33:82).
ms	at *Y*	*male sterile:* anthers shriveled, not usually exserted; 17 84 220 223.
ms2	9-67	*male sterile:* like *ms*; 17 84 165.
ms5	5-near *v3*	*male sterile:* 17 84.
ms6	(allele *po*)	
ms7	7-near *g1*	*male sterile:* 17 84 165.
ms8	8-14	*male sterile‡:* like *ms*; 17 84.
ms9	1-near *P*	*male sterile:* like *ms*; 17 84.
ms10	10-near *bf2*	*male sterile:* like *ms*; 17 84.
ms11	—	*male sterile:* 17 84 165.
ms12	1-	*male sterile:* like *ms*; 17 84.
ms13	5S	*male sterile:* like *ms*; 17 84.
ms14	1-near *as*	*male sterile:* like *ms*; 17 84.
ms17	1-23	*male sterile:* like *ms*; 84.
Ms21	—	*male sterile:* dominant; pollen grains developing in presence of *Ms21* are defective and nonfunctional if *sks*, normal if *Sks*; 156 298.
ms-si	(allele *si*)	
Mst	10-63	*modifier of R-st* (was *M^{rr}*): affects expression of *R-st*; 8 92.
na	3-86	*nana‡:* short, erect dwarf; no response to gibberellins; 248 366.
na2	5	*nana:* like *na*; 84.
(NCS)	—	*nonchromosomal stripe:* maternally inherited light green leaf striping; 324.
nec	8-	*necrotic* (was *nec*-6697*; allele *sn*, *sienna*-7748*): chlorotic seedling that stays rolled, wilts and dies (E.G. Anderson).
nec2	1S	*necrotic* (was *nec*-8147*): green seedling develops necrotic lesions at 2-3 leaf stage; lethal (E.G. Anderson).
nec3	5-near centromere	*necrotic* (was *nec*-E409*): seedlings emerge with tightly rolled leaves that turn brown and die without unrolling; manually unrolled leaves tan with dark brown crossbands (Neuffer, M.G., MNL 47:150).
nl	10-24	*narrow leaf‡:* leaf blade narrow, some white streaks; 84.
(NOR)	6S	*nucleolus organizer:* codes for ribosomal RNA; 246.
o	4-near *gl3*	*opaque:* endosperm starch soft, opaque; 206.
o2	7-16	*opaque‡:* like *o*; high lysine content; 39 68 94 101 185 190 202 204 206.
o4	(allele *fl*)	
o5	7-near *gl*	*opaque:* like *o* (Robertson, D.S., MNL 41:94).
o6	—	*opaque:* like *o*; lethal seedling (Nelson, O.E., MNL 46:203).
o7	10-80	*opaque:* like *o*; high lysine content; 190.
Og	10-16	*old gold‡:* variable bright yellow stripes on leaf blade; 84.
oy	10-12	*oil yellow‡:* seedling oily greenish-yellow; 84.
P	1-26	*pericarp color‡:* red pigment in cob and pericarp; 4 16 34 80 92 111 114 179 228.
(P)	10-57	*plant color component at R:* anthocyanin pigmentation in seedling leaf tip, coleoptile, anthers; 9 26 30 35 74 85 136 293 346 347.
pa	1-58	*pollen abortion:* 42.
pb	at *Y*	*piebald‡:* very light, irregular green bands on leaf; 84.
pb4	6-	*piebald:* like *pb*; 84.

Table 1—continued.

Symbol	Location	Name, phenotype, literature
pd	—	*paired rows:* single vs. paired pistillate spikelets; *pd* is found in teosinte also; 97.
pe	—	*perennialism:* vegetatively totipotent in combination with *gt* and *id;* 319.
pg2	3-	*pale green:* seedling light yellowish green; mature plant green and somewhat weak; 84.
pg11	6-37	*pale green‡:* like *pg2,* but mature plant pale and vigorous; duplicate factor with *pg12;* 269.
pg12	9-64	*pale green:* see *pg11;* 269.
pg14	3S	*pale green‡* (was *pg^m*): pale green leaves with normal green sectors, controlled by *En* (*Spm*); 92 231-233 238.
Phos4	—	*alkaline phosphatase:* electrophoretic mobility; no hybrid bands; 37.
Phos8	—	*alkaline phosphatase:* electrophoretic mobility; 38.
Pl	6-48	*purple plant‡:* sunlight-independent purple pigment in plant; 82 84.
pm	3-near *ts4*	*pale midrib‡:* midrib and adjacent tissue lighter green; 84.
Pn	7-112	*papyrescent‡:* long, thin, papery glumes on ear and tassel (Galinat, W.C., and P.C. Mangelsdorf, MNL 31:67).
po	6-4	*polymitotic‡:* microspore division without chromosome division; 84 245.
pr	5-46	*red aleurone‡:* changes purple aleurone to red; 50 78 82 92 117 138 179 258 259 362.
ps	5-19	*pink scutellum‡* (=*vp7*): viviparous; endosperm and scutellum pink, seedling white with pink flush; 174 279 280 282 284.
Pt	6-59	*polytypic ear‡:* proliferation of pistillate tissue to produce irregular growth on ear and tassel; 253.
Pu	—	*purple plumule:* 84.
Pu2	—	*purple plumule:* 84.
Px	—	*peroxidase:* electrophoretic mobility; no hybrid bands; null allele is known; 36 37 115.
Px2	—	*peroxidase:* electrophoretic mobility; 163.
Px3	—	*peroxidase:* electrophoretic mobility; 163.
Px4	—	*peroxidase:* electrophoretic mobility; 163.
Px5	—	*peroxidase:* presence-absence; 163.
Px6	—	*peroxidase:* presence-absence; 163.
Px7	—	*peroxidase:* electrophoretic mobility; null allele is known; 163.
py	6-68	*pigmy‡:* leaves short, pointed; fine white streaks; 84.
pyd	9S	*pale yellow deficiency:* pale yellow seedling; deficiency for short terminal segment of chromosome arm; 175.
R	10-57	*colored‡:* red or purple color in aleurone and/or anthers, leaf tip, brace roots, etc.; 8-10 26 28-30 33 35 57 74 78 81 82 85 92 95 113 133 134 136 138 182 186 198 258 259 293 305 320 321 337 340 342 344-347 360-363 372 373.
R2	2-49	*colored‡:* duplicate factor with *R* for aleurone color (either *R* or *R2* is required); allelic to *B,* affects plant color; 363.
ra	7-32	*ramosa‡:* ear branched, tassel conical; 220 223 253.
ra2	3-26	*ramosa‡:* irregular kernel placement; tassel many-branched, upright; 223.
ra3	—	*ramosa:* (R.A. Brink).
rd	1L	*reduced:* semi-dwarf plant; 109 207.
rd2	6L	*reduced* (was *spl*)*:* like *rd,* but not as extreme; 109.
Rf	3-45	*restorer:* restores fertility to *cms-T;* complementary to *Rf2;* 18 19 76 154.
Rf2	9-near *wx*	*restorer:* see *Rf;* 18 19 76 154.
Rf3	2L	*restorer:* restores fertility to *cms-S;* 19 40 76 154 326.
Rg	3-48	*ragged‡:* chlorotic tissue between veins of older leaves, causing holes and torn appearance; 184.
rgd	6-0	*ragged seedling‡:* seedling leaves narrow, thread-like, have difficulty in emerging (Kramer, H.H., MNL 31:120).
rhm	6S	*resistance to Helminthosporium maydis:* chlorotic-lesion reaction with race 0; 330.
Rp	10-0	*rust resistant‡:* resistant to *Puccinia* spp.; 276 277 374.
Rp3	3-49	*rust resistant:* resistant to *Puccinia sorghi;* 374.
Rp4	4-27	*rust resistant:* resistant to *Puccinia sorghi;* 374.
rp7	2-11+	*rust susceptibility* (was *rp_x*)*:* susceptible to *Puccinia sorghi* (Cornu, A., MNL 35:134).
Rs	—	*rough sheath:* 84.
rs2	1-near *as*	*rough sheath:* 84.
rt	3-40	*rootless‡:* secondary roots few or absent; 84.

Table 1—continued.

Symbol	Location	Name, phenotype, literature
S	4S	*colored scutellum:* colored scutellum in presence of *s5* and two of *S2, S3, S4,* if aleurone color is present (*A A2 Bz Bz2 C C2 R*); 84 332.
S2	—	*colored scutellum:* see *S*; 84 332.
S3	—	*colored scutellum:* see *S*; 84 332.
S4	—	*colored scutellum:* see *S*; 84 332.
S5	—	*inhibitor of scutellum color:* see *S*; 84 332.
(S)	10-57	*seed color component at R:* anthocyanin pigmentation in aleurone; 9 26 30 35 74 85 136 293 346 347.
sen	3-	*soft endosperm:* endosperm starch soft, opaque; duplicate factor with *sen2* (Stierwalt, T.R., and P.L. Crane, MNL 48:139).
sen2	7-	*soft endosperm:* see *sen* (Stierwalt, T.R., and P.L. Crane, MNL 48:139).
sen3	1-	*soft endosperm:* like *sen*; duplicate factor with *sen4* (Stierwalt, T.R., and P.L. Crane, MNL 48:139).
sen4	—	*soft endosperm:* see *sen3* (Stierwalt, T.R., and P.L. Crane, MNL 48:139).
sen5	2-	*soft endosperm:* like *sen*; duplicate factor with *sen6* (Stierwalt, T.R., and P.L. Crane, MNL 48:139).
sen6	5-	*soft endosperm:* see *sen5* (Stierwalt, T.R., and P.L. Crane, MNL 48:139).
Sg	—	*string cob:* reduced pedicels; 97 98.
sh	9-29	*shrunken:* inflated endosperm collapses on drying, forming smoothly indented kernels; 39 54 55 68 94 101 179 190 192 204 301.
sh2	3-111.2	*shrunken:* inflated, transparent, sweet kernels collapse on drying, becoming angular and brittle; 68 94 101 150 166 190 202 204 229 318 371.
sh3	(allele *bt*)	
sh4	5-	*shrunken* (= *sh-fl*)*:* collapsed, chalky endosperm; 190 204 229 318.
sh5	(allele *bt*)	
si	6-19	*silky* (= *ms-si*)*:* multiple silks in ear; sterile tassel with silks; 84.
sk	2-56	*silkless:* pistils abort, no silks; 84.
Sks	2-near *v4*	*suppressor of sterility* (was S^{Kys})*:* pollen grains developing in presence of *Ms21* are defective and nonfunctional if *sks,* normal if *Sks*; 156 298.
sl	7-50	*slashed:* leaves slit longitudinally by necrotic streaks; 84.
sm	6-58	*salmon:* silks salmon color with *P-RR*, brown with *P-WW*; 84.
so	—	*orange scutellum:* duplicate factor with *so2*; 84 269 332.
so2	—	*orange scutellum:* see *so*; 84 269 332.
sp	4-66	*small pollen:* pollen grains small, not competitive with normal; 167 327.
sp2	10-near *du*	*small pollen:* like *sp*; 276.
Spm	various	*suppressor-mutator:* transposable factor, regulates responsive element at *a-m1, c2-m, pg14-m,* etc.; 92 178-180 203 212 234-237.
sr	1-0	*striate:* many white striations or stripes on leaves; 84.
sr2	10-92	*striate‡:* white stripes on leaf and sheath (see *Ej*) (Joachim, G., and C.R. Burnham, MNL 27:66).
sr3	10S	*striate‡:* virescent and striate to striped (Glover, D.V., MNL 42:151).
st	4-55	*sticky chromosome‡:* small plant, striate leaves, pitted kernels resulting from sticky chromosomes; 299 358.
su	4-71	*sugary‡:* endosperm wrinkled and translucent when dry; sweet at milk stage; 39 68 69 94 98 101 190 202 204 318 334.
su2	6-57	*sugary:* endosperm glassy, translucent, sometimes wrinkled; 39 68 94 101 202 204 318.
sy	—	*yellow scutellum:* 84 332.
T	various	*reciprocal translocation‡:* general symbol for exchange of parts between two nonhomologous chromosomes.
Ta	—	*transaminase:* electrophoretic mobility; hybrid bands occur; 163.
tb	1-124	*teosinte branched;* many tillers; nodes with slender branches ending in unbranched tassel (Burnham, C.R., MNL 35:87).
td	5-	*thick tassel dwarf:* (E.G. Anderson).
tn	5-	*tinged:* small plant, pale green leaf tip; 84.
tn2	(allele *oy*)	
Tp	7-46	*teopod‡;* many tillers, narrow leaves, many small partially podded ears, tassel simple; 221 370.
Tp2	10-41	*teopod‡:* like *Tp* (Peterson, H., MNL 33:41).
tr	—	*two-ranked ear:* distichous vs. decussate phyllotaxy in ear axis; 97.
ts	2-74	*tassel seed:* tassel pistillate and pendant; if removed, small ear with irregular kernel placement develops; 223.
ts2	1-24	*tassel seed‡:* like *ts,* but branches pendant rather than whole tassel; 220 223.

Table 1—continued.

Symbol	Location	Name, phenotype, literature
Ts3	1-119	*tassel seed:* dominant; upright staminate branches terminate in pistillate clusters; 223.
ts4	3-55	*tassel seed‡:* tassel compact, upright, with pistillate and staminate florets; 223.
Ts5	4-56	*tassel seed:* dominant; nearly normal tassel with scattered, short silks; 220 223.
Ts6	1-158	*tassel seed‡:* tassel pistillate to mixed, compact; ear with irregular kernel placement; 220 223.
Tu	4-107	*tunicate‡:* kernels enclosed in long glumes; tassel glumes large, coarse; 169 220 223.
ub	—	*unbranched‡:* tassel with one spike; 219.
v	9-66	*virescent‡:* yellowish white seedling, greens rapidly; 71 283.
v2	5-87	*virescent‡:* like v, but greens slowly; 71 283.
v3	5-25	*virescent‡:* light yellow seedling, greens rapidly; 71 283.
v4	2-83	*virescent‡:* like v2; 71 283.
v5	7-24	*virescent‡:* like v, but older leaves have white stripes; 71 283.
v8	4-near *Tu*	*virescent:* like v2; 283.
v12	5-	*virescent:* like v3; 283.
v13	—	*virescent:* 283.
v16	8-0	*virescent:* like v2; 283.
v17	4-	*virescent:* like v, but greening from base to tip; 283.
v18	10-	*virescent:* like v; 52 53 283.
v19	1-	*virescent:* like v; 283.
v21	8L	*virescent* (was v*-A552, v*-E25): grainy virescent, greening from tips and margins inward (Beckett, J.B., and M.G. Neuffer, MNL 47:147).
v22	1-near *an*	*virescent* (was v*-8983): like v (E.G. Anderson).
v23	4-near *su*	*virescent* (was v*-8914): like v (E.G. Anderson).
va	7L	*variable sterile:* male sterile with some fertile anthers; 84.
Vg	1-85	*vestigial‡:* glumes very small, cob and anthers exposed; 220 223 224.
Vm	10-near *du*	*virescent mutable* (was V^m-1817): reduced chlorophyll, dominant; endosperm etched; homozygous lethal; 233.
vp	3-128	*viviparous:* embryo fails to become dormant; chlorophyll and carotenoids unaffected; anthocyanins in aleurone suppressed; 279 280 375.
vp2	5-18	*viviparous‡:* embryo fails to become dormant; white endosperm, white seedling; anthocyanins unaffected; 279-282.
vp5	1-1	*viviparous‡:* like vp2; 279-282 375.
vp7	(= *ps*)	
vp8	1-154	*viviparous:* embryo fails to become dormant; chlorophyll, carotenoids, and anthocyanins unaffected; small, pointed-leaf seedling; 279.
vp9	7-25	*viviparous‡:* like vp2; 12 278-282 284.
w	6-near *py*	*white:* white seedling; 67 282.
w2	10-73	*white:* white seedling; 67 282.
w3	2-111	*white‡:* like vp2; 7 12-14 67 159 202 278-282.
w11	9-near *wx*	*white:* like w; 67 282.
w13	3-near *lg2*	*white:* white seedling, mutable (this symbol was incorrectly used for y10 on the linkage map in reference 217) (Peterson, P.A., MNL 40:64); 282.
w14	6-near *py*	*white* (was w*-8657): like w; 67 281 282.
w15	6-near *y*	*white* (was w*-8896): like w; 14 67 282.
Wc	9-104	*white cap:* kernel with white crown and pale yellow endosperm; 84.
wd	9S	*white deficiency‡:* white seedling; deficiency for distal half of first chromomere of short arm; 175 282.
wi	6-	*wilted:* chronic wilting, delayed differentiation of metaxylem vessels; 252.
wl	—	*white leaf base:* leaf base variably white in seedling and older plants; 84.
ws	—	*white sheath:* light yellow leaf sheaths; duplicate factor with ws2; 84.
ws2	—	*white sheath:* see ws; 84.
ws3	2-0	*white sheath‡:* white leaf sheath, culm, husks; 84.
wt	2-60	*white tip:* tip of first leaf white and blunt (Sprague, G.F., et al., MNL 39:164).
wx	9-59	*waxy‡:* amylopectin (stained red by iodine) replaces amylose (blue staining) in endosperm and pollen; 25 39 68 92 94 101 131 176 177 179 200-204 229 318 334 376.
Y	6-17	*yellow‡:* carotenoid pigments in endosperm; some alleles affect green pigments in seedlings; 168 278 280-283.

Table 1—continued.

Symbol	Location	Name, phenotype, literature
y8	7-18	*yellow:* light yellow endosperm (Jenkins, M.T., MNL 21:33).
y9	10-near *nl*	*yellow:* pale yellow endosperm, slightly viviparous; green to pale green seedlings and plants (Robertson, D.S., MNL 44:81); 282.
y10	3-75	*yellow* (was *w*-7748*): pale yellow endosperm; white seedling; 280 282.
yd	6-	*yellow dwarf:* 84.
yd2	3-near *lg2*	*yellow dwarf:* (Robertson, D.S., MNL 48:70).
yg	5-near *v2*	*yellow-green:* yellow-green seedling and plant; 84.
yg2	9-7	*yellow-green‡:* like *yg*; 175.
ys	5-55	*yellow stripe‡:* yellow tissue between leaf veins, reflects iron deficiency symptoms; 22.
ys3	3-near *Lg3*	*yellow stripe:* like *ys* (Wright, J.E., MNL 35:111).
z	—	*zeta-carotenic:* carotenoids photosensitive; embryo viviparous; accumulates zeta-carotene; 88 124 282.
zb	—	*zebra:* yellowish crossbands on older leaves; 84.
zb2	—	*zebra:* crossbands on seedling leaves; 84.
zb3	5-near *v2*	*zebra:* crossbands on older leaves; 84.
zb4	1-19	*zebra‡:* regularly spaced cross-bands on earlier leaves; enhanced by cool temperatures; 84 119.
zb6	4-84	*zebra:* regularly spaced cross-bands on older leaves; enhanced by cool temperatures (Hayes, H.K., and M.S. Chang, MNL 12:8).
zl	1-28	*zygotic lethal:* sporophyte fails; 84.
zn	10-see map	*zebra necrotic‡:* necrotic tissue appears between veins in transverse leaf bands on half-grown or older plants (Horovitz, S., MNL 22:42).
zn2	—	*zebra necrotic:* like *zn*; 108.

Factors are known that affect any tissue type — in the kernel including the aleurone tissue, starchy portion of the endosperm, embryo axis, scutellum, and pericarp; in the seedling and plant including the roots, mesocotyl, coleoptile, coleorhiza, leaves, sheaths, culm, tassel, anther, pollen grain, husks, ear, ovules, silks, and cob. Some factors influence form or stature, surfaces, textures, and colors, including red, purple, yellow, green, brown, white, blue-fluorescence, and yellow-fluorescence. The numerous isozyme loci display detailed specificity of function according to tissues and conditions.

The accompanying linkage map (Fig. 1) delineates those loci that have been positioned by recombination analysis. Genetic locations of centromeres are included, although in most instances these are subject to considerable uncertainty. Genetic locations are shown for breakpoints of B-A translocations, which generate deficiencies for the segment distal to the breakpoint. Cytological features are given in Fig. 2.

Dosage Effects, Pleiotropisms, Cell Autonomy, and Duplicate Factors

Noteworthy examples of dosage effects are found at several loci. Dosage effects for *Y* with *y* in the endosperm, where triploidy permits four different constitutions (triplex, duplex, simplex, and nulliplex for *Y*), can often be distinguished visually; correspondingly, a linear relationship between *Y* dosage and vitamin A content has been established (Mangelsdorf and Fraps, 1931). Comparable visual dosage effects are found for anthocyanin intensity in the aleurone tissue under control of some alleles at the *A, C, C2,* and *In* loci; for floury endosperm with *Fl2* and *fl2*; and for frequency of mutation of several loci (e.g., *a, R-st*). A nonlinear relationship of dosage to mutation frequency is found for *Dt* (Nuffer, 1955; Rhoades, 1936) (see Fig. 3); for *Ac*, however, increased frequency is accompanied by a delay in the time of

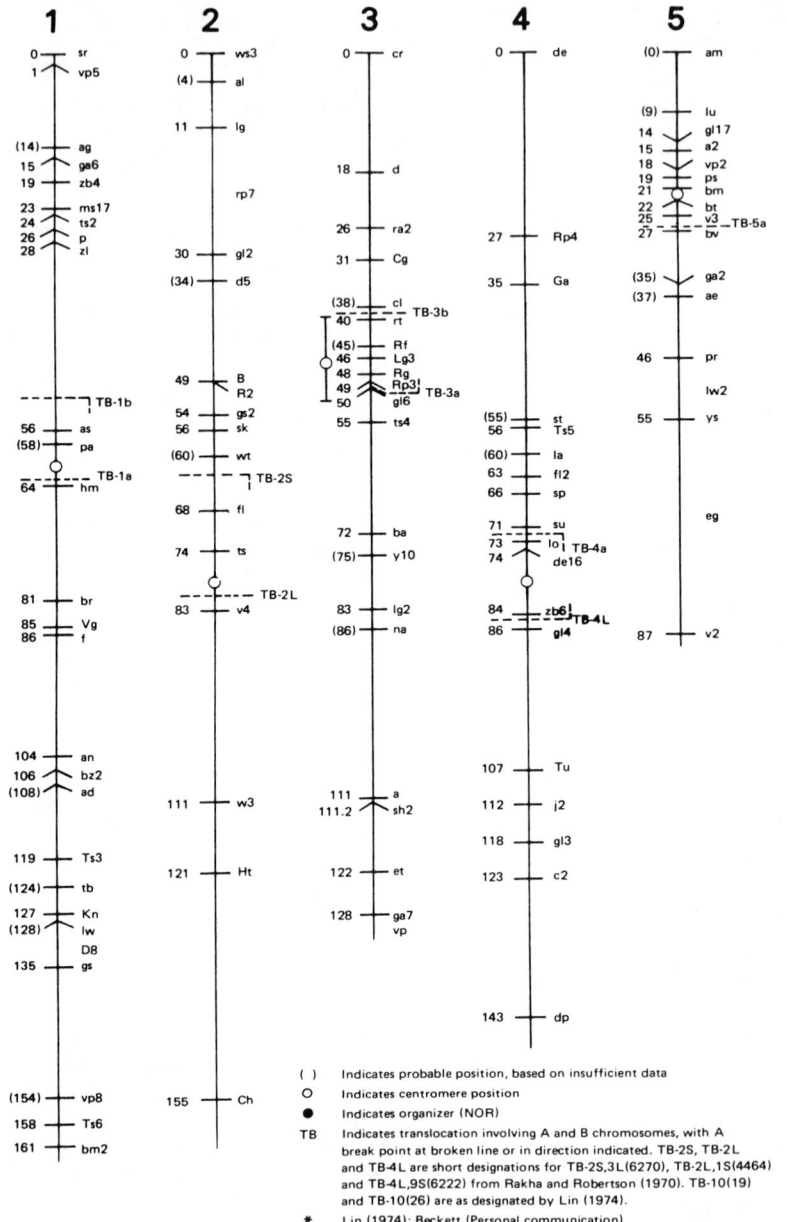

Fig. 1. Linkage map of maize.

THE GENETICS OF CORN

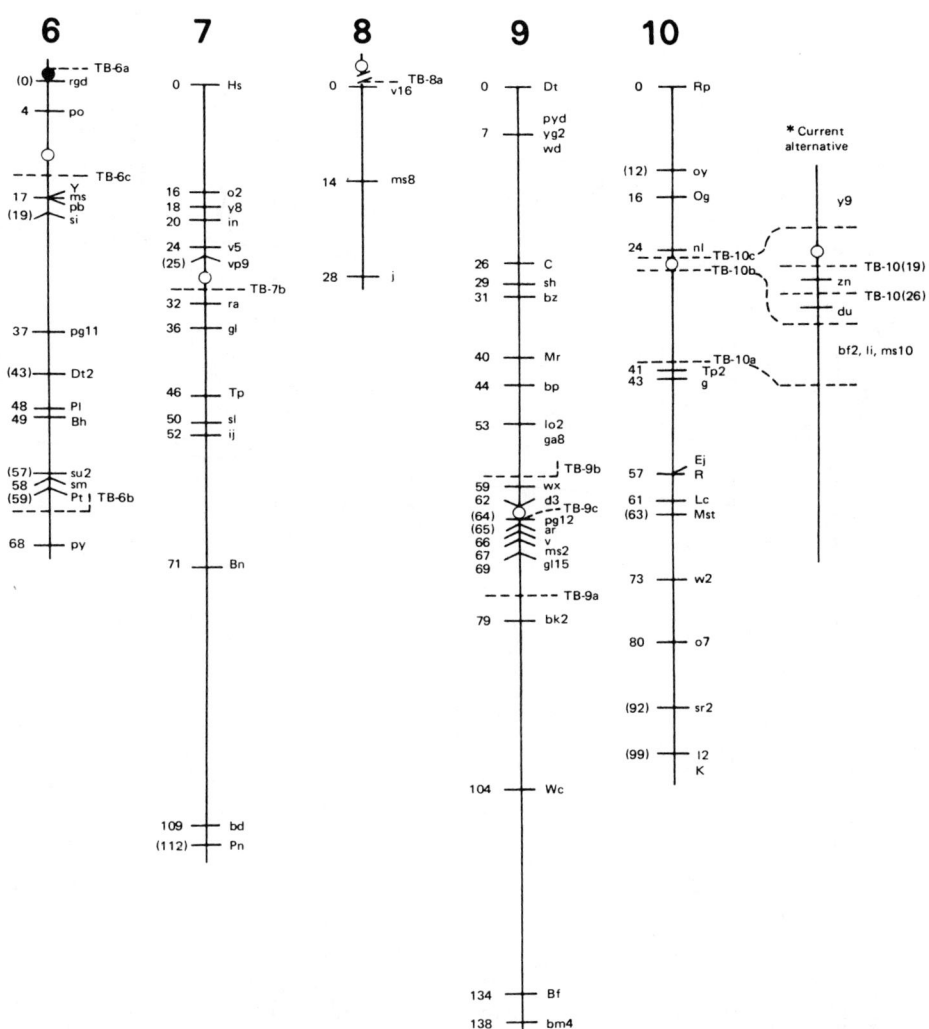

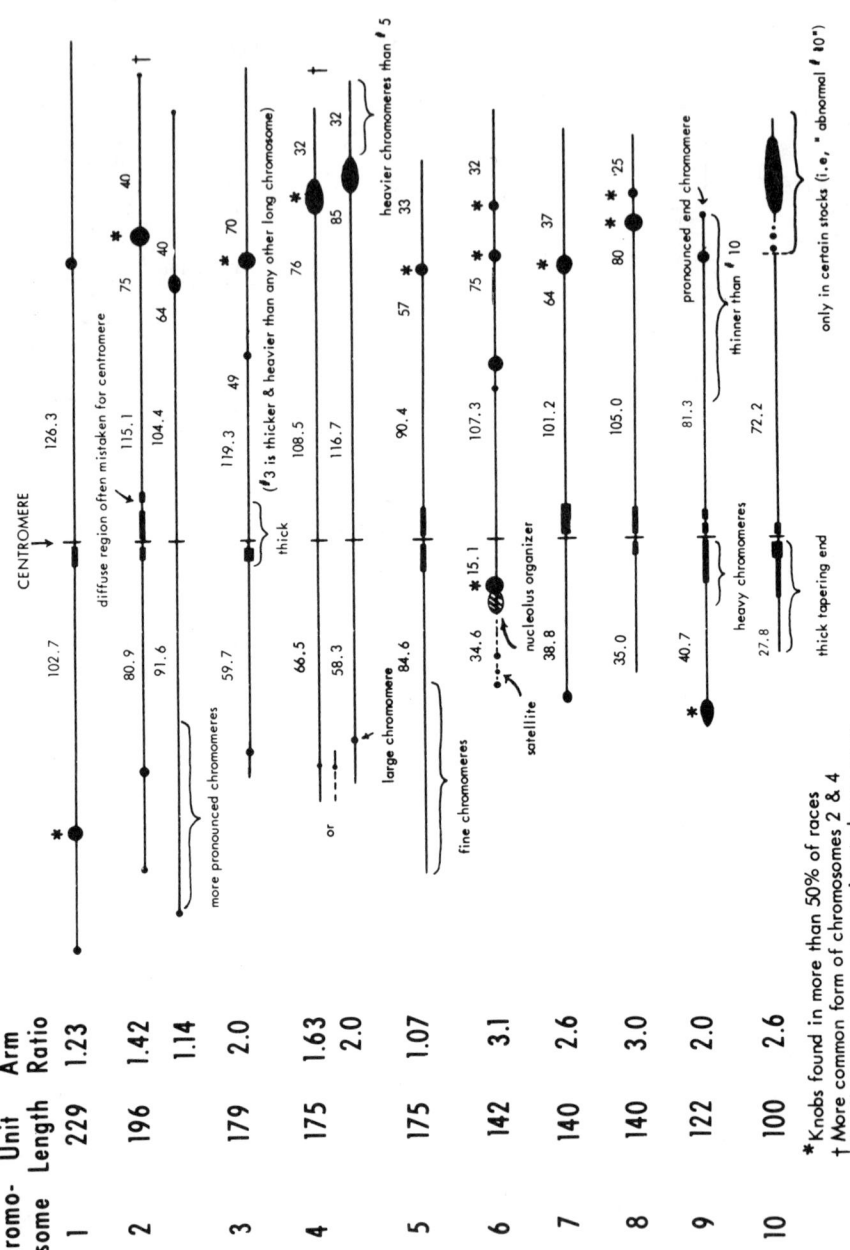

Fig. 2. Cytological map of maize chromosomes, drawn to scale with major distinguishable characteristics (based on published and unpublished data of McClintock, Longley, and Kato).

mutational events (Brink and Nilan, 1952; McClintock, 1951). Schwartz (1965) has discussed the possibility that, for some factors, activity may be determined by derivation (e.g., the sex lineage through which a factor is transmitted) and has suggested that derivation may be the cause of the effects observed for *Fl*, where two doses of *Fl* (transmitted maternally) result in flinty and two doses of *fl* (transmitted maternally) result in floury endosperm. The *Fl fl* relationship has classically been assumed to reflect dosage and thresholds of expression. Experimental discrimination by Kermicle (1970b) between the dosage vs. activation hypotheses for expressions of *R* has established that pollen-transmitted *R* is in a less active state than maternally transmitted *R* (Fig. 4). Differences between factor activities on transmission

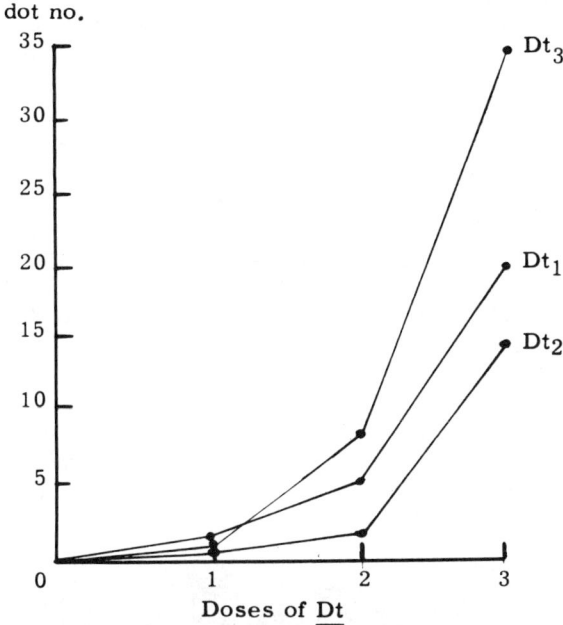

Fig. 3. Exponential dosage effects of *Dt, Dt2,* and *Dt3* on the frequency of *A* dots on *a a a* kernels. (From Nuffer, 1955).

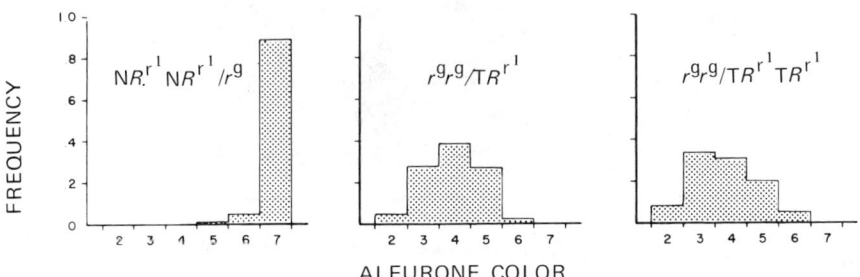

Fig. 4. Differences in level of expression of male-transmitted vs. female-transmitted *R'*. Kernels receiving two doses of *R'* from the male (right) show the low levels typical of kernels receiving one dose from the male (center) rather than the high level typical of kernels receiving two doses from the female (left). (From Kermicle, 1970b.)

through the two sexes were suggested by Schwartz (1965) to explain his observations on *E2* esterase. On the other hand *E* (Schwartz, 1960a) and several other isozyme factors, notably *Adh, Amy, Cat,* and *E10,* show band distributions that are consistent with dosage of alleles determining different mobilities but producing similar numbers of enzyme units/allele in the endosperm (Fig. 5).

Inasmuch as pleiotropism is always arbitrarily defined awaiting further genetic or physiological knowledge, a listing of factors that show the phenomenon reveals more the state of knowledge or the particular bias of the compiler than any inherent validity as a listing. Among notable pleiotropisms are those affecting chlorophylls, carotenoids, and seed dormancy, including *al, cl, lw, lw2, lw3, lw4, ps, vp, vp2, vp5, vp8, vp9,* and *z*. The andromonoecious dwarfs, *an, d, d2, d3, d5,* and *D8,* all show both reduced stature and sex reversal. A few other factors show double effects that might be considered to be disconnected in origin: *cp2, de16, de17, et, gl11, gl17, su,* and *Vm*.

For the majority of known factors it appears that the effects of their action are virtually cell-autonomous — that is, that the occurrence of sectoring for the function, either through chromosome loss or through mutation, leads to little or no cross-feeding of adjacent tissues, indicating that no easily diffusible product is responsible for the effects of that factor. From both the literature and the authors' experience, virtual autonomy at the cell level is clearly characteristic of the following factors: *A, A2, Ac, B, Bk2, Bm, Bt, Bz, Bz2, C, C2, cm, cms-S, Ds, Dt, Dt2, Dt3, Gl, Gl2, ij, j, j2, Kn, Lg, Lg2, Og, P, Pg14, Pr, R, Rf3, Sh, Sh2, Spm, Su, Wd, Wt, Wx, Y, Yg2*. The cross-feeding found for the anthocyanin system, discussed in the section on these factors, is limited in extent to only one or a few cells. Very few factors are found to be nonautonomous. Presumably many or all of the hormone-responding types (e.g., the gibberellin-responding dwarfs) are nonautonomous; a mutable allele of *an, an-m,* appears to be a specific example (Neuffer, unpublished).

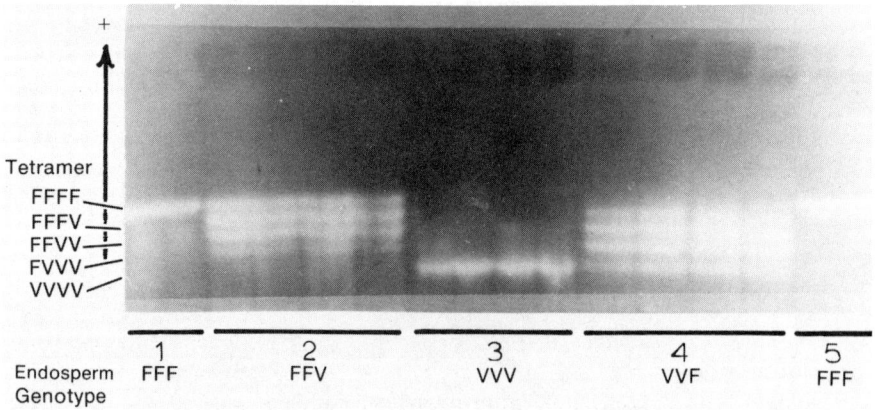

Fig. 5. Dosage effects of *Cat-F* and *Cat-V* alleles. Two monomers that differ in electrophoretic mobility on starch gels form five types of tetramers. The cross of a *Cat-F* ♀ × a *Cat-V* ♂ yields offspring (2) with strong bands for the FFFF and other F-bearing tetramers but little or no VVVV, which is consistent with a gene (and monomer) dosage for *Cat-F* twice that for *Cat-V*, while the reciprocal cross shows precisely the reverse (4). (From Scandalios, 1965.)

Effects attributable to duplicate factors are frequently found but are usually difficult enough to analyze that many are not carried through to full demonstration; stocks of several appear to have been lost. Occurrences of duplicate factors were tabulated by Sprague (1932a), and the possible significance of duplicate factors in identifying gross duplicated segments in the corn genome has been discussed by Rhoades (1951). Among the factors in the locus list, the following show 15:1 (duplicate factor) interactions: certain alleles at the B locus with certain ones at R; $E5$-I with $E5$-II; lu with $lu2$; $lw3$ with $lw4$; $pg11$ with $pg12$; $S2$, $S3$, and $S4$; sen with $sen2$; $sen3$ with $sen4$; $sen5$ with $sen6$; so with $so2$; and ws with $ws2$.

GENETIC SYSTEMS

Substantial genetic information has developed on the effects of unit factors that control certain traits, especially characteristics of the kernel, major pigments as well as some other biochemical constituents, and plant form. Other significant areas of analysis include studies on enzyme variations, on reaction to pests, and on several unique or unorthodox systems of inheritance that have been explored in some depth.

Form and Texture of the Kernel

The corn kernel is one of the most remarkable platforms on which genetic expressions can be studied. Because the kernel is of considerable size and is borne in compact clusters of several hundred individuals, most segregating traits affecting the kernel can be recognized at a glance, and ratios can be estimated by scanning or counting with little or no need for magnification or other aids or even for shelling and sorting. Part of the reason for this visual convenience is the transparentness of the outer maternal tissue, the pericarp, often dubbed the "hull" in popped corn. The substantial triploid endosperm and the shield-shaped germ (the cotyledon and the embryo), both derived as the new generation through meiosis and the sexual process, are conspicuous and are surveyed directly without dissection. With as little as $10\times$ magnification it is possible to resolve the individual cells of the aleurone tissue (the one-cell-thick outer layer of the endosperm) and to recognize mutations or other differential expressions on a field of 150,000 to 250,000 cells representing 15 to 20 mitotic divisions that occurred in sequence and in contiguity. This field of aleurone cells is the "petri dish" of many of the genetic analyses that are conducted. The anthocyanin pigments in the aleurone tissue and carotenoid pigments throughout the endosperm, discussed in separate sections, are particularly distinctive visible traits in the aleurone tissue. Only slightly less conspicuously, the starchy internal portion of the endosperm can change in translucency, detectable with the naked eye or with the aid of a diffuse, light-transmitting surface or similar device.

Many genetic factors affect endosperm form and texture so drastically that the kernel becomes distorted or collapsed such that the change in form itself is recognizable. However, the inheritance of the "dent" form that typifies field corn of the US Corn Belt is complex and has been difficult to study as a genetically controlled expression. Because the dent form tends to interfere with discernment and classification of simpler traits, most strains selected for qualitative genetic work have either the solid, hard (flinty) endosperm type characteristic of popcorns or the round to flat-topped type of the flints of the northeastern USA and of certain meal corn strains, in which the expressions of most endosperm traits are recognized more reliably.

A somewhat arbitrary grouping of kernel morphology types, following common laboratory practice and terminology, separates them into "brittle," "shrunken," "defective," "etched" and "pitted," "sugary" and "glassy," "opaque" and "floury," "viviparous," and "germless."

In the *brittle* type of kernel the endosperm is extremely collapsed. Mutants at any of the four loci *bt, bt2, sh2,* or *sh4* give kernels with closely similar phenotypes. The endosperm in *bt, bt2,* and *sh2* before drying is like a fluid-filled sac (in *sh2* greatly distended, balloon-like, and partially transparent) that develops very little starch. During the drying phase of maturation the kernel shrinks and collapses into a crumpled, angular structure with marked concavities, brittle edges, and a rough surface texture (Fig. 6). In a study of phenotypic interactions of several endosperm factors, Garwood and Creech (1972) found that combinations of *bt, bt2,* and *sh2* with each other and with *ae, du, fl, h, o2, sh, su, su2,* and *wx* are all essentially the same in appearance — that is, the brittle phenotype overrides other, less drastic, visible effects. Although the phenotypic similarity among the four brittle types implies that the end consequences of each mutation are the same (namely gross failure of normal starch deposition), their individual biochemical bases are complex and difficult to define (Nelson and Burr, 1973; Ozbun et al., 1973; Shannon and Creech, 1972). Weaver et al. (1972) reported altered electrophoretic mobility for ADP glucose pyrophosphorylase in *bt2* endosperms and suggested that *Bt2* may be the structural gene (coding locus) for this enzyme. Because the sugar content of *sh2* kernels remains high longer than that of *su* kernels, Laughnan (1953) proposed that *sh2* be used as an alternative factor for sweet corn. The "super-sweet" varieties now in use, which Galinat (1971b) described as having "found a niche with those who like honey-sweet sweet corn," are homozygous for *sh2* in place of *su.* Creech (1968) and Brown et al. (1971) have summarized data on carbohydrate contents and starch granule properties during development in a number of genotypes, including *bt, bt2, sh2,* and many combinations involving these factors; lipid contents have been presented by Flora and Wiley (1972). Alleles identified at the *Sh2* locus include the *a-x* deficiencies isolated by Stadler and Roman (1948).

The *shrunken* kernel characteristic of *sh* is less severely collapsed than the brittle type. The endosperm before drying is fluid-filled and distended,

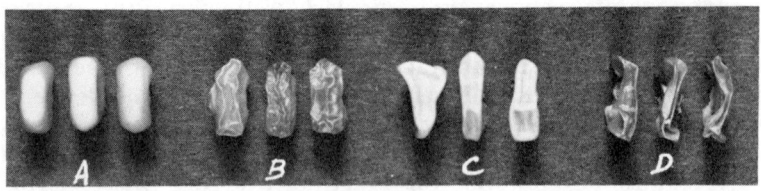

Fig. 6. Phenotypes of the sweet corn factors. (A) normal, *Su Sh2;* (B) sugary, *su Sh2;* (C) super-sweet, *Su sh2;* (D) sugary-supersweet, *su sh2.* (From Laughnan, 1953.)

and has a reduced quantity of solids, which are confined to the periphery surrounding a cavity. Upon drying the kernel usually collapses variably from the top, sides, or bottom, producing smooth, soft-line indentations that may leave a few residual pockets and seams detectable by dissection; whenever collapse does not occur a hollow cavity remains, rounded in outline and distinct from the small, angular cavity found in dent forms. The surface texture of dry *sh* kernels is smooth; the light transmission properties are only slightly different from normal. Interactions with other factors (Garwood and Creech, 1972) show nothing unusual, suggesting that the effects of *sh* may stand alone relative to other factors. The discovery by Schwartz (1960b) of a major protein in *Sh* that is absent in *sh* endosperms (Fig. 7), and the demonstration of electrophoretic variants for this protein in some induced *sh* mutants (Chourey, 1971; Chourey and Schwartz, 1971) indicate that *Sh* is the structural gene for this protein. Alleles identified at the *Sh* locus include the protein-positive mutants and electrophoretic variants (one of which, *sh-S*, complements the others phenotypically) as well as the protein-negatives reported by Chourey and Schwartz, and two probable deficiencies for the *Sh Bz* interval found by Mottinger (1970) following X-ray mutagenesis.

Defective kernel types, represented by the phenotypes of *cp, cp2, de, de16, de17, mn,* and *mn2,* generally are poorly filled. Commonly the pericarp tends to separate from its contents and becomes crumpled or wrinkled. Plant phenotypes are sometimes affected as well. Although 16 or more *de* loci are listed in Emerson et al. (1935), the original series is no longer available in authentic strains. Simply inherited defective types, often incapable of germination, are common in mutagenized material, but little analysis has been done with this class of mutants and most occurrences have remained unexamined. Morphological and histological studies on *mn* are presented by Lowe and Nelson (1946); on *de17* by Brink and Cooper (1947).

Etched and *pitted* kernels are frequent anomalies of development, and tests for inheritance of this type often are futile exercises. A few mutants are established, however, and their nature suggests that there is interesting potential in this morphological class. The factor *st,* sticky chromosomes, in connection with its cytological effects causes irregular plant growth and sectoring, and irregular endosperm development resulting in fissures and pits at

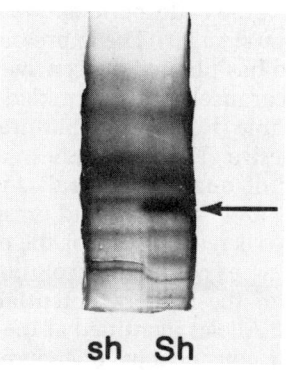

Fig. 7. Major protein band in starch gel electrophoresis found in *Sh* endosperm extracts (right, at arrow) but not in *sh* (left). (From Schwartz, 1960b.)

Fig. 8. Portion of ear homozygous for *et* (etched). In kernels with colored aleurone tissue, the fissured endosperm may be especially distinct. (From Neuffer et al., 1968.)

maturity. An allele, *st-e*, affects the endosperm only at high temperatures (Schwartz, 1958). Two other factors, *et* (etched, a recessive; see Fig. 8) and *Vm* (virescent-mutable, a dominant), develop fissured endosperms regularly. The associated chlorophyll-defective seedlings (virescent with *et* and white to pale-green with *Vm*) indicate that the kernel effects may derive from underlying causes that relate to more general developmental defects. Cox (1966) has identified a dosage-dependent modifier that results in early embryonic elimination of homozygous *et* kernels. Pitted areas develop occasionally in kernels homozygous for intense purple aleurone color (with *in*) but these appear to be composed of disordered, burst cells rather than of fissured areas.

Sugary and *glassy* are phenotypic terms that identify the appearance of a group of kernel types in which the endosperm shows increased transparency, not unlike crude glass or crystal sugar. The expression of the sugary factor, *su*, is familiar to anyone who has planted garden sweet corn. Dry *su* kernels have a glassy, gumlike appearance, and a wrinkled, irregular form that is characteristic and unmistakable (Fig. 6). The mature endosperm just before drying is distended and cohesive. The *su* varieties grown for table use have been selected for a number of quantitative traits, including tenderness of pericarp, desirable flavor and aroma, and resistance to pests, but homozygosity for the *su* factor is responsible for the primary effects of sweetness and creamy texture in sweet corn. The evolution of sweet corns under human selection and some of the modern potentials and implications are discussed by Galinat (1971b). Alleles identified at the *Su* locus include *su-am* (amylaceous), which has no conspicuous phenotypic effect except in interaction with *du; su-st*, starchy endosperm recessive to *su* (Dahlstrom and Lonnquist, 1964), and *su-66*, which is intermediate in expression (Creech,

Table 2—Expressions and effects in combination of some of the factors
determining opaque, floury, and glassy endosperms†

Constitution	Su Wx	Su wx	su Wx	su wx
All dominant	normal	opaque	wrinkled, glassy	wrinkled, glassy to opaque
ae	tarnished	collapsed, translucent (new)	translucent (new)	collapsed, translucent (new)
du	tarnished	collapsed, opaque (new)	wrinkled, glassy	wrinkled, glassy
fl	opaque, floury	opaque, floury	wrinkled, glassy	
h	opaque, floury	opaque, floury		
o2	opaque, floury	opaque, floury	wrinkled, glassy	
su2	tarnished	opaque	wrinkled, glassy	wrinkled, glassy
ae du	translucent (new)	collapsed, opaque (new)	wrinkled, translucent	collapsed, translucent (new)
ae su2	tarnished	translucent (new)	wrinkled, translucent (new)	
du su2	translucent (new)	collapsed, opaque (new)	wrinkled, glassy	

†Adapted from Garwood and Creech, 1972.

1968). Kernels homozygous for *su2*, though similar to *su* in appearance, generally are less deviant from normal in all respects. Two other glassy types, *ae* and *du*, tend to be a little collapsed but only slightly wrinkled. The term "tarnished" has been used to identify the dull glassiness of *su2, ae,* and *du* kernels. The expressions and interactions given in Table 2 identify certain new or unique expressions in the endosperm of combinations involving this group of factors, specifically for *ae* when homozygous recessive with *su, du,* or *wx,* and for *du* with *su2* and *wx.* Although interactions suggest complementing functional effects, carbohydrate patterns during kernel development are very similar for *su, su2, ae,* and *du* singly (Creech, 1968); presumably these patterns contribute similarly to the commonalities in phenotype but derive from different causes. Data on the enzymes and on polysaccharide structures (Nelson and Burr, 1973; Shannon and Creech, 1972) suggest that these factors are somehow instrumental in the regulation of branching of starch and other polysaccharides. Specialized industrial applications for the starch from select hybrids homozygous for *ae* have become significant because of their higher content of amylose, the straight-chain unbranched starch type (Senti, 1967).

Factors that can result in *opaque* appearance or *floury* texture in the endosperm include *o, o2, o5, o6, o7, fl, fl2, h,* and *wx.* At first glance the opaque appearance of kernels homozygous for any of these is the same, but close inspection reveals discrete distinctions between some of them. Whereas most of these are chalk-like in appearance and texture, *wx* kernels display a uniform, marble-like opacity, and a hardness similar to that of normal kernels. Cut with a blade, *wx* endosperm chips away evenly, leaving a smooth, opaque surface, while normal endosperm (in the corneous side portions) breaks unevenly and leaves an irregular, transparent surface. The starch in the cut surface of non-waxy endosperm, whether flinty, floury, opaque, glassy, or brittle, will stain blue, turning quickly to black, with iodine-potassium iodide solution, while that of homozygous *wx* (waxy) will stain reddish-brown, turning soon to dark brown. A drop of dilute solution (light amber solution prepared as needed by dilution from a stock solution of, for example, 10 g KI and 5 g I_2 or less in 100 ml of water) is sufficient, but the working solution occasionally needs replenishment because of its instability.

Precise work such as staining of pollen requires more exacting preparations and procedures (Nelson, 1968). Alleles that have been identified at the Wx locus include wx-a (Argentine), which has intermediate starch constitution and staining properties, and a number of recombinationally distinct stable and mutable alleles (Nelson, 1959, 1968). There is considerable industrial use of the pure amylopectin (branched-chain starch) from waxy strains as a gelling agent. Floury kernels (all of the opaque and floury factors other than wx) are more chalk-like than wx in appearance and in hardness in all or most of the endosperm, the tissue crumbling away loosely when cut. Some of the floury types (fl, $fl2$, and h) are difficult to classify in segregating material, especially in diverse backgrounds, even with the aid of an arrangement for observing transmission of light through the endosperm. The expressions of fl and $fl2$ are dosage-dependent, though to different degrees: whereas with two or three doses of Fl the kernels are flinty and with two or three doses of fl they are floury, each dosage level for $fl2$ is distinguishable (with difficulty) from each other level. In contrast, dominance is characteristic of 0, 02, 05, 06, 07, H, and Wx. Anthocyanin pigmentation in the aleurone tissue is reduced and irregular in o kernels. A group of duplicate recessive factors (15:1 interaction) of the floury class, designated sen through $sen6$, has been identified recently (T.R. Stierwalt and P.L. Crane, 1974, MNL 48:139-140). The flour and meal corns, used by natives of the Americas for hand grinding and for parching, were strains that had been selected in use to carry fl or one of the other factors of this class. Changes in protein composition are found for $o2$ (Mertz et al., 1964), $fl2$ (Nelson et al., 1965), and $o7$ (Misra et al., 1972), with important potential impact toward designed improvement in the nutritional balance (for humans and nonruminant domestic animals) of amino acids, especially lysine and tryptophan, in corn proteins. The mechanism by which these single-gene changes result in alterations in the balance of the whole spectrum of proteins stored in the endosperm has attracted intensive experimentation, which is summarized by Nelson and Burr (1973). That the effect is indirect is suggested by the studies of Misra et al. (1972), which show that not only $o2$, $o7$, and $fl2$ have these protein changes but also the carbohydrate-modifying factors su, sh, $sh2$, $sh4$, bt, and $bt2$. Except for wx, which interacts with ae and du (see Table 2), the members of this group do not show unique phenotypes in interactions (Garwood and Creech, 1972). Evidence is summarized by Nelson and Burr (1973) supporting the view that Wx is the structural gene for a starch granule-bound (i.e., not soluble) nucleoside diphosphate sugar-starch glucosyl transferase (accepting UDPG or ADPG), an enzyme that is given only a modest role in starch accumulation according to the analyses of Ozbun et al. (1973). The phenotypic interactions of Ae, Du, and Wx (Garwood and Creech, 1972) indicate that they work in concert in the maximal accumulation of starch.

Viviparous kernels constitute a diverse class in which the embryos, instead of maturing into the normal dormant state, tend to grow into plantlets, while still on the maturing ear. Even if a plantlet does not become discernible, moderate to severe distortion and irregular translucency in the endosperm are common, presumably due to partial digestion during premature germination. Alleles at the loci *al*, *ps*, *vp*, *vp2*, *vp5*, *vp8*, *vp9*, *w3*, *y9*, and *z* display these effects to varying degrees. Robertson (1955, 1961) has characterized most of the viviparous factors; vivipary in at least four (*ps, vp, vp5, vp8*) is determined by the embryo genotype independently of the endosperm genotype, and effects on endosperm color (anthocyanins and carotenoids) are determined specifically by the endosperm genotype. Diverse allelic forms have been identified at the *Al, Vp9,* and *W3* loci (Robertson, 1971); the notable relationships of this group of factors to chlorophylls and carotenoids are considered in the section on factors affecting these pigments. The physiology of *vp* and *vp5* endosperms and embryos during premature plantlet formation has been found to be parallel to that in normal germination (Wilson et al., 1973). A condition simulating vivipary is induced following infection of developing ears by a particular strain of *Diplodia maydis* (Calvert et al., 1969).

Germless kernels, like viviparous kernels, are often distorted in form or altered in endosperm translucency. Occasional kernels without a normally developed embryo occur as seemingly spontaneous accidents in development and cannot be studied further. Segregations for germless kernels, for which the symbol *gm* has been used, are frequent following mutagen treatments, but little has been done with this class of mutants.

Anthocyanins and Related Pigments

According to East and Hayes (1911), segregations for red and blue colors in corn kernels had attracted enough attention among plant hybridizers before 1900 to lead to some anticipatory counting of proportions; ultimately these segregations contributed strongly to the rediscovery of Mendel's laws, especially in the work of Correns. The red and blue pigmentations have been used repeatedly in genetic studies, primarily because they are nonvital and are easy to observe. Moreover, because the genetics of pigment control is more fully elaborated than that of any other character or constituent, and because virtually every tissue can form anthocyanins under specified genetic conditions, these markers permit development of versatile and flexible research designs. The information that follows is intended to provide a survey of the genetics of these factors and a guide to the basic literature on the state of research into their action and nature. Genetic control of certain red and brown pigments of the pericarp, cob, and silks appears to overlap, or at least to have factors in common with that of anthocyanins, even though the pigments are not strictly anthocyanins and their biochemistry is unknown. The brown-midrib factors, which probably are concerned with biochemical pathways closely related to those of the anthocyanins, also are considered briefly in this section.

Adequate description of the genetics of the pigment factors requires consideration of three facets: 1) the loci involved, including the allelic diversity found at each locus; 2) the tissues affected, which can be diverse yet very explicit, and 3) the interactions among the effects of the different loci.

At least 14 distinct loci affect the qualitative, quantitative, and distributional elaboration of anthocyanin pigments and their relatives; only the smallest linkage group, number 8, lacks a member of this class. The 14-plus loci are listed in alphabetical order, according to the tissues over which they have control, in the accompanying chart (Table 3). Each locus has its own particular effects: some, like *A*, are essential to anthocyanin formation

Table 3—Anthocyanins and related pigments: the genotypic constitutions presently known to be required for pigmentation in certain tissues

Tissue															
Kernel colors															
Aleurone layer	A	A2		B-Peru†	Bz	Bz2	C§	C2§	In§				Pr§		R†§
Scutellum	A	A2		B-Peru†	Bz	Bz2	C	C2					Pr§		R†(S and s5 and two of S2, S3, S4)
Plumule	A	A2			Bz		C	C2						Pu Pu2	
Seedling colors															
Coleoptile (exposed)	A	A2		B†	Bz	Bz2		C2				Pl			R-r†‡
Coleoptile (unexposed)	A	A2		B†	Bz	Bz2		C2							R-r†‡
Leaf tip and margin	A	A2		B	Bz	Bz2		C2							R-r†
Auricle	A	A2			Bz	Bz2		C2							
Leaf blade	A	A2			Bz	Bz2		C2		Lc					
Roots (unexposed)	A	A2		B†	Bz	Bz2		C2				Pl			R-r†‡
Sheath (exposed)	A	A2		B†	Bz	Bz2		C2							R-r†‡
Plant colors															
Sheath, husks, culm (exposed)	A	A2	a3†	B†	Bz	Bz2		C2§							
Sheath, husks, culm (unexposed)	A	A2	a3†	B†	Bz	Bz2		C2				Pl			
Auricle	A	A2		B†	Bz	Bz2		C2		Lc†					
Leaf blade	A	A2		B†	Bz	Bz2		C2		Lc†					
Tassel glume base	A	A2		B†	Bz	Bz2		C2							R-r†‡
Tassel glume face	A	A2	a3†	B§	Bz	Bz2		C2							R-r†
Anthers (exposed)	A	A2		B†	Bz	Bz2		C2				Pl			R-r†
Anthers (unexposed)	A	A2			Bz	Bz2									R-r†
Silks red	A										P-RR				sm
salmon	A										P-WW				sm
brown	A														
Cob purple (unexposed)	A	A2		B§	Bz	Bz2		C2§			P-WR§	Pl			
brick-red	A									ch	P-RR				
Pericarp lacquer-red	a†§			Bp							P-RR				
brown	A			Bp											
chocolate				bp†						Ch					
purple (unexposed)	A	A2		B†	Bz	Bz2				Lc†		Pl			R-r†‡

†Within one line of the table, only one of these factors is required (see text). ‡Some other alleles can substitute for the one listed.
§See text for effects of the allelic series at the indicated locus.

Table 4—Aleurone color expressions of the known factors controlling anthocyanins and related pigments. Aleurone color is given for kernels homozygous for the limiting factors specified in the left column (in the presence of all other factors) with the four possible combinations of the modifying factors *Pr, pr* and *In, in*. The pericarp color given is expressed in otherwise clear pericarp tissue over aleurone tissue of the indicated genotype

Limiting factor	Pr In	Pr in	pr In	pr in
All color factors present	purple	deep purple; pericarp brown	red	deep red
a	colorless	faint brown; pericarp brown	colorless	colorless; pericarp faint brown
a2	colorless	faint brown; pericarp brown	colorless	colorless; pericarp faint brown
bz	purple bronze	brownish purple; pericarp brown	red bronze	pink; pericarp yellow brown
bz2	purple bronze	brownish purple; pericarp brown	red bronze	pink; pericarp yellow brown
bz bz2	near colorless	not seen	colorless	not seen
c	colorless	colorless	colorless	colorless
C-I†	colorless; few purple dots	pale; few deep purple dots	colorless; few red dots	red pale; few deep red dots
c2‡	colorless	pale purple	colorless	pale red
r§	colorless	colorless	colorless	colorless

†Expression indicated is in one dose (*C-I C C*); two doses (*C-I C-I C*) or three doses (*C-I C-I C-I*) generally completely colorless; *C-I c c* and *C-I C-I c* are completely colorless. In certain backgrounds *C* also has dosage relationship to *c* (*C c c* pale, *C C c* dark purple, *C C C* very dark purple.) ‡*c2* has clear dosage effect. Homozygous recessive kernels (*c2 c2 c2*) are colorless, those with one dose of *C2* (*C2 c2 c2*) are pale, two-dose kernels are dark pale, and three-dose kernels are fully colored.
§In most backgrounds *R r r* kernels are mottled while *R R r* and *R R R* kernels are fully colored; see text for effects of certain alleles at the *B* locus ("*R2*".)

in virtually every tissue; while others, like *C* and *Pl*, are essential only to one or a few tissues. Broad and parallel control over pigmentation in a common set of tissues is found for the factors *A2, Bz,* and *Bz2*, which complement each other and *A* in most tissues. A few factors, such as *B* and certain alleles at *R*, have duplicate-like effects on pigmentation in particular tissues; these are identified by daggers in the chart. In designing experiments the chart may be used either to identify which tissues can be used to follow the effects of a selected locus or to identify for a selected tissue those factors for which genetic variation is known. As an aid in diagnosis of genotype, the chart may be used to identify which loci are presently known to be required for pigmentation in each tissue; thus, given an observation that pigment is present in one or more specific tissues of an individual or strain under study it is possible to use the chart to identify genotypic constitution for the known factors. Descriptions of the expressions of the type alleles at each locus, in aleurone tissue and in plant tissues, are given in abbreviated form in Tables 4 and 5. More detailed studies of expressions, alleles, structure, and action have been carried out on many of the loci, and the briefs below outline some of the supplemental information.

    A   Expressions of the allele types at the *A* locus reported and analyzed by Laughnan (1948, 1955) and Nuffer (1961), with a few additions, are given in Table 6. Existing and experimentally isolated variants include alleles with several intermediate levels of expression; the faintest, *a-b,* can be distinguished from *a* in the aleurone tissue only with difficulty even by testing for anthocyanin with dilute acid (e.g., 4% hydrogen chloride, applied to cut

Table 5—Plant color expressions of the known factors controlling anthocyanins and related pigments. Color of plant tissues is given for plants bearing the limiting factors specified in the left column (in the presence of all other factors) with the four possible combinations of the factors B, b and Pl, pl

Limiting factor	B Pl	B pl	b Pl	b pl
All dominant including R-r or r-r	purple†	sun-red in exposed tissues; pink anthers†	green plant; purple anthers, glumes, base, and brace roots	green plant; pink anthers, red glumes, base, and brace roots
pr	deep maroon†	sun-red in exposed tissues; pink anthers†	green plant; maroon anthers, glumes, base, and brace roots	green plant; pink anthers, glumes, base, and brace roots
a	brown†	faint brown, especially glume base; sun-dependent	green; brown base and brace roots	green
$a2$	brown; brown tissues deteriorate†	faint brown, especially glume base; sun-dependent	green; brown base and brace roots	green
bz	red brown; some deterioration†	weak red brown; sun-dependent	green; red-brown anthers, base, and brace roots	green
$bz2$	red brown; some deterioration†	weak red brown; sun-dependent	green; red-brown anthers, base, and brace roots	green
$c2$	purple auricles and glumes; faint elsewhere	sun-red auricles and glumes	green	green
R-g or r-g	purple plant† green anthers	sun-red plant;† green anthers	green	green

†Expressed in leaf sheaths, auricles and blades, culm, cob, husks, glumes, anthers, coleoptile, seedling base, and brace roots. Same applies where indicated in B pl column only if tissue is exposed to sunlight.

cells, resulting in a characteristic bright pink color change). With P-RR, pericarp color determined by A-st and some of the other alleles is lacquer-red (non-anthocyanin pigments); A-b and certain other alleles determine brown color dominant to that of A-st; a and some others determine recessive brown. Laughnan (1952) has shown that A-b consists of separable components α(A-d) and β(like A-st). Dots of color in a aleurone tissue and stripes in a plant tissues are mutations of a to A and intermediate alleles caused by Dt (Rhoades 1941); Dt can also induce changes in the A to a direction (Nuffer, 1961). Other mutable alleles are described in the section on mutability. The anthocyanins in purple aleurone tissue, cyanidin and pelargonidin glucosides, accompanied in husks by traces of peonidin glucosides (Chen, 1973; Harborne and Gavazzi, 1969; Kirby and Styles, 1970; Sando et al., 1935; Styles and Ceska, 1972), are absent in a tissues; but glycosides of closely related flavonols, quercetin and kaempferol, are increased in a husks and aleurone tissue (Kirby and Styles, 1970; Laughnan, 1950).

$A2$ As $a2$ kernels age in storage, brown pigments tend to accumulate, perhaps through the oxidation of phenolic compounds paralleling that in the development of brown pigments in the husks. In husks and sheaths, $a2$ B Pl tissues develop necrotic areas that may expand until the entire plant collapses; however, if either B or Pl is absent necrosis does not develop. Mutable variants at $A2$ are described in the section on mutability. In addition to flavonols (Kirby and Styles, 1970), leucoanthocyanidins accumulate in $a2$ aleurone tissues (Reddy and Reddy, 1971).

Table 6—Types of alleles at the *A* locus and their effects on aleurone, plant, and pericarp colors

Allele type		Aleurone color with all color factors present	Plant color with *B Pl*	Pericarp color with *P-RR*	Pericarp with *P-WW* *b Pl r-ch*	Notes
A-st		purple	purple	red	cherry	Common N. Am. allele
	(β)	purple	purple	red	cherry	Derived from *A-b*
A-b	(αβ)	purple	purple	dominant brown	cherry	Collected from Ecuador
A-b:P	(βα′)	purple	purple	dominant brown	cherry	Collected from Peru
A-r		purple	purple	red	cherry	From *a* by *Dt* action
A-rb		purple	purple	red-brown	cherry	From *a* by *Dt* action
A-br		purple	purple	recessive brown	cherry	From *a* by *Dt*; dosage effect
A-lt		dilute purple	deep red-brown	recessive brown	red-brown	From *a* by *Dt*; dosage effect
A-w		dilute purple	deep red-brown			From *a* by *Dt* or UV; dosage effect
A-d	(α)	dilute purple	deep red-brown	dominant brown	red-brown	Derived from *A-b*; competes with most *A* alleles
a-p	(α′)	pale purple	red-brown	dominant brown	brown	Collected from Peru; competes with most *A* alleles
a-b		faint purple	red-brown	dominant brown	brown	
a		colorless	brown	recessive brown	brown	Standard; responds to *Dt*
a-s		colorless	brown	recessive brown	brown	Stable with *Dt*; from *a* by *Dt* action
a-X		colorless	brown	recessive brown	brown	From *A* by x-rays; deficiencies including *Sh2*

After Laughnan (1948, 1955) and Nuffer (1961).

a3   Information on the genetics of this recessive intensifier of plant color is limited; other factors essential to the phenotype may include specific alleles at *B* or *R,* and the factor could possibly be allelic to *A.*

B   The alleles listed by Emerson (1921) are *B, B-w, b-s,* and *b;* authentic stocks of the two intermediate alleles, *B-w* and *b-s,* have not been maintained. The allele types studied by Coe (1966b) have the following properties:

     B   Intense pigmentation in most tissues; paramutable (changed by *B′*)
     B-b   Intense glume base and culm, weak elsewhere; nonparamutable
     B-v   Variegated, mutating to *B* and *b-v;* paramutable
     b-v   Green plant mutants from *B-v;* paramutable
     B′   Intense glume base and culm, weak elsewhere; paramutagenic
     b   No anthocyanin (with green-plant alleles at *R*); nonparamutable

"Paramutagenic" alleles cause "paramutable" alleles to be reduced in expression (and to become paramutagenic), as discussed in a separate section.
The following two alleles, unique in that they confer aleurone color in place of *R,* are described by Styles et al. (1973):

     B-Peru   Strong aleurone and scutellum color, supplanting *R*; faint plant color
     B-Bolivia   Variable aleurone color, supplanting *R*; strong plant color

Bz   The aleurone tissue of *bz* kernels in different backgrounds can range from virtually colorless to dark brown or purple-brown approaching *Bz* in phenotype; dilute acid (4% hydrogen chloride) applied to cut cells results in bright pink color change characteristic of anthocyanins, developing from gray-brown vacuolar pigments. Brown wall pigments are evident in *bz* aleurone cells in some backgrounds. Plant tissues contain strong anthocyanin coloring during development, browning and soon becoming necrotic. Anthers of *bz* plants are brilliantly yellow-fluorescent under ultraviolet light, regardless of genotype for any other known factors. Mutants from *Bz* isolated by Mottinger (1970, 1973) following X-irradiation include variants that are probably small deletions (at least one including *Sh*) and some that are unstable. Other mutable alleles are discussed in the section on mutability. In addition to giving a detailed description of the expression of *bz*, Rhoades (1952) explores the implications of McClintock's (1951) observation of anthocyanin "borders" at the interface between colorless *(C-I Bz)* and bronze *(C bz)* sectors in aleurone tissue. Luteolinidin (3-deoxycyanidin) has been reported in hydrolyzed extracts from *bz* plant tissues by Styles and Ceska (1972). The flavonol quercetin and its glucoside, isoquercitrin, are present in *Bz* pollen, but the glucoside and the enzyme that catalyzes its formation from quercetin are absent in *bz* pollen (Larson and Coe, 1968); presumably the accumulation of quercetin, which is bright yellow-fluorescent in contrast to its glucoside, is responsible for the striking yellow fluorescence of *bz* anthers.

Bz2 Closely similar to *bz* (but lacking the yellow-fluorescent anther); originally found as a mutable allele, *bz2-m*, from which stable alleles have been isolated. The mutant *an-6923*, isolated from material irradiated in nuclear bomb tests, is allelic to both *an* and *bz2* and is homozygous viable.

C The existence of three types of alleles, inhibitor *(C-I)*, color-determining *(C)*, and colorless *(c)*, has long been known; expressions of combinations of these are sketched in a footnote to Table 4 and have been detailed further by Coe (1962). A large test for recombination between *C-I* and *c* yielded no recombinants (Coe, 1964). Kirby and Styles (1970) demonstrated light-dependent pigment synthesis during germination of *c* kernels, which Chen (1973; Hsu-Chen and Coe, 1973) has found to be specific to particular conditional alleles, designated *c-p* (positive), in contrast to *c-n* (negative). The allele carried in at least one of the long-standing *c sh* pedigrees of the Maize Genetics Coop. is *c-n*. The dominant factor *Bh* causes variable, diffuse blotches of color in the aleurone tissue of *c* kernels.

C2 Recessive *(c2)* plant tissues develop substantial pigment, but the aleurone tissue contains no detectable pigment except in the presence of homozygous intensifier *(in)* or sugary *(su)*. The dominant dilute color factor of Brink and Greenblatt (1954), *C2-Idf*, appears to be allelic.

In In addition to intensifying anthocyanin pigmentation specifically in the aleurone layer, the recessive allele, *in*, also conditions a metallic gold to brown sheen in the pericarp over aleurone tissue homozygous for *in*. A dominant dilute factor, *In-D*, appears to be allelic.

P The alleles studied by Anderson (1924) are symbolized with two letters, the first pertaining to the pericarp color and the second to the cob color, with *R* designating lacquer-red pericarp or brick-red cob; *O* orange; *W* white (clear) pericarp or white cob; *C* white-cap pericarp; and *V* variegated, as follows: *P-RR*, *P-OR*, *P-WR*, *P-OW*, *P-CW*, *P-CR*, *P-WW*, and *P-VV*. The allele *P-MO* (mosaic), an exception to the symbol series, is mutable and generally similar in appearance to *P-VV*, but mutates less frequently to uniform red. The mutability at this locus is described in the section on mutability. Silk color is affected by the allele of *P* present with *sm*. The symbol *p* is used interchangeably with *P-WW*.

Pr In plant tissues, the effects of *pr* (redder color than *Pr*) are not distinct enough to permit unambiguous classification. In aleurone tissues, however, the expressions are usually clear-cut. *Pr* determines mostly cyanidin glucoside with small amounts of pelargonidin glucoside, while *pr* determines

Table 7—Types of alleles at the *R* locus and their effects on aleurone and other colors

Allele type	Aleurone color with all color factors present	Aleurone color in one dose from male	Embryo or scutellum color	Seedling and plant color with *b Pl*†	Notes‡
R-r	purple	mottled		purple	Paramutable; (P) (S)
R-g	purple	mottled		colorless	Paramutable; (p) (S) or (S)
r-r	colorless	colorless	colorless	purple	(P) (s) or (P)
r-g	colorless	colorless	colorless	colorless	(p) (s) etc.
R-ch	dilute purple	mottled		deep purple	Paramutable; (P) (S) and others
R-mb	marbled	marbled		colorless	Paramutagenic
R-nj	navajo§	navajo§	purple§	purple	
R-sc	purple	purple	purple		From *R-st*; some are paramutagenic
R-scm	purple	purple	purple	colorless	From *R-mb*; some are paramutagenic
R-st	stippled	stippled	stippled	colorless	Paramutagenic
r-ch	colorless	colorless		deep purple	
r-x	colorless	colorless		colorless	Deficiency from *R-r*

†Expressed as purple to red pigment (disregarding green) in roots, mesocotyl, coleoptile, lower leaf sheaths, tassel glume face, anthers, and silks, often but not always in concert in all of these tissues for a particular allele (see text); in pericarp with *P-WW* color is expressed as red streaks (for *R-r* or *r-r*) or cherry-red to purple (for *R-ch* or *r-ch*).   ‡"Paramutagenic" alleles cause "paramutable" alleles to be reduced in expression, as discussed in a separate section; (P) and (S) represent components for plant color and seed color, respectively (see text).   §Color restricted to the crown portion of the aleurone tissue; one specific allele, *R-nj:Cudu*, determines exceptionally strong color in the embryo (see text).

reverse proportions of these two pigments (Chen, 1973; Harborne and Gavazzi, 1969; Kirby and Styles, 1970). The homologous flavonols, respectively quercetin and kaempferol, have been identified in hydrolyzed extracts of certain *Pr* and *pr* genotypes by Kirby and Styles (1970), and the parallel leucoanthocyanidins have been identified by Reddy (1964) in *a2* kernels.

R   More diverse varieties of expression are known at *R* than at any other locus; variants are found to affect at least 12 tissue types differentially. Expressions of the types of alleles are given in Table 7, which is drawn primarily from the descriptions given by Brink (1958) and by Styles et al. (1973). The first four types, designated by Emerson (1921), categorize alleles according to effects on aleurone color (*R* vs. *r*) and on plant color (*-r* vs. *-g*, respectively designating red or green plant parts, noted primarily in the seedling leaf tip and the anthers). Several symbols overlap these categories; for example, *R-ch* and *r-ch* determine plant color that would be designated *-r*, but they also condition unusually strong (cherry) pericarp color that has led to the special *-ch* designation instead. The pattern alleles, *R-mb*, *R-st*, and *R-nj*, are each specific in the distribution of pigmented areas in the aleurone tissue: *R-mb* (marbled) confers large blotches of color with well-defined borders; *R-st* (stippled) confers small, sharply defined colored sectors (Fig. 9); *R-nj* (navajo crown) confers color with diffuse, graded borders in the crown portion of the kernel. The designation *-sc* is applied to self-colored alleles (i.e., those that are uniformly colored rather than mottled when present in one male-derived dose); mutants with self-colored expression arise by mutation from *R-st* and *R-mb* (Ashman, 1960; Weyers, 1961). That the mottling expression (irregular, scattered color) typical of *r r R* kernels is determined by transmission through the male (regardless of dosage) has been shown by Kermicle (1970b; see Fig. 4). According to the exacting studies carried out by Fogel (1946) and summarized by Stadler (1951), *R-r* alleles can be arranged in a series according to very fine increments in level of expression in the various plant tissues, but an increment of difference in one tissue does not necessarily presage a parallel increment in another. Independent mutation of

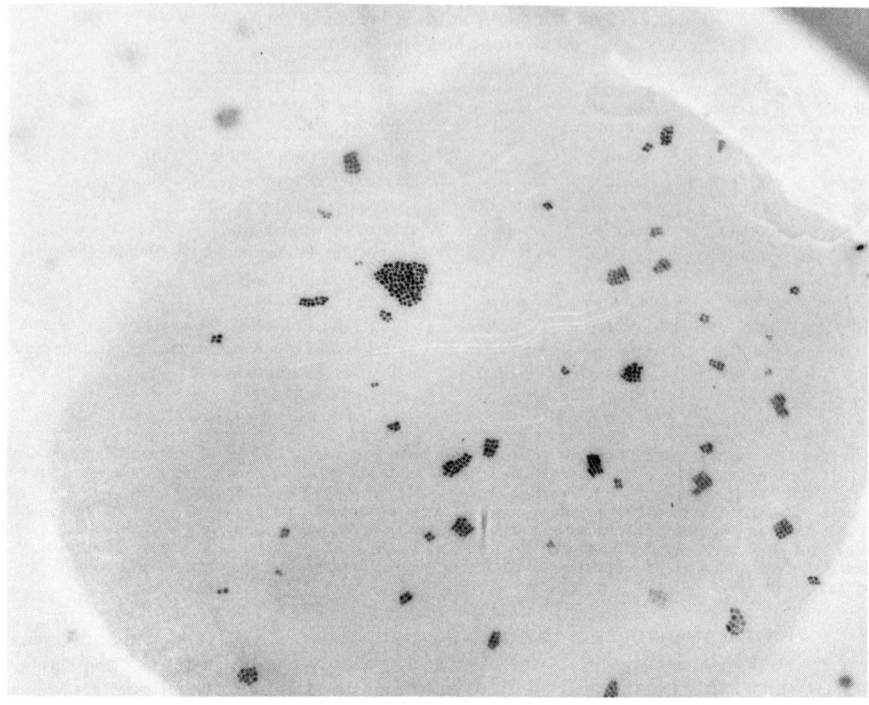

Fig. 9. Region of stippled *(R-st)* aleurone tissue, showing sectors (clones) of cells resulting from mutations of *R-st* to *R-sc* (self-colored) during development (×25).

plant color and seed color expression (Stadler, 1951; see Fig. 10), and correlations with crossing over (Emmerling, 1958; Stadler and Emmerling, 1956; Stadler and Nuffer, 1953) have established compoundness at the locus, designated by (P) for plant color and (S) for seed color. Considering the incremental differences among alleles, and the demonstration by Sastry (1970) that *R-ch* alleles undergo independent mutation of expressions in at least four tissues, the locus may be even more complex in structure. The picture is confounded, however, by the existence of *"R-g"* alleles (colored aleurone, green anthers) that condition color in certain plant tissues (coleoptile, mesocotyl, roots) and mutate to colorless in both aleurone and plant tissues in one step (Bray and Brink, 1966; Stadler, 1951). The specific expressions of *R-nj:Cudu* in the endosperm (navajo crown) and the embryo provide differential expression that identifies kernels in which exceptional ("noncorresponding") embryos can be selected that are potentially haploids or diploids of maternal or paternal origin (Greenblatt and Bock, 1967; Kermicle, 1969; Nanda and

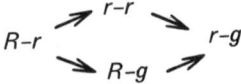

Fig. 10. Mutational sequence for *R-r*, in which *r-g* (lacking both plant and aleurone color) arises either through *r-r* (lacking aleurone color) or through *R-g* (lacking plant color), but not in a single step from *R-r*. (After Stadler, 1951.)

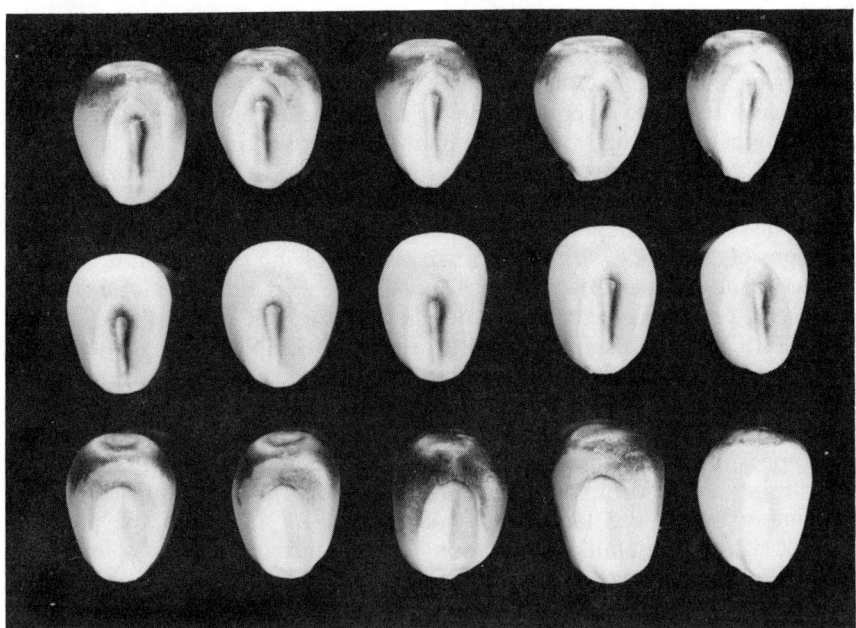

Fig. 11. Expression of *R-nj* and its use as a marker to screen for exceptional events at fertilization. The kernels shown are from the cross of *r r* × *R-nj r*. Top row: kernels with the anthocyanin pigmentation characteristic of the presence of *R-nj* in the endosperm (color restricted to the crown) and in the embryo. Center and bottom rows: exceptional kernels in which a sperm carrying *R-nj* entered the embryo, while one carrying *r* entered the endosperm, and vice versa; both types result from heterofertilization in this cross. The kernels in the bottom row demonstrate the expression by which haploid embryos can be recognized in the dry kernel in crosses of *r r* × *R-nj R-nj* (for maternal haploids) or *R-nj R-nj* × *r r* (for paternal haploids). (Courtesy of J.L. Kermicle.)

Chase, 1966; see Fig. 11). A deficiency, *r-x1*, was derived by L.J. Stadler in 1944 from *R-r* following X-irradiation (the deficiency termed *r-x2* is the same one as *r-x1*, accidentally misidentified). Unlike *X-1* (Stadler, 1933), which was a cytologically visible terminal deficiency, *r-x1* is intercalary and is cytologically undetectable; it is transmitted only through the female and lacks aleurone and plant color effects. It has been found to induce non-disjunctional events in the megagametophyte that result in high frequencies of aneuploid eggs, generating monosomes, trisomes, and more complex aneuploids (Weber, 1973). Among factors mapping in the *R* region whose functional relationships are curiously confounded are *Lc* (leaf color), *M-st* (modifier of *R-st*), *Ej* (extender of *j* and other striping factors), and *cm* (chloroplast mutator).

Field corn lines of the U.S. Corn Belt are regularly *r-r* or *r-g*; usually *c* and *pl*; usually *A, A2, Bz, Bz2*; and *B-b* or *b*. Sweet corn and popcorn lines are more diverse in aleurone color factors, often carrying *C-I*; super-sweet hybrids carry *a*, derived from the source of *sh2* and closely linked to it.

Interactions of the anthocyanin factors in aleurone, scutellum, and plant tissues are known well enough to warrant elaboration; only parts of these interactions are given or implied in the foregoing tables.

In the aleurone tissue the formation of full purple pigment requires nine factors: A, $A2$, Bz, $Bz2$, C, $C2$, In, Pr, and either R or certain rare alleles at the B locus (Tables 3 and 4). Complementation characterizes the interaction of seven of the nine factors: If any one or more of a, $a2$, c, $c2$, or r is homozygous recessive no color develops; if bz or $bz2$ is homozygous very little purple pigment develops and brown pigment is formed instead (as long as A, $A2$, C, $C2$, and R are present), while if bz and $bz2$ are both homozygous no color develops. East and Hayes (1911) identified the 9:7 ratio resulting from complementary interactions in F_2 from $C\ c\ R\ r$, the 9:3:4 from epistatic interactions of $C\ c$ or $R\ r$ with $Pr\ pr$, and ratios involving the color inhibitor, C-I; Emerson (1918) identified the 27:37 complementary ratio from $A\ a\ C\ c\ R\ r$. Absence of complementary interaction is characteristic of c with C-I, which were shown by Hutchison (1922) to be allelic. A complementary interaction forms distinct borders of anthocyanin at the interface between sectors of C-I Bz (colorless) and C bz (bronze) tissue (McClintock, 1951; see Fig. 12). Rhoades (1952) observed that the most intensely pigmented cells in these borders appear to be the edgemost C bz cells. The seven complementary factors determining anthocyanin in the aleurone tissue form two groups based on their interaction (Coe, 1957) with homozygous in, in which the pericarp develops a brassy metallic sheen even if the aleurone tissue is colorless or bronze due to homozygosity for a, $a2$, bz, or $bz2$, but not if c, $c2$ or r is homozygous recessive or if C-I is present. Kernels homozygous for $c2\ su$ or $c2$ in develop faint purple color. Reddy and Coe (1962) carried out com-

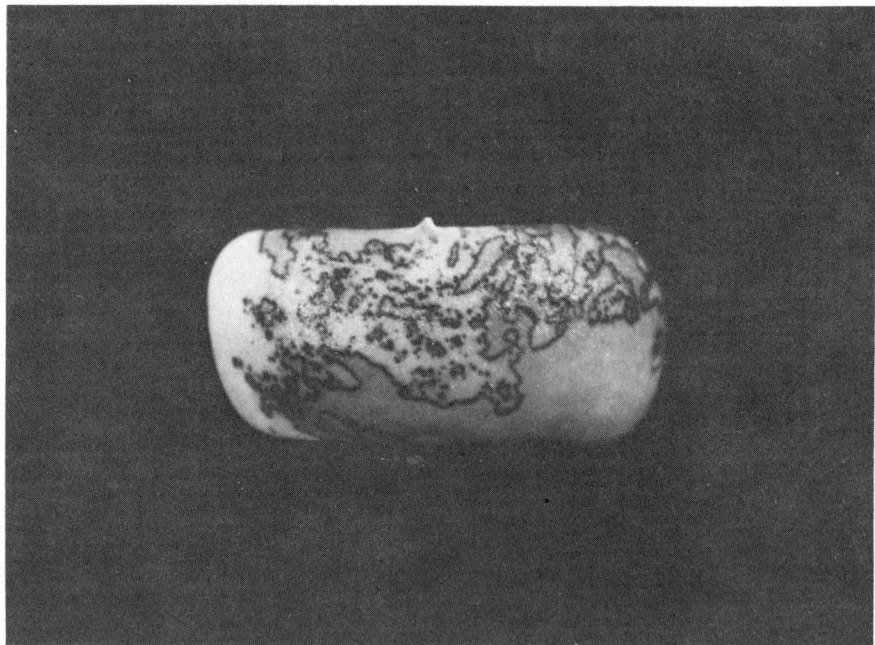

Fig. 12. Borders of anthocyanin formed by complementary interaction of products diffusing between cells of aleurone tissue carrying the inhibitor C-I with Bz (background of upper portion), and cells carrying C with bz (lower right, and the lightly colored central regions of small sectors). (From McClintock, 1951.)

plementation tests with tissue fragments for all pairwise combinations of the seven complementary factors, singly recessive, and *C-I*; interactions were found that identified one member of a pair as donor and the other as receiver of transferable substances (possibly precursors), suggesting a sequence of action of the factors (according to the order, receiver block first, donor block second) as follows: *(C-I), C, (C2), R, (In), A, A2, Bz, Bz2*, in which parentheses indicate equivocal data. Subsequently, Reddy (1964) tested for the leucoanthocyanidins characteristic of *a2* aleurone tissue in a series of double recessive combinations involving *a2*, and found that leucoanthocyanidin formation requires *C, C2, R,* and *A,* but not *Bz* or *Bz2,* as expected from the proposed sequence of action of the factors. The leucoanthocyanidin quantity is influenced by *in*, and the hydroxylation pattern by *pr*, indicating that the action of *In* and *Pr* could precede the action of *A2*. Kirby and Styles (1970) report that the hydroxylation of flavonols that accumulate in *a* aleurone tissue is influenced by *Pr*, indicating that the action of *Pr* could precede that of *A*. One enigma is the observation of Rhoades (1952) that *C bz* cells are most intensely pigmented at borders with *C-I Bz*, which is not as expected by simple consideration of the proposed sequence of action. Whether the many other compatible observations reflect the actual order of action of the factors awaits further biochemical-genetic analysis. Scutellum color requires all of the factors for aleurone color, along with certain constitutions for a scutellum-specific series of factors (Sprague, 1932a). *S* is indispensable, as is homozygous *s5* (i.e., *S5* inhibits color), and color further requires a dominant allele from any two of *S2, S3,* and *S4*, or all three. The resulting interactions generate either 9:7 or 15:1 ratios by segregation for the same two factor pairs, depending upon whether the third pair is homozygous dominant or homozygous recessive. The scutellum color factors have not yet been interrelated functionally with the others.

In plant tissues (e.g., husks) the formation of full purple pigment requires seven factors: *A, A2, B, Bz, Bz2, C2,* and either *Pl* or exposure to sunlight (Tables 3 and 5). With homozygous *b*, plant tissues generally contain neither purple pigments nor their brown alternatives, although limited pigment in some tissues is conditioned by certain *R* alleles whose effects duplicate the action of *B*. With homozygous *pl*, purple or brown pigments develop in exposed tissues only. For the formation of anthocyanin *A, A2, B*, and *Pl* are complementary, while *Bz, Bz2*, and *C2* are complementary for the formation of full pigmentation (substantially less anthocyanin, soon largely replaced by brown pigments, is characteristic of the recessives *bz, bz2*, or *c2*). Emerson (1921) identified the interactions of *A, B, Pl,* Pr, and *R*, including the sun-dependent synthesis of pigment with *pl*. Laughnan (1951) found that the phenotypes of the double-recessive combinations *a bz* and *a a2* resemble that of recessive *a* alone, and suggested that the action of *A* could precede that of *A2* and *Bz*. Because the *a2 bz* combination, on the other hand, resembles neither *a2* nor *bz* alone, the sequence of action of *A2* and *Bz* could not be designated. Recent work along similar lines with *Bz2* and the other factors (Coe, unpublished) indicates that the action of *A2* precedes that of *Bz2*, and that the sequence of action indicated from the above-mentioned studies with aleurone tissue (*C2, A, A2, Bz, Bz2*) is not contradicted. According to developmental studies by Styles et al. (1973), various alleles at *R*

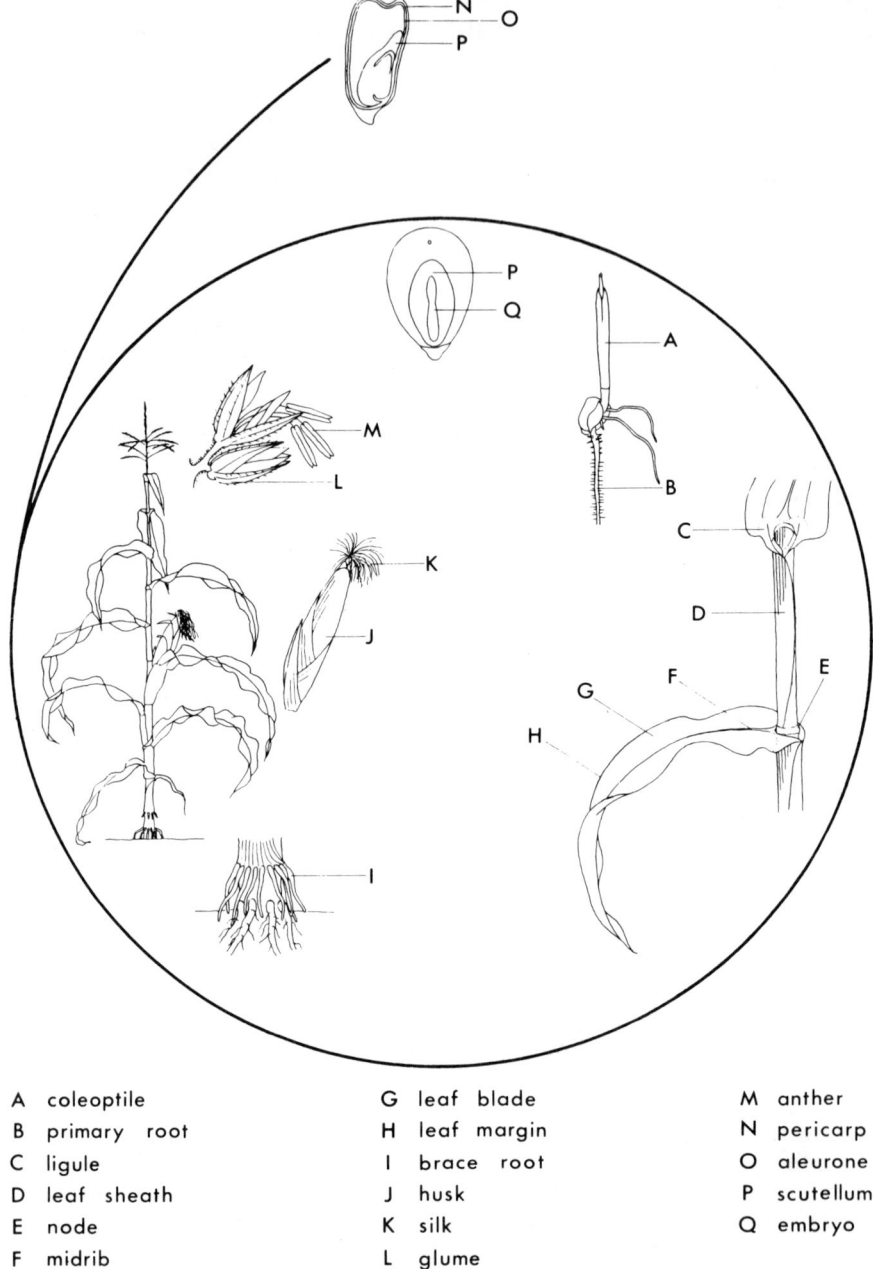

A	coleoptile	G	leaf blade	M	anther
B	primary root	H	leaf margin	N	pericarp
C	ligule	I	brace root	O	aleurone
D	leaf sheath	J	husk	P	scutellum
E	node	K	silk	Q	embryo
F	midrib	L	glume		

Fig. 13. Diagrammatic life cycle of maize (left) illustrating tissues in which anthocyanin can occur, and representatations of the effects of a group of *R* and *B* alleles (right). For a given allele, tissues that develop pigment are identified by keyed letters: intensity is indicated by bold lines (moderate to strong pigmentation) or dotted lines (weak or variable pigmentation). (From Styles et al., 1973.)

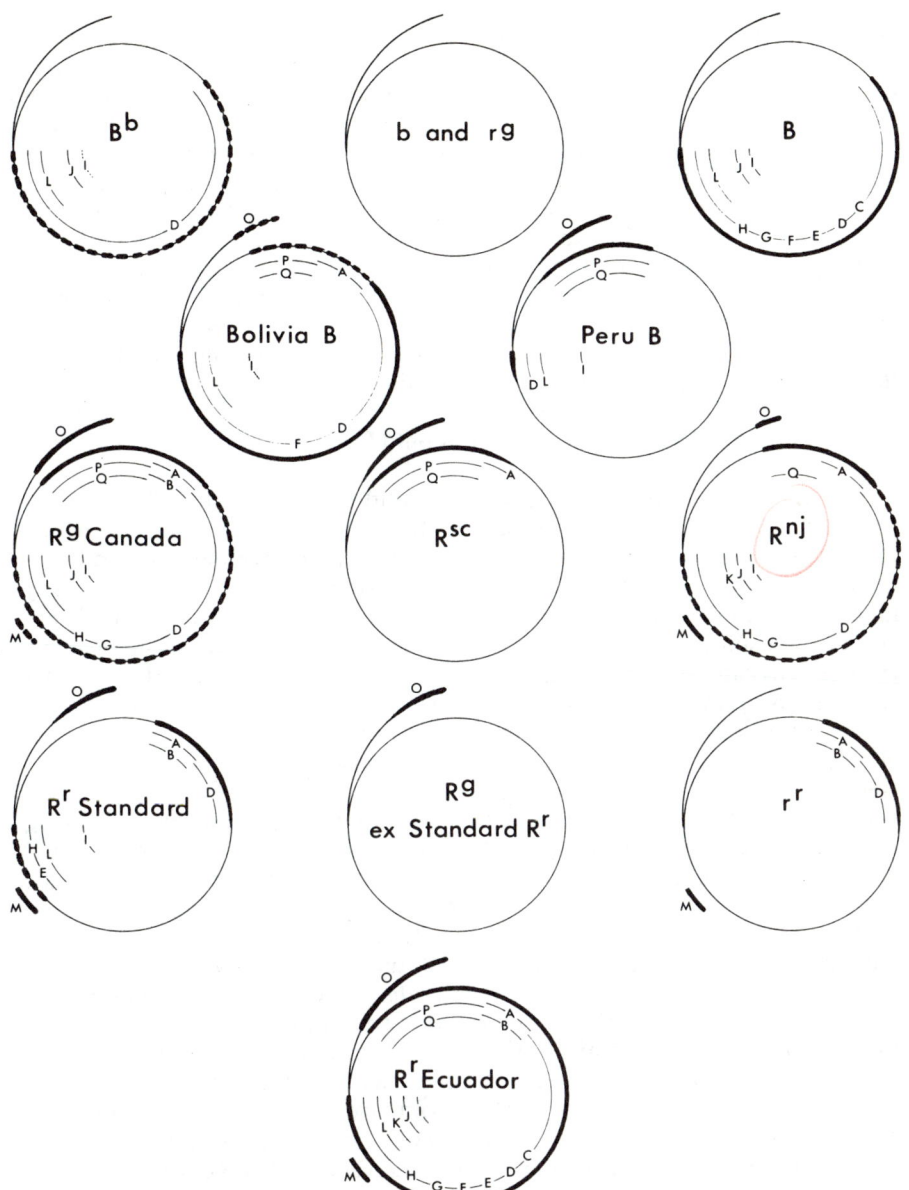

and B differ not only in rate of activity but also in time of onset and termination of activity during the life cycle (Fig. 13): duplicate interactions occur very specifically in some tissues for some pairs of alleles at the two loci, but not in every tissue or for every set of alleles.

The brown-midrib factors, bm, $bm2$, $bm3$, and $bm4$, condition brown pigmentation that is prominent in the leaf midrib but also present in the leaf sheath, husks, culm, cob, and other parts to varying degrees. The brown pigments, visually very similar to those accumulated in a B Pl plant tissues, are found in the cell walls. McClintock (1938) has demonstrated that tissues homozygous deficient for *Bm* have the same expression as the homozygous recessive for *bm*. Studies of lignin contents and properties (Gee et al., 1968; Kuc and Nelson, 1964; Kuc et al., 1968) show that lignin content is lower for all four brown-midrib types and for their double-recessive combinations. The color of alkali lignin (the de-esterified "core"), which is greyish-tan from normal plants, is reddish-brown from *bm* plants, and specific differences are found among the four types in phenolic acids released during de-esterification as well as in other measures of the constituents and properties of lignin. Muller et al. (1971) report data on normal, *bm*, *bm3*, and the double recessive, relevant to the possibility that improved digestibility of cattle silage might result from the lowered lignin content.

Chlorophyll and Carotenoid Pigmentation

In the endosperm, embryo, seedling, and mature plant, 117 separately designated loci are known to influence the photosynthetic pigments, which are under such diverse varieties of genetic, developmental, and environmental control that systematized descriptions of part of the diversity may tend to misrepresent the whole. For example, although carotenoids are visually conspicuous in the endosperm they are screened from view by chlorophylls in the seedling and older plant; consequently, altered carotenoids in leaf tissues may be unrecognized unless chemical studies have been carried out. Some alterations in carotenoids, on the other hand, are closely associated with conspicuous changes in chlorophyll synthesis, function, or stability; some are associated with embryo dormancy as well, yet some factors that affect dormancy have no notable effects on the pigments. Further, alleles at one losuc may differ markedly among themselves in their effects and modifying factors (genetic or environmental) may after specific expressions dramatically. Thus, because the multiple dimensions of analysis that could be applied to these factors, including detailed genetic analysis, have not always been applied, a coherent and consistent system of relationships among the factors cannot be extracted, and the categorizations that follow are necessarily both artificial and oversimplified. [Extensive genetic, physiological, and biochemical information is presented in a survey (Robertson, 1975) that appeared following completion of the present manuscript.]

The keyed tabulation in Table 8 compares effects of the factors with each other. The yellow-orange carotenoid color of the endosperm (with *Y* and other factors) may be reduced to white or pale-yellow by homozygosity for any of 15 factors, a few of which affect endosperm morphology.

Table 8—Chlorophyll and carotenoid factors: Comparative effects (see key below) on endosperm, embryo, seedling, and plant

Genotype§	Endosperm (color with *Y*)	Embryo dormancy	Seedling Color	Seedling Pattern	Mature plant Color	Mature plant Pattern
al	W or PY	D or R	G to W	AC	G to W	AC
al-Brawn	W or PY		PG	A	G to PG	A
ar	Y	D	W	VC		FC
bl			G			U
cl clm	W or PY	D	W		X	
cl Clm-2	W or PY	D	PG		X	
cl Clm-3	W or PY	D	G		G	
cm				SW		SW†
cp2	M	D	PG	U	X	

THE GENETICS OF CORN

Table 8—continued.

Genotype§	Endosperm (color with Y)	Embryo dormancy	Seedling Color	Seedling Pattern	Mature plant Color	Mature plant Pattern
dp	Y	D	W	V	G	
Ej				U		U
et	YM	D	W	V	G	
f	Y	D	W	VC		F
g	Y	D	YG	H	YG	H
g2			YG		YG	
g5		D	Y		YG	
gs		D	G			LP
gs2		D	G			LP
hcf		D	GU		M	
hcf2		D	GU		X	
hcf3		D	GU		X	
ij	Y	D	W	V		SW†
j	Y	D	G			SW‡
j2	Y	D		SW		SW‡
l	Y	D		U		U
l2		D	Y		X	
l3		D	Y		X	
l4		D	Y		X	
l6		D	Y		X	
l7		D	Y		Y	
l8		D	Y		X	
l10	Y	D	Y		X	
l11		D	Y		X	
l12		D	Y		X	
li			G			F
lu lu2			YG	H	YG	H
lw	W or PY	D	W		X	
lw2	W or PY	D	W		X	
lw3 lw4	W or PY	D	W		X	
ly	O	D	O		X	
(NCS)				SP		SP
nec		D	PG	N	X	
nec2		D	G	N	X	
nec3		D	T	N	X	
Og		D	G			SY‡
oy oy	Y	D	Y		YG	
oy-t oy-t	Y	D	YG		YG	
Oy-y Oy-y	Y	D	PY		X	
Oy Oy-y	Y	D	YG		YG	
oy Oy-y	Y	D	Y		X	
oy-t Oy-y	Y	D	Y		X	
pb	W or PY	D	PG	ZH	PG	Z
pb4	Y	D	YG	VC		ZU
pg2			PG	C	G	M
pg11 pg12	Y	D	YG		YG	

Continued next page

†Sectorial loss of chlorophyll is maternally inherited. ‡Sectorial loss of chlorophyll appears not to be inherited.
§Homozygous unless otherwise indicated.

Key:
- A albescent; tending to white in later-developing tissues
- C cool temperature (e.g., 20 C) heightens expression
- D dormant embryo
- F finely streaked with white linear markings on leaves
- G green
- H high temperature (e.g., 35 C) heightens expression
- L linearly marked with lighter color between vascular strands
- M morphologically affected; see descriptions in the locus list and text
- N necrotic (tissues die in part or all of a plant)
- O orange-pink (pink endosperm, pink-albino seedling)
- P pale
- R rampant embryo (not dormant); endosperm roughened or opaque
- S striped in morphogenetic pattern
- T tan
- U unique effects on phenotype; see descriptions in the locus list and text
- V virescent seedling; white or yellow portions may remain as such or may become green, but later-developing tissues are green
- W white
- X dies
- Y yellow
- Z zebra-banded across the width of the leaf

Table 8 — continued

Genotype	Endosperm (color with Y)	Embryo dormancy	Seedling Color	Seedling Pattern	Mature plant Color	Mature plant Pattern
pg14		D	PG	U	PG	U
pm						FU
ps	O	R	O		X	
py	Y	D		FM		FU
pyd	Y	D	PY		X	
Rg	Y	D	G			U
so so2	U					
sr		D		F or S		F or S
sr2		D		F or S		F or S
sr3		D		F or S		F or S
st	YM	D		FM		FM
st-e	MH	D				
su	YM	D	PG	VC	G	
sy	U					
tn			PG	V	G	
v	Y	D	W	VC	G	
v2	Y	D	W	VC	G	
v3	Y	D	PY	VC	G	
v4	Y	D	Y	VC	G	
v5		D	YG	VC	G	F
v8		D	W	V	X	
v12		D	W	VC	G	
v13		D	W	V	G	
v16	Y	D	W	VC	G	
v17		D	W	VU		
v18		D	PY	V	G	
v19		D	W	VC	G	
v21		D	W	VU		
v22		D	W	V	G	
v23		D	W	V	G	
Vm	M		PG		X	
vp	YUM	R	G		G	
vp2	W or PY, M	R	W		X	
vp5	W or PY	R	W		X	
vp8	YM	R	G	M	G	M
vp9	W or PY	R	W		X	
vp9-4889	PY	D	PG	C	X	
w	Y	D	W		X	
w2	Y	D	W		X	
w3	W or PY	R	W		X	
w3-8686	W or PY	D	PG	C	X	
w11		D	W		X	
w13	Y	D	W	SG	X	
w14	Y	D	W		X	
w15	Y	D	W		X	
Wc	PY	D	G		G	
wd	Y	D	W		X	
wl				U	G	U
ws ws2						U
ws3	Y					U
wt	Y	D	W	VM	G	
y	W or PY	D	G		G	
y-8549	W or PY	D	PG	ZH	PG	Z
y-wmut	W and Y	D	PG, G	ZH	PG, G	Z
y8	PY		G		G	
y9	PY	R	PG	H	G	
y10	W or PY	D	W		X	
y10-8624	W and Y	D	W, G		X	
yd			Y	M		
yd2			Y	M		
yg			YG		YG	
yg2	Y	D	YG	C	YG	
ys			G			LY
ys3		D	G			LY
z		R	PG		X	
zb						Z
zb2				Z		
zb3						Z
zb4				ZC		
zb6						ZC
zn						ZN
zn2						ZN

Embryo dormancy is affected by nine factors. Seedling color ranges from green to pale green, yellow-green, yellow, pale yellow, tan, orange-pink, or white; seedling coloration patterns include albescent, virescent, striped, zebra-banded, finely streaked, and necrotic. Plant color, when the type is viable, ranges from green to pale green, yellow-green, or yellow; plant patterns include albescent, striped, zebra-banded, fine streaked, and linearly marked. A few of the factors affect the morphology of seedlings and plants. Temperature has been found to influence the expression of several of the factors, and many others may be similarly influenced but have not been tested systematically. The varying effects of light intensity on expression have been examined in only a few types (not identified in the table).

Pertinent to the photosynthetic process itself, the mutants *hcf, hcf2,* and *hcf3,* recently identified and studied by Miles and Daniel (1974), are of particular interest. In these mutants, the light energy absorbed in the chloroplast is released as red fluorescence (from chlorophyll) in unusually high proportion. In electron micrographs *hcf2* chloroplasts show substantial alterations in structure. Experimental studies implicate *hcf* with photosystem I because of its limited ability to reduce NADP, while *hcf2* and *hcf3* are implicated with photosystem II.

The pigmentation factors can be categorized (making reasonable assumptions for missing data in Table 8) into combinational and pattern classes with little overlapping as follows:

	Endosperm	**Embryo**	**Seedling**	**Plant**
YDPG	yellow	dormant	pale or yellow or virescent	green
YDPP	yellow	dormant	pale or yellow-green	pale or yellow-green
YDPX	yellow	dormant	pale or yellow or tan or white	dies
WDGG	white or pale	dormant	green	green
WDPX	white or pale	dormant	pale or white	dies
YRGG	yellow	viviparous	green	green
WRPP	white or pale	viviparous	green to white	green to white
WRWX	white or pale	viviparous	white	dies
S			striped	
F			finely streaked	
L			linearly marked	
Z			zebra-banded	

Factors of the YDPG class condition reduced chlorophyll in the seedling, but their effects are not continued into the mature plant. This expression is termed *virescent* (becoming green — Demerec, 1924). The term is often used without distinction between the broad sense (the organism at first having reduced chlorophyll and later becoming green) and the narrow sense (cells with reduced chlorophyll becoming green). Mutants of this class include all of the *v* group except *v8,* and *ar, et, f, ij, pg2, su, tn,* and *wt.* Many of these are only conditionally viable, depending upon temperature or other conditions, and could be termed conditional YDPX mutants; this is specifically the case for *ar, et, f, pg2, v2, v4, v12, v16,* and *v19.* Three members of the class, *ar, f,* and *v5,* are members of the finely streaked (F) class in mature plants; *ij* is a member of the striped (S) class. The effects of *ij* on maternally inherited characters are discussed elsewhere in this paper. The white leaf tips on *wt* seedlings do not change to green; in fact, the first leaf appears not to expand fully, and remains somewhat blunted. Although *su* (sugary endosperm) is not ordinarily included among the chlorophyll factors, pale yellow-virescent seedlings very commonly develop from homozygous kernels. The studies of Demerec (1924) established that high temperatures stimulate development of green color in seedlings of *v, v2, v3, v4, v5,* or *f;* Eyster (1933) showed that *ar* seedlings have the same temperature dependence, and that initial development of plastids and green pigment occurs adjacent to the vascular bundles. In *v18* seedlings Chollet and Paolillo (1972) have found that plastid morphology before greening is irregular and that photosynthetic capacities are greatly reduced; Chollet and Ogren (1972) have studied enzymatic activities developing during greening.

The YDPP factors condition reduced chlorophyll both in the seedling and in the mature plant, effects that are usually termed *pale-green (pg* group), *yellow-green (yg* group), or *golden (g* group). While *pg2* becomes green and is thus included in the YDPG class, the other extant pale-green types, namely the *pg11 pg 12* duplicate factors (Rhoades, 1951) and *pg14* (Peterson, 1960), are pale in the mature plant; the same is true of *g, g2, g5, l7,* duplicate factors *lu* and *lu2, oy, yd, yd2, yg,* and *yg2.* Conditional viability is characteristic of *g5, l7,* some alleles of *oy,* and *yg2.* The expression of *g* is heightened by high temperatures (Emerson et al., 1935); the trait can be classified in seedlings in a moderately warm sandbench or flat by slicing off the tops at ground level and observing the cut tips for golden vs. green color. The lutescent trait, controlled by *lu lu2,* is also heightened by high temperatures (Shortess et al., 1968). McClintock (1944) found that the *wd* terminal deficiency is deficient for the *Yg2* locus. Alleles at the *oy* locus (Neuffer and Beckett, 1973, MNL 47:149-150) form a series (see Table 8) with two distinguishable recessives, *oy* and *oy-t,* and a dominant, *Oy-y;* homozygosity for *Oy-y* or heterozygosity of the dominant with either *oy* or *oy-t* results in inviability. Cox and Dickinson (1971), in testing a group of chlorophyll-deficient mutants, found that etiolated leaf segments of *l7,* among others, fail to accumulate normal levels of starch when incubated on glucose in the dark, suggesting that *l7* (and four others of the YDPX class) may be blocked in functions essential to starch accumulation.

Chlorophyll-deficient lethals (YDPX class) include most members of the *luteus* group (the *l* factors except *l* and *l7*), the *necrotics (nec* group), most of the *white* group (*w* factors, except that *w3* is WRPX while its allele *w3-8686* is WDPX), and *cp2, pyd, v8, Vm,* and *wd.* Under restrictive conditions, usually cool temperatures, many members of the YDPG and YDPP classes are lethal, as indicated in the preceding paragraphs. McClintock (1944) identified *pyd* and *wd* cytologically as terminal deficiencies, *pyd* distal to *yg2* and *wd* proximal, such that *pyd* and *yg2* are "allelic" to *wd* but not to each other (see Fig. 14). Tests of several mutants of this class identified very low capability for starch accumulation in *l4,* and reduced capability in *l3, l12,* and *w11,* while *w, w2, w14,* and *w15* showed qualitatively normal starch accumulation in etiolated leaf segments supplied with glucose and incubated in the dark (Cox and Dickinson, 1971). The ultrastructure of the plastids and the pigment contents in *l3, l4, l11, w3-8686,* and *w15* are pictured and tabulated in detail by Bachmann et al. (1973).

White or pale yellow endosperm factors that do not affect embryo dormancy or chlorophyll pigmentation (WDGG class) are known at the three loci *Wc, y,* and *y8. Wc* conditions dominant white crown and pale yellow endosperm in the presence of *Y* (white with homozygous *y*); a somewhat lighter allele, *Wc-wh,* is known but homozygotes for either allele are faint yellow, not white. *Y* conditions orange-yellow endosperm in the absence of *Wc,* while homozygous *y* conditions pure white endosperm in most backgrounds. Dosage levels for *Y* are often visually distinguishable, and Mangelsdorf and Fraps (1931) established that these visible differences are paralleled by increases in vitamin A content in proportions closely approximating 0:1:2:3 for the four dosage levels. In addition to the allelic alternatives at *Y* that affect both carotenoids and chlorophyll, *y-8549* (was *pastel-8549*) and *y-wmut* (Robertson and Anderson, 1961), the factors *l10, l12, ms, pb, si,* and *w15* are also closely associated and have curiously confounded effects. Richardson et al. (1962) have found that plastid pigments of *y-8549* seedlings develop to nearly normal levels at cool temperatures but are substantially lower than normal at warm temperatures. The *y8* factor is a simple recessive that conditions pale yellow in the presence of *Y*.

The white-endosperm albino lethals (WDPX class) include *cl,* the *lemon-white* group (*lw, lw2,* and duplicate factors *lw3 lw4*), *ly,* certain of the alleles of *Y* mentioned previously, and *y10.* Robertson et al. (1966) have shown that *cl* seedlings retain the capacity to manufacture at least as much chlorophyll as normal seedlings. A range of alleles at the dominant modifier

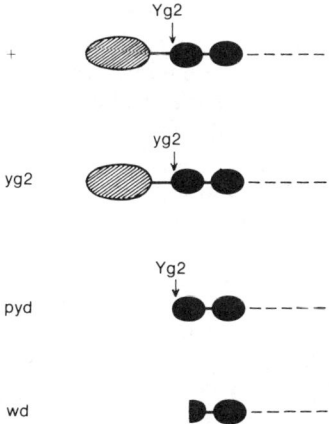

Fig. 14. Cytological constitutions and allelic relationships of three chlorophyll factors located at the end of the short arm of chromosome 9. The large, terminal heterochromatic knob and adjacent short chromatin thread, present with *Yg2* and *yg2,* are absent in *pyd.* The distal part of the first distinct chromomere is also absent in *wd.* Because *pyd* carries *Yg2,* and *wd* does not, the *Yg2* locus must be in the chromomere segment between the two breakpoints. If the cytology were not known, the allelic relationships would be difficult to interpret. (From McClintock, 1944.)

Seedling phenotypes of the homozygotes and combinations:
+/+,+/yg2,+/pyd,+/wd,yg2/pyd   green
yg2/yg2,yg2/wd   yellow-green
pyd/pyd,pyd/wd   pale yellow
wd/wd   white

locus, *Clm*, conditions greening of *cl* seedlings to essentially normal color and viability. While the chlorophyll synthesizing capacity remains almost constant regardless of modifiers, the carotene and xanthophyll levels are increased, suggesting that normal protection against photo-destruction of chlorophyll may be lacking in *cl* and that the modifier corrects this defect (Robertson et al., 1966). Similarly kernels and seedlings homozygous for *ly* accumulate precursors and alternative carotenoids, including the orange-red pigment lycopene, in place of β-carotene, and show photosensitivity of chlorophylls (Horvath et al., 1972; Walles, 1971). Cox and Dickinson (1971) report normal levels of starch accumulation by leaves of *cl, lw,* and *lw3 lw4* but reduced levels for *lw2*. Pigment contents and ultrastructure for *cl* (with and without *Clm* modification) and for *lw* are pictured and tabulated in detail by Bachmann et al. (1973).

Dormancy-affecting factors, termed *viviparous,* include some that have no effects on chlorophyll or carotenoid pigments, *vp* and *vp8* (YRGG class); some that affect kernel carotenoids and chlorophyll but are nonlethal, *al* and *y9* (WRPP); and some that affect both pigment types and are lethal, *ps, vp2, vp5, vp9, w3,* and *z* (WRWX). Robertson (1955) has characterized *ps, vp, vp2, vp5, vp8, vp9,* and *w3*. Uniquely, *vp* partially or completely blocks or reverses anthocyanin synthesis in the aleurone tissue; none of the others has this effect. Effects of the viviparous factors on endosperm form and texture are described in the section on kernel factors. Although *al* embryos tend to be viviparous, they are usually dormant; if this variability relates to the bright-light sensitivity and low-light enhancement of pigment content found by Sander et al. (1968), *al* is an important physiological test system for relating dormancy to pigment contents. The *ps* factor, like the dormant *ly* with which it may be allelic, accumulates lycopene (Robertson et al., 1966) and may be deficient in the stabilization of chlorophyll. Intensive studies on *w3* and *w3-8686* (Anderson and Robertson, 1960; Bachmann et al., 1967, 1969, 1973; Richardson et al., 1962) have characterized the effects of these alleles on pigment stabilization and plastid ultrastructure, evidently resulting from deficiency of colored carotenoids (Fig. 15); Cox and Dickinson (1971) find normal starch accumulation by *w3* leaves. The zeta-carotene factor, *z,* accumulates precursors of β-carotene at very high levels in etiolated seedlings, and shows low photostability of chlorophylls (Horvath et al., 1972); the double-mutant combination homozygous for *z* and *ly* does not accumulate lycopene (Faludi-Daniel et al., 1967). Because some members of this class have very similar effects, allelism may be found in some instances when they are tested inter se.

Leaf striping in the lengthwise pattern characteristic of cell divisions and elongation is conditioned by several nuclear factors (S class), including the recessives *cm, ij, j,* and *j2* and the dominant, *Og,* and a maternally inherited property, (NCS). Although the two japonica factors, *j* and *j2,* and the dominant, *Og,* appear not to transmit any altered nuclear or cytoplasmic state through the germ line to the next generation, the other two nuclear factors, *cm* and *ij,* do exhibit maternal transmission of altered plastid characters. Rhoades (1943) demonstrated that *ij* induces irreversible changes in plastid character. Shumway and Weier (1967) found that aberrant plastids in *ij* plants lack normal internal structures and components, including

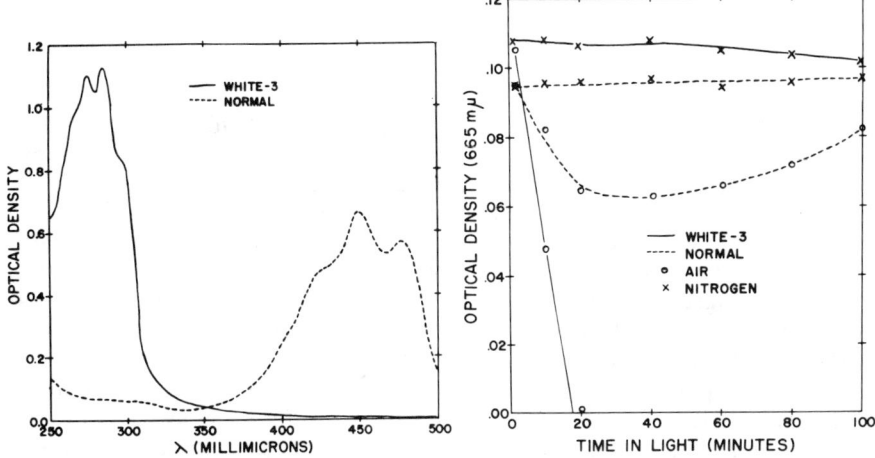

Fig. 15. Carotenoids and chlorophylls in *w3* and normal seedlings. Left: The absorption spectra of hexane-extracted carotenoids. Right: The photostability of chlorophylls, measured by the quantity of chlorophyll present in dark-grown seedlings following timed exposures to light in air and in a nitrogen atmosphere. (From Anderson and Robertson, 1960).

ribosomes. Stroup (1970) found that *cm* conditions maternally inherited white mutations similar to those induced by *ij*. The factor *Ej* increases the extent of variegation in *j* and other S and F types.

F class factors display fine linear markings on older leaves that may or may not be clonal or follow morphogenetic patterns. This expression is characteristic of certain virescent (YDPG) factors, *ar, f,* and *v5;* the *striate* factors, *sr, sr2,* and *sr3;* and *li, pm, py,* and *st.*

The linearly marked (L) factors include *green-stripe, gs* and *gs2,* and *yellow-stripe, ys* and *ys3.* Their pattern of darker color at the vascular strands with lighter color between can be taken to indicate limited uptake or transport of some nutrient, in view of the demonstration by Bell et al. (1958) that ferrous iron can correct the *ys* phenotype through the roots, and that either ferric or ferrous iron can correct the phenotype in floated leaf sections.

Zebra banded leaves (Z class) are characteristic of *pb, pb4, zb, zb2, zb3, zb4, zb6, zn,* and *zn2.* In addition, two alleles at the *Y* locus, *y-8549* and *y-wmut,* tend to develop zebra-banding, possibly related to the expression of *pb,* which is located at *Y.* Zebra-banding sometimes is seen in *nec3* and *al* as well. The cause of this particular pattern type is presumably related to the effects of diurnal light or temperature patterns, as suggested for *zb4* by Hayes (1932) and for *al* by Sander et al. (1968).

Among other factors that affect chlorophyll and carotenoid pigmentation, the first-named luteus factor, *l,* when homozygous confers yellow pigmentation in tissues that would otherwise be white, such as in albino seedlings or japonica plants; *wl,* the *ws ws2* duplicate factors, and *ws3* affect localized chlorophyll development in the leaf base or sheath; and *bl* conditions disease-like yellow-necrotic lesions. In the scutellum of the kernel, Sprague (1932a) has shown that the duplicate factor pair, *so so2,* and the *sy* factor condition, respectively, orange and yellow color.

Plant Form

An impressive range of variation in the morphology of vegetative and reproductive structures is found among races, lines and varieties. While casual observation may suggest that this variation could only be controlled by the overlapping effects of very large numbers of factors whose effects are expressed in small increments, straightforward genetic analysis has established over 70 loci conditioning easily classifiable, genetically simple morphological variations in plant stature and in the form and structure of vegetative and reproductive parts. Even more numerous studies on continuous or measured variation in agronomic and other traits, analysis of which is not attempted here, suggest that unit factors with very much smaller effects could be extracted systematically and could be manipulated by choice.

Reduced plant stature controlled by single factors ranges from moderate differences that can be classified only in very uniform backgrounds to extremely shortened or morphologically deviant types that are unmistakable. Semidwarf or compact plants that show only minor morphological alterations are conditioned by *br, br2, br3, bv, cr, ct, ct2, mi, na, na2, rd, rd2,* and *td.* Kempton (1921) reported on the characteristics of various heritable and nonheritable brachytic plant types, including that conditioned by *br,* in which the plants have shortened internodes, especially below the ear, but conventional leaf size and attitude at maturity. Kempton suggested that the brachytic form might be advantageous in agronomic use because of its sturdiness of culm and relatively modest interference with grain yield. That semidwarf plant types may have a yield advantage, especially in high-density plantings, is supported by data from Nelson and Ohlrogge (1957, 1961) on *br2, ct,* and *rd.* The shortening of *na* plants apparently results from lowered auxin levels, shown by van Overbeek (1935) to result from higher rates of destruction of auxin in *na* than in normal seedlings. Leaf initiation rates in developing *br2* embryos were found by Stein (1955) to be slightly lower than normal, and Scott and Campbell (1969) determined that *br2* plants develop

Fig. 16. Growth response of dwarf seedlings to a single application of gibberellic acid ranging from 0.001 to 100 nanograms per plant. (From Phinney and West, 1960).

THE GENETICS OF CORN

fewer internodes below the ear and shorter internodes throughout the plant than normal. The andromonoecious dwarfs form a distinctive group (*an, d, d2, d3, d5,* and *D8*) in which the ears bear perfect (monoclinous) flowers, although the anthers in the ear generally do not shed pollen and remain among the kernels. With *d, d2, d3, d5,* or *D8* the plants are characteristically shortened and compressed, forming rosettes that have broad, crinkled, foreshortened leaves and more tillers than normal; with *an* or with an intermediate allele at *d* the shortening of stature is less extreme. Stein (1955) found the leaf initiation rate in the embryo and the final leaf number in the plant to be reduced in *d* relative to normal. Phinney (1956) and Phinney and West (1960) determined that each of the recessive andromonoecious dwarfs

Fig. 17. Response of dwarf plants to regular applications of gibberellic acid. Top: untreated plants at maturity. Bottom: treated plants that have been supplied the hormone in aqueous solution (1 to 10 nanograms every 2 to 3 days, to a total of 300 nanograms) from the seedling stage to maturity. (Courtesy of B.O. Phinney.)

responds to small amounts of gibberellins with a prompt, brief spurt of growth (Fig. 16); if hormone is supplied regularly in appropriate quantities normal stature, normal habit, and normal ears (without anthers) develop (Fig. 17). The dominant dwarf, *D8*, shows no response whatever, and others of the reduced-stature types that have been tested (e.g., *br, bv, na, na2,* and *py*) show little or no response. An exhaustive survey of 143 gibberellins and allied compounds (Brian et al., 1967) revealed common response levels by *d, d2, d3,* and *d5* for most compounds. Differential responses for several compounds suggest possible individual specificities among these mutants in interconversion or in biosynthetic pathways. Several other factors condition reduced stature in association with other effects, such as defective endosperm (*de16, de17*), glossy seedlings (*gl11, gl17*), and other abnormalities (*Clt, py, rgd, yd*), and may be reduced as a result of indirect metabolic effects or limited nutrition.

Vegetative form and structure is affected, apart from stature, in diverse ways. Leaf shape is affected by foreshortening in the andromonoecious dwarfs and *py*; by narrowing in *Cg, nl, Tp,* and *Tp2*; by crinkling or outgrowths in *cr* and *Kn*; and by tearing, growth failure, or slashing in *Rg, rgd,* and *sl* plants. For *Kn*, Gelinas et al. (1969) have determined that the outgrowths, occurring along vascular strands, are anatomically similar to the intercalary meristem at the transition between leaf sheath and blade. Gelinas and Postlethwait (1969) find higher levels of compounds that inhibit the auxin inactivating enzyme, IAA oxidase, in *Kn* seedlings. Nickerson and Hewitson (1963) report that knots are increased on *Kn* leaves by an auxin, indole butyric acid, and decreased by gibberellic acid. Mericle (1950) has studied the course of cell and tissue breakdown in leaves of *Rg* plants. Leaf adherence is found in *ad* and *Ce* plants. Leaf attitude is affected by many of the reduced-stature factors and by *lg, lg2,* and *Lg3*, while phyllotaxy is altered by *Cg, Tp, and Tp2*. The basis for upright leaf attitude in *lg, lg2,* and *Lg3* plants is a change in the transition zone between leaf sheath and blade, in which the smooth, auriculate tissue is supplanted and the filmy, collar-like ligule is reduced or absent. Wilting is affected by *gs* and by *wi*; Postlethwait and Nelson (1957) have found *wi* plants to be characterized by imperfectly developed vascular elements. Sheath surface is affected by *Hs, Rs,* and *rs2*. Basal branching increases the number of tillers in the andromonoecious dwarfs and in *Cg, gt, tb, Tp,* and *Tp2* plants. Prostrate culms lacking normal geotropic response develop with *la*, in which van Overbeek (1938) found high auxin content in upper portions of nodes and internodes, in contrast to the auxin distribution in normal plants placed in a horizontal position. In the coleoptile, Hertel et al. (1969) have found that seedlings homozygous for *ae*, in which the starch grains are smaller than normal, have less efficient geotropic responses, and they have suggested that a statolith model for these responses is thus sustained. Root branching is greatly reduced in *rt* plants, to the extent that support is necessary in order to grow the plants to maturity. Treatments of *rt* plants with tri-iodo benzoic acid tend to increase the formation of brace roots (Nickerson, 1969, MNL 43:179-181, and preceding reports). Brittle stalks, leaves, tassels, silks, cobs, and other parts are conditioned by *bk2*, effects of which can be recognized as early as the 5-leaf stage; plants are difficult to bring to maturity in windy or stormy locations, even with support.

THE GENETICS OF CORN

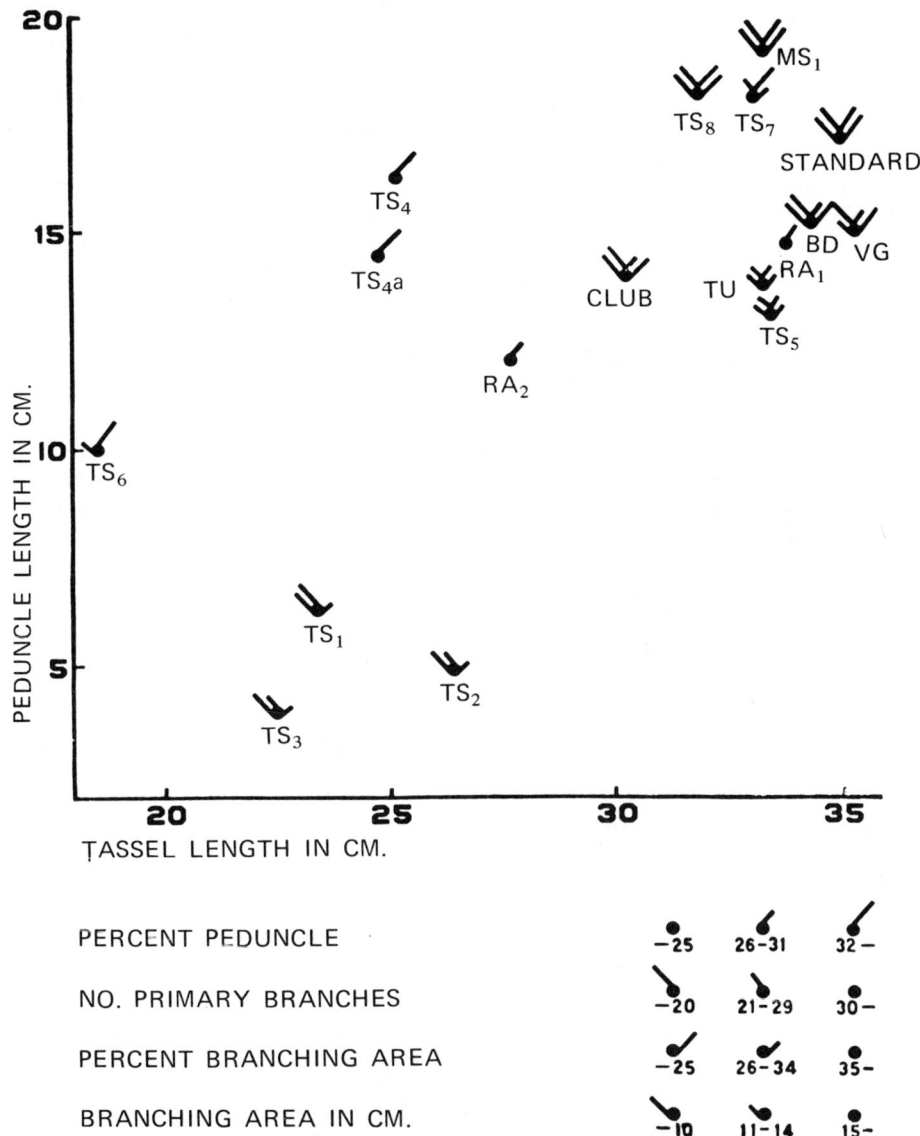

Fig. 18. Pictorial representation of morphological measurements from the tassels of several genetic types. (From Nickerson and Dale, 1955.)

Reproductive structures are subject to genetically and physiologically determined changes in the degree of proliferation and branching, and to alterations of sexual determination. Although in conventional Corn Belt strains sex alterations are not common, many of the indigenous strains in the Americas and elsewhere tend to have pistil-bearing parts or even small ears in the tassel, particularly on tillers, and stamen-bearing parts in or on the conventional ear, including fertile tassel-like extensions from the tip. Genetic factors that have been identified as conditioning sex alteration include the andromonoecious dwarfs, *an, d, d2, d3, d5,* and *D8*, in which perfect flowers bearing anthers develop in the ear, all of which except *D8* can be normalized with gibberellins, as previously indicated. The tassel-seed factors, *ts, ts2, Ts3, ts4, Ts5,* and *Ts6,* result in monoclinism or partial to total sex inversion in the tassel, described in comparative morphological terms by Nickerson and Dale (1955); in addition to inversion, the spikelets tend to show proliferation, branching, and similar changes (Fig. 18 and 19). Proliferation and branching without inversion is also common; for example, tassels and ears of *ra* and *ra2,* and ears of *bd* plants are many-branched. Nickerson and Dale (1955) and Postlethwait and Nelson (1964) have shown that initiation of spikelets is replaced by initiation of branches, which then bear spikelets (Fig. 20). Nickerson (1960a) found that *ra* tassels, but not ears, could be normalized by gibberellin treatments, while several other tassel types could not. Polytypic (*Pt*) ears and tassels develop proliferated branches, spikelets, glumes, or pistils (Postlethwait and Nelson, 1964). Form of ear is affected also by *is, pd, Sg,* and *tr* (Galinat, 1971a), while multiplied silk initiation is brought about by *si* and silk growth failure by *sk.* Glumes become exceptionally extended under the influence of *Pn* or *Tu* but do not extend normally with *Vg.* Mangelsdorf and Galinat (1964) find that *Tu* can be subdivided by crossing

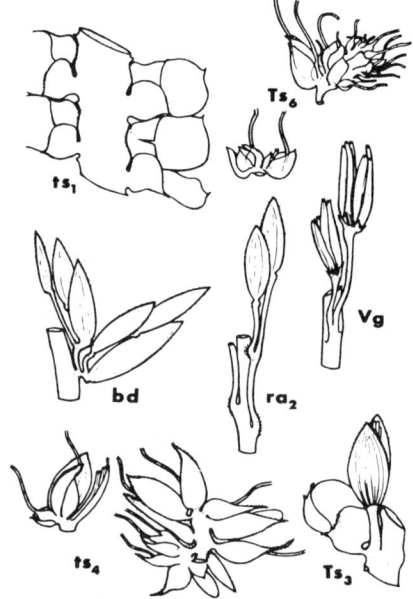

Fig. 19. Drawings to same scale of spikelet pairs found on lowermost primary tassel branches of certain mutants showing characteristic features: *ts,* adaxial view of the thick, ribbon-like branch showing cupules and a developed second floret (silks removed). *Ts6,* pedicellate (right) and sessile (below) spikelets with hyaline awnless glumes; the pedicellate spikelet forms numerous small naked pistils. *bd,* both spikelet axes form extra spikelets. *ra2,* an extra internode is found between the primary axis and the point of departure of the sessile spikelet. *Vg,* glumes do not develop, so the stamens are left naked. *ts4,* sessile spikelet (left) has one male and two female florets; the pedicellate spikelet (right) forms numerous small spikelets, with many of the thin glumes ending in soft awn-like tips. *Ts3,* sessile female and pedicellate male spikelets in one spikelet pair; the pedicellate axis adheres to the edge of the cupule for part of its length (silks removed). (From Nickerson and Dale, 1955.)

THE GENETICS OF CORN

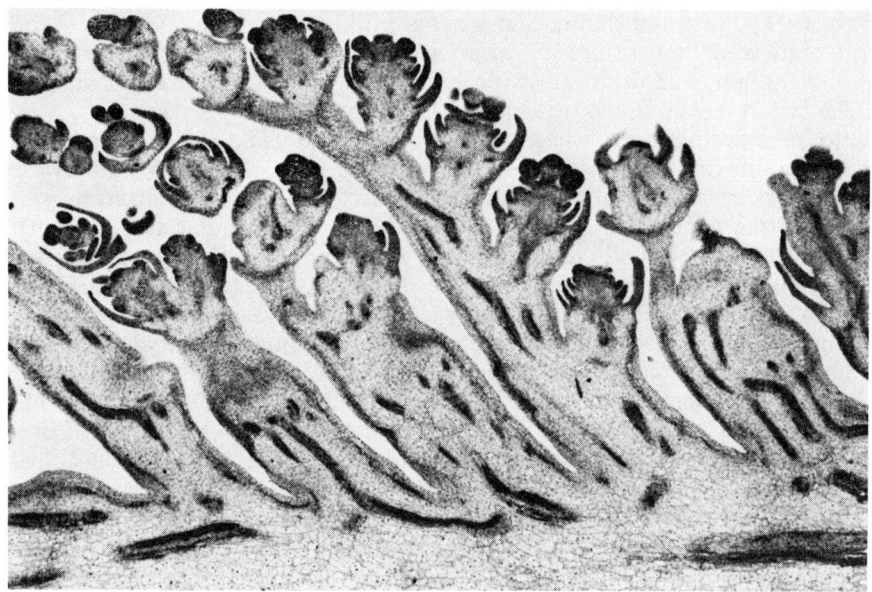

Fig. 20. Longitudinal section of a ramosa (*ra*) ear, showing branches bearing normal florets. (From Postlethwait and Nelson, 1964.)

Fig. 21. Normalization of *Tp* and *Cg* plants with gibberellic acid *(GA)*. Left to right: Ear and tassel of *Tp/-* plant treated regularly with GA; *Tp/-* control; ear and tassel of *Cg/+* plant treated regularly with GA; *Cg/+* control. (From Nickerson, 1960b).

over into two distinguishable components with intermediate effects, *tu-l* and *tu-d,* similar to "half-tunicate" variants, *tu-h,* found as mutations. Glumes in *eg* tassels open to an unusually wide angle. Branching at the level of initiation of the inflorescence is conditioned by *tb, Tp,* and *Tp2.* The effects of *Tp* include free tillering, narrow leaves, greatly increased leaf number, long bracts or spathes in the inflorescences, and variable monoclinism (Weatherwax, 1929); Nickerson (1960b) found that *Tp* plants could be normalized by gibberellin treatments (Fig. 21). *Cg* plants, though very similar to *Tp* plants, characteristically branch very heavily at the tiller level, have many narrow leaves, and form bracts in the poorly developed ears; vegetative propagation is easily accomplished by division or by adventitious shoot and root formation (Galinat and Andrew, 1953). Singleton (1951) demonstrated that *Cg* plants are reversibly normalized by autumn conditions, presumably daylength or temperature, when grown in the greenhouse; Nickerson (1960b) found that *Cg* plants can be normalized by gibberellin treatments (Fig. 21). Angeles (Stebbins, 1967) showed that mitotic frequency in the flank of the apical meristem of *Cg* plants is much higher than in normal, leading to increased leaf initiations. Initiation or completion of the inflorescences fails under the influence of *ba, ba2,* and *ub.* The tassel does not expand normally with *ad, Clt, ct2,* or *td.*

Initiation and development of flowering has been shown to be affected by a recessive factor, *id* (Singleton, 1946). Galinat and Naylor (1951) found that a definite size, age, or number of leaves is prerequisite to floral initiation, which is short-day dependent in *id* plants (Fig. 22); vegetative reversal occurs if long days follow floral initiation. Shaver (1967) defined three recessive factors, *gt* (grassy tillers), *pe* (perennialism), and *id* as necessary for perennial habit in derivatives from hybrids between corn and perennial teosinte.

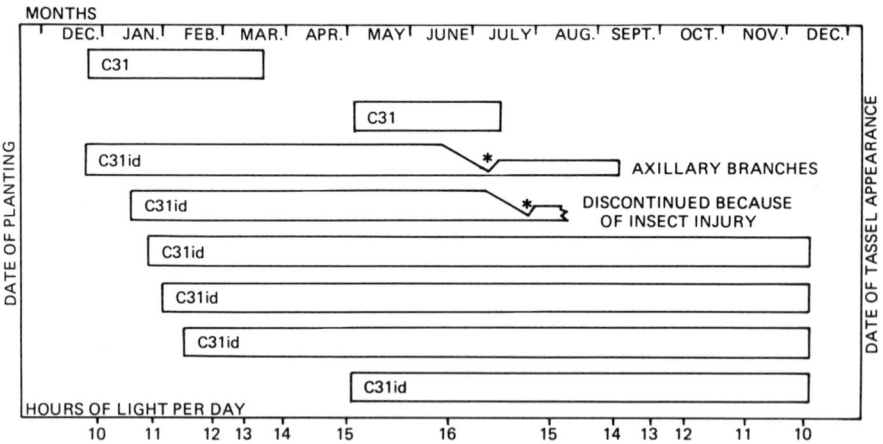

Fig. 22. Flowering dates of normal and *id* plants from plantings made over several months, demonstrating daylength dependence for floral induction (bottom bars vs. top two), vegetative reversal of flowering by increasing daylength (third from top), and growth-period requirements in *id* plants (bottom six bars). (From Galinat and Naylor, 1951.)

THE GENETICS OF CORN

Leaf Cuticle Waxes

In normal seedlings the first five or six leaves form a bluish-white waxy bloom on the upper and lower epidermis. In *glossy* seedlings (*gl* group) little or no bloom forms; in bright light a "glossy" effect results, in which the surface of the leaf reflects as if polished, in contrast to the dull appearance of normal leaves. If a normal leaf is rubbed lightly with the fingers, the bloom rubs away and the surface becomes glossy. Since leaves with bloom are water

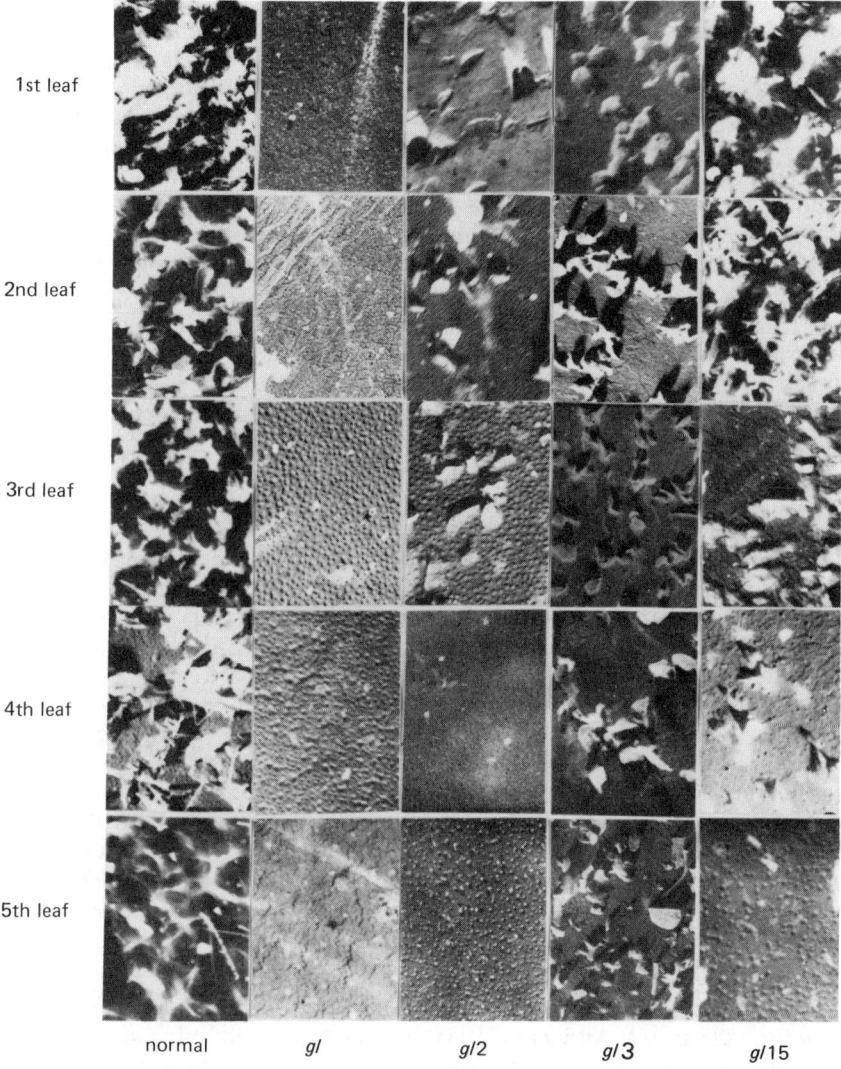

Fig. 23. Electron micrographs ($\times$ 10,000) of shadowed surfaces of the first five leaves of normal and glossy plants, showing wax projections of the cuticle. (From Bianchi and Marchesi, 1960).

repellent whereas glossy leaves retain water droplets, water can be sprinkled or sprayed on the seedlings for clearer discrimination (Hayes and Brewbaker, 1928). Bianchi and Marchesi (1960) point out that the water-repellency of normal leaves results in a reflective interface between water and the leaf surface, so retained droplets and immersed leaves have a mirror-like reflectance with normal seedlings but not with glossy. Although most of the glossy types are classifiable beginning with the first two leaves, in *gl15* seedlings the first two leaves usually have normal appearance; the third leaf is often variable, but usually glossy; and the fourth and later leaves are glossy. In contrast, normal plants begin to produce leaves that have glossy patches at about the sixth leaf, and later leaf blades are entirely glossy; however, the leaf sheaths on normal plants continue to develop enough bloom to permit classification for two of the group, *gl* and *gl15*, if not for the others. Conditions of seedling development that may detract from the development of bloom on normal leaves include suboptimal light, infections, injury, and some weak genetic types; some of the glossy factors are associated with weak seedlings and may be glossy for indirect reasons. The electron microscopic studies of Bianchi and Marchesi (1960; see Fig. 23) demonstrate that the bloom is composed of minute projections from the cuticle on the first five leaves of normal seedlings, that *gl* seedlings have virtually no projections, *gl2* a few on the first three leaves but none on the fourth and fifth, *gl3* intermediate amounts on the first five leaves and *gl15* was gl^H) normal amounts on the first two leaves, decreasing for the next three leaves. A mutable at the *Gl* locus, *gl-m*, has been identified. Intracistron recombination at *Gl* has been reported by Salamini and Lorenzoni (1970).

Kernel Lipids

The inbred line R84, which has a lower ratio of oleate to linoleate in the kernel than does Illinois High Oil, carries a single recessive factor, *ln*, which conditions the difference in ratio without greatly altering the total fatty acid content (Poneleit and Alexander, 1965; see Fig. 24). Differences between these two lines and a third, C103, are more complex (de la Roche et al., 1971). Considering the dramatic effects on lipid content and proportions shown by a number of mutants that affect kernel form and texture (Flora and Wiley, 1972), great complexity of direct and indirect genetic influences on lipids can be anticipated. For example, *bt2* kernels have double the normal lipid content and a higher proportion of oleate, while *o2* kernels have half the lipid content and slightly lower oleate relative to normal. Plewa and Weber (1973b) report that kernels monosomic for chromosome 2 have higher oleate and lower linoleate than normal.

Fluorescent Seedlings and Anthers

Routine screening of seedling cultures under an ultraviolet lamp led to recognition (Teas and Anderson, 1951) of a factor conditioning strong blue-white fluorescence of the seedling leaves (rather than the typical dull reddish-purple) and blue-fluorescent anther walls. Because the factor shows dominant expression in the anther walls (though recessive in the seedling), the designation *Bf* is used. Teas and Anderson found in *Bf* anthers and

THE GENETICS OF CORN

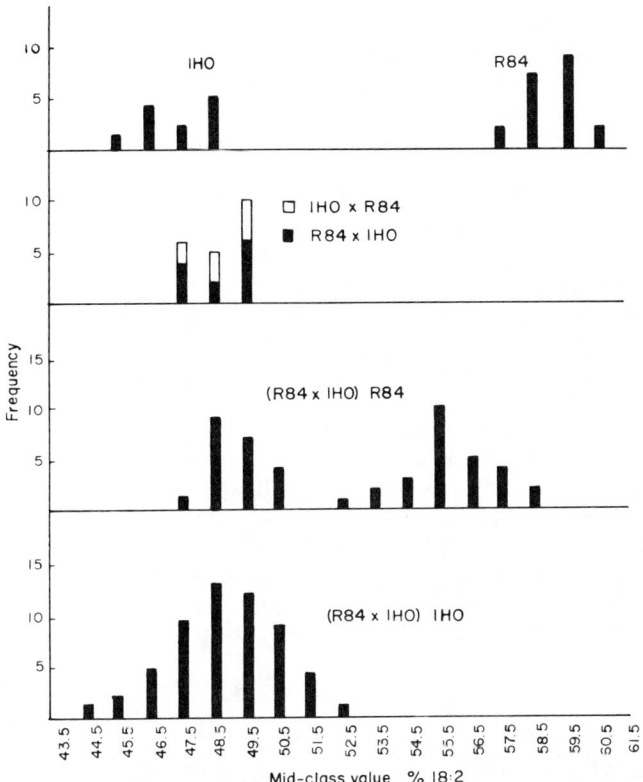

Fig. 24. Frequency distributions of individual kernels according to linoleic acid content (percent total oil) for lines R84 (*ln ln*) and IHO (++), their reciprocal hybrids, and both backcrosses. (From de la Roche et al., 1971.)

seedlings three chromatographically distinctive blue-fluorescent compounds, one of which could be identified as anthranilic acid and all of which supported growth of tryptophan-requiring (anthranilate-utilizing) *Lactobacillus* and *Neurospora crassa*. A second factor, *bf2*, has phenotypic effects similar to *Bf* in seedling and anther fluorescence. Brilliantly yellow-fluorescent anthers are conditioned by *bz*; this recessive expression is independent of any of the other known factors that influence anthocyanins and related pigments. Red fluorescence in seedlings of *hcf*, *hcf2*, and *hcf3*, blocked in photosynthesis, is discussed in the section on chlorophyll pigmentation.

Enzyme Variations in Electrophoresis

Since the finding by Schwartz (1960a) that three variants of one of the esterases in young endosperms are determined by three alleles at a single locus, genetically controlled variation in electrophoretic properties of enzymes has been reported at over 30 loci. The physiological and theoretical implications of several of these variations have been reviewed by Nelson (1967) and by Nelson and Burr (1973).

Variants differing in mobility during electrophoresis normally will differ in charge properties of the enzyme molecule, due to differences in one to a few amino acids. In most cases the variants found in corn are codominant — that is, heterozygotes produce both of the variants rather than only one. A finding of codominance is generally considered strong evidence that the locus is coding for the enzyme and is the "structural gene."

With E, the heterozygote shows not only codominance but also a unique band not present in either homozygous type. The hybrid between a strain homozygous for E-F (an allele determining a fast-migrating band) and one homozygous for E-S (slow) shows an intermediate band as well as both parental bands. Schwartz (1960a) demonstrated codominance, hybrid bands, and gene dosage effects in the endosperm, all consistent with the random formation of dimeric enzymes (FF, FS, SS) from F and S units coded by each allele. Similar identification of hybrid bands has led to the same interpretation of dimeric structure for the esterase controlled by $E3$ (Schwartz, 1964), for the alcohol dehydrogenase controlled by Adh (Schwartz and Endo, 1966), and for the transaminase controlled by Ta (MacDonald and Brewbaker, 1972). A tetrameric structure has been interpreted for the catalase controlled by Cat, in which three hybrid bands are found along with the parental bands in heterozygotes — i.e., homozygous Cat-F × Cat-S produces endosperms with FFFF, FFFS, FFSS, FSSS, and SSSS units, apparently in proportions according to gene dosage (Beckman et al., 1964a; see Fig. 5). Definitive evidence for tetrameric structure of catalase was adduced by Scandalios (1965) from the formation of hybrids in vitro following

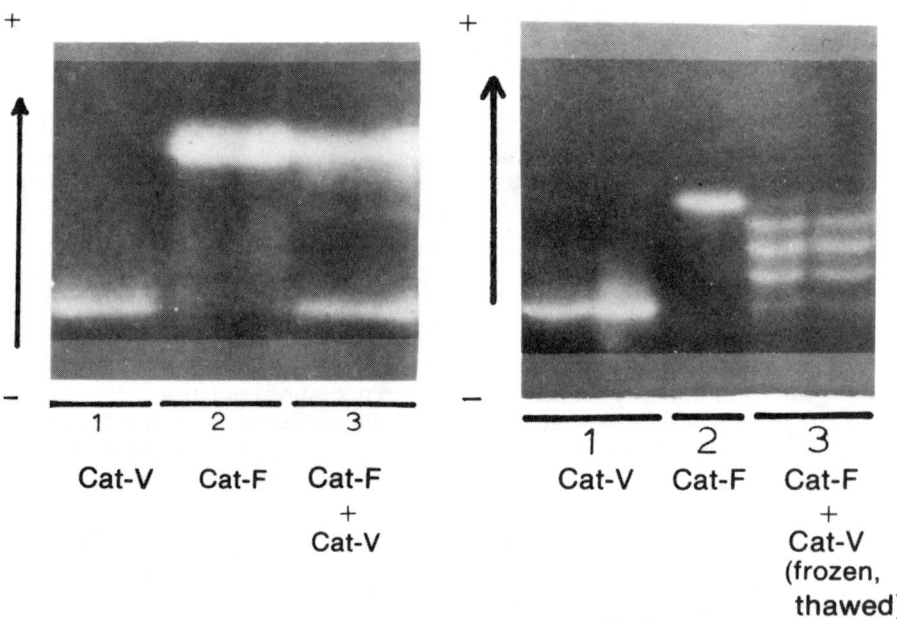

Fig. 25. Electrophoresis patterns demonstrating reassociation of catalase subunits in vitro to form hybrid tetramers.
Left: 1) Samples from two kernels of Cat-V strain; 2) Samples from two kernels of Cat-F strain; 3) Two samples of a mixture of the two strains.
Right:1) Samples from two kernels of Cat-V; 2) Sample from one kernel of Cat-F; 3) Two samples of a mixture of the two strains, dissociated and reassociated by freezing and thawing in $1M$ NaCl. Compare these patterns with those formed by hybrids (Fig. 5). (From Scandalios, 1965.)

Table 9—Alleles designated at loci showing discrete variants in electrophoretic properties of enzymes

Locus	Active alleles in order of electrophoretic mobility†	Complex alleles	Null
Adh-	Ct Cm F S1108 S296 S	FC	0
Adh2-	N P		
Amy-	A B		
Amy2-	A B		
Ap-	S I F		
Ap2			
Ap3			
Bt2	Bt2 bt2		
Cat-	F K M S V		
Cat2			
Cx-	F S		N(null)
E-	F F' L N R S S' T W		
E2	E2		e2
E3-	F S		
E4-	F F' E (or N) N' D (or S) C		N(null)
E5-I-	1 2		
E5-II-	1 2		
E6-	1		null
E7-	1		null
E8-	1		null
E9-	1 2		null
E10-	1 2		
E12-	F S		
E16-	F N S		
Ep-	A B		
Lp-	F S		
Lp2-	F S		
Phos4-	F N S		
Phos8-	F S		
Px-	F (or 1) N (or 2) S (or 3)		null
Px2-	1 2		
Px3-	5 4 1 2 3	6	
Px4-	1 2 3		
Px5-	1		null
Px6-	1		null
Px7-	1 2		null
Ta-	2 1		

†Direction of migration is dependent upon specific details of technique; the order given does not necessarily identify slow to fast or anodal to cathodal order of migration.

dissociation and reassociation of the subunits (Fig. 25). Similar evidence supports formulation of a dimeric structure for alcohol dehydrogenase (Fischer and Schwartz, 1973; Fig. 26).

Reported alleles at the loci identified by electrophoresis are listed in Table 9. Exceptions to codominance exist at a few loci: *E2, E6, E7, E8, Px5,* and *Px6* have so far shown only presence vs. absence of bands of catalytic activity. Lacking catalytically active variants differing in mobility, there is a strong likelihood that such loci may be indirectly regulating the synthesis of the enzyme rather than coding for it (MacDonald and Brewbaker, 1972). Several loci have been found to condition variations in electrophoretic

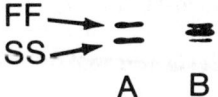

Fig. 26. Electrophoresis patterns demonstrating reassociation of alcohol dehydrogenase subunits in vitro to form hybrid dimers.
(A) Mixture of preparations from *Adh-F* and *Adh-S* strains.
(B) Mixture as in A, dissociated and reassociated by freezing and thawing in 1M NaCl. (From Fischer and Schwartz, 1973.)

mobility and to have null alleles as well (*Adh, Cx, E4, E9, Px, Px7*), which is in keeping with the range of variants anticipated for a coding locus. Presence of an immunologically cross-reacting material (CRM) that lacks catalytic activity has been demonstrated for some null alleles of *Adh* (Schwartz, 1971b, 1973), showing that an inactive molecular product is formed in these instances. One system, *E5-I* and *E5-II*, appears to determine the presence of a fast set of bands, in contrast to a slow set, through a duplicate-factor (15:1) interaction (MacDonald and Brewbaker, 1974).

It should be emphasized that several of the listed isozyme loci are somewhat more provisional in status than are most other genetic factors; those for which some supporting genetic analysis has been published (or communicated in the *Maize Newsletter*) are included in the tabulations of loci and alleles (Tables 1 and 9). For certain loci, differences exist in current interpretation and terminology or in strains or techniques among two or more laboratories; this is particularly the case for alcohol dehydrogenase, but also for *E, E4,* and *Px*.

Schwartz (1960a) suggested that the occurrence of hybrid enzymes for the *E* esterase might be relevant to hybrid vigor — specifically that a hybrid molecular complex (heterodimer) might be more active than its parental counterparts (homodimers), conferring potential vegetative advantage on the hybrid organism. Relevant to the populational aspect of this suggestion, however, Brown (1971) has examined the distribution of alleles at *E, E3, E4, E12, Adh,* and *Px* in strains that have been subjected to long-term selection for high and low oil and protein, and has found Hardy-Weinberg (i.e., random) distributions for each, including the three (*E, E3, Adh*) that form hybrid bands. That *Adh* function is important to survival and growth in germination under flooding (anaerobic) conditions has been demonstrated by the survival of *Adh-S* seedlings under conditions in which *Adh-O* (null) seedlings die (Schwartz, 1969), and by faster growth rates in flooded *Adh-S* than in flooded *Adh-F* seedlings (Marshall et al., 1973). For the enzyme alcohol dehydrogenase, a heterodimer (Cm F) composed of one unit (Cm) that is relatively insensitive to alkaline pH but low in activity coupled with one (F) that is sensitive but highly active results in a complemented, stable and active enzyme (Schwartz and Laughner, 1969); with two other alleles, the specific activity of FF, FS, and SS dimers falls in that order while the stability to heat of FF and FS is similar and exceeds that of SS (Felder and Scandalios, 1971); negative complementation (i.e., loss of stability) is found for the heterodimer between a temperature-sensitive subunit (S1108) and a typical one (Ct) following dialysis (Schwartz, 1971a), in spite of stabilization of the heterodimer in the absence of dialysis.

Longo and Scandalios (1969) have shown that variant isozymes of malic dehydrogenase localized in the mitochondria are under nuclear control.

Reaction to Pests and to Adverse Conditions

Simple genetic control of reaction to pathogens, insects, herbicides, and flooding has been identified. Numerous analyses, not detailed here, have established multiple-factor control of reaction to viruses, rusts, smut, blights and other infectious diseases, to the European corn borer and the earworm, and to various adverse conditions; in some instances studies have succeeded

THE GENETICS OF CORN

in localizing one to several differential factors to chromosomes (e.g., Scott et al., 1966), but unit factors have not been designatable.

Monogenic control by *Rp* of reaction to a particular form of the rust fungus *Puccinia sorghi* was localized by V.H. Rhoades (1935) to chromosome 10. Studies summarized in a report by Wilkinson and Hooker (1968) have identified several related determinants closely linked to *Rp*, including one conditioning reaction to a different species, *Puccinia polysora* (Ullstrup, 1965), and two independent loci with several alleles, *Rp3* and *Rp4*. Storey and Howland (1967) summarize work designating several distinct loci conditioning reaction to *P. polysora*. In the absence of adequate means by which to test for allelism according to the standard cis-trans test criteria, and lacking data either on proximity of several of the factors or on exchange with outside markers, we have deemed it prudent to identify listed designations only to *Rp, Rp3, Rp4,* and *rp7,* each of which is unambiguously distinct in genetic location. Reports in the literature have specified symbols *Rp2, Rp5,* and *Rp6*, and *Rpp* through *Rpp11*.

Among *Helminthosporium* species, reaction to race I of *H. carbonum* is determined by either of two loci. *Hm* confers full resistance (although some alleles are intermediate) and *Hm2* (in the presence of homozygous recessive *hm hm*) confers lower resistance that becomes progressively stronger as the plants develop (Nelson and Ullstrup, 1964). Relative to *H. turcicum* the factor *Ht* confers hypersensitive resistance in the form of chlorotic lesions (in contrast to wilt-type lesions with *ht*) to 166 world-wide isolates of the fungus (Hooker et al., 1965). Calub et al. (1974) find that while leaf diffusates from

Fig. 27. Two monogenic dominants that mimic the lesions formed with *Helminthosporium* species. The large-lesion type (left), case no. 28, forms watery spots that tend to enlarge and develop as in sensitive reactions to fungal invasion, while the small-lesion type (right), case no. 48, forms local lesions characteristic of hypersensitive reactions.

infected *Ht* plants are inhibitory to spore germination in vitro, diffusates from strains with multiple-factor resistance are substantially more inhibitory. In the presence of *Bx*, a dominant factor that determines occurrence of a cyclic hydroxamate ("DIMBOA") in the roots and other parts, *H. turcicum* infection levels are substantially reduced either in *ht ht Bx Bx* or *Ht Ht Bx Bx*, relative to their *bx bx* counterparts (Couture et al., 1971). Relationships may exist between DIMBOA and resistance to pathogens, insect pests, and herbicides. For *H. maydis* the relationship of cytoplasmic constitution to race T and its pathotoxin is discussed in the section on extrachromosomal inheritance; for race O, Smith and Hooker (1973) have identified a recessive factor, *rhm*, conferring resistance. Neuffer (1973) reports two newly induced, monogenic dominant traits (Fig. 27) that either mimic closely the lesion-forming reaction with *Helminthosporium* species, or confer new reactions to one or more ubiquitous pathogens.

The recessive factor *ag*, identified by Horovitz and Marchioni (1942) and mapped to chromosome 1, results in definite resistance to leaf feeding by grasshoppers. Authentic stocks are no longer extant.

A recessive factor in one particular strain of the inbred line GT112 determines susceptibility to the widely used herbicide Atrazine (2-chloro 4-ethylamino 6-isopropylamino-S-triazine). The factor has not been symbolized but has been located to chromosome 8 (Scott and Grogan, 1969).

Flooding has been shown to be differentially injurious to plants with specific genotypes for factors determining alcohol dehydrogenase (Schwartz, 1969; Marshall et al., 1973).

Gametophyte Formation and Development

The dramatic fecundity of most reasonably vigorous strains is the basis for the ease with which sterilities, deviant ratios, and gametophyte characters can be recognized. Diverse sorts of genetic alterations have been identified in the processes from the beginning of formation of sporogenous tissues through the functioning of the gametes. Rather than specific cytogenetics of these factors, only a brief summary of phenotypic expressions and characterizations will be attempted here.

Types that fail to generate normal, functional pollen grains but are female-fertile, termed *male steriles*, include the *ms* group of recessive, nuclear factors and the dominant *Ms21*, and the three types of cytoplasmic male steriles (*cms-C, cms-S, cms-T*), which are discussed in the section on extrachromosomal inheritance. The fertility restoration factors, *Rf* and *Rf2* for *cms-T* and *Rf3* for *cms-S*, could be defined as nuclear factors whose recessive alleles condition male sterility in the appropriate cytoplasm. Beadle (1932) determined that the stages of failure of the male steriles through *ms16* ranged from the earliest phases of meiosis to late phases of pollen grain maturation. The course of microgametophyte development in *ms2, ms7*, and *ms11* breaks down between the 5th and 10th day after meiosis (Madjolelo et al., 1966). Among other recessive factors affecting pollen but not ear fertility are divergent spindle (*dv*), in which the pollen is substantially reduced in viability through chromosome losses but is not completely aborted (Clark, 1940), and desynaptic (*dy*), which has similar phenotypic effects (Nelson and Clary, 1952). The fact that the recessive male sterility factors are expressed only in homozygous plants could be interpreted to mean that the gene action of the dominant alleles, whenever it may be initiated, may affect the pollen through "maternal" (i.e., mother plant) processes rather than through autonomous effects in the gametophyte. Examples of the autonomous type are brought forward in a succeeding paragraph. The potential for application of nuclear male sterility factors (controlled through the use of cytogenetic devices) in hybrid seed production has been emphasized (Phillips et al., 1971).

Factors that influence both male and female fertility in homozygous recessive plants include *am, as, el, ig,* and *po,* each affecting parts of the meiotic process or development of the gametophytes. B chromosomes and the K-10 knob influence events in meiosis and the gametophyte more subtly, affecting crossing over, chromosome dynamics, transmission, and chromosome breakage without grossly affecting fertility.

In the male gametophyte proper, virtually any substantial deficiency results in functional failure or abortion of the pollen grain, and to a lesser degree the same is true of the female gametophyte. From these facts it might be supposed that a large number of loci would influence vital functions required for normal development of the gametophyte; that relatively few factors of this kind have been identified may indicate that most such functions are required in both gametophytes and that mutants consequently are not maintainable. In the female, the two factors *lo* and *lo2* cause abortion of ovules carrying a gametophyte with either factor. In the male, *pa* determines the abortion of pollen grains carrying the factor (Burnham, 1941), while *sp* and *sp2* condition reduced size of the pollen grain and an extreme disadvantage in competition with normal pollen grains. Mangelsdorf (1932) succeeded in separating *sp* pollen grains from normal ones by sieving and completed normal pollinations with the separated samples. *Sks* is a gametophyte-specific factor that conditions normal development and functioning of *Sks* pollen grains produced on plants bearing the dominant male sterility factor *Ms21* (Schwartz, 1951; Leng and Bauman, 1955).

Several factors have been found to affect biochemical properties of the pollen grain. The most commonly used of these for genetic studies or for classroom demonstrations is the starch variation conditioned by the *wx* locus. Pollen grains carrying *Wx* accumulate enough amylose starch to stain black with I_2-KI solution, while *wx* pollen grains contain amylopectin starch alone and stain light brown, with the result that heterozygous *Wx wx* plants produce pollen composed of equal numbers of easily distinguishable grains. Slightly lowered transmission (about 48% instead of 50%) of *wx* relative to *Wx* is due to slower germination and establishment of *wx* pollen grains, followed by equal growth rates, according to direct observations on live silks (Sprague, 1933; see Fig. 28); the differences are even more pronounced for pollen grains produced in *su su* plants. Nelson (1959, 1968) has used the *Wx wx* pollen character to survey large numbers of meiotic products as an aid to the study of genetic fine structure at the *wx* locus. With homozygous *wx* the effects of the *ae* locus also can be distinguished in the pollen, and fine structure of the *ae* locus has been studied accordingly (Moore and Creech, 1972). The occurence in pollen from *Bz Bz, Bz bz,* and *bz bz* plants of a 2:1:0 ratio for the enzymatic activity associated with this locus (Larson and Coe, 1968)

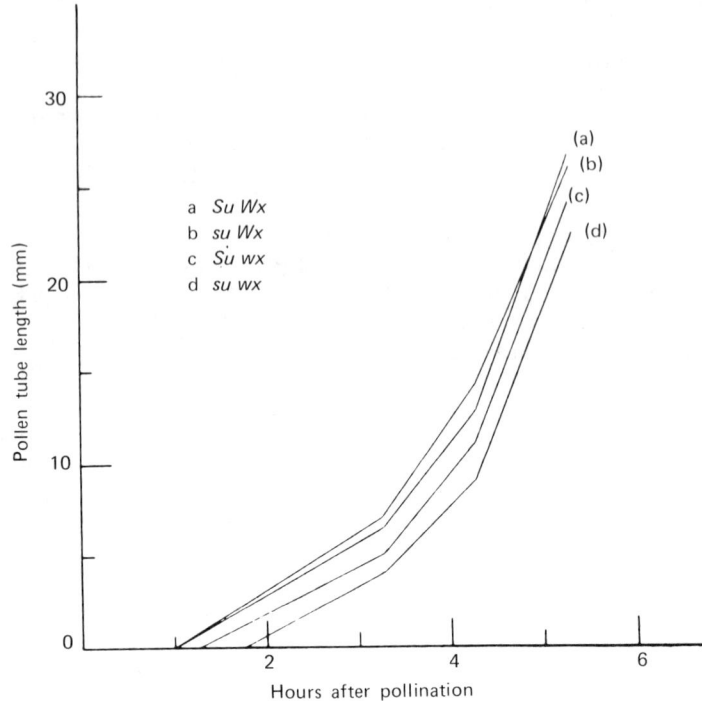

Fig. 28. Growth of pollen tubes of four genotypes. Although pollen germination is delayed in *Su wx* (c) and *su wx* (d), all four types grow at equal rates. (From Sprague, 1933).

suggests that the *Bz* factor may be functional in individual pollen grains in determining the glucosylation of the flavonol quercetin. Reports of anthocyanin development in the pollen appear to be the result of absorption of these pigments from exceptionally intense anther walls under unusual environmental conditions rather than of pigment synthesis in the pollen grain itself.

Effects of factors of the *Ga* group, which affect the functioning of pollen grains individually, are detected only by differential transmission of marker factors linked to them. For the original factor on chromosome 4, pollen tubes carrying *Ga* function equally in competition with *ga* on *ga* silks but function in 90 to 100% of fertilizations on *Ga* silks (Mangelsdorf and Jones, 1926; Emerson, 1934). A stronger allele, *Ga-s*, is more effective in competition and is restrictive against *ga* in silks, to the point of exclusion (Schwartz, 1950). House and Nelson (1958) demonstrated by ^{32}P labelling of pollen that *ga* tubes grow slowly and stop in a few hours on *Ga-s* silks. Jimenez and Nelson (1965) found an additional factor on chromosome 4, *Ga9*, which is differentially competitive on *Ga* silks but is indifferent to *Ga9* silks. The *Ga2* factor on chromosome 5 is dependent on *Ga2* silk constitution (Burnham, 1936), although Longley (1961) reported (in another source) behavior of a chromosome 5 factor that was consistent with effects of a silk factor genetically distinct from the pollen factor. The *Ga8* factor on chromosome 9 also is dependent on *Ga8* in the silks (Bianchi et al., 1969).

Regulatory Systems Controlling Gene Expression and Mutation

The analysis by McClintock and others of certain types of genetically controlled variegation and its relationship to chromosome breakage, gene regulation, and patterns of development has been one of the major milestones in genetics. Inherited somatic variegations expressed by the *P* locus were described in detail by Emerson (1914); and the first clear example of a heritable factor (*Dt*) controlling mutation at another locus (*A*) was reported by Rhoades (1938, 1941). The elemental nature of mutation-controlling factors first became clear when McClintock (1950, 1951) discovered and analyzed the transposable elements *Ds* and *Ac*. A more complex system with similar properties, the *I En* system of Peterson (1953), was later shown (Peterson, 1965) to be equivalent to the *Spm* system of McClintock (1956). Brink and Nilan (1952) and Barclay and Brink (1954) showed that the *P-VV* variegation of Emerson was an expression of the *Ac Ds* system operating at the *P* locus, and the mutable factor *bz2-m* and its mutator, *M*, were found to be of the same system (Nuffer, 1955).

A listing of regulatory systems with their designated controlling elements and some typical affected loci is given in Table 10. Presented with pertinent definitions in Table 11 are symbolic representations of the operational relationships between the structural gene *A* and the elements and components of the *Dt, Ac Ds,* and *Spm* systems. The three most intensively studied systems will be described in conjunction with the tables; these descriptions are heavily dependent upon the authoritative and precise review of the characteristics of these systems presented by McClintock (1965). [A current, thorough, and thought-provoking review (Fincham and Sastry, 1974) appeared following completion of the present manuscript.]

The elements of regulatory systems exhibit some properties that are Mendelian in that the elements are generally associated with the chromosomes and are transmitted, despite frequent exceptions, from parent to progeny as genetic material. Further, their only phenotype is their effect on the action of known genetic markers or on chromosomes. They also express some distinctly non-Mendelian properties exemplified by the transposition

Table 10—Listing of regulatory systems and some affected loci†

System	Regulatory element	Controlling element	Designation	Reference
Dt	Dt	—	a a-m1(Dt)	Rhoades, 1938; Nuffer, 1961
Ac Ds	Ac	Ds	a-m3 a-m4 a2-m4 c-m1 c-m2 c-m4 sh-m1 sh-m2 bz-m1 bz-m2 bz-m4 wx-m1 wx-m5 wx-m6 wx-m7 wx-m9	McClintock, 1965
	Mp = Ac	Ds	P-VV	Brink and Nilan, 1952, Barclay and Brink, 1954
	M = Ac	Ds	bz2-m	Nuffer, 1955
Spm	Spm	Spm	a-m1(Spm) a-m2 a-m5 c2-m1 c2-m2 a2-m1 a2-m5 pr-m2 pr-m3 c-m5 wx-m8	McClintock, 1965
	En = Spm	I	pg14-m a-m(En) a2-m w13-m	Peterson, 1953, 1970
R-st	—	I-R	R-st	Ashman, 1960
R-mb	—	—	R-mb	Weyers, 1961
Mr	Mr	—	R-m	Neuffer et al., 1968
R-v, r-v	—	—	R-v, r-v cherry	J. K. Stadler, 1955
β-m	—	—	∝β-m ∝-m β-m	Neuffer, 1963
c-m	—	—	c-m	Schwartz, 1960c

†Dash signifies that element not identified.

Table 11—Definitions and symbolic representations for known regulatory systems

Controlling element (*Ds, I*): Responding unit of a regulatory element system; controls the action of the structural gene at which it is located.

Regulatory element (*Ac, Dt, Spm*): Regulating unit of a regulatory element system; regulates the activity of a controlling element.

Structural gene (*A*): Functional genetic unit that determines a single genetic trait.

1)	*A*	Dominant standard allele at the *A* locus on chromosome 3; used as an example to designate the active structural gene and its response to controlling elements (See text for description of *A* action).
2)	*a*	Recessive, inactive allele at the *A* locus; usually refers to the recessive allele studied by Rhoades that responds to *Dt*.
3)	*a-m*	Recessive, mutable allele under the control of one of the controlling elements as represented in (11) through (18); reverts to some form of *A* in response to a specific regulatory element.
4)	*a-s*	Recessive, inactive allele of *A*; does not respond to *Dt* specifically or to any regulatory element.
5)	*Ac*	*Activator:* Transposable regulatory element that initiates and regulates activity of *Ds* controlling elements; operational equivalents are *Mp* (of *P*) and *M* (of *bz2-m*).
6)	*Ds*	*Dissociation:* A transposable controlling element that responds to regulation by *Ac* to control activity of a structural gene and/or to break chromosomes at the site of its insertion.
7)	*Dt*	*Dotted:* Regulatory element that initiates and regulates a presumed controlling element at the *A* locus to cause frequent reversions of *a-m* to various levels of *A* and vice versa.
8)	*En*	*Enhancer:* Peterson's designation for the regulatory element of the *Spm* system; operationally equivalent to *Spm*.
9)	*I*	*Inhibitor:* Peterson's designation for the controlling element of the *Spm* system; used in combination with McClintock's terminology in this publication.
10)	*Spm*	*Suppressor-mutator:* A complex, transposable, regulatory element with two major activities or components (McClintock, 1965): Component-1 suppression (control of the action of the structural gene where it resides through regulation of the controlling element) and Component-2 mutation (change of state of activity and transposition).
11)	*A·Ds*	*a-m:* Structural gene *A* suppressed by the controlling element *Ds* and therefore phenotypically null (*a*) in the absence of *Ac*, but responding to *Ac* regulation; e.g., *a-m4* of McClintock.
12)	*A·Ds, Ac*	Like (11) but with *Ac* present at a separate locus, causing *Ds* to respond and to produce changes in *A* function that result in sectors of color of varying sizes (time of event) and intensity, at varying frequencies; this is a two-unit system (see 13).
13)	*A·Ac*	Like (12) but with *Ac* present at the *A* locus, causing behavior as a single-unit "autonomous" system in which *Ac* appears to possess the properties of both a controlling element and a regulatory element; the possibility of a *Ds* element actually being present is not ruled out.
14)	*A·I·Spm*	*a-m:* Structural gene *A* autonomously under control of the *Spm* system with a controlling element *I* and a fully active *Spm* element inserted at the locus; phenotypically null, because the active suppressor (component-1) causes the controlling element to suppress *A* action, with frequent reversions (dots) to pale and full color because of changes in state or transposition initiated by the active mutator (component-2).
15)	*A·I*	Gene *A* partially affected by the presence of *I* without an active *Spm*; uniform, stable phenotype ranging from nearly colorless to nearly colored depending on the state of the locus at inception of this condition.
16)	*A·I, Spm*	Same as (15) but with a fully active *Spm* present at another location; phenotypically colorless kernel with pale and colored reversions (like 14).
17)	*A·I·Spm* (active-1)	Gene *A* suppressed by *I* that is regulated by an active component-1 but stable because of the absence of an active mutator component; phenotype is stable and colorless, but with the addition of a fully active *Spm* the phenotype will be colorless with pale and colored dots (like 14).
18)	*A·I·Spm* (active-2)	Gene *A* partially affected by *I* with an inactive suppressor (component-1) and an active mutator (component-2) present but not expressed because suppression must precede mutation. This would constitute a stable configuration that appears like (15) but is not responsive to active *Spm*.

of elements from site to site on the chromosomes, by changes in number and state, and by the regulatory effect they have on each other and on known Mendelian factors.

The *Dt* system, discovered in Black Mexican sweet corn by Rhoades (1936), has been found in unrelated varieties from diverse sources (Nuffer, 1955). It consists of an activating element, *Dt*, which causes apparent mutational changes to occur at certain alleles of the *A* locus, and an implied responding element, which controls the action of *A* to express these changes. Other loci than *A* may be affected but they have not been identified. The initial change as reported by Nuffer (1961) is usually from *A* to *a* or to *a-m* — recessive, inactive alleles that are stable in the absence of *Dt* but have the property of reverting back to several different levels of *A* activity if *Dt* is present. The reversions occur in somatic cells at various stages, giving colored sectors in tissues where anthocyanin develops. In the aleurone tissue these sectors appear as dots, hence the designation *dotted*. Each Dt-controlled *A*→*a* event may produce a different *a-m* with regard to the type and frequency of response to *Dt*, while each *a-m*→*A* event may produce a different *A* allele with regard to its anthocyanin producing ability and its response to *Dt* (Nuffer, 1961; see Fig. 29). Changes also occur from *a-m* to *a-s* (stable, not responsive to *Dt*). Increases in dosage of the responding allele (*a* or *a-m*) show a linear increase in the frequency of dots with no conspicuous change in the time, type, or location of events; increases in dosage of *Dt* show an exponential increase in frequency with no conspicuous change in time, type, or location (Rhoades, 1941; Nuffer, 1955; see Fig. 3). *Dt* elements have been located on the short arm of chromosome 9 and the long arms of chromosomes 6 and 7 (Nuffer, 1955). New *Dt* elements have been induced by chromosome breakage, either by chromosome manipulation (Doerschug, 1973; McClintock, 1950, 1965; Fig. 30) or by ultraviolet or x-ray treatment (Neuffer, 1966).

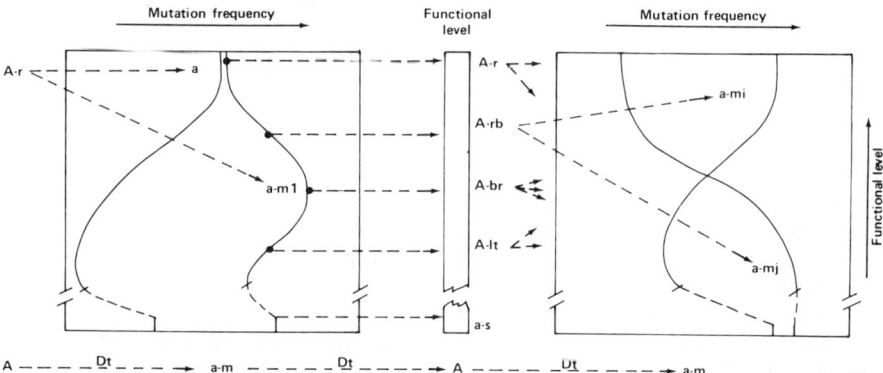

Fig. 29. Diagram representing pedigrees of mutational events induced at the *A* locus by *Dt*. Each particular mutable allele (*a* or *a-m*) is distinctive in the frequencies (profile) with which reversions to specific functional levels of *A* occur, and each of these reversions will in turn exhibit an equally distinctive profile for the production of additional mutable alleles. (After Nuffer, 1961.)

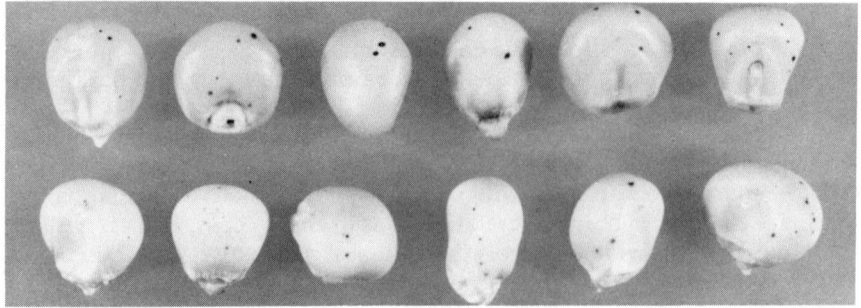

Fig. 30. Induction of *Dt* activity by chromosome breakage.
Upper row: Kernels homozygous for *a*, carrying standard *Dt*.
Lower row: Kernels homozygous for *a*, in which activity imitating *Dt* has been initiated by chromosome breakage.
(From McClintock, 1965.)

The *Ac Ds* system, discovered by McClintock (1950), is the foundation on which the present level of understanding of controlling element systems is built. McClintock (1951) showed in a classic demonstration that the system appeared under conditions of chromosome breakage and rearrangement in the short arm of chromosome 9. However, at least two variations of this system were found to be present in genetic lines that had not been subjected knowingly to chromosome breakage (Barclay and Brink, 1954; Nuffer, 1955). The system consists of two types of elements. The *controlling element*, designated *Ds*, has several characteristic properties: 1) transposability, the property of moving from site to site in the chromosome complement; 2) control of activity of genes affected by the presence of *Ds*, such that the gene may be either active, partially active, or inactive, depending on the state of *Ds*; 3) control of chromosome integrity at the site, such that the chromosome may or may not break or disassociate; 4) ability to change in state (frequency, timing, or quality of events). The *regulatory* element, designated *Ac*, acts as a regulator of *Ds* activity in that *Ds* transposes, causes chromosome breakage, or changes its state of activity only in the presence of *Ac*. In the absence of *Ac*, *Ds* cannot change its state but is locked in the condition present at the time of *Ac* removal. *Ac* generally has an unusual dosage response, in that additional *Ac* units have the effect of delaying the timing of events that are regulated by *Ac*. The *Ac* element can be located elsewhere in the genome, not necessarily at *Ds*. Additional units apparently arise because the *Ac* element can move (transpose) from site to site in the chromosome complement and can be copied additional times in the cell cycle as replication advances past the new sites of insertion (Greenblatt, 1968a). As a consequence cell lineages or progenies occur with more than one *Ac* element (McClintock, 1951). Except for the dosage relationship, *Ac* units apparently act independently. *Ac* transposes, changes in state of activity, suppresses the action of adjacent loci, and possibly causes chromosome breakage. Loci controlled directly by an *Ac* element at the site are said to constitute an autonomous one-unit system in contrast to loci controlled by a *Ds* element regulated by *Ac*, which are said to constitute a two-unit system. There is some question as to whether or not *Ac* can by itself break chromosomes and control gene loci. It may be that apparent one-unit systems actually are composed of two units (*Ac* and *Ds*) tightly associated so that they act together and give the appearance of a single *Ac* unit with these properties, as is the case with the *Spm* system to be described later. A representative pedigree of observations and events typical of those that are found in studies with the *Ac Ds* system will serve to characterize the behavior of the units.

If one begins by crossing a stock carrying a *Ds* element in a particular chromosome with one carrying an *Ac* element, *Ac* will activate *Ds* and any of the following events may occur at various cell stages, depending on the state of *Ds* and on the number and state of *Ac*. If *Ds* is not located at a recognizable gene locus, its phenotype (with *Ac*) will result only from its ability to cause chromosome breakage and the loss of genes on the arm distal to its location. Under these conditions and with a characteristic frequency, *Ds* will leave its original site and move (transpose) to a new site. The new site will most often be near the original one, with more distant locations and other chromosomes being progressively less likely. When *Ds* moves to a new site its activity at the old site usually disappears, and it may exhibit all or only some of its former characteristics. If the new site is at a recognizable gene locus, *Ds* may suppress the action of the gene, resulting in an inactive recessive mutant; and if *Ac* is removed (by segregation or other events) a stable recessive allele will be the visible result. If *Ac* remains in the lineage, further transpositions, chromosome dissociations (if this property was transferred with *Ds*), and changes in state of *Ds* will occur. The result of further transposition will be the removal (complete or partial) of suppression of the gene by *Ds*, permitting expression of the gene function and phenotype in subsequent cell lineages, and the possible appearance of *Ds* at additional new sites. Transpositions may occur at all stages of development or at limited stages, depending on the state of the *Ds* element and the number of *Ac* elements in the lineage (the higher the number of *Ac* elements, the later the events at *Ds*). If a *Ds*-suppressed gene is one that controls anthocyanin pigmentation in the kernel and plant, such as *A,* the phenotype will be a colorless kernel (suppression of *A*) with frequent dots (sectors) of color representing the recovery of *A* action, and a green or brown plant with sectors of red or purple in appropriate tissues. The result of dissociation at the new location may be the same as at the original site except that a different chromosome segment will be lost. Changes in the state of *Ds* can be of several types: the *Ds* element can become less or more responsive to *Ac* regulation, resulting in fewer or more transpositions (dots or sectors); the *Ds* element can become responsive only to certain special conditions, resulting in events (sectors) only in certain tissues or events only of certain types (size or color intensity); the *Ds* element may also change its degree of suppression so that the gene may be partially expressed (giving, in the case of *A,* an allele with faint anthocyanin in the aleurone and in plant tissue).

The *Spm* system, discovered and described independently by Peterson (1953, 1965) and McClintock (1956, 1965), possesses many of the characteristics of the *Ac Ds* system but is more complex. We have taken the liberty in this summary of using primarily the symbols of McClintock, supplemented by those of Peterson and our own when such admixture is effective. The *Spm*

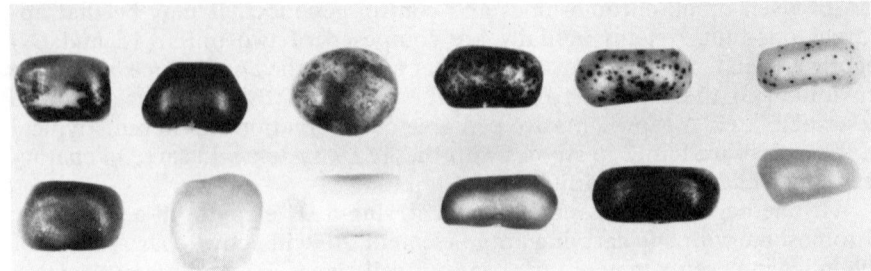

Fig. 31. Series of different states derived from *a-m1* of the *Spm* system, each expressed in the presence (upper row) or absence (lower row) of active *Spm*. The initial state from which the others were derived is exhibited by the first kernel (on the left) in each row. (From McClintock, 1965.)

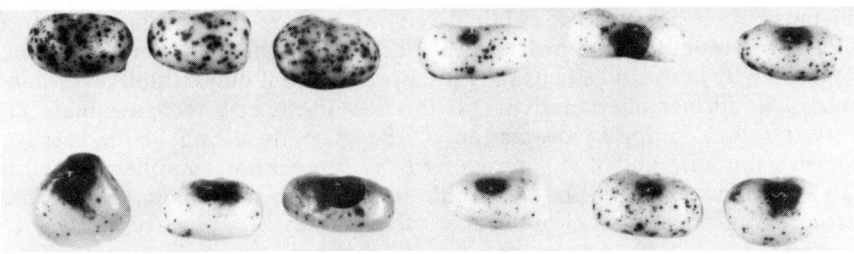

Fig. 32. Patterns produced by *a-m1* of *Spm* in the presence of an *Spm* active in both components 1 and 2 (first three kernels on the left in upper row) and in the presence of an *Spm* with component 1 preset to be inactive in the crown (resulting in uniform pigment synthesis) but active in the lower parts of the kernel (resulting in mutational events on a colorless background). (From McClintock, 1965.)

system is composed of a controlling element (*I* in Peterson's terminology) that affects the activity of a structural gene where it is inserted, and a complex regulatory element designated *Spm*. *Spm* has two components of action: component-1 is a suppressor that regulates *I*; component-2 is a mutator that regulates transposition and changes in state of the controlling element and of component-1.

Component-1, when active, causes the controlling element to suppress the action of any structural gene at the site of the inserted element. The result is an apparently nonfunctional gene (in the case of *A*, a colorless kernel). When component-1 is inactive, the controlled gene may exhibit partial to full activity (in the case of *A*, from near-colorless to fully colored kernels; see Fig. 31), depending on the position of insertion or on the state of the gene at the time of insertion of the controlling element. Changes in state of component-1 from activity to inactivity are mutation-like events regulated by component-2. When component-1 is associated with a controlling element that has lost its ability to respond to component-2 (i.e., to mutate), cyclical changes in the phase of activity of component-1 can be observed. These changes, from active to inactive and back again, seem to depend on preset timing or on responses to given conditions and result either in delay of expression to later cell or plant generations or in differential expressions in different tissues. Delayed expression is seen in cases when a nonfunctional gene under control of these systems suddenly becomes functional in later cell or whole plant generations (McClintock, 1968); tissue differences are seen as patterns of expression that do not follow cell lineages — for example, colored crown or base of the kernel (McClintock, 1965; Peterson, 1966; see Fig. 32).

Component-2 regulates mutational changes of the other units in the system; when active, it causes abrupt changes of varying kinds. It can trigger the transposition of the controlling element to other sites in the genome, often leaving the original gene modified in its expression or in its subsequent response to further *Spm* activity (in the case of *A,* the effect of component-2 activity is a colorless kernel with colored sectors or dots of varying size and intensity). Component-2 can cause changes in the controlling element such that it no longer responds to component-2 but only to component-1 (colorless kernel without dots). It can also cause permanent loss of activity of component-1 either by transposition or by change to an inactive state, to produce a state of the gene that is no longer responsive to *Spm* activity. In this case the gene may be left in its original state — i.e., fully functional or at some intermediate level, but completely stable.

Component-2 regulates all transpositions within the system and can itself undergo transposition; under its direction, part or all of the system may be transposed to a new site in the genome. Component-2 can change to a different level of activity, which will also alter the activity or expression of the gene, the controlling element, and component-1. If component-2 assumes an inactive state, component-1 and the controlling element will be locked into the condition present at the time of the change; however, the addition of an *Spm* element with an active component-2 will restore activity to the system. An inactive component-2 may revert to the active form (at a rate characteristic for each inactive case), producing new states of component-2 expression.

Additional interactions and effects are known. Component-2 is effective only when component-1 is active. Release of a gene from its controlling element frequently does not leave it in its original condition. The time of occurrence of changes in gene action and some aspects of the frequency of changes are expressions of the state of the locus; when two different states are present as alleles, each responds to the system independently of the other. Dominant modifiers of the system have been identified that alter the frequency of mutation-inducing responses of the controlling element in the presence of an active component-1, supplant inactive component-2 by supplying component-2 activity, and are transposable. Increased doses of fully active *Spm* or the modifier do not affect the frequency of mutation-inducing responses of the controlling element.

Many aspects of the regulatory systems are still not fully understood. Though much is known about the behavior of the elements in these systems, very little is known about their nature; they appear to be foreign to (but often intimately associated with) the genetic material. Nelson (1968) placed the controlling elements of *wx-m8* of the *Spm* system and *wx-m1* and *wx-m6* of the *Ac Ds* system within the cistron of the *wx* locus. Neuffer (1966) was unable to remove the controlling elements from the *A* locus by crossing over, for several mutable alleles that are a part of the *Dt* and *Ac Ds* systems;

however, McClintock (1968) found that with certain other alleles the controlling elements could be removed by crossing over. Kermicle (1970a) found that many of the changes in mutability of the *R-st* system are associated with exchange. It is assumed that some aspects of these systems are present but inactive in the genomes of many organisms (McClintock 1968), but the only well-documented way to bring them to observable activity is through chromosome breakage (Doerschug, 1973; McClintock, 1965; Neuffer, 1966). In any case, the discovery and analysis of these systems has added a new dimension to the study of inheritance.

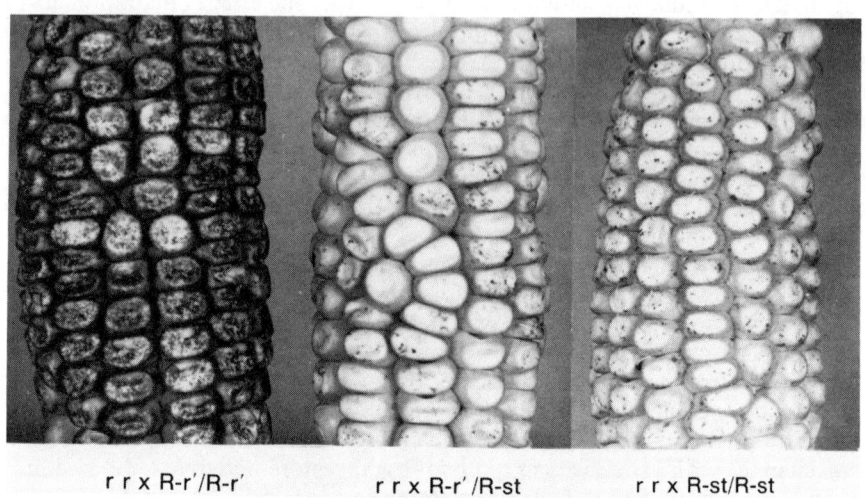

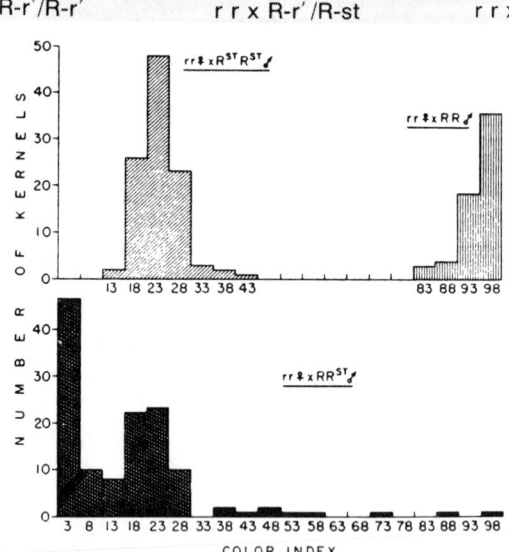

Fig. 33. Paramutation of *R-r*. Top: ears from testcrosses of sibling plants of the three genotypes derived from the self-fertilization of *R-r/R-st*, showing (left) the medium mottled expression of *R-r'/R-r'* (*R-r* once-exposed to *R-st* and partially returned to standard *R-r* level in passage through homozygosity); (center) the dramatically reduced expression of *R-r'/R-st* (*R-r* twice-exposed to *R-st*); and (right) the standard expression of *R-st*. Bottom: graph showing the color grade distributions of the kernels from the three ears pictured above. (From Brink, 1956.)

THE GENETICS OF CORN

Paramutation

Brink (1956) discovered a form of unorthodox inheritance at the R locus in which the functional capacity of an allele is heritably altered by its genetic history, and Coe (1959) found a similar system at the B locus. As defined by Brink (1973), paramutation is "an interaction between alleles that leads to directed, heritable change at the locus." A full, referenced survey of the extensive genetic analyses that have been done on these systems and on related ones in other organisms is given in the review by Brink (1973). The basis in conventional genetic or physicochemical terms for the paramutation phenomena at both R and B is not known, but there may be some relationships between the systems of controlling elements, paramutation, and aberrant ratio.

In brief, paramutation at the R locus consists in an inherited reduction in the functional capacity of a sensitive (paramutable) allele after it has passed through a heterozygote with a paramutagenic allele, which remains unchanged (Fig. 33). Paramutable alleles are those which are subject to mottling (R-r, R-g, R-ch); paramutagenic alleles include R-st, R-mb, certain of their mutational derivatives, and a number of other naturally occurring alleles. When the paramutant allele, R', is maintained for several generations in the homozygous state, it progressively regains part of its original functional capacity; when maintained heterozygous with r or with a deficiency, R'

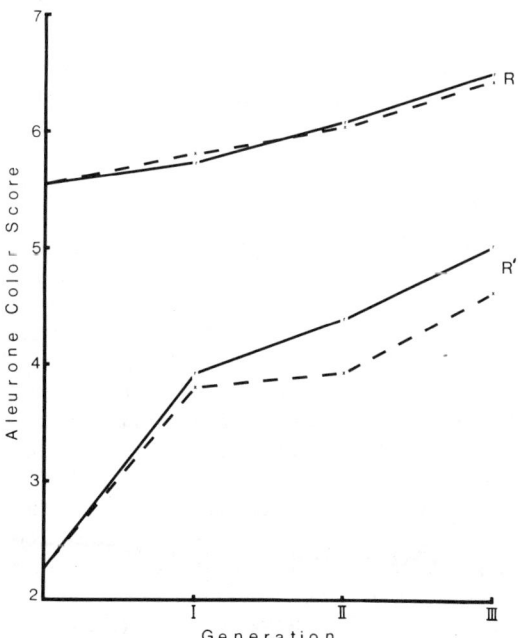

Fig. 34. Recovery in level of function of paramutant R' (lower lines), and parallel enhancement of R (upper lines) when maintained heterozygous with r (solid lines) or with the deficiency r-$x1$ (broken lines). (From Styles and Brink, 1969).

regains nearly all of its original level of function (Styles and Brink, 1969; see Fig. 34). Since parallel changes in level of function of R are found without prior exposure to a paramutagenic allele, R alleles may be considered metastable, capable of change (in small increases or decreases) in particular passages without the necessity of an "inducing" allele (Styles and Brink, 1966), although the effects of R-st in decreasing R function are much more dramatic and are unidirectional. The spontaneous mutation (primarily recombination) rate of paramutant R' is usually lower than that of its non-paramutant counterpart (Bray and Brink, 1966). Mutagens such as X-rays tend to reduce the paramutability of R, to partially restore function in R', and to reduce the paramutagenicity of R-st (Shih, 1969; Shih and Brink, 1969). R' is progressively reduced, by repeated passages with a paramutagenic allele, to near-absence of function (Mikula, 1961). R-st has been shown to be fractionable by recombination into several units, including multiple determiners of its paramutagenicity (Ashman, 1965; Kermicle, 1970a, 1974).

Paramutation at the B locus consists in a uniform, inherited change in the level of functional capacity of B after it has passed through a heterozygote with a paramutagenic allele, B', or with certain other alleles that have been exposed to B' previously (Coe, 1966b). Once changed, B' becomes uniformly paramutagenic and retains a constant level of functional capacity, not subject to further reduction or to mutagen alterations. X-ray induced losses of the B' chromosome segment during development in $B'B$ individuals reveal B to be unchanged during most or all of the life cycle. The color level of tissues of $B'B$ plants, including the tassels and cobs, is darker than that on $B'B'$ and $B'b$, showing that B is not reduced all the way to B' level of function during the course of vegetative growth, even though all gametes formed on $B'B$ plants are uniformly reduced. Mutations to B' occur spontaneously in B B plants at variable frequencies.

Aberrant Ratio

In progenies derived from pollen of multiple-dominant plants artificially infected with barley stripe mosaic virus (BSMV) crossed onto an uninfected multiple-recessive ear parent, Sprague and McKinney (1966) found unorthodox distortions of segregation ratios whose cause remains perplexing. The distorting effect, termed Aberrant Ratio (AR), is specifically virus initiated and is systematically inherited, raising the suspicion that the conventional genetic system has been supplemented or altered by the infectious agent.

AR is found in rather high frequency (1 in 200) in F_2 progenies from crosses of multiple-recessive (a su pr wx) ear parents with pollen from infected multiple-dominant plants. Ratio distortions ranging from small deviations close to the expected (i.e., 60:40) to extremes as deviant as 95:5 can be found for any one of the factors in either direction (low A or low a). Deviations are equal via male and female, in contrast to the behavior of other known systems that affect either segregation pattern or transmission by conventionally explained mechanisms. The degree of distortion is inherited, showing occasional changes in direction or degree. AR is dependent upon testcross parents (e.g., a a) derived from the AR source, inasmuch as testcrosses to controls show no distortion (Sprague and McKinney, 1971). AR at one locus does not affect ratios of other segregating factors in the same progenies, and there appears to be no interaction between AR at one locus and AR at another in the same pedigree. In a 3-point testcross, Sprague and McKinney (1971) established that AR at the C locus (low C) did not affect two other linked factors, $yg2$ and sh, and that the pattern of crossing over fit expectations calculated on the basis that C expression is modified to that of c. Progeny tests of kernels from low A and low a strains show results agreeing with the assumption that A expression is modified to a in low A strains and a to A in low a strains, with occasional reversion in each instance.

Extrachromosomal Inheritance

Variants have been recognized that affect either chlorophyll pigmentation or male fertility and show non-Mendelian inheritance in being transmitted entirely through the maternal parent. In a thorough, analytical review Duvick (1965) surveys these variants, including some that appear to affect plant vigor and specific agronomic traits as well as chlorophyll, and concentrates particularly on cytoplasmically inherited male sterility and its properties as related to seed production. The practical utilization of one of the variants affecting male fertility has been one of the triumphs of genetics as applied to production practices, and its specific disease reaction has been an instructive lesson in crop ecology.

The chlorophyll variant whose inheritance was reported by Anderson (1923) showed stripes of green and pale green tissue in the leaves. Striped plants transmitted striped or entirely pale green character to progeny through the maternal parent only, along with variable proportions of entirely green individuals. Pale green seedlings were lethal, and green plants remained stable in progeny tests. Since "ear map" plantings (see section on developmental studies) from striped individuals showed clustering of types, with striped seedlings most often occurring near the border between green and pale green (Fig. 35), striping may have been due to sorting-out of two or more differing determinants, presumably plastids. A very similar case, independent in origin and darker in phenotype, was studied by Demerec (1927); tests for infectious transmission of the variant were unsuccessful. Nonchromosomal stripe (NCS), analyzed by Shumway and Bauman (1967), showed a similar phenotype (light green striped), similar clustering in ear

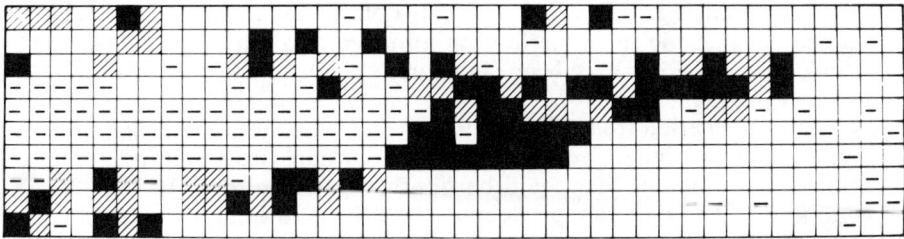

Fig. 35. Ear map distribution of a maternally inherited chlorophyll deficiency, demonstrating sectorial clustering of pale green (solid) and striped (hatched) seedlings with green (open) ones. Ovaries and kernels that could not be tested are indicated by a dash. (After Anderson, 1923.)

maps, and failure of infectious transmission, but entirely light green plants were not reported and some of the green plants yielded abnormal progeny; wide variation in plant and seed vigor, distinct from the chlorophyll effects, was observed. In crosses of NCS plants with unrelated males, slightly lower frequencies of NCS progeny occurred than in pollinations within the NCS strain.

The recessive nuclear factor *ij* was shown by Rhoades (1946), in a now-classic study, to condition the loss of chlorophyll in sectors in the plant, inherited maternally in subsequent progeny. Although some striping tendency continues to be inherited, suggesting that two or more differing unitary determinants are involved, the majority of striped individuals derived by the *ij* effect transmit entirely green or entirely white character to progeny. Shumway and Weier (1967) found that plastids in white tissue of homozygous *ij* plants have abnormal granal and lamellar organization and lack ribosomes. Rhoades (1950) later reported induction of cytoplasmic male sterility by *ij* (which would bear importantly on the nature of determinants), but Duvick (1965) deduces evidence indicating that the sterility may have been derived from pre-existing *cms-S* in the *iojap* strains. Stroup (1970) identified a second factor, *chloroplast mutator (cm)*, very similar to *ij* in its effects.

A variant from Peru affecting male fertility was shown by Rhoades (1931, 1933) to be transmitted only through the maternal parent, with some irregularity in phenotype (i.e., "fertile" maternal progeny when tested yielded male-sterile products). Whether this variant is related in any way to subsequently discovered ones has not been tested, and pedigreed material is no longer extant. Duvick (1965) identifies and details this and later discoveries of cytoplasmic male sterility, mostly categorized into two groups, *cms-S* (USDA type) and *cms-T* (Texas type), on the basis of their pattern of behavior (fertility restoration) when crossed to a series of discriminating pollinator strains carrying restorer factors. Restoration of fertility has been found to be under the control of conventional nuclear factors. Restoration of full fertility to *cms-T* is controlled by two complementary dominant factors, *Rf* and *Rf2*. In addition to these factors for full fertility, Beckett (1966) has found that partial restoration of fertility to *cms-T* is conferred by several other, ill-defined factors, at least some of which are distinct from *Rf* and *Rf2*. Restoration of fertility to *cms-S*, conferred by *Rf3*, is distinctive in that pollen grains carrying *Rf3* develop normally, while *rf3* grains abort if they are derived on *cms-S* plants (Buchert, 1961). Pollen from plants of *Rf3 rf3 (cms-S)* constitution is half aborted and half normal. Attempts to induce or to reverse cytoplasmic male sterility by mutagenesis, heat treatments, or infectious transmission have been singularly unfruitful (Duvick, 1965; Shumway and Bauman, 1966).

Over the period from 1950 to 1970, *cms-T* came into increasing use as a major aid in crossing to produce hybrid seed corn by virtue of its reliability as a male-sterile parent that could be returned genetically to the male-fertile condition in the final grain production stage. In 1970 a nation-wide epidemic of *Helminthosporium maydis* race T (a new race specifically virulent with *cms-T*) spread across host fields estimated to contain *cms-T* in 75 to 90% of plantings, causing severe losses in yield on a national scale (National

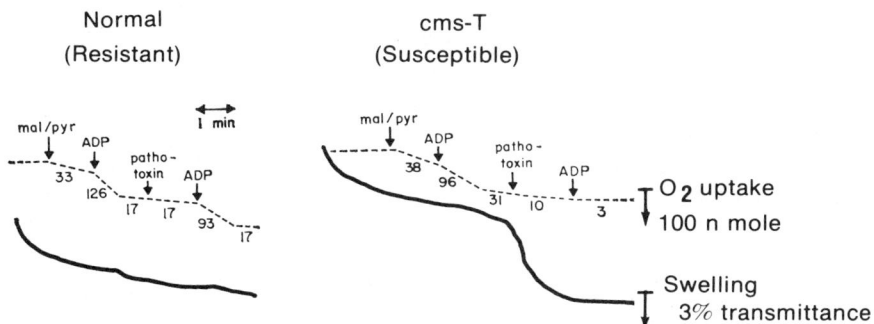

Fig. 36. Changes in the dynamic properties of mitochondria from normal (left) and *cms-T* (right) cytoplasm after exposure to the pathotoxin from *Helminthosporium maydis* race T. Following the initiation of oxygen uptake with substrate (mal/pyr), which can then be stimulated by ADP, the addition of the pathotoxin causes dramatic swelling in *cms-T* mitochondria and interferes with subsequent ADP stimulation. (From Miller and Koeppe, 1971.)

Academy of Sciences, 1972). Current seed production developments concentrate on increasing the diversity of cytoplasmic and nuclear sources (Duvick, 1972).

Beckett (1971) defined, among 30 cytoplasmically male-sterile collections, three groups with respect to their fertility patterns in crosses with 18 inbred lines: S group (20 sources); T group (four sources); and a new class, C group (two sources). The remaining four sources either gave ambiguous experimental results or were inadequately tested. The same 30 sources, tested for reaction to race T of *H. maydis,* were all resistant except for the four identified as *cms-T* by the fertility criteria (Smith et al., 1971). Duvick (1972) notes that altogether 108 collections of cytoplasmic male steriles from widely diverse parts of the world have been tested and that most (if not all) are categorizable as *cms-T, -S,* or *-C*. In addition to the disease susceptibility and fertility restoration patterns cited above and the agronomic properties discussed by Duvick (1965), *cms-T* has been found to exhibit extreme sensitivity of its mitochondria to the pathotoxin produced by the fungus and to several other agents (Gengenbach et al., 1973; Miller and Koeppe, 1971; see Fig. 36), and sensitivity of pollen tubes to the pathotoxin (Laughnan and Gabay, 1973b). Shah and Levings (1973) report no differences in properties of chloroplast DNA between *cms-T* and normal.

Although *cms-S* has been considered to be rather stable (Duvick, 1965), Singh and Laughnan (1972) identified exceptional fertile progeny in certain specific strains. Crossing analyses indicated that the exceptions resulted from instances either of mutation to normal cytoplasm or, conceivably, of transfer of normal cytoplasm through the pollen. Analysis of further events in the same strains (Laughnan and Gabay, 1973a) reveals the occurrence of tassel and ear chimeras for cytoplasmically inherited changes, demonstrating that these are mutation-like events occurring in the cytoplasm rather than transfers through the pollen. More rare events, not found by Singh and Laughnan, are characteristically like mutations to nuclear restorers. Laughnan and Gabay suggest that the two kinds of mutational events may involve establishment of the same fertility-conferring element either in the cytoplasm or in the nucleus.

METHODS OF GENETIC ANALYSIS

The art of designing and conducting experiments in any organism is dependent upon traits and properties peculiar to that organism, although most specific methodologies can be adapted to other organisms by modifications appropriate to their biological differences. Brief consideration is given in this section to specific methods applied in genetic analysis in corn.

Analysis of Genetic Constitution and Behavior

Efficient experimental strategies are aided by the availability of a variety of techniques associated with pollination, the ease of multiple testing of the genotype of single individuals, a broad foundation of specific biological and genetic information, and extensive knowledge of anomalies in genetic transmission.

While strains differ widely in ear habit and potential, self or crosspollinations of one or more ears on a plant and multiple crosses with single or repeated collections of pollen can be chosen as convenient. Normally only one ear per plant is used, but two ears on one stalk can often be obtained if both ear shoots are well developed and both pollinations are completed with a 24-hour period; pollination of the lower ear a day ahead of the upper ear sometimes gives better results. Tillered plants may produce additional ears receptive over a longer period of time. A single plant can be testcrossed as both pollen and ear parent, in an exactly reciprocal manner if desired. Several manipulative techniques depend on the common practice of cutting back the end of the husks and silks to an even surface the day before pollination, allowing the silks to grow overnight into a convenient brush of receptive strands 1 to 2 cm long. Two or more pollen types can be applied to a single ear while separating the silk brush into segments with a card or other stiff object; ears with sharply defined sections usually result. Relative growth rates of pollen tubes can be differentiated according to the silk length traversed before fertilization (e.g., Kempton, 1936; Mulcahy, 1971). Pollen can be separated according to size by sieving (e.g., Mangelsdorf, 1932). Pollen carrying wx has been shown to be less sensitive than Wx to environmental exposure (Kempton, 1936). Liquid media have been identified in which pollen can be suspended and can be exposed to selective agents in solution before pollination (Coe, 1966a and unpublished), but differential selection or enrichment for specific pollen genotypes in liquids has not been reported. [Schwartz and Osterman (Genetics 83:63-65, 1976) have shown that ADH-negative pollen grains can be recovered from large populations of ADH-positives following exposure of pollen to vapors of allyl alcohol, which alcohol dehydrogenase converts to a toxic product.]

With the manipulability described above, the testing of genetic constitution of single individuals can be very extensive and discriminating. Commonly, "tester" lines are planted solely for pollinations intended to identify the constitution of experimental plants. Emerson (1918) designated tester stocks singly recessive for each of the aleurone color factors a, c, and r. Crosses with a "c tester" strain, which is homozygous recessive for c but dominant for A, $A2$, Bz, $Bz2$, $C2$, and R, will give colorless seeds only when crossed with an unknown carrying c or one of the color inhibitor factors. The progeny from a typical crossed ear consist of up to several hundred kernels, which can be expeditiously scored or observed for segregation ratios, recombinants, mutations, or variations in expression. On an ear with as many as 300 kernels, two proportions differing by as little as 10% can be distinguished with 95% confidence, and genotypes expected as infrequently as one in 64 will be included with 99% confidence (Hanson, 1959). The ease of making testcrosses and of scoring segregating ears makes feasible studies requiring the recognition and classification of ratio aberrancies, such as those caused by preferential segregation with the abnormal neocentric knob K-10 (Rhoades and Dempsey, 1966a), by gametophyte factors (Emerson, 1934), or by Aberrant Ratio (Sprague and McKinney, 1971), as well as the detection of low-frequency mutational events such as the chromosome losses induced by B chromosomes (Rhoades and Dempsey, 1972, 1973a). Some types of experimental work are facilitated by carrying along recessive or otherwise concealed factors, traced entirely by test-crosses, in the strains under study. Tester-based analysis is particularly essential for the effective study of regulatory systems. With repeated crosses to a series of uniform lines that amount to testers, it becomes possible to make exacting discriminations among cytoplasm types affecting male fertility and among the nucleus types interacting with them (Beckett, 1966, 1971).

Information is available on several kinds of biological events that may cause unexpected results or may be used as devices for experimental designs. Heterofertilization, the occurrence of embryos and endosperms of differing genetic constitution (non-correspondence), results from the fertilization of the egg and the polar nuclei by genetically differing sperm from different pollen tubes (Sprague, 1932b). In pollinations with artificially mixed pollen samples or with segregating pollen marked with endosperm and embryo markers it is possible to identify kernels derived from embryo sacs that have been penetrated by at least two pollen tubes and to analyze fertilizations or other events accordingly (e.g., Sarkar and Coe, 1971a; see Fig 11). In recombination analyses involving an endosperm trait with an embryo or plant trait, segregating markers transmitted through the male must be verified by progeny tests whenever heterofertilization might confound the results. Heterofertilization frequencies of 1 to 2% are common, and Sprague (1932b) found up to 25% heterofertilization in some strains. Another type of heterofertilization event, due to the regular nondisjunction of the centromere of B chromosomes in the male gametophyte, is recognizable by noncorrespondence and deficiencies if a section of an A chromosome is attached to the B centromere by translocation (Roman and Ullstrup, 1951). In spontaneous sectors trisomic for chromosome 3, tetraspory (all four products of meiosis contributing to the eight nuclei of the embryo sac) rather than monospory is found, causing noncorrespondence and anomalous classes (Neuffer, 1964; see Fig. 37). Gynogenesis or androgenesis (maternal or paternal haploidy or diploidy) can be recognized in the embryos of dry seeds by genetic marking (Coe and Sarker, 1964; Greenblatt and Bock, 1967; Kermicle, 1969; Nanda and Chase, 1966; see Fig. 11). In crosses that involve a particular strain, Coe's stock 6, the frequency of gynogenetic haploids is sub-

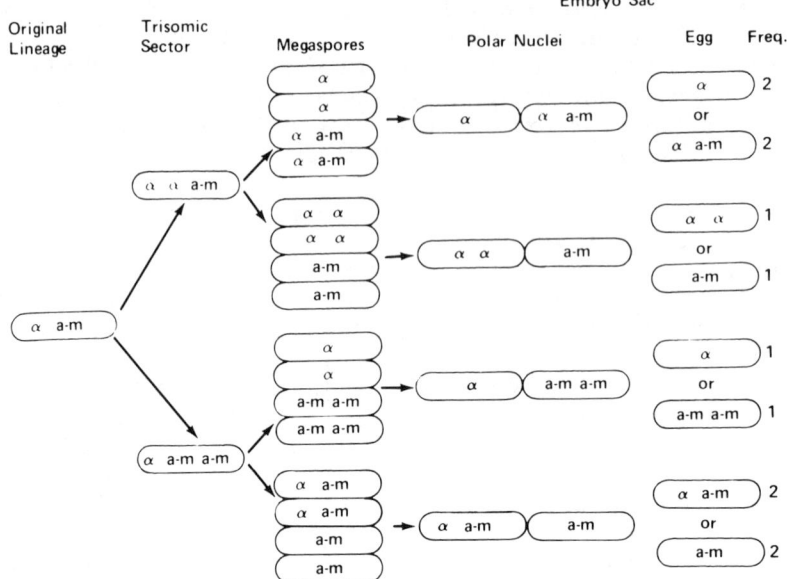

Fig. 37. Genetic consequences of exceptional tetrasporic embryo sac development in sectors trisomic for chromosome 3. Spontaneous trisomy in a heterozygous diploid leads to genetic noncorrespondence between polar nuclei and eggs in both chromosomal and genetic constitutions. Numbers indicate the relative frequencies of the various distributions. (After Neuffer, 1964.)

stantially increased over normal (Sarker and Coe, 1966), while the frequency of androgenetic haploids is greatly increased in seed progeny from plants bearing *ig* (Kermicle, 1969). Monosomy may occur spontaneously or under the influence of *r-x1* (Weber, 1973) and can cause noncorrespondence. Newly occurring polysomy also can cause noncorrespondence and other anomalous classes. Gynogenetic diploids from crosses of 2n × 2n or 2n × 4n appear to be derived by the fusion of meiotic products to form diploid sporophytes rather than by the doubling of reduced products, and thus may not be homozygous (Sarkar and Coe, 1971b, c). Tetraploids arising from crosses of 2n × 4n, however, may be derived by doubling of reduced products to give diploid gametes through the ear parent, while tetraploids arising from 4n × 2n appear to be derived by fusion of meiotic products in the male (Sarkar and Coe, 1971b). Plants homozygous for *el* produce diploid eggs whose origin appears to be from incompletely reduced meiotic products (Rhoades and Dempsey, 1966b). The genetics of autotetraploids, taking into account all known phenomena that affect chromosome behavior and transmission, is detailed by Doyle (1973).

Several aspects of genetically controlled anomalies in transmission are identified in the section on gametophyte formation and development. Transmission can be affected by factors that result in lethality in the ovule *(lo, lo2)* or pollen *(pa; cms-S* with *rf3)*, including most chromosomal deficiencies; by factors that influence pollen grains or tubes differentially *(ga* group; *sks* with *Ms21; sp; sp2; wx)*; by preferential assortment to functional

THE GENETICS OF CORN

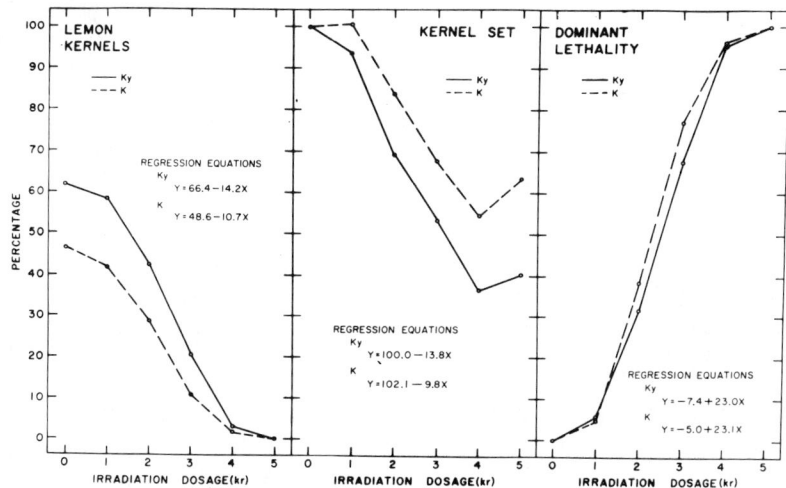

Fig. 38. Effects of increasing levels of gamma irradiation on the function of pollen grains. The pollen of two strains (K and Ky) dominant for lemon endosperm, irradiated and mixed with untreated white endosperm pollen, yields decreasing percentages of lemon kernels (left) and decreasing total kernel set (center), which indicates increasing dominant lethality in the fertilized kernels (right) rather than pollen lethality. (From Pfahler, 1967b.)

megaspores (K-10 and knobs); by differential fertilization (B chromosomes); or by lethality of the zygote (et with modifier; zl). More generally, pollen mixtures show differences in fertilization percentage among strains, depending on the pollen sources and ear parents (Pfahler, 1965, 1967a). Gamma irradiation, however, appears not to alter fertilization ability in competition but results in dominant lethality in the zygotes (Pfahler, 1967b; see Fig. 38). Mulcahy (1971) has tested for differences in average weight of kernel following competition in pollen mixtures, and has found kernel weight to be correlated strongly with faster tube growth. Murakami et al. (1972) have found that pollen of F_1 hybrids succeeds in achieving higher fertilization percentages than the parent lines, in competition with a constant pollen type in mixtures.

Unsuccessful attempts to achieve genetic transformation with preparations containing DNA have been reported (Coe and Sarkar, 1966; Kivilaan and Blaydes, 1974).

Locating Factors to Chromosome and to Map Position

A factor pair that is not yet placed to chromosome can be located in one of the 10 linkage groups by the use of marker stocks for each chromosome. In addition to the conventional method, which can become tedious, a number of efficient special methods have been contrived, including the use of chromosomal aberrations as markers, aberrations associated with convenient gene markers, hemizygosity, transmission defects, and trisomy.

In conventional tests with marker stocks the unplaced factor is made heterozygous with each of a series of markers chosen to survey the linkage map and the F_1 plants are testcrossed (if possible) or self-fertilized. There is no standard set of markers; a common procedure is to use whatever strains bearing contrasting markers may be on hand and the results are often accordingly haphazard. The marker strains should carry two well-separated markers on the same chromosome or segment; usually it is not practical or efficient at this stage to include large numbers of closely linked markers with the intention of locating the factor and mapping it precisely in one cycle. Markers that affect the seed or young seedling are generally preferred because classifications can be conducted in the laboratory months ahead of field plantings and because expenditures of time and materials are substantially lower. Although dominant markers are efficient in that coupling (cis) heterozygotes produce a mathematically more informative array in the F_2 (Hanson, 1959), few suitable dominants are available; most are older-plant traits and many have low productivity combined with variable penetrance. Data that are obtained in testcross or F_2 progenies with markers can be tested for association in the assortment of two factor pairs at a time with the 2×2 chi-square test, which is applied without concern for the linkage phase (cis vs. trans) or for deviations in ratios. If necessary, agreement with the expected linkage phase and with expected individual ratios can be judged by inspection or by appropriate chi-square tests. Any genetic interactions between a marker and the unplaced factor will confound the statistical test and can suggest linkage spuriously. Interactions should be reconciled by development of appropriate expected proportions and testing of observed values against these proportions. Data involving interactions are inefficient for the estimation of recombination percentages, as can be seen in the graphs presented by Allard (1956).

Chromosomal aberrations with easily recognized phenotypic expressions are reliable markers for linkage tests with unplaced factors. Reciprocal translocations (interchanges) are most often applied in this way (see Burnham, 1966), since clear-cut semisterility is characteristic of heterozygous plants for most interchanges. In pollen, semi-sterility (50% empty pollen grains) is recognized easily on a black, opaque surface with a 30× or higher pocket microscope in the field; on the ear, irregular seed set is observed, resulting from the abortion of half the ovules. Homozygous interchanges, almost without exception, have normal fertility that is phenotypically indistinguishable from the standard chromosome constitution, so that the procedure normally followed is to testcross rather than to self-fertilize the F_1 (translocation/mutant) and to analyze the data as a backcross in coupling (cis) for a recessive mutant or in repulsion (trans) for a dominant. Because each interchange has breakpoints (semisterility points) in two chromosomes, overlapping sets of translocations are selected that will narrow the conclusion to a single chromosome; clues to map position are often detected with well-chosen sets. The semi-sterility linkage method is so routine for simple factors that it is not often reported in print; for complex traits, references and a paradigm are given in the report by Scott et al. (1966).

Anderson (1956) and Burnham (1966) detail the use of the gene-marked translocation technique, in which a series of interchanges is selected to include a chromosome bearing a convenient gene marker associated by the translocation with each of the other chromosomes. The common set maintained in the stocks of the Maize Cooperation employs *wx* with 18 selected translocations. Crosses of the mutant strain are made with each of the *wx* translocation strains and the F_1 plants are testcrossed or self-fertilized. Data are tested for association between the segregation for wx and the segregation of the mutant. This method is limited or inadequate if the mutant has an expression that is confounded with that of the gene marker. The method depends upon both the physical linkage of interchanged chromosome segments and the 'pseudolinkage' resulting from selective survival of complete chromosome segments.

Methods for locating a factor to a chromosome that make use of hemizygosity (heterozygosity with a deficiency) are the most rapid and generally the most reliable. A recessive factor can be located through observations in the immediate generation if it is made hemizygous and is "uncovered" (i.e., shows recessive expression). Dosage-dependent factors also can be recognized when hemizygous and can be located in the immediate generation. Dominant factors require a second-generation progeny test, which is usually definitive because of failure in transmission of the deficiency or inviability of the homozygote (i.e., in the critical test, when the dominant factor is hemizygous, no recessives will segregate from the hemizygote). Several procedures that make use of hemizygosity can be applied, including B-A translocations, monosomics, maintainable deficiencies, and induced deficiencies.

The most effective method by which deficiencies of known location and extent can be generated is by the use of B-A translocations (Roman and Ullstrup. 1951), in which the unique behavior of the supernumerary B chromosome results in transmissible sperm nuclei deficient for whatever portion of an arm of the A chromosome lies beyond the breakpoint of the translocation. A B-A translocation uncovering a substantial part of each arm is now available for 17 of the 20 arms of the A complement, with which at least 70% of the known map length can be made 'hypoploid' (hemizygous) at will (J.B. Beckett, personal communication). Locations of each of these translocations are indicated on the accompanying linkage map. The procedure is quite simple, in spite of the underlying intricacies on which it is based. Pollen from plants bearing the desired B-A translocation (homozygous or heterozygous) is crossed onto plants carrying the factor to be located (each pollen parent should also be crossed onto a known recessive tester that will validate the occurrence of deficient sperm). Hypoploids are usually distinct in morphology and range in frequency from about 10% to 45%; in the critical test (the one in which the factor of interest becomes hemizygous) the supposed hemizygotes can be tested for errors, contamination, haploidy, or confounding of expressions.

Plewa and Weber (1973a) have demonstrated recently the use of monosomics to identify chromosomes carrying dosage-dependent factors affecting fatty acid constitution. Monosomics are generated at high frequencies under the effects of the $r-x1$ deficiency (Weber, 1973). Because each monosomic is morphologically distinctive it becomes possible through this system to locate any factor to chromosome by correlating the monosomy with the genetic results of hemizygosity. Viable recessives and dosage-dependent factors will be placed by "uncovering" and others by failure of segregation from the critical monosomic.

A few small, maintainable deficiencies have been identified (Rhoades and Dempsey, 1973b), some as simple deletions and some as products derived from chromosomal aberrations; these can be used to localize factors very precisely in the chromosome, but they are very limited in extent and therefore are limited in usefulness. In addition to the simple deficiencies cited by Rhoades and Dempsey, which uncover *a, sh2, bm, pyd, wd, yg2, c, sh, bz, r,* and *sr2*, deficiencies derived from chromosomal aberrations include ones that uncover *f* and *fl2* (Gopinath and Burnham, 1956), *po* and the organizer and satellite of chromosome 6 (Phillips et al., 1971a), and *ar* (Turcotte, unpublished).

Locating of factors to chromosome and segment often has been done by inducing a series of losses of the dominant allele and determining cytologically at the pachytene stage the particular segments of chromosome that have been lost. This approach was applied initially by McClintock (1931) to locate several factors affecting plant characters. The deficient chromosomes usually are not transmitted through either gametophyte and so are not maintained. The method may seem ineffective for endosperm traits, where the cytology cannot be conducted; however, it is fully effective if the deficiency is induced in the pollen before the two sperm nuclei are separated, so that they correspond in carrying the deficiency to both the endosperm and the embryo: *Bz2, Dt2,* and *Dt3* were located in this fashion (Nuffer, 1955). Generally the other methods described in this section, especially the B-A translocations, are more easily applied than are induced deficiencies because ad hoc cytology is not required. Phenotypically, the frequency of spontaneous or induced loss of a chromosome segment that includes an unlocalized endosperm factor can provide an estimate of the physical distance of the factor from its centromere (Fabergé, 1957).

Defective transmission, especially through the pollen, is characteristic of duplicated or deficient segments. Linkage of the defective transmission effect with factors in the same chromosome but outside the aberrant region permits tests for location of unplaced factors. Rhoades and Dempsey (1973b) point out that map distance from a simple deficiency to a gene determined in this way is as accurate as that obtained by gene-to-gene mapping. More complex deficiencies and duplications will provide less precise mapping data but greater efficiency in identifying linkage, inasmuch as recombination tends to be reduced when synapsis is complex. For example, when a series of tests with B-A translocations fails to uncover an unplaced recessive, the tests can be extended by application of the defective transmission effect through self-fertilization or testcrossing of hypoploid F_1 plants (Roman and Ullstrup, 1951). Factors located in the euploid segment of the chromosome will show sharply deviant ratios due to linkage with the deficient segment.

All of the 10 trisomes of maize are available and are identified rather easily in uniform stocks. Trisomic inheritance for a factor, given information identifying which particular chromosome is trisomic, identifies the chromosome on which the factor is located. The trisomic method was used initially by McClintock and Hill (1931). Little application has been made of other equivalent techniques such as tertiary trisomes, or of telocentrics and isochromosomes, in locating factors to chromosome.

Mapping is relatively straightforward once a factor is located to chromosome, because the linkage groups are extensively established and many factors are identified approximately to cytological positions. Mapping, per se, is conventionally conducted by preparing a hybrid with two, three, or more heterozygously marked points and testcrossing whenever possible, preferably with the homozygous tester as pollen parent in order to avoid problems arising from heterofertilization. Elementary genetics textbook examples of multiple-point testcross data analysis are typical of conventional mapping arithmetic and logic. When simple testcrosses are not available, special procedures such as the product moment method (tables and formulas in Immer, 1930) or maximum likelihood (Allard, 1956) are applied to partial testcross, F_2, F_3, or other data types. For systematic tabulation and reporting of data the concise but complete format used by Emerson et al. (1935) is normally followed.

Developmental Studies

Analysis of patterns and mechanisms of development with the aid of genetic methods or materials has received only moderate attention. In maize, as with any species in which diverse genetic materials are available, there exists substantial potential for the analysis of developmental processes through the use of genetic devices. Approaches that have been applied include morphological-physiological analysis of mutant types, clonal analysis either by radiation-induced events or by genetically elaborated changes, and the study of differential genic interactions or expressions.

Morphological-physiological analysis can be exemplified by the several studies cited previously in the section on plant form. For example, in plants of *Cg* genotype the homologies of specific vegetative and reproductive structures have been deduced from studies of the structural gradations between extreme *corngrass* and normal or near-normal individuals induced by varying physiological treatments (Galinat and Andrew, 1953). Also in *Cg*, Angeles (Stebbins, 1967) was able to relate localized mitotic dynamics to the initiation of leaf primordia in the shoot apex. The anatomical and morphological descriptions of development in the embryo and endosperm by Randolph (1936) and in the vegetative and reproductive parts by Kiesselbach (1949) form the foundations upon which experiments of these kinds can be based.

In plants from seeds irradiated in atomic tests, Anderson et al. (1949) identified clonal sectors bearing chromosomal aberrations in the tassel by classifying for aborted pollen, branch by branch. The average sector was estimated to include about two branches, representing one of the seven or eight cells in the dormant meristem of the embryo that are destined to give rise to the tassel. X-ray induced sectors of loss of dominant *Yg2* in seeds heterozygous for the marker (Stein and Steffensen, 1959a,b) permit discrimination among preformed, predetermined, and undetermined leaves in the dormant embryo, and identification of differential lateral and horizontal divisions and growth depending upon cell positions in the leaf. Similar sectors induced in marked heterozygotes for *yg2* or for *wd* provide the data by which Steffensen (1968; see Fig. 39) has developed a detailed reconstruction of the anatomy and elaboration of the shoot apex.

Clonal sectors occurring during development of the ear can be recognized following mutational changes of *P-VV* to red pericarp (Emerson,

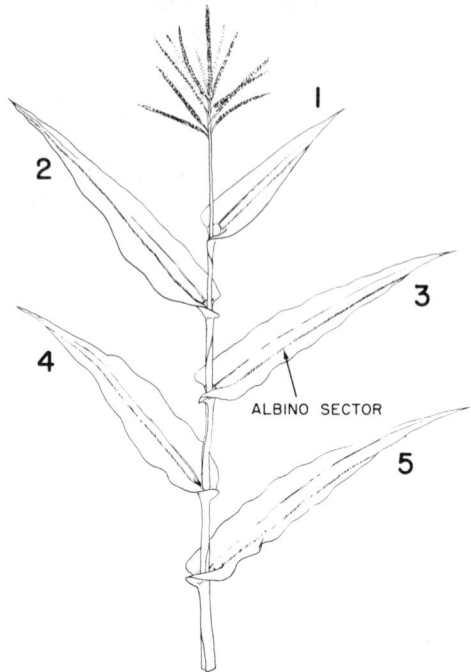

Fig. 39. Developmental relationships revealed by clonal elaboration into sectors. Following irradiation of seeds heterozygous for *wd*, proliferation of the cells of the meristem leads to albino sectors with systematic relationships that reflect the pattern of development of an individual cell of the irradiated embryo meristem. (From Steffensen, 1968.)

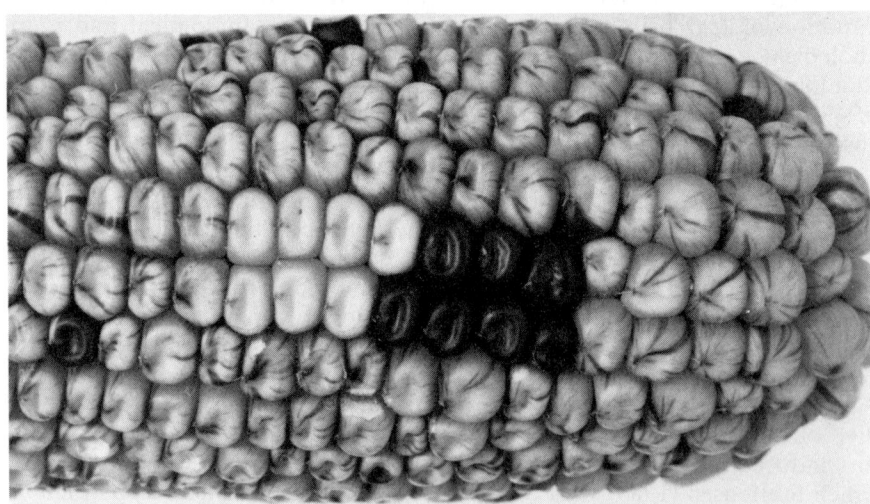

Fig. 40. Contiguous sister clones ("twin spots") in the ear, showing the mirror-image relationship characteristic of the development of sister cells bearing the reciprocal products of a mutational event at *P-VV*. (From Brink and Nilan, 1952.)

THE GENETICS OF CORN

1914). The striking potential of sister clones of cells developing in contiguity following reciprocal mutational events is shown in the study of twin mutations described and pictured by Brink and Nilan (1952; see Fig. 40) and Greenblatt and Brink (1962). Anderson (1923) extended the technical use of ear clonality to seedling traits, planting kernels according to their positions on the ear and generating "ear maps" of the sectorial distribution of a cytoplasmically varying chlorophyll trait (Fig. 35). In the developing plant, genetically determined chlorophyll sectoring (under the influence of the several leaf-striping factors discussed in the section on plant form) generates clonal sectors suited for use in morphogenetic analysis. During development of the endosperm, systematic cell divisions and enlargement lead to conspicuous clonal events in the outer layer (aleurone tissue) following the mutation of a to A under the influence of Dt, and of R-st to R-sc (Fig. 4), among others. Striking figures showing the form of elaboration of the inner endosperm as marked by events occurring with mutable wx are pictured by McClintock (1965; see Fig. 41) and Peterson (1974).

Differential gene action in isozyme formation in different tissues has been implied in a number of studies (e.g., Scandalios and Felder, 1971). The complexities of specific allelic vs. nonallelic control, temporal changes in cell milieu that may effect molecular alterations, and the dynamics of molecular synthesis and breakdown over extended time intervals, however, present as many challenges as insights into developmental processes.

With the aid of factors controlling anthocyanin pigmentation (several alleles at R and B), Styles et al. (1973) have initiated a promising form of multidimensional analysis of gene action in controlling biosynthesis of a visible product, in which they find that particular alleles may differ from

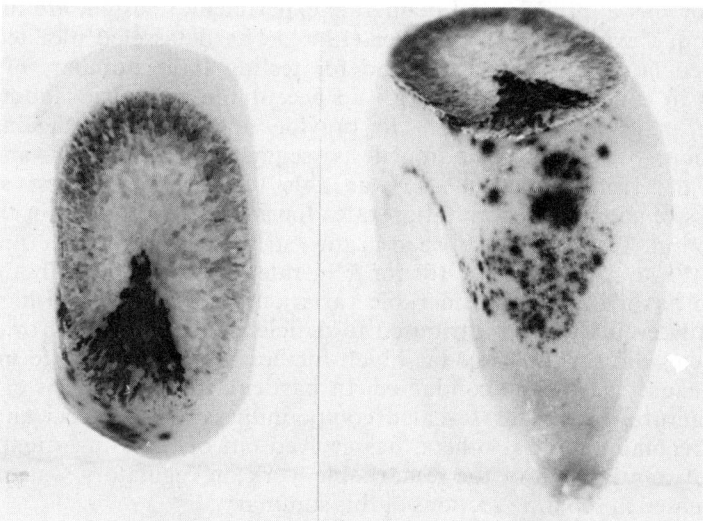

Fig. 41. Developmental patterns in the endosperm revealed by mutational changes proliferated into clones. The above kernel, shown in two views, began its growth with inactive Ac plus wx-$m7$ and a-$m3$ (both of which are mutable and can respond to active Ac). During early development Ac became active in one cell, resulting in frequent reversions to Wx (dark-staining areas in the inner endosperm) and to A (pigmented spots in the aleurone layer). The wedge-like clone characterizes the morphogenetic pattern in the endosperm. (From McClintock, 1965.)

others in their time of onset of activity and its cessation, and in their level of action (Fig. 13).

Mutation Studies

Mutation can be defined in the broader sense as a change in some characteristic of a cell or organism that may be transmitted to succeeding generations. Included among mutational changes are effects ranging from changes in whole genomes to changes in a single nucleotide pair in the DNA molecule; we shall limit our discussion to methods of analysis that involve the gene itself and its relationships with its immediate neighbors on the chromosome. A given study may take any of several approaches, including analysis of spontaneous events or of events induced by mutagenic agents; of events at random loci or at selected loci; of the kind and nature of events or of mutational mechanisms. In the following pages, methods that may be specific to a particular approach are taken up with that approach, while more general procedures and designs that facilitate identification, categorization, and quantitation of events are taken up in the latter part of the section.

Stadler (1946) alluded to the materiality of spontaneous mutation studies in higher organisms as follows: "The study of spontaneous mutation is laborious at best, but it seems an indispensable preliminary to the interpretation of mutations induced experimentally, and it may be the only approach now open for the study of gene evolution." Much effort had already been and would continue to be spent on studying spontaneous events at random loci, but a more productive approach has been to study intensively the mutational behavior at a few selected loci. The special technical suitability of maize for this approach, and deliberate experimental design, are discussed with unique explicitness by Stadler (1946). The detasseled plot technique (described later) provided a method for testing large numbers of female gametes in controlled matings with an acceptable amount of labor. Using multiple endosperm characters to provide adequate tests against contamination, phenocopies, androgenesis, segmental deficiencies, and other sources of error, Stadler (1942) was able to provide the first accurate measures of spontaneous mutation rates for a number of genes in the same background. The rates established in this early work range from 0 in 1.5×10^6 for Wx to 273 in 0.55×10^6 for R (a rate of 492×10^{-6}). Though later research has identified considerable variation at these loci and has shown that many events can be attributed to deficiency or to crossing over, these frequencies did set the stage on which further investigations into mutation and its causes have been conducted. In particular, the diagnosis of genetic fine structure, which has revealed compoundness at some loci and intracistron recombination at others, has evolved out of these investigations, as has a substantial part of the remarkable work on regulatory systems; both are discussed in separate sections of this summary.

Artificially induced mutation was first identified through the studies of Stadler (1928) with x-rays. Losses of specific dominant markers were recognized as mosaic endosperms in kernels of recessive stocks that had been pollinated by dominant strains and x-rayed on the ear at or near the time of fertilization (this work, although some months earlier into print, was concurrent with studies also by Stadler demonstrating that heritable mutations were induced at random loci in barley, *Hordeum vulgare* L.) Irradiation at the fertilization stage is particularly difficult in that it requires either a portable x-ray unit with elaborate shielding, or growth of plants in portable containers that can be transported to the irradiation equipment. Consequently most irradiations are applied to either dormant seeds or pollen unless the specific purpose compels otherwise. For pollen, x-ray or gamma ray dosages in the range of 1,000 to 2,000 r are generally applied either to freshly shed pollen samples (e.g., Amano and Smith, 1965; Mottinger, 1970; Pfahler, 1971) or to flowering tassel branch segments, freshly picked, from which pollen is collected as the anthers open on the morning of treatment (e.g., Nuffer, 1957). Pfahler (1967b) has shown that gamma-irradiated pollen grains (up to 5,000 r) are fully competitive with untreated ones in achieving fertilization but that irradiation leads to failure in kernel development (i.e., dominant lethality), increasingly with increasing dosages (Fig. 38). Although mutagenesis was not a part of their study, House and Nelson (1958) introduced ^{32}P into pollen by injecting a phosphate solution into a well bored in the stem just below the ear-bearing node at the time of first pollen shedding. For dormant dry seeds, dosages are generally applied in the range of 5,000 to 30,000 r (e.g., Pfahler, 1970), while for seeds soaked 24 hours in water or for growing plants, dosages in the range of 1,000 to 3,000r are typically suitable (e.g., Stein and Steffensen, 1959a).

Ultraviolet light was demonstrated to be mutagenic by Stadler and Sprague (1936), who also found that shorter wavelengths were more effective, per unit of energy applied, than longer wavelengths. Stadler and Uber (1942) demonstrated that irradiation of pollen from two sides was substantially more effective than from one side, inasmuch as the constituents of the pollen absorb strongly; because the sperm nuclei are eccentrically placed, more homogeneous treatment results from two-sided than from one-sided irradiation with the same total dose. Treatments accordingly are best done with a two-sided (e.g., Nuffer, 1957) or a tumbling system (e.g., Pfahler, 1973).

Early work with chemical mutagens was limited by the difficulties encountered in attempting to reach the germ cells and in trying to make reliable quantitative tests. Gibson et al. (1950) applied mustard gas vapor to pollen, at saturation or diluted through a measuring and mixing apparatus, and found increasing effects with greater exposure times and concentrations. Kreizinger (1960) induced mutations by application of diepoxybutane to flowering tassels, either cut and held in solution to take up the mutagen through the cut ends or supplied with solution by means of a wick in a hole bored in the stalk of the tassel. Neuffer and Ficsor (1963) induced mutations with ethyl methane sulfonate (EMS) solutions introduced into cotton packed around the maturing tassel 3 to 5 days before flowering. Amano and Smith (1965) induced mutations with EMS by holding pre-soaked seeds in solutions or by immersing cut root ends of young seedlings. The problems of quantitation inherent in all of these early methods have been largely overcome through treatments applied to pollen in paraffin oil. Following the finding by Coe (1966a; MNL 40:108) that pollen could be suspended in paraffin oil for extended time periods before pollination, Neuffer (1968; MNL 45:146) perfected quantitative methods by which EMS or nitrosoguanidine (NG), or potentially any agent, can be applied to pollen. Fresh pollen is mixed with paraffin oil carrying the mutagen in suspension (freshly prepared and thoroughly dispersed by vigorous stirring), stirred intermittently for a measured period, and applied to receptive silks with a small brush. Treatment standards for concentration and exposure time can be established by in vitro germination tests of treated pollen on agar media (Neuffer, 1972; MNL 45:146). For material treated by this method the general mutation rate can be very high; for example, in one study with a control rate of 0.7%, 26% of surviving pollen grains treated with EMS, and 54% of those treated with NG, underwent at least one mutation.

Most severe precautions must be taken to avoid contact with mutagens such as EMS and NG by the investigator, or inadvertent contact of others with treated materials or contaminated objects; especially in oil, these potent mutagens-carcinogens can be particularly dangerous, penetrating, and difficult to sanitize.

Some of the pioneering studies on the nature of radiation-induced mutations have been conducted with corn. Heritable x-ray induced changes in both barley and corn (Stadler, 1930) were initially thought to be changes of the gene from one form to another (gene mutation), but continuing investigation by Stadler and his associates showed that, in corn, x-rays produce only chromosomal aberrations or losses of genetic material. Stadler and Sprague (1936) and Stadler (1939) identified significant differences in the types and frequencies of mutational events produced by ultraviolet light and x-rays. When mature pollen carrying dominant markers is treated and crossed onto a stock carrying recessive markers, x-ray treated material produces ears with many whole-kernel losses and smaller numbers of fractional losses, most of which are larger than 15/16 of the kernel surface (Fig. 42). Ultraviolet light, on the other hand, produces a somewhat smaller number of whole-kernel losses and a larger number of fractionals, which distribute around a mean of one-half kernel in size. These differences were interpreted by Stadler (1939) to indicate that x-rays cause chromosome-type changes while ultraviolet causes more discrete chromatid changes. The very large fractional losses produced by x-rays were suggested by Stadler to represent chromosome breakage events in which the fragment rejoins and regains mitotic distribution after a number of divisions, leading to "recovery spots" in limited portions of the endosperm. In addition, where x-rays produce high frequencies of endosperm and embryo losses, ultraviolet produces only 1/10 as many embryo losses/endosperm loss. This difference may result in part from the different morphogenetic patterns of the two structures, and perhaps in part from differential repair of x-ray and ultraviolet-induced changes in rapidly dividing young endosperm cells vs. the more slowly dividing embryo cells. Stadler and Roman (1948) obtained 415 cases of change of the A locus in material exposed to x-rays. Only two of these approached the standards of behavior that would be required to classify them as changes of A to a (i.e., gene mutation), and each of these (a-$X1$ and a-$X2$), like an obvious deficiency case *(a-X3)* studied in comparison, involved more than one genetic trait, affected crossing over in the region, had altered

THE GENETICS OF CORN

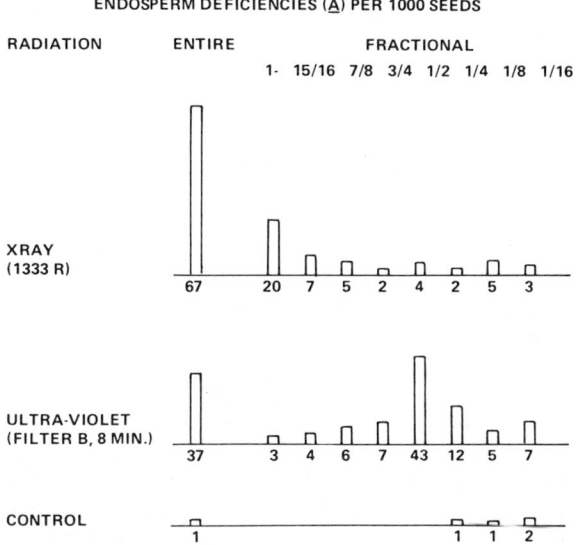

Fig. 42. Comparison of the effects of x-rays and ultraviolet light on the frequencies of entire and fractional losses in the endosperm. (From Stadler, 1939.)

transmission, and was inviable as a homozygote. To determine whether or not x-rays and ultraviolet light could change or remove a marker without affecting its close neighbors, Nuffer (1957) treated pollen carrying three closely linked markers (α β Sh2, all within 0.3 map units) with these two mutagenic agents. Recognition of the critical cases (α Sh2) was facilitated by the inclusion of a-m in the ear parent and Dt in the pollen parent. The absence of a single case of the critical type from x-ray treatment in a population of 8,739 kernels, compared with 10 cases in 8,888 from ultraviolet, indicates that x-rays do not often produce such discrete changes. With EMS, the limited results of a similar study by Neuffer and Ficsor (1963) indicate that this mutagen can cause discrete losses of β alone. Mottinger (1970) has identified mutants of Bz following x-ray treatment which, like the a-X mutants of Stadler and Roman (1948), appear to be small deletions affecting more than the one locus. Seeking the equivalent of back mutations, which might be viewed as instances of discrete change, Stadler (1944) attempted to determine whether x-rays produce dominant mutation of a, the recessive that reverts to A in the presence of Dt but not in its absence. Stadler made crosses to produce kernels of a a A and a a a constitutions (both lacking Dt) on the same ears and treated the ears with x-rays 73 to 81 hours after pollination, a time selected to give loss sectors for A of approximately the same size as the colored dots that occur on a Dt kernels. Treatment with 800 r of x-rays produced A a a kernels with an average of 14 to 34 loss sectors, while the a a a kernels from treatments of 800 and 1,600 r had not a single colored dot in a population that would have had 900,000 losses of A. The same population would have produced 400,000 dots of a to A if Dt had been present in three doses. In a similar experiment using highly mutable alleles, a-m1 of the Dt system and a-m1 of the Spm system (see the section on regulatory elements),

Neuffer (1966) found that neither x-rays nor ultraviolet produces changes from a-m to A directly, but both mutagens induce new Dt or Spm activity, which produces in turn the reversions from a-m to A. The occurrence of new regulatory activity is known to result from chromosome breakage. All of the above evidence leads to the conclusion that, in corn, x-rays do not produce gene mutation at any appreciable frequency in comparison with chromosomal aberrations and gene losses. This is not in agreement with findings in other organisms, where the preponderance of evidence indicates that radiation-induced gene mutations are frequent. This difference has not yet been explained. Ultraviolet light, on the other hand, appears to produce discrete changes.

It is of historical interest to note that Stadler (1939) and Stadler and Uber (1942) reported the first genetic experiments that clearly pointed to nucleic acids as the basic genetic material. Pollen carrying dominant markers was subjected to a series of specific wavelengths of ultraviolet light and crossed onto recessive strains. Taking into account the differential absorption of each wavelength by the wall and contents of the pollen grain and the eccentric position of the nuclei, so as to estimate the effectiveness per unit of energy reaching the nucleus, they determined that the most effective wavelengths for producing mutations and losses are those around 254 to 265 nm. Since this action spectrum corresponds with the absorption spectrum for nucleic acids (Fig. 43), Stadler and Uber cautiously but correctly stated that "nucleic acid is intimately associated with the function of the genetic material."

Among the more convenient general procedures and methods for study of mutation at selected loci is the detasseled plot technique developed as a genetically marked system by Stadler (1946). The method usually consists in planting two to four rows of ear stock carrying the dominant alleles of the loci to be studied alternately with one or two rows of pollen stock carrying the recessive alleles, ordinarily in an isolated field. The ear rows are detasseled and the ears are pollinated naturally by the pollen rows, as in the classical method of hybrid corn production. The ears so produced have kernels that are nominally heterozygous for the genes in question and may be examined for recessive mutants. This method has the advantage of producing large numbers of test kernels (populations of the order of 10^6 are feasible) with relatively modest effort. Since the female gametes are held in place in regular kernel rows with fixed relationships, it is also possible to account directly for problems such as contamination and sectoring.

Quantitation of mutational events by the somatic sector technique has been used by several investigators, including Steffensen (1968) and Ficsor (1965). It consists in preparing seeds heterozygous for genes expressed in the seedling or growing plant, treating the seeds with a mutagenic agent, growing them, and examining the resulting plants for mutant sectors. The size of a sector depends on the stage at which the treatment becomes effective and the extent of growth following mutation; the population tested depends on the number of primordial cells present at the time of treatment. The advantages of this method are that experiments are quickly and easily conducted once the seeds are produced, and that large populations of cells are tested in moderate-scale studies. One disadvantage is that progeny tests of the sectors

THE GENETICS OF CORN

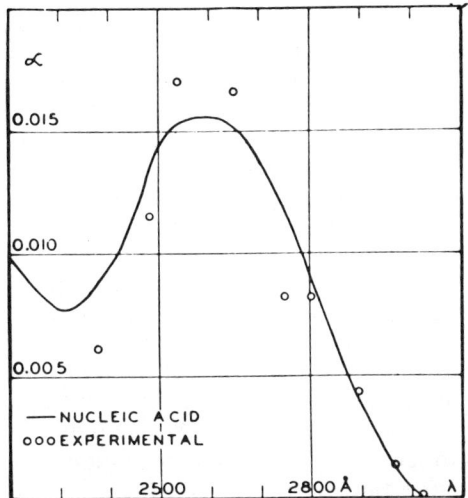

Fig. 43. Comparison of the mutagenic effectiveness of a series of wavelengths of ultraviolet light (open circles) with the absorption spectrum of DNA (smooth curve). (From Stadler and Uber, 1942).

are unobtainable except where they occur in the germ line; even then, a sector may include multiple copies of a single mutational event rather than independently arising mutations.

An extensively used method for studying mutation at random loci in species with perfect flowers involves seed treatment. It involves treating mature seeds with a mutagenic agent, growing an M_1 (first generation from treatment), selfing or outcrossing the M_1 plants, then planting and selfing an M_2 to be classified in M_3 for the segregation of recessive mutants. The advantage of this method is the ease of the initial treatment and the simplicity of subsequent operations. However, it has two major disadvantages for corn: first, it requires extensive testing of the M_2 to save and identify individual mutants, and second, confusion arises about individual mutant identities and about frequencies because treatment gives rise to sectors that may transmit copies of each mutational event. The relative efficiency of different procedures for the extraction of mutational events following seed treatment is considered in the latter part of this section.

Truly large and decisive populations can be obtained by use of pollen as the treated stage. By counting the pollen in individual anthers and multiplying by the number of anthers in a tassel, we have estimated that a modest-sized tassel will generate 15×10^6 pollen grains. With proper treatment and conservative distribution, the pollen from a few tassels can produce very large numbers of offspring. Pollen has additional advantages: 1) The gametophyte is a small, self-contained, and uniform but easily invaded structure; 2) The three nuclei mutate independently, so that one sperm may be affected without impairing the function of the other or of the tube nucleus, thus permitting the transmission of quite deleterious mutants; 3) Contaminations and spontaneous endosperm mutations that occur prior to the treatment time can be recognized because they will affect all three nuclei and will be transmitted as cases having a genetically corresponding endosperm and embryo, while mutations induced in mature pollen will appear in either the endosperm or the embryo, but not in both; 4) Each kernel arising from treated pollen is potentially heterozygous for a unique induced mutant that can be expressed immediately (if dominant) or in the progeny of the M_1 (if recessive). Thus when mature pollen carrying dominant endosperm markers is treated with a mutagenic agent and crossed on ears carrying recessive alleles for the same markers, losses or mutations will be expressed immediately in the endosperm; the embryo will carry similar but independent events that cannot be seen until the next generation. The initial endosperm events will include primary events seen as whole kernel mutants having the recessive phenotype and secondary events expressed as sectors of recessive phenotype. The numbers, sizes, and characteristics of these sectors have been effectively used to analyze the actions of such treatments as x-rays and ultraviolet light (Stadler, 1939; Nuffer, 1957), chromosome breakage (McClintock, 1941), and chemical mutagens (Neuffer and Ficsor, 1963).

One important factor in the design of mutation experiments that is often overlooked or misinterpreted is the specific numerical efficiency of the method used. This can best be seen by comparing the consequences of different handling methods, including seed vs. pollen treatments, sample sizes, and selfing vs. crossing. It has been established (Anderson et al., 1949; Steffensen, 1968) that in the mature seed there are four to eight primordial cells that will develop into male gametes. It has not been clearly established whether the same primordial cells produce the female gametes, in which case the gametes would be mutationally concordant, or whether different primordial cells are involved. The authors have noted, however, that sectors for loss of *A, B,* or *Pl* induced by treatment at the dry seed stage do not include both the tassel and the ear. Treatments applied at the seed stage may affect any one of the primordial cells independently and produce a sector generating half normal and half mutant gametes. If we were to assume eight primordial cells and concordance, the population treated would be 16 genomes times the number of seeds treated. For those mutants having no selective effects, 1/16 of the gametes from an M_1 plant undergoing a mutational event would carry the mutant. The M_1 plants may be either crossed to a standard strain or selfed, with the results indicated in Table 12 and discussed below.

If one treats 100 kernels, each carrying eight diploid primordial germ cells, the resulting M_1 plants will carry 1,600 treated genomes. If the treatment produces mutants in 25% of the genomes (a frequency somewhat higher than is characteristically found for treatment at the seed stage, but used here in order to make equivalent comparisons with pollen treatment), then 400 mutants will have been produced. Crossing the M_1 plants as female by a standard strain will produce 100 ears that will preserve all the mutants. The frequency with which the gametes and kernels will carry a particular mutant will be 1 in 16. For the recessive mutants, a planting of these seeds with a subsequent self-fertilization will be required to express any recessive mutants. According to Hanson (1959), a sample of 47 individuals is required in order to have 95% certainty of obtaining one individual occurring at a frequency of 1 in 16. Therefore, to detect 95% (380) of the 400 mutants in the 100 ears will require planting 47 kernels from each ear, a total of 4,700 plants to be selfed. Thus, an input of 4,800 plants (100 M_1 plus 4,700 M_2) will produce 380

Table 12—Comparison of the efficiency of different methods of treatment and handling for mutagenesis

Treatment and handling	M_1 plants	Treated genomes	Mutants produced at 25% effectiveness	M_2 selfed	Mutants detected	Plants grown $M_1 + M_2$	Efficiency (mutants/plant)
Seed; M_1 crossed; large M_2	100	1,600†	400	4,700‡	380	4,800	0.08
Seed; M_1 selfed; large M_2							
Concordance	100	1,600	400	2,300	380	2,400	0.16
Nonconcordance	100	3,200†	800	4,700	760	4,800	0.16
Seed; M_1 selfed; minimal M_2							
Concordance	1,000	16,000	4,000	1,000	500	2,000	0.25
Nonconcordance	1,000	32,000†	8,000	1,000	500	2,000	0.25
Pollen; M_1 selfed	2,000	2,000	500	—	500	2,000	0.25

†Assuming eight nonconcordant primordial cells each for the tassel and ear.
‡Required to detect 95% of mutants; 7,200 would be required to detect 99%.

mutants, an efficiency of 0.08 mutant/cultured plant.

Selfing the M_1 plants is more efficient than crossing them by a standard strain. Assuming concordance, a double sample (from the ear and the tassel) would be provided by each selfed M_1 ear, and only 23 M_2 kernels, or 2,300 M_2 plants, would be required to detect 380 mutants; thus an input of 2,400 plants would give an efficiency of 0.16. With nonconcordance (the true situation), a self will provide one test each from two eight-celled primordial sets (ear and tassel) for a total of 3,200 treated genomes and 800 mutants. These will require 4,700 individuals to detect 760 mutants, resulting in an efficiency of 0.16, which is the value for concordance. The tendency to try to save all of the mutants by taking a very large sample from the M_2 ears is counterproductive; for example, a sample of 52 kernels (5,200 plants) would be required to save 99.9% of the mutants (efficiency 0.075). An additional problem arises from maximal sampling, namely the duplication and confounding of mutants. A single mutant event in a primordial cell will be duplicated many times through the cell divisions before gamete formation, and many copies of the same mutant will be produced. For this reason, only one mutant of a particular type can be accepted in the progeny of each M_1 plant as a unique mutant. This is of considerable consequence, inasmuch as recessive alleles at many loci have similar, confoundable phenotypes.

A more efficient sampling method, developed by Rédei (1974), is to treat large numbers of seeds and grow a large M_1 and then take a minimal sample from each M_1 for the M_2 to be selfed. For corn the minimal sample would consist of a single seed from each M_1 ear. Following previous assumptions, treatment of 1,000 seeds will affect 16,000 genomes and produce 1,000 plants that will carry 4,000 mutants. Planting one seed from each M_1 will test 2,000 gametes and save 500 of the mutants. In terms of total input, an investment of 2,000 selfed plants (1,000 M_1 and 1,000 M_2) will yield 500 mutants for an efficiency of 0.25. Furthermore, since only one sample of two gametes is taken from each M_1 plant the problem of duplicate copies is eliminated; if by chance (1 in 256, again assuming concordance) both gametes contribute the same mutant, it would be homozygous immediately and would be recognizable as such. With nonconcordance there are two independent samples with no chance of homozygosity for the single event.

An equally efficient approach with some additional advantages over that just presented is to treat pollen; an investment of less than 100 plants for treatment plus a test by selfing of 2,000 M_1 plants will produce 500 mutants for an efficiency of 0.25 (slightly less because of the plants needed for the original crosses with treated pollen). Pollen treatment has several advantages: 1) Each mutant seen is an independent event, so that except for normal attrition all mutants produced are saved — this allows easy comparison of mutation rates and of relative frequencies of different mutant types; 2) Variations among different lines or between the sexes in primordial cell number are not a concern in the estimation and comparison of mutation rates; 3) Dominant mutants are easily recognized as such and are ready for immediate testing.

Analysis of Genetic Fine Structure

The occurrence of high rates of "mutation" that could not be explained in terms of established modes of inheritance engendered an early interest in detailed structural analysis at A and R; some other loci have been similarly explored, either in their own right genetically or out of morphological, evolutionary, or biochemical interest. Although cytological proof is lacking, the genetic evidence at both A and R points overwhelmingly to the presence of varying numbers of segments (tandem repeats) that are related but distinguishable in function and are capable of synapsis and crossing over inter se. The primary observations supporting this compoundness are: 1) The regular occurrence of stepwise events (inferring that two or more units are "mutating" independently); 2) The association of events in homozygotes with recombination between flanking markers (inferring that synapsis and crossing over are occurring between differentiated components); and 3) The demonstrable experimental reconstruction of compound forms. However, compound structure should not necessarily be assumed for a locus that has not been studied with the same thoroughness; some other loci that have been studied appear to be composed of single functional units. The operations and devices that have been employed thus far in studies of genetic fine structure include some that are common to any such study and some that are cunningly unique.

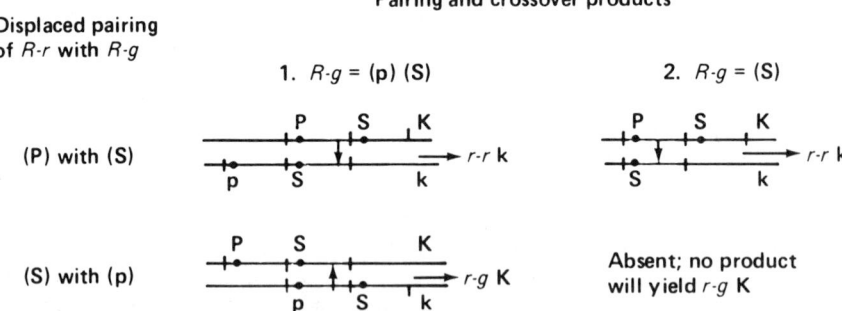

Fig. 44. Determination of the presence (1) or absence (2) of a synaptic element (p) in *R-g*. For an *R-g* carrying (p), the two types of displaced pairing will yield both *r-r* k and *r-g* K by crossing over. For an *R-g* lacking the element, only *r-r* k products will be observed. (After Emmerling, 1958.)

Laughnan (1949) pointed out that stepwise events at the A locus, in which frequent changes occur from A-b (strong color) to A-d (dilute) and from A-d to a (colorless), but not from A-b directly to a, suggest that A-b consists of two independently changing units differing in functional capacity. Similar patterns of events at the R locus (Stadler, 1951), changes from R-r to R-g or r-r but not directly to r-g (Fig. 10), bear the same implications. For A-b Laughnan designated two components, (α) and (β); Stadler designated (P) and (S) components for R-r. Following theoretical development of a restrictive set of expected events, Emmerling (1958) was able to distinguish experimentally two classes of R-g derivatives from R-r (Fig. 44), the first able to yield colorless (r-g) by crossing over in heterozygotes with R-r [inferring that synapsis could occur between an inactive plant color unit (p) in R-g and the (S) unit of R-r], and the second unable to yield such derivatives [inferring that no remnant plant color element synaptically homologous to (S) is present]. Laughnan (1961b) tested in a similar way for the presence of a synaptically effective but functionally inactive unit (beta-null) in A-d without identifying this type. Ashman (1965) found stepwise events involving R-st, for which the elements (Sc), (I-R), and (Nc) have been designated (Kermicle, 1974), and Sastry (1970) has postulated on similar grounds that R-ch alleles contain four or more independently mutating components.

Marking of the region under study to reveal association of "mutational" events with recombination of external markers has been a standard practice in each effective study of fine structure. Recombination occurring largely or entirely in a single direction is found for A-d events from A-b/a (Laughnan, 1949) and A-b/A-b:P (Laughnan, 1955); for r-r events from R-r/R-g (Stadler and Nuffer, 1953); for Wx events from wx-$90/wx$-C (Nelson, 1962, 1968); for r-g, r-nc (near colorless) and compound events from R-st/R-r (Ashman, 1965); and for Gl events from heteroallelic (gl-a/gl-b) hybrids (Salamini and Lorenzoni, 1970). Recombination in either direction is associated with tu-d and tu-l events from Tu/tu (Mangelsdorf and Galinat, 1964). As expected for a compound locus, recombination in either direction is associated with events arising in homozygotes of R-r (Stadler and Emmerling, 1956). From the proportion of r-r to R-g events arising by crossing over in R-r/R-r, Dooner and Kermicle (1971) have devised quantifications of the relative position of

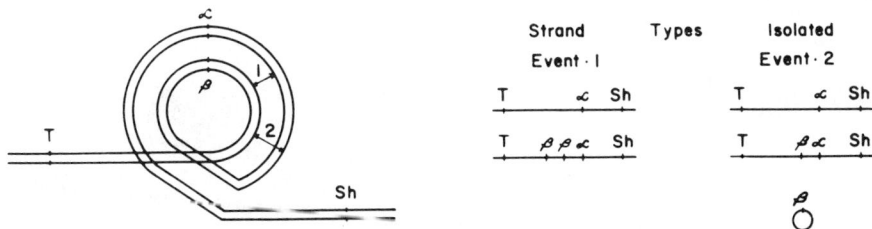

Fig. 45. Diagrammatic representation of auto-association. Left: adjacent, homologous β and α segments of a single chromosome, paired with each other at meiosis to form a double loop. Right: strands isolated following exchange event 1 (involving sister chromatids) and exchange event 2 (involving a single chromatid). Both events yield a "noncrossover" α strand; however, the first event produces a complementary strand carrying two β members and one α, while the second event yields a parental-type strand carrying β and α plus an acentric ring representing the β member. (From Laughnan, 1961b).

the sites differentiating (P) from (S) in their respective duplicated segments, of the frequency of oblique synapsis, and of the genetic length of the duplications.

Nonrecombinant events are at least as informative as recombinant events. Laughnan (1952, 1961a) determined not only that some A-d events arise without crossing over from A-b/a but, further, that A-d events arise from A-b heterozygous with a carried in an inversion, or even from hemizygous A-b, in each case at a higher frequency than from A-b/a. These observations among others point to the occurrence of auto-association (synapsis of tandem repeats within the same strand) and crossing over, to yield losses or gains in numbers of units, as the basis for "noncrossover" events at A (Laughnan, 1961b; see Fig. 45). The occurrence of a class of noncrossover R-g events from R-r, in which no synaptic unit homologous to (S)

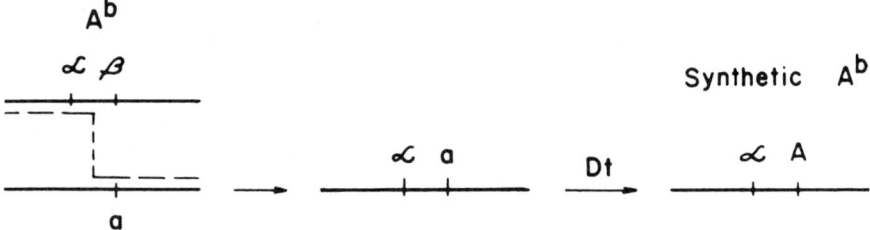

Fig. 46. Steps in reconstruction of a synthetic A-b. The compound αa is first derived by crossing over and then mutates to the synthetic A-b (αA) by the action of Dt. (From Laughnan, 1961b.)

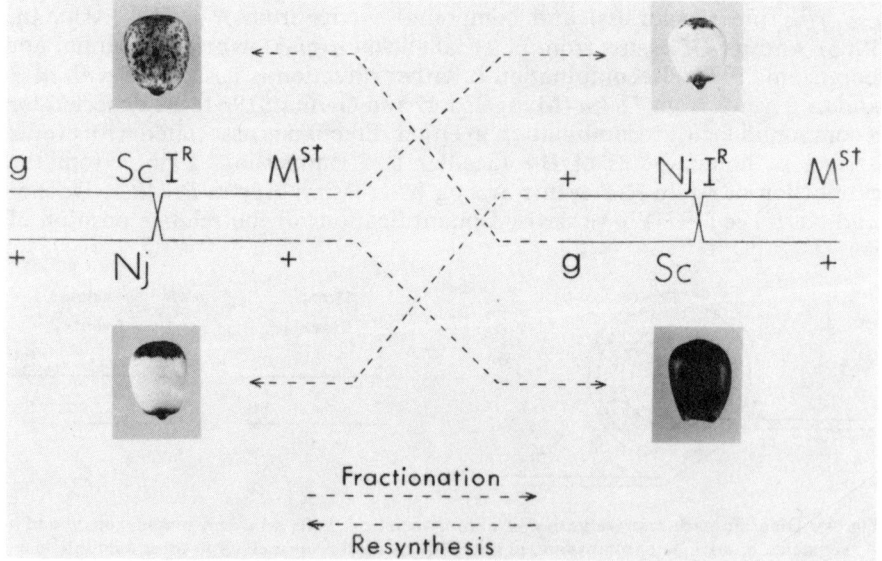

Fig. 47. Fractionization of R-st, formation of a new compound, and subsequent resynthesis of R-st. The reciprocal products from crossing over between R-st and R-nj (left) are the component Sc (R-sc) and a new compound consisting of the elements Nj and IR (R-nj;st), which yield a resynthesized R-st by crossing over. (From Kermicle, 1974.)

seems to be present, may very well be due to the same mechanism (Emmerling, 1958). Kermicle (1970a) reports higher than normal frequencies of *R-sc* events from *R-st* hemizygotes. Bianchi and Tomassini (1965) studied the frequency of *Wx* events in plants hemizygous for *wx* and found a barely significant increase in events, which appear to include substantial numbers of premeiotic events that may spuriously infer auto-association.

Experimental construction or reconstruction of compounds by prediction is among the more impressive evidences supporting the existence of compoundness. Occurrence of a new compound with the properties of both the (α) component and a (response to *Dt*) was found by Laughnan (1952) to be associated with crossing over in *A-b/a* hybrids. The reconstruction of a synthetic *A-b* from (α)*a* by *Dt*-induced mutation of *a* to *A* (Fig. 46), in which *A* is the equivalent of (β), produced a compound that once again yielded (α) products in association with crossing over (Laughnan, 1961b). Mangelsdorf and Galinat (1964) reconstructed *Tu* from its products, *tu-d* and *tu-l*. Coe (1964) attempted unsuccessfully to construct compounds of *C-I* with *C* and suggested that either compoundness or synaptic homology of compound parts was lacking. Reconstruction of *R-st* from its products was predicted and verified by Ashman (1970). A new compound in which the mutability element (I-R), originally in *R-st* in cis relationship with a seed color component, becomes associated with a plant color component and causes it to

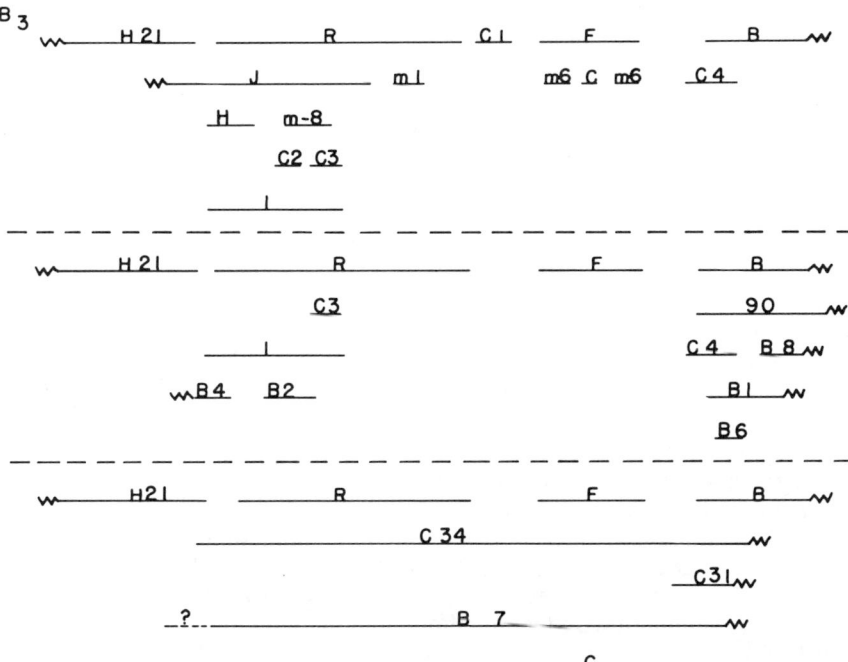

Fig. 48. Location of alleles within the *wx* locus. The three horizontal divisions of the figure depict the locations of alleles that have been tested in crosses among themselves. If the lines representing two mutants overlap, the mutants do not recombine; if the lines do not overlap, the mutants do recombine. The serration terminating some of lines indicates that for that particular mutant there is no mutant to the left (or right) with which it recombines, leaving no means of delimiting the mutant on that side. (From Nelson, 1968.)

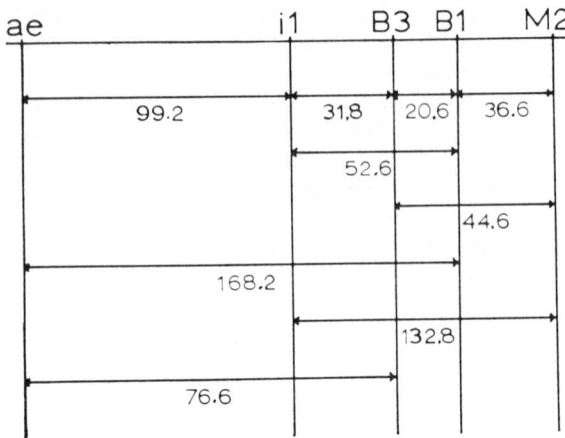

Fig. 49. Map of five mutations at the *ae* locus. The map distances were determined from the frequencies of wild type pollen grains produced by plants heterozygous for two alleles at a time. The numbers are frequencies per 10^5 pollen grains, doubled to account for reciprocal events. (From Moore and Creech, 1972.)

become mutable, has been shown to arise in association with crossing over (Kermicle, 1974). Construction through crossing over of a new compound, *R-nj:st*, that in turn yields predicted products by crossing over (Kermicle, 1970a, 1974; see Fig. 47) adds a dimension to the analysis of the diversity of components at the *R* locus.

Comparative rates of events in themselves provide data by which the order of sites and the relative distances between them can be specified. At the *Wx* locus, Nelson (1959, 1962, 1968) has mapped 24 alleles (Fig. 48) by determining whether *Wx* products arising from heteroallelic hybrids (*wx-a/wx-b*) exceed those arising from homoallelic constitutions; the data by which this mapping could be done were obtained by observations in large populations of pollen grains. Yu and Peterson (1973) identified alterations in the rates of such events with changes in distance from the centromere to the *Wx* locus. Pollen data have also been used to map five alleles at the *Ae* locus (Moore and Creech, 1972; see Fig. 49). As indicated previously, Dooner and Kermicle (1971) have mapped the relative position of the (S) and (P) elements in duplicated segments by quantifying the proportions of different classes of events at the *R* locus.

Association of events with the stage of meiosis is one of the major clues to recombinational rather than mutational causes. Laughnan (1952) has reasoned that the occurrence of noncrossover *A-d* events singly rather than in clusters indicates a meiotic process rather than a mutational one. Emmerling (1958) reported one large sector of *r-r* on an ear of *R-r/R-r*, but the vast majority of events at the *R* locus are isolated (i.e., meiotic). In studies of the distributions of events in tassels of homozygous and hemizygous *wx* plants, Bianchi and Tomassini (1965) found clustering, indicating that the events in these constitutions frequently arise through mutation. In order to study the fine structure of the mutable allele *R-st*, Kermicle (1970a) concentrated on events influenced by crossing over through narrowing consideration of events to those occurring in or near the meiotic phase.

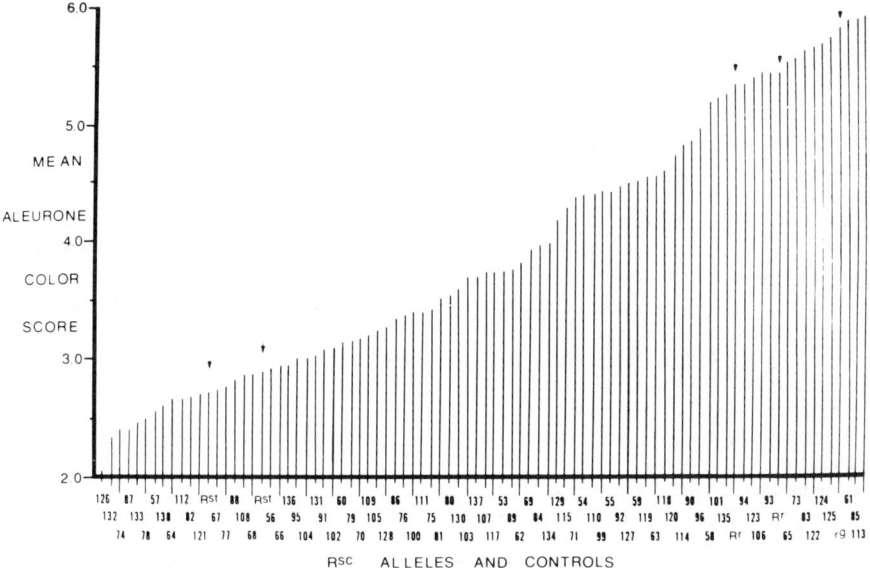

Fig. 50. Continuous variation in paramutagenic effect of a series of *R-sc* mutants derived from *R-st*. Each numbered line represents the average color score of a standard *R-r* exposed to a particular *R-sc* and then testcrossed to *r-g*. The lines marked with arrows represent controls (standard *R-r* exposed to *R-st*, *R-r* and *r-g*). (From McWhirter and Brink, 1962.)

Quantitative variations in the degree of paramutagenicity and paramutability, analyzed by recombination studies, support the concept that multiple inserted elements are responsible (Brink, 1973). McWhirter and Brink (1962) found that *R-sc* events derived from *R-st* vary continuously in paramutagenicity from essentially none up to the level of *R-st* (Fig. 50); recombination analysis reveals that paramutagenic material is distributed unevenly in the *R* region around the (I-R) element and that events occurring with crossing over affect the paramutagenic material differently from event to event (Kermicle, 1974). Bray and Brink (1966) have shown that the frequencies of events arising from various *R-r* and *R-g* alleles following paramutation are reduced, suggesting that the materials responsible for the paramutant state may interfere with crossing over.

ACKNOWLEDGMENTS

Cooperative investigations of ARS-USDA and the Missouri Agricultural Experiment Station. Journal series No. 7183. We have been fortunate to have the generous help of J.B. Beckett, R.J. Lambert, E.B. Patterson, and M.M. Rhoades in studying and critiquing the manuscript in its entirety. Selected parts were reviewed by W.C. Galinat, J.L. Kermicle, J.R. Laughnan, O.E. Nelson, N.H. Nickerson, D.S. Robertson, and M.S. Zuber, whose suggestions have been invaluable. Many of the photographs and drawings were made available through the kind consent (and frequently the prints) of the respective authors cited with the figures. We appreciate the spirited aid of Karen Sheridan in suggesting editorial and technical revisions and of Sheila McCormick for systematic attention to the references and many other tasks; Paul Bolen, Kenneth Leto, and Marion Murray also provided suggestions, comments, and revisions. The care, patience, and good cheer of Maxine Roberts and Genevieve Monroe in carrying the demanding typing load is gratefully appreciated.

LITERATURE CITED

1. Allard, R.W. 1956. Formulas and tables to facilitate the calculation of recombination values in heredity. Hilgardia 24:235-278.
2. Amano, E., and H.H. Smith. 1965. Mutations induced by ethyl methanesulfonate in maize. Mutat. Res. 2:344-351.
3. Anderson, E.G. 1923. Maternal inheritance of chlorophyll in maize. Bot. Gaz. 76:411-418.
4. ----. 1924. Pericarp studies in maize. II. The allelomorphism of a series of factors for pericarp colors. Genetics 9:442-453.
5. ----. 1956. The application of chromosomal techniques to maize improvement. Brookhaven Symp. Biol. 9:23-36.
6. ----, A.E. Longley, C.H. Li, and K.L. Retherford, 1949. Hereditary effects produced in maize by radiations from the Bikini atomic bomb. I. Studies on seedlings and pollen of the exposed generation. Genetics 34:639-646.
7. Anderson, I.C., and D.S. Robertson. 1960. Role of carotenoids in protecting chlorophyll from photodestruction. Plant Physiol. 35:531-534.
8. Ashman, R.B. 1960. Stippled aleurone in maize. Genetics 45:19-34.
9. ----. 1965. Mutants from maize plants heterozygous $R^r\ R^{st}$ and their association with crossing over. Genetics 51:305-312.
10. ----. 1970. The compound structure of the R^{st} allele in maize. Genetics 64:239-245.
11. Ayonoadu, U., and H. Rees. 1968. The influence of B-chromosomes on chiasma frequencies in Black Mexican sweet corn. Genetica 39:75-81.
12. Bachmann, M.D., D.S. Robertson, and C.C. Bowen. 1969. Thylakoid anomalies in relation to grana structure in pigment-deficient mutants of *Zea mays*. J. Ultrastruct. Res. 28:435-451.
13. ----, ----, ----, and I.C. Anderson. 1967. Chloroplast development in pigment deficient mutants of maize. I. Structural anomalies in plastids of allelic mutants at the w_3 locus. J. Ultrastruct. Res. 21:41-60.
14. ----, ----, ----, and ----. 1973. Chloroplast ultrastructure in pigment-deficient mutants of *Zea mays* under reduced light. J. Ultrastruct. Res. 45:384-406.
15. Baker, R.L., and D.T. Morgan, Jr. 1969. Control of pairing in maize and meiotic interchromosomal effects of deficiencies in chromosome 1. Genetics 61:91-106.
16. Barclay, P.C., and R.A. Brink. 1954. The relation between modulator and activator in maize. Proc. Nat. Acad. Sci. USA. 40:1118-1126.
17. Beadle, G.W. 1932. Genes in maize for pollen sterility. Genetics 17:413-431.
18. Beckett, J.B. 1966. Inheritance of partial male fertility in maize in the presence of Texas sterile cytoplasm. Crop Sci. 6:183-184.
19. ----. 1971. Classification of male-sterile cytoplasms in maize *(Zea mays* L.) Crop Sci. 11:724-727.
20. Beckman, L., J.G. Scandalios, and J.L. Brewbaker. 1964a. Catalase hybrid enzymes in maize. Science 146:1174-1175.
21. ----, ----, and ----. 1964b. Genetics of leucine aminopeptidase isozymes in maize. Genetics 50:899-904.
22. Bell, W.D., L. Bogorad, and W.J. McIlrath. 1958. Response of the yellow-stripe maize mutant (ys_1) to ferrous and ferric iron. Bot. Gaz. 120:36-39.
23. Bianchi, A., and G. Marchesi. 1960. The surface of the leaf in normal and glossy maize seedlings. Z. Vererbungsl. 91:214-219.
24. ----, M.R. Parlavecchio, and F. Restaino. 1969. Behaviour and linkage relationship of gametophyte factors in chromosome 9 of maize. Genet. Agrar. 22:345-361.
25. ----, and C. Tomassini. 1965. Reversion frequency of *waxy* pollen type in normal and hypoploid maize plants. Mutat. Res. 2:352-365.
26. Bray, R.A., and R.A. Brink. 1966. Mutation and paramutation at the R locus in maize. Genetics 54:137-149.

27. Brian, P.W., J.F. Grove, and T.P.C. Mulholland. 1967. Relationships between structure and growth-promoting activity of the gibberellins and some allied compounds, in four test systems. Phytochemistry 6:1475-1499.
28. Brink, R.A. 1956. A genetic change associated with the R locus in maize which is directed and potentially reversible. Genetics 41:872-889.
29. ----. 1958. Paramutation at the R locus in maize. Cold Spring Harbor Symp. Quant. Biol. 23:379-391.
30. ----. 1973. Paramutation. Annu. Rev. Genet. 7:129-152.
31. ----, and D.C. Cooper. 1947. Effect of the de_{17} allele on development of the maize caryposis, Genetics 32:350-368.
32. ----, and I.M. Greenblatt, 1954. Diffuse, a pattern gene in *Zea mays.* J. Hered. 45:46-50.
33. ----, J.L. Kermicle and N.K. Ziebur. 1970. Derepression in the female gametophyte in relation to paramutant R expression in maize endosperms, embryos, and seedlings. Genetics 66:87-96.
34. ----, and R.A. Nilan. 1952. The relation between light variegated and medium variegated pericarp in maize. Genetics 37:519-544.
35. ----, E.D. Styles, and J.D. Axtell. 1968. Paramutation: directed genetic change. Science 159:161-170.
36. Brown, A.H.D. 1971. Isozyme variation under selection in *Zea mays.* Nature 232:570-571.
37. ----, and R.W. Allard. 1969a. Inheritance of isozyme differences among the inbred parents of a reciprocal recurrent selection population of maize. Crop. Sci. 9:72-75.
38. ----, and ----. 1969b. Further isozyme differences among the inbred parents of a reciprocal recurrent selection population of maize. Crop Sci. 9:643-644.
39. Brown, R.P., R.G. Creech, and L.J. Johnson. 1971. Genetic control of starch granule morphology and physical structure in developing maize endosperms. Crop Sci. 11:297-302.
40. Buchert, J.G. 1961. The stage of the genome-plasmon interaction in the restoration of fertility to cytoplasmically pollen-sterile maize. Proc. Nat. Acad. Sci. USA 47:1436-1440.
41. Burnham, C.R. 1936. Differential fertilization in the *Bt-Pr* linkage group of maize. J.Am. Soc. Agron. 28:968-975.
42. Burnham, C.R. 1941. Cytogenetic studies of a case of pollen abortion in maize. Genetics 26:460-468.
43. ----. 1966. Cytogenetics in plant improvement. p. 139-187 *In* K.J. Frey (ed.) Plant breeding. Iowa State Univ. Press, Ames.
44. Calub, A.G., B.J. Long, and G.M. Dunn. 1974. Production of inhibitory compounds in corn inbreds with monogenic and multigenic resistance to *Helminthosporium turcicum.* Crop Sci. 14:303-304.
45. Calvert, O.H., M.S. Zuber, and M.G. Neuffer. 1969. Vivipary in *Zea mays* induced by *Diplodia maydis.* Phytopathology 59:239-240.
46. Carlson, W.R. 1973. Instability of the maize B chromosome. Theor. Appl. Genet. 43:147-150.
47. Chao, S.E., and J.G. Scandalios. 1969. Identification and genetic control of starch-degrading enzymes in maize endosperm. Biochem. Genet. 3:537-547.
48. ----, and ----. 1971. Alpha-amylase of maize: Differential allelic expression at the *Amy-1* gene locus, and some physiochemical properties of the isozymes. Genetics 69:47-61.
49. ----, and ----. 1972. Developmentally dependent expression of tissue specific amylases in maize. Mol. Gen. Genet. 115:1-9.
50. Chen, Shu-Mei Hsu. 1973. Anthocyanins and their control by the C locus in maize. Ph. D. Thesis, Univ. Missouri, Columbia, Diss. Abst. 35(2):883 B. Microfilm No. 74-18497.
51. Chilton, M., and B.J. McCarthy. 1973. DNA from maize with and without B chromosomes: a comparative study. Genetics 74:605-614.
52. Chollet, R., and W.L. Ogren. 1972. Greening in a virescent mutant of maize. II. Enzyme studies. Z. Pflanzenphysiol. 68:45-54.
53. ----, amd D.J. Paolillo, Jr. 1972. Greening in a virescent mutant of maize. I. Pigment, ultrastructural, and gas exchange studies. Z. Pflanzenphysiol. 68:30-44.

54. Chourey, P.S. 1971. Interallelic complementation at the Sh_1 locus in maize. Genetics 68:435-442.
55. ----, and D. Schwartz. 1971. Ethyl methanesulfonate-induced mutations of the Sh_1 protein in maize. Mutat. Res. 12:151-157.
56. Clark, F.J. 1940. Cytogenetic studies of divergent meiotic spindle formation in Zea mays. Am. J. Bot. 27:547-559.
57. Coe, E.H., Jr. 1957. Anthocyanin synthesis in maize — A gene sequence construction. Am. Nat. 91:381-385.
58. ----. 1959. A regular and continuing conversion-type phenomenon at the B locus in maize. Proc. Natl. Acad. Sci. USA. 45:828-832.
59. ----. 1962. Spontaneous mutation of the aleurone color inhibitor in maize. Genetics 47:779-783.
60. ----. 1964. Compound versus bifunctional nature of the C locus in maize. Genetics 50:571-578.
61. ----. 1966a. Liquid media suitable for suspending maize pollen before pollination. Proc. Mo. Acad. Sci. 3:7-8.
62. ----. 1966b. The properties, origin, and mechanism of conversion-type inheritance at the B locus in maize. Genetics 53:1035-1063.
63. ----, and K.R. Sarkar. 1964. The detection of haploids in maize. J. Hered. 55:231-233.
64. ----, and ----. 1966. Preparation of nuclear acids and a genetic transformation attempt in maize. Crop. Sci. 6:432-435.
65. Couture, R.M., D.G. Routley, and G.M. Dunn. 1971. Role of cyclic hydroxamic acids in monogenic resistance of maize to *Helminthosporium turcicum*. Physiol. Plant Pathol. 1:515-521.
66. Cox, D.K. 1966. An investigation of aberrant transmission associated with the etched locus in maize, Ph.D. Thesis, Univ. Illinois at Urbana-Champaign. Diss. Abst. 26(12):6986. Microfilm No. 66-04160.
67. Cox, E.L., and D.B. Dickinson. 1971. Identification of maize seedling mutants lacking starch accumulation capacity. Biochem. Genet. 5:15-25.
68. Creech, R.G. 1968. Carbohydrate synthesis in maize. Adv. Agron. 20:275-322.
69. Dahlstrom, D.E., and J.H. Lonnquist. 1964. A new allele at the sugary-1 locus in maize. J. Hered. 55:242-246.
70. de la Roche, I.A., D.E. Alexander, and E.J. Weber. 1971. Inheritance of oleic and linoleic acids in Zea mays L. Crop Sci. 11:856-859.
71. Demerec, M. 1924. Genetic relations of five factor pairs for virescent seedlings in maize. Cornell Univ. Agric. Exp. Stn. Mem. 84.
72. ----. 1927. A second case of maternal inheritance of chlorophyll in maize. Bot. Gaz. 84:139-tants and double mutants of maize. Phytochemistry 7:1435-1436.
73. Doerschug, E.B. 1973. Studies of dotted, a regulatory element in maize. I. Inductions of dotted by chromatid breaks. II. Phase variation of dotted. Theor. Appl. Genet. 43:182-189.
74. Dooner, H.K., and J.L. Kermicle. 1971. Structure of the R^r tandem duplication in maize. Genetics 67:427-436.
75. Doyle, G.G. 1973. Autotetraploid gene segregation. Theor. Appl. Genet. 43:139-146.
76. Duvick, D.N. 1965. Cytoplasmic pollen sterility in corn. Adv. Genet. 13:1-56.
77. ----. 1972. Potential usefulness of new cytoplasmic male steriles and sterility systems. Proc. 27th Annu. Corn Sorghum Res. Conf., Am. Seed Trade Assoc.
78. East, E.M., and H.K. Hayes. 1911. Inheritance in maize. Conn. Agric. Exp. Stn. Bull. 167.
79. Efron, Y. 1970. Tissue specific variation in the isozyme pattern of the AP_1 acid phosphatase in maize. Genetics 65:575-583.
80. Emerson, R.A. 1914. The inheritance of a recurring somatic variation in variegated ears of maize. Am. Nat. 48:87-115.
81. ----. 1918. A fifth pair of factors, A a, for aleurone color in maize, and its relation to the C c and R r pairs. Cornell Univ. Agric. Exp. Stn. Mem. 16.
82. ----. 1921. The genetic relations of plant colors in maize. Cornell Univ. Agric. Exp. Stn. Mem. 39.

83. ----. 1934. Relation of the differential fertilization genes, *Ga ga,* to certain other genes of the *Su-Tu* linkage group of maize. Genetics 19:137-156.
84. ----, G.W. Beadle, and A.C. Fraser. 1935. A summary of linkage studies in maize. Cornell Univ. Agric. Exp. Stn. Mem. 180.
85. Emmerling, M.H. 1958. An analysis of intragenic and extragenic mutations of the plant color component of the R^r gene complex in *Zea mays.* Cold Spring Harbor Symp. Quant. Biol. 23:393-407.
86. Eyster, W.H. 1933. Plastid studies in genetic types of maize: argentia chlorophyll. Plant Physiol. 8:105-121.
87. Fabergé, A.C. 1957. The possibility of forecasting the relative rate of induced loss for endosperm markers in maize. Genetics 42:454-472.
88. Faludi-Daniel, A., F. Lang, A. Nagy, and B. Faludi. 1967. The inheritance of carotenoid types in maize. Acta Agron. Acad. Sci. Hung. 16:1-6.
89. Felder, M.R., and J.G. Scandalios. 1971. Effects of homozygosity and heterozygosity on certain properties of genetically defined electrophoretic variants of alcohol dehydrogenase isozymes in maize. Mol. Gen. Genet. 111:317-326.
90. ----, ----, and E.H. Liu. 1973. Purification and partial characterization of two genetically defined alcohol dehydrogenase isozymes in maize. Biochim. Biophys. Acta 318:149-159.
91. Ficsor, G. 1965. Chemical mutagenesis in *Zea mays.* Ph. D. Thesis, Univ. Missouri, Columbia. Diss. Abst. 27(6B):1728. Microfilm No. 65-14482.
92. Fincham, J.R.S., and G.R.K. Sastry. 1974. Controlling elements in maize. Annu. Rev. Genet. 8:15-50.
93. Fischer, M., and D. Schwartz. 1973. Dissociation and reassociation of maize alcohol dehydrogenase: allelic differences in requirement for zinc. Mol. Gen. Genet. 127:33-38.
94. Flora, L.F., and R.C. Wiley. 1972. Effect of various endosperm mutants on oil content and fatty-acid composition of whole kernel corn (*Zea mays* L.). J. Am. Soc. Hort. Sci. 97:604-607.
95. Fogel, S. 1946. The allelic variability and action of the gene *R* in maize. Ph.D. Thesis, Univ. Missouri, Columbia. Diss. Abst. 9(2):94. Microfilm No. 00-01276.
96. Freeling, M., and D. Schwartz. 1973. Genetic relationships between the multiple alcohol dehydrogenases of maize. Biochem. Genet. 8:27-36.
97. Galinat, W.C. 1971a. The origin of maize. Annu. Rev. Genet. 5:447-478.
98. ----. 1971b. The evolution of sweet corn. Mass. Agric. Exp. Stn. Res. Bull. 591:1-20.
99. ----, and R.H. Andrew. 1953. Stolon-producing corn. Agron. J. 45:122-123.
100. ----, and A.W. Naylor. 1951. Relation of photoperiod to inflorescence proliferation in *Zea mays* L. Am. J. Bot. 38:38-47.
101. Garwood, D.L., and R.G. Creech. 1972. Kernel phenotypes of *Zea mays* L. genotypes possessing one to four mutated genes. Crop Sci. 12:119-121.
102. Gee, M.S., O.E. Nelson, and J. Kuc. 1968. Abnormal lignins produced by the *brown-midrib* mutants of maize. II. Comparative studies on normal and *brown-midrib-1* dimethylformamide lignins. Arch. Bioch. Biophys. 123:403-408.
103. Gelinas, D.A., and S.N. Postlethwait. 1969. IAA oxidase inhibitors from normal and mutant maize plants. Plant Physiol. 44:1553-1559.
104. ----, ----, and L.F. Bauman. 1966. Developmental studies in the *Zea mays* mutant clumped tassel *(Ct).* Am. J. Bot. 53:615.
105. ----, ----, and O.E. Nelson. 1969. Characterization of development in maize through the use of mutants. II. The abnormal growth conditioned by the knotted mutant. Am. J. Bot. 56:671-678.
106. Gengenbach, B., D.E. Koeppe and R.J. Miller. 1973. A comparison of mitochondria isolated from male-sterile and nonsterile cytoplasm etiolated corn seedlings. Physiol. Plant. 29:103-107.
107. Gibson, P.B., R.A. Brink, and M.A. Stahmann. 1950. The mutagenic action of mustard gas on *Zea mays.* J. Hered. 41:232-238.
108. Giesbrecht, J. 1965. A second zebra-necrotic gene in maize. J. Hered. 56:118, 130.

109. Glover, D.V. 1970. Location of a gene in maize conditioning a reduced plant stature. Crop Sci. 10:611-612.
110. Gopinath, D.M., and C.R. Burnham. 1956. A cytogenetic study in maize of deficiency-duplication produced by crossing interchanges involving the same chromosomes. Genetics 41:382-395.
111. Greenblatt, I.M. 1968a. The mechanism of modulator transposition in maize. Genetics 58:585-597.
112. ----. 1968b. Paramutation at the diffuse locus in maize. Proc. XII Int. Cong. Genet. 1:254.
113. ----, and M. Bock. 1967. A commercially desirable procedure for detection of monoploids in maize. J. Hered. 58:9-13.
114. ----, and R.A. Brink. 1962. Twin mutations in medium variegated pericarp maize. Genetics 47:489-501.
115. Hamill, D.E., and J.L. Brewbaker. 1969. Isoenzyme polymorphism in flowering plants. IV. The peroxidase isoenzymes of maize (*Zea mays*). Physiol. Plant. 22:945-958.
116. Hanson, W.D. 1959. Minimum family sizes for the planning of genetic experiments. Agron. J. 51:711-715.
117. Harborne, J.B., and G. Gavazzi. 1969. Effect of *Pr* and *pr* alleles on anthocyanin biosynthesis in *Zea mays*. Phytochemistry 8:999-1001.
118. Harris, J.W. 1968. Isozymes of the E_4 esterase in maize. Genetics 60:186-187.
119. Hayes, H.K. 1932. Heritable characters in maize. XLIII. Zebra seedlings. J. Hered. 23:415-419.
120. ----, and H.E. Brewbaker. 1928. Glossy seedlings in maize. Am. Nat. 62:228-235.
121. Hertel, R., R.K. de la Fuente, and A.C. Leopold. 1969. Geotropism and the lateral transport of auxin in the corn mutant amylomaize. Planta 88:204-214.
122. Hooker, A.L., R.R. Nelson, and H.M. Hilu. 1965. Avirulence of *Helminthosporium turcicum* on monogenic resistant corn. Phytopathology 55:462-463.
123. Horovitz, S., and A.H. Marchioni. 1942. Herencia de la resistencia a la langosta en el maiz "Amargo." Anales Inst. Fitotecnico Santa Catalina 2 (1940): 27-52. (Publ. No. 23.)
124. Horváth, G., J. Kissimon, and A. Faludi-Dániel. 1972. Effect of light intensity on the formation of carotenoids in normal and mutant maize leaves. Phytochemistry 11:183-187.
125. House, L.R., and O.E. Nelson, Jr. 1958. Tracer study of pollen-tube growth in cross-sterile maize. J. Hered. 49:18-21.
126. Hsu-Chen, S.M., and E.H. Coe, Jr. 1973. C_1 locus control of anthocyanin synthesis in maize. Genetics 74:s120-121.
127. Hutchinson, C.B. 1922. The linkage of certain aleurone and endosperm factors in maize, and their relation to other linkage groups. Cornell Univ. Agric. Exp. Stn. Mem. 60:1421-1473.
128. Immer, F.R. 1930. Formulae and tables for calculating linkage intensities. Genetics 15:81-98.
129. Jimenez T., J.R., and O.E. Nelson. 1965. A new fourth chromosome gametophyte locus in maize. J. Hered. 56:259-263.
130. Kempton, J.H. 1921. A brachytic variation in maize. USDA Bull. 925.
131. ----. 1936. Modification of a Mendelian ratio in maize by pollen treatments. J. Agric. Res. 52:81-121.
132. Kermicle, J.L. 1969. Androgenesis conditioned by a mutation in maize. Science 166:1422-1424.
133. ----. 1970a. Somatic and meiotic instability of *R*-stippled, an aleurone spotting factor in maize. Genetics 64:247-258.
134. ----. 1970b. Dependence of the *R*-mottled aleurone phenotype in maize on mode of sexual transmission. Genetics 66:69-85.
135. ----. 1971. Pleiotropic effects on seed development of the indeterminate gametophyte gene in maize. Am. J. Bot. 58:1-7.
136. ----. 1974. Organization of the paramutational components of the *R* locus in maize. Brookhaven Symp. Biol. 25:262-280.

137. Keisselbach, T.A. 1949. The structure and reproduction of corn. Nebr. Agric. Exp. Stn. Res. Bull. 161.
138. Kirby, L.T., and E.D. Styles. 1970. Flavonoids associated with specific gene action in maize aleurone, and the role of light in substitution for the action of a gene. Can. J. Genet. Cytol. 12:934-940.
139. Kivilaan, A., and D.F. Blaydes, 1974. Attempts to achieve genetic transformation in plants. Mich. State Univ. Agric. Exp. Stn. Res. Rep. 246:1-5.
140. Kleese, R.A., and R.L. Phillips. 1972. Electrophoretic mutants as useful markers for chromosome aberrations. Genetics 72:537-540.
141. Kreizinger, J.D. 1960. Diepoxybutane as a chemical mutagen in *Zea mays*. Genetics 45:143-154.
142. Kuc, J., and O.E. Nelson. 1964. The abnormal lignins produced by the *brown-midrib* mutants of maize. I. The *brown-midrib-1* mutant. Arch. Biochem. Biophys 105:103-113.
143. ----, ----, and P. Flanagan. 1968. Degradation of abnormal lignins in the *brown-midrib* mutants and double mutants of maize. Phytochemistry 7:1435-1436.
144. Larson, R.L., and E.H. Coe, Jr. 1968. Enzymatic action of the Bz_1 anthocyanin factor in maize. Proc. XII Int. Cong. Genet. 1:131.
145. Laughnan, J.R. 1948. The action of allelic forms of the gene A in maize. I. Studies of variability, dosage and dominance relations. The divergent character of the series. Genetics 33:488-517.
146. ----. 1949. The action of allelic forms of the gene A in maize. II. The relation of crossing over to mutation of A^b. Proc. Natl. Acad. Sci. USA 35:167-178.
147. ----. 1950. The action of allelic forms of the gene A in maize. III. Studies on the occurence of isoquercitrin in brown and purple plants and its lack of identity with the brown pigments. Proc. Natl. Acad. Sci. USA 36:312-318.
148. ----. 1951. Reaction sequence in anthocyanin synthesis in maize. Genetics 36:559-560.
149. ----. 1952. The action of allelic forms of the gene A in maize. IV. On the compound nature of A^b and the occurrence and action of its A^d derivatives. Genetics 37:375-395.
150. ----. 1953. The effect of the $sh2$ factor on carbohydrate reserves in the mature endosperm of maize. Genetics 38:485-499.
151. ----. 1955. Structural and functional aspects of the A^b complexes in maize. I. Evidence for structural and functional variability among complexes of different geographic origin. Proc. Natl. Acad. Sci. USA 41:78-84.
152. ----. 1961a. An evaluation of the multiple crossover and sidechain hypotheses based on an analysis of alpha derivatives from the A^b-P complexes in maize. Genetics 46:1347-1372.
153. ----. 1961b. The nature of mutation in terms of gene and chromosome changes. p. 3-29 *In* Mutation and plant breeding, N.A.S.-N.R.C. Publ. 891.
154. ----, and S.J. Gabay. 1973a. Mutations leading to nuclear restoration of fertility in S male-sterile cytoplasm in maize. Theor. Appl. Genet. 43:109-116.
155. ----, and ----. 1973b. Reaction of germinating maize pollen to *Helminthosporium maydis* pathotoxins. Crop Sci. 13:681-684.
156. Leng, E.R., and L.F. Bauman. 1955. Expression of the "Kys" type of male sterility in strains of corn with normal cytoplasm. Agron. J. 47:189-191.
157. Levings, C.S., III, and C.W. Stuber. 1971. A maize gene controlling silk browning in response to wounding. Genetics 69:491-498.
158. Lin, B.Y. 1974. TB-10 breakpoints and marker genes on the long arm of chromosome 10. Maize Genet. Coop. Newsl. 48:182-184. (Cited by permission.)
159. Liu, H.Z., and H.L. Everett. 1965. An analytic study of the w-3 genetic lesion in *Zea mays* L. Plant Physiol. 40:433-436.
160. Longley, A.E. 1961. A gametophyte factor on chromosome 5 of corn. Genetics 46:641-647.
161. Longo, G.P., and J.G. Scandalios. 1969. Nuclear gene control of mitochondrial malic dehydrogenase in maize. Proc. Natl. Acad. Sci. USA 62:104-111.
162. Lowe, J., and O.E. Nelson, Jr. 1946. Miniature seed — a study in the development of a defective caryopsis in maize. Genetics 31:525-533.

163. MacDonald, T., and J.L Brewbaker. 1972. Isoenzyme polymorphism in flowering plants. VIII. Genetic control and dimeric nature of transaminase hybrid maize isoenzymes. J. Hered. 63:11-14.
164. ----, and ----. 1974. Isoenzyme polymorphism in flowering plants. IX. The E_5-E_{10} esterase loci of maize. J. Hered 65:37:42.
165. Madjolelo, S.D.P., C.O. Grogan, and P.A. Sarvella. 1966. Morphological expression of genetic male sterility in maize (Zea mays L). Crop Sci. 6:379-380.
166. Mains, E.B. 1949. Heritable characters in maize. Linkage of a factor for shrunken endosperm with the a_1 factor for aleurone color. J. Hered. 40:21-24.
167. Mangelsdorf, P.C. 1932. Mechanical separation of gametes in maize. J. Hered. 23:288-295.
168. ----, and G.S. Fraps. 1931. A direct quantitative relationship between vitamin A in corn and the number of genes for yellow pigmentation. Science 73:241-242.
169. ----, and W.C. Galinat. 1964. The tunicate locus in maize dissected and reconstituted. Proc. Natl. Acad. Sci. USA 51:147-150.
170. ----, and D.F. Jones. 1926. The expression of Mendelian factors in the gametophyte of maize. Genetics 11:423-455.
171. Marshall, D.R., P. Broue, and A.J. Pryor. 1973. Adaptive significance of alcohol dehydrogenase isozymes in maize. Nature (London) New Biol. 244:16-17.
172. McClintock, B. 1931. Cytological observations of deficiencies involving known genes, translocations and an inversion in Zea mays. Mo. Agric. Exp. Stn. Res. Bull. 163.
173. ----. 1938. The production of homozygous deficient tissues with mutant characteristics by means of the aberrant mitotic behavior of ring-shaped chromosomes. Genetics 23:315-376.
174. ----. 1941. The association of mutants with homozygous deficiencies in Zea mays. Genetics 26:542-571.
175. ----. 1944. The relation of homozygous deficiencies to mutations and allelic series in maize. Genetics 29:478-502.
176. ----. 1950. The origin and behavior of mutable loci in maize. Proc. Natl. Acad. Sci. USA 36:344-355.
177. ----. 1951. Chromosome organization and genic expression. Cold Spring Harbor Symp. Quant. Biol. 16:13-47.
178. ----. 1956. Controlling elements and the gene. Cold Spring Harbor Symp. Quant. Biol. 21:197-216.
179. ----. 1965. The control of gene action in maize. Brookhaven Symp. Biol. 18:162-184.
180. ----. 1968. Genetic systems regulating gene expression during development. Devel. Biol. Suppl. 1:84-112.
181. ----, and H.E. Hill. 1931. The cytological identification of the chromosome associated with the R-G linkage group in Zea mays. Genetics 16:175-190.
182. McWhirter, K.S., and R.A. Brink. 1962. Continuous variation in level of paramutation at the R locus in maize. Genetics 47:1053-1074.
183. Melville, J.C., and J.G. Scandalios. 1972. Maize endopeptidase: genetic control, chemical characterization, and relationship to an endogenous trypsin inhibitor. Biochem. Genet. 7:15-31.
184. Mericle, L.W. 1950. The developmental genetics of the Rg mutant in maize. Am. J. Bot. 37:100-116.
185. Mertz, E.T., L.S. Bates, and O.E. Nelson. 1964. Mutant gene that changes protein composition and increases lysine content of maize endosperm. Science 145:279-280.
186. Mikula, B.C. 1961. Progressive conversion of R-locus expression in maize. Proc. Natl. Acad. Sci. USA 47:566-571.
187. Miles, C.D., and D.J. Daniel. 1974. Chloroplast reactions of photosynthetic mutants in Zea mays. Plant Physiol. 53:589-595.
188. Miller, O.L., Jr. 1963. Cytological studies in asynaptic maize. Genetics 48:1445-1466.
189. Miller, R.J., and D.E. Koeppe. 1971. Southern corn leaf blight: susceptible and resistant mitochondria. Science 173:67-69.

190. Misra, P.S., R. Jambunathan, E.T. Mertz, D.V. Glover, H.M. Barbosa, and K.S. McWhirter. 1972. Endosperm protein synthesis in maize mutants with increased lysine content. Science 176:1425-1427.
191. Moore, C.W., and R.G. Creech. 1972. Genetic fine structure analysis of the *amylose-extender* locus in *Zea mays* L. Genetics 70:611-619.
192. Mottinger, J.P. 1970. The effects of X rays on the bronze and shrunken loci in maize. Genetics 64:259-271.
193. ----. 1973. Unstable mutants of bronze induced by pre-meiotic X-ray treatment in maize. Theor. Appl. Genet. 43:190-195.
194. Mouli, C., and N.K. Notani. 1970. Absence of a detectable change in *Ds* at the A_1 locus in maize following mutagenic treatments. Can. J. Genet. Cytol. 12:436-442.
195. Mulcahy, D.L. 1971. A correlation between gametophytic and sporophytic characteristics in *Zea mays* L. Science 171:1155-1156.
196. Muller, L.D., R.F. Barnes, L.F. Bauman, and V.F. Colenbrander. 1971. Variations in lignin and other structural components of brown midrib mutants of maize. Crop Sci. 11:413-415.
197. Murakami, K.I., M. Yamada, and K. Takayangi. 1972. Selective fertilization in maize, *Zea mays* L. I. Advantage of pollen from F_1 plant in selective fertilization. Jpn. J. Breed. 22:203-208. (Japanese with English tables and summary).
198. Nanda, D.K., and S.S. Chase. 1966. An embryo marker for detecting monoploids of maize *(Zea mays* L.). Crop Sci. 6:213-215.
199. National Academy of Sciences. 1972. Genetic vulnerability of major crops. Washington, D.C.
200. Nelson, O.E. 1959. Intracistron recombination in the *Wx/wx* region in maize. Science 130:794-795.
201. ----. 1962. The waxy locus in maize. I. Intralocus recombination frequency estimates by pollen and by conventional analyses. Genetics 47:737-742.
202. ----. 1967. Biochemical genetics of higher plants. Annu. Rev. Genet. 1:245-268.
203. ----. 1968. The *waxy* locus in maize. II. The location of the controlling element alleles. Genetics 60:507-524.
204. ----, and B. Burr. 1973. Biochemical genetics of higher plants. Annu. Rev. Plant Physiol. 24:493-518.
205. ----, and G.B. Clary. 1952. Genic control of semi-sterility in maize. J. Hered. 43:205-210.
206. ----. E.T. Mertz, and L.S. Bates. 1965. Second mutant gene affecting the amino acid pattern of maize endosperm proteins. Science 150:1469-1470.
207. ----, and A.J. Ohlrogge. 1957. Differential responses to population pressures by normal and dwarf lines of maize. Science 125:1200.
208. ----, and ----. 1961. Effect of heterosis on the response of *compact* strains of maize to population pressures. Agron. J. 53:208-209.
209. ----, and A.J. Ullstrup. 1964. Resistance to leaf spot in maize. Genetic control of resistance to race I of *Helminthosporium carbonum* Ull. J. Hered. 55:194-199.
210. Neuffer, M.G. 1963. Transposition of mutability among components of a compound allele of the A_1 locus in maize. Proc. XI Int. Cong. Genet. 1:44-45.
211. ----. 1964. Tetrasporic embryo-sac formation in trisomic sectors of maize. Science 144:874-876.
212. ----. 1966. Stability of the suppressor element in two mutator systems at the A_1 locus in maize. Genetics 53:541-549.
213. ----. 1968. Chemical mutagens effective on maize pollen. Proc. XII Int. Cong. Genet. 1:118.
214. ----. 1972. *In vitro* germination of pollen as a measure of effectiveness of chemical mutagens in maize. Am. Soc. Agron. Abstr. (1972):26.
215. ----. 1973. Induced dominant disease mimics in maize. Genetics 74:s194.
216. ----, and E.H. Coe, Jr. 1974. Corn (maize). p. 3-30 *In* R.C. King (ed.) Handbook of genetics. Vol. 2. Plenum, New York.

217. ----, and ----. 1976. Linkage map and annotated list of genetic markers in maize. p. 833-847 In G.D. Pasman (ed.) Handbook of biochemistry and molecular biology, 3rd ed. CRC Press, Cleveland.
218. ----, and G. Ficsor. 1963. Mutagenic action of ethyl methanesulfonate in maize. Science 139:1296-1297.
219. ----, L. Jones and M.S. Zuber. 1968. The mutants of maize. Crop Sci. Soc. Am., Madison, Wis.
220. Nickerson, N.H. 1960a. Studies involving sustained treatment of maize with gibberellic acid. II: Responses of plants carrying certain tassel-modifying genes. Ann. Mo. Bot. Gard. 47:243-261.
221. ----. 1960b. Sustained treatment with gibberellic acid of maize plants carrying one of the dominant genes Teopod and Corn-grass. Am. J. Bot. 47:809-815.
222. ----. 1962. Studies involving sustained treatment of maize with gibberellic acid. III: Responses of *Zea mays* plants carrying the gene "lazy." Am. Midl. Nat. 67:125-131.
223. ----, and E.E. Dale. 1955. Tassel modifications in *Zea mays*. Ann. Mo. Bot. Gard. 42:195-212.
224. ----, and W.M. Hewitson. 1963. Morphological and anatomical differences in maize induced by treatment with growth-regulating substances. Am. J. Bot. 50:622.
225. Nuffer, M.G. 1955. Dosage effect of multiple *Dt* loci on mutation of *a* in the maize endosperm. Science 121:399-400.
226. ----. 1957. Additional evidence on the effect of X-ray and ultraviolet radiation on mutation in maize. Genetics 42:273-282.
227. ----, 1961. Mutational studies at the A_1 locus in maize. I. A mutable allele controlled by *Dt* Genetics 46:625-640.
228. Orton, E.R. 1966. Frequency of reconstitution of the variegated pericarp allele in maize. Genetics 53:17-25.
229. Ozbun, J.L., J.S. Hawker, E. Greenburg, C. Lammel, and J. Preiss. 1973. Starch synthetase, phosphorylase, ADPglucose pyrophosphorylase, and UDPglucose pyrophosphorylase in developing maize kernels. Plant Physiol. 51:1-5.
230. Palmer, R.G. 1971. Cytological studies of ameiotic and normal maize with reference to premeiotic pairing. Chromosoma 35:233-246.
231. Peterson, P.A. 1953. A mutable pale green locus in maize. Genetics 38:682-683.
232. ----. 1960. The pale green mutable system in maize. Genetics 45:115-133.
233. ----. 1963. Influence of mutable genes on induction of instability in maize. Proc. Iowa Acad. Sci. 70:129-134.
234. ----. 1965. A relationship between the *Spm* and *En* control systems in maize. Am. Nat. 99:391-398.
235. ----. 1966. Phase variation of regulatory elements in maize. Genetics 54:249-266.
236. ----. 1968. The origin of an unstable locus in maize. Genetics 59:391-398.
237. ----. 1970. The *En* mutable system in maize. III. Transposition associated with mutational events. Theor. Appl. Genet. 40:367-377.
238. ----. 1974. Unstable genetic loci as a probe in morphogenesis. Brookhaven Symp. Biol. 25:244-261.
239. Pfahler, P.L. 1965. Fertilization ability of maize pollen grains. I. Pollen sources. Genetics 52:513-520.
240. ----. 1967a. Fertilization ability of maize pollen grains. II. Pollen genotype, female sporophyte and pollen storage interactions. Genetics 57:513-521.
241. ----. 1967b. Fertilization ability of maize pollen grains. III. Gamma irradiation of mature pollen. Genetics 57:523-530.
242. ----. 1970. Reproductive characteristics of *Zea mays* L. plants produced from gamma-irradiated kernels. Radiat. Bot. 10:329-335.
243. ----. 1971. *In vitro* germination and pollen tube growth of maize (*Zea mays* L.) pollen. V. Gamma irradiation effects. Radiat. Bot. 11:233-237.

244. ----. 1973. *In vitro* germination and pollen tube growth of maize (*Zea mays* L.) pollen-VII. Effects of ultraviolet irradiation. Radiat. Bot. 13:13-18.
245. Phillips, R.L., C.R. Burnham, and E.B. Patterson. 1971a. Advantages of chromosomal interchanges that generate haplo-viable deficiency-duplications. Crop Sci. 11:525-528.
246. ----, R.A. Kleese, and S.S. Wang. 1971b. The nucleolus organizer region of maize (*Zea mays* L.): Chromosomal site of DNA complementary to ribosomal RNA. Chromosoma 36:79-88.
247. Phinney, B.O. 1956. Growth response of single-gene dwarf mutants in maize to gibberellic acid. Proc. Natl. Acad. Sci. USA 42:185-189.
248. ----, and C.A. West. 1960. Gibberellins and the growth of flowering plants. Symp. Soc. Study Dev. Growth 18:71-92.
249. Plewa, M.J., and D.F. Weber. 1973a. The use of monosomics to detect genes conditioning lipid content in *Zea mays* L. embryos. Can. J. Genet. Cytol. 15:313-320.
250. ----, and ----. 1973b. The effect of monosomy on linoleic acid concentration in *Zea mays* embryos. Genetics 74:s214.
251. Poneleit, C.G., and D.E. Alexander. 1965. Inheritance of linoleic and oleic acids in maize. Science 147: 1585-1586.
252. Postlethwait, S.N., and O.E. Nelson, Jr. 1957. A chronically wilted mutant of maize. Am. J. Bot. 44:628-633.
253. ----, and ----. 1964. Characterization of development in maize through the use of mutants. I. The Polytypic (*Pt*) and ramosa-1 (*ra*$_1$) mutants. Am. J. Bot. 51:238-243.
254. Pryor, T., and D. Schwartz. 1973. The genetic control and biochemical modification of catechol oxidase in maize. Genetics 75:75-92.
255. Rakha, F.A., and D.S. Robertson. 1970. A new technique for the production of A-B translocations and their use in genetic analysis. Genetics 65:223-240.
256. Randolph, L.F. 1936. Developmental morphology of the caryopsis in maize. J. Agric. Res. 53:881-916.
257. Reddy, A.R., and G.M. Reddy. 1971. Chemico-genetic studies of leucoanthocyanidin in a maize mutant. Curr. Sci. 40:335-337.
258. Reddy, G.M. 1964. Genetic control of leucoanthocyanidin formation in maize. Genetics 50:485-489.
259. ----, and E.H. Coe, Jr. 1962. Inter-tissue complementation: A simple technique for direct analysis of gene-action sequence. Science 138:149-150.
260. Rédei, G.P. 1974. Economy in mutation experiments. Z. Pflanzenzücht. 73:87-96.
261. Rhoades, M.M. 1931. Cytoplasmic inheritance of male sterility in *Zea mays*. Science 73:340-341.
262. ----. 1933. The cytoplasmic inheritance of male sterility in *Zea mays*. J. Genet. 27:71-93.
263. ----. 1936. The effect of varying gene dosage on aleurone colour in maize. J. Genet. 33:347-354.
264. ----. 1938. Effect of the *Dt* gene on the mutability of the a_1 allele in maize. Genetics 23:377-397.
265. ----. 1941. The genetic control of mutability in maize. Cold Spring Harbor Symp. Quant. Biol. 9:138-144.
266. ----. 1943. Genic induction of an inherited cytoplasmic difference. Proc. Natl. Acad. Sci. USA 29:327-329.
267. ----. 1946. Plastid mutation. Cold Spring Harbor Symp. Quant. Biol. 11:202-207.
268. ----. 1950. Gene induced mutation of a heritable cytoplasmic factor producing male sterility in maize. Proc. Natl. Acad. Sci. USA 36:634-635.
269. ----. 1951. Duplicate genes in maize. Am. Nat. 85:105-110.
270. ----. 1952. The effect of the bronze locus on anthocyanin formation in maize. Am. Nat. 86:105-108.
271. ----, and E. Dempsey. 1966a. The effect of abnormal chromosome 10 on preferential segregation and crossing over in maize. Genetics 53:989-1020.
272. ----, and ----. 1966b. Induction of chromosome doubling at meiosis by the elongate gene in maize. Genetics 54:505-522.

273. ----, and ----. 1972. On the mechanism of chromatin loss induced by the B chromosome of maize. Genetics 71:73-96.
274. ----, and ----. 1973a. Chromatin elimination induced by the B chromosome of maize. I. Mechanism of loss and the pattern of endosperm variegation. J. Hered. 64:12-18.
275. ----, and ----. 1973b. Cytogenetic studies on a transmissible deficiency in chromosome 3 of maize. J. Hered. 64:125-128.
276. ----, and V.H. Rhoades. 1939. Genetic studies with factors in the tenth chromosome in maize. Genetics 24:302-314.
277. Rhoades, V.H. 1935. The location of a gene for disease resistance in maize. Proc. Natl. Acad. Sci. USA 21:243-246.
278. Richardson, L.B., D.S. Robertson, and I.C. Anderson. 1962. Genetic and environmental variation: effect on pigments of selected maize mutants. Science 138:1333-1334.
279. Robertson, D.S. 1955. The genetics of vivipary in maize. Genetics 40:745-760.
280. ----. 1961. Linkage studies of mutants in maize with pigment deficiencies in endosperm and seedling. Genetics 46:649-662.
281. ----. 1971. Investigations of the bipartite structure of the white-albino maize mutants. Mol. Gen. Genet. 112:93-103.
282. ----. 1975. Survey of the albino and white-endosperm mutants of maize: their phenotypes and gene symbols. J. Hered. 66:67-74.
283. ----, and I.C. Anderson. 1961. Temperature-sensitive alleles of the Y_1 locus in maize. J. Hered. 52:53-60.
284. ----, M.D. Bachmann, and I.C. Anderson. 1966. Role of carotenoids in protecting chlorophyll from photodestruction-II. Studies on the effect of four modifiers of the albino cl_1 mutant of maize. Photochem. Photobiol. 5:797-805.
285. Roman, H., and A.J. Ullstrup. 1951. The use of A-B translocations to locate genes in maize. Agron. J. 43:450-454.
286. Salamini, F., and C. Lorenzoni. 1970. Genetical analysis of glossy mutants of maize. III. Intracistron recombination and high negative interference at the gl_1 locus. Mol. Gen. Genet. 108:225-232.
287. Sander, C., L.J. Laber, W.D. Bell, and R.H. Hamilton. 1968. Light sensitivity of plastids and plastid pigments present in the albescent maize mutant. Plant Physiol. 43:693-697.
288. Sando, C.E., R.T. Milner, and M.S. Sherman. 1935. Pigments of the mendelian color types in maize. Chrysanthemin from purple-husked maize. J. Biol. Chem. 109:203-211.
289. Sarkar, K.R., and E.H. Coe, Jr. 1966. A genetic analysis of the origin of maternal haploids in maize. Genetics 54:453-464.
290. ----, and ----. 1971a. Analysis of events leading to heterofertilization in maize. J. Hered. 62:118-120.
291. ----, and ----. 1971b. Anomalous fertilization in diploid-tetraploid crosses in maize. Crop Sci. 11:539-542.
292. ----, and ----. 1971c. Origin of parthenogenetic diploids in maize and its implications for production of homozygous lines. Crop Sci. 11:543-544.
293. Sastry, G.R.K. 1970. Paramutation and mutation of R^{ch} in maize. Theor. Appl. Genet. 40:185-190.
294. Scandalios, J.G. 1965. Subunit dissociation and recombination of catalase isozymes. Proc. Natl. Acad. Sci. USA 53:1035-1040.
295. ----, and M.R. Felder. 1971. Developmental expression of alcohol dehydrogenases in maize. Dev. Biol. 25:641-654.
296. ----, E.H. Liu, and M.A. Campeau. 1972. The effects of intragenic and intergenic complementation on catalase structure and function in maize: a molecular approach to heterosis. Arch. Biochem. Biophys. 153:695-705.
297. Schwartz, D. 1950. The analysis of a case of cross-sterility in maize. Proc. Natl. Acad. Sci. USA 36:719-724.
298. ----. 1951. The interaction of nuclear and cytoplasmic factors in the inheritance of male sterility in maize. Genetics 36:676-696.

299. ----. 1958. A new temperature-sensitive allele at the sticky locus in maize. J. Hered. 49:149-152.
300. ----. 1960a. Genetic studies on mutant enzymes in maize: synthesis of hybrid enzymes by heterozygotes. Proc. Natl. Acad. Sci. USA 46:1210-1215.
301. ----. 1960b. Electrophoretic and immunochemical studies with endosperm proteins of maize mutants. Genetics 45:1419-1427.
302. ----. 1960c. Analysis of a highly mutable gene in maize. A molecular model for gene instability. Genetics 45:1141-1152.
303. ----. 1962. Genetic studies on mutant enzymes in maize. III. Control of gene action in the synthesis of pH 7.5 esterase. Genetics 47:1609-1615.
304. ----. 1964. A second hybrid enzyme in maize. Proc. Natl. Acad. Sci. USA 51:602-605.
305. ----. 1965. Regulation of gene action in maize. Proc. XI Int. Cong. Genet. 2:131-135.
306. ----. 1969. An example of gene fixation resulting from selective advantage in suboptimal conditions. Am. Nat. 103:479-481.
307. ----. 1971a. Subunit interaction of a temperature-sensitive alcohol dehydrogenase mutant in maize. Genetics 67:515-519.
308. ----. 1971b. Dimerization mutants of alcohol dehydrogenase of maize. Proc. Natl. Acad. Sci. USA 68:145-146.
309. ----. 1973. Comparisons of relative activities of maize Adh_1 alleles in heterozygotes — analyses at the protein (CRM) level. Genetics 74:615-617.
310. ----, and T. Endo. 1966. Alcohol dehydrogenase polymorphism in maize — simple and compound loci. Genetics 53:709-715.
311. ----, L. Fuchsman, and K.H. McGrath. 1965. Allelic isozymes of the pH 7.5 esterase in maize. Genetics 52:1265-1268.
312. ----, and W.J. Laughner. 1969. A molecular basis for heterosis. Science 166:626-627.
313. Scott, G.E., and C.M. Campbell. 1969. Internode length in normal and brachytic-2 maize inbreds and single crosses. Crop Sci. 9:293-295.
314. ----, F.F. Dicke, and G.R. Pesho. 1966. Location of genes conditioning resistance in corn to leaf feeding of the European corn borer. Crop Sci. 6:444-446.
315. ----, and C.O. Grogan. 1969. Location of a gene in maize conditioning susceptibility to atrazine. Crop Sci. 9:669-670.
316. Senti, F.R. 1967. High-amylose corn starch: its production, properties, and uses. p. 499-522 In R.L. Whistler and E.F. Paschall (eds.) Starch: Chemistry and technology. Academic Press, New York.
317. Shah, D.M., and C.S. Levings, III. 1973. Characterization of chloroplast and nuclear DNAs of Zea mays L. Crop. Sci. 13:709-713.
318. Shannon, J.C., and R.G. Creech. 1972. Genetics of storage polysaccharides in Zea mays L. Ann. N.Y. Acad. Sci. 210:279-289.
319. Shaver, D.L. 1967. Perennial maize. J. Hered. 58:270-273.
320. Shih, K.L. 1969. Effects of pretreatment with X rays on paramutability or paramutagenicity of certain R alleles in maize. Genetics 61:179-189.
321. ----, and R.A. Brink. 1969. Effects of X-irradiation on aleurone pigmenting potential of standard R^r and a paramutant form of R^r in maize. Genetics 61:167-177.
322. Shortess, D.K., J.E. Wright, and W.D. Bell. 1968. The lutescent mutant in maize: I. Inheritance patterns and environmental effects. Genetics 58:227-235.
323. Shumway, L.K., and L.F. Bauman. 1966. The effect of hot water treatment, x-ray irradiation and mesocotyle grafting on cytoplasmic male sterility of maize. Crop Sci. 6:341-342.
324. ----, and ----. 1967. Nonchromosomal stripe of maize. Genetics 55:33-38.
325. ----, and T.E. Weier. 1967. The chloroplast structure of iojap maize. Am. J. Bot. 54:773-780.
326. Singh, A., and J.R. Laughnan. 1972. Instability of S male-sterile cytoplasm in maize. Genetics 71:607-620.

327. Singleton, W.R. 1940. Influence of female stock on the functioning of small pollen male gametes. Proc. Nat. Acad. Sci. USA 26:102-104.
328. ----. 1946. Inheritance of indeterminate growth in maize. J. Hered. 37:61-64.
329. ----. 1951. Inheritance of corn-grass, a macromutation in maize, and its possible significance as an ancestral type. Am. Nat. 85:81-96.
330. Smith, D.R., and A.L. Hooker. 1973. Monogenic chlorotic-lesion resistance in corn to *Helminthosporium maydis.* Crop Sci. 13:330-331.
331. ----, ----, S.M. Lim, and J.B. Beckett. 1971. Disease reaction of thirty sources of cytoplasmic male-sterile corn to *Helminthosporium maydis* race T. Crop Sci. 11:772-773.
332. Sprague, G.F. 1932a. The inheritance of colored scutellums in maize. USDA Tech. Bull. 292:1-43.
333. ----. 1932b. The nature and extent of hetero-fertilization in maize. Genetics 17:358-368.
334. ----. 1933. Pollen tube establishment and the deficiency of waxy seeds in certain maize crosses. Proc. Natl. Acad. Sci. USA 19:838-841.
335. ----, and H.H. McKinney. 1966. Aberrant ratio: an anomaly in maize associated with virus infection. Genetics 54:1287-1296.
336. ----, and ----. 1971. Further evidence on the genetic behavior of AR in maize. Genetics 67:533-542.
337. Stadler, J.K. 1955. An analysis of instability at the R locus in maize. Ph.D. Thesis, Univ. Missouri, Columbia. Diss. Abst. 16(3): 435. Microfilm No. 00-14972.
338. Stadler, L.J. 1928. Genetic effects of x-rays in maize. Proc. Nat. Acad. Sci. USA 14:69-75.
339. ----. 1930. Some genetic effects of x-rays in plants. J. Hered. 21:3-19.
340. ----. 1933. On the genetic nature of induced mutations in plants. II. A haplo-viable deficiency in maize. Mo. Agric. Exp. Stn. Res. Bull. 204:1-29.
341. ----. 1939. Genetic studies with ultraviolet radiation. Proc. VII Int. Cong. Genet. 269-276.
342. ----. 1942. Some observations on gene variability and spontaneous mutation. Spragg memorial lectures on plant breeding (third series). Michigan State College, East Lansing.
343. ----. 1944. The effect of X-rays upon dominant mutation in maize. Proc. Natl. Acad. Sci. USA 30:123-128.
344. ----. 1946. Spontaneous mutation at the R locus in maize. I. The aleurone-color and plant-color effects. Genetics 31:377-394.
345. ----. 1951. Spontaneous mutation in maize. Cold Spring Harbor Symp. Quant. Biol. 16:49-63.
346. ----, and M.H. Emmerling. 1956. Relation of unequal crossing over to the interdependence of R^r elements (P) and (S). Genetics 41:124-137.
347. ----, and M.G. Nuffer. 1953. Problems of gene structure. II. Separation of R^r elements (S) and (P) by unequal crossing over. Science 117:471-472.
348. ----, and H. Roman. 1948. The effect of X-rays upon mutation of the gene A in maize. Genetics 33:273-303.
349. ----, and G.F. Sprague. 1936. Genetic effects of ultra-violet radiation in maize. Proc. Natl. Acad. Sci. USA 22:572-591.
350. ----, and F.M. Uber. 1942. Genetic effects of ultraviolet radiation in maize. IV. Comparison of monochromatic radiations. Genetics 27:84-118.
351. Stebbins, G.L. 1967. Gene action, mitotic frequency and morphogenesis in higher plants. Dev. Biol. Suppl. 1:113-135.
352. Steffensen, D.M. 1968. A reconstruction of cell development in the shoot apex of maize. Am. J. Bot. 55:354-369.
353. Stein, O.L. 1955. Rates of leaf initiation in two mutants of *Zea mays,* dwarf-1 and brachytic-2. Am. J. Bot. 42:885-892.
354. ----, and D.M. Steffensen. 1959a. The activity of X-rayed apical meristems: a genetic and morphogenetic analysis in *Zea mays.* Z. Vererbungsl. 90:483-502.
355. ----, and ----. 1959b. Radiation-induced genetic markers in the study of leaf growth in *Zea.* Am. J. Bot. 46:485-489.

356. Stevens, B.J. 1970. Nucleolar development in the divergent spindle mutant of *Zea mays.* Cong. Int. Microsc. Electron. C.R. 7th. p. 241-242.
357. Storey, H.H., and A.K. Howland. 1967. Resistance in maize to a third East African race of *Puccinia polysora* Underw. Ann. Appl. Biol. 60:297-303.
358. Stout, J.T., and R.L. Phillips. 1973. Two independently inherited electrophoretic variants of the lysine-rich histones of maize *(Zea mays).* Proc. Natl. Acad. Sci. USA 70:3043-3047.
359. Stroup, D. 1970. Genic induction and maternal transmission of variegation in *Zea mays*. J. Hered. 61:139-141.
360. Styles, E.D., and R.A. Brink. 1966. The metastable nature of paramutable *R* alleles in maize. I. Heritable enhancement in level of standard R^r action. Genetics 54:433-439.
361. ----, and ----. 1969. The metastable nature of paramutable *R* alleles in maize. IV. Parallel enhancement of *R* action in heterozygotes with *r* and in hemizygotes. Genetics 61:801-811.
362. ----, and O. Ceska. 1972. Flavonoid pigments in genetic strains of maize. Phytochemistry 11:3019-3021.
363. ----, ----, and K.T. Seah. 1973. Developmental differences in action of *R* and *B* alleles in maize. Can. J. Genet. Cytol. 15:59-72.
364. Teas, H.J., and E.G. Anderson. 1951. Accumulation of anthranilic acid by a mutant of maize. Proc. Natl. Acad. Sci. USA 37:645-649.
365. Ullstrup, A.J. 1965. Inheritance and linkage of a gene determining resistance in maize to an American race of *Puccinia polysora.* Phytopathology 55:425-428.
366. van Overbeek, J. 1935. The growth hormone and the dwarf type of growth in corn. Proc. Natl. Acad. Sci. USA 21:292-299.
367. ----. 1938. "Laziness" in maize due to abnormal distribution of growth hormone. J. Hered. 29:339-341.
368. Walles, B. 1971. Chromoplast development in a carotenoid mutant of maize. Protoplasma 73:159-175.
369. Ward, E.J. 1973. Nondisjunction: localization of the controlling site in the maize B chromosome. Genetics 73:387-391.
370. Weatherwax, P. 1929. The morphological nature of Teopod corn. J. Hered. 20:325-330.
371. Weaver, S.H., D.V. Glover, and C.Y. Tsai. 1972. Nucleoside diphosphate glucose pyrophosphorylase isoenzymes of developing *normal, brittle-2*, and *shrunken-2* endosperms of *Zea mays* L. Crop Sci. 12:510-514.
372. Weber, D.F. 1973. A test of distributive pairing in *Zea mays* utilizing doubly monosomic plants. Theor. Appl. Genet. 43:167-173.
373. Weyers, W.H. 1961. Expression and stability of the marbled allele in maize. Genetics 46:1061-1067.
374. Wilkinson, D.R., and A.L. Hooker. 1968. Genetics of reaction to *Puccinia sorghi* in ten corn inbred lines from Africa and Europe. Phytopathology 58:605-608.
375. Wilson, G.F., A.M. Rhodes, and D.B. Dickinson. 1973. Some physiological effects of viviparous genes vp_1 and vp_5 on developing maize kernels. Plant Physiol. 52:350-356.
376. Yu, M.H., and P.A. Peterson. 1973. Influence of chromosomal gene position on intragenic recombination in maize. Theor. Appl. Genet. 43:121-133.

Chapter 5
The Cytogenetics of Corn

WAYNE R. CARLSON
The University of Iowa
Iowa City, Iowa

Corn is at once an outstanding cytological and genetical instrument, thus providing ideal ground for the study of cytogenetics. In the past two decades we have advanced our understanding of earlier work and struck out in new directions with studies of intragenic (allelic) recombination, the initiation of synapsis, functioning of the nucleolus organizer, genetic activity of the B chromosome, and gross duplications within the corn genome. This chapter emphasizes subjects of current importance but attempts to provide, in addition, a general picture of corn cytogenetics.

I. CHROMOSOME MORPHOLOGY

The 10 chromosomes comprising the monoploid or gametic number in corn are morphologically distinguishable by (1) their relative lengths, (2) distinctive chromomere patterns, (3) deep-staining knobs in characteristic positions, (4) location of the centromere (determining arm ratios), and (5) the degree of heteropycnosis in the chromomeres adjacent to the centromeres. These diagnostic features are most clearly evident at pachynema when the chromosomes are found as elongated paired threads; it is this stage which has been so widely used in cytogenetic studies. Corn cytogenetics received its first great impetus when McClintock in 1929 reported that the 10 chromosomes could be recognized by relative lengths, arm ratios, and knob positions at prophase of the first microspore division. The longest member of the haploid complement was designated as chromosome 1, the second longest as chromosome 2, etc. Later, from her studies of the morphology of the pachytene chromosomes, which set the stage for a remarkable series of advances, it became apparent that the relative lengths of the 10 chromosomes were in general the same at pachynema and at the first microspore prophase. There are, however, two changes in rank. Chromosome 5 is slightly longer than is chromosome 4 at pachynema and chromosome 8 is somewhat longer than is chromosome 7. As a consequence of its possessing the nucleolus-organizing region, chromosome 6 is always found associated with the nucleolous. It is the most readily identified of all the 10 chromosomes.

This chapter is a revision of the 1955 chapter by M.M. Rhoades. Occasional segments of the original chapter have been retained intact.

Features which are used to identify the chromosomes generally do not vary from strain to strain. However, the number, position, and size of heterochromatic knobs is not constant. Some lines have no heterochromatic knobs while others may have one or more on each chromosome. Certain positions on the chromosome accommodate series of different-sized knobs. Once the karyotype of a particular strain has been determined, knobs become quite useful for routine chromosome classification. An advantageous result of the variation in knob constitution between different strains has been the widespread utilization of knobs as cytological markers which can be treated in a similar manner to genetic markers.

Knobs consist of extra chromatin and are not derived from the regular complement of genes. The length of a chromosome is increased by the presence of a knob. Since knobs remain in a relatively contracted condition throughout the mitotic cycle (Morgan, 1943) their relative contribution to the length of a condensed metaphase chromosome is much greater than to that of a prophase chromosome. The absence of knobs in certain strains demonstrates that no vital function is associated with them. Also, nonessential functions have not been attributed to knobs, and they may be genetically inert. Evidence that heterochromatic regions are not transcribed has been accumulating in recent years.

Two distinct classes of heterochromatin, termed facultative and constitutive, have been identified. Facultative heterochromatin is transitory: the chromatin has alternate heterochromatic and euchromatic states. Studies of facultative heterochromatin have shown that conversion from a euchromatic to a heterochromatic condition results in gene inactivation (Berlowitz, 1965; Brown and Nur, 1964; Lyon, 1961, 1962, 1971; Nur, 1967). The corn heterochromatin is the constitutive type which always remains condensed. Constitutive heterochromatin sometimes consists of redundant (repetitive) DNA (see "Duplications" in Chromosomal Rearrangements section of this chapter). A single nucleotide sequence, generally too small to code for a useful protein, may through repeated duplication make up all of the DNA present in a large heterochromatic region. Constitutive heterochromatin may, therefore, be devoid of useful information. Nevertheless, it is difficult to accept the idea that knobs could arise and persist in a population if they did not at some time provide one or more useful functions. Perhaps the latent ability of knobs to attach spindle fibers (see "Abnormal Chromosome 10 in Supernumerary Elements" section of this chapter) and the general association of centromeres with heterochromatin derives from a past or present function of constitutive heterochromatin in chromosome movement. Novitski (1955) demonstrated a role for centromeric heterochromatin in chromosomal disjunction in *Drosophila*.

Strains of corn from different parts of the world exhibit an amazing range of phenotypes and knob constitutions, but there is little evidence that gross chromosomal rearrangements are maintained in corn populations. That corn has a remarkably stable karyotype is indicated by the failure of Cooper and Brink (1937) to find reciprocal translocations in 55 strains from Latin America and also by the inability of Rhoades and Dempsey (1953) to find evidence of either translocations or long inversions in their study of the chromosomal homology of 90 exotic races. However, a few short inversions

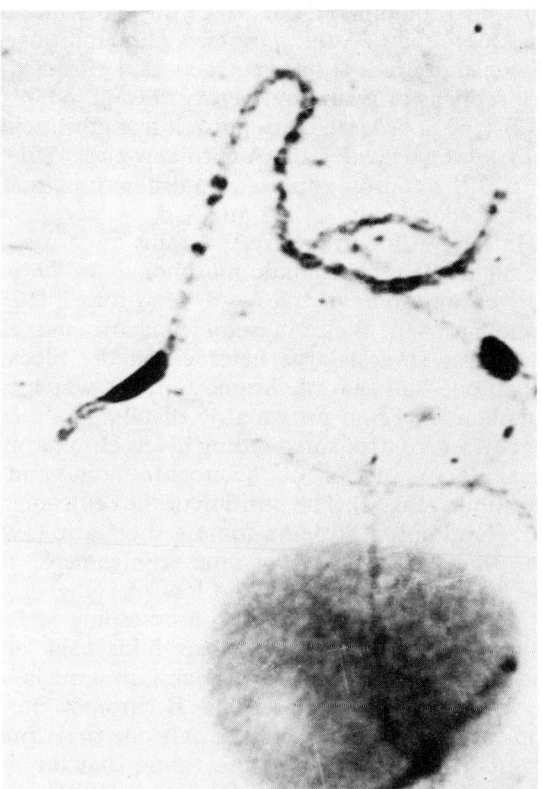

Fig. 1. Abnormal chromosome 10 in pachynema. Note the three chromomeres in 10L and the supernumerary segment with the large heterochromatic knob.

have been reported in the literature. The most striking exception to the uniformity of corn chromosomes is found with chromosome 10. There are two major forms of chromosome 10 (Longley, 1937). A rare and abnormal kind of chromosome 10 is shown in Fig. 1. Abnormal 10 differs from the usual or normal chromosome in chromomere pattern and overall length. The three chromomeres present in the distal one-sixth of the long arm of abnormal 10 are missing in normal 10. Abnormal 10 also possesses a large, chiefly heteropycnotic region attached to the end of the long arm which is not present in normal 10. Abnormal 10 — normal 10 bivalents are heteromorphic and the prominent knob of abnormal 10 is unpaired. The origin of the extra terminal segment of abnormal 10 can only be conjectured. The unusual genetic properties of abnormal 10 are reviewed in section X, 1.

In addition to the complement of 10 chromosomes found in all corn plants, the supernumerary B chromosomes may also be present. Kuwada (1911, 1915, 1919, 1925), who was the first to make extensive chromosome counts in corn, reported that the somatic number was not the same in different races. He found that sweet varieties often had more than 20 chromosomes, whereas most field corn varieties had 20 chromosomes. Later studies by Fisk (1925, 1927), Longley (1927), Reeves (1925), Kiesselbach and

Petersen (1925), and Randolph (1928a, b) confirmed Kuwada's finding that some strains had more than 20 chromosomes, although this was the number most frequently encountered. It became clear that chromosomes present in excess of 20 are replicas of a supernumerary chromosome which Randolph designated as a B-type in order to distinguish it from the members of the normal complement, which he called the A chromosomes. Wide variation exists in the number of B chromosomes present in different plants. This numerical instability is discussed in sections IX, 1 and X, 2.

The B chromosomes found in diverse strains of corn and in teosinte all have a unique, highly heterochromatic morphology which is quite different from that of any chromosome of the A set. McClintock (1933) described in detail the morphology of the B chromosome in pachynema. The chromosome is small and contains several large heterochromatic blocks. Its length is slightly more than one-half that of chromosome 10, which is the shortest of the normal complement. From proximal to distal ends, the B chromosome consists of the centromere with surrounding heterochromatin, a euchromatic segment, several large blocks of heterochromatin and a very short euchromatic terminus. (Fig. 2). The position of the centromere was described as terminal, but Randolph reported a minute short arm (1941) and Carlson (1970a) found that misdivision of the centromere to produce an isochromosome was accompanied by the loss of some of the centromeric heterochromatin. Apparently, the B chromosome has a diminutive heterochromatic short arm which frequently folds back on the long arm, giving the appearance of a telocentric. The chromosome is best classified as subtelocentric. In somatic metaphases, the B chromosome is readily identified as the shortest chromosome and the only one that appears telocentric. Genetic data support the cytological observation that the B chromosome is not homologous with A chromosomes. Randolph (1941) in a test of 46 loci distributed among 17 different arms of the A chromosomes failed to find any indication that the B chromosome possessed any of these loci. The origin of the B chromosome from any member of the regular complement is therefore unlikely.

B chromosomes appear to be genetically inert or subinert, since they are not essential for normal growth and development. No phenotypic differences are observed in plants with low numbers of B chromosomes, but if the number is greater than 10 to 15, several abnormalities appear which increase in intensity with greater numbers of B chromosomes. These include reduction in fertility, decreased vigor, defective kernels, aborted pollen etc. Seed is rarely produced on plants with 25 or more B chromosomes. The lack of functional significance of B chromosomes made possible Randolph's recovery (1941) of reduced derivatives of the B. The modified B chromosomes of Randolph arose spontaneously and were designated C, D, E, and F chromosomes in order of decreasing size. Longley (1956) suggested that unequal crossing over between or within B chromosomes may explain the origin of Randolph's diminutive chromosomes. The B chromosome and abnormal chromosome 10 are discussed further in section X.

As was noted earlier, studies of chromosome morphology in corn are almost invariably performed with pachytene material. No other stage of division, meiotic or mitotic, provides the detailed picture available in

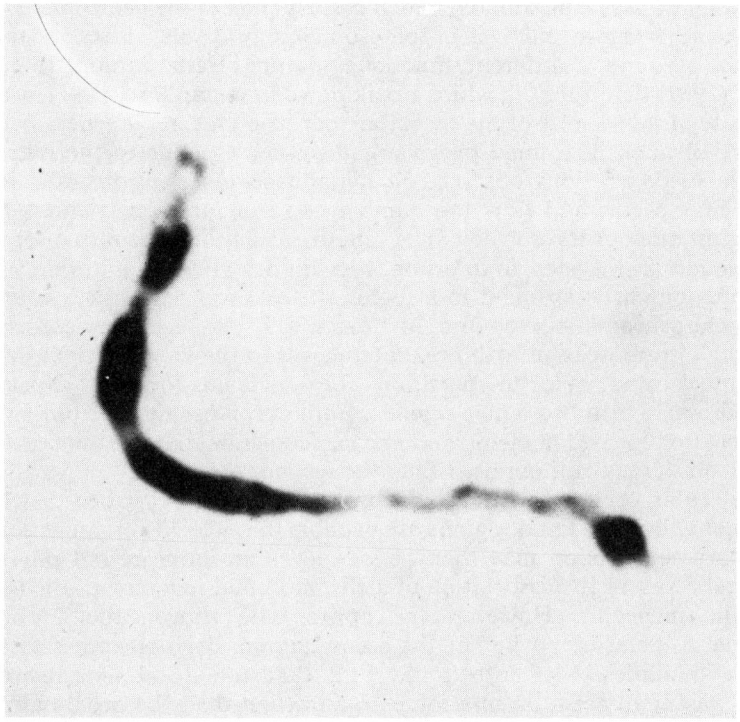

Fig. 2. The B chromosome in pachynema. The smaller, terminal heterochromatic region surrounds the centromere.

pachynema. However, for reasons of convenience and as a result of some improvements in technique it is sometimes preferable to work with mitotic prophase or metaphase cells. The classification of B chromosomes and abnormal 10 can be readily accomplished in somatic cells. In addition, Chen (1969) has shown that all the corn chromosomes can be identified in mitotic metaphase cells through the use of centromeric position, overall length, and specific morphological features such as the satellite on 6S (Fig. 3). Filion and Walden (1973) have tested the limits of somatic karyotype analysis in maize with a series of reciprocal translocations. Somatic analysis was found to be much less sensitive than meiotic analysis in detecting changes in chromosome length. Nevertheless, the amount of information that can be extracted from somatic cells has increased considerably in recent years. The work of Horn (section VII) on homologous associations of somatic chromosomes in corn is a clear example of the benefits derived from improvements in somatic karyotype analysis.

II. NUCLEOLUS ORGANIZER REGION

In somatic metaphase cells, the chromosomes assume a densely staining configuration in which euchromatin and heterochromatin become indistinguishable. Each chromosome retains a short, light-staining region called the primary constriction, which corresponds to the centromere. In addition, one or more pairs of chromosomes may display a secondary constriction of quite a different functional nature. Heitz demonstrated with *Crepis* (1931) that nucleoli, which break down in metaphase, are reformed in telophase in the vicinity of the secondary constrictions. As a general rule, the number of nucleoli formed by an organism corresponds to the number of secondary constrictions observed in metaphase. As nucleoli grow in size, fusion may occur, and thus the number of secondary constrictions gives a maximum number for nucleoli. The functional significance of the secondary constriction is still open to question. McClintock (1934) found that nucleoli arise only on chromosome 6 in a region of the short arm which contains a large heterochromatic knob and, just distal to it, the secondary constriction. Utilizing a translocation with breakpoints within the heterochromatic knob, McClintock was able to partition the nucleolus-forming capacity of chromosome 6 into two active regions. Both chromosomes derived from the translocation formed nucleoli, and one can conclude that the knob is the site of nucleolus organization rather than the secondary constriction. In addition, the nucleolus organizer (NOR) of chromosome 6 is repetitive, since both proximal and distal knob regions are capable of nucleolus organization. The secondary constriction may then be viewed as an unimportant physical or chemical result of the association of a chromosomal region with the NOR or with the nucleolus. However, the appropriate translocations were not available to determine whether the secondary constriction might also have NOR capabilities. McClintock did find that the distal heterochromatic region was more active in nucleolus organization than the proximal region, which could be due to the distal association with a secondary constriction. Also, Lima de Faria and Sarvella (1962) found in pachynema an association of the nucleolus with the distal part of the knob plus the "gap" (secondary constriction) in 83.2% of cells and attachment of the distal and median knob segments plus the gap to the nucleolus in 12.6% of cells. An association of the knob but not the gap with the nucleolus was found in only 2.5% of cells. The findings suggest an organizer capacity in the distal knob region and perhaps in the secondary constriction. More conclusive evidence of the role of the secondary constriction in nucleolus organization was provided recently by R.L. Phillips (unpublished data). A series of translocations was utilized in which one of the exchange points of each translocation was within the heterochromatic NOR or the adjacent euchromatic "gap." In each case, the chromosome with the proximal region of 6S was found to form a nucleolus, confirming McClintock's results. However, the size of the nucleolus formed, in comparison to the total nucleolar volume, was quite small for chromosomes that did not include at least some of the nucleolar gap region. The unpublished results suggest that the nucleolar gap is quite important in organization of the nucleolus, although small nucleoli can be formed in its absence.

Without further evidence, one might conclude that the role of NOR's is to form a nucleolus from materials synthesized elsewhere in the cell. However, recent work has shown that NOR's synthesize RNA and that the nucleolus provides for a secondary processing of the RNA. The first work leading in this direction was that of Lin (1955). Lin utilized a translocation between chromosome 6 and a B chromosome. The interchange point of 6 was

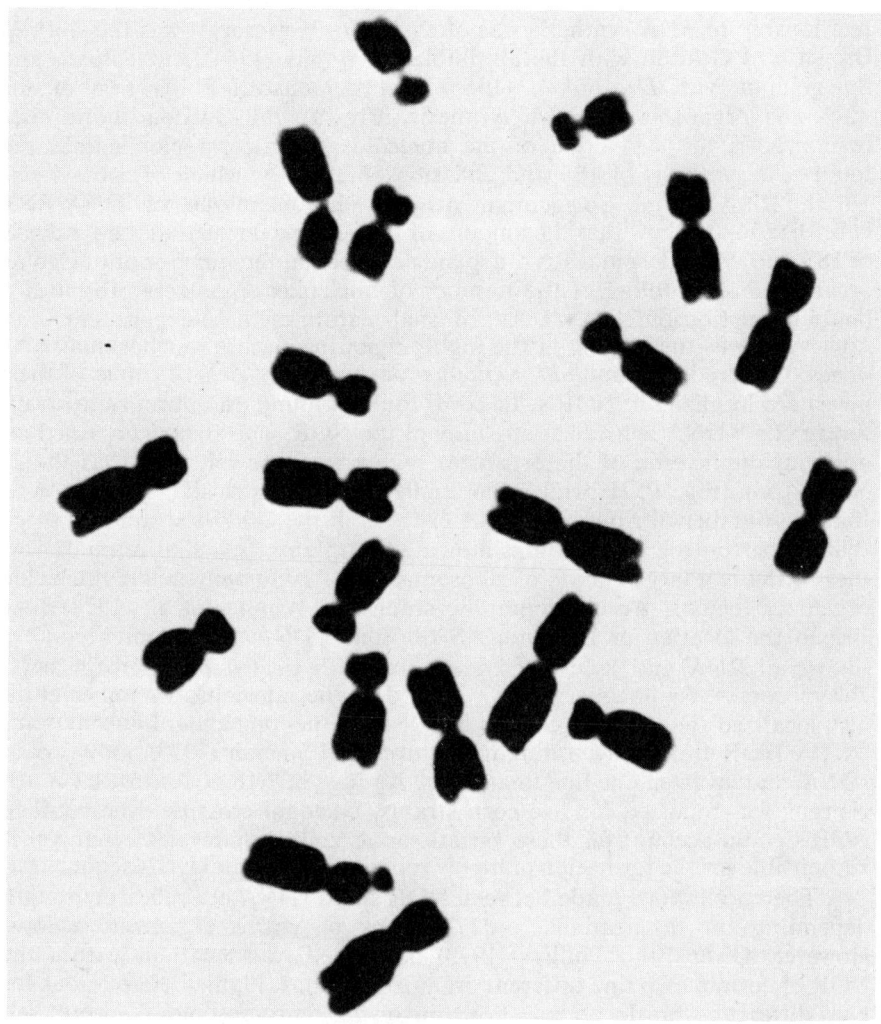

Fig. 3. Metaphase of mitosis. Chromosome 6 is identified by the presence of two constrictions. Other chromosomes display characteristic arm ratios and total lengths.

within the heterochromatic organizer so that two NOR's were produced by the translocation. The B^6 chromosome of the interchange carried the distal half of the NOR plus the very short remainder of 6S. This chromosome, which carried little more genetic activity than that of the NOR, was extracted from the translocation and maintained as a supernumerary in plants with two normal 6's. By inbreeding and selection the number of B^6's could be increased, so that a range of plants with 0 to 5 B^6's plus two 6's was obtained. Lin used microspectrophotometry to determine the amount of RNA present in the nucleoli of plants with various numbers of B^6's and the amount of nucleolar RNA and concluded that the nucleolus organizer has a synthetic as well as an organizing activity. The functioning of the nucleolus and the organizer was also studied in a number of other organisms and some relationship to RNA synthesis was often found. However, it was the work of Brown and Gurdon with the amphibian *Xenopus* (1964) and Ritossa and Spiegelman with *Drosophila* (1965) that demonstrated the role of the nucleolus organizer in RNA synthesis. Brown and Gurdon found that homozygosis for a deletion of the nucleolus organizer region eliminated totally the synthesis of 18S and 28S ribosomal RNA, which of course, was lethal. Ritossa and Spiegelman utilized the technique of DNA-RNA hybridization to show that the amount of the genome devoted to the synthesis of 18S and 28S ribosomal RNA depends entirely on the number of nucleolus organizers. A doubling of the number of nucleolus organizers produced a doubling of ribosomal DNA. An unusual feature of these experiments and later work was the finding of the highly repetitive nature of ribosomal RNA genes. Hundreds *(Drosophila, Xenopus)* or thousands *(Zea)* of copies of these genes are localized in NOR's. In corn, Phillips found an appropriate strain for an RNA-DNA hybridization study of the NOR. The strain contained an adjacent duplication of the organizer region and was referred to as the 2-NOR strain (Fig. 4). Hybridization studies by Phillips et al. (1971b) showed that the number of ribosomal DNA cistrons in the 2-NOR strain was twice that in the control. The findings indicate that plants as well as animals synthesize the two large classes of ribosomal RNA exclusively in the nucleolus organizer regions. Autoradiographic studies of Wimber et al. (1974) confirmed the location of 18S and 28S ribosomal DNA. They iodinated (^{125}I) ribosomal RNA and hybridized it in situ with pachytene microsporocyte chromosomes. Radioactivity was confined to the nucleolus. (Wimber et al. also localized the smaller 5S class of rRNA to the long arm of chromosome 2). The NOR of corn, according to Ramirez and Sinclair (1973), may vary in rDNA content from one line to another. A range of 1.18×10^4 to 1.82×10^4 cistrons was reported for five corn strains. Unequal crossing over between NOR's could account for these variations, as could whatever mechanism is responsible for the formation of highly repetitive ribosomal DNA sequences.

The correlations made between NOR's and rDNA are relatively rough, depending on duplications and deletions of entire organizer regions. However, Givens and Phillips (1973) have used translocations within the NOR of corn to examine different organizer regions. Plants heterozygous for two different translocations, both involving chromosome 6, were self pollinated to produce duplications of the heterochromatic segment of the NOR or duplications of the euchromatic segment. Hybridization studies localized most of the rDNA to the heterochromatic region and relatively little to the euchromatic region. The intriguing possibility arises here that the euchromatic gap region of the organizer may control nucleolus organization while the heterochromatic region contains the rDNA. Alternately, Givens and Phillips have suggested that the heterochromatin of the organizer is inactive and the euchromatic segment contains the only active rDNA cistrons.

III. THE EUPLOID SERIES

Although the normal gametic complement consists of 10 different chromosomes and diploids have two sets or 20 chromosomes, many modifications of this number have been found. In a euploid series, where entire sets are added or subtracted, the number of sets may be less than

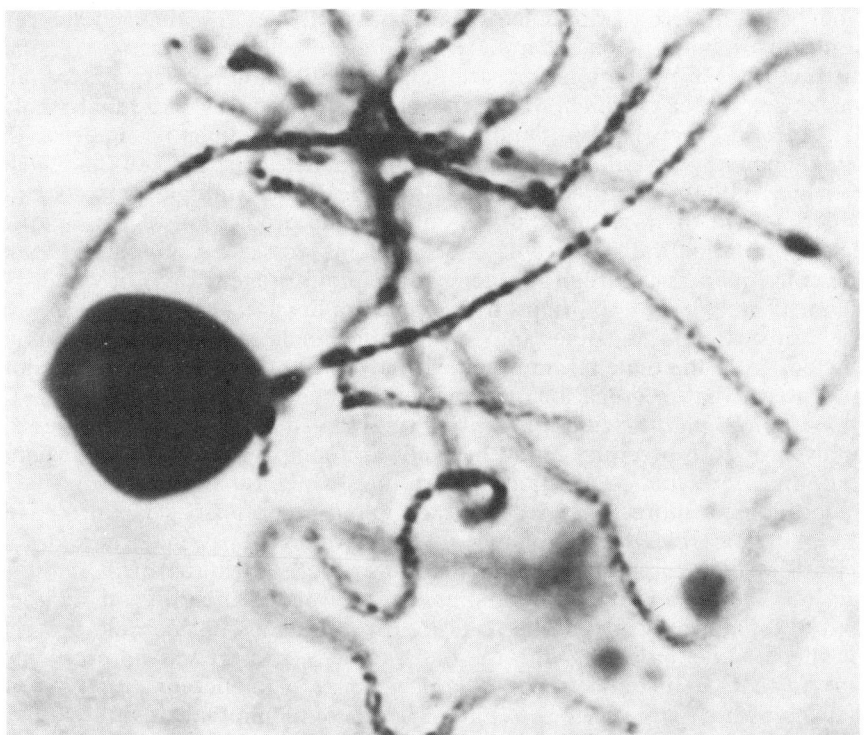

Fig. 4. The 2-NOR strain of Phillips in pachynema. Note the two large dark-staining regions associated with the nucleolus. Reprinted with permission from Chromosoma 36:83.

(monoploid) or more than (polyploid) the diploid number. Monoploid or haploid plants, which have one set of chromosomes, arise spontaneously in crosses at different rates, but generally in the range of 1 per 1,000 to 2,000 kernels. Although monoploid sporophytes are smaller than their diploid sibs, there is no striking external morphological difference between them. That this is characteristic of all members of the euploid series is understandable, since all have a balanced chromosomal complement and differ solely in the number of times the monoploid complement is replicated. Cell and nuclear size are smallest in monoploids and become progressively larger as the degree of ploidy increases. By far the majority of monoploids that arise are maternal in origin. Maternal monoploids arise from kernels with a normal (triploid) endosperm and a haploid embryo. The origin of maternal monoploids is not understood, but one sperm of the pollen grain effects fertilization of the polar nuclei while the other is, in some way lost. Chase (1969) and Sarker and Coe (1966) showed by gene dosage tests that the fusion of both sperm from a pollen grain with the polar nuclei is not the source of maternal monoploids. Androgenetic monoploids are produced very infrequently, with a rate of roughly 1 per 200,000 kernels. In this instance both sperm function, but a maternal nucleus is apparently lost.

The production of monoploids has been shown to be under genetic control and therefore subject to selection. Chase (1949, 1951) found that the incidence of maternal monoploid plants was affected by the genetic constitution of both the male and female parents. Different strains display distinct rates of monoploid formation. Coe (1959) found a particular genetic stock (Coe's stock 6) which, when self-pollinated, produces maternal haploids at a rate in excess of 3%. Outcrosses of stock 6 as either the male or the female parent gave enhanced monoploid frequencies (Sarkar and Coe, 1966). Kermicle (1969) identified a mutant gene *ig* (indeterminate gametophyte) which acts in the female parent to enhance the production of monoploids. The *ig* mutation has pleiotropic effects on embryo sac development. One of the consequences is a high frequency of haploid progeny (3%) of which the majority are *paternal* in origin. A practical application of this system may be the transfer of inbred lines into cytoplasmic male sterile backgrounds. A nucleus from the male is combined with cytoplasm from the female without any intermediate segregation of genes between the two lines (Goodsell, 1961).

Haploid plants possess two properties that may be of value in genetic analysis or corn breeding. First, haploids are homozygous for all their genes and allow for rapid production of pure lines which otherwise require many generations of inbreeding. Second, monoploids express all genes present in the plant regardless of dominance or recessiveness. They may therefore be useful in screening new mutations. However, both attributes of monoploids can only be utilized if 1) monoploids can be isolated in sufficient numbers and 2) diploid progeny can be recovered from them. Considerable progress has been made in dealing with the first problem, and the second problem is not as acute as it might seem. The occurrence of monoploids may be enhanced by use of Coe's Stock 6, but a 3% rate of monoploidy is still relatively difficult to work with when large numbers of haploids are desired. More useful are various genetic procedures which have been developed to screen populations for haploids. The most successful of these employ endosperm and scutellum (embryo) color markers, in which selection can be applied prior to germination. The *C-I* marker (Coe and Sarker, 1964) and the *R-nj: cudu* marker (Greenblatt and Bock, 1967; Nanda and Chase, 1966) have been successfully employed in haploid selection. During selection, kernels in which the scutellum lacks the appropriate dominant marker from the male parent but obviously carries this marker in the endosperm are putative haploids. Efficiency of these procedures is very high. Sarker et al. (1972) have recently combined the *C-I* selection procedure with the stock 6 genetic background to provide a higher rate of recovery of haploids per number of seeds screened. As a result of selection procedures, haploids can be produced in reasonable numbers for a variety of experiments.

The second problem in dealing with haploids is the recovery of diploid progeny. As Chase (1969) has pointed out the theoretical expectations for fertility in haploids are much too small in comparison to observed seeds sets and pollen fertility. Random migration of 10 chromosomes to the poles should seldom give a single viable kernel on haploid ears, but seed set of 30 to 40 or more has been found. According to Chase, monoploids occasionally undergo chromosome doubling during development so that diploid sectors frequently appear in the tassel or ear. The frequent appearance of diploid sectors is believed to result from a competitive advantage of the diploid over the monoploid condition. Diploid sectors are observable as clusters of seeds on the ear or localized segments of pollen fertility on the tassel. Approximately 1 in 10 haploid plants can be self-fertilized to produce pure lines as a result of diploid sectors in the male and female inflorescences. More frequent recovery of seeds can be obtained by using a diploid male parent. The work of Alexander (1964) and Weber and Alexander (1972) on crossing over in haploid corn (section VII) depended on the ability to recover seed from haploids.

Moving up the euploid series, the triploids are the first polyploids encountered. Triploids which arise in diploid populations usually come from the fertilization of a diploid or unreduced egg by a haploid sperm, but the male parent may occasionally contribute the diploid number of chromosomes (Rhoades, 1936). Triploids also occur regularly in the hybridization of diploid with tetraploid lines. The triploids found in diploid matings may arise spontaneously in normal material with a low frequency or may be of frequent occurrence in lines with genically induced disturbances of meiotic processes. Beadle (1930) reported many triploids in the cross of *asynaptic (as)* corn by haploid pollen where the formation of unreduced eggs was brought about by lack of pairing of the homologous chromosomes and nondisjunction. Rhoades and Dempsey (1966a) found triploids in the progeny of plants homozygous for the *elongate (el)* gene where suppression of one of the meiotic divisions occurs.

In triploids the chromosomes usually are associated at metaphase I as trivalents and less frequently occur as bivalents and univalents. A great majority of the spores of triploid plants are aneuploid with chromosome numbers ranging from 11 to 19 as a consequence of the random or near-random segregation of each set of three homologous chromosomes. Many of these aneuploid spores are unable to develop into viable gametophytes, and triploids characteristically have a high percentage of aborted ovules and pollen grains. When triploids were used as the pollen parent in crosses with diploids, McClintock (1929b) found a marked selection in favor of pollen grains with 10, 11, or 12 chromosomes; those with higher numbers were unable to compete successfully in achieving fertilization. The severe competition found among the male gametophytes was not evident in the reciprocal cross, where there was a wide range of chromosome numbers in the functional eggs. The different primary trisomes isolated by McClintock in the progenies of diploid by triploid crosses and their subsequent correlation with specific linkage groups marked the beginning of corn cytogenetics.

The infertility of triploid corn is an unavoidable consequence of the segregation of sets of three homologous chromosomes. Tetraploids, however, are potentially fertile since 2-2 disjunction of homologous chromosomes can occur. Consequently, tetraploid lines of corn have been developed and maintained. Randolph (1932) utilized heat treatment of young ears during early divisions of diploid embryos to produce tetraploid (as well as octoploid) plants. Alexander (1957) synthesized various tetraploid lines by crossing diploids homozygous for the elongate mutation as female to tetraploid males. Tetraploid corn resembles the diploid in height and growth habit but has broader leaves, sturdier stalks, larger tassels, and ears and kernels of increased size.

Randolph (1935) found that because of irregularities in meiotic segregation not all of the offspring of tetraploid parents have 40 chromosomes. The number of chromosomes ranged from 37 to 42, although

40 was the most frequent class. Apparently, multivalent association is less reliable than bivalent pairing in producing proper chromosomal disjunction. Gilles and Randolph (1951) determined the relative frequency of quadrivalent and bivalent associations in a tetraploid strain at the beginning and at the end of a 10-year period during which the tetraploid line had been maintained by selecting the more vigorous and fertile plants. There were fewer quadrivalents and more bivalents at the end of the 10-year period than at the beginning. It appears that in selecting for vigor and fertility there was a concomitant selection for a gene or genes influencing the type of chromosome association.

Levels of polyploidy higher than the tetraploid were obtained by Rhoades and Dempsey utilizing the *elongate* mutant (1966). They obtained a whole series of polyploids including 3N, 4N, 5N, 6N, and 7N plants. Nevertheless, experimental work in polyploids is restricted mainly to the tetraploid since it is by far the simplest polyploid to maintain and utilize experimentally. Recombination in autotetraploids has been studied by a number of workers and was recently reviewed by Doyle (1973).

IV. ANEUPLOIDY

In contrast to the euploid series, where each chromosome of the complement is present the same number of times, are the genically unbalanced aneuploid types. At the diploid level the addition to the normal complement of one or more chromosomes gives hyperploids. Plants with a single extra chromosome are known as trisomes. There are several classes of trisomes including primary, secondary tertiary and telo-trisomes.

Primary trisomes are those in which one member of the monoploid set is in triplicate while the remaining nine are in duplicate. All of the 10 possible primary trisomes have been isolated in corn. These have been derived chiefly from n + 1 gametes produced by triploids, although they have also arisen from meiotic nondisjunction in tertiary trisomes. All of the primary trisomes are smaller and less vigorous than their disomic sibs. Some can be recognized by characteristic phenotypic differences. This is true for trisomes 1, 2, 3, 4, 5, 7, and 8 but particularly for trisomes 5, 7, and 8. Plants trisomic for chromosome 5 have thicker, broader leaves with blunter tips, a more compact and stockier tassel, and a shorter stature than do disomes. The general appearance of plants trisomic for chromosome 5 is so pronounced that an accurate classification can be made in a segregating family for trisomic and disomic individuals. Plants trisomic for chromosome 7 can also be readily recognized by their stiff, narrow leaves with a leathery texture. Trisomes for chromosome 8 are more slender and flower several days earlier than their disomic sibs.

The primary trisomes isolated by McClintock (1929b) from a spontaneously occurring triploid were extremely useful in cytogenetical studies. The association of 6 of the 10 linkage groups with specific chromosomes followed immediately. The way in which this was done is illustrated by McClintock and Hill's (1931) demonstration that the shortest chromosome of the complement carried the genes in the *R-G* linkage group. Plants trisomic for chromosome 10 were crossed with plants carrying recessive mutant genes in the different linkage groups. The F^1 trisomic plants were either selfed or backcrossed to recessive individuals. In the ensuing F_2 or backcross progenies most markers gave 3:1 or 1:1 ratios, respectively, and it was evident that these genes were not located in the trisomic chromosome being tested. Aberrant ratios, on the other hand, were found for the r marker, linking it to chromosome 10.

Trisomic genetic ratios are subject to variation, depending on several factors. If a trisomic plant of AAa constitution is pollinated by recessive pollen, a 5:1 ratio of $A:a$ is expected if 50% of the ovules are n +1. Since less than half of the eggs are n + 1, owing to elimination of univalent chromosomes during meiosis, the observed ratio is less than 5:1, depending upon the frequency of univalents, but it is always strikingly different from 1:1 disomic ratios. Not only does the observed trisomic ratio depend upon the frequency of n + 1 ovules, but it is also influenced by the amount and kinds of crossing over between the gene and the centromere. If the locus of A is far enough removed from the centromere so that frequent multiple crossovers occur between it and the centromere, the assortment of the six chromatids, four with the dominant A allele and two with the recessive a, will approach randomness, and a ratio of approximately 4:1 instead of 5:1 would be produced if there were equal numbers of n and n + 1 gametes. Conversely if the gene is near the centromere, in terms of genetic units, there is no effect of crossing over on the ratio of dominant to recessive gametes.

Another potential influence on testcross results is the occurrence of preferential pairing between two homologues to the exclusion of the third. Bivalent-univalent configurations then might regularly involve A/A bivalents and a univalents. This nonrandom pairing would tend to enhance the frequency of the A phenotype in progenies through the production of Aa (n + 1) and A (n) gametes. Doyle (1967) examined preferential pairing in trisomics of chromosome 3, using the A color marker. He constructed plants with only one dominant allele ($A/a/a$). Preferential pairing between the two a-carrying chromosomes should enhance the frequencies of Aa and a gametes with respect to the A-type gamete. Doyle eliminated consideration of the Aa gametes by crossing the trisomics as male parents to an aa tester. Only balanced gametes survive the pollen competition. In the absence of preferential segregation, one-third of the gametes were expected to be A and two-thirds a. In the control plants carrying very similar chromosomes 3, the frequency of A gametes was, as expected, 33%. However, combinations of the standard chromosome 3 (a) with homologues (A) of different racial origin or structure produced evidence of preferential pairing. To the extent that preferential pairing occurred an increase in the frequency of a gametes resulted. Doyle found preferential pairing in a number of situations. Surprisingly, he also found under certain presumably rare circumstances a preferential A/a pairing. A possible explanation for the latter finding was given recently (Doyle, 1974).

In a secondary trisome the extra chromosome is not a replica of one of the members of the monoploid set but is an isochromosome consisting of two identical arms. If the order of loci in chromosome 5 beginning with the distal end of the short arm is (a b c d e) centromere (f g h i j k), the isochromosome derived from the short arm is (a b c d e) centromere (e d c b a) in constitution, and the other isochromosome would be (k j i h g f) centromere (f g h i j k). Since each of the 10 chromosomes can give rise to two isochromosomes, a total of 20 secondary trisomes is possible, but only isochromosomes for the short arm of chromosome 5 (Rhoades, 1933a; 1940) the long arm of chromosome 10 (Emmerling, 1958a) and both arms of chromosome 6 (Maguire, 1962) have been identified. In addition, an isochromosome of the B^9 chromosome (section X, 2) has been found (Carlson, 1970a). Isochromosomes arise by breakage through the centromere (misdivision) and joining across of chromatids belonging to the same chromosome arm. Formation of isochromosomes may occur directly from members of the chromosomal complement or indirectly by misdivision of a telocentric chromosome, which itself was produced by centromeric misdivision. In general, misdivision occurs when a centromere is incapable of carrying out its function of orienting and moving the chromosome poleward during meiosis or mitosis. A prime source of misdivision products is the univalent at meiosis which lacks the pairing partner needed for proper orientation. Isochromosomes may also be generated by irradiation (Maguire, 1962) but these chromosomes may arise from breakage adjacent to the centromere rather than within it. A slight difference in proximal chromatin content may exist between the arms of such an "isochromosome."

In tertiary trisomes the extra chromosome is composed of parts of two nonhomologous chromosomes. Whereas the number of primaries is 10 and that of secondaries is 20, there is no limit to the number of tertiary trisomes, since their constitution depends upon the contributions from the two parental chromosomes. This will vary widely, depending upon the position of the breakage points. The numerous tertiary trisomes in corn have been derived from plants heterozygous for translocations in which nondisjunction of the ring of 4 chromosomes at anaphase I of meiosis results in a functional 11-chromosome gamete. Little cytogenetic work has been done with tertiary trisomes, but they have been used in the placement of genes within the cytological chromosome (Dempsey and Smirnov, 1964).

Telotrisomes contain an extra telocentric chromosome derived from one of the twenty chromosome arms in maize. Thus far, only the 5S telocentric (Rhoades, 1933a, 1940) and the 6L telocentric (Doyle, 1972a) have been recovered in the telotrisomic condition (5, 5, telo 5; 6, 6, telo 6). Rhoades et al. (1967) found a number of telocentrics in the progeny of plants carrying B chromosomes (section X, 2) but the telocentrics were present in partial monosomic chromosome combinations such as 3, telo 3 rather than in trisomics. Telocentric chromosomes arise by centromeric misdivision, damage to the centromere through irradiation, etc., or, in the special case of Rhoades and coworkers, by breakage of a dicentric bridge in the region of the centromere.

In addition to the trisomes another basic class of aneuploids is the 2n-1 monosomes. These individuals have one of their chromosomes in single dose and the rest in duplicate. Monosomes are difficult to produce in corn since even if they arise accidentally during plant development, the condition is not transmissible. The n-1 gametophyte lacks genes necessary for survival and only balanced gametes are produced. Recently, Kante Satyanarayama (in Weber, 1973) found that a particular mutation *(r-xl)* induces nondisjunction during the megagametophyte divisions with the frequent production of n + 1 and n - 1 cells. Since eggs do not require a balanced set of genes for survival, monosomes can be recovered in the progeny. Weber (1970, 1973) has utilized the system to produce 9 of the 10 possible monosomes. In addition, an occasional double monosome and even triple monosomes have been produced, in which two or three different chromosomes are present in one dose.

The detrimental effects of aneuploidy in corn on the sporophyte and gametophyte are often severe, and the selection and maintenance of aneuploids are difficult. However, in a number of organisms, the tolerance to aneuploidy is greater than in corn, allowing for more extensive studies of changes in chromosome number. In addition to the basic aneuploid types discussed here, a number of complex aneuploids have been found with other organisms, and these are discussed by Khush (1973) in his comprehensive review of the subject.

V. MEIOSIS IN CORN

Meiosis in corn is not unusual and has been described by Rhoades (1950, 1955). The photographs of Rhoades (1955) are reproduced here (Fig. 5, 6, 7) but the stages of meiosis will not be reviewed. Instead, a few observations on meiosis will be made.

The meiotic divisions have been studies in microsporocytes with carmine (or orcein) smear preparations. Limited studies of megasporogenesis reveal that meiosis here is essentially the same as in microsporocytes. The only notable difference between micro and megasporogenesis is that in the former four functional microspores are produced, whereas in the latter, only the inner (basal) member of the linear set of four megaspores gives rise to a functional gametophyte.

Premeiotic divisions in the sporogenous cells of the anther are characterized by the marked elongation or attenuation of the chromosomes. The prophase chromosomes in these mitotic divisions are much longer than those in somatic cells. It is as if the chromosomes were being made ready for the onset of meiosis at which time the leptotene chromosomes as extremely long and slender threads. During meiosis, synapsis is accomplished in zygonema. Zygotene chromosomes are drawn into a tight knot next to the nucleolus, and emergence of the chromosomes from this chromatin mass marks the onset of pachynema. The chromosomes in zygonema are difficult to observe, as is the process of synapsis. Pairing of homologues is accompanied by formation of the synaptinemal complex between paired segments. The corn synaptinemal complex is typical in structure and time of appearance (Underbrink et al. 1967; Gilles, 1973). When regions of non-homologous pairing occur, the synaptinemal complex may also link these regions (Ting, 1971, 1973; Gilles, 1973). In addition to synapsis, certain chromosomal regions may fuse with similar regions on other chromosomes. Centromeric fusions and heterochromatic fusions appear concurrently with synapsis but are quite distinct phenomena. The subjects of homologous and nonhomologous synapsis and chromosome fusions are discussed in section VII.

The most useful stage for cytogenetic work is pachynema, which provides a detailed view of chromosomal rearrangements, heterochromatic regions, centromeric positions, etc. However, changes in chromosome number of multivalent associations are most readily revealed in diakinesis. Identification of dicentric chromosomes is made in anaphase I and anaphase II when chromatin bridges form between disjoining centromeres. The quartet stage found at the end of meiosis is used to analyze chromosomal segregation in multivalents. Whereas individual chromosomes cannot be recognized in a quartet of cells, the nucleolus acts as a marker for the presence or absence of chromosome 6. When chromosome 6 is involved in a translocation, meiotic segregation may transmit 0, 1, or 2 nucleolus organizers to the microspores and classification of the nucleoli present in members of a quartet reveals the type of meiotic segregation that occurred (section IX, 1).

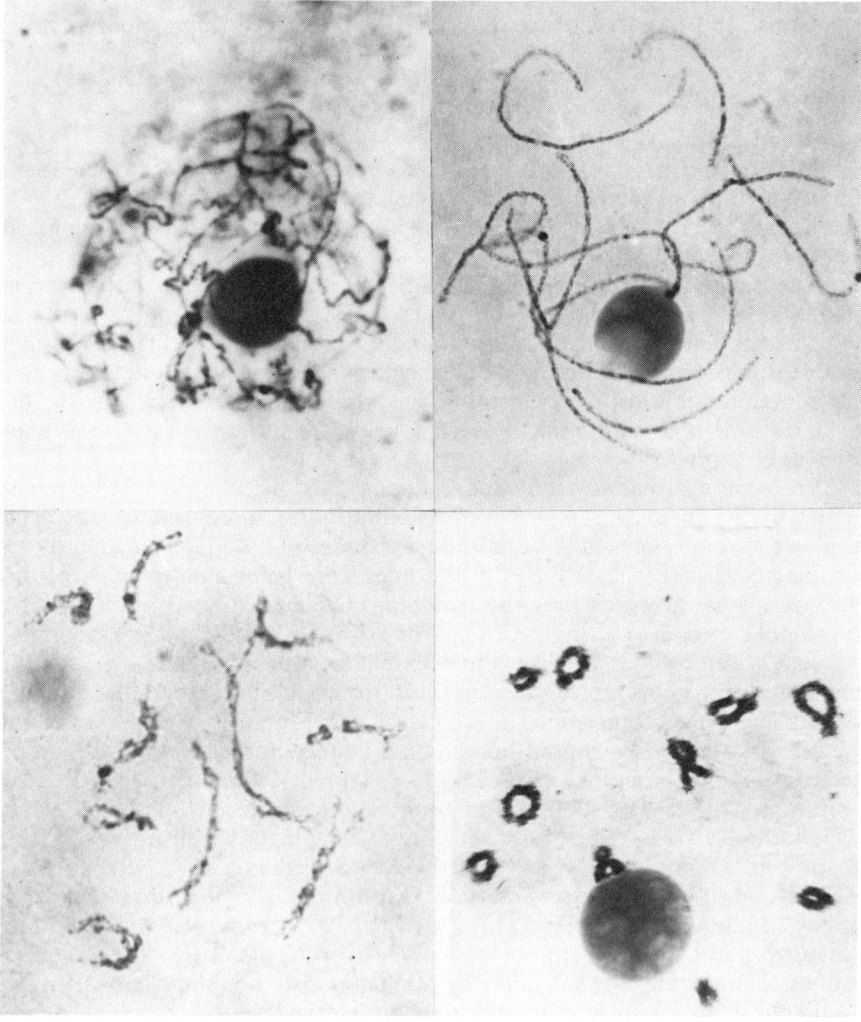

Fig. 5. Leptonema (upper left), pachynema (upper right), diplonema (lower left), and diakinesis (lower right) in corn.

THE CYTOGENETICS OF CORN 241

VI. MUTATIONS OF MEIOSIS, MITOSIS, AND CHROMOSOME STRUCTURE

Mutations that involve the structure or behavior of chromosomes are difficult to isolate for at least two reasons. First, the mutations often have no external phenotype and must be identified cytologically. Second, the mutant phenotype may be lethal, since chromosome division or chromosome integrity are affected. Nevertheless, a number of mutations which affect the chromosomes have been found in corn. The reader is referred to the following studies:

1) *asynaptic (as)* — Baker and Morgan, 1969; Beadle, 1930, 1933; Miller, 1963; Nel, 1971; Rhoades, 1947; Rhoades and Dempsey, 1949.
2) *divergent spindle (dv)* — Clark, 1940.

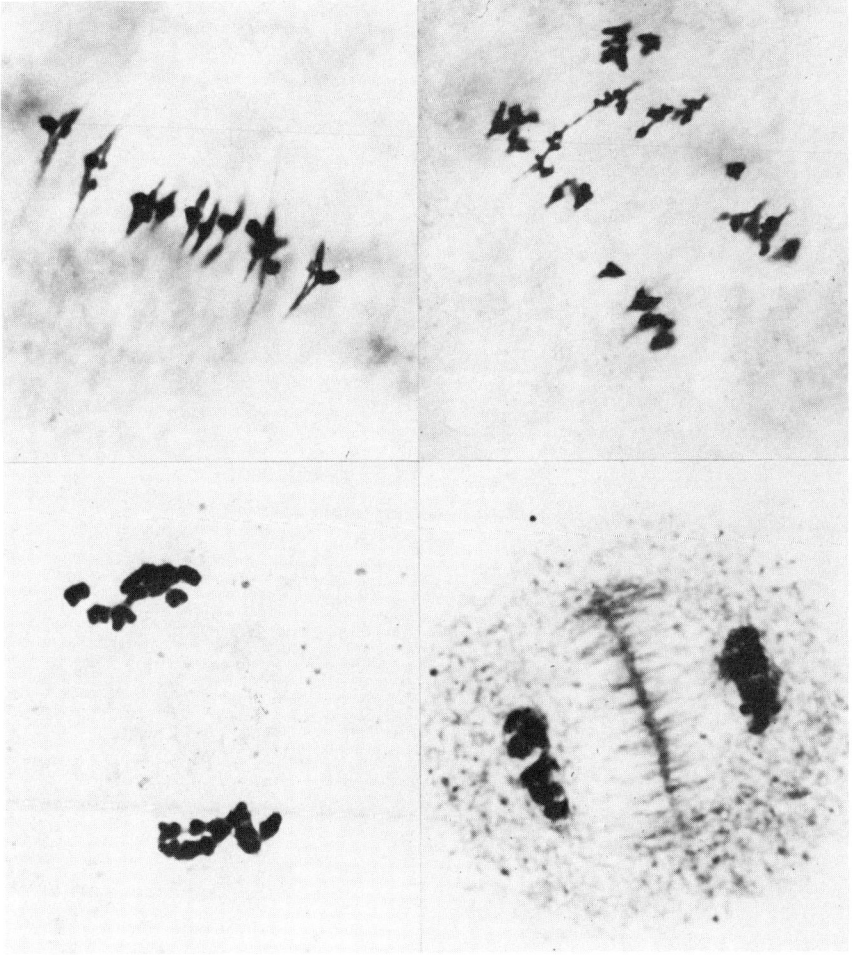

Fig. 6. Metaphase I (upper left), anaphase I (upper right, telophase I (lower left), and interphase II (lower right) in corn.

3) *polymitotic* (*po*) — Beadle, 1931.
4) *sticky* (*st*) — Beadle, 1932a, 1937; Schwartz, 1958; Stout and Phillips, 1973.
5) failure of cytokinesis — Beadle, 1932b.
6) *ameiotic* (*am*) — Palmer, 1971; Rhoades, 1956a.
7) *elongate* (*el*) — Rhoades and Dempsey, 1966a.
8) *r-x1* Satyanarayama — see Weber, 1973.
9) modified B-type chromosomes — Carlson, 1973b.
10) lysine-rich histone locus (*alh*) — Stout and Phillips, 1973.

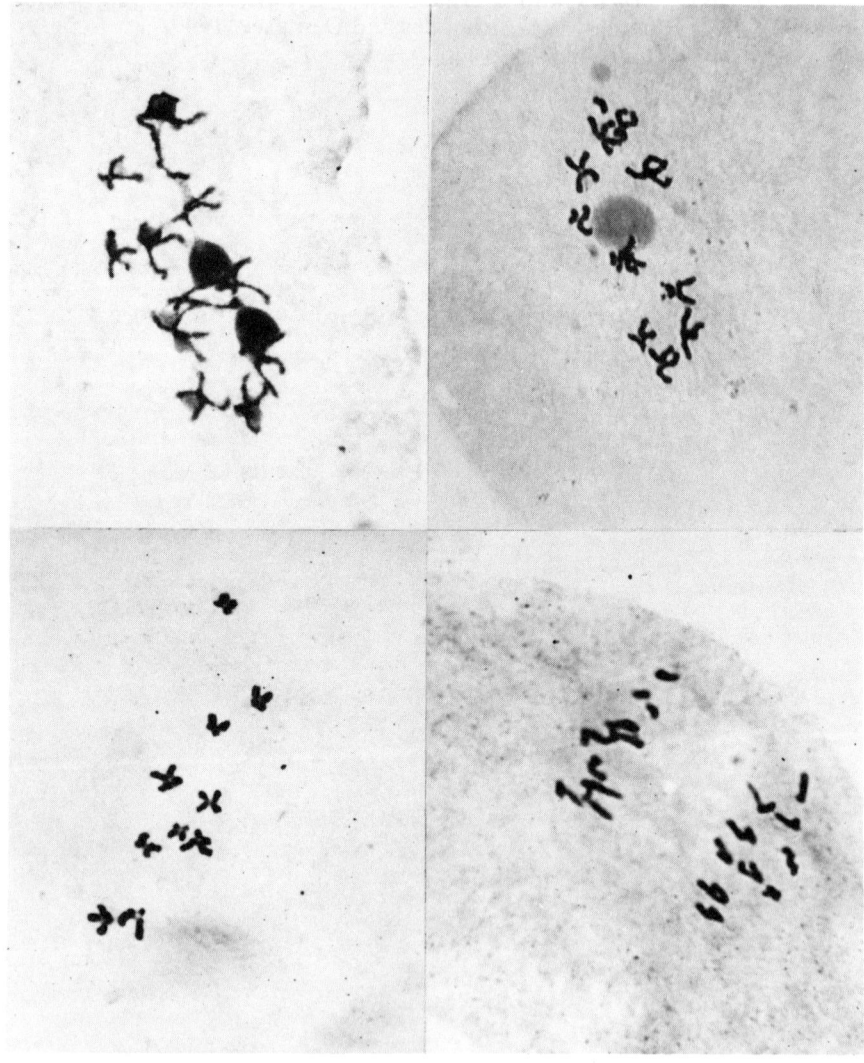

Fig. 7. Early and late prophase II (upper left and right), metaphase II (lower left) and anaphase II (lower right) in corn.

Certain of the mutations provide useful cytogenetical tools, for example in the production of diploid eggs by elongate or asynaptic plants. However, the mutant phenotypes do not provide much insight, as yet into the vital functions which they control. It may be necessary to develop systematic procedures for the isolation of large numbers of mutants before a meaningful description can be made of the genes which control meiosis, mitosis, and chromosomal organization (Carlson, 1973b). In addition, analysis of the functioning of specific mutations will require the increased utilization of biochemical techniques such as electrophoresis (Stout and Phillips, 1973).

VII. SYNAPSIS

The pairing of homologous chromosomes in meiosis is a prerequisite of crossing-over in zygonema-pachynema and disjunction in anaphase I. The subject of synapsis was reviewed recently by Moens (1973). Pairing is controlled by a number of genes, as evidenced by the variety of synaptic mutants that have been identified in different organisms (Riley and Law, 1965). Apparently the occurrence of synapsis and the maintenance of synapsis are separate events, since both asynaptic and desynaptic mutants have been found. Synapsis will be considered here in terms of initiation, elongation, and maintenance of pairing.

Theoretical explanations of the process of pairing have often foundered on the problem of initiating chromosome synapsis. If chromosomes are randomly distributed in the leptotene nucleus, the association of homologues may occur over a distance of several microns. Presumably a pairing force that operates over a considerable distance is required, but the identification of such a force has proved difficult. Several workers have suggested that "long-range" pairing forces do not exist. However, the view that homologues meet and pair by random chance seems unlikely in view of the striking regularity of synapsis. Current explanations of synaptic initiation often assume a premeiotic alignment of homologues that eliminates the need for "long-range" pairing forces. These ideas will be discussed later.

Following the initiation of synapsis the chromosomes are held together and elongation of synapsis from the point of initial contact can occur. In addition, secondary sites of pairing may occur as the chromosomes are drawn together by elongation of the initial pairing site. Pairing is initiated in zygonema and maintained through pachynema until the chromosomes open out in diplonema.

Throughout the period of synapsis a structure called the synaptinemal complex is present (Moses, 1968; Westergaard and von Wettstein, 1972). The synaptinemal complex consists of two lateral elements and a central element, as revealed by electron microscopy. In leptonema, each chromosome possesses a single lateral element, which probably is attached between sister chromatids. In zygonema the lateral elements of homologues pair up and a central element forms between them to complete the synaptinemal complex. When the chromosomes open out in diplonema the complex is shed, but remnants of it may remain at the newly formed chiasmata. By metaphase the chromosomes are completely free of the synaptinemal complex and the brief existence of this structure during chromosome pairing suggests an important role for it in synapsis and crossing over. Whether the synaptinemal complex is required for the initiation, elongation or maintenance of synapsis is not known. Recent evidence from maize demonstrates that the synaptinemal complex may form in regions of nonhomologous pairing (Gilles, 1973; Ting, 1971, 1973). This finding indicated a certain nonspecificity in the pairing of the lateral elements of chromosomes and suggests that the synaptinemal complex may not carry the information needed for the initiation of synapsis. Nevertheless, the synaptinemal complex could be involved in zipper synapsis following initiation or in the maintenance of synapsis. A role for the synaptinemal complex in recombination is also difficult to assign. The synaptinemal complex appears in most organisms to be a prerequisite to crossing over. In general, absence of the synaptinemal complex is accompanied by a lack of crossing over. An exception was found by Grell et al. (1972) who showed that crossing over occurs in *Drosophila ananassae* males even though the males do not have synaptinemal complexes present in meiosis. The finding thus far is unique to *Drosophila ananassae*, and it should be noted that the rate of crossing over in males is only about one-third the rate found in females.

Having considered the basic elements of synapsis, we will now return to certain problems for a more detailed discussion. Studies of the initiation of chromosome synapsis have centered on two questions.

1.) How are homologous chromosomes brought together for initial contact?

2) Are there specific sites on homologous chromosomes that initiate synapsis?

The second question is related to the first, since the identification of specific initiation sites on the chromosomes, such as the centromeres or the telomeres, has a bearing on the mechanism or existence of "long range" pairing.

The idea that homologous chromosomes may be aligned prior to meiosis in preparation for pairing is not new (Smith, 1942). Pre-alignment of homologues provides an explanation for the regularity of bivalent associations in meiosis and also explains the general absence of interlocking bivalents that might arise from the synapsis of randomly arranged chromosomes in the zygotene nucleus. The test for prealignment of chromosomes is the demonstration of non-meiotic association of homologous chromosomes. Non-meiotic "pairing" of homologues may occur in somatic or premeiotic cells. While somatic pairing does not lead to meiotic pairing, the presence of somatic pairing in an organism may infer a similar alignment in premeiotic cells. Somatic association of homologues has been detected in *Triticum aestivum* (Feldman et al., 1966), *Zea mays* (Horn, 1970, 1971; Miles, 1968, 1970); and several other organisms. Horn has found that most of the chromosomes in corn root tip metaphases are somatically associated. In general, somatic "pairing" is loose and must be determined in a quantitative manner. If the spatial separation between two homologous chromosomes in a cell is significantly less than that expected for a random distribution of chromosomes, the homologues are said to be "paired." While somatic pairing is not intimate, the association that is seen at metaphase may be a remnant of pairing that occurs in interphase and is disrupted by the breakdown of the nuclear membrane. Also, the loose somatic "pairing" may be tightened up in the important premeiotic stages.

If somatic "pairing" plays a role in meiotic synapsis then the two events should share some common properties. Miles (1968) found in corn that chromosome 6 (identified somatically by the secondary constriction) is not associated with its homologue in root tip metaphases. However, a significant association of the chromosomes 6 was found in the presence of abnormal chromosome 10 (section X, 1). Since abnormal chromosome 10 also enhances meiotic synapsis and crossing over, a relationship between somatic pairing and meiotic pairing may exist. More direct evidence for the alignment of chromosomes before meiosis should be present in the mitotic divisions that precede meiosis. Premeiotic pairing, sometimes very intimate, has been observed in *Triticum aestivum* (Feldman et al., 1966), *Haplopappus gracilis* and *Rhoeo discolor* (Brown and Stack, 1968), *Plantago ovata* (Stack and Brown, 1969) and other organisms. However, Palmer (1971) reported that the two chromosomes 6 in corn are not obviously associated during prophase of the last premeiotic mitosis. Maguire (1972a), on the other hand, has reported close premeiotic associations in corn. Palmer relied on the secondary constriction of chromosome 6 for identification of homology while Maguire compared similarities in length and centromeric position. The contradiction in results has not been resolved. Certainly the most direct evidence for a prealignment of chromosomes that aids meiotic pairing should be available from cytological studies of leptonema. Unfortunately, in most organisms, including corn, this is a very difficult stage to analyze. Favorable preparations of leptonema were obtained by Walters (1970) with *Lilium longiflorum,* but no evidence of pre-alignment of homologues was observed. Premeiotic association, therefore, does not appear to be of universal occurrence as would be expected for a mechanism controlling the initiation of synapsis. In organisms such as the Ascomycetes, which undergo a zygotic meiosis, the opportunity for premeiotic pairing is absent altogether. While pre-alignment of chromosomes may play a role in the prevention of interlocking bivalents for some organisms, it does not seem to be an indispensable element of synapsis.

Another possible factor in the initiation of synapsis is the interaction of chromosomes with the nuclear envelope. In some organisms the nuclear membrane plays an important role in chromosome separation during cell division (Pickett-Heaps, 1969). It has been suggested that the nuclear membrane of most organisms controls chromosome movement only during the initiation of synapsis. Movement of chromosomes through contact with the nuclear envelope could be an essential step in bringing homologues together. The role of the nuclear envelope in synapsis has been studied primarily because of the finding that the synaptinemal complex is attached to the nuclear envelope in a number of organisms (Moens, 1973). Particularly in animals and some fungi, both ends of the synaptinemal complex of each chromosome are regularly attached to the nuclear envelope, with no other chromosomal attachments to the membrane. This means that in pachynema, all chromosomes are attached by their telomeres to the nuclear envelope. In a few cases, unpaired meiotic chromosomes in zygonema have also been shown to be attached through their axial cores (lateral elements) to the nuclear membrane. The attachment of unpaired chromosomes and paired chromosomes to the membrane indicates that during synapsis and chromosome movement, telomeres may remain attached to the nuclear envelope.

While the attachment of telomeres to the membrane may be general for animals, this is not always the case in plants (Moens, 1973). In corn, serial sections of five pachytene nuclei were used to reconstruct the 10 bivalents and identify attachments to the nuclear envelope (Gilles, 1973). In general, chromosomes were attached to the nuclear envelope by the telomeres, but some free ends were also seen. Gilles suggested that during synapsis chromosome ends may be regularly attached to the nuclear envelope, but later following synapsis they occasionally are released from the membrane and are free in pachynema. An unusual feature of the telomeric attachments in corn was the lack of synaptinemal complex involvement. The SC usually terminated 2,000 to 4,000 A away from the nuclear envelope, and contact was made through chromatin.

If the initiation of synapsis is mediated by the nuclear envelope, it follows that synapsis should begin at or near the regions attached to the nuclear envelope — the telomeres. However, direct identification of initial pairing sites is difficult, since the early zygotene stage is short and usually difficult to observe. In corn, the problem has been approached through the use of opposite arm interchanges (Burnham et al., 1972; Tabata, 1962). Burnham and coworkers utilized translocations of chromosome 1 and chromosome 5. Different 1 to 5 translocations with breaks in opposite arms were combined in a single plant, and the pairing patterns were observed in pachytene. If one translocation had breaks in 5L and 1L, the other translocation would have breaks in 5S and 1S. Combinations of 5L, 1S with 5S, 1L were also made. Since two translocations are involved, two "crosses" may form during synapsis; but in the case of opposite arm interchanges, the crosses are linked in a "double-cross" configuration. In the absence of complete synapsis of all homologous regions, which often happens in structural heterozygotes, "pairs" of chromosomes may also form. The "pairs are necessarily heteromorphic and, depending upon the type of synapsis that occurs, the "pairs" may be homologous in the region of the centromere and not at the telomeres or vice versa. If pairing regularly initiates in proximal regions of the chromosome, pairs with non-homologous ends are expected; whereas homologous ends are expected if synapsis begins at or near the telomeres. Most of the translocations examined formed heteromorphic bivalents in which one chromosome was noticeably longer than the other. Pairing of homologous ends produced "buckling" in the centromeric area, whereas initiation of pairing at the centromere gave overlapping chromosome ends. Using a wide variety of 1 to 5 translocations with breakpoints in many different positions, Burnham et al. found that pairs with nonhomologous ends were virtually nonexistent. Among 449 cells with pairs only one had pairs with nonhomologous ends. Obviously pairing does not begin at the centromere. When all four break points of the two translocations were at least 0.3 of the respective arm lengths away from the centromere, the "pairs" formed still had homologous ends, indicating a lack of pairing initiation for a considerable distance on either side of the centromere. Further analysis of pairing configurations in opposite-arm interchanges led to an approximation of the frequencies of pairing initiation at different sites on corn chromosome arms. The highest frequency of initiation is subterminal, and initiation in the proximal two-thirds of the chromosome arm occurs rarely, if at all. The results establish the existence of preferred sites of pairing initiation and are consistent with a nuclear membrane mediated pairing. In a more direct observation of chromosome pairing, Moens (1969) found in *Locusta migratoria* that formation of the synaptinemal complex begins near the nuclear membrane. Although the experiments presented are suggestive, much work obviously remains to be done before the initiation of synapsis can be fully understood.

After the initiation of pairing, secondary sites of pairing are established along the chromosome, and zipper synapsis between the sites completes pairing. As Sved (1966) has observed, pairing may initiate distally and subsequent pairing may be a race between the zipper synapsis from established regions of pairing and the formation of separate paired regions. The existence of multiple secondary pairing sites is especially obvious in inversion heterozygotes. Zipper synapsis from the distal ends of a bivalent without secondary pairing initiations would not produce the typical inversion loop. On the other hand, the occasional (sometimes frequent) lack of loop formation in inversion heterozygotes is an indicator of the absence of secondary site pairing in the inversion loop.

Another problem encountered in understanding synapsis is the occurrence of preferential pairing. When more than two potential synaptic partners, as in polyploids or aneuploids, are present in an organism, a preferential association of specific chromosome pairs may occur. The most familiar instance of preferential pairing occurs in polyploids containing distinct but related genomes. For example, the corn/perennial teosinte tetraploid hybrid contains two complete diploid genomes. Pairing tends to occur within genomes (and not between genomes) so that bivalent formation is favored at the expense of quadrivalents, although some quadrivalents do occur (Shaver, 1962). Preferential pairing may also occur within a species. Thus, in corn evidence from polyploids and aneuploids (Doyle, 1963, 1967; Rhoades, 1957) indicates that chromosomes which are more closely related in structure or racial origin tend to pair to the exclusion of less closely related homologues. Here the discrimination between potential pairing partners is remarkable, since seemingly minor differences between homologous chromosomes block synapsis (see also section IV).

At the opposite end of the spectrum from preferential pairing is nonhomologous pairing. Nonhomologous pairing is rare when an appropriate synaptic partner is present, except for certain structural heterozygotes. The position of the "cross" in some translocation heterozygotes is constant from one pachytene cell to the next (homologous pairing), whereas other translocations display a variability in the region of exchange (nonhomologous pairing). With inversion heterozygotes, complete synapsis may occur in the absence of a loop configuration, indicating nonhomologous pairing. Depending upon the particular inversion being studied, the frequency of nonhomologous pairing may be high, moderate, or low. The improper synapsis found in structural heterozygotes may result from a rapid, nonspecific elongation phase of synapsis which follows initial homologous pairing at one or a few localized sites.

Nonhomologous pairing occurs frequently when chromosomes or chromosomal segments lack homologues. In haploids, monosomics, and plants with heteromorphic bivalents, chromosomal regions are present singly. In pachynema, these regions may synapse either with segments of the same chromosome (foldback pairing) or another chromosome (heterologous pairing). Plants which are monosomic or contain heteromorphic bivalents can only engage in foldback pairing, which they do frequently (McClintock, 1933). Haploids may participate either in foldback or heterologous pairing. McClintock (1933) found that nonhomologous synapsis was frequent in haploids, since most chromosomes appeared double at pachynema. Although foldback pairing was frequently observed by McClintock, the relative levels of foldback and heterologous pairing could not be determined. The chromosomes of haploids tend to clump together during pachynema, making analysis difficult. Ting (1966) reported that individual chromosomes of a haploid are occasionally separated from the mass of chromatin during pachynema, and the single chromosomes invariably displayed foldback pairing. Later studies of Ting (1971, 1973) demonstrated the presence of normal synaptinemal complexes within regions of foldback pairing in haploids. Similarly, Gilles (1973) found synaptinemal complexes in foldback regions of abnormal 10/normal 10 heteromorphic bivalents (section X, 1).

The appearance of synaptinemal complexes during foldback pairing, raises the question of whether crossing over may occur between nonhomologous chromosomal sites. Most evidence suggests that it does not. Nonhomologous synapsis in structural heterozygotes is associated with a reduction in crossing over (Burnham, 1934; Rhoades, 1968; and section IX, 3). Extensive attempts failed to select illegitimate crossovers from a structural heterozygote which frequently engaged in nonhomologous synapsis. (Weber, 1968). However, cytological observations with haploids have shown that bivalents are present at a low rate from diakinesis to anaphase I (McClintock, 1933; Snope, 1967; Ting, 1966). Chromosome associations at these late stages usually result from chiasma formation. In addition, Alexander (1964) isolated several products of crossing over from the progeny of haploids. Depending on the direction of pairing, foldback synapsis and exchange may yield inversion chromosomes while crossing over between different chromosomes may produce reciprocal translocations. Other products of non-homologous exchange, such as dicentrics and deficiencies, will generally be inviable. Alexander screened the diploid progeny of monoploid corn for plants with semisterile pollen. He recovered a number of reciprocal translocations. (As is discussed in section IX, 1 and 2, both translocations and inversions give semisterile pollen when heterozygous). The finding suggested that exchange occurs at a low rate, between non-homologous chromosomes. Subsequent characterization of the translocations, however, provided some of the best evidence against the occurrence of non-homologous crossing over. Twenty-two translocations recovered from monoploids were examined cytologically by Weber and Alexander (1972). Twelve of the translocations were independent isolations of the same aberration with exchange occurring between 6L and 7L at approximately the same cytological sites. The existence of preferred sites of exchange suggests the presence of duplicated segments within the maize genome that engage in homologous exchange. Two other translocations with breaks in 2L and 6L also appear to be repeats arising from recombination between homologous regions. The remaining translocations were dissimilar and could represent exchange between short or relatively divergent duplicated segments. Alternately, they could arise from non-homologous crossing over or non-meiotic, breakage-exchange. However, the finding that 14/22 translocations isolated arose by homologous crossing over in very limited regions of the corn genome indicates the rarity of truly non-homologous exchange.

Non-homologous associations are rarely observed in pachynema of structurally normal diploid plants with the exception of fusions between heterochromatic knobs and between centromeres. Knob fusions and centromeric fusions superficially resemble homologous synapsis. However, fusions are atypical in that they involve associations between two or more bivalents rather than two chromosomes. It would seem unlikely that crossing over occurs between non-homologous chromosomes involved in fusions, since the high rate of fusion in pachynema would produce frequent translocations in the progeny, which is not observed. Weber and Alexander (1972) found that centromeric fusions occur in monoploid as well as diploid corn, but the reciprocal translocations which they recovered from the progeny of monoploids did not involve breaks at centromeric sites. Gilles (1973) reported that the synaptinemal complexes which enter fusions retain their identity and traverse the region of fusions without disruption. Thus centromeric and heterochromatic fusions are phenomena that are superimposed upon synapsis, do not affect it, and do not initiate crossing over.

Another aspect of the association of non-homologues is the phenomenon of distributive pairing. Grell (1962) proposed that achiasmate pairing occurs between nonhomologues in *Drosophila* females when homologous pairing partners are not available. The pairing is maintained until anaphase, when the nonhomologues disjoin. The hypothesis accounts for the nonrandom distribution of univalents during meiosis in *Drosophila*. However, the absence of chiasmata suggests that the chromosomes are held together in first division of meiosis by an unusual mechanism. Whether distributive pairing is a unique property of *Drosophila* chromosomes was investigated by Michel (1966), Michel and Burnham (1969), and Weber (1966, 1969). Univalents were generated in several ways, but common to the investigations was the use of double trisomic plants. When bivalent-univalent formation occurs in double trisomics, two non-homologous univalents are available for distributive pairing. Pairing of nonhomologues at diakinesis and metaphase I was noted frequently by Michael (1966) and Michel and Burnham (1969), but rarely by Weber (1966, 1969). Chromosome distribution was analyzed at anaphase I or prophase II into 10 to 12 (non-disjunction) vs. 11 to 11 (disjunction) classes. Again, discordant results were obtained. While the different findings might be attributed to variations in genetic background of the plants being tested, the double trisomic condition with the frequent occurrence of trivalents rather than bivalents and univalents is not the best situation for analyzing pairing and disjunction of non-homologues. Satyanarayama (unpublished; cited by Weber, 1973) recently found a method in corn for producing large numbers of monosomic plants, including some double monosomes, which contain two univalent, nonhomologous chromosomes (section IV). Weber (1970, 1973) has utilized the monosome-generating system to construct several double monosomes (Fig. 8). The double monosomes have two univalents present in all meiotic cells and provide an ideal test of distributive pairing. Weber found a uniformly low rate of pairing at diakinesis (avg., 3.7% pairs) and metaphase I (avg., 2.2%) for five double monosome combinations. All associations (even atypical ones) were counted and the frequency of pairing may therefore be an overestimate. Disjunction of the univalents was determined in both anaphase I and prophase II and found to be random. Thus distributive pairing does not occur in corn and appears to be unique to *Drosophila*.

VIII. CROSSING OVER IN CORN

In this account of crossing over, we will discuss work with corn that has helped elucidate the process of crossing over. In light of considerable evidence we will assume that crossing over occurs during synapsis in meiotic prophase. The molecular mechanism of recombination is presently unknown, but recent discussions may be found in Fincham and Day (1971), Radding (1973), Sobell (1973) and Stadler (1973).

1. Cytological Demonstration of Genetic Crossing Over

Genetic data on crossing over are explicable on the assumption that an actual exchange of segments occurs between the paired chromosomes, and in 1931, Stern working with *Drosophila* and Creighton and McClintock with corn obtained proof of the correlation between cytological and genetic crossing over. The cytogenetic demonstration in corn was as follows. Certain strains have a large, easily recognizable knob on the end of the short arm of chromosome 9, whereas other strains have a chromosome 9, which is knobless. In the reciprocal translocation T8-9a, the interchange point in chromosome 9 is in the long arm. The presence or absence of both the knob and the translocation point afford two cytologically detectable markers on chromosome 9 which are constant features of those chromosomes which possess them and whose inheritance in subsequent generations can be followed with precision. The recessive genes, *yg2, c,* and *wx* lie in the short arm of 9 between the translocation point and the terminal knob, the *yg2* locus is close to the knob, while *wx* is nearest the translocation point. Individuals were obtained heterozygous for the terminal knob, the translocation point, and the recessive genes located between the two cytological markers. These plants were testcrossed by structurally normal individuals carrying the recessive alleles and having knobless chromosomes 9. The ensuing progenies were classified for genetic crossovers between the *yg2, c,* and *wx* loci; both crossover and noncrossover plants were examined cytologically at meiosis to determine if a physical exchange of chromatin occurred between the two cytological markers when a genetic crossover within this region took place. Creighton and McClintock (1931, 1935) found that genetic crossovers in regions between the two heteromorphic points were invariably accompanied by an exchange of segments between the two chromosomes. Stern (1931) in his extensive study in *Drosophila* found a similar correlation. Brink and Cooper (1935), using a different cytogenetic set up from that employed

THE CYTOGENETICS OF CORN

by Creighton and McClintock, reported additional evidence of this correlation in corn.

2. Chromatid Crossing Over

In corn as in *Drosophila*, it has been possible through genetic studies to demonstrate that the chromosomes are divided into chromatids at the time of crossing over and that each exchange involves only two of the four chromatids. The test of chromatid crossing over requires recovery of more than one of the four chromatids coming from a single bivalent. Chromatid exchange produces two crossover and two non-crossover chromatids, whereas crossing over at the chromosomal level produces all crossover chromatids. The recovery of both crossover and non-crossover chromatids from a bivalent in corn and *Drosophila* demonstrated that crossing over occurs at the chromatid level. In corn trisomic chromosome 5 plants allowed recovery of two of the four chromatids involved in a chiasma and proof of chromatid exchange (Rhoades, 1933b). More recently, the meiotic mutants *asynaptic* and *elongate* have been used in tests of chromatid crossing over, since they produce diploid (2N) eggs from diploid plants and allow recovery of two chromatids from the same meiocyte. (Nel, 1971; Rhoades and Dempsey 1966a). Recovery of two of the four meiotic chromatids is referred to as half-tetrad analysis. In *Neurospora* all four products of meiosis may be recovered in the ascus (tetrad analysis) providing a complete picture of recombination. Again, recombination is found to occur between chromatids rather than chromosomes.

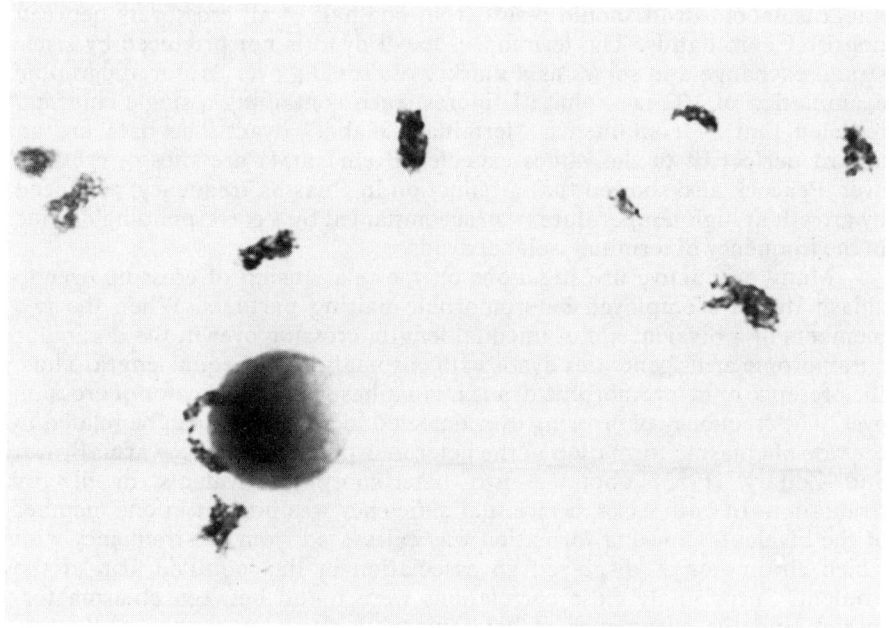

Fig. 8 — Double monosomic cell in diakinesis.

3. Relation Between Crossing Over and Chiasmata

At the end of pachynema, homologous chromosomes separate. Chiasmata prevent the complete separation of homologues and maintain the bivalent structure until anaphase I. It is obvious in favorable diplotene cells that chiasmata are points of reciprocal chromatid exchange between chromosomes. The general assumption is that crossing over between nonsister chromatids, probably in pachynema, is later revealed in the form of chiasmata. Convincing proof that chiasmata are sites of physical exchange between nonsister chromatids has been difficult to obtain. An early experiment with corn (Beadle, 1932c) correlated crossing over and chiasmata in a heterozygous translocation. The cytological situation was favorable for an accurate classification of chiasmata within a genetically marked chromosome segment. The expected ratio of 2:1 between chiasmata and crossing over was found. (The 2:1 relationship is expected on the basis that one-half of the chromatids in a tetrad are unaffected by a chiasma). However, the design of this experiment has been criticized (Rhoades, 1955) and subsequent work has also often been questioned (Fu and Sears, 1973).

Nevertheless, autoradiographic studies of Peacock (1970), 1971) have provided strong evidence of the relationship between chiasmata and crossing over. Grasshopper spermatocytes were examined in metaphase I following incorporation of H^3-thymidine at the last premeiotic interphase. In the absence of exchange, each dyad of the metaphase I tetrad contains one labelled and one unlabelled chromatid. However, crossing over or sister strand exchange produces chromatids which are partially labelled and partially unlabelled. Peacock observed that a particular class of metaphase I tetrad with a "terminal isolabel" dyad should result from one-half of all crossovers between nonsister chromatids. The terminal isolabel dyad is not produced by sister strand exchange and serves as a marker of crossing over. Autoradiographic examination of 102 metaphase I figures, each containing a single chiasma, revealed that 57 contained a "terminal isolabel" dyad. The data are an almost perfect fit to the values expected if chiasmata are sites of crossing over. Peacock also showed that a reduction in chiasma frequency, produced by growth at high temperature, was accompanied by a corresponding decline in the frequency of terminal isolabel dyads.

Many productive investigations on the relationship of crossing over to chiasmata have employed heteromorphic pairing partners. When the two members of a bivalent are of unequal length, crossing over in the dissimilar chromosome arms generates dyads with chromatids of unequal length. Thus, the presence of heteromorphic dyads in anaphase I is an indicator of crossing over. The frequency of crossing over detected in anaphase I can be related to the rate of chiasma formation in the heteromorphic chromosome arm. Brown and Zohary (1955) obtained two heteromorphic bivalents in lily by irradiation. In both cases, a terminal deficiency was present in one member of the bivalent. Chiasma formation was calculated from the frequency with which chromosomes displayed an association in the modified arm of the bivalent. Surprisingly close correlations were found between chiasma formation and the occurrence of heteromorphic dyads in anaphase. Kayano (1960) carried out a similar experiment in *Disporum sessile*. One of the sources of dissimilarity between pairing partners in this case was a reciprocal translocation. In a heterozygous translocation, crossing over between the centromere and the "cross" yields a heteromorphic dyad. The correlation made here was between chiasmata in the interstitial region (section IX, 1) and heteromorphic dyads at anaphase. Again, an excellent correlation was found. The only objection to this work is the absence of a direct correlation between chiasmata and crossing over, since the presence of heteromorphic dyads is not a genetic identification of recombination. However, we know that disjunction of the centromere is reductional in anaphase I (McClintock, cited by Rhoades, 1955), so that heteromorphic dyads could not arise through a reassortment of centromeres in the bivalent and equational disjunction. They must result from physical exchange between chromatids. In light of the experiments which relate genetical and cytological exchange (Creighton and McClintock, 1931; Stern, 1931) the findings of Brown and Zohary (1955) and of Kayano (1960) provide strong evidence that chiasmata are sites of genetic crossing over.

4. Chiasma Interference, Chromatid Interference, and the Upper Limit of Recombination

Chiasma interference is the well-known inhibitory effect of one chiasma on the formation of an adjacent chiasma. It is measured genetically by the coefficient of coincidence. In regions which are very short in terms of map units, interference is strong. As the distance between two potential chiasmata increases, interference declines to the point where it no longer operates. The coefficient of coincidence varies from 0.0 (complete interference) to 1.0 (no interference) but does not generally exceed 1.0 (negative interference). Studies of intragenic (allelic) recombination often yield interference values greater than 1, but the crossovers found within genes may result from gene conversion rather than conventional exchange (Fincham and Day, 1971; Stadler, 1973). The relationship between gene conversions and chiasmata is still not fully understood, but we will assume that chiasma interference is always positive.

Chiasma interference limits the number of chiasmata that may occur along the length of a chromosome arm. (Interference is generally not found to operate across the centromere). An upper limit to the number of chiasmata per arm produces a distribution of chiasmata between as well as within bivalents which is nonrandom. For example, in rye, almost all bivalents have one or two chiasmata with a deficiency in the frequency of bivalents with higher or lower numbers of exchanges. Jones (1967, 1974) identified a strain of rye in which chiasma interference was apparently absent and found a random distribution of chiasmata between bivalents which conformed to the theoretically expected Poisson distribution. Interestingly, the random distribution of chiasmata resulted in a significant frequency of univalents. The absence of interference did not, in this case, produce an increased chiasma frequency, but only a redistribution of chiasmata among the chromosomes so that bivalents with greater and fewer exchanges than normal were found.

Alterations in chiasma interference can produce localized changes in rates of crossing over. The B chromosome of corn, which increases crossing over in certain chromosomal regions (section X, 2), is known to operate in part by decreasing interference and thereby increasing the rate of multiple crossovers (Hanson, 1969). Rhoades (1968) found that a segment of chromosome 3, which does not normally engage in chiasma formation (section VIII, 6) nevertheless affects the rate of crossing over in regions adjacent to it. A piece of chromosome 3 was transferred by rearrangement to the short arm of chromosome 9 between the Sh and Wx loci. Rhoades found with the modified chromosome 9 homozygote that regions distal to the inserted segment were enhanced in recombination. The finding was explained as an affect on chiasma interference in which the inserted segment acts as a spacer between proximal and distal regions of 9S. Interference was shown to be low for double exchanges which spanned the chromosome 3 insertion.

Chromatid interference refers to the influence of those chromatids involved in one chiasma on the chromatids selected for exchange at an adjacent site. Within a bivalent there are three possible kinds of double chiasmata. Two strand doubles are those in which the same two chromatids are involved in both chiasmata. In three strand doubles, one of the chromatids involved in the first chiasma does not participate in the second exchange. There are two types of three strand doubles depending on the chromatids participating. In four strand exchanges, the two noncrossover chromatids at the first chiasma exchange at the second site. Each type of double exchange produces a unique set (tetrad) of chromatids which can be identified genetically with appropriate markers. If selection of chromatids is random and not influenced by adjacent chiasmata, the proportions of two, three, and four strand double exchanges will be 1:2:1. Any deviation from randomness can be classified as chromatid interference. Positive chromatid interference would give an excess of four strand doubles while negative interference would enhance the frequency of two strand exchanges. Identification of the different classes of double chiasmata is relatively difficult in most organisms since the four products of meiosis cannot be recovered intact. Thus tetrad analysis in corn and *Drosophila* is ruled out. However, half tetrad analysis has been utilized in corn and *Drosophila* to provide evidence of random chromatid crossing over (Beadle and Emerson, 1935; Emerson and Beadle, 1933; Rhoades and Dempsey, 1966a). A combined cytological and genetical study of chromatid interference was performed by Rhoades and Dempsey (1953), utilizing the fact that a four strand double exchange in an inversion loop can be identified by its unique cytological product: a double dicentric bridge with two acentric fragments. Results again indicated a random selection of chromatids.

Not all organisms have a completely random assortment of two strand, three strand, and four strand double exchanges. Extensive work with *Neurospora* tetrads has demonstrated a slight excess of two strand doubles and a tendency toward two strand doubles has been reported for some other organisms (Fincham and Day, 1971). Nevertheless, little data deviate sharply from the 1:2:1 ratio, and randomness may generally be assumed.

The upper limit of genetic recombination can be calculated if one assumes no chromatid interference. If in a given bivalent there is one exchange between two loci, 50% of the chromatids will be crossovers and 50%

THE CYTOGENETICS OF CORN

will be noncrossovers, since only two of the four chromatids are involved in the exchange of segments. When two chiasmata occur between two loci, the genetic results are more complex. Two strand doubles between two loci yield two double crossover and two noncrossover chromatids. Three strand doubles produce one noncrossover, two single-crossover, and one double crossover chromatid: whereas in four strand double exchanges all four chromatids are single crossovers. The resulting strands from all double exchanges assuming no chromatid interference, are found in a ratio of 1 noncrossover: 2 single crossovers: 1 double crossover. Since the double crossover chromatids have the two loci in the parental association they are not recombinants, as is true for the noncrossover strands. All of the single crossover chromatids are recombinations, so the ratio of chromatids with parental recombinations to new or recombinations is 1:1. A similar ratio is found when more than two exchanges occur (Emerson and Rhoades, 1933). It follows that the percentage of recombination between any two genes no matter how far apart they are on the linkage map can never exceed 50 with random chromatid crossing over. That 50% is the upper limit of recombination in corn is illustrated by the following data from Emerson and Rhoades.

Map Distances		Recombination Values	
P	0	P-as	25
as	25	P-f	41
f	58	P-an	45
an	75	P-$bm2$	49
$bm2$	128		

Although P and $bm2$ are 128 map units apart, they show approximately 50% recombination. It should also be noted that the approach to 50% recombination is gradual and does not occur as soon as the map distance reaches 50%. Even in relatively long map lengths, a chiasma may occasionally fail to occur, and 50% recombination is not reached until all cells have one or more chiasmata.

5. Sister-strand Exchange

Breakage and exchange between sister strands of a chromosome is not detectable by most cytological or genetical techniques. However, sister exchange in a ring chromosome produces a dicentric ring chromosome which is readily seen as a double bridge in anaphase of mitotic divisions. Dicentric rings were first noted somatically by McClintock, who suggested (1938b) that they could arise from sister exchange. That sister exchange also occurs meiotically was demonstrated by Schwartz's (1953a, b) work with a ring chromosome 6. Schwartz studied the cytological consequences of crossing over in corn heterozygous for a normal chromosome 6 and a ring-shaped chromosome 6 which was deficient for part of the satellite region in 6S and for a small distal segment of the long arm. Due to the shortness of 6S, only crossovers in the long arm between the ring and rod chromosomes 6 were considered. Schwartz determined the anaphase configurations expected for a single chiasma, and all possible double chiasmata. Conventional crossing over can produce dicentric bridges in AI and occasionally in AII. The expected configurations were found plus an unexpected class of double bridges in AII (10%). The double bridges at AII arise from sister strand crossing over in the ring chromosome (plus no crossing over or a two strand double exchange between nonsister chromatids). In addition, the number of double bridges in AI was expected to equal the number of cells with a single bridge in anaphase II, assuming no sister strand exchange. An excess of AII single bridges was found which also can be explained by sister strand exchange. The findings of Schwartz were recently confirmed by Miles (1970), who analyzed crossing over in a ring-rod heterozygote of chromosome 10.

The significance of sister strand exchange remains a question. It does not seem to be a vital phenomenon. Most of the evidence indicates that sister strand crossing over does not occur in *Drosophila*. If sister strand crossing over in meiosis and mitosis result from a common event, the mechanism of sister exchange may be quite different from that producing genetic recombination, which is restricted to meiosis. Sister exchange and genetic crossing over could be temporally separated during meiosis and, for example, sister strand crossing over might occur as an anomaly of DNA replication rather than as a component of the recombination mechanism. On the other hand, Schwartz's data indicate that sister strand crossing over is much more frequent in meiosis than mitosis, so that a relationship may exist between sister exchange and conventional recombination.

6. Distribution of Crossovers in the Cytological Chromosomes

In *Drosophila* a comparison of the cytological and genetical maps indicates that crossing over per unit of physical length near the centromere is less frequent than in distal regions. This reduction in crossover frequency has been attributed to the influence of the centromere, (Beadle, 1932d; Mather, 1939) since it has been demonstrated that an increase in recombination values occurs when a proximal segment is shifted away from the centromere. Conversely, regions normally near the distal ends undergo a reduction in crossing over when brought close to the centromere. There is evidence in corn that crossing over in distal segments is much more frequent per unit of physical length than in more proximally located regions. Some data suggest that the corn centromere has an inhibitory effect on chiasma formation (Rhoades and Dempsey, 1953; Yu and Peterson, 1973).

Perhaps the most striking evidence of a disproportionate frequency of crossing over in distal regions of corn chromosomes comes from studies with the short arm of chromosome 9. The total length of the genetic map of the short arm is between 60 and 70 units. The *wx* locus lies approximately in the middle of the short arm of the pachytene chromosome. Nevertheless, the distal half of the short arm extending from the *wx* locus to that of *Dt*, which is in or near to the terminal knob, has a map length of 59 units (Rhoades, 1945). The *yg2* locus is situated in the terminal chromomere of the short arm (McClintock, 1944). Disregarding the heteropycnotic terminal knob, the only part of the short arm distal to *yg2* is the threadlike strand which connects the last chromomere and the knob. Although physically very near one another the *Dt* and *yg2* loci are seven map units apart.

Evidence for the unequal distribution of crossovers along the chromosomes also comes from studies with chromosome 2. Morgan (1950) found that in In2b one breakpoint was in the middle of the short arm of chromosome 2 with the other in 2L. The map length of the short arm is about 79 crossover units, and it was found that 66 of these units are distal to the inversion breakpoint, although they must be contained in only one-half of the short arm.

The higher rate of crossing over in distal regions may be related to the process of synapsis. Burnham et al. (1972) provided evidence that synapsis in corn chromosomes begins in subterminal regions and never in the vicinity of the centromere. The finding may reflect a greater synaptic ability in distal vs. proximal regions, which produces a higher rate of crossing over. Alternatively, distal regions may be synapsed for longer periods of time than proximal regions, allowing more time for recombination. A third possibility is that chiasmata form first in distal regions in the absence of chiasma interference, while proximal chiasmata form later and are severaly limited by interference. Clues to the proximal vs. distal control of chiasma formation may be present in experiments with agents which affect the rate of crossing over. Proximal chromosomal regions seem particularly sensitive to changes in the rate of recombination. In corn, abnormal chromosome 10 has a striking ability to enhance recombination for all chromosomes of the complement, and its primary effect is in proximal chromosomal regions. (Miles, 1970; Nel, 1973; Rhoades and Dempsey 1957, 1966b). Crossing over in proximal regions of *Drosophila* chromosomes also can be altered by a number of agents, and the subject is reviewed by Nel (1971, 1973).

A striking example of nonrandom chromosomal distribution of chiasmata which is not related to proximity to the centromere comes from recent findings of Rhoades (1968) with a segment of chromosome 3 that was inserted into the short arm of chromosome 9. The normal location of the segment is not near the centromere of chromosome 3, since the proximal breakpoint is about 0.6 of the long arm away. However, the segment, consisting of 10% of 3L, seldom if ever engages in chiasma formation. Its removal from chromosome 3 does not reduce recombination distances for flanking markers in the deficient 3, nor does insertion of the segment in chromosome 9 increase the genetic length of 9S. The absence of chiasmata at both the normal and transposed locations of this segment suggests an autonomous control over recombination. The latent ability of the region to undergo chiasma formation was demonstrated by the specific inducement of recombination in the presence of B chromosomes (Rhoades 1968).

The purposes of variability in recombination potential are speculative. Restriction of recombination may be useful in keeping coadapted blocks of genes together (Dobzhansky, 1970). Suppression of crossing over might also be favored in regions of repetitive DNA (section IX, 3) where presumably oblique synapsis and unequal crossing over could occur. However, some differences in rates of recombination may be related only to the mechanism of synapsis and not to fitness.

7. Crossing Over in Male and Female Flowers

It is of both practical and theoretical interest to know whether rates of crossing over differ during micro and mega sporogenesis. The work of Burnham (1950a, 1953, 1958, 1961), Clark (1956), Collins and Kempton (1927), Emerson and Hutchison (1921), Eyster (1921, 1922), Nel (1973), Phillips (1969), Rhoades and Rhoades (1939), Rhoades (1941), Stadler (1926), and others have led to some general conclusions.

1. Many regions, including all those tested on chromosomes 2 and 10 have similar rates of crossing over in male and female gametes.
2. When rates of crossing over differ between the sexes, it is almost always higher in the male gametes.
3. Differences in rates of recombination between male and female flowers may appear in one genetic environment or chromosomal arrangement but not another.

Interpretations of differences in male vs. female crossing over are complicated by the variety of results obtained under different circumstances. Burnham (1950a) and Clark (1956) found that recombination in a translocation heterozygote may show differences in crossing over between the male and female flowers, but no influence of sex is found in the homozygous arrangement (normal or translocation chromosomes). Phillips (1969) found a variety of effects of interchanges on crossing over in reciprocal crosses and also found that the genetic background influences variation in crossing over between male and female flowers. Data of Nel (1973) suggest that changes in recombination produced by supernumerary elements may depend in part on sex. He found that the B chromosome enhances crossing over more in the male than the female and that abnormal chromosome 10 has the reverse effect. These experiments suggest caution in relating cytological work in the male to genetic results in the female, even in chromosomal regions where previous tests show no differences in rates of recombination.

8. Allelic Recombination

Recombination studies in corn and most eukaryotes have been largely confined to intergenic crossing over, since intragenic (allelic) crossing over requires excessively large population sizes. However, Benzer's classic experiments on allelic crossing over in the virus T4 prompted Nelson (1957) to question whether the gene in eukaryotes could also be subdivided by recombination. He suggested use of the corn gametophyte rather than the sporophyte to obtain the large sample size needed. Since the waxy (wx) gene is expressed in the pollen and can be readily classified, this gene was proposed for testing. Nelson gathered recessive alleles of wx and tested recombination between them in heteroallelic combinations. Pollen was collected from plants carrying heteroalleles and classified on slides (by I-KI staining) into recessive (brown) and dominant (black) phenotypes. Plants with two *waxy* alleles produced mainly recessive pollen, but an occasional dominant resulted from recombination. The dominant was seen as an isolated dark-staining pollen grain in a field of lighter staining pollen. Rapid classification of large populations was possible and Nelson was able to subdivide the *wx* locus and map sites within it (1959, 1962). The validity of the approach was demonstrated with large-scale conventional tests of crossing over between *wx* heteroalleles (Nelson, 1962, 1968).

More recently (1968), Nelson utilized the waxy system to map the position of several controlling elements within the *wx* locus. Controlling elements have some of the properties of bacterial episomes (Peterson, 1970) and upon insertion into the chromosome may repress gene activity at the site of insertion. Nelson obtained three *wx* alleles, each of which arose from the dominant *Wx* allele following insertion of a controlling element. The position of each controlling element within the gene map was determined by recombination tests with other *wx* alleles. Surprisingly, each of the controlling elements was found at a different position within the locus and none of the controlling elements was positioned at one end of the gene. The finding makes less likely the suggestion that controlling elements regulate gene activity by repressing RNA synthesis. By analogy to bacterial regulatory systems, one expects controlling elements to be inserted at one end of the *wx* locus if they control transcription.

Yu and Peterson (1973) adapted the waxy system to test the effect of the centromere on crossing over. The centromere has an inhibiting effect on intergenic crossing over in *Drosophila,* and perhaps in corn (section VIII, 6). Much evidence now suggests that intergenic and allelic crossing over are closely related phenomena, so that a change in recombination in the waxy locus may well reflect a general modification of recombination. Allelic recombination was selected for analysis since a region of constant genetic length, the waxy locus, could be tested for recombination at many distances from the centromere. Recessive alleles were transferred into translocation chromosomes by crossing over and homozygous translocations, heteroallelic for recessive waxy alleles, were constructed. Recombination between waxy heteroalleles was determined by pollen classification in several different translocation positions. The results suggested that crossing over increases with distance of a gene from the centromere and is unaffected by proximity to the distal end of a chromosome arm.

Gametophytic classification of genes is an interesting concept and highly rewarding in the case of the waxy locus. However, it is limited by the number of genes that can be classified in the pollen. Only one other gene, amylose extender (*ae*), has been shown to have a suitable pollen phenotype (Creech and Kramer, 1961). Moore and Creech (1972) recently detected allelic recombination at the *ae* locus.

Conventional genetic techniques have supplemented gametophytic analyses to demonstrate allelic recombination in corn. The work of Salamini and Lorenzoni (1970) with glossy (*gl*) mutants is of importance, since evidence for gene conversion was found. Basically, gene conversion occurs when a heterozygote, *Aa,* yields $3A$ and $1a$ or $3a$ and $1A$ chromatids from a meiotic tetrad. An allele on one chromatid is said to convert the allele on another chromatid to its own type. Gene conversion occurs over much shorter distances than the gene itself, and conversion involves a mutant site plus some adjacent nucleotides. As a result, recessive heteroallelic combinations of genes (*gl-a/gl-b*) can give wild type (*Gl*) or the double mutant (*gl-a,b*) by gene conversion. The event for a single chromatid (i.e., *Gl*) appears to be a simple intragenic crossover. However, tetrad analysis reveals the nonreciprocal nature of the event. For the converted site within the *gl* locus, there is a ratio of 3 to 1. Also, unlike conventional crossovers, the *Gl* or double mutant individuals need not be recombinant for genetic markers on either side of the locus. Salamini and Lorenzoni studied recombination between different *gl* heteroallelic combinations in plants of the constitution *O2 gl-a S1/o2 gl-b sl*. Wild type *Gl* recombinants were selected and the chromosomes were analyzed for crossing over between *o2* and *sl*. In more than one-half of the cases, intragenic recombination was not accompanied by crossing over of the flanking markers, even though the *o2* and *sl* loci are only 20 map units apart. Nelson (1962) found similar results in conventional genetic analyses of crossing over between *wx* heteroalleles. The finding of intragenic crossing over in the absence of intergenic exchange is strong evidence for gene conversion, although the critical demonstration of conversion requires tetrad analysis. Gene conversion is critical to current theories of recombination which unite allelic and intergenic recombination (Fincham and Day, 1971; Stadler, 1973) and tetrad analysis in the fungi and *Drosophila* as well as the work described here indicate that gene conversion is a general phenomenon.

IX. CHROMOSOMAL REARRANGEMENTS
1. Translocations

Translocations are exchanges of segments between nonhomologous chromosomes. Simple translocations, where a fragment of one chromosome becomes terminally attached to an unbroken end of another chromosome, have never been found in corn or in any other organism. Broken ends of chromosomes are said to be "sticky" and will unite only with other broken ends, thus restricting the types of chromosomal rearrangements that occur. Reciprocal translocations arise when breaks occur in two different chromosomes, producing four chromosome fragments each with a broken end. Rejoining between appropriate fragments can produce two interchanged monocentric chromosomes each with blocks of genes from the two parental nonhomologous chromosomes. These interchanged chromosomes behave normally throughout the somatic mitoses and are completely stable. In plants homozygous for the interchange there are two new pairs of chromosomes whose behavior is in no way different from that of the other eight pairs.

The first translocations found in corn were of spontaneous origin. However, the great majority of those which have been studied were induced by ionizing radiation. Longley (1961) lists over 1,000 different translocations whose points of interchange have been cytologically determined, and nearly all of these were induced. Longley concluded from his extensive work with translocations that breakpoints in the chromosomes were much more frequent in heterochromatic regions, particularly in segments adjacent to the centromere. Longley considered that heterochromatic regions are especially susceptible to chromosome breakage. Jancey and Walden (1972) also analyzed the exchange points of corn translocations. They found a nonrandom distribution of breakpoints, with exchanges near the centromere being preferred. In addition, a nonrandom pattern of chromosome association was

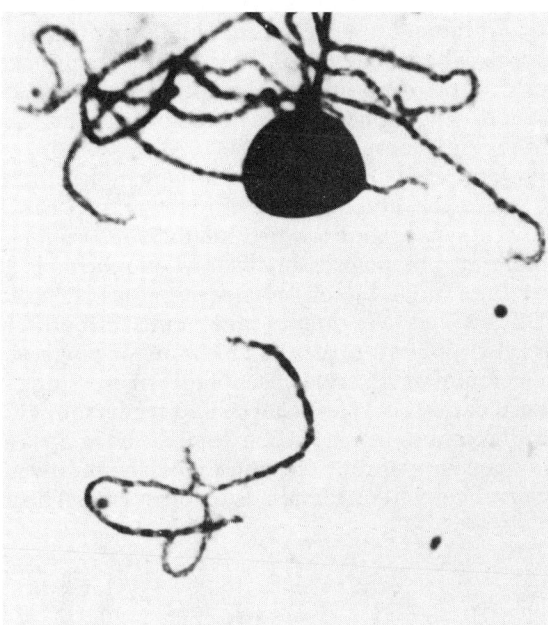

Fig. 9. Heterozygous translocation in pachynema.

found with the translocations. For example, when the exchange point of one chromosome of a translocation was within the centromere, the second chromosome also frequently had its breakpoint through the centromere. Nonrandom joining of broken chromosomal regions suggests that a specific organization exists within the interphase nucleus which orients chromosomes with respect to each other and restricts the opportunities for exchange.

Detection of the exchange points of a translocation can be made in pachynema. In plants heterozygous for a translocation, two normal and two interchanged chromosomes are present. A four-armed synaptic configuration is produced at pachynema by the pairing of homologous regions (Fig. 9). The center of the cross configuration denotes the breakpoints if strictly homologous pairing occurs. However, Burnham (1932a, 1934) and McClintock (1933) have shown that the position of the cross is not constant. Shifting of the position of the cross from the true position involves the pairing of nonhomologous regions.

The nonhomologous associations frequently found in translocation figures makes it difficult to ascertain the exact point of interchange, unless the breakpoints occurred in regions of strikingly dissimilar chomomere pattern. The points of interchange determine the lengths of the four arms of the multivalent figure at pachynema. Two opposite arms of the cross have centromeres, but the other two are without centromeres. The segments between the centromeres and the center of the cross are known as interstitial regions. If both points of interchange were close to the centromeres, crossovers would infrequently occur in the interstitial region. Translocations of this type generally form open rings of four at diakinesis (Fig. 10), since there will be one or more crossovers in each of the four arms.

However, if an additional crossover occurs in an interstitial region, a "figure 8" complex results. The frequency of these figure 8 complexes is determined by the number of crossovers in the interstitial segments; it will be high in those translocations where one or both of the breakpoints are some distance removed from the centromere.

One of the four arms of the cross configuration will be short in those translocations where the break in one chromosome is near the end. There may be a failure of synapsis of the two homologous regions comprising the short arm or there may be no crossing over when pairing is achieved. Chains (Fig. 11) rather than rings of four characterize translocations of this type (Burnham, 1932b). When the interchange points in both chromosomes are nearly terminal two "pairs" rather than chains or rings of four are found because of the lack of pairing and consequently of crossing over in both short arms of the translocation complex (Clarke and Anderson, 1935).

Following diakinesis, translocation figures move onto the MI spindle. Considering an open ring of four chromosomes, the multivalent may assume either a zigzag arrangement or remain as an open ring. The zigzag or twisted

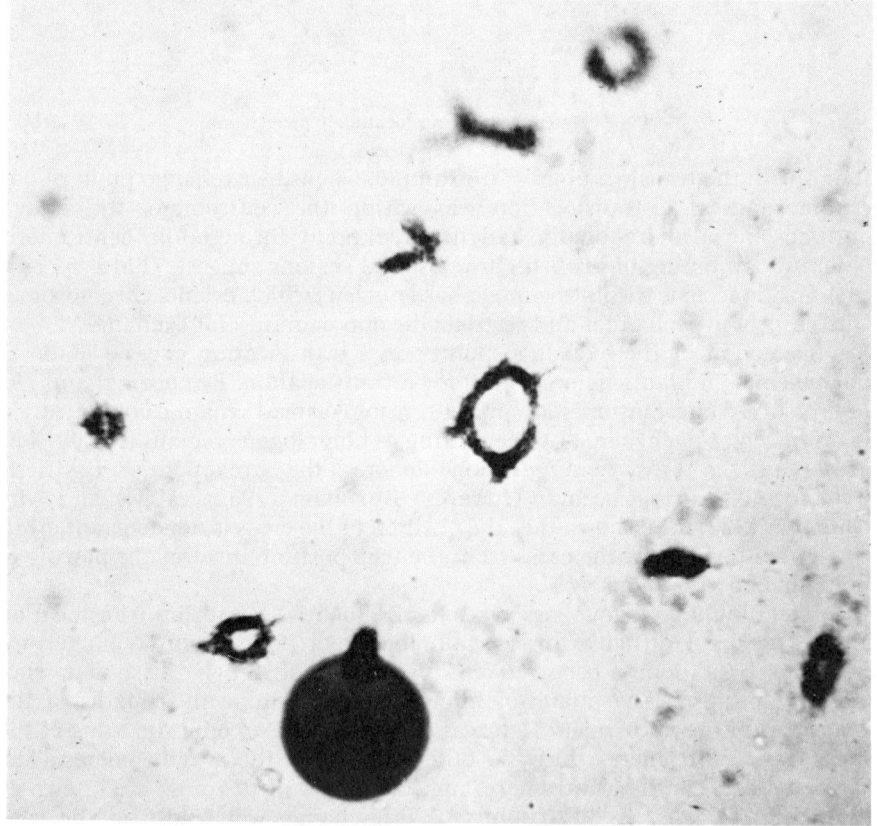

Fig. 10. Heterozygous translocation in diakinesis. Chiasmata in each of the four arms of the "cross" produced this open ring configuration.

ring orientation leads to alternate chromosomes of the ring passing to the same pole at anaphase. If, for example, we have a translocation involving chromosomes 2 and 5 with one break near the centromere in 2L and one in 5S also close to the centromere, the ring of four will consist of two normal chromosomes, 2 and 5, and two interchanged chromosomes, 2^5 and 5^2. The 2^5 chromosome has a centromere from chromosome 2, while the 5^2 chromosome obtained its centromere from chromosome 5. With alternate disjunction the normal chromosome 2 and chromosome 5 go together and the two interchanged chromosomes 2^5 and 5^2 pass to the opposite pole at anaphase I. The quartet of spores formed at the end of the two meiotic divisions will consist of two microspores with normal chromosomes and two with the interchanged chromosomes. All four will develop into functional male gametophytes, since none is deficient or duplicate for any genes.

However, instead of the ring of four assuming the zigzag orientation on the spindle it may become attached to the spindle as an open ring. This will result in adjacent members of the ring passing to the same pole. In adjacent-1 segregation homologous centromeres pass to opposite poles, i.e., chromosome 2 goes with 5^2 while chromosome 5 goes with 2^5. All four spores of the resulting quartet will be deficient and duplicate; consequently they will abort. With adjacent-2 disjunction, homologous centromeres pass to the same pole. In the above example this would mean that chromosome 2 went with 2^5 and 5 with 5^2. As in the case of adjacent-1 segregation all four microspores would abort because of their deficient-duplicate constitution.

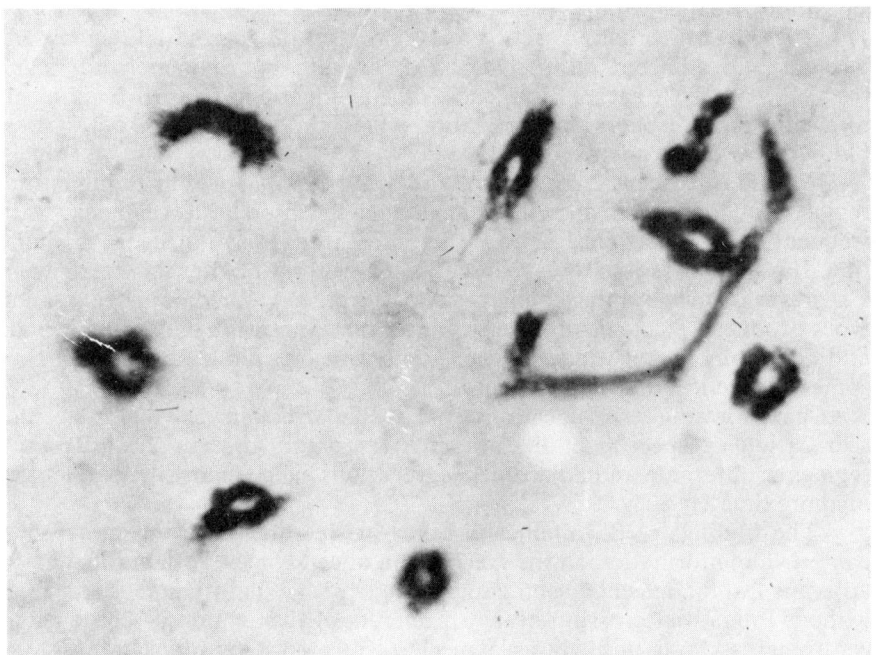

Fig. 11. Heterozygous translocation in diakinesis. Failure of exchange in one of the four arms of the "cross" gave this chain of 4 figure.

If alternate and the two kinds of adjacent segregation occurred with equal frequency, two-thirds of the pollen would be abortive. However, since approximately 50% of the pollen is viable, the number of pollen mother cells with an alternate orientation is equal to the sum of those with the two types of adjacent segregation. Ovule sterility is also about 50%, indicating that orientation of the ring of four is essentially the same in megasporocytes as in microsporocytes. It has recently been shown (Endrizzi, 1974) that two types of alternate orientation exist, which may be visualized as transformations of adjacent 1 and adjacent 2 rings into a zigzag arrangement by a twist of the rings. Both types of zigzag rings give the same segregational products and the two classes of rings are generally not distinguishable cytologically. However, in terms of orientation they are separate forms. If the four possible types of orientation (adjacent 1, adjacent 2, two alternate classes) occur with equal frequency, the 50% gametic sterility is explicable.

The use of chromosome 6 with its nucleolar organizer has made it possible to distinguish between alternate, adjacent 1, and adjacent 2 segregation. This is owing to the fact that different kinds of segregation lead to recognizably different quartets of microspores in terms of the number of nucleoli present. When all cells of a quartet receive one nucleolus organizer, as normally occurs, each cell develops one nucleolus. However, segregation from a translocation involving chromosome 6 may lead to unbalanced spores in which cells with two nucleoli and zero nucleoli are found within a quartet. Burnham (1949, 1950b), following the earlier work of McClintock (1934), made an extensive study of nucleolar segregation with chromosome 6 translocations. He reached the following conclusions:

1) In those translocations with no crossing over in the interstitial regions, adjacent-1 and adjacent-2 segregations occur with approximately the same frequency (ca. 25% each). Both lead to the production of abortive deficient-duplicate spores. Alternate segregation which produces viable spores takes place in 50% of the microsporocytes.

2) There is no adjacent-2 segregation following crossing over in an interstitial region. In those translocations having long interstitial segments with frequent crossovers (figure 8 rings) there was little if any adjacent-2 disjunction. It was not possible to determine the percentage of alternate to adjacent 1 segregations when interstitial crossovers occurred, since both segregation types produced cytologically identical quartets. Interestingly, both alternate and adjacent 1 segregations yield quartets with two viable and two abortive spores. Gametic lethality is 50% as a result.

3) Chain-forming translocations have low frequencies of adjacent-2 segregation irrespective of the amount of crossing over in the interstitial segments. Alternate and adjacent 1 segregations each occur with frequencies of approximately 50%.

The findings of Burnham and others with translocation heterozygotes can provide information on the mechanism of centromeric orientation. Since adjacent 1 and adjacent 2 segregations occur with equal frequencies in rings without interstitial crossing over, a preferential disjunction of homologous centromeres seems unlikely as a mechanism of orientation. However, the marked tendency of translocation figures to disjoin, with two chromosomes going to each pole, indicates that a mechanism regulating chromosome movement is in effect in multivalent figures as well as bivalents. In this regard, Nicklas and Koch (1969) found with grasshopper bivalents that stable centromeric orientation to the poles requires tension on the centromere. Tension is produced by the two centromeres of a bivalent pulling in opposite directions. In the absence of tension, as in unipolar orientation, spindle attachments are unstable and reorientations occur until a bipolar arrangement is achieved. The regular 2 to 2 disjunction of translocation heterozygotes may occur to satisfy a requirement for tension on all centromeres of the figure. Also, the absence of adjacent 2 segregation from chains of 4 may result from failure to meet the need for centromeric tension on all the chromosomes. Nicklas has suggested (1967) that a second centromeric property which insures bipolar orientation is the tendency of centromeres to attach spindle fibers in a particular direction. When bivalents are detached by microsurgery from their spindle fiber attachments, they reorient in the direction in which they are approximately "facing." In other words, centromeres have "faces" which in meiotic bivalents are oriented in opposite directions. Perhaps the inhibition of adjacent-2 segregation from translocation heterozygotes by interstitial chiasmata depends on the juxtaposition of homologous centromeres and their tendency to "face" in opposite directions.

Although several different chromosomal combinations are produced by segregation from a translocation heterozygote, genetic recombination is influenced only by the two classes of gametes which do not abort. In the absence of interstitial chiasmata, functional spores come only from alternate segregation, which produces gametes with two normal or two interchange chromosomes in equal amounts. When a plant heterozygous for a translocation is crossed with a chromosomally normal plant, two kinds of zygotes are produced: those homozygous normal and those heterozygous for the translocation. The homozygous normal plants have no pollen or ovule sterility, whereas heterozygotes have 50% aborted pollen and ovules. Since the occurrence of aborted pollen and ovules marks of the presence of the interchange chromosome, inheritance of the translocation may be followed in linkage studies as if it were a dominant gene for semisterility. The center of the pachytene cross configuration represents the translocation "gene" and is designated by the symbol T. Genes lying in the four arms of the cross will be linked with T (semisterility) and with each other. When only the recombination values between T and the loci of a single arm are considered, the usual linear map is obtained, with T located at one end of the map. However, when recombination values between genes in all four arms are determined, the combined linkage data do not give a linear order but form a cross-shaped map with T at the center of the cross.

Recombination values for regions adjacent to the translocation point are usually lower in translocation heterozygotes than in normal plants. This reduction in crossing over is produced by imperfect synapsis in pachynema. In many pachytene figures, it is observed that homologous regions are loosely paired at the exchange points. This asynapsis can account for some of the reduced crossing over. Also, nonhomologous pairing is frequent in the center of the cross configuration, as shown by variation in the position of the exchange point (Burnham 1932a, 1934; McClintock 1933). Since crossing over rarely, if ever, occurs between nonhomologous regions (section VII), an inhibition of crossing over due to nonhomologous pairing is expected. While heterozygosity for a translocation frequently inhibits crossing over markedly in the region of exchange, there are exceptions to this rule (Anderson et al., 1955; Maguire, 1968; Roberts, 1972). Roberts found in *Drosophila* that median or medial-distal breaks in the chromosome arms had the greatest depressive effect on recombination, whereas translocations near the distal tip had little affect on crossing over. Anderson found a proximal region of chromosome 4 that was unaffected in recombination by the presence of the translocation cross figure. Maguire (1968) found, in a *Zea-Tripsacum* interchange, little reduction in crossing over near the exchange point.

The T marker of translocation heterozygotes offers a number of advantages over conventional genetic markers in studying the inheritance and linkage relations of genes determining agronomic characters. In particular, regions of chromosomes devoid of known mutant genes can be marked by translocations and analyzed for agronomically important genes. Also, the presence of the interchange chromosomes does not modify the morphology of the plant, and the frequent inhibition of recombination at the translocation point increases the effectiveness of linkage tests. Traits such as ear length, days-to-silking, smut resistance, etc., have been linked to specific chromosomes with translocations by many workers.

Translocations have also proved useful in the cytological placement of new gene mutations. The assignment of a gene to the correct chromosome may be accomplished in corn in a number of ways:
1) a complete chromosomal series of trisomic plants crossed to the mutant will identify the correct chromosome in backcrosses by the unusual genetic ratio found with a particular trisomic
2) recombination of the mutant with a number of marker genes spanning all the linkage groups will place the gene on a chromosome and with a linkage group
3) the T marker of a translocation can be used in the same way as marker genes
4) certain methods, that will be described, utilize translocations to make chromosomal segments hemizygous. If a recessive mutant is combined with a chromosome lacking the dominant allele, the mutant will be expressed and thereby located.

Of the methods described, procedures 1 and 2 are seldom used today. Trisomic plants are weak and somewhat difficult to maintain in stocks. Classification of trisomics, which is essential to their efficient use, is not always clear-cut when morphological characters are utilized, and cytological classification is time-consuming. Conventional marker genes also have drawbacks in the assignment of genes to chromosomes. Easily classified marker genes that are located so as to cover the genetic length of a chromosome are not always available. The remaining more practical procedures 3) and 4) involve the use of translocations.

Let us consider linkage of a newly arisen mutant gene to the translocation marker T. The advantages of using the T marker in recombination tests include the fact that it is linked to regions of two chromosomes rather than one. Thus one marker allows a linkage test for two chromosomes, with the slight disadvantage that once a gene is linked to T further tests are needed to determine which of the two chromosomes carries the locus. A discrimination between the two chromosomes can be made with standard linkage tests or with linkage data from other translocation crosses. For example, a gene that is linked to the T marker of T1-3 and T2-3 must be located on the common chromosome 3. A further advantage of translocations in assigning loci to chromosomes derives from the frequent reduction of crossing over in the vicinity of T. The inhibition of crossing over enhances the linkage of T with genes in the four arms of the translocation and reduces the chance that genes some distance from the break point will not be linked to T.

The main problem with the utilization of translocations in linkage tests is the necessity of classifying T. The classification requires growing plants to maturity and checking the pollen with a hand microscope for semisterile vs normal pollen. Alternately, plants are grown to maturity and allowed to open pollinate. Reduced seed set marks the T/N plants. Both procedures are tedious and a simple alternative is available. If a translocation point occurs near a known marker gene that is readily classified, the marker can be used as a substitute for T (Anderson, 1956). Due to poor synapsis at the exchange point the locus may show very little recombination with T and serve as an accurate substitute for it. If the marker is expressed in the endosperm (as well as the gene being tested for linkage) germination of seeds is not even needed. A series of 20 translocations, with each translocation linked to the *waxy* locus of chromosome 9, has been developed. The series is maintained by the corn stock center at the University of Illinois. The translocations were selected to provide good coverage of the corn genome. With this series, the advantage of testing linkage for two chromosomes in one cross is lost, since chromosome 9 is common to all translocations. However, the advantage of the system lies in the fact that a very easily classified endosperm marker, *waxy,* can be used to test the linkage of genes to all chromosomes of the complement.

Hemizygous methods for placing mutant genes on the chromosome have the considerable advantage of requiring only one generation of crossing. Hemizygosity is established in the F_1, and if the mutant phenotype is expressed, the gene is located to the hemizygous region. One procedure for establishing hemizygosity utilizes translocation heterozygotes which yield viable unbalanced gametes. In those translocations where one break is near the end of one chromosome, adjacent-1 disjunction yields spores deficient for the segment distal to the break and duplicate for the part of the other chromosome. If the deficient segment is not too long so that genes essential for gametophyte development are not included within it, these deficient-duplicate spores can develop into functional female gametophytes. (The male gametophyte is much more sensitive to genic imbalance than the female and pollen transmission is unlikely. Since many deficient-duplicate pollen grains appear normal or nearly so with the ordinary microscope, inviability may frequently result from pollen competition rather than pollen abortion). When the egg of a deficient-duplicate gametophyte is fertilized by sperm carrying a recessive allele, the F_1 individual will display the recessive character, if the locus lies within the deficient segment. Patterson (1952, 1959) demonstrated the validity of this approach by placing several loci on the cytological maps of chromosomes 2 and 9. However, only short terminal regions of the chromosomes can be analyzed in this way and the procedure is therefore of limited usefulness.

In addition to generating hemizygosity, translocations with viable deficient-duplicate meiotic products give abnormal genetic ratios in testcrosses. One-half of the adjacent 1 gametes are viable, and genes associated with the surviving deficient-duplicate gamete are elevated in frequency with respect to alleles linked to the abortive chromosomal combination (Burnham, 1932b). Mutant genes may be tested for linkage to translocations with viable deficient-duplicate gametes by construction of the translocation/mutant heterozygote and testcrossing to the mutant homozygote. In the absence of linkage, a 1:1 genetic ratio is expected, whereas complete linkage of the mutant gene to the translocation exchange point gives a 2:1 ratio. (The 2:1 ratio assumes that translocations with viable unbalanced gametes form chains rather than rings of four at meiosis, due to the shortness of one interchanged segment. Chains of four give 50% adjacent 1 segregation). The procedure does not rely on hemizygosity for linkage of mutant genes, and therefore requires an extra generation beyond the F_1. However, the technique is not limited to short, terminal chromosomal regions and can be used to mark the entire genome with an appropriate set of translocations (Phillips et al., 1971a). The system has an advantage over the use of standard translocations tests of linkage, since only the mutant gene needs to be classified and not the T marker (or associated gene) as well.

Undoubtedly the most useful and promising method of gene placement involves translocations between the B chromosome and members of the normal chromosomal complement (A chromosomes). The A-B translocations are capable of generating hemizygosity for regions of the genome without the limitations imposed by gametophytic inviability. In order to understand this method of gene placement, we must first consider the unique property of B chromosome variability in pollen transmission.

Studies of B chromosome transmission through the male and female parents were made by Longley (1927), Randolph (1941), and Blackwood (1956). Transmission is normal through the female but quite unusual in the male parent. Randolph found that the great majority of offspring from the cross OB female $\times$ 2B male had zero or two B chromosomes, rather than the expected one B. A cytological examination of plants with one or two B chromosomes did not reveal any unusual behavior of the B chromosome during or prior to microsporogenesis. Randolph suggested that the B chromosome is subject to a post meiotic irregularity in division. Unfortunately, cytological observation of post-meiotic divisions is very difficult, and genetic analysis is not possible since the B chromosome is devoid of marker genes. However, Roman (1947) was able to elucidate the behavior of B chromosomes in the pollen mitoses by involving them in translocations with A chromosomes and using the mutant genes provided by the latter as genetic markers.

Roman isolated eight different A-B translocations. The first to be analyzed was TB-4a, and its properties have since been found to be characteristic for other A-B translocations. In TB-4a, the point of interchange in chromosome 4 is in the short arm approximately 0.25 of the length of the arm away from the centromere. The breakpoint in the B is at or near the junction of the proximal euchromatic region with the distal heterochromatin. The two new chromosomes formed by the translocation are A^B and B^4 chromosomes, designated 4^B and B^4. The 4^B chromosome contains 4L, part of 4S, and the noncentric portion of the B distal to the breakpoint. The B^4 chromosome includes the B centromere, proximal euchromatic segment of the B, and the distal region of 4S. The Su locus of chromosome 4 is in the region of 4S that was transferred to the B^4 chromosome. When plants homozygous for the interchanged chromosomes, with the Su allele in both B^4 chromosomes, were used as the female parent in crosses with recessive su pollen parents, all of the resulting kernels were Su in phenotype. The reciprocal cross, however, gave ears with many kernels showing the su phenotype. These su kernels lacked the B^4 chromosome in the endosperm, but the embryos of such kernels carried two B^4 chromosomes. Likewise many of the Su kernels had embryos lacking the B^4 chromosome, which presumably was present in double dose in the endosperm. This dissimilarity in constitution of the embryo and endosperm showed that nondisjunction of the B^4 chromosome occurred at the second pollen mitosis. The 4^B chromosome, on the contrary, behaved normally in pollen transmission. Nondisjunction of the B^4 did not occur 100% of the time, since some Su kernels gave rise to embryos with one B^4. (Rates of nondisjunction are quite variable from one experiment to the next, but frequencies below 50% appear to be rare).

Significantly, it is the B^4 and not the 4^B chromosome which undergoes mitotic nondisjunction. Among Roman's different A-B translocations the B was broken at various places along its length, but it was consistently found that the B^4 chromosome engages in nondisjunction while the A^B is stable. Thus, a proximal region, probably the centromere or adjacent heterochromatin, must be involved in nondisjunction. However, as we discuss in section X, 2, at least one other region of the B chromosome is active in controlling nondisjunction.

The A-B translocations are unusual in several respects, in addition to their variable transmission through the pollen. The B chromosome is of no noticeable value to the plant and can be eliminated. Therefore, the heterozygous A-B translocation need contain only three rather than four chromosomes. Crosses between a homozygous A-B translocation plant (female) and a normal plant lacking B chromosomes produce heterozygotes (A, A^B, B^4) which form trivalents rather than quadrivalents in meiosis. The A and A^B chromosomes disjoin regularly from the trivalent. However, as Robertson has shown for TB-9b (1967), the B^4 chromosome does not regularly disjoin from its pairing partner, the A chromosome. Instead, the B^4 moves to the pole at random with respect to the A chromosome, generating four meiotic products in equal amounts: A; AB^4; A^B; A^BB^4. The balanced spores A^BB^4, and A are transmitted while the A^B spore, which is deficient for genes on the A chromosome, aborts. The A B^4 combination is duplicate for a segment of A and deficient for part of the B chromosome, but the deficiency is not important and this is basically a duplication carrying spore. It is transmitted readily through the female parent and less well through the male. In the pollen, transmission of AB^4 spores depends on the extend of the A duplication. Large duplications inhibit greatly the ability of pollen to compete successfully for fertilization. However, the A B^4 pollen from several translocations is known to be transmissible, although at a reduced rate.

When the complication of nondisjunction of the B^A is added to the anomaly of a dispensable chromosome participating in a translocation, the types of individuals that can be generated with one A-B translocation are considerable. To avoid confusion, a translocation carrying marker genes on both the A^B and B^A chromosomes is desirable in experimental work, and for this purpose, Roman's TB-9b is ideal (Carlson, 1969a).

A-B translocations have the unique ability to transmit sperm with a duplication or a deficiency for specific A chromosomal regions. The pollen is not affected by genic imbalance in the sperm since the euploid tube nucleus controls pollen activity. Transmission of deficient sperm in crosses of A-B translocations provides a hemizygous technique for locating genes on the chromosomes. (Roman and Ullstrup, 1951). Recessive mutants may be crossed as female parents to a series of different A-B translocations. Genes located in the B^A segment of a translocation will be hemizygous in some of the progeny, allowing a test for the presence of the mutant gene in this region. In order to exploit this technique fully, at least 20 different A-B translocations are needed, two for each chromosome, with the points of interchange close to the centromere and in different arms. All mutant genes could be placed in one of the chromosome arms with such a series of translocations. Unfortunately, the appropriate series of A-B translocations proximal breaks is not available. However, in recent years, the translocations produced by Roman have been added to by Beckett (1967, 1968, 1972) and Rakha and Robertson (1970) so that a series of translocations covering approximately two-thirds of the known genetic linkage map of corn has been assembled (Neuffer and Beckett, 1971). Neuffer and Beckett (1971, 1972, 1973) have demonstrated the usefulness of this series of translocations by locating a large number of new mutants to the appropriate chromosome arm.

To improve the series of A-B translocations available, new translocations must be generated. These may be obtained by X-irradiation, but Rakha and Robertson (1970) have devised a novel scheme for producing new A-B translocations by crossing over. The idea is to combine an A-B translocation and an A-A translocation in one plant. The two translocations must involve a common A chromosome so that crossing over can occur between them. If the B^A chromosome is chosen so that it can pair with the homologous A^A chromosome proximal to the breakpoint of the A-A interchange, crossing over will transfer an interchange segment from a distal region of A^A to the B^A chromosome. The new B^A chromosome will carry parts of two A chromosomes, and the region newly joined to the B^A chromosome can be analyzed for mutant genes. Further details of the complex translocations recovered in this manner are described by Rakha and Robertson (1970).

In addition to locating recessive mutant genes, a number of other possible roles for A-B translocations have been suggested (Laughnan and Coe, 1953; Bianchi et al., 1961). Of particular interest is the suggestion of Robertson (1964) and Peterson and Wernsman (1964) that segments of A chromosomes may be isolated and transferred intact from one line to another without being modified in the process by crossing over. Such a procedure has value in analyzing the genetic contents of regions of the A chromosomes and in maintaining favorable linked gene combinations during breeding.

Basically the procedure is as follows:
1) cross a homozygous A-B translocation as female parent to an inbred line, for example M14. All the progeny are heterozygous for the translocation and semisterile in the pollen, due to the abortion of A^B gametes.
2) Backcross the plants as female parents to M14 males and select heterozygotes by pollen sterility. The procedure is repeated until an M14 line with the A-B translocation is produced.
3) Cross the M14 translocation stock as male to another line, perhaps W22. Select hypoploid (AA^B) plants which lack the B^A and display a high rate (50%) of pollen sterility.
4) Backcross hypoploid plants as female to the M14 translocation stock. Only A-containing gametes are transmitted by the hypoploid. Again select hypoploids and repeat the process until the W22 line has been converted to essentially an M14 line.

When the process of backcrossing is complete the segment of the A chromosome which was hemizygous in each cross, due to hypoploidy, is still an intact W22 segment. This region had no opportunity for crossing over with an M14 segment since it was always maintained as a hemizygote. When backcrossing is complete the final hypoploid may be self-pollinated, and a normal plant, lacking the A-B translocation is produced. This plant carries M14 genes at all loci except in the formerly hypoploid segment, where W22 genes are present.

2. Inversions

Compared with translocations, little work has been done with inversions in corn, even though they constitute one of the more interesting types of aberrations. Convincing genetic evidence of inversions in *Drosophila* was found by Sturtevant (1931). The first cytogenetical studies of inversions were made by McClintock (1931, 1933) working with corn. Later investigations of inversions in corn were reported by Anderson (1936), Chao (1959), Clark (1942), McClintock (1938a), Morgan (1950), O'Mara (1942), Rhoades and Dempsey (1953), Rhoades (1956b, 1963), Rinehart (1970), and Russell and Burnham (1950).

There are two kinds of simple inversions. In the paracentric type the inverted segment lies within a single arm of the chromosome, whereas in the pericentric type the centromere is included within the inverted segment. Paracentric inversions generally have no karyotypic effect in somatic cells, whereas pericentric inversions may produce marked changes in the position of the centromere. McClintock (1931, 1933) showed that in both types a loop-shaped configuration is produced at pachynema by homologous synapsis in normal/inversion heterozygotes (Fig. 12). However, the subsequent cytological behavior of the two inversion types is strikingly different, since the dicentric bridges and acentric fragments produced by crossing over in heterozygous paracentric inversions are not found in pericentric inversions.

A typical paracentric inversion at pachynema is diagrammed in Fig. 13. The proximal region is that segment between the centromere and the beginning of the loop; the inversion loop consists of the inverted segment, and the distal portion is the terminal uninverted region. Listed in Table 1 are the anaphase configurations coming from various types of exchanges.

Table 1—Cytological and genetical products of crossing over within a bivalent heterozygous for a paracentric inversion

Configuration	Type of Tetrad	Microspores produced		
		Viable non-crossover	Viable double crossover	Abortive
Anaphase I				
1. No bridge, no fragment	a. No Exchange	4	0	0
	b. 2-strand double in loop	2	2	0
2. 1 bridge, 1 fragment	a. Single exchange in loop	2	0	2
	b. 3-strand double in loop	1	1	2
3. No bridge, 1 fragment	a. 3-strand double with 1 exchange in loop and one in proximal region	2	0	2
4. 2 bridges, 2 fragments	a. 4-strand double in loop	0	0	4
Anaphase II				
1. No bridge	a. From all above types except #3			
2. 1 bridge	a. From #3 above			
3. 2 bridges (1 in each sister cell)	a. Certain triple exchanges where two are in loop			

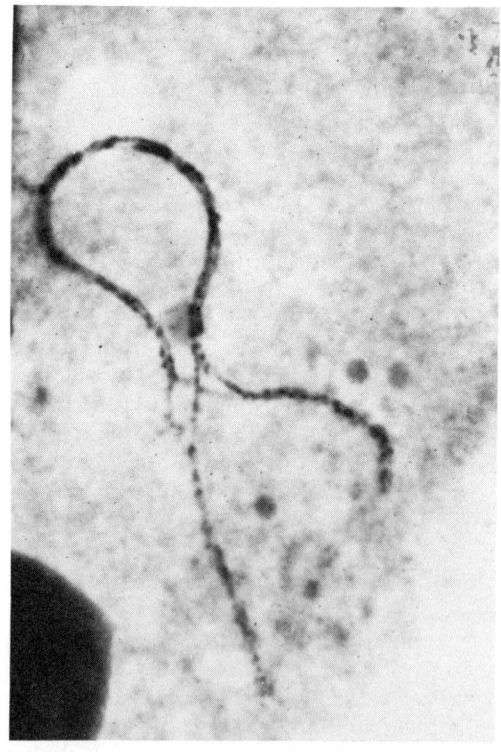

Fig. 12. Heterozygous pericentric inversion in pachynema.

In anaphase I, the dicentric bridge with an associated acentric fragment is characteristic of a paracentric inversion. It results from a single chiasma in the inversion loop or a three-strand double in the loop. A single exchange in the loop may also produce a no bridge, one fragment configuration if a second exchange occurs between the centromere and the inversion loop. A measure of double exchanges is provided by the two bridge, two fragment class produced by a four strand double exchange. Assuming that double chiasmata occur in a ratio of 1:2:1 for two strand/three strand/four strand events, all doubles can be calculated from the two bridge, two fragment data. Anaphase II of meiosis has relatively few bridges in comparison to anaphase I. Bridges come primarily from three strand doubles where one exchange is proximal to the loop and one exchange is within the loop.

Listed in Table 1 are the expected frequencies of non-crossover, double crossover, and abortive microspores produced by each tetrad type. Theoretically the same expectations should apply to the megaspores. However, following meiosis in the female three of the four products die and only the basal megaspore survives. If certain chromatids are preferentially included in the basal megaspore, the products of crossing over in an inversion heterozygote could differ between the male and the female. The idea comes from findings with *Drosophila*. As in corn, only one of the meiotic products gives rise to an egg nucleus. Beadle and Sturtevant (1935) suggested that chromatids involved in a dicentric bridge are confined to the central nuclei of the *Drosophila* tetrad due to the failure of bridge breakage. Only non-crossover or double crossover balanced chromatids are transferred to the inner egg nucleus. Support for the hypothesis was provided by Beadle and Sturtevant (1935) and Sturtevant and Beadle (1936). In corn, delayed breakage of the anaphase I bridge could potentially orient noncrossover chromatids to the outer poles of the linear tetrad (selective orientation). Since the surviving (basal) megaspore develops from one of the outer poles, the result would be a decrease in gametophytic lethality and a difference in the effects of inversion heterozygosity between the male and female. Indeed, Morgan (1950) with inversion 4a, and Russell and Burnham (1950), with inversion 2c, reported considerable pollen sterility, but low egg lethality in inversion heterozygotes.

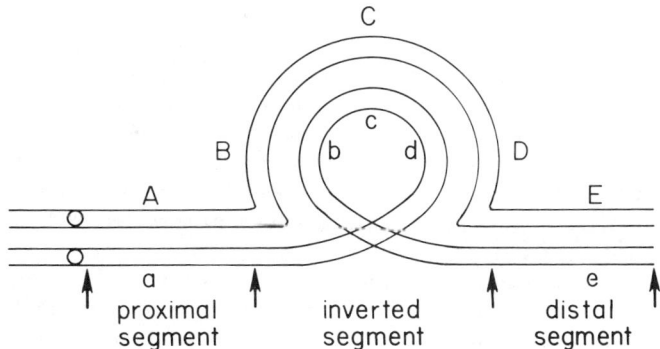

Fig. 13. Diagram of pairing in a heterozygous paracentric inversion. The proximal segment is that between the centromere and the inversion loop while the distal segment includes the region from the chromosome terminus to the inversion loop.

However, Rhoades and Dempsey (1953, and unpublished data), with inversion 3a, did not find evidence of greater female than male fertility. One explanation for the contradictory findings assumes that selective orientation may occur in some genetic backgrounds or with certain chromosomes but not in all situations. Alternatively, rates of crossing over in the male and female inflorescences may control differences in male and female fertility. In certain backgrounds, the rate of crossing over may be significantly lower in the female, producing a lower frequency of dicentric bridges and higher fertility. Recently a number of inversion heterozygotes were checked for ovule and pollen sterility by Rinehart (1970). In general, female fertility was found to be markedly higher than fertility in the male and some inversions gave evidence of complete female fertility. Possible differences in recombination between the male and female do not appear to be important in these results. For example, it was shown in some cases that the addition of abnormal chromosome 10 to the female parent lowered fertility somewhat, but not to the level of N10/N10 males. Abnormal chromosome 10 (section X, 1) markedly increases rates of recombination in regions of structural heterozygosity, as are found in Inv/N bivalents (Rhoades, 1968). Unexpectedly, Rinehart found that inversion 3a heterozygotes have greater female than male fertility. This is in contradiction to the results of Rhoades and Dempsey and suggests that varying genetic backgrounds may contribute to the presence or absence of selective orientation. The work of Rinehart indicates that selective orientation occurs in corn although not always with the efficiency seen in *Drosophila*.

The production of dicentrics and their loss through inviability or selective orientation accounts for the ability of paracentric inversions to "suppress" crossing over. In fact, crossing over may occur frequently within an inversion loop but the chromatids involved in a chiasma are lost. Only the relatively rare double crossover chromatids (from two strand or three strand double chiasma) are balanced and survive. Pericentric inversions also inhibit crossing over since deficient-duplicate chromatids are produced. No dicentric bridge is involved, however, and selective orientation is not possible. Consequently, egg and pollen sterility occur at similar rates with pericentric inversions. The production of inviable gametes in both paracentric and pericentric inversions varies with the amount of crossing over within the inversion loop. A maximum rate of 50% pollen sterility occurs with long inversions, and this value corresponds to the maximum rate of recombination which can occur between any two points in a chromosome.

Inversions of any kind are rare in corn populations for obvious reasons. However, the harmful effects of inversions are directly related to the chiasma frequency within the inversion loop. As a result, short inversions or inversions in regions with low recombination frequencies have occasionally been found. A paracentric inversion composed of the distal two-thirds of the short arm of chromosome 8 has been observed in certain races of corn (McClintock, 1959). Variant arm ratios for chromosomes 2 and 4 have been attributed to pericentric inversions. In the absence of direct evidence it may be a mistake to routinely assign differences in centromeric position to inversions, since centromeric transpositions do occur (Jackson, 1973). However, McClintock found (1933) an occasional inversion loop in heterozygotes for two forms of chromosome 4.

Several experimental applications for inversions have been found. Most obvious is their widespread use as suppressors of crossing over. In addition, realization of the role of the centromere in the movement of chromosomes came from observations on the acentric fragments produced by inversion crossing over. The failure of these fragments to form chromosomal fibers at metaphase and their irregular movement on the spindle during anaphase, which led to their failure to be included in daughter nuclei, demonstrated the control of centromeres over chromosomal disjunction. Rhoades and Dempsey (1953) combined cytological and genetical data on crossing over in a paracentric inversion to provide evidence for the absence of chromatid interference in maize. Dobzhansky and Rhoades (1938) suggested that paracentric inversions can be utilized in the cytological placement of genes affecting agronomic characters. Sprague (1941) and Chao (1959) successfully employed the procedure in locating certain agronomic traits. McClintock (1938a, 1939, 1941b) studied the fate of dicentric bridges generated by crossing over in paracentric inversions and certain more complex rearrangements. This latter work was reviewed by McClintock (1951) and Faberge (1958) and will be discussed in some detail.

McClintock discovered that the chromosome ends produced by breakage of a dicentric bridge are unstable (often referred to as "sticky" or "raw" ends). The instability follows two paths. A broken chromosome may rejoin with another broken chromosome to form a dicentric. The fate of such dicentrics will be discussed later. Alternatively, a broken chromosome may replicate to form daughter chromatids. Replication, however, is improper in the presence of a freshly broken chromosome, and the daughter chromatids are joined together at the broken end. Thus, replication of a broken chromosome generates a new dicentric, establishing a breakage-fusion-bridge cycle which continues through successive divisions. The cycle is referred to as "chromatid type" since fusions are between chromatids rather than whole chromosomes. When dicentrics are produced by crossing over in inversion heterozygotes, bridges appear in anaphase I and in the subsequent male and female gametophytic mitoses. Bridges do not occur in anaphase II, since disjunction of the dyads is not preceded by DNA replication. If gametophytes with broken chromatids are viable, the chromatid type breakage-fusion-bridge cycle may continue after fertilization. However, dicentrics produced by paracentric inversions are usually not transmitted through the gametophyte. Inversion heterozygotes produce dicentric bridges plus an acentric fragment, and the acentric carries all genes distal to the inversion loop. Unless the deficiency produced by loss of the acentric fragment is very small (Rhoades and Dempsey, 1953), the chromatids derived from a dicentric bridge are inviable. McClintock, therefore, developed two aberrations (a tandem duplication and a complex inversion) in which crossing over yields a dicentric bridge but no acentric fragment. With appropriate positioning of the breakpoint in the dicentric bridges of the male or female a functional gametophyte can be produced. Upon transmission of a broken chromosome through the gametophyte, the breakage-fusion-bridge cycle continues in the endosperm and is marked by a sectored loss of dominant genes from the unstable chromosome during development. Depending on the point of bridge breakage in the endosperm mitoses, one or more markers may be lost at one time. In the sporophyte, the cycle does not occur and stable chromosomes are found. Apparently a function present in the embryo but lacking in the endosperm or gametophyte causes a "healing" of broken chromosomes.

The chromatid type breakage-fusion-bridge cycle may be established by the action of controlling elements as well as through recombination between structurally dissimilar chromosomes. A discussion of controlling elements is given elsewhere. We wish to note only that the excision of a controlling element (comparable to a bacterial episome) from a chromosome may cause breakage of the chromosome. Subsequent fusion at the broken chromosome end initiates the breakage-fusion-bridge cycle.

One might expect the breakage-fusion-bridge cycle to produce very complex deficient-duplicate chromosomes as a result of random bridge breakage, but this is not always true. McClintock noted (1941b) that the bridges involved in a cycle tend to break at the same place in successive divisions. The interpretation was that fusion of sister chromatids following breakage may be weak or incomplete. A cytological study of corn endosperm by Schwartz and Murray (1957) indicated that bridge formation does not always occur during a chromatid type breakage-fusion-bridge cycle. In agreement with McClintock, they suggested that incomplete fusion of chromatids occurs frequently, leading to normal rather than dicentric anaphase configurations. In addition, Miles (1971) found that certain chromosomes derived from dicentric bridges always separate at the original breakage site during endosperm development. Miles utilized a ring-rod heterozygote of chromosome 10 to generate dicentric bridges. The R allele marked the ring 10 and r was on the rod 10. Since ring chromosomes are unstable in both endosperm and plant development, they display a variegated phenotype (section IX, 5). In testcrosses of the ring-rod heterozygote one expects variegated color on the endosperm (R on unstable ring) or no color (r on rod 10). In addition, crossing over between the ring and rod yields dicentric bridges in which R may be associated with a rod chromosome with a freshly broken end. One expects this rod to be unstable in the endosperm (variegated color), but stabilized by "healing" in the sporophyte (solid color). In a testcross of rr ♀ × R-ring 10/r-rod 10 one expects all R progeny to have a variegated endosperm phenotype, generated by a ring chromosome or an unstable rod chromosome. However, an unexpected class of progeny was found in which the endosperm phenotype was solid-colored rather than variegated. Six kernels of this type were germinated and all the plants were shown to contain a stable rod (R) chromosome derived from a dicentric bridge. The unusual aspect of this class is the apparent absence of a breakage-fusion-bridge cycle in the endosperm, even though a freshly broken chromosome was undoubtedly present. Miles suggested that breakage at certain chromosomal sites leads only to weak fusion of sister chromatids in successive generations such that all dicentric bridges break at the same position and a stable phenotype results in the endosperm. The idea receives support from the finding that all six chromosomes derived from solid colored kernels arose from breaks in heterochromatic regions. Two of the chromosomes had been broken in centromeric heterochromatin and four resulted from breakage either through or adjacent to a heterochromatic knob carried on the ring 10. Thus heterochromatin may provide an extreme example of the tendency for bridge-breakage to occur at the same point in succeeding divisions.

In addition to the chromatid breakage-fusion-bridge cycle, McClintock demonstrated the existence of a chromosome cycle. If broken chromosomes are introduced into the zygote from both the male and female parents, the broken ends may fuse with each other forming a dicentric. The dicentric bridge now spans two different centromeres rather than connecting one centromere. In anaphase two bridges may be produced, or no bridges, depending on centromeric orientation. If two bridges appear, breakage of the bridges in anaphase sends two chromosomes with "sticky" ends to each pole. The two broken chromosomes of each daughter cell fuse with each other to re-form a dicentric. This chromosome cycle occurs in both the sporophyte and the endosperm, and variegation for sporophytic markers can be observed. Apparently, fusion between broken chromosomes occurs in the sporophyte before the "healing" can stabilize the chromosomes.

The chromatid cycle and the chromosome cycle are distinct phenomena. However, Schwartz and Murray (1957) provided evidence that a chromatid cycle can be converted to a chromosome cycle by nondisjunction. If the bridge produced by a chromatid cycle fails to break and the dicentric migrates to one pole, a chromosome type breakage-fusion-bridge cycle is established in successive divisions.

A special example of the chromosome cycle is found with ring chromosomes. Occasionally ring chromosomes give rise to double dicentric rings by sister strand crossing over (section VIII, 5). Breakage of the double ring in anaphase gives a rod chromosome with two broken chromosome ends. Fusion of the two chromosome arms restores the ring chromosome condition.

3. Duplications

The duplication of base sequences, genes, chromosome segments or whole chromosome sets has played an important and, as yet, not fully documented role in evolution. Although the duplication and evolution of certain genes has been traced, which is true for the hemoglobin and myoglobin genes, the extent of gene duplication within any species is usually difficult to ascertain. Once duplication occurs, the functions of duplicate loci may diverge with evolution until the relationships between them are no longer apparent.

The identification of duplications within the genome has relied mainly on the occurrence of crossing over between supposedly nonhomologous regions. Compound loci that consist of tandem repeats have been found at the *R* locus (Dooner and Kermicle, 1971; Emmerling, 1958b; Stadler and Nuffer, 1953) and the *A* locus (Laughnan, 1949, 1952a, b). Oblique synapsis and crossing over can occur between alleles of a tandem repeat with the consequent production of new alleles. The "mutants" produced by oblique synapsis are associated with recombination for flanking markers and thus can be identified. The situation is similar to the classic case of oblique synapsis at the *Bar* locus in *Drosophila* except that in corn the duplications are naturally occurring and seem to involve duplication of a single cistron.

Laughnan (1955, 1961) has found evidence with the *A* locus that in the absence of a pairing partner, a tandem repeat may pair on itself and undergo intrastrand crossing over. Evidence for intrachromosomal crossing over in *Drosophila* at the *Bar* duplication has also been found (Peterson and Laughnan, 1961, 1963; Gabay and Laughnan, 1973). This unique phenomenon of "auto-association" may potentially allow some discrimination between theories of chromosomal synapsis.

Schwartz and Endo (1966) identified a compound locus in maize through electrophoresis. An allele of the *Adh* locus (*alcohol dehydrogenase*) was found which produces two isozymic variants of the enzyme. The two isozymic variants are completely linked in inheritance, indicating the presence of an adjacent duplication.

Gross redundancy between chromosomes of the corn genome has also been found. In monoploid corn, the opportunity exists for crossing over between duplicate segments of the genome, if the regions are long enough occasionally to generate chiasmata. Among the products of such "nonhomologous" crossing over are translocations. Weber and Alexander (1972) demonstrated the existence of at least two duplications in the corn genome, through cytological analysis of translocations recovered from monoploid plants. This work is discussed in section VII.

Certain duplications have been identified in corn by nucleic acid hybridization techniques. The number of cistrons present for a particular gene can be determined by RNA-DNA hybridization, if the RNA product of the gene is available. Phillips, et al. (1971b) have shown that the genes which code for 18S and 28S ribosomal RNA are present in thousands of copies in the nucleolus organizer region of corn (section II). A high level of redundancy for ribosomal RNA genes has been found to occur generally, and apparently reflects a need for greater levels of ribosome synthesis than single cistrons can provide.

The discovery of high levels of gene duplication with hybridization techniques has changed prevalent concepts of genetic organization and, at the very least, suggests that chromosomes are capable of internal, precise duplication of genes on a grand scale. In addition to the redundancy of ribosomal DNA, the genomes of eukaryotes contain surprisingly high levels of repetitive DNA sequences (Britten and Davidson, 1971; Britten and Kohne, 1968; Flamm, 1972; Walker, 1968; Walker and Hennig, 1970; Waring and Britten, 1966). When the DNA of a species is denatured and then allowed to renature, the rate of formation of double helices can reveal the existence of redundancy within the genome. Redundant regions that are highly duplicated renature rapidly, moderately repetitive DNA renatures less rapidly, and unique sequences form double helices very slowly. Highly repetitive DNA is clustered at specific chromosomal sites, often the centromeric heterochromatin, and it may be that constitutive heterochromatin and highly repetitive DNA are synonymous (Arrighi, et al., 1970; Gall, 1973; Jones, 1970; Pardue and Gall, 1970). Theories on the functioning of highly redundant DNA have been developed by Yunis and Yasmineh (1971), Walker (1971), and Mazrimas and Hatch (1972). Moderately repeated DNA sequences appear to be interspersed with unique sequences along the length of eukaryotic chromosomes (Davidson et al., 1973). Britten and Davidson

(1969, 1971) have suggested a role for moderately redundant DNA in gene regulation. Repetitive DNA does not appear to function as do typical genes, since messenger RNA is synthesized primarily by nonrepetitive sequences (Davidson et al., 1973; Klein et al., 1974). Repetitive DNA is present in corn, as in other eukaryotes, and was recently examined by Chilton and McCarthy (1973) (section X, 2).

In addition to naturally occurring duplications, procedures are available for synthesizing specific duplications. Beadle and Sturtevant (1935) showed with *Drosophila* that crossing over between overlapping inversions of the same chromosome could be utilized to produce duplications. The procedure was applied in corn to generate duplications in chromosome 3 (Rhoades and Dempsey, 1956b). Gopinath and Burnham (1956) utilized corn translocations to produce gametes containing duplications without any accompanying deficiencies. If two translocations are combined which involve the same two chromosomes, and if they have exchange points in the appropriate positions on the chromosome arms, one of the products of meiotic segregation will be a viable duplication-type spore. The duplication should be egg transmissible and, although it will not be competitive in the pollen, it should give viable pollen grains as seen in a field microscope. For a number of appropriate translocation intercrosses, Gopinath and Burnham found that pollen and ovule lethality was reduced below 50%, as expected. Duplications should, therefore, be found in the progeny when the two-translocation hybrids are crossed as the seed parent.

Another approach to inducing duplications is that of Doyle (1971, 1972). Doyle irradiated kernels that were heterozygous (in repulsion) for the closely linked markers a and $sh2$ of chromosome 3. The A $sh2/a$ $Sh2$ plants were grown and testcrossed, and progeny with the A Sh phenotype were selected. The A Sh phenotype may result from crossing over, but since the markers are closely linked and the kernels were irradiated, the phenotype could also result from some anomaly that places the A $Sh2$ and a $Sh2$ segments in the same gamete. Nondisjunction of chromosome 3, for example, could produce a trisomic plant with the A Sh phenotype. Also, breakage and exchange between homologous chromosomes, which produces a duplication could account for the phenotype. Doyle found that most A Sh cases arose from crossing-over or trisome formation, but two duplications were identified.

Duplications have been utilized experimentally in corn in a number of ways. Only one example will be reviewed. F.C. Beard isolated an insertional translocation or transposition in which approximately 10% of the long arm of chromosome 3 was inserted into the short arm of chromosome 9. Since both the deficient chromosome 3 and the duplication (transposition) chromosome 9 were recovered, the two chromosomes constitute a balanced chromosomal rearrangement. However, the Tp9 chromosome may be combined with normal chromosomes 3, producing a duplication. Rhoades (1968) found that the transposed segment of 3L disrupts synapsis in the short arm of chromosome 9. When Tp9/N9 heterozygotes were examined cytologically, a buckle corresponding to the unpaired segment of 3L was expected to be present in pachynema. While a "buckle" was often present, the position of the unpaired region varied considerably along the length of 9S, indicating that the 3L segment was not regularly excluded from pairing. In some cases, the Tp9/N9 heterozygote displayed complete pairing in 9S despite the difference between the chromosomes in arm length. Obviously, nonhomologous pairing is quite frequent in Tp9/N9 heterozygotes. The unusual pattern of synapsis is accompanied by a drastic reduction in recombination along the length of 9S. Several conclusions can be drawn from the findings with Tp9/N9 heterozygotes. First, there is a good correlation between nonhomologous pairing and the absence of crossing over, as expected. Second, pairing between homologous regions to the exclusion of nonhomologous or less homologous regions (i.e. preferential pairing) does not always occur. Third, crossing over probably does not precede and control synapsis, as suggested by Maguire (1972b), but follows homologous pairing. Other findings with this interesting duplication are discussed in sections VII; VIII, 6; X, 2.

4. Deficiencies

Deficiencies arise in a number of ways, in addition to chromosome breakage and loss of a segment. Adjacent segregation from translocation heterozygotes produces deficient (as well as duplicate) gametes. Crossing over in an inversion heterozygote results in the loss of chromosomal regions distal to the inversion loop. The breakage of dicentric bridges, generated in a number of ways, may result in chromatin loss. Plants with abnormal chromosome numbers, such as monosomics or monoploids, produce gametes deficient for whole chromosomes.

Despite the variety of ways in which deficiencies can be produced, they have played a relatively minor role in corn cytogenetics. Both the male and the female gametophyte are highly sensitive to the loss of most chromosomal regions and even small deficiencies are usually not transmissible. As a result, few studies on the properties of deficient chromosomes have been possible. However, specific deficiencies may be maintained in culture for study if the deleted chromosomal region is present in a supernumerary ring chromosome (McClintock, 1938b). Sporadic loss of portions or all of the ring chromosome during development produces homozygous deficient sectors in the plant (section IX, 5). In addition, deficiencies may be selected and studied within the period of one generation under appropriate circumstances. Baker and Morgan (1969) pollinated chromosome 1 tester plants with x-ray-treated pollen carrying dominant markers. In the progeny, seedlings with a recessive phenotype were identified. Cytological studies of these plants revealed that deletions on chromosome 1 often accounted for the recessive phenotype. The purpose of the experiment was to produce plants which were hemizygous for the *asynaptic* locus. It was shown that two doses of the wild type allele (As) are needed for normal meiotic synapsis. Hemizygous $As/$-plants showed varying degrees of asynapsis. Since As/as heterozygotes have normal chromosome pairing, it was concluded that the as allele is not an amorphic but rather a hypomorphic allele.

A number of other experiments on deficiencies have depended on the finding that gametophytes which lack short, specific chromosomal segments are sometimes viable. McClintock (1944) recovered viable terminal deficiencies of chromosome 9 which were formed by breakage of dicentric bridges. Rhoades and Dempsey isolated, from Inv 3a heterozygotes, some terminally deficient chromosomes generated by crossing over in the inversion loop (1953). More recently, Rhoades and Dempsey (1973a) reported on a viable terminal deficiency in 3L which arose in an unknown manner. Phillips et al., (1971a) listed a number of translocations from which deficient gametes may be recovered following adjacent 1 segregation. In general deficient gametes that are transmissible survive only through the female and are not competitive in the efficient pollen screen, although exceptions (McClintock, 1944) are known.

Deficiencies that are transmissible through the gametophyte are sometimes too small for cytological detection. This problem was encountered in experiments designed to elucidate the mutagenic properties of X-rays. The majority of "mutations" produced by X-rays result from gross chromosomal changes, including deletions, rather than intragenic changes. The work of Stadler and Roman (1948), Neuffer (1957) and Mottinger (1970) failed to demonstrate evidence of x-ray induced point mutations in corn. Among the "mutants" that were recovered from x-ray treatment, the most difficult to analyze were small chromosomal changes which were apparently short deficiencies rather than point mutations. In this work, several procedures for identifying short deficiencies were developed. When a putative intragenic mutation occurs at a locus not known to be vital to the organism, the association of gametophytic or recessive sporophytic lethality with the mutation suggests the loss of vital loci adjacent to the gene being studied. Also, a reduction in crossing over in regions spanning the mutation is indicative of chromatin loss rather than point mutation. The concomitant loss of adjacent markers with mutation of a locus is strongly suggestive of a deficiency. In addition, an inability to select back mutations to wild type is characteristic of chromatin loss. However, no infallible test for short chromosomal deficiencies is available, and mutations that appear by one criterion to be deficiencies may not be confirmed by other criteria. For example, an apparent deficiency of both the sh and bz loci of chromosome 9 found by Mottinger (1970) eliminated recombination in the sh-bz region as expected, but had no detrimental effect on gametophytic or sporophytic viability. On the other hand, a deficiency involving the A locus of chromosome 3 $(a$-$x1)$ had detrimental effects on gametophytic and sporophytic viability but showed no consistent effect on recombination (Stadler and Roman, 1948).

5. Ring Chromosomes

The unusual properties of ring chromosomes were first investigated by McClintock. She obtained two rings involving segments from the proximal region of the short arm of chromosome 5 (McClintock, 1932, 1938b, 1941a). Both of the rings were induced by breaking a normal chromosome 5 with x-rays. In each case one break occurred in the centromere, which was fractured into two portions, and the second point of breakage was in the short arm. Fusion of broken ends resulted in the formation of a ring chromosome and a deficient rod chromosome. The centromeres of both rod and ring chromosomes were able to function normally. Plants were obtained with two deficient rods and one or more ring chromosomes. Since the chromatin of the ring chromosome was composed of that missing from the deficient rod chromosome, the young sporophytes initially were normal in appearance but soon manifested sectors of various kinds of mutant tissue. Cytological observations revealed that the ring chromosome was not present unaltered in all cells but that some cells had larger rings with certain segments in duplicate. Others had smaller rings deficient for certain portions originally present in the ring, and in some cells the ring chromosome had been completely eliminated.

Modified ring chromosomes arise in the following way. In some mitotic prophases, as the probable result of sister strand exchange, the two halves of a divided ring chromosome form either a continuous dicentric double sized ring or two interlocked rings instead of two separate ring chromosomes. Stalling of these figures in the center of the spindle during anaphase can exclude the ring chromosomes from daughter cells. Breakage of the chromosomes during anaphase and poleward migration produces, monentarily, rod chromosomes. However, fusion occurs between the two broken ends of these rod chromosomes, giving rise to ring chromosomes again. Ring chromosomes essentially participate in a chromosome type breakage-fusion-bridge cycle (section IX, 2). When a double dicentric ring pulls apart in anaphase, breakage of the two bridges may be unequal so that duplications and deficiencies arise in the daughter chromosomes. The constitution of ring chromosomes newly derived from a double bridge depends upon the position where the double sized ring breaks. Unequal breakage leads to one ring with duplicated segments and one deficient for the segments in duplicate in the other ring.

The somatic instability of ring chromosomes in corn occurs in both the endosperm and the sporophyte and produces a characteristic variegated phenotype. The sporadic loss of genetic markers on a ring chromosome may be observed when the ring is placed in a genetic background in which only the ring chromosome carries the dominant allele for specific genetic markers. Ring chromosome instability has provided experimenters with a very useful cytogenetic tool. Homozygous deficiencies can be produced in the sporophyte with the aid of ring chromosomes (McClintock, 1938b, 1941a, and section IX, 4). Sister strand crossing over can be visualized with ring chromosomes (section VIII, 5). Changes in chromosome structure from a rod to a ring or vice versa (section X, 1) can be identified with the aid of the variegated phenotype of ring chromosomes.

The harmful effects of sister-strand exchange on ring stability occur in the meiotic as well as somatic cells of corn (Schwartz, 1953c; and section VIII, 5). In addition, crossing over between rod and ring chromosomes in meiosis yields single dicentric bridges. Following bridge breakage, genes from a ring chromosome may be transferred to a rod chromosome with a freshly broken end. The rod chromosome undergoes a breakage-fusion-bridge cycle in gametophytic divisions and is genetically imbalanced. It is not surprising, therefore, that ring chromosomes are laboratory products which are never found in natural eukaryotic populations.

X. SUPERNUMERARY ELEMENTS

Supernumerary elements are extra chromosomes or chromosomal segments that are not derived by duplication from the normal complement, have no vital function, and may be present or absent in different members of the species. Although heterochromatic knobs may fall under this definition, we will limit this discussion to two supernumerary elements identified in corn: the extra segment of abnormal chromosome 10 and the B chromosome. There are a number of similarities between the B chromosome and the extra region of abnormal chromosome 10. Both contain large heterochromatic regions as well as euchromatic segments. Both have specific and dramatic effects on recombination among A chromosomes. In addition, the two supernumerary elements maintain themselves in corn populations, at least in part, by accumulation mechanisms. The accumulation mechanisms act to increase the frequency of these elements in the gametes of plants through modifications of centromeric behavior. Despite these similarities, Ting's (1958) suggestion that abnormal 10 arose by translocation between a B chromosome and normal 10 seems unlikely. Both Randolph (1941) and Rhoades (1942) found that the B and abnormal 10 were cytologically dissimilar. Snope (1967) found no pairing between abnormal 10 and a B chromosome in haploid plants. Furthermore, the modifications of the long arm of chromosome 10 which have given rise to the term "abnormal" include not only the addition of a supernumerary terminal segment but also changes in the chromomeric pattern of 10L adjacent to the supernumerary region. It is doubtful that the altered chromomeric pattern of 10L resulted from a B-10 translocation, since a distal region of the normal 10 would be lost in the process and presumably vital genes would be missing. More likely is the interpretation that part of the modified region of abnormal chromosome 10 is not supernumerary, but homologous to part of normal chromosome 10. Finally, the accumulation mechanisms of the B chromosome and abnormal 10 are strikingly different (see below). The evolution of accumulation mechanisms by the B and abnormal 10 resembles the convergent evolution of different species in which similar problems of adaptation produce superficially similar responses.

The origin of the supernumerary elements in corn is unknown, although they apparently are relics of chromosomes that once were involved in the ancestry of the plant. Certainly at one time these chromosomal regions served a useful function. However, according to one idea (Ostergen, 1945) the development of a supernumerary condition may be accompanied by the loss of useful genetic functioning. Both the B chromosome and abnormal 10 may be maintained in populations by their accumulation mechanisms rather than selection. Supernumerary elements in other words may be parasitic. The alternate idea, that supernumeraries may serve some useful genetic function is appealing but difficult to substantiate. In maize, the only obvious effect of B chromosomes is a detrimental one when B's are present in unusually high numbers. However, it is well established that both the B chromosome and abnormal 10 change the distribution of chiasmata in A chromosome bivalents and may "unlock" genetic regions in which chiasmata seldom, if ever, occur. They, therefore, tend to increase genetic variation subject to selection and this may be beneficial to the population. The B chromosomes in rye (Jones and Rees, 1967) and grasshopper (John and Hewitt, 1965) have also been shown to alter chiasma distribution or frequency. Recombination effects are a promising area of investigation for possible beneficial effects of supernumeraries. In addition, some studies indicate that supernumeraries may confer a selective advantage on individuals which carry them. The reader is referred to Rees and Ayonoadu (1973) for an introduction to the problem. In the following discussions, the unusual properties of abnormal 10 and the B chromosome will be described.

1. Abnormal Chromosome 10

Abnormal chromosome 10 is found in some strains of corn from Latin America and the southwestern USA (Longley, 1937, 1938). The chromosome is identical to normal 10 in the morphology of the short arm and most of the long arm. However, in the distal one-sixth of the long arm there are three large and distinct chromomeres in abnormal 10 not found in the corresponding section of normal 10. The most distal of these chromomeres lies very near the end of the long arm of normal 10. In addition, abnormal 10 has a distal region that is not present in normal 10. Beginning just distal to the last prominent chromomere, there is a light-staining region comprising somewhat less than half of the length, at pachynema, of the extra segment. Next, a large heterochromatic knob is present and the extra segment is terminated by a short euchromatic segment with small and diffuse chromomeres (Fig. 1). Crossing over in normal 10/abnormal 10 heterozygotes is rare in the distal region of the long arm with a dissimilar chromomeric structure, indicating some genetic divergence in this region (Rhoades, 1942; Kikudome, 1959). Ting has found another rare abnormal chromosome 10, which Kikudome (1961) described. The chromosome is similar to the more common abnormal 10 but has a unique chromosomal morphology. We will be concerned here with the chromosome found by Longley.

Three phenomena occur in plants carrying abnormal chromosome 10 in heterozygous or homozygous condition. These are: preferential segregation, neocentromere formation, and an enhancement of recombination. Rhoades and Dempsey (1966b) presented a comprehensive study of these phenomena and the subject was earlier reviewed by Rhoades (1952, 1955).

Preferential segregation is the accumulation mechanism of abnormal chromosome 10. Rhoades (1942) found in abnormal 10/normal 10 heterozygotes that abnormal 10 was recovered through the female in about 70% of the progeny rather than the 50% expected. Longley (1945) later found that abnormal 10 was able to induce preferential segregation in other bivalents if the chromosomes were heterozygous for a knob. Chromosomes with knobs, including abnormal 10, are preferentially transmitted through the female in competition with knobless homologues when abnormal chromosome 10 is present. Genes which are linked to the knobs are also recovered in frequencies above 50%. As the distance between a gene and a knob increases, crossing over tends to randomize the association between them and preferential segregation of the gene declines. Genes which are closely linked to knobs can be utilized as accurate markers of preferential segregation in place of cytological classification.

Kikudome (1959) found that the frequency of preferential segregation for a chromosome depends on the size of its knob. Three chromosomes with small, medium, and large knobs at the distal end of 9S were made heterozygous with a knobless 9. In the presence of abnormal 10 the knobbed chromosomes, designated K^S, K^M, and K^L, were recovered in 59%, 65%, and 69%, respectively of the progeny. In addition, Kikudome found that pairwise

THE CYTOGENETICS OF CORN

combinations of chromosomes with knobs of different sizes resulted in preferential recovery of the chromosome with the larger knob. Perhaps large knobs are more efficient in neocentromere formation than small knobs (see below).

The phenomenon of preferential segregation is an end product of neocentromere formation, according to the hypothesis of Rhoades (1952). Neocentromeres are regions in addition to the normal centromeres to which groups of spindle fibers become attached. Rhoades and Vilkomerson (1942) found that abnormal 10 induces neocentromere formation on certain

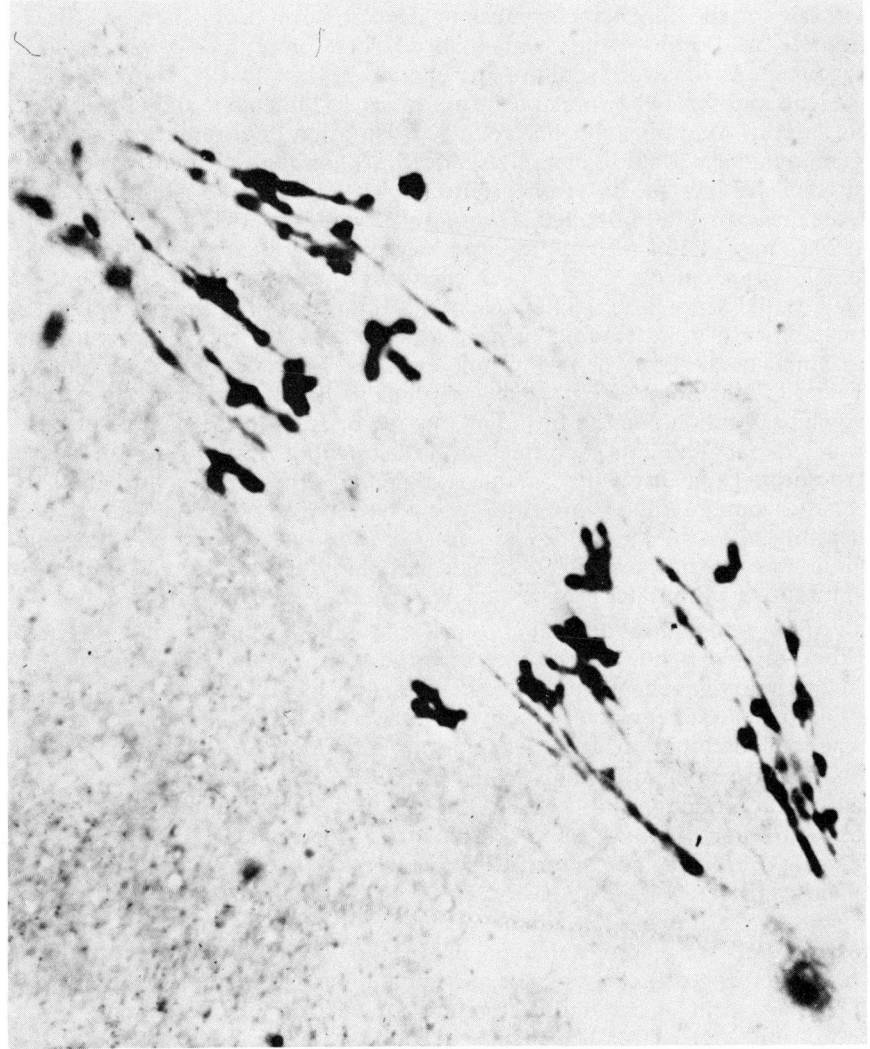

Fig. 14. Neocentromeres in anaphase I. Chromosome arms with neocentromeres are extended toward the poles in advance of the conventional centromere. Abnormal chromosome 10 was present in this cell.

chromosomes at both meiotic divisions. The neocentromeres draw the chromosomes precociously to the poles, producing anaphase I dyads and anaphase II monads whose neocentric arms reach the poles in advance of the centromere (Fig. 14). The number of chromosomes which possess neocentromeres is the same as the number of knob-bearing chromosomes and it is believed that neocentromeres arise at or near each heterochromatic knob. Rhoades' hypothesis, that neocentromeres account for preferential segregation, depends on the linear arrangement of the two meiotic spindles during megasporogenesis. Consider a bivalent consisting of one knobbed and one knobless chromosome. In the presence of abnormal 10, neocentromeres will arise on the knobbed chromosome. If a crossover occurs between the centromere and knob of the bivalent, dyads heteromorphic for the knob are produced. As the dyads separate in anaphase I, the knobbed chromatids take the lead and throughout anaphase are closer to the pole than is the knobless chromatid. Assuming that the relative orientation of chromatids is retained from anaphase I until metaphase II, the knobbed chromatids at second division will face the outer or terminal poles, while the knobless chromatids face inward. The knobbed chromatids would then pass to the outer megaspores, one of which, the basal megaspore, will survive to produce the female gametophyte. On this scheme, only heteromorphic dyads undergo preferential segregation. The lack of 100% preferential segregation in testcrosses could be attributed in part to an occasional lack of crossing over proximal to the knob or to multiple exchanges between the knob and centromere which do not yield heteromorphic dyads. Alternatively, the extent to which orientation persists from one meiotic division to the next could determine in part the rate of preferential segregation. Support for Rhoades' hypothesis came from the demonstration that crossing over proximal to a heterochromatic knob is essential for preferential segregation (Rhoades and Dempsey, 1966b). The transposition 9 chromosome (section IX, 3) virtually eliminates crossing over in 9S, when it is combined with a normal 9. Concomitantly it eliminates preferential segregation for the terminal knob of 9S.

The value of abnormal 10 and also the B chromosome in maize cytogenetics depends on the fact that these supernumerary elements have their primary genetic effects on basic mechanisms of heredity. In addition to its influence on centromeric activity, abnormal 10 has a striking effect on genetic recombination. In general, abnormal 10 appears to increase recombination in proximal regions (Miles, 1970; Nel, 1971; Rhoades and Dempsey, 1966b) although the Wx to Gl-15 region of chromosome 9, which spans the centromere, is unaffected by it (Nel, 1971). Abnormal 10 also promotes crossing over in regions of structural heterozygosity (Rhoades and Dempsey, 1966b) and knob heterozygosity, possibly by reducing asynapsis in these regions. An example is Kikudome's (1959) work with chromosome 9 bivalents heterozygous for a terminal knob in the short arm. Kikudome found that knob heterozygosity causes a sharp decline in crossing over in the distal half of 9S. The drop in recombination was greatest in the presence of a large knob, while a small knob had little effect and a medium-sized knob had an intermediate effect. Recombination in the wd to wx region was 26.9% in K^S/k heterozygotes, 17.7% for K^M/k plants and 12.7% for K^L/k. When abnormal 10 was added to the above chromosomal combinations, crossing-over was raised to 31.5% for K^S/k, 26.8% for K^M/k and 30.3% for K^L/k. In each chromosomal combination, approximately the same level of recombination was found in the presence of abnormal 10. The finding suggests that abnormal 10 acts in regions of low recombination, perhaps to restore a "normal" level of crossing over. Consistent with this interpretation, the regions adjacent to the centromere discussed earlier have much lower rates of crossing over per cytological length than do distal regions (section VIII, 6).

Why abnormal 10 should enhance crossing over in the corn genome is not known. Possibly the phenomenon is another element of abnormal 10's accumulation mechanism, since crossing over between a knob and its centromere is a prerequisite to preferential segregation. Alternatively, abnormal 10 may serve the function in corn populations of increasing genetic variability. Regardless, the chromosome provides a rare and useful tool in the study of recombination.

Abnormal chromosome 10 controls or influences neocentromere formation, preferential segregation, and recombination. Since these phenomena are exclusively associated with abnormal 10 in segregating progeny, the controlling genes must be located somewhere within the distinctive segment which terminates 10L. A procedure for localizing these genetic factors was developed by Emmerling (1959). She utilized a modified abnormal chromosome 10 which had been converted to a ring by X-irradiation. The ring arose from distal breaks in the chromosome arms, one near the tip of 10S and the other approximately in the middle of the large 10L knob. The ring chromosome, therefore, contained all of the distinctive regions of abnormal 10 except for the distal half of the knob and the short euchromatic tip. The ring was transmissible through the gametophyte, and rod 10/ring abnormal 10 heterozygotes were constructed. Crossing over between rod and ring chromosomes generates dicentric bridges without accompanying acentric fragments. With proper positioning of the breakpoint of the bridge, viable rod chromosomes can be recovered which contain regions derived from the ring chromosome. As discussed in section IX, 2, the freshly broken end of a chromosome recovered from a dicentric is stabilized in the sporophyte. In Emmerling's work, the rod 10 was marked by the r-g allele and the ring abnormal 10 by the R-r allele. The R locus is distal in 10L and accurately marks the abnormal segment of the ring chromosome. Ring chromosomes are somatically unstable and the purple plant color of R-r is variegated on plants carrying the ring chromosome. Rod derivatives of dicentric bridges were identified by noting the occasional appearance of a stable purple plant phenotype in the progeny, indicating the transfer of R-r from a ring to a rod chromosome. Two rod abnormal 10 derivatives, designated K^s and K^o, were analyzed. Breakage of the dicentric within the abnormal 10 knob produced K^s with a small knob, while breakage just proximal to the knob generated the K^o knobless chromosome. In both cases, the three chromomeres of abnormal chromosome 10 plus the euchromatin between the chromomeres and the knob were present. The K^o and K^s chromosomes induced neocentromeres, although perhaps less strongly than a typical abnormal 10. Both chromosomes also induced preferential segregation, but at a drastically reduced rate. Preferential segregation of a knob on chromosome 9 was barely above the 50% level (53%) in the presence of either K^o or K^s. The findings demonstrate that a site(s) essential to high levels of preferential segregation is located within the heterochromatic knob or distal euchromatic tip of abnormal 10. In addition, the residual amount of preferential segregation found in the K^o and K^s chromosomes suggests that regions proximal to the heterochromatic knob contribute to preferential segregation. Interpretation of the findings concerning neocentremeres is more difficult, since the extent of neocentromere formation cannot be quantified. However, if neocentromere formation and preferential segregation are synonymous, one must assume that the level of neocentromeric activity in the presence of K^o and K^s is lower than that produced by abnormal chromosome 10.

Miles (1970) reconstructed the ring abnormal 10 chromosome of Emmerling from a duplicated rod chromosome which itself arose by bridge-breakage from the ring. The duplicated rod was capable of crossing over between its two arms and formation of the ring K10 (procedure of Schwartz, 1953a). From the derived ring K10, Miles followed the procedure of Emmerling and recovered a number of derivative rod abnormal 10 chromosomes. Her findings suggest that all the genes controlling preferential segregation are located in the knob. Rod chromosomes with different sized knobs were compared and the size of the knob was found to be directly related to the extent of preferential segregation. If Emmerling's K^o chromosome carried a small piece of the knob, which is possible, all regions controlling preferential segregation may be located in the knob. Evidence was also obtained by Miles which suggests a relationship between neocentromere formation and preferential segregation. It is possible to derive from the ring abnormal 10/rod normal 10 bivalent a rod chromosome with the abnormal segment in the short arm. When such a chromosome was recovered, it was found that neither neocentromere formation nor preferential segregation were induced by it, even though all abnormal 10 regions present in the ring chromosome were present in the rod. Since several derivatives of the ring chromosome display both phenomena when the abnormal segment is in 10L, the finding was unexpected. It was also shown that transfer of the abnormal segment from 10S to 10L was accompanied by a restoration of neocentromere formation and preferential segregation. The genetic regions controlling these events were suppressed when present in 10S, apparently by some type of position effect. Since both neocentromere formation and preferential segregation were influenced by the position effect it seems likely that they originate from a common genetic site(s). Finally, Miles localized the effect of abnormal 10 on recombination. She found that crossing over is controlled by a locus (or loci) proximal to the knob. Among the rod derivative chromosomes studied, no effect of differences in size of the heterochromatic knob on recombination was found. However, loss of the euchromatin between the knob and the modified chromomeric region of K10 eliminated the recombination effect.

While the picture developed by Emmerling and Miles may seem complete, a cautionary note should be added. Some rather paradoxical results, particularly with respect to preferential segregation of genes on the derivative rod chromosomes, have been noted. In addition, the correlation of neocentric activity with preferential segregation has never been a precise quantitative relationship (Rhoades and Dempsey, 1966b). These problems probably are explicable within the context of Rhoades' hypothesis, but further investigation is warranted.

2. The B Chromosome

A considerable number of organisms have supernumerary or B chromosomes (Battaglia, 1964). Several B chromosomes in different species possess accumulation mechanisms. The corn B chromosome behaves normally during development of the female gametophyte, but is transmitted in excess numbers through the pollen. Roman (1947, 1948) analyzed pollen transmission of B chromosomes with the use of A-B translocations (section IX, 1). Two quite unusual phenomena were discovered. First, the B chromosome frequently undergoes nondisjunction at the second pollen mitosis. If one B chromosome is present in the generative nucleus, sperm containing zero and two B chromosomes are produced. Second, when sperm with zero B's and two B's compete during fertilization, the sperm with two B's fertilizes the egg approximately two-thirds of the time and the polar nuclei one-third of the time. The latter phenomenon, preferential fertilization, results in an increase in the number of B's in the progeny above that present in the parent. Together, nondisjunction and preferential fertilization of the egg constitute the B chromosome accumulation mechanism.

Mitotic nondisjunction is an unusual event, seldom amenable to study. However, B chromosomes undergo nondisjunction regularly at the second pollen mitosis, providing a powerful tool in the study of chromosome movement. Nondisjunction of the B occurs in as many as 98% of pollen grains and is seldom observed at frequencies below 50%. Since nondisjunction occurs only at the second pollen mitosis, genetic regulation of the event seems necessary. Roman (1949) showed, with the A-B translocation B-4a, that two separable regions of the B are involved in nondisjunction. The B^4 chromosome with the proximal half of the B undergoes nondisjunction at the second pollen mitosis, while the 4^B chromosome does not. However, if the B^4 chromosome is separated from the 4^B and maintained as a supernumerary in $44B^4$ plants, the chromosome is no longer capable of nondisjunction. Apparently, the distal half of the B on 4^B must be present at the second pollen mitosis for nondisjunction of the B^4. After Roman's findings, work with several other A-B translocations with different breakpoints in the B chromosome produced consistent results: B^A chromosomes undergo nondisjunction, but only in the presence of A^B (see Ward, 1973a) or an intact B chromosome (Carlson, 1974a). Recently, Ward (1973a) utilized TB-8a with a near terminal exchange point in the B to demonstrate that the distal euchromatic tip of the B (on 8^B) is required for nondisjunction of the B^8 chromosome.

Localization of a proximal factor in nondisjunction was obtained by Carlson through an analysis of the products of misdivision of the B^9 chromosome of TB-9b. Misdivision of B^A centromeres occurs occasionally, and an isochromosome of the B^9 was isolated (Carlson, 1970a). The isochromosome lacks the minute short arm of a B^9, but contains the long arm in double dose. The chromosome retains the ability to undergo nondisjunction at the second pollen mitosis, indicating that the short arm of the

B is not required for nondisjunction. Misdivision of the B^9 isochromosome to form telocentrics also occurs and several telocentrics were recovered. Four telocentric derivatives of the isochromosome were found that lack the ability to undergo nondisjunction (Carlson, 1973b). Since the B^9 telocentrics are not induced to undergo nondisjunction in the presence of extra B chromosomes (unpublished data), the function missing from them is a *cis* rather than a *trans* activity. The telocentric B^9's differ little from a normal B^9 except for the lack of a short arm and possible absence of part of the centromere. Since the short arm of the B is not needed for nondisjunction, as shown by the B^9 isochromosome, the important modification of the B^9 telocentrics may be damage to the centromere. Alternatively, telocentrics *per se* may be incapable of nondisjunction due, perhaps, to lack of heterochromatin on both sides of the centromere.

The findings of Ward with TB-8a and Carlson with B^9 misdivision products may be interpreted as localization of the proximal and distal regions of the B which Roman found to control nondisjunction. However, there is no reason to believe, except for the sake of simplicity, that only two sites are involved in nondisjunction. Multiple sites could exist between the centromere and the distal tip of the B which are necessary for nondisjunction. These sites could be identified, theoretically, by the analysis of point mutations or gross chromosomal rearrangements of the B. In this regard, a number of spontaneously occurring aberrations of B-type chromosomes have been found over a period of time. Randolph (1941) isolated reduced derivatives of the B chromosome. Longley found (1956) a rearrangement of the B-9a translocation chromosomes. Carlson (1973a) and Ghidoni (1973) reported unstable B^A chromosomes which Ghidoni characterized as ring chromosomes. The repeated discovery of B chromosome rearrangements indicated the feasibility of selecting "mutants" of the mitotic nondisjunction system (Carlson, 1973b). As a result, different mutations (deletions, point mutations) of the A-B translocation TB-9b were collected with a common phenotype: the inability to carry out nondisjunction at the second pollen mitosis. Twelve TB-9b mutants have thus far been found, with cytological examination of the translocations and localization of mutant sites in a preliminary stage.

The mechanism of nondisjunction is unclear, but one hypothesis was developed from certain experimental results. Rhoades, Dempsey, and Ghidoni (1967) and Rhoades and Dempsey (1972, 1973b) found that B chromosomes can block the disjunction of A chromosomes at the second pollen mitosis. The B chromosomes are effective only in a particular "high-loss" genetic background, and only act upon A chromosomes which carry heterochromatic knobs. It appears that B chromosomes prevent the splitting of heterochromatic knobs in mitotic anaphase. As a result, dicentric bridges are formed at the second pollen mitosis and chromosome breakage often results. Frequently, one sperm receives a complete A chromosome while the other receives a deficient, often telocentric chromosome. Typical nondisjunction seldom is found. It was proposed that the unusual interaction of B's and heterochromatic knobs found in the "high-loss" background was a reflection of the usual pattern of B chromosome nondisjunction. A gene(s) on the B chromosome may cause specific heterochromatin to become "sticky" at the second pollen mitosis. The stickiness could result from a delay in DNA replication. It was proposed that centromeric heterochromatin is the nondividing site on the B and that the proximal location of centromeric heterochromatin produces nondisjunction rather than the dicentric bridge formation found for A chromosomes.

In addition to nondisjunction, preferential fertilization is an essential element of the B accumulation mechanism. The phenomenon has been analyzed with these questions in mind.
1) What is the rate of preferential fertilization?
2) Are there genes on the B chromosome which control preferential fertilization?
3) Does the presence of B chromosomes in a sperm cell alter the fertilization behavior of the sperm and in what way?

Rates of preferential fertilization are relatively constant, in contrast to rates of nondisjunction. Most workers using A-B translocations report fertilization of the egg by $A^B B^A B^A$ sperm in approximately two-thirds of pollinations while the A^B sperm succeeds one-third of the time. A striking exception to this rule was found when TB-9b plants were crossed as males to a particular inbred female. The female parent blocked preferential fertilization in all crosses and the $9^B B^9 B^9$ sperm succeeded no more often than the 9^B sperm in fertilization of the egg. Crosses of the same TB-9b males to other inbred lines produced normal rates of preferential fertilization, so that inhibition of preferential fertilization in the anomalous set of crosses could be attributed to the female parent (Carlson, 1969b). The mutant line should be of value when studies of the biochemical nature of preferential fertilization are performed.

Whether the B chromosome carries genes which control preferential fertilization is not known. In A-B translocations, the sperm carrying two B^4's preferentially fertilizes the egg. Thus, genes controlling preferential fertilization could be localized in a proximal region of the B chromosome. Alternatively, a difference in chromosome number or chromatin content between sperm with zero and two B^4's could trigger preferential fertilization in the absence of any specific genetic activity. If sperm differing in the number of A chromosomes could be generated, the possible nonspecificity of preferential fertilization could be tested.

In this discussion of preferential fertilization, we have assumed that B chromosomes in some way confer on sperm the ability preferentially to fertilize the egg. An alternative idea is that the two sperm of a pollen grain differ from each other in fertilization capacity regardless of the presence or absence of B chromosomes. During nondisjunction B chromosomes may find their way into the sperm which has an advantage in fertilizing the egg. If true, B-type chromosomes should migrate to a particular pole during nondisjunction at the second pollen mitosis. However, the B^4 and B^9 chromosomes of TB-4a and TB-9b segregate randomly with respect to each other during nondisjunction (Carlson, 1969b), giving no evidence of a specific polar orientation. Also, if B's do not affect the fertilization capacities of sperm, one expects that preferential fertilization with an A-B translocation will be unaffected by the addition of normal B chromosomes. However, when plants carrying TB-9b and 6 to 8 extra B chromosomes were crossed as male parents to a tester, preferential fertilization by sperm carrying two B^9's was completely absent. It is assumed that B chromosomes entered both sperm of the pollen grains and nullified any advantage of the B^9-containing sperm (Carlson, 1969b). The finding also demonstrates that the B chromosome accumulation mechanism breaks down when more than a few B chromosomes are present in a plant, thus setting an upper limit to its effectiveness. The B chromosomes, therefore, do influence the process of fertilization. They apparently do so by conferring a competitive advantage on sperm in fertilization of the egg (Carlson, 1970b).

In addition to the unique genetic functions associated with the B chromosome accumulation mechanism, B chromosomes also have the ability to influence recombination in certain regions of the genome. Hanson (1961, 1962, 1969) first reported an effect of B chromosomes on recombination. He found a slight enhancement of crossing over in proximal regions of chromosome 3 and chromosome 9 together with a decline in interference. The B chromosomes showed a dosage effect, with higher numbers giving higher rates of recombination. Nel (1973) found that B's increase crossing over for proximal regions of chromosome 5, the effect also being dosage dependent. Ayonoadu and Rees (1968) provided cytological confirmation of the influence of B's on crossing over. They found an increase in mean chiasma frequency in response to B chromosomes. Rhoades (1968) discovered that a segment of chromosome 3 which normally engages in little or no recombination undergoes frequent crossing over in the presence of B's. The chromosome 3 region was involved in a chromosomal aberration (transposition) which made analysis possible (section VIII, 6). Perhaps a number of untested regions of the corn genome behave in a similar way. Ward (1973b) used the transposition studied by Rhoades as a sensitive indicator of the B chromosome influence on recombination. Ward attempted to identify the site(s) on the B which control recombination. Several A-B translocations were used to separate and test different regions of the B. However, no specific recombinogenic locus was found and it was concluded that a number of genes scattered along the chromosome affect crossing over.

Whether or not the B chromosome and abnormal chromosome 10 have homologous genetic systems for influencing crossing over is not known. Both chromosomes may preferentially enhance recombination in proximal rather than distal chromosomal regions. However, B chromosomes show a dosage-dependent effect on recombination, whereas abnormal 10 does not (Rhoades and Dempsey, 1966b). Also, Nel (1973) found for regions on chromosome 5 that abnormal 10 was more active in raising the rate of crossing over during megasporogenesis than microsporogenesis and B chromosomes had the opposite effect. Abnormal 10 dramatically enhances crossing over in regions of structural or knob heterozygosity (section X, 1), but B chromosomes do not show a similar effect in In3a heterozygotes (Rhoades and Dempsey, unpublished data).

The B chromosome lends itself to the study of yet another basic genetic problem. Heterochromatic regions have been shown in some cases to contain highly repetitive DNA which may or may not differ from bulk DNA in G-C content (section IX, 3). The numerical variability of B chromosomes together with their high heterochromatin content allows analysis of the DNA in heterochromatin. Rinehart (1966) determined the buoyant density of DNA from plants without B's or with an average of 4.8 B's. The same value, corresponding to 42% GC was found in plants with or without B chromosomes. Similar results were obtained by Pitout and van Schaik (1968) and Pryor (1973). Rinehart's extractions and those of Pitout and van Schaik were made with isogenic Black Mexican lines differing only in the presence or absence of B's. The Black Mexican strain has no A chromosome knobs of significant size, making the test a sensitive one. Chilton and McCarthy (1973) analyzed buoyant densities and the renaturation kinetics of DNA from plants with zero B's and five B's. They failed to find any differences in DNA composition attributable to B chromosomes. In hybridization studies, the redundant (rapidly annealing) DNA from B-containing plants was found to cross-react completely with DNA of O-B plants. Chilton and McCarthy suggest from this finding that B chromosomes and A chromosomes are closely related, which does appear to be true for redundant sequences. The contrary conclusion of Randolph (1941) that B's and A's do not have genes in common probably relates to unique rather than repetitive sequences.

ACKNOWLEDGMENTS

The author and editor are deeply indebted to Dr. M.M. Rhodes for his contributions to this chapter and for the courtesy of his figures (1, 2, 5, 6, 7, 9, 10, 11, 14). Fig. 3, 4, 8 and 12 were the courtesy of J.D. Horn, R.L. Phillips, D.F. Weber, and D.T. Morgan, respectively.

LITERATURE CITED

Alexander, D.E. 1957. The genetic induction of autotetraploidy: a proposal for its use in corn breeding. Agron. J. 49:40-43.

----. 1964. Spontaneous reciprocal translocation during megasporogenesis of maize haploids. Nature 201:737-738.

Anderson, E.G. 1936. Induced chromosomal alterations in maize. p. 1297-1310, In Biological Effects of Radiation, Vol. II, McGraw-Hill, New York.

----. 1956. The application of chromosomal alterations in maize. p. 1297-1310, In Biological Effects of Radiation, Vol. II, McGraw-Hill, New York.

----, Kramer, H.H., and Longley, A.E. 1955. Translocations in maize involving chromosome 4. Genetics 40:500-510.

Arrighi, F.E., T.C. Hsu, P. Saunders, and G.F. Saunders, 1970. Localization of repetitive DNA in the chromosomes of *Microtus agrestis* by means of in situ hybridization. Chromosoma 32:224-236.

Ayonoadu, U., and H. Rees, 1968. The influence of B-chromosomes on chiasma frequencies in Black Mexican sweet corn. Genetica 39:75-81.

Baker, R.L., and D.T. Morgan, Jr. 1969. Control of pairing in maize and meiotic interchromosomal effects of deficiencies in chromosome 1. Genetics 61:91-106.

Battaglia, E. 1964. Cytogenetics of B chromosomes. Caryologia 17:245-299.

Beadle, G.W. 1930. Genetical and cytological studies of Mendelian asynapsis in *Zea mays*. Cornell Univ. Agric. Exp. Stn. Mem. 129: 1-23.

----. 1931. A gene in maize for supernumerary cell divisions following meiosis. Cornell Univ. Agric. Exp. Stn. Mem. 135:1-12.

----. 1932a. A gene for sticky chromosomes in *Zea mays*. Z Induk, Abstamm. Vererbungsl. 63:195-217.

----. 1932b. A gene in *Zea mays* for the failure of cytokinesis during meiosis. Cytologia 3:142-155.
----. 1932c. The relation of crossing over to chromosome association in *Zea-Euchlaena* hybrids. Genetics 17:481-501.
----. 1932d. A possible influence of the spindle fiber on crossing over in *Drosophila*. Proc. Natl. Acad. Sci. USA 18:160-165.
----. 1933. Further studies of asynaptic maize. Cytologia 4:269-287.
----. 1937. Chromosome aberration and gene mutation in sticky chromosome plants of *Zea mays*. Cytologia (Fujii Jubilee volume) p. 43-56.
----, and Emerson, S. 1935. Further studies of crossing over in attached X chromosomes in *Drosophila melanogaster*. Genetics 20:192-206.
----, and A.H. Sturtevant. 1935. X chromosome inversions and meiosis in *Drosophila melanogaster*. Proc. Natl. Acad. Sci. USA 21:384-390.
Beckett, J.B. 1967. Two new B-type translocations. Maize Genet. Coop. News Letter 41:139.
----. 1968. A B-type translocation involving the short arm of chromosome 3. Maize Genet. Coop. News Letter 42:132.
----. 1972. An expanded set of B-type translocations in maize. Genetics 71:s3-4.
Berlowitz, L. 1965. Correlation of genetic activity, heterochromatinization, and RNA metabolism. Proc. Natl. Acad. Sci. USA 53:68-73.
Bianchi, A., G. Bellini, M. Contin, and E. Ottaviano. 1961. Non-disjunction in presence of interchanges involving B-type chromosomes in maize, and some phenotypical consequences of meaning in maize breeding. Z. Vererbungslehre 92:213-232.
Blackwood, M. 1956. The inheritance of B chromosomes in *Zea mays*. Heredity 10:353-366.
Brink, R.A., and D.C. Cooper. 1935. A proof that crossing over involves an exchange of segments between homologous chromosomes. Genetics 20:22-35.
Britten, R.J. and E.H. Davidson. 1969. Gene regulation for higher cells: a theory. Science 165:349-357.
----, and ----. 1971. Repetitive and non-repetitive DNA sequences and a speculation on the origins of evolutionary novelty. Quart. Rev. Biol. 46:111-138.
----, and D.E. Kohne. 1968. Repeated sequences in DNA. Science 161:529-540.
Brown, D.D., and J.B. Gurdon. 1964. Absence of ribosomal RNA synthesis in the anucleolate mutant of *Xenopus laevis*. Proc. Natl. Acad. Sci. USA 51:139-146.
Brown, S.W., and U. Nur. 1964. Heterochromatic chromosomes in the coccids. Science 145:130-136.
----, and D. Zohary. 1955. The relationship of chiasmata and crossing over in *Lilium formosanum*. Genetics 40:850-873.
Brown, W.V., and S.M. Stack. 1968. Somatic pairing as a regular preliminary to meiosis. Bull. Torrey Bot. Club 95:369-378.
Burnham, C.R. 1932a. The association of non-homologous parts in chromosomal interchange in maize. Proc. 6th Intern. Congr. Genet. 2:19-20.
----. 1932b. An interchange in maize giving low sterility and chain configurations. Proc. Natl. Acad. Sci. USA 18:434-440.
----. 1934. Chromosomal interchanges in maize: Reduction of crossing-over and the association of non-homologous parts. Am. Naturalist 68:81-82.
----. 1949. Chromosome segregation in maize translocations in relation to crossing over in interstitial segments. Proc. Natl. Acad. Sci. USA 35:349-356.
----. 1950a. Crossing over differences in reciprocal backcrosses in corn. Maize Genet. Coop. News Letter 24:56-57.
----. 1950b. Chromosome segregation in translocations involving chromosome 6 in maize. Genetics 35:446-481.
----. 1953. Crossing over differences in male and female. Maize Genet. Coop. News Letter 27:64-65.
----. 1958. Crossing over in ♂ vs. ♀. Maize Genet. Coop. News Letter 32:93.
----. 1961. Differences in recombination in ♂ and ♀. Maize Genet. Coop. News Letter 35:86.

----, J.T. Stout, W.H. Weinheimer, R.V. Kowles, and R.L. Phillips. 1972. Chromosome pairing in maize. Genetics 71:111-126.

Carlson, W.R. 1969a. A test of homology between the B chromosome of maize and abnormal chromosome 10, involving the control of nondisjunction in B's. Mol. Gen. Genet. 104:59-65.

----. 1969b. Factors affecting preferential fertilization in maize. Genetics 62:543-554.

----. 1970a. Nondisjunction and isochromosome formation in the B chromosome of maize. Chromosoma 30:356-365.

----. 1970b. A test for involvement of the polar nuclei in preferential fertilization. Maize Genet. Coop. News Letter 44:91-92.

----. 1973a. Instability of the maize B chromosome. Theor. Appl. Genet. 43:147-150.

----. 1973b. A procedure for localizing genetic factors controlling mitotic nondisjunction in the B chromosome of maize. Chromosoma 42:127-136.

----. 1974a. B chromosomes induce nondisjunction in $9B^9$ pollen. Maize Genet. Coop. News Letter 48:81-83.

----. 1974b. A mutant of chromosomal behavior in mitosis and meiosis. Maize Genet. Coop News Letter. 48:76-80.

Chao, C.Y. 1959. Heterotic effects of a chromosomal segment in maize. Genetics 44:657-677.

Chase, S.S. 1949. Monoploid frequencies in a commercial double cross hybrid maize, and in its component single cross hybrids and inbred lines. Genetics 34:328-332.

----. 1951. The monoploid method of developing inbred lines. Proc. 6th Annu. Hybrid Corn Industry-Research Conf. 6:29-34.

----. 1969. Monoploids and monoploid derivatives of maize (Zea mays L.) Bot. Rev. 35:117-167.

Chen, Chi-Chang. 1969. The somatic chromosomes of maize. Can. J. Genet. Cytol. 11:752-754.

Chilton, M., and B. McCarthy. 1973. DNA from maize with and without B chromosomes: a comparative study. Genetics 74:605-614.

Clark, E.M. 1956. A comparison of crossing over in pollen and ovules in translocations involving the short arm of chromosome 9. Dissert. Abstr. 17:725.

Clark, F.J. 1940. Cytogenetic studies of divergent meiotic spindle formation in Zea mays. Am. J. Bot. 27:547-559.

----. 1942. Cytological and genetic studies of sterility in inbreds of similar and diverse genetic origin. Conn. Agric. Exp. Stn. New Haven, Bull 465:705-726.

Clarke, A.E., and E.G. Anderson. 1935. A chromosomal interchange in maize without ring formation. Am. J. Bot. 22:711-716.

Coe, E.H. 1959. A line of maize with high haploid frequency. Am. Nat. 93:381-382.

----, and Sarkar, K.R. 1964. The detection of haploids in maize. J. Hered. 55:231-233.

Collins, G.N., and J.H. Kempton. 1927. Variability in the linkage of two seed characters in maize. USDA Agric. Bull. 1468.

Cooper, D.C., and R.A. Brink. 1937. Chromosome homology in races of maize from different geographical regions. Am. Nat. 71:582-587.

Creech, R.G., and H.H. Kramer. 1961. A second region in maize for genetic fine structure studies. Am. Nat. 95:326-328.

Creighton, H.B., and B. McClintock. 1931. A correlation of cytological and genetical crossing-over in Zea mays. Proc. Natl. Acad. Sci. USA 17:492-497.

----, and ----. 1935. The correlation of cytological and genetical crossing-over in Zea mays. A corroboration. Proc. Natl. Acad. Sci. USA 21:148-150.

Davidson, E.H., B.R. Hough, C.S. Amenson, R.J. Britten, D.E. Graham, B.R. Neufeld, M.J. Smith, R.B. Goldberg, G.A. Galau. 1973 Sequence organization in animal DNA's. p. 9-30. In: Molecular cytogenetics (ed.) B.A. Hamkalo and J. Papaconstantinou. Plenum, New York.

Dempsey, E., and V. Smirnov. 1964. Cytological location of gl-15. Maize Genet. Coop. News Letter. 38: 71-73.

Dobzhansky, T. 1970. Genetics of the evolutionary process. Columbia Univ. Press, New York.

----, and M.M. Rhoades. 1938. A possible method for locating favorable genes in maize. J. Am. Soc. Agron. 30:668-675.

Dooner, H.K., and J.L. Kermicle. 1971. Structure of the R-r tandem duplication in maize. Genetics 67:427-436.

Doyle, G. 1963. Preferential pairing in structural heterozygotes of Zea mays. Genetics 48:1011-1027.

----. 1967. Preferential pairing in trisomics of Zea mays. In Chromosomes today, vol. 2. Plenum, New York.

----. 1971. Tandem duplications from translocations between homologous chromosomes in maize. Genetics 68:s16(abstr.)

----. 1972a. A telocentric trisome and its potential use in the production of commercial hybrid corn using genic male sterility. Maize Genet. Coop. News Letter 46:142-146.

----. 1972b. A tandem duplication and an intrachromosomal displaced duplication induced by irradiation. Maize Genet. Coop. News Letter 46:138-142.

----. 1973. Autotetraploid gene segregation. Theor. Appl. Genet. 43:139-146.

----. 1974. Nonrandom pairing in trisomics with two standard chromosomes 3 and one chromosome 3 treated with ethyl methane sulfonate or nitrosoguanidine. Maize Genet. Coop. News Letter 48:126-128.

Emerson, S., and G.W. Beadle. 1933. Crossing-over near the spindle fiber in attached-X chromosomes of Drosophila melanogaster. Z. Indukt. Abstamm. Vererbungsl. 65:129-140.

----, and Hutchinson, C.B. 1921. The relative frequency of crossing over in microspore and in megaspore development in maize. Genetics 6:417-432.

----, and M.M. Rhoades. 1933. Relation of chromatid crossing over to the upper limit of recombination percentages. Am. Nat. 67:374-377.

Emmerling, M.H. 1958a. Evidence of non-disjunction of abnormal chromosome 10. J. Hered. 49:203-207.

----. 1958b. An analysis of intragenic and extragenic mutations of the R-r gene complex in Zea mays. Cold Spring Harbor Symp. Quant. Biol. 23:393-407.

----. 1959. Preferential segregation of structurally modified chromosomes in maize. Genetics. 44:625-645.

Endrizzi, J.E. 1974. Alternate-1 and alternate-2 disjunctions in heterozygous reciprocal translocations. Genetics 77:55-60.

Eyster, W.H. 1921. The linkage relations between the factors for tunicate ear and starchy-sugary endosperm in maize. Genetics 6:209-240.

----. 1922. The intensity of linkage between the factors for sugary endosperm and for tunicate ears, and the relative frequency of their crossing over in microspore and megaspore development. Genetics 7:597-601.

Faberge, A.C. 1958. Relation between chromatid-type and chromosome-type breakage-fusion-bridge cycles in maize endosperm. Genetics 43:737-749.

Feldman, M., T. Mello-Sampayo, and E.R. Sears. 1966. Somatic association in Triticum aestivum. Proc. Natl. Acad. Sci. USA. 56:1192-1199.

Filion, W.G., and D.B. Walden. 1973. Karyotype analysis: the detection of chromosomal alterations in the somatic karyotype of Zea mays L. Chromosoma 41:183-194.

Fincham, J.R.S., and P.R. Day. 1971. Fungal genetics. 3rd ed. Blackwell Sci. Publ., Philadelphia.

Fisk, E.L. 1925. The chromosomes of Zea mays. Proc. Natl. Acad. Sci. USA. 11:352-356.

----. 1927. The chromosomes of Zea mays. Am. J. Bot. 14:54-75.

Flamm, W.G. 1972. Highly repetitive sequences of DNA in chromosomes. Int. Rev. Cytol. 32:1-51.

Fu, T.K., and E.R. Sears. 1973. The relationship between chiasmata and crossing over in Triticum aestivum. Genetics 75:231-246.

Gabay, S.J., and J.R. Laughnan. 1973. Recombination at the Bar locus in an inverted attached-X system in Drosophila melanogaster. Genetics 75:485-495.

Gall, J.G. 1973. Repetitive DNA in Drosophila. p. 59-74. In: B.A. Hamkalo and J. Papaconstantinou (eds.) Molecular Cytogenetics. Plenum, New York.

Ghidoni, A. 1973. Changes in the structure of the B^4 chromosomes in maize. Theor. Appl. Genet. 43:151-161.

Gilles, A., and L.F. Randolph. 1951. Reduction of quadrivalent frequency in autotetraploid maize during a period of 10 years. Am. J. Bot. 38:12-17.

Gilles, C.B. 1973. Ultrastructural analysis of maize pachytene karyotypes by three dimensional reconstruction of the synaptonemal complexes. Chromosoma 43:145-176.

Givens, J.F., and R.L. Phillips. 1973. Localization and estimation of the number of active versus inactive rDNA cistrons within the nucleolus organizer region of Zea mays. Genetics 74:594-595.

Goodsell, S.F. 1961. Male sterility in corn by androgenesis. Crop Sci. 1:227-228.

Gopinath, D.M., and C.R. Burnham. 1956. A cytogenetic study in maize of deficiency-duplication produced by crossing interchanges involving the same chromosomes. Genetics 41:382-395.

Greenblatt, I.M., and M. Bock. 1967. A commercially desirable procedure for detection of monoploids in maize. J. Hered. 58:9-13.

Grell, R.F. 1962. A new hypothesis on the nature and sequence of meiotic events in the female of Drosophila melanogaster. Proc. Natl. Acad. Sci. USA. 48:165-172.

----, Bank, H., and G. Gassner. 1972. Meiotic exchange without the synaptinemal complex. Nature (London) New Biol. 240:155-157.

Hanson, G.P. 1961. Alteration of recombination frequencies in A- by B- chromosomes. Maize Genet. Coop. News Letter 35:61-62.

----. 1962. Crossing over in chromosome 3 as influenced by B-chromosome. Maize Genet. Coop. News Letter 36:34-35.

----. 1969. B chromosome-stimulated crossing over in maize. Genetics 63:601-609.

Heitz, E. 1931. Die Ursache der gesetzmassigen Zahl, Lage, Form, und Grosse pflanzlicher Nukleolen. Planta 12:775-844.

Horn, J.D. 1970. Somatic association as a general phenomenon in maize. Maize Genet. Coop. News Letter 44:198-200.

----. 1971. Somatic association: the effects of various methods of arresting spindle fiber development. Maize Genet. Coop. News Letter 45:209-210.

Jackson, R.C. 1973. Chromosomal evolution in Haplopappus gracilis: a centric transposition race. Evolution 27:243-256.

Jancey, R.C., and D.B. Walden, 1972. Analysis of pattern in distribution of breakage points in the chromosomes of Zea mays L. and D. melanogaster meigen. Can. J. Genet. Cytol. 14:429-442.

John, B., and G.M. Hewitt. 1965. The B-chromosome system of Myrmeleotettix maculatus (Thunb.). II. The statics. Chromosoma 17:121-138.

Jones, G.H. 1967. The control of chiasma distribution in rye. Chromosoma 22:69-90.

----. 1974. Correlated components of chiasma variation and the control of chiasma distribution in rye. Heredity 32:375-387.

Jones, K.W. 1970. Chromosomal and nuclear location of mouse satellite DNA in individual cells. Nature 225:912-915.

Jones, R.N. and H. Rees. 1967. Genotypic control of chromosome behavior in rye. XI. The influence of B chromosomes on meiosis. Heredity 22:333-347.

Kayano, H. 1960. Chiasma studies in structural hybrids. III. Reductional and equational separation in Disporum sessile. Cytologia 25:461-467.

Kermicle, J.L. 1969. Androgenesis conditioned by a mutation in maize. Science 166: 1422-1424.

Khush, G.S. 1973. Cytogenetics of aneuploids. Academic Press, New York.

Kiesselbach, T.A., and N.F. Petersen. 1925. The chromosome number of maize. Genetics 10:80-85.

Kikudome, G.Y. 1959. Studies on the phenomenon of preferential segregation in maize. Genetics 44:815-831.

----. 1961. Cytogenetic behavior of a knobbed chromosome 10 in maize. Science 134:1006-1007.

Klein, W.H., W. Murphy, G. Attardi, R.J. Britten, and E.H. Davidson. 1974. Distribution of repetitive and nonrepetitive sequence transcripts in HeLa mRNA. Proc. Natl. Acad. Sci. USA 71:1785-1789.

Kuwada, Y. 1911. Meiosis in the pollen mother cells of *Zea mays*. Bot. Mag. (Tokyo) 25:163-181.

----. 1915. Ueber die Chromosomenzahl von *Zea mays*. Bot. Mag. (Tokyo) 29:83-89.

----. 1919. Die Chromosomenzahl von *Zea mays*. Ein Beitrag zur Hypothese der Individualitat der Chromosomen und zur Frage uber die Herkunft von *Zea mays*. J. Coll. Sci. Imperial Univ. Tokyo. 39:1-148.

----. 1925. On the number of chromosomes in maize. Bot. Mag. (Tokyo) 39:227-234.

Laughnan, J.R. 1949. The action of allelic forms of the gene *A* in maize. II. The relation of crossing over to mutation of *A-b*. Proc. Natl. Acad. Sci. USA. 35:167-178.

----. 1952a. The action of allelic forms of the gene *A* in maize. IV. On the compound nature of *A-b* and the occurrence and action of its *A-d* derivatives. Genetics 37:375-395.

----. 1952b. The *A-b* components as members of a duplication in maize. Genetics 37:598. (Abstr.).

----. 1955. Intrachromosomal association between members of an adjacent serial duplication as possible basis for the presumed gene mutations from *A-b* complexes. Genetics 40:580 (Abstr.).

----. 1961. The nature of mutations in terms of gene and chromosome changes. p. 3-29. *In:* Mutation and plant breeding, NAS-NRC Publ. 891.

----, and E.H. Coe. 1953. A-B interchanges as a method of screening for new mutants in specific segments. Maize Genet. Coop. News Letter 27:55-56.

Lima-de-Faria, A., and P. Sarvella. 1962. Variation of the chromosome phenotype in *Zea, Solanum,* and *Salvia*. Chromosoma 13:300-314.

Lin, M. 1955. Chromosomal control of nuclear composition in maize. Chromosoma 7:340-370.

Longley, A.E. 1927. Supernumerary chromosomes in *Zea mays*. J. Agric. Research 35:769-784.

----. 1937. Morphological characters of Teosinte chromosomes. J. Agric. Res. 54:835-862.

----. 1938. Chromosomes of maize from North American Indians. J. Agric. Res. 56:177-196.

----. 1945. Abnormal segregation during megasporogenesis in maize. Genetics 30:100-113.

----. 1956. The origin of diminutive B-type chromosomes in maize. Am. J. Bot. 43:18-22.

----. 1961. Breakage points for four corn translocation series and other corn chromosome aberrations maintained at the California Institute of Technology. USDA. ARS 34-16.

Lyon, M.F. 1961. Gene action in the X chromosome of the mouse (*Mus musculus* L.) Nature 190:372-373.

----. 1962. Sex chromatin and gene action in the mammalian X-chromosome. Am. J. Human Genet. 14:135-148.

----. 1971. Possible mechanisms of X chromosome inactivation. Nature New Biol. 232:229-232.

Maguire, M.P. 1962. Pachytene and diakinesis behavior of the isochromosomes 6 of maize. Science 138:445-446.

----. 1968. The effect of synaptic partner change on crossover frequency in adjacent regions of a trivalent. Genetics 59:381-390.

----. 1970. Non-random metaphase I orientation of the chromosomes of a trivalent. Genetica 41:361-368.

----. 1972a. Premeiotic mitosis in maize: Evidence for pairing of homologues. Caryologia 25:17-23.

----. 1972b. The temporal sequence of synaptic initiation, crossing-over and synaptic completion. Genetics 70:353-370.

Mather, K. 1939. Crossing over and heterochromatin in the X chromosome of *Drosophila melanogaster*. Genetics 24:413-435.

Mazrimas, J.A., and F.T. Hatch. 1972. A possible relationship between satellite DNA and the evolution of kangaroo rat species (genus *Dipodomys*). Nature New Biol. 240:102-105.

McClintock, B. 1929a. Chromosome morphology in *Zea mays*. Science 69: 629.

----. 1929b. A cytological and genetical study of triploid maize. Genetics 14: 180-222.

----. 1931. Cytological observations of deficiencies involving known genes, translocations, and an inversion in *Zea mays*. Missouri Agric. Exp. Stn. Res. Bull. 163: 1-30.

----. 1932. A correlation of ring-shaped chromosomes with variegation in *Zea mays*. Proc. Natl. Acad. Sci. USA 18: 677-681.

----. 1933. The association of non-homologous parts of chromosomes in the mid-prophase of meiosis in *Zea mays*. Z. Zellforsch. mikroskop. Anat. 19:191-237.

----. 1934. The relation of a particular chromosomal element to the development of the nucleoli in *Zea mays*. Z. Zellforsch. Mikrosk. Anat. 21: 294-328.

----. 1938a. The fusion of broken ends of sister half-chromatids following chromatid breakage at meiotic anaphases. Missouri Agric. Exp. Stn. Res. Bull. 290: 1-48.

----. 1938b. The production of homozygous deficient tissues with mutant characteristics by means of aberrant mitotic behavior of ring-shaped chromosomes. Genetics 23:315-376.

----. 1939. The behavior in successive nuclear divisions of a chromosome broken at meiosis. Proc. Natl. Acad. Sci. USA 26:405-416.

----. 1941a. The association of mutants with homozygous deficiencies in *Zea mays*. Genetics 29:542-571.

----. 1941b. The stability of broken ends of chromosomes in *Zea mays*. Genetics 26:234-282.

----. 1944. The relation of homozygous deficiencies to mutations and allelic series in maize. Genetics 29:478-502.

----. 1951. Chromosome organization and genic expression. Cold Spring Harbor Symp. Quant. Biol. 16:13-47.

----. 1959. Chromosome constitutions of Mexican and Guatemalan races of maize. Carnegie Inst. Wash. Yearb. 59:461-472.

----, and H.E. Hill. 1931. The cytological identification of the chromosome associated with the *R-G* linkage group in *Zea mays*. Genetics 16:175-190.

Michel, K. 1966. Non-homologous pairing in double trisomics in maize. Maize Genet. Coop. News Letter 40:105.

----, and C.R. Burnham. 1969. The behavior of nonhomologous univalents in double trisomics in maize. Genetics 63:851-864.

Miles, J.H. 1968. Somatic association of homologues induced by abnormal chromosome 10. Maize Genet. Coop. News Letter 42:77-78.

----. 1970. Influence of modified K10 chromosomes on preferential segregation and crossing over in *Zea mays*. Ph.D. thesis, Indiana University.

----. 1971. Probable weak fusion of chromatids during a breakage-fusion-bridge cycle. Maize Genet. Coop. News Letter 45:136-139.

Miller, O.L. 1963. Cytological studies in asynaptic maize. Genetics 48:1445-1466.

Moens, P. 1969. The fine structure of meiotic chromosome polarization and pairing in *Locusta migratoria* spermatocytes. Chromosoma 28: 1-25.

----. 1973. Mechanisms of chromosome synapsis at meiotic prophase. Int. Rev. Cytol. 35:117-134.

Moore, C.W., and R.G. Creech. 1972. Genetic fine structure analysis of the amylose-extender locus in *Zea mays* L. Genetics 70:611-619.

Morgan, D.T., Jr. 1943. The formation of chromocenters in interkinetic nuclei of maize by knobs and B chromosomes. J. Hered. 34:195-198.

----. 1950. A cytogenetic study of inversions in *Zea mays*. Genetics 35:153-174.

Moses, M. 1968. Synaptinemal complex. Ann. Rev. Genet. 2:363-412.

Mottinger, J.P. 1970. The effect of X-rays on the bronze and shrunken loci in maize. Genetics 64:259-271.

Nanda, D.K., and S.S. Chase. 1966. An embryo marker for detecting monoploids of maize (*Zea mays* L.) Crop Sci. 6:213-215.

Nel, P.M. 1971. Studies on the genetic control of recombination in *Zea mays*. Ph.D. thesis. Indiana University.

----. 1973. The modification of crossing over in maize by extraneous chromosomal elements. Theor. Appl. Genet. 43:196-202.

Nelson, O.E. 1957. The feasibility of investigating genetic fine structure in higher plants. Am. Nat. 91:331-332.

----. 1959. Intracistron recombination in the *Wx/wx* region in maize. Science 130:794-795.

----. 1962. The *waxy* locus in maize. I. Intralocus recombination frequency estimates by pollen and by conventional analysis. Genetics 47:737-742.

----. 1968. The *waxy* locus in maize. II. The location of the controlling element alleles. Genetics 60:507-524.

Neuffer, M.G. 1957. Additional evidence on the effect of X-ray and ultraviolet radiation on mutation in maize. Genetics 42:273-282.

----, and J.B. Beckett. 1971. Location of new mutants by A-B translocation method. Maize Genet. Coop News Letter 45:144-146.

----, and ----. 1972. Location of new mutants by the A-B translocation method. Maize Genet. Coop News Letter 46:130.

---, and ---. 1973. New mutants located by A-B translocation method. Maize Genet. Coop. News Letter 47:148-149.

Nicklas, R.B. 1967. Chromosome micromanipulation II. Induced reorientation and the experimental control of segregation in meiosis. Chromosoma 21:17-50.

----, and C.A. Koch. 1969. Chromosome micromanipulation III. Spindle fiber tension and the reorientation of maloriented chromosomes. J. Cell Biol. 43:40-50.

Novitski, E. 1955. Genetic measures of centromere activity in *Drosophila melanogaster*. J. Cell Comp. Physiol. 45 (Suppl. 2): 151-169.

Nur, U. 1967. Reversal of heterochromatinization and the activity of the paternal chromosome set in the male mealy bug. Genetics 56:375-389.

O'Mara, J.G., 1942. A cytogenetic study of *Zea* and *Euchlaena*. Missouri Agric. Exp. Stn. Res. Bull. 341:1-16.

Ostergren, G. 1945. Parasitic nature of extra fragment chromosomes. Bot. Notiser 2:157-163.

Palmer, R.G. 1971. Cytological studies of ameiotic and normal maize with reference to premeiotic pairing. Chromosoma 35:233-246.

Pardue, M.L., and J.G. Gall. 1970. Chromosomal localization of mouse satellite DNA. Science 168:1356-1358.

Patterson, E.B. 1952. The use of functional duplicate-deficient gametes in locating genes in maize. Genetics 37:612.

----. 1959. Report on maize cooperative. Maize Genet. Coop. News Letter 33:131.

Peacock, W.J. 1970. Replication, recombination, and chiasmata in *Goniaea australasiae (Orthopetera: Acrididae)*. Genetics 65:593-617.

----. 1971. Cytogenetic aspects of the mechanism of recombination in higher organisms. Stadler Genet. Symp. 1, 2:123-152.

Peterson, H.M., and J.R. Laughnan. 1961. Nonrecombinant derivatives at the Bar locus in *Drosophila melanogaster*. Genetics 46:889.

----, and ----. 1963. Intrachromosomal exchange at the Bar locus in *Drosophila*. Proc. Natl. Acad. Sci. USA 50:126-133.

Peterson, P.A. 1970. Controlling elements and mutable loci in maize: their relationship to bacterial episomes. Genetica 41:33-56.

----, and E.A. Wernsman. 1964. A monosomic-type approach in maize breeding. Crop Sci. 4:533-535.

Phillips, R.L. 1969. Recombination in *Zea mays* L. II. Cytogenetic studies of recombination in reciprocal crosses. Genetics 61:117-127.

----, C.R. Burnham, and E.B. Patterson. 1971a. Advantages of chromosomal interchanges that generate haplo-viable deficiency-duplications. Crop. Sci. 11:525-528.

----, R.A. Kleese, and S.S. Wang. 1971b. The nucleolus organizer region of maize (*Zea mays* L.): chromosomal site of DNA complementary to ribosomal RNA. Chromosoma 36:79-88.

Pickett-Heaps, J.D. 1969. The evolution of the mitotic apparatus: an attempt at comparative ultrastructural cytology in dividing plants. Cytobios 1:257-280.

Pitout, M.J., and N. van Schaik. 1968. Deoxyribonucleic acid from Black Mexican sweet corn. S. Afr. J. Sci. 64:25-29.

Pryor, T. 1973. Characterization of DNA from maize lines of different heterochromatic constitution. Maize Genet. Coop. News Letter 47:28-30.

Radding, C. 1973. Molecular mechanisms in genetic recombination. Ann. Rev. Genet. 7:87-111.
Rakha, F.A., and D.S. Robertson. 1970. A new technique for the production of A-B translocations and their use in genetic analysis. Genetics 65: 223-240.
Ramirez, S.A., and J.H. Sinclair. 1973. Intraspecies variation of ribosomal gene redundancy in *Zea mays*. Maize Genet. Coop. News Letter 47:67-70.
Randolph, L.F. 1928a. Chromosome numbers in *Zea mays* L. Cornell Univ. Agric. Exp. Stn. Mem. 117.
----. 1928b. Types of supernumerary chromosomes in maize. Anat. Rec. 41:102.
----. 1932. Some effects of high temperature on polyploidy and other variations in maize. Proc. Natl. Acad. Sci. USA 18:222-229.
----. 1935. Cytogenetics of tetraploid maize. J. Agric. Res. 50:591-605.
----. 1941. Genetic characteristics of the B-chromosomes in maize. Genetics 26:608-631.
Rees, H., and U. Ayonoadu. 1973. B chromosome selection in rye. Theor. Appl. Genet. 43:162-166.
Reeves, R.G. 1925. Chromosome studies of *Zea mays*. Proc. Iowa Acad. Sci. 32:171-177.
Rhoades, M.M. 1933a. A secondary trisome in maize. Proc. Natl. Acad. Sci. USA 19: 1031-1038.
----. 1933b. An experimental and theoretical study of chromatid crossing over. Genetics 18:535-555.
----. 1936. Note on the origin of triploidy in maize. J. Genet. 33:355-357.
----. 1940. Studies of a telocentric chromosome in maize with reference to the stability of its centromere. Genetics 25:483-520.
----. 1941. Different rates of crossing over in male and female gametes of maize. J. Am. Soc. Agron. 33:603-615.
----. 1942. Preferential segregation in maize. Genetics 27:395-407.
----. 1945. On the genetic control of mutability in maize. Proc. Natl. Acad. Sci. USA 31:91-95.
----. 1947. Crossover chromosomes in unreduced gametes of asynaptic maize. Genetics 32:101 (Abstr.)
----. 1950. Meiosis in maize. J. Hered. 41:58-67.
----. 1952. Preferential segregation in maize. p. 66-80. *In:* Heterosis. Iowa State College Press, Ames.
----. 1955. The cytogenetics of maize. p. 123-219. *In*: G.F./Sprague (ed.) Corn and corn improvement. Academic Press, New York.
----. 1956a. Genic control of chromosomal behavior. Maize Genet. Coop. News Letter 30:38-42.
----. 1956b. Studies with overlapping inversions. Maize Genet. Coop. News Letter 30:42-47.
----. 1957. Preferential pairing in structurally heterozygous triploids. Maize Genet. Coop. News Letter 31:75-76.
----. 1963. Cytogenetic studies with inversion 3c. Maize Genet. Coop. News Letter 37:56-58.
----. 1968. Studies on the cytological basis of crossing-over. p. 229-241. *In:* W.J. Peacock and R.D. Brock (eds.) Replication and recombination of genetic material.
----, and E. Dempsey. 1953. Cytogenetic studies of deficient-duplicate chromosomes derived from inversion heterozygotes in maize. Am. J. Bot. 40:405-424.
----, and ----. 1957. Further studies on preferential segregation. Maize Genet. Coop. News Letter 31:77-80.
----, and ----. 1966a. Induction of chromosome doubling at meiosis by the *elongate* gene in maize. Genetics 54:505-522.
----, and ----. 1966b. The effect of abnormal chromosome 10 on preferential segregation and crossing over in maize. Genetics 53:989-1020.
----, and ----. 1972. On the mechanism of chromatin loss induced by the B chromosome of maize. Genetics 71:73-96.
----, and ----. 1973a. Cytogenetic studies on a transmissable deficiency in chromosome 3. J. Hered. 64:125-128.
----, and ----. 1973b. Chromatin elimination induced by the B chromosome of maize. J. Hered. 64:13-18.

----, ----, and A. Ghidoni. 1967. Chromosome elimination in maize induced by supernumerary B chromosomes. Proc. Natl. Acad. Sci. USA 57:1626-1632.
----, and V.H. Rhoades. 1939. Genetic studies with factors in the tenth chromosome in maize. Genetics 24:302-314.
----, and H. Vilkomerson. 1942. On the anaphase movement of chromosomes. Proc. Natl. Acad. Sci. USA 28:433-436.
Riley, R., and C.N. Law. 1965. Genetic variation in chromosome pairing. Adv. Genet. 13:57-114.
Rinehart, K.V. 1966. Maize DNA composition: analysis of plants with and without B chromosomes. Maize Genet. Coop. News Letter 40:56-58.
----. 1970. A study of selective orientation in heterozygous paracentric inversions of maize. Ph.D. thesis. Indiana University, Bloomington.
Ritossa, F.M., and S. Spiegelman. 1965. Localization of DNA complementary to ribosomal RNA in the nucleolus organizer region of *Drosophila melanogaster*. Proc. Natl. Acad. Sci. USA 53:737-745.
Roberts, P. 1972. Differences in synaptic affinity of chromosome arms of *Drosophila melanogaster* revealed by differential sensitivity to translocation heterozygosity. Genetics 71:401-415.
Robertson, D.S. 1964. Transfer of intact segments of maize chromosomes. A possible method. J. Hered. 55:107-114.
----. 1967. Crossing over and chromosomal segregation involving the B^9 element of the A-B translocation B-9b in maize. Genetics 55:443-449.
Roman, H. 1947. Mitotic nondisjunction in the case of interchanges involving the B-type chromosome in maize. Genetics 32:391-409.
----. 1948. Directed fertilization in maize. Proc. Natl. Acad. Sci. USA 34:36-42.
----. 1949. Factors affecting mitotic nondisjunction in maize. Rec. Genet. Soc. Am. 18:112.
----, and A.J. Ullstrup. 1951. The use of A-B translocations to locate genes in maize. Agron. J. 43:450-454.
Russell, W.A., and C.R. Burnham. 1950. Cytogenetic studies of an inversion in maize. Sci. Agric. 30:93-111.
Salamini, F., and C. Lorenzoni. 1970. Genetical analysis of glossy mutants of maize. III. Intracistron recombination and high negative interference at the *gl* locus. Molec. Gen. Genet. 108:225-232.
Sarkar, K.R., and E.H. Coe, 1966. A genetic analysis of the origin of maternal haploids in maize. Genetics 54:453-464.
----, S. Panke, and J.K.S. Sachan. 1972. Development of maternal-haploidy-inducer lines in maize (*Zea mays* L.) Indian J. Agric. Sci. 42:781-786.
Schwartz, D. 1953a. The behavior of an X-ray-induced ring chromosome in maize. Am. Nat. 87:19-28.
----. 1953b. Evidence for sister-strand crossing over in maize. Genetics 38:251-260.
----, 1958. A new temperature sensitive allele at the *sticky* locus in maize. J. Hered. 49:149-152.
----, and T. Endo. 1966. Alcohol dehydrogenase polymorphism in maize — simple and compound loci. Genetics 53:709-715.
----, and C. Murray. 1957. A cytological study of breakage-fusion-bridge cycles in maize endosperm. Proc. Int. Gen. Symp. 277-279.
Shaver, D.L. 1962. A study of meiosis in perennial teosinte, in tetraploid maize and in their tetaploid hybrid. Caryologia 15: 43-57.
Smith, S.G. 1942. Polarization and progression in pairing. II. Premeiotic orientation and the initiation of pairing. Can. J. Res. 20:221-229.
Snope, A.J. 1967. The relationship of abnormal chromosome 10 to B-chromosomes in maize. Chromosoma 21:243-249.
Sobell, H. 1973. Symmetry in protein-nucleic acid interaction and its genetic implications. Adv. Genet. 17:411-490.
Sprague, G.F. 1941. The location of dominant favorable genes in maize by means of an inversion. Genetics 26:170.

Stack, S.M., and W.V. Brown. 1969. Somatic and premeiotic pairing of homologues in *Plantago ovata.* Bull Torrey Bot. Club. 96:143-149.

Stadler, D.R. 1973. The mechanism of intragenic recombination. Ann. Rev. Genet. 7:113-127.

Stadler, L.J. 1926. Variability of crossing over in maize. Genetics 11:1-37.

----, and M.G. Nuffer. 1953. Problems of gene structure. II. Separation of *R-r* elements (S) and (P) by unequal crossing over. Science 117:471-472.

----, and Roman, H. 1948. The effect of X-rays upon mutation of the gene *A* in maize. Genetics 33:273-303.

Steinitz-Sears, L.M. 1966. Somatic instability of telocentric chromosomes in wheat and the nature of the centromere. Genetics 54:241-248.

Stern, C. 1931. Zytologisch-genetische Untersuchungen als Beweise fur die Morgansche Theorie des Faktorenaustauschs. Biol. Zentr. 51:547-587.

Stout, J.T., and R.L. Phillips. 1973. Two independently inherited electrophoretic variants of the lysine-rich histones of maize. Proc. Natl. Acad. Sci. USA. 70:3043-3047.

Sturtevant, A.H. 1931. Known and probable inverted sections of the autosomes of *Drosophila melanogaster.* Carnegie Inst. Wash. Publ. 421:1-27.

----, and Beadle, G.W. 1936. The relations of inversions in the X chromosome of *Drosophila melanogaster* to crossing over and disjunction. Genetics 21:554-604.

Sved, J.A. 1966. Telomere attachment of chromosomes. Some genetical and cytological consequences. Genetics 53:747-756.

Tabata, M. 1962. Chromosome pairing in intercrosses between stocks of interchanges involving the same two chromosomes in maize. I. Diakinesis configurations in relation to breakage positions. Cytologia 27:410-417.

Ting, Y.C. 1958. On the origin of abnormal chromosome 10 in maize (*Zea mays* L.) Chromosoma 9:286-291.

----. 1966. Duplications and meiotic behavior of the chromosomes in haploid maize (*Zea mays* L.). Cytologia 31:324-329.

----. 1971. Further studies on the synaptonemal complex of haploid maize. Genetics 68:s67. (Abstr.)

----. 1973. Synaptonemal complex of haploid maize. Cytologia 38:497-500.

Underbrink, A.G., Y.C. Ting, and A.H. Sparrow. 1967. Note on the occurrence of a synaptinemal complex at meiotic prophase in *Zea mays* L. Can. J. Genet. Cytol. 9:606-609.

Walker, P.M.B. 1968. How different are the DNA's from related animals? Nature 219:228-232.

----. 1971. Origin of satellite DNA. Nature 229:306-308.

----, and W. Hennig. 1970. Variations in the DNA from two rodent families (*Cricetidae* and *Muridae*). Nature 225:915-919.

Walters, M.S. 1970. Evidence on the time of chromosome pairing from the preleptotene spiral stage in *Lilium longiflorum* "Croft." Chromosoma 29:375-418.

Ward, E.J. 1973a. Nondisjunction: localization of the controlling site in the maize B chromosome. Genetics 73:387-391.

----. 1973b. The heterochromatic B chromosome of maize: the segments affecting recombination. Chromosoma 43:177-186.

Waring, M., and R.J. Britten. 1966. Nucleotide sequence repetition: a rapidly reassociating fraction of mouse DNA. Science 154:791-794.

Weber, D.F. 1966. A test for distributive pairing. Maize Genet. Coop. News Letter 40:50-51.

----. 1969. A test for nonhomologous recombination in *Zea mays.* Genetics 60:235.

----. 1969. A test of distributive pairing in *Zea mays.* Chromosoma 27:354-370.

----. 1970. An attraction between nonhomologous univalent chromosomes and further tests of distributive pairing in *Zea mays.* Genetics 64:s65.

----. 1973. A test of distributive pairing in *Zea mays* utilizing doubly monosomic plants. Theor. Appl. Genet. 43:167-173.

----, and D.E. Alexander. 1972. Redundant segments in *Zea mays* detected by translocations of monoploid origin. Chromosoma 39:27-42.

Westergaard, M., and D. von Wettstein. 1972. The synaptinemal complex. Ann. Rev. Genet. 6:71-110.

Wimber, D.E., P.A. Duffey, D.M. Steffensen, and W. Prensky. 1974. Localization of the 5S RNA genes in *Zea mays* by RNA-DNA hybridization in situ. Chromosoma 47:353-359.

Yu, M.H., and P.A. Peterson. 1973. Influence of chromosomal gene position on intragenic recombination in maize. Theor. Appl. Genet. 43:121-133.

Yunis, J.J., and W.G. Yasmineh. 1971. Heterochromatin, satellite DNA, and cell function. Science 174:1200-1209.

Chapter 6 Corn Breeding

G.F. SPRAGUE
University of Illinois
Urbana, Illinois

S.A. EBERHART
Funk Seeds International, Inc.
Bloomington, Illinois

From the beginning of corn's domestication until at least the beginning of this century, corn breeding involved only the simplest form of mass selection. Because each ear was harvested separately, any variation in ear size, color, or kernel characteristics was readily apparent, and ears judged most desirable were saved to plant the ensuing crop. Mass selection was followed in turn by other procedures: ear-to-row breeding, the development of inbred lines and their use in hybrids, and more recently, population improvement. At each stage in this sequence, the methods used were based on the then current understanding of the biological and genetic principles involved.

I. EARLY HISTORY OF CORN BREEDING

A. Mass Selection

Most of the progress achieved through mass selection was realized before the advent of experiment stations or sound field experimentation. As a consequence, information on the effectiveness of this method is almost completely lacking. Judging by the wide range in plant and ear characteristics among the corn races and varieties of the world, we surmise that the selection practiced to bring about such differences must have been effective. Little can be said, however, about rates of either progress or genetic change.

Data on mass selection suggest that rates of change may have been quite variable, depending upon the attribute under selection, with yield the least amenable to change. A few reports will serve to illustrate the divergent results obtained.

Mr. H.E. Bidwell of Minnesota reported to the editor of the *Agricultural Annual* (1868) on selection for prolificacy. A part of the letter is quoted as follows:

A man in Tennessee gave me a good idea which I think worth publishing. He said, "Five years ago my corn yielded but one ear to each stalk, on an average, although I had long practiced selecting my seed corn from stalks bearing two ears. It occurred to me that the ears on the two-eared stalks were fertilized by adjoining plants bearing one ear only. I therefore resolved to raise my seed corn by itself, giving it the best of soil and culture, and before the silks appeared, breaking off the male flowers (tassels or spindles) from those having but one ear. You see the results — entire fields bearing nearly uniformly two ears to the stalk.

Table 1—Effect of selection for long and short ears in the variety Clarage†

Year	Length of ear (in inches)						Yield (bu/acre)		
	Seed			Crop			Long	Short	Diff.
	Long	Short	Diff.	Long	Short	Diff.			
1907	9.43	7.11	2.32	7.56	7.12	0.44	64.95	65.38	-0.43
1908	—	—	—	—	—	—	68.22	67.77	+0.45
1909	8.90	6.60	2.30	7.92	6.87	1.05	85.49	82.58	+2.91
1910	9.50	6.90	2.60	6.25	5.58	0.67	31.03	36.74	-5.71
1911	8.50	6.20	2.30	6.85	6.51	0.34	64.81	69.28	-4.47
1912	8.90	6.60	2.30	7.20	6.28	0.92	62.13	55.90	+6.23
1913	8.90	6.30	2.60	7.58	6.29	1.29	68.96	63.61	+5.35
1914	9.10	6.10	3.00	—	—	—	61.68	62.29	-0.61
8 year avg.	9.03	6.54	2.49	7.23	6.44	0.78	63.41	62.94	+0.47

†Adapted from Table XV, Williams and Welton Corn Experiments, Ohio Agric. Exp. Stn. Bull. 282, 1915.

Selection in this situation would be expected to be somewhat more effective than the selection normally practiced because it involved a greater than average control of the male parent.

Data on the effectiveness of mass selection for several ear characters were presented by Williams and Welton (1915), Table 1. These data indicate that the selection for ear length was not effective in separating the original population into two distinct groups, nor did the selection practiced have any important correlated effect on yield. Before one concludes that mass selection was ineffective, however, several aspects of the experimental design should be considered. Some of the factors that minimized the apparent effectiveness would include:

1) Small selection differential. The average difference in ear length for the two contrasting types was only 6.3 cm (2.5 in.). Neither long nor short ears were chosen unless they conformed to corn show standards. The effect of environment on phenotype would operate to reduce further the selection differential.
2) Lack of parental control. The contrasting selections were grown in adjacent plots to increase the accuracy of yield comparisons. This system of planting would facilitate interpollination between the two groups, thus minimizing any long-term trends.

The data available do not provide any measure of the importance of these two factors. It seems, however, that these two factors might have been sufficiently important to have masked any effects resulting from the selection practiced.

If one examines the data on corn yield trials published by experiment stations from about 1860 to 1900, it will be noted that many varieties were included for only a short period and then discarded because of poor yielding ability. Varities such as 'Reid' did not come into prominence until after winning at the Illinois State Fair at Peoria in 1891 and the World Columbian Exposition in 1893. This variety had its origin in a chance hybrid between a strain known as 'Gordon Hopkins' and a yellow flint known as 'Little Yellow.' From 1890 to 1920, this variety became very widely distributed, and local strains were developed by numerous breeders and corn show enthusiasts. In the report of the corn yield test for 1920, sponsored by the Iowa Corn and Small Grain Growers Association, Professor H.D. Hughes, then in charge of the Farms Crops Section, Iowa State University, was quoted as follows:

> There are hundreds of strains of Reid's Yellow Dent Corn in Iowa, and there is as much difference in the yielding power of different strains of this one variety as between distinct varieties.

It would not be logical, however, to attribute all the observed differences in yields to the mass selection practiced in isolating the many substrains. Accidental hybridization leading to increased genetic variability cannot be ruled out as a possible factor contributing to the observed strain differences. There seems little justification, however, for the assumption that mass selection was not one of the causal factors involved.

The corn shows played a prominent role in the recognition and distribution of varieties from about 1890 to 1920. A scorecard for judging corn was prepared by Orange Judd for the Illinois State Fair of 1891. The distribution of a variety was largely dependent upon winning at a local, state, or national show. Prize-winning single or 10-ear samples often sold for considerable sums. Gradually, it came to be appreciated that a winning sample was a measure of the showman's ability and patience in sorting over a large volume of ears to choose a sample corresponding to the showcard ideal rather than a measure of the breeding worth of the particular sample. As this idea grew, numerous experiments were conducted to measure the relationship between various ear and kernel characteristics and yielding ability. These experiments have been summarized by Richey (1922). In general, long, smooth, heavy ears with less than average number of kernel rows and shelling percentage produced greater average yields than did contrasting types.

Two varieties that later became important in hybrid development were 'Lancaster Sure Crop' and 'Krug.' Lancaster Sure Crop was developed in Pennsylvania and Krug in Illinois. The selection practiced in each variety was along utility lines, such as ear length and weight, rather than for attributes deemed important under the corn show scorecard.

Techniques for mass selection have been refined by Gardner (1961), and progress from the more recent mass selection programs are discussed in Section VII — Recurrent Selection.

B. Varietal Hybridization

Varietal hybridization has played a dual role in corn breeding. Varietal hybrids provided foundation material from which many of the standard varieties were selected and stabilized by mass selection. In addition, it supplied some of the earliest information on yield heterosis in corn and thus, indirectly, encouraged subsequent work on inbreeding and hybridization.

The religious ceremonies of many Indian tribes provided for the semicontrolled mixing of types characterized by different endosperm colors. Hybridization in the hands of the Indians may have played a much more important role in the development of corn varieties and types than is commonly recognized. Wellhausen et al. (1952) suggested that varietal hybridization was one of the important factors in the present diversity of corn types found in Mexico. Varietal hybridization was sufficiently common in early colonial

times to merit some attention. Beverley (1705) stated:
> There are four sorts of Indian corn, two of which are early ripe and two, late ripe; all growing in like manner . . . The late ripe corn is diversified by the shape of the grain only — one looks as smooth and full as the early ripe corn, and this they call Flint corn; the other has a larger grain and looks shrivell'd with a dent on the back of the grain, as if it had never come to perfection; and this they call She-corn. This is esteem'd by the planters, as the best source for increase, yet I can't see, but that this also produces the Flint corn accidentally among the others.

These observations suggest that the accidental or intentional intercrossing of these two types was not a new situation. Lorain (1825) some 100 years later describes the merits of mixing and intercrossing of the Gourd seed ("She") and flint types.

It usually is assumed that the dent types of corn had their origin from crosses between the Northern Flint and Gourd seed types. Anderson and Brown (1952) suggested that this crossing was most intensive from about 1800 to 1870. The report of Beverley indicated that this mixing process was well under way by 1700, although probably less systematically than occurred in later years. It seems logical to assume that it began immediately upon the introgression of the Gourd seed type into the area previously dominated by the flint types. This may have been about 1000 A.D., because archeological evidence for the southeastern quarter of the USA reveals the presence of only the flint types prior to that date.

Controlled experiments on varietal hybridization from which yield data were obtained were first reported by Beal (1877). Additional results were cited in his reports for 1880 and 1881-82. Individual hybrids tested outyielded the parents by amounts ranging from approximately 10 to 50%. Beal's interest in hybridization studies was further strengthened by Darwin's studies (1877) on the same subject. Beal was interested in establishing cooperative experiments on corn hybridization. The report of Ingersoll (1882) was clearly a direct outgrowth of this cooperative attempt. After the work of Beal and Ingersoll, an intensive investigation of varietal hybrids was begun at many different stations.

The individual experiments in this early work must be interpreted with considerable caution. Many of the experiments were based on single plots. Morrow and Gardner (1893) presented some data on variation in yield among replicated plots of the same variety. In the variety 'Murdock,' having a mean yield of 57.6 bu, the range in yield among the four replicates was 11.9 bu. (Conversions to metric units have not been made because original data were reported in nonmetric units, and errors might be introduced by the conversion.) In the variety 'Leaming,' the observed range was 18.9 bu, with an average yield of 70.1 bu. Although the individual contrasts between parents and hybrids were poorly determined, the results from all experiments considered as a group followed a rather consistent pattern. Richey (1922) has summarized the data for 244 comparisons. Of this number, 82.4% produced more, and 17.6% produced less than the average of the parents. Slightly over one-half of the crosses produced yield in excess of the higher-yielding parent.

The result took on increased significance when they were summarized with respect to the characteristics of the parents. With very few exceptions,

Table 2—Comparative performance of varietal hybrids involving diverse endosperm types (Hayes and Olson, 1919)

	Yield as a percentage of Minn. 13	
Pistillate parent	Pistillate parent	Cross
Flour		
Blue Soft	96.7	132.5
Flints		
Smutnose	110.8	127.6
King Phillip	100.0	119.9
Longfellow (NK)	100.5	119.6
Longfellow (Burwells)	104.9	114.1
Mercer	91.8	97.3
Dents		
Northwestern	105.9	115.7
Chowen	99.0	114.9
Rustler	112.1	112.4
Minnesota No. 23	96.7	110.9
Silver King	100.9	106.7
Murdock	80.3	101.6

Table 3—Summary of Dent × Dent hybrids compared with the higher yielding parent (Smith and Brunson, 1928)

Variety × Reid Yellow Dent	Gain		Loss	
	bu/acre	%	bu/acre	%
Champion White Pearl	1.89	3.9	—	—
Boone County White	1.61	3.3	—	—
Leaming (Chester's)	1.58	3.7	—	—
Calico	0.38	0.9	—	—
Silvermine	—	—	0.18	0.4
Crimson	—	—	0.42	1.0
Reid Yellow Dent	—	—	0.53	1.2
Golden Eagle	—	—	0.84	1.8
Riley's Favorite	—	—	1.73	4.1
Leaming (Maxey's)	—	—	3.93	8.9

flint × dent or flour × dent crosses produced yields substantially greater than the higher-yielding parent. In the dent × dent combinations, the superiority of the hybrids was less consistent. In general, the crosses exhibited the greatest superiority when the parents were high yielding and differed appreciably in type. Some idea of the difference in performance of variety hybrids is afforded by the data in Table 2 (Hayes and Olson, 1919) and Table 3 (Smith and Brunson, 1928).

The data in Table 2 represent averages for parents and crosses involving from 1 to 4 years of tests. All yields are espressed as a percentage of 'Minn. 13,' which was the staminate parent of all crosses. With but one exception, all crosses produced yields exceeding those of the higher-yielding parent. These results indicate a high degree of genetic diversity among the parents tested.

The data in Table 3, representing averages covering a 4 or 5-year period, present a decided contrast. All these hybrids involved 'Reid' as one parent and a locally adapted dent as the other parent. All comparisons were made with the higher yielding of the two parents. Only four of the crosses produced yields in excess of the higher-yielding parent, and in each instance, the increase was small and nonsignificant. Thus, within the Corn Belt dents as a group, genetic diversity was quite limited.

Many of the early reports on varietal hybrids give instructions on how to produce hybrid seed. Some state that the method was being used by farmers. There is no information on the extent of its use, however, and it seems probable that varietal hybrids were never planted on more than a small fraction of the USA corn acreage.

C. Ear-to-row Selection

The ear-to-row system of breeding was devised and put into use at the Illinois Station in 1896 (Hopkins, 1899). In its simplest form, the method involves the selection of a number of ears on the basis of some criteria of interest and subsequent evaluation of these ears by a progeny test. Ears for the next cycle of selection would be chosen from those progenies exhibiting the highest means for the criteria of interest (e.g., yield, chemical composition). The Illinois studies differed from this simple model in that ears were grown in progeny rows, but selections were based on a truncated segment of the entire distribution, regardless of the progeny-rows in which the extreme deviates may have occurred. Thus, the method employed was a form of *mass selection* and not ear-to-row breeding as used later by other workers or as used today in half-sib progeny tests. In spite of these operational differences, the results are most conveniently reviewed in this section.

The first attributes to receive attention in the Illinois experiments were the oil and protein percentages of the grain. The results from the two series were rather similar, and we shall discuss only the results obtained with oil percentages. One hundred and sixty-three ears of the variety 'Burr White' were individually analyzed for oil percentage of the shelled grain. A group of 24 ears having the highest oil percentage was chosen as foundation material for the "high" strain. Twelve ears having the lowest oil percentage were used to form the "low" strain. Each group was grown in an isolated plot, with each of the parent ears being planted as a separate entry. At harvest, 10 ears were saved from each row. The four ears from each row judged most suitable for seed corn were analyzed individually. The 24 ears highest in oil or the 12 ears lowest in oil, regardless of the row from which they had come, were used to plant the ear-to-row plot the ensuing year. Modifications of this scheme were introduced at various times (Winters, 1929). Several progress reports were presented (Dudley et al., 1974b; Leng, 1961, 1962; Woodworth et al., 1952). Results will be presented in detail in Chapter 10. Striking separations in both the oil and protein selections occurred during the 70-year-history of this experiment.

Smith (1909) reported data on the effects of ear-to-row selection on several plant attributes, including plant and ear height. Approximately 24 ears were selected for the "high-ear" group and a similar number for the "low-ear" group. These were planted ear-to-row, the two groups being placed end to end. Among the progeny, the first criterion was productiveness, and the second, height of ear. At the end of a six-year selection period, the "high-ear" strain was approximately twice the ear height of the "low-ear" strain.

The selection practiced was effective in increasing the spread in both plant and ear heights between these two contrasting groups. Also, there was a close relationship between plant and ear heights. Furthermore, height of plant or ear was closely associated with time of flowering, the shorter types being the earlier. In 1907 and 1908, the low-ear strain was approximately 1 week earlier in flowering than the high-ear strain. Thus, although no physical isolation of the two types was practiced, an effective physiological isolation was rapidly achieved owing to the association between plant or ear height and time of flowering. Selection was continued in the high and low-ear strains for some years, but no further reports on progress were presented.

The results of ear-to-row selection for chemical and various plant attributes (such as height, ear inclination, and leaf area) indicated this breeding system to be effective. It was natural, therefore, that the system should be extended to an even more important characteristic, yield of grain. The early results from such studies indicated that the method was effective. As more data became available, however, the results were generally disappointing. The failure to obtain continuing favorable results was ascribed to various causes, and modifications were introduced to overcome the assumed limitations. Some of the modifications suggested and tried were detasseling of alternate halves of alternate rows, use of duplicate plots for each ear-row, and the compositing of remnant seed from the high-yielding rows. None of these modifications seemed to increase the effectiveness of the method materially. Some of these modifications have been reintroduced with some success, e.g., the modified ear-to-row recurrent selection scheme (Lonnquist, 1964). The ineffectiveness, therefore, probably resulted from the inadequate field-plot techniques of the period.

An experiment reported by Smith and Brunson (1925) provides more data on ear-to-row selection for yield than does any other report available. A total of 990 ears was selected from a local variety and planted ear-to-row. At the same time, a composite involving these same ears was grown in an isolated plot to provide a standard for the measurement of progress. In addition, this composite seed lot was used as a check in every 10th row in the ear-row evaluation plot. The check plots ranged in yield from 3.0 to 13.3 kg/plot. The 40 highest and 40 lowest yielding ear rows, on the basis of their performance relative to adjacent checks, were chosen to form the high-yielding and low-yielding strains. In subsequent years, 4 ears were saved from each of the 10 highest or lowest progenies to continue the strain. In each strain, the most vigorous appearing plants were tagged before harvest, and the ears finally selected to represent either the high or low-yield strain were chosen from these tagged plants. After 5 years, a "special" high-yield plot was established in a separate breeding plot. The foundation material was remnant seed of the nine highest ear-row selections from the high plot. Eight of these were detasseled and used as females, and the highest-yielding row was used as a male. Eighty ears were selected from this mating plot for the next year's ear-row test plot. All the ear-row tests throughout the experiment were based on single rows. Because each of the three selections was grown in a separate area, yields were not directly comparable. Separate tests were established to provide direct comparability. Either two or four replications were used in these comparisons.

Selection for yield was ineffective in each of the three populations. In the low-yield strain, the selection intensity for yield was doubtlessly reduced through the use of the most vigorous plants. Inbreeding effects would have been appreciable in all three populations and would have been at a maximum in the "special" high-yield plot because of the reduced numbers and the method of formation. On the basis of current knowledge, the limitations imposed through use of single plots were probably as great a barrier to progress as the inbreeding depression effect. When both of these limitations were minimized, reasonable rates of progress were achieved (Section VII). Studies of the type just described were interpreted as indicating that mass or ear-to-row selection was ineffective for modifying yield, and alternative breeding procedures were sought.

D. Early Inbreeding Experiments

The first inbreeding experiments in corn were reported by Darwin (1877). These studies were continued for only a single generation and presumably had little effect on the future of corn breeding. Inbreeding experiments were begun at Illinois sometime before 1900, but the studies were never reported in detail. Shamel (1905), who had worked in the Illinois program from 1898 to 1902, mentioned that marked reduction occurred in yield in material that had been self-fertilized for four generations. His conclusions for a practical method of corn breeding were as follows:

> It would seem that the improvement of our crops can be most rapidly effected with permanent beneficial results by following the practice of inbreeding or crossing, to the degree in which these methods of fertilization are found to exist in the kind of plant under consideration.

The first inbreeding experiments that led to an interpretation of inbreeding depression and the restoration of vigor upon crossing were reported by G.H. Shull. These studies were initiated because of his interest in the quantitative character, number of kernel rows per ear, rather than corn breeding per se. In his first report (1908), Shull concluded:

> 1) that in an ordinary field of corn, the individuals are very complex hybrids, 2) that the deterioration which takes place as a result of self-fertilization is due to the gradual reduction of the strain to a homozygous condition, and 3) that the object of the corn breeder should not be to find the best pure line but to find and maintain the best hybrid combination.

In a discussion of the pure-line method, Shull (1909) states:

> The process may be considered under two heads, (1) finding the best pure lines, and (2) the practical use of the pure lines in the production of seed corn.
> (1) In finding the best pure lines, it will be necessary to make as many self-fertilizations as practicable, and to continue these year after year until the homozygous state is nearly or quite attained. Then all possible crosses are to be made among these nearly pure strains and the F_1 plants coming from such crosses are to be grown in the form of an ear-to-row test, each row being the product of a different cross . . . (2) After having found the right pair of pure strains for the attainment of any desired result in the way of yield and quality, the method of producing seed corn for the general crop is a very simple though somewhat costly process.

Techniques for producing and testing lines have undergone extensive modification since this report in 1909, but the two steps outlined still form the basis for most of the corn breeding programs of today.

East (1908, 1909) presented data on the effects of inbreeding and crossbreeding of maize. He was of the opinion that, although the pure-line method of breeding was sound theoretically, it was not commercially feasible. As a more practical substitute, he advocated, as did Collins (1910) and many others, the use of varietal hybrids. (In a letter to George Allee of Newell, Iowa in 1915, Shull indicated also that varietal hybridization might be more feasible than the production of F_1 hybrids between pure lines.)

Inbreeding work on corn was begun at a number of stations after the reports by Shull and East. The investigators seemed much more impressed by the reduction in vigor upon inbreeding than by the increased vigor following hybridization. This is not too surprising in view of the relatively small number of lines available and the practice of maintaining any line that could be propagated. Jones (1918) presented a summary of the results of the studies on inbreeding and crossbreeding of corn conducted at the Connecticut Station. His suggestion of double-cross production from high-yielding single crosses removed hybrid corn from an intriguing theoretical possibility to that of a widely accepted commercial development. [Shull tested double-cross hybrids in 1915, but the results were not published until many years later (Shull, 1952)].

II. DEVELOPMENT OF INBRED LINES

Maize is normally cross-pollinated, and therefore, artificial control is necessary to effect inbreeding. Control is accomplished by covering the developing ear shoot, before silks have emerged, by a small glassine or parchment "shoot" bag. After emergence of the silks, the appropriate tassels are bagged, and the next day, the pollen within the bag is transferred to the silks (stigmas) of the desired plant. Many different methods have been developed for this transfer (Sprague, 1942), and each method is satisfactory in the hands of an experienced operator.

A. Direct Isolation Methods

Inbred lines were initially extracted from open-pollinated varieties. As elite inbred lines were identified, these lines were then used as parental material to produce synthetic varieties as well as F_2 and backcross populations from which other inbred lines could be developed. Several methods have been proposed and used in the development of inbred lines.

1. Standard Method

Self-fertilization is the most common form of inbreeding used. In this situation, pollen from a single plant is applied to the silks of the same plant. Inbreeding normally is accompanied by selection. Plants to be inbred are selected for vigor, freedom from disease, or other desired attributes. Because many desirable attributes are not apparent at the time of pollination, the plants are reselected at harvest time, and any plants that have developed undesirable weaknesses of any sort are discarded. Ears that are saved from the self-pollinated plants surviving this second selection are planted ear-to-row the following season. Selection is practiced both among and within progenies. Only the more desirable plants within the superior progenies are used for further inbreeding. After three or four generations of inbreeding, the lines surviving the selection process become separated into distinct and relatively true-breeding types. The main reduction in vigor occurs during the first few generations of self-fertilization (East and Jones, 1919).

The most extensive studies on changes in quantitative traits upon inbreeding have been presented by Hallauer and Sears (1973). They concluded that "the relation between the mean performance and levels of inbreeding can be described by a genetic model based on the cumulative effects of loci with dominance for most traits; i.e., inbreeding depression resulted from an increase in the frequency of homozygous recessive deleterious loci." A linear model accounted for more than 92% of the total variability for plant and ear height and for yield. A quadratic model was significant for ear-leaf width, kernel-row number, ear diameter, and kernel depth, but even for these traits the relation was essentially linear.

Jones (1939) reported that reductions in plant height ceased after five generations and reductions in yield after 20 generations. In some instances, sib lines separated after different numbers of generations of self-fertilization were clearly different. Such differences were assumed to be the result of spontaneous mutations rather than delayed segregation from relic heterozygosity.

Studies designed to measure mutation ratios of quantitative genes have been reported by Sprague, et al. (1960) and Russell et al. (1963). In the first study, the parental material was doubled-monoploid stocks presumed to be completely homozygous. The second test involved a series of inbred lines that had been self-pollinated for 10 to 20 generations, maintained in each generation as ear-row progenies. In each experiment, two self-pollinated ears were saved at random to represent each line of descent. At the completion of the inbreeding cycle, the sublines, each descended in a dichotomous manner from a single parental ear, were compared in individual replicated tests. Data were taken on plant and ear traits. The criterion for mutation was a significant difference between parent and progeny or between siblings within a given generation. In the monoploid test, the average mutation rate was 4.5 per 100 gametes tested. In the long-time inbred test, the rate for corresponding traits was 2.8 mutations per trait per 100 gametes tested. In each instance, mutation rates seemed to be much higher than the average rate for qualitative traits, 10^{-5}.

Russell and Vega (1972) compared genetic stability in 11 long-time inbred lines. Progeny from these lines were evaluated for both seed and plant characters after 1, 3, 5, 7, 9, and 11 generations of additional self-pollination. Significant differences were detected in approximately 40% of the comparisons. Major changes were observed in 2 of the 11 lines, but most changes observed, although significant, were too small to be practical.

Genetic evidence for the number and character of mutations involved was not feasible because of lack of specific information on number of loci involved in each genetic trait. With mutation rates of this order, long-time stability of inbred lines is dependent upon a continuous selection for type.

2. Single-Hill Method

This modification of the standard method was suggested by Jones and Singleton (1934). It differs from the standard procedure by the substitution of a single three-plant hill for a progeny row in each generation of inbreeding.

The method permits the handling of a larger number of items in a given space and, therefore, a potentially larger zygotic sample. The method minimizes the opportunity for selection within progenies. Yermanos (1952) presented some data on the relative test-cross performance of lines isolated by the standard and single-hill methods, both sets of lines having their origin from the same S_1 ears. The results were not critical because comparisons involved equal numbers of lines rather than equal areas. The results indicated a slight superiority in yield for single-hill lines, but a much greater degree of susceptibility to both root and stalk lodging. This method has been used for the development of random lines for theoretical studies, but has never been used extensively in the development of commercial lines.

3. Pedigree Selection

This procedure represents no significant departure from the standard method, except that the parental material is normally derived from some planned hybrid. This procedure has been used most extensively in the Minnesota program. Inbred parents are chosen, which in combination possess a series of desirable attributes. Parents are than crossed, and a series of new lines is derived by the standard techniques of self-fertilization and selection. Reports by Hayes and Johnson (1939) and Johnson and Hayes (1940) provide information on the results obtained.

A somewhat similar breeding scheme was used by Lindstrom (1939). This scheme was designated as cycling and involved the isolation of new lines from first-generation backcrossed material. Through successive repetitions or cycles, it was hoped that progressive improvements could be achieved for both lines and their hybrid combinations. Preliminary results seemed encouraging, but no critical data are available.

4. Homozygous Diploids

The production of homozygous individuals through the use of viable haploids has been suggested as a desirable breeding procedure by several workers. Two very different procedures have been suggested.

Burnham (1946) proposed the development of an Oenothera-type of multiple translocation stock involving each of the chromosome pairs. Such a stock could then be used as follows: (1) cross multiple translocation to some heterozygous source; (2) examine pollen of F_1 plants and self-pollinate those exhibiting a high degree of sterility; (3) grow F_2 progeny and examine pollen for abortion, saving only those having normal pollen; (4) outcross plants with normal pollen to a standard (i.e., carrying no translocations); (5) grow test crosses and examine pollen, discarding all plants with partly sterile pollen. Test crosses exhibiting normal pollen would be the desired type, and the normal plant under test would be the desired homozygote.

Thus far, a multiple translocation stock has not been developed. Translocations in corn do not exhibit any marked degree of directed segregation, so a multiple translocation stock, if found, would be very difficult to maintain.

Einset (1943) cited unpublished data by Randolph in which a system of dominant markers was used to detect monoploids. Chase (1949, 1952a,

1952b) used this scheme and reported on the frequency of haploids obtained under a number of different sources. As an average for the stocks studied, monoploids occurred with a frequency of about 1 per 1,000, and among such monoploids, selfed progeny were obtained in a frequency of about 1 in 10. Thus, an average of 1 homozygous strain could be expected for each 10,000 seedlings tested.

This system was tried on a very extensive scale by one of the large hybrid seed companies. Its use was discontinued after a period of years, suggesting that they found the system to have no real advantage over the standard method. The method may have real value, however, in special circumstances. Goodsell (1961) was able to obtain androgenous monoploids in a series of inbred lines. This permitted the incorporation of cytoplasmic sterility in a single generation, as opposed to the six to eight generations required by conventional backcrossing. Kermicle (1969) reported the use of the mutant indeterminate gametophyte (ig) to obtain androgenetic monoploids. The major obstacle is the low frequency of restoration to the diploid level to have a useful homozygous line.

Improvement of Existing Lines

Other methods have been devised to improve existing elite lines. A line may be superior in most respects, but deficient in one or two traits. The goal is to improve the deficient traits while retaining all favorable traits.

1. Convergent Improvement and Backcrossing

The concept of convergent improvement was developed by Richey (1927). Later Richey and Sprague (1931) presented data from a series of experiments designed to supply information on (1) types of gene action involved in yield heterosis and (2) the value of convergent improvement as a breeding method. Considered solely from a breeding standpoint, convergent improvement involves the reciprocal addition to each of two inbred lines of the dominant favorable genes lacking in one parent and present in the other. Backcrossing to one line is continued for usually three or more generations. Selection is practiced during the backcrossing period to retain as many as possible of the factors contributed by the nonrecurrent parent. Backcrossing and selection are performed in parallel, each of the original lines serving as the recurrent parent in one series. The experimental data indicated improvements in the yield of both the recovered lines and the corresponding single-cross combination. Murphy (1942) presented additional data supporting these conclusions.

Simple backcrossing has been used in the improvement of lines to a much greater degree than has convergent improvement. Many lines currently in extensive use have been derived by simple backcrossing. This method offers a convenient procedure for the introduction of relatively simply inherited characteristics into an established line. If the trait of interest is not completely dominant, alternate selfing and backcrossing may be desirable to ensure retention of the introduced gene complex to a maximum degree. The experience of many corn breeders suggests that backcross improvement can be undertaken profitably with only the very best of the inbred lines available.

2. Gamete Selection

This breeding scheme was proposed by Stadler (1944), and limited data bearing on the usefulness of the method were presented by Pinnell et al. (1952), Lonnquist and McGill (1954), and Giesbrecht (1964).

The theoretical advantage of the method arises from the fact that, when zygotes of a given level of desirability occur with a frequency of x, then gametes of equal desirability will occur with a frequency of $\sqrt{x}$. The method consists in the crossing of an elite line with a random sample of pollen from an open-pollinated variety. F_1 plants grown from such seed differ from each other only in the gametic complement contributed by the variety. The individual F_1 plants are self-fertilized and outcrossed to a tester. Crosses of the elite line with the same tester serve as a check. Any testcross giving a performance superior to the elite line $\times$ tester combination is presumed to have received a superior gamete from the variety. Stadler (1944) stressed the desirability of continued sampling of open-pollinated varieties and suggested that gametic sampling offered a more efficient sampling procedure than the direct isolation of lines by standard inbreeding procedures.

Gamete selection has one obvious disadvantage; the superior gametes identified cannot be isolated as homozygous zygotes. The final product may be considered as an improved elite line, rather than a new line. Richey (1947) suggested that the corn breeder is interested not only in superior gametes, which in combination yield superlative zygotes, but in satisfactory zygotes as well; a satisfactory zygote was defined as one from which a superlative zygote may be isolated by further inbreeding and selection. Hallauer (1970) suggested a zygote selection method to develop superior new inbred lines. An elite inbred line would be used as the tester in this procedure, and the final product would be a new inbred line rather than an improvement of an existing line.

Hayes, et al. (1946) suggested that synthetics and single- and double-cross hybrids may be used equally as well as open-pollinated varieties in the gamete selection system. Results presented by Pinnel et al. (1952) indicated that new lines that give performance superior to the elite parent or parents can be obtained from application of this scheme.

The use of the backcrossing and gamete selection schemes, as well as pedigree selection, imposes a possible restriction that was not, initially, fully recognized. Unless special precautions are taken, the long-continued use of any of these breeding schemes leads to an increasing degree of relationship among the surviving lines with possible adverse effects on the germplasm base (Sprague, 1972). In the North Central Region, the existing lines were assigned to one of two separate groups, A or B, in the 1940's. The component lines of a hybrid, to be used as foundation material for the isolation of new lines, were to be chosen so that the two parental inbreds were members of the same group. Thus, continued repetition of any of the breeding systems might have lead to increasing relationship within a group, but might have maintained the diversity between groups. Unfortunately, lines were assigned to groups at random rather than by a system based on genetic diversity between groups, and many new lines have been extracted from crosses between lines in different groups.

III. EVALUATION OF INBRED LINES

The development of inbred lines poses no problem equal in complexity to those involved in the evaluation of lines. The final merit of any line is judged by its performance in hybrid combinations.

A. Effect of Selection During Inbreeding on Yield of Hybrids

Selection during inbreeding may serve more than one purpose. First, it would eliminate lines with limited or no commercial value. Second, it would ensure that propagation is confined to the most vigorous plants, thereby increasing potential commercial usefulness. Third, the visual selection practiced during inbreeding may lead to improved hybrid performance. In this section, we shall be concerned only with the third possibility.

Richey and Mayer (1925) presented data indicating that no general advantage in crossbred performance was derived after three generations of inbreeding. The differences between hybrids made after the second and third generations of inbreeding were somewhat erratic but, as an average, favored the more highly inbred parents. Davis (1934) reported that high-combining S_2 lines, as measured by inbred-variety crosses, also exhibited high combining ability when tested as S_4 and S_5 lines.

Jenkins (1935) compared a series of inbreds derived from the varieties 'Lancaster' and 'Iodent' from the first through the eighth generations of inbreeding, excepting the seventh. For each line in each generation studied after S_1, comparisons involved a selected and discarded sibling. All lines were evaluated as inbred-variety crosses, Krug being used as the common tester. Jenkins concluded:

Selection was ineffective in isolating strains whose crosses differed from those of their parents in productiveness, or in any other character studied. The inbred lines acquired their individuality as parents of topcrosses very early in the inbreeding process and remained relatively stable thereafter.

This interpretation was questioned by Richey (1945). By use of Jenkins's data and by various combinations of generations, he concluded that selection had been effective as measured by the performance of $(S_2 - S_3)$ vs. $(S_6 - S_8)$. He failed, however, to point out that, with the same manipulations, other contrasts could be developed in which selection seemed to have decreased combining ability. Singleton and Nelson (1945) presented data that they interpreted as indicating an improvement in combining ability during inbreeding.

Sprague and Miller (1952) presented information from two separate experiments in which six inbred lines were crossed in all possible combinations within each generation of inbreeding from S_1 through S_5. The first set of six lines was derived from Stiff Stalk Synthetic, and in the second set, each line was from a separate source. The mean square associated with "generations" was partitioned into a linear and residual component. Had selection been effective, the linear component, a measure of general combining ability effects, should have been significant. No statistically significant trend was observed in either experiment. Variation due to specific combining ability was significant, but the magnitude was erratic in both experiments. The selection practiced was effective in increasing resistance to stalk breaking in one experiment and ineffective in the other.

Payne and Hayes (1949) presented data contrasting testcross performance of a series of F_2 plants and their F_3 progenies. They concluded that the testing of the F_2's was of doubtful practical value. Their results, however, indicate clearly that the F_2's with the highest combining values produced a proportionately larger percentage of high-combining F_3 progenies than did the F_2's with low combining values.

Brown (1967) reported data on a series of 1,160 random lines from the varieties 'Reid,' Lancaster, 'Krug,' and 'Midland.' These had survived six generations of selfing with no selection in any generation other than ability to reproduce. These were compared with 200 lines derived from the breeding program having been subjected to the normal selection procedures. Both sets of lines were evaluated in topcrosses. The topcross yield data indicated no significant difference between the two groups for yield or percentage of standing stalks. He concluded "that visual selection as practiced during inbreeding has little, if any direct influence on yield in hybrid combinations."

Russell and Teich (1967) found that visual selection within and among progenies derived from the single cross M14 × C103 resulted in a positive gain for combining ability. Gains were somewhat greater for the subpopulation grown and selected at the higher population densities.

The weight of evidence presented suggests that selection practiced during inbreeding has less effect on combining ability for yield than was once thought. Selection, however, is highly effective in modifying lines with respect to general vigor, maturity, and insect or disease resistance; each of these characteristics may plan an important role in suitability of lines for the production of commercial hybrids.

B. Relationship Between Inbreds and Hybrid Progeny

A clearer understanding of the relationship of hybrid performance and parental inbred lines would be of value in developing elite parental lines.

1. Correlation Between Traits of Inbreds and Hybrids

Numerous studies have reported the relationship between various attributes of an inbred line and the same attribute or yield of its hybrid progeny (Jorgenson and Brewbaker, 1927; Kiesselbach, 1922; Nilsson-Leisner, 1927; Richey, 1924; Richey and Mayer, 1925). The most detailed and comprehensive correlation studies were those reported by Jenkins (1929). These studies involved correlations among characters of the inbred line and various characters of their crossbred progeny. The correlations involving yield as one variable are of greatest interest. Within inbred lines, positive and significant correlations were obtained between yield and plant height, ear length, ear diameter, shelling percentage, and number of ears per plant. Significant negative correlations were obtained for yield and chlorophyll grade, date of silking, and maturity of the harvested ears. Within the parent-progeny series and also within the F_1 hybrid series, significant and positive correlations were obtained between yield and dates of silking and tasseling, plant height, ear length and diameter, number of ears, and number of nodes per plant. The highest multiple correlation involving yield was 0.42.

Hayes and Johnson (1939) reported correlations between the yield of inbred-variety crosses and various attributes of the inbred parents. Yield was positively and significantly correlated with each of the 12 attributes studied. The multiple correlation between yield and the 12 attributes of the inbred parents was 0.67. Where direct comparisons are possible, this was a somewhat closer relationship than found by Jenkins. In most of these studies, simple correlations involving yield as one variable were too low to have useful predictive value.

These correlation studies were conducted before the concept developed by Panse (1940) that phenotypic correlations involved two components, one genetic and one environmental. Robinson et al. (1951) presented both phenotypic and genetic correlations involving a series of eight attributes in crosses among F_2 plants derived from three single-crosses. In only a few instances did the estimates of association differ appreciably. Ears per plant was the only attribute found to have a high positive correlation with yield. These authors illustrate how genetic correlations may be used in the construction and evaluation of selection indices.

Obilana and Hallauer (1974) evaluated 247 S_6 unselected lines of BSSS ('Stiff Stalk Synthetic') developed by single-seed descent from 250 S_o plants. Genetic and phenotypic correlations between yield and 15 plant, tassel, and ear traits were calculated for these inbred lines. The genetic and phenotypic correlations between yield and plant and tassel traits were either small or zero. The correlations between yield and the ear components generally were higher, with kernel depth having the highest genetic correlation (0.76) with yield. Generally, high positive genetic correlations of kernel-row number (0.56), ear length (0.58), and ear diameter (0.62) with yield also were obtained. For primarily one-eared lines, the relatively high correlations of the ear traits with yield and among the ear traits showed that each of the ear traits contributed to yield and that adjustments made in any one ear trait were reflected in the other ear traits.

2. Genetic Diversity Among Varieties and Derived Lines

It is difficult to provide any precise definition of genetic diversity. Strains of the same varietal origin are presumed to have a greater degree of genetic similarity than different varieties; Corn Belt varieties, in turn, are more closely related to each other than to some of the Latin American races. Some of these genetic dissimilarities may be revealed by ancestral relationship, geographical distribution, or differences in gross morphology. Differences in heterosis, however, have provided a more direct and more useful measure of genetic diversity to the corn breeder. Moll et al. (1962) reported heterosis of 21 to 39% for crosses involving varieties from the USA and Puerto Rico, whereas the heterosis in variety crosses among the local varieties was only 9 to 22%. Although the expression of heterosis as percentage of the midparent increased the difference in this experiment, the heterosis (q/ha) was still least among the local varieties.

Some of the early work on genetic diversity obtained from varietal hybrids was summarized in Section II. More recent data are available from several studies. Hallauer (1972) presented data on the yield of nine synthetics and their F_1 hybrids. Two of these involved exotic germplasm derived from the varieties Tuxpeno and Eto. Constant parent heterosis for the group ranged from 5% for BSCB9 ('Corn Borer Synthetic #9') to 30% for BSSS. The highest yielding combinations approached the yield level of the commercial single-cross standards. Paterniani and Lonnquist (1963) presented data on 12 races of corn of diverse origin and their F_1 hybrids grown in Brazil. The heterotic response, measured from the mid-parent, ranged from -11 to 101%. The correlation between mid-parental values and F_1 yields was 0.677. Troyer and Hallauer (1968) evaluated a diallel of 10 early varieties and their crosses from Canada, Japan, Poland, Russia, and the USA. The average heterosis was 43%, with the greatest heterosis for crosses of 'Syldecka' from Poland with the three varieties from Russia. The average heterosis of crosses among the six varieties collected from Canada was lower than any of the other comparisons.

Wu (1939) showed that lines derived from the same single-cross consistently produced lower yields when tested in single cross combinations than did lines having only one or no parents in common. Hayes and Johnson (1939) reported similar data. Lines were derived from a series of single crosses. The derived line then would have different degrees of relationship: both parents, one parent, or no parents in common. In crosses between lines unrelated in origin, 28 of 43 single crosses were equal or superior in yielding ability to the standard double-cross checks. In the groups having one or both parents in common, only 6 of 15 and 1 of 6 crosses, respectively, were equal or superior to the double-cross checks. Johnson and Hayes (1940) presented additional data indicating a significant difference in single-cross performance of related and unrelated lines. Eckhardt and Bryan (1940) and Cowan (1943) also stressed the importance of genetic diversity in the production of high-yielding hybrids.

Hoegemeyer (1974) evaluated 144 intrapopulation and 192 interpopulation single crosses involving S_7 lines from BS10 and BS11. Although the average superiority of interpopulation crosses was only 3.0 and 12.2% greater than that of the BS11 and BS10 intrapopulation crosses, respectively, 28 of the 30 best crosses were interpopulation crosses.

It has been common experience that the best hybrids involve inbred lines derived from different varietal sources. Lines derived from Stiff Stalk Synthetic have combined well with lines involving Lancaster germplasm (B37 × Oh43, A632 × A619, N28 × Mo17, and B73 × Mo17).

C. Early Testing

The breeding methods discussed thus far defer any testcross evaluation of inbred lines until a high degree of homozygosity has been attained. Then, F_1 crosses among or between groups of lines were evaluated. Such a system was satisfactory with limited numbers, but it became burdensome as ever-increasing numbers of lines became available. A search was made for simpler and more efficient testing procedures.

The method of early testing was suggested by Jenkins (1935) and Sprague (1939). Under this scheme, inbreeding and testcross evaluation proceed concurrently. Experimental data bearing on the efficiency of the method have been presented by Sprague (1946, 1952) and Lonnquist (1950). These data plus additional information from ear-to-row tests indicate that individual selected ears from an open-pollinated variety exhibit marked differences in the yielding ability of their progeny. An identification of these high-yielding genotypes early in the course of inbreeding might permit the heavy discarding of inferior and the concentration of effort on the desirable material at the stage of inbreeding when selection would be expected to be most effective.

Jones (1922) presented data on inbred-variety crosses, but his interest was in relative performance rather than in a method of evaluating lines. From studies on inbred-variety crosses Lindstrom (1931) noted that certain lines were very prepotent for ear-type, disease resistance, and uniformity of maturity and suggested the commercial use of such hybrids. Davis (1927) presented data on inbred-variety crosses and used this procedure to estimate the combining ability of S_2 lines. An additional report was presented in 1934. The most extensive report on the performance of inbred-variety crosses was presented by Jenkins and Brunson (1932). Correlations between the yields of topcrosses and the mean performance of these same lines in a series of single crosses ranged from 0.53 to 0.90. The pooled correlation, involving 77 lines, was 0.75. On the basis of these studies, they concluded that it would be safe to discard the lower combining 50% of the lines under test without serious risk of losing valuable material. The remaining 50% would be tested further in single-cross combinations.

Johnson and Hayes (1936) presented additional data on this system of early testing. Eleven lines from the variety 'Golden Bantam' were compared as topcrosses and in a diallel single-cross test. The lines exhibiting low combining ability in topcrosses tended to be below average in their mean single-cross performance. Conversely, the lines exhibiting the highest means in single-cross combinations were also above average in topcross combinations. They concluded that inbred-variety evaluation provided a rapid and satisfactory method for the preliminary evaluation of inbred lines.

The results presented by Sprague (1942), Lonnquist (1950), and Wellhausen (1952) indicated that early testing aided in the detection of lines having high combining ability. Two public lines used most extensively in commercial seed production in 1970, B14, and B37, were derived through early testing procedures (Sprague, 1946).

Several workers have raised objections to the early testing procedure. For the most part, these objections stem from economic, rather than genetic, considerations. Where genetic considerations were involved, they were based on the presumed effectiveness of phenotypic selection during inbreeding (Section III, A).

The several schemes of recurrent selection (Section VII) were a logical outgrowth of the work on early testing. These recurrent schemes, especially full-sib reciprocal recurrent selection (Hallauer, 1967, 1973; Lonnquist and Williams, 1967) and reciprocal recurrent selection-inbred tester (Russell and Eberhart, 1975), can be extremely efficient sources of improved lines each cycle of selection. When a comprehensive breeding system is used (Section

VIII), the development of inbred lines becomes a part of the system. The type of tester and the method used in the final identification of superior hybrids are determined by the recurrent selection procedure. This system is very efficient because the performance trials for population improvement serve as early testing for inbred line development.

IV. THE PREDICTION OF DOUBLE-CROSS AND THREE-WAY CROSS PERFORMANCE

The number of combinations and permutations of n items increases very rapidly with an increase in n. For this reason, it is not possible to produce and evaluate all possible three-way or double-cross combinations among even a small number of lines. Ten inbred lines can be combined to produce 45 single, 340 three-way, and 630 double-cross combinations, disregarding reciprocals. Obviously, some form of prediction would be highly useful.

The first studies on double-cross prediction were conducted by Jenkins (1934). Eleven lines were used and data obtained on (1) inbred-variety crosses of these lines, (2) 53 of the possible single-cross combinations, and (3) 42 of the possible double-cross combinations. The performance data were then used to contrast the relative effectiveness of four different methods of prediction. These were (a) the mean performance of the six possible combinations among any set of four lines, (b) the average performance of the four nonparental singles, (c) the mean performance of a set of four lines over a series of single-crosses, and (d) the average top-cross performance of a group of four lines. The agreement between predicted and observed performance of the 42 double-crosses was measured by means of correlations. There was little difference between methods a, b, and c, with method d providing the poorest estimate of the four tried. The effectiveness of method b has been reported by several investigators (Doxtator and Johnson, 1936; Anderson, 1938; Hayes et al., 1943). A number of procedures have been developed to facilitate routine prediction computations (Millang and Sprague, 1940; Combs and Zuber, 1949; Clem, et al., 1956).

In Jenkin's method b, the double-cross hybrid performance is predicted from single-cross means as follows:
$$DC_{AB \cdot CD} = \tfrac{1}{4}(SC_{AC} + SC_{AD} + SC_{BC} + SC_{BD}).$$
A similar formula can be used for three-way cross hybrids:
$$TWC_{AB \cdot C} = \tfrac{1}{2}(SC_{AC} + SC_{BC}).$$
When the single-crosses have been grown in a diallel mating design, the general combining (g_i) and specific combining (s_{ij}) effects from the single-cross model can be used for double-cross and three-way-cross prediction, also (Otuska, et al., 1972):
$$\widehat{SC}_{ij} = m + g_i + g_j + s_{ij}$$
$$\widehat{DC}_{ik \cdot kl} = m + \tfrac{1}{2}(g_i + g_j + g_k + g_l) + \tfrac{1}{4}(s_{ik} + s_{il} + s_{jk} + s_{jl})$$
$$\widehat{TWC}_{ij \cdot k} = m + \tfrac{1}{2}(g_i + g_j) + g_k + \tfrac{1}{2}(s_{ik} + s_{jk})$$

Jenkin's four methods of prediction involve somewhat different assumptions with respect to type of gene action involved. All assume some role for additive gene action; i.e., a gene contributed by a line produces its characteristic effect, regardless of order of pairing. In addition, method b permits the recognition of nonadditive effects arising from dominance and various types of epistasis.

The problems of double-cross prediction have received considerable attention from quantitative geneticists (Cockerham, 1967; Eberhart, 1964; Eberhart, et al., 1964; Rawlings and Cockerham, 1962). It has been established that Jenkins's method c (average performance of a set of four lines over a series of single crosses) will be the best predictor only when all gene effects are additive. Similarly, method b (average of four nonparental singles) is best when the genetic variance is due only to dominance, but b is unbiased with either additive or dominance effects (no epistasis). In any given situation, genetic variances will be unknown and almost certainly will involve a mixture of effects. Otsuka et al. (1972), Stuber, et al. (1973), and Stuber and Moll (1974) showed that lower correlations of observed and predicted (method b) double-cross performance resulted from genotype $\times$ environmental interaction than from epistasis unless trials were grown in a large number of environments. Hence, the selection procedure for developing three-way or double-cross hybrids is more efficient if predicted means are obtained from single-cross trials, and only the elite three-way or double-cross hybrids are actually evaluated.

V. TYPE OF GENE ACTION

The breeder is interested in establishing the genetic worth or breeding value of inbred lines, hybrids, or populations. In a strict Mendelian sense; however, breeding values cannot be expressed as genotypic values. Although Mendelian inheritance is involved, alternative definitions are required that are consistent with the measurement and evaluation procedures used. A general elaboration of the definitions may be found in any standard quantitative genetics text (Allard, 1960; Falconer, 1960; Kempthorne, 1957), and the details of the procedures and specific models may be found in the appropriate references.

Knowledge concerning the various types of gene action and their relative magnitude in conditioning the various attributes of importance is basic to the maximum efficiency of a breeding program. Fortunately, a larger body of such information is available for corn than for any other single crop. As a broad generality, the genetic variability of the more important traits is due largely to additive genetic variance (Compton, et al., 1965; Gardner, 1963; Lindsey, et al., 1962; Lonnquist, 1961; Robinson and Comstock, 1955; Sprague, 1966). Nonadditive variance due to dominance and epistasis may exist, but is commonly smaller in magnitude.

Estimates of genetic variances can be obtained from any mating designs involving one or more types of families derived from a breeding population in equilibrium where there is no selection of plants to be used as parents. Designs involving sets of families in which the parents had different levels of inbreeding in each set of families will increase the number of covariances among relatives and, theoretically, epistatic variances can be estimated, as well as the additive and dominance variance components.

Sprague and Tatum (1942) divided gene action involved in combining ability into two categories, general and specific. Later developments in statistical genetic theory have shown that, when unselected inbred lines from breeding populations are evaluated in diallel crosses, the variance component arising from general combining ability effects is half the additive variance (plus additive by additive types of epistasis), and the variance component arising from specific combining effects is dominance variance (plus epistasis).

Most current corn breeding methods make use of heterosis (Gowen, 1952). Heterosis may result from partial to complete dominance, overdominance, epistasis, or some combination of these (Comstock and Robinson, 1952). If partial to complete dominance predominates, a possibility exists for the eventual development of stable, high-yielding homozygous genotypes. If overdominance or overdominant types of epistasis predominates, however, the highest yielding genotypes must be heterozygotes.

The most extensive data on degree of dominance has been presented by Gardner (1963), Moll et al. (1964), and Moll and Robinson (1967). In F_2 populations derived from a cross of two inbred parents, testcross progenies were developed by crossing random F_2 plants to each of the parental lines. Analysis of the resulting performance data can be used to provide estimates of additive and dominance variances. The ratio of $\sqrt{2\sigma_D^2/\sigma_A^2}$ then provides a measure of the average level of dominance because the gene frequency at all segregating loci should be 0.5. Ratios exceeding 1.0 may be due to either overdominance or linkage disequilibrium. Genetic variance estimates for yield and certain other traits were observed to exceed 1.0. When the F_2 populations were advanced to F_n and the same procedure repeated, however, estimates for average degree of dominance were materially reduced, often below 1.0, indicating linkage disequilibrium as a more likely source of bias than overdominance.

Overdominance, or pseudo-overdominance from very tight linkages, has been detected in fixed models by use of nearly isogenic stocks. Backcross subline derivatives of inbred B14 involving three loci — R_p^d, Rf_1 and wx — were used by Russell and Eberhart (1970). The 27 possible genotypes were developed and then were compared in replicated experiments. Significant main effects, additive and dominance, were observed for yield and the seven other traits measured, with dominance being of lesser importance. First-order interaction effects (a × a, d × d, a × d) were generally larger than second-order interaction effects (a × a × a, etc.), but only a few cases of significance were noted for the interaction effects. A second experiment (Russell, 1971) involving the inbred Hy and the three loci, br_2, rf_1, and wx, produced roughly comparable results. The frequency of significant additive and dominance effects at the individual loci and epistatic effects among loci was greater for all traits in Hy than in B14. These results, while indicating the presence of overdominance or pseudo-overdominance and certain types of epistasis, are valid for only the specified genotypes. They indicate no requirement for a change in current approaches to population improvement.

Three somewhat different approaches have been used for estimating epistatic genetic variances. The first method utilizes genetic variances and covariances derived from Design I and Design II analyses (Cockerham, 1954, 1956). In studies reported by Eberhart et al. (1966), additive gene action accounted for the major portion of the genetic variance observed for the seven traits under investigation in the 'Jarvis' and 'Indian Chief' varieties. Dominance was somewhat greater for yield than for other traits, particularly in the variety Jarvis. It could not be shown that epistasis contributed significantly to the total genetic variance. Epistatic variability was not detected in the interpopulation variance component analyses for the same varieties by Stuber et al. (1966).

The second approach (Stuber and Moll, 1969) involved F_1 crosses in a Design II experiment and derived S_1 progenies from each F_1. Sets of F_1 and their derived S_1 progeny were grown in the same split-block of a replicated design. This model permits estimation of additive, dominance, and certain types of epistatic effects. The pooled analysis indicated that no more than 10% of the total genetic variability could be accounted for by epistasis. In a similar approach, sets of two, three and four-line crosses involving unselected Jarvis and Indian Chief lines and corresponding sets of crosses involving lines derived after three cycles of reciprocal recurrent selection (Stuber and Moll, 1971) were evaluated. Comparison of mean squares provided no evidence that three cycles of reciprocal recurrent selection for grain yield had changed the magnitude of variation due to epistasis in the traits evaluated. However, several sets exhibited significant epistasis for certain traits.

The general conclusion to be drawn from these and similar experiments involving random mating populations (Chi et al., 1969; Silva, 1974; Wright et al., 1971) is that epistasis is of limited importance in random-mating, equilibrium populations. Epistasis, however, may be of importance in unique hybrid combinations.

The third general approach makes use of selected parents (fixed models). Within this group, several alternative procedures have been used. Gamble (1962a) has used a generation mean analysis involving P_1, P_2, F_1, F_2, F_3, F_1P_1, F_1P_2, and BCS_1 generations involving a series of inbred lines. Significant epistasis was found for certain traits in some hybrid combinations. Darrah and Hallauer (1972) conducted similar studies involving sets of "good" and "poor" inbred lines and first and second-cycle inbreds. Although second-cycle inbreds demonstrated significantly more epistasis than did first-cycle inbreds, correlation analyses indicated that the nonepistatic model was sufficient to describe the gene action, because all r values exceeded 0.9.

Bauman (1959), Gorsline (1961), Sprague et al. (1962), and Sprague and Thomas (1967) utilized differences among types of hybrids in estimating epistasis. A significant difference between $A \times C$, $B \times C$ vs. $(A \times B) \times C$ would be indicative of the presence of net epistasis. Epistatic variance, however, cannot be estimated by this procedure. These studies are in general agreement in indicating that epistasis exists in certain genetic combinations. Even under these favorable conditions for a qualitative estimate, generally epistasis seems to play only a minor role in yield heterosis. This minor role, however, may be of significance when certain specific genotypes are involved.

Results from recurrent selection studies (Section VII) also provide information on gene action. When expected gain has been computed with the assumption of negligible epistatic variance, observed gain has been similar to expected gain (Moll and Stuber, 1971, 1975). Overdominant gene action does not appear to be more important than complete dominance. Russell et al. (1973) and Horner et al. (1973) reported that the use of an inbred tester in recurrent selection studies for specific combining ability improved the general combining ability as much as the specific combining ability with the inbred tester. [In the following discussion, the method of selection is indicated in parenthesis (see Tables 7 and 8), and Cn specifies the numbers of cycles of recurrent selection completed. The source populations were Stiff Stalk Synthetic (BSSS) and Corn Borer Synthetic #1 (BSCB1).]

The population cross BSSS (HT) C7 × BSCB1 (R) C5 was as high yielding as BSSS (R) C5 × BSCB1 (R) C5 (Eberhart et al, 1973a; Genter and Eberhart, 1974), even though BSSS (R) was selected for specific combining ability with BSCB1 (R) in a reciprocal recurrent selection study, and BSSS (HT) was improved in a recurrent selection for general combining ability study with the Iowa 13 double cross as a tester. Russell and Eberhart (1975) reported that the cross of the two BSSS substrains, BSSS (R) C5 × BSSS (HT) C6, yielded as much as BSSS (R) C5 × BSCB1 (R) C5 or BSSS (HT) C6 × BSCB1 (R) C5. Although Moll and Stuber (1971) reported greater improvement in the Jarvis × Indian Chief variety cross from reciprocal recurrent selection than from full-sib selection within varieties, these results are not inconsistent with expectations from a model involving complete dominance.

A broad generalization covering all studies involving types of gene action indicates the almost universal importance of additive effects. Dominance effects rank second in importance with epistatic effects of generally minor importance. This general relationship is consistent with a large number of breeding and evaluation procedures in common use — topcross testing, hybrid prediction, and the genetic advances made with the several recurrent selection schemes.

VI. GENOTYPE BY ENVIRONMENTAL INTERACTIONS

If the relative performance of hybrids were little influenced by environment, then tests conducted in a single environment would suffice to provide adequate information for a reliable ranking of the hybrids. Experience indicates that genotype × environmental interactions are usually of significance, not only for hybrids, but also for families from breeding populations, and these interactions are caused by many factors.

When hybrids or families (genotypes) are grown at different locations for several years, estimates of the variance components provide information on the relative importance of the genotype × location, genotype × year, and genotype × location × year interactions (Comstock and Moll, 1963). Moll (personal communication) has summarized the relative magnitude of these components to the genotypic variances from three experiments in North Carolina (Moll and Robinson, 1967) and from three experiments in Iowa and Nebraska (Lonnquist and Gardner, 1961; Matzinger et al., 1959; Rojas and Sprague, 1952) as shown in Table 4. They have interpreted the large genotype × location × year and the small genotype × location and genotype × year variances in North Carolina to mean that the differential response of genotypes in different environments does not seem to be associated with particular locations or years. Permanent location factors, such as soil types, did not seem as important as the variable factors, such as amount and distribution of rainfall, which caused moisture and temperature stresses in these experiments.

The larger amount of genotype × year interactions in the Iowa and

Table 4—Estimates of Genotype × environmental interaction variance components expressed as percentage of the genotypic variance

Variety Component	Moll and Robinson (1967)			Rojas and Sprague (1952)	Matzinger et al. (1959)	Lonnquist and Gardner (1961)
	Jarvis		Variety crosses			
	Single crosses	Half-sib families				
Genotype × location	6	-11	-12	25	27	48
Genotype × year	9	-11	-17	132	95	47
Genotype × year × location	119	70	47	26	28	53
Number of locations	4	5	3	3	3	3
Number of years	3	5	2	2	3	2
Number of genotypes	64	60	15	45	45	66

Nebraska experiments also suggests that the changes in environment from year to year due to weather patterns are important, but these yearly differences tend to affect the environments at the various locations in a similar manner. Gamble (1962b) also reported that the genotype × year interaction was the primary source of interaction in a generation mean study in Iowa.

Stratification of environments and evaluating genotypes within regions of similar ecological conditions has been used to reduce the genotype × environmental interaction, particularly with respect to length of the growing season. Allard and Bradshaw (1964) classified the variation due to these more permanent features as predictable variation in contrast to the unpredictable variation due to weather, insect infestations, and disease infections.

Eberhart and Russell (1966) recommended the development of indexes based on environmental factors such as rainfall, temperature, and soil fertility. Indexes to measure disease and insect damage might be included also if the genotypes have different levels of resistance. Solar radiation (Duncan, et al., 1973) and growing-degree units (Wang, 1960) are other factors that may be involved. The differential response of genotypes to these indexes could be used to gain a clearer understanding of the cause of the genotype × environmental interaction. Finlay and Wilkinson (1963) and Eberhart and Russell (1966, 1969) used the mean of all genotypes at a location as an environmental index to give a general measure of stress factors affecting yield and found that the differential linear responses to this index removed part of the interaction. Eberhart and Russell (1966, 1969) reported that the variation in the environmental response of the corn single crosses evaluated was caused primarily by additive genetic effects.

In Eastern Africa, the differential temperature response among varieties, as measured by an altitudinal index, removed a substantial amount of the genotype × environmental interaction (Eberhart, et al. 1973b). Corsi and Shaw (1971) and Sopher et al. (1973) have developed procedures to compute a moisture stress or a drought stress index that should be extremely useful in separating the moisture stress factors from the several types of stress factors confounded in the environmental index.

When indexes to characterize the environments are not available, the genotypes must be evaluated in a sufficiently large number of environments to estimate an average response. The elite genotypes are then expected to give superior performance on the average. But if the key factors causing the genotype × environmental interactions can be determined and appropriate indexes can be developed, the response of each genotype to these indexes can be determined. Performance predicted for each genotype for a specific environment is often of greater value than a large series of trials in environments that have not been indexed (Eberhart, et al., 1973b).

With the index technique, the number of environments must be greater than the number of indexes, and the environments must be sufficiently different for the several indexes to preclude a nonsingular "design matrix" of the indexes.

Sprague and Federer (1951) examined a series of topcross, single-cross, and double-cross experiments repeated over locations or years to obtain information on the hybrid and hybrid × location or hybrid × year variance component estimates as a guide to refinements in testing procedures. Both the hybrid × location and hybrid × year interactions were significant. They reported that the interaction components were higher for the single crosses than for the double crosses included in the experiments. Eberhart and Russell (1969) also reported larger genotype × environmental interactions for single crosses than for double crosses, and Wright et al. (1971) obtained similar results with three-way cross and single-cross interactions with environments. The smaller genotype × environmental interaction is expected for three-way and double-cross hybrids because each hybrid is a mixture of genotypes. Allard and Bradshaw (1964) used the term "population buffering" to distinguish this type of stability from the genetic stability of "individual buffering" of stable genotypes.

When unrelated single crosses are developed from crossing random lines from a population, Cockerham (1961) pointed out that the genetic variance among hybrids involves additive, dominance, and epistatic variances ($\sigma_A^2 + \sigma_D^2 + \sigma_I^2$). But the variances among three-way and double crosses are $\frac{3}{4}\sigma_A^2 + \frac{1}{2}\sigma_D^2$ and $\frac{1}{2}\sigma_A^2 + \frac{1}{4}\sigma_D^2$, respectively (disregarding epistasis). Hence, the genotype × environmental interactions would be expected to approach corresponding ratios ($\sigma_{AE}^2 + \sigma_{DE}^2$) : ($\frac{3}{4}\sigma_{AE}^2 + \frac{1}{2}\sigma_{DE}^2$) : ($\frac{1}{2}\sigma_{AE}^2 + \frac{1}{4}\sigma_{DE}^2$), where σ_{AE}^2 and σ_{DE}^2 are the additive × environmental and dominance × environmental interactions, respectively.

Eberhart and Russell (1966, 1969) proposed that the deviation mean square obtained from the regression of yield on the environmental index could be used as a measure of stability for each hybrid. Certain single crosses were as stable as double crosses, but the inheritance of this type of stability seemed complex. Single-cross hybrids that have a high level of barrenness under moisture or temperature stresses tend to be unstable. This stability parameter does not seem to be useful for double-cross, three-way cross, or variety-cross hybrids because the population buffering due to the mixture of genotypes in a complex hybrid masks the differences due to individual buffering of the component genotypes. Moll and Stuber (1974) have summarized several approaches to the problem of genotype × environmental interactions including other stability parameters.

Population improvement (Section VII) seems to offer the methodology for developing hybrids with greater stability. If improved breeding populations can be developed with high levels of resistance to diseases and insects and with a high degree of tolerance to temperature and moisture stresses, hybrids involving lines from these improved populations should interact less with varying environmental conditions than the current hybrids.

VII. RECURRENT SELECTION

Recurrent selection was suggested by Hayes and Garber (1919) as a method of improving corn varieties, and East and Jones (1920) and Jenkins (1940) published detailed descriptions of this breeding scheme. Subsequent empirical results have shown recurrent selection to be effective in improving breeding populations (Burton et al., 1971; Darrah et al., 1972; Eberhart et al., 1973a; Gardner, 1969; Horner et al., 1973; Lonnquist, 1961; Moll and Stuber, 1971; Russell et al., 1973; Webel and Lonnquist, 1967).

When a character is controlled by a few genes and their effects are not masked by environmental variation, a large number of individuals can be grown and evaluated, and the individuals homozygous for all favorable alleles for that character can be selected and used. But most characters of interest to a corn breeder do not meet these requirements. Lindstrom (1939) considered limitations that a breeder must overcome under four subdivisions: (1) large number of genes, (2) masking effects of the environment, (3) complicated system of gene interaction, and (4) inadequate methods of isolating and evaluating lines. There is no definite information on the number of genes influencing grain yield in corn, but the consensus is that the number is large. "Student" (1934) estimated that oil percentage of the corn kernel is conditioned by at least 20 to 40 loci and possibly 200 to 400. If comparable estimates were derived from data now available from the Illinois Selection Experiments (Dudley et al., 1974), the estimate would likely be substantially greater than Student's minimum estimate. However, if no more than this minimum number of loci were involved in yield and if one were dealing with a population heterozygous for these 20 loci, it would require an area of approximately 36,450,000 ha (90 million acres) to provide a population of a size to give an even chance for the occurrence of an individual homozygous for the 20 favorable alleles. Linkage would impose still further restrictions.

The identification of a completely dominant individual, if it did occur, would pose insurmountable difficulties. Therefore, in populations of a size feasible for the breeder to handle, the opportunity for detecting individuals that surpass the mean of the population by more than two or three standard deviation units is rather remote. A point is soon reached where an increase in population size cannot be expected to yield commensurate returns. Lonnquist (1951) pointed out that, when gene frequency is increased from a relatively low level, the proportion of individuals of the desired genotype is greatly increased. For instance, when the gene frequency at each locus is 0.9, there is an even chance of 15 homozygous desirable individuals per 1,000, rather than 1 individual in 36,450,000 ha (gene frequency of 0.5 at each locus when 20 loci are segregating). The only solution, therefore, seems to be a system that provides an opportunity for a gradual increase in the frequency of desirable alleles in the population being sampled.

The importance of the masking effect of the environment as a limitation to selection on a single-plant basis requires no elaboration (Section V), but the masking effect of environment can be reduced to a manageable level with some type of family selection when families are evaluated in sufficient environments with adequate replication.

The types of gene action involved in yield and other attributes necessary or desirable in commercial hybrids were reviewed in Section IV. Results from numerous experiments indicate that additive genetic variation, for yield and most traits of interest, is sufficient to obtain progress from recurrent selection.

Many recurrent selection methods and techniques have been proposed to improve breeding populations. They require the selection of plants with superior phenotypes in the breeding population and the intermating of these selected individuals to form a new population. These recurrent selection procedures will gradually increase the frequency of favorable alleles. Selection can be based on the phenotype of an individual (mass selection) or on the mean phenotype of families. When families are used, three phases are involved: (1) forming families, (2) evaluating these families and selecting those that are superior, and (3) intercrossing plants produced from remnant seed of the selected families (or selfed seed of the parents) to form the improved breeding population for the next cycle of improvement.

The recurrent selection methods can be divided into two main categories, depending on the objectives of the population improvement programs (Moll and Stuber, 1974). Intrapopulation improvement will tend to maximize improvement of the population itself and the inbred lines derived from it, whereas interpopulation improvement will maximize improvement in the population cross and hybrids between lines from the two different populations for characters controlled by genes with a relatively high level of dominance. Rapid improvement of a population formed from the advanced generation of the population cross also can be expected with interpopulation selection.

A. Intrapopulation Improvement

Several alternative procedures have been developed and studied in some detail. These include (1) mass selection, (2) modified ear-to-row selection, (3) half-sib selection, (4) full-sib selection, (5) S_1 or S_2 selection, and (6) testcross selection with a broad-base tester or with an inbred-line tester.

The mathematical formula to predict gain from selection is probably the most valuable tool provided to the plant breeder by statistical geneticists. Lush (1945) pointed out that expected gain could be predicted by multiplying the selection differential (the mean of selected individuals minus the overall mean) by heritability (the proportion of the phenotypic variance that is additive genetic variance). Because the selection differential can be expressed as $k \cdot \sigma_p$ with a normal distribution,

$$Gy = HD/y = (k \cdot \sigma_{g'}^2)/(y \cdot \sigma_P)$$

$$Gy = \frac{c \cdot k \cdot \sigma_{g'}^2}{y \cdot \sqrt{\frac{\sigma_e^2 + \sigma_{ge}^2 + \sigma_g^2}{rm} \quad \overline{m}}} \quad \text{and}$$

$$\sigma_e^2 = \frac{\sigma_u^2 + (\sigma_G^2 - \sigma_g^2)}{n} + \sigma^2$$

Table 5—Form of the analysis of variance of families evaluated in replicated trials in two or more environments

Source	df	Mean square	Expected mean square†
Environments	(m-1)		
Reps/environments	m(r-1)		
Families	(f-1)	MS$_3$	$\sigma_e^2 + r\sigma_{ge}^2 + rm\sigma_g^2$
Environments by families	(m-1)(f-1)	MS$_2$	$\sigma_e^2 + r\sigma_{ge}^2$
Pooled Error	m(r-1)(f-1)	MS$_1$	σ_e^2
Plants within plots	(n-1) rmf	MS$_4$	σ_w^2

†σ_g^2 is the genetic variance among families, σ_{ge}^2 is the genotype × environmental variance, $\sigma_e^2 = (\sigma_w^2/n + \sigma^2)$, where σ^2 is the plot-to-plot environmental variance and $\sigma_w^2 = \sigma_u^2 + (\sigma_G^2 - \sigma_g^2)$, where σ_u^2 is the within plot environmental variance and σ_G^2 is the total genetic variance among individual plants in the source population.

Table 6—Genetic interpretations of covariances among half-sibs, among full-sibs, and among individuals in families developed by self-fertilization when parents are not inbred and when epistasis is negligible

Type of covariance	Genetic interpretation†
Half-sib	$(¼)\sigma_A^2$
Full-sib	$(½)\sigma_A^2 + (¼)\sigma_D^2$
Selfed‡	$\sigma_{A'}^2 + (¼)\sigma_D^2$

†When the parents are unrelated inbred plants, the coefficients of σ_A^2 and $\sigma_{A'}^2$ increase by (1 + F) where F is the coefficient of inbreeding of the parents. The coefficient of the σ_D^2 changes also. ‡$\sigma_{A'}^2$ fo selfed progenies is not equal to σ_A^2 for non-inbred progenies unless dominance effects are negligible o unless the gene frequency is 0.5 at all segregating loci in the population.

where Gy is the expected gain per year, H is heritability on an individual plant or family mean basis depending upon the unit selected, D is the selection differential, c is determined by parental control, k is the standardized selection differential for a normal distribution and is a function of the selection intensity (Falconer, 1960), σ_g^2 is the portion of the genetic variance among individuals or families due to additive effects, σ_P is the square root of the phenotypic variance of the individual or family mean, y is the number of years per cycle, σ_e^2 is the experimental error, σ_{ge}^2 is the genotype × vironment interaction, σ_g^2 is the genetic variance among individuals or families, σ_G^2 is the total genetic variance among individuals, σ_u^2 is the within-plot environment variance, σ^2 is the plot-to-plot environmental variance, n is the number of plants per plot, r is the number of replications per environment, and m is the number of environments (locations).

The analysis of variance and expected mean squares for family selection are given in Table 5. The genetic variance among families (σ_g^2) can be represented as a covariance among relatives and expressed in terms of additive (σ_A^2) and dominance (σ_D^2) variance. The expectations are given in Table 6 assuming negligible epistasis.

Cockerham (1956, 1963) and Horner et al. (1969) showed that genetic variances among relatives increase with inbreeding. If the parental plants are inbred and unrelated, the coefficient of σ_A^2 increases by (1 + F) in both the numerator and phenotypic variance, where F is the coefficient of inbreeding of the parents. The coefficient of σ_D^2 in the phenotypic variance also changes slightly.

Expected gain for six methods of intrapopulation improvement can be derived from these relations and are given in Table 7. The expected gain

Table 7 — Expected genetic gain per cycle (Gc) under different intra population schemes with non-inbred parents

Method	Expected gain (Gc)†	Crop seasons per cycle
Mass Selection (M) (a) One sex	$\dfrac{k \cdot (1/2)\sigma_A^2}{\sqrt{\sigma_u^2 + \sigma_{AE}^2 + \sigma_{DE}^2 + \sigma_A^2 + \sigma_D^2}}$	1
(b) Both sexes	$\dfrac{k \cdot \sigma_A^2}{\sqrt{\sigma_u^2 + \sigma_{AE}^2 + \sigma_{DE}^2 + \sigma_A^2 + \sigma_D^2}}$	1 or 2
Modified ear-to-row (E)‡	$\dfrac{k \cdot (1/8)\sigma_A^2}{\sqrt{\dfrac{\sigma_e^2}{rm} + \dfrac{1/4\sigma_{AE}^2}{m} + 1/4\sigma_A^2}}$	1
Half-sib (H)	$\dfrac{k \cdot (1/4)\sigma_A^2}{\sqrt{\dfrac{\sigma_e^2}{rm} + \dfrac{1/4\sigma_{AE}^2}{m} + 1/4\sigma_A^2}}$	2
Full-sib (F)	$\dfrac{k \cdot (1/2)\sigma_A^2}{\sqrt{\dfrac{\sigma_e^2}{rm} + \dfrac{(1/2\sigma_{AE}^2 + 1/4\sigma_{DE}^2)}{m} + (1/2\sigma_A^2 + 1/4\sigma_D^2)}}$	2
Testcross Population as tester (HT)	$\dfrac{k \cdot (1/2)\sigma_A^2}{\sqrt{\dfrac{\sigma_e^2}{rm} + \dfrac{1/4\sigma_{AE}^2}{m} + 1/4\sigma_A^2}}$	3 or 4
S_1 selection (S)§	$\dfrac{k \cdot \sigma_{A'}^2}{\sqrt{\dfrac{\sigma_e^2}{rm} + \dfrac{(\sigma_{AE'}^2 + 1/4\sigma_{DE}^2)}{m} + (\sigma_{A'}^2 + 1/4\sigma_D^2)}}$	

†σ_u^2 is the within-plot environmental variance, σ_{AE}^2 and σ_{DE}^2 are the additive by environmental and dominance by environmental interactions, σ_A^2 and σ_D^2 are the additive and dominance variance, k is the standardized selection differential, n is the number of plants per plot, r is the number of replications per environment, m is the number of environments. ‡If the mass selection within ear rows is for the primary trait only, an additional component should be added:

$$\dfrac{k(3/8)\sigma_A^2}{\sqrt{\sigma_u^2 + (3/4)\sigma_{AE}^2 + \sigma_D^2 + (3/4)\sigma_A^2 + \sigma_D^2}}$$

§The definition of additive genetic variance components changes slightly with inbreeding.

refers to the change per cycle in the population per se, except for S_1 selection and testcross selection, where expected gain refers to the change in the mean of a random set of S_1 lines and the mean testcross performance of the improved population with the tester, respectively. Because significant positive estimates of epistatic variances have not been obtained (Chi et al., 1969; Eberhart et al., 1966; Silva, 1974; Stuber et al., 1966; Wright et al., 1971), the epistatic bias might be expected to be small and has not been considered.

Operationally, the recurrent schemes fall into three somewhat different groups. In mass selection, half-sib selection, and modified ear-to-row selection, no controlled pollinations are necessary, but adequate isolation is essential. With mass selection, choice of parents is based on individual plant performance. Gardner (1961) suggested selection within blocks as a means of reducing the effect of environmental variation. When mass selection is practiced on the maternal plants only (one sex), gain is reduced because of the lack of parental control for the pollen source (c = 1/2). If the trait is expressed before pollination, undesirable plants can be eliminated to give parental control of both sexes (c = 1). Alternatively, pollen can be collected from selected plants, bulked, and used to pollinate other selected plants; or the selected plants can be selfed, and this seed can be planted in isolation the following season for recombination. Obviously selecting both sexes will give twice as much gain as selection for one sex only and will often justify the extra ex-

Table 8—Expected gain for the population cross per cycle of selection under different interpopulation selection scheme with noninbred parents†

Method	Expected gain (Gc)	Crop seasons per cycle
Reciprocal recurrent selection (R)‡	$\dfrac{k \cdot (1/4)\sigma^2_{A(1)}}{\sqrt{\dfrac{\sigma^2_{e(1)}}{rm} + \dfrac{1/4\sigma^2_{AE(1)}}{m} + 1/4\sigma^2_{A(1)}}} + \dfrac{k \cdot (1/4)\sigma^2_{A(2)}}{\sqrt{\dfrac{\sigma^2_{e(2)}}{rm} + \dfrac{1/4\sigma^2_{AE(2)}}{m} + 1/4\sigma^2_{A(2)}}}$	3
Full-sib reciprocal recurrent selection (FR)	$\dfrac{k \cdot (1/2)\sigma^2_A}{\sqrt{\dfrac{\sigma^2_e}{rm} + \dfrac{(1/2\sigma^2_{AE'} + 1/4\sigma^2_{DE'})}{m} + (1/2\sigma^2_{A'} + 1/4\sigma^2_{D'})}}$	3

†Variance components are defined for the population cross and differ from the corresponding components within a population (Table 7) because the gene frequencies in both populations are involved.
‡When the number of tester plants (t) is finite, the phenotypic variance for each population is increased by

$$\left(\frac{1}{t}\right)\left[\frac{(1/4\sigma^2_{AE} + 1/4\sigma^2_{DE})}{m} + (1/4\sigma^2_A + 1/4\sigma^2_D)\right]$$

pense involved. For characters such as yield with large genotype × environmental interactions, the phenotypic variance will normally be very large for mass selection in comparison with any method of family selection where trials have been grown in different environments. But expense also will be much less for mass selection than for family selection. Gardner (1968, 1969) has reported a good rate of progress (2.7% per year) for grain yield in 'Hays Golden,' and Lonnquist (1967) reported even greater yield improvement from mass selection for prolificacy. Adaptive mass selection in two adapted and eight semi-exotic populations in Nebraska was effective in increasing yield and prolificacy, especially in the semi-exotic populations where yield was increased an average of 5% per year (Mathema, 1971). But plant and ear height were increased as well.

Hallauer and Sears (1972) were able to decrease the interval from planting to silking by 20 days (3.8 days per year) in the 'Eto' variety with five cycles of mass selection in Iowa. A correlated response in decreased plant height of 75 cm was observed. Heritability estimates (broad sense on an individual plant basis) were 59% and 58% for silking date and plant height.

Seventy generations of selection for high protein, low protein, high oil, and low oil content in the 'Burr's White' variety resulted in means 12, 8, 27, and 10 standard deviations, respectively, beyond the original mean (Dudley et al., 1974). These changes were obtained by evaluating ears from approximately 6,000 plants in each of the four populations. In contrast, only 2.6 × 10^{-12} individuals in a normally distributed population are expected to exceed the mean by seven or more standard deviations. Mild selection combined with frequent opportunity for recombination in this recurrent selection program has been extremely efficient in obtaining the desired changes. Yield has declined, however, in these populations.

Hallauer and Sears (1969), Darrah et al. (1972), and Brown (see Genter and Eberhart, 1974) obtained little improvement in yield and usually an increase in plant and ear height. Much higher plant densities per hectare were used in these later mass selection experiments.

When the family selection methods are used, the superior progenies are identified on the basis of replicated field tests grown in several environments. A limit on the possible number of replications and environments is imposed by the number of kernels on the selected ears for all methods of family selection, except for testcross selection.

Modified ear-to-row selection (Lonnquist, 1964) is a type of half-sib selection in that selected ears are planted ear-to-row in a replicated test in different environments. One trial is grown in isolation and the ear rows are detasseled so that pollen is provided by a bulk sample of all entries. Superior plants are mass selected within the highest yielding ear-rows (based on family means over environments).

The half-sib selection procedure differs from the modified ear-to-row in that remnant seed of the high-performing families is recombined to form the new base population, rather than seed from the isolated trial that has been pollinated from the bulk planting for the modified ear-to-row. Recombination adds one generation to the cycle time for half-sib selection, but this can be done in a winter nursery so that a cycle per year is also possible. Recombination and family formation are accomplished simultaneously in the winter nursery. Gain per cycle is expected to be twice as great for half-sib selection (with winter nursery) as for modified ear-to-row (no mass selection for the primary trait), because only one sex is selected in the latter ($c = \frac{1}{2}$).

Mass selection for the primary trait within half-sib families is an important aspect of modified ear-to-row selection (Compton and Comstock, 1975), because gain per year from half-sib selection (with no winter nursery) is expected to be as high as for modified ear-to-row when mass selection within families is deleted, and costs should be half as much because yield trials are grown every other year. The value of an extra generation of recombination for modified ear-to-row selection cannot be considered in the formula, however. Empig et al. (1972) provided a formula for the extra increment of expected gain that should be added for within-family mass selection during recombination when the mass selection within half-sib families is for the primary trait. The suggestion for agronomic selection during recombination and family formation by Lonnquist (1964) was an extremely valuable idea, and it is appropriate for all methods of family selection.

Modified ear-to-row selection has been extremely effective in improving the mean yield of the population (Paterniani, 1967; Webel and Lonnquist, 1967). Bahadur (1974) reported 5.3% increase in yield per year in 'Hays Golden' from 10 cycles of modified ear-to-row selection in Nebraska. But undesirable increases were noted for plant lodging (4.3% per year), grain moisture at harvest (1.4% per year) and ear height (2.9% per year). Dudley et al. (1974) selected for increased kilograms of lysine per hectare in four opaque-2 populations for three cycles in Illinois. Yield was increased 2.4 to 4.8% per year.

Full-sib selection requires only two generations per cycle when plant-to-plant crosses are made between plants from different selected families, because recombination and family formation are accomplished simultaneously: i.e., season 1, recombination-family formation; and season 2, performance trials. Although gain will be proportional to $(1/2)\ \sigma_A^2$, the phenotypic variance is larger than for half-sib selection. When plants within each full-sib family are selfed in the nursery the same season as the performance trials are conducted and a bulk of seed from several elite selfed plants is used to represent each selected line for recombination, gain from selection is increased, because the parents of the full-sib families are inbred for the subsequent cycle, $\sigma_{g'}^2 = (1 + F)\sigma_A^2$.

Information on observed progress from full-sib selection with two seasons per cycle is limited. Bolton (personal communication) completed five cycles of full-sib selection in Ukiriguru Composite A in Tanzania in 5 years. Moll and Stuber (1971) reported 2.5 to 4.0% gain per cycle in five varieties in North Carolina. To obtain estimates of additive and dominance variance components, however, they used a modified selection scheme in which each plant used as male was crossed to four plants used as females. Hence, three seasons per cycle were required.

When population improvement is obtained through S_1 selection, the improvement in the population per se cannot be predicted exactly unless dominance effects are negligible because the definition of σ_A^2 changes slightly with inbreeding (Cockerham, 1963; Empig et al., 1972). This formula gives a good estimate of expected gain for the population, only if the gain for the population is approximately proportional to the gain in the mean S_1 line performance.

The yield of BSK(S) was increased 16.3% by four cycles of S_1 selection (Burton et al., 1971). The average testcross yields with four single cross tester increased 10.6%. VCBS(S)C4 (developed from the 'Virginia Corn Belt Southern' variety by four cycles of S_1 selections) outyielded the original variety by 20% in a regional trial (Genter and Eberhart, 1974), and the improvement was expressed in the population crosses also.

Although Hull began S_1 selection in 1928 (Hull, 1945), he discontinued it after one cycle. S_2 selection (Hull, 1952) was initiated in 'Fla. 767' for a comparative study with testcross selection, and progress from five cycles of selection was reported by Horner et al. (1973). Testcross grain yield with Fla. 767 and 'Fla. 3W' (a broad-base tester) had been improved 1.1 and 0.9 q/ha per cycle, respectively. But no improvement in yield was detected in the population per se.

Testcross selection is a form of half-sib selection. Three seasons are required per cycle, because each phase requires a separate season. In testcross selection, a plant is selfed and crossed to several random plants from the population as tester. Alternatively, plants can be selfed one season, and the S_1 lines can be interplanted with the tester the next season and detasseled to obtain the testcross seed. But four, rather than three, seasons per cycle are required with the latter procedure. After testcross performance is evaluated and the superior families are selected, remnant selfed seed of the corresponding male plant is used for recombination. Progress for testcross selection with the population as tester is proportional to $(1/2) \sigma_A^2$ instead of $(1/4) \sigma_A^2$ (Table 7), but three (or four) seasons per cycle are required instead of the two with regular half-sib selection. When the number of tester plants (t) is not infinite, the phenotypic variance, σ_P^2, is increased by

$$\frac{1}{t}[\frac{(\frac{1}{4}\sigma_{AE}^2 + \frac{1}{4}\sigma_{DE}^2)}{m} + (\frac{1}{4}\sigma_A^2 + \frac{1}{4}\sigma_D^2)].$$

When testcross selection is used, Rawlings and Thompson (1962) recommended that the tester have a low gene frequency for the dominant allele at each locus. Allison and Curnow (1966) pointed out that the original source population or a low-yielding substrain developed from it should be good testers.

Hull (1945) recognized the advantage of using a homozygous line as a tester and proposed an appropriate recurrent selection program. His objective, however, was an improved topcross or three-way cross hybrid with the tester line as the male parent. Results from this Florida program (Horner et al., 1973) with Fla. 767 and from Iowa (Russell et al., 1973) with Alph (BS12) have demonstrated that testcross selection with an inbred tester improved the general combining ability, as well as the specific combining ability. Although the terms "recurrent selection for general combining ability" and "recurrent selection for specific combining ability" (Hull, 1952) have been used commonly for a broad base and an inbred line tester, respectively, these recent results do not seem to justify a continued usage. Horner et al. (1973) reported greater improvement in the testcross yields for the Fla. 767 substrain improved with an inbred tester than for the substrain improved with the population per se as tester. With the assumption of complete dominance at all loci and negligible epistasis, statistical theory would predict a superiority for the use of an inbred line from the population as tester. The genetic variance among testcrosses would be greater, because an inbred line would have a gene frequency of 0.0 or 1.0 at all loci for the favorable allele. The determination of the theoretical relationship is difficult, but preliminary results from Kenya (Darrah et al., 1972) give a ratio of 30/17 for σ_g^2 in KCA(HI) with an inbred tester vs. KCA(HT) with the population as tester. Horner et al., (1973) reported a ratio of 5.3/2.0 for a non-related line vs. the population tester. The inbred line tester should be changed periodically, however, because no selection pressure can be applied at loci with a frequency of 1.0 for the favorable allele in the tester line.

Most intrapopulation programs involving family performance testing have been restricted in effective population size. Consequently, inbreeding has increased rapidly, and many alleles have reached fixation because of random drift, or possibly linkage, rather than as a response to selection. The superiority of the CO × Cn crosses over the Cn substrains and the heterosis between substrains developed from a common source suggest that the effect of inbreeding has been serious (Burton et al., 1971; Eberhart et al., 1973a; Horner et al., 1973).

B. Interpopulation Improvement

Reciprocal recurrent selection was proposed by Comstock et al. (1949). The procedure is a type of half-sib selection (testcross selection with a broad-base tester) in which the reciprocal population is used as the tester. The expected gain from selection refers to the improvement of the population cross, and it is the sum of expected gain for each population with its tester (Table 8).

The contribution to the improvement in the population cross due to each parental population can be estimated when crosses are made to a constant reciprocal population (Co or Cn) in addition to the Cn × Cn crosses when progress from selection is evaluated. Russell et al. (1973) found that Alph

Table 9—Average yield gains per cycle with different intrapopulation recurrent selection schemes

Selection scheme	Population	No. of cycles	Avg. gain per cycle %	Reference
Mass Selection (M)	Hays Golden	13	2.7	Gardner, 1969
	Tropical	3	11.1	Johnson, 1963
	Krug	6	1.6	Hallauer & Sears, 1969
	Iowa Ideal	5	1.4	Hallauer & Sears, 1969
	KCA	5	1.6	Darrah, 1975
	NC	5	2.2	Mathema, 1971
Ear-to-Row (E)	SSSSo$_2$	3	4.8	Dudley et al., 1974a
	SSSSflo$_2$	3	3.4	Dudley et al., 1974a
	D.O.o$_2$	3	2.4	Dudley et al., 1974a
	D.O.fl$_2$	3	4.6	Dudley et al., 1974a
	KCA	6	2.2	Darrah, 1975
	Hays Golden	10	5.3	Bahadur, 1974
Full-sib (F)	Jarvis	6	3.5	Moll & Stuber, 1971
	Indian Chief	6	2.8	Moll & Stuber, 1971
	Jarvis (F) × Indian Chief (F)	6	2.5	Moll & Stuber, 1971
	(Jarvis × Indian Chief)syn-4	6	2.8	Moll & Stuber, 1971
	CI21 × NC7	10	4.0	Moll & Stuber, 1971
S$_1$ selection (S)	BSK	4	3.9	Burton et al., 1971
	VLE	2	1.1	Genter, 1973
	VCBS	4	5.4	Genter & Eberhart, 1974
	VLE(S) × VLE(HT)	2	4.1	Genter, 1973
	BSK(S) × BSK(HT)	4	6.9	Burton et al., 1971
	VCBS(S) × VCBS(HT)	3,4	6.5	Genter & Eberhart, 1974
S$_2$ selection (S2)	Fla. 767(S2) × Fla.767	5	2.2	Horner et al., 1973
	Fla.767(S2) × Fla.3W syn	5	1.9	Horner et al., 1973
Testcross (HT)	Fla.767(HT) × Fla.767	5	2.3	Horner et al., 1973
	Fla.767(HT) × Fla.3W syn	5	2.5	Horner et al., 1973
	VLE	2	-0.7	Genter, 1973
	VCBS	3	7.3	Genter & Eberhart, 1974

contributed more than (WF9 × B7)syn-2, whereas Eberhart et al. (1973a) reported approximately equal contributions from BSSS(R) and BSCB1(R).

The improvement in the Jarvis × Indian Chief population performance has been greater with six cycles of reciprocal recurrent selection (3.5% per cycle) than for full sib selection in each parental population (2.5% per cycle) in North Carolina (Moll and Stuber, 1971). In Kenya, the yield of the KII(R) × Ec573(R) variety-cross hybrid has been increased 2.3 q/ha per cycle (7.4%) with reciprocal recurrent selection (Darrah, 1975).

Procedures for full-sib reciprocal recurrent selection have been described by Hallauer (1967, 1973) and Lonnquist and Williams (1967). Jones et al. (1971) pointed out that, because only half as many families must be evaluated as for reciprocal recurrent selection to maintain the same effective population size in each population, the selection intensity can be doubled. This doubling of the selection intensity offsets the increase in phenotypic variance for full-sib versus half-sib families. Progress is faster and the procedure is simpler when at least one population produces two ears per plant. Lonnquist (1967) Hallauer and Sears (1972), and Torregroza et al. (1973) demonstrated that prolificacy is highly heritable, however, and it can be increased rapidly in a population.

Because empirical results (Darrah et al., 1972; Horner et al., 1973) and statistical genetic theory indicate greater variation among families when an inbred line is used as the tester, a modification of reciprocal recurrent selection with an inbred line from the reciprocal population as a tester can be expected to produce greater progress from selection than where the reciprocal

Table 10—Average yield gains with different interpopulation recurrent selection schemes

Selection	Population	No. of cycles	Avg. gain cycle %	Reference
Reciprocal	Jarvis (R)	6	2.3	Moll & Stuber, 1971
	Indian Chief (R)	6	1.2	Moll & Stuber, 1971
	Jarvis(R) × Indian Chief(R)	6	3.5	Moll & Stuber, 1971
	BSSS(R)	5	0.4	Eberhart et al., 1973a
	BSCB1(R)	5	0.9	Eberhart et al., 1973a
	BSSS(R) × BSCB1(R)	5	4.5	Eberhart et al., 1973a
	K II(R)	3	0.3	Darrah, 1975
	Ec 573 (R)	3	5.6	Darrah, 1975
	K II(R) × Ec 573(R)	3	7.4	Darrah, 1975
Testcross	Krug(HT) × (WF9 × M14)	3	3.2	Lonnquist, 1961
	BSK(HT)	4	1.9	Burton et al., 1971
	BSK(HT) × (WF9 × W22)	4	2.4	Burton et al., 1971
	BSK(HT) × (B14A × M14)	4	3.2	Burton et al., 1971
	BSSS(HT)	7	1.4	Eberhart et al., 1973a
	BSSS(HT) × Ia13	7	2.6	Eberhart et al., 1973a
	BSSS(HT) × BSCB1 (R)	7,5	3.8	Eberhart et al., 1973a
	F5B(HT) × (F44 × F6)	2	2.7	Horner et al., 1965
Testcross	BS12(I1I)	5	4.8	Russell et al., 1973
	BS12(HI) × B14A	5	4.5	Russell et al., 1973
	(WF9 × B7)syn-2(HI)	5	4.0	Russell et al., 1973
	(WF9 × B7)syn-2(HI) × B14A	5	1.8	Russell et al., 1973
	BS12(HI) × (WF9 × B7)syn-2	5	7.4	Russell et al., 1973
	Fla.767(HI) × Fla.767	5	4.0	Horner et al., 1973
	Fla.767(HI) × Fla.3W syn	5	4.9	Horner et al., 1973

population is used as tester (Eberhart et al. 1973a; Russell and Eberhart, 1975). Even though the same inbred line, B14, was used as a tester to improve Alph and (WF9 × B7)syn-2, gain in Alph × (WF9 × B7)syn-2 (4.06 q/ha/cycle or 7.4%) was nearly the sum of gain for B14 ×(WF9 × B7)syn-2 (1.32 q/ha/cycle) and B14 × Alph (3.09 q/ha/cycle) (Russell et al., 1973). As progress is made, inbred lines from subsequent cycles can be substituted as the tester whenever they become available.

A critical comparison of relative efficiency of the many alternative recurrent selection systems poses problems. Progress in plant breeding is dependent upon two separate elements: (1) the genetic characteristics of the base populations and (2) the efficiency of the selection procedures practiced. The base populations in use range from open-pollinated varieties of long standing to newly formed composites.

Alternative schemes are not directly comparable in progress achieved, because different base populations and different selection intensities were used, and the experiments varied in their efficiency in reducing environmental effects and genotype × environmental interactions.

The results from several long-term studies have been assembled in Tables 9 and 10. Comparisons have been made in terms of percentage of gain per cycle, but should be translated to gain per year when seasons per cycle differ.

VIII. A COMPREHENSIVE BREEDING SYSTEM

The application of mathematical and statistical theory to basic genetic principles and the resulting theoretical developments in statistical genetics have provided the foundation for a more scientific approach to corn breeding. Breeders have devised procedures and techniques based on the statistical genetic theory and have conducted experiments to evaluate empirically the adequacy of the models and theory. Although a few minor discrepancies have been observed, the statistical genetic models have been satisfactory.

A careful consideration of the various aspects of corn breeding in light of the information from statistical genetic theory and the empirical experiments can suggest many ways in which the effectiveness and efficiency of corn breeding can be improved over the traditional inbreeding and hybridization methods that have evolved over the years (Dudley and Moll, 1969; Eberhart et al., 1967; Empig et al., 1972; Moll and Stuber, 1974).

Such a comprehensive breeding system has three distinct phases: (1) the development of two or more breeding populations from diverse sources so that the population-cross mean(s) will be at the highest level possible and the populations will have maximum additive genetic variation within each, (2) continuous population improvement by an effective recurrent selection program, and (3) the development of superior hybrids from each cycle of selection by an efficient and systematic procedure. Where open-pollinated varieties are required for commercial use, either a parental population or the advanced generation of the population cross may be used.

A. Developing Breeding Populations

Experience and statistical genetic theory indicate that, when several varieties with similar yields are composited into a breeding population, the mean yield of the new population will be higher than the average of the parental varieties when dominance effects are important (Darrah et al., 1972; Eberhart, 1971; Hallauer and Sears, 1969). Because gene frequencies in the new population will be averages for the parental varieties (weighted according to the contribution of each source variety), the frequencies at many loci will move from 0 or 1 to intermediate values. Hence, both the mean and variance can be expected to be somewhat higher than in the source varieties, especially if the source varieties have been maintained with small effective population sizes. If bias from epistatic effects is negligible, the formula developed by Wright (1922) for the F_2 population from homozygous lines is appropriate when the source varieties are in linkage equilibrium:

$$\text{Population mean} = \overline{VC} - \frac{\overline{VC} - \overline{V}}{n},$$

where $\overline{VC}$ is the mean of the possible crosses among the n source varieties and $\overline{V}$ is the mean of the varieties.

The cross between the breeding populations A and B also can be predicted with a general formula similar to the double-cross prediction formula:

$$A \times B = \tfrac{1}{mn}(VC_{11} + VC_{12} + ... + VC_{1n} + VC_{21} + ... + VC_{2n} + ... + VC_{m1} + ... + VC_{mn}),$$

where VC_{ij} are crosses of the ith variety (i = 1 to m) in population A and the jth variety (j = 1 to n) in population B.

Neal (1935) and Kinman and Sprague (1945) obtained information on predicted and observed means that substantiates the general utility of the formula for populations developed from single- and double-cross hybrids. Hayes and Garber (1919) and Sprague and Jenkins (1943) presented data on populations involving a larger number of parental lines. The mean of the breeding population should stabilize after one generation of random mating (syn-2 or F_2), except for linkage. Data from Kiesselbach (1933) and Sprague and Jenkins (1943) show very little change from the syn-2(F_2) to the syn-4(F_4). Castro et al. (1968) obtained no evidence for epistasis in a diallel involving five varieties, the variety crosses, the variety crosses selfed and the variety crosses random mated (syn-2). The agreement of observed and predicted syn-2 means was excellent.

These two formulas for the population-cross mean provide the basis for an objective development of breeding populations. The variety-cross diallel can be evaluated in a series of environments (Castro et al., 1968; Eberhart, 1971; Gardner, 1961; Hallauer and Malithano, 1976; Lonnquist and Gardner, 1961) and this information can be used to assign varieties to the appropriate breeding population or to select two of the varieties as the breeding populations. Although the primary objectives are to maximize (1) the population-cross mean and (2) the genetic variance within breeding populations, high yielding parental populations also are desirable. The variety diallel does not provide information on genetic variances within source varieties or breeding populations developed from compositing varieties, but populations developed by compositing several varieties should vary as much as or more than the source varieties because genetic variances are highest at intermediate gene frequencies.

Before an extensive variety-cross evaluation program is planned, however, the expected superiority of the cross between the breeding populations developed from such a program over the cross from populations developed by less extensive procedures must be considered. The use of performance information available and a knowledge of diversity because of geographic isolation or parentage may be nearly as effective. Funds and time spent on recurrent selection may be much more productive than the use of resources in extensive diallel evaluation trials. Although the 'Krug' × 'Reid' variety crosses were not as high yielding as 'Barber Reid × 'Golden Republic,' Lonnquist and Gardner (1961) reported that $K_{II(A)} \times R_{II}$ (developed by two cycles of recurrent selection) was the highest yielding of the 66 variety crosses. After several cycles of recurrent selection in BS12 and BSSS, the variety cross BS12(HI)C5 × BSSS(R)C6 was 24% higher yielding than the best cross among unimproved varieties (Hallauer and Malithano, 1976), and it approached the yields of the single-cross checks.

Breeding populations that will yield as high as or higher than populations involving only local germplasm have been developed from local and exotic material. PHWI(MT)C7 and (Indian Chief × 'Diente de Caballo')(S)C2 gave higher yields than any variety involving only germplasm from the USA in a regional trial grown at seven locations (Eberhart, 1971). Each of these populations involved approximately half local and half West Indies germplasm (Goodman, 1965; Moll et al., 1962). Hallauer and Malithano (1976) reported very high yields from BS16 developed from 'Eto' by six cycles of mass selection for earlier flowering (Hallauer and Sears, 1972). A small proportion of Corn Belt germplasm was used in developing Eto (Chavarriaga, 1966).

Two Corn Belt composites and several semi-exotic populations have been developed in Nebraska (Mathema, 1971). After one cycle of adaptive

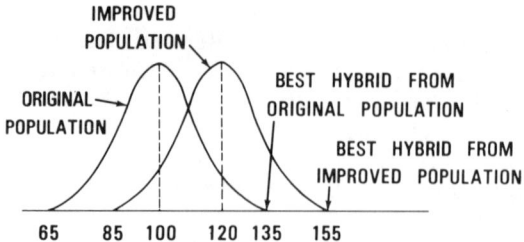

Fig. 1. An expected distribution for a finite number of single crosses from the original and an improved population.

mass selection, NMX(MA)C1 ('Corn Belt' × 'Mexican') yielded 17% more than NC (Corn Belt) and NZB (Corn Belt × 'Brazilian'), and NCB (Corn Belt × 'Carribbean') yielded as much as NC. Improvement per year averaged 2.9 q/ha (5.0%) for the eight semi-exotic populations and 1.9 q/ha (3.1%) for two Corn Belt populations. Although plant and ear heights are high and lodging is often excessive in the original populations involving exotic germplasm, these traits respond to selection (Genter and Eberhart, 1974; Hallauer and Malithano, 1975; Hallauer and Sears, 1972).

The improvement in performance obtained from the introduction of exotic germplasm can be even greater in areas where corn was not indigenous. Harrison (1970), Bolton (1971), and Eberhart et al. (1973b) reported highest yields in East Africa from breeding populations that involve germplasm recently introduced from Central and South America.

Information is somewhat limited on the comparison of genetic variances for diverse populations. Goodman (1965) reported greater genetic variance in the West Indies composite than in a comparable Corn Belt composite without the exotic germplasm. Moll and Robinson (1967) reported greater genetic variance in (Jarvis × Indian Chief)syn-4 than in either parental variety. The genetic variance in KCA ('Kitale II' × Ec573)syn-3 was larger than in Kitale II, but it was similar to the genetic variance in Ec573 (Darrah et al., 1972). Progress from four cycles of modified ear-to-row selection was also greater in Ec573 and KCA than in Kitale II.

Genetic variance in 'Krug Yellow Dent' and (Krug Yellow Dent × 'Tabloncillo') syn-3 were compared by Shauman (1971). The estimates of additive genetic variance were larger for yield and other closely related traits in the semi-exotic population than for Krug Yellow Dent.

B. Improving Breeding Populations

The improvement of the population cross between breeding populations through recurrent selection is the key to the more effective development of improved hybrids (Fig. 1). With the traditional inbreeding and hybridization system, most of the research efforts are directed to the identification of superior hybrid combinations, and limited resources remain available for the improvement of the source materials. Recent empirical results and statistical genetic theory clearly indicate that, when research is directed toward the improvement of the breeding populations, a corresponding improvement will be obtained in hybrids derived from these populations.

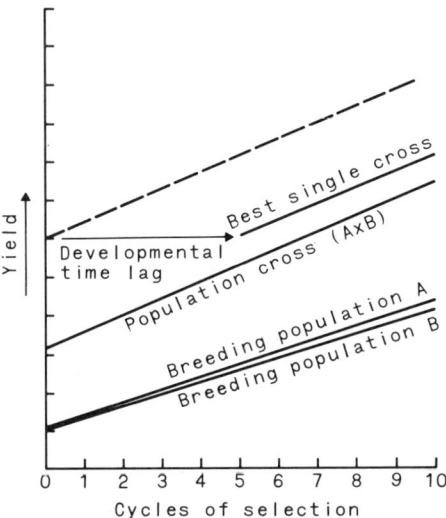

Fig. 2. **Expected improvement of breeding populations, the population cross, and the best single cross with reciprocal recurrent selection.**

Because complex traits such as yield are controlled by many genes with small effects, changes in genetic variances will change very slowly as the gene frequencies gradually change. No changes in genetic variances have been detected in five to seven cycles of selection by Moll and Stuber (1971), Penny and Eberhart (1971), Darrah et al. (1972), or Russell et al. (1973). Genetic variation among S_1 lines from 'Hays Golden' mass selection populations, C9 and I9, seemed to be less than among S_1 lines from Hays Golden (Harris et al., 1972), but in earlier studies no changes had been detected (Gardner, 1969; Lonnquist, 1966). The expected relations among breeding materials are shown diagrammatically in Fig. 2 for a limited number of cycles (assuming that changes in genetic variance components are negligible). Because the rate of improvement of the best single cross will be parallel to the improvement in the population cross, the objective of breeding programs should be to maximize improvement of the population cross within the funds and facilities available.

The rate of improvement (gain from selection) for interpopulation selection was given in Table 10. Gain will increase when breeding populations with maximum genetic variance are used; but the initial mean performance also must be high. Suitable plot techniques to minimize the experimental error are essential (good cultural practices and appropriate experimental designs). The genotype × environmental interaction can often be reduced with separate breeding programs in major ecological zones. Prolificacy may increase the stress tolerance and thereby reduce the genotype × environmental interaction (Hallauer, 1972; Russell and Eberhart, 1968).

The formula for predicting progress clearly shows that gain can be increased by reducing the phenotypic variance with increased numbers of replications and environments. If the genotype × year interaction is due to unpredictable weather factors, lengthening the recurrent selection cycle to permit testing over years is seldom efficient because of the reduction in gain

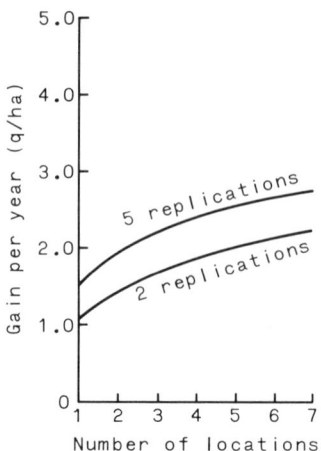

Fig. 3. Expected gain in Jarvis (R) × Indian Chief (R) with 2 years per cycle.

per year. Increasing the number of locations for the yield evaluation trials is usually advisable, and in certain instances the number of replications per location might be increased. The formula can be used, however, to allocate resources objectively for maximizing gain. Eberhart (1970) used variance components estimated from a breeding-methods study in Kenya to provide information with several intrapopulation selection methods. Twenty to 25 plants/plot with two replications grown in four environments were shown to be satisfactory for yield evaluation trials in KCA. Three replications in three locations, however, were predicted to give equal gains. Russell (1973) compared expected gain for several intrapopulation selection methods in BSSS in Iowa. The relative expected gains were very similar to those for KCA in Kenya, but expected gain per cycle was less.

The estimates of variance components from Jarvis (crossed to Indian Chief) and Indian Chief (crossed to Jarvis) given by Moll and Robinson (1967) can be used to predict gain from reciprocal recurrent selection in North Carolina (the number of tester plants was assumed to be infinite). Two replications in four locations are expected to give nearly as much gain as five replications in two locations (Fig. 3). Winter nurseries for recombination or family formation can increase progress per year by 50% because a cycle can be completed in 2 years, and breeders in certain tropical areas may be able to obtain three seasons per year to triple gain (Fig. 4).

Because the additive genetic variance among families increases with inbreeding, gain per cycle will increase if unrelated S_1 plants are crossed and selfed, instead of S_0 plants. But if the length of cycle is increased, gain per year may be similar (Fig. 5). Yield evaluation trials will be grown less often, however, and this saving may permit a higher selection intensity. Because S_2 lines will be recombined when S_1 plants are selfed and crossed to the tester, five seasons per cycle are suggested to provide an additional cycle of recombination: (1) selfing, (2) forming testcross families and selfing, (3) trials evaluating the yield potential of testcross families, (4) recombining selected S_2 lines, and (5) recombining crosses between S_2 lines. If five seasons are

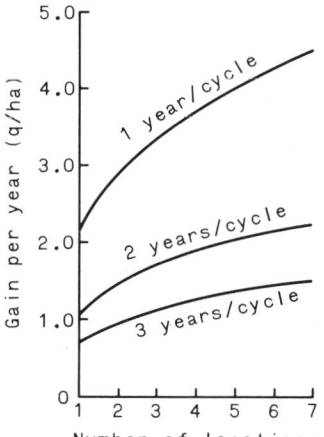

Fig. 4. Expected gain in Jarvis (R) × Indian Chief (R) with two replications per location as influenced by years per cycle.

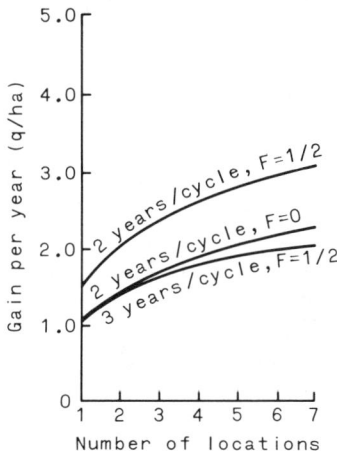

Fig. 5. Expected gain in Jarvis (R) × Indian Chief (R) with two replications per location, as influenced by levels of inbreeding.

available to complete a cycle in 2 years (F = ½) but three seasons cannot be obtained to complete a cycle in 1 year (F = 0), the extra generation of inbreeding is warranted.

Empirical results indicate the effectiveness of reciprocal recurrent selection (Darrah et al., 1972; Eberhart et al., 1973a; Moll and Stuber, 1971). The merits of full-sib reciprocal recurrent selection (Hallauer, 1967, 1973; Jones et al., 1971; Lonnquist and Williams, 1967) and reciprocal recurrent selection — inbred tester (Eberhart et al., 1973a; Russell and Eberhart, 1975) should be considered and compared with reciprocal recurrent selection in breeding-method studies.

Unfortunately, appropriate variance components are not available to compare expected gain from selection for interpopulation selection methods. The Jarvis and Indian Chief data, however, can be used in the formula given in Table 8 to provide general approximations. Assuming negligible epistatic

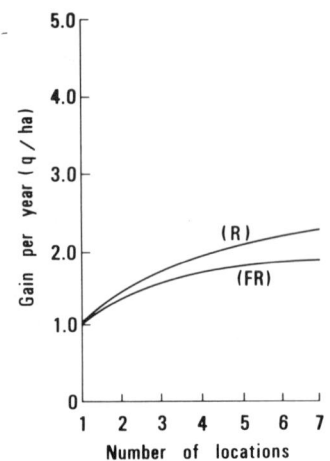

Fig. 6. Expected gain in Jarvis × Indian Chief with two replications per location for reciprocal recurrent selection (R), and fall-sib reciprocal recurrent selection (FR).

variance, additive variance with reciprocal recurrent selection (R), $\sigma^{2'}_{A(i)}$ was estimated as $4\sigma^2_{m(i)}$ and $\sigma^{2'}_{D(i)}$ as $4(\sigma^2_{f(i)} - \sigma^2_{m(i)})$. Because no estimates are available for full-sib reciprocal recurrent selection (FR), $\sigma^{2'}_A$ was estimated as $2(\sigma^2_{m1} + \sigma^2_{m2})$ and $\sigma^{2'}_d$ as $2[(\sigma^2_{f1} - \sigma^2_{m1}) + (\sigma^2_{f2} - \sigma^2_{m2})]$. Corresponding relationships were used for the genotype × environmental components. Because only half as many families need be evaluated with full-sib reciprocal recurrent selection to maintain a similar effective population size (Jones et al., 1971), the selection intensity should be twice as great to give comparable numbers of yield plots as for reciprocal recurrent selection (R). Computations were made for two replications and noninbred parents.

With selection intensities of 5% (k = 2.05) for full-sib reciprocal recurrent selection and 10% (k = 1.75) for reciprocal recurrent selection, expected gain for these methods are similar (Fig. 6). Jones et al. (1971) obtained greater gain from full-sib than from half-sib reciprocal recurrent selection in a computer simulation study, but they used unrealistically low selection intensities, 10/40 (k = 1.24) and 10/20 (k = 0.77).

With complete dominance (overdominance and epistasis negligible), reciprocal recurrent selection — inbred tester may be expected to give greater gain from selection than either reciprocal recurrent selection or full-sib reciprocal recurrent selection. The recurrent selection for specific combining ability studies initiated by Sprague and Miller (1950) are similar to reciprocal recurrent selection — inbred tester except for the use of the same line as a tester for both populations. When progress in Alph and (WF9 × B7)syn-2 with B14 as a tester was evaluated (Russell, et al., 1973), the gain in the population cross, Alph(HI) × (WF9 × B7)syn-2(HI), was 4.09 q/ha or 7.4% per cycle compared to 2.73 or 4.6% per cycle for BSSS(R) × BSCB1(R) (Eberhart et al., 1973a). The differences in source populations confounds the comparison, but these empirical results conform to expectations. Expanded research on this selection method may be very productive.

Gain from selection can be increased substantially for any selection method by increasing the selection intensity. The constant related to selection intensity, k, increases from 1.16 for a 30% selection intensity to 1.40, 1.75, 2.05, and 2.42 for 20, 10, 5 and 2% respectively. Funds and facilities usually limit the number of families that can be evaluated; hence, a compromise must be made between selection intensity and effective population size. The number of lines recombined should be as large as possible to increase the effective population size (N_e), because expected total advance and "half-life" of recurrent selection progress are proportional to effective population size (Robertson, 1960). A 10% selection intensity with an effective population size of 30 to 45 might be a reasonable compromise (Rawlings, 1970). Baker and Curnow (1969) suggested replicated selection programs, each with an effective population size of 16 or less, rather than one large program. This procedure seems to have an advantage, however, only in very long-term experiments. The heterosis between two substrains of BSK (Burton et al., 1971) suggests that gain from the four cycles of intrapopulation selection would have been 24% more if the effective population size had been increased from 20 to 40.

Li (1955) pointed out that when individuals (S_0 plants or S_1 lines representing the parental plants) are recombined systematically to ensure that all individuals contribute exactly the same number of gametes to the next generation, the effective size, N_e, is twice the number, N, of individuals (or S_1 lines recombined). With unequal number of males and females, the formula is given as

$$N_e = \frac{4 N_0 N_1}{N_0 + N_1}$$

where N_0 and N_1 are the number of males and females.

When the parents of the lines to be recombined are inbred such as with S_2 lines, the effective population size is reduced:

$$N_e = 2 \left(\frac{N}{1 + F_P} \right)$$

where F_p is the coefficient of inbreeding of the parental plants of the lines. Hence, with homozygous lines, N_e equals N.

The effective population size is important, because genetic drift causes undesirable alleles to be fixed in the homozygous condition before selection can increase the gene frequency of all desirable alleles to 1.0. Falconer (1960) expressed the inbreeding coefficient as a function of the effective population size:

$$F_t = \tfrac{1}{2N} + (1 - \tfrac{1}{2N}) F_{t-1} = 1 - (1 - \Delta F)^t,$$

where t corresponds to the cycle of selection and ΔF is $\tfrac{1}{2N}$. When the possibility of self-fertilization is excluded, $\Delta F - \tfrac{1}{2N+1}$ is a preferable approximation.

With recurrent selection, parental lines rarely, if ever, contribute equally to subsequent generations. Breeders usually initiate the next cycle of selection immediately after the first intercrossing among selected parents to reduce the generation interval. An examination of pedigrees over cycles of selection with this procedure will reveal that parents of preceding cycles have not contributed equally. In extreme cases the effective population size, N_e, may approach N rather than $2N$.

Table 11—Expected levels of inbreeding (F) for the population with varying effective population sizes

Cycle of selection	Effective population size (N_e)†				
	20	30	40	60	80
1	0.05	0.03	0.02	0.02	0.01
5	0.22	0.15	0.12	0.08	0.06
10	0.39	0.28	0.22	0.15	0.12
20	0.62	0.48	0.39	0.28	0.22
40	0.86	0.73	0.63	0.48	0.39

†$N_e = 2(\frac{N}{1 + F^*})$, where N is the number of lines recombined and F^* is the coefficient of inbreeding of the parental plants for the lines ($F^* = 0$ for S_1 lines, 0.5 for S_2 lines, etc.).

The relative rates of inbreeding for varying effective population sizes (N_e) are shown in Table 11. Of course, these levels are appropriate only when a realistic estimate of N_e can be obtained.

A homozygous line with favorable alleles at all loci would be the final goal if all loci showed partial to complete dominance. This goal cannot be achieved, however, if the undesirable allele is fixed at many of the loci because of genetic drift. Although mutations or introgression of germplasm from other selected breeding populations can restore part of the lost variability, minimizing random inbreeding during selection will result in higher mean performance and maintain a greater amount of genetic variability. For loci exhibiting overdominance or overdominant types of epistasis, retaining the alleles until selection can fix the appropriate allele in the reciprocal populations will result in greater gain than if the same allele is fixed in both populations by random drift.

Eberhart et al. (1973a) reported 5.2 q/ha improvement in BSSS(HT)C7 over BSSS C0. If 30 lines had been selected from 300, rather than 10 from 100, each cycle, however, expected inbreeding in BSSS(HT)C7 would have been 0.11, instead of 0.29. Assuming a yield depression from inbreeding similar to the estimate obtained by Hallauer and Sears (1973) for BSSS (0.449 q/ha per % of inbreeding), BSSS(HT)C7 would have been expected to yield an additional 8.1 q/ha, which would have given a gain of 1.90 q/ha per cycle instead of the 0.74 gain reported with an effective population size of 20. Similar computations indicate a probable gain per cycle of 1.47 q/ha for BSSS(R) for 30/300 compared with observed gain of 0.24 with 10/100 lines.

Horner et al. (1973) estimated an inbreeding depression of 0.63 q/ha per 1% inbreeding in Fla. 767 for five cycles of testcross selection-inbred tester. He concluded that the improvement in Fla. 767 would have been considerably greater if a larger effective population size had been used.

When funds and facilities permit the evaluation of large numbers of families, higher selection intensities (1% to 5%) may be desirable in the breeding population, if a corresponding germplasm pool is maintained as a reserve for each breeding population. Mass selection with a large effective population size (300 to 500 plants selected) and moderate selection intensity (20 to 40%) would allow recombination each year and minimize the loss of desirable alleles because of genetic drift. The combined effects of linkage, intense selection, and small effective population size merit additional studies. Meanwhile, the use of germplasm pools might circumvent such problems. After each cycle of intense interpopulation selection, the elite lines could be introgressed into the appropriate germplasm pool. If genetic variation

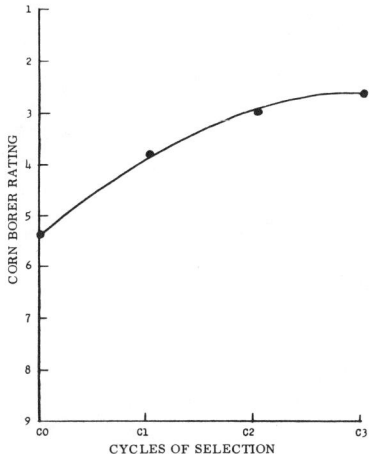

Fig. 7. Average improvement for corn borer resistance in five corn varieties by recurrent selection with S_1 selection (from Penny et al., 1967).

decreases after several cycles of intense selection, each breeding population could be crossed to its corresponding germplasm reserve and backcrossed to the breeding population, to restore genetic variation with a minimum loss of mean performance.

Several of the improved population crosses produce yields similar to those of many commercial hybrids (Darrah et al., 1972; Eberhart et al., 1973a; Genter and Eberhart, 1974; Russell and Eberhart, 1975). Nevertheless, they are not commercially acceptable because of excessive root and stalk lodging and insect or disease susceptibility. Because many of these other agronomic traits are highly heritable, selection pressure can be exerted in the nurseries during recombination and family formation, as suggested by Lonnquist (1964) and Penny and Eberhart (1971). Both half-sib and full-sib selection, as well as mass selection, can be used for these agronomic traits in the breeding nurseries, so that only elite families are evaluated for yield. As long as yield is not highly correlated with the factors causing susceptibility to disease or insects (evaluated in the absence of diseases or insect infestations), selection of resistant plants for selfing and crossing in the nursery should not reduce genetic variability or gain from selection for yield.

Although few examples for the improvement of disease or insect resistance of corn from recurrent selection are available, it has been extremely effective when used. Penny et al. (1967) used S_1 selection to reduce the average rating for leaf feeding by the European corn borer (*Ostrinia nubilalis* (Hubner) from 5.4 to 2.5 (1 to 9 scale) in only three cycles of recurrent selection (Fig. 7). Resistance to first brood seems to be under multigenic control (Scott et al., 1966).

Mass selection was used by Zuber et al. (1971) to reduce the percentage of ears with kernel damage due to the feeding of earworms (*Heliothis zea* (Boddie)) from 80.8% in Synthetic C to 58.7% in cycle 10 and from 64.5% in Synthetic S to 38.2% in cycle 10, with an average reduction of 2.8% per cycle. The C10 of these populations equalled or exceeded the check hybrid (Dixie

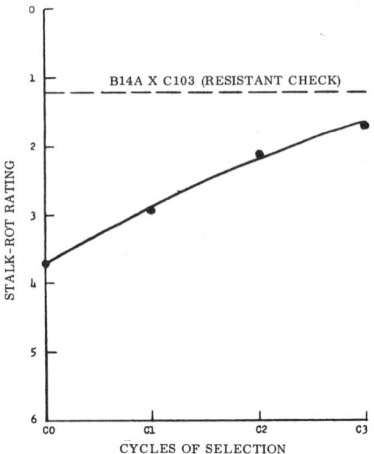

Fig. 8. Varietal improvement for stalk-rot resistance in the Lancaster corn variety by recurrent selection with S_1 selection (from Jinahyon and Russell, 1969).

18 with 55.9% kernel damage) in ear worm resistance.

Jenkins et al. (1954) decreased the susceptibility for leaf blight (*Helminthosporium turcicum* (Pass.) of four F_2 populations (a resistant and susceptible parent) from an average rating of 3.3 to 2.1 (0 to 5 scale) by three cycles of mass selection. In a susceptible × intermediate-resistance cross (CI. 2 × T8), the rating was changed from 4.0 to 3.0, and a few progenies were more resistant than either parent. Reciprocal translocations identified six chromosome arms associated with the resistance from NC34 in the (R4 × NC34) F_2 (Jenkins et al., 1957).

Three cycles of S_1 selection were used to improve the stalk-rot resistance (*Diplodia zeae* (Schw.) Lev. in the variety Lancaster (Jinahyon and Russell, 1969). The mean rating (0 to 5 scale) changed from 3.7 in the C0 to 1.7 in the C3, in comparison with 1.2 for the resistant single cross check, B14A × C103 (Fig. 8). Field stalk lodging showed a decrease from 40% to 26%, and grain yield remained unchanged. El-Rouby and Russell (1966) reported that nine different chromosome arms in two highly resistant lines, B14 and C103, carried a gene or genes that condition resistance to *D. zeae*.

In four cycles of mass selection (both sexes) for stalk crushing strength, the response was essentially linear in MoSQA and MoSQB in Missouri (Zuber, 1973). The average crushing strength was changed from 354 hg (780 lbs.) to 588 kg (1,295 lbs.) in the high-selection substrain (16% per cycle) and to 256 kg (564 lbs.) in the low-selection strain. Average field stalk lodging was changed from 3.1 to 1.7 and 13.7% in the high and low substrains, respectively. No changes were observed for yield, grain moisture at harvest, shelling percentage, test weight, or ear height grade.

Morphological traits with high heritabilities also can be changed easily by recurrent selection. The leaf angle of a corn variety, 'Puerto Rico Group I,' was changed to 10 to 12% per cycle with four cycles of bidirectional mass selection in Thailand (Ariyanayagam et al., 1974).

Resistant breeding populations with average yield and high-yielding populations without insect and disease resistance have little value as source materials of elite inbred lines. Hence, simultaneous recurrent selection for higher yields and desirable agronomic traits is required. When these traits are manifested at different stages of growth, multistage selection (Young, 1964) can be an efficient procedure. If traits cannot be selected at different growth stages or at different phases in the recurrent selection cycle, a selection index (Smith, 1936) is expected to be more efficient than tandem selection (Hazel and Lush, 1942). Subandi et al. (1973) suggested that a multiplicative type of index for the selection of erect plants, higher yields, and fewer dropped ears may be useful. Indexes have been used for cold tolerance (Mock and Eberhart, 1972) and ear worm resistance (Widstrom, 1974). When heritabilities are high (as with family selection methods) and genetic correlations are low, however, a base index (with the relative economic weights as the phenotypic weights) is expected to be nearly as efficient as the conventional index in improving the aggregate genetic worth (Falconer, 1960; Suwantaradon and Eberhart, 1975).

Pesek and Baker (1969, 1970) proposed a modified index in which desired gains are used to obtain phenotypic weights that are expected to result in reaching the desired gains simultaneously for all traits. This procedure has merit when the relative economic weights cannot be obtained, but it may be considerably less efficient than the base or conventional index when the relative economic weights are known for all traits. Even with a selection index, gain from selection for any given trait normally decreases as additional traits are added in the index. Hence, traits to be included in the index and the desired gains should be chosen objectively.

S_1 selection is especially amenable to multistage selection. When the recombination phase is obtained in isolation from a planting of bulked seed of selected lines from the previous cycle of selection, a large number of half-sib families can be available. Selection among and within the half-sib families for highly heritable traits at pollination time can be followed by selection among selfed plants in the selected half-sib families for other highly heritable traits at harvest. Hence, all S_1 families in the yield trials will have exceeded prescribed standards for important agronomic traits.

The use of a five-season cycle with S_2 selection also has certain advantages. During the first season of selfing, mass selection can be used to select parental plants with desirable traits that have a high heritability. Further selection among S_1 lines for agronomic traits with lower heritabilities can be achieved the following season, as well as another cycle of mass selection within S_1 lines. Gain from mass selection within lines will be reduced, however, because only half the additive genetic variance will remain within S_1 lines. If only one selfed plant in each selected S_1 line is retained for S_2 yield trials, genetic variance among selfed families will be increased by $1+F(1+\frac{1}{2})$ over S_1 selection. When 3 years are required for S_2 selection and 2 years for S_1 selection, the expected gain per year for yield is often similar for S_1 and S_2 selection and greater gain is possible for S_2 selection for the other agronomic traits. The value of two cycles of recombination from S_2 selection in breaking repulsion phase linkages vs. a single season of recombination normally obtained with S_1 selection cannot be derived easily, but this extra season of recombination without selection may be of considerable value. Reciprocal recurrent selection-inbred tester will be as amenable as S_1 or S_2 selection for multistage selection in prolific populations. The scheme can be similar, except that one ear of each selected plant can be pollinated by the inbred tester, and the other ear can be selfed. In nonprolific material, however, the testcrosses must be made on the inbred tester, and this procedure will require slightly more labor in making sufficient selfs and testcrosses to permit the same selection intensity within families at harvest.

C. Developing Hybrids from Improved Populations

The rate of improvement of hybrids will be proportional to the improvement of the population cross between breeding populations (Fig. 2). Hence, the efficient development of new hybrids after each cycle of population improvement will be an important phase of the breeding program.

The procedures in the traditional inbreeding and hybridization system, as outlined in Sections II and III, can be used to develop superior hybrids from the improved breeding populations, but greater efficiency will be obtained by use of the yield evaluation phase of the interpopulation selection as the early testing for inbred line development. Full-sib reciprocal recurrent selection was proposed initially because of the expected efficiency from the use of all types of gene action in developing single-cross hybrids (Hallauer, 1967, 1973; Lonnquist and Williams, 1967). Hoegemeyer (1974) showed that the hybrids of the selected reciprocal pairs of inbred lines give superior yields because of specific combining ability, but these lines also have good general combining ability with the other selected lines from the reciprocal population.

When M full-sib families have been evaluated in a full-sib reciprocal recurrent selection program, the m S_1 lines from population A and n S_1 lines from population B might be chosen as superior lines (based on agronomic traits as lines per se as well as for yield as full-sib families) for the development of new single cross hybrids. The elite lines from population A should be evaluated in hybrid combination with the elite lines from population B in an AB factorial mating design. When the best hybrid is selected from the $m \cdot n$ hybrids evaluated, the true selection intensity will be closer to $1/M^2$, rather than $1/m \cdot n$. Similarly, in reciprocal recurrent selection or reciprocal recurrent selection-inbred tester, M testcross families will have been evaluated for general combining ability in population A and N families in population B. When the elite $m \cdot n$ hybrids among inbred lines derived from the selected lines are evaluated in an AB factorial mating design, selection of the best hybrid will result in a selection intensity approaching $1/M \cdot N$, rather than $1/m \cdot n$.

When random inbred lines are obtained from a population in equilibrium, the expected superiority of the best hybrid over the population mean can be obtained with a formula similar to the one given under Section VIII, A. The variance among hybrids, σ_g^2, is $(\sigma_A^2 + \sigma_D^2)$, $(3/4) \sigma_A^2 + (1/2) \sigma_D^2$, and $(1/2) \sigma_A^2 + (1/4)\sigma_D^2$ for single crosses, three-way crosses, and double crosses (Cockerham, 1961), assuming negligible epistatic variance. Gain will be proportional to σ_g^2, rather than $\sigma_{g'}^2$, because the superior hybrid will be selected without further recombination. Expected superiorities from an $A_i \times B_j$ single cross and $(A_i \times A_k)(B_j \times B_l)$ double cross are given in Table 12 for the

Table 12—Expressed superiority† (expressed as percentage of the population-cross mean) of hybrids developed from breeding populations

Selection intensity	BSSS(R) × BSCB1(R)‡		Jarvis (R) × Indian Chief (R)§		BS10(FR) × BS11(FR)¶
	SC	DC	SC	DC	DC
4.0	20.4	12.4	15.0	9.4	11.7
1.0	25.0	15.2	18.5	11.5	14.3
0.1	32.0	19.5	23.6	14.8	18.3
0.005	40.5	24.6	29.9	18.8	23.2
Reps (r)	3	3	5	5	2
Locations (m)	6	6	2	2	9

† $\Delta = k \cdot (\sigma_A^2 + \sigma_D^2) / \sqrt{\frac{\sigma^2}{rm} + \frac{(\sigma_{AE}^2 + \sigma_{DE}^2)}{m} + (\sigma_A^2 + \sigma_D^2)}$ for SC and

$\Delta = k(1/2\sigma_A^2 + 1/4\sigma_D^2) / \sqrt{\frac{\sigma^2}{rm} + \frac{(1/2\sigma_{AE}^2 + 1/4\sigma_{DE}^2)}{m} + (1/2\sigma_A^2 + 1/4\sigma_D^2)}$ for DC.

‡Variance components from Penny and Eberhart (1971). §Variance components from Bari (1973).
¶Variance components from Hallauer (1973).

BSSS (R) × BSCB1 (R), the Jarvis (R) × Indian Chief (R), and the double cross from the BS10 (FR) × BS11 (FR) populations.

Bari (1973) reported that the best single cross between unselected Jarvis and Indian Chief S_{13} lines exceeded the mean of the set of 132 single crosses evaluated by 86 g/plant (0.190 lb./plant) or 29%. In comparison, the best single cross between unselected Jarvis (R)C6 and Indian Chief (R)C6 S_5 lines exceeded the mean of the 133 single crosses evaluated by 64 g/plant (0.141 lb/plant) or 19%. Because no changes in genetic variances had been detected, expected gain was 59 g/plant for both sets of single crosses. Topcrosses of S_1 lines from 'Hays Golden' C9 and I9 populations yielded 17% more than comparable topcrosses involving S_1 lines from the original Hays Golden variety (Harris et al. 1972).

Russell and Eberhart (1975) reported that the best cross between selected S_2 lines from the fifth cycle of reciprocal recurrent selection outyielded the parental population cross, BSSS(R)C5 × BSCB1(R)C5, by 35%. This hybrid was significantly higher yielding than B37 × Oh43 and was equal in root and stalk strength. Although 25 hybrids were evaluated in this phase, 100 testcrosses were evaluated for each population in the reciprocal recurrent selection program. The best BS10(FR) × BS11(FR) single cross yielded 25.6 q/ha (31.8%) more than the average of the interpopulation single crosses involving unselected lines (Hoegemeyer, 1974).

A topcross hybrid 'Florida 200,' (F44 × F6) × F5B(HT)C3, was released for commercial use (Horner et al., 1957) and was replaced by 'Florida 200A,' (F44 × F6) × F5B(HT)C5 (Horner et al., 1965, 1972). The improvement of the male parent, F5B(HT)C5, resulted in a topcross hybrid with a 5.4% higher yield than Florida 200. Furthermore, 200A had a higher percentage of erect plants, more ears per plant, and lower ear height.

In Kenya, the variety cross hybrid H611 (Kitale II(R) × Ec573(R)) has been used commercially. An improved version, H611C [Kitale II (HT)C1 × Ec573(R)C2], gave 26% greater yield than the original H611 (Darrah et al., 1972). The improvement in Ec573(R) has also resulted in 24% higher yields of the topcross hybrid H613C [(F × G) × Ec573(R)C2] than for the original topcross H613B [F × G) × Ec573CO]. Because reduced hybrid seed costs from H613C resulted from higher yields on the F × G seed parent and reduced labor requirements with more uniform tasseling, H613C accounted for 64% of the seed sales in Kenya for 1973 (Darrah, 1975).

Not only does the reciprocal recurrent selection-inbred tester method give promise of faster improvement in the population cross (Fig. 6 and Russell et al., 1973), but also it may make the use of topcross (inbred-variety) hybrids competitive with single crosses. When the inbred testers are elite inbred lines suitable for use as parents, the $A(RI)Cn \times B_j$ and $B(RI)Cn \times A_i$ topcrosses may have commercial potential. The $BS12(HI)C5 \times B14A$ topcross gave good performance in regional trials involving elite local checks (Genter and Eberhart, 1974). Once the suitability of such a topcross is established, the latest cycle of selection, $C(n + 1)$, can be substituted immediately for the Cn population. But an additional 4 to 8 years will be required to develop and identify new lines for improved single crosses. Because of the developmental time lag, the $C(n + 4)$ topcross may be available at the same time that a single cross produced from a new line out of the Cn population can be released.

When the disease and insect resistance of the breeding populations have been improved sufficiently in the interpopulation selection program, all hybrids developed from these improved breeding populations should have satisfactory resistance.

LITERATURE CITED

Allard, R.W. 1960. Principles of plant breeding. John Wiley and Son. New York. 485 p.

----, and A.D. Bradshaw. 1964. Implications of genotype environment interactions in applied plant breeding. Crop Sci. 4:503-507.

Allison, J.C.S., and R.N. Curnow. 1966. On the choice of tester parent for the breeding of synthetic varieties of maize (*Zea mays* L.). Crop Sci. 6:541-544.

Anderson, D.C. 1938. The relation between single and double-cross yields in corn. J. Am. Soc. Agron. 30:209-211.

Anderson, E., and W.L. Brown. 1952. Origin of corn belt maize and its genetic significance in heterosis. p. 124-148. *In*: J.W. Gowen (ed.) Heterosis. Chapter 8, Iowa State College Press. Ames.

Ariyanayagam, R.P., C.L. Moore, and V.R. Carangal. 1974. Selection for leaf angle in maize and its effect on grain yield and other characters. Crop Sci. 14:551-555.

Baker, L.H., and R.N. Curnow. 1969. Choice of population size and use of variation between replicate populations in plant breeding selection programs. Crop Sci. 9:555-560.

Bari, A. 1973. Comparison of the distribution of single cross yields before and after reciprocal recurrent selection. Ph.D. Thesis. N.C. State Univ. at Raleigh. 116 p.

Bauman, L.F. 1959. Evidence of non-allelic gene interactions in determining yield, ear height and kernel row number in corn. Agron. J. 51:531-534.

Beal, W.J. 1877. Report of the professor of botany and horticulture. Annu. Rep. Mich. Board Agric. p. 41-59.

----. 1880. Indian corn. Annu. Rep. Mich. Board Agric. pp. 279-289.

----. 1881-82. Report of the professor of botany and horticulture. Annu. Rep. Mich. Board Agric. p. 98-153.

Beverley, R. 1705. The history and present state of Virginia. Printed by R. Parker at the Unicorn under the Piazza's of the Royal Exchange.

Bolton, A. 1971. Territorial maize variety trials in Tanzania 1966-70. E. Afr. Agric. For. J. 37:109-124.

Brown, W.L. 1967. Results of non-selective inbreeding in maize. Der Zuchter 37:155-159.

Burnham, C.R. 1946. An "Oenothera" or multiple translocation method of establishing homozygous lines. J. Am. Soc. Agron. 38:702-707.

Burton, J.W., L.H. Penny, A.R. Hallauer, and S.A. Eberhart. 1971. Evaluation of synthetic populations developed from a maize population (BSK) by two methods of recurrent selection. Crop Sci. 11:361-367.

Chase, S.S. 1949. Monoploid frequencies in a commercial double-cross hybrid and in its component single-cross hybrids and inbred parents. Genetics 34:328-332.

----. 1952a. Production of homozygous diploids of maize from monoploids. Agron. J. 44:263-267.

----. 1952b. Monoploids in maize. p. 389-399. In: J.W. Gowen (ed.) Heterosis. Chapter 25, Iowa State College Press. Ames.

Chavarriaga, E. 1966. Maiz Eto una variedad producida en Colombia Instituto Colombian Agropecuario. Agric. Trop. (1) 30 p.

Chi, R.A., S.A. Eberhart, and L.H. Penny. 1969. Covariances among relatives in a maize variety (*Zea mays* L.). Genetics 63:511-520.

Clem, Mary, C.C. Moser, and G.F. Sprague. 1956. Simplified punched card procedures for predicting double cross performance. Agron. J. 48:319-320.

Cockerham, C.C. 1956. Analysis of quantitative gene action. Brookhaven Symp. Biol. 9:53-68.

----. 1961. Implications of genetic variances in a hybrid breeding program. Crop Sci. 1:47-52.

----. 1963. Estimation of genetic variances. p. 53-94. In: W.D. Hanson and H.F. Robinson (eds.) Statistical genetics and plant breeding. Natl. Acad. Sci.-Natl. Res. Counc. Publ. 982. Washington, D.C.

----. 1967. Prediction of double crosses from single crosses. Der Zuchter 37:160-169.

Collins, G.N. 1910. The value of first generation hybrids in corn. USDA Bur. Plant Ind. Bull. 19:1-45.

Combs, J.B., and M.S. Zuber. 1949. Further use of punched card equipment in predicting the performance of double-crossed corn hybrids. Agron. J. 41:485-486.

Comstock, R.E., H.F. Robinson, and P.H. Harvey. 1949. A breeding procedure designed to make maximum use of both general and specific combining ability. Agron. J. 41:360-367.

Corsi, W.C., and R.H. Shaw. 1971. Evaluation of stress indices for corn in Iowa. Iowa State J. Sci. 46:79-85.

Cowan, J.R. 1943. The value of double-cross hybrids involving inbreds of similar and diverse origin. Sci. Agric. 23:287-296.

Darrah, L.L. 1975. Maize genetics, Record of Research Annual Report, 1971. E. Afr. Agric. For. Res. Organ. Annu. Rep. (In press).

----, S.A. Eberhart, and L.H. Penny. 1972. A maize breeding study in Kenya. Crop Sci. 12:605-608.

----, and A.R. Hallauer. 1972. Genetic effects estimated from generation means from diallel sets of maize inbreds. Crop Sci. 12:615-621.

Darwin, C. 1877. The effects of cross and self fertilization in the vegetable kingdom. D. Appleton and Co., New York. 482 p.

Davis, R.L. 1927. Report of the plant breeder. P.R. Agric. Exp. Stn. Rep. 1927. p. 14-15.

----. 1934. Maize crossing values in second generation lines. J. Agric. Res. 48:339-357.

Doxtator, C.W., and I.J. Johnson. 1936. Prediction of double-cross yields in corn. J. Am. Soc. Agron. 28:460-462.

Dudley, J.W., D.E. Alexander, and R.J. Lambert. 1974. Genetic improvement of modified protein maize. CIMMYT-Purdue Int. Symp. on Protein Quality in Maize. Mexico, 1972.

----, R.J. Lambert, and D.E. Alexander. 1974. Seventy generations of selections for oil and protein concentration in the maize kernel. p. 181-212. In: J.W. Dudley (ed.) Seventy generations of selection for oil and protein in maize. Crop Sci. Soc. Am. Madison, Wis.

----, and R.H. Moll. 1969. Interpretation and use of estimates of heritability and genetic variances in plant breeding. Crop Sci. 9:257-262.

Duncan, W.G., D.L. Shaver, and W.A. Williams. 1973. Insolation and temperature effects on maize growth and yield. Crop Sci. 13:187-191.
East, E.M. 1908. Inbreeding in corn. Annu. Rep. Conn. Agric. Exp. Stn. for 1907. p. 419-428.
----. 1909. The distinction between development and heredity in inbreeding. Am. Nat. 43:173-181.
----, and D.F. Jones. 1919. Inbreeding and outbreeding. J.B. Lippincott Co., Philadelphia. 285 p.
Eberhart, S.A. 1964. Theoretical relations among single, three-way and double-cross hybrids. Biometrics 20:522-539.
----. 1970. Factors effecting efficiencies of breeding methods. Afr. Soils. 15:669-680.
----, Seme Debela, and A.R. Hallauer. 1973a. Reciprocal recurrent selection in the BSSS and BSCB1 maize populations and half-sib selection in BSSS. Crop Sci. 13:451-456.
----, R.H. Moll, H.F. Robinson, and C.C. Cockerham. 1966. Epistatic and other genetic variances in two varieties of maize. Crop Sci. 6:275-280.
----, L.H. Penny, and M.N. Harrison. 1973b. Genotype by environment interactions in maize in Eastern Africa. E. Afr. Agric. For. J. 39:61-71.
----, and W.A. Russell. 1966. Stability parameters for comparing varieties. Crop Sci. 6:36-40.
----, and W.A. Russell. 1969. Yield and stability for a 10-line diallel of single-cross and double-cross maize hybrids. Crop Sci. 9:357-361.
----, W.A. Russell, and L.H. Penny. 1964. Double-cross hybrid prediction when epistasis is present. Crop Sci. 4:363-366.
Eckhardt, R.C., and A.A. Bryan. 1940. Effect of method of combining the four inbred lines of a double-cross of maize upon the yield and variability of the resulting double-crosses. J. Am. Soc. Agron. 32:347-353.
Einset, J. 1943. Chromosome length in relation to transmission frequency of maize trisomes. Genetics 28:349-364.
El-Rouby, M.M., and W.A. Russell. 1966. Locating genes determining resistance to *Diplodia maydis* in maize by using chromosome translocations. Can. J. Genet. Cytol. 8:233-240.
Empig, L.T., C.O. Gardner, and W.A. Compton. 1972. Theoretical gains for different population improvement procedures. Nebr. Agric. Exp. Stn. Bull. M26 (revised). 22 p.
Falconer, D.S. 1960. Introduction to quantitative genetics. Ronald Press Co., New York. p 365.
Gamble, E.E. 1962a. Gene effects in corn (*Zea mays* L.). I. Separation and relative importance of gene effects for yield. Can. J. Plant Sci. 42:339-348.
----. 1962b. Gene effects in corn (*Zea mays* L.). III. Relative stability of the gene effects in different environments. Can. J. Plant Sci. 42:628-634.
Gardner, C.O. 1961. An evaluation of effects of mass selection and seed irradiation with thermal neutrons on yield of corn. Crop Sci. 1:241-245.
----. 1963. Estimates of genetic parameters in cross fertilizating plants and their importance in plant breeding. p. 225-252. *In*: W.D. Hanson and H.F. Robinson (eds.) Statistical genetics and plant breeding. Natl. Acad. Sci.-Natl. Res. Counc. Publ. 982. Washington, D.C.
Genter, C., and S.A. Eberhart. 1974. Performance of original and advanced maize populations in their diallel crosses. Crop Sci. 14:881-885.
Giesbrecht, J. 1964. Gamete selection in two early maturing corn varieties. Crop Sci. 4:19-21.
Goodman, M.M. 1965. Estimates of genetic variance in adapted and exotic populations of maize. Crop Sci. 5:87-90.
Goodsell, S. 1961. Male sterility in corn by Androgenesis. Crop Sci. 1:227-228.
Gorsline, G.W. 1961. Phenotypic epistasis for ten quantitative traits in maize. Crop Sci. 1:55-58.
Hallauer, A.R. 1967. Development of single-cross hybrids from two-eared maize populations. Crop Sci. 7:192-195.
----. 1970. Zygote selection for the development of single-cross hybrids in maize. Adv. Front. Plant Sci. 25:75-81.
----. 1972. Third phase in the yield evaluation of synthetic varieties of maize. Crop Sci. 12:16-18.
----. 1973. Hybrid development and population improvement in maize by reciprocal full-sib selection. Egypt. J. Genet. Cytol. 2:84-101.

----, and D. Malithano. 1976. Evaluation of maize varieties for their potential as breeding populations. Euphytica 25:117-127.
----, and J.H. Sears. 1968. Second phase in the evaluation of synthetic varieties of maize for yield. Crop Sci. 8:448-451.
----, and J.H. Sears. 1969. Mass selection for yield in two varieties of maize. Crop Sci. 9:47-50.
----, and J.H. Sears. 1972. Integrating exotic germplasm into corn belt maize in breeding programs. Crop Sci. 12:203-206.
----, and J.H. Sears. 1973. Changes in quantitative traits associated with inbreeding in a synthetic variety of maize. Crop Sci. 13:327-330.
Harris, R.E., C.O. Gardner, and W.A. Compton. 1972. Effects of mass selection and irradiation in corn measured by random S_1 lines and their testcrosses. Crop Sci. 12:594-598.
Harrison, M.N. 1970. Maize improvement in East Africa. In C.L.A. Leakey (ed.) Crop improvement in East Africa. Common Agric. But. Farnham Royal.
Hayes, H.K., and R.J. Garber. 1919. Synthetic production of high protein corn in relation to breeding. J. Am. Soc. Agron. 11:308-318.
----, and P.J. Olson. 1919. First generation crosses between standard Minnesota corn varieties. Minn. Agric. Exp. Stn. Tech. Bull. 183:5-22.
----, and I.J. Johnson. 1939. The breeding of improved selfed lines of corn. J. Am. Soc. Agron. 31:710-724.
----, R.P. Murphy, and E.H. Rinke. 1943. A comparison of the actual yields of double-crosses of maize with their predicted yield from single-crosses. J. Am. Soc. Agron. 35:60-65.
----, E.H. Rinke, and Y.S. Tsiang. 1946. Experimental studies on convergent improvement and backcrossing in corn. Minn. Agric. Exp. Stn. Tech Bull. 172:3-40.
Hazel, L.N., and J.H. Lush. 1942. The efficiency of three methods of selection. J. Hered. 33:393-399.
Hoegemeyer, T.C. 1974. Selection among and within full-sib families for the development of single crosses in maize (Zea mays L.). Ph.D. dissertation. Iowa State Univ. Ames, Iowa.
Hopkins, C.G. 1899. Improvement in the chemical composition of the corn kernel. Ill. Agric. Exp. Stn. Bull. 55:205-240.
Horner, E.S., W.H. Chapman, M.C. Lutrick, and H.W. Lundy. 1969. Comparison of selection based on yield of topcross progenies and S_2 progenies in maize (Zea mays L.). Crop Sci. 9:539-543.
----, H.W. Lundy, M.C. Lutrick, and W.H. Chapman. 1973. Comparison of three methods of recurrent selection in maize. Crop Sci. 13:485-489.
----, ----, ----, and R.W. Wallace. 1965. Florida 200A, an improved yellow field corn hybrid for north and west Florida. Fla. Agric. Exp. Stn. Circ. S-165.
Hull, F.H. 1945. Recurrent selection and specific combining ability in corn. J. Am. Soc. Agron. 37:134-145.
Jenkins, M.T. 1929. Correlation studies with inbred and crossbred strains of corn. J. Agric. Res. 39:677-721.
----. 1934. Methods of estimating the performance of double-crosses in corn. J. Am. Soc. Agron. 26:199-204.
----. 1935. The effect of inbreeding and of selection within inbred lines of corn upon the hybrids made after successive generations of selfing. Iowa State J. Sci. 3:429-450.
----, and A.M. Brunson. 1932. Methods of testing inbred lines of corn in crossbred combinations. J. Am. Soc. Agron. 24:523-530.
----, A.L. Robert, and W.R. Findley, Jr. 1954. Recurrent selection as a method for concentrating genes for resistance to Helminthosporium turcicum leaf blight in corn. Agron. J. 46:89-94.
----, ----, and ----. 1957. Genetic studies of resistance to Helminthosporium turcicum in maize by means of chromosomal translocations. Agron. J. 49:197-201.
Jinahyon, S., and W.A. Russell. 1969. Evaluation of recurrent selection for stalk-rot resistance in an open-pollinated variety of maize. Iowa State J. Sci. 43:229-237.
Johnson, E.C. 1963. Mass selection for yield in a tropical corn variety. Am. Soc. Agron. Abstr. p. 82.

Johnson, I.J., and H.K. Hayes. 1936. The combining ability of inbred lines of Golden Bantam sweet corn. J. Am. Soc. Agron. 28:246-252.

----, and H.K. Hayes. 1940. The value in hybrid combinations of inbred lines of corn selected from single-crosses by the pedigree method. J. Am. Soc. Agron. 32:479-485.

Jones, D.F. 1918. The effects of inbreeding and cross-breeding upon development. Conn. Agric. Exp. Stn. Bull. 207:5-100.

----. 1922. The productiveness of single and double-cross first generation hybrids. J. Am. Soc. Agron. 14:241-252.

----. 1939. Continued inbreeding in maize. Genetics 24:462-473.

----, and W.R. Singleton. 1934. Crossed sweet corn. Conn. Agric. Exp. Stn. Bull. 361:489-536.

Jones, L.P., W.A. Compton, and C.O. Gardner. 1971. Comparison of full-sib and half-sib reciprocal selection. Theor. Appl. Genet. 41:36-39.

Jorgenson, L., and H.E. Brewbaker. 1927. A comparison of selfed lines of corn and first-generation crosses between them. J. Am. Soc. Agron. 19:819-830.

Kempthorne, O. 1957. An introduction to genetic statistics. John Wiley and Son, New York. p. 545.

Kermicle, J.L. 1969. Androgenesis conditioned by a mutation in maize. Science 166:1422-1424.

Kiesselbach, T.A. 1922. Corn investigations. Nebr. Agric. Exp. Stn. Res. Bull. 20. p. 5-151.

----. 1933. The possibilities of modern corn breeding. Proc. World Grain Exhibit and Conf., Canada 2:92-112.

Kinman, M.L., and G.F. Sprague. 1945. Relation between number of parental lines and theoretical performance of synthetic varieties of corn. J. Am. Soc. Agron. 37:341-351.

Leng, E.R. 1961. Predicted and actual response during long-term selection for chemical composition in maize. Euphytica 10:368-378.

----. 1962. Results of long-term selection for chemical composition in maize and their significance in evaluating breeding systems. Z. Pflanzenzuecht. 47:67-91.

Lindstrom, E.W. 1939. Analysis of modern maize breeding principles and method. Proc. 7th Int. Genet. Congr. Edinburgh, Scotland. p. 191-196.

Lonnquist, J.H. 1950. The effect of selection for combining ability within segregating lines of corn. Agron. J. 42:503-508.

----. 1951. Recurrent selection as a means of modifying combining ability in corn. Agron. J. 43:311-315.

----. 1964. Modification of the ear-to-row procedure for the improvement of maize populations. Crop Sci. 4:227-228.

----. 1967. Mass selection for prolificacy in maize. Der Zuchter 37:185-187.

----, and C.O. Gardner. 1961. Heterosis in intervarietal crosses in maize and its implication in breeding procedures. Crop Sci. 1:179-183.

----, and D.P. McGill. 1954. Gamete selection from selected zygotes in corn breeding. Agron. J. 46:147-150.

----, and N.E. Williams. 1967. Development of maize hybrids through selection among full-sib families. Crop Sci. 7:369-370.

Lorain, J. 1825. Nature and reason harmonized in the practice of husbandry. Carey and Lea, Philadelphia. 563 p.

Lush, J.L. 1945. Animal breeding plans. Iowa State University Press, Ames, Iowa. p. 442.

Matzinger, D.F., G.F. Sprague, and C.C. Cockerham. 1959. Diallel crosses of maize in experiments repeated over locations and years. Agron. J. 51:346-350.

Millang, A., and G.F. Sprague. 1940. The use of punched card equipment in predicting the performance of corn double-crosses. J. Am. Soc. Agron. 32:815-816.

Moll, R.H., and H.F. Robinson. 1967. Quantitative genetic investigations of yield in maize. Der Zuchter 37:192-199.

----, M.F. Lindsey, and H.F. Robinson. 1964. Estimates of genetic variances and level of dominance in maize. Genetics 49:411-423.

----, W.S. Salhuana, and H.F. Robinson. 1962. Heterosis and genetic diversity in variety crosses of maize. Crop Sci. 2:197-198.

----, and C.W. Stuber. 1971. Comparisons of response to alternative selection procedures initiated with two populations of maize (*Zea mays* L.). Crop Sci. 11:706-711.

----, and ----. 1974. Quantitative genetics — empirical results relevant to plant breeding. Adv. Agron. 26:277-313.

Morrow, G.E., and F.D. Gardner. 1893. Field experiments with corn. Ill. Agric. Exp. Stn. Bull. 25:173-203.

Murphy, R.P. 1942. Convergent improvement with four inbred lines of corn. J. Am. Soc. Agron. 34:138-150.

Neal, N.P. 1935. The decrease in yielding capacity in advanced generations of hybrid corn. J. Am. Soc. Agron. 27:666-670.

Nilsson-Leisner, G. 1927. Relation of selfed lines of corn to F_1 crosses between them. J. Am. Soc. Agron. 19:440-454.

Obilana, A.T., and A.R. Hallauer. 1974. Estimation of variability of quantitative traits in BSSS by using unselected maize inbred lines. Crop Sci. 14:99-103.

Otsuka, Y., S.A. Eberhart, and W.A. Russell. 1972. Comparison of prediction formulae for maize hybrids. Crop Sci. 12:325-331.

Panse, V.G. 1940. The application of genetics to plant breeding. II. The inheritance of quantitative characters and plant breeding. J. Genet. 40:283-302.

Paterniani, E. 1967. Selection among and within half-sib families in a Brazilian population of maize (*Zea mays* L.). Crop Sci. 7:212-216.

----, and J.H. Lonnquist. 1963. Heterosis in interracial crosses of maize (*Zea mays* L.). Crop Sci. 3:504-507.

Payne, K.T., and H.K. Hayes. 1949. A comparison of combining ability in F_2 and F_3 lines of corn. Agron. J. 41:383-388.

Penny, L.H., and S.A. Eberhart. 1971. Twenty years of reciprocal recurrent selection with two synthetic varieties of maize (*Zea mays* L.). Crop Sci. 11:900-903.

----, G.E. Scott, and W.D. Guthrie. 1967. Recurrent selection for European corn borer resistance in maize. Crop Sci. 7:407-409.

Pesek, J., and R.J. Baker. 1969. Desired improvement in relation to selection indexes. Can. J. Plant Sci. 49:803-804.

----, and ----. 1970. An application of index selection to the improvement of self-pollinated species. Can. J. Plant Sci. 50:267-276.

Pinnell, E.L., E.H. Rinke, and H.K. Hayes. 1952. Gamete selection for specific combining ability. p. 378-388. *In:* J.W. Gowen (ed.) Heterosis. Chapter 24. Iowa State College Press. Ames.

Rawlings. J.O. 1970. Present status of research on long- and short-term recurrent selection in finite populations — choice of population size. Meeting Working Group Quant. Genet. Se 22. Int. Union Forestry Res. Org. 2nd (Raleigh, NC). p. 1-15.

----, and C.C. Cockerham. 1962. Analysis of double-cross hybrid populations. Biometrics 18:229-244.

----, and D.L. Thompson. 1962. Performance level as criterion for the choice of maize testers. Crop Sci. 2:217-220.

Richey, F.D. 1922. The experimental basis for the present status of corn breeding. J. Am. Soc. Agron. 14:1-17.

----. 1924. Effects of selection on the yield of a cross between varieties of corn. USDA Bull. 1209:1-19.

----. 1927. The convergent improvement of selfed lines of corn. Am. Nat. 61:430-449.

----. 1945. Isolating better foundation inbreds for use in corn hybrids. Genetics 30:456-471.

----. 1947. Corn breeding, gamete selection, the Oenothera method and related miscellany. J. Am. Soc. Agron. 39:403-412.

----, and L.S. Mayer. 1925. The productiveness of successive generations of self-fertilized lines of corn and of crosses between them. USDA Bull. 1354:1-18.

----, and G.F. Sprague. 1931. Experiments on hybrid vigor and convergent improvement in corn. USDA Tech. Bull. 267. 22 p.

Robertson, A. 1960. A theory of limits in artificial selection. Proc. Royal Soc. 153, B:234-249.

Robinson, H.F., R.E. Comstock, and P.H. Harvey. 1951. Genetic and phenotypic correlations in corn and their implication to selection. Agron. J. 43:282-287.

Rojas, B.A., and G.F. Sprague. 1952. A comparison of variance components in corn yield trials. III. General and specific combining ability and their interaction with locations and years. Agron. J. 44:462-466.

Russell, W.A. 1971. Types of gene action at three gene loci in sublines of a maize inbred line. Can. J. Genet. Cytol. 13:322-334.

----. 1973. Improvement of maize populations for sources of inbred lines. VII Meeting of Eucarpia Maize & Sorghum Section.

----, and S.A. Eberhart. 1970. Effects of three gene loci in the inheritance of quantitative characters in maize. Crop Sci. 10:165-169.

----, and ----. 1975. Hybrid performance of selected maize lines from reciprocal recurrent selection and testcross selection programs. Crop Sci. 15:1-4.

----, ----, and U.A. Vega O. 1973. Recurrent selection for specific combining ability for yield in two maize populations. Crop Sci. 13:257-261.

----, G.F. Sprague, and L.H. Penny. 1963. Mutations affecting quantitative characters in long-time inbred lines of maize. Crop Sci. 3:175-178.

----, and A.H. Teich. 1967. Selection in *Zea mays* L. by inbred line appearance and test-cross performance in low and high densities. Iowa Agric. Home Econ. Exp. Stn. Res. Bull. 552:919-946.

Scott, G.E., F.F. Dicke, and G.R. Pesho. 1966. Location of genes conditioning resistance of corn to leaf feeding of the European corn borer. Crop Sci. 6:444-446.

Shamel, A.D. 1905. The effect of inbreeding in plants. USDA Yearb. Agric. p. 377-392.

Shull, G.H. 1908. The composition of a field of maize. Am. Breed. Assoc. Rep. 4:296-301.

----. 1909. A pure line method of corn breeding. Am. Breed. Assoc. Rep. 5:51-59.

Silva, J.C. 1974. Estimation of genetic and environmental variances and covariances in the BSSS maize (*Zea mays* L.) population. Ph.D. dissertation. Iowa State Univ. Ames, Iowa p. 155.

Singleton, W.R., and O.E. Nelson. 1945. The improvement of naturally cross-pollinated plants by selection in self-fertilized lines. IV. Combining ability of successive generations of inbred sweet corn. Conn. Agric. Exp. Stn. Bull. 490:458-498.

Smith, L.H. 1909. The effect of selection upon certain physical characters of the corn plant. Ill. Agric. Exp. Stn. Bull. 132.

----, and A.M. Brunson. 1925. An experiment in selecting corn for yield by the method of the ear-row breeding plot. Ill. Agric. Exp. Stn. Bull. 271:561-583.

----, and ----. 1928. Experiments in crossing varieties as a means of improving productiveness in corn. Ill. Agric. Exp. Stn. Bull. 306:374-386.

Sopher, C.D., R.J. McCracken, and D.D. Mason. 1973. Relationships between drought and corn yields on selected south Atlantic coastal plain soils. Agron. J. 65:351-354.

Sprague, G.F. 1939. An estimation of the number of top-crossed plants required for adequate representation of a corn variety. J. Am. Soc. Agron. 31:11-16.

----. 1942. Production of hybrid corn. Iowa Agric. Exp. Stn. Agric. Ext. Serv. Bull. p. 48.

----. 1946. Early testing of inbred lines of corn. J. Am. Soc. Agron. 38:108-117.

----. 1952. Early testing and recurrent selection. p. 400-417. *In* J.W. Gowen (ed.) Heterosis. Chapter 26. Iowa State College Press. Ames, Iowa.

----. 1972. Genetic vulnerability to diseases and insects in corn and sorghum. Proc. 26th Annu. Corn and Sorghum Research Conf. p. 96-104.

Sprague, G.F., and W.T. Federer. 1951. A comparison of variance components in corn yield trials. II. Error, year × variety, location × variety and variety components. Agron. J. 43:535-541.

----, and M.T. Jenkins. 1943. A comparison of synthetic varieties, multiple crosses, and double-crosses in corn. J. Am. Soc. Agron. 35:137-147.

----, and P.A. Miller. 1950. A suggestion for evaluating current concepts of the genetic mechanisms of heterosis in corn. Agron. J. 42:161-162.

----, and ----. 1952. The influence of visual selection during inbreeding on combining ability in corn. Agron. J. 44:329-331.

----, W.A. Russell, and L.H. Penny. 1960. Mutations effecting quantitative traits in the selfed progeny of double-monoploid maize stocks. Genetics 45:855-866.

----, ----, ----, T.W. Horner, and W.D. Hanson. 1962. Effect of epistasis on grain yield in maize. Crop Sci. 2:206-208.

----, and L.A. Tatum. 1942. General vs. specific combining ability in single crosses of corn. J. Am. Soc. Agron. 34:923-932.

----, and W.L. Thomas. 1967. Further evidence of epistasis in single and three-way cross yields of maize. Crop Sci. 7:355-356.

Stadler, L.J. 1944. Gamete selection in corn breeding. J. Am. Soc. Agron. 36:988-989.

Stuber, C.W., and R.H. Moll. 1969. Epistasis in maize (Zea mays L.). I. F_1 hybrids and their S_1 progeny. Crop Sci. 9:124-127.

----, and ----. 1971. Epistasis in maize (Zea mays L.). II. Unselected populations. Genetics 67:137-149.

----, and ----. 1974. Epistasis in maize (Zea mays L.). IV. Crosses among lines selected for superior intervariety cross performance. Crop Sci. 14:314-317.

----, ----, and W.D. Hanson. 1966. Genetic variances and interrelationships of six traits in a hybrid population of Zea mays L. Crop Sci. 6:455-459.

----, and W.P. Williams, and R.H. Moll. 1973. Epistasis in maize (Zea mays L.). III. Significance in prediction of hybrid performance. Crop Sci. 13:195-200.

"Student." 1934. A calculation of the minimum number of genes in Winters selection experiments. Ann. Eng. 6:77-82.

Subandi, W.A. Compton, and L.T. Empig. 1973. Comparison of the efficiencies of selection indices for three traits in two variety crosses of corn. Crop Sci. 13:184-186.

Torregroza, M., F. Arboleda, J.A. Rivers, C. Diaz, and E. Arias F. 1973. Evaluacion de la seleccion masal por prolificidad en 2 poblaciones. XIII. Reunion Anual del Programa Cooperativo Centro-americano para el Mejoramiento de los Cultivos Alimenticios.

Troyer, A.F., and A.R. Hallauer. 1968. Analysis of a diallel set of early flint varieties of maize. Crop Sci. 8:581-584.

Wang, Jen Yu. 1960. A critique of the heat unit approach to plant response studies. Ecology 41:785-790.

Webel, O.D., and J.H. Lonnquist. 1967. An evaluation of modified ear-to-row selection in a population of corn (Zea mays L.). Crop Sci. 7:651-655.

Wellhausen, E.J. 1952. Heterosis in a new population. p. 418-450. In: J.W. Gowen (ed.) Heterosis. Chapter 27. Iowa State College Press. Ames, Iowa.

----, L.M. Roberts, and X.E. Hernandez. 1952. Races of maize in Mexico. Bussey Inst. Harvard. p. 9-223.

Widstrom, N.W. 1974. Selection indexes for resistance to corn earworm, based on realized gains in corn. Crop Sci. 14:673-675.

Williams, C.G., and F.A. Welton. 1915. Corn experiments. Ohio Agric. Exp. Stn. Bull. 282:69-109.

Winters, F.L. 1929. The mean and variability as affected by continuous selection for composition in corn. J. Agric. Res. 39:451-476.

Woodworth, C.M., E.R. Leng, and R.W. Jugenheimer. 1952. Fifty generations of selection for protein and oil in corn. Agron. J. 44:60-65.

Wright, J.A., A.R. Hallauer, L.H. Penny, and S.A. Eberhart. 1971. Estimating genetic variance in maize by use of single and three-way crosses among unselected inbred lines. Crop Sci. 11:690-695.

Wright, S. 1922. The effects of inbreeding and cross-breeding on guinea pigs. III. Crosses between highly inbred families. USDA Tech. Bull. 1121:1-61.

Wu, S.K. 1939. The relationship between the origin of selfed lines and their value in hybrid combinations. J. Am. Soc. Agron. 31:131-140.

Yermanos, D.M. 1952. Comparative efficiency of three methods of isolating lines of corn. M.S. Thesis. Iowa State College. Ames, Ia.

Young, S.S.Y. 1964. Multi-stage selection for genetic gain. Heredity 19:131-145.

Zuber, M.S. 1973. Evaluation of progress in selection for stalk quality. Proc. 28th Annu. Corn Sorghum Research Conf. Am. Seed Trade Assoc. p. 110-122.

----, M.L. Fairchild, A.J. Keaster, V.L. Fergason, G.F. Krause, E. Hilderbrand, and P.J. Loesch, Jr. 1971. Evaluation of 10 generations of mass selection for corn earworm resistance. Crop Sci. 11:16-18.

Chapter 7 Breeding Special Industrial and Nutritional Types

D.E. ALEXANDER
University of Illinois
Urbana, Illinois

ROY G. CREECH
Mississippi State University
Mississippi State, Mississippi

Corn is no longer "just corn." Plant breeders have produced diverse types that already have, or that ultimately can be of, commercial consequence. These variants involve modifications of quality and quantity — quality of protein, quality of starch, quality of oil, as well as their relative amounts. Thus corn has been, or can be, bred to more precisely meet the specific nutritional requirements of animals or to meet particular industrial needs — or to meet the whims of human preference. The high sugar vegetable corns and popcorn represent the latter.

The development of special nutritional and industrial types could not have occurred had the theorist and the empiricist not been working with corn. The discovery and maintenance of mutants by the geneticist and the eventual use of those mutants exemplify science at its best.

HIGH OIL CORN

The concentration of oil, henceforth referred to as oil content, exhibits wide variation in corn. Typically ears from open-pollinated, Corn Belt cultivars range from 3.5 to 6.0% oil, although higher and lower values may also be encountered. Great extremes in the species also are possible. The 'Illinois Low Oil' strain contains less than 0.75% oil, and the mean oil percent of the 'Ilinois High Oil' strain reached 18.2% in its 75th generation of selection, and ears as high as 20.6% have been encountered (J.W. Dudley, personal communication).

Realized heritabilities in early generations of selection in heterogeneous populations often are 0.6 to 0.7. The opportunity to increase oil content is inescapably obvious — great variation exists and much of it is heritable.

However, a fundamental question exists whether corn should be bred for higher oil content. This is best advanced by the often-repeated phrase "corn is a starch crop." However one cannot ignore the possibility that higher oil corn may be more valuable as animal feed or as an industrial crop. Furthermore it potentially may be a more effective trapper of energy than ordinary corn, i.e., calorie yield/ha may be higher.

Impetus came to oil breeding programs as a consequence of the development of wide-line nuclear magnetic resonance spectroscopy as a quantitative analytical tool (Bauman et al., 1963; Alexander et al., 1967). The scheme permits nondestructive analysis of samples as small as a single kernel. Compared to gravimetric analysis cost, the method is inexpensive. Large numbers of samples may be analyzed in a relatively short time since as little as 2 seconds may be required. Breeding for high oil content therefore becomes an appealing effort because progress can be rapid and relatively inexpensive.

Why should high oil corn be bred? We shall attempt to answer this question from the viewpoints of 1) a swine feeder, 2) a wet miller, and 3) a corn breeder.

High Oil Corn as a Feed

The value of corn as a feed is a function of its chemical composition and of the relative digestibility of these components. We shall ignore all components except oil, protein, and carbohydrates in our consideration of the question. We also shall consider all carbonaceous compounds as a homogeneous group on the grounds that changes in oil level will have no effect on the proportions of sugars, kinds of starch, and fiber. Carbohydrates are defined in this context as being that which remains after oil and protein have been removed.

Nordstrom et al. (1972) fed 7% oil corn to growing-finishing pigs and found that the amount of grain required for each unit of gain in live weight was 5 to 6% less than in diets involving ordinary corn. It may be of some practical interest also to point out that less soybean meal was required in the high oil corn diets. The amount of reduction was substantial, i.e., 22% less for the 16% protein diet and 41% less for the 13% diet.

Increased intake of oil by swine has adverse effects on bacon quality. The bacons become "soft," presumably because of direct incorporation of unsaturated fatty acids into depot fats. However, Nordstrom (1972) found that 7% oil corn had little effect on quality although diets containing 16% oil through addition of corn oil produced carcasses that were unacceptable for conventional processing.

Experiments conducted by Lynch et al. (1972) at the Illinois Station suggest that the advantage of higher oil corns in swine diets were negligible. The diets under comparison contained 4% and 5.9% oil and it might be expected that the anticipated caloric advantage of the higher oil strain was not large enough to be detected by the trial.

High Oil Corn as an Industrial Crop

Corn oil is more valuable than corn starch. As long as this remains true, higher oil corns shall be potentially attractive to wet millers in particular. This statement, however, greatly simplifies the complexities of the price structure of wet milling products. If, however, we consider the value of the components of corns of different oil content, some insight into the relative potential values can be derived.

An attempt has been made to compute the increased values of higher oil corns, relative to corn of 4.5% oil content, as a function of oil and starch

Table 1—Increase in value (USA $) per quintal of high oil corns over 4.5% oil corn as functions of oil and starch prices. (Values in parentheses are based on pounds, bushels)†

Oil	Oil	Starch		
		15.4 (0.07)	26.4 (0.12)	39.7 (0.18)
%		$/kg		
6.0	0.55 (0.25)	0.43 (0.11)	0.32 (0.08)	0.16 (0.04)
	0.77 (0.35)	0.67 (0.17)	0.55 (0.14)	0.39 (0.10)
	0.99 (0.45)	0.90 (0.23)	0.79 (0.20)	0.63 (0.16)
7.0	0.55 (0.25)	0.71 (0.18)	0.51 (0.13)	0.28 (0.07)
	0.77 (0.35)	1.10 (0.28)	0.91 (0.23)	0.67 (0.17)
	0.99 (0.45)	1.50 (0.38)	1.30 (0.33)	1.06 (0.27)
8.0	0.55 (0.25)	1.42 (0.26)	0.71 (0.18)	0.39 (0.10)
	0.77 (0.35)	1.54 (0.39)	1.26 (0.32)	0.94 (0.24)
	0.99 (0.45)	2.13 (0.54)	1.85 (0.47)	1.50 (0.38)
9.0	0.55 (0.25)	1.30 (0.33)	0.94 (0.24)	0.51 (0.13)
	0.77 (0.35)	2.01 (0.51)	1.69 (0.43)	1.22 (0.31)
	0.99 (0.45)	2.72 (0.69)	2.36 (0.60)	1.93 (0.49)
10.0	0.55 (0.25)	1.58 (0.40)	1.41 (0.29)	0.59 (0.15)
	0.77 (0.35)	2.44 (0.62)	2.01 (0.51)	1.54 (0.39)
	0.99 (0.45)	3.31 (0.84)	2.87 (0.73)	2.40 (0.61)

†Assumptions: 1) dry matter computed on basis of no. 2 corn (15.5% H_2O) and 2) increase in oil content equals decrease in starch content.

values (Table 1). At least two variables have been deliberately neglected: 1) the effect of oil levels on relative level of recovery of gluten feed and of gluten meal, by-products of different market value, and 2) the cost of providing identity-maintained corn to the miller.

Clearly, higher oil corns can be considerably more valuable, from a wet milling standpoint, than ordinary corns containing 4.5% oil. One would anticipate that an advantage would continue to exist because oil is more valuable than starch.

Effects of Selection on Oil Content

Dudley et al. (1974) reported that oil content in corn can range from 0.1 to 19.6% as shown in individual ears of the Illinois Low Oil and Illinois High Oil strains. Furthermore ample evidence exists that oil content is highly heritable. Since the NMR method permits rapid and nondestructive analysis, it is somewhat puzzling that breeders have not considered breeding for high oil content worthy of attention.

Sprague et al. (1952) were able to increase oil content of Iowa Stiff Stalk Synthetic from 4.2 to 7.0% in two cycles of recurrent selection. Hume et al. (1914) reported some success in selecting for oil content in 'Minnesota 13,' an open-pollinated cultivar. However the work apparently did not continue beyond the first cycle.

Attempts were made in the 1930's and 1940's by Woodworth, Jugenheimer, and others to recover high oil versions of several inbreds by outcrossing to Illinois High Oil and subsequently backcrossing and selfing. These efforts failed to produce hybrids equal in performance to lower oil types. Perhaps the failure can be attributed to the fact that rapid improvement in yield of ordinary hybrids occurred simultaneously, hence the earlier US13 types were obsolete before oil hybrids of similar parentage could reach commercial production. It is more likely, however, that the relationship among recovered high oil lines was great enough to depress yield in the

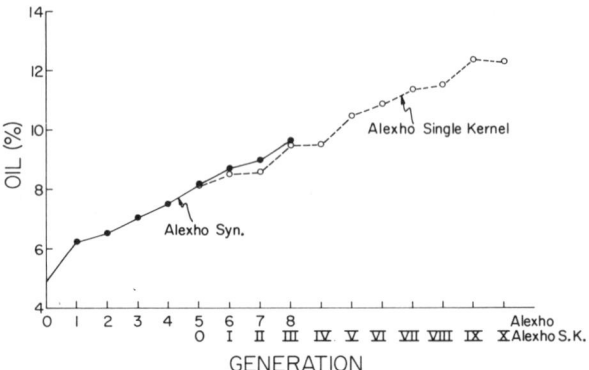

Fig. 1. Effect of selection for higher oil content in Alexho Synthetic.

derived hybrids. Selection in several unrelated gene pools has subsequently been carried out (Alexander, 1971). Two will be described.

A base population consisting of 56 open-pollinated cultivars was random-mated several generations. Several hundred plants were selfed and ears harvested from the better 300 to 500. Individual ears were analyzed and the higher 50 to 70 were recombined. Eight cycles of selection were carried out. At cycle five, a single kernel intrafamily scheme was inaugurated. Twenty kernels from each of 77 ears were analyzed by NMR and the higher three were planted ear-to-row. At anthesis, the better plants from each row were pollinated with a mixture of pollen collected from selected plants from all rows. This procedure, because of its success in increasing oil content, has been continued and the original scheme abandoned. A plot of the progress encountered is shown in Fig. 1. Progress over 15 generations has averaged 0.5%/cycle.

Single kernel selection, similar to that described above, has been carried out in 'Reid Yellow Dent,' an open-pollinated cultivar. From 100 to 300 kernels from each of 400 ears are analyzed each cycle. Progress is summarized in Table 2 (Alexander, unpublished data). Progress over three generations of selection has averaged about 0.8%/cycle, although gains appear to be smaller in all cycles beyond the first.

Performance of High Oil Hybrids

Lines developed by selfing and sibbing from Alexho Synthetic and Alexho Single-Kernel Synthetic have been tested in combinations with Iowa Stiff Stalk Synthetic lines and particularly with R802A, a backcross recovery of B37 and Illinois High Oil. Performance of the better hybrids is shown in Table 3 (Alexander, unpublished data).

Calorie Content of High Oil Corn

Since oil is higher in calorie content than an equal amount of carbohydrate or protein, a smaller amount of high oil corn will be isocaloric to a larger amount of ordinary corn. An equation that assists in computing this

Table 2—Progress in selecting for higher oil content in Reid Yellow Dent by single-kernel method

Cycle	$\bar{X}$ oil content	$\bar{X}$ oil content of selected kernels	Selection differential	h^2
		%		
0	4.50	6.18	1.85	0.68
I	5.75	7.34	1.59	0.38
II	6.35	8.46	2.11	0.38
III	7.16	9.00	1.84	
IV	7.80			

Table 3—Performance of high oil and standard hybrids, Urbana, 1973†

	Yield‡	Oil	H$_2$O in grain	Protein	Lodged
	quintals/ha		%		
R802A × R805	97 (154)	8.3	22	12.5	0
R802A × R806	108 (172)	8.5	23	12.7	1
R801^{HT} × 10120-8	107 (171)	7.0	22	12.8	5
R805 × Mo17	109 (173)	6.0	24	12.1	1
R806 × N28	110 (175)	6.7	26	10.3	2
Mo17 × N28	118 (188)	3.3	25	11.1	1
Commercial hybrid	101 (161)	6.3	26	11.6	3
Four commercial hybrids§	108 (172)	4.2	24	10.4	4

†L.S.D. = 11.3 quintals/ha (18 bu/acre) C.V. = 8.7% ‡bu/acre in parentheses.
§Mean of four popular commercial hybrids.

value is:

$$\text{Isocaloric amount of high oil corn} = \frac{452.15}{9.4\,(\%\text{ oil}) + 4.15\,(\%\text{ carbohydrate}) + 5.65\,(\%\text{ protein})}$$

where 452.15 is equal to the calorie content of 100 g of corn containing 4.5% oil and 10% protein and the denominator describes the composition of the high oil corn.

For example, if one computes the isocaloric amounts of 8 and 4.5% oil corn, and each is assumed to contain 10% protein, then 95.79% as much of the higher oil corn will contain as many calories as the lower oil corn.

Although bomb calorimeter data are of value, computations on the digestible energy of corns of different oil content may be more pertinent to animal feeders. Digestible energy, in this context, is defined as being the difference between bomb-calorimeter values and energy in feces, i.e., that energy utilized by swine weighing 11 to 34 kg (25 to 75 pounds). The equation becomes:

$$\text{Isocaloric digestible energy of high oil corn} = \frac{422.50}{9\,(\%\text{ oil}) + 4\,(1\text{-}\%\text{ oil})}$$

where 422.5 is equal to the digestible energy of 100 g of corn containing 4.5% oil and 10% protein and the denominator describes the digestible calorie value of the particular high oil corn.

Digestible energy values for corns of differing oil content are shown in

Table 4—Digestible energy† of corns of different oil content‡

Oil	Relative Calories	Total kcal/kg
%		
2.5	98.3	4,149
4.5	100.0	4,220
7.0	102.1	4,309
8.0	103.0	4,345
9.0	103.8	4,380
10.0	104.6	4,416

†Digestible energy—Gross energy, as determined by bomb calorimeter, minus energy in feces (derived from feeding trials with swine of 11 to 34 kg (25 to 75 lbs) live weight.
‡Assumptions: a. Oil = 7.62 kcal/g
b. Carbohydrate = 4.06 kcal/g
c. Protein = 4.06 kcal/g

Table 5—Isocaloric amounts of corns of differing oil contents

Oil	Bomb calorimeter	Digestible energy†
%		
2.5	102.0	102.2
4.5	100.0	100.0
7.0	96.87	97.1
8.0	95.79	96.0
9.0	94.79	94.9
10.0	93.81	93.9

†As determined from feeding experiments with 11 to 34 kg (25 to 75 lbs) swine.

Table 4. Isocaloric digestible energy and calorimeter energy values for corns of different oil contents are shown in Table 5. Bomb calorimeter values agree closely with digestible energy values obtained from feeding trials with young swine.

Oil Quality

Oil quality is usually considered to be a function of its melting point and of its keeping qualities, i.e., freedom from rancidity after heating and storage. However, in this context, we shall consider it as a function of the relative amounts of unsaturated and saturated fatty acids. A high quality oil in this sense, is one having a relatively high proportion of linoleic acid and lesser amount of of oleic, palmitic, and stearic acids. Linolenic acid, however, is an undesirable component of cooking oils, even though it is an unsaturate, because upon oxidation, it produces a varnish-like product.

Lofland and Quackenbush (1954) found that iodine number, a measure of unsaturation, and oil content were inversely related. The correlation among 392 samples, ranging from 1.13 to 13.80% oil was -0.69. Quackenbush et al. (1963) found that iodine number ranged from 111 to 151 in 125 inbreds well known to corn breeders of that time. Oil percent in the same group of lines ranged from 1.2 to 5.7%. Analysis of their data shows that the coefficient of correlation between iodine number and oil percent is +0.13, a value not significantly different from zero. Alexander (unpublished data) found significant negative correlations between percent linoleic acid and percent oil in cycles of synthetics selected for oil content. For original population and 11 cycles in Alexho Synthetic, $r = -0.60$ and for 6 cycles in Synthetic D.O., $r = -0.81$.

Alexander (unpublished data) found a significant negative correlation of -0.18 between percent linoleic acid and percent oil content among 342 inbreds, primarily of U.S. origin. The range in linoleic was 32.6 to 71.6% and in oil, 2.0 to 10.2%. Although the data from several sources suggest that a biological relationship exists between high oil content and low levels of unsaturates, exceptions exist. R802A, a derivative of (B37 × IHO)B37, combines a reasonably high oil content with a desirable level of linoleic acid, i.e., 7% oil and approximately 60% linoleic acid. Jellum (1970) found that linoleic acid percentage ranged from 19 to 71% in oils of exotic cultivars.

Genetic constitution has a greater effect on fatty acid composition and on quantity of oil than the environment. Homozygous stocks monitored for many years at the Illinois Station are remarkably stable and show little variation. Poneleit and Alexander (1965) proposed that a single gene (ln) was responsible for the difference in level of linoleic acid of R84, an inbred related to Illinois High Oil, and Illinois High Oil. De la Roche (1970) extended the study and proposed that a single locus controlled the 20 percentage point difference in level of linoleic acid of inbreds C103 and R84.

The biosynthesis of the unsaturated fatty acids oleic and linoleic is thought to proceed in a pathway:

$$12{:}0 \to 14{:}0 \to 16{:}0 \to 18{:}0 \to 18{:}1 \to 18{:}2$$

One would conclude, a priori, that the transformation of oleic (18:1) to linoleic (18:2) would be effected by a relatively simple genetic mechanism, i.e., less complex than the inheritance of quantity of oil, for example. Furthermore one would anticipate a reciprocal relationship in relative amounts of the two. Both of these expectations are borne out by the genetic studies cited before.

Kannangara et al. (1973) have suggested that linolenic acid (18:3) is not derived from linoleic acid (18:2) but is synthesized through a separate pathway:

$$12{:}0 \to 12{:}1 \to 12{:}2 \to 12{:}3 \to 14{:}3 \to 16{:}3 \to 18{:}3$$

Corn oil is relatively low in linolenic acid although some inbreds are known to carry as much as 1.5% of it. It might be expected that mutation at loci responsible for any of the transformational steps could block completely or partially restrict its synthesis, or conversely, to increase the relative amount of it.

Although little or no effort has been made to establish gene pools that carry favorable alleles for high levels of unsaturates, or to consciously breed hybrids with high quality oils, little doubt can exist that corns may be bred with substantially higher or lower levels. It is interesting to note that linoleic acid content of commercial corn oil produced in the US prior to 1964 averaged 58.7% but by 1968 had increased to 61.9%. Presumably the change came about through increased production of hybrids genetically predisposed to synthesize a greater proportion of the more desirable fatty acid. Should breeders ignore this attribute, it is conceivable that the introduction of new inbreds into commercial production could bring shifts to higher, or lower, levels.

Inbreds developed in the northern US tend to be higher in linoleic acid than those originating from southern cultivars. For example, the mean for 169 northern lines was 58%, for 63 southern lines, 48% (Alexander, unpublished data).

Almost all fatty acids in corn oil are esterified with glycerol to form triglycerides, although free fatty acids exist in trace amounts in freshly extracted oil. Stereochemical analysis of several plant storage lipids has shown that the distribution of the different acids is not at random among the three positions, and that the more unsaturated ones tend to occupy the middle (number 2) position. De la Roche (1970) and de la Roche et al. (1971) have shown that the distributions are different in different corn inbreds and further, that the distributions are under genetic control.

The peculiarities of triglyceride stereospecificity is of some practical consequence:
1. Oxidation of double bonds is less likely to occur during heating and perhaps under unfavorable grain storage conditions in fatty acids carried at the 2-position than for those at the 1- and 3-positions.
2. Fatty acids at the 1- and 3-positions are removed by lipase in the intestine and the monoglyceride passes into the blood stream. Some of the monoglycerides are acted upon by enzymes of the animal and new triglycerides are produced. Depot fats composed of triglycerides derived from plant triglycerides high in unsaturates would therefore be expected to have altered physical qualities such as a lowered melting point.

Although peripheral to the main thrust of this report, the amount and kinds of tocopherol (vitamin E) carried in corn is of some consequence. First of all, vitamin E is an important conponent of animal diets. To illustrate: in swine, a deficiency of vitamin E and selenium is associated with white muscle disease. Second, the keeping qualities of stored corn, i.e., the avoidance of oxidation of double bonds of the unsaturates, should be materially improved by the presence of adequate amounts of the tocopherols, particularly the gamma isomer.

Quackenbush et al. (1963) found that vitamin E ranged from 0.03 to 0.33% of the oil of a group of 125 corn belt inbreds. Levy (1973) found that the inheritance of vitamin E in corn was polygenic and that maternal effects, probably cytoplasmic in nature, affected levels of the gamma isomer. He also found a correlation of +0.60 between the amount of linoleic acid and vitamin E/g of oil in a series of 18 dent inbreds.

PROTEIN: QUANTITY AND QUALITY

Although corn is ordinarily thought of as being a carbonaceous feed, it annually produces more protein than the soybean (*Glycine max*(L.) Merr.] in the US. Unfortunately ordinary corn protein is less valuable in the diets of nonruminants than soybean protein largely because of a deficiency of lysine and tryptophan.

Protein concentration in the grain, hereafter referred to as protein content, is influenced importantly by cultural conditions as well as by heredity. The magnitude of effects are very different, however. Earley and DeTurk (1948) found that protein content of a commercial hybrid planted at a rate of 56,800 plants/ha (23,000 plants/acre) on soil not fertilized with nitrogen was 8.8%. Protein content was 10.8% on plots on which 269 kg/ha (240 lbs./acre) of N had been added. Many studies have been conducted with planting rates, fertility levels, hybrids, etc., as variables, with similar conclusions. Genetic constitution, however, can more profoundly affect protein level.Dudley et al. (1974) reported that the Illinois High Protein strain reached 26.6% protein in its 70th generation of selection, 12 S.D.'s above the mean of the original population. The mean protein content of the Illinois Low Protein strain was 4.4% after 70 generations of selection.

Modified Protein Corn

A remarkable discovery, made by Mertz et al. (1964), has added a new and important dimension not only to maize breeding but to other monocotyledonous species as well. That discovery, of course, was that a single gene (opaque-2) affected the synthesis of zein and that the amino acid profile of the endosperm was modified in a manner that vastly improved its biological quality. Not only were the relative amounts of lysine and tryptophan increased, but the iso-leucine-leucine ratio was narrowed. Since 1964, mutants with similar gross effects have been found in barley (*Hordeum vulgare* L.) and sorghum [*Sorghum bicolor* (L.) Moench] as well as a second gene (floury-2) in corn (Nelson et al., 1965). It can be anticipated with some confidence that low-prolamine mutants will be found in other diploid monocots provided that a reasonable effort is mounted to identify them.

The opaque-2 mutant in particular has some promise in practical corn agriculture. Limited marketing of opaque-2 hybrids, dubbed "high lysine" hybrids in the US, has been carried out in Brazil, Colombia, and the US. Vigorous research programs have been underway, particularly in the eastern European corn-producing countries and in the USSR. It can be expected that commercial production will rise, particularly in those countries that are unable to produce adequate supplies of protein supplements for animal feeding or that depend heavily on corn as a human food.

Adoption of opaque-2 hybrids in commercial agriculture in the US depends primarily on three factors: 1) agronomic performance relative to widely used dent hybrids, 2) the relative costs of corn and soybean meal, and 3) the availability of hybrid seed. A swine feeder would be disinclined to adopt new methods, i.e., the production of new kinds of hybrids and the development of strategies for feeding the new corns, if corn and soybean meal were of equal cost. On the other hand, if soybean meal were relatively more expensive than corn, he would be more likely to adopt high lysine corn. Production of seed is not likely to occur if hybrid performance is unsatisfactory. Seed producers also may be reluctant to invest in the research necessary to develop the new hybrids since they already produce dent hybrids that meet existing demands.

Even in the face of several compelling negative factors, commercial production of opaque-2 hybrids has proceeded in the US with reasonable progress. In 5 years (1970-75), US production of opaque-2 corn rose from essentially zero to more than an estimated 1.3 million metric tons in 1975.

Performance of Modified-protein Corns

Lambert et al. (1969) compared field performance of opaque-2 hybrids and their normal counterparts. They found that the opaque-2 versions were significantly lower in yield (8%) and w/hundred kernels (5%). Opaque-2 hybrids also were higher in grain moisture (20%), oil content (13%), and protein content (2%). However two opaque-2 hybrids of a group of 10 were higher in yield than their normal counterparts. Plots were harvested by a plot combine similar to harvestors used by farmers. The proportion of damaged kernels was much greater in the opaque-2 hybrids (89% greater) than in the normal counterparts.

Bauman (1975) reported that some opaque-2 hybrids were competitive with commercial hybrids. Brown (1975) concluded that although recently developed opaque-2 hybrids were much superior to earlier experimental hybrids, they were still inferior in yield and possessed undesirable kernel characteristics. He further proposed that these problems might be overcome by the breeding of modified endosperm types with higher test weight and presumably with greater resistance to insect and disease damage as well as to machine harvest and handling damage.

Virtually all breeders have encountered segregants in homozygous o_2 backgrounds that possess islands or bands of vitreous endosperm. In fact, some are phenotypically similar to dent or flint corns and are difficult to distinguish from them. Epistatic effects involving the o_2 allele and the fl_2, su_2, and wx have been reported. Glover et al. (1975) reported that the kernel density of the double mutant $o_2 su_2$ was nearly equal to that of ordinary dents, and perhaps of even more significance, that the lysine/protein ratio of the endosperm was at least as desirable as that of opaque-2 itself.

Although high-quality protein in a vitreous phenotype is unquestionably desirable in minimizing damage during harvest and handling, and may be preferred in some markets, there is no conclusive evidence that it is essential for high yield. Dudley et al. (1971) found that the correlation between kernel density and yield was not different from zero in opaque-2 synthetics, but that a significant positive correlation (0.25 and 0.30) existed in two floury-2 synthetics. Bauman (1975) has questioned the necessity for developing vitreous types for the temperate zones.

Zuber (1975) reported in a preliminary way that recurrent selection for higher lysine content in three open-pollinated dent cultivars resulted in an increase in percentage of lysine in the grain. However when the original populations and the cycles were grown in the same year, the differences largely disappeared except in one cultivar (Zuber, personal communication). Bauman (personal communication) also found minimal variation of lysine concentration in the whole grain of dent synthetic cultivars.

Opportunities unquestionably exist to further increase the proportions of proteins that carry more desirable amino-acid patterns than those of ordinary opaque-2 maize. Glover et al. (1975) pointed out that zein concentration in the endosperm can be reduced by appropriate genetic manipulation. For example, zein fraction II in o_2/o_2 stocks composes approximately 30% of the protein, but in the double mutant $o_2 su_2$, only 18.5% of the protein is zein. Low levels of zein also have been found in opaque-2 homozygotes. For example, the percentage of zein in whole kernel protein is approximately 22% in R802, a high oil opaque-2 inbred (Alexander, unpublished data). Although not directly comparable to the endosperm data of Glover, the R802 data suggest that further improvement in protein quality of the whole grain can be expected should appropriate breeding schemes be directed towards that end.

Feeding Value of Opaque-2 Corn

Mertz (1966) reported that ". . . for the young rat, opaque-2 maize proteins are well balanced, with only a minor deficiency in lysine and tryptophan." Pickett (1966) found that opaque-2 corn, appropriately supplemented with vitamins and minerals, was adequate to support normal growth of finishing pigs [54.5 kg (120 lbs.)]. Rogler (1966) found that opaque-2 corn alone did not produce normal growth in chicks, but when supplemented with methionine, produced better gains and feed conversion than normal corn at sub-optimal protein levels. Bressani (1966) concluded that for children the nutritive value of opaque-2 protein was about 90% that of skim milk. Clark (1966) concluded that 300 g of opaque-2 corn/day by an adult male would provide 93% of the protein requirement but only 40% of the calorie requirement. In contrast, consumption of 500 to 600 g of ordinary corn would provide at least 80% of the required calories but only 70% of the protein requirement. Ingestion of more than 300 g of opaque-2 corn would be expected to further improve N retention.

A great deal of research carried out since these early reports confirms that opaque-2 corn grain can be an important source of high quality protein in the diets of nonruminants (Bressani, 1975; Jensen et al., 1969; Maner, 1975; Pradilla et al., 1975).

CARBOHYDRATES

Characteristics of Endosperm Carbohydrates

Many studies have been conducted on the type and quantity of carbohydrates in corn kernels. Much of the early work involved open-pollinated cultivars and was concerned with change in sugar content during kernel development. Post harvest transformations in sweet corn kernels were also investigated extensively because of their significance in the palatability of the processed product.

Straughn (1907) reported that the sugar content of mature kernels of different cultivars of sweet corn were similar; however, he noted that the degree of wrinkling of the kernels was associated with amount of sugar. Highly wrinkled kernels were slightly higher in sugar than less wrinkled kernels. The finding that sugar content of corn in the edible stage changed rapidly during the first few hours after removal from the plant was of special interest. Straughn and Church (1909) and Appleman and Arthur (1919) established that the loss in sugar was caused by both respiration and transformation into polysaccharides, chiefly starch. Doty et al. (1945) conducted similar experiments on 39 inbred and hybrid strains of sweet corn over a 4-year period. They concluded that the genetic constitution of the plant affected the conversion of sucrose to polysaccharides during storage. Appleman and Eaton (1921) found a progressive decrease in total sugars, an increase in fat, and a very high increase in starch during the development of kernels from open-pollinations. Culpepper and Magoon (1924, 1927) found that amount of sugars increased up to 15 days after pollination and then decreased slowly during the latter stages of kernel development. Reducing sugars were highest at the initial stages and decreased throughout growth and development. Sucrose increased rapidly until the 15th day and then decreased slowly during the next 15 days. In this case, most of the sugar was sucrose. Polysaccharides increased continuously during the development of the ear. Sweet corn cultivars, unlike standard dent corn, were high in watersoluble polysaccharides (WSP).

Hassid (1945) presented a review of the chemical structure and properties of the two components of starch, amylose and amylopectin, and animal glycogen. Parker (1935), Hassid and McCready (1941), and Sumner and Somers (1943) described the water-soluble polysaccharide that had been observed in sweet corn. Although problems occurred in sample preparations, apparently due to presence of degradative enzymes, they were able to characterize the WSP fraction of sweet corn and demonstrate that it possessed a branched structure similar to, but more highly branched than, amylopectin. This highly branched fraction was termed phytoglycogen because of its similarity to animal glycogen.

Lampe and Myers (1925), Evans (1941), Andrew et al. (1944), Earley (1952) and Ingle et al. (1965) conducted studies on the changes which occur in the carbohydrate constituents during kernel development. The findings were essentially the same as those reported by Culpepper and Magoon in 1924, even though the materials and techniques were different. The data show a relationship between moisture content, reducing sugars, sucrose, starch, and total dry weight, which follow a pattern during kernel development. Moisture decreased constantly and starch and total dry weight showed a constant increase, beginning about 2 weeks after fertilization. Reducing sugar content was high in unfertilized ovules and remained high until about 10 days after fertilization, at which time it began to decrease slowly. Sucrose content decreased slightly immediately after fertilization and then increased rapidly, reaching its highest concentration about 18 days after fertilization. After 18 days sucrose began to decrease and reached a low level in mature kernels.

Earley (1952) demonstrated that starch was present in ovules prior to fertilization. Bernstein (1943) suggested that the small amount of starch in corn kernels one week after fertilization was in the pericarp.

Peat et al. (1956) extracted phytoglycogen from seed corn kernels with 0.01 M mercuric chloride to inhibit carbohydrases likely to degrade phytoglycogen. When mercuric chloride was used, the sugars present in the extract were glucose, fructose, sucrose, raffinose, and a trace quantity of maltose. Without the inhibitor the fraction contained glucose, fructose, maltose, maltotriose, and higher maltodextrins, but no sucrose or raffinose. These results showed the importance of avoiding autolytic changes during extraction.

Bond and Glass (1963) identified seven low molecular weight carbohydrates at various stages of germination of maize kernels. Raffinose was present during the 2-day steeping period, but it disappeared after 24 hours in the germination chamber. Fructose, glucose, and sucrose were present during all stages of germination. Glucose concentration increased during germination to a level about 57 times that at the beginning, a level almost as great as that of sucrose. Myoinositol and glycerol were present at the beginning but disappeared during early phases of germination. Maltose was present after 48 hours.

Hydrolysis of translocated sugar has recently been shown to be necessary prior to its movement into the storage cells of corn endosperm (Shannon, 1972). This was suggested because of the relatively high proportion of radioactivity in the endosperm monosaccharides soon after treatment of the whole plant with $^{14}CO_2$. Hydrolysis appeared to occur in the pedicel region. One hour after $^{14}CO_2$ treatment began, almost all the ^{14}C-sugar of the cob was sucrose. However in the endosperm most of the radioactivity was in the monosaccharide fraction. In the pedicel region, radioactivity existed in both sucrose and the monosaccharides. After entry into the endosperm tissue, sucrose was rapidly resynthesized from the monosaccharides prior to its utilization in starch synthesis.

Tsai et al. (1970) reported on the reducing sugars, sucrose, and starch content of dent corn endosperms in the 8 and 28 days post-pollination period. There was an increase in the total extractable carbohydrates over time. Eight to 10 days after pollination the reducing sugars comprised over 80% of the extractable carbohydrates. The amount of sucrose had increased to a maximum 12 days after pollination. There was little starch present through 12 days, but it rapidly increased thereafter. Although the percentage of soluble sugars declined with time, the maximum quantity of both reducing sugars and sucrose (per endosperm) was reached 16 days after pollination.

Endosperm Mutants

Many mutations in corn affect endosperm components. Several alter the type and quantity of carbohydrates, including starch. Collins (1909) described a waxy mutant that was different from normal corn in sugar and starch content. Weatherwax (1922) demonstrated by iodine staining that starch in the endosperm consisted solely of a "rare" form of carbohydrate called "erythrodextrin," currently known as amylopectin. This starch stained red with iodine, in contrast to "amylodextrin," currently termed amylose, which stained blue with iodine. Bates et al. (1943) and Sprague et al. (1943) confirmed that waxy endosperm of corn contains starch consisting of nearly all amylopectin, a branched-chain polysaccharide. The presence of amylopectin in cereal had been previously demonstrated by Parnell (1921). Other waxy mutants have been reported by Bear (1944) and Mangelsdorf (1947). The waxy mutants produce starch that is practically all amylopectin (Bates et al., 1943). Andres and Bascialli (1941) reported two waxy mutations in a cultivar of Argentine flint that possessed a small amount of amylose. These waxy mutations are allelic to standard wx.

It is now generally accepted that the starch granule is usually a mixture of two polysaccharides (Greenwood, 1956; Whelan, 1961). Amylopectin is the major component in most starches, ranging from 75 to 85%, and is of high molecular weight in a magnitude of 1×10^7. Amylopectin consists of chains of α-D-(1-4) and α-D-(1-6)-glucosidic linkages which form a branched molecule. Amylose is primarily linear with α-D-(1-4)-linked glucose residues.

Two sugary genes, designated sugary-1 (su_1) and sugary-2 (su_2), were described by East and Hayes (1911) and Eyster (1934). Mangelsdorf (1947) reported two gene mutations, designated dull (du) and amylaceous sugary (su_{am}), that also altered the carbohydrate properties of maize endosperm.

Amylaceous sugary was later found to be allelic to the su_1 gene. Andrew et al. (1944) reported that the recessive allele wx increased sugars and water-soluble polysaccharides in either su_1 or Su_1 backgrounds. Cameron (1947) reported that the genes su_1 and du interact to increase the proportion of amylose to approximately 65%. In contrast, the amylose content of standard dent corn is about 25%. These two genes also increased the WSP fraction and decreased the amount of starch.

Kramer and Whistler (1949) established that the su_2 gene increased the amylose content to about 35%. Dvonch et al. (1951) and Dunn et al. (1953) reported that the genes du, su_1, and su_2 interact to increase the amylose content of starch to about 77%; however, no increase in absolute yield of amylose was obtained because the amount of endosperm starch decreased. Cameron and Cole (1959) reported that the gene combinations $su_1 du$ and $su_1 su_2$ partially inhibited the accumulation of starch, compared to su_1 alone.

Vineyard and Bear (1952) described a gene mutation, termed amylose-extender (ae), that substantially increased amylose content without drastically decreasing the total starch content. Endosperm starch of ae stocks, in many backgrounds, contains approximately 60% amylose. Deatherage et al. (1954) reported the development of an ae hybrid that was high in starch yield but 60% of the starch was amylose. Kramer et al. (1958), using various combinations of the genes ae, du, su_1, su_2, and wx, found that amylose contents ranged from zero to over 70%.

Zuber et al. (1960) suggested that the range in amylose content of 54 to 71% in maize inbreds possessing the ae gene was due to modifier genes. They also reported that starch content decreased as amylose content increased.

The gene shrunken-1 (sh_1) is an endosperm mutation that reduces the amount of starch in the endosperm to a very significant degree, resulting in a collapsed shrunken kernel. The mutation was first described by Hutchinson (1921). Burnham (1944) described a mutant somewhat similar to sh_1 but which later was shown by Mains (1949) to not be allelic to sh_1. This gene was designated shrunken-2 (sh_2).

Laughnan (1953) studied the effects of the genes sh_2 and su_1 on the distribution of endosperm carbohydrates. It was found that almost 20% of the dry weight of the shrunken kernels was composed of sugars, which was about a 10-fold increase over standard dent kernels. Most of the sugar consisted of sucrose, which made up 16% of the dry weight of the endosperm. A corresponding decrease in the starch was observed. Total carbohydrate production was lowest in the double recessive, $sh_2 su_1$. In comparison with su_1 kernels, the $sh_2 su_1$ kernels showed a decrease in the accumulation of WSP. Laughnan concluded that sh_2 precedes su_1 in the synthesis of starch and WSP. This conclusion was based on the assumption that each gene controlled a specific biochemical step in starch and WSP synthesis. He pointed out that by using additional mutations in specific combinations it should be possible to gain valuable information about starch synthesis in corn.

Cameron and Teas (1953) have shown that two other mutations, brittle-1 (bt_1) and brittle-2 (bt_2) reduce endosperm starch substantially without the accumulation of WSP. Both genes caused increases in reducing sugars and sucrose and reductions in total starch. Creech et al. (1963) and Creech (1965) reported the effects of the genes ae, du, sh_2, su_1, su_2, and wx, singly and in combinations, on qualitative and quantitative changes in carbohydrates in

Table 6—Quantities of various carbohydrates† and total dry matter‡ in whole kernels of several corn genotypes at four stages of development§¶

Genotype	Kernel age	Reducing sugars	Sucrose	Total sugar	WSP#	Starch	Total carbohydrates††	Dry matter
	days				%			
normal	16	9.4	8.2	17.6	3.7	39.2	60.5	15.7
	20	2.4	3.5	5.9	2.8	66.2	74.9	27.1
	24	1.6	2.6	4.8	2.8	69.2	76.1	37.2
	28	0.8	2.2	3.0	2.2	73.4	78.6	43.8
ae	16	8.6	21.9	30.6	5.7	20.8	57.2	18.4
	20	4.8	13.9	18.7	4.2	37.6	60.5	26.0
	24	3.1	8.3	11.4	3.7	48.9	64.0	34.0
	28	1.9	7.4	9.4	4.4	49.3	62.9	37.5
du	16	8.8	15.5	24.2	4.1	25.1	53.4	16.2
	20	4.8	10.5	15.3	2.7	44.6	62.6	25.6
	24	2.8	6.1	9.0	2.4	56.5	67.9	33.5
	28	1.3	6.7	8.0	1.9	59.9	69.8	38.9
sh_2	16	6.9	21.4	28.3	5.6	22.3	56.1	16.8
	20	4.9	29.9	34.8	4.4	18.4	57.6	20.3
	24	4.4	24.9	29.4	2.4	19.6	51.4	22.9
	28	3.6	22.1	25.7	5.1	21.9	52.8	26.3
su_1	16	9.2	16.5	25.7	14.3	23.3	65.3	19.9
	20	5.4	10.2	15.6	22.8	28.0	66.5	25.6
	24	3.6	9.5	13.1	28.5	29.2	70.8	30.5
	28	3.9	4.4	8.3	24.2	35.4	69.6	37.6
su_2	16	7.4	10.5	16.7	3.6	39.3	59.6	17.5
	20	3.5	9.2	12.7	3.1	50.7	61.8	24.9
	24	1.9	2.6	4.5	2.5	63.9	70.9	34.9
	28	1.4	1.9	3.3	1.9	64.6	69.8	43.6
wx	16	10.1	9.6	19.7	3.5	34.1	57.2	14.9
	20	3.5	5.2	8.7	2.3	53.3	64.6	23.9
	24	2.5	4.5	7.0	2.8	61.9	71.5	33.1
	28	1.6	1.7	3.3	2.2	69.0	74.5	37.3
$ae\ su_1$	16	6.9	12.6	19.6	3.7	18.3	41.5	19.3
	20	3.7	8.3	12.0	3.6	29.3	44.9	24.8
	24	2.2	5.3	7.6	3.6	37.2	48.4	31.5
	28	2.1	5.3	7.4	3.2	34.4	45.1	33.9
ae wx	16	6.1	23.8	29.9	4.2	19.7	53.9	18.3
	20	3.8	23.2	27.0	4.6	26.6	58.2	23.5
	24	3.9	17.9	22.4	5.6	37.1	64.9	25.0
	28	3.2	12.3	15.4	4.6	39.5	59.5	28.3
du wx	16	7.3	25.5	32.8	5.5	21.3	59.6	21.2
	20	4.1	15.8	19.9	12.2	34.3	66.4	25.7
	24	3.8	11.6	15.4	11.4	37.9	64.7	30.4
	28	3.0	9.5	12.5	11.6	45.4	69.5	34.8
$sh_2\ su_1$	16	8.9	24.1	33.1	5.0	7.2	47.3	20.5
	20	8.1	25.4	33.5	4.9	11.7	50.1	23.8
	24	7.1	19.1	27.8	4.6	14.4	46.9	25.2
	28	5.7	20.1	24.5	4.9	15.7	45.4	24.6

corn endosperm during kernel development from 16 through 28 days after pollination. Part of these data are presented in Table 6. Significant differences between genotypes at different stages of kernels development were noted for all carbohydrate and dry matter analyses. Total sugars, reducing sugars, and sucrose contents were negatively correlated with dry matter and starch content. Creech and McArdle (1966) reported on carbohydrate profiles of mature corn kernels of the same genotypes used in the earlier work (Creech, 1965). In general, data on the carbohydrates of mature kernels were similar to the findings for 28-day kernels for all the genotypes investigated, which included several combinations of *ae, du, sh*$_2$*, sμ*$_1$*, su*$_2$*,* and *wx*.

Anderson et al. (1962) reported that high-amylose corn (*ae*) had more protein and less endosperm starch than standard dent corn. They attributed the lower test weight of *ae* grain to its lower starch content.

Table 6—Continued

Genotype	Kernel age	Reducing sugars	Sucrose	Total sugar	WSP#	Starch	Total carbohydrates††	Dry matter
	days				%			
$su_1 su_2$	16	4.9	16.8	21.8	33.7	11.9	67.5	20.1
	20	2.8	11.2	14.1	31.5	20.1	65.6	28.5
	24	2.4	9.6	12.0	31.0	20.5	63.5	31.1
	28	2.5	10.4	12.8	36.9	18.9	68.6	35.4
$ae\ du\ su_1$	16	12.8	24.6	37.3	9.6	23.6	70.5	17.3
	20	9.2	18.0	27.2	12.4	30.9	70.5	22.6
	24	4.7	15.5	21.3	16.1	32.7	70.0	25.8
	28	4.6	10.6	15.3	18.2	38.0	71.5	27.6
$ae\ du\ wx$	16	6.8	39.9	46.7	4.2	15.9	66.7	18.5
	20	4.1	34.6	38.7	3.6	26.6	68.9	24.6
	24	3.6	30.7	34.3	4.5	31.1	69.9	25.8
	28	4.4	23.7	28.1	4.9	32.0	65.1	24.5
$ae\ su_1 su_2$	16	8.5	23.2	31.7	6.6	23.8	62.0	20.3
	20	3.5	9.7	13.2	10.4	41.6	65.3	27.1
	24	2.7	7.9	10.6	10.6	39.6	61.1	31.5
	28	2.4	8.6	11.0	11.0	41.0	65.9	34.1
$ae\ su_1\ wx$	16	8.0	28.2	36.2	4.5	22.0	62.7	16.3
	20	5.2	21.9	27.0	8.4	30.7	66.0	21.7
	24	3.5	15.0	18.5	12.2	38.5	69.1	25.8
	28	2.8	11.1	13.9	12.4	38.3	64.5	26.2
$du\ su_1\ wx$	16	5.9	16.8	21.7	24.4	14.7	60.8	22.0
	20	3.2	10.2	13.4	36.1	21.4	70.9	27.9
	24	2.9	7.8	10.7	38.4	17.5	66.6	33.4
	28	2.3	6.7	9.0	47.5	15.9	72.3	35.3
$du\ su_2\ wx$	16	9.2	25.7	34.9	4.6	17.2	53.4	15.2
	20	5.2	19.5	24.7	14.8	24.7	64.2	20.8
	24	3.0	10.6	13.3	14.3	33.9	61.5	27.9
	28	2.7	8.9	11.6	16.7	38.1	64.5	30.7
Golden Cross Bantam (su_1) Sweet Corn	16	8.1	15.4	23.6	7.8	28.7	60.1	16.1
	20	3.2	5.5	8.7	27.0	35.5	71.3	26.8
	24	1.9	3.9	5.9	33.3	38.5	77.7	33.5
	28	1.6	1.9	3.6	34.8	33.9	72.3	37.0
L.S.D., genotypes within ages								
5%		3.2	10.4	10.9	10.4	14.2	15.3	2.9
1%		4.2	13.9	14.5	13.9	18.8	20.4	3.9
L.S.D., ages within genotypes								
5%		2.4	6.0	5.8	4.8	7.6	7.5	2.9
1%		3.2	7.9	7.7	6.3	10.1	9.9	3.8

†Percent of dry matter. ‡Percent of fresh weight. §Three replications.
¶Adapted from Creech (1965). #WSP, water-soluble polysaccharides. ††Sum of weights of reducing sugar, sucrose, WSP, and starch/dry wt.

Interactions

Combinations of genes affecting kernel properties are being studied by an increasing number of field corn and sweet corn breeders in order to obtain types with altered chemical properties. Garwood and Creech (1972) described kernel phenotypes of 94 corn genotypes possessing 1 to 4 mutated genes (Table 7). They were studied in the following backgrounds: the dent inbreds W64A and W23, sweet corn inbreds S3-61, I453, and I5125, and derivatives of the dent single cross W23 × L317. Use of the information in Table 7 can minimize the number of confirmatory tests needed to develop particular homozygotes. For example, to develop a double recessive, single recessive plants are crossed and the F_1 is self-pollinated. Kernels homozygous for the gene that is hypostatic are selected. Segregation for the double mutant can easily be observed following self-pollination of the selected class and the double recessive can be selected. Alternatively, the F_1 may be crossed with a plant homozygous for the hypostatic gene, the mutant kernels selected, and following self-pollination of the selected class, segregation for the double mutant can be observed. Verification of the newly developed genotype by a confirmatory test is recommended. If phenotypes of the single recessives and multiple recessive are indistinguishable, in contrast to the epistatic situation previously described, testing will be required when the second self-pollination is made. For example, to combine bt_1 and bt_2, single recessive plants are crossed, the F_1 is self-pollinated, and the "brittle" kernels selected. Progeny from the selected kernels is self-pollinated and also crossed to both bt_1 and bt_2 tester stocks. The bt_1 bt_2 genotypes may be identified only by confirmatory tests. Kramer (1961) gave several examples of the isolation of triple recessive genotypes. His procedure calls for crosses of double recessive types having one gene in common to combine the desired genes with subsequent isolation by self-pollination and selection as described in the previous paragraphs.

Triple recessives can also be developed from F_1 heterozygous for three loci by using three cycles of self-pollination and selection. In cases of epistasis, selection order is important. For example, to isolate ae su_1 wx, waxy kernels are selected from the ear of the self-pollinated F_1 plant. The waxy kernels are grown, the F_2 plants are self-pollinated, and sugary kernels (su_1 wx) are selected and planted from ears also segregating for ae. The F_3 plants are self-pollinated and the ae su_1 wx kernels are obtained from segregating ears. Other orders of selection will not give visible segregations in all generations (Garwood and Creech, 1972).

Investigations on the role of the genes ae and su_1 in phytoglycogen accumulation in endosperm were conducted by Ayers and Creech (1969) by quantitating and characterizing the WSP from 16 genotypes resulting from crosses among lines homozygous for normal (Ae and Su_1), ae, su_1, and ae su_1. Only the endosperms homozygous for su_1 contained phytoglycogen. Increased doses of ae decreased the amount of phytoglycogen from 38.8 to 4.2% of the dry weight at the 20-day post-pollination stage. The β-amylolysis limit phytoglycogen from the double mutant suggested that it may be more loosely branched than the polysaccharide from su_1 endosperm.

Carbohydrate properties of single and multiple combinations of alleles are helpful in elucidating their interactions. For example, the gene sh_2 apparently has an effect on the ratio of sugars and polysaccharides. The su_1 gene is associated with a significant increase in WSP, as reported by Culpepper and Magoon (1924); however, the gene ae is epistatic to su_1 and the double recessive ae su_1 possesses little or no WSP. The nature of these interactions at the biochemical level is not known.

The data of Kramer et al. (1958) Zuber et al. (1958), Creech et al. (1963), Creech (1965), and of previously cited earlier workers on the roles of the genes ae, du, sh_2, su_1, su_2, and wx, indicate that starch synthesis is genetically complex. The ae gene, in addition to changing the amylose content on a percentage basis (Vineyard and Bear, 1952), also causes a substantial increase in

Table 7—Kernel phenotypes of 94 genotypes of corn†

Genotype‡	Gene expressed§	Kernel phenotype
ae	ae	Tarnished, translucent, or opaque; Sometimes semi-full
$ae\ bt_1$	bt_1	Shrunken, opaque to tarnished
$ae\ bt_2$	bt_2	Shrunken, opaque to tarnished
$ae\ du$	C	Translucent, not as full as ae alone; S.C.¶: Etched, translucent, or tarnished
$ae\ fl_1$	ae	Tarnished, occasional small floury sectors
$ae\ h$	ae	Tarnished
$ae\ o_2$	o_2	Opaque; tarnished sectors observed in W23 × L317 background
$ae\ sh_1$	sh_1	Collapsed, opaque
$ae\ sh_2$	sh_2	Shrunken, opaque to tarnished
$ae\ su_1$	C	Not quite as full as ae, translucent (tarnished in S.C.), may have opaque caps
$ae\ su_2$	ae	Translucent or opaque, etched base
$ae\ wx$	C	Semi-full to collapsed, translucent or glassy, may have opaque caps; S.C.: Slightly fuller, etched, translucent to glassy#
$ae\ du\ sh_2$	sh_2	Shrunken, translucent to glassy
$ae\ du\ su_1$	su_1	Wrinkled, translucent; S.C.: Slightly wrinkled, translucent
$ae\ du\ su_2$	C	Semi-collapsed, translucent
$ae\ du\ wx$	C	Shrunken, opaque to tarnished; S.C.: Semi-collapsed, opaque
$ae\ sh_2\ su_1$	sh_2	Shrunken, opaque
$ae\ sh_2\ su_2$	sh_2	Shrunken, opaque to tarnished
$ae\ sh_2\ wx$	sh_2	Shrunken, opaque
$ae\ su_1\ su_2$	C	Partially wrinkled, translucent to tarnished
$ae\ su_1\ wx$	C	Semi-collapsed, opaque to translucent; S.C.: Etched, semi-full, translucent
$ae\ su_2\ wx$	C	Etched, semi-full or wrinkled, translucent
$ae\ du\ su_1\ wx$	C	S.C.: Etched, semi-full, translucent to tarnished
bt_1	bt_1	Shrunken, opaque to tarnished
$bt_1\ bt_2$	I	Shrunken, opaque to tarnished
$bt_1\ du$	bt_1	Shrunken, opaque to tarnished
$bt_1\ fl_1$	bt_1	Shrunken, opaque to tarnished
$bt_1\ h$	bt_1	Shrunken, opaque to tarnished
$bt_1\ o_2$	bt_1	Shrunken, opaque
$bt_1\ sh_1$	bt_1	Shrunken, opaque to tarnished
$bt_1\ sh_2$	I	Shrunken, opaque††
$bt_1\ su_1$	bt_1	Shrunken, opaque to tarnished
$bt_1\ su_2$	bt_1	Shrunken, tarnished
$bt_1\ wx$	bt_1	Shrunken, opaque to tarnished
bt_2	bt_2	Shrunken, opaque to tarnished
$bt_2\ du$	bt_2	Shrunken, opaque to tarnished
$bt_2\ fl_1$	bt_2	Shrunken, opaque to tarnished
$bt_2\ h$	bt_2	Shrunken, opaque to tarnished
$bt_2\ o_2$	bt_2	Shrunken, opaque
$bt_2\ sh_1$	bt_2	Shrunken, opaque to tarnished
$bt_2\ sh_2$	I	Shrunken, opaque to tarnished
$bt_2\ su_1$	bt_2	Shrunken, translucent to tarnished
$bt_2\ su_2$	bt_2	Shrunken, opaque to tarnished
$bt_2\ wx$	bt_2	Shrunken, opaque to tarnished
du	du	Opaque to tarnished; S.C.: Semi-collapsed, translucent with some opaque sectors
$du\ fl_1$	fl_1	Opaque
$du\ h$	du	Slightly tarnished to opaque††
$du\ o_2$	o_2	Opaque, tarnished sectors observed in W23 × L317 background

Genotype	Gene expressed§	Kernel phenotype
$du\ sh_1$	sh_1	Collapsed, opaque
$du\ sh_2$	sh_2	Shrunken, opaque to tarnished
$du\ su_1$	su_1	Wrinkled, glassy (duller than su_1 alone); S.C.: Extremely wrinkled, glassy‡‡
$du\ su_2$	C	Translucent, etched
$du\ wx$	C	Semi-collapsed, opaque; S.C.: Shrunken, opaque
$du\ sh_2\ su_1$	C	Shrunken, translucent
$du\ sh_2\ su_2$	sh_2	Shrunken, opaque
$du\ su_1\ su_2$	su_1	Wrinkled, glassy
$du\ su_1\ wx$	su_1	Wrinkled, glassy
$du\ su_2\ wx$	C	Semi-collapsed, opaque, etched
fl_1	fl_1	Opaque§§
$fl_1\ h$	I	Opaque
$fl_1\ o_2$	o_2	Opaque
$fl_1\ sh_1$	sh_1	Collapsed, opaque
$fl_1\ sh_2$	sh_2	Shrunken, opaque
$fl_1\ su_1$	su_1	Wrinkled, glassy
$fl_1\ su_2$	su_2	Tarnished, etched at base
$fl_1\ wx$	wx	Opaque
h	h	Opaque
$h\ o_2$	o_2	Opaque
$h\ sh_1$	sh_1	Collapsed, opaque
$h\ sh_2$	sh_2	Shrunken, opaque to tarnished
$h\ su_2$	su_2	Tarnished, etched at base
$h\ wx$	wx	Opaque
o_2	o_2	Opaque
$o_2\ sh_1$	sh_1	Collapsed, opaque
$o_2\ sh_2$	sh_2	Collapsed, opaque
$o_2\ su_1$	su_1	Wrinkled, glassy
$o_2\ su_2$	C	Tarnished, tarnished with opaque sectors, opaque
$o_2\ wx$	o_2	Opaque
sh_1	sh_1	Collapsed, opaque
$sh_1\ sh_2$	sh_2	Shrunken, opaque to tarnished
$sh_1\ su_1$	C	Extremely wrinkled, glassy with opaque sectors
$sh_1\ su_2$	Both	Collapsed, tarnished
$sh_1\ wx$	sh_1	Collapsed, opaque to tarnished (wx segregation visible in sh_1sh_1)
sh_2	sh_2	Shrunken, opaque to tarnished
$sh_2\ su_1$	sh_2	Shrunken, opaque to tarnished
$sh_2\ su_2$	sh_2	Shrunken, opaque to translucent
$sh_2\ wx$	sh_2	Shrunken, opaque
su_1	su_1	Wrinkled, glassy: S.C.: Not as extreme
$su_1\ su_2$	su_1	Wrinkled, glassy
$su_1\ wx$	su_1	Wrinkled, glassy to opaque (wx segregation often visible in su_1su_1)
$su_1\ su_2\ wx$	su_1	Wrinkled, glassy
su_2	su_2	Slightly tarnished to tarnished, often etched at base
$su_2\ wx$	wx	Opaque, often etched
wx	wx	Opaque

†Adapted from Garwood and Creech (1972). ‡Gene symbols and the meaning of terms describing kernel phenotype are given in the text. §If one gene is responsible for the phenotype, that gene is listed. Slight modifications in this phenotype, if the gene is not completely epistatic, are indicated in the description. Interaction (C) signifies a complementary expression giving a new phenotype differing from the phenotypes of the stocks possessing the individual genes. Indistinguishable (I) is used to indicate that the individual mutations and the double recessive appear identical or nearly identical.
¶S.C. means the phenotype observed in sweet corn inbreds. #Dosage effects are observed as follows with wx homozygous: $Ae\ ae\ ae$ is tarnished, semi-full and $Ae\ Ae\ ae$ is like homozygous wx (5).
††Tentative observation. Allele tests are incomplete. These loci are closely linked. ‡‡Dosage effects are observed as follows with du homozygous: $Su_1\ su_1\ su_1$ is semi-full, glassy and $Su_1\ Su_1\ su_1$ is like homozygous du. §§$Fl_1\ fl_1\ fl_1$ is floury while $Fl_1\ Fl_1\ fl_1$ is normal (Neuffer et al., 1968).

sugars (Creech, 1965) and a reduction in starch (Kramer et al., 1958; Zuber et al., 1960; Creech et al., 1963; Creech, 1965; Creech and McArdle, 1966). In addition, *ae* combined with *wx* and *du wx* causes dramatic increases in sugars and reductions in starch. This supports the view that *ae, du,* and *wx* may be involved in separate starch synthesis pathways.

The amylose content data of Kramer et al (1958), shown in Table 8, combined with the data on effects of *ae* and *wx* on sugars and polysaccharide content (Creech, 1965), also suggested that the dominant allele *Ae* may be involved in the formation of branched-chain polysaccharides (amylopectin) and that the dominant allele *Wx* may be involved in the synthesis of straight-chain polysaccharides (amylose). Nelson and Rines (1962) reached a similar conclusion concerning *Wx*. They reported that starch granules from *wx* endosperms were deficient in uridine diphospho-glucose (UDPG) transferase activity, but that it was present in starch granules of *Wx* endosperms. Since *wx* endosperm lacked both amylose starch and starch granule-bound UDPG transferase activity, Nelson and Rines (1962) suggested that UDPG transferase is necessary for amylose synthesis. This enzyme caralyzes the transfer of glucose from UDPG to an α-D-(1-4) linkage on the nonreducing end of the polysaccharide acceptor. Nelson and Tsai (1964) later reported that 17 different waxy mutants transform glucose to starch at about one-tenth the rate of similar preparations from kernels of non-waxy maize. The source of the activity was indicated to be in starch granules isolated from the embryo. The endosperm, which is the site for most starch synthesis, apparently was devoid of starch granule bound transferase activity. Badenhuizen and Chandorkar (1965) suggested that the low UDPG transferase in *wx* corn starch granules may be due to the absence of straight-chain or amylose starch. They isolated starch granules varying in amylose content from developing kernels and assayed them for amylose content and UDPG transferase activity. There was little association between amylose content and UDPG-transferase activity in kernels of the two amylose strains.

Sandstedt et al. (1962) investigated the pancreatic digestibility of isolated corn starches. The data are presented in Table 9. The *ae* gene appears to be associated with both high amylose (Vineyard and Bear, 1952) and high resistance to enzyme action. Kramer et al. (1958) and Brown (1966) reported that *ae* starch had a very high birefringence end-point temperature (BEPT). However *du* and su_2 endosperms, although higher in amylose than normal endosperm, seemed to be associated with high digestibility. They suggested that the answer may be in the structure of the starch granule, i.e., in differences in the bonding of the starch molecules and/or in possible anomalous linkages between molecules. Brown (1966) suggested that the su_2 starch was less crystalline than normal starch but this does not account for the results with *wx*. The su_2 starch granules also had a lower BEPT than normal granules. Sandstedt suggested that improving the digestibility of starch for commercial strains of corn may be important in improving the nutritional value of corn, especially for nonruminants.

Four corn genotypes, normal, su_2, o_2, and the double mutant $o_2 su_2$, were evaluated for protein quality by four methods — amino acid analysis, chemically available lysine, rat protein efficiency ratio, and meadow vole protein efficiency ratio (Brinkman et al. 1974). The tests gave similar results which indicated that the improved protein characteristics associated with the

Table 8—Amylose content of mature kernels of 24 genotypes of corn†

Gene combination	Amylose in starch
	%
normal	27
ae	61
du	38
su_1	29
su_2	40
wx	0
ae du	57
ae su_1	60
ae su_2	54
ae wx	15
du su_1	63
du su_2	47
du wx	0
su_1 su_2	55
su_1 wx	0
su_2 wx	0
ae du su_1	41
ae du su_2	48
ae su_1 su_2	54
ae su_1 wx	13
du su_1 su_2	73
du su_1 wx	0
du su_2 wx	0
su_1 su_2 wx	0

†Adapted from Kramer et al. (1958).

Table 9—Comparison of the susceptibility of various starches to pancreatic digestion†

Starch source	Digestion
	%
normal	69
ae	24
du	80
su_1	71
su_2	88
wx	85
ae du	42
ae su_2	22
du su_2	76

†Adapted from Sandstedt et al. (1962).

o_2 gene were retained in the double mutant containing the su_2 gene. The high digestibility of starch of su_2 did not show an advantage in the double mutant. Kernel texture was improved in the double mutant in that it was harder. Glover et al. (1975) reported that the kernel density of o_2 su_2 was nearly equal to that of ordinary dents.

Breeding Waxy Corn

Waxy corn was brought to the US from China in 1908 and maintained as a genetic curiosity. Its name derives from the waxy appearance of the endosperm exposed in a cleanly cut cross section. Waxy starch differs from ordinary starch in that it consists almost exclusively of amylopectin, a branched chain molecule.

An extensive cross-breeding program was initiated in 1937 in an effort to introduce the wx allele into regular high-yielding hybrid corns. The work was conducted primarily at Iowa State University by G.F. Sprague and colleagues. As reported by S.A. Watson in Chapter 13 on utilization, an estimated 300 to 360 million metric tons of waxy corn is produced annually in the USA. Generally no severe problems have been encountered in the transfer of the gene wx to commercial inbreds and new experimental inbreds.

Breeding High Amylose Corn

Bear (1958) reported having found a recessive allele in 1950 which changed the amylose-amylopectin ration from 1:3 to 1:1 in mature corn endosperm without materially reducing total starch production. The gene was designated amylose extender (ae), as previously reported in this chapter.

Because the physical-chemical properties of corn starch having a 1:1 amylose to amylopectin ratio were desirable for certain industrial purposes, hybrid development was initiated using the ae gene (Bear, 1958). Early breeding efforts consisted primarily of using the backcross method to incorporate the gene into existing inbreds. Bear et al. (1958) found wide differences in amylose content among derived inbreds. The differences suggested that a modifier complex existed and therefore selection during backcrossing was deemed necessary. Alternate backcrossing and selfing were employed during the conversion of Corn Belt inbreds to high amylose content with selection each generation for maximum amylose content (Helm et al., 1967).

A consistent problem in high amylose corn has been a reduction in amount of starch, which may in fact be a reduction in amylopectin starch, without a compensating increase in amount of amylose. Moisture content of the grain also is higher relative to comparable dents. These problems have been largely overcome, or reduced to a tolerable level, and amylo-maize is well established as a specialty corn in the seed trade. To our knowledge, no other mutant exists that will produce higher amylose content (or lower amylopectin content) than the standard ae gene. Alleles at the ae locus have been discovered and characterized; however, none have effected a superior starch to that of standard ae.

Breeding Sweet Corn with Higher Sugar Content

Most of the commonly used US sweet corn hybrids are single crosses involving the su_1 gene in combination with modifiers that increase sugar content or otherwise improve quality. East and Hayes (1911) described the su_1 gene in sweet corn. Laughnan (1953) studied the effects of the genes sh_2 and su_1 on the distribution of endosperm carbohydrates. He found that almost 20% of the dry weight of sh_2 kernels was composed of sugars. The $sh_2 su_1$ kernels showed a decrease in accumulation of water-soluble polysaccharides. Laughnan concluded that sh_2 might be very useful in improving the quality of sweet corn. The hybrid, Illini X-tra Sweet, and similar hybrids, possess the gene sh_2. The major objection to these hybrids lies in their seed quality which is generally inferior to that of standard sweet hybrids. Selection has been successful in improving germination. Creech and McArdle (1966) suggested that ae, du and wx might also be utilized to improve the quality and maintenance of quality of harvested sweet corn. The genes ae, du, and wx have been transferred to several sweet corn inbreds by R.G. Creech at the Pennsylvania State University and these inbreds are being released to the seed trade. It has not yet been established whether the $ae\ du\ wx$ combination will be accepted by the seed trade.

POPCORN

Beadle (1972) reported that in 1939 he had popped teosinte and that the kernels had exploded out of their hard fruit cases and were indistinguishable from ordinary popcorn. If teosinte (*E. mexicana*) is the wild progenitor of modern corn, then the earliest domesticated corn must have been a poppable type. Teosinte kernels are flinty and quite similar in texture to the popcorns.

According to Brunson (1955) popcorns may be classified into two primary types: the rice corns that have pointed kernels and the pearl types that have smooth rounded crowns. He also recognized three sub-types: Jap Hulless with short thick ears and long kernels, Lady Finger with many small ears, and the eight-row types that are in reality modified flints.

In commercial channels, cultivars were rapidly displaced by hybrids, primarily of the eight-row types in the US in the early 1940's.

Popcorn, and of course sweet corn, is consumed after simple in-the-home processing. Quality, therefore, is of prime importance and the major emphasis of breeders has been to maintain both high quality and satisfactory agronomic performance in hybrids. The prime quality feature is popping expansion. Standard evaluation schemes have been evolved which measure the volume of the popped corn and relates this value either to the volume of the raw grain or to its weight. The latter system is used by the Purdue Agricultural Experiment Station. Their data show that popular hybrids commonly produce more than 40 hl popped corn/kg of grain (1,100 cubic inches/pound).

Popping expansion is affected both by genotype and by environment. Abundant evidence exists that the trait is polygenic and that selection can effect change. Heritability estimates range from 70% (Grisson, 1951) to 90% (Clary, 1954). Weaver and Thompson (1957) reported that a 15-generation modified mass selection scheme increased popping expansion from 22.2 to 35.8 volumes in narrow-base white hulless population. Johnson and Eldredge (1953) found that no striking relationship existed between yield and popping expansion and that it was possible to add useful alleles from dents to popcorn by backcrossing. Recovery of high popping expansion was sufficiently frequent to suggest that a comparatively simple genetic mechanism was responsible. Subsequent breeding work suggests, however, that there is a negative correlation between yield and popping expansion (R.B. Ashman, personal communication).

Popping expansion is affected by moisture content of the grain, degree of damage of pericarp and endosperm, and by drying method. Huelsen and Bemis (1955) in a series of studies were able to show that popping expansion varied in relation to moisture level, but further, that hybrids differed in the ranges of moisture which would produce optimum expansion. Several hybrids had an acceptable range of five percentage points but others less suitable for commercial production had a narrower acceptable range. In general, moisture contents ranging from 13.5 to 15.5% produce maximum expansion values. They also found that rate of drying had negligible effect on popping expansion of well-matured corn, although rapid drying of corn containing 25% moisture or more materially reduced it.

Popping expansion is also affected by the amount of damage inflicted on the grain during harvest and handling. Although optimum moisture content will vary as a consequence of machinery adjustment and other variables, popcorns can be harvested with minimal damage by combine or field shellers at 15 to 18% moisture. However optimum grain quality is encountered by whole ear harvest followed by natural drying or by low temperature drying.

Popcorn production in the US has ranged from 24,000 metric tons of ear corn in 1934 to 2,560,000 metric tons in 1971. Yield has improved over the years from approximately 1.8 metric tons/ha in 1920 to more than 3.3 metric tons in 1973. Yields unquestionably improved, in part, as a consequence of the introduction of hybrids and of their subsequent improvement. Three-way crosses are more widely used than single crosses primarily because kernels of the latter type are usually too small for successful commercial planting. Double crosses are less popular than three-ways because they lack uniformity, an attribute of some importance in quality. "T" cytoplasm, once widely used in hybrid production, has been abandoned and conventional mechanical and manual detasseling is used.

LITERATURE CITED

1. Alexander, D.E. 1971. Progress in breeding maize for oil content. p. 74-78. *In* I. Kovacs (ed.) Proc. 5th Meeting of the Maize and Sorghum Section of Eucarpia. Akademiai Kiado, Budapest, Hungary.
2. ----, L. Silvela S., F.I. Collins, and R.C. Rodgers. 1967. Analysis of oil content of maize by wide-line NMR. J. Am. Oil Chem. Soc. 44:555-558.
3. Anderson, R.A., D.E. Uhl, W.L. Deatherage, and E.L. Griffin. 1962. Composition of the component parts of two hybrid high-amylose corns. Cereal Chem. 39:282-286.
4. Andres, J.M., and P.C. Bascialli. 1941. Caracters hereditarios aislados en maices cultivados en la Argentina. Univ. Buenos Aires Inst. Genet. 2:1.
5. Andrew, R.H., R.A. Brink, and N.P. Neal. 1944. Some effects of the waxy and sugary genes on endosperm development in maize. J. Agric. Res. 69:355-371.
6. Appleman, C.O., and J.M. Arthur. 1919. Carbohydrate metabolism in green sweet corn during storage at different temperatures. J. Agric. Res. 17:137-152.
7. ----, and S.V. Eaton. 1921. Evaluation of climatic temperature efficiency for the ripening process in sweet corn. J. Agric. Res. 20:795-805.
8. Ayers, J.E., and R.G. Creech. 1969. Genetic control of phytoglycogen accumulation in maize (*Zea mays* L.). Crop Sci. 9:739-741.
9. Badenhuizen, N.P., and K.R. Chandorkar. 1965. UDPG-alpha-glucan glucosyltransferase and amylose content of some starches during their development and under various external conditions. Cereal Chem. 42:44-54.
10. Bates, L.L., D. French, and R.E. Rundle. 1943. Amylose and amylopectin content of starches determined by their iodine complex formation. J. Am. Chem. Soc. 65:142-148.
11. Bauman, L.F. 1975. Germ and endosperm variability, mineral elements, oil content, and modifier genes in opaque-2 maize. p. 217-227. *In* E.T. Mertz (ed.) High quality protein maize. Dowden, Hutchinson and Ross, Inc., Stroudsburg, Penn.
12. ----, T.F. Conway, and S.A. Watson. 1963. Heritability of variations in oil content of individual corn kernels. Science 139:498-499.
13. Beadle, G.W. 1972. The mystery of maize. Field Mus. of Nat. Hist. Bull. 43 (10):2-11.
14. Bear, R.P. 1944. Mutations for waxy and sugary endosperm in inbred lines of dent corn. J. Am. Soc. Agron. 36:89-91.
15. ----. 1958. The story of amylomaize hybrids. Chemurgic Digest 17:5.

16. ----, M.L. Vineyard, M.M. MacMasters, and W.L. Deatherage. 1958. Development of "amylomaize" — corn hybrids with high amylose starch. II. Results of breeding efforts. Agron. J. 50:598-602.
17. Bernstein, L. 1943. Amylases and carbohydrates in developing maize endosperms. Am. J. Bot. 34:517-526.
18. Bond, A.B., and R.L. Glass. 1963. The sugars of germinating corn (*Zea mays*). Cereal Chem. 40:459-466.
19. Bressani, R. 1966. Protein quality of opaque-2 maize in children. p. 34-39. *In* E.T. Mertz (ed.) Proc. High Lysine Corn Conf., Corn Refiners Assn., 1001 Connecticut Ave. N.W., Washington, DC 20036.
20. ----. 1975. Improving maize diets with amino acid and protein supplements. p. 38-57. *In* E.T. Mertz (ed.) High quality protein maize. Dowden, Hutchison and Ross, Inc., Stroudsburg, Penn.
21. Brown, R.P. 1966. Genetic control of starch granule development in maize endosperm. M.S. Thesis. Penn. State Univ., University Park.
22. Brinkman, G.L., J.S. Shenk, R.G. Creech, and D.L. Garwood. 1974. Comparisons between rats, voles and chemical methods for the determination of protein quality of four maize (*Zea mays* L.) genotypes. Nutr. Rep. Int. 10:61-68.
23. Brown, W.L. 1975. Worldwide seed industry experience with opaque-2 maize. p. 256-264. *In* E.T. Mertz (ed.) High quality qrotein maize. Dowden, Hutchinson and Ross, Inc., Stroudsburg, Penn.
24. Brunson, A.M. 1955. Popcorn p. 423-441. *In* G.F. Sprague (ed.) Corn and corn improvement. Academic Press, Inc., New York.
25. Burnham, C.R. 1944. Midrib color. Maize Genet. Coop Newsl. 18:15.
26. Cameron, J.W. 1947. Chemico-genetic bases for the carbohydrates in maize endosperm. Genetics 32:459-485.
27. ----, and D. Cole. 1959. Effects of the genes su_1, su_2, and du on carbohydrates in developing maize kernels. Agron. J. 51:424-427.
28. ----, and H.J. Teas. 1953. Carbohydrate relationships in developing endosperms of brittle-1, brittle-2, and related maize genotypes. Proc. Int. Congr. Genet. 9:822-823.
29. Clark, H.E. 1966. Opaque-2 corn as a source of protein for adult human subjects. p. 40-44. *In* E.T. Mertz (ed.) Proc. High Lysine Conf. Corn Refiners Assn., 1001 Connecticut Ave. N.W., Washington, DC 20036.
30. Clary, G.B. 1954. A study of the inheritance of popping expansion in popcorn. Ph.D. Thesis. Purdue Univ.
31. Collins, G.N. 1909. A new type of Indian corn from China. US Bur. Plant Inds. 161. 31 p.
32. Creech, R.G. 1965. Genetic control of carbohydrate synthesis in maize endosperm. Genetics 52:1175-1186.
33. ----, and L.J. McArdle. 1966. Gene interaction for quantitative changes in carbohydrates in maize kernels. Crop Sci. 6:192-194.
34. ----, ----, and H.H. Kramer. 1963. Genetic control of carbohydrate type and quality in maize kernels. Maize Genet. Coop. Newsl. 37:111-120.
35. Culpepper, C.W., and C.A. Magoon. 1924. Studies upon the relative merits of sweet corn varieties for canning purposes and the relation of maturity of corn to the quality of the canned product. J. Agric. Res. 28:403-443.
36. ----, and ----. 1927. A study of the factors determining quality in sweet corn. J. Agric Res. 34:413-433.
37. Deatherage, W.L., M.M. MacMasters, M.L. Vineyard, and R.P. Bear. 1954. A note on starch of high amylose content from corn with high starch content. Cereal Chem. 31:50-52.
38. De la Roche, I.A. 1970. Genetics and biochemistry of maize lipids. Ph.D. Thesis. Univ. Illinois, Urbana.
39. ----, I.A., E.J. Weber, and D.E. Alexander. 1971a. Genetic aspects of triglyceride structure in maize. Crop Sci. 11:871-874.
40. ----, ----, and ----. 1971b. Stereo-specific analysis of maize tryglycerides. Lipids 6 (8):525-530.

41. Doty, D.M., G.M. Smith, J.R. Roach, and J.T. Sullivan. 1945. The effect of storage on the chemical composition of some inbred and hybrid strains of sweet corn. Indiana Agric. Exp. Stn. Bull. 503. 31 pp.
42. Dudley, J.W., R.J. Lambert, and D.E. Alexander. 1971. Variability and relationships among characters in Zea mays L. synthetics with improved protein quality. Crop Sci. 11:512-514.
43. ----, ----, and ----. 1974. Seventy generations of selection for oil and protein concentration in the maize kernel. p. 181-212. In J.W. Dudley (ed.) Seventy generations of selection for oil and protein concentration in maize. Crop Sci. Soc. of Am. Special Pub.
44. Dunn, G.M., H.H. Kramer, and R.L. Whistler. 1953. Gene dosage effects on corn endosperm carbohydrates. Agron. J. 45:101-104.
45. Dvonch, W., H.H. Kramer, and R.L. Whistler. 1951. Polysaccharides of high-amylose corn. Cereal Chem. 28:270-280.
46. Earley, E.B. 1952. Percentage of carbohydrates in kernels of station reid yellow dent corn at several stages of development. Plant Physiol. 27:184-190.
47. ----, and E.E. de Turk. 1948. Corn protein and soil fertility. What's new in the production, storage and utilization of hybrid seed corn. Am. Seed Trade Assoc. Washington, D.C.
48. East, E.M., and H.K. Hayes. 1911. Inheritance in maize. Conn. Agric. Exp. Stn. Bull. 167. 142 pp.
49. Evans, J.W. 1941. Changes in the biochemical composition of the corn kernel during development. Cereal Chem. 18:468-473.
50. Eyster, W.H. 1934. Genetics of Zea mays. Bibliogr. Genet. 11:187-392.
51. Garwood, D.L., and R.G. Creech. 1972. Kernel phenotypes of Zea mays L. genotypes possessing one to four mutated genes. Crop Sci. 12:119-121.
52. Glover, D.V., P.L. Crane, P.S. Misra, and E.T. Mertz. 1975. Genetics of endosperm mutants in maize as related to protein quality and quantity. p. 228-240. In E.T. Mertz (ed.) High quality protein maize. Dowden, Hutchinson and Ross, Inc., Stroudsburg, Penn.
53. Greenwood, C.T. 1956. Aspects of the physical chemistry of starch. Adv. Carbohydr. Chem. 11:335-393.
54. Grisson, D.B. 1951. Heritability and association of characters affecting popping volume in dent corn-popcorn crosses. M.S. Thesis. Iowa State College.
55. Hassid, W.Z. 1945. Recent advances in the molecular constitution of starch and glycogen. Fed. Proc. Fed. Am. Soc. Exp. Biol. 4:227-234.
56. ----, and R.M. McReady. 1941. The molecular constitution of glycogen and starch from the seed of sweet corn. J. Am. Chem. Soc. 63:1632-1635.
57. Helm, J.L., V.L. Fergason, and M.S. Zuber. 1967. Development of high-amylose corn (Zea mays L.) by the backcross method. Crop Sci. 7:659-662.
58. Huelsen, W.A., and W.P. Bemis. 1955. Artificial drying and rehydration of popcorn and their effects on popping expansion. Illinois Agric. Exp. Stn. Bull. 616.
59. Hume, A.N., M. Champlin, and H. Loomis. 1914. Selecting and breeding corn for protein and oil in South Dakota. South Dakota Agric. Exp. Stn. Bull. 153.
60. Hutchinson, C.B. 1921. Heritable characters of maize. VII. Shrunken endosperm. J. Hered. 12:76-83.
61. Ingle, J., D. Beitz, and R.H. Hayeman. 1965. Changes in composition during development and maturation of maize seeds. Plant Physiol. 40:835-839.
62. Jacobson, B.S., C.G. Kannangara, and P.K. Stumpf. 1973. The elongation of medium chain trienoic acids to α-linolenic acid by spinach chloroplast stroma system. Biochem. Biophys. Res. Commun. 52:1100-1108.
63. Jellum, M.D. 1970. Plant introductions of maize as a source of oil with unusual fatty acid compositions. J. Agric. Food Chem. 18:365.
64. Jensen, A.H., D.H. Baker, D.E. Becker, and B.G. Harmon. 1969. Comparison of opaque-2 corn, milo and wheat in diets for finishing swine. J. Animal Sci. 29(1):16-19.
65. Johnson, I.J., and J.C. Eldredge. 1953. Performance of recovered popcorn inbred lines derived from outcrosses to dent corn. Agron. J. 45:105-110.
66. Kannangara, C.G., B.S. Jacobson, and P.K. Stumpf. 1973. In vivo biosynthesis of α-linolenic acid in plants. Biochem. Biophys. Res. Commun. 52:648-655.

67. Kramer, H.H. 1961. Genes and gene combinations affecting corn endosperms. Proc. 16th Annu. Hybrid Corn Indus.-Res. Conf. 16:57-63.
68. Kramer, H.H., R.P. Bear, and M.S. Zuber. 1958. Designation of high amylose gene loci in maize. Agron. J. 50:229.
69. ----, P.L. Pfahler, and R.L. Whistler. 1958. Gene interactions in maize affecting endosperm properties. Agron J. 50:207-210.
70. ----, and R.L. Whistler. 1949. Quantitative effects of certain genes on the amylose content of corn endosperm starch. Agron. J. 41:409-411.
71. Lambert, R.J., D.E. Alexander, and J.W. Dudley. 1969. Relative performance of normal and modified protein (opaque-2) maize hybrids. Crop Sci. 9:242-243.
72. Lampe, L., and M.T. Meyers. 1925. Carbohydrate storage in the endosperm of sweet corn. Science 61:290-291.
73. Laughnan, J.R. 1953. The effect of the sh_2 factor on carbohydrate reserves in the mature endosperm of maize. Genetics 38:485-499.
74. Levy, R.D. 1973. Genetics of vitamin E in corn grain. Ph.D. Thesis. Univ. Illinois, Urbana.
75. Lofland, H.B., and F.W. Quackenbush. 1954. Distribution of fatty acids in corn oil. J. Am. Oil Chem. Soc. 31:412-414.
76. Lynch, P.B., D.H. Baker, B.G. Harmon, and A.H. Jenson. 1972. Feeding value for growing-finishing swine of corns of different oil contents. J. Anim. Sci. 35(5):1108.
77. Mains, E.B. 1949. Heritable characters in maize: linkage of a factor for shrunken endosperm with the a_1 factor for aleurone color. J. Hered. 40:21-24.
78. Maner, J.H. 1975. Quality protein maize in swine nutrition. p. 58-81. In E.T. Mertz (ed.) High quality protein maize. Dowden, Hutchison and Ross, Inc. Stroudsburg, Penn.
79. Mangelsdorf, P.C. 1947. The inheritance of amylaceous sugary endosperm and its derivatives in maize. Genetics 32:448-458.
80. Mertz, E.T. 1966. Growth of rats on opaque-2 maize. In E.T. Mertz (ed.) Proc. High Lysine Corn Conf. Corn Refiners Assoc., 1001 Connecticut Ave., N.W. Washington, D.C. 20036.
81. ----, L.S. Bates, and O.E. Nelson. 1964. Mutant gene that changes protein composition and increases lysine content of maize endosperm. Science 145:279-280.
82. Nelson, O.E., E.T. Mertz, and L.S. Bates. 1965. Second mutant gene affecting the amino acid pattern of maize endosperm proteins. Science 150:1469-1470.
83. ----, and H.W. Rines. 1962. The enzymatic deficiency in the waxy mutant of maize. Biochem. Biophys. Res. Commun. 9:297-300.
84. ----, and C.Y. Tsai. 1964. Glucose transfer from adenosine diphosphate-glucose to starch in preparations of waxy seeds. Science. 145:1194-1195.
85. Neuffer, M.G., G.L. Jones, and M.S. Zuber. 1968. The mutants of maize. Crop Sci. Soc. of Am. Madison, Wis.
86. Nordstrom, J.W., B.R. Behrends, R.J. Meade, and E.H. Thompson. 1972. Effects of feeding high oil corns to growing-finishing swine. J. Anim. Sci. 35:357.
87. Parker, M.W. 1935. Physical and chemical properties of the soluble polysaccarides in sweet corn. Plant Physiol. 10:713-725.
88. Parnell, F.R. 1921. Note on the detection of segregation by examination of the pollen of rice. J. Genetics 11:209-212.
89. Peat, S., W.J. Whelan, and J.R. Turvey. 1956. The soluble polyglucose of sweet corn (Zea mays). J. Chem. Soc. p. 2317-2322.
90. Pickett, R.A. 1966. Opaque-2 corn in swine nutrition. p. 19-22. In E.T. Mertz (ed.) Proc. High Lysine Corn Conf. Corn Refiners Assoc. 1001 Connecticut Ave. N.W., Washington, DC 20036.
91. Poneleit, C.G., and D.E. Alexander. 1965. Inheritance of linoleic and oleic acids in maize. Science 147 (3665):1585-1586.
92. Pradilla, A.G., D.D. Harpstead, D. Sarria, and F.A. Linares. 1975. Quality protein maize in human nutrition. p. 27-37. In E.T. Mertz (ed.) High quality protein maize. Dowden, Hutchinson and Ross, Inc., Stroudsburg, Penn.
93. Quackenbush, F.W., J.G. Firch, A.M. Brunson, and L.R. House. 1963. Carotenoid, oil, and tocopherol content of corn inbreds. Cereal Chem. 40:250-259.

94. Rogler, J.C. 1966. A comparison of opaque-2 and normal corn for the chick. p. 23-25. *In* E.T. Mertz (ed.) Proc. High Lysine corn Conf. Corn Refiners Assoc., 1001 Connecticut Ave. N.W., Washington, DC 20036.
95. Sandstedt, R.M., D. Strahan, S. Ueda, and R.C. Abbot. 1962. The digestibility of high-amylose corn starches compared to that of other starches. The apparent effect of the *ae* gene on susceptibility to amylose action. Cereal Chem. 39:123-131.
96. Shannon, J.C. 1972. Movement of C^{14}-labeled assimilates into kernels of *Zea mays* L. I. The pattern and rate of sugar movement. Plant Physiol. 49:198-202.
97. Sprague, G.F., B. Brimhall, and R.H. Hixon. 1943. Some effects of the waxy gene in corn on properties of the endosperm starch. J. Am. Soc. Agron. 35:817-822.
98. ----, P.A. Miller, and B. Brimhall. 1952. Additional studies of the relative effectiveness of two systems of selection for oil content of the corn kernel. Agron. J. 44:329-331.
99. Straughn, M.N. 1907. Sweet corn investigations. Maryland Agric Exp. Stn. Bull. 120. p. 37-78.
100. ----, and C.G. Church. 1909. The influence of environment on the composition of sweet corn. USDA Bur. Chem. Bull. 127. 60 p.
101. Sumner, J.B., and G.F. Somers. 1943. The water-soluble polysaccharrides of sweet corn. Arch Biochem. Biophys. 4:7-9.
102. Tsai, C.U., F. Salamini, and O.E. Nelson. 1970. Enzymes of carbohydrate metabolism in the developing endosperm of maize. Plant Physiol. 46:299-306.
103. Vineyard, M.L., and R.P. Bear. 1952. Amylose content. Maize Genet. Coop. Newsl. 26:5.
104. Weatherwax, P. 1922. A rare carbohydrate in waxy maize. Genetics 7:568-572.
105. Weaver, B.L., and A.E. Thompson. 1957. Fifteen generations of selection for improved popping expansion in White Hulless popcorn. Ill. Agric. Exp. Stn. Bul. 616.
106. Weber, E.J., and D.E. Alexander. 1975. Breeding for lipid composition in corn. J. Am. Oil Chem. Soc. 52(9):370-373.
107. Whelan, W.J. 1961. Recent advances in the biochemistry of glycogen and starch. Nature 190:954-957.
108. Zuber, M.S., W.L. Deatherage, C.O. Grogan, and M.M. MacMasters. 1960. Chemical composition of kernel fractions of corn samples varying in amylose content. Agron. J. 52:572-575.
109. ----, C.O. Grogan, W.L. Deatherage, J.E. Hubbard, W.E. Schulze, and M.M. MacMasters. 1958. Breeding high amylose corn. Agron. J. 50:9-12.
110. ----, and J.L. Helm. 1975. Approaches to improving protein quality in maize without use of specific mutants. p. 241-252. *In* E.T. Mertz (ed.) High quality protein maize. Dowden, Hutchinson and Ross, Inc., Stroudsburg, Penn.

Chapter 8 Diseases of Corn

A.J. ULLSTRUP
Purdue University
West Lafayette, Indiana

Since the publication of the first edition of this volume in 1955 some new, undescribed diseases of corn (*Zea mays* L.) have appeared in the USA. A new race of a well-known pathogen, *Helminthosporium maydis* Nisik. & Miy., became destructive. Some diseases that were of little economic importance have become more prevalent in certain seasons during this span of about 20 years. At least two fungus parasites appear to have been introduced into this country. An unusual parasite of corn — a seed plant, called witchweed — was first observed in the Carolinas in 1956. Apparently this plant was introduced from the Eastern Hemisphere, but the time and the means of introduction are unknown.

In the earlier edition of "Corn and Corn Improvement," it was pointed out that estimates of annual disease losses in corn ranged from 2 to 7% in the USA, and that compared to other crops corn was relatively healthy. This is generally true today. Statements were also made that in the USA the diseases of corn are seldom, if ever, destructive over wide areas, and that production has not been seriously limited by disease when weather and soil conditions were favorable for growth of the crop. These latter statements must now be modified following the catastrophic epidemic of southern corn leaf blight in 1970. This disease *did* occur in severe proportions over a wide area, i.e., much of the corn-growing regions east of the Mississippi River, and in some sections west of this boundary. The weather in the 1970 growing season *was* favorable for the growth of corn, but it also was optimum for rapid reproduction of a race of the pathogen causing southern corn leaf blight.

The question may arise as to why new diseases have appeared and why some older, well-known diseases, that have generally been of minor importance, have become more prevalent in some seasons during the past several years. Some of the pathogens of *new* diseases may have arisen by mutation; some were probably inadvertently introduced from outside the USA; unusually favorable environmental conditions may have played a role in the noted increase in severity and prevalence of some corn diseases.

Awareness of corn diseases has been stimulated by the epidemic of 1970. It demonstrated that corn, too, is vulnerable to severe disease under some circumstances. Since 1970 the crop has been under closer scrutiny, perhaps, than ever before. Farmers, seed producers, extension personnel, and researchers are looking more closely at corn from the standpoint of its diseases. More research on the pathology of corn is being pursued now than in earlier years. This side effect of the epidemic of 1970 obviously is not entirely responsible for the increase in the numbers and the severity of corn disease — some of these diseases appeared prior to 1970. The epidemic simply helped draw attention to the diseases of corn and their potential effects.

Diseases may affect corn production in several ways: Reduction in stand due to seedling diseases is one way losses may be incurred. Such decreases in stand, however, are not directly reflected in yield depression. The remaining healthy plants will tend to compensate stand reductions by taking advantage of the availability of greater amounts of nutrients and soil moisture. Leaf diseases reduce functional photosynthetic area and thus lower yield. Here, yield reduction is related to the time when a severe level of disease is attained. The earlier this occurs the greater the yield loss. Stalk rots and root rots cause lodging, and if onset is before full maturity the grain weight is reduced. Lodged corn is difficult to harvest by machine, and often appreciable numbers of ears are left in the field. Ear rots reduce feeding quality of the grain and, in at least one ear disease, the infected grain is toxic to swine.

In this chapter, discussion will be limited to those corn diseases occurring in the USA, with emphasis mainly on those caused by bacteria, fungi, and viruses.

I. GENERAL CONSIDERATIONS

1. Factors Affecting Disease Development

Fluctuations in severity and prevalence are characteristic of most infectious plant diseases. This variation depends on three important factors, namely, the environment, the presence or absence of an often variable disease-inciting agent, and the relative resistance of the host. A fourth factor, that of the presence or absence of a vector and the effect of the environment on it, comes into play in certain diseases. Epidemics occur when these three factors, and sometimes a fourth, interact. The absence, or the failure, of one factor to function may upset this interaction and check the diseases. Thus the environment may be optimum for a particular disease to flourish, the pathogen present, but if the host is resistant, disease development will be minimal or entirely suppressed. Similarly, the pathogen may be present, the host susceptible, but if the environment is unfavorable the disease will be of little, or no, consequence. An example of the role of the fourth factor, the vector, operative only in some diseases, would be where the environment was optimum for disease, the host susceptible and a virulent pathogen present, but if the vector population was much reduced, the level of disease would be low. Shifts in relationships among these factors determine the variations commonly observed in the prevalence and severity of plant diseases.

a. *Environment.* Temperature, rainfall, and humidity have a marked influence on plant diseases. Fertility, drainage, and pH of the soil may play indirect, but nonetheless certain, roles. Although the environmental requirements may differ for different diseases of corn, many become prevalent when moisture supplies are ample and the temperature within a moderate range (20 to 30 C) such as are near optimum for growth of the host. Water on the surface of the plant is a requisite for germination of spores of the fungi that attack corn. It is also necessary for penetration of the host by some of the pathogenic bacteria. Some diseases become severe when the temperature is relatively high, others increase when the temperature is in a moderate range. A few diseases thrive when corn is under the stress of low soil temperatures (8 to 10 C). Nutrition of the host affects the development of some diseases, particularly those in which the pathogen possesses a relatively *low level of parasitism* and becomes aggressive only when the host is under stress. At least one disease occurs only where drainage is inadequate and the soil becomes waterlogged during a period from a few days after planting until seedlings are 10 to 15 cm high.

b. *Host Resistance.* The inherent resistance or susceptibility of the host may be a determining factor in the occurrence of disease. Inbred lines and hybrids exhibit marked differences in resistance to certain diseases. Resistance to many of the common diseases is determined by a number of genes. The mode of inheritance of resistance is then said to be polygenic, or quantitative. Resistance to some diseases is governed by a single dominant gene, and more rarely by a recessive gene. In some diseases the mode of inheritance of resistance may be either monogenic or polygenic. Resistance to each different race of certain corn pathogens is determined by a specific gene.

Many of the present-day hybrids have some resistance to a few of the more economically important diseases. When these and other less prevalent diseases become severe, however, such hybrids may sustain some damage. The progress that has been made during the past 25 years in breeding for disease resistance in corn has been appreciable. The results obtained thus far offer much promise for future endeavor in this area.

c. *The Disease-Inciting Agent.* Most diseases of corn are caused by fungi, some by bacteria, and a few by viruses. At least one corn disease is caused by a mycoplasma-like entity, one by a higher seed plant, and several by nematodes. A majority of the known diseases of corn occur in the USA, but fortunately a few of the most severe downy mildews and certain of the virus diseases have not been found in this country. The absence of some of these from the USA may be due to an unfavorable environment having a direct effect on the pathogen or an indirect effect on the vector, providing the latter is necessary for the pathogen to establish infection and to be transmitted. Some diseases of corn not found in this country may have been excluded because of effective quarantines.

Variation in pathogenicity is often observed in plant disease organisms, but some appear to be relatively constant. Such variability may arise through mutation, through combination and resegregation attending the sexual processes in heterothallic fungus pathogens, or through reassortment of nuclei of different genetic potentials in thalli of heterocaryotic fungi. The occurrence of specialized pathogenic races within species of disease-producing organisms may be accounted for on these theoretical bases. In some instances new physiologic races appear relatively frequently and as a consequence complicate and compound the efforts of plant breeders. Most of the pathogens of corn appear to be relatively stable, but occasionally new races, or strains, as they are referred to in the case of viruses, arise. The pathogens of the two corn rusts in the USA each appear to be composed of a number of physiologic races that can be differentiated by their pathogenicity toward certain genotypes of the host.

2. Control of Corn Diseases

There are several means by which corn diseases may be controlled or held in check so that they cause minimal losses. By far the most important method of controlling corn diseases is through the use of resistant hybrids. No hybrid is resistant to all diseases and whether one could be developed, or is even necessary, is questionable. Hybrids resistant to the major diseases occurring over wide areas can be developed through modern plant breeding procedures. New and diverse sources of resistance to diseases must be found and such resistance incorporated into otherwise agronomically desirable inbred lines.

The use of fungicides to control corn diseases is limited because of extensive culture of the crop and the low monetary value per unit of product. Fungicides are applied to seed to protect them from rot and, subsequently, seedling blight. Foliar applications of fungicides are made on high-income corn such as winter crops of sweet corn grown in the South for shipment to northern markets. The price per unit is sufficiently high to warrant expenditures for fungicides. In some areas failure to protect sweet corn from the ravages of certain leaf diseases means little or no harvest of marketable ears. Seed corn, which demands a high price, can also be economically protected from leaf diseases where the latter are severe and reduce quality of the product. Proper protection demands routine spraying since there is at present no satisfactory method of long-range forecasting of prevalence and severity of most corn diseases. At present, routine spraying of corn grown for animal feed generally is not economically feasible. The manner in which corn grows — the progressive development, until flowering, of new leaf tissue that must be covered with a fungicide as it becomes exposed — means frequent and costly applications of the chemical.

Certain cultural practices have been shown to minimize some corn diseases. Maintaining well-balanced soil fertility is one of these practices. Observations indicate that the severity of the stalk rots can be reduced when major plant-food deficiencies are corrected and soil fertility restored to proper balance. There is need for more research designed specifically to determine the basic relationships between plant nutrition and corn diseases. At present a general statement cannot be made that all diseases of corn can be controlled by increasing soil fertility.

Crop rotation and sanitation measures, such as clean plowing and destruction of refuse harboring pathogens of corn diseases, have been recommended. Crop rotation is generally advisable to improve tilth and conserve fertility. Some rotation sequences have shown to be less favorable than others for development of certain corn diseases. Under some circumstances crop rotation is not always practical, e.g. in river-bottom land. With the increasing popularity of minimum-till for seed-bed preparation, questions arise relating to survival of some disease-inciting organisms that are on or within plant refuse on the soil surface. Observations clearly suggest that some corn pathogens do remain viable in the surface refuse and can act as a source of primary inoculum in initiating epidemics. Under conditions where no-till or minimum-till is practiced, diseases often become prevalent earlier than where clean plowing is done and the pathogens are buried, along with debris, beneath the soil surface where they cannot function in establishing disease. If it has been shown by past experience that the hazard of disease is slight, then minimum-tillage practices will save time and money. On the other hand if some diseases, in which the pathogens survive in surface debris, have caused losses, then clean culture might be used to advantage as a partial control measure.

II. SEED ROT AND SEEDLING BLIGHTS

Seed rot and seedling diseases are not often of major economic importance. With improvements in the mechanics of processing seed corn some of the factors that predispose the seed and seedlings to these diseases have been minimized. The general use of fungicides as seed-treating materials has been effective in reducing losses.

The symptoms of seed rot are not especially specific to the pathogen involved and consist simply in complete decay of the seed before, or at about the time, germination commences. Seedling blight, which may occur prior to, or after, emergence may present a number of symptoms. Brown, water-soaked lesions on the roots, complete rotting of root tips to give the appearance of root pruning, or brown lesions on the mesocotyl are characteristic below-ground symptoms of seedling infection by a number of pathogenic fungi (Fig. 1). Progressive wilting and eventual dying beginning at the tips of seedling leaves is an above-ground symptom often associated with mesocotyl infection by *Pennicillium oxalicum* Thom. (Johann et al., 1931) Conspicuous brown, sunken lesions on the mesocotyl are symptoms that follow infection by *Gibberella zeae* (Schw.) Petch, [Syn. = *G. roseum* f. sp. *cerealis* (Cke.) Snyd. & Hans.; *G. saubinetti* (Mont.) Sacc.]; (Asexual stage = *Fusarium roseum* f. sp. *cerealis*, Graminearum) [Syn. = *F. graminearium* Schw.] and *Diplodia maydis* (Berk.) Sacc. [Syn. = *D. zeae* (Schw.) Lev.]. A distinct virescence of seedling leaves following infection by *Aspergillus flavus* Link and *A. tamarii* Kita has been described (Koehler and Woodworth, 1938). Infection by these fungi is uncommon. Stunting is a symptom common to most seedling blights. Because of the variation and overlapping of symptoms of seed rots and seedling blights it is difficult to indicate which pathogen is involved.

One seedling condition, occasionally observed, is characterized by failure to emerge. When uncovered the seedlings are found twisted, gnarled and etiolated. No pathogen has been isolated, and all tissues appear sound. The cause is unknown. It is not related to crusting of the soil. Shallow spring plowing, early and deep planting so that seed is in contact with overturned debris are cultural practices that have been associated with the condition.

A number of species of fungi are capable of infecting corn seed and seedlings. Some, and probably the most important, are strictly soil inhabitants; others are soil invaders, but are mostly aerial in habit and are carried on and within seed. Among the former are several species of *Pythium*. In an assay of some Wisconsin soils the following were found in descending order of prevalence to be parasitic on corn seed and seedlings: *Pythium irregulare* Buisman, *P. debaryanum* Hesse, *P. ultimum* Trow., *P. rostratum* Butler, *P. paerocandrum* Drechsler, and *P. vexans* de Bary (Hoppe and Middleton,

1950). The population of species varied quantitatively and qualitatively with soil type and crop sequence. In view of these findings it may be expected that these and other species occur in greater or lesser prevalence in different localities, depending on environmental conditions.

The morphology of members of the genus *Pythium* is comparatively simple. Mycelium is made up of fine, hyaline filaments that are irregularly branched and coenocytic when young. Sporangia are either spheroidal of filamentous and germinate by the release of motile zoospores from an evanescent vesicle. Oogonia, in which usually one oospore develops, are

Fig. 1. Seed rot and seedling blight.

spherical, ellipsoidal or limoniform, and possess smooth or spiny walls. Germination is not often observed, but when it does occur it is accomplished by release of zoospores. All species appear to be homothallic. A detailed study of this genus has been published with respect to taxonomy, host range, and geographical distribution (Middleton, 1943).

Typical corn-ear pathogens, *Gibberella zeae, Diplodia maydis* and *Fusarium moniliforme* Sheld. (Sexual stage = *Gibberella moniliforme* (Sheld.) Snyd. & Hans.); [Syn. = *G. Fujikuroi* (Saw.) Wr.], have been isolated from diseased seedlings. These are most destructive when they have become well established in the seed before harvest. After such seed is planted these fungi may become rapidly invasive. There is some disagreement regarding the virulence of *F. moniliforme*. Some report it to be vigorously parasitic on corn seedlings (Djakaminhardja et al., 1970; Edwards, 1935; Futrell and Kilgore, 1969; Valleau, 1920; Voorhees, 1933b). Others consider it to be weakly parasitic and invading senescent or debilitated tissue (Leonian, 1932; Limber, 1927; McKeen, 1951a). In unpublished research the writer has found this fungus to be widely distributed and readily isolated from tissue that is weakened or dead. The following organisms, and a few less frequently encountered fungi, have been associated with seed rot and seedling blight of corn. *Nigrospora oryzae* (Berk & Br.) Petch [Syn. = *Basisporium gallarum* Moll]; *Rhizoctonia zeae* Voorhees; *Rhizoctonia bataticola* (Taub.) Butl.; [Syn. = *Macrophomina phaseoli* (Maubl) Ashby; *Macrophomina phaseolina* (Tassi) G. Goid.; *Sclerotium bataticola* Taub.], *Trichoderma* spp. These generally are less virulent and not so prevalent as the *Pythium* spp., and the more common corn ear-rot fungi.

Most of the pathogens become aggressive when the seedlings are under stress of low temperatures that sometimes prevail after planting. *Penicillium oxalicum*, however, is an exception and can cause disease at soil temperatures optimum for germination and growth of seedlings (Diachun, 1939, Johann et al., 1931).

The relative maturity or degree of "finish" of the seed is an important factor relating to seed decay and seedling blight. Mature, well-finished seed is much more resistant to invasion by pathogenic fungi than poorly finished seed. Soundness of the seed is another factor that may determine the prevalence of these diseases (Tatum, 1954; Tatum and Zuber, 1943). Mechanical abrasions afford easy access by the pathogens to the interior of the seed from which they can invade the embryo or the seedling. Age of seed and storage conditions are other critical factors relating to seedling diseases. Old seed and seed stored at relatively high temperatures and humidities is highly susceptible to invasion by seedling-disease pathogens.

Control of seed rot and seedling diseases can be practiced by several means. One nearly universally used control measure is treating seed with a fungicide especially formulated for this purpose. These fungicides are complex organic compounds that are especially toxic to *Pythium* spp. These materials are protectants against invasion by soil-borne fungi; they are not volatile and hence have little or no effect in eradicating pathogenic fungi within the seed. Organic mercury compounds, which are no longer used because of their high mammalian toxicity, are volatile and more effective in controlling internally borne fungi. Present-day seed corn fungicides impose little, or no, hazard to handlers of treated corn. Such corn, however, should not be fed to livestock.

Biological control of seed decay and seedling blights of corn has been investigated. This involves coating the seed with spores of fungi antagonistic to the pathogens, but harmless to the seedling (Chang and Kommedahl, 1968). It has not been put to practical use.

Other means of control involve correction or elimination of those factors that predispose seed and seedlings to these diseases, i.e., avoidance of excessive mechanical damage by careful seed processing; use of well-matured seed; storage of seed at relatively low temperature and humidity; and delaying planting until soils are warm and favorable for rapid germination and growth.

"Cold testing" of seed corn is a practice that may be used to determine 1) effectiveness of seed-treating materials, 2) the germination of seed under adverse conditions of such as those encountered in the field, and 3) the resistance of genotypes to seed rot and seedling blight. Cold testing consists in placing seed in contact with moist field soil for about 10 days at 10C. Seeds are then removed to a temperature of 25 to 30C to determine their germinability. Placing seed in contact with soil is essentially an inoculation with soil-borne pathogens, primarily species of *Pythium*. Muck soil is often used for this purpose since one or more species of *Pythium* are usually present in abundance.

Little is known of the basis for resistance to seed rot and seedling diseases. Research suggests many genes are involved (Hoppe, 1929; McIndoe, 1931). Some progress has been made in selecting genotypes in which seeds are capable of germinating at low temperatures. This character in itself appears to be directly related to resistance to these diseases. Kernels with horny endosperm are generally more resistant to seed and seedling pathogens than kernels with soft-starchy or sugary endosperm.

III. STALK AND ROOT ROTS

Diseases of corn stalks are widespread in the USA, and although there is much interannual variation in severity, some stalk rot occurs in every field every year. Various estimates have been made of the losses due to stalk rot, but over an extended period of years it is perhaps of greater economic importance than any other single corn disease. Stalk rot of corn is a complex condition involving several pathogenic fungi and bacteria.

When the onset of stalk rot occurs before physiological maturity (30 to 35% kernel moisture) premature dying takes place. Then appreciable losses are sustained because kernels fail to fully develop (Hooker and Britton, 1962). When stalk rot develops later the yield of grain is not reduced, but diseased plants may lodge badly. Ears on lodged plants are difficult to harvest with machinery and many are left in the field unless gleaned by livestock.

The distinction between stalk and root rots is not clear-cut. Most of the common fungus parasites of the stalks will also attack roots. Some fungi seem to be confined to the stalk, and a few may be considered strictly root parasites.

A number of fungi and bacteria have been isolated from stalks and roots. Some, but not all, are similar in their levels of parasitism in that they are comparatively weakly parasitic and become invasive only after stalk or root tissues are approaching maturity or are under other stresses. The pathogens could be classified as those that are strongly parasitic and those that are only weakly so.

Corn stalks and roots mature, become senescent and eventually die. The beginning of this process occurs several weeks after silking and continues. It is during this time that stalks and roots are invaded by parasitic fungi. When young corn stalks are inoculated with certain of these fungi little or no invasion of the tissue takes place until several weeks post-silking; then rotting commences and progresses rapidly if the host is susceptible and slowly if it is resistant. The time of onset and rate of senescence appear to determine, at least in part, how soon and how rapidly the tissues are invaded. This would seem to provide some evidence that some of the stalk-rot fungi are weak parasites and become invasive when the host is under stress of the aging process. Imposing a stress such as removal or destruction of leaf or root tissue — either naturally by means of leaf diseases, insect injury or hail damage, or by artificial clipping and pruning — accelerates spread of these fungi in the stalks and roots. Another example suggesting that some of these fungi are not aggressive parasites is the resistance of the stalks of *indeterminate* (*id*) mutants. This mutant fails to flower, remaining juvenescent and free from stalk rot for its entire life, only to succumb eventually to freezing. Its siblings, differing only by a single gene from the mutant, flower and bear ears normally. The normal plants are invaded by certain stalk-rotting fungi as senescence progresses.

The influence of host nutrition on the severity of some stalk rots has been investigated (Abney and Foley, 1971; Foley and Wernham, 1957; Martens and Arny, 1967a; Otto and Everett, 1956). There is no complete agreement relative to the effects of different nutrients except that under a balanced nutritional regime stalk rot is less severe than where an imbalance of nutrition occurs. Excessively high levels of N coupled with low amounts of K favor premature dying, stalk rotting, and stalk breakage. Some evidence indicates that KCl is superior to other forms of K in minimizing the severity of stalk rot (Martens and Arny, 1967a; Younts and Musgrave, 1958; Zuber and Grogan, 1958). The basic physiological effect on the Cl-ion is not fully understood, but it appears to be related to the N metabolism of the host.

Death of pith cells of the internodes of the stalk and of cells in the nodal plates appears to precede invasion of the stalks by the several fungi responsible for stalk rot. Cell death is indicative of senescence, and early onset is correlated with susceptibility (Pappelis, 1965; Pappelis and Smith, 1963; Pappelis and Williams, 1966).

Early work (Holbert et al., 1935) demonstrated that removal of leaves increased susceptibility while removal of ears or prevention of fertilization increased resistance. It is a common observation that leaf damage caused by disease, insects, or hail predisposes corn stalks to rotting; and that barren plants usually are free from stalk rot and remain upright among the ear-bearing plants that are lodged.

Mechanical strength of stalks is closely related to lodging. Evidence has shown that thick rind and high crushing strength of the stalk are directly correlated with resistance to lodging (Zuber, 1962), but not necessarily associated with resistance to stalk rot. Mechanical strength coupled with stalk rot resistance is a goal in breeding for standability. One objective in the development of superior hybrids is to obtain those in which physiological maturity is reached before the onset of stalk rot. This means a full-season hybrid that continues to deposit dry matter in the kernels for a maximum period of time but which does mature before killing frosts. After physiological maturity is reached and fungus parasites begin invading the aging stalk and root tissues, standability is then dependent on mechanical strength.

Resistance to stalk rot and lodging appear to be inherited in a quantitative manner with a number of genes involved (Jinahyon and Russell, 1969; Sprague, 1954). Significant progress has been made in the development of genotypes of corn resistant to stalk rot and lodging through the use of different breeding procedures. This is attested by comparison of the resistance of such popular current inbred lines as B14, B37, Oh43 and H60 with CI540, R4, L317, Oh28 and Os420 that were used in hybrids 30 to 35 years ago. Some of the advances made in stalk rot resistance have been negated by such cultural practices as increasing plant populations per hectare, sometimes beyond the nutritive capacity of the soil to support and by the addition of excessive amounts of N, a practice sometimes creating an imbalance in nutrients.

Resistance to stalk rot has been positively correlated with high total sugar content of the stalks (Craig and Hooker, 1961; Martens and Arny, 1967; Messiaen, 1957). Total sucrose content was found to be highest about 1 week before silking, when resistance is fully expressed, and then declined with maturity. Removal of ears resulted in a higher sucrose content than found in ear-bearing plants (Sayre et al., 1931). Barren plants are generally resistant to stalk rot. Sugar content per se is probably not directly responsible for resistance, but rather is indicative of physiological activity of the tissue.

An ether extract made from stalks soon after pollination was more inhibitory to in vitro growth of *Diplodia maydis, Gibberella zeae* and *Nigrospora oryzae* than extracts made later in the season when resistance declines. However, extracts from plants that were defoliated to increase susceptibility did not enhance growth of the fungus, nor was there a reduction of growth on extracts from plants in which pollination was prevented in order to increase resistance (Johann and Dickson, 1945). It has also been pointed out that ether extracts of stalks made at silking of both resistant or susceptible genotypes were inhibitory to growth of *G. zeae*. The inhibitory factor disappeared in both extracts with age, but faster in that of the susceptible host (Barnes, 1959).

An aglucone, 2, 4-dihydroxy-7-methoxy-1, 4 benzoxazin-3-one (DIMBOA), which is formed by the action of an enzyme on a glucoside precursor when corn tissue is injured, has been reported to be responsible for resistance in corn stalks to infection by *G. zeae*, to infection of leaves by *Helminthosporium turcicum* Pass. (Klun, 1969; Molot, 1969 a, b, c; Molot and Anglade, 1968) and to the first brood of the European corn borer. DIMBOA is chemically unstable and decomposes into 6-methoxy-2-

benzoxazolinone (MBOA). The concentration of MBOA, which is stable, is correlated with the levels of the unstable DIMBOA. Bioassays of the glucoside, of DIMBOA and of MBOA, indicate that only DIMBOA is biologically active.

The several stalk rots of corn are usually named in part after the particular parasite involved, such as gibberella stalk rot, diplodia stalk rot, pythium stalk rot, or less specifically, bacterial stalk rot. At least one of these diseases, charcoal rot, is named after the black appearance of infected stalk tissue.

Two monographic treatises dealing with the several aspects of stalk rots of corn have been published in recent years (Christensen and Wilcoxson, 1966; Koehler, 1960).

1. Diplodia Stalk Rot

Diplodia stalk rot is a common disease throughout the central Corn Belt and is probably responsible for much of the stalk lodging and premature killing of plants. The disease fluctuates from year to year, being much less prevalent when weather is abnormally cool during the maturation of corn.

The disease does not ordinarily become apparent until several weeks after silking. Affected plants die suddenly and leaves take on a light grayish-green color resembling frost injury. Lower internodes lose their bright green color and solidity, become brown or straw-colored, spongy, and easily crushed (Fig. 2A). The pith rapidly disintegrates leaving only the vascular bundles intact. Plants that are killed prematurely produce chaffy, poorly finished grain. The amount of stalk breakage following attacks by this disease is always most severe when high wind and driving rains prevail prior to harvest. When invasion and killing is delayed until after physiological maturity, little, if any, yield reduction is sustained, but the hazard of stalk breakage remains. Roots attacked by the pathogen are discolored and disintegrated, and so is the stalk tissue. A definite sign of diplodia stalk rot is the presence of pycnidia of the fungus on the stalks (Fig. 2B). These spore-bearing structures are black, subepidermal, and most abundant near the nodes. In some years they are numerous in the fall, while in other years they do not appear and mature until the following summer.

Diplodia stalk rot is incited by *Diplodia maydis*. The fungus produces two types of spores, the most common of which are brown to olivaceous in color, generally two-celled, straight to slightly curved, and measure 25 to 30 $\times$ 5 to 6 μ. The second, less frequently seen, are scolicospores, which are hyaline, needle or threadlike, and 25 to 35 $\times$ 1 to 2 μ in size (Johann, 1939). Both are borne in pycnidia and, under warm, wet conditions, are exuded in a mucous-like matrix in long chains from these spore-bearing structures. When these spore chains dry they fragment and are then dispersed by wind currents. No sexual stage of the fungus is known.

Infection of stalks may take place at the nodes following germination of spores that lodge behind leaf sheaths (Durrell, 1923). It has also been suggested that infection occurs in the seedling stage. The pathogen, which may be in the soil or within the seed, invades the mesocotyl or crown of the young plant. Such seedlings do not die, but remain parasitized for the remainder of the growing season. After pollination, when physiological activities gradually lessen with the approach of maturity, the quiescent mycellium becomes active and lower nodes and stalks are then invaded (McNew, 1937). It is questionable if infection of stalks usually arises from infected seed. The generally high quality of hybrid seed, and the very low percentage of infection found in such seed does not correlate with the amount of stalk rot observed each year.

Fig. 2. Stalk rot: A, diplodia stalk rot; B, pycnidia of *Diplodia maydis* embedded in stalk near lower nodes; C. perithecia of *Gibberella zeae* on the surface of a diseased stalk; D, charcoal rot showing dark gray to black discoloration.

Diplodia stalk rot appears to become more prevalent when rainfall in August and September is ample or excessive. This seems to be most evident if precipitation in June and July has been deficient.

Control of diplodia stalk rot can be effected, at least in part, by growing hybrids resistant to the disease. Resistance to this disease is inherited in a quantitative manner with a number of genes involved. Genes governing resistance are dominant or partially dominant to susceptibility (El-Rouby and Russell, 1966).

Resistance to leaf diseases will also tend to minimize severity of stalk rot by providing functional leaf tissue at least until physiological maturity is reached.

Maintenance of balanced fertility by judicious application of necessary elements and adjustment of planting rates consistent with soil fertility are practices that will help reduce severity of this disease.

2. Gibberella Stalk Rot

Gibberella stalk rot is widely distributed, but occurs more frequently in the northern and eastern areas of the USA. When growing seasons are relatively cool, such as in 1972, the range of this disease will extend farther southward into the central Corn Belt, where it may become the dominant stalk disease.

Symptoms of this disease are similar to those of diplodia stalk rot. Early onset of the disease is expressed by sudden wilting and change in color of leaves from bright green to dull green and rapidly turning to straw color. The exterior of lower internodes of infected plants are straw colored as opposed to green of healthy plants, and the pith is disintegrated leaving only the vascular bundles in evidence. A pink-to-reddish discoloration often, but not invariably, is evident in this disease. This distinguishes it from diplodia stalk rot.

Gibberella zeae is the inciting agent of this disease. The perithecia of this fungus are blue-black, spherical, and borne on the surface of infected corn stalks, especially around lower nodes (Fig. 2C). These structures are often seen in the fall, but generally the spores contained within them do not mature until the following late spring or summer after warm, moist weather prevails. Within the perithecia numerous asci develop. These are long, sac-like hyaline structures, each containing eight ascospores. These spores are hyaline, long-oval in shape, one to three septate, and measure 20 to 30 × 3 to 5 μ in size. Mature ascospores are forcably ejected from the ostiole at the tip of the neck of the perithecia during wet periods and then dispersed by wind currents to susceptible tissue, where under favorable conditions, they germinate and establish infection. Spores of the asexual stage (*Fusarium graminearum*) are hyaline, slightly curved, with pointed ends, three to five septate, 30 to 60 × 4 to 6 μ in size and borne on the mycelium.

Factors influencing prevalence of the disease are essentially the same as those affecting diplodia stalk rot. Gibberella stalk rot and diplodia stalk rot are both diseases of senescense and, being so, are not only adversely affected by the same factors, i.e. imbalance in soil nutrients, destruction of functional leaf area, but amenable to control by use of resistant, full-season hybrids.

Resistance to gibberella stalk rot is closely correlated with resistance to diplodia stalk rot. Resistance to these stalk rots is quantitative, with a number of genetic factors involved, many of them probably linked or associated with those governing time and rate of maturation.

3. Fusarium Stalk Rot

Fusarium stalk rot is widespread and occurs in the USA wherever corn is grown. It seems to be more prevalent, however, in the drier, warmer regions of the country.

Symptoms of the disease are not greatly different from those of diplodia and gibberella stalk rots. Early development of the disease is characterized by premature death of plants. The interior of the lower internodes of stalks is discolored, and pith is disintegrated. The red coloration and the black perithecia sometimes found in gibberella stalk rot do not occur in this disease, nor are pycnidia in evidence, whereas they are in diplodia stalk rot. A pale pink to whitish mycelium sometimes develops at the nodes of infected plants and within the disintegrated internodes.

The disease is caused by *Fusarium moniliforme* (Sexual stage = *Gibberella moniliforme*). *Fusarium moniliforme* (Sheld.) var. *subglutinans* Wr. & Rg. (Sexual stage = *Gibberella fujikuroi* (Saw.) Wr. var. *subglutinans* Ed.) is often isolated from corn plants exhibiting symptoms of fusarium stalk rot. This fungus differs from *F. moniliforme* in that its microconidia are borne in false heads instead of chains.

Macroconidia of *F. moniliforme* are sparse. They are hyaline, 15 to 60 × 2.5 to 5 μ in size, curved near the tips, and bear three to five septa. Microconidia are abundant 5 to 12 × 2 to 3 μ in size, and single-celled. The sexual stage, *G. moniliforme*, is exceedingly rare in nature. Perithecia are blue-black and globose. Asci within the perithecia are elongated, sac-like and measure 75 to 100 × 10 to 16 μ. Each ascus contains eight ascospores which are straight and taper toward tips, one-septate, and 12 to 17 × 4.5 to 7 μ in size.

Macroconidia of *F. moniliforme* var. *subglutinans* are slightly less curved and usually bear three septa. Microconidia are essentially indistinguishable from those of *F. moniliforme* except that they are borne in false heads, never in chains. The sexual stage, *G. moniliforme* var. *subglutinans,* is also found sparsely in nature, and is morphologically similar to *G. moniliforme*.

Both *F. moniliforme* and the variety *subglutinans* are widespread throughout the world. They appear especially prevalent in drier areas and are frequently isolated from plant debris and from debilitated or senescent tissue. They are generally not vigorously parasitic, but often rapidly colonize tissue initially invaded by more active parasites.

4. Charcoal Rot

Charcoal rot is found in the drier areas where corn is grown (Hoffmaster et al., 1943; Livingston, 1945). In 1932 the disease was described on corn and other crops in the Sacramento Valley of California (Mackie, 1932). In some years the disease has appeared in the central Corn Belt (Semeniuk, 1942;

Tehon and Boewe, 1939).

The disease, like diplodia, gibberella, and fusarium stalk rots, is one of senescence. Symptoms first become apparent when plants approach maturity. The outside of the lower internodes becomes a straw colored to dark brown. The pith eventually becomes disintegrated leaving the vascular bundles in evidence. A distinguishing characteristic of the disease is the presence of numerous, minute, black sclerotia of the pathogen scattered over the vascular bundles and the inside walls of the stalk (Fig. 2D). This gives the tissue a gray-black cast and it is from this symptom that the name of the disease is derived.

Relatively high soil temperatures (38 to 42 C) and low soil moisture favored development of charcoal rot of corn in Nebraska (Livingston, 1945).

Charcoal rot is caused by *Macrophomina phaseolina* (Tassi) G. Goid. [Syn. = *Macrophomina phaseoli* (Maubl.) Ashby; *Rhizoctonia bataticola* (Taub.) Butl.; *Sclerotium bataticola* Taub.]. The fungus has a wide host range, particularly among crop plants in warm-temperature and tropical areas. Some races of the fungus produce spores; isolates from corn, however, are sterile and do not form spores. Spores are oval, single-celled, hyaline, measure 16 to 32 × 7 to 10 μ in size and are borne in black, globose pycnidia. Sclerotia are generally smooth, black and range from 0.05 to 0.20 mm in size.

Little is known concerning inheritance of resistance to charcoal rot of corn such as might be applicable to control of the disease. The type of parasitism appears to be similar in some respects to the three stalk rots discussed above, and on these bases control of the disease will probably be effected by methods similar to those that have been used successfully to reduce the severity of these stalk rots.

5. Pythium Stalk Rot

Pythium stalk rot, while widely distributed where corn is grown under warm, humid conditions, is not of major economic importance because it usually occurs in restricted places in fields and does not become severe over extensive areas.

The disease is different from those stalk rots already described in that invasion by the pathogen may occur in actively growing stalks well before silking. Pythium stalk rot is not a disease of senescence; and in this respect the casual agent is a much more vigorous parasite, being independent of stresses on the host plant to effect invasion and to establish disease.

The conspicuous symptom of the disease is the soft, brown rot usually of a single internode near the soil line (Fig. 3A). Infected plants soon fall over, but do not break. Plants will often remain alive and the prostrate plant may even develop roots at internodes above the rotted portion.

Pythium stalk rot occurs following periods of abnormally warm, moist weather. Frequently the disease is found in river bottom land where such conditions prevail and air drainage is poor.

Pythium stalk rot is caused by *Pythium aphanidermatum* (Eds.) Fitz. [Syn. = *P. butleri* Subr.], a rapidly growing fungus favored by relatively high temperatures. Mycelium is hyaline, non-septate, except adjacent to fruiting structures. The asexual spore-bearing structure (sporangiophores) are filamentous, branched or unbranched measuring 50 to 1,000 $\times$ 4 to 20 μ and give rise to sporangia containing kidney-shaped, biciliate zoospores 12 x 7.5 μ in size. Zoospores, when released, move to infection courts of the host and there establish infection. The sexual spores (oospores) are terminal on the mycelium, spherical and 17 to 19 μ in size. The oospores are encased in a thick wall providing resistance to the extremes of environment. They germinate by production of a germ tube. The fungus is strictly a soil inhabitant.

Fig. 3. Stalk rot: A, pythium stalk showing rotting of a single internode; B, anthracnose of the stalk showing gray-brown discoloration of the pith; C, bacterial stalk and top rot (courtesy Arthur Kelman, University of Wisconsin); D, bacterial stalk rot.

The minor importance of the disease has not warranted any concerted effort to control the disease. Differences in resistance among inbred lines has been reported (Elliott, 1943).

6. Anthracnose

This disease attacks both stalks and leaves of corn. It has been generally of minor importance except where a few localized epidemics have occurred (Dale, 1963, Morgan and Kantzes, 1971; Warren et al., 1973; Williams and Willis, 1963). The disease was severe in parts of Virginia and eastern North Carolina in 1972 and 1973, particularly on no-till plantings.

Early onset of the stalk rot phase of anthracnose presents some symptoms similar to those of diplodia stalk rot, i.e. rapid wilting and dying of the leaves accompanied by a grayish-brown discoloration of the pith (Fig. 3B). Rotting of the pith begins at the crown of the plant and may extend upward for several internodes. The exterior of infected stalks is light brown and often marked by narrow, black streaks running parallel to the long axis of the stalk. These are composed of fruiting structures of the pathogen. The severity of the stalk rot phase of anthracnose is often intensified by the leaf blight caused by the same fungus.

Anthracnose is caused by *Colletotrichum graminicola* (Ces.) Wils. Conidia are hyaline, one-celled, crescent-shaped with pointed ends, borne singly at the tips of short conidiophores, and measure 26 to 31 × 5 to 5.5 μ. Conidiophores, which develop in acervuli are erect, unbranched, hyaline, non-septate, and 5 to 13.5 × 1.5 to 3.5 μ in size. Acervuli are saucer-shaped, brown to black, and are produced on necrotic, infected tissue and on plant debris. Black spines (setae) develop around the spore-bearing area and are one of the identifying characters of the organism and the disease on corn. They average 5.3 μ at the base and 114.3 μ long and have one to four septa. They can be identified by 10 to 20 × magnification within the narrow, black streaks on the surface of infected stalks. A sexual stage of the pathogen has been observed when pure cultures of *C. graminicola* were grown on sterilized corn leaves (Politis and Wheeler, 1972). The morphology of perithecia, asci and ascospores indicate that they are of the *Glomerella* type, but a specific term has not yet been assigned to this stage. A treatise on the mycological aspects of the genus and the relationships of the species has been published (von Arx, 1957).

Races of the fungus that infect sorghum appear unable to attack corn and those that infect the latter do not cause disease in sorghum.

Differences in resistance to anthracnose are evident among inbred lines and hybrids. With the apparent increase in prevalence of the disease in recent years, efforts to transfer such resistance to a broader array of corn germplasm will probably be initiated. Sweet corn is more susceptible to the disease than dent corn, although the latter may suffer severe stalk breakage.

7. Bacterial Stalk and Top Rot

This disease, which involves both stalk and leaf tissue, is found in areas where corn is grown with overhead irrigation (Hoppe and Kelman, 1969;

Kelman et al., 1957). There is no evidence of spread of the disease to healthy plants beyond the area of the overhead irrigation suggesting that the water is the source of inoculum.

The first symptom is a wilting of the leaves in the whorl of the plant, followed rapidly by a soft rot in the uppermost parts of the stalk (Fig. 3C). As the breakdown of tissue progresses plants droop and when lower internodes become rotted the plants fall over. Diseased leaves can be readily pulled from the whorl owing to rotting at their bases.

The cause of the disease has been identified as *Erwinia chrysanthemi* Burk et al. (corn pathovar). The bacterium has been isolated in North Carolina and Wisconsin from corn showing symptoms of bacterial stalk and top rot. What appears to be the same pathogen has also been isolated from diseased corn in Egypt. The bacterium appears to be distinct in its cultural, physiological and pathogenic characteristics from *Erwinia chrysanthemi* Burkholder et al. and *E. carotovora* (Jones) Holland. The organism is a Gram-negative rod with peritrichous flagella. In addition to its ability to cause disease in corn, it also incites a soft rot in potato, squash, onion, cucumbers, carrots, and cabbage. The bacterium is rapidly invasive causing extensive rotting and collapse of corn stalk and leaf tissue within 48 to 72 hours.

Control of the disease can be obtained by eliminating the overhead system of irrigation, or by disinfecting the water by addition of small, non-phytotoxic amounts of chlorine (Thompson, 1965).

Limited observations indicate some differences in resistance to the disease among inbred lines of corn. The disease has not yet become of sufficient economic importance to warrant intensive search for sources of resistance and to transfer such resistance to agronomically acceptable genotypes.

8. Bacterial Stalk Rot

Bacterial stalk rot of corn is generally of minor economic importance. It occurs mainly during warm wet weather and where flooding or surface irrigation is practiced.

A bacterial stalk rot and the causal agent was described several years ago in Arkansas (Rosen, 1922). Later what appeared to be the same disease was reported from West Virginia (Stanley and Orton, 1932).

A conspicuous symptom of the disease is a soft rot, usually of a single internode, at or near the soil line (Fig. 3D). The rotted tissue is brown and may have a putrid odor. Infected plants may frequently show a sudden wilting and drying of the leaves and, because of the breakdown of the stalk tissue, fall over. Vascular bundles remain intact.

Erwinia dissolvens (Rosen) Burkh. [Syn. = *Phytomonas dissolvens* Rosen] is the inciting agent. It is a short, Gram-negative rod 0.5 to 0.9 × 1.5 to 1.2 μ in size, motile by a single polar flagellum or non-motile. Colonies on the commonly used agar substrates are white glistening and fast-growing. The bacteria penetrate through stomata, hydathodes, or wounds.

Another bacterial stalk rot with essentially the same symptoms was described in California (Ark, 1940). The inciting agent, *Pseudomonas lapsa* (Ark) Starr & Burkh. [Syn. = *Phytomonas lapsa* Ark], is a small, Gram-negative rod 0.5 to 1.5 µ long with one to four polar flagella. On nutrient agar, colonies are grayish-blue, glistening, and smooth. On 1% glucose-nutrient agar, a green fluorescent pigment is produced by the organism. The organism has been demonstrated to retain viability on dried corn seed for over a year (Ark 1941). The hazard of the organism being carried on commercial seed corn is probably negligible since it is not systemic and would unlikely come into direct contact with seed. The bacterium has been isolated from beetles (*Diabrotica sp.*) suggesting that this insect in the adult or larval form (corn root worm) may act as a vector.

Erwinia carotovora f. sp. *zeae* Sabet has been shown to be the cause of a stalk rot in tropical areas outside the USA. That disease and its causal organism described from Egypt (Sabet, 1954) is probably the same as described above under "Bacterial Stalk and Top Rot."

The bacterial stalk rots, with the exception of that caused by *Pectobacterium chrysanthemi* pathotype *zeae* are similar to each other in symptoms. The causal agents are also related with respect to most morphological and physiological characters. The symptoms incited by these bacteria bear close resemblance also to those of pythium stalk rot.

Other bacterial pathogens have been associated with stalk rots of corn — *Pseudomonas alboprecepitans* Rosen and *Erwinia stewartii* (E.F. Smith) Dye — but these will be treated under "Leaf Diseases" because the most predominant and conspicuous symptoms incited by these are on leaves.

Little is known about the control of bacterial stalk rots in the USA or of the relative resistance to the diseases. The infrequency of their occurrence has precluded investigations leading to practical control measures.

9. Black Bundle Disease

Black bundle disease was intensively studied several years ago, and at that time considered to be general in the corn-growing areas of the USA (Reddy and Holbert, 1924). In Illinois the incidence was estimated at about 6.5% with as high as 70% infected plants in some fields (Koehler and Holbert, 1930).

Symptoms become evident after kernels have reached the dough stage or later. One of the first symptoms is the purpling of stalks and leaves; such stalks are barren or bear only nubbins. Multiple ear shoots and tillering are other symptoms. Positive indication of the disease is the blackening of vascular bundles. Any one or combinations of these symptoms may appear in a single plant.

Cephalosporium acremonium Cda. is considered to be the inciting agent of the disease. Spores of the fungus are one-celled, hyaline, oval-shaped, 4.5 × 1.5 µ in size, and borne in groups held together in a slimy matrix on the tips of short, simple conidiophores. It was believed that the fungus is an active parasite that invades seedlings and progresses systemically through the vascular system. Artificial inoculation with pure cultures reproduced symptoms associated with the disease under natural conditions (Reddy and Holbert, 1924). Evidence has been presented contrary to these findings (Harris, 1936). In only a few instances could the fungus be found associated with discolored vascular bundles. Discoloration of vascular bundles could be induced by water stress or by an unbalanced nutrient supply. In the field, drought and P deficiency accentuated blackening of the vascular bundles, and the latter condition was evident in some inbred lines regardless of treatment. It was concluded from this work that the fungus is a weak parasite which occasionally gained entrance to the vascular system through wounds.

When plants are naturally barren or the condition is artificially induced, they frequently show purpling of stalks and leaves, and vascular bundles are discolored. These symptoms result from barrenness and seem not to be caused by a pathogen. Planting seed lots in which a high percentage (60%) of the kernels showed infection by *C. acremonium* resulted in healthy, productive plants without the symptoms described above.

At the present time with the use of generally high-yielding hybrids few of the symptoms in the syndrome of black bundle disease are seen. Some barren plants are observed, but generally this can be traced to causes other than infection by *C. acremonium*. Further critical research on this disease may resolve the apparent confusion that exists as to causes and effects.

Several other fungi have been associated with stalk rot. These are seldom, if ever, of major economic importance and appear to be more frequently encountered as secondary invaders, attacking senescent plants or those under environmental stresses. Among these are *Nigrospora oryzea* (Berk, Br.) Petch, which causes a black streaking on the exterior of the stalk and black discoloration of the pith.

Phaeocytostroma ambigua (Mont.) Petr. [Syn. = *Phaecocytosporella zeae* Stout] is responsible for tan spots on the lower parts of the stalk. Spores are borne in black, long-necked pycnidia found on the surface of stalks. Spores are single-celled, oval, brown, and 9 to 15 $\times$ 4 to 6 μ in size.

Pyrenochaeta terrestris (Hansen) Gorenz, Walker & Larsen, is found at basal portions of the stalk and imparts a red color to the invaded tissue. This can often be confused with infection caused by *G. zeae*. Spores are hyaline, ovoid, 1.5 to 6.0 $\times$ 1.8 to 2.5 μ in size and borne on simple conidiophores in black, globose pycidia.

10. Root Rots

Pythium root rot of corn is widely distributed (Branstetter, 1927; Valleau et al., 1926) but apparently of minor economic consequence. The disease is found mostly in poorly drained or physically compacted soils where the oxygen supply is deficient.

Roots are yellowish to brown and flaccid. Invasion of the pathogen may take place at root ruptures and eventually progress through the entire root system. Plants may be attacked at any stage of development and root lodging follows in severe cases of the disease.

A number of species of *Pythium* have been associated with root rot of corn. *Pythium arrhenomanes* Drechs, *P. debaryanum* Hesse and *P. graminocolum* Sub. are among these. They are differentiated from one another on the bases of size, shape, and arrangement of asexual and sexual structures (Middleton, 1943). These like many species of the genus have relatively wide host ranges.

Specific control measures have not been developed for pythium root rot, although differences in resistance to the disease has been observed among inbred lines (Elliott, 1942).

Most of the typical stalk rotting fungi discussed above, *D. maydis, G. zeae, F. moniliforme, M. phaseolina,* and others may invade and effectively weaken senescent corn roots.

Phaeocytostroma ambigua and *Pyrenochaeta terrestris* have been shown to cause root rot in seedlings and have been isolated from mature plants in the fields (Craig and Koehler, 1958). These fungi are not often observed and probably are weakly parasitic after invasion of more aggressive pathogens. When massive artificial inoculations are made on plants under stress, or where such fungi are introduced into sterilized soil and no competition with other fungi exists, they may cause extensive necrosis.

IV. EAR ROTS

Corn is susceptible to a number of ear-rotting pathogens. Five of these are widely distributed and have at times reduced yield and quality of grain (Stevens and Wood, 1935). Reduction in nutritive value of infected corn has been demonstrated (Mitchell and Beadles, 1940). Usually, however, ear rots have not caused severe losses over extensive areas. Surveys made of the relative prevalence and distribution of ear rots over a period of years have shown wide interannual variation and areas of greatest incidence (Hoppe, 1938-1943; Koehler, 1959).

1. Diplodia Ear Rot

Diplodia ear rot is widely distributed over the U.S. Corn Belt, but individual epidemics appear to be limited in extent (Ullstrup, 1964). In these observations the incidence of the disease was slightly over 60% where plants were adjacent to the source of inoculum, but decreased rapidly to a trace 122 m (500 ft.) away in a leeward direction.

One of the earliest symptoms of diplodia ear rot is the bleaching of husks at the butt of the ear. Infection may take place from silking until maturity but the time of greatest susceptibility appears to be during the 3 to 4 weeks after full silk (Raleigh, 1930; Ullstrup, 1949). Early infection often results in complete rotting of ears (Fig. 4A). Such ears are lightweight, husks tightly adherent to each other, owing to growth of the fungus between them, and kernels grayish-brown. Black pycnidia, which contain spores of the pathogen may develop on and between husks, on kernels, and cob. White mycelium may also be seen on the husks, the kernels, and the ear shank.

When infection occurs late in the development of the ear there may be no outward appearance of rotting. When such ears are carefully examined, growth of mycelium will often be evident at the tips of the kernels and on the cob. Infection almost invariably takes place at the butt of the ear and progresses toward the tip.

Diplodia ear rot is caused by *Diplodia maydis*; its morphology and life cycle have been described in the section on "Diplodia Stalk Rot". The fungus usually completes one cycle during a growing season. Pycnidia and their contained pycnidiospores may develop on infected corn ears or stalks during late summer or fall, or they may form on corn debris the following summer. If the spores mature in the fall, ears are usually then beyond the optimum stage for

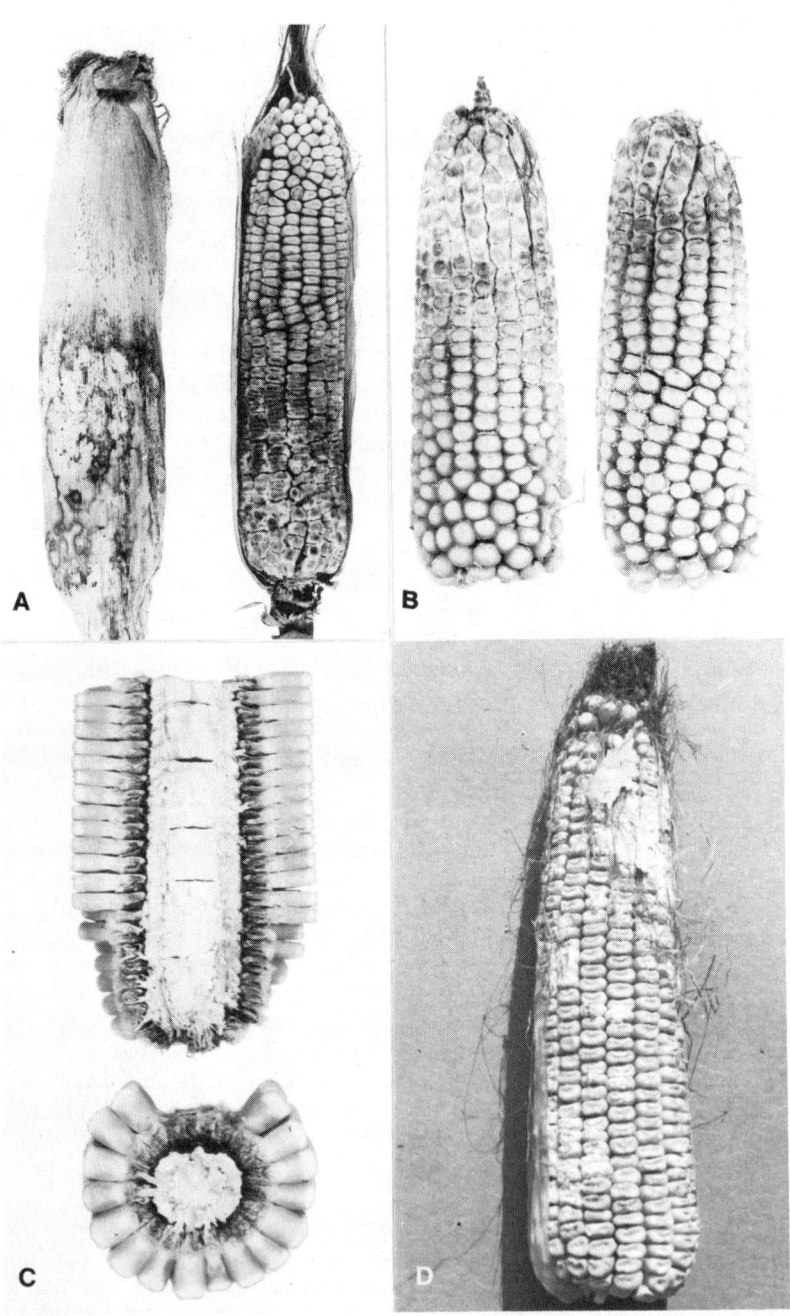

Fig. 4. Ear and kernel rots: A, diplodia ear rot; B, gibberella ear rot; C, nigrospora cob rot; D, fusarium kernel rot.

infection, or mature, so that a second cycle of the fungus is not initiated. It is probable that because of the single cycle that acute epidemics of diplodia ear rot are limited to a few hundred yards in extent. In contrast, a pathogen with repeating spore cycles can build up a large mass of inoculum capable of spreading over a wide area.

Several factors may influence the incidence of diplodia ear rot, among these weather conditions play an important role. Frequent rainfall from full silk to 4 to 5 weeks thereafter is especially favorable for development of the disease (Durrell, 1923; Koehler, 1950).

Husk coverage and time of ear declination may contribute to susceptibility. Open-husked genotypes and those in which ears are held in an upright position for a long time have been reported as susceptible (Koehler, 1951). There is some disparity of opinion on the factor of husk coverage. Tight husk coverage was believed to be positively associated with susceptibility (Boewe, 1936). When ears are infected early and completely overrun by the fungus they do tend, because of their light weight, to be held upright, and husks are tightly adherent to one another. These conditions of husk coverage and ear declination may be effects of infection rather than causes of susceptibility. More observations are needed, extending over a wide array of genotypes under both artificially induced epidemics of the disease and under disease-free conditions.

The nature of resistance has not been extensively explored, but wide differences in resistance to diplodia ear rot are observed (Ullstrup, 1949; Koehler, 1953). Anatomical characteristics of corn kernels seem not to be associated with the relative resistance of corn genotypes to infection of ears by *D. maydis* (Johann, 1935).

While diplodia ear rot is widespread and occasionally acute over restricted areas, it has not yet become of sufficient economic importance to stimulate searches for resistance and the incorporation of this trait into adapted genotypes. Methods of artificial inoculation of corn ears with *D. maydis* have been developed which differentiate between resistant and susceptible corn germplasm (Ullstrup, 1970a). This method could be a useful tool in a breeding program designed to develop resistant inbred lines and hybrids.

Genes conditioning resistance to diplodia ear rot appear to be quantitatively inherited (Wiser et al., 1960).

2. Gibberella Ear Rot

Gibberella ear rot, or red ear rot, occurs throughout the Corn Belt, but is somewhat more prevalent in the northern parts of this general region. The disease is favored by cool, humid weather. Within recent years — 1965 and again in 1972 — gibberella ear rot has been responsible for economically important epidemics (Tuite et al., 1974; Ullstrup, 1966).

The distinguishing symptom of this disease is the appearance of a pinkish to red mold at the tips of ears which progresses toward the butt. The rot involves all kernels as it develops and is unlike fusarium ear rot in which infected kernels are scattered over the ear (Fig. 4B). Usually only the upper half of the ear is involved, seldom is the entire ear rotted as in diplodia ear rot. Husks are often pink at their tips and tightly adherent. Occasionally black incrustations are evident on the upper parts of the husks. They are composed of masses of perithecia which usually do not mature (produce viable spores) until the following summer when ample moisture is present.

Gibberella ear rot is caused by *G. zeae,* the morphology of which has been described in the section on "Gibberella Stalk Rot."

The epidemics of gibberella ear rot are generally more widespread (Tuite et al., 1974; Ullstrup, 1967), than those of diplodia ear rot which seem to be much more localized and immediately adjacent to primary sources of inoculum (Ullstrup, 1964). In *G. zeae* it is possible for a number of spore cycles of the asexual stage, *Fusarium graminearum,* to occur during the growing season. This may bring about an increase in the volume of inoculum. Where wheat and barley are grown, in addition to corn, epidemics in these crops of head blight, caused by *G. zeae,* can bring about further build-up and spread of inoculum available for infection of corn ears.

In addition to losses sustained directly from rotted ears, corn infected with *G. zeae* is toxic to swine. The animals may refuse to eat such infected corn or, if consumed, react by vomiting. Thus, nutritional intake is lessened and animals fail to gain weight. Death may ensue in severe cases of such enteric disturbances (Curtin and Tuite, 1966). Toxicity is expressed when 3 to 5% of the corn is infected. It is not presently known whether two toxins are involved or if the two reactions, refusal and vomiting, in the animals are caused by one toxin.

An additional toxin appears to develop in corn stored for several months at temperatures of about 15 C or below. The precise time and temperature required for the toxin to form is not known. Swine respond with marked endocrine upsets. Young gilts come into heat, older females may suffer uterine prolapse, or, if they are pregnant, may abort. Boar pigs may show overt development of the mammary system. This toxin has been identified and characterized (Caldwell et al., 1969; Stob et al., 1962).

Swine are the only farm animals in which a major problem occurs in feeding corn or other grains infected with *G. zeae.* Ruminants and poultry seem not to be affected by these toxins. No antidotes or amendments to feed have been devised to eliminate toxicity in feed. Dilution with sound corn to lower the percentage of infected kernels below 3% has been effective in some instances.

Marked differences in resistance to gibberella ear rot are observed among inbred lines and hybrids, but no specific efforts have been made to develop inbred lines of high resistance. The gradations from relatively high resistance to susceptibility suggest that resistance to this disease, as in diplodia ear rot, is governed by a number of genes.

No control measures have been devised for gibberella ear rot. If and when the need arises for highly resistant genotypes, genetically controlled resistance will probably be the method of choice.

3. Nigrospora Cob Rot

Nigrospora cob rot, although widespread, is rarely of any economic importance.

A distinguishing symptom of this disease is the shredding of the cob

(Fig. 4C). This may occur at the tip or the butt but more frequently at the latter site. This characteristic is especially conspicuous when, on shelling, infected cobs shatter. Unlike diplodia and gibberella ear rots, no marked symptoms are apparent in unhusked ears; there is no evidence of tight adherence of husks. Kernels are usually poorly finished, pinched at their tips and show much soft starch on their faces. Kernels are easily pressed into the disintegrated cob, in which the chaff is chocolate-brown rather than red. In infected white-cob corn, chaff is yellowish to tan. In severely infected ears, kernels will show masses of black spores of the pathogen on their tips. These spore masses are evident also in the pith of such ears.

Nigrospora oryzae is the cause of this disease. Spores of the fungus are black, oval to spherical in shape, 10 to 20 μ in diam. and borne on short simple conidiophores on the mycelium. No sexual stage of the fungus is known. The fungus overseasons on corn debris infected the prior year (Standen, 1944).

Nigrospora oryzae is weakly parasitic and attacks ears of plants that have been subjected to some stress — drought, cold, poor nutrition, or other diseases. Genotypes susceptible to leaf diseases and stalk rot often show a greater incidence of nigrospora cob rot (Durrell, 1925). Secondary ears in which there is no seed-set will frequently be overrun by *N. oryzae*. The tissue in these ears matures early and dies rapidly, thus affording a good substrate for this weakly parasitic fungus.

Specific control practices have not been devised for this disease. The low level of parasitism shown by this fungus suggests that full-season, adapted hybrids grown under recommended nutritional regimes are the most practical measures for control of nigrospora cob rot.

4. Fusarium Kernel Rot

Fusarium kernel rot is worldwide in distribution and can be found every year. It is especially prevalent in the drier parts of the Corn Belt (Hoppe, 1942) and in California (Smith and Madsen, 1949). The disease is also reported as the most important on corn ears in the southeastern USA (Boling and Grogan, 1965). Severe levels of fusarium ear rot have been observed also where soil was frequently flooded and humidity of the air was generally high (Hoppe, 1942).

Symptoms consist of groups of infected kernels scattered randomly over the ears (Fig. 4D). Whitish-pink to lavender mold is typical of the infection. Genotype and moisture content of the grain influence the color of infected kernels. Infection is generally more frequent at the tip of the ear and is often associated with the boring and channeling of ear worms.

Spores of the pathogen may enter through the silk channel at the tip of the ear and initiate infection on immature kernels (Koehler, 1942). The isolation of the fungus from nodal and internodal stalk tissue suggests the fungus may traverse the vascular system (Foley, 1959). If it can be established that the pathogen is fully systemic, infection of kernels may be postulated to arise from inoculum transported through the xylem elements.

Fusarium ear rot is caused by *Fusarium moniliforme* and a closely related *F. moniliforme* var. *subglutinans*. Minor morphological differences separate the respective, rarely occurring, sexual stages of the two pathogens (Ullstrup, 1936). Morphological characteristics of the asexual stages of each is described under the section on "Fusarium Stalk Rot."

Fusarium moniliforme is widely dispersed throughout the world. The fungus apparently is adapted to a wide range of environments. It can easily be isolated from many types of diseased tissue where it often invades after a primary pathogen. There are indications that it is not an aggressive, highly specialized parasite. Seed samples with no visible disease symptoms, when plated on potato dextrose agar (PDA), showed 60% internal infection. When remnant samples were planted in the field they produced healthy seedlings, plants that matured normally, bearing ears with no more kernel infection than those grown from control seed lots with little or no internal infection. The fungus rapidly overruns kernels in which the seed coat is broken as a consequence of "silk-cut" or "popped-kernel."

No specific control practices are known for the disease other than selections against susceptibility. The nature of susceptibility is not understood, but sweet corn is generally much more susceptible than dent corn. This is often brought out in ears segregating for *starchy* and *sugary* kernels; the latter will almost invariably be infected with *F. moniliforme* and the starchy kernels nearly free from visible symptoms. The difference may be due to more readily available substrate for the fungus or morphological characters such as thin, easily penetrated seed coat. Some high lysine inbred lines and hybrids are susceptible to this disease (Ullstrup, 1971).

Resistance is inherited in a quantitative manner and one study with a limited array of genotypes suggests that increasing frequency of genes determining resistance in populations, might be accomplished most expediently by reciprocal recurrent selection (Boling and Grogan, 1965).

5. Gray Ear Rot

This disease has been widespread over the USA, but prevalence has been low. In a few isolated cases the incidence was observed as high as 10% in seed fields where double-cross hybrids were being made (Ullstrup, 1946). An unusual aspect in the epidemiology of the disease has been its virtual disappearance in the past 20 years. This is based on field observations, and on absence of the pathogen in hundreds of samples of corn from the east-central Corn Belt that are analyzed each year, in the Department of Botany and Plant Pathology at Purdue University, for the presence of internally borne fungi. The reasons for the apparent absence of the disease are obscure. There have been neither radical changes in the environment, nor any marked trends toward resistance within host germplasm. The array of genotypes observed to be susceptible was broad and none among them was highly resistant.

The first symptoms of the disease resemble those of diplodia ear rot and the two can be easily confused, White, felty mycelium is frequently seen between the husks, where its growth "cements" them together, and between kernels. Ears infected early remain in an upright position and become slate-gray. The cob and kernels of such early-infected ears become completely permeated by the mycelium which, with age, turns from white to gray or black. When infected ears are broken, minute black sclerotia of the fungus can be distinguished on cob and beneath the seed coat of kernels (Fig. 5A). The

DISEASES OF CORN

presence of these black sclerotia and dark-gray color of the ears in advanced stages of the disease distinguish it from diplodia ear rot.

The cause of gray ear rot, *Macrophoma zeae* Tehon and Daniels, was first described on corn leaves in Illinois (Tehon and Daniels, 1927). Large leaf lesions result from infection and both pycnidia of *M. zeae* and perithecia of its sexual stage develop together in such lesions (Fig. 5B). Pycnidia are oval,

Fig. 5. Ear and kernel rots: A, gray ear rot showing sclerotia beneath the seed coat and within the cob; B, pycnidia and perithecia of the pathogen of gray ear rot embedded in necrotic tissue of a leaf lesion; C, penicillium ear rot; D, cladosporium kernel rot; E, trichoderma ear rot; F, "blue eye" showing the gray-green development of *Penicillium* sp. in the germ.

black, carbonous, embedded in mesophyll tissue, 65 to 120 μ in diam. Spores are hyaline to pale-green, nonseptate, oval to fusiform, and 17 to 31 × 6.5 to 8.5 μ in size. The sexual stage, *Physalospora zeae* Stout, was described in 1930 (Stout, 1930), also on corn leaves. The relationship between the sexual and asexual stages was suspected, but not established. Perithecia, embedded in the necrotic tissue of lesions, are black, carbonous, globose, and 75-235 μ in diam. Asci are stalked, straight to curved, double-walled, 85 to 150 × 13 to 22 μ, and contain eight ascospores. Paraphyses are obscure, hyaline and filamentous. Ascospores are hyaline to pale olivaceous, one-celled, ellipsoid, and 19 to 25 × 6.5 to 8 μ in size. A third type of fruiting body is sometimes observed. These are small, black, flask-shaped structures with one or two, long, tapering beaks, and contain minute single-celled, hyaline spores. These spores are exuded in a viscous matrix from ostioles at the apices of the beaks. Germination has not been observed and their function unknown. Attempts to "spermatize" cultures of the fungus with these spores, in order to initiate the sexual stage, failed. Leaf lesions in which the three fruiting structures develop are tan to gray-brown and range in size from 10 to 45 × 2.5 to 5 cm. The three spore-bearing bodies develop also on stalks immediately below the tassel following infection beneath the uppermost leaf sheath. These fruiting bodies have never been observed on kernels. The relationship between asexual stage, *M. zeae,* and the sexual stage, *P. zeae,* and proof of pathogenicity has been established (Ullstrup, 1946). The pathogen appears to survive overwinter in leaf lesions and in the tissues of the tassel peduncle.

Factors affecting prevalence of the disease are not known, but judging its "once-geographical distribution," warm, wet weather in August and September seems favorable.

The disease made its appearance for so short a time that reactions of inbred lines and hybrids could not be adequately examined. No control for the disease could be recommended. Cultural practices or soil fertility seemed not to have marked effects on its incidence.

6. Other Ear and Kernel Rots

A number of other ear and kernel rots of corn have been described, but most of these generally are of little economic importance because they have restricted distribution, occur only rarely, and then not in severe epidemic form. None of these has been of sufficient importance to warrant the development of specific control measures and generally no pronounced differences in resistance to these ear rots has been observed among genotypes.

Penicillium ear rot is occasionally encountered. Bluish-green, powdery mold growth appears on and between the kernels, and especially on the glumes of the cob (Fig. 5C). *Penicillium oxalicum* Currie and Thom is the most common fungus associated with this disease, although other species of *Penicillium* occasionally cause ear rot. Seedlings grown from infected kernels are generally attacked by the pathogen and often die.

Aspergillus ear rot appears as a black, powdery mold on and between kernels. *Aspergillus niger* Van Tiegh. is the fungus causing this ear rot. Other species — *A. flavus* Link and members of the *A. glaucus* group — can cause ear rot. Rots incited by the latter two are generally greenish-yellow and green respectively. *Aspergillus niger* appears to be relatively more prevalent in drier parts of the United States (Hoppe, 1938-1943 incl.; Melchers, 1956; Taubenhouse, 1920).

Physalospora ear rot caused by *Physalospora zeicola* Ell and Ev. is found in some of the Gulf States (Eddins and Voorhees, 1933). Black or dark brown, felty mycelium grows over most of the ear. Infection usually begins at the butt and progresses toward the tip of the ear. Mildly infected ears may show only a few black kernels.

Perithecia of the fungus are black, globose, and develop most frequently on corn stalks and husks. Asci are double-walled, contain 8 ascospores and measure 95 to 140 × 10 to 13 μ. Ascospores are hyaline, oval non-septate, granular and 20 to 23 × 8 to 9 μ in size. Spores of the asexual stage, *Diplodia frumenti* Ell. and Ev., are 20 to 28 × 11 to 15 μ, oval, light to dark brown, two-celled and striate. Spores are borne in black, globose pycnidia 170 to 470 μ in diam. which emerge from fissures in the exterior of infected corn stalks. *Diplodia frumenti* is differentiated from *Macrophoma zeae* on the basis of size, shape, color, and number of septations of the pycnidiospores. Those of the latter are one-celled, and do not possess striations on their surfaces.

In the USA rhizoctonia ear rot appears to be confined to the Gulf States. Symptoms are characterized by the salmon-pink color of the mycelial growth over the ear. In later stages of development the color may change to dull gray; white to dark brown sclerotia 0.1 to 0.5 mm in diam. form on the husks. *Rhizoctonia zeae* Voorhees is the causal agent of the ear rot (Voorhees, 1934). No spore forms have been associated with this fungus. It thrives at relatively high temperatures (33 C) and survives as dormant mycelium and sclerotia in debris.

Cladosporium kernel rot appears as black striations on the faces of the kernels or as a greenish-black mold over the entire kernels (Fig. 5D). When the seed coat on the kernel cap is broken the greenish black mycelium of the pathogen develops on the exposed starch (Hoppe, 1961; Koehler, 1959). The fungus is often associated with insect injury on the kernels, and is sometimes coincidental with early frost damage to corn ears.

The causal fungus, *Cladosporium herberum* (Pers.) Link ex S.F. Gray [Syn. = *Hormodendrum cladosproioides* (Fres.) Sacc.], produces olive-brown, one- to three-septate, cylindrical spores 5 to 23 × 8 to 15 μ in size. These are borne on sparsely branched, erect conidiophores that range up to 250 μ long.

Rhizopus ear rot appears more commonly following abnormally hot, dry growing seasons. The mycelial growth is white to pale gray with minute black spore-bearing structures (sporangia) dispersed throughout. Little is known about the disease and the species of *Rhizopus* involved has not been identified.

An ear rot occurring in Mississippi caused by *Gonatobotrys zeae* Futrell and Bain has recently been described (Futrell and Bain, 1968). The mycelium on kernels is white to yellowish. Spores are single-celled, hyaline, and borne singly on conidiophores.

Southern diplodia ear rot is found occasionally in the southern states. The symptoms are similar to those incited by *D. maydis*. The causal agent *Diplodia macrospora* Earle differs from *D. maydis* in size of the pycnidiospores. Those of the latter are approximately 24 × 5.3 μ while those of

D. macrospora average 68 × 9 μ (Eddins, 1930). Pycnidia are black, flask-shaped and 150 to 450 μ in diam. While the fungus primarily attacks ears it also causes small leaf lesions and a stalk rot. Pycnidia in which the long cylindrical spores are borne develop on stalks, husks and in leaf lesions.

Trichoderma ear rot usually occurs when corn is under some stress. During the southern corn leaf blight epidemic in 1970 this ear rot was occasionally seen where plants were rapidly killed by Race T of *Helminthosporium maydis*. The disease is recognized by bright green powdery growth of the pathogen on and between the kernels (Fig. 5E). *Trichoderma viride* Pers, is the causal fungus.

Cephalosporium acremonium is frequently isolated from corn kernels. No specific rotting of the ear is evident, but individual kernels infected with *C. acremonium* may show pale vertical striations. This symptom, however, is not specific for infection by this fungus.

Other fungi cause diseases of corn ears, but are associated more frequently with the diseases they incite on other parts of the plant. Examples of these are Race O and especially Race T of *Helminthosporium maydis*, Races I and II of *H. carbonum* Ullstrup, and an, as yet, unnamed species of *Helminthosporium* recently reported in the Midwest (Hooker et al., 1973). These will be discussed under the section on "Leaf Diseases." It should be pointed out that *H. turcicum* Pass. was once identified as causing an ear rot of corn (Sherbakoff and Mayer, 1937). Later it was reported that another species, not *H. turcicum*, was involved (Sherbakoff, 1951). There is no record of *H. turcicum* causing a typical ear rot, although the fungus might occasionally be isolated from kernels.

V. STORAGE ROTS

Any of the typical ear rot pathogens discussed above may continue to grow and invade kernel and cob tissues in storage. These pathogens are sometimes referred to as "field fungi" as distinct from "storage molds." Most of the latter do not cause typical ear diseases in the field. Usually the storage molds are able to grow at lower temperatures and on kernels with lower moisture contents than required by most ear rot fungi (Quasem and Christensen, 1960).

Storage rots reduce nutritive value and market grade of corn (Christensen and Kaufman, 1969). In addition some storage molds can, as a consequence of their growth on corn, produce substances — mycotoxins — that are highly toxic to humans and animals (Christensen, 1971; Berg, 1972).

Evidence of rotting of ear corn in storage is the growth of mycelium and production of masses of fungus spores between the kernels, particularly at their tips where they are attached to the cob. In shelled corn storage molds may sometimes develop sufficiently to cover the surface of the corn in the bin and cause "cakeing" of the grain. One rather distinctive symptom of deterioration in stored corn is referred to in the grain trade as "blue-eye" (Fig. 5F). This is so named because of the growth of species of *Penicillium* in the germ and beneath the seed coat. This occurs when moisture accumulates in the upper layers, owing to storage of grain with high moisture, or as a result of leakage of rain water into the bin.

The storage molds are mainly species of the genera *Penicillium* and *Aspergillus*. To a lesser extent, and under specific conditions, species of *Cladosporium, Paecilomyces, Mucor, Rhizopus* and *Fusarium* may also cause problems when corn is improperly stored.

Storage molds can be controlled by drying shelled corn to 12% moisture or less and maintaining it at this level. Ear corn can be stored without danger in well-ventilated cribs if the moisture content of the grain is in the range of 18 to 22%.

VI. LEAF DISEASES

The prevalence and severity of some leaf diseases fluctuates from year to year depending largely on environmental conditions. A given set of environmental factors favoring one of these diseases may not necessarily be optimum for the development of others. Moderate temperatures and ample moisture in the form of rain or heavy dew are ideal conditions for most leaf diseases caused by fungi. Some, such as eyespot and yellow leaf blight, seem to thrive at lower temperatures than does southern corn leaf blight. Bacterial wilt seems little affected by moisture, but moderate to relatively high temperatures accentuate severity of symptoms.

Within the past 15 years three *new* leaf diseases of corn have appeared in the USA — eyespot, yellow leaf blight and southern corn leaf blight caused by Race T. Anthracnose and gray leaf spot, which in the past have caused little trouble, have increased in prevalence and severity in some areas.

When leaf diseases become severe, they not only depress yield by reducing the photosynthetic area of leaves, but also lower the yield and quality of silage (Allinson and Washko, 1972). Severe leaf infection also predisposes plants to stalk rots; and at least one disease of ears — nigrospora cob rot — seems to become more prevalent following severe leaf infection.

1. Northern Corn Leaf Blight

Northern corn leaf blight is found in most of the humid climates where corn is grown. In the USA it occurs mostly in the eastern half of the Corn Belt to the Atlantic Coast and southward. Localized epidemics have been found outside this general region when the environment was favorable.

The epidemiology of the disease is not fully understood, but based on observations and its general geographic distribution it appears to be favored by comparatively moderate temperatures 22 to 28 C and heavy dew. Following a year when the incidence and severity of northern corn leaf blight is suppressed by dry weather there is then a gradual build up in prevalence in subsequent years, providing environmental factors are near optimum. This fluctuation occurred in the early 1940's when in 1943 the disease was widespread and locally severe in the eastern parts of the Corn Belt. In 1944 the disease was negligible because of hot, dry weather. In subsequent years there was a gradual increase in prevalence until 1951 when it again attained a peak in severity. In 1952 little of the disease could be found in Illinois, Indiana, and Ohio, again because of dry hot weather, but in Wisconsin, where temperatures were relatively moderate, the disease was abundant. During any given season a gradual increase in prevalence and severity is observed. This may be due, in part, to successive infections which increase the amount of inoculum. There is some evidence, too, that susceptibility increases along with gradual maturation after pollination.

Reduction in grain yield depends on severity and time of establishment of the disease. In artificially induced epidemics of northern corn leaf blight susceptible hybrids showed losses of 40 to 68% when the disease became established by the time plants were in full silk, and attained a high level severity 2 to 4 weeks later. In years when the disease did not become severe on susceptible hybrids until 6 to 7 weeks after silking no measurable losses in yield were sustained. In resistant hybrids losses were nonsignificant. The reduction in yield was found to be closely related to the number of resistant inbreds in double-cross hybrids (Ullstrup and Miles, 1957).

Symptoms of northern corn leaf blight appear as long, elliptical, grayish-green or tan lesions ranging up to 16 to 18 $\times$ 3 to 4 cm in size. Within 48 to 72 hours following infection of the leaves by the pathogen, small chlorotic flecks appear at points of penetration. These are barely visible except where a large number of penetrations occur over a small area of leaf surface. Lesions first become evident 8 to 10 days after infection is established and then appear as small wilted areas 10 to 15 $\times$ 2 to 3 mm in size. Lesions continue to enlarge up to 15 to 18 $\times$ 3 to 4 cm in dimension (Fig. 6A). The time required for lesions to develop depends on temperature; rapid development occurs at temperatures of about 25 to 30 C; under lower temperatures a longer time is required. Spores develop in the necrotic tissue of the lesions during damp weather; often the spores are arranged in a concentric zonate pattern. Observations made both under natural and artificial epidemics indicate that kernels are not attacked by the pathogen. Although no rotting of the ears has been seen it is possible that the causal agent could be isolated from kernels.

The reaction of plants with monogenic resistance conferred by the gene *Ht* is characterized by relatively small, narrow necrotic areas up to 30 to 40 $\times$ 2 to 8 mm in size and surrounded by a chlorotic halo (Fig. 6B). The size of the necrotic lesion and the width of the chlorotic halo vary with the genetic background into which the *Ht* gene is introduced.

The inciting agent of northern corn leaf blight is *Helminthosporium turcicum* Pass. Spores are spindle-shaped, often slightly curved on one side, olive-gray, average 105 $\times$ 20 μ and have one to nine septations. A conspicuous and identifying feature of the spores of *H. turcicum* is the protruding hylum. The hylum of most of the other species of this genus, which attack corn, is internal or contained within the wall of the cell attached to the conidiophore. A sexual stage has been developed in vitro following the pairing of appropriate mating types of the fungus on suitable media (Luttrell, 1958). Perithecia are black, ellipsoidal, and measure about 360 to 720 μ in height by 350 to 500 μ in diameter. Asci are long, cylindrical, measure approximately 27 $\times$ 200 μ and bear 1 to 8 ascospores. Ascospores are fusoid, hyaline, average 3 septa, and are about 15 $\times$ 62 μ in size. Walls of acospores are constricted at the septa. The sexual stage has not been found under field conditions.

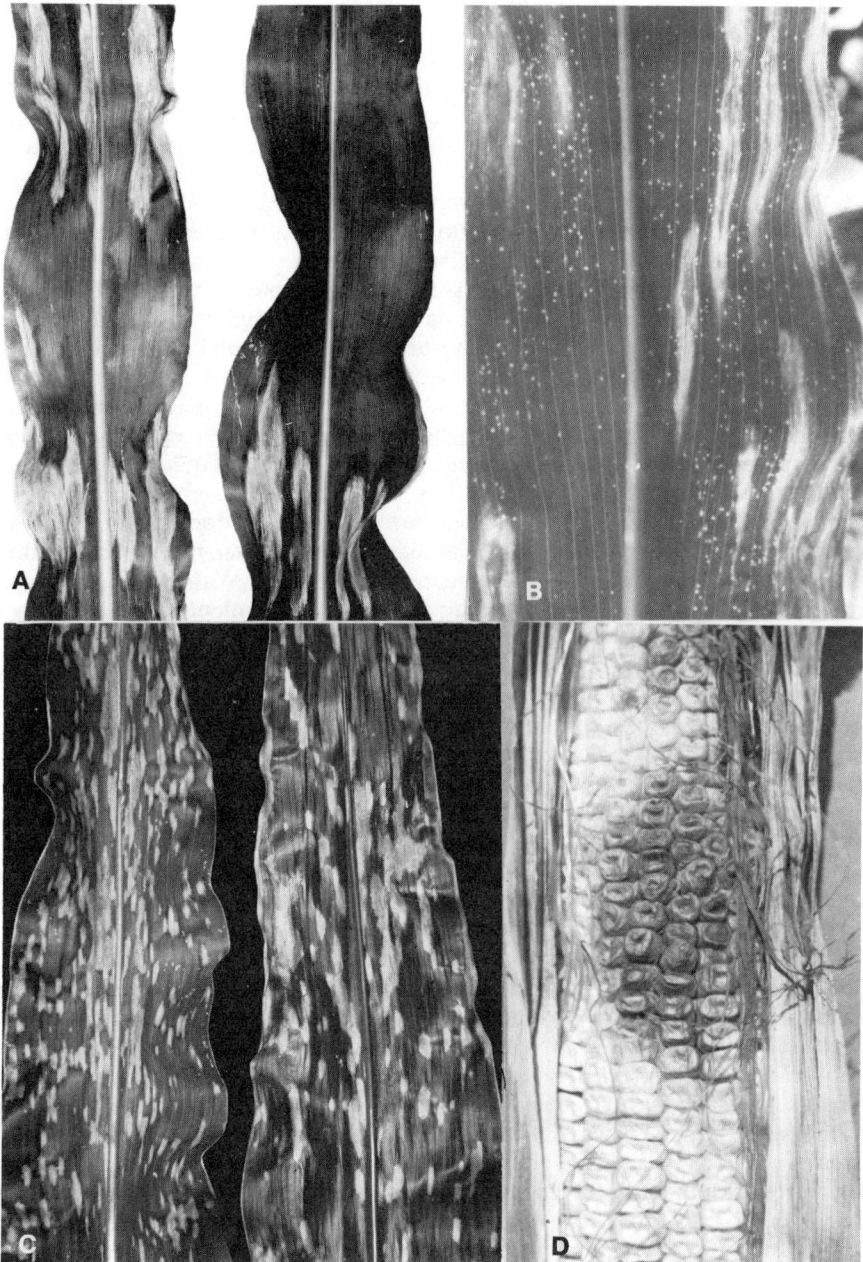

Fig. 6. Leaf diseases: A, northern corn leaf blight showing a typical susceptible reaction; B, northern corn leaf blight showing the resistant reaction conditioned by the monogenic dominant *Ht*; C, southern corn leaf blight on *Tcms* corn caused by Race T of *H. maydis*; D, ear rot on *Tcms* corn caused by Race T of *H. maydis*.

Under field conditions *H. turcicum* causes diseases on corn, sorghum (*Sorghum bicolor* (L.) Moench); sudangrass (*S. sudanensis* Stapf); johnsongrass (*S. halepense* (L) Pers.) and teosinte (*Euchlaena mexicana* Schrad.). A number of other grasses have been mentioned, in the literature, from which *H. turcicum* has been isolated (Hooker et al., 1965), but epidemics are not observed on these hosts. In "Diseases of Cereals and Grasses" (Sprague, 1950), only the four species cited above, exclusive of teosinte, are listed as hosts.

Physiologic races have been reported within *H. turcicum*. In one study isolates from sudangrass, sorghum ('Atlas') johnsongrass and corn were tested for their ability to incite disease on these hosts. Two isolates from sudangrass and one from Atlas sorghum were pathogenic on Gooseneck sorghum, but not on corn. Four isolates from corn infected sudangrass, but two did not. All cultures from corn were pathogenic only on this host (Lefebvre and Sherwin, 1945). Further studies, in which single-ascospore cultures produced in the laboratory from matings of isolates collected on different hosts were used, confirmed existence of physiologic races based on host genera and species (Rodriguez and Ullstrup, 1962). Thus far there is no evidence of differential pathogenicity of *H. turcicum* within genotypes of corn. Physiologic specialization has been reported (Robert, 1960), but this referred to differences in virulence among isolates. These races were "pathic" races, not "hostic" races as these terms were defined (Waggoner and Wallin, 1952). A given isolate expressed the same level of virulence over an array of corn genotypes; and virulence was not constant, but changed with time. Resistant inbred lines, whether having polygenic or monogenic resistance have remained so — they have not become susceptible during the past 20 years that some of these have been in existance. The possibility always remains of new physiologic races arising that are capable of overcoming the resistance of some genotypes now used to control northern corn leaf blight. (What appears to be a new physiologic race, capable of attacking corn with the *Ht* gene has recently been observed in Hawaii.) Experiments have shown that some monascosporic cultures derived from crosses of isolates of *H. turcicum* were able to incite susceptible type lesions on resistant inbred lines of corn (Nelson et al., 1965). The crossing and backcrossing was done in the laboratory under conditions where the sexual stage can be developed. Since the perithecial stage has not been found under field conditions there may be some doubt if sexual combination and segregation occurs in that environment such as to have important and practical significance.

Helminthosporium turcicum can survive for at least one year in dried leaves, but spores apparently lose their germinability in this time. It has been suggested that the dormant mycelium, in corn debris infected the previous year, functions under field conditions to produce spores that initiate infection in the early summer (Robert and Findley, 1952). Tests conducted over several years failed to demonstrate overwintering of the fungus in Wisconsin (Hoppe, 1962). In Nebraska chlamydospores, which developed within old conidia, germinated to form spores in the growing season (Boosalis et al., 1967). Essentially the same observations were made in Minnesota where chlamydospore-like structures, borne singly or in chains on dormant hyphae, germinated to form spores which incited infection (Asare-Nyako, 1964). It has been pointed out that the pathogen can overwinter in Pennsylvania, but that the genotype of the corn and the particular "race" (isolate) may influence survival. Furthermore, these "races" have different abilities to overwinter saprophytically on debris of different genetic constitution (Nelson and Scheifele, 1970).

Observations of localized epidemics that occur suddenly in fields where no appreciable amount of debris was present, and often later than might be expected if the inoculum source was from debris, suggest that "spore showers" may account for the onset of the disease (Hoppe and Arny, 1966).

Resistance in corn to infection by *H. turcicum* is believed to be due to phytoalexins (Lim, et al., 1970); and also to cyclic hydroxamic acids (Couture, et al., 1971). Resistance to infection and subsequent invasion of host tissue by *H. turcicum* in genotypes bearing the gene *Ht*, has been shown to be due to two phytoalexins, A1 and A2. These are blue fluorescent compounds but differ chemically from each other. The compounds are not found in healthy tissue, but are elaborated by the host only in response to penetration. When a drop of spore suspension is placed on a leaf of a resistant genotype the phytoalexins are detectable in the droplet (diffusate) 3 days later. Isolates of *H. turcicum* from johnsongrass, to which corn is resistant, did not elicit production of phytoalexin in infection droplets placed on corn leaves. Virulent cultures from corn, when used in the infection droplet, caused elaboration of a higher concentration of phytoalexin than did those of less virulence.

A second substance that may be responsible for resistance in corn to infection by *H. turcicum* is a cyclic hydrosamate: 2,4-dehydroxy-7-methoxy-1,4 benzoxazine-3-one (DIMBOA). This compound does not exist as such in intact corn tissue, but is elaborated in response to penetration, or other injury, as a result of enzymatic action on a glycoside precursor which is nontoxic. The latter compound inhibited germination of spores of *H. turcicum* at concentrations of 1 to 10 ppm. The compound is also implicated as the basis for resistance to infection of corn stalks by *Gibberella zeae* and the first brood of European corn borer, *Ostrinia nubilalis* Hubn. (Anglade and Molot, 1967a; Anglade and Molot, 1967b; Molot, 1969a, b, c).

Genetic resistance in corn to infection by *H. turcicum* is determined in some genotypes by several genes (Hughes and Hooker, 1971; Jenkins and Robert, 1952; Jenkins and Robert, 1959; Jenkins and Robert, 1961; Jenkins, Robert and Findley, 1952). This polygenic resistance was the first to be found and exploited for control of northern corn leaf blight. Polygenic resistance appears to control numbers of lesions that develop on plants and has little effect on lesion size.

A second type of resistance to the pathogen was found to be controlled by a monogenic dominant designated as *Ht* (Hooker, 1961b; Hooker, 1963). Monogenic resistance appears to control lesion size and not lesion numbers. Since the publication of the first finding of this type resistance in Ladyfinger popcorn and in the inbred line GE440, a number of other sources of the same type of monogenic resistance have been identified (Hooker et al., 1964; Ullstrup, 1963).

Control of northern corn leaf blight is done largely through the incorporation of both the polygenic and monogenic types of resistance into inbred lines.

Chemical control of the disease on commercial sweet corn and in seed fields is a common practice in southern Florida where winter crops of these are grown and where in most years northern corn leaf blight can severely limit production in plants not protected with fungicide sprays. Application of the fungicides must be done frequently during the growing season, but with these high-value crops the practice is economically feasible. Application of fungicides on plants in seed fields or on commercial corn is rarely done in the Corn Belt.

Northern corn leaf blight has been shown to be less severe when the supply of potassium, in the form of KC1, is adequate than where a deficiency existed (Hooker et al., 1963). The relationship of potassium to this disease is most pronounced on impoverished soils. Weather conditions and host resistance are more important in determining severity of this disease than host nutrition.

2. Southern Corn Leaf Blight

Southern corn leaf blight is widely distributed over the world, and although it has been identified as far north as Quebec in Canada, it is found mainly in warm-temperate and tropical regions.

In the USA the disease was not of major economic importance until 1970. It can be found nearly every year when moist, humid weather prevails, especially in seed fields when susceptible inbreds are grown. Southern corn leaf blight has generally not been so prevalent or severe in the Corn Belt as northern corn leaf blight or bacterial leaf blight.

In 1970 the epidemic of southern corn leaf blight was caused by a new physiologic race of the pathogen that was exceptionally virulent on corn with Texas male-sterile cytoplasm (*Tcms*). About 90% of the corn grown in the USA at that time was of this type. In the 1970 growing season, a highly virulent pathogen was present, a favorable environment prevailed — warm and humid weather over most of the eastern third of the country — and a uniformly susceptible host was extensively grown. The three important factors necessary for the development of a plant disease were at hand. The growing season of 1971 began with these three factors still present — a susceptible host, a virulent pathogen, and an environment favorable for disease development. In midsummer, however, temperatures lowered and the disease, although it did appreciable damage, subsided and did not develop as anticipated (Felch and Barger, 1971). The effects of the disease were widespread and had many ramifications (Hooker, 1972; Tatum, 1971; Ullstrup, 1972).

Symptoms of this disease are distinct from those of northern corn leaf blight, but are similar to, and can be confused with, the symptoms of yellow leaf blight and anthracnose. Lesions are tan, oblong to spindle-shaped, range up to 40 to 50 × 6 to 12 mm, and often show a faint concentric pattern of light and dark necrotic tissue. Often lesions are parallel-sided due to limitation of lateral spread, of pathogen, by the leaf veins (Fig. 6C). As with a number of leaf pathogens, lesions are generally smaller when larger numbers are present and larger when only a few are on the leaf. The leaf lesions caused by both races are generally similar. Those incited by Race T are somewhat more spindle-shaped and often smaller because of the larger numbers on a leaf.

Race T of the pathogens attacks all parts of corn plant with Texas-male-sterile cytoplasm. In addition to parasitizing leaves, ears may be penetrated through the tip, the butt or directly through the husk (Fig. 6D). Lesion on leaves of *Tcms* corn are surrounded with chlorotic areas of varying size. Growth of the fungus over the kernels imparts a gray-black color to the ears. Stalks are also attacked particularly at the nodes where the fungus invades from behind the leaf sheaths. Sporulation of Race T is especially profuse on leaf sheaths, and husks as well as on leaf lesions of *Tcms* corn. On corn with normal cytoplasm lesions are small — 2 to 4 mm in diam. and tan to brown. They may develop on leaf sheaths and husks as well as leaf blades.

Leaf lesions caused by Race O on normal cytoplasm corn tend to be more parallel-sided than those on *Tcms* corn caused by Race T. Race O rarely attacks ears even under severe epidemics of the disease. There are two recorded instances of ear rot of sweet corn caused by Race O (Burton, 1968; Robert, 1956).

Southern corn leaf blight is caused by *Helminthosporium maydis* Nisik. & Miy. Conidia of the asexual stage are long, spindle-shaped, tapering to rounded ends, curved, pale, olivaceous-brown, average $90 \times 15\,\mu$ in size and have 3 to 13 septations. These spores have greater curvature and are lighter in color than those of other species of the genus parasitic on corn. The asexual stage was described in Japan in 1926 (Nisikado and Miyake, 1926).

In 1925 the sexual stage of *H. maydis, Ophiobolus maydis* Drechs., was described in the USA. The description was based on its development, in a moist chamber, on infected leaves that had been collected in Florida and the Philippines (Drechsler, 1925). Perithecia are globose, black, 0.4 to 0.6 mm in diam. and beaked. Asci are hyaline, straight to slightly curved, cylindrical, 160 to 180 $\times$ 24 to 28 μ in size and bear up to eight, but typically four, ascospores. Ascospores are pale, smoky-colored, 130 to 340 $\times$ 6 to 7 μ in size with five to nine septations and are in a helicoid arrangement in the ascus. The sexual stage has been produced only under laboratory conditions when cultures of compatible mating type are grown together on appropriate substrates (Nelson, 1957b). A report of the occurrence of the sexual stage under field conditions in Florida (Schenck, 1970) was later shown to be another fungus (Schenck, 1972). The closest approach to occurrence of the sexual stage in the field was that when the perithecial stage was originally described in 1925. Infected leaves *collected in a field* were kept in a moist chamber in the laboratory. The development of mature ascospores under such conditions suggests that both mating types of the fungus must have been present in the infected leaves.

In 1934 a new genus, *Cochliobolus*, was erected and the pathogen transferred to it (Drechsler, 1934). The sexual stage of the fungus is now recognized under the binomial, *C heterostrophus* Drechs.

The pathogen is composed of two physiologic races differentiated on the basis of their pathogenicity on normal and *Tcms* corn (Hooker, et al., 1970a; Hooker et al., 1970b; Smith et al. 1970). These were designated as Race O and Race T. The former being that race upon which the original description was based and which was extant in the USA prior to 1969. There was no knowledge of the existance in *H. maydis,* of physiologic races possessing differential pathogenicity until late in 1969. During that growing season the hyper-susceptibility of *Tcms* corn was observed in a few locations in Illinois (Scheifele et al., 1970). Indiana, Iowa, Kentucky, Missouri, Florida, and Alabama (Scheifele, 1971). An isolation of Race T from the 1968 crop, collected from a sealed corn crib has been recorded (Foley and Knaphus, 1971). It has been pointed out that Race T may have existed in the USA prior to its recognition in the fall of 1969 (Nelson et al., 1970).

The first observation of the extreme susceptibility of *Tcms* corn was reported from the Philippines in 1961 (Mercado and Lantican, 1961). This was interpreted not as being due to a different race of the pathogen, but to warm, wet conditions that prevail during the rainy season in the Philippines and which are so favorable for reproduction and spread of the pathogen.

The two races of *H. maydis* are indistinguishable on the basis of the morphological characters in both the sexual and the asexual forms. Race T has been reported to produce "sclerotia" on malt extract agar; Race O does not (Locci and Locci, 1972).

Race T produces a host-specific pathotoxin which, when added to isolated mitochondria of *Tcms* corn, increased respiration, and caused an irreversible swelling of mitochondrial membranes. These changes did not take place when the toxin was added to mitochondria from corn with normal cytoplasm (Miller and Koeppe, 1971). Race O is reported to produce a host-specific toxin (Smedegard-Petersen and Nelson, 1969), but other studies could not confirm this (Comstock and Scheffer, 1972).

Infection of *Tcms* corn by Race T is accompanied by increased electrolyte and peroxidase leakage from the tissue, compared to that found when normal cytoplasm corn is infected. Infection of leaves of *Tcms* corn by Race O induced much less electrolyte and peroxidase leakage (Garraway, 1973).

The origin of Race T in the USA is somewhat obscure. It may have been introduced from the Philippines where it, or a race having a similar level of virulence on *Tcms* corn, has been observed (Aala, 1964; Mercado and Lantican, 1961; Villareal and Lantican, 1964; Villareal and Lantican, 1965). Surveys were conducted in different parts of the USA for *H. maydis* with respect to virulence and mating type. Race T was found in 12% of leaves collected in the South, but none recovered from leaves from Iowa or New York. All of the Race T isolates from Iowa were mating type A and 9 of 10 from New York were of this type. Mating types A and a were recovered in about equal numbers from the South. Mating types in the midwestern and northeastern populations in 1971 were about the same as in 1970. The southern population changed drastically from 1970 when 80% was type A. These findings are interpreted to indicate Race T to be of recent origin in the USA and other parts of the world. It may have arisen in the South by mutation from Race O which is more prevalent there than in other parts of the USA (Leonard, 1973).

Resistance to Race O is quantitative in some genotypes with incomplete dominance (Pate and Harvey, 1954). A second type of resistance has been observed in some Nigerian inbred lines in which the reaction to infection is a small, circular, chlorotic lesion supporting little sporulation (Craig and Daniel-Kalio, 1968). This type of resistance was reported to be recessive and controlled by two linked, recessive genes (Craig and Fajemisin, 1969). Later studies indicated the resistance to be controlled by a monogenic recessive (Smith and Hooker, 1973).

Resistance to Race T is primarily cytoplasmically controlled, but modification of reaction to infection may be influenced by nuclear genes.

Control of southern corn leaf blight caused by Race T of *H. maydis* has been accomplished by abandoning the use of *Tcms* corn in commercial production and reverting to normal cytoplasm corn. Control of the disease caused by Race O, if this is necessary, can be done by transfering the gene(s) governing resistance to agronomically acceptable genotypes.

Some control of the disease, caused by Race T, during the epidemics of 1970 and 1971 was effected by foliar application of the fungicides. This was expensive and necessitated spraying at 7 to 10-day intervals not only to protect the newly expanding foliage but to replenish the washed off or inactivated fungicide.

3. Bacterial Wilt

This disease, variously known as bacterial wilt, Stewart's wilt, Stewart's disease or Stewart's leaf blight, was described first in the USA in 1897 (Stewart, 1897). The disease is found mainly in the northeastern part of the country from Iowa eastward to the Atlantic Coast. It is found outside this general region, but less frequently.

Unlike the leaf blights caused by fungi, bacterial wilt does not require damp weather and heavy dew for rapid spread and development. Conditions favoring rapid transpiration, such as hot, dry weather, and rapid air movement, increase the severity of the symptoms in infected plants. Winter temperatures appear to have a marked influence on the incidence of bacterial wilt. It has been demonstrated that in summers following mild winters the prevalence of the disease is generally greater than when the growing season is preceded by severe winters (Haenseler, 1937; Stevens, 1934; Stevens and Haenseler, 1941). When the sum of the mean temperatures for December, January, and February, in degrees Fahrenheit is 100 or more the disease may be expected to be severe on susceptible genotypes the following growing season. When this sum is about 85 or less the incidence is expected to be low. The relationship of winter temperatures to the incidence of this disease has been the basis for predicting its prevalence (Stevens, 1936; Stevens and Haensler, 1941). The forecasts have proven to be fairly accurate. In 1953 a severe outbreak of bacterial wilt was predicted on the basis of temperatures during the winter of 1952-53. The prevalence of the disease in the summer of 1953 was the greatest in recent years (Boewe, 1953).

The effect of low winter temperatures is not directly on the inciting agent, but rather on the insect vector. It has been demonstrated that the corn flea beetle *(Chaetocnema pulicaria* Melsh.), the toothed flea beetle *(C. denticulata* Ill.), and the spotted cucumber beetle, *(Diabrotica undecimpunctata howarti* Barb.) act as vectors in the field. Other insect vectors have been identified under experimental conditions (Pepper, 1967), but *C. pulicaria* is by far the most important in the survival and spread of the pathogen in the field (Elliott and Poos, 1934). The survival of the pathogen is thus related to overwintering of the vector. If large numbers of this insect survive, as they do in mild winters, the prevalence of bacterial wilt is proportionately greater than when few vectors overwinter as occurs when temperatures are abnormally low.

While the pathogen has a wide host range among grasses (Ivanoff, 1935; observations suggest that hosts other than corn are not important in the epidemiology of Stewart's wilt.

Symptoms of the disease on systemically infected susceptible corn appear as long, pale-green or yellowish streaks of varying and irregular width which become progressively necrotic (Fig. 7A). Severely infected plants become stunted, wilt rapidly and may often show cavities within the pith of the stalk. In such plants bacterial masses will ooze from the cut surface of vascular bundles. The pathogen may traverse the entire vascular system and pass into the kernels. This occurs only occasionally in the most severely infected inbred lines of sweet corn. Such infected kernels may serve to disseminate the pathogen. The probability of this occurring in dent corn or in resistant sweet corn is low.

On dent corn the disease is generally less severe although a few inbred lines may be as susceptible as some sweet corn. Infection on leaves of dent corn is more restricted and usually does not become fully systemic. Lesions originate at points where flea beetles have injured the leaf in their feeding. The manifestation of the disease on dent corn is sometimes referred to as the leaf-blight, or late-infection phase (Fig. 7B). The latter term alludes to the increasing prevalence of restricted lesions on leaves during late summer. A considerable amount of leaf killing may occur on susceptible genotypes as a result of coalescence of lesions.

The causal organism, *Erwinia stewartii* (E.F. Smith) Dye, has been known by a number of binomials since its original description in 1897 (Stewart, 1897). These synonyms are: *Pseudomonas stewarti* E.F. Smith; *Bacterium stewarti* E.F. Smith; *Aplanobacter stewarti* (E.F. Smith) McCulloch; *Bacillus stewarti* E.F. Smith) Holland; *Phytomonas stewarti* (E.F. Smith) Bergey et al.; *Xanthomonas stewarti* (E.F. Smith) Dowson.

Erwinia stewartii is a non-motile, non-flagellate, non-spore-forming, capsule-forming, Gram-negative rod 0.4 to 0.8 $\times$ 0.9 to 2.2 μ in size. Colonies on glucose-nutrient agar range in diameter from 1.8 to 12 mm, and in color from creamy to orange, but are typically buff-yellow. Colony appearance has been correlated with virulence (Wellhausen, 1937a; Lindstrom, 1938; Ivanoff et al., 1938). Large, smooth, spreading, watery, mucoid, colonies were virulent; avirulence was correlated with smaller, rouch, raised and non-mucoid colonies.

Successive passage of the organism through resistant inbred lines of corn has been shown to increase the virulence of the initial culture for corn (Wellhausen, 1937a). Serial passage through susceptible inbred lines decreased virulence for corn. Virulence could not be increased beyond a certain level in resistant corn nor decreased beyond a certain point in susceptible corn. The bacterium appeared to reach a stage of equilibrium with the host and the environment after which further serial passage had no effect. Serial passage through highly resistant grass hosts, unrelated to corn, decreased virulence for corn, but increased virulence for the specific host through which it was passed.

DISEASES OF CORN

Wide differences in resistance to the disease occur among different types of corn. In general sweet and flint corns are most susceptible, popcorns less so, and dent corns the most resistant. One study showed that, among open-pollinated varieties of sweet, flint and dent corns, there was a high positive correlation between plant height and resistance, and between lateness and

Fig. 7. Leaf diseases: A, early, systemic infection by *Erwinia stewartii;* B, late infection phase or leaf blight phase (Stewart's disease); C, yellow leaf blight; D, eyespot.

resistance within each group (Ivanoff, 1936). The same general correlations were found among inbred lines of sweet corn when inoculated with the pathogen (Ivanoff and Riker, 1936). Dominance of resistance was observed generally among single-cross hybrids of susceptible × resistant parents. Among sweet corns Evergreen types are most resistant, Country Gentleman types slightly less so, and the Bantam types most susceptible.

Resistance to bacterial wilt appears to be determined by two major, one, minor dominant, supplementary and independently inherited genes (Wellhausen, 1937b). An association was observed between red cob color and resistance and white cob color and susceptibility suggesting some degree of genetic linkage between genes determining cob color and disease resistance.

Control of bacterial wilt has been obtained by selection for resistance among inbred lines following artificial inoculation. This achievement was one of the early examples of disease control in corn through the use of genetically conditioned resistance. Prior to 1932 the disease was at times a limiting factor in production of sweet corn. In 1933 Golden Cross Bantam was released (Smith, 1933). This single cross was developed specifically for resistance to bacterial wilt, but it also possessed the attributes of high quality and excellent yield. The hybrid was not immune to the disease, but it did have enough resistance to reduce the hazard of disease to tolerable levels.

Some control may be effected by reducing flea beetle populations with insecticides, but this practice must begin when plants are in the seedling stage and continued for several weeks until near the time of tasseling.

It has been demonstrated that corn seedlings deficient in potassium are more susceptible to infection than those deficient in either phosphorus or nitrogen. High applications of nitrogen increased the severity of the disease (Spencer and McNew, 1938). In field-grown plants it has been observed that deficiencies of potassium and excesses of nitrogen have the same influence as demonstrated in the seedling stage.

4. Yellow Leaf Blight

Yellow leaf blight was recognized first in Wisconsin in 1967 (Arny et al., 1970), and at about the same time in Ontario, Canada (Gates and Mortimer, 1969). It is now known to occur in the northern reaches of the Corn Belt and in most of the northeastern states. The distribution of the disease suggests that cool, wet weather favors development and spread of the pathogen. In 1968 the disease was manifested as an unusual seedling blight in central Pennsylvania (Scheifele and Nelson, 1969). The southern range of the disease has not been definitely established.

Yellow leaf blight is characterized by lesions on lower leaves that are rectangular to oblong-elliptical in shape, and 7 to 10 × 15 to 20 mm in size. These are buff-colored in their necrotic centers and have a brown border. The necrotic lesions are characteristically surrounded by a chlorotic halo of varying width (Fig. 7C). On upper leaves the lesions are narrower and confined by the veins. Dark pycnidia may be found in the necrotic lesions. Pycnidia develop mainly on the upper surface of lesions and are partially embedded in the necrotic tissue. Lesions of yellow leaf blight are similar to, and easily confused with, those of southern corn leaf blight or anthracnose if fruiting structures of the inciting agent are not present. Identity can be resolved by placing the diseased leaves in moist chambers for 3 to 4 days to permit sporulation of the pathogen.

The causal agent of yellow leaf blight is *Phyllosticta maydis* Arny and Nelson. Pycnidia are dark reddish-brown, subglobose, 60 to 150 μ in diam., embedded in the necrotic tissue, with ostioles protruding from the lesion surface. Pycnidiospores are non-septate, oblong to ellipsodial, hyaline, and typically with two oil droplets near each end. Spore size varies, but most fall within the range of 12 to 15 × 4 to 6 μ. Cross walls are not evident in freshly emerged spores, but on germination one or more may develop. When pycnidia are moistened spores are exuded in cirri (Arny and Nelson, 1971).

When the pathogen was first observed it was believed by some to be *Phyllosticta zeae* Stout (Scheifele and Nelson, 1969); others identified it as *Aschochyta zeae* Stout. Further studies showed that spores of *P. zeae* were shorter (4.5 to 7 × 2 to 3.5 μ) than those of the pathogen in question. Pycnidiospores of *A. zeae,* as originally described, measure about 8.5 to 13.5 × 3 to 4.5 μ and were indicated as being "obscurely uniseptate" (Stout, 1930). Evidence seems now established that the inciting agent of yellow leaf blight is neither *P. zeae* nor *A. zeae,* but a different fungus, *P. maydis.*

Mycosphaerella zeae-maydis Mukunya and Boothroyd, the sexual stage of the fungus, was first found on overwintered corn debris. It also developed in pure cultures of *Phyllosticta maydis* derived from single ascospores, thus demonstrating that the fungus is homothallic (Mukunya and Boothroyd, 1973). Pseudothecia of *M. zeae-maydis* are dark brown, globose, 86 - 192 μ in diameter, and embedded in overwintered leaf tissue, with ostioles exposed. Asci are cylindrical or clavate, straight or curved, and have thick hyaline walls. Mature asci measure about 45.5 × 11.3 μ, and bear 8 biseriately arranged ascospores. Ascospores are straight or curved, hyaline, about 16 × 5 μ in size, tapering toward rounded ends, two-celled and with a marked constriction at the cell wall. The upper cell of each spore is larger than the lower.

Phyllosticta maydis overwinters in corn debris on the surface of the soil. It is from this source that primary inoculum is derived in the early summer.

Wide variations in resistance to the pathogen has been observed among inbred lines and hybrids, and genotypes with Texas male-sterile cytoplasm and the "P-type" of male sterility were, with few exceptions, much more susceptible than their respective normal-cytoplasm versions (Arny et al., 1970; Ayers et al., 1970; Scheifele and Nelson, 1969; Scheifele et al., 1969). All types of male sterility did not confer susceptibility as did the Texas male-sterile cytoplasm and "P-type" male-sterile cytoplasm. Genotypes with the "C" and "S-type" male sterility reacted generally as their normal-cytoplasm versions (Nelson et al., 1971).

The behavior of genotypes with Texas male-sterile cytoplasm to infection by *P. maydis* is similar to that observed when such corn is infected by Race T of *H. maydis.* Texas male-sterile cytoplasm confers susceptibility, but it appears that disease reaction can be modified by genes.

Corn is the most important host of *P. maydis,* but limited observations have shown that sudangrass and *Setaria* spp. to be susceptible (Arny et al., 1970).

Control of yellow leaf blight involves the elimination of Texas male-sterile cytoplasm corn. This has been done as a consequence of the epidemic of southern corn leaf blight. If need for resistance to yellow leaf blight is indicated, sufficient resistant germplasm is available for use in development of inbred lines and hybrids that combines resistance and desirable agronomic attributes.

The survival of the pathogen in corn debris indicates that clean plowing would help delay early development of the disease (Sutton et al., 1972). Some control has been observed in small plot experiments using fungicides, but this has been considered too expensive except for protection of plants in seed fields (Arny et al., 1970).

5. Eyespot

Eyespot of corn was first observed in the United States in 1968 (Ullstrup et al., 1969). The disease has not been of major economic importance, but in parts of Minnesota, Wisconsin, Iowa, Illinois, Indiana, Ohio and Michigan small localized epidemics have caused appreciable leaf killing. Eyespot and the causal agent were described first in Japan in 1959 (Narita and Hiratsuka, 1959) where the disease is called "brown spot."

Early symptoms consist of small, oval to circular, translucent lesions 1 to 4 mm in size. With age the centers of the lesions become tan and are surrounded by a reddish-brown to purple margin with a narrow, yellow halo (Fig. 7D). This pattern of coloration and the size of the lesion suggests the name "eyespot." The same type of lesions developed on leaf sheaths and husks. At first the lesions may occur in zones or patches on the leaves owing to the collection of spores in the leaf whorl where they germinate and penetrate if moisture is present. Maturing leaves appear to be more susceptible to infection than young leaves. The early symptoms can be confused with some of the noninfectious "physiological" spottings that are genetically controlled. Young lesions are also similar to those seen in Curvularia leaf spot.

Eyespot is caused by *Kabatiella zeae* Narita and Hiratsuka. Spores are slightly curved, hyaline, pointed at their ends, nonseptate, and with average measurements of $27 \times 3.6\,\mu$. The spores are produced successively on the tips of short, simple conidiophores which emerge from stomata in the necrotic parts of lesions. No sexual stage of the fungus has been observed. The pathogen overwinters as dormant mycelium in old corn debris on the soil surface, and it is from this that primary inoculum arises. Cool, moist weather favors growth and sporulation of the pathogen; and it is in those parts of the world where such weather often prevails that the disease has been observed (Arny et al., 1971).

Some inbred lines and hybrids have shown tolerance to the disease, but reactions to it among a broad array of genotypes has not been examined.

No specific control of eyespot has been devised, but considering that corn debris, infected the previous year, is the source of primary inoculum, clean plowing and rotation would appear to delay early onset of the disease.

6. Anthracnose

Anthracnose of corn has not been of major economic importance in the United States, even though widespread over the eastern part of the country (Sprague, 1950). In recent years, however, there have been some localized epidemics causing increases in stalk rot, leaf killing and loss in yield (Dale, 1963; Morgan and Kantzes, 1971; Warren et al., 1973; Williams and Willis, 1963). In eastern North Carolina the disease was severe in 1972 and 1973. In Europe (Böning and Wallner, 1936; Messiaen, 1956; Zwillenberg, 1959), Asia (Chowdhury, 1936; Pupipat and Mehta, 1970; Quebral, 1958) and Africa (Krüger, 1965) the disease causes greater losses.

The disease is favored by ample rainfall and heavy dew coupled with moderate to warm temperatures 24 to 30 C.

Symptoms on leaves appear as oblong or broad spindle-shaped lesions that are tan to brown and measure up to 40×12 mm (Fig. 8A). When lesions are abundant the size is much smaller. Lesions often develop first near the tips of leaves. Diffuse chlorotic areas often form around lesions suggesting that a toxin may be produced by the pathogen. Lesions form on leaf sheaths, and husks as well as leaf blades. The fruiting structures of the fungus — "acervuli" — are present in the necrotic parts of the lesions and are often arranged in concentric circles. The acervuli, if not present, do form within 24 to 48 hours after placing lesions in a humid chamber. Setae, the black spines present on the acervuli, are diagnostic characters that distinguish lesions of anthracnose from those of southern corn leaf blight or yellow leaf blight — diseases with which anthracnose may be superficially confused. Symptoms on stalks have been descrived under the section on "Stalk Rots."

Anthracnose of corn is caused by *Colletotrichum graminicola*. This fungus has a wide host range among wild and cultivated grasses. *Sorghum* spp. appear generally more susceptible than corn. Physiologic races are present within the species. Isolates of the pathogen from corn have been shown to cause disease in that host, but not in *Sorghum* spp. and isolates from the latter unable to incite disease in corn (Dale, 1963; Messiaen, 1956). In India, however, isolates from corn and from sorghum could incite disease in both genera (Chowdhury, 1936). The pathogen is capable of overwintering on corn debris and it is from such sources that primary inoculum originates.

Colletotrichum graminicola has long been known only in an asexual form. Recently a sexual stage has been reported occurring in pure cultures under laboratory conditions (Politis and Wheeler, 1972). The sexual stage has not been observed in the field. The asexual stage of *C. graminicola* is described under the section on "Stalk Rots." The pathogen may gain entrance to ears and sporulation has been observed on kernels. Survival of the pathogen on and within infected seed has been demonstrated.

Many grasses in addition to corn and *Sorghum* spp. are susceptible to *C. graminicola,* but the number of discrete "hostic" races that may be involved in the extensive host range has not been determined, other than the two races differentiated on corn and *Sorghum* spp.

7. Goss's Bacterial Wilt

This disease was first found in 1969 in south central Nebraska. Since that time it has been recognized in other areas of that state; and in 1971 it was discovered in western Iowa (Wysong et al., 1973). It has at times been referred to as "Nebraska Wilt" or leaf freckles and wilt, but the above has been selected as the official name of the disease.

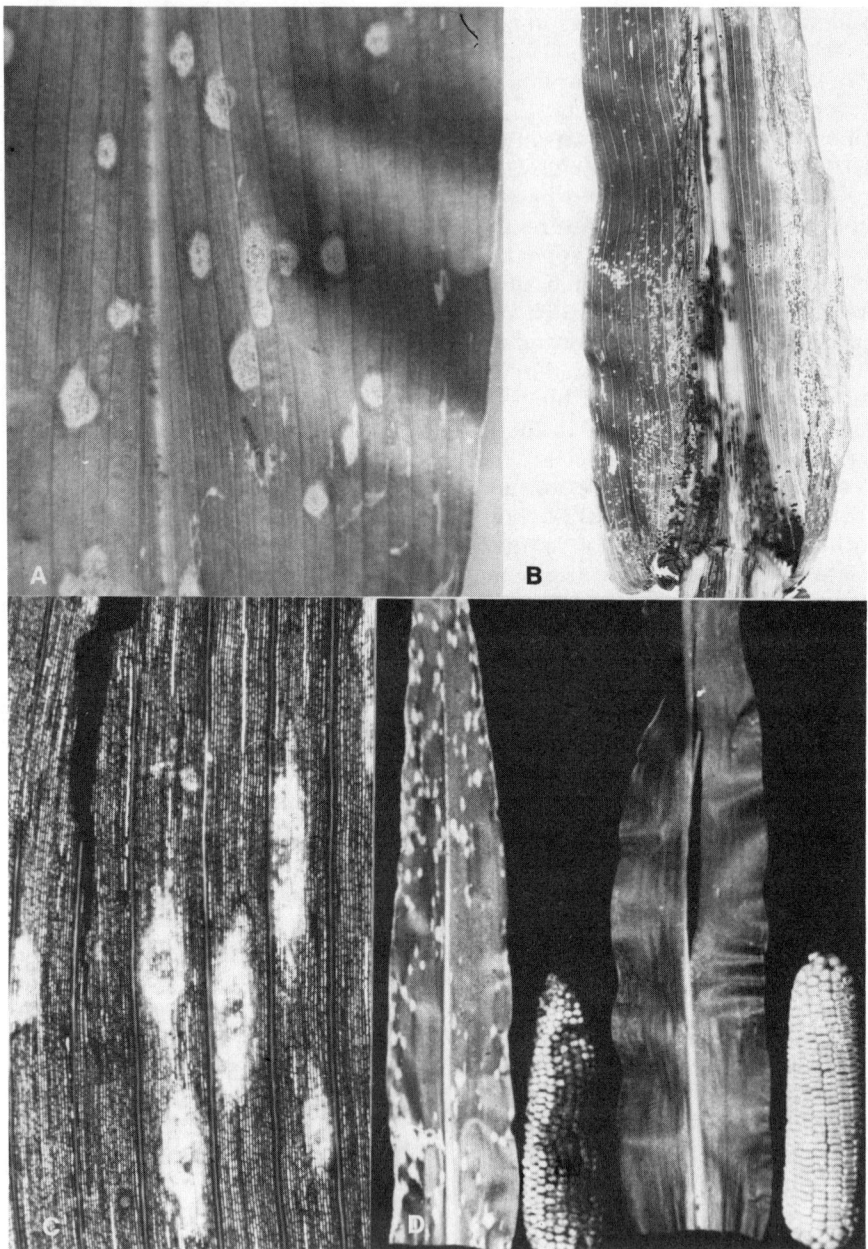

Fig. 8. Leaf diseases: A, anthracnose on leaves; B, brown spot or physoderma disease; C, northern leaf spot (about 4 ×) (courtesy, A.L. Hooker, University of Illinois); D, helminthosporium leaf spot showing symptoms on both the leaf and ear contrasted with resistant reaction.

The environmental conditions favoring spread and development of the disease are not fully understood.

Symptoms of this disease are somewhat similar to those of Stewart's disease. Long greenish-yellow lesions or stripes with wavy margins are the predominant symptoms, but discrete, water-soaked spots adjacent to the veins is one characteristic that differentiates this disease from Stewart's wilt. Plants may become infected at any stage of development. Occasionally watersoaking and the exudation of the pathogen may result in seedling leaves sticking together or, in older plants, prevent emergence of the tassel. A brown, root and stalk rot may also develop with discoloration of lower internodes. When stalks of infected plants are cut in cross-section an orange-colored bacterial exudate issues from the ends of the vascular bundles. The pathogen is primarily a vascular parasite, but parenchymatous tissue may also be invaded. Severely infected plants show marked leaf firing and stalk rotting. The pathogen, being fully systemic in the host may traverse the entire vascular system and invade kernels. These may be a means by which the pathogen is introduced into new areas.

"Goss's Wilt" is caused by *Corynebacterium nebraskense* Vidaver and Mandel. The bacterium is a non-motile rod measuring 1.6 to 2.0 $\times$ 0.5 μ and is Gram-positive and non-flagellate. The bacterium secretes an orange pigment in the host. On artificial substrates such as nutrient-dextrose agar with thiamine, colonies are mucoid, convex, and apricot colored. Thiamine is necessary for growth and production of the orange endopigment.

The bacterium apparently required wounding of the host tissue to establish infection. No vector for the pathogen has been found. *Corynebacterium nebraskense* may overwinter on diseased debris, and it is from this source that primary inoculum derives.

The host range of the bacterium is relatively narrow — corn, *Sorghum* spp., teosinte, (*Euchlaena mexicana* Schrad.) eastern grama grass (*Tripsacum dactyloides* L.) and sugarcane (*Saccharum officinarum* L.) ('Hinahina') were found susceptible to artificial inoculation. Under field conditions only sweet and dent corn have been observed susceptible. None of the common grass or broadleafed weeds in fields where infection of corn was severe showed evidence of the disease (Schuster, 1972).

The bacterium varies in pathogenicity and may lose virulence in artificial culture.

In tests for resistance among inbred lines of corn most of over 100 subjected to artificial inoculation were susceptible, a few showed a degree of resistance and none was immune (Schuster et al., 1972).

8. Brown Spot

Brown spot, or physoderma disease, is prevalent in the South Atlantic and the Gulf states. Yield losses as high as 25% have been recorded in some fields in Mississippi, with 100% lodging (Broyles, 1959). It has been observed as far west as Kansas and northward into South Dakota. In some years brown spot has been locally severe in bottom lands along the Ohio River in Ohio and southeastern Illinois (Burns and Shurtleff, 1973).

The disease is favored by warm, wet weather and this is reflected in its geographical range.

Symptoms of the disease are mainly on leaves below the ear or leaf sheaths and stalks, but in severe infection all leaves may show some disease. First indications of infection appear as small, yellowish spots that eventually turn brown. The yellowish to brown spots often occur in bands across the leaf blade. The spots coalesce to form larger brown blotches at the base of the leaves and on the adjoining leaf sheaths (Fig. 8B). The blotches on the leaf sheaths may be confused with purple sheath spot. Stalks infected with the pathogen usually break at the nodes. Cells of infected tissue eventually break down exposing pockets of dusty, golden-brown spores of the pathogens.

Brown spot is caused by *Physoderma maydis* Miyabe [Syn. = *P. zeae maydis* Shaw], an obligate parasite of relatively simple structure and life history. Sporangia, produced in great abundance in invaded host cells, are smooth, thick-walled, golden-brown 20 to 30 × 18 to 24 μ in size, globe-shaped, except for a flattened area on one side. When sporangia germinate a cap or lid opens to release a thin-walled endosporangium, which in turn ruptures to liberate a number of zoospores. The zoospores are thin-walled, mostly uniciliate, but occasionally are multiciliate (Ojerholm, 1934), 5 to 7 × 3 to 4 μ in size. Cilia are 3 to 4 times longer than the zoospores. These motile zoospores eventually come to rest on the host and establish infection. Penetration takes place mostly during daylight and only in meristematic tissue (Broyles, 1956b, 1962). The periodicity of infection into newly developed meristem tissue at the leaf base results in formation of bands of yellowish spots across the leaf blades. The affinity of the pathogen for meristematic tissue also accounts for the sharp break at the nodes of infected stalks. Infection may take place from the seedling stage to about the time of silking. After zoospores invade the host cells large vegetative cells develop in a matrix of fine hyphal strands. The enlarged cells develop into sporangia and ultimately the parasitized host cells die, releasing the mature sporangia. It has been shown that zoospores, after coming to rest on the host surface, produce irregular, extramatricular sporangia which, when mature, eject numerous zoospores through an apical pore. These secondary zoospores are similar to those produced by the golden-brown, resting sporangia except, for somewhat smaller size. After discharge another and often a third sporangium may form within the original (Sparrow, 1934). This succession of zoospore production by sporangia developed from the zoospores originally released from the golden-brown, resting sporangia accounts for secondary spread of the pathogen on individual host plants and adjacent plants (Broyles, 1956a).

The minimum temperature for germination is about 18 C, the optimum 28 and the maximum 36 C. The optimum pH for germination is 8.5 to 9.0 with the minimum and maximum respectively about 3.0 and near 10. Sporangia do not germinate in complete darkness. Germination percentage increased according to the logarithm of light intensity from 0 to 12 foot-candles. Intensities up to 393 ft-c did not significantly increase germination. Germination is highest in blue light, low in green and none in red or yellow light. Oxygen is necessary for light reaction and subsequent germination (Hebert and Kelman, 1958). Direct sunlight is lethal to sporangia as well as a

30-day exposure to 80 C. Viability of sporangia is maintained after one month exposure to either 0 or 70 C; and at room temperatures they may survive for 2 years (Voorhees, 1933a).

The host range of *P. maydis* is relatively narrow, only corn and teosinte have been found susceptible (Eddins, 1933). Marked differences have been found among inbred lines under conditions of both natural infection and artificial inoculation (Harvey, Thompson and Hebert, 1955). Some qualitative differences in symptomology have been observed. Few inbreds appear to be highly resistant, but a sufficient number show a level of resistance sufficient to use in breeding programs.

Resistance is determined by a number of genes with additive effect being most important. Dominance and epitstatic effects have been found somewhat less important in determining resistance (Moll et al., 1963; Thompson 1969; Thompson et al., 1963).

Control by use of fungicides has some promise for protection against the disease in seed fields (Broyles, 1956b) but this may not be economical in *produce* corn.

It has been demonstrated that resting sporangia overwinter in debris left on the surface of the soil and they constitute the primary source of inoculum in the spring and early summer (Burns and Shurtleff, 1973). Clean plowing would destroy the potential of such material and hence delay onset of the disease and reduce its intensity during the growing season.

9. Northern Leaf Spot

Northern leaf spot was first recognized in 1972 in Illinois, Wisconsin, Indiana and Michigan (Hooker et al., 1973). The disease may have been observed before, but perhaps mistaken for another leaf blight. Its appearance in the northern parts of the Corn Belt during 1972, which was unusually cool and wet, suggests that the disease is favored by such environmental conditions.

In the field lesions appear on leaf blades, sheaths and husks. Leaf lesions are generally oval, tan and occasionally showing a concentric pattern of coloration within the necrotic area. Lesions range in size up to 8×12 mm. The size, shape, and color of the margin of lesions varies with the genotype of the host and age of lesions.

Symptoms on leaves are distinct from those produced by Races O and T of *H. maydis*, *H. turcicum* or by Race I of *H. carbonum*. They are more nearly intermediate in size and shape between those incited by Race I and Race II of *H. carbonum* (Fig. 8C). Symptoms on ears are characterized by black mold growth on and between the kernels. Invasion of the ear may be through the tip, butt or sides. The black rot is visually indistinguishable from that incited by Race I or II of *Helminthosporium carbonum*.

The pathogen is a member of the genus *Helminthosporium,* but a species name has not yet been assigned. Spores are elongated, tapering, slightly curved, rounded at ends, golden-yellowish when young, becoming olivaceous with age, and average about $66 \times 13\,\mu$ in size. The number of septations ranges from 3 to 12. The spores are thus intermediate in size between those of *H. maydis* and those of *H. carbonum*. They are less curved than the spores of *H. maydis,* but more so than those of *H. carbonum.* Morphologically the spores are similar to those of *H. setariae* Saw.

The fungus does not produce a toxin as does Race T of *H. maydis* or Race I of *H. carbonum.* The pathogen is equally pathogenic on normal cytoplasm corn and corn with Texas male-sterile cytoplasm. The *Ht* gene, which governs resistance to infection by *H. turcicum* has no effect on resistance to this pathogen. Inbred lines or hybrids susceptible or resistant to Race I of *H. carbonum* show no differential reaction to the fungus. Some of the commonly used inbred lines, B14, B37, W64A, and N26, appear susceptible following inoculation; the inbred line Oh545 is somewhat more resistant.

The pathogen is not highly virulent on hybrids and thus the disease may become a problem mainly on inbred lines and not in farmers' fields. Control, if this should become necessary, will probably be best accomplished through genetic resistance. As with other species of *Helminthosporium,* the fungus may overwinter in corn debris and the latter will thus become a source of primary inoculum. Clean plowing should be helpful in delaying early infection.

10. Helminthosporium Leaf Spot

This disease was recognized first in Indiana in 1938 (Ullstrup, 1944). Since that time it has been found on a few inbred lines of corn in the eastern one-third of the USA. Moderate temperatures and ample moisture favor the disease.

Lesions produced by Race I of the pathogen are tan, oval, with a definite zonate pattern in the necrotic portions, and measure 15 × 25 mm. All parts of the plant are susceptible (Fig. 8D). In damp weather sporulation is profuse in the lesions on leaf sheaths. The pathogen penetrates any part of the ear and imparts a black charred appearance. Lesions incited by Race II of the fungus, found mostly on lower leaves, are brown, elongated, and 5 × 25 mm in size. Symptoms on ears, following infection by Race II of the pathogen, are indistinguishable from those caused by Race I.

Helminthosporium carbonum Ullstrup, the causal agent of this disease, appears to be composed of two races. Spores of the asexual stage are elongated, slightly curved, olive-brown, rounded at the ends, with a range in size of 25 to 100 × 7 to 8 μ. The hylum does not protrude as in *H. turcicum,* but appears to be within the confines of the wall. The number of septa ranges from 2 to 12 with an average of about 6. Spores show less curvature than those of *H. maydis* and are distinctly darker in color than those of either the latter or *H. maydis* and are distinctly darker in color than those of either the latter or *H. turcicum.*

The sexual stage, *Cochliobolus carbonum* Nelson can be produced in vitro when compatible mating types are grown in association on suitable substrates (Nelson, 1959). The sexual stage has not been found occurring naturally under field conditions. Mature perithecia are black, globose to ellipsoidal, beaked, and 355 to 550 μ in height by 320 to 430 μ in diam. Asci are cylindrical to clavate, hyaline, average 197 × 23 μ in size, and contain 1 to 8 ascospores arranged in a helicoid manner. Ascospores are filiform, hyaline, 5 to 9 septate and measure about 245 × 8 μ. The thread-like ascospores are forceably discharged from the ostiole of the perithecium. The sexual stage has served as a model for a number of genetic studies on the fungus (Nelson 1960, 1961, 1964, 1970; Nelson and Ullstrup, 1961; Scheffer et al., 1967).

Race II of *H. carbonum* is weakly parasitic and shows little or no differential pathogenicity on inbred lines. The reaction of corn seedlings to infection by Race II is characterized by minute, chlorotic-necrotic flecks similar to those incited by Race I on resistant inbred lines.

A host-specific toxin, produced both in vivo and in vitro by Race I of *H. carbonum*, has been isolated and characterized (Pringle, 1971; Pringle and Scheffer, 1967; Scheffer and Ullstrup, 1965).

A number of grasses, when artificially inoculated with Race I, react to penetration and infection (Nelson and Kline, 1971). Under natural conditions in the field, however, such reactions, if they do occur are benign and have not been of practical consequences. These grasses, in the field, are not afflicted with a disease incited by Race I of *H. carbonum*. Corn appears to be the only host for this pathogen under conditions of natural infection.

Resistance to Race I of *H. carbonum* is general — only a few inbred lines are susceptible. Those presently known to be susceptible are Pr, K61, K44, Mo21a, and N31. Resistance to Race I is conditioned by a single, dominant gene located on chromosome 1 (Roman and Ullstrup, 1951).

11. Southern Leaf Spot

This disease occurs on corn, *Sorghum* spp., *Pennisetum glaucum* (L.) R., Br., and *Eragrostis ciliansis* (All.) Link. The disease is of no economic importance and occurs mainly in southern USA (Lefebvre and Johnson, 1941; Young et al., 1947).

Lesions on corn are yellowish to tan with a brown margin and necrotic centers, spindle-shaped, and measure 2 to 5 × 1 to 2 mm. These may coalesce and form larger necrotic areas.

The pathogen of the leaf disease is *Helminthosporium rostratum* Drechs. Spores are 32 to 184 × 14 to 22 μ, straight to slightly curved, dark olivaceous, 3 to 15 septate, and possess a protruding hilum. A distinctive feature of the spores is the thickened walls of the proximal and distal cells.

The disease has not been sufficiently important to warrant development of control practices.

12. Zonate Leaf Spot

This disease was first observed on sweet sorghum (*Sorghum bicolor*) in 1940 in Louisiana (Bain and Edgerton, 1943). The disease is much more prevalent and conspicuous on *Sorghum* spp. than on corn. Zonate leaf spot occurs more frequently in tropical and warm temperate areas, but may be encountered occasionally as far north as Indiana. The disease is of negligible economic importance on corn in the USA.

Symptoms are expressed as distinct spots or blotches with a characteristic zonate pattern of purplish-brown, irregular, rings or circles, alternating with tan, nectoric tissue (Fig. 9A). The lesions are first apparent as small yellowish, water-soaked spots that eventually enlarge up to 4 to 5 cm in diam. On sorghum the lesions are more conspicuous and show more reddish-purple, rather than brown pigmentation, as in corn.

The pathogen inciting zonate leaf spot is *Gloeocercospora sorghi* Bain and Edgerton. The fruiting body is a sporodochium which originates from a stomate and lies on the surface of the leaf. These structures are visible as minute, salmon-colored bodies in the necrotic tissue. Conidiosphores within the sporodochia measure 5 to 10 μ, are hyaline, and simple or branched.

Fig. 9. Leaf diseases: A, zonate leaf spot; B, holcus spot (courtesy, M.G. Boosalis, University of Nebraska); C, bacterial stripe showing lesions on lower leaves; D, chlorosis of upper leaves following infection of lower leaves. Upper chlorotic leaves are not infected.

Conidia are thread-like, few to many-septate, curved to straight, 20 to 195 × 3 μ in size and borne in a pinkish, mucous matrix. As the lesions mature, black, round to elliptical sclerotia form within the disintegrating tissue. In artificial culture the fungus requires thiamine and biotin for growth (Malca and Ullstrup, 1960).

The pathogen, in addition to infecting corn, causes disease on several species of sorghum, including sudangrass and johnsongrass, on sugarcane and on *Agrostis* sp. (Bain and Edgerton, 1943).

13. Holcus Bacterial Spot

This disease was first recognized in Iowa in 1916, although at that time little was known as to the causal agent or other aspects of the disease. In 1926 a report of studies was made in which the pathogen, its host range and control measures were discussed (Kendrick, 1926). The name of the disease is derived from *Holcus* which, at that time, was the generic term of *Sorghum*, and also a host of the pathogen.

Holcus bacterial spot has not been of economic importance in the United States. In 1966 it became widespread on corn, 55 to 70 cm tall, over an extensive area in eastcentral Nebraska. Infection apparently followed a heavy rain and windstorm that watersoaked leaf tissue with water contaminated with the pathogen. A hot, dry period suppressed further development of the disease (Weihing and Vidaver, 1967).

Early symptoms appear as round or irregular dark green watersoaked spots. With age the lesions may become 10 mm in diam. with a brown necrotic center and a reddish-brown margin (Fig. 9B). When viewed in transmitted light a yellowish halo is seen around older lesions. Infection often begins at the edges of leaves where the pathogen may enter through hydathodes (Kendrick, 1926). The lesions may sometimes be confused with the necrotic spots induced by the herbicide "Paraquat" when this is used around corn and drift of the spray falls accidently on corn.

The causal agent of holcus bacterial spot is *Pseudomonas syringae* v. Hall [Syn. = *P. holci* (Kendrick) Bergey et al.; *Xanthomonas holcicola* (Elliott) Starr and Burkholder]. The organism is a short rod, averaging 0.7 to 2.1 μ in size, Gram-negative and motile with one to several polar flagella. Colonies are round, smooth, glistening grayish-white on nutrient-dextrose agar. The optimum temperature for growth is between 25 to 30 C. *Pseudomonas syringae* has an exceedingly wide host range among both monocots and dicots. Strains of the bacterium may exist which differ from each other on the basis of groups of hosts attacked.

The host range of the pathogen causing holcus bacterial spot includes, in addition to corn, sudangrass (*S. sudanense*), johnsongrass (*S. halipensis*), broomcorn (*S. bicolor* var. *technicum*), *S. bicolor* and Setaria spp. Most of the common cultivated and wild grasses, including wheat (*Triticum aestivum* L.) and oats (*Avena sativa* L.) are not susceptible.

The pathogen can survive in soil and overwinter in debris of corn and sorghum infected during the previous growing season. Splashing rain and wind may carry the bacteria to the leaves of host plants where, in the presence of a film of water, they enter stomata and hydathodes and establish infection.

Holcus bacterial spot has not been sufficiently prevalent or severe to warrant development of control measures.

14. Bacterial Stripe

This disease, of little importance on corn, is more prevalent on *Sorghum* sp. It becomes widespread when driving rains are frequent during early stages of growth of corn.

Typical symptoms on corn leaves appear as elongated, parallel-sided, amber to olive-colored, oil-soaked, translucent lesions (Fig. 9C). Lesions first appear on lower leaves and, with conditions favorable for spread of the pathogen, gradually develop on upper leaves, but seldom on those above the ear. With age the lesions become brown and necrotic. Young lesions are gorged with the pathogen. A secondary symptom which appears only on certain of the most susceptible inbred lines — Oh28 and L304A — is the pronounced bleaching of interveinal tissue as new leaves grow from the whorl (Fig. 9D). This chlorosis may increase so that the entire leaf blade is white. The chlorotic tissue is not infected. Another characteristic of the disease syndrome is the buckling and distortion of the stalk owing to the failure of the leaves to unfurl while internodes continue to grow.

On sorghum, leaf symptoms are much the same in general pattern of the lesions, but their color is purplish-red. No secondary chlorosis develops in the sorghum.

The pathogen, originally described in 1905 (Smith, 1905), is *Pseudomonas andropogonis* (E.F. Sm.) Stapp [Syn. = *Bacterium andropogoni* E.F. Sm.; *Phytomonas andropogoni* (E.F. Sm.) Bergey et al.]. It is a short rod with rounded ends, occurring singly or in pairs, 1.5 to 2.5 × 0.5 to 0.8 μ in size, Gram-negative, and motile by a single, polar flagellum (Ullstrup, 1960).

The organism survives in the soil, then in early summer may come in contact with leaves through rain splash and wind. In the presence of a film of water on leaf surfaces, cells of the bacterium enter stomata and establish infection.

The disease in corn is mainly one of inbred lines; seldom is it seen on hybrids. As a consequence it is of minor economic importance and no effort has been made to control the disease.

15. Bacterial Leaf Blight and Stalk Rot

Bacterial leaf blight was first reported in the USA in 1929 (Johnson et al., 1929). Since that time it has been reported in a number of southern and central states (Johnson et al., 1949; Pady, 1944; Slagg, 1944). In 1973 it occurred in a localized area in North Carolina. It appears to be confined to the United States, although a similar disease has been reported in Australia (Ludbrook, 1942). The disease has been of minor importance except for occasional, localized outbreaks.

Symptoms of the disease are distinct from other bacterial diseases of corn in this country. Leaf lesions appear as well-defined, narrow necrotic stripes ranging from small spots 1 to 2 mm long to those 400 mm or more in length. At first lesions are water-soaked and translucent; later becoming necrotic with straw-colored centers and brown margins. Lesions may coalesce to form extensive necrotic areas. Shredding of leaves in the necrotic portions is a characteristic of the disease (Fig. 10A). Stalks usually become infected just above the joint of ear attachment, where a dark brown discoloration occurs. Severely diseased plants are stunted, tops may die as a result of stalk infection and sometimes several, small, barren, diseased ears will develop in place of a single healthy ear.

Pseudomonas alboprecipitans Rosen [Syn. = *Bacterium alboprecipitans* Rosen; *Phytomonas alboprecipitans* (Rosen) Bergey et al.] is the causal organism of bacterial leaf blight and stalk rot. The specific epithet refers to the white sediment formed by the pathogen when grown on several agar and liquid substrates. It is rod-shaped, Gram-negative, about $0.6 \times 1.0 \mu$ in size occurring singly or in chains up to 10 cells long and motile by a single, polar flagellum. On nutrient-dextrose agar colonies are small, smooth, and pale-white in reflected light. Under field conditions the bacterium attacks corn and *Setaria lutescens* (Weigel) F.T. Hubb, the first host on which it was described. Later work established that the organisms on corn was identical with the pathogen described on *S. lutescens* (Johnson et al., 1949).

Because of the minor importance of the disease inbred lines and hybrids have not been evaluated with respect to their resistance.

16. Diplodia Leaf Spot

This disease is occasionally encountered on corn in some of the southern states. It has been of no economic importance.

Lesions appear first as minute yellowish water-soaked spots which may enlarge and coalesce into necrotic areas 4 to 5×1.0 to 1.2 cm in size. Mature lesions are tan to brown with darker brown margins and develop mainly near the base of the leaf blade.

Diplodia macrospora Earle is the pathogen causing the disease. It is more important as the cause of an ear rot, and is described under the section on "Ear Rots."

17. Purple Sheath Spot

Purple sheath spot is a common disease, but appears not to have any important effect on corn. It was often confused, in 1970 and 1971, with the sheath infections caused by Race T of *H. maydis*. The disease may often be mistaken for brown spot.

Irregular purplish-brown blotches of varying size develop on the outer surface of the sheaths (Fig. 10B). Beneath these spots the inner surface of the sheath is discolored and necrotic. The disease becomes conspicuous after silking and is almost universally present on corn. A number of fungi and bacteria have been isolated from necrotic sheaths. It has been assumed that these organisms life saprophytically on debris — pollen and soil particles — that collect behind the sheaths and become mildly parasitic and invade the maturing tissue (Durrell, 1920). Some inbred lines and hybrids show more purple sheath spot than others.

18. Gray Leaf Spot

Gray leaf spot has been observed in Illinois (Tehon and Daniels, 1925), Kentucky and Tennessee (Hyre, 1943), South Carolina, (Kingsland, 1963), and in other southern states (Chupp, 1953). In 1973 the disease was prevalent in parts of Virginia and Kentucky. The disease has been of minor economic

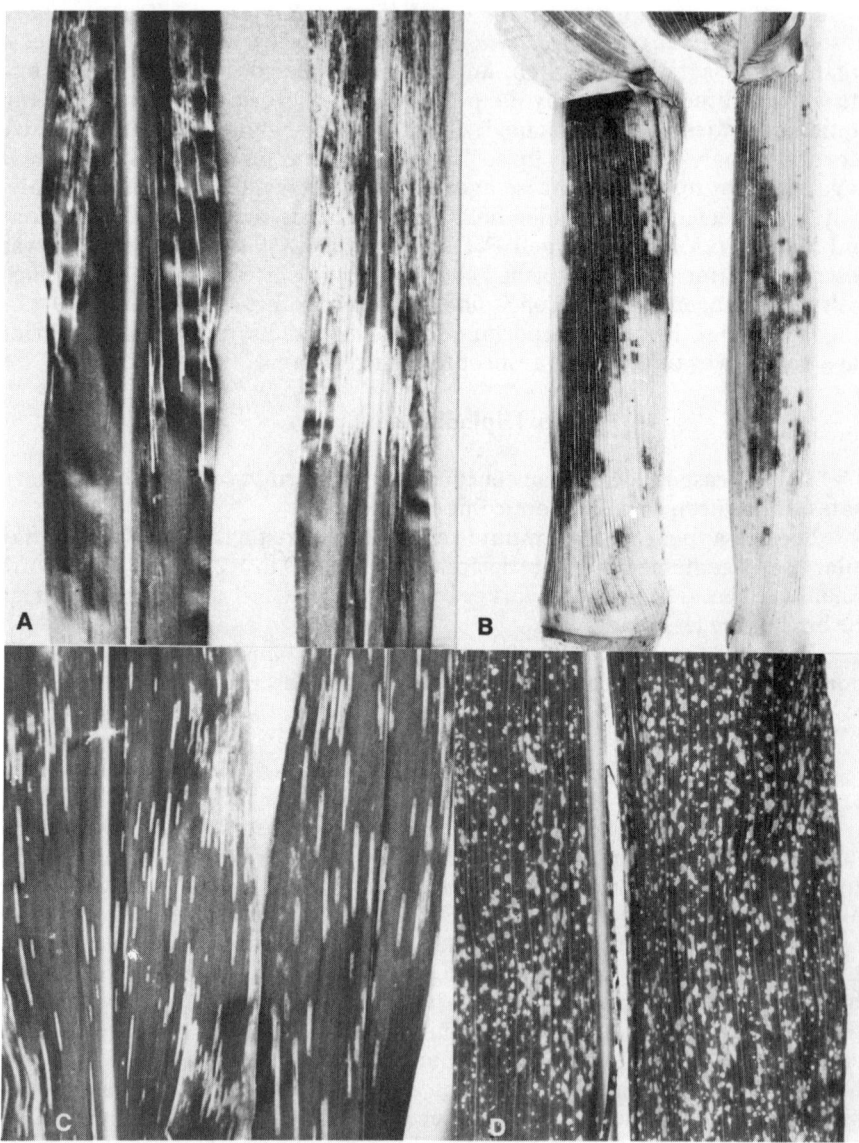

Fig. 10. Leaf diseases: A, bacterial leaf blight and stalk rot. Typical leaf symptoms showing long narrow lesions and shredding of tissue; B, purple sheath spot; C, gray leaf spot (courtesy, C. Roane, Virginia Polytechnic Institute): D, curvularia leaf blight.

importance and little work has been done on it. From its general distribution the disease appears to be favored by warm, wet weather.

Symptoms are characterized by small, brown, elongated lesions delimited in lateral spread by leaf veins. Mature lesions are tan to brown, later becoming gray, measure up to 0.5×1 to 2.5 cm, and bounded by a narrow maroon border (Fig. 10C).

Gray leaf spot may be caused by two closely related fungi — *Cerospora zeae-maydis* Tehon and Daniels and *C. sorghi* Ell. & Ev. The conidia of the former are borne on pale conidiophores emerging in fascicles from stomata. Conidia are hyaline, cylindrical, straight to curved, tapering, 3 to 9 septate and average 25 to 30×4 to $8\,\mu$ in size. Conidia of *C. sorghi* are hyaline, cylindrical, straight to curved, multiseptate measure 30 to 70×2 to $4\,\mu$ in size, and borne on dark brown conidiophores. The two species are distinguished on the bases of the pale conidiophores and wider conidia in *C. zeae-maydis*.

In addition to corn *C. sorghi* has been identified as a pathogen on *Sorghum* and *Panicum* spp. *Cerospora zeae-maydis* has been reported only on corn (Chupp, 1953). The symptoms produced by each, on corn, are essentially the same. Both species are relatively weak parasites that attack corn as it begins to mature. *Cerospora zeae-maydis,* and probably *C. sorghi* overwinter on corn debris.

19. Curvularia Leaf Spot

This disease occurs occasionally in southern United States, but it has not reached economic importance. It is favored by warm, moist weather.

Symptoms appear first as minute, circular to oval, straw-colored lesions which may be water-soaked in early stages. When first evident spots are 1 to 2 mm in diam. and with age seldom attain a size greater than 5 to 7 mm (Fig. 10D). Mature lesions are straw-colored to brown with a reddish-brown margin. Lesions may be confused with some of the chlorotic-necrotic, noninfectious, "physiological" spottings that are observed on some inbred lines.

In the USA *Curvularia maculans* (Bancroft) Boedijn has been identified as the causal agent of the disease (Nelson, 1956). This species has been relegated to synonymy with *C. eragrostidis* (P. Henn) J.A. Meyer (Ellis, 1966). Spores are straight, elliposoid or barrel-shaped, almost always 3-septate, with the middle septum darker and thicker than the other two. Terminal cells of the spores are much lighter in color than the two middle cells. The size of spores average $27.4 \times 14.5\,\mu$. The members of the genus *Curvularia* bear some superficial similarity to those of *Helminthosporium*. Conidia of the former are generally shorter and broader than those of the latter, and the terminal cells of the spores of *Curvularia* are almost always lighter in color than the centrally located cells. A monographic treatment of the genus together with a description of species has been made (Ellis, 1966).

A number of other species have been identified as pathogens on corn in tropical areas (Mabadeje, 1969; Vasal et al., 1970). In the USA Curvularia leaf spot has not been of sufficient importance to justify development of control practices. Nothing is known concerning the reaction of inbred lines to infection, or the mode of inheritance of resistance.

20. Chocolate Spot

This is a widespread disease that has only recently been described. It occurs on lower leaves and especially where plants grow on soils deficient in potassium. There is no evidence thus far that the disease causes any appreciable loss in yield or contributes to stalk lodging.

Symptoms are characterized by dark brown, oval to oblong, necrotic lesions, measuring up to 15×5 mm in size. A distinct yellow halo usually surrounds the necrotic area.

Chocolate spot is caused by *Pseudomonas syringae,* but apparently by a different strain than that inciting holcus bacterial spot. The pathogen is described under the section on that disease. Inoculation trials in the greenhouse were successful only when the test plants were grown in a regime of potassium deficiency (Arny et al., 1972).

VII. DOWNY MILDEW

Corn is susceptible to eight downy mildew diseases, three of which occur in the United States. The downy mildews of corn are far more destructive in the tropical parts of the Eastern Hemisphere particularly, the Philippines, Indonesia, Thailand, and India. Each disease is caused by a specific pathogen, all of which are obligate parasites and systemic in their hosts.

These diseases have been named according to the country where they were first found: Philippine downy mildew, *Sclerospora philippinensis* Weston; Javanese downy mildew, *S. maydis* (Raciborski) Butler, or the host on which they were first described: sugarcane downy mildew, *S. sacchari* Miyaki, sorghum downy mildew, *S. sorghi* (Kulk.) Weston and Uppal, spontaneum downy mildew, *S. spontanea* Weston (from wild sugarcane, *Saccharum spontaneum* L.); or from a conspicuous and identifying symptom: crazy top, *Sclerophthora macrospora* (Sacc.) Thirum., Shaw, Naras., brown stripe downy mildew, *Sclerophthora rayssiae* var. *zeae* Payak and Renfro, and green ear disease, *Sclerospora graminicola* (Sacc.) Schroet (after the symptom on pearl millet, *Pennisetum glaucum* R.Br).

The identification of individual downy mildew diseases of corn may present some difficulties. Some of these diseases have several symptoms in common, e.g., stunting, chlorotic striping, narrowing of leaves, phyllody of floral parts, and barrenness. Not all of these downy mildews are alike in expression of all symptoms. Brown stripe downy mildew, for example, does not show phyllody, and crazy top does not show eventual necrosis of chlorotic stripes. The pathogens of some of these diseases are only slightly different from one another. This fact at times requires critical examination of these fungi in order to properly separate them. Finally, as in other diseases, environment and host genotype may interact to modify symptoms. These conditions should be taken into consideration in making diagnoses of the downy mildews of corn.

Fig. 11. Downy mildews: A, crazy top showing phyllody of the tassel; B, sorghum downy mildew showing narrowing, chlorosis and striping of leaves, and stunting of the plants; C, "down" on corn leaf due to production of conidia and conidiophores of the pathogen; D, contrast in reaction to *Sclerospora sorghi* between resistant and susceptible corn genotypes.

1. Green Ear Disease

This disease, named after the symptom incited on pearl millet is extremely rare on corn. It was observed in Iowa in 1925 and again in 1927 (Melhus et al., 1928). The disease was also found in Wisconsin in 1921 (Weston, 1929). In the USA *Setaria viridis* (L.) Beauv. is relatively more frequently infected than corn.

Because of the rarity of the disease, descriptions of the symptoms on corn are meager. Gray blotching of leaves, mottling and chlorotic striping are early symptoms on young plants. Infected plants become dark green and stocky owing to shortened internodes. Production of sporangia is sparse on corn and oospores were not observed on this host (Melhus et al., 1928).

The pathogen, *Sclerospora graminicola* (Sacc.) Schroet., produces ovoid sporangia 13 to 37 × 11 to 25 μ in size, on branched conidiophores. Sporangia germinate by production of biciliate, reniform zoospores. Oospores are formed on some grass hosts, including *S. viridis*, but not on corn or sorghum. These globose sexual spores range in diameter from 19 to 45 μ and are thick-walled. The pathogen has a wide host range in the *Gramineae*, especially in the family *Panicae*.

Although it has been mentioned that the pathogens of downy mildew diseases of corn are obligate parasites, this pathogen, *S. graminicola*, has been shown to make limited growth on a complex, artificial medium. The fungus was established in vitro on callus tissue of pearl millet and from this living substrate it grew onto the agar medium. Abnormal reproductive structures developed, but after two transfers, at 20-day intervals, it declined rapidly in vigor (Tiwari and Arya, 1969).

2. Crazy Top

This is a downy mildew disease of corn that occurs sporadically in all parts of the USA (Semeniuk and Mankin, 1964; Ullstrup, 1952; Ullstrup and Sun, 1969). It is of little economic importance, but occasionally it has caused damage in individual fields. The disease appears only following waterlogging or flooding of the soil.

There are many symptoms in the disease syndrome, but rarely do all appear at one time in a field of infected plants. One early symptom is excessive tillering and stunting. Infected plants may also show chlorotic striping, but this is less intense and less frequently seen than in some of the other downy mildew diseases. Leaves on severely infected plants may be narrow, leathery in texture, and finely rugose on the lower side. Leaf sheaths also show a finely rugose surface in severely infected plants. Occasionally the leaf whorls of diseased plants will be tightly rolled and curved. Elongation of ear shanks is another symptom that is sometimes observed. The most conspicuous symptom, and from which the disease derives its name (Koehler, 1939), is the condition of phyllody in the tassel (Fig. 11A). In such tassels normal floral parts are replaced by leaves, resulting in a bizarre, bushy-appearing tassel. Often such tassels will show much common smut infection. This is because of the abundance of meristematic tissue in these bushy tassels which renders them susceptible to infection by *Ustilago maydis* (DC.) Cda. In severe infection the floral structures may be completely suppressed. The great majority of infected plants are barren, the female inflorescence being replaced by husks. Rarely a small, poorly filled ear may develop on an infected plant.

The pathogen is systemic in its hosts and may traverse the entire plant including the seed. Systemic infection of seed has been demonstrated (Ullstrup, 1970b; Ullstrup and Sun, 1969) but because the pathogen is easily destroyed by drying seed at 40 C, and soon loses viability in stored seed, there is little danger of the fungus being carried in commercial seed corn. Furthermore, only a few infected plants bear seed and these on poorly filled nubbin ears that would be discarded before final processing of seed corn.

The fungus is an obligate parasite and attempts to establish it on artificial substrates in the absence of living host tissue have failed (Ullstrup, 1970). The same pathogen has been isolated from rice onto very simple agar media (Katsura, 1952) and then inoculated into rice seedlings, but symptoms were not reproduced in such infected seedlings (Akai, 1959).

Sclerophthora macospora (Sacc.) Thirum., Shaw and Naras. [Syn. = *Sclerospora macrospora* Sacc.] has a wide host range among cultivated and wild grasses (Semeniuk and Mankin, 1964; Ullstrup, 1955a; Whitehead, 1958).

For a number of years the etiological agent of crazy top was not known (Koehler, 1939; Semeniuk et al., 1946), probably because the findings of earlier Italian investigations on this disease were published in journals not readily available in the USA (Cugini and Traverso, 1902; D'Ippolito and Traverso, 1902; Gabboto, 1918). In 1952 the constant association of the fungus with corn showing characteristic symptoms was recorded (Ullstrup, 1952). This evidence was circumstantial at that time since rigid proof of pathogenicity of the fungus was not determined by inoculation, reproduction of symptoms and reisolation. These conditions have now been satisfied as far as can be where an obligate parasite is involved (Semeniuk and Mankin, 1964; Ullstrup, 1970b).

Control of the disease lies in correction of soil drainage since the requisite for disease development is waterlogged soil for a period soon after planting until plants are 15 to 20 cm (6 to 8 in.) tall.

3. Sorghum Downy Mildew

Sorghum downy mildew was recognized first on sorghum in India in 1907, but the causal organism was believed to be *Sclerospora graminicola* because of similarity of symptoms produced by the latter on other grasses, and because the oospores found in diseased sorghum resembled those of *S. graminicola* in millet. Later the pathogen was considered a variety of the latter and assigned the name *S. graminicola* var. *andropogonis-sorghi* Kulkarni. Further research (Weston and Uppal, 1932) on the pathogen resulted in renaming it *Sclerospora sorghi* (Kulk.) Weston and Uppal.

The disease is now recognized on corn in Asia and Africa, and in 1963 it was first observed in Texas (Reyes et al., 1964). Since that time it has spread over a number of southern states (Futrell and Frederiksen, 1970) and in 1973 it was found in southern Indiana (Warren et al., 1974). The means by which the fungus was introduced into the USA is not known. There is no report of its occurrence in the Western Hemisphere prior to 1961 when it was found on sorghum (Reyes et al., 1964). In its northward progress into the Corn Belt it has not caused serious damage on corn. Sorghum and sudangrass have sustained more injury and it is on these crops where it is generally first seen.

The symptoms, like crazy top, are varied within the disease syndrome. Initial symptoms of systemically-infected seedlings appear as stunting and chlorotic striping. The first diseased leaf often shows a sharp demarcation between diseased and healthy tissue imparting a "half-diseased leaf" appearance. Leaves on diseased plants are narrower and more erect than those of healthy plants (Fig. 11B). Phyllody of tassels is a frequent symptom. Stalks are often brittle, may show a marbled grayish-brown discoloration of the pith, and proliferation of brace roots at several nodes above those of healthy plants (Warren et al., 1974). Leaf shredding is common in sorghum, but only in some genotypes of corn does this occur (Frederiksen and Bockholt, 1969). Sporulation of the pathogen is abundant on both surfaces of infected sorghum leaves, but less so on corn. It appears as a downy covering and is composed of conidiophores and conidia (Fig. 11C). Local infections, resulting from infection by conidia, appear as elongated chlorotic stripes mostly on lower leaves. Most infected plants are barren, but occasionally some will set some seed on nubbin ears. Such ears frequently develop an aborted tassel at their tips, and ear shanks may show a marked increase in length and number of nodes. Seed produced on systemally infected plants may be internally infected, but the possibility of transmission through commercial seed is remote since the fungus does not survive long in diseased seed (Jones, et al., 1972).

Conidia of *S. sorghi* are hyaline, suborbicular, 15 to 29 × 15 to 27 μ in size, and germinate generally by germ tube. Conidia are borne on elaborately branched conidiophores that may measure 80 to 150 μ from the septum of the basal cell to the first branch. The branch system is comprised of primary, secondary and tertiary branches that terminate in sterigmata about 13 μ long, on which individual conidia are borne. Oospores are spherical, 35 to 37 μ in size, with a yellowish wall 1 to 3 μ thick (Weston and Uppal, 1932). They are produced in abundance in *Sorghum* spp., but rarely in corn tissue.

Conidial formation and germination is favored by moderate temperatures (ca. 21 C) and saturated atmosphere. Conidia are carried by wind currents to corn leaves where they germinate, penetrate by germ tube and establish local infection. Oospores which survive in the soil, possibly for a number of years, may germinate either by germ tube or indirectly by zoospores. These penetrate roots of young plants and initiate systemic infection. The mode of germination by oospores and the subsequent infection process is not fully understood.

Sclerospora sorghi, in addition to infecting corn, also causes disease in *Sorghum* spp. and teosinte. It is not known if specialized, *hostic* races exist within the species.

Practical control of the disease in corn may be done through transferring genes for resistance to agronomically adapted genotypes. An appreciable number of inbred lines of corn have been tested for resistance under conditions of natural infection in Texas (Frederiksen et al., 1971; 1973). Most have been found susceptible, but a few have shown sufficient resistance to be used as parental material in breeding operations (Fig. 11D).

Since the pathogen over-seasons as oospores in old sorghum debris and soil, crop rotation and deep plowing to bury plant residues are promising cultural practices as a means of control. Chemical control also holds some promise (Matocha et al., 1974).

Studies on inheritance of resistance in corn to sorghum downy mildew, although not exhaustive, indicate resistance to be dominant to intermediate in effect and that only a few genes may be involved in its determination (Bockholt and Frederiksen, 1972).

VIII. VIRUS AND MYCOPLASMA DISEASES

Several years ago it was pointed out that only four virus diseases of corn were present in the USA; at that time corn stunt was included among these (Ullstrup, 1955b). These diseases were unimportant and none was present in the Corn Belt. A few years later the potential threat of virus diseases of the corn crop of this country was emphasized, and rightly so, because of the unusual variability of these disease incitants —Pound, 1960).

At present eight virus diseases of corn occur under field conditions in this country. One of these, corn mosaic, occurs in Hawaii and other tropical areas of the world, but thus far not within the contiguous 48 states.

Corn stunt found first in Texas and originally classified as a virus (Kunkel, 1946, 1948), is now believed to be incited by a mycoplasma-like organism (Granados et al., 1968; Maramorosch, et al., 1968). This is based on the presence of mycoplasma-like bodies in infected host tissue as well as in that of one vector; and on remission of symptoms following treatment of the host with tetracycline (Grenados, 1969).

Several other virus diseases of corn occur outside the United States and even greater numbers of viruses, normally inciting disease in other hosts, have been experimentally transmitted to corn (Ohio Agr. Res. & Dev. Center, 1971; Thornberry, 1966).

1. Sugarcane Mosaic

In the USA this disease has been found on corn grown in close proximity to fields of susceptible sugarcane. Studies made years ago to measure the effects of the disease on corn indicated only a slight decrease in yield and quality (Stoneberg, 1927). With the introduction of resistant sugarcane varieties the disease has become of little economic importance on corn.

On corn, symptoms vary widely with genotype. Generally, they appear as ring spots of ovoid shape and assuming a mosaic pattern of light green to yellowish streaks of varying width. Variations in symptoms are also associated with strains of the virus. There is no necrosis and stunting of growth is not pronounced (Brandes, 1920; Brandes and Klaphaak, 1923).

This disease, as with many other virus diseases, has been known by other names (Smith, 1972); *Saccharum virus 1*, Smith; *Marmor sacchari,* Holmes; *Saccharum virus 1,* Brandes; *Sugarcane Yellow Stripe Disease* (Virus), Wakker and Went; *Sugarcane Mottling Disease,* Stevenson; *Grass Mosaic Virus,* Brandes and Klaphaak; *Sugarcane Virus 1*, J. Johnson.

The infectious particle is rod-shaped 620×15 mμ; 760×10 mμ in situ (Gold and Martin, 1955; Herold and Weibel, 1963). The virus is sap transmissible and at least five aphid species have been demonstrated to be vectors. The virus is not seed transmitted.

At least seven strains have been reported within this virus — A, B, C, D, E, F, and H. Strains C, D, and F are unstable, Differentiation of strains is based primarily on symptoms developed on varieties of sugarcane (Abbott and Tippett, 1966). Some strains of the virus appear to be related to strain B of maize dwarf mosaic and to isolates of the latter from corn in California.

The host range of the virus is wide; among these are nine cultivated hosts including corn, sorghum, sudangrass, and pearl millet. A number of common wild grasses are susceptible.

Because the disease is not of appreciable economic importance in corn no efforts have been made to control it other than to avoid planting corn close to susceptible sugarcane. Little is known about the inheritance of resistance, but disease reaction appears to be determined by several genes. Marked differences in resistance have been observed among genotypes of corn.

2. Maize Leaf Fleck

This disease has been observed only once, and only in the northeastern area of the San Francisco Bay region (Stoner, 1952).

Small, pale, roughly circular spots in the interveinal areas of oldest leaves are the first symptoms. These develop 6 to 8 weeks after artificial inoculation of 14-day-old seedlings. These spots enlarge to about 2 mm in diam. Marginal firing and yellowing follows flecking, and proceeds from *Rhopalosiphum prunifoliae* for their lifetime. Investigations on serology and strains of the virus have not been undertaken.

The morphology of the infectious particle is unknown. The virus has not been mechanically transmitted and is not seed- or soil-borne. Three aphid species are known vectors and the virus is persistent in *Myzus persicae* and *Rhopalosiphum prunifoliae* for their lifetime. Investigations on serology and strains of the virus have not been undertaken.

3. Bromegrass Mosaic

This disease has been recorded on corn in South Dakota; this is the only occurrence under natural conditions (Stoner et al., 1967).

Descriptions of symptoms on field-grown corn are meager because the disease is of such rare occurrence. Chlorotic stripes 1 to 1.5 cm in width running parallel to leaf veins are the characteristic early symptoms. Artificially inoculated seedlings show lenticular chlorotic lesions 1 to 1.5×3 to 4 mm in size 3 to 4 days after inoculation. After 5 days systemic symptoms appear which are characterized by mottling, streaking, and wilting. Seedlings of some genotypes succumb to inoculation — Ky201, B37, Oh28, H59 and P39-5 — others are highly resistant — Pr1, 33-16, N6, and W22. The disease is also known by the binomial, *Marmor graminis* McKinney.

The infectious particles are of 2 types; one a spheroid of about 23 mμ in size, the other is of the same general shape but about 31 mμ in diam. (Ford et al., 1970). The serological relationships of the virus have not been examined other than to show identity of the South Dakota isolate to the type. No strains of the virus have been noted.

The virus is readily sap transmissible to corn by the usual mechanical means. The vector operative under natural conditions is not known, but nematodes have been suspected to perform this function.

The host range of bromegrass mosaic virus is wide, including many cultivated and wild grasses. Wheat, barley, oats, sorghum, and rye are among the economic hosts. In addition, many forage grasses are susceptible. Several dicots are among suscepts, some of which produce local lesions in response to infection (Ford et al., 1970).

Because the disease is of no importance on corn, control practices have not been explored. Nothing is known concerning the mode of inheritance of genes governing resistance in corn.

4. Cucumber Mosaic

This disease is found on corn only where it is planted near celery or other hosts of the virus. Cucumber mosaic has been reported on corn only in Florida (Wellman, 1934).

Primary symptoms consist of chlorotic spots around punctures made by the vectors. The chlorosis spreads toward the base of the leaf. Systemic symptoms appear as long elliptical streaks arranged parallel to the veins. Some mottling may accompany systemic development. As plants mature marked stunting occurs along with buff-colored blotches, necrosis and shredding of leaves. The disease is also known as "Celery Virus 1" Wellman.

The infectious particle is ovoid to spherical and about 35 mμ in diameter (Sill et al., 1952). Numerous strains of the virus are known, but celery virus 1 is the strain identified with the naturally occurring disease on corn (Wellman, 1934; Price, 1935).

The virus is readily sap transmissible, and under field conditions a number of aphid species act as vectors. Seed transmission has not been established.

5. Corn Mosaic

In the United States corn mosaic is found only in Hawaii. What appears to be the same disease occurs also in the Caribbean area, South America, and Africa.

The first symptoms of the disease are expressed as elongated white flecks on one side of the midrib near the base of the leaf. These flecks enlarge, coalesce and become conspicuous stripes (Fig. 12A). Withering, red to purple coloration of leaves, and necrosis of chlorotic tissue develop with age (Kunkel, 1927). Stunting, interveinal necrosis, reduced ear size, and barrenness accompany severe infection (Brewbaker and Aquilizan, 1965). Shortening of husks resulting in exposure of the ear is a characteristic symptom. Such exposed ears are vulnerable to bird damage and rotting by fungi. Necrosis of phloem tissue has been reported (Cook, 1936).

The disease has been known by a number of synonyms: *Zea virus 1* Smith, *Marmer zeae* Holmes, *Corn mosaic virus* Kunkel, *Corn leaf stripe virus* Stahl, and *Corn virus 1* J. Johnson.

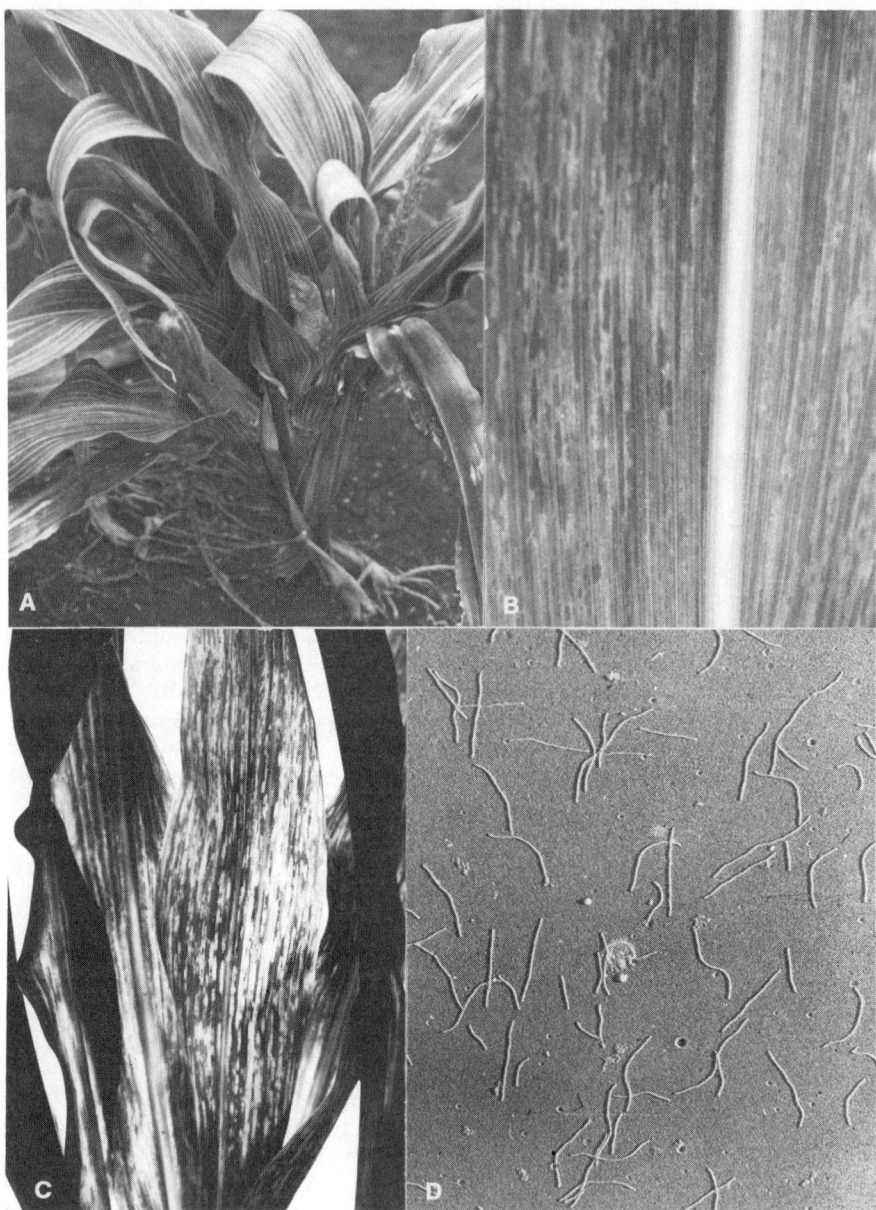

Fig. 12. Virus diseases: A, maize mosaic (courtesy, J.L. Brewbaker, University of Hawaii); B, Wheat streak mosaic; C, maize dwarf mosaic; D, long flexuous rods characterize the infectious particle of maize dwarf mosaic virus (electron micrograph by C.E. Bracker, Purdue University).

The infectious particle is bullet-shaped, about 241 to 255 × 72 to 90 mµ, and apparently possessing an envelope with thread or knob-like protrusions (Herold and Munz, 1967). The existence of strains within the virus has not been investigated, and no serological studies have been performed. The virus is transmitted by the leaf-hopper *Peregrinus maidis* Ashm. in which it is persistent (Kunkel, 1927; Stahl, 1927). The virus is not sap transmissible, but seed transmission has been reported (Cook 1936).

The disease is of economic importance in Hawaii. Sweet corn is generally susceptible although resistance has been found (Brewbaker and Aquilizan, 1965). Resistance to the disease has been shown to be governed by a single gene, but dominance is not evident (Brewbaker and Aquilizan, 1965).

Some control of the disease may be obtained by use of insecticides to reduce the leafhopper population, but this is not practical on a large scale where corn is grown for animal feed.

6. Wheat Streak Mosaic

Wheat streak is found on corn in the United States and Canada (Finley, 1957; How, 1963; Paliwal, Slykhuis and Wall, 1966; Tunac and Nagel, 1969). The disease has not been of major importance on hybrid corn, but in seed-production fields susceptible inbred lines have been severely damaged (Finley, 1957; How, 1963). The disease appears to be widespread, but with little, if any, visible evidence of debilitating effects. The virus can be transmitted from symptomless corn plants.

Small, chlorotic, oval to elliptical spots at tips of older leaves accompanied often by general chlorosis of older leaves are characteristic symptoms (Williams et al., 1967). Long, chlorotic streaks showing concentric or spindle-shaped patterns may often be observed on younger leaves (Fig. 12B). Stunting is marked in inbred lines, but not often apparent in hybrids. On husks a reticulate pattern of dark green on a light green background is sometimes observed. In some genotypes this reticulation becomes red to purple as husks mature and become straw-colored. Necrosis is not observed. Symptom expression in inbred lines is strikingly influenced by genotype, and among these inbred lines are great differences in resistance. Red streaking of kernels was at one time believed to be a part of the syndrome on corn (Tunac and Nagel, 1969; Williams et al., 1967), but this has since been shown to be caused by a salivary phytotoxin secreted by the vector of wheat streak mosaic and independent of the virus (Nault et al., 1967). Synonyms of the disease are *Yellow streak mosaic virus* McKinney, and *Green streak mosaic virus* McKinney.

The infectious particle is a flexuous rod about 670 to 750 × 15 mµ in size (Williams et al., 1967). The virus is sap transmissible. The natural, and only, vector is a mite, *Aceria tulipae*, (Slykhius, 1955; Williams et al., 1967). The virus is not seed borne, but it can be isolated from corn kernels before full maturity. There is evidence of physiologic strains within the vector with respect to ability to transmit the virus (Del Rosario and Sill, 1965).

A number of cultivated and wild host are susceptible to the virus; among these are wheat, barley, oats, rye, teosinte, sand lovegrass, and several species of *Setaria* and *Panicum* (Williams et al., 1967).

Nothing is known concerning mode of inheritance of resistance to the disease, but among inbred lines marked differences in this attribute are evident.

7. Maize Dwarf Mosaic

Maize dwarf mosaic is widely distributed in the United States from the Eastern Seaboard (Roane and Troutman, 1965) to the West Coast (Shepherd et al., 1964) and from the northern states (Smith, 1966) to the Gulf States (Bond and Pirone, 1966). The disease has been of appreciable economic importance in the USA; this is especially true where corn is grown close to infestations of johnsongrass. From about 1963 to 1967, and in some subsequent years, corn in many areas in the south-central Corn Belt and eastward to the Atlantic Coast sustained serious economic losses.

Typical mottling of young leaves characterized by light and dark green areas is the primary symptom (Fig. 12C). The pattern and intensity of the mosaic symptom varies with the genotype of the host. Some inbred lines and hybrids show a fine stippling, in others the mottling is intense with sharp contrasts between shades of green, and in size of the individual spots or "mottles." As plants mature mottling tends to fade and general chlorosis and reddening of leaves becomes evident. Stunting and proliferation of adventitious buds are apparent. These symptoms closely resemble those of corn stunt, and maize chlorotic dwarf. Badly infected plants are predisposed to root rots and stalk rots caused by fungi, but this is not peculiar to maize dwarf mosaic. Poor fill and barrenness accompany severe infection (Dale, 1964; Janson et al., 1965; Shepherd et al., 1964; Williams and Alexander, 1964). Severity of disease in individual plants is positively correlated with early infection. Plants infected relatively late in their development may show little damage.

The infectious virus particle is a long, flexuous rod approximately 750 to 800 × 15 mμ in size (Fig. 12D). The serological relationships of the virus are not entirely clear. Some studies indicate Strain B (non-johnsongrass strain) to be related to Strain H of sugarcane mosaic virus. Other results indicate all strains of maize dwarf mosaic virus could be placed in one group on a serological basis and distinct from strains that do not infect johnsongrass. In these same studies, sugarcane virus Strains A, B, C, and E, and an Australian collection fell into a third group. Strain H of sugarcane mosaic virus was serologically distinct from all others (Snazelle et al., 1971). The many similarities between maize dwarf mosaic virus and sugarcane mosaic virus suggest the former is a strain of the latter.

Strain A was identified first in Ohio and is widespread over the country (Bancroft et al., 1966; Williams and Gordon, 1967). Strain B was reported first in Pennsylvania and is differentiated from Strain A on the basis of serology and its failure to infect johnsongrass (Gordon and Williams, 1970; Mackenzie et al., 1966). Recently four additional strains were distinguished on the basis of host reaction, symptoms, and certain physical properties

(Louie and Knoke, 1970). These four strains appear to be closely related to Strain A.

Maize dwarf mosaic virus is readily sap transmissible (Dale, 1964; Williams and Alexander, 1964), and at least twelve aphid species are capable of acting as vectors (Bancroft et al., 1966; Daniels and Toler, 1969; Messieha, 1967). The virus is non-persistent and stylet-borne in its vectors. Evidence has shown that inbred lines of corn may vary in susceptibility to maize dwarf mosaic depending on method of inoculation. The inbred lines I11A and W70 were more susceptible to mechanical inoculation than when inoculated with viruliferous aphids (*Schizaphis graminum*). The inbred line Oh7B was more susceptible to aphid inoculation than by mechanical means, while Oh07 was equally susceptible by both methods. Experiments on seed transmission have been contradictory (Bancroft et al., 1966; Sehgal, 1966; Shepherd and Holdeman, 1965; Williams et al., 1968). If seed transmission does occur, it probably does so rarely.

Resistance to maize dwarf mosaic tends to be partially dominant in crosses between resistant and susceptible inbred lines. Relatively few genes appear to govern resistance which simplifies development of resistant inbred lines combining this attribute with acceptable agronomic traits (Dollinger et al., 1970; Johnson, 1971; Loesch and Zuber, 1967).

8. Maize Chlorotic Dwarf

This disease was recognized first in Ohio and in Kentucky (Bradfute, Louie and Knoke, 1972; Pirone et al., 1972) but has since been identified also in Indiana and a number of southern states. Since the disease has only recently been found comparatively little is known about it.

Symptoms appear as chlorotic blotches, marginal chlorosis, splitting of margins of young leaves at right angles to the long axis, and stunting. The stunting results from shortening of all internodes to give an appearance of a miniature plant. Vein clearing is a distinguishing symptom. Symptoms resemble those of corn stunt, and of maize dwarf mosaic in its later stages of development after the mosaic pattern has faded. For this reason the three diseases can be confused, especially where their geographical ranges overlap. Double infection by this virus and maize dwarf mosaic virus is especially debilitating.

The infectious particle is isometric, and about 27 to 30 mμ in diam. In this respect it differs from the long, flexuous rods of maize dwarf mosaic and from the mycoplasma-like bodies associated with corn stunt. The particles appear individually and in aggregates in the cytoplasm and central vacuole of phloem cells (Bradfute et al., 1972).

The virus particles are transmitted by the leafhopper, *Graminella nigrifrons* Forbes in a non- or semi-persistent manner (Nault et al., 1973). It has not been mechanically transmitted. Johnsongrass is one of the hosts commonly infected by the virus (Pirone et al., 1972).

The disease resembles tungro disease of rice in infectious particle characteristics and its non-persistent association with a leaf-hopper vector.

The disease may be of the same described as a strain of corn stunt in Ohio (Choudhury and Rosenkranz, 1973; Rosenkranz, 1969).

9. Corn Stunt

Corn stunt was reported first in the lower Rio Grande Valley in 1945 (Altstatt, 1945). What was believed to be the same disease (Kunkel, 1948) was observed earlier in the San Joaquin Valley of California (Frazier, 1945). It was later pointed out that the disease in California was probably not stunt, but sugarcane mosaic on corn (Frazier et al., 1966). The disease is found in most southern states and the Southwest where it can cause appreciable damage. In Mexico, Central and South America corn stunt is often a limiting factor in corn production. In Latin America the disease is known as "achaparramiento."

In the field a most striking symptom is the proliferation of tillers and elongation of ear shanks at nearly every node (Fig. 13A). This gives the plants a bushy appearance. If infection takes place early all internodes are shortened; if infection is later only upper internodes will be reduced in length. Small, circular, chlorotic spots develop at the bases of the leaves on young plants. These often coalesce to form stripes that are discrete or diffuse (Kunkel, 1948; Altstatt, 1945) (Fig. 13B). Mottling or mosaic manifestations are not characteristic. Reddish-purple strips and general chlorosis appear as plants become older (Fig. 13C). At this stage symptoms are much the same as on mature plants infected with maize dwarf mosaic or maize chlorotic dwarf. Pronounced proliferation of roots is sometimes observed (Kunkel, 1948). Numerous ear shoots and poorly filled ears, or barrenness, are common symptoms in plants infected early (Altstatt, 1945).

Symptoms will vary with time of infection in relation to host development, with genotype of the corn, with the strain of the causal agent, and probably with the environment.

Corn stunt was originally believed to be caused by a virus (Kunkel, 1946; 1948) because it possessed many attributes often associated with plant diseases of viral nature. In 1968 some evidence was presented suggesting that the causative agent of corn stunt might be a mycoplasma-like organism (Granados et al., 1968; Maramorosch et al., 1968; Shikata et al., 1969). The fact that complete blockage or remission of symptoms could be accomplished by treatment with tetracyclines, which are effective against mycoplasma-incited diseases, strengthened this position (Granados, 1969). Electron micrographs of tissues of corn showing symptoms of the disease, and of tissues of infective leafhoppers have revealed helical filaments that are not in healthy corn or non-infective vectors (Chen and Granados, 1970; Davis et al., 1972). *Spiroplasma* has been tentatively assigned as a trivial term for the organism (Davis and Worley, 1973). The helical morphology, contractile movements of filaments of the organism associated with corn stunt distinguish it from other described mycoplasmas. Each filament is bounded by a unit membrane but no cell wall, sheath, or secondary membrane. Filaments are 0.2 to 0.25 $\times$ 3 to 15 μ, with regular wavelength, often with spherical bodies attached — 0.4 to 0.6 μ in diam. The size, and lack of cell wall, axial fibrils, flagellar structure and an envelope, suggest affinities with other mycoplasmas. The helical structures are most numerous in phloem cells of diseased plants. The organism can be maintained in a cell-free broth for several weeks. Recent findings have shown the corn stunt organism to be serologically related to

DISEASES OF CORN

Spiroplasma citri, the mycoplasma responsible for citrus stubborn disease (Tulley et al., 1973).

The corn stunt organism can be transmitted under field conditions by *Dalbulus maidis* DeL. & W. (Kunkel, 1946), *D. elimatus* Ball (Neiderhauser and Cervantes, 1950), *Graminella nigrifrons* Forbes (Granados et al., 1966), *Deltocephalus sonorus* Ball (Granados et al., 1968) and *Baldulus tripsici*

Fig. 13. Corn stunt: A, chlorosis, stunting, multiple ear shoots; B, striations at leaf bases; C, chlorosis, reddish stripes and blotches; D, shortening of internodes, multiple ear shoots.

Kramer & Whitcomb (Grenados and Whitcomb, 1971). The vectors are not equally efficient in their transmission and they have different, but overlapping, geographical ranges. After feeding on diseased corn tissue an incubation period is required by the vectors before it can transmit the organism. This period varies with the species of leafhopper. There is a time lapse of over 3 weeks between inoculation of plants by the insects and appearance of symptoms.

Two strains of the corn stunt organism were recognized in Mexico (Maramorosch, 1955). Later a Louisiana strain (Granados et al., 1966) and a strain from Ohio were identified (Rosenkranz, 1969). The strain found in Ohio and transmitted by *G. nigrifrons* may be what is now believed to be the incitant of maize chlorotic dwarf.

Resistance to corn stunt appears to be governed by relatively few genes with strong additive effects. Dominance is generally lacking and epistasis has no strong influence. Heterosis per se confers tolerance to the disease (Grogan and Rosenkranz, 1968). More recent studies using a larger number of inbred lines and their hybrid combinations, have confirmed these results. In this study high, negative correlations were observed between yield and severity of disease (Nelson and Scott, 1973). Control of corn stunt, as suggested in these investigations can be accomplished by development of resistant inbred lines.

IX. CORN SMUTS

Corn is susceptible to two smut diseases. Common smut, boil smut, or blister smut, as the disease is known in some parts of the world, is widely distributed in many, but not all, areas where corn is grown. Head smut, of lesser importance in the USA, is found mainly in the western states and rarely east of the Great Plains. It appears to be more prevalent in Asia, Africa, and Australia.

1. Common Smut

The incidence of common smut in the USA varies from year to year and from one locality to another. In the Corn Belt estimates of losses from common smut range from a trace to 6%. It is doubtful if losses exceed 2% over wide areas. In sweet corn losses may be much higher. A monographic treatment of the many aspects of this disease has been published in 1963. (Christensen, 1963).

The number, size and location of smut galls determine the amount of yield depression that may be sustained (Immer and Christensen, 1928b; Johnson and Christensen, 1935; Smith, 1936). In determinations of the effect of smut galls formed where tassels were removed from plants of the seed-parent in seed fields, galls under 5 cm (2 in.) in diam. reduced yield about 9%, those 5 to 7.6 cm (2 to 3 in.), 14% and those over 7.6 cm in diam. depressed yield 40% in individual plants. It was further observed that smut infection resulted in smaller kernels and ears (Menzies and Stanberry, 1947). In other tests smut affected yield only when it induced barrenness, and plants in which the tassel was infected with smut actually yielded more than smut-free plants (Garber and Hoover, 1928).

DISEASES OF CORN

Symptoms of common smut are conspicuous and easily recognized. All above-ground parts of the corn plant are susceptible, although the apical meristem can be infected and galls produced beneath the soil surface when plants are young (Ullstrup and Britton, 1968) (Fig. 14A). Galls are formed and continue to develop only in meristematic tissue. Where cells mature rapidly, gall formation is arrested and they remain small, often hard, and few

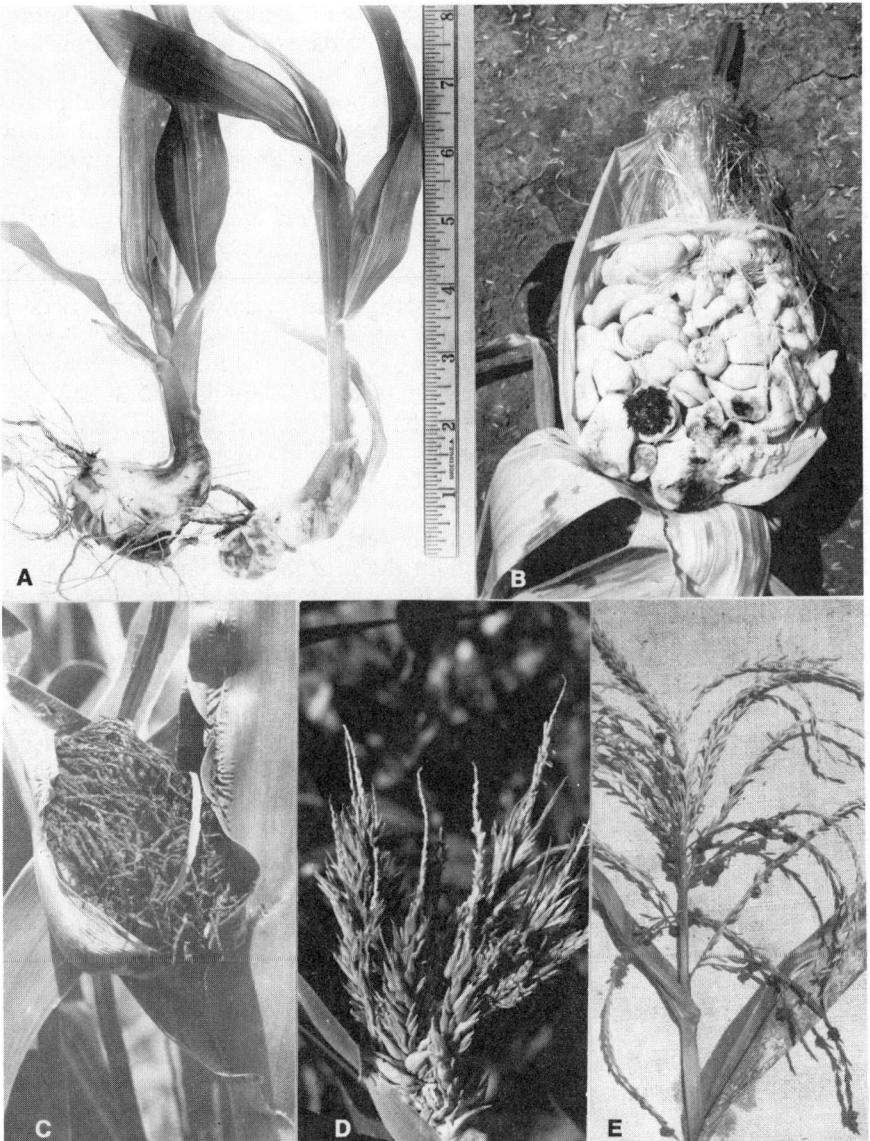

Fig. 14. Corn smuts: A, common smut showing galls involving growing point of young seedlings; B, common smut on ear; C, head smut on ear; D, head smut on tassel showing phyllody (proliferation of floral parts); E, false smut, only the tassel is infected.

teliospores of the fungus are produced within them. Galls are at first glistening and white (Fig. 14B). The interior is composed of soft white tissue that may show streaks of black where formation of teliospores of the pathogen has been initiated. Within a short time, first the interior and then the entire gall becomes a mass of black, powdery teliospores of the fungus. Early infection may cause stunting and even killing of young plants, but this is rarely seen in the field.

Ustilago maydis (DC.) Cda. [Syn. = *U. zeae* (Schw.) Ung.] is the inciting agent of common corn smut. Teliospores (chlamydospores) are brownish-black, thick-walled, heavily echinulate, oval to spherical and 8 to 12 μ in diam. Germination is attended by the formation of a septate promycelium on which small hyaline, thin-walled, ellipsoid sporidia are borne. The optimum temperature for germination is within the range of 26 to 34 C, the minimum about 8 C and the maximum 36 to 38 C (Jones, 1923). Growth of the fungus on artificial media is composed of mycelial strands and sporidia; the latter reproduce by budding. Teliospores are not formed outside the host on artificial media. The fungus is composed of numerous races that may differ in cultural behavior or pathogenicity or both. These arise by mutation or as a result of combination and resegregation of genetic material following hybridization between compatible mating types (Christensen, 1929; Stakman et al., 1933). Sporidia are usually uninucleate and haploid and those of opposite mating type fuse to form binucleate infection hyphae. These fusion hyphae are dikaryotic and are capable of establishing infection (Bowman, 1946; Sleumer, 1932). The infection hyphae survive in the host, but on most artificial substrates they die and the colony remains composed of budding sporidia of the two compatible types. On certain types of agar substrate the fungus can be induced to form diploid thalli (Puhalla, 1968). Occasionally the diploid nucleus in the teliospore is not reduced on germination and a diploid thallus develops. Single sporidia from such cultures are pathogenic (solopathogenic). Haploid sporidia are non-pathogenic (Chilton, 1940; Christensen, 1931).

Penetration of meristematic tissue may take place by the binucleate infection hyphae resulting from sporidial fusion, or by infection hyphae arising directly from germinating teliospores (Walter, 1934). Invasion of the host may be by direct penetration of the thin-walled meristematic cells or through wounds that occur through various agencies — hail (Coffman et al., 1926), by insects (Pan, 1959; Pepper and Haenseler, 1944; Yakovleva, 1963), by detasselling operations in seed fields (Menzies and Stanberry, 1947), or by cultivating machinery. The parasitic mycelium stimulates host cells to increase in size and numbers. The gall is thus composed of hypertrophied and hyperplasic host and fungus tissue. Although some evidence has been presented to suggest systemic infection by *U. maydis* (Davis, 1936), most investigators believe that galls result from local infections (Itzerott, 1938; Piemeisel, 1917; Scurti, 1949). Because most galls disperse teliospores when the host is mature or near maturity, and possess little meristematic tissue, there is little, if any, secondary infection unless young plants are available. Most smut galls are a result of infection by primary inoculum that survived from the previous growing season. The host range of *U. maydis* is restricted to corn and teosinte.

The epidemiology of common corn smut is not fully understood. Some observations suggest that high humidity favors smut infection (Piemeisel, 1917), but most reports indicate dry, warm weather (Coffman et al., 1926; Immer and Christensen, 1928a; Platz, 1929; Potter and Melchers, 1925). During a period of cool, wet weather when growth of corn was much retarded, the incidence of smut on young plants was unusually high (Ullstrup and Britton, 1968). The slower growth rate, following early planting, may account for more smut in plantings made at that period (Piemeisel, 1917).

The addition of phosphate (84 kg/ha) with calcium carbonate and with calcium silicate reduced the incidence of smut respectively from 12.1% to 7.4% and from 13.1% to 5.8%. Further increases of phosphate did not further reduce the prevalence (Tamimi and Hunter, 1970). Potassium applied at the rate of 27.5 kg/ha as KC1 increased the incidence of smut in Wisconsin. The increase in susceptibility was accompanied by a decrease in reducing sugars and an increase in nitrogen in the pith (Martens and Arny, 1966). It has been frequently observed that heavy application of barnyard manure increases the incidence of common smut. It is not known if this is a result of a change in the physiology of the host to render it more susceptible or that the enrichment of the soil favors the pathogen.

The most effective means of controlling smut is through selection and breeding for resistance. It was pointed out in 1942 that most of the inbred lines used in the Corn Belt, at that time, had a reasonable degree of resistance to smut (Stringfield and Bowman, 1942). This applies at present. Those that are somewhat less resistant have other compensating attributes. In crosses between resistant and susceptible inbred lines, resistance tends to be intermediate (Hoover, 1932; Immer, 1927). In studies on the genetics of resistance it has been shown that this attribute is determined by a relatively large number of genes (Burnham and Cartledge, 1939; Hoover, 1932; Saboe and Hayes, 1941).

Little is known of relationships of cultural practices such as "minimum tillage" vs. clean plowing to the incidence of common smut. Destruction of smut galls may reduce inoculum in a small garden plot, but where corn is extensively grown this is obviously impractical. The avoidance of heavy applications of barnyard manure, of planting corn on old feed lots, or of excessively high additions of nitrogen to the soil, may be helpful in reducing the incidence of smut.

The few reports on the relationship of insect control to corresponding reduction in the incidence of smut would seem worthy of further investigations (Pan, 1959; Pepper and Haenseler, 1944; Yakovleva, 1963).

2. Head Smut

Head smut of corn was first observed in the United States in 1890 in eastern Kansas (Norton, 1895). The disease is not important on corn in this country, although there are a few reports of its occurrence east of the Great Plains (Potter, 1914), it is confined mainly to California, Washington, Oregon and Idaho. In the latter state it has been increasing in recent years in the sweet corn, seed-producing areas (Simpson, 1966). The disease is widespread and in some countries it is of appreciable economic importance. Head smut of sorghum, incited by a race of the same pathogen is much more important.

The conspicuous and identifying symptoms of head smut of corn become evident when tassels and ears appear. These organs may be partly or completely converted to a mass of black, powdery spores (Fig. 14C). Each sorus is covered with a membrane that soon ruptures to expose the spore mass. Tassels or ears, or both, may become infected in a single plant. Occasionally sori will form on leaves. Infected plants usually show some stunting and in some instances it may be severe. Proliferation (phyllody) of floral parts in the tassel and ear often occurs (Fig. 14D).

Sphacelotheca reiliana (Kühn) Clint. [Syn. = *Sorosporium reiliana* (Kuhn) McAlpine; *Ustilago reiliana* Kuhn] is the causal agent of head smut of corn. Teliospores are reddish-brown to black, echinulate, and measure 9 to 12 μ in diam. These germinate over a range of temperatures, the optimum being 21 to 28 C. A septate promycelium is formed bearing small, hyaline, thin-walled, single-celled, haploid sporidia. Penetration of the seedling takes place by direct penetration by the promycelium or following fusion of sporidia of compatible mating type to form binucleate infection hyphae. The parasitic mycelium develops systematically in the host, and sori composed of spore masses are formed in the floral parts when they appear. The morphology and life cycle is similar to that of *U. maydis*. Seedling infection, systemic development of the mycelium, and the formation of sori almost exclusively in tassels and ears distinguishes *S. reiliana* from *U. maydis*.

Two races of the pathogen have been recognized: one that attacks only corn and the second which is confined to several sorghum species (Halisky, 1963; Reed et al., 1927). Some studies however have demonstrated that the sorghum race may also infect some genotypes of corn (Al-Sahaily and Mankin, 1960). It has been suggested that head smut on corn and sorghum should be distinguished as varieties rather than as races (Al-Sahaily et al., 1963b). These workers found that isolates from different sorghum species collected in different areas could be divided into 4 races on 55 varieties of sorghum and into 5 races when sweet corn (cv North Star) was included in the host differentials. It has also been shown that hybrids between isolates from corn and from sorghum were pathogenic on both hosts (Al-Sahaily et al., 1963a).

Little is known of inheritance of resistance to the disease, but some striking differences are observed among inbred lines of sweet corn. (Simpson and Fenwick, 1968, 1971). Among 21 races of Mexican corn, statistically significant differences in the amount of smut were observed (Fuentes, 1963).

Some control has been effected by in-furrow soil treatments with different fungicides (Fenwick and Simpson, 1967; Simpson and Fenwick, 1968; Simpson and Fenwick, 1971). This appears to be economically practical where sweet corn seed is produced. These studies have shown that inoculum is soil-borne in the form of teliospores from the previous year, and not internally seed-borne. Seed treatment, however, can protect seed from infection (Simpson and Fenwick, 1971).

3. False Smut

This disease is not a smut; it is only so-called because of a superficial resemblance to common smut and because it develops on the tassels of corn. In the United States the disease has been reported only in Louisiana (Haskell

DISEASES OF CORN

and Diehl, 1929). It has been observed in some tropical areas, and a similar disease caused by the same pathogen is found on rice.

The disease is confined to the tassels were hard, irregular to subspherical galls, 4 to 15 mm in diam. are formed. These are the sclerotia of the pathogen (Fig. 14E). The surface is irregular, olive-green to black and velvety. The interior of the sclerotium is whitish. The exterior is made up of black spores that are spherical to ovate, which, when mature, have a warty surface and are 4 to 7 μ in diameter.

Ustilagenoidea virens (Cke.) Tak. is the pathogen causing false smut.

X. CORN RUSTS

Corn is susceptible to three rusts, each caused by a distinct pathogen. Common rust and southern corn leaf rust occur in the United States. Tropical corn rust has not been reported on corn in this country, but the pathogen has been recorded on teosinte in Florida (Cummins, 1971). None of the corn rusts has been of major economic importance in this country; occasionally common corn rust has been prevalent over wide areas, and in a few instances locally severe (Wallin, 1951). Southern Corn rust has become severe in rare instances, but not over broad regions. It has been of major economic importance in West Africa (Rhind et al., 1952). There are few quantitative data reporting the effect on yield of common corn rust (Hooker, 1962) and none of southern corn rust in the USA.

1. Common Corn Rust

This disease is widespread over the USA wherever corn is grown. The prevalence and severity vary widely from year to year. Generally, the infections appear relatively late in the growing season so that no important damage results.

Common rust is recognized by the circular to elongate, cinnamon-brown, powdery pustules (uredinia) scattered over both surfaces of the leaves (Fig. 15A). As plants mature the pustules become brownish-black owing to replacement of the urediospores by black teliospores. The pustules may develop on any above-ground part of the plant, but are most abundant on leaves.

Puccinia sorghi Schw. is the inciting agent of common corn rust. It is an obligate parasite. Urediospores are cinnamon-brown, globoid to ellipsoid, finely to moderately echinulate, 26 to 32 × 23 to 29 μ in size, with three to four equatorial pores (Fig. 15B). Urediospores are dispersed by wind and, under favorable conditions, will continue to initiate infections on corn throughout the growing season. Each spore contains two haploid nuclei as does each cell of the mycelium which develops in the leaf tissue following infection. As the host matures teliospores develop in the pustules. These spores are brownish-black, oblong to ellipsoid, rounded or obtuse at both ends, and are 29 to 54 × 16 to 23 μ in size (Fig. 15C). Teliospores are two-celled, slightly constricted at the septum, and the spore wall thickened at the apex. Each spore is attached to a pedicel once or twice the length of the spore. The cells of teliospores are binucleate, but before germination the two haploid nuclei fuse to form the truly diploid phase of the fungus. Normally, teliospores germinate in the spring and early summer when moisture and temperature are favorable to form a basidium on which small, hyaline, thin-walled basidiospores develop. With the onset of germination miotic division of the diploid nuclei takes place and four haploid nuclei are formed, each of which moves into one of the basidiospores. Basidiospores are incapable of parasitizing corn, but can infect a number of species of wood sorel (*Oxalis* spp.), the alternate host (Mains, 1934; Smith, 1926). Following penetration of the leaves of *Oxalis* spermagonia (pycnia) are formed on the upper surface. These are inconspicuous, few in number and resemble pimple-like eruptions. Minute hyaline spores, spermatia, develop within the spermagonia and are exuded in a syrupy matrix. The spermatia, which are uninucleate and haploid, fuse with receptive structures (paraphyses) that protrude from the ostioles of the spermagonia. The single nucleus of the spermatium passes into a paraphysis and pairs, but does not fuse, with the latter. Fusion takes place only between spermatia and paraphyses originating from spermagonia of opposite mating type. By successive division and migration of nuclei, diploidization proceeds through the mycelium associated with a "fertilized" spermagonium. Without fusion between spermatia and paraphyses the mycelium within the *Oxalis* leaf remains haploid and further development of the fungus is stopped. Following diploidization the cluster cup, or aecial, stage of the parasite develops on the lower surface of the leaf. Aecia bear numerous binucleate aeciospores which are globoid to ellipsoid, finely verrucose, pale-yellow and measure 18 to 26 $\times$ 13 to 19 μ. Aeciospores are carried by air currents to corn leaves where, with favorable moisture and temperature, they establish infection. Following infection urediospores are again produced. Urediospores may overwinter in southern USA and serve as initial inoculum on corn in the spring and summer thus bypassing the necessity of the alternate host, *Oxalis,* in the life cycle of the fungus. Where this does occur urediospores may be carried progressively northward as the growing season advances and furnish the bulk of primary inoculum of common corn rust in the Corn Belt. Although *Oxalis* spp. have been found in the Corn Belt bearing aecial infection (Smith, 1926) the infrequent occurrence of this species, and the even rarer event of their infection by *P. sorghi,* suggests that the alternate host plays a minor role in the epidemiology of this disease. The heterothallic nature of common corn rust was discovered in 1931 (Cummins, 1931) and later verified by cytological studies (Allen, 1933, 1934). In the latter investigations alternative methods of formation of the dikaryon mycelium (diploidization) were pointed out. One is the possible anastomosis between haploid hyphae of opposite mating type within the *Oxalis* leaf, or by fusion of germ tubes of germinating spermatia with haploid hyphae often present in the stomatal openings of the leaf.

Relatively moderate temperatures and ample moisture favor spread and development of common corn rust. Studies have shown the minimum temperature for germination of urediospores is near 4 C, the optimum over a broad range of 17 to 25 C, and the maximum under 31 C (McKeen, 1951b; Smith, 1926; Weber, 1922). Germination at 25 C was found to be 100% in a saturated atmosphere but only 3% at 97.5% relative humidity (Smith, 1926).

DISEASES OF CORN

Wide differences in resistance to *P. sorghi* are observed in both the seedling and adult-plant stages. Specific resistance is expressed in the seedling stage as minute chlorotic flecks following infection by the pathogen. This type of resistance is specific for one or more physiologic races of the pathogen. Some genotypes of corn are resistant to a large number of races, others to only a few. Inbred line selections from the variety Cuzco appear to carry resistance to all races of *P. sorghi* presently within the USA. In some

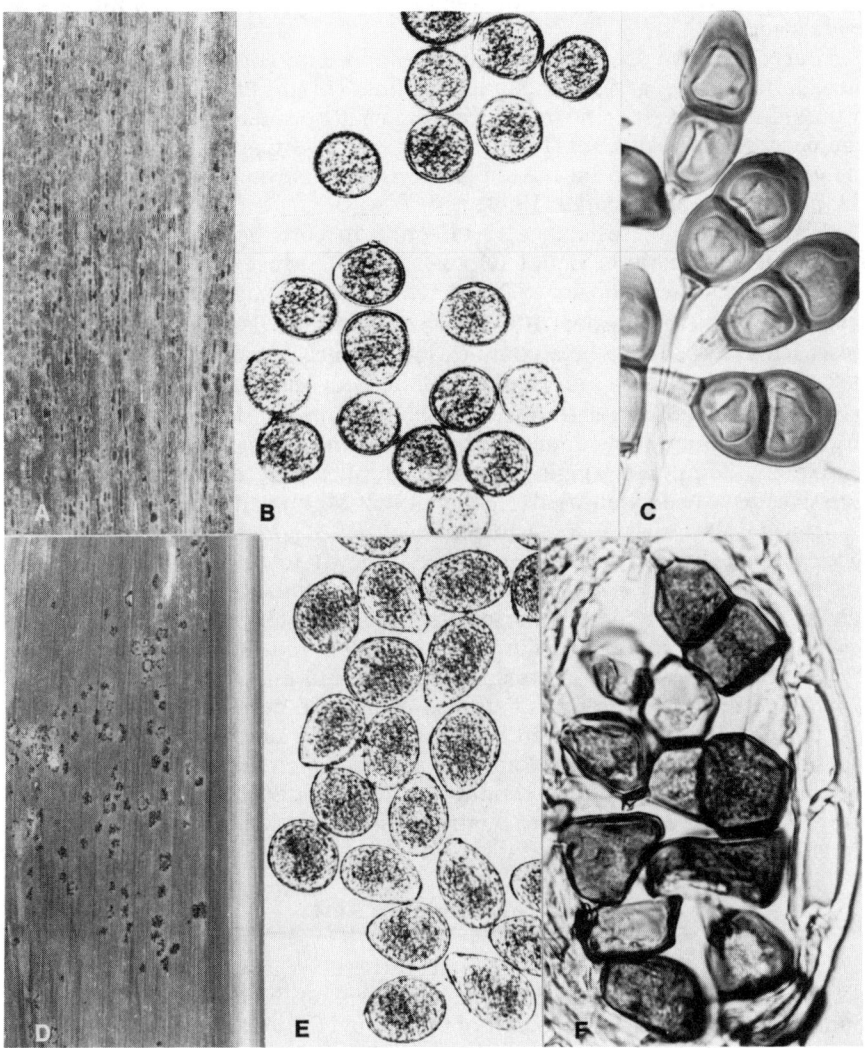

Fig. 15: Corn rusts: A, uredinia of common rust; note linear shape of pustules; B, urediospores of *Puccinia sorghi*; C, teliospores of *P. sorghi* (courtesy, G.B. Cummins, Purdue University); D, uredinia of southern corn rust, note that pustules are smaller and more circular than those of common corn rust; E, urediospores of *P. polysora*, these are larger and more ovoid than those of *P. sorghi*; F, teliospores of *P. polysora* (courtesy, G.B. Cummins, Purdue University), note epidermis retained over pustule.

types of resistance chlorotic halos surround small pustules. A second type of resistance found in corn functions in the mature plant. This is generalized resistance; it is conditioned by a large number of genes; it is expressed in numbers of lesions and effective against most, if not all, races. Most corn grown in the United States has some mature plant resistance, but some have very little. This can be observed in years when rust is relatively abundant. The rust reaction of mature field grown plants is relatively consistent from year to year since natural inoculum in the field is generally made up of a mixture of several races.

Early work on seedling resistance to common corn rust demonstrated the wide range in reaction among inbred lines (Mains et al., 1924). This type of resistance, to a specific race, was shown to be determined by a single dominant gene (Mains, 1931). In subsequent investigations the location of this gene was found to be on the short arm of chromosome 10 (Rhoades, 1935; Rhoades and Rhoades, 1939).

The genetics of resistance to infection in corn by *P. sorghi* has been studied in considerable detail (Hooker, 1969). Most of the race-specific resistance has been shown to be determined by single dominant genes. Several loci on chromosome 10 have been identified (Hooker, 1969). Among these, Rp_1a, appears to be a complex locus consisting of at least 14 allels or "pseudoallels" (Saxena and Hooker, 1968). Loci on other chromosomes have been located (Malm and Hooker, 1962; Russell and Hooker, 1962). In addition to the many single dominant genes governing rust resistance, single incompletely dominant, single recessive, duplicate recessive and triplicate recessives have been identified (Hooker, 1962; Malm and Hooker, 1962).

Besides the resistance to infection by *P. sorghi* found in corn, other species of the tribe *Maydeae* have been shown to be resistant. Perennial teosinte (*Euchlaena perennis* Hitch.) and some annual teosinte (*E. mexicana* Schrad.) exhibit a typical hypersensitive reaction (chlorotic-necrotic flecks) when inoculated in the seedling stage (Malm and Beckett, 1962). These sources could be employed in rust resistance breeding.

While up to the present there has been little need to control common rust in dent corn in the United States, it has been suggested, and wisely so, that both a diversity of single dominant genes — such as the "Cuzco gene" — and mature-plant resistance should be incorporated into our present and future hybrids in order to guard against possible epidemics and the advent of new destructive races of the pathogen (Hooker, 1969).

2. Southern Corn Rust

Southern corn was described first in the USA (Cummins, 1941), but it has not caused any appreciable damage to corn in this country. The disease is found sparingly in the eastern half of the United States as far north as Illinois (Hooker, 1961a)., Wisconsin, and Massachusetts. It is more prevalent in the South, and in 1973 some localized epidemics occurred in some southeastern states. Until the late 1940's the disease was confined to the Western Hemisphere, but since that time it has been identified in West Africa, where it caused appreciable damage, in other African countries, Thailand, Malaysia, and the Philippines.

Southern corn rust can be confused with common corn rust, but there are some characteristics that distinguish the two from each other. The uredinia of southern corn rust are generally smaller, more circular in outline, and lighter in color — tending more toward orange-red — than those of common rust (Fig. 15D). Telia are brownish-black, circular to elongate, and distinguished from those of common corn rust by retention of the leaf epidermis over the pustules for a long time. In this respect the telia resemble those of crown rust of oats. Telia often appear in a circle around individual uredinial pustules.

Puccinia polysora Underw. is the inciting agent of southern corn rust. Urediospores are yellowish-golden, mostly ellipsoid or ovoid, echimulate, with four to five equatorial pores and 29 to 36 $\times$ 23 to 29 μ in size (Fig. 15E). Teliospores are angular to ellipsoid or oblong, two-celled, the cell wall over the apex not thickened, as in *P. sorghi*, chestnut-brown, smooth, pedicel yellow or brownish, and up to 30 μ long; one-celled teliospores are often observed (Fig. 15F). The aecial stage of the pathogen is not known. Continued propagation of the pathogen depends on the repeating uredinal stage, on its overseasoning, or on wind dispersal of spores into areas where it cannot overseason.

Southern corn rust thrives under warm, humid environments. In East Africa this rust is found at the lower altitudes where temperatures are generally high, whereas common corn rust is prevalent in the highlands where the temperatures are moderate (Nattrass, 1953). The geographical distribution of southern corn rust follows this general pattern of being more prevalent in warmer areas.

In addition to corn, this pathogen attacks teosinte, *Erianthus alopecuroides* (L.) Ell., *Tripsacum dactyloides* L., *T. lanceolatum* Rupr., *T. laxum* Nash., and *T. pilosum* Scrib & Merr.

The fungus is composed of a number of physiological races differentiated on the basis of parasitism on various corn genotypes. Three races have been identified in Kenya, East Africa (Ryland and Storey, 1955; Storey and Howland, 1957; Storey et al., 1958; Storey and Howland, 1961; Storey and Howland, 1967). Seven races have been identified in the United States (Robert, 1962; Ullstrup, 1965) and two in Nigeria (Lallmahomed and Craig, 1968). Single genes governing resistance have been mostly dominant, but some are incompletely so (Storey and Howland, 1957; Storey and Howland, 1967). Eleven of such genes designated Rpp_1 to Rpp_{11} have been identified. The gene Rpp_9 is closely linked to the complex locus $Rp_1{}^d$, on chromosome 10 governing resistance to *P. sorghi*.

Control of the disease which has been a problem in West and East Africa has been accomplished through the use of these single dominant and incompletely dominant genes (Storey et al., 1958).

3. Tropical Corn Rust

Tropical corn rust is confined to the American Tropics from Mexico to Columbia. The pathogen has been reported from Florida on teosinte (Cummins, 1971). Its hosts, in addition to corn and teosinte, are *Tripsacum dactyloides, T. laxum, T. lanceolatum* and *T. pilosum*.

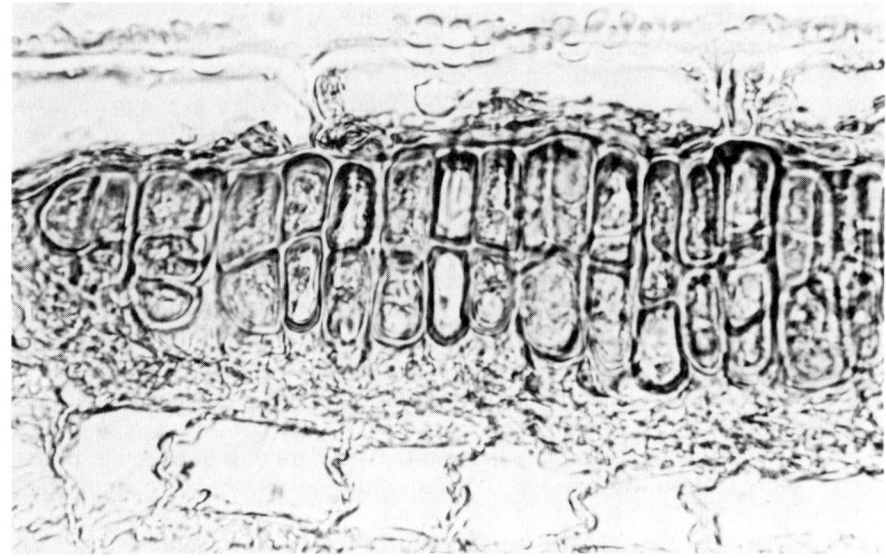

Fig. 16. Corn rusts: teliospores of *Physopella zeae*, the pathogen of tropical rust. Note the columnar arrangement of teliospores beneath epidermis (courtesy, G.B. Cummins, Purdue University).

The pathogen of tropical corn rust is *Physopella zeae* Cumm. and Ramachar [Syn. = *Angiospora zeae* Mains]. Uredinia are amphigenous, yellowish, and generally circular to oblong. Urediospores are elliptical or ovoid, colorless or pale yellow, echinulate, 18 to 24 × 12 to 18 μ in size. Pores in the equatorial region number about 5, but are obscure. Telia are black or brown and covered by the epidermis. Teliospores are single-celled, cuboid or oblong, in chains of two to four spores, golden to light, chestnut-brown and measure 12 to 18 × 10 to 14 μ (Fig. 16). The aecial stage of the pathogen is unknown.

Because the disease has been of minor importance, little is known of the reaction of inbred lines of corn or of the genetics of resistance. Of 139 corn inbred lines and collections tested in Guatamala 7 were found resistant (Schieber and Dickson, 1963).

XI. NEMATODE DISEASES

Nematode diseases of corn are not recognized as being of economic importance in the Corn Belt of the USA — although they have been observed as far north as New York (Edmunds et al., 1967; Miller, et al., 1963), Wisconsin (Griffin, 1964), and Minnesota (Taylor and Schleden, 1959). These diseases are more troublesome in the southern and southeastern regions of the country (Holdeman, 1955; Nelson, 1955; Perry, 1956; Young, 1964). Nematodes are favored by warm soil temperatures, a long growing season, and porous, sandy soils with sufficient moisture to facilitate their migration.

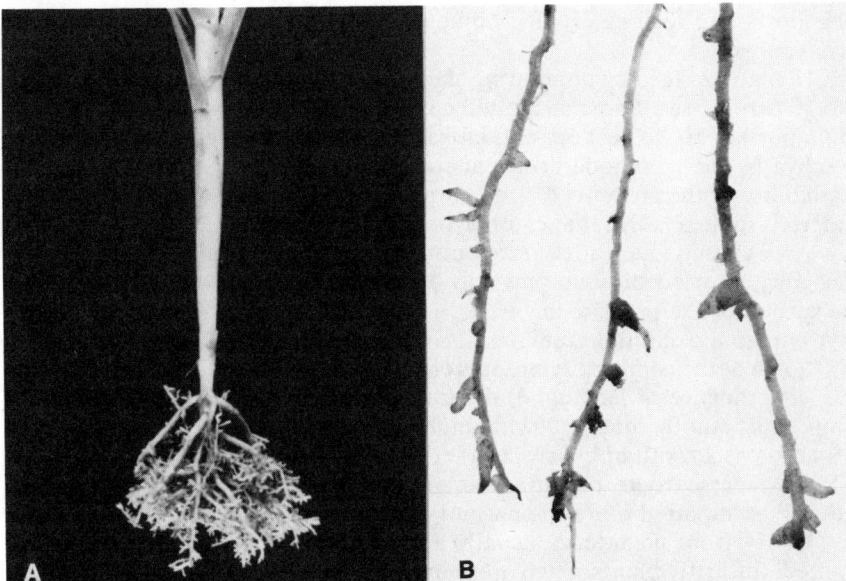

Fig. 17. Nematode diseases: A, short, deformed roots following attack by the stubby root nematode, *Trichodorus christiei*; B, symptoms caused by the root knot nematode, *Meloidogyne incognita* (courtesy, K.R. Barker, North Carolina State University).

Nematodes are small, wormlike in appearance and are sometimes referred to as eelworms. The plant parasitic species range in length from about 300 to 1,000 μ, but a few may be longer. In general they are eel-shaped, smooth, unsegmented, round in cross section and without legs or other appendages. Because of their small diameter individual nematodes cannot be seen without magnification.

The general life cycle of most plant parasitic nematodes is similar. Eggs hatch into larvae which are similar to adults. Nematodes pass through four larval stages and each stage ends with a molt; the first molt may occur in the egg. At the end of the fourth molt the larvae differentiates into adult males and females. These may then mate and the female produce fertile eggs. In the absence of males, females may produce eggs pathogenetically, or they may produce their own sperm to fertilize the eggs. The life cycle from egg to egg usually requires 3 or 4 weeks, but this period may be longer under unfavorable environments. Some nematodes cannot parasitize host plants until after the first or second larval stage. Infective stages must obtain nutrients from a favorable host or die. In the absence of a suitable host the entire population may die in a few months, but in some species eggs survive in the soil for several years. Among the several hundred species that attack plants none parasitizes man or animals.

Over 40 species of nematodes have been found feeding on, or associated with, corn roots. As a result of feeding the root system is debilitated and its function impaired. Plant parasitic nematodes have been grouped according to their method of feeding: Endoparasitic nematodes spend part of their life histories within root tissue. Ectoparasitic types live outside of the roots of their hosts and feed externally. Some nematodes are both internal and external parasites.

Nematodes feed by puncturing the host cells with a stylet that may make 8 to 10 thrusts/sec. Some of the cell contents aggregate around the stylet tip and a portion of the host cell contents is ingested by the nematode. Injection of saliva by the nematode brings about changes in the host cell and aids in availability of the contents for the parasite. Some species apparently secrete materials in their saliva that cause hypertrophy of host cells.

As a consequence of the large number of species of nematodes that attack corn, a variety of symptoms may be found. Symptoms not only vary with the species of the parasite involved, but with stage of host development, soil environmental conditions, and numbers of nematodes.

Some of the common symptoms observed are: 1) Stunting, 2) root lesions as a consequence of feeding, 3) deformation of roots such as production of many short stubby roots following injury to root tips (Fig. 17A), gall or knot formation as a result of hyperplasia induced by feeding of the nematode (Fig. 17B), 4) coarse roots devoid of small rootlets, 5) wilting resulting from reduced or impaired root systems, and 6) chlorosis.

Injury from nematodes usually appears in fields as irregular areas of stunted, unthrifty plants. Such areas may increase in size in succeeding years if the nematodes are not controlled.

Control may be practiced, in part, by the use of resistant hybrids (Baldwin and Barker, 1970a; Baldwin and Barker, 1970b; Nelson, 1957a). Resistance in corn to nematode infection is not as clear-cut as the resistance expressed toward many fungal and bacterial diseases.

Soil application of nematacides has long been used to reduce nematode injury in many horticultural crops, but in corn this method is somewhat limited because of economic considerations. The relatively low unit value of the harvested product, except for seed corn or sweet corn, may limit use of nematacides. Some marked increases in yield of corn have been demonstrated in some areas (Cooper et al., 1959; Griffin, 1964; Young, 1964). In other experiments fumigants reduced populations, but yield increases did not always follow (Edmunds et al., 1967).

Crop rotation with non-host plants has been effective in reducing the injury from nematodes (Tarte, 1971). This necessitates knowing the species of the nematode involved and their host ranges so that the proper rotations can be established.

XII. WITCHWEED

Witchweed is a seed plant parasitic on corn and a large number of other plants, most of which are members of the *Gramineae,* and the *Cyperaceae*, but a few are dicots (Nelson, 1958; USDA, 1957a). The parasite was first found in North Carolina in 1956 and in South Carolina in 1957. There is some evidence that it may have been present in these areas as early as 1946 (Garriss and Wells, 1956; Holdeman, 1957). The areas where witchweed was found are in the eastern parts of these states and separated from each other by the state line (USDA, 1957b). These reports are the first records of the parasite outside of the Eastern Hemisphere.

DISEASES OF CORN

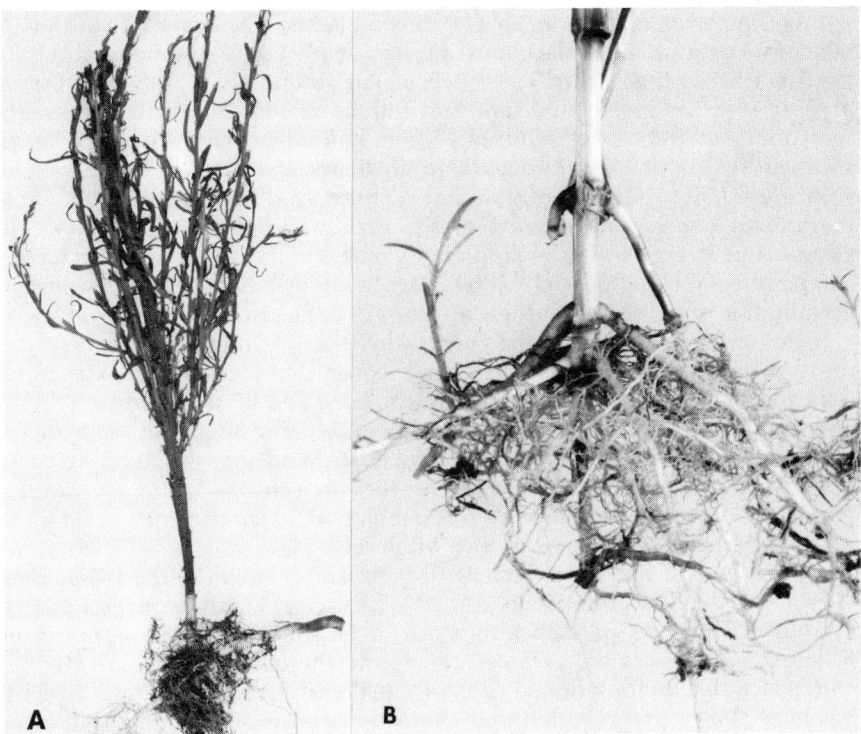

Fig. 18. Witchweed: A, witchweed; B, witchweed plant parasitizing corn roots (courtesy, H.R. Garriss, North Carolina State University).

Corn plants parasitized by witchweed present a generally unthrifty appearance. Plants are stunted, chlorotic, as though deficient in nitrogen, and wilted, even though soil moisture is ample. Yield losses vary from a trace to complete, depending on severity of infection. Severely infected plants die before producing seed.

Striga lutea Lour. [Syn. = *S. asiatica* (L.) Kuntz], commonly called witchweed, is the parasite involved in this disease. There are a number of species of *Striga*, most of which are parasitic and grow mainly in tropical or subtropical areas. Thus far only *S. lutea* has been observed in the United States. It is an inconspicuous plant about 20-25 cm in height (Fig. 18A). Leaves are bright green, narrow, strap-shaped, and slightly hairy. Stems are square and leaves are borne oppositely. Flowers are small, brick-red or scarlet, but a few may be yellowish red or almost white, and borne in the axils of the leaves. Flowers are tubular with a two-lipped corolla. Seeds of witchweed are exceedingly small — less than 0.25 mm long — and because of their minute size, are easily dispersed by wind and water. One plant may produce up to about 500,000 seeds and these may remain dormant in the soil for as long as 20 years. Seeds require an after-ripening period of 15 to 18 months before germination.

Germination appears to consist of two phases. The first involves the uptake of water and takes place most rapidly near 21 C. The time required for the first phase ranges from 2 to 4 weeks. The second phase, in which the enbryo emerges, is most important and unique in that a stimulant from the roots of host plants is required. Some non-hosts may also secrete the stimulant. This phase requires 24 to 48 hours and is most rapid at temperatures of 30 to 32 C. The stimulant is absorbed during the first 2 hours of the second phase. The identity of the germination stimulant is not fully known, but it appears to be similar to coumarin-like derivatives or related compounds (Worsham et al., 1964). Other known chemicals can stimulate germination of the seed, but these appear not to be involved.

Following germination the young witchweed seedling must contact the roots of a host plant or die. Small conical or bell-shaped structures (haustoria) develop on the white roots of *S. lutea* and these are pressed tightly against those of the host. By chemical or enzymatic processes, cells of the host roots are dissolved at the point of contact and the vascular systems of host and parasite appear to connect in the "nucleus" — a spongy, swollen part of the haustorium. Through this contact, of which there are great numbers on parasitized host roots, the witchweed plant obtains nutrients and water for growth and development (Rogers and Nelson, 1962.) It has been shown that the flow of nutrients is from host to parasite. After the seedling is established it grows parasitically beneath the soil for 4 to 6 weeks during which time it causes the greatest damage to the host plant (Fig. 18B). On emergence the seedling develops chlorophyll and becomes semi-parasitic. It has been shown that the chlorophyll in the parasite is functional, but when the plant is placed in the dark it will continue to grow by obtaining photosynthates from its host (Rogers and Nelson, 1962). Flowers of witchweed are produced about 1 month after emergence and the seed 1 month later.

Control of witchweed can be practiced by several means. The use of "catch" crops is one of these and involves planting of a host crop — corn, sorghum, or sudangrass — and plowing down the entire planting before the parasite produces seed. A second cultural practice is to plant "trap" crops such as soybeans, flax, cowpeas or sunflowers, which are not true hosts, but do secrete the germination stimulant. The seed germinates, but being unable to establish a parasitic relationship, dies. In this method true weed hosts must be eradicated from the field. The nature of resistance of the trap crops is not understood. A third method of control is the use of herbicides such as 2,4-D to destroy the witchweed. Trap crop and catch crop methods should be continued for 4 to 5 years and can be used in conjunction with herbicide application to any surviving witchweed.

Extreme care must be used in the movement of crops, machinery or containers out of an infested area. The parasite appears to have been contained within the infested areas through use of cultural practices and quarantine.

XIII. NON-INFECTIOUS DISEASES

There are a number of non-infectious diseases of corn that sometimes are cause for concern among farmers and seed producers. Generally they are localized in extent, but occasionally, cold, heat or drought injuries may extend over an appreciable area. Some of these are primarily under genetic control, others are triggered by the environment, while still others are induced by cultural practices.

1. Genetically Controlled Abnormalities

Some of the non-infectious diseases or abnormalities are directly related to the genetic constitution of the corn. These are most often encountered in breeding nurseries of seed producers where in the course of developing inbred lines, plants may become homozygous for such traits. This is so because many of these characters are recessive and not apparent in the heterozygous condition.

"Silk-cut" is one abnormal condition that appears in some inbred lines in different seasons. It is characterized by a horizontal crack over the face of the kernel (Fig. 19A). Silk-cut is something of a misnomer because it has nothing to do with the silk. It appears to come about after a dry period, before full maturity, that is followed by wet weather which may bring about a resurgence of deposition of dry matter in the kernel. This brings about a stress on the seed coat which breaks as a consequence. The crack in the seed coat exposes the underlying starch providing a substrate for the growth of fungi, predominantly *Fusarium moniliforme*. It was believed at one time that this fungus was the cause of silk cut, but observations have shown that its presence in the fissures is an effect and not a cause.

A similar abnormality, again more prevalent in inbred lines, is referred to as "popped-kernel" (Fig. 19B). The folklore attached to this condition was that high temperature was responsible. The factors bringing about popped-kernel appear to be the same as those operative in silk-cut, except that the rupture of the seed coat takes place over the crown of the kernel. The condition occurs irregularly and seems more prevalent among genotypes with hard, flinty kernels.

Occasionally certain inbred lines and a few hybrids will exhibit tightly rolled whorl leaves, sometimes to the extent that the roll will twist and bend. This occurs irrespective of environmental conditions and distinct in appearance, and in its cause, from "buggy-whip." Eventually the leaves unfurl and the tassel emerges.

A leaf abnormality peculiar to certain genotypes is expressed as a crinkling of the leaf blade. This condition is not related to the environment, but rather to the genetic make-up of the inbred line.

A large number of leaf spots, blotches and stripings of various kinds may be found in inbred lines when they become homozygous for these conditions. Some of these are determined by single, usually recessive, genes (Emerson, 1923; Giesbrecht, 1965). Rarely a leaf spot will be conditioned by a dominant gene and appear in the heterozygous condition. These spots of varied pattern may be chlorotic at first and become necrotic later in the development of the plant (Fig. 19C). They may be chlorotic and remain so to maturity, or they may appear first as necrotic and remain so. Some of the blotches become more abundant as the plant matures, and may be lethal (Ullstrup and Troyer, 1967). (Fig. 19D). A number of the spots on corn leaves show a concentric pattern resembling some of the symptoms of

Fig. 19. Non-infectious diseases: A, "silk-cut;" B, popped kernel; C, leaf spotting resembling some infectious diseases; D, lethal leaf spot, plants are completely covered with lesions and die before seed set.

infectious nature, and may often be confused with these. It has been suggested that these leaf spots may be cytoplasmically conditioned or that they are incited by viruses (Atanasoff, 1964, 1965). Discrete genetic ratios, as has been demonstrated in studies on inheritance of these spottings, would not be expected if determined by the cytoplasm. The suggestion that they are of viral origin seems untenable since they occur under varying environments and have not been reproduced by sap transmission through the intervention of a vector or by grafting. Until such symptoms can be reproduced by such methods their viral nature is not proven.

2. Induced Injuries and Abnormalities

Some of the abnormalities that may be confused with, or suspected of, being infectious diseases, are brought about through certain cultural practices. Most of these result from the improper use of some of the herbicides. Some of the effects come about through the interaction of the environment and the herbicide, even when the proper dosage is maintained. Insecticides and the presently used fungicides applied to seed or foliage seldom cause injury.

The herbicide, 2,4-D, may cause brittleness of the stalk, at lower internodes and fasciation, distortion and upturning of the brace roots (Fig. 20A). Under some conditions the leaf whorl will become tightly furled and similar to a genetically-controlled condition associated with some inbred lines.

A most unusual malformation associated with 2,4-D is "buggy-whip." In this, little or no leaf blade is formed and the upper portions of the plants including leaves and tassel are tightly enclosed in a tube-like structure that is thick and heavily lignified (Fig. 20B). The tube appears to be an extension of the leaf sheath and lateral growth of the midrib. In some plants a small amount of leaf blade may develop near the tip or at various levels along the whip. The name of this abnormality is derived from its superficial resemblance to a tapering buggy whip. On many plants the tube prevents the emergence of the tassel which remains enclosed. In other instances growth pressure may split the tube and release the tassel. Where this condition prevails the incidence of smut is often increased, particularly at the juncture of the leaf sheath and node or where the leaf sheath joins the midrib. This may result from the disturbances in normal growth patterns which prolongs the meristematic stage and delays maturation of cells at these growing points. Buggy-whip occurs relatively infrequently. It appears when extended periods of cool, damp, overcast weather follows application of the herbicide (Ullstrup, 1969).

Some of the other systemic herbicides, mainly the carbamate types, may induce stunting and various distortions when applied improperly.

Paraquat, a contact herbicide, will incite small necrotic spots on corn leaves that bear resemblance to some of the infectious leaf blights (Fig. 20C). This herbicide is not used on corn, but sometimes drift from the spray being used to eradicate roadside weeds, or to defoliate soybeans, will cause this

Fig. 20. Non-infectious diseases: A, fasciation of brace roots, caused by 2,4-D (Courtesy, J.L. Williams, Purdue University); B, "buggy whip," also caused by 2,4-D; C, paraquat injury.

Fig. 21. Non-infectious diseases: A, cold injury showing silvery-gray blotches on leaves: B, interveinal dieing in mature leaves due to heat and drought.

reaction on corn leaves.

Ammonia burn sometimes occurs when this gas is used as a fertilizer and escapes from an applicator as it is raised at the ends of corn rows. This appears as straw-colored, irregular areas on the leaves. Its identity can usually be established by appearance on the "turn rows" in the field.

Cold injury, in contrast to freezing, occurs when temperatures in the spring and early summer drop to 2 to 7 C, and this followed by a clear sunny day. The leaves have a silvery-gray appearance for several weeks (Fig. 21A). The basis for the reaction is not understood, but no typical necrosis takes place.

Drought injury appears as general stunting owing to shortening of internodes. This often imparts a "feather duster" appearance to the plants. Dieing of leaf tissue between the veins appears to result from a combination of excessively dry, hot weather (Fig. 21B).

"Top-firing," in which the upper leaves are killed and assume a straw-color, follows extremely high temperatures. If this occurs near the time of emergence of tassels, these, including the pollen are destroyed. Marked differences are noted among inbred lines and hybrids with respect to tolerance to heat.

3. Nutritional Deficiencies

The most commonly observed nutritional deficiencies observed in corn in the USA are those involving the three major nutrients — nitrogen, phosphorus and potassium. Occasionally some of the "trace elements" will be in short supply and the typical deficiency symptoms of these become apparent.

Nitrogen deficiency results in spindly, chlorotic and stunted plants. Marked chlorosis develops first as a V-shaped pattern that starts at tips of the lowermost leaves, moves toward the leaf base and progressively upward on the plant (Fig. 22A). Chlorotic areas may eventually become necrotic. Nitrogen starvation is apt to appear on wet soil when temperatures are below

Fig. 22. Nutritional deficiencies: A, nitrogen deficiency; B, potassium deficiency; C, phosphorus deficiency; D, magnesium deficiency; E, zinc deficiency.

optimum for growth of corn. Corn grown on soils low in organic matter, which permits excessive leaching, often shows nitrogen starvation.

Phosphorus deficiency is characterized by retarded growth (Fig. 22C), and purpling or reddening of leaf tissues. This is often most acute in seedlings and young plants. Ears on P-deficient plants are poorly filled at their tips and may be constricted on one side. Extremely high or too low pH of the soil may restrict P availability, even though a sufficient amount of the element is present.

Potassium starvation appears as yellowing and dieing of margins of the leaves (Fig. 22B). The necrosis begins at the tips of the lowermost leaves and moves upward on the plant. Ears on such plants show little kernel development at their tips.

Magnesium deficiency, when severe, will appear on young plants as chlorotic striping between the veins with marked stunting and eventual necrosis of leaf tissue (Fig. 22D). The symptoms are most often seen where corn is grown on acid, sandy soil and also in areas where calcareous limestone has been used over an extended period to correct acidity. The use of dolomitic limestone can correct the deficiency in most cases. When the condition is severe, foliar or soil application of magnesium sulfate is recommended.

Zinc starvation is often seen in young plants and appears as elongated pale stripes between the midrib and margin. The chlorotic areas may eventually die and become straw-colored (Fig. 22E). As plants grow older and roots penetrate deeper into soil horizons, where zinc is in ample supply, the newly developed leaves will not show deficiency symptoms. Acute Zn deficiency can be corrected by foliar, or soil, application of zinc salts. The deficiency often appears on corn growing on knolls in a field where the pH of the soil may be too high.

Manganese is rarely seen, but may occur on soils where the pH is excessively high. Muck and sandy soils high in organic matter are sometimes deficient in this element. Pale yellowish streaking of leaf tissue between the veins is characteristic of manganese starvation. Foliar sprays or soil applications containing Mn salts can correct the problem.

Detailed accounts and illustrations of these and other nutrient deficiencies in corn have been published (Krantz and Melsted, 1964).

LITERATURE CITED

(Cutoff date for literature citations, September 1, 1973.)

Aala, F.T. 1964. The corn leaf blight disease: a problem in the production of corn seed involving cytoplasmic male sterility. Philippine J. Plant Indust. 29:115-122.

Abbott, E.V., and R.L. Tippett. 1966. Strains of sugarcane mosaic. USDA, ARS Tech. Bul. 1340.

Abney, T.S., and D.C. Foley. 1971. Influence of nutrition on stalk rot development in Zea mays. Phytopathology 61:1125-1129.

Allen, Ruth F. 1933. The spermatia of corn rust, Puccinia sorghi. Phytopathology 23:923-925.

----. 1934. A cytological study of heterothallism in Puccinia sorghi. J. Agric. Res. 49:1047-1068.

Akai, S. (Ed.). 1959. Studies on the downy mildew disease of rice plants II: 1-181. Plant Protection Sec. Bur. Agric. Admin., Min. Agric. Forestry, Japan.

Allinson, D.W., and W.W. Washko. 1972. Influence of a disease complex on yield and quality components of silage corn. Agron. J. 64:257-258.

Al-Sahaily, I.A., and C.J. Mankin. 1960. Reaction of corn and sorghum to corn and sudangrass head smuts. Plant Dis. Rep. 44:113-114.

----, C.J. Mankin and G. Semeniuk. 1963a. Pathogenicity to sorghum and corn of a hybrid between head smut fungi on each of these crops. Phytopathology 53:360-361.

----, ----, ----. 1963b. Physiologic specialization of *Sphacelotheca reiliana* to sorghum and corn. Phytopathology 53:723-726.

Altstatt, G.E. 1945. A new corn disease in the Rio Grande Valley. Plant Dis. Rep. 29:533-534.

Anglade, P., and P.M. Molot. 1967a. Biochemical factors of combined resistance in maize to certain insects and pathogenic fungi. Biochem. Meded. Rijsfac. Landb. Wetensch. Gent. No. 3/4:327-328.

----, and ----. 1967b. Clarification of the relationship between the susceptibility of maize lines to *Ostrinia nubialis* and *H. turcicum*. Ann. Epiphyt. 18:279-284.

Ark, P.A. 1940. Bacterial stalk rot of field corn caused by *Phytomonas lapsa*. Phytopathology 30:1 (Abstr.).

----. 1941. Persistence of *Phytomonas lapsa* on seed of field corn. Plant Dis. Rep. 25:202.

Arny, D.C., and R.R. Nelson. 1971 *Phyllosticta maydis* sp. nova, the incitant of yellow leaf blight of maize. Phytopathology 61:1170-1172.

----, S. Saad, W.H. Hughes, and G.L. Worf. 1972. Is sick corn related to nutrition? Better Crops Soils 56:20-21.

----, E.B. Smalley, A.J. Ullstrup, G.L. Worf, and R.W. Ahrens. 1971. Eyespot of maize, a disease new to North America. Phytopathology 61:54-57.

----, G.L. Worf, R.W. Ahrens, and M.F. Lindsey. 1970. Yellow leaf blight of maize in Wisconsin: Its history and the reactions of inbreds and crosses to the inciting fungus (*Phyllosticta* sp.). Plant Dis. Rep. 54:281-285.

Arx, J.A. von. 1957. Die Arten der Gattung Colletrotrichum. Phytopathol. Z. 29:413-468.

Asare-Nyako, A. 1964. Germination of *Helminthosporium turcicum* conidia overwintered in Minnesota soil. Phytopathology 54:886 (Abstr.).

Atanasoff, D. 1964. Viruses and cytoplasmic heredity. Z. Pflanzensuchtung 51:197-214.

----. 1965. Leaf fleck disease of maize and its possible relationship to cytoplasmic inheritance. Phytopathol. Z. 52:89-95.

Ayers, J.E., R.R. Nelson, Carol Koons, and G.L. Schiefele. 1970. Reactions of various maize inbreds and single crosses in normal and male-sterile cytoplasm to the yellow leaf blight organism, (*Phyllosticta* sp.). Plant Dis. Rep. 54:277-280.

Bain, D.C., and C.W. Edgerton. 1943. The zonate leaf spot, a new disease of sorghum. Phytopathology 33:220-226.

Baldwin, J.G., and K.R. Barker. 1970a. Host suitability of selected hybrids, varieties and inbreds to corn to populations of *Meliodogyne* spp. J. Nematol. 2:345-350.

----, and ----. 1970b. Histopathology of corn hybrids infected with root knot nematode, *Meliodogyne incognita*. Phytopathology 60:1195-1198.

Bancroft, J.B., A.J. Ullstrup, Mimi Messieha, C.E. Bracker, and T.E. Snazelle. 1966. Some biological and physical properties of a midwestern isolate of maize dwarf mosaic virus. Phytopathology 56:474-478.

Barnes, J.M. 1959. Extraction and bioassay of an antifungal substance from inbreds and hybrid corn differing in susceptibility to *Gibberella zeae*. Phytopathology 49:533 (Abstr.).

Berg, G.L. (Ed.). 1972. Master manual on molds and mycotoxins. Celanese Chemical Co., New York.

Bockholt, A.J., and R.A. Frederiksen. 1972. Breeding corn for resistance to sorghum downy mildew. Am. Soc. Agron. Abstr. 64:18.

Boewe, G.H. 1936. The relation of ear rot prevalence in Illinois corn fields to ear coverage by husks. State of Ill. Nat. Hist. Survey Div. Contr. Sec. of Appl. Bot. and Plant Pathol. Publ. No. 273.

----. 1953. Stewart's disease prospects for 1953. Plant Dis. Rep. 37:311-312.

Boling, M.B., and C.O. Grogan. 1965. Gene action affecting host resistance to fusarium ear rot of maize. Crop Sci. 5:305-307.

Bond, W.F., and T.P. Pirone. 1966. Distribution and identification of mechanically transmitted corn viruses in Louisiana. Plant Disease Rep. 50:325-326.

Böning, K., and F. Wallner. 1936. Welke, fusskrankheit und andere shadigungen an mais durch *Colletotrichum graminicolum* (Ces.) Wilson. Phytopathol. Z. 9:99-110.

Boosalis, M.G., D.R. Sumner, and A.S. Rao. 1967. Overwintering of conidia of *Helminthosporium turcicum* on corn residues and in the soil in Nebraska. Phytopathology 57:990-996.

Bowman D.H. 1946. Sporidial fusion in *Ustiliago maydis*. J. Agric. Res. 72:233-243.

Bradfute, O.E., R. Louie, and J.K. Knoke. 1972. Isometric virus-like particles in maize with stunt symptoms. Phytopathology 62:748 (Abstr).

Brandes, E.W. 1920. Mosaic disease of corn. J. Agric. Res. 19:517-522.

----, and P.J. Klaphaak. 1923. Cultivated and wild hosts of sugarcane or grass mosaic. J. Agric. Res. 24:247-261.

Branstetter, B.B. 1927. Corn root rot studies. Missouri Agric. Exp. Stn. Bull. 113.

Brewbaker, J.L., and F. Aquilizan. 1965. Genetics of resistance in maize to a mosaic stripe virus transmitted by *Peregrinus maidis*. Crop Sci. 5:412-415.

Broyles, J.W. 1956a. Observations on secondary spread of *Physoderma maydis* on corn. Phytopathology 46:8 (Abstr.).

----. 1956b. Observation on time and location of penetration in relation to amount of damage, and chemical control of *Physoderma maydis*. Phytopathology 46:8 (Abstr.).

----. 1959. Incidence of brownspot of corn in Mississippi in 1957, and estimations of its effect on yield. Plant Dis. Rep. 43:19-21.

----. 1962. Penetration of meristematic tissues of corn by *Physoderma maydis*. Phytopathology 52:1013-1016.

Burnham, C.B., and J.L. Cartledge. 1939. Linkage relations between smut resistance and semisterility in maize. J. Am. Soc. Agron. 31:924-933.

Burns, E.E., and M.C. Shurtleff. 1973. Observations on *Physoderma maydis* in Illinois: Effects of tillage practices in field corn. Plant Dis. Rep. 57:630-633.

Burton, C.L. 1968. Southern corn leaf blight on sweet corn ears in transit. Plant Disease Rep. 52:847-851.

Caldwell, R.W., John Tuite, Martin Stob, and Robert Baldwin. 1969. Zearalenone production by Fusarium species. Appl. Microbiol. 20:31-34.

Chang, I-pin, and Thor Kommedahl. 1968. Biological control of seedling blight in corn by coating kernels with antagonistic microorganisms. Phytopathology 58:1395-1401.

Chen, Tseh-An, and R.R. Granados. 1970. Plant-pathogenic mycoplasma-like organisms: Maintenance in vitro and transmission to *Zea mays* L. Science. 167:1633.

Chilton, St. John P. 1940. Delayed reduction of the diploid nucleus in the promycelia of *Ustilago zeae*. Phythopathology 30:622-623.

Choudhury, M.M., and E. Rosenkranz. 1973. Differential transmission of Mississippi and Ohio corn stunt agents by *Graminella nigrifrons*. Phytopathology 63:127-133.

Chowdhury, S.C. 1936. The disease of *Zea mays* caused by *Colletotrichum graminicolum*. Indian J. Agric. Sci. 6:833-843.

Christensen, C.M. 1971. Mycotoxins. Critical Reviews in Environmental Control 2:57-80. The Chemical Rubber Co. Cleveland, Ohio.

----, and H.H. Kaufman. 1969. Grain storage. The role of fungi in quality loss. Univ. of Minn. Press, Minneapolis, Minn.

Christensen, J.J. 1929. Hybridization in *Ustilago zeae*. Minnesota Agric. Exp. Tech. Bul. 65.

----. 1931. Studies on the genetics of *Ustilago zeae*. Phytopathol. Z. 4:129-188.

----. 1963. Corn smut caused by *Ustilago maydis*. Monograph No. 2. Am. Phytopathol. Soc.

----, and R.D. Wilcoxson. 1966. Stalk rot of corn. Monograph No. 3. Am. Phytopathol. Soc.

Chupp, Charles. 1953. A monograph of the fungus genus Cercospora. (Privately published) Ithica, New York.

Coffman, F.A., W.H. Tisdale, and J.F. Brandon. 1926. Observations on corn smut at Akron, Colorado. J. Am. Soc. Agron. 18:403-411.
Comstock, J.C., and R.P. Scheffer. 1972. Production and relative host specificity of a toxin from *Helminthosporium maydis* Race T. Plant Dis. Rep. 56:247-251.
Cook, M.T. 1936. Phloem necrosis in the stripe disease of corn. Phytopathology 26:90.
Cooper, W.E., J.C. Wells, J.N. Sasser, and T.G. Bowery. 1959. The efficiency of preplant and postplant applications of 1,2-dibromo-3-chloropropane for control of the sting nematode, *Belonolarmus longicaudatus*. Plant Dis. Rep. 43:903-908.
Couture, R.M., D.G. Routley, and G.M. Dunn. 1971. Role of cyclic hydroxamic acids on monogenic resistance of maize to *Helminthosporium turcicum*. Physiol. Plant Pathol. 1:515-521.
Craig, Jeweus, and L.A. Daniel-Kalio. 1968. Chlorotic lesion resistance to *Helminthosporium maydis* in maize. Plant Disease Rep. 53:134-136.
----, and J.M. Fajemisin. 1969. Inheritance of chlorotic lesion resistance to *Helminthosporium maydis* in maize. Plant Dis. Rep. 53:742-743.
----, and A.L. Hooker. 1961. Relation of sugar trends and pith density to diplodia stalk rot in dent corn. Phytopathology 51:376-382.
----, and B. Koehler. 1958. *Pyrenochaeta terrestris* and *Phaeocytosporella zeae* on corn roots. Plant Dis. Rep. 42:622-623.
Cugini, G., and G.B. Traverso. 1902. La *Sclerospora macrospora* Sacc. parasitica della *Zea mays* L. Stn. Sper. Agric. Ital. 35:46-49.
Cummins, G.B. 1931. Heterothallism in corn rust and the effect of filtering pycnial exudate. Phytopathology 21:751-753.
----. 1941. Identity and distribution of three rusts of corn. Phytopathology 31:356-357.
----. 1971. The rust fungi of cereals, grasses and bamboos. Springer-Verlag, New York.
Curtin, T.M., and John Tuite. 1966. Emesis and refusal of feed in swine associated with *Gibberella zeae*-infected corn. Life Sci. 5:1937-1944.
Dale, J.L. 1963. Corn anthracnose. Plant Disease Rep. 47:245-249.
----. 1964. Isolation of a mechanically transmissible virus from corn in Arkansas. Plant Disease Rep. 48:661-663.
Daniels, N.E., and R.W. Toler. 1969. Transmission of maize dwarf mosaic by the greenbug, *Schizaphus graminum*. Plant Disease Rep. 53:59-61.
Davis, G.N. 1936. Some of the factors influencing the infection and pathogenicity of *Ustilago zeae* (Beckm.) Unger on *Zea mays* L. Iowa Agric. Exp. Stn. Bull. 199.
Davis, G.N. 1936. Some of the factors influencing the infection and pathogenicity of *Ustilago zeae* (Beckm.) Unger on *Zea mays* L. Iowa Agric. Exp. Stn. Bull. 199.
----, R.F. Whitcomb. T. Ishijima, and R.L. Steere. 1972. Helical filaments produced by a mycoplasma-like organism associated with corn stunt disease. Science 176:521-523.
Del Rosario, Maria Salome E., and W.H. Sill. 1965. Physiological strains of *Aceria tulipae* and their relationship to transmission of wheat streak mosaic virus. Phytopathology 55:1168-1175.
Diachun, S. 1939. The effect of some soil factors on penicillium injury of corn seedlings. Phytopathology 29:231-241.
D'Ippolito, G. and G.B. Traverso. 1902. La *Sclerospora macrospora* Sacc. parasitica della inflorescenze virescenti di *Zea mays* L. Le Stag. Agric. Ital. 35:46.
Djakamihardja, Sjamsudin, G.E. Scott, and M.C. Futrell. 1970. Seedling reaction of inbreds and single crosses of maize to *Fusarium moniliforme*. Plant Disease Rep. 54:307-310.
Dollinger, E.J., W.R. Findley, and L.E. Williams. 1970. Inheritance of resistance to maize dwarf mosaic in maize (*Zea mays* L.). Crop Sci. 10:412-415.
Drechsler, Charles. 1925. Leaf spot of maize caused by *Ophiobolus heterostrophus* n. sp. the ascigerous stage of a Helminthosporium exhibiting bipolar germination. J. Agric. Res. 31:701-726.
----. 1934. Phytopathological and taxonomic aspects of Ophiobolus, Pyrenophora, Helminthosporium and a new genus, Cochliobolus. Phytopathology 24:953-983.

Durrell, L.W. 1920. A preliminary study of the purple sheath spot of corn. Phytopathology 10:487-495.

----. 1922. The nodal infection of corn by *Diplodia zeae.* Iowa Acad. Sci. Proc. 29:346-347.

----. 1923. Dry rot of corn. Iowa Agric. Exp. Stn. Res. Bull. 77:345-376.

----. 1925. Basisporium dry rot of corn. Iowa Agric. Exp. Stn. Res. Bull. 84:135-160.

Eddins, A.H. 1930. A new diplodia ear rot of corn. Phytopathology 20:733-742.

----. 1933. Infection of corn plants by *Physoderma zeae-maydis* Shaw. J. Agric. Res. 46:241-253.

----, and R.K. Voorhees. 1933. *Physalospora zeicola* on corn and its taxonomic and host relationships. Phytopathology 23:63-72.

Edmunds, J.E., C.W. Boothroyd, and W.F. Mai. 1967. Soil fumigation with D-D for control of *Pratylenchus penetrans* in corn. Plant Disease Reptr. 51:15-19.

Edwards, E.T. 1935. Studies on *Gibberella fujikuroi* var. *subglutinans,* the hitherto undescribed ascigerous stage of *Fusarium moniliforme* var. *subglutinans* and its pathogenicity on maize in New South Wales. Dept. Agric. N.S. Wales Sci. Bull. 49.

Elliott, Charlotte. 1942. Relative susceptibility to pythium root rot of twelve dent corn inbreds. J. Agr. Res. 64:711-723.

----. 1943. Pythium stalk rot of corn. J. Agric. Res. 66:21-39.

----, and F.W. Poos. 1934. Overwintering of *Aplanobacter stewarti.* Science (N.S.) 80:289-290.

El-Rouby, M.M. and W.A. Russell. 1966. Locating genes determining resistance to *Diplodia maydis* in maize by using chromosome translocations. Can. J. Genet. Cytol. 8:233-240.

Ellis, M.B. 1966. Dematiaceous Hyphomycetes VII. Curvularia, Brachysporium, etc. Mycological Papers No. 106:1-57, C.M.I. Kew.

Emerson, R.A. 1923. The inheritance of blotch leaf in maize. Cornell Agric. Exp. Stn. Mem. 70:3-16.

Felch, R.E., and G.L. Barger. 1971. Epimay and southern corn leaf blight. Weekly Weather and Crop Bull., Oct. 5, 1971. Environmental Data Ser. NOAA. U.S. Dep. Commerce. Stat. Rep. Ser. USDA.

Fenwick, H.S., and W.R. Simpson. 1967. Suppression of corn head smut by in-furrow application of pentachloronitrobenzene. Plant Dis. Rep. 51:626-628.

Finley, A.M. 1957. Wheat mosaic disease of sweet corn in Idaho. Plant Dis. Rep. 41:589-591.

Foley, D.C. 1959. Systemic infection of corn by *Fusarium moniliforme.* Phytopathology 49:538 (Abstr.).

----, and George Knaphus. 1971. *Helminthosporium maydis* race T in Iowa in 1968. Plant Dis. Rep. 55:855-857.

----, and C.C. Wernham. 1957. The effect of fertilizers on stalk rot of corn in Pennsylvania. Phytopathology 47:11-12 (Abstr.).

Ford, R.E., Helen Fagbenle, and W.N. Stoner. 1970. New hosts and serological identity of bromegrass mosaic from South Dakota. Plant Dis. Rep. 54:191-195.

Frazier, N.W. 1945. A streak disease of corn in California. Plant Dis. Rep. 29:212-213.

----, J.H. Freitag, and Q.L. Holdeman. 1966. Evidence that corn stunt virus may not occur in California Plant Dis. Rep. 50:318-320.

Frederiksen, R.A., and A.J. Bockholt. 1969. *Sclerospora sorghi,* a pathogen of corn in Texas. Plant Dis. Rep. 53:566-569.

----, ----, and A.J. Ullstrup. 1971. Reaction of selected midwestern corn inbred lines to *Sclerospora sorghi.* Plant Dis. Rep. 53:995-998.

----, ----, and ----. 1973. Reaction of selected corn inbred lines to *Sclerospora sorghi* II. Plant Dis. Rep. 57:42-43.

Fuentes, S. 1963. Resistance to head smut of Mexican races of corn. Phytopathology 53:24 (Abstr.).

Futrell, M.C., and D.C. Bain. 1968. *Gonatobotrys zeae* sp. nov., a new pathogen on corn. Phytopathology 58:728 (Abstr.).

----, and R.A. Frederiksen. 1970. Distribution of sorghum downy mildew (*Sclerospora sorghi*) in the USA. Plant Dis. Rep. 54:311-314.

----, and Marcia Kilgore. 1969. Poor stands of corn and reduction of root growth caused by *Fusarium moniliforme.* Plant Dis. Rep. 53:213-215.

Gabboto, L. 1918. La peronosporia de mais. Cultivatore 65:331-333.

Garber, R.J., and M.M. Hoover. 1928. The relation of smut infection to yield in maize. J. Am. Soc. Agron. 20:735-746.

Garraway, M.O. 1973. Electrolyte and peroxide leakage as indicators or susceptibility of various maize inbreds to *Helminthosporium maydis* races O and T. Plant Dis. Rep. 57:518-522.

Garriss, H.R., and J.C. Wells. 1956. Parasitic herbaceious annual associated with corn disease in North Carolina. Plant Dis. Rep. 40:837-839.

Gates, L.F., and C.G. Mortimer. 1969. Three diseases of corn (*Zea mays*) new to Ontario: crazy top, a Phyllosticta leaf spot, and eyespot. Can. Plant Dis. Surv. 49:128-131.

Geisbrecht, John. 1965. A second zebra-necrotic gene in maize. J. Hered. 56:118-119.

Gold, A.H., and A.J. Martin. 1955. Electron microscopy of particles associated with sugarcane mosaic. Phytopathology 45:694 (Abstr.).

Gordon, R.R., and L.E. Williams. 1970. The relationship of a maize virus isolate from Ohio to sugarcane mosaic virus strains and the B strain of maize dwarf mosaic virus. Phytopathology 60:1293 (Abstr.).

Granados, R.R. 1969. Chemotherapy of the corn stunt disease. Phytopathology 59:1556 (Abstr.).

----, R.R. Gustin, Karl Maramorosch, and W.N. Stoner. 1968. Transmission of the corn stunt virus by the leafhopper, *Deltocephalus sonorus* Ball. Contrib. Boyce Thompson Inst. 24:57-60.

----, Karl Maramorosch, and E. Shikata. 1968. Mycoplasma: Suspected etiologic agent of corn stunt. Proc. Natl Acad. Sci. USA 60:841-844.

----, ----, Travis Everett, and T.P. Pirone. 1966. Leafhopper transmission of a corn stunt virus from Louisiana. Phytopathology 56:584 (Abstr.).

----, and Robert Whitcomb. 1971. Transmission of corn stunt mycoplasma by the leafhopper *Baldulus tripsaci.* Phytopathology 61:240-241.

Griffin, G.D. 1964. Association of nematodes with corn in Wisconsin. Plant Dis. Rep. 48:458-459.

Grogan, C.O., and E. Rosenkranz. 1968. Genetics of host reaction to corn stunt virus. Crop Sci. 8:251-254.

Haenseler, C.M. 1937. Correlation between winter temperature and incidence of sweet corn wilt in New Jersey. Plant Dis. Rept. 21:298-301.

Halisky, P.M. 1963. Head smut of sorghum, sudangrass and corn caused by *Sphacelotheca reiliana* (Kühn) Clinton. Hilgardia 34:287-304.

Harvey, P.H., D.L. Thompson, and T.T. Hebert. 1955. Reaction of inbred lines of corn to brown spot. Plant Dis. Rep. 39:973-976.

Harris, M.R. 1936. The relationship of *Cephalosporium acremonium* to the black-bundle disease of corn. Phytopathology 26:965-980.

Haskell, R.J., and W.W. Diehl. 1929. False smut of maize, Ustilaginoidea. Phytopathology 19:589-592.

Hebert, T.T., and A. Kelman. 1958. Factors influencing the germination of resting sporangia of *Physoderma maydis.* Phytopathology 48:102-106.

Herold, F., and K. Munz. 1967. Morphological studies of maize mosaic virus. Phytopathology 57:8 (Abstr.).

----, and J. Weibel. 1963. Electron microscopic demonstration of sugarcane mosaic virus particles in cells of *Saccharum officinarum* and *Zea mays*. Phytopathology 53:469-471.

Hoffmaster, D.E., J.H. McLaughlin, W.W. Ray, and K.S. Chester. 1943. The problem of dry rot caused by *Macrophomina phaseoli.* Phytopathology 33:1113-1114 (Abstr.).

Holbert, J.R., P.E. Hoppe, and A.L. Smith. 1935. Some factors affecting infection with and spread of *Diplodia zeae* in the host tissue. Phytopathology 25:1113-1114.

Holdeman, Q.L. 1955. The present known distribution of the sting nematode, *Belonolaimus gracilis,* in the coastal plain of the southeastern United States. Plant Dis. Rep. 39:5-8.

----. 1957. Striga reported in South Carolina. Plant Dis. Rep. 41:133-134.

Hooker, A.L. 1961a. Occurrence of *Puccinia polysora* in Illinois. Plant Dis. Rep. 45:236.

----. 1961b. A new type of resistance in corn to *Helminthosporium turcicum.* Plant Dis. Rep. 45:780-781.

----. 1962. Corn leaf diseases. Proc. 17th Annu. Hybrid Corn Ind.-Res. Conf. p. 24-36.

----. 1963. Inheritance of chlorotic-lesion resistance to *Helminthosporium turcicum* in seedling corn. Phytopathology 53:660-662.

----. 1969. Widely based resistance to rust in corn. *In*: J.A. Browning (Ed.) Disease consequences of intensive and extensive culture of field crops. Iowa State Univ. Spec. Rep. No. 64.

----. 1972. Southern leaf blight of corn — present status and future prospects. J. Environ. Qual. 1:244-249.

----, and M.P. Britton. 1962. The effects of stalk rot on corn yields in Illinois. Plant Dis. Rep. 46:9-13.

----, H.M. Hilu, D.R. Wilkinson, and C.G. Van Dyke. 1964. Additional sources of chlorotic lesion resistance to *Helminthosporium turcicum* in corn. Plant Dis. Rep. 48:777-780.

----, P.E. Johnson, M.C. Shurtleff, and W.D. Pardee. 1963. Soil fertility and northern corn leaf blight infection. Agron. J. 55:411-412.

----, A. Mesterhazy, D.R. Smith, and S.M. Lim. 1973. A new helminthosporium leaf blight of corn in the northern corn belt. Plant Dis. Rep. 57:195-198.

----, R.R. Nelson, and H.M. Hilu. 1965. Avirulence of *Helminthosporium turcicum* on monogenic resistant corn. Phytopathology 55:462-463.

----, ----, S.M. Lim, and J.B. Beckett. 1970a. Reaction of corn seedlings with male-sterile cytoplasm to *Helminthosporium maydis.* Plant Dis. Rep. 54:708-712.

----, ----, ----, and M.D. Musson. 1970b. Physiological races of *Helminthosporium maydis* and disease resistance. Plant Dis. Rep. 54:1109-1110.

Hoover, M.M. 1932. Inheritance studies of the reaction of selfed lines of maize to smut (*Ustilago zeae*). West Virginia Agric. Exp. Stn. Bull. 253.

Hoppe, P.E. 1929. Inheritance of resistance to seedling blight of corn caused by *Gibberella saubinetii.* Phytopathology 19:79-80.

----. 1938 - 1943. Relative prevalence and geographical distribution of various rot fungi in the 1937 corn crop (and subsequent years). Plant Dis. Rep. 22:234-241, 1938; 23:142-148, 1939; 24:210-213, 1940; 25:148-152, 1941; 26:145-149, 1942; 27:199-202, 1943.

----. 1942. Fusarium ear rot in sweet corn. Plant Dis. Rep. 26:458.

----. 1961. A new kernel rot disease of corn in Wisconsin. Plant Dis. Rep. 45:99.

----. 1962. Does the corn leaf blight fungus survive Wisconsin winters? Plant Dis. Rep. 46:444-445.

----, and D.C. Arny. 1966. Factors affecting the survival of *Helminthosporium turcicum* in corn leaf tissue. Plant Dis. Rep. 50:377-380.

----, and A. Kelman. 1969. Bacterial top and stalk rot disease of corn in Wisconsin. Plant Dis. Rep. 53:66-70.

----, and J.T. Middleton. 1950. Pathogenicity and occurrence in Wisconsin soils of Pythium species which cause seedling disease in corn. Phytopathology 40:13.

How, Shao Chung. 1963. Wheat streak virus on corn in Nebraska. Phytopathology 53:279-280.

Hughes, G.R., and A.L. Hooker. 1971. Gene action conditioning resistance to northern corn leaf blight in maize. Crop Sci. 11:180-184.

Hyre, R.A. 1943. *Cercospora zeae-maydis* on corn in eastern Tennessee and Kentucky. Plant Dis. Rep. 27:553-554.

Immer, F.R. 1927. The inheritance of reaction to *Ustilago zeae* in maize. Minnesota Agric. Exp. Stn. Tech. Bull. 51.

----, and J.J. Christensen. 1928a. Influence of environmental factors on the seasonal prevalence of corn smut. Phytopathology 18:589-598.

----, and ----. 1928b. Determination of losses due to smut infections in selfed lines of corn. Phytopathology 18:599-602.

Itzerott, D. 1938. On the germination and growth of *Ustilago zeae* with special reference to infection. Phytopathol. Z. 1:155-180.

Ivanoff, S.S. 1935. Studies on the host range of *Phytomonas stewarti* and *P. vascularum.* Phytopathology 25:992-1002.

----. 1936. Resistance to bacterial wilt of open-pollinated varieties of sweet, dent and flint corn. J. Agric. Res. 53:917-926.
----, and A.J. Riker. 1936. Resistance to bacterial wilt of inbred strains and crosses of sweet corn. J. Agric. Res. 53:937-954.
----, ----, and H.A. Dettwiler. 1938. Studies on the cultural characteristics, physiology and pathogenicity of strain types of *Phytomonas stewarti.* J. Bacteriol. 35:235-253.
Janson, B.F., L.E. Williams, W.R. Findley, E.J. Dollinger, and C.W. Ellett. 1965. Maize dwarf mosaic: new corn virus disease in Ohio. Ohio Agric. Exp. Stn. Ext. Circ. 460.
Jenkins, M.T., and A.L. Robert. 1952. Inheritance of resistance to leaf blight of corn caused by *Helminthosporium turcicum.* Agron. J. 44:136-140.
----, and ----. 1959. Evaluating the breeding potential of inbred lines of corn resistant to the leaf blight caused by *Helminthosporium turcicum.* Agron. J. 51:93-96.
----, and ----. 1961. Further genetic studies of resistance to *Helminthosporium turcicum* Pass. in maize by means of chromosomal translocations. Crop Sci. 1:450-455.
----, ----, and W.R. Findley, Jr. 1952. Inheritance of resistance to helminthosporium leaf blight in populations of F_3 progenies. Agron. J. 44:438-442.
Jinahyon, S., and W.A. Russell. 1969. Evaluation of recurrent selection for stalk-rot resistance in open-pollinated varieties of maize. Effects of recurrent selection for stalk-rot resistance on other agronomic characters in an open-pollinated variety of maize. Ia. State J. Sci. 43:229-251.
Johann, Helen. 1935. Histology of the caryopsis of yellow dent corn, with reference to resistance and susceptibility to kernel rots. J. Agric. Res. 51:855-883.
----. 1939. Scolecospores in *Diplodia zeae* Phytopathology 29:67-71.
----, and A.D. Dickson. 1945. A soluble substance in corn stalks that retards growth of *Diplodia zeae* in culture. J. Agric Res. 71:89-110.
----, J.R. Holbert, and J.G. Dickson. 1931. Further studies on penicillium injury to corn. J. Agric. Res. 43:757-790.
Johnson, A.G., L. Cash, and WA. Gardner. 1929. Preliminary report on a bacterial disease of corn. Phytopathology 19:81-82 (Abstr.).
----, A.L. Robert, and Lillian Cash. 1949. Bacterial leaf blight and stalk rot of corn. J. Agric. Res. 78:719-732.
Johnson, G.R. 1971. Analysis of genetic resistance to maize dwarf mosaic disease. Crop Sci. 11:23-24.
Johnson, I.J., and J.J. Christensen. 1935. Relation between number, size and location of smut infections to reduction in yield of corn. Phytopathology 25:223-233.
Jones, E.S. 1923. Influence of temperature on the spora germination of *Ustilago zeae.* J. Agric. Res. 24:593-597.
Jones, B.L., J.C. Leeper, and R.A. Frederiksen. 1972. *Sclerospora sorghi* in corn: its location in carpellate flowers and mature seeds. Phytopathology 62:817-819.
Katsura, K. 1952. Some information on the downy mildew disease of the rice plant. Ann. Phytopathol. Soc. Japan. 16:170.
Kelman, A., L.H. Person, and T.T. Hebert. 1957. A bacterial stalk rot of irrigated corn in North Carolina. Plant Dis. Rep. 41:798-802.
Kendrick, J.B. 1926. Holcus bacterial spot of *Zea mays* and *Holcus* species. Iowa Agric. Exp. Stn. Bull. 100:301-334.
Kingsland, G.C. 1963. Cercospora leaf blight of corn. A case history of a local epiphytotic in South Carolina. Plant Dis. Rep. 47:724-725.
Klun, J.A. 1969. Relation of chemical analysis for DIMBOA and visual resistance rating for first-brood corn borer. Proc. 24th Annu. Corn and Sorghum Res. Conf. p. 55-60.
Koehler, B. 1939. Crazytop of corn. Phytopathology 29·817-820.
----. 1942. Disease infection in hybrid seed corn. Internat'l. Crop Imp. Assn. Ann. Rept. (1941) 23:26-31.
----. 1950. Stalk and ear rot. Rep. Fifth Hybrid Corn Industry-Res. Conf. p. 33-46.
----. 1951. Husk coverage and ear declination in relation to corn ear rots. Phytopathology 41:22 (Abstr.).

----. 1953. Ratings of some yellow corn inbreds for ear rot resistance. Plant Dis. Rep. 37:440-444.
----. 1959. Corn ear rots in Illinois. Illinois Agric. Exp. Stn. Bull. 639.
----. 1960. Cornstalk rots in Illinois. Illinois Agric. Exp. Stn. Bull. 658.
Koehler, B., and J.R. Holbert. 1930. Corn diseases in Illinois, their extent and control. Illinois Agric. Exp. Stn. Bull. 354.
----, and C.M. Woodworth. 1938. Corn seedling virescences caused by *Aspergillus flavus* and *A. tamarii*. Phytopathology 28: 811-823.
Krantz, B.A. and S.W. Melsted. 1964. Nutrient deficiencies in corn, sorghum and small grains. In: H.B. Sprague (ed.) Hunger signs in crops David McKay Co., New York.
Krüger, W. 1965. *Colletotrichum graminicolum* (Ces.) Wilson on maize in South Africa. S. Africa J. Agric. Sci. 8:881-886.
Kunkel, L.O. 1927. The corn mosaic distinct from sugarcane mosaic. Phytopathology 17:41 (Abstr.).
----. 1946. Leafhopper transmission of corn stunt. Proc. Natl. Acad. Sci. USA 32:246-247.
----. 1948. Studies on a new corn virus disease. Ark. Gesamte Virusforschung. 1:24-46.
Lallmahomed, G.M. and Jeweus Craig. 1968. Races of *Puccinia polysora* in Nigeria. Plant Dis. Rep. 52:136-138.
Lefebvre, C.L., and A.G. Johnson, 1941. Collection of fungi, bacteria and nematodes of grasses in the United States. Plant Dis. Rep. 25:556-579.
----, and H.S. Sherwin. 1945. Races of *Helminthosporium turcicum*. Phytopathology 35:487 (Abstr.).
Leonard, K.J. 1973. Association of mating type and virulence in *Helminthosporium maydis* and observations of the race T populations in the United States. Phytopathology 63:112-115.
Leonian, L.H. 1932. The pathogenicity and variability of *Fusarium moniliforme* from corn. West Virginia Agric. Exp. Stn. Bull. 248.
Limber, D.F. 1927. *Fusarium moniliforme* in relation to diseases of corn. Ohio J. Sci. 27:232-246.
Lim, S.M., A.L. Hooker, and J.D. Paxton. 1970. Isolation of phytoalexins from corn with monogenic resistance to *Helminthosporium turcicum*. Phytopathology 60:1071-1075.
Lindstrom, E.W. 1938. Genetic investigations of bacterial wilt in corn. Iowa Agric. Exp. Stn. Rep. (1937-38):46-47.
Livingston, J.E. 1945. Charcoal rot of corn and sorghum. Nebraska Agric. Exp. Stn. Res. Bull. 136.
Locci, R., and J.R. Locci. 1972. Possible means of differentiating, on malt agar, between physiologic races of "T" and "O" of *Helminthosporium maydis*. Riv. Patol. Veg. 8 (4):232-238.
Loesch, P.J., Jr., and M.S. Zuber. 1967. An inheritance study of resistance to maize dwarf mosaic in corn (*Zea mays* L.). Agron J. 59:423-425.
Louie, R., and J.K. Knoke. 1970. Evidence for strains of maize dwarf mosaic virus in southern Ohio. Phytopathology 60:1301 (Abstr.).
Ludbrook, W.V. 1942. Top rot of maize, sweet corn and sorghum. J. Council Sci. Indust. Res. 15:213-219.
Luttrell, E.S. 1958. The perfect stage of *Helminthosporium turcicum*. Phytopathology 48:281-287.
Mabadeje, S.A. 1969. Curvularia leaf spot of maize. Trans. Brit. Mycol. Soc. 52:267-271.
Mackenzie, D.R., C.C. Wernham, and R.E. Ford. 1966. Differences in maize dwarf mosaic virus isolated in northeastern United States. Plant Dis. Rep. 50:814-818.
Mackie, W.W. 1932. A hitherto unreported disease of maize and beans. Phytopathology 22:637-644.
Mains, E.B. 1931. Inheritance of resistance to rust, *Puccinia sorghi*, in maize. J. Agric. Res. 43:419-430.
----. 1934. Host specialization of *Puccinia sorghi*. Phytopathology 24:405-411.
----, J.F. Trost, and G.M. Smith. 1924. Corn resistant to rust, *Puccinia sorghi*. Phytopathology 14:47 (Abstr.).

Malca, I., and A.J. Ullstrup. 1960. Vitamin requirements of *Gleocercospora sorghi*. Bul. Torrey Bot. Club 87:271-275.

Malm, N.R., and J.B. Beckett. 1962. Reaction of plants in the tribe Maydeae to *Puccinia sorghi* Schw. Crop Sci. 2:360-361.

----, and A.L. Hooker. 1962. Resistance to rust, *Puccinia sorghi* Schw., conditioned by recessive genes in two corn inbred lines. Crop Sci. 2:145-147.

Maramorosch, K. 1955. The occurrence of two distinct types of corn stunt in Mexico. Plant Dis. Rep. 39:896-898.

----, E. Shikata, and R.R. Granados. 1968. Mycoplasma-like bodies in leafhoppers and diseased plants. Phytopathology 58:886 (Abstr.).

Martens, J.W., and D.C. Arny. 1966. The effect of potassium fertilizer on the incidence of smut in a susceptible corn inbred. Plant Dis. Rep. 50:12-13.

----, and D.C. Arny. 1967a. Effects of potassium and the chloride-ion on root necrosis, stalk rot and pith condition in corn (*Zea mays* l.) Agron. J. 49:499-502.

----, and ----. 1967b. Nitrogen and sugar levels in pith tissue in corn as influenced by plant age and potassium and chloride-ion fertilization. Agron. J. 59:332-334.

Matocha, P.,R.A. Frederiksen, and L. Reyes. 1974. Control of Sorghum downy mildew in grain sorghum by soil incorporation of potassium azide. Indian Phytopathol. 27:322-324.

McIndoe, K.G. 1931. The inheritance of the reaction of maize to *Gibberella saubinetii*. Phytopathology 21:615-639.

McKeen, W.E. 1951a. A preliminary study of corn seedling blight in southern Ontario. Can J. Bot. 29:125-137.

----. 1951b. An exceptionally early outbreak of corn rust in Canada. Plant Dis. Rep. 35:367.

McNew, G.L. 1937. Crown infection of corn by *Diplodia zeae*. Iowa Agric Exp. Stn. Res. Bull. 216:191-222.

Melchers, L.E. 1956. Fungi associated with Kansas hybrid seed corn. Plant Dis. Rep. 40:500-506.

Melhus, I.E., F.H. Van Haltern, and D.E. Bliss. 1928. A study of *Sclerospora graminicola* on *Setaria viridis* and *Zea mays*. Iowa Agric. Exp. Stm. Res. Bull. 111:297-338.

Menzies, J.D., and C.O. Stanberry. 1947. The effect of terminal smut galls on yield and seed grade of detasseled hybrid corn. Phytopathology 37:363 (Abstr.).

Mercado, A.C., and R.M. Lantican. 1961. The susceptibility of cytoplasmic male sterile lines of corn to *Helminthosporium maydis* Nisikado and Miy. Philippine Agric. 45:235-243.

Messiaen, C.M. 1956. Sur quelque anthracnoses des plantes cultivees. Ann. Epiphytes 3:285-299.

----. 1957. Richesse en sucre des tiges de maiz et verse parasitaire. Rev. Pathol. Veg. Entom. Agric. 26:209-213.

Messieha, Mimi. 1967. Aphid transmission of maize dwarf mosaic virus. Phytopathology 57:956-959.

Middleton, J.T. 1943. The taxonomy, host range and geographical distribution of the genus Pythium. Mem. Torrey Bot. Club 20. No. 1.

Miller, R.E., C.W. Boothroyd, and W.F. Mai. 1963. Relationship of *Pratylenchus penetrans* in roots of corn in New York. Phytopathology 53:313-315.

Miller, R.J., and D.E. Koeppe. 1971. Southern corn leaf blight: susceptible and resistant mitochondria. Science 173:67-69.

Mitchell, H.H., and J.R. Beadles. 1940. The impairment in nutritive value of corn grain damaged by specific fungi. J. Agric. Res. 61:135-142.

Moll, R.H., D.L. Thompson, and P.H. Harvey. 1963. A quantitative genetic study of the inheritance of resistance to brown spot (*Physoderma maydis*) of corn. Crop Sci. 3:389-391.

Molot, P.M. 1969a. Studies on maize resistance towards helminthosporium and fusarium diseases. I. Part played by the chemical composition of the plant. Ann. Phytopathol. 1:55-74.

----. 1969b. Studies on maize resistance towards helminthosporium and fusarium diseases. II. Factors of resistance. Ann. Phytopathol. 1:353-366.

----. 1969c. Studies on maize resistance towards helminthosporium and fusarium diseases. III. Behavior of phenolic compounds. Ann. Phytopathol. 1:367-383.

----, and P. Anglade. 1968. Resistance of maize inbreds to leaf-blight disease and to European corn borer in relation with a compound likely identical with 6-methoxy-2(3)-benzoxazolinone. Ann. Epiphytes 19:75-95.
Morgan, O.D., and J.G. Kantzes. 1971. Observations of *Colletotrichum graminicola* on T corn and blends in Maryland. Plant Dis. Rep. 55:755.
Mukunya, D.M., and C.W. Boothroyd. 1973. *Mycosphaerella zeae-maydis* the sexual stage of *Phyllosticta maydis.* Phytopathology 63:529-532.
Narita, T., and Y. Hiratsuka. 1959. Studies on *Kabatiella zeae* n. sp., the causal fungus of a new leaf spot disease of corn. Ann. Phytopathol. Ser. Japan 24:147-153.
Nattrass, R.M. 1953. Occurrence of *Puccinia polysora* in East Africa. Nature 171:527.
Nault, L.R., M.L. Briones, L.E. Williams, and B.D. Barry. 1967. Relation of the wheat curl mite to kernel red streak of corn. Phytopathology 57:986-989.
----, W.E. Styer, J.K. Knoke, and H.N. Pitre. 1973. Semipersistent transmission of a leafhopper-borne maize chlorotic dwarf virus. J. Econ. Entomol. 66:1271-1273.
Nelson, L.R. and G.E. Scott. 1973. Diallel analysis of resistance of corn (*Zea mays* L.) to corn smut. Crop Sci. 13:162-164.
Nelson, R.R. 1955. Nematode parasites of corn in the Coastal Plain of North Carolina. Plant Dis. Rep. 39:818-819.
----. 1956. A new disease of corn caused by *Curvularia maculans.* Plant Dis. Rep. 40:210-211.
----. 1957a. Resistance in corn to *Meloidogyne incognita.* Phytopathology 47:25-26.
----. 1957b. Heterothallism in *Helminthosporium maydis.* Phytopathology 47:191-192.
----. 1958. Preliminary studies on the host range of *Striga asiatica.* Plant Dis. Rep. 42:376-382.
----. 1959. *Cochliobolus carbonum,* the perfect stage of *Helminthosporium carbonum.* Phytopathology 49:807-810.
----. 1960. The genetics of compatibility in *Cochliobolus carbonum.* Phytopathology 50:158-160.
----. 1961. Evidence of gene pools for pathogenicity in species of *Helminthosporium.* Phytopathology 51:736-737.
----. 1964. Genetic inhibition of perithecial formation in *Helminthosporium carbonum.* Phytopathology 54:876-877.
----. 1970. Genes for pathogenicity in *Cochliobolus carbonum.* Phytopathology 60:1335-1337.
----, J.E. Ayres, and J.B. Beckett. 1971. Reactions of various corn inbreds in normal and different male-sterile cytoplasms to the yellow leaf blight organism (*Phyllosticta* sp.). Plant Dis. Rep. 55:401-403.
----, ----, H. Cole, and D.H. Peterson. 1970. Studies and observations on the past occurrence and geographical distribution of isolates of race T of *Helminthosporium maydis.* Plant Dis. Rep. 54:1123-1126.
----, and D.M. Kline. 1971. The pathogenicity of fifty-two isolates of *Cochliobolus carbonum* to twenty-two gramineous species. Plant Dis. Rep. 55:325-327.
----, A.L. Robert, and G.F. Sprague. 1965. Evaluation of genetic potentials in *Helminthosporium turcicum.* Phytopathology 55:418-420.
----, and G.L. Scheifele. 1970. Factors affecting the overwintering of *Trichometasphaeria turcica* on maize. Phytopathology 60:369-370.
----, and A.J. Ullstrup. 1961. The inheritance of pathogenicity in *Cochliobolus carbonum.* Phytopathology 51:1-2.
Niederhauser, J.S. and Javier Cervantes. 1950. Transmission of corn stunt in Mexico by a new insect vector, *Baldulus elimatus.* Phytopathology 40:20 (Abstr.).
Nisikado, Y., and C. Miyake. 1926. Studies on two helminthosporium diseases of maize caused by *Helminthosporium turcicum* Pass. and *Ophiobolus heterostrophus* Drechs. *Helminthosporium maydis* Nisikado et Miyake. Ber. Ohara Inst. Landwirtsch. Forsch. Kurashiki, Japan. 3:221-266.
Norton, J.B.S. 1895. *Ustilago reiliana* on corn. Bot. Gaz. 5:463.
Ohio Agric. Res. Dev. Center. 1971. A list of references: maize virus diseases and corn stunt. 34 p.
Ojerholm, E. 1934. Multiciliate zoospores in *Physoderma zeae-maydis.* Bul. Torrey Bot. Club 61:13-18.

Otto, H.J. and H.L. Everett. 1956. Influence of nitrogen and potassium fertilizer on the incidence of stalk rot of corn. Agron J. 43:301-305.

Pady, S.M. 1944. Bacterial leaf blight and top rot of corn in Nebraska and Kansas. Plant Dis. Rep. 28:826.

Paliwal, Y.C., J.T. Slykhuis, and R.E. Wall. 1966. Wheat streak mosaic virus in corn in Ontario. Can. Plant Dis. Surv. 46:8.

Pan, S.F. 1959. Swedish fly and blister smut on maize. [in Russian]. Zashch. Moscow 4:25-26 *(In* Rev. Appl. Mycol. 38:739).

Pappelis, A.J. 1965. Relationship of seasonal changes in pith condition ratings and density to gibberella stalk rot of corn. Phytopathology 55:623-626.

----, and F.G. Smith. 1963. Relationship of water content and living cells to the spread of *Diplodia zeae* in corn stalks. Phytopathology 53:1100-1105.

----, and J.H. Williams. 1966. Patterns of cell death in elongating corn stalks. Trans. Illinois State Acad. Sci. 59:195-198.

Pate, J.B., and P.H. Harvey. 1954. Studies on the inheritance of resistance in corn to *Helminthosporium maydis* leaf spot. Agron. J. 46:442-445.

Pepper, E.H. 1967. Stewart's bacterial wilt of corn. Am. Phytopathol. Soc. Monograph 4.

Pepper, B.B., and C.M. Haenseler. 1944. Control of European corn borer and ear smut on sweet corn with dusts and sprays. New Jersey Agric. Exp. Stn. Circ. 486.

Perry, V.G. 1956. Nematodes affecting corn in Florida, Alabama, Maryland and Wisconsin. Phytopathology 46:23 (Abstr.).

Piemeisel, F.J. 1917. Factors affecting parasitism of *Ustilago zeae.* Phytopathology 7:294-307.

Pirone, T.P., O.E. Bradutfe, P.H. Freytag, M.C. Young, and C.G. Poneleit. 1972. Virus-like particles associated with a leafhopper-transmitted disease of corn in Kentucky. Plant Dis. Rep. 56:652-656.

Platz, G.A. 1929. Some factors influencing the pathogenicity of Ustilago zeae (Beckm.) Unger. Iowa State Coll. J. Sci. 3:177-214.

Politis, D.J., and H. Wheeler. 1972. The perfect stage of *Colletotrichum graminicola.* Plant Dis. Rep. 56:1026-1027.

Potter, A.A. 1914. Head smut of sorghum and maize. J. Agric. Res. 2:339-371.

----, and L.E. Melchers. 1925. Study of the life history and ecological relations of the smut of corn. J. Agric Res. 30:161-173.

Pound, G.S. 1960. Plant pathology and plant breeding. Proc. 15th Annu. Hybrid Corn Indust.-Res. Conf. p. 7-14.

Price, W.C. 1935. Classification of celery mosaic virus. Phytopathology 25:947-953.

Pringle, R.B. 1971. Amino acid composition of the host-specific toxin of *Helminthosporium carbonum.* Plant Physiol. 48:756-759.

----, and R.P. Scheffer. 1967. Isolation of the host-specific toxin and a related substance with non-specific toxicity from *Helminthosporium carbonum*. Phytopathology 57:1169-1172.

Puhalla, J.E. 1968. Compatibility reactions on solid media and interstrain inhibition in *Ustilago maydis.* Genetics 60:461-473.

Pupiat, Udom, and U.R. Mehta. 1970. Stalk rot of maize caused by *Colletotrichum graminicolum.* Indian Phytopathology 22:346-348.

Quasem, S.A. and C.M. Christensen. 1960. Influence of various factors on the deterioration of stored corn by fungi. Phytopathology 50:703-709.

Quebral, F.C. 1958. Anthracnose of corn. Philippine Agric. 41:250-263.

Raleigh, W.P. 1930. Infection studies of *Diplodia zeae* (Schw.) Lev. and control of seedling blights of corn. Iowa Agric. Exp. Stn. Res. Bull. 124:93-121.

Reddy, C.S., and J.R. Holbert. 1924. The black-bundle disease of corn. J. Agric. Res. 27:177-205.

Reed, G.M., Marjory Swaby, and L.A. Kolk. 1927. Experimental studies on head smut of corn. Bull. Torrey Bot. Club 54:295-310.

Reyes, L.D., T. Rosenow, R.W. Berry, and M.C. Futrell. 1964. Downy mildew and head smut diseases of sorghum in Texas. Plant Dis. Rep. 48:249-253.

Rhind, D., J.M. Waterston, and F.C. Deighton. 1952. Occurrence of *Puccinia polysora* in West Africa. Nature 169:631.

Rhoades, M.M., and V.H. Rhoades. 1939. Genetic studies with the factors in the tenth chromosome in maize. Genetics 24:302-314.

Rhoades, V.H. 1935. The location of a gene for disease resistance in maize. Proc. Natl. Acad. Sci. USA 21:243-246.

Roane, C.W., and J.L. Troutman. 1965. The occurrence and transmission of maize dwarf mosaic in Virginia. Plant Dis. Rep. 49:665-667.

Robert, A.L. 1956. *Helminthosporium maydis* on sweet corn ears in Florida. Plant Dis. Rep. 40:991-995.

----. 1960. Physiologic specialization in *Helminthosporium turcicum*. Phytopathology 50:217-220.

----. 1962. Host ranges and races of the corn rusts. Phytopathology 52:1010-1012.

----, and W.R. Findley. 1952. Diseased corn leaves as a source of infection in artificial and natural epidemics of *Helminthosporium turcicum*. Plant Dis. Rep. 36:9-10.

Rodriguez, A.E., and A.J. Ullstrup. 1962. Pathogenicity of single-ascospore isolates of *Trichometasphaeria turcica*. Phytopathology 52:599-601.

Rogers, W.E., and R.R. Nelson. 1962. Penetration and nutrition of *Striga asiatica*. Phytopathology 52:1064-1070.

Roman, H., and A.J. Ullstrup. 1951. The use of A-B translocations to locate genes in maize. J. Agron. 43:450-454.

Rosen, H.R. 1952. The bacterial pathogen of corn stalk rot. Phytopathology 12:497-498.

Rosenkranz, Eugen. 1969. A new leafhopper-transmissible corn stunt disease in Ohio. Phytopathology 59:1344-1346.

Russell, W.A., and A.L. Hooker. 1962. Location of genes determining resistance to *Puccinia sorghi* Schw. in corn inbred lines. Crop Sci. 2:477-480.

Ryland, A.K., and H.H. Storey. 1955. Physiological races of *Puccinia polysora* Underw. Nature 176:655-656.

Sabet, K.A. 1954. A new bacterial disease of maize in Egypt. Empire J. Exp. Agric. 22:65-67.

Saboe, L.C., and H.K. Hayes. 1941. Genetic studies of reactions to smut and of firing in maize by means of chromosomal translocations. J. Am. Soc. Agron. 33:463-470.

Saxena, K.M., and A.L. Hooker. 1968. On the structure of a gene for disease resistance in maize. Proc. Natl Acad. Sci. USA 61:1300-1305.

Sayre, J.D., V.H. Morris, and F.D. Richey. 1931. The effect of preventing fruiting and of reducing the leaf area on the accumulation of sugars in the corn stem. J. Am. Soc. Agron. 23:751-753.

Scheffer, R.P., and A.J. Ullstrup. 1965. A host-specific toxic metabolite from *Helminthosporium carbonum*. Phytopathology 55:1037-1038.

----, R.R. Nelson, and A.J. Ullstrup. 1967. Inheritance of toxin production and pathogenicity in *Cochliobolus carbonum* and *Cochliobolus victoriae*. Phytopathology 57:1288-1291.

Scheifele, G.L. 1971. Geographical distribution of *Helminthosporium maydis* race T for 1969 and 1970. Plant Dis. Rep. 55:302-306.

----, and R.R. Nelson. 1969. The occurrence of Phyllosticta leaf spot of corn in Pennsylvania. Plant Dis. Rep. 53:186-189.

----, ----, and Carol Koons. 1969. Male sterility cytoplasm conditioning susceptibility of resistant inbred lines of maize to yellow leaf blight caused by *Phyllosticta zeae*. Plant Dis. Rep. 53:656-659.

----, W. Whitehead, and C. Rowe. 1970. Increased susceptibility to southern leaf spot (*Helminthosporium maydis*) in inbred lines and hybrids of maize with Texas male-sterile cytoplasm. Plant Dis. Rep. 54:501-503.

Schenck, N.C. 1970. Perithecia of *Cochliobolus heterostrophus* on corn leaves in Florida. Plant Dis. Rep. 54:1127-1128.

----. 1972. *Phaeosphaeria herpotricha* on southern corn leaf blight-infected plants in Florida. Plant Dis. Rep. 56:276.

Scheiber, Eugenio, and J.G. Dickson. 1963. Pathogenicity studies with *Physopella zeae* on corn. Phytopathology 53:25 (Abstr.).

Schuster, M.L. 1972. Leaf freckle and wilt, a new corn disease. Proc. 27th Annu. Corn Sorghum Res. Conf. p. 176-191.

----, W.A. Compton, and Betty Hoff. 1972. Reaction of corn inbred lines to the new Nebraska leaf freckle and wilt bacterium. Plant Dis. Rep. 56:863-865.

Scurti, J. 1949. Action at a distance in neoformations caused by *Ustilago maydis*. Nuovo G. Bot. N.S. 56:740-741. (Rev. Appl. Mycol. 30:34).

Seghal, O.P. 1966. Host range, properties and partial purification of a Missouri isolate of maize dwarf mosaic virus. Plant Dis. Rep. 50:862-866.

Semeniuk, G. 1942. Charcoal rot of maize, new to Iowa. Proc. Iowa Acad. Sci. 49:256.

----, and C.J. Mankin. 1964. Occurrence and development of *Sclerophthora macrospora* on cereals and grasses in South Dakota. Phytopathology 54:409-416.

----, I.E. Melhus, J.R. Wallin, C.L. Gilly, and Muriel O'Brien. 1946. A dwarfing and witches broom on corn in Iowa. Phytopathology 36:40 (Abstr.).

Shepherd, R.J., D.H. Hall, and D.E. Purcifull. 1964. Occurrence of a new virus disease of corn in California. Plant Dis. Rep. 48:749.

----, and Q.L. Holderman. 1965. Seed transmission of a johnsongrass strain of sugarcane mosaic virus in corn. Plant Dis. Rep. 49:468-469.

Sherbakoff, C.D. 1951. A correction of the identification of *Helminthosporium* causing black ear rot of corn. Phytopathology 41:378.

----, and L.S. Meyer. 1937. Black ear rot of corn. Phytopathology 27:207.

Shikata, E., K. Maramorosch, and K.C. Ling. 1969. Presumptive mycoplasma etiology of yellows diseases. FAO Plant Protection Bull. 17:121-128.

Sill, Jr., W.H., W.C. Burger, M.A. Stahman, and J.C. Walker. 1952. Electron microscopy of cucumber virus 1. Phytopathology 42:420-422.

Simpson, W.R. 1966. Head smut of corn in Idaho. Plant Dis. Rep. 50:215-217.

----, and H.S. Fenwick. 1968. Chemical control of corn head smut. Plant Dis. Rep. 52:726-727.

----, and H.S. Fenwick. 1971. Suppression of corn head smut with carboxin seed treatments. Plant Dis. Rep. 55:501-503.

Slagg, C.M. 1944. Diseases of corn in Nebraska and Kansas. Plant Dis. Rep. 28:1041-1042.

Sleumer, H.O. 1932. On the sexuality and cytology of *Ustilago zeae*. Z. Bot. 25:209-263.

Slykhius, J.T. 1955. *Aceria tulipea* Kiefer (Acarina: Eriophyidae) in relation to the spread of wheat streak mosaic. Phytopathology 45:116-128.

Smedegard-Petersen, V., and R.R. Nelson. 1969. The production of host-specific pathotoxins by *Cochliobolus heterostrophus*. Can. J. Bot. 47:951-957.

Smith, D.R. and A.L. Hooker. 1973. Monogenic chlorotic-lesion resistance in corn to *Helminthosporium maydis*. Crop Sci. 13:330-331.

----, ----, and S.M. Lim. 1970. Physiologic races of *Helminthosporium maydis*. Plant Dis. Rep. 54:819-822.

Smith, E.F. 1905. Bacteria in relation to plant disease. Carnegie Inst. Wash. Publ. 27. Vol. 5. p. 1-2. Washington, D.C.

Smith, F.L. 1936. The effect of corn smut on the yield of grain in the San Joaquin Valley of California. J. Am. Soc. Agron. 28:257-265.

----, and C.B. Madsen. 1949. Susceptibility of inbred lines of corn to fusarium ear rot. Agron. J. 41:347-348.

Smith, G.M. 1933. Golden cross bantam sweet corn. USDA Circ. No. 268.

Smith, K.M. 1972. A textbook of plant virus diseases. Third Ed. Academic Press, New York.

Smith, M.A. 1926. Infection and spore germination studies with *Puccinia sorghi*. Phytopathology 16:69 (Abstr.).

Smith, N.A. 1966. A virus disease of Michigan field corn. Michigan State Ext. Circ. BPP66-4.

Snazelle, T.E., J.B. Bancroft, and A.J. Ullstrup. 1971. Purification and serology of maize dwarf mosaic and sugarcane viruses. Phytopathology 61:1059-1063.

Sparrow, F.K. 1934. The occurrence of true sporangia in the Physoderma disease of corn. Science (N.S.) 79:563-564.

Spencer, E.L., and G.L. McNew. 1938. The influence of mineral nutrition on the reaction of sweet-corn seedlings to *Phytomonas stewarti*. Phytopathology 28:213-223.

Sprague, G.F. 1954. Breeding for resistance to stalk rot. Proc. Ninth Annu. Hybrid Corn Indust.-Res. Conf. p. 38-43.

Sprague, Roderick. 1950. Diseases of cereals and grasses in North America. Ronald Press, New York.

Stahl, C.F. 1927. Corn stripe disease in Cuba not identical with sugarcane mosaic. Trop. Plant Res. Found. Bul. 7:12p.

Stakman, E.C., L.J. Tyler, and G.E. Hafstad. 1933. The constancy of cultural characters and pathogenicity in variant lines of *Ustilago zeae*. Bull. Torrey Bot. Club 60:565-572.

Standen, J.H. 1944. Chemical and physical characteristics of maize cobs in relation to the growth of *Nigrospora oryzae*. Phytopathology 34:315-323.

Stanley, A.R., and C.R. Orton. 1932. Bacterial stalk rot of sweet corn. Phytopathology 22:26.

Stevens, N.E. 1934. Stewart's disease in relation to winter temperature. Plant Dis. Rep. 18:141-149.

----. 1936. Was there an outbreak of bacterial wilt of corn in central Illinois in 1891 and 1892? Trans. Illinois Acad. Sci. 29:73-75.

----, and C.M. Haenseler. 1941. Incidence of bacterial wilt of sweet corn, 1935-1940: Forecast and performance. Plant Dis. Rep. 25:152-157.

----, and J.I. Wood. 1935. Losses from corn ear rots in the United States. Phytopathology 25:281-283.

Stewart, F.C. 1897. A bacterial wilt of sweet corn. N.Y. (Geneva) Agric. Exp. Stn. Bull. 130:423-439.

Stob, M.R., R.S. Baldwin, J. Tuite, F.N. Andrews, and K.G. Gillette. 1962. Isolation of an anabolic uterotrophic compound from corn infected with *Gibberella zeae*. Nature 196:1318.

Stoneberg, H.F. 1927. The productiveness of corn as influenced by the mosaic disease. USDA Tech. Bull. 10.

Stoner, W.N. 1952. Leaf fleck, an aphid-borne persistent virus disease of maize. Phytopathology 42:683-689.

----, R.D. Gustin, and M.L. McComb. 1967. A virus infecting maize in South Dakota. Plant Dis. Rep. 51:705-709.

Storey, H.H., and A.K. Howland. 1957. Resistance in maize to the tropical American rust fungus, *Puccinia polysora* Underw. I. Genes Rpp_1 and Rpp_2. Heredity 11:289-301.

----, and ----. 1961. The tropical rust disease of maize caused by *Puccinia polysora* Underw. East African Agric. Forest Res. Org. Annu. Rep. p. 56.

----, and ----. 1967. Resistance in maize to a third race of *Puccinia polysora*. Ann. Appl. Biol. 60:297-303.

----, ----, J.S. Hemingway, J.D. Jameson, B.S.T. Baldwin, H.C. Thorp, and G.E. Dixon. 1958. East African work on breeding maize resistant to the tropical American rust, *Puccinia polysora*. Empire J. Expt. Agric. 26:1-17.

Stout, G.L. 1930. New fungi found on Indian corn. Mycologia 22:271-287.

Stringfield, G.H., and D.H. Bowman. 1942. Breeding corn hybrids for smut resistance. J. Am. Soc. Agron. 34:486-494.

Sutton, J.C., A. Bootsma, and T.J. Gillespie. 1972. Influence of some cultural practices on yellow leaf blight of maize. Can. Plant Dis. Surv. 52:89-92.

Tamimi, Y.N., and J.E. Hunter. 1970. Effect of P, $CaCO_3$ and $CaSiO_3$ fertilization upon incidence of corn smut. Agron. J. 62:496-498.

Tarte, R. 1971. The relationship between preplant populations of *Pratylenchus zeae* and growth and yield of corn. J. Nematol. 3:330-331.

Tatum, L.A. 1954. Seed permeability and "cold-test reaction" in *Zea mays*. Agron. J. 46:8-10.

----, 1971. The southern corn leaf blight epidemic. Science 171:1113-1116.

----, and M.S. Zuber. 1943. Germination of maize under adverse conditions. J. Am. Soc. Agron. 35:48-59.

Taubenhaus, J.J. 1920. A study of the black and yellow molds of ear corn. Texas Agric. Exp. Stn. Bull. 270.

Taylor, D.P., and E.G. Schleden. 1959. Nematodes associated with Minnesota crops. II. Nematodes associated with corn, barley, oats, rye and wheat. Plant Dis. Rep. 43:329-333.

Tehon, L.R., and G.H. Boewe. 1939. Charcoal rot in Illinois. Plant Dis. Rep. 23:312-325.

----, and E.Y. Daniels. 1925. Notes on the parasitic fungi of Illinois. II. Mycologia 17:240-249.

----, and ----. 1927. Notes on the parasitic fungi of Illinois III. Mycologia 19:110-129.

Thompson, D.L. 1965. Control of bacterial stalk rot of corn by chlorination of water in sprinkler irrigation. Crop Sci. 5:369-370.

----. 1969. Quantitative genetic estimates for brown spot resistance in corn. Crop Sci. 9:246-247.

----, J.O. Rawlings, and R.H. Moll. 1963. Inheritance and breeding information pertaining to brown spot resistance in corn. Crop Sci. 3:511-514.

Thornberry, H.H. 1966. Plant pests of importance to North American agriculture. Index of plant virus diseases. USDA Agric. Handb. No. 307. 446 p.

Tiwari, M.M., and H.C. Arya. 1969. *Sclerospora graminicola* — axenic culture. Science 163:291-293.

Tuite, J., G. Shanor, G. Rambo, J. Foster, and R.W. Caldwell. 1974. The gibberella ear rot epidemics of corn in Indiana in 1965 and 1972. Cereal Sci. 19.

Tulley, J.G., R.F. Whitcomb, J.M. Bove, and P. Saglio. 1973. Plant Mycoplasmas: Serological relation between agents associated with citrus stubborn and corn stunt diseases. Science 182:827-829.

Tunac, J.B., and C.M. Nagel. 1969. Reaction of dent corn inbreds and hybrids to *Aceria tulipae* and wheat mosaic virus. Plant Dis. Rep. 53:662-664.

Ullstrup, A.J. 1936. The occurrence of *Gibberella fujikuroi* var. *subglutinans* in the United States. Phytopathology 26:685-693.

----. 1944. Further studies on a species of Helminthosporium parasitizing corn. Phytopathology 34:214-222.

----. 1946. An undescribed ear rot of corn caused by *Physalospora zeae*. Phytopathology 36:201-212.

----. 1949. A method for producing artificial epidemics of diplodia ear rot. Phytopathology 39:93-101.

----. 1952. Observations on crazy top of corn. Phytopathology 42:675-680.

----. 1955. Crazy top of some wild grasses and the occurrence of the sporangial stage of the pathogen. Plant Dis. Rep. 39:839-840.

----. 1955. Diseases of corn. *In* G.F. Sprague (ed.) Corn and corn improvement. Academic Press, New York.

----. 1960. Bacterial stripe of corn. Phytopathology 50:906-910.

----. 1963. Sources of resistance to northern corn leaf blight. Plant Dis. Rep. 47:107-108.

----. 1964. Observations of two epiphytotics of diplodia ear rot of corn in Indiana. Plant Dis. Rep. 48:415-418.

----. 1965. Inheritance and linkage of a gene determining resistance in maize to an American race of *Puccinia polysora*. Phytopathology 55:425-428.

----. 1966. A widespread epiphytotic of gibberella ear rot in the USA. Int. Symp. Plant Pathol. New Delhi, India. p. 46.

----. 1969. "Buggy-whip" of corn and its relation to the incidence of common smut. Plant Dis. Rep. 53:250-252.

----. 1970. Methods for inoculating corn ears with *Gibberella zeae* and *Diplodia maydis*. Plant Dis. Rep. 54:658-662.

----. 1970. Crazy top of maize. Indian Phytopathol. 23:250-261.

----. 1971. Hyper-susceptibility of high-lysine corn to kernel and ear rots. Plant Dis. Rep. 55:106.

----. 1972. The impacts of the southern corn leaf blight epidemics of 1970-1971. Ann. Rev. Phytopathol. 10:37-50.

----, and M.P. Britton. 1968. An unusual epiphytotic of common corn smut in Indiana and Illinois. Plant Dis. Rept. 52:922-923.

----, and S.R. Miles. 1957. The effects of some leaf blights of corn on grain yield. Phytopathology 47:331-336.
----, and M.H. Sun. 1969. The prevalence of crazy top of corn in 1968. Plant Dis. Rep. 53:246-250.
----, E.B. Smalley, G.L. Worf, and R.W. Ahrens. 1969. Eyespot: a serious new disease of corn in the United States. Phytopathology 59:105 (Abstr.).
----, and A.F. Troyer. 1967. A lethal leaf spot of maize. Phytopathology 57:1282-1283.
USDA 1957a. Witchweed (*Striga asiatica*) a new parasitic plant in the United States. Plant Disease and Identification Sec. Special Publ. No. 10.
----. 1957b. Witchweed. ARS #22-41.
Valleau, W.D. 1920. Seed corn infection with *Fusarium moniliforme* and its relation to root and stalk rots. Kentucky Agric. Exp. Stn. Bull. 226.
----, P.E. Karraker, and E.H. Johnson. 1926. Corn root rot, a soil-borne disease. J. Agric. Res. 33:453-476.
Vasal, S.K., C.L. Moore, and Udom Pupipat. 1970. Inheritance of resistance to curvularia leaf spot in maize. SABRAO Newsl. Mishima 2:81-89. (Plant Breed. Abstr. 41:7522, 1971).
Villareal, R.L., and R.M. Lantican. 1964. The effect of "T" cytoplasm on yield and other agronomic characters of corn. Philippine Agric. 48:144-147.
----, and ----. 1965. Cytoplasmic inheritance of susceptibility to helminthosporium leaf spot in corn. Philippine Agric. 49:294-300.
Voorhees, R.K. 1933a. Effect of certain environmental factors on the germination of sporangia of *Physoderma zeae-maydis*. J. Agric. Res. 47:609-615.
----. 1933b. *Gibberella moniliforme* on corn. Phytopathology 23:368-378.
----. 1934. Sclerotial rot of corn caused by *Rhizoctonia zeae* n. sp. Phytopathology 24:1290-1303.
Waggoner, P.E., and J.R. Wallin. 1952. Variation in pathogenicity among isolates of *Phytophthora infestans* on tomato and potato. Phytopathology 42:645-648.
Wallin, J.R. 1951. An epiphytotic of corn rust in the north central region of the United States. Plant Dis. Rep. 35:207-211.
Walter, J.M. 1934. The mode of entrance of *Ustilago zeae* into corn. Phytopathology 24:1012-1020.
Warren, H.L., D. Scott, and R.L. Nicholson. 1974. Occurrence of sorghum downy mildew on maize in Indiana. Plant Dis. Rep. 58: 430-432.
----, R.L. Nicholson, A.J. Ullstrup, and E.G. Sharvelle. 1973. Observations of *Colletotrichum graminicola* on sweet corn in Indiana. Plant Dis. Rep. 57:143-144.
Weber, G.F. 1922. Studies on corn rust. Phytopathology 12:89-97.
Weihing, J.L., and A.K. Vidaver. 1967. Report of holcus leaf spot (*Pseudomonas syringae*) epidemic on corn. Plant Dis. Rep. 51:396-397.
Wellman, F.L. 1934. Infection of *Zea mays* and various other *Gramineae* by the celery virus in Florida. Phytopathology 24:1035-1037.
Wellhausen, E.J. 1937a. Effect of the genetic constitution of the host on the virulence of *Phytomonas stewarti*. Phytopathology 27:1070-1089.
----. 1937b. Genetics of resistance to bacterial wilt in maize. Iowa Agric. Exp. Stn. Res. Bull. 244:69-114.
Weston, Jr. W.H. 1929. The occurrence of *Sclerospora graminicola* on maize in Wisconsin. Phytopathology 19:391-397.
----, and H.N. Uppal. 1932. The basis for *Sclerospora sorghi* as a new species. Phytopathology 22:573-586.
Whitehead, M.D. 1958. Pathology and pathological histology of downy mildew, *Sclerophthora macrospora*, on six graminicolous hosts. Phytopathology 48:485-493.
Williams, L.E., and L.J. Alexander. 1964. An unidentified virus isolated from corn in southern Ohio. Phytopathology 54:912 (Abstr.).
----, and D.T. Gordon. 1967. Preliminary studies on the identity of a mosaic virus from corn in Ohio. Phytopathology 57:836 (Abstr.).
----, ----, L.R. Nault, L.J. Alexander, O.E. Bradfute, and W.R. Findley. 1967. A virus of corn and

small grains in Ohio and its relation to wheat streak mosaic virus. Plant Dis. Rep. 51:207-211.

----, W.R. Findley, E.J. Dollinger, and R.M. Ritter. 1968. Seed transmission studies of maize dwarf mosaic in corn. Plant Dis. Rep. 52:863-864.

----, and G.M. Willis. 1963. Disease of corn caused by *Colletotrichum graminicolum.* Phytopathology 53:364-265.

Wiser, W.J., H.H. Kramer, and A.J. Ullstrup. 1960. Evaluating inbred lines of corn for resistance to diplodia ear rot. Agron. J. 52:624-626.

Worsham, A.D., D.E. Moreland, and G.C. Klingman. 1964. Characterization of the *Striga asiatica* (witchweed) germination stimulant from *Zea mays* L.J. Exp. Bot. 15:556-567.

Wysong, D.S., A.K. Vidaver, H. Stevens, and D. Stenberg. 1973. Occurrence and spread of an undescribed species of *Corynebacterium* pathogenic on corn in the western corn belt. Plant Dis. Rep. 57:291-294.

Yakovleva, N.P. 1963. Tissue and age specialization of the causal agent of blister smut. J. Agric. Sci. Moscow 2:45-51. (Rev. Appl. Mycol. 43:3203, 1964).

Young, G.Y., C.L. Lefebvre, and A.G. Johnson. 1947. *Helminthosporium rostratum* on corn, sorghum and pearl millet. Phytopathology 37:180-183.

Young, P.A. 1964. Control of corn nematodes with Volex and D-D. Plant Dis. Reptr. 48:122-123.

Younts, S.E., and R.P. Musgrave. 1958. Chemical composition, nutrient absorption, and stalk rot incidence of corn as affected by chloride in potassium fertilizer. Agron. J. 50:426-429.

Zuber, M.S. 1962. A mechanical method for evaluating stalk lodging. Proc. 17th Annu. Hybrid Corn Indust.-Res Conf. p. 15-23.

----, and C.O. Grogan. 1958. A new technique for measuring stalk rot incidence of corn as affected by chloride in potassium fertilizer. Agron. J. 59:426-429.

Zwillenberg, H.H.L. 1959. *Colletotrichum graminicola* on maize and various other plants. Phytopathol. Z. 34:417-425.

Chapter 9 The Most Important Corn Insects

F. F. DICKE
Pioneer Hi-Bred International, Inc.
Johnston, Iowa

During the past two decades there have been substantial changes in crop production practices. Such changes as the decline in oat acreage, the ascendancy of soybean acreage, the upsurge in the use of commercial fertilizer, improved corn hybrids, and effective pesticides have had an impact on insect populations. Collective improvements and generally favorable climatic conditions have made it possible to about double the average per-acre yield of corn in the USA in this period.

Certain insects such as the chinch bug and grasshopper (locusts), which in the past have been serious pests, in recent years have been at relatively low populations in corn-producing areas. Environmental conditions characterized by sustained below-normal rainfall and above-normal temperatures, which are regarded favorable for epizootics of these pests, have not prevailed since the drought years of 1930-1936. Furthermore, effective insecticides are available to cope with an insect problem should it reoccur. Indeed, some investigators believe these problems are a thing of the past.

Nevertheless, the corn crop is subject to a complex of insect attack from the time it is planted until it is utilized as food or feed. Other crops, particularly small grains, forage grasses, and legumes are sources of insects that attack corn and are also sources of the prey that help keep the population complex in balance. This ecological relationship has to be considered a part of the corn insect problem.

It is customary to classify the injurious species according to their habitat or to the plant structures that they commonly infest. The stage of plant development is frequently an important factor in determining what structures are attacked and in the food habits and damage potential of the insects.

In general this chapter has been arranged in sections based on a seasonal host-insect biological relationship as follows: (1) research trends, (2) soil insects, (3) insects attacking the leaf, stalk, and ear, (4) lesser recognized groups, (5) insects in relationship to corn diseases, (6) stored grain insects, (7) insect resistance in corn, and (8) chemical control status. In addition, a list of common and scientific names of insects mentioned in this chapter are given in an appendix.

Specific identification of insects is essential where infestation is encountered because of biological differences between species. When questions of identity arise it is desirable to collect specimens and preserve them in a 70% alcohol solution for future reference. An attempt has been made to briefly describe the injurious forms and the typical injury to the plant structures involved. Literature citations have been more or less limited to publications that include supporting references.

Many of the early records on distribution, abundance, and biology of corn insects were assembled by USDA entomologists and state entomologists of several states, especially New York, Illinois, and Missouri. The most complete special reports are those of Forbes (1893, 1904) in Illinois.

A large number of insect species have been recorded. In discussing the sources of American corn insects, Neiswander (1926) listed fewer than 400 species. Many of these species do not have a proved host relationship. With the expansion of corn culture throughout the world the number of economic species would be considerably increased and heritable protective traits are increasingly important. Especially in tropical areas, insect populations directly and indirectly exert severe stress on corn culture. An effort has been made, therefore, to discuss protective traits, particularly those that are highly developed on the ear of corn originating in tropical areas of the Western Hemisphere.

RESEARCH TRENDS

Records of the identity, prevalence, and biology of corn insects in America prior to the establishment of official entomology in 1854 are scattered and fragmentary. The complex before the days of systematic records, however, was similar to that of today. The major groups infesting different parts of the plant and the grain were well represented. Some species have extended their range, whereas new species have been introduced from foreign sources.

The epizootics of many of the important species show a periodicity pattern, although not necessarily in regular cycles. Populations may remain relatively stable for several years and then suddenly increase. The limiting factors in such periodicity are environmental. A high rate of survival in some insects is favored by dry and warm conditions during critical periods of development and activity; survival of other species is favored by cool, moist conditions. The variable pattern of the environment, therefore, affects the general corn-insect complex from year to year.

Much of the research to control corn insects has centered around applied biological and ecological methods. A good deal of emphasis was placed on natural enemies. Efforts were mainly directed toward methods of limiting populations through adjustments in plowing or seedbed preparation and planting time, crop rotation, and sanitation. The effect of compatible cultural practices to reduce insect abundance in corn is perhaps of greater significance than is commonly appreciated. Through usage many such practices have become established procedure. Others, such as indiscriminate burning to destroy overwintering forms, have been discarded with the adoption of better agronomic and soil conservation methods.

During the period 1930-1950 there was practically a complete turnover from open-pollinated to hybrid varieties of corn. With this conversion the development of new inbred lines and hybrid combinations has become a continuing process. Extensive breeding, with improved techniques and exposure to varying environmental conditions, have resulted in new hybrids with improvements in yields, standing ability, and resistance to insects and diseases.

In recent years there has been a significant adjustment in research activities on the corn insect complex. A group of factors have contributed to stimulating an increase in host and insect relationship studies, namely: (1) the upsurge in the use of commercial fertilizers, particularly nitrogen, which resulted in increased monoculture of corn and corn rootworm problems; (2) development of strains of rootworms resistant to chlorinated hydrocarbon insecticides, which intensified the search for new insecticides and resistant corn hybrids; (3) dispersal of the western corn rootworm, southwestern corn borer, and western bean cutworm to new areas; (4) the widespread increase in virus and spiroplasma diseases in corn and the vector-host relationships; (5) exploration research on male sterility through radiation and genetic procedures; (6) insect pathology and pheromone research; and (7) a significant increase in a regional and interdisciplinary approach in research. The overall outlook appears to be promising for improvements in the practical control of corn insects under a wide range of environmental conditions.

SOIL INSECTS

Soil insects are those that inhabit the soil while they are in some way causing plant injury. The soil forms may be injurious to the roots or other subterranean parts of the plant, or, as with some cutworms and beetles, they may sever or feed on plant parts above the ground. Because they are hidden in the soil, many of these species ordinarily escape attention when in the immature form but may be well known in the adult stage. Plant injury symptoms aboveground are commonly the first indications of an infestation. With a knowledge of their adaptations and habits one can usually detect where certain species are apt to be most abundant. The economic species of soil insects have been treated more in groups than those feeding on the aerial parts of the plant, perhaps because of the large number of species involved. Thus, cutworms, wireworms, and white grubs have been grouped, although there may be deviations in biology among the species within a group.

Northern and Western Corn Rootworms

Description, Injury, and Distribution

The larvae of the northern corn rootworm *Diabrotica longicornis* are slender, white to pale yellow, with a yellowish-brown head, and about 10 mm long when full grown. The young larvae may be found feeding on root hairs and on or in young lateral roots, causing them to turn reddish or brown. The infestation by the larvae follows a more or less definite pattern related to the development of the crown roots. Apple and Patel (1963) presented data that showed the larvae feeding on progressively higher rings of crown roots as they emerged at the nodes and as the season advanced. The feeding habits of immature larvae on roots of seedlings are shown in Fig. 1. Later stages may be found on the young crown root buds near the leaf sheath attachment. The pupae, naked and white, are located in earthen cells in the soil near the plants. The adults of northern rootworm are about 5 mm long and pale green. During the latter part of the summer the beetles are injurious to the silks and tips of ears of corn and may also be found on the flowers of other plants. When abundant on newly emerged silk they may interfere with full pollination. The northern species has been predominantly a pest in the Central States, but also occurs in the eastern and southern states (Forbes, 1893; Webster, 1913). Chiang (1973) reviewed the literature on the bionomics of the northern and western species.

Habits and Biology

The northern corn rootworm has a single generation annually. The adults emerge in the summer and are prevalent in corn fields until fall. They feed gregariously on the silks and also on pollen of corn and other plants. There is a tendency for the beetle population to shift to the later silking fields in which oviposition is likely to be more concentrated. The eggs are deposited in the soil mostly in corn fields in the latter part of the summer. This habit is an important factor in the fate of the progeny the following year. When corn follows corn, a favorable environment is provided for the development of the next generation. After hatching late in the spring and early summer, the young larvae establish themselves on or in young roots of corn and other minor grass hosts, or perish. Their feeding on or in the lateral roots results in poor root growth or "root pruning" as shown in Fig. 2, with varying degrees of regeneration of fibrous secondary roots. The injured roots are commonly infected with root rots, a condition that may be implicated in varietal responses (Bigger, 1932; Forbes, 1893; Palmer and Kommedahl, 1969).

Under heavy infestation, plants are poorly anchored and are very susceptible to root lodging. Because of interference with machine harvesting and field loss, lodging may be as important in the rootworm problem as direct loss of yield. According to Tate and Bare (1946), in Nebraska, continuous crops of corn and moist soil conditions favor the development and survival of larvae.

Diabrotica virgifera, the western corn rootworm, is closely related to the northern corn rootworm. In early publications it was usually referred to as the Colorado corn rootworm. Periodically it was a pest of corn in Colorado, western Nebraska and Kansas (Gillette, 1912; Tate and Bare, 1946). This species began to dominate the rootworm population with its rapid spread eastward in the early 1960's. The distribution now (1976) extends to parts of Indiana, Michigan, and Ohio. It is generally accepted that the high populations during this period were associated with the development of populations of both the western and northern species resistant to the chlorinated hydrocarbon insecticides, principally aldrin, and increased monoculture of corn. The latter was made possible by rapid increase in the use of commercial fertilizers, particularly nitrogen.

The larva of the western corn rootworm is similar in appearance and feeding habits to the northern species (see Fig. 1). Specific characters for identification of mature larvae of three species of *Diabrotica* have been described by Mendoza and Peters (1964). The adult female is greenish and is

THE MOST IMPORTANT CORN INSECTS

identified by horizontal stripes on the wing covers; the males are dark and usually are lacking in the wing stripes. The life history is practically the same as that of the northern species and the same control practices are advocated. The beetles feed vigorously on the leaves of "whorl stage" corn and likewise attack the florets of emerging tassels. The leaves of some genotypes are more susceptible than others.

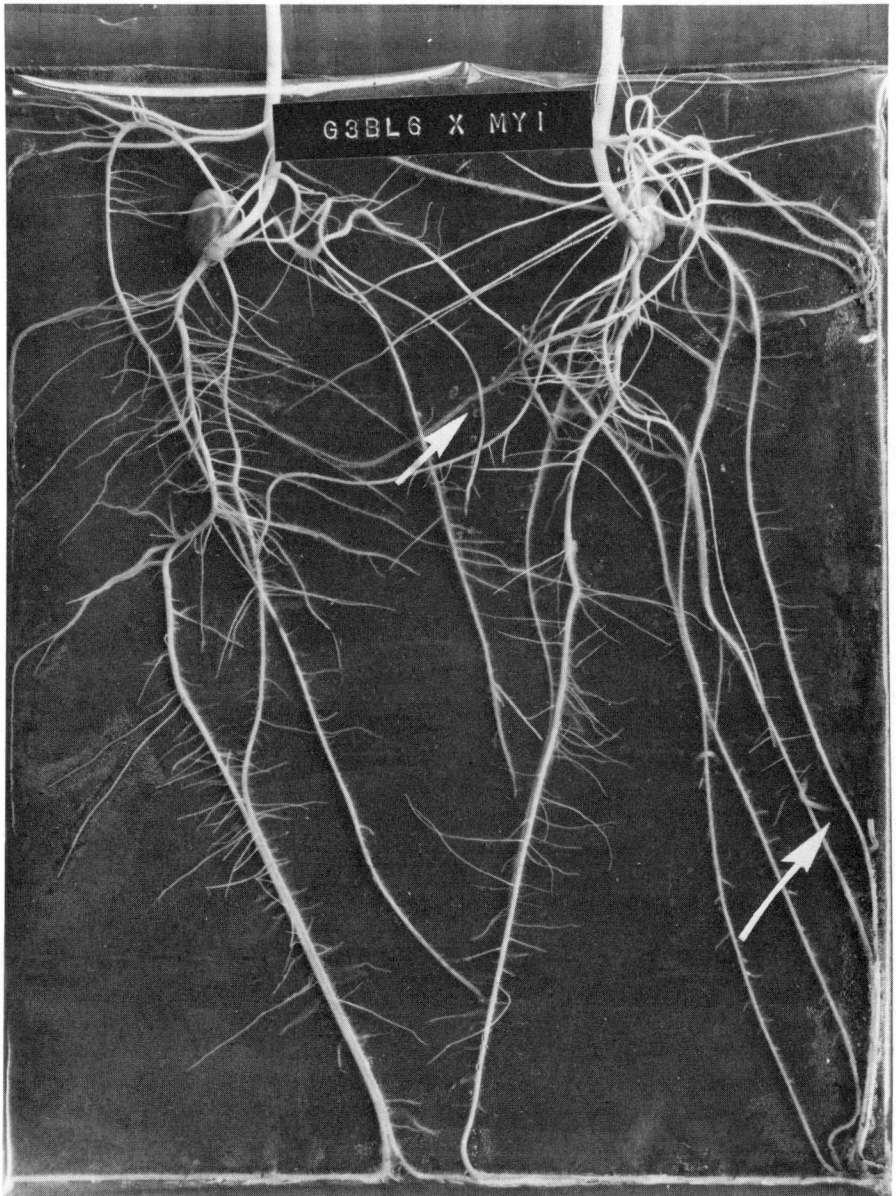

Fig. 1 — Immature larvae of Diabrotica feeding on lateral roots of corn growing in plastic bags.

Cultural Practices

The northern and western corn rootworm problem is associated with cropping practices. For many years rotation of crops, with not more than two successive crops of corn, has been advocated in time of threatening rootworm populations (Forbes, 1893). Bigger (1932) showed that a 4-year rotation gave almost complete relief, whereas a short rotation using an intervening crop of

Fig. 2 — Root pruning by northern and western corn rootworms, and regrowth trait.

oats and sweet clover did not adequately reduce rootworm injury, an indication that substantial egg deposition took place in oat stubble ground in late summer. The work of Branson and Ortman (1970) would support this conclusion. They have made extensive studies of the host range of larvae of both the northern and western species. Corn was found to be the favored host for completing the life cycle. Limited populations were recovered from green and yellow foxtail, foxtail millet, wheat, barley, spelt, intermediate wheatgrass, and pubescent wheatgrass.

Tate and Bare (1946), working in Nebraska, found less lodging after fall plowing, listed planting, and timely irrigation than after other cropping practices. Excessive irrigation at tasseling time and thereafter increased lodging. In further tests in Nebraska, Hill et al. (1948) reported increased yields and reduction in lodging under irrigation when corn was grown in rotation with sweet clover, when it was fertilized with nitrogen, or when manure was applied. Nitrogen did not decrease rootworm populations but stimulated root recovery.

The practice of short rotation, particularly with an intervening crop of soybeans, has become a common practice in the Corn Belt. Where monoculture of corn is practiced, several effective insecticides are available to insure against appreciable yield losses. Chemical control practices will be discussed in a later section.

Southern Corn Rootworm

Description, Injury, and Distribution

The southern corn rootworm, also known as the spotted cucumber beetle, injures corn in both the larval and the adult stages. The young larvae are slender, white to yellowish, becoming greenish-yellow as they mature. The full grown larva is about 12 mm long and has a brownish head shield and brown dorsal shield on the ninth abdominal segment. The larvae and pupae resemble those of the northern and western species and are easily confused with them. The adult is about 6 mm long, green, with 11 or 12 black spots on the wings. Symptoms of larval injury are holes through the base of small plants, tillering following growing point injury and root injury similar to that described for the other species (Forbes, 1893; Webster, 1913).

The southern corn rootworm has many hosts among the grasses, cucurbits, and legumes. It is widely distributed east of the Rocky Mountains, including southern Canada, but is most serious as a pest in the southeastern states and in the lower and middle Mississippi Valley. In the middle Corn Belt the beetles appear on alfalfa (*Medicago sativa* L.) or young corn in the latter part of May. There is as yet no evidence that any form overwinters in this area. It is presumed that the populations of beetles are migrants from more southern areas.

Habits and Biology

In warm climates the southern corn rootworm overwinters as an adult under debris, but may be active or feed on green plants on warm days in the winter. It is not known to have an inherent diapause system in any of its stages of development. In the spring the beetles fly to legumes, cucurbits, or grasses, where they feed and deposit their eggs in the soil. In southern areas, according to Isely (1929) and Arant (1929), the main source of the injury to young corn is from early-stage larvae that were present in the soil prior to seedbed preparation and corn planting. However, oviposition has been observed after corn emergence. In areas where this species is a pest, cool and wet springs are favorable for building up populations in winter legumes and on lowlands where favorable hosts are prevalent.

The first brood of larvae feed in or on the young roots and also burrow through the plants near the base. This feeding stunts or kills the growing point of plants and frequently induces tillering. Luginbill (1918) reported the first brood of larvae as the most important on corn in South Carolina. However, there appears to be an overlapping of generations, and Isely (1929) concluded that in Arkansas the number of generations was indeterminate. In the northern areas there are thought to be one or two generations. Forbes (1893) reported root pruning and subsequent root lodging after the bud-feeding stage and presence of larvae from June to August. On older corn, larval feeding was found in the base of the stalk, on the young brace roots, and on the lower leaf sheaths.

The injury in the northern states is easily confounded with that of the northern and western species. In southern areas the injury may be mistaken for that of *Diatraea* spp., the sugarcane beetle, or wireworms. The adults commonly feed on the leaves of corn from the seedling to later whorl stages of growth and on the silks. The feeding on the whorl leaves may be confused with that of some of the lepidopterous budworms.

Cultural Practices

Various cultural practices have been advocated for reducing rootworm injury, such as crop rotation, thick planting, fertilization, green manure crops, clean tillage, and adjustment of planting dates. Good soil and fertilization practices minimize the rootworm injury by stimulating root recovery and development. Isely (1929) recommended the elimination of wild grasses, on which the larval infestation begins, about a month before the planting of corn. Similarly, Arant (1934) and Eden and Arant (1953) suggested turning under winter legumes on or before 15 April and planting early in May in areas with a climate similar to that in Alabama. Cultural practices are not known to have an effect on populations in Corn Belt areas where the infestations originate from a migrant population.

Cutworms — Many Species

Description, Injury, and Distribution

Cutworms are the larvae of noctuid moths. The typical cutworm found attacking corn has a plump, curled-up appearance. The color of the larvae varies with the species from a light-glassy to a grayish-black or brown. The larvae feed at night and their presence in the soil is indicated by plants cut off at or below the surface of the ground. The moths are usually gray to brown.

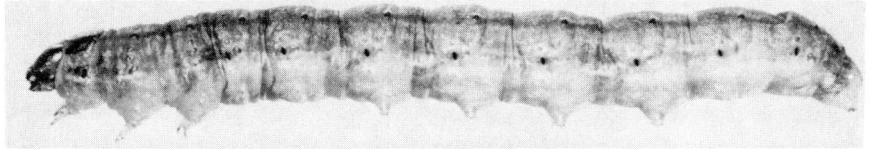

Fig. 3 — Larva of the black cutworm.

They fly at night and congregate at lights, where they may be readily trapped to determine their abundance in a locality. Among the most important species on corn in the USA are the black cutworm, *Agrotis ipsilon,* the glassy cutworm, *Crymodes devastator,* the dingy cutworm, *Feltia subgothica,* and the claybacked cutworm, *Agrotis gladiaria*. These species are widely distributed in the USA and Canada, but vary in abundance in different areas. The black cutworm, shown in Fig. 3, is worldwide in distribution and is of general importance on other crops (Forbes 1904; Stanley, 1936; Walkden, 1950).

Habits and Biology

The incidence of cutworm larvae in corn fields is usually dependent on the populations present in the previous season and the practices followed for seedbed preparation. The moths deposit their eggs in hay or sod land early in the fall. The larvae generally develop on grass or clover and pass the winter about half-grown. Typical areas for cutworm infestation in corn are in grassy river or creek bottoms to which moths are attracted for oviposition.

The black cutworm, however, overwinters in the pupal stage in southern areas. It has two or more generations, depending on the bioclimate, whereas the other three species have one generation. Immature larvae are usually present in the field when the crop is planted. The larvae occupy burrows in the soil, feeding on plant parts that they have severed and pulled in from the surface. The girdling of plants is therefore a means to an end, and results in a loss of plants out of proportion to the food requirements of the larvae. Advanced larval stage may burrow into plants near the soil surface, causing plants to produce tillers or die. Loss of stand may be substantial even with a low population (Crumb, 1929; Forbes, 1904).

Cultural Practices

Late summer or early fall plowing to eliminate the young larvae has been recommended for many years. This practice, however, is not desirable in hilly areas, where grass sod serves to prevent erosion of the soil through the winter. Under such conditions spring plowing is the logical practice. Most farmers replant when stands have been depleted by cutworm infestation.

Wireworms — Several Species

Description, Injury, and Distribution

Wireworms are the larvae of the common click beetles. They are slender, shiny, rather sluggish, buff to reddish-brown, heavily chitinized, and usually from 12 to 40 mm long. Figure 4 shows a typical wireworm specimen.

Fig. 4 — Wireworm larva.

Typical symptoms of their presence in corn are poor stands, dying seedlings, and tillering of young plants. In addition to corn, wireworms are injurious to small grains, forage grasses, and vegetable crops. The widely distributed species on corn are *Agriotes mancus,* the wheat wireworm; *Melanotus cribulosus; Aeolus mellillus; Melanotus communis;* and *Horistonotus uhlerii,* the sand wireworm (Forbes, 1893; Glen et al., 1943; Tenhet and Howe, 1939; Thomas, 1940).

Habits and Biology

The wireworm problem in corn is inherited from the previous crop of pasture or forage grasses and, in some cases, small grain. A generation is completed in 1 to 3 or more years, depending on the species and climatic factors. A typical generation develops as follows: The larvae pupate in earthen cells in middle summer. The adults emerge in late summer and early fall. In some species the pupae remain in the cells in the soil and the adults emerge in the spring. The eggs are deposited in grass or small-grain fields and hatch in about 2 weeks. The young larvae feed on the roots until fall, when they burrow deeper into the soil for hibernation. They resume feeding on roots or sod during the following summer, go again into hibernation and complete their feeding in the following spring. When corn is planted on infested ground, the immature larvae migrate to the germinating seed or young plants, burrow through the seed, and feed on the developing roots and crown. Loss of stand, therefore, is the initial effect, particularly if large populations are present.

Tenhet and Howe (1939) found a 1-year cycle of the sand wireworm in South Carolina, and Jewett (1942) reported a full and a partial second generation of *Aeolus mellilus* in Kentucky, with the most injury from the overwintered larvae in the spring. *A. mancus, M. communis,* and *M. cribulosus* are long-cycle species that sometimes injure corn (Hawkins, 1936;

Hyslop, 1915). In south Florida *M. communis* has a single generation annually. The adults emerge in the summer and deposit their eggs on grass areas. Fall-planted corn is subject to attack under these conditions. There appears to be general agreement that the wireworms attacking corn and wheat (*Triticum aestivum* L.) are favored by a moist, clay soil with liberal amounts of organic matter. However, *H. uhlerii,* an important species in the South, thrives in light, sandy soils low in organic matter. In the Pacific Northwest several species of wireworms attack corn and other crops (Lane, 1941).

Cultural Practices

Until recent years cultural practices were considered the only means of relief for wireworm infestations. These practices varied with the species involved. Hawkins (1936) in Maine reported that a short rotation and the use of crimson clover or buckwheat as a green manure crop reduced reinfestation. Clean cultivation and summer plowing or fallowing, regular crop rotation, and soil drainage have also been recommended, the objective being to eliminate susceptible vegetation and to disturb the insect activities as much as possible.

Billbugs

Description, Injury, and Distribution

The billbugs are snout beetles. Most of them are reddish or brown, but they may appear grayish when covered with mud. They vary from about 5 to 25 mm in length. The larvae are white with a brownish head, rather plump, and legless. Injury on young corn plants shows up as a series of transverse oblong or round holes in the leaves and some destruction of the growing point which results in tillering, stunting, or loss in stand.

As pests of corn, billbugs are widely distributed. The species most injurious to corn in the USA are the maize billbug (*Sphenophorus maidis*); timothy billbug, *S. zeae;* bluegrass billbug, *S. parvula;* and the southern corn billbug, *S. callosa* (Forbes, 1904; Kelly, 1911; Satterthwait, 1919).

Habits and Biology

Most of the billbugs normally inhabit moist or swampy soil areas, where the larvae develop on various species of grasses such as rushes, sedges, reeds, timothy, and bluegrass (Poa spp.). Infestations on corn, therefore, are likely to occur when this crop is planted in or near ground that had these conditions and plant associations in previous seasons. In some species only the adults are injurious, whereas in others, notably the maize and southern corn billbugs, the larvae also infest corn. Only a single generation occurs annually in the species commonly recorded on corn, the adults overwintering among the roots or in the roots or culms of their hosts.

In the spring the adults feed on bulbous parts of rhizomes of grasses. They begin to infest corn during the seedling and post-seedling stages of growth, eating small holes into the plants at about the soil surface, thereby piercing the rolled leaves in the bud. As the leaves unroll they show the typical transverse series of holes, which increase in size with the growth of the leaves. Severe injury in seedlings may cause distortion of the young leaves or even the death of the plants. In a study of the habits of *S. callosa* on corn in North Carolina, Metcalf (1917) found eggs deposited in plants near or below the soil surface and also among the roots. The eggs were laid in May or June. Larvae were found feeding in the plants and externally on the roots. This feeding resulted in the death of young plants and stunting when older plants were attacked. The injury caused by the larvae in the lower part of the stem was more serious than that caused by the adults. Similar observations were reported by Cartwright (1929) from South Carolina for *S. maydis*.

Cultural Practices

There appears to be general agreement that drainage of swampy areas and the elimination of the common wild host grasses associated with them will lessen the incidence of infestations on corn grown in or near such an environment. Rotation of crops, early planting, clean cultivation, and use of fertilizer are the normal practices that reduce injury and stimulate plant growth to overcome it (Cartwright, 1929; Metcalf, 1917; Satterthwait, 1919).

Webworms

Description, Injury, and Distribution

The adults of webworms are crambid or close-wing moths (*Crambus* spp.). The larvae are slender, yellow to pink with regular dark spots, and about 25 mm in length when full grown. When injury on young plants is first noticed, the larvae, then about 12 mm long, are located in silken web-lined burrows in the soil near the plants. Typical attack at or below the ground level shows gnawed places pits, or sometimes cut off plants, somewhat resembling cutworm injury. Above the soil surface the infested plants have distorted leaves with ragged eaten places in them. Sod webworms are occasionally injurious to corn in the Corn Belt following spring plowing of sod land. The most common species injurious to corn are the corn root webworm (*Crambus caliginosellus*), striped webworm (*C. mutabilis*), and larger sod webworm (*C. trisectus*) (Ainslie, 1922; Forbes, 1905).

Habits and Biology

According to Ainslie (1922), the corn root webworm has a single generation annually, whereas the striped webworm and larger sod webworm have at least two full generations. All three species pass the winter as partly grown larvae. The webworm moths drop their eggs in flight on bluegrass, timothy, and other grass sod. Thus, the development of larvae is closely tied up with grass sod and perhaps the legumes commonly found with it. Corn is a host only where it follows sod. The eggs from which the overwintering larvae develop are deposited in the summer. The young larvae establish themselves within their flimsy cocoons or webs in the sod, feed on foliage during late summer and early fall, and establish themselves in burrows in the soil before winter. They are about 12 mm long. With warm weather in the spring, feeding is resumed. It is at this time that there may be some migration to corn along edges of fields. The larvae complete their development in about 1 month. When corn is planted on spring-plowed sod land infested with webworms, the seedling stage is attacked, resulting in loss of stand (Ainslie, 1922-1927).

Cultural Practices

Cultural practices for controlling sod webworms have not been evaluated for many years. Late summer or early fall plowing has been recommended for eliminating the partially grown larvae before they enter hibernation. However, fall plowing would not be an acceptable practice on soils that are subject to erosion. Since most of the larvae become full grown while the corn is in the seedling stage, serious crop injury does not usually occur in corn that is replanted (Ainslie, 1922).

White Grubs

Description, Injury, and Distribution

White grubs are the larvae of brown beetles commonly known as May or June beetles. The larvae are sluggish, white, with a brown head and rather prominent legs. They have two slightly curved, narrowly spaced rows of spines on the ventral side of the last abdominal segment. Larvae feed on the roots of corn, causing stunted, wilted, or dying plants beginning with the seedling stage of growth.

There are many species of white grubs, which vary in abundance in different sections of the country. Chamberlin and Callenbach (1943) found the following species to be most numerous on cereal and forage crops in Wisconsin: *Phyllophaga rugosa, P. hirticula, P. fusca,* and *P. tristis.* According to Luginbill and Painter (1953), these species are widely distributed and may be involved in crop damage.

Habits and Biology

The primary source of white grubs in cornfields is soil that has been in grass and in some instances in soybeans. Bluegrass and timothy sods are particularly favorable for their development. The beetles ordinarily prefer to feed on the foliage of trees, but they also feed on weeds and crop plants.

Most of the important species of May beetles require 3 years to complete a generation. Of the four species listed only *P. tristis* has a 2-year cycle. The adults emerge from the soil in the spring and lay their eggs a few centimeters below the surface. The eggs hatch in about 3 weeks. The young larvae feed through the first season on living roots and decaying vegetable matter at about plow-sole level or in the top soil. Late in the fall they burrow deeper into the soil and remain inactive during the winter. The second-year larvae feed throughout the growing season on the roots of plants and are the most injurious stage on corn when it is planted on infested sod ground. In the fall the larvae again go deep into the soil to overwinter. The feeding period in the third year is relatively short. Pupation takes place in cells in midsummer. The adults emerge in the early fall but remain in the earthen cells over winter to issue from the soil in the spring. There is an overlapping of broods which results in beetle flights each year. For 3-year-cycle species these broods are designated as A, B, and C. Brood A is by far the most abundant and is the one present in outbreak years. Brood B is usually unimportant, but brood C may be injurious in certain areas.

When corn is planted in infested ground the grubs congregate near the base of the plants, where they feed on and sever the young roots. Likely places of infestation are on high ground near wooded areas where the adults concentrate on their feeding. It is therefore not uncommon to find the important infestations associated with the poorer soils in an area (Davis, 1922; Luginbill and Chamberlin, 1953).

Cultural Practices

In areas where white grubs are periodically a problem, Luginbill and Chamberlin (1953) have proposed certain cropping practices utilizing crops that are least susceptible to attack. They suggest avoiding the common grasses as much as possible and using alfalfa and clovers as the dominant pasture and hay crops.

An effective method of eliminating grubs from infested fields to be planted to corn is through pasturing by hogs. This practice is not desirable on permanent pastures because of the serious uprooting of the sod. Crows, blackbirds, skunks, and opossums are effective predators of white grubs.

Corn Root Aphid

Description, Injury, and Distribution

The corn root aphid (*Aphis maidiradicis*) is bluish-green and about the size of a pinhead. It sucks the sap from the roots of corn and cornfield weeds. Typical external evidence of injury are dwarfing and yellowing and reddening of plants before they are knee-high, and ant burrows around the plants.

This aphid is widely distributed in the USA but is most abundant in the Corn Belt (Davis, 1949; Forbes, 1915).

Habits and Biology

The corn root aphid is dependent on ants for survival, principally the cornfield ant. This aphid has both sexual egg laying and parthenogenetic viviparous forms. The sexual form deposits the eggs in the fall, whereupon the ants transfer the eggs to their nests and nurture them during the winter. When the eggs hatch in the spring, the ants transfer the young aphids to the roots of cornfield weeds and later, after corn is planted, to the roots of corn. The ants feed on the sweetish fluids excreted from the cornicles of the aphids. During the spring and summer the aphids reproduce viviparously, some of the offspring being winged forms. The winged aphids emerge from the soil and fly to other fields where with the attending ants new colonies are produced (Davis, 1949). Successive crops of corn, satisfactory spring host plants, and a favorable environment for the cornfield ant are important factors in building up populations.

Cultural Practices

The corn root aphid is of economic importance only on corn (Davis, 1949). Therefore, rotation of crops will hold down both aphid and ant populations. In the presence of both insects Bigger and Bauer (1939) found that plowing shortly before planting reduced aphid infestations. Early plowing and tillage to keep favored wild hosts under control before planting corn was the most desirable practice. In connection with the other cultural practices Davis advocated thorough cultivation and maintenance of a high fertility level to stimulate plant growth.

Seed-corn Maggot

Description, Injury, and Distribution

The adult of the seed-corn maggot (*Hylemya platura*) is a grayish fly similar to the housefly but with a more prominent thorax. The larvae are pearly white, almost conical, tapering anteriorly, and have black hook-like mandibles. When full grown it is about 7 mm long.

The larvae mine into germinating seed in the soil and subsequently attack young seedlings. The species is worldwide in distribution. In addition to corn, it is an important pest of many vegetable crops (Forbes, 1893; Hawley, 1922).

Habits and Biology

The habits and biology of the seed-corn maggot have been studied mostly on vegetable crops. The adults are attracted to decaying vegetable or animal matter on which the eggs are deposited. The maggots attack the sprouting seed and seedlings. In the warmer parts of the country adults are abroad throughout the year. Reid (1940) reported that diapause was not observed in North and South Carolina, where he trapped adults every month of the year. He found puparia in the fields and also observed oviposition in the winter months. Hawley (1922) in New York reported collecting adults on wheat stubble in the spring and assumed that the insect spends the winter in the puparium. It is generally agreed that the maggots become injurious on seeds and seedlings under cool, moist soil conditions.

Cultural Practices

As a pest of corn the infestations are most prevalent in early plantings on soils with liberal amounts of plowed-under decaying plant material. Under such conditions, Hawley (1922) suggested that shallow planting and seed of good viability, to obtain germination as quickly as possible is a means of avoiding stand failure.

Sugarcane Beetle

Description, Injury, and Distribution

The adult of the sugarcane beetle (*Euetheola rugiceps*), sometimes called the rough-headed cornstalk beetle, is black, about 12 mm long, and similar in appearance to the common dung beetles. The main injury on corn is in the seedling and postseedling stages of growth. The beetles feed on sprouting seed and gouge and gnaw through the leaf sheaths into the stem near the crown. Evidence of beetle injury are wilting or drying of the inner leaves of the bud and tillering of the plants that survive injury.

This beetle is an important pest in southern states, where outbreaks sometimes occur. In addition to corn, it is also a pest of rice and sugarcane (Baerg, 1942; Ingram et al., 1951; Phillips and Fox, 1924).

Habits and Biology

The larvae develop in soils with liberal amounts of decaying vegetable matter consisting mainly of grasses, commonly *Paspalum* spp., bermudagrass (*Cynodon* spp.), and johnsongrass. Phillips and Fox (1924) observed infestations in such plant associations under moist or poorly drained soil conditions.

The beetles deposit their eggs in the soil from spring to early summer, depending on temperatures. The incubation period requires about 2 weeks. Larval development covers from 6 to 8 weeks, and the pupal period about 2 weeks. The beetles emerge in late summer and early fall, feed on the wild grasses and overwinter in the soil. With a substratum attractive to the beetles for oviposition there is little movement from infested fields.

Cultural Practices

Phillips and Fox (1924) reported that systematic rotation, keeping land well cultivated, and good drainage created an unfavorable environment for this beetle. They also advocated early planting and fertilization to stimulate plant growth, but Baerg (1942) did not concur in early planting. Ingram et al. (1951), in connection with this pest on sugarcane, recommended elimination of the sod breeding areas and substitution of adapted legumes.

Grape Colaspis

Description, Injury, and Distribution

The grubs of grape colaspis (*Colaspis brunnea*) are white with brownish heads and thoracic shields and about 3 mm in length when full grown. Compared with the common white grubs, they are not as slender, have less prominent legs, and lack the two rows of spines on the ventral side of the last abdominal segment.

The adults are small, pale brown beetles, which have many hosts but commonly feed on clover, bean, grape, and strawberry foliage.

The species is widely distributed east of the Rocky Mountains. It has been reported as injurious to young corn mainly in the North Central States (Bigger, 1928; Forbes, 1905; Lindsay, 1943; Petty and Apple, 1966).

Habits and Biology

Grubs found on young corn originate from infestations in ground planted to clovers or soybeans or alfalfa. The beetles emerge in midsummer and deposit their eggs in the soil. Incubation requires 1 to 2 weeks. The young larvae commonly become established and feed on the roots of clovers during the latter part of the summer and early fall. The winter is passed as immature larvae. With the appearance of warm weather in the spring the larvae feed on roots, completing their development late in the spring.

Bigger (1928) reported the larvae to be most abundant on red clover, sweet clover, soybeans, and timothy, in the order named. The most severe injury on corn occurred when red clover ground was plowed late in the spring. Similarly, Lindsay (1943) in season-history studies in Iowa also reported damage to corn following red clover.

Seedcorn Beetles

Description, Injury, and Distribution

The injurious forms of the seedcorn beetles are the adults. The adult of *Agonoderus lecontei* is oblong, about 6 mm long, and dark with two brown stripes on its wing covers. The adult of *Clivina impressifrons,* the slender seedcorn beetle, is about 6mm long, shiny dark red with a constricted articulation between the first and second thoracic segments. In characteristic injury the germinating seed is attacked, and one observes holes into or hollowed out kernels with dead or stunted sprouts. Both species are widely distributed in the USA and Canada (Forbes, 1893; Phillips, 1909).

Habits and Biology

The two species hibernate as adults. They are attracted to light in large numbers, especially on warm evenings in the spring. The presence of beetles throughout the season indicates that there are one or more overlapping generations annually. Little is known of the habits of the immature stages. It is believed that the larvae are predaceous on other soil insects and may also inhabit sod. The adults of *Agonoderus lecontei* are sometimes pests in lawns and golf greens. Both species occur in wet grassy ground (Forbes, 1893; Hamilton, 1935; Phillips, 1909).

Cultural Practices

In soils inhabited by these beetles Phillips (1909) suggested delaying planting until there is assurance of quick germination. Recent experience has shown that deep planting should be avoided.

INSECTS ATTACKING THE LEAF, STALK, AND EAR

Most of the insects injurious on the leaf, stalk, and ear may at some time feed on any of the major structures of the corn plant. There is a degree of similarity among the species in that the parts attacked depend to a large extent on the stage of plant development and on the life history of the insect. Some species may develop on the foliage and partially on the tassel in one generation, but in a subsequent generation mostly on ear structures. This group, predominantly lepidoptera, includes some of the most important corn insects.

The Corn Earworm

Description, Injury, and Distribution

The newly-hatched larvae of corn earworm (*Heliothis zea*) are light gray with conspicuous small dark hairs. The full grown larvae range in color from red and brown to green with a striped appearance. They may be located among the whorl leaves or on the emerging tassel but are most frequently found feeding in the tip of the ears. Figure 5 shows a full grown larva in the tip of the ear and inserts of the moth and eggs. The moths vary in color from a light olive green to buff. The eggs, laid singly on the leaves, emerging tassel, and silks are off-white, dome-shaped, with ribs converging at the top. The insect is worldwide in distribution. Because it also damages cotton, tomatoes, legumes, and other crops, entomologists rate it as one of our most injurious insects.

Habits and Biology

For many years the corn earworm has received a great deal of attention by entomologists. Quaintance and Brues (1905) assembled the early information on the biology, distribution, and control of the insect as a pest of cotton. Many of the studies reported by them are useful in an understanding of the earworm problem on corn.

The corn earworm hibernates as a pupa in a burrow prepared in the soil by the larva. Moths emerge from the pupae from early spring to early summer, depending on the bioclimate (Barber, 1936a; Phillips and Barber, 1929). Early in the season the moths oviposit on the corn leaves and emerging tassel, but as silks appear, a high proportion of the eggs are deposited on fresh silks. Depending on the stage of plant development, the young larvae become established on the leaves in the whorl, on the florets of the tassel, or on the silks.

As the larvae hatch on the silks, or migrate there from other parts of the plant, feed, and grow they gradually infest the tip of the ear. When the husks are loose, feeding extends along the side of the ear. This type of feeding, illustrated in Fig. 6, is especially prevalent on field corn in the more northern areas when the corn is in the dough stage. If the ear has a long husk, extension, and silk channel, the larvae sometimes mature on the silk. (Varietal resistance is discussed in a later section.) When feeding is completed, the lar-

THE MOST IMPORTANT CORN INSECTS 519

va leaves the ear, usually from the tip end or through an exit hole made through the husk. In the summer, pupation takes place within a few days and moths emerge in 2 or 3 weeks, or they may have a period of aestivation. The hibernating pupae become established in late summer and early fall. In

Fig. 5 — Full-grown corn earworm on ear tip with inset of moth and eggs.

warm climates there may be as many as five or six generations annually. In the central Corn Belt area there may be one or two generations, or perhaps there may be some infestations that originate from migrant moths from more southern areas (Barber, 1936b; Blanchard, 1942; Ditman and Cary, 1931; Garman and Jewett, 1914).

The attraction of the moths to fresh silk for oviposition causes them to move from the less attractive older silks to the more attractive new silks. This

Fig. 6 — Typical grain injury caused by the earworm on loose-husked variety.

tends to limit larval development to that corn which has not reached the dough stage for much of the growing season. Callahan (1957) presented evidence of oviposition response of moths to various wavelengths of light. On late maturing corn there is a more prolonged oviposition, and more of the larvae mature on dough stage corn. Because the larvae are cannibalistic, usually only one larva completes its development in an ear. Under high levels of infestation there may be a succession of larvae infesting the ears, particularly in the late summer and early fall. Eggs laid after the first week in September in Virginia are considered to be of little significance in the development of overwintering pupae (Dicke, 1939; Phillips and Barber, 1929; Phillips and Barber, 1933).

As corn becomes unattractive, the moths move to other hosts for oviposition, particularly to cotton, tomatoes, and legumes. The importance of cotton or tomatoes in building up populations is not well-known. Isely (1935) found that the reproductive capacity was highest when the insect was reared on corn and lowest when reared on tomatoes. In records covering 25 years, he observed that this insect was most serious where the acreages of corn and cotton were about equal.

Hibernation and overwintering of the earworm has received a great deal of attention because it has a bearing on the general abundance of the insect, at least early in the growing season. The conditions that determine whether or not larvae will develop into aestivating or hibernating pupae, the extent of successful hibernation, and seasonal abundance have been investigated in different parts of the country. The maturity of the corn on which the larvae feed may determine whether or not the pupae will hibernate and may affect the rate of overwinter survival. Phillips and Barber (1929) found that larvae maturing on dough stage corn late in the summer developed into pupae that most successfully survived the winter. In hibernation studies conducted in southeast Georgia under cage conditions, Barber (1941) reported that an average of 51% of the individuals that entered the soil for pupation in late summer and early fall survived to the following summer.

Ditman (1938) studied the water and fat relations of prepupae and pupae that developed from larvae reared on silk and dough stage corn. He concluded that the reduced percentages of water, the increased percentage of fat, and the reduced saturation of fat are intensified by a diet of dough stage corn, and that these factors are associated with the ability of the earworm to withstand the temperature hazards of hibernating conditions. Some pupae from larvae reared on dough stage corn were able to survive and produce moths after an exposure to temperatures of -6 to -10 C for 10 days. In further studies Ditman et al. (1940) reported that diapause in the pupae may result when the larvae are exposed to low temperatures.

Experimental evidence shows that dry soil conditions favor survival of aestivating as well as hibernating pupae (Barber, 1941; and Barber and Dicke, 1939). From a study conducted near the District of Columbia it was concluded that a mean temperature of 0 C or less from December through February reduced the abundance of the earworm the following season (Dicke, 1939). Blanchard (1942) summarized the results of cooperative hibernation experiments conducted in 15 central and northeastern states during the period 1935-39. Survival of hibernating pupae was recorded in all of the states except Iowa. In Ohio and Indiana, larvae entering the soil in October and early November under cool soil temperatures showed a low rate of pupation. Eichmann (1940) reported a similar experience in Washington. The information at hand indicates that under conditions favorable for development and with a mild winter the earworm may hibernate farther north than has been supposed.

Natural Enemies

There is a considerable mortality of hibernating pupae throughout the period from fall to summer. In soils where earthworms are abundant mortalities may be high when the pupae become embedded in the droppings deposited in the burrows. Diseases, particularly fungus, take a small toll. There is a rather low rate of parasitism and predation on larvae infesting the ear. *Trichogramma minutum* frequently parasitizes a high proportion of the eggs, the amount of parasitism reaching a peak in the fall. Among the predators *Orius insidiosus* is responsible for exhausting a considerable proportion of the eggs deposited on silks (Phillips and Barber, 1929; Phillips and Barber, 1933).

Cultural Practices

One of the earliest recommendations for control of the corn earworm was plowing. It was recommended many years before definite evidence of its effectiveness for destroying hibernating pupae was available. Experiments in Virginia (Barber and Dicke, 1937) showed that fall plowing resulted in a high mortality of overwintering pupae. Spring plowing and fall disking were less effective.

The effect of planting date on the amount of earworm injury has been studied at several points. In Virginia tests, early planted corn produced the best yield and had a low rate of earworm damage. Similar observations in Georgia have confirmed these results (Barber, 1936a; Phillips and Barber, 1934).

European Corn Borer

Description, Injury, and Distribution

The newly-hatched larvae of the European corn borer (*Ostrinia nubilalis*) are about 2 mm in length, light-greenish after early feeding, with a brown to black head. The larvae take on a more opaque greenish gray or pink cast in later stages and have light longitudinal lines along the body. The full grown larva is about 25 mm long and varies in color, usually grayish to light brownish, and frequently has a reddish tinge.

The pupae are 12 to 20 mm long, slender and brown; they may be located on the foliage, in the ears, or in the stalk, and in surface debris in the spring. The moths are buff to brownish, frequently with reddish wing veins and a wing expanse of about 25 mm. Detailed descriptions of all the stages were published by Vinal and Caffrey (1919). Characteristic injury indicating the presence of larvae are shot-hole and elongated lesions on the leaf blades, in the midribs and behind the sheath. In later development broken tassels, and holes and burrows in the stalk become prominent.

The presence of the European corn borer in the USA was first reported by Vinal (1917) in Massachusetts. Shortly afterward infestations were found in New York and Pennsylvania, 1919; Ontario, Canada, 1920; and Ohio, 1921. Separate points of introduction from foreign sources or spread through transportation were indicated in this rapid westward establishment. Severe damage occurred in Kent and Essex counties, Ontario, in 1925 and 1926, and caused an appreciable reduction of corn acreage (Caffrey and Worthley, 1927; Stirrett, 1938). Movement westward and southward was gradual between 1927 and 1936. After the drought year of 1936 the spread into southern states and across the North Central region proceeded rapidly and the insect reached the Rocky Mountain area in northern Colorado in 1950. Brindley et al. (1975) reported the presence of infestations in south Georgia and Alabama. The general abundance of the borer touched a peak in 1949 when there were high populations over much of the central Corn Belt.

Habits and Biology

There are numerous published accounts on the biology and habits of the European corn borer. The results of experiments and observations reported from some areas may seem at variance with other areas. These differences are primarily due to differences in the environment and the population complexes produced in it. The fact that the species is of economic importance in such diverse environments as those prevailing in Canada and northern USA, and Israel and Egypt is positive evidence that a versatile genetic mechanism is an important factor in developing these adaptations. In an extensive interregional study of populations Brindley et al. (1975) reported that diapause is controlled by multigenetic factors that respond to temperature and photoperiod.

The first extensive report on life history studies in the USA was made by Vinal and Caffrey (1919). Most of this work was done on sweet corn and with an insect that was predominantly two-brooded. Their descriptions of the biology are in general still applicable to a two-generation infestation similar to that encountered in Massachusetts. Under a single generation, which was predominant for many years in the Great Lakes area, similar data were brought together by Crawford and Spencer (1922), Caffrey and Worthley (1927), and Huber et al. (1928).

The European corn borer hibernates as a full grown larva in cornstalks or plant debris. Time of pupation and moth emergence depends on the weather conditions and to some extent on whether or not the prevailing population is single or multiple-brooded. For most of the infested area moth emergence begins in May or June. Most of the eggs are deposited between dusk and midnight. The important factors limiting moth activity are low temperatures and high winds. The earliest planted corn is most attractive to the moths of the first brood for oviposition, whereas late-planted or late-maturing corn is more attractive to the moths of subsequent broods. The stage of development of corn is therefore an important factor in the rate of oviposition and figures prominently in determining the need for insecticides (Barber, 1925; Huber et al., 1928; Stirrett, 1938). In a given area the stage of development of the corn crop varies widely during the first-brood, egg-laying period. The vegetative or whorl stage of growth increases progressively in attractiveness and susceptibility to larval establishment and survival as plant growth increases.

The most significant increase in larval survival occurs with the exposure of the tassel. A high susceptibility to larval survival persists throughout the reproductive stages of the plant but recedes significantly about 2 weeks after silking. This condition prevails in early market sweet corn during the first brood oviposition period in varying degrees and in late maturing field and sweet corn during the second brood oviposition period. The pattern of larval survival during the vegetative and reproduction stages of growth is shown in Fig. 7.

In the vegetative or whorl stage of growth the primary place of larval establishment is near the upper limits of the surface moisture level within the spirally rolled leaves. The larvae are predominantly sheath and midrib feeders during the third and fourth stages, whereas the later stages invade the stalk, shank, and the ear.

When corn is in the reproductive stage during either first or second brood oviposition, the young larvae become established largely on the structures associated with the inflorescence, in the florets, on pollen accumulations at the axils of the leaves, or on ear structures. The pattern of feeding by the middle and late stage larvae is about the same for both broods, with perhaps a higher concentration of young larvae in the upper region of the plant during the second brood infestation. The full grown larvae have a habit of establishing themselves in the lower part of the stalk (Batchelder, 1949; Caffrey and Worthley, 1927; Dicke, 1954).

Biological Control

The USDA established its program of foreign parasite introduction from Europe in 1919 and from the Orient in 1927. With the establishment of certain of the parasites in the older infested areas, federal and state agencies participated in recovery and redistribution programs with the most effective species. Baker et al., (1949) brought together the significant information on extensive research activities on biological control in the USA and Canada. Arbuthnot (1953), in reviewing the status of parasites in the USA reported that *Lydella thompsoni* (formerly *L. stabulans*), *Simpiesis viridulus, Eirborus tetebrons* (formerly *Horogones punctorius)* and *Macrocentrus grandii* to be widely established, and *L. thompsoni* to be abundant over most of the infested area. The latter species has become rare in recent years (personal communication from Leslie C. Lewis, ARS-USDA). The most common predators are several species of lady beetles, the hemipteron *Orius insidiosus,* and the downy and hairy woodpeckers. Contributions on a broad spectrum of biological research on the European corn borer have been reviewed by Brindley and Dicke (1963) and Brindley et al. (1975).

THE MOST IMPORTANT CORN INSECTS

Cultural Practices

The reduction of population levels through the disposal of infested crop residues was advocated in central Europe as early as 1897 (Babcock and Vance, 1929). Similar practices were later adopted in the USA and Canada (Caf-

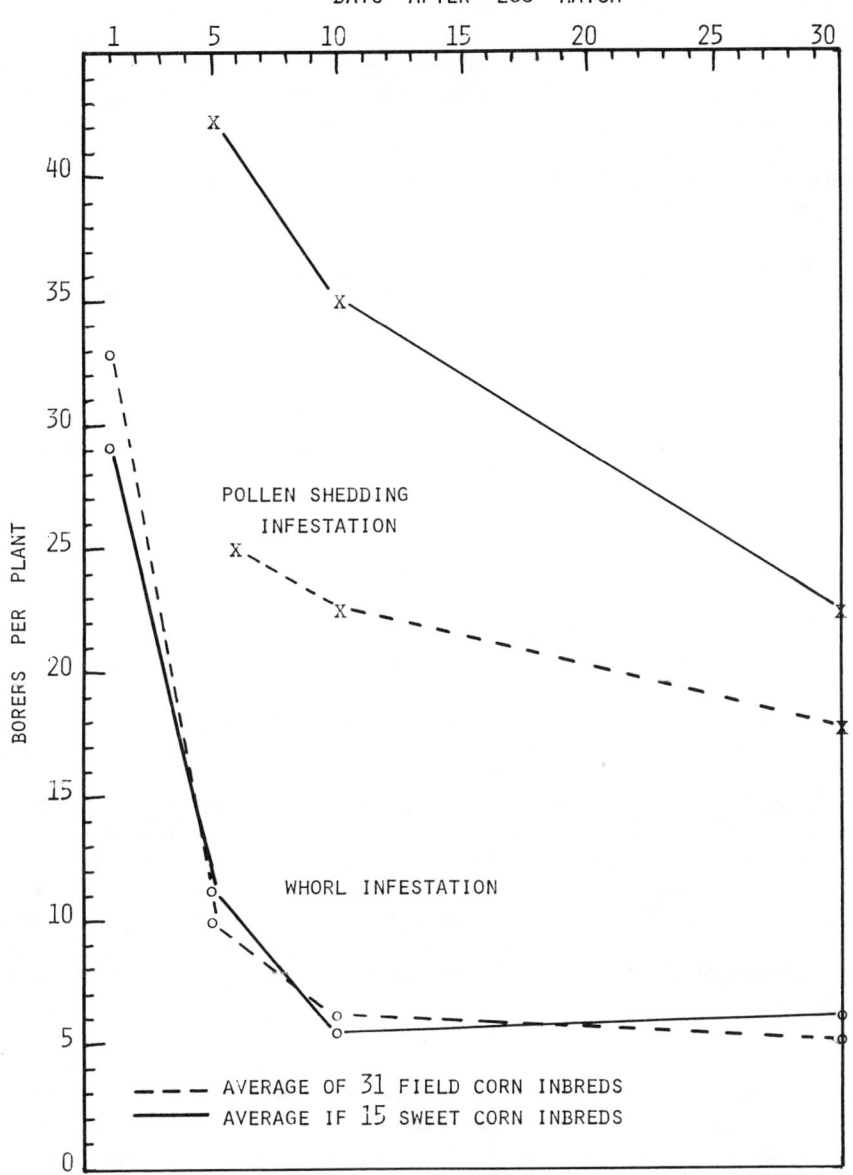

Fig. 7 — Comparison of larval survival of the European corn borer during the whorl stage period (first brood) and the pollen shedding period (second brood).

frey and Worthley, 1927; Vinal and Caffrey, 1919). Much of the larval population of a given season is destroyed by the usual farm tillage and harvesting practices. The most effective reductions of the overwintering populations can be obtained through ensiling and stalk chopping, followed by plowing under the crop residue (USDA, 1962).

Based on a 10-year study on the effect of farm practices on the reduction of corn borer populations from October to June in Illinois, Bigger and Petty (1953) presented data to show that the general practice of disking stalk fields for oats is an important factor contributing to the continued threat for corn borer damage. Eighty percent of the larval population in the spring after tillage were found in the disked fields. They advocated a right adjustment of corn picker rolls and heavy pasturing of stalk fields as supplementary to good plowing as the most effective cultural control practice. The effect of minimum or no-tillage practices on population survival has yet to be determined.

In the early biological studies of the European corn borer it was established that under univoltine (single-brooded) infestation, the earliest planted corn receives the highest rate of egg deposition and infestation and the latest planted corn the least. In the areas that were subject to univoltine infestation, delayed planting was encouraged. This practice was never well received because over a period of years early planted corn has a yield advantage and is less subject to attack by the second brood and to stalk breakage.

Corn Leaf Aphid

Description, Injury, and Distribution

The corn leaf aphid (*Rhopalosiphum maidis*) is small, dark bluish-green in color. Its wingless colonies first become conspicuous on emerging leaves and tassel tip (Fig. 8). The winged form becomes abundant in large crowded colonies on the leaves and tassels at about pollen shedding and silking time. Before tassel emergence the early nymphal (all females) infestation occurs typically on the moist part of the leaves in the whorl, where the insects suck nutrients from the phloem elements (Fig. 9). When the infestations begin relatively early in the season on mid-whorl corn and build up large colonies by the time of tassel eclosion, anthesis is impeded and may result in varying degrees of barrenness. The species is worldwide in distribution and has been identified as a vector of virus disease in corn and other species of grasses. The role of the corn leaf aphid in virus transmission is discussed in a later section (Dicke, 1969; Wildermuth and Walter, 1932).

Habits and Biology

As far as known the corn leaf aphid overwinters successfully only in warmer climates. Wildermuth and Walter (1932) reported corn, sorghum, and barley as preferred cultivated hosts with abundant populations on winter barley in the southwestern states. Forbes (1893) recognized the species as distinct from the corn root aphid and failed to find an oviparous form or any winter survival of the viviparous forms in Illinois. In Wisconsin, Orlob and

Medler (1961) did not find any forms in the winter but found colonies with mature apterous forms on barley in mid-May. The source of infestation in the northern areas is therefore believed to stem from populations of the winged form carried in by southerly air currents. In southern Indiana, Saugstad and Everly (1967) found reproducing populations on winter barley during the winter months of 1965-1966. In airplane collections in eastern Kansas, Taylor and Berry (1968) captured the corn leaf aphid and other cereal crop aphid species at an altitude of 670 m. At Johnston, Iowa, the earliest record on corn from 1966 to 1969 was winged forms on 3 June 1966 (Dicke 1969). The winged or alate form has an important role in the initial establishment of colonies. This form has a habit of locating the colonies on the semi-etiolated spiral of newly emerging leaves. The feeding activities of large colonies in the

Fig. 8 — Corn leaf aphid infestation on top leaves and lower part of tassel.

vascular system results in an accumulation of carbohydrates in the leaves and abnormal synthesis of anthocyanin and reddening of the leaves (McCulloch, 1921; and Wildermuth and Walter, 1932).

In detailed biological studies Randell (1970) found that crowded colonies of first and second instar nymphs produced a high proportion of

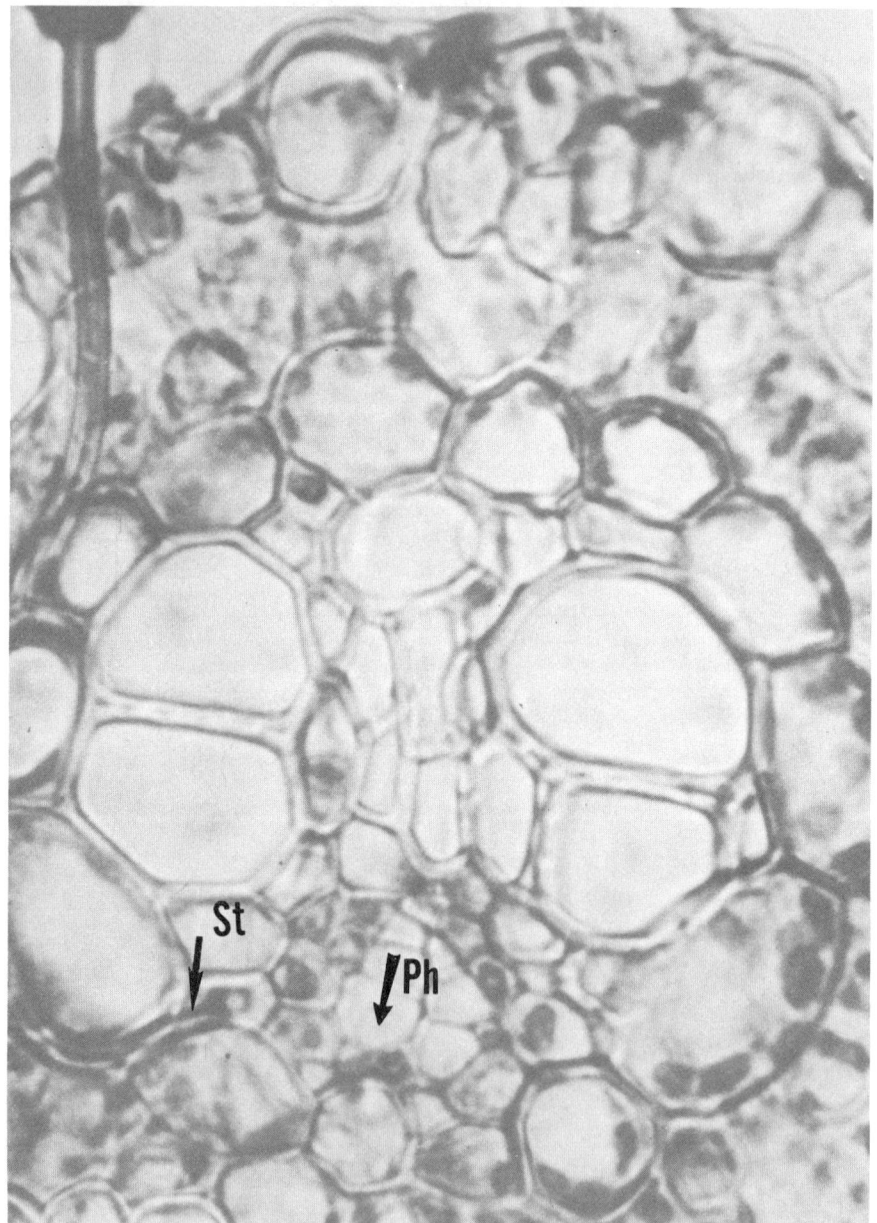

Fig. 9 — Aphid with stylet in place to feed in vascular elements. St, stylet; Ph, phloem.

winged forms which he interpreted to be a postnatal effect.

Climatic conditions play an important role in determining the rate of colony development and yield reduction. Triplehorn (1959), in a comparison of yield losses in Ohio in 1957 and 1958, reported severe reduction in yield mainly through barrenness, under stress of high infestation and soil moisture in 1957; whereas in 1958 under high infestation with abundant moisture supply, the yield losses were light. Everly (1960a) reported similar results on high yield loss under soil moisture stress.

Extremes of temperature are unfavorable for aphid development. The aphids seem to reach their maximum reproductive capacity in periods of cool temperatures which are unfavorable for some of their important predators and parasites. Wildermuth and Walter (1932) observed this relationship in the southwestern states and Walter and Brunson (1940) reported similar observations in Indiana. *Lysiphlebus testaceipes,* a small hymenopterous parasite, the common lady beetles, *Hippodamia convergens* and *Ceratomegilla fuscilabris,* and the lacewing flies (*Chrysopa* spp.) are the dominant natural enemies (Wildermuth and Walter, 1932).

Fall Armyworm

Description, Injury, and Distribution

The young larvae of the fall armyworm *(Spodoptera frugiperda)* are slightly greenish with black heads. They are commonly found feeding on the leaves in the whorl of young corn plants. With the emergence of the tassel they feed on the tassel buds and surrounding leaves. Later the ear is favored.

The mature larvae vary from greenish to a grayish-brown, have a light-colored inverted 'Y' on the head, and dorsal lines running lengthwise of the body. The moth has variegated gray forewings, which are held at an angle over the back when at rest.

The fall armyworm is an important pest of corn and other grasses in the South Atlantic and Gulf States. It is frequently a limiting factor in profitable corn production in tropical America unless protective insecticides are used. Periodically it spreads to the Middle Atlantic States and the southern part of the North Central region, where it becomes prevalent in the later maturing corn in August and September. It has been observed as far north as Canada (Forbes, 1893; Luginbill, 1950).

Habits and Biology

The fall armyworm overwinters in subtropical and tropical America and migrates northward each year by moth flight as the weather warms up. Diapause or aestivation is not known to occur in any of its life stages. On corn the eggs are usually deposited on the under side of the leaves in large masses covered with the scales and hair of the moth. In the whorl stage of growth the young larvae feed on the tender parts of the whorl leaves. Upon tassel eclosion the florets are a preferred site of feeding. In successive stages of growth the larvae feed on leaf sheath and ear parts. The stage of plant development is therefore an important factor in the feeding habits of the larvae. When the larvae become established in the whorl, the unfurled leaves show irregular elongated feeding areas as shown in Fig. 10.

With the appearance of the ear the larvae feed on the husks, in the shank, and on the silks, and also attack the kernels. The injury at the tip of the ear can easily be confused with that caused by the corn earworm. On becoming full grown the larvae leave the plant and prepare short tunnels and elliptical cells in the soil in which to pupate.

It requires about a month to complete a generation. In the South there may be as many as six generations annually, whereas in the more northern areas there may be only a single brood of larvae resulting from migrant moths (Burkhardt, 1952; Dicke and Jenkins, 1945; Luginbill, 1928, 1950).

The fall armyworm has several important hymenopterous and dipterous parasites as well as hemipterous and coleopterous predators. Some of the common entomogenous fungi have also been recorded in several southern localities (Luginbill, 1928; and Vickery, 1929).

Corn Borers

Description, Injury, and Distribution

The larval forms of the southern cornstalk borer (*Diatraea crambidoides*) and the southwestern corn borer (*D. grandiosella*) and the injuries they cause are similar. The young larvae are dull white to yellowish with black heads, about 2 mm long, with prominent transverse, closely spaced dark spots, giving the general appearance of a ring around each segment. With successive molts the dark spots are further apart, becoming pale shortly before pupation. In the winter form the larvae practically lose their spots, leaving the black spiracles very prominently displayed.

The larvae of the sugarcane borer (*D. saccharalis*) and Neotropical corn borer (*D. lineolata*) are superficially very similar. Ordinarily they can be differentiated from the southern cornstalk borer by their brown spots. Typical injury by the larvae of these species in the whorl stage of corn growth is at first on the unrolled leaves. As the leaves unfold the midribs and sheaths are the main points of attack, which is followed by burrowing in the ear parts and the stalks. Holes across the leaves and broken midribs are characteristic of later stages of leaf injury similar to that shown for the European corn borer in Fig. 25.

Severe injury to young plants may kill the growing point, causing dead heart and subsequent tillering. Holes are made in the stalks by the older larvae. The symptoms of the second or third brood feeding on corn that has tasseled are sheath and ear part injury, and holes and burrows in the stalk. Girdled stalks in the lower internodes are more typical for the southwestern corn borer than for the other species. The similarity of plant injury sometimes confuses their identity with the corn earworm and fall armyworm.

The moths of the four species are of pale to medium straw color with white hindwings and a spread of about 25 to 30 mm. They are generally distinguished by the spots on the forewings. The southern cornstalk borer moths have a prominent black spot in the middle of the wing and seven similar spots along the margin. These spots are 'V'-shaped in the moth of the

THE MOST IMPORTANT CORN INSECTS

Fig. 10 — Typical whorl stage damage by the fall armyworm.

sugarcane borer but not conspicuous in the moth of the southwestern corn borer.

The southern cornstalk borer is generally distributed along Atlantic coastal areas and Gulf States. It occurs as far west as Texas and north to Kansas. It has been ranked among the important pests of the corn crop in the South Atlantic coastal region. The southwestern corn borer has spread from the southwestern states and has northern limits of survival in Colorado, Kansas, Missouri, and Illinois. It is a hazard in corn production in parts of Texas and the southeastern states. The sugarcane borer is of most importance on corn in sugarcane areas and is frequently a pest in Central and South America and in rice producing areas of the Gulf States. The Neotropical corn borer is generally distributed in Central and northern South America and in the Caribbean Islands. In the USA it occurs in southern Texas (Ainslie, 1919; Leiby, 1920; Davis et al. 1933; Cartwright, 1934; Ingram and Bynum, 1941; Wilbur et al., 1950; USDA, 1966).

Habits and Biology

The habits and biology of the four species of *Diatraea* are sufficiently similar in the corn crop to discuss them together. The southern cornstalk borer and the southwestern corn borer hibernate as full grown larvae in the stalk near the base or crown, sometimes erroneously called the tap root. Fig. 11 shows the usual position of the hibernating larva of the southwestern corn borer. Pupation and moth emergence occurs in May or June, depending on the bioclimate. The larvae of the sugarcane borer overwinter in various parts of the stalk, as shown in Fig. 12, but seldom below the level of the ground. The overwintered larvae pupate in the spring, the pupal stage requiring about 3 weeks. The time of the first brood pupation and moth emergence varies with the bioclimate from April to June.

The eggs of the four species are normally deposited in masses on the leaves or in the sheaths in fish-scale fashion. The young larvae feed in the whorl of unfolding leaves, following a pattern similar to several of the other lepidopterous caterpillars occurring on young corn. They invade the growing point and cause dead heart or they burrow in the stalk. There is a tendency for the larvae to leave the upper part of the stalk and to reenter at lower points. Larvae of Neotropical corn borer have adapted to aestivate during the dry season and pupate after rains resume (USDA, 1966).

On late planted corn, in the whorl stage of growth, the habits are the same as those of the first brood. Whereas on the older tasseled corn there is

Fig. 11 — Typical position of the overwinter hibernating larva of the southwestern corn borer in the base of the stalk.

little feeding on the leaf blade but on the sheath, husks of the primary ear, and ear shoots (Fig. 13). Again the older larvae feed and burrow in the stalk. A detailed study of the feeding habits of the larvae of the southwestern corn borer on two stages of corn growth has been reported on by Davis et al. (1972).

Phillips et al. (1921) reported two generations for *Diatraea crambidoides* in Virginia, and Cartwright (1934) recorded mostly two or three overlapping generations in South Carolina, which is similar to the observations on *D.*

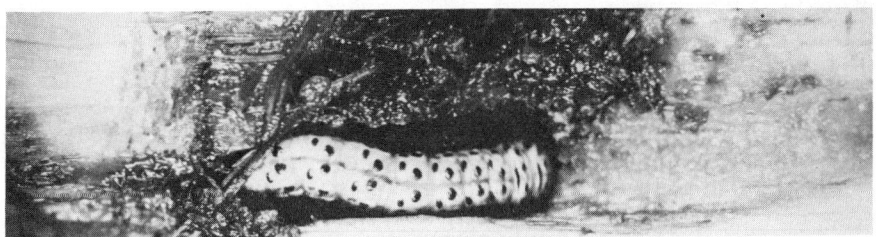

Fig. 12 — Sugarcane borer burrowing in stalk.

Fig. 13 — Husk penetration and ear injury by the southwestern corn borer.

grandiosella by Walton and Bieberdorf (1948) and Wilbur et al. (1950). When three generations occurred, each of the first two produced some hibernating larvae. Henderson and Davis (1970) directed insecticide tests against second and third generation infestation of the southwestern species. In the sugarcane borer, Ingram and Bynum (1941) report from two to four generations. Of particular significance in the second and third brood infestations of the southwestern corn borer is the severe breakage caused by larvae girdling the rind of the stalk, usually below the ear. This has not been reported as serious with the other species.

Cultural Practices

Procedures advocated for reducing the populations of the southern cornstalk borer and the southwestern corn borer are very similar. They are aimed at eliminating as many of the overwintering larvae as possible by exposure of the corn stubble, which they inhabit below the soil surface, to freezing temperatures on the soil surface. A high percentage of the larvae are killed by subfreezing temperatures. The stubble or stalks should be plowed or disked out in the fall and allowed to remain on the surface until March. High mortalities among the hibernating larvae also result when stalks are scattered and broken up in such operations. Moths fail to reach the soil surface when the stalks are plowed under to a depth of 10 cm (Cartwright, 1934; Davis et al., 1933, 1972; Phillips et al., 1921; Wilbur et al., 1950).

The southwestern corn borer has been most troublesome in the areas where grain sorghums are better adapted than corn. Wilbur et al. (1950) therefore advocated the substitution of sorghums for corn. This has become a common practice in Oklahoma (Walton and Bieberdorf, 1948). In a study of planting dates a partial escape from infestation by the first brood was observed by these workers and also by Davis et al. (1933). This practice has been further discussed by Davis et al. (1973). In South Carolina Cartwright (1934) observed lower infestations of the southern cornstalk borer in late planted corn, but discouraged this practice in areas where billbugs and the lesser cornstalk borer (*Elasmopalpus lignosellus*) are important factors. He also observed that properly fertilized corn will outgrow injury and produce fair yields.

Ingram and Bynum (1941) regarded the source of the sugarcane borer as a pest of corn to be sugarcane debris and rice stubble. A high percentage of the overwintering larvae may be destroyed by burning the sugarcane trash. In rice stubble the hibernating larvae are materially reduced by grazing, burning, flooding, and dragging.

Armyworm

Description, Injury, and Distribution

The young larvae of the armyworm (*Pseudaletia unipuncta*) are pale green with alternate brown to reddish and light longitudinal stripes, which become more conspicuous in successive instars. The anterior prolegs of the first two instars are poorly developed. In these stages the larvae are similar to measuring worms. Late instar larvae vary in color from green to light brown and red. The head is pale brown with a mottled area on each side of the median suture. When feeding in the corn ear, the mature larvae may easily be mistaken for the fall armyworm or the corn earworm.

The moths of the armyworm are brownish with a white spot near the center of the forewing. They are nocturnal in their flight habits, but are attracted to light in large numbers during outbreak years.

Injury on young corn is at first on the leaves and in the whorl; later the older larvae feed more on the edge of leaves. The latter injury is common when migrations occur from adjoining small grain. A typical infestation shows plants with elongated lesions on and along the edge of leaves.

The armyworm is widely distributed over the USA and Canada as well as in foreign countries. Outbreaks are frequently local and sporadic, but high general populations have at times occurred over large sections of the eastern parts of the USA and Canada (Forbes, 1904; Walton and Packard, 1951; ARS, 1961).

Habits and Biology

The primary host plants of the armyworm are wheat, oat, barley, rye, and forage grass. Forbes (1904) reported that the winter is passed as partially grown larvae; this fact was confirmed by Davis and Satterthwait (1916) and Knight (1916). The overwintering larvae complete their development early in the spring and then burrow into the soil to pupate. The first brood of moths are abroad from April to June, depending on the bioclimate. They commonly deposit their eggs in bead-like clusters behind the sheaths and on leaves of small grain near the base of the plants. On hatching the larvae hide among the leaves during the day and feed on the upper leaves at night.

The presence of an infestation does not become conspicuous until the later instars, or about the time the heads of the grain appear. Migration to contiguous corn occurs from May to July, depending on the locality under conditions of overpopulation, and may extend a considerable distance into a field. Some infestations have been observed on early planted corn which resulted directly from first brood oviposition. Forbes (1904) and Davis and Satterthwait (1916) estimated three complete generations in Illinois and Indiana, respectively. Knight (1916) observed two broods in New York. At about the roasting-ear stage of corn the larvae are not uncommon at the tip of the ear, causing injury typical of the corn earworm.

Musick and Tuttle (1973) observed a high infestation of the armyworm in corn planted by the no-tillage method (see Fig. 14). A maximum reduction of severe damage was obtained by an application of carbofuran granules placed in the furrow at the rate of 2.4 ounces per 1,000 feet of row.

During an armyworm outbreak large numbers of larvae fall prey to parasites, predators, and diseases. Knight (1916) recorded two species of tachinid flies, four of *Hymenoptera,* and five of predaceous beetles in his observations in western New York. Among other predators the most common feeders are the blackbirds, sparrows, domestic fowls, skunks, and toads (Walton and Packard, 1951).

Lesser Cornstalk Borer

Description, Injury, and Distribution

The full grown larva of the lesser cornstalk borer (*Elasmopalpus lignosellus*) is about 25 mm in length, and greenish with conspicuous brown longitudinal stripes. The body colors of the young larvae vary from pale yellow to pale green. The larvae have a distinctive habit of jerking and skipping when disturbed.

Typical injury in young plants is characterized by tunnels into the plant at or slightly below the surface of the soil (Fig. 15). Heavy populations may cause loss of stand, stunting, and tillering. In older plants the injury in the stems is of a girdling nature near the ground, and consequently plants are subject to stalk breakage. The presence of earthen-like, silk-webbed tubes attached at the entrance of the tunnels into the plant is evidence of an infestation. These tubes are occupied by the larvae when they are not feeding in the plant (Kirk, 1960; Luginbill and Ainslie, 1917).

Fig. 14 — Armyworm infestation under no-till planting conditions.

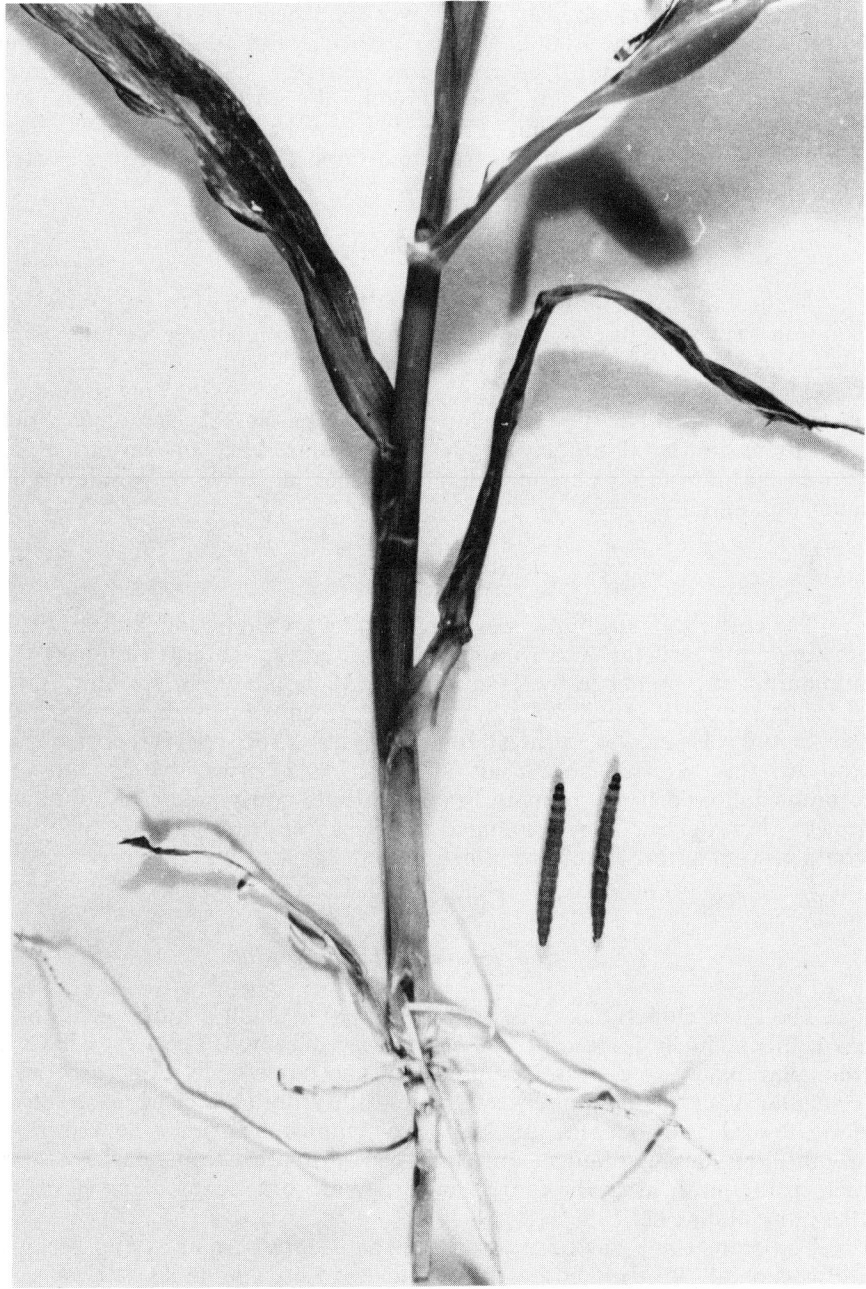

Fig. 15 — Corn plant showing typical injury by the lesser corn stalk borer with full-grown larvae on right.

The species is widely distributed over the warmer parts of North and South America and the islands of the West Indies. In addition to corn, the more important crops attacked in the southern states are sugarcane, sorghum, cowpeas, and beans. Among weeds, nut grasses (*Cyperus* spp.) are sources of seasonal populations. The moths have been recorded in the northern states, but infestations of economic importance have been reported only as far north as Arkansas (Isely and Miner, 1944).

Habits and Biology

According to Luginbill and Ainslie (1917), only the larvae survive the winter in their stalk burrows, and pupation and moth emergence take place late in the winter and in early spring. At Columbia, South Carolina, the first-brood larvae on corn were nearly full grown the first week of June. Based on rearing experiments, they concluded that the insect developed four generations in that latitude. The records of outbreaks on various crops showed that infestations are most likely to occur under weedy conditions on sandy upland soils.

Cultural Practices

The cultural control practices suggested by Luginbill and Ainslie (1917) are aimed primarily at destroying the overwintering population through the elimination of crop debris by clean plowing in late fall or early winter. Early planting and fertilization are advocated as a means of producing vigorous growth and tolerance to an infestation. Isely and Miner (1944) reported that corn was the preferred host in an outbreak in Arkansas in 1943. This infestation followed heavy rains in May and June, and resulted in abundant growth of crabgrass. They concluded that clean cultivation before planting would prevent outbreaks of this insect.

Chinch Bug

Description, Injury, and Distribution

The adult chinch bug (*Blissus leucopterus*) is about 4 mm in length. It has a black, finely pubescent body with whitish wings. The wings have a triangular black area near the middle of the outer edge, and there is a triangular black scutellum between them at the anterior. The egg is reddish, elongate-oval, and about 0.7 mm long. The nymphs, the primary source of injury through sucking plant sap, are light red at first, becoming darker red in each instar, until after the fourth molt they are practically black (Forbes, 1904; Luginbill, 1922).

The chinch bug has been recorded from Central America and Mexico and is generally distributed over the major corn growing areas of USA and southern Canada. Leonard (1966) reviewed the status of the several species of *Blissus* in relation to distribution and biology. He concluded that *Blissus hirtus* is a subspecies of *B. leucopterus* but does not infest corn. Although *B. leucopterus* has many hosts, corn figures prominently in population build-up. Outbreaks have occurred periodically for more than 150 years. The most serious losses on corn have occurred in the upper Mississippi Valley, Illinois being in about the center of the region of periodic outbreaks. Shelford and Flint (1943) made a study of the factors involved in the fluctuation of chinch bug populations in the upper Mississippi Valley from 1823 to 1940. They concluded that population levels were correlated with meterorological conditions. Populations sufficient to cause crop damage were associated with above-normal temperatures and below-normal rainfall between March and October. No adverse effect could be attributed to below-normal winter temperatures.

Habits and Biology

So far as known the chinch bug feeds exclusively on plants belonging to the grass family. They attack wheat, barley, rye, oat, corn, and sorghum. The adult chinch bugs overwinter deep in clumps of sod of perennial grasses and in accumulations of leaves or debris, concentrating in fence rows and southern or western exposure of wooded areas. They fly to such cover in the fall from corn, sorghum, or other hosts (Forbes, 1904; Headlee and McColloch, 1913). In the spring the bugs fly to the small grains. In Illinois, Benton and Flint (1938) found wheat, rye, and barley most attractive to the bugs in the spring, and oats least attractive. Early planted corn may also be a host to the spring brood, especially in the Southwest.

The eggs are deposited around the base of the plants and behind the sheaths of the lower leaves, where they hatch in 1 to 2 weeks. The bugs obtain their food by sucking the sap from the plants, thereby causing serious injury to small grain under heavy infestations. Painter (1928), in a histological study of the tissues pierced by the sucking mouth parts of nymphs and adults, established that the stylets were thrust into the vascular bundles. He concluded that the food was obtained chiefly from the phloem and that the conductive tissue elements became clogged through the feeding process.

Small grain begins to ripen before much of the first brood reaches the adult stage. This condition results in the well-known, on-the-foot migration from small-grain fields to adjoining corn, where the nymphs complete their development. The adults become abundant a few weeks later and distribute themselves by flight on corn, sorghum, or other grasses. The second brood bugs complete their development on these crops in late summer and early fall. A third brood has been observed in the southern areas. The flight to their overwintering cover occurs in the late summer and fall, being heaviest at temperatures over 38 C following cool nights. Among the important natural control agents that have been recorded are two fungus diseases, *Beauveria bassiana* and *Entomophthora aphidis,* which are most effective when rainfall is abundant. A minute egg parasite, *Eumicrosoma benefica,* is also considered to be of importance.

Cultural Practices

Various cropping practices have long been advocated when a chinch bug problem is threatened. The spectacular migration from small-grain fields to adjoining corn led to the recommendation to avoid planting corn next to small grain whenever possible. The cultural practices that produce good, vigorous stands of small grain and legumes have been observed to be unfavorable for the development of chinch bugs (USDA, 1954).

Grasshoppers

Description, Injury, and Distribution

Many species of grasshoppers Melanoplus spp. and *Schistocera americana* attack corn but only several of these cause serious crop losses during epidemic years. Of foremost importance on corn is the differential grasshopper, *M. differentialis.* This species is about 38 mm long, dorsally dark-greenish to olive-brown to yellow underneath. Its distribution extends from the Great Plains States to the Atlantic Coast and north to the southern part of Canada.

Of almost equal importance to corn is the slightly smaller two-striped grasshopper (*M. bivittatus*). It is distinguished by two yellow stripes which extend back from the eyes along the sides to the end of the abdomen. The body is generally olive-brown above to yellow underneath. *M. bivittatus* is generally distributed over the USA and north into the Canadian provinces.

Of lesser importance on corn are the migratory grasshopper and the red-legged grasshopper (*M. sanguinipes* and *M. femur-rubrum*). These are about the same size and general appearance, being about 25 mm and reddish-brown to gray dorsally to greenish-yellow beneath. The red-legged grasshopper has highly colored red hind tibiae. Both species are distributed throughout most of North America and are more specific on small grain, clover, alfalfa, and pasture.

In the southern states the American grasshopper, *S. americana*, is a large species 50 to 75 mm long and of reddish-brown color and is occasionally a pest on corn (Blatchley, 1920; Parker and Connin, 1964; Shotwell, 1958).

Habits and Biology

Eggs of these species are deposited sometime during the late summer and early fall in pods 25 to 50 mm below the surface of the soil. Depending upon the species, the pods are 25 to 40 mm long and contain 15 to 100 eggs/pod. The differential and two-striped grasshoppers concentrate their egg pods among the roots of weeds or grass along fence rows, roadsides, ditch banks, and other margins of cropped fields and pastures. Egg pods of migratory and the red-legged species are deposited in more open areas of small grain stubble and bare spots in alfalfa and clover fields. The American grasshopper deposits its pods in old weedy pastures, stump, and abandoned lands typical among cropped fields in the southeastern states (Blatchley, 1920; Parker and Connin, 1964; Shotwell, 1941).

The eggs of grasshoppers found on corn hatch in the spring, sometime between 1 April and 1 July, depending upon weather conditions and species. The young nymphs at first are less than 6 mm long and develop in stages by five to seven molts at intervals of 4 to 8 days. Most of the injury to corn is done after midsummer by adults or nymphs in the last two stages of development which migrate to corn from newly harvested small grain, hay fields, and roadsides (Shotwell, 1941, 1952).

Grasshoppers show a distinct preference for corn silks and eat them down to the cob, a practice which may interfere with pollination. A population of 15 to 30 m^2 will devour the leaves to the midrib or cut them off at the axils, leaving the bare stalks. Grasshoppers thrive under moderately dry climatic conditions. Fungus and bacterial diseases are effective in population control in areas that have a rainy climate (Parker and Connin, 1964).

Cultural Practices

Tillage practices, where they are compatible with good cropping procedures, are helpful in destroying egg pods of cropland species. This can be accomplished either in the fall or spring by disking or moldboard plowing. Laying soil open to wind erosion should be avoided (USDA, 1964).

Corn Flea Beetle

Description, Injury and Distribution

The adult is the primary injurious form of the corn flea beetle (*Chaetocnema pulicaria*). It is oblong, very dark, shiny green, and averages about 2 mm long. The feeding lesions on the leaves form a pattern of narrow, closely spaced lines which normally run parallel with the veins. On the average the lesions are about 6 mm in length. They are the points of origin of bacterial wilt, which in its more advanced stages, may obscure the beetle lesions in susceptible varieties. The role of this beetle as a vector of the disease is discussed elsewhere (Poos and Elliot, 1936).

Large populations of the corn flea beetle occur periodically in sections from the Atlantic states to as far west as Colorado. Infestations on field and sweetcorn are sometimes severe. Forbes (1905), in discussing the importance of this insect in several states, indicated that wilting was associated with the injury and that the beetles were especially fond of sweetcorn.

Other species of flea beetles are commonly found on corn and cause similar injury. The toothed flea beetle, *C. denticulata,* which is metallic green and about twice as large as *C. pulicaria*, is more closely associated with the wild grasses that persist in cornfields. This species is also capable of transmitting Stewart's disease but with less virulence (Elliott and Poos, 1940; Poos and Elliott, 1936). In the southwestern states, *C. ectypa* is generally distributed as a pest of corn (Wildermuth, 1917). Evidence of its ability to harbor and transmit Stewart's disease has not been established.

Habits and Biology

The adults of the corn flea beetle feed on many different plants. Forbes (1905) and Elliott and Poos (1940) list a wide host range but the cereal and forage grasses, wheat, oat, barley, rye, bluegrass, timothy, and orchardgrass are probably the most important. Poos (1955) for several years investigated the biology of the species and its relationship to the transmission of bacterial wilt.

In the vicinity of Washington, D.C., the insect was found to overwinter as an adult, mostly in the top 25 mm of the soil and primarily in bluegrass sod. On days when the air temperature near the surface of the soil reached 35 to 48 C for a period of a few hours, the adults become active during the winter. In the spring they became generally dispersed. New broods of adults were observed at various times during the season. Emergence of a brood was concentrated after a good rainfall following dry soil conditions. This was observed as early as 9 June and as late as 8 August, making it evident that at least two generations of this insect were completed each season. Under laboratory and greenhouse conditions a generation was completed in about a month.

Small larvae were found developing on 21 species of grains and grasses, including corn (both field and sweet), barley, oat, wheat, orchardgrass, timothy, redtop, and Italian ryegrass. Larvae would not feed on alfalfa and adults lived only about 10 days when confined to this host plant. Evidence was obtained indicating that a generation developed on wheat and timothy in 1940 before corn was generally available in the field.

Japanese Beetle

Description, Injury, and Distribution

The injurious form of the Japanese beetle (*Popillia japonica*) on corn is the adult. It is about 12 mm long, oval, with a metallic green body and brownish wing covers. The beetle can readily be identified by the six conspicuous tufts of white hair along the outer edge of each wing cover. Typical injury is scatter-grain, particularly at the tip of the ear, due to continuous beetle feeding on emerging silks. As a pest of corn this beetle is most important in the Middle Atlantic and Southern New England states. The Japanese beetle was discovered in New Jersey in 1916. In approximately 60 years it has extended its range as far south as the Gulf States, west to western areas of the North Central region. Colonies have become established by hitchhiking beetles on conveyances moving out of the heavily infested areas. The larvae are sod feeders and have their optimum survival in sandy soil. The polyphagous nature of the beetles readily permits situations for successful establishment of colonies in grassy areas (Fleming, 1970, 1972).

Habits and Biology

The emergence period of the beetles extends from June to July and there is but one brood annually. Egg deposition is predominantly in grass sod. On hatching the larvae feed on fine plant roots and organic matter in the soil until fall and again in the spring. Pupation occurs in May and early June. Favorable situations for the development of larvae are lawns or short grass pastures. Host selection is odoriferous in nature. The adults feed on many species of plants, corn in the silking stage being one of the preferred hosts. The beetles have a gregarious feeding habit and when large populations concentrate in a field with emerging silks, pollination and seed set is interfered with to varying degrees. The scatter-grain is therefore the result of continued attack and shearing of emerging silk (Coon, 1951; Fleming, 1972; Hawley and Metzger, 1940).

LESSER RECOGNIZED GROUPS

Thrips

Several species of thrips are almost invariably present on corn, particularly in the early stage of growth. The writer has recorded varietal resistance under high populations of the common grass thrips, *Anaphothrips obscurus,* and lesser populations of *Frankliniella tenuicornis* in northern Ohio. Everly (1960b) observed the same species and *F. fusca* and *Aeolothrips bicolor* as injurious to seedling corn. Thrips injury is characterized by a silvery mottled appearance of the lower leaves which is sometimes more intense along the edges of the leaves. The young larvae are primarily located in the moist, rolled leaves of the bud, whereas the adults commonly occur under the fully exposed leaves. Sources of corn resistant to *F. occidentalis* and *F. williamsi* have been evaluated by Granados (1970). Elliot and Poos (1940) reported on tests with *A. obscurus, Aeolothrips fasciatus* and *Heliothrips femoralis* as possible vectors of bacterial wilt. A single isolation was obtained from *A. obscurus.*

Leafhoppers

Many species of leafhoppers have been recorded on corn. Neiswander (1926) lists 59 species. The significance of their injury, with the exception of several species, is not well established. The lantern fly, *Peregrinus maidis,* is a pest of corn in tropical and semitropical areas. The nymphal instars concentrate feeding activity on the bud leaves. This species and several others are discussed in a subsequent section as virus and spiroplasma vectors. Leafhoppers are nearly always abundant on small grain, pasture and other grasses. Some of the same species commonly occur on corn. A pale green mirid, *Trigonotylus brevipes,* feeding on the leaf blades has been observed to be abundant on early to mid-whorl stages of growth for several years in Ohio and Iowa.

Western Bean Cutworm

The western bean cutworm (*Loxagrotis albicosta*) has been a pest of field beans for many years and has been recorded as injurious to corn in several western states. In recent years it has extended its range eastward in Nebraska and Kansas. Injurious infestations have occurred on corn in the Platte River Valley, mainly in the Grand Island area. Raun et al. (1968) reported on extensive losses to about 121,600 ha of corn, primarily in central Nebraska.

The full grown larva, about 40 mm long, tunnels into the soil where it passes the winter in a cell. Pupation takes place in early summer followed by moth emergence shortly thereafter. The moth is brown with creamy white stripes near the front of the forewing. The wing expanse is about 40 mm and the body is brown to tan in color. The eggs are deposited in masses in July on the upper surface of the leaves. The stage of corn growth during this period ranges from late whorl to tasseling. Upon hatching the brownish larvae feed on the whorl leaves and on the developing tassel. The pattern of larval feeding progresses to the leaf sheath, husks and developing ear. Injury to the ear can be severe (Fig. 16). There is a single generation annually.

The full grown larvae leave the plant to enter the soil from late summer to early fall. An end result of ear infestation is not only the actual grain devoured but also a considerable amount of moldy grain (Douglas et al., 1957).

The spread eastward of the bean cutworm has been rapid. In its normal habitat it has preferred field beans as a host. On corn the highest populations have occurred in irrigated areas. Whether or not the large area of nonirrigated corn beyond the present distribution limits will restrict further spread remains to be seen.

Corn Blotch Leaf Miner

The corn blotch leaf miner (*Agromyza parvicornis*) attracts attention in some localities each year. This insect belongs to a large family of *Diptera* in which larvae develop in between the epidermal layers of leaves. As the larvae develop the "mines" increase in size and appear as irregular, dead, or blighted spots. The full grown larvae are about 6 mm long and pale green in color. They pupate in the soil. The adult flys emerge from 2 to 3 weeks later. Egg deposition is on the leaves. The hatching larvae enter the leaf directly as they leave the eggshell.

The species has several host plants among the Gramineae but appears to have a preference for corn. Under high populations blotched leaves may occur from the seedling to the top leaves of the plant. After several generations during the growing season the insect enters diapause as a puparium, although continuous breeding has been reported from Florida; Phillips (1914) lists 20 species of parasites reared at several locations.

Infestations are usually not noticed until the damage is almost complete. For this reason chemical control practices have been believed to be of doubtful value.

Spider Mites

Mites are small, light to dusky arthropods (four pairs of legs) that vary in size from less than 1 mm for the newly hatched to about 2 mm when full grown. The eggs are minute, spherical and off-white in color. The normal feeding site is under the leaves and the injury first appears as small stippled areas in the leaf. As the numbers increase the chlorophyl breaks down and leaves take on a grayish color. Mites are vigorous spinners of silk, which forms a protective webbing over the feeding areas on the leaves. Disruption of the silk webbing through rains is an important factor in natural control of populations.

Infestations of spider mites have become commonplace in the drier western corn growing areas, particularly on irrigated ground. When populations begin to build up on young corn, severe stunting may occur.

THE MOST IMPORTANT CORN INSECTS

Ehler (1973) reported on the identity of mites collected on corn and sorghum in three areas of Texas. The species most frequently collected on corn was the Banks grass mite, *Oligonychus pratensis,* which has a wide distribution over the Great Plains area. The other species identified were the two-spotted spider mite, *Tetranychus urticae,* the carmine spider mite, *T. cinnabarinus,* and *O. stickneyi.* He emphasized the importance of proper identification in field observations and experimentation.

Pink Scavenger Caterpillar

Description, Injury, and Distribution

The injurious form of the pink scavenger caterpillar (*Sathrobrota rileyi*) is a pink to rose-colored larva about 10 mm long when full grown. The larvae begin to infest the ear in the field and continue to feed on the ears after harvest in storage. The moth has mottled yellow, brown, and black forewings

Fig. 16 — Illustration of husk penetration and ear injury by advanced stage larvae of the western bean cutworm.

and long-fringed narrow hindwings. It is smaller than the Angoumois grain moth with which its populations may be mixed. In typical injury the ear is webbed with grass between the rows of grain and in the larval cavities. The injury and the silk webbing are not unlike that of the Indian-meal moth. The species is tropical and subtropical in distribution and is a pest of corn, sorghum, and cotton in the South Atlantic and Gulf States (Chittenden, 1916).

Habits and Biology

According to the records published by Chittenden (1916) the pink scavenger caterpillar worm infests corn in the field that has been infested with earworms or has had the husk cover broken in some other way. He observed ears infested about milk stage or a little later. It is indicated that the breeding is continuous through the fall and after harvest in storage. The larvae feed also on the husk and cob but have a habit of feeding in the grain on the embryo and the endosperm. Douglas, et al. (1962) found a positive correlation between the infestation of ears by the corn earworm and the pink scavenger caterpillar. Good husk cover gave partial protection against infestation. Wiseman et al. (1970) in studies in Georgia, similarly concluded that husks of selected hybrids provided significant protection against the earworm and the pink scavenger caterpillar and that there was a close connection between the damage incurred from the two insects.

Garden Symphylan

The garden symphylan (*Scuttigerella immaculata*) is a "centipede-like" soil arthropod which has been frequently observed as a pest of corn in recent years. The eggs are deposited in runways in moist soil. Newly hatched larvae are white with six pairs of legs. There are six molts and the number of legs increase to 12 pairs by the last molt. Adults are about 6 mm long, white, sometimes grayish in the abdominal segments. They have prominent, protruding antennae.

This "centipede" has many hosts among the vegetable crops. It feeds on roots and underground stems, asparagus and sprouting seeds being favored hosts (Wymore, 1931).

INSECTS IN RELATIONSHIP TO CORN DISEASES

In crop improvement primary consideration has to be given to yield, quality, and harvestability under a range of environmental conditions. The experimental experience through which our goals are achieved is a continuing process. An inbred line or variety may be reasonably homozygous or uniform but the environment to which it is exposed is in a continual state of flux. To counteract the plasticity of the environment, a plant breeder has to compromise with genetic factors that will, in his judgment, produce the best end result. In order to achieve varietal improvements we need to have knowledge of the components that contribute to these end results.

The role of insects in the transmission or dissemination of pathogenic and saprophytic organisms is an important part of the interdisciplinary corn development program. These problems are encountered each year to varying degrees from the time of germination to plant maturity and storage of the grain.

To begin with, the soil is a complex environment abounding in microorganisms, particularly fungi, that may infest plant parts injured by soil-inhabiting arthropods. Then, proceeding through the vegetative stages of growth there are many species of insects that become involved directly or indirectly in the fungous, bacterial, virus, and spiroplasma diseases of the corn plant.

To obtain an understanding of the insect-related (or mite) diseases of corn, it is obvious that basic research on biological relationships between the host and its parasites is essential. Information developed through such studies builds the stepping-stones of methodology for evaluating the many qualities needed to withstand the hazards present in the environment. In recent years there has been an increasing cooperative effort in this direction by state, federal, and private agencies. It is the purpose of this section to review some of the contributions made on the status of the more important insect-related diseases.

Fungous Diseases

In much of the Corn Belt, parasitic and saprophytic fungi probably pose more problems in corn production than any of the other groups of organisms. The primary roots of the seedling, the radical and seminal roots, are commonly infected with *Fusarium* spp. after they have served their function and become senescent. Thus, inoculum is normally present in close proximity of any later root injury that may occur.

Diabrotica rootworms are generally prevalent in much of the corn growing areas of North and Central America. Periodically populations reach high levels. The habits of the larvae encompass feeding on root hairs, small rootlets, burrowing in tender portions of roots, and removal of rings of young crown roots. Palmer and Kommedahl (1969) have reported on the *Fusarium* species associated with roots of corn infested with *D. longicornis*. They isolated *F. oxysporum, F. moniliforme, F. roseum f.* sp. *cerealis graminearum,* and *F. tricinctum. Fusarium roseum* and *F. moniliforme* were pathogenic, whereas *F. oxysporum* and *F. tricinctum* were secondary invaders and grew only on tissues damaged by the rootworm. Both larvae and adults were vectors of *Fusarium* spp. Eggs and frass also harbored these organisms.

On sweetcorn, Pepper and Haenseler (1944) observed less corn smut (incited by *Ustilago maydis*) where the European corn borer was controlled with insecticides, indicating that the organism gained entrance in the plants at the sites of larval injury. Smut infection may be a cause of barrenness and poor ear development in corn.

The microorganisms associated with what are generally called stalk rots are of significant importance in determining the quality, yield and harvestability of maize. Interference with the translocation of nutrients in the vascular system predisposes the stalk for development of common stalk rot organisms, such as *Diplodia maydis* and *Gibberella zeae*. In a study of losses caused by varying levels of infestation by the European corn borer, Patch et al. (1941) reported increased stalk breakage below the ear as the level of larval population increased. Christensen and Schneider (1950) in Minnesota, investigated the organisms that developed in the stalk and ear after infestation with the European corn borer. They found that species of *Fusarium* were the most common pathogens isolated from injured points in the stalk, shank, and ear. Among the fungi isolated from rotted internodes of corn infested with European corn borer larvae were *Alternaria* spp., *Aspergillus* spp., *Mucor* spp., *Rhizopus* spp., *Penicillium* spp., *Diplodia maydis*, and *Gibberella zeae*. Substantially the same group of fungi were isolated from internal parts of living and dead larvae. They indicated that weak pathogens may play an important role in disease development of injured stalks.

The most important insects associated with ear injury (kernel) are the corn earworm and the European corn borer. In years of general infestation by either species, ear fungi takes a substantial toll in kernel damage and quality. Koehler (1942) studied ear diseases in relation to corn earworm infestation. Ear rot incited by *F. moniliforme* was increased by corn earworm damage. *Gibberrella zeae* infection increased somewhat less, and *Nigrospora* spp. and *Cephalosporium* sp. increased only slightly through earworm infestation. Earworm injury did not increase *D. maydis* infection.

Recent investigations by Lillehoj et al. (1975) have revealed the presence of alflatoxin, produced by *Aspergillus flavus*, on corn produced in diverse locations. In Georgia, Widstrom et al (1975) reported a relationship between the concentration of aflatoxin and the incidence of ear infestation by the European corn borer, *O. nubilalis*, the fall armyworm *S. frugiperda*, and the corn earworm, *H. zea*, in the order listed.

Bacterial Diseases

Bacterial leaf blight incited by *Erwinia stewartii* (sometimes called Stewart's bacterial blight) has been recognized mainly as a disease of sweetcorn. However susceptible cultivars of fieldcorn can be readily found. According to the *Compendium of Corn Diseases* (American Phytopathological Society, 1973), this disease has been reported from Europe and Asia. All of these reports are apparently from warm climate areas. Likewise, in the Western Hemisphere, bacterial leaf blight is annually present in southern states and the southern fringe of the Central Corn Belt east to the Atlantic Coast. In 1972 and 1973 this blight was prevalent as far west as Kansas. Stevens (1936) monitored bacterial leaf blight in relation to winter temperature for a series of years and predicted increased severity of the disease in seasons following mean temperature for December, January, and February that totaled 68 C or more. The relationship of insects to this disease was observed by Rand and Cash (1933). They were able to transfer the organism from diseased plants to healthy plants under cage conditions with the corn flea beetle and the southern corn rootworm.

The importance of the corn flea beetle in the incidence of bacterial leaf blight was investigated extensively by Elliot and Poos (1934) and Poos and Elliot (1936), particularly in connection with the overwintering of the organism. They found that at least 19% of the overwintered corn flea beetles carried the organism. In tests of some 40 species of insects as possible vectors they concluded that only the corn flea beetle and the toothed flea beetle were vectors of the pathogen under field conditions. From May to September in 1934 they found 40% of the corn flea beetle harboring the pathogen. Under experimental conditions, the southern corn rootworm transferred the pathogen from diseased to healthy corn. An association between winter temperatures and abundance of the corn flea beetle and incidence of bacterial leaf blight was indicated in 1934 in New York and the New England states. In that year the corn flea beetle was not found abundantly north of central Pennsylvania and bacterial leaf blight was much reduced over that of 1932 and 1933.

The favored overwintering habitat of the corn flea beetle is in grass sod and in well-drained areas. During warm periods in early spring they can be collected from a variety of grasses and small grain, particularly wheat. Although the biology of larval and pupal development is somewhat fragmentary, the evidence at hand is that the first generation may be completed on these grasses although some of the overwintering population of beetles may feed upon and infect early planted corn. A short life-cycle is indicated and populations may reach high concentrations by midsummer, when elongated feeding scars and wilt lesions can be readily observed. Poos (1955) recovered larvae of the corn flea beetle from soil surrounding 21 species of Gramineae belonging to 17 genera. Poos (1939) reported that it was strongly indicated that Kentucky bluegrass, orchardgrass, and wheat, which do not show symptoms, provide inoculum for the vector before corn is available in the field. The pathogen was recovered from these species after beetles exposed to diseased corn plants fed upon them. Little information has been developed on the host-vector relationship of this wilt disease since these reports were published. After many years of experience and observation on this problem the writer suspects that symptomless hosts such as winter wheat and corn flea beetle relation thereto could be one of the factors involved in the low incidence of the corn flea beetle and bacterial wilt in central Iowa where very little wheat is produced.

Virus and Spiroplasma Diseases

Most of the insect vectors of virus diseases in corn belong to the aphid and leafhopper families. Until some 10 years ago virus diseases were not regarded as threatening problems of corn production in the USA. For many years virus diseases encompassed those maladies whose pathogens could be transmitted by biological or mechanical techniques but could not be observed by the maximum magnification of available microscopes. The electron microscope has made it possible to describe the configuration of these pathogens, which has aided in the identification and classification process. This has also resulted in separating many pathogens from the viruses and establishing an additional hierarchy of classification currently designated as spiroplasms. Most of the Cicadelid leafhopper-transmitted pathogens now fall into the spiroplasma group and are causal agents of yellows disease.

A sugarcane-mosaic pathology program identified the corn leaf aphid as a vector of sugarcane mosaic in corn (Brandes, 1920). About the same time Kunkel (1921) reported the corn planthopper (a Fulgorid) as a possible vector of corn mosaic, later referred to as corn stripe. This disease was later determined to be distinct from sugarcane mosaic (Stahl, 1927; Kunkel, 1927). Kunkel's studies were confirmed by Carter (1941) and McEwen and Kawanishi (1967).

Streak disease of corn has had a long history in parts of Africa. H.H. Storey's many publications on this disease and its vectors extend from 1924 to 1967. He made detailed studies of the vector relationships of *Cicadulina mbila, C. zeae,* and *C. storeyi* in epidemiology of what was identified as a virus disease. In his many years of experimentation he was successful in transmitting the pathogen only with these species of leafhoppers (Storey, 1939).

In 1963 a new regime in virus investigations began primarily in corn and sorghum, as a result of pockets of virus disease outbreaks in Mississippi, Louisiana, and Ohio. Stoner (1968) compiled reports from many investigators representing practically all of the corn or sorghum producing states, and provinces in Canada. He summarized the results of investigations in progress from 1963 to 1967. The Ohio Agricultural Research and Development Center (1973) issued a list of the many references pertaining to Maize Virus Diseases and Corn Stunt.

Samples from Mississippi led to a determination of Corn Stunt Virus (Maramorosch, 1963) with *Dalbulus maidis* as the vector. Many years of investigation by Kunkel and Maramorosch with leafhopper vectors had established that CSV (now classified as a spiroplasma) was a persistent virus with an incubation period of about 2 weeks.

Working with an Ohio source of disease, Stoner et al. (1964) established that the corn leaf aphid was a nonpersistent vector of the virus infecting corn in southern Ohio. They were able to mechanically transmit the pathogen and to observe that the disease was different from known virus diseases of corn. The disease, as it prevailed in the southern area, became known as corn stunt and that found in the Ohio river areas, maize dwarf mosaic virus (MDMV). It was early recognized that johnsongrass was a factor in the incidence of this disease in the field. This grass is a favored host of the corn leaf aphid. Fig. 17 shows a typical mosaic virus infested plant. Shephard (1965) reported on the properties of mosaic virus of corn and johnsongrass and the relation to sugarcane mosaic virus. Detailed studies on the properties of the virus and the vector relationship were reported by Bancroft et al. (1966) and Messieha (1967). It was shown that the virus was transmissable by 11 species of aphids, namely *Dactynotus sp., Brevicoryne brassicae, Rhopalosiphum padi, Rhopalosiphum maidis, Aphis maidiradicis, Myzus persicae, Rhopalomyzus poae, Therioaphis maculata, Acyrthosiphon pisum, Aphis gossypii,* and *Aphis craccivora.* The acquisition period of the virus was short (10 to 20 sec.) and that the retention period by the aphids was only 10 to 20 min. Effective sources of host-vector relationship were shown in 10 species of Gramineae. In tests of many grasses, Ford (1967) determined that 10 species of perennial grasses found in Iowa were susceptible to MDMV when the grasses were mechanically inoculated.

THE MOST IMPORTANT CORN INSECTS

A series of papers on vector-host, and pathogen relationships have been published from several institutions. All of this research has some sequential bearing on advances in this area of research. Several workers have contributed to recent developments (Choudhury and Rosenkranz, 1973; Davis et al., 1972; Granados et al., 1966, 1968; Maramorosch et al., 1968; Nault and Bradley, 1969; Nault et al., 1973; Pitre, 1968; Pitre and Hepner, 1967; Rosenkranz, 1969).

An isometric virus causing chlorosis and stunting in corn was recovered from corn, johnsongrass, sweet sorghum, giant foxtail and yellow foxtail by *Graminella nigrifrons* and transmitted to corn. This virus has since been named maize chlorotic dwarf virus (MCDV) (Bradfute et al., 1972; Nault et al., 1973). The more recent information gathered by investigators working

Fig. 17 — Inbred line infected with mosaic virus.

with *G. nigrifrons* has identified johnsongrass as a reservoir for two viruses, MCDV as only transmissible biologically by *G. nigrifrons* and MDMV as aphid-borne and mechanically transmissible.

Periodically wheat streak mosaic virus (WSMV) has been prevalent in corn, but the disease caused by this pathogen has not been as virulent as in wheat. Slykhuis (1953), in transmission work, established that *Eriophyes tulipae,* an eriophyd mite, was an efficient vector. Oldfield (1970) published a review of the literature on mite transmission of plant viruses. The vector, *E. tulipae,* is widely distributed and has been recorded on a variety of hosts. On corn, *E. tulipae* causes kernel red streak which is mostly concentrated on the tip end of the ear. This kernel condition was at first thought to be the result of WSMV infection. Later, it was demonstrated that non-viruliferous as well as viruliferous mites were able to induce the red streaks through their feeding activity on the kernels (Nault et al., 1967).

The classification of pathogens causing yellows diseases in plants, usually transmitted by leafhoppers, as virus was questioned by Doi et al. (1967). Their studies of electron microscope sections of mulberry infected with dwarf disease showed microorganisms in the phloem elements that they described as mycoplasma-like agents. Ishiie et al. (1967) reported that some tetracycline antibiotics had a suppressive and therapeutic effect on mulberry dwarf infected plants. Similarly, Granados (1969b) reported such effects with the corn stunt pathogen in corn and in the leafhopper vector, *Dalbulus elimatus.* These events set in motion the reexamination of many diseases, classified as viruses, transmitted by leafhoppers. An extensive literature has developed in this area of pathology. Maramorosch et al. (1968, 1970) and Granados (1969b) described mycoplasma-like structures in diseased plants and in insect vectors. Whitcomb and Davis (1970) have published a review of literature on this matter. Recently, the generic name *Spiroplasma* has been proposed for the spiral-like or helical pathogens that have been referred to as mycoplasma-like agents for several years (Davis and Worley, 1973; Whitcomb et al., 1973). Corn stunt, transmitted by leafhoppers of *Dalbulus* spp., falls into this classification (Davis, 1973). Granados (1969a) obtained suppressive effects with tetracycline antibiotics on corn stunt symptoms and in the ability of the leafhopper vector *(D. eliminatus)* to transmit the pathogen.

A number of virus or spiroplasma diseases, commonly referred to as noneconomic, found in small grain and other grasses, have been transmitted to corn. A list of these diseases is given in American Phytopathology Society (1973). The potential of such causal pathogens and their vectors should not be discounted. Carter (1973) has assembled a large mass of information on insects in relation to plant diseases.

STORED-GRAIN INSECTS

The most injurious insects found in stored corn also infest other grain, cereal products, and stored food or feed. The problems of controlling these insects on corn are frequently related to those on other grain and products in which culture infestations are unknowingly maintained. Cotton et al. (1953) have brought together the significant information on the causes of outbreaks of stored-grain insects. To a considerable extent this information applies to the heavy grain-producing region of the North Central states but much of it also applies to other regions.

Stored-grain insect infestations in corn are most intense in warm climates and tropical areas where the problem begins in the field and continues on through the storage period. Varieties with good husk cover are helpful in protection against heavy field infestation. The following section gives only some of the more general information on a group of the most important forms encountered in corn storage and the conditions under which they occur.

Rice Weevil and Maize Weevil

Description, Injury and Distribution

The rice weevil (*Sitophilus oryzae*) and the maize weevil (*Sitophilus zeamaize*, also called the larger rice weevil) are dark reddish-brown snout beetles usually about 4 mm long, with light spots on each wing cover. The thorax is densely pitted, legs are prominent, and the wings are well developed. The larvae which feed in corn and small grain are white, legless, thick-bodied grubs.

These weevils are warm-climate insects and infest the grain in the field and continue reproducing in storage until the grain is used. In the lower temperate zone and in tropical areas they are classed among the most important economic insects. In the USA field infestations of the rice weevil extend into the southern part of the North Central states region following mild winters. The primary source of infestations in stored corn by this species in colder areas is from the transportation of infested grain or from small local cultures located in favorable spots for surviving the winter (Dean, 1913; Cotton, 1920; Kuschel, 1961; Widstrom et al., 1972a).

Habits and Biology

It is typical for the *Sitophilus* adults to eat small cavities in the grain, in which the eggs are deposited. Under favorable temperature and humidity, egg deposition by the rice weevil may extend over 4 or 5 months. On hatching the larvae feed in the grain on both germ and endosperm. Pupation takes place in the grain in the cavity made by the larva during its feeding. During the summer months Cotton (1920) observed that it required about 35 days to complete a generation. There were as many as seven generations annually in the South. This number is dependent on the bioclimate and would be considerably less in northern areas. Figure 18 shows a mature ear infested with rice weevil (also some Angoumois grain moth) and typical grain with adult emergence holes.

The adult beetles are strong fliers, which is an important factor in the dissemination and establishment of field infestations where there is normally an overwinter survival in stored grain. Cotton (1920) observed that the infestations in corn in the field began after the grain became firm. This corresponds closely to the time earworms have emerged from the ear and left openings favorable for weevil access. The relationship between poor husk protection and increased injury by the corn earworm and rice weevil in the field and in storage was reported by several workers (Cartwright, 1930; Cotton, 1920; Eden, 1952; Floyd and Powell, 1958; Hinds, 1914; Kyle, 1918; Wiseman et al., 1970).

The relationship between injury of ears in storage and field infestations by the rice weevil is further supported by the results on the control of earworms by insecticides. Douglas and Smith (1953) reported almost complete absence of the rice weevil in ear corn until harvest in November in local fields where they had practically eliminated the corn earworm and fall armyworm with a DDT-mineral oil spray.

Strains of corn that gave protection against infestations by the corn earworm and the rice weevil were developed by Kyle (1918). In recent years programs have emphasized this approach in areas where rice weevils are chronic pests. Eden (1952) confirmed that rice weevil injury decreased as the husk extension and layers increased but observed that the effect of the two characters was independent. In tests of kernel hardness in 20 varieties and hybrids adjusted to 12% moisture he calculated that there was a 1% decrease in damage for each $16/in.^2$ of pressure required to penetrate the kernels with a 1.5 mm hollow punch. A variation in pericarp thickness was not correlated with the amount of damage.

VanDerSchaaf (1969) tested 337 samples of corn obtained from the International Germplasm Seed Bank, Chapingo, Mexico. Of 96 of the more resistant samples tested by the no-choice method, 20 contained various degrees of resistance when tested by a free-choice method. He concluded that strains 214 (coastal tropical flint), 222 (Chaudelle), and 291 (unknown race) were substantially resistant.

In a test of 10 hybrids against an insect complex, Wiseman et al. (1970) ranked the corn earworm, pink scavenger caterpillar, and the maize weevil as the insects of most economic importance on corn in their studies in Georgia. Some hybrids were resistant to insect damage by virtue of husks and/or kernels.

In an appraisal of methods of measuring corn kernel resistance to the maize weevil in 41 inbred lines, Widstrom et al. (1972) determined that among six criteria the total progeny reared rated better or comparable to any of the other methods evaluated. The criteria were: (1) grain weight loss, (2) total weevil progeny, (3) % damaged kernels, (4) % of parent mortality, (5) % of progeny mortality and (6) weight/weevil.

The Granary Weevil

Description, Injury, and Distribution

Granary weevil (*Sitophilus granarius*) is closely related to and easily confused with the rice weevil. The light spots on the wing covers, typical of the rice weevil, are lacking, however, and the wings are reduced to a vestige. The larva inhabits the grain like the rice weevil. It has a wide host range among cereal grains and products, in which it has become a cosmopolitan pest (Dean, 1913).

Fig. 18 — Rice weevil infestation originating in the field.

Habits and Biology

The granary weevil has a life history very similar to that of the rice weevil. Observations in corn storage bins have shown that it is not as widespread a pest of stored corn as the rice weevil. It is somewhat more tolerant to colder climates than the rice weevil (Decker, 1941) and is not considered a serious pest of corn. In Kansas, Dean (1913) has reported four or five broods annually. Because of their inability to fly, the beetles become disseminated only through the transport of grain and other infested products stored under conditions favorable for their multiplication and survival.

Flour Beetles

Description, Injury, and Distribution

The flour beetles (*Tribolium* spp.) are flat, reddish-brown, and about 4 mm long. The larvae are about 6 mm long when full grown and have a forked termination of the last abdominal segment. Although they are usually associated with flour and meal storage they are commonly found in stored corn. Like most of the other stored-product insects they are worldwide in distribution and thrive on a large variety of seed and food products. In the extreme northern states and Canada survival is believed to occur only in heated buildings. The most common species are *T. castaneum* and *T. confusum* (Good, 1936). In the late summer *T. castaneum* becomes abundant in shelled corn stored in the commercial corn area.

Habits and Biology

Both adults and larvae feed on stored corn but prefer broken kernels to undamaged grain. Because they shun light they are seldom noticed when populations are low. The adults are long-lived, but the developmental stages are short. The oviposition period may extend for as long as a year. The number of generations produced annually depends on the bioclimate. Continuous breeding is believed to occur in the Gulf states where *T. castaneum* may be associated with infestations of other grain insects in the field. Dissemination by flight for a short distance is considered possible only for *T. castaneum*. For this reason this species is the predominant one in farm-stored grain. *T. confusum* has been extensively used as a test insect for nutrition research.

Flat Grain Beetle

Description, Injury and Importance

The flat grain beetle *(Cryptolestes pusillus)* is flat, reddish-brown, and about 1.5 mm long, with long antennae extending 'V'-shaped from the head. The adults feed mostly on grain dust or broken kernels, whereas the larvae burrow into the germ. This species has been noticeable for its presence in large numbers in out-of-condition corn in storage. Infestations show grain

THE MOST IMPORTANT CORN INSECTS

with the germ eaten out and association with primary infestations by the rice weevil and the cadelle beetle (Cotton, 1947; and Cotton et al. 1953).

Habits and Biology

The flat grain beetle is attracted to and breeds in grain that is in poor condition. Decker (1941) and Cotton et al. (1953) report the species as among the most abundant form in shelled corn stored in steel bins. It has, however, not been regarded as a serious pest in sound corn. A closely related species *C. ferrugineus* is smaller in appearance and habits to *C. pusillis*. Cotton (1947) states that it is more resistant to cold weather and more commonly found in stored grain in northern states.

Sawtoothed Grain Beetle

Description, Injury, and Distribution

The sawtoothed grain beetle (*Oryzaephilus surinamensis*) is readily identified by the six tooth-like projections on each side of the thorax. These also occur on the pupa. It is dark brown and about 2.5 mm long. The larva is slender and yellowish-white, with three pairs of legs. Both adult and larva are free feeders upon or in grain, injuring more seeds than they consume. The species is worldwide in distribution and infests many grains, seeds, and foodstuffs.

Habits and Biology

The adults of this beetle are long-lived, whereas the immature forms develop relatively fast requiring only about 3 weeks under favorable conditions. The egg deposition period ranged from 6 to 10 months, according to Back and Cotton (1926). On completing development, the larva transforms into a pupa in a light cocoon bound together with foodstuff fragments. Because of the long and variable adult life, it is difficult to determine the number of generations. In the vicinity of Washington, D.C., Back and Cotton (1926) estimated from four to five generations and believed breeding to be continuous in subtropical and tropical climates. The winter is passed in the adult stage. Under low humidity in heated buildings, development of larvae was much inhibited.

Cadelle

Description, Injury and Distribution

The cadelle beetle (*Tenebroides mauritanicus*) is one of the largest of the common stored-grain pests. The adult is dark reddish-brown, oblong, flat, and about 8 mm long with a conspicuous constriction just ahead of the wings. The mature larva is about 20 mm long, white, with a black head and two caudal spines. The head and thoracic segments are narrower than the abdominal segments. Both adults and larvae prefer the germ for food.

The larva has a habit of burrowing into wooden structures before pupation, a habit which is an important factor in control practices in some areas. Other grain insects usually follow up its attack. The species is worldwide in distribution. It has a wide host range among grains and stored products (Back and Cotton, 1926). It is not one of the abundant species in shelled corn stored in steel bins.

Habits and Biology

Among the important stored-grain insects the cadelle is probably the most hardy under winter temperatures in the more northern corn-growing areas. The overwintering population is made up of both adults and larvae. The infestations in the spring result from the beetles and larvae that survive the winter. In life history studies, Back and Cotton (1926) reported oviposition throughout the spring and summer months. Continuous breeding occurs in warm climates.

Angoumois Grain Moth

Description, Injury, and Distribution

The injurious form of the Angoumois grain moth (*Sitotroga cerealella*) is a small white caterpillar with yellowish head which feeds within the grain. Pupation occurs in the cavity hollowed out by the larva. Typical indication of an infestation are kernels with round holes or with small channels stopping just below the seed coat. The moths are buff-colored with pointed hair-fringed wings measuring about 12 mm when spread (Flg. 19). As a pest of grain the Angoumois grain moth is ranked next to the rice weevil in economic importance. It is distributed throughout the world, infesting grain in storage and in the field in the lower temperate and tropical areas. It only occurs in outbreak numbers in the southern part of the central Corn Belt after a series of warm winters allow populations to build up. In the southern area it is never as troublesome in corn as the rice weevil (Cotton et al., 1953; Simmons and Ellington, 1933).

Habits and Biology

In areas where the Angoumois grain moth is a pest, corn serves as an important source for building up infestations in other grains, particularly in the field and in storage. As with the rice weevil corn becomes infested in the field, the ears with loose husks and exposed tips being most susceptible to attack. In warm climates there is a continuing infestation in cribbed corn. In stored shelled corn the infestation is concentrated near the top of the bin. Simmons and Ellington (1933) reported on the life history and host activity of the insect in Maryland. In that area they observed winter survival primarily in the full-grown larval stage in grain in storage and in wheat straw piles, baled straw, and in litter. Light infestations were present in corn with exposed ear tips at harvest time. Cartwright (1930) presented data that showed an inverse relationship between damage and husk extension beyond the tip of the ear.

THE MOST IMPORTANT CORN INSECTS

Fig. 19 — Angoumois grain moth showing long hair at the end of the wings, and kernels with emergence holes.

The effect of infestation by the Angoumois grain moth on germination and stand was investigated by Everly et al. (1963). Infested popcorn was found to be unsuitable for seed, and germination was not increased by an application of a fungicide. Unless dent corn was highly infested, germination and final stand reduction was not serious. Treatment of dent corn seed with a fungicide decreased loss in stand. Infestation in seed reduced seedling vigor in dent corn by 50 and 90% in popcorn.

Different levels of amylose content in the endosperm and their effect on the biology of the Angoumois grain moth were studied by Peters et al. (1972a). They found highly significant differences in the weight of moths when larvae were reared in strains of corn with endosperm differing in levels of amylose, but there was no difference in the number of moths emerged per kernel. A decreasing rate of infestation resulted when moths collected from certain strains were used to infest noninfested ears. In further tests Peters et al. (1972b) reported that amylose and fat content of corn kernels was negatively correlated with weight of moths. However, kernel hardness had a significant effect on the weight of male moths. Size of female moths appeared to be reduced by the size and weight of kernels of a strain of popcorn.

Indian-Meal Moth

Description, Injury, and Distribution

The fullgrown larvae of the Indian-meal moth (*Plodia interpunctella*) are about 12 mm long and vary in color from whitish to green or pink. When at rest the moths are easily recognized by their brown color and a light, transverse band on the forepart of the wings (Fig. 20). In typical injury the

Fig. 20 — Indian meal moth showing the typical whitish cross-band on the wings.

embryo of the corn kernel is devoured, and silken webs are spun over the infested grain or meal. It breeds in ear, shelled, and ground corn and on a wide variety of grain and cereal products. It is widely distributed in grain-growing countries (Dean, 1913).

Habits and Biology

In the southern areas the Indian-meal moth breeds in ear corn through most of the year. In the North Central States the moths fly to the bins of shelled corn in midsummer and lay their eggs on the surface grain. In some years this insect is so abundant that the entire surface of the bins are covered with a sheet of silk. In 1972, damaging infestations of bulk and bagged seed corn in commercial warehouses were observed in Iowa. Survival of large numbers of larvae exposed in an unheated building was observed at Johnston, Iowa, in January 1974 following a week of near and below -32 C. In this case the larvae pupated shortly after being brought into the laboratory, indicating a nondiapause condition. What appeared to be normal moths emerged. In chemical control this species has shown high resistance to the commonly used malathion seed treatment.

Other Insects

Among other insect pests of stored grain the foreign grain beetle (*Ahasverus advena*), the hairy fungus beetle, (*Typhaea stercorea*) and the larger black flour beetle (*Cynaeus augustus*) are often quite abundant in shelled corn stored in the North Central States. They are attracted to corn that is slightly out of condition, but seldom cause appreciable damage to corn in good condition.

Control of Stored Grain Insects

The practices recommended for controlling infestations of insects in stored corn or small grain vary with the environmental conditions under which the crop is produced, the handling of the crop and storage facilities, and the species present. In the environment, temperature and moisture are the important limiting factors in the prevailing abundance of these pests in different regions.

Taking climatic conditions into consideration, Cotton (1938) evaluated the storage problems in four general regions. In Region 1, which comprises the northern corn-growing areas, crib-stored corn is free of these problems overwinter. In shelled corn the sources of infestation are from old grain and feed situated in places favorable for winter survival. Farm storage is considered to be relatively safe. Region 2 comprises the central part of the commercial corn-growing area. Monthly inspection of grain during the warm months is advised. In Region 3, which consists of a zone from Pennsylvania to Virginia, the southern part of the North Central region and the northern part of the southern region, grain corn storage becomes more hazardous because of more favorable conditions for winter survival in susceptible products stored in protected places. In the southern part of this region, field infestations of the Angoumois grain moth and rice weevil — beginning in wheat before harvest and continuing in corn later in the season — make crib and bin storage on the farm hazardous. Under these conditions monthly inspection, except in the winter, is advised. In the southern corn-growing areas, designated as Region 4, conditions are favorable for the development of storage pests and for their winter survival. Field and storage infestations at varying levels are chronic under farm conditions. Immediate treatment when grain is placed in storage, and monthly inspection for possible retreatment is essential in this area.

Control Practices

Systematic application of preventive measures and chemical treatment are effective in keeping stored grain pests in check. Practices for farm-stored grain have been discussed by Cotton et al. (1953); and Walkden et al. (1954). The following are effective preventive measures.

Store in weather-tight rodent-proof bins, preferably of steel.

Clean out all bins before loading with grain;

Spray walls and floors of bins and adjacent woodwork of farm buildings with an approved and registered material;

Clean up and dispose of litter, waste grain, and feed that have accumulated in and around farm buildings.

Store only dry grain (12% moisture or less).

Insecticides registered for treatment of stored grain insects are labeled with ample information about the insects, dosage, use of the material, and handling hazards. Methods and materials are not static. When a grain insect problem is encountered it is advisable to seek the advice of a state extension service or research personnel who are familiar with species identity and current control practices.

INSECT RESISTANCE IN CORN

Insect resistance factors in corn are heritable and have become established through an evolutionary process of mutation and natural selection by man. Painter (1951) defined three general types of insect resistance in plants: (1) antibiosis, (2) tolerance, and (3) preference or nonpreference. In relative antibiosis, the host plant has a varying degree of effect on the development and survival of insect species. Relative tolerance is the varying effect an insect has on its host; whereas, preference as a variable, involves the degree of avoidance of the host for oviposition or food. These three types of resistance may, to some degree, occur in the same host variety.

Because of its wide variability corn inherently produces adaptations that permit it to survive in very different environments. Husk protection qualities on the ear are well developed in southern and tropical areas where the hazards from earworms and graininfesting insects can be severe. In more northern areas, where these hazards are minor or do not exist, the selection pressure has been for shorter seasoned varieties with relatively loose husks adapted for fast drying, ease of hand or machine picking, and safe storage of grain. The bracts surrounding the seed of some sorghum varieties give protection against the maize weevil. Biological as well as physical hazards play an important role in the varietal development process in a particular environmental area.

Breeding for insect resistance in corn and other crops has received an increasing amount of attention in recent years. Perhaps some of this emphasis was generated by the development of strains of insects resistant to insecticides. By and large the stress of epidemics has a beneficial effect on efforts for varietal improvement. Probably the most important factor in recent advances is the interdisciplinary approach to varietal development. In these programs entomologists assume major responsibility for biological relationship studies between host and pest and for exploration in collections of varieties for locating resistance factors. Plant breeders have primary responsibility for utilizing resistance factors and determining the nature of inheritance. A by-product of biological relationship research are methods of evaluating relative resistance. Efficient methods to evaluate infestation are needed by plant breeders.

Insect Rearing Techniques

The USDA and state experiment stations have for many years cooperated to improve hybrids for resistance to a number of corn insects. This work was initially done under natural infestations. The oldest continuous program to employ extensive artificial infestation is that on the European corn borer which has been in progress over 45 years. The source of infestation was egg masses produced in the laboratory by moths that emerged from infested cornstalks stored in cages (Dicke, 1932; Dicke, 1954; Guthrie et al., 1965; Patch, 1947).

In the first edition of this monograph George Sprague stated, "The lack of regular infestations has proved to be a serious deterrent to studies on inheritance of resistance or breeding for resistance." There are now numerous institutions that have research projects on insect nutrition, insect pathology, antibiotics, and artificial diets, involving many species of insects. The Diptera, Lepidoptera, and Coleoptera have been intensely involved in these developments. Artificial diets with an agar base are now in common use for mass-producing several species of corn insects for plant resistance and other biological research. Among corn insects, most of the studies on nutrition and the development of artificial diets has centered around lepidopterous pests. Soil insects, such as rootworms (Coleoptera) and sucking forms, particularly Homoptera (leafhoppers and aphids), are more difficult to rear on artificial diets. Some species of aphids, however, have been reared on artificial media (Auclair, 1963). It is appropriate to review some of these contributions.

The first work on an artificial diet for nutrition studies on the European corn borer and the chemical nature of plant resistance was reported by Bottger (1940, 1942). Bottger collaborated with J.D. Sayre, who assisted in making estimates for nutrient requirements based on the nature of the tissue on which the larvae develop. In this attempt a single generation was completed. Stanley Beck, of the University of Wisconsin, and his students have made many contributions on the nutritional aspects of artificial diets and also on the chemical nature of resistance (Beck and Stauffer, 1950, 1957; Beck et al., 1949; Chippendale and Beck, 1964).

The advances made in artificial diets have been largely dependent on modifications of sources of protein, such as casein, wheatgerm, and soyameal, and in minerals, sugar, and vitamins. An important ingredient in diets is the antimicrobial agent to reduce contaminating microorganisms to a minimum (Guthrie, 1971; Lynch and Lewis, 1971; Ouye, 1962). Many contributions on artificial diets with several cotton insects have been made by Erma Vanderzant. Recently she assembled a review of artificial diets of insects (Vanderzant, 1974). Artificial infestation techniques for establishing infestations to measure differential earworm damage have been in use for many years (Blanchard et al., 1942). Bennett and Josephson (1962) used an artificial diet to rear earworm larvae for infesting corn ears.

Methods for large-scale production of the corn earworm and fall armyworm on artificial diets have been developed at the Southern Grain Insect Laboratory, Tifton, Georgia, for plant resistance and other research (Burton, 1969; Perkins et al, 1973).

Artificial diets have been developed for rearing the southwestern corn borer for plant resistance and other biological research (Davis et al., 1973; Jacob and Chippendale, 1971; Reddy and Chippendale, 1972).

Since the development of strains of the western and northern corn rootworms resistant to chlorinated hydrocarbon insecticides, particularly aldrin (Ball and Weekman, 1963; Bigger, 1963; Blair et al., 1963; Patel and Apple, 1966) there has been a pronounced effort devoted to finding germplasm with protective resistance factors against these two species as well as the southern corn rootworm. Although small-scale methods for artificially infesting corn plots have been used, applications for the usual-sized breeding and testing programs have not been adequate. However, delayed plantings of corn for attracting ovipositing beetles in the late summer (the northern and western species overwinter in the egg stage) has given adequate infestations in some areas the following season to permit effective breeding work.

Corn Rootworm

The studies on rootworm resistance or tolerance in corn have been largely restricted to three species, *Diabrotica virgifera* (western), *D. longicornis* (northern), and *D. undecimpunta howardi* (southern). The development of larvae in relation to root structures is similar for the western and northern species in that the newly hatched larvae prefer to feed on roothair and small rootlets (see Fig. 1), and outer cortex followed by feeding on tender crown roots as they emerge behind the leaf sheath attachment. The southern species is more commonly found feeding on tender root tips and burrowing in the center of the young roots.

Up to the present, resistance or tolerance factors have dealt with the ability to regenerate new roots after injury, volume and fibrousness of roots, angle from vertical, and relative resistance to primary and weak root pathogens. Figure 21 illustrates different root types and responses. For practical purposes agronomists have evaluated rootworm tolerance by differential root lodging. The stage of plant maturity and root development are important factors in relative differences in plant response.

To obtain detailed data on root injury it is necessary to dig and clean roots by pressure washing. Estimates of injury can be readily made by a rating system. A system in common use for evaluating relative varietal resistance qualities is the 9-class system. Figure 22 shows high susceptibility in lodged row (class 9) and excellent tolerance in row to the right (class 1). Vertical-pull devices have been successfully used to obtain comparative tolerance ratings by Ortman et al. (1968) and Zuber et al. (1971). These researchers found a highly significant correlation among root-pulling resistance, root volume, and root classification, indicating that any of the three characteristics would be suitable for tolerance evaluations. Mo22 × T8 had the smallest pull difference between the treated and untreated plots. Fitzgerald et al. (1968) evaluated roots of commercial hybrids through mechanical damage and concluded that root pruning could be useful in selecting for root regrowth potential in breeding for tolerance or resistance to rootworms in the absence of an infestation.

Some of the early records on rootworm tolerance or resistance were reported from Ohio and Illinois (Anonymous, 1937-1938; Bigger et al., 1938, 1941). In these tests, involving the southern corn rootworm, the inbreds Oh56 and 38-11 were least damaged. In a 1952 test of 200 topcrosses in Iowa, the parent lines 38-11, B2, and B14, or derivatives of crosses involving these lines, regenerated their roots and had good lodging resistance under a high level of northern corn rootworm infestation.

With the rapid dispersion of the western corn rootworm, much of the recent resistance research has been done under the population dominance of this species in Iowa, Missouri, and South Dakota. Eiben and Peters (1965) reported the following inbred lines as being above average in performance under rootworm infestation: SD10, Hy, B57, B67, B9A, B45, and B46. Fitzgerald and Ortman (1964) tested many inbred lines, single crosses, synthetic variety derivatives, and plant introductions (PI) at different locations in South Dakota and Iowa. Several inbreds were consistent in a high level of performance under infestation and transmitted these qualities to single

Fig. 21 — Characteristic root types representing six inbred lines, from right to left: (1) strong, spreading, deep system with excellent lodging resistance; (2) vertical, deep, weak resistance to lodging under rootworm attack; (3) intermediate in root numbers and spread, disease resistant; (4) relatively high root numbers with dark lower portion disease-infested; (5 and 6), relatively poor root system, susceptible to lodging under rootworm attack.

crosses. They also found wide differences in leaf feeding on inbred lines by western corn rootworm beetles. Sifuentes and Painter (1964), in Kansas tests, estimated that resistance to leaf feeding by western corn rootworm beetle had a monogenic type of inheritance.

In a study of the performance of 22 inbred lines, using nine different measurements, Ortman and Gerloff (1970), concluded that correlations with field performance are best for a root growth index and root pulling resistance. Tolerance may also be expressed in relative degree of stunting (Fig. 23).

In a survey of 2,000 plant introductions, Wilson and Peters (1973) selected 441 entries for a retest. Among the retest entries the following showed the best response to infestations: PI 177606, PI 239099, PI 214288, PI 177245, PI 303923, and PI 257625. The resistance was believed to be mostly an expression of tolerance.

The resistance in sorghum roots to western corn rootworm larval attack has been attributed to a cyanogenic glucocide. Cyanide is liberated by the enzymatic action of a hydrolytic glucosidase (Branson et al., 1969). In a study of root resistance to larval feeding of the western corn rootworm, reciprocal crosses between corn and *Tripsacum dactyloides* gave mixed results. The roots of *T. dactyloides* were highly resistant to larval attack. *Z. mays* × *T. dactyloides* was susceptible, whereas the reciprocal cross was resistant. Branson and Guss (1972) explained these results as indicating that the resistance was either inherited through the cytoplasm or that the resistance genes were on the lost *Tripsacum* genome in the *Z. mays* × *T. dactyloides* cross.

Corn Earworm

Through a long period of ear selection under stress of earworm populations, protective husk qualities became fairly well fixed in tropical and southern open-pollinated varieties. In contrast these qualities were not well developed in the Corn Belt varieties. Evidence of this is apparent in the results of some varietal tests made by Garman and Jewett (1914). Hinds (1914) pointed out that the extension and tightness of husks provided protection against earworm injury and subsequent field infestation by the rice weevil. Collins and Kempton (1917) demonstrated that characters for resistance to ear infestation could be transferred from field corn to sweetcorn and that husk extension and the number of husk layers were associated with protection against earworm injury. Kyle (1918) reported selections with such characters in Georgia. In a comparison of varieties with different types of husk cover and extension, Phillips and Barber (1931) concluded that ears protected by a long, tight husk had less earworm damage and suffered less from grain beetles, grain moths, and birds. The varieties of southern origin gave the best protection against earworm injury.

With the establishment of many corn breeding programs and development of hybrids in the 1930's by federal and state agencies, evaluations of in-

Fig. 22 — Difference in susceptibility to corn rootworm infestation; center row very susceptible; right row with good resistance.

THE MOST IMPORTANT CORN INSECTS

bred lines and hybrid combinations for resistance to the earworm and associated field infestation of grain insects became a cooperative, routine procedure. Information on relative resistance of numerous inbred lines and hybrid combinations came from many locations in southern and Corn Belt areas and have been reported by Painter and Brunson (1940), Blanchard

Fig. 23 — Inbred line showing stunting effect from rootworm attack (foreground); rear of row treated with insecticide.

(1941); Richey (1944); Dicke and Jenkins (1945); Douglas (1948); Blanchard and Douglas (1953), and Douglas and Eckhardt (1957). The chemical nature of resistance factors in corn silk, which is a primary structure for establishment of newly hatched larvae, has been a subject of investigation. Walter (1957) presented evidence of a lethal factor in silks of sweetcorn. Nutritional factors in silks as a source of relative resistance have not been conclusive. However, reducing total sugars were reported to be higher in the susceptible single crosses by Knapp et al., (1965). In a comparison of silks of the resistant inbred Ab18 and the susceptible inbred Ab34, McCain et al. (1963) found no qualitative differences in free amino acids, but quantitative analyses were not made.

The area of preference as a feeding response in larvae of the corn earworm on plant parts of corn and other hosts has received extensive attention in recent years. Starks et al. (1965) in preference experiments with fourth-instar larvae showed that water extracts of silks and kernels possessed a feeding stimulus. In pursuing this phase of host plant resistance, McMillian and Starks (1966) found a differential larval feeding response in water extracts from primary and secondary hosts, corn silks and kernels being preferred over extracts from sorghum, cotton, tomato, and tobacco plant parts. Such differential feeding responses were also found between corn lines (Starks and McMillian, 1967).

In a study of husk qualities among hybrids, significant protection against several insect species resulted in increased yields. The corn earworm and the pink scavenger caterpillar were found to be more damaging than the maize weevil (Wiseman et al., 1970). The various facets in Fig. 24 shows the protective effect of husk qualities in a comparison where the exposure to a corn insect complex was on slit and nonslit husks. Earworm resistance in corn and other related biological problems have been presented in a comprehensive review by McMillian and Wiseman (1972).

European Corn Borer

The early tests on varietal resistance showed that the early varieties of sweet and flint corn supported higher populations of larvae than the longer seasoned dent varieties (Crawford and Spencer, 1922; Caffrey and Worthley, 1927). Huber et al., (1928) found differences in the larval population between varieties of corn and the ability of some varieties to tolerate damage. In tests of F_3 and F_4 lines extracted from crosses of 'Maize Amargo' and local Michigan varieties, Marston (1931) reported higher populations of larvae on the local varieties than on the recovered lines.

With the advent of extensive breeding programs by federal and state agencies in field, sweet, and popcorn there was continued testing of inbred lines and hybrid combinations of them. The results of these tests were reported by Patch et al. (1942), Patch and Everly (1945), Schlosberg and Baker (1948), Meyers et al. (1937), and in numerous special reports of federal and state agencies.

It was established that resistance and tolerance qualities of certain lines were transmitted to their crosses and that there was a significant correlation between stalk breakage in the absence and presence of an infestation.

Fig. 24 — Illustration of the effect of protective husk qualities under earworm infestation; five ears on the left with unslit husks and light injury; five ears on right on which husks were slit to expose the ear, resulting in severe infestation.

Hybrids had less stalk breakage than open-pollinated varieties and lost less yield as a result of a first-brood infestation (Patch, 1937; Patch et al., 1941). In a comparison of the development of first-brood larvae on susceptible and resistant crosses, Patch (1943) reported slower development and smaller larvae on the resistant cross. Patch (1943) also reported that with tassel appearance, pollen shedding, and silking, plants became highly susceptible to larval survival. This condition applies to early market sweetcorn during the first-brood infestation and to varying degrees in late planted sweetcorn and field corn during the second-brood period. Everly (1947) reported that inbred lines varied in attractiveness to moths for oviposition and that this trait was heritable in crosses.

A comparison of the biological relationship involved between the survival of larvae during the whorl stage of plant development (first brood) and that of the flowering stage of plant development (second brood) (see Fig. 7) and the implication in breeding procedures was reported by Dicke (1950, 1954) and Guthrie et al. (1960). Effective sources of resistance to a first-brood infestation and moderate levels of resistance or tolerance to a second-brood infestation have been developed in progressive stages. Figures 25 and 26 show contrasting differences between a high level of susceptibility (Fig. 26) and a high level of resistance to a first-brood infestation (whorl stage of growth). Note the small feeding punctures on leaves in Fig. 26. Figure 27 shows severely broken inbred line in contrast to a resistant sister line on the right. Methods for evaluating resistance with studies on inheritance and corn breeding procedures have received continued emphasis. The more recent advances in resistance research have been reported by Penny and Dicke (1956), Guthrie and Stringfield (1961), Chiang and Holdaway (1960), Zuber and Dicke (1964), Pesho et al., (1965), Scott et al. (1967), Penny et al. (1967), Guthrie et al. (1970), Guthrie and Dicke (1972), and Jennings et al. (1974).

The basis of resistance has been investigated by histological and chemical procedures. Bell (1956), working with whorl stage, first-brood infestations, determined that the primary feeding point by first instar larvae is

Fig. 25 (left) — Inbred line highly susceptible to first brood (whorl stage infestation) European corn borer. Fig. 26 (right) — Inbred line showing high level of resistance to first brood European corn borer (whorl stage infestation).

on the upper leaf surface and that the bulliform cells are the initial point of attack. He observed that the outer walls of the bulliform cells were thinner than the inner walls and that staining reaction of the walls differed between WF9 (susceptible, and L317 (resistant).

The chemical nature of resistance has been studied for many years. From extract concentrates of whorl stage corn incorporated in an artificial diet, it was determined that toxic factors present in resistant corn increased mortality of newly hatched larvae and also inhibited their growth. An isolate in pure form was known as resistance factor A and another nonpurified substance was designated factor B. Factor (RFA) was characterized as MBOA (6-methoxy-benzoxazolinone) (Beck and Smissman, 1960, 1961; Beck and Stauffer, 1957; Loomis et al., 1957). Later studies revealed that MBOA is an end product of DIMBOA (2,4-dihydroxy-7-methoxy-1, 4-benzoxaxine-3-one), and that the latter, a labile substance, is actually the chemical factor that is present in resistant strains of corn in the whorl stage of development (Wahlroos and Virtanen, 1959; Klun and Brindley, 1966; Klun et al., 1967). In a study of a group of inbred lines, Klun and Robinson (1969), found a significant correlation between the concentration of DIMBOA, as measured by the end product MBOA, and the resistance rating of the inbreds when grown and infested in the whorl stage of growth in the field. These results

Fig. 27 — Broken row in foreground shows high susceptibility to second brood European corn borer (late summer) infestation and lodging; background, comparable row, a sister line, with good resistance.

were further confirmed by Klun et al. (1970) in a test that showed a high correlation between concentration of DIMBOA in 11 inbreds and a set of diallel crosses, and their resistance ratings. The effects for general and specific combining ability were also highly significant.

Corn Leaf Aphid

Differences in varietal response to the corn leaf aphid have been recorded over a period of years. McColloch (1921) observed that there was a tendency for the later maturing varieties to be most heavily infested although there were exceptions. Walter and Brunson (1940) found no plant characters that could be consistently correlated with resistance. A large proportion of the inbreds and hybrids tested indicated that large and compact tassels were attractive to aphids. With some exception, hybrids were less heavily infested than the parental lines.

Snelling et al., (1940) tested a large number of inbred lines at different places in Illinois and observed little barrenness in plants with no infestation, whereas in aphid-infested plants there was a high rate of barrenness. Huber and Stringfield (1942) presented data showing a correlation between resistance to corn leaf aphid and the European corn borer in a group of lines tested in single-cross combinations. Coon et al. (1948) reported a significant correlation between the carotene content of the seed and the degree of aphid infestation, the high carotene parent lines being the more susceptible. Such correlation seem to depend on the reactions and resistance balance among the lines under test. From tests of many inbred lines and hybrids differences in relative degree of resistance to aphid development and infestations were found (Dicke, 1969; Everly, 1967; Neiswander and Triplehorn, 1961; Rhodes and Luckmann, 1967).

Fall Armyworm

An increasing amount of effort has been devoted to searching for sources of resistance to fall armyworm infestations in recent years. Some inbred lines are less subject to ear attack through the husks and through the silk channel than others. In areas where the fall armyworm is important the corn earworm is usually dominant. Therefore, it is necessary to consider the resistance of corn to both insects. In tests conducted in Virginia it was apparent that northern inbred lines in general are much more subject to fall armyworm attack on the husks, ear, and in the shank than some lines having southern or tropical corn in their parentage.

In greenhouse tests with seedlings of several hybrids, Wiseman et al. (1967a) were able to adapt a numerical rating system for detecting significant differences in plant damage 3 and 5 days after first instar larvae were introduced. In field tests in Kansas with 81 Latin American lines, Wiseman et al. (1967b) found 'Cuba Honduras 46-J, 'Eto Amarillo' and some 'Antigua' lines to be least damaged. Widstrom et al. (1970) concluded from a test of 36 inbred lines that among 14 characters considered, the mechanical protection of tight husks reduced damage considerably.

In a study of eight inbred lines and their F_1 progeny, Widstrom et al. (1972) concluded that general combining ability was highly significant in conditioning resistance. Dominance and/or specific combining ability added little to resistance but heterosis contributed substantially to F_1 progenies. They suggested recurrent selection based on selfed progeny performance as a procedure for developing resistant plant populations.

Southwestern Corn Borer

As a result of the rapid spread of the southwestern corn borer eastward over the southern states, a concerted effort has been made to find sources of resistance that would effectively reduce stalk breakage and eardropping connected with the infestation of the second and third generation. For breeding purposes, York and Whitcomb (1963) developed a synthetic variety in which eight inbred lines were blended. In a screening program with single crosses, Davis et al. (1973) found two lines of 'Antigua Gpo2' origin that showed some resistance to first-brood leaf feeding. Several southern single crosses showed some degree of resistance to second-brood attack.

Chinch Bugs

Flint (1921) reported the open-pollinated varieties 'Champion White Pearl' and 'Golden Beauty' as resistant or tolerant to the attack of second-brood chinch bugs. In further tests in Illinois, Holbert et al. (1934) obtained data demonstrating that inbred lines resistant to second-brood bugs transmitted these qualities to top, three-way, and double cross hybrids. In a comparison of hybrids and open-pollinated varieties Holbert et al. (1935) under an infestation of second-brood bugs found 'Illinois Hybrid 391' to be superior in standing ability and yield. In tests under second and third-brood infestations conducted in Kansas and Oklahoma, Painter et al. (1935) reported F_1 hybrids to be more resistant on the average than their inbred parents, which they thought to be due to one or both of two conditions, namely, specific inherited resistance and tolerance or escape from injury associated with heterosis. They considered heterosis to be an important factor. Hybrids in general were superior to open-pollinated varieties but a wide range of resistance was found in both groups. The pattern on corn appeared to be similar to that found on the sorghums. With populations at relatively low levels, little emphasis has been placed on breeding for resistance to chinch bugs in recent years.

Grasshoppers — Many Species

General oubreaks of grasshoppers occur periodically, and, similar to chinch bugs have been associated with a continued period of dry weather conditions. High populations are prevalent in some parts of the world each year. Little information has become available on resistance in corn since the drought years of 1930-36. Brunson and Painter (1938) reported on differential feeding of grasshoppers on corn and sorghum in Kansas. They estimated the percent defoliation in open-pollinated varieties, inbred lines in uniform top crosses, and in double and three-way hybrids. Extreme contrasts in grasshopper injury were found. In their tests, the varieties and inbred lines of corn showing the greatest resistance originated in areas where grasshoppers are a natural element of the environment. Because of generally low levels of "crop hopper" populations, exploration and breeding for resistance has occupied a low priority in varietal development programs for many years.

CHEMICAL CONTROL STATUS

The importance of the development of strains of insects resistant to insecticides should not be underestimated in varietal development. Resistance to arsenicals, for example, was well documented for the codling moth (*Laspeyeresia pomonella*) before organochlorine, organophosphorous, and carbamate compounds became available. Strains of the housefly resistant to DDT were recognized a few years after DDT came into general use. Ball and Weekman (1963) reported on resistance of field populations of the western corn rootworm to aldrin. For the same species, Ball (1973) reported on the potential resistance to diazinon and phorate.

Brown (1968) summarized the information on cross-resistance of the different types of compounds and on the genetic nature of resistance in various species. He reported that of 224 species of insects and acarines known to have developed resistance, 127 attack field or forest crops or stored products.

Chemical companies have made the major contributions in research and development of new chemicals and in providing monetary grants to state and federal agencies for evaluating formulations. Nonpersistent materials have largely replaced the persistent chlorinated hydrocarbon materials. There has been increasing extension of legalistic regulation in recent years which requires detailed research on a broad spectrum of biological activity of insecticides for establishing use labels or thresholds. In spite of the high cost of development and official registration, new compounds have been introduced by the chemical industry with considerable frequency, particularly organophosphorus and carbamate types, of which some have broad-spectrum systematic activity in plants. A great deal of research effort is being devoted to sex attractants (pheromones) and hormones. These materials have been synthesized for many species and practical application is receiving increased attention.

Probably the most significant development in the application of chemical control of corn insects during the past quarter century was the development of granular carriers and solvents for incorporating the toxicant in granular formulations. Farrar (1953) pointed out the advantages of granular materials in avoiding drift, uniformity of distribution, ease of handling, adaptation and for controlling soil and whorl inhabiting insects. He reported on impregnating such clay materials as attapulgite and bentonite with heptachlor and dieldrin, and modifications on aerial and ground equipment for distributing granular formulations. Farrar tested particle size as well as other physical properties. He reported good control results on whorl infestation of the corn earworm and a group of grass and soil inhabiting insects.

Although such materials as bran, cornmeal, and sawdust had been used for many years as carriers of insecticides for control of cutworms, grasshoppers, etc., the impregnation of granular clay materials generated a great deal of research activity that extended into a wide aspect of handling pesticides and equipment to apply them. Because of the simplicity of procedures, this development met with immediate favorable response from growers.

Continuing programs investigating insect biology in relation to chemical control, including new materials and equipment, are maintained by federal and state agencies and private companies on a group of major corn insects. The information developed is regularly shared among investigators and extension personnel through reports and direct contact. This pooled experience has accelerated obtaining adequate information on many chemicals applied under different environmental conditions.

The efficacy of insecticidal control of the European corn borer became established among growers of high-market-value, early-planted, susceptible sweetcorn. Application schedules were worked out for liquid and dust formulations consisting of such toxicants as rotenone, nicotine, and fluosilicates. These materials were quickly replaced by DDT and other chlorinated hydrocarbon formulations between 1945-1950. At the same time equipment was developed, both aerial and ground, that was adapted to field corn production. By 1950 the use of these materials in the central Corn Belt had increased rapidly, particularly for the soil insect complex. It is possible only to review some of the developments from 1950 to the present.

As a continuing research program, detailed biological and comparative studies of formulations have been conducted with many chemicals for control of the European corn borer. These studies emphasized equipment, development of the infestation on the corn plant, dosages, and timing of applications, relative toxicty to the plant and insect, and position and amount of residual toxicant at intervals after application. Over a period of some 20 years DDT has been the standard of comparison and maintained good performance under all conditions over all of these years. Several other organochlorine chemicals have been approximately equal to DDT. Because of persistent residue problems, these materials have given way to several organophosphorus and carbamate materials which have shown satisfactory performance. The following references represent the efforts that have gone into this comprehensive research: Berry, 1972; Berry et al., 1972; Cox et al., 1956a, 1956b; Dicke, 1954; Edwards and Berry, 1972; Fahey et al., 1953, 1956, 1965; Klun and Robinson, 1971; Lovely et al., 1956; McWhorter et al., 1972; Munson et al., 1970.

Chemical control research on the corn earworm was pursued extensively in southern areas, particularly on sweetcorn. DDT formulations received wide acceptance with multiple applications. Resistance to DDT gradually developed. The less persistent formulations of toxaphane, carbaryl, parathion, and gardona have been in common use for a period of years (Anderson and Nakakihara, 1968; Harrison, 1962; James and Greene, 1972; Lingren and Bryan, 1965; McMillian et al., 1972.

The dispersion of the southwestern corn borer over southern corn growing areas has encouraged chemical control investigations and the biological factors involved. Davis et al. (1972) studied the movements and feeding habits of the larvae in the whorl and tassel stages of growth to provide information on timing and placement of insecticides. Significant reductions in girdling and infestation were obtained with various formulations of azodrin, carbofuran, endrin, and endosulfon by Henderson and Davis (1970). The systemic effect of carbofuran on the southwestern corn borer was reported by Whitcomb et al., (1966) and Keaster and Fairchild (1968). In connection with carbofuran tests on the southwestern corn borer, Keaster and Fairchild (1968) observed a reduction in the incidence of corn virus disease indicating reduced vector activity. Pitre (1968) presented similar results on control of the leafhopper (*Graminella nigrifrons*) and the effect on corn stunt disease. The role of the leafhopper as a vector of maize chlorotic dwarf virus is discussed in a previous section.

The most extensive search for new chemicals and methods and equipment for applying them during the past 10 years has been on soil insects, particularly the *Diabrotica* rootworm complex. The factors involved in this development have been previously discussed. Farmers growing "corn after corn" have become convinced that applying a rootworm insecticide is good insurance against significant yield and harvest loss. In the presence of a rootworm infestation the injury effect is linked with environmental stress, temperatures and moisture, factors that are not always predictable.

There has been an active regional participation by state and federal agencies in the Corn Belt in the chemical control program. Large numbers of formulations have been tested each year and published or processed. Reports are available to growers with information on their acceptance and registry status.

All institutions, public and private, strongly emphasize that users of chemicals adhere closely to the instructions provided on each container label. These instructions are specific for target species; however, nontarget pests are not included.

APPENDIX

Common and Scientific Names* of Corn Insects

American grasshopper, *Schistocerca americana* (Drury).
Angoumois grain moth, *Sitotroga cerealella* (Oliver).
army cutworm, *Euxoa auxiliaris* (Grote).
armyworm, *Pseudaletia unipuncta* (Haworth).
Banks grass mite, *Oligonychus pratensis* (Banks).
billbugs, *Sphenophorus spp.* (snout beetles).
blackfaced leafhopper, *Graminella nigrifroms* (Forbes).
black flour beetle, *Tribolium audux* Halstead.
bluegrass billbug, *Sphenophorus parvulus* Gyllenhal.
cadelle beetle, *Tenebroides mauritanicus* (Linnaeus).
carmine spider mite, *Tetranychus cinnabarinus* (Boisduval).
chilo, *Chilo zonellus* Swinhow**.
chinch bug, *Blissus leucopterus leucopterus* (Say).
claybacked cutworm, *Agrotis gladiaria* (Morrison).
codling moth, *Laspeyresia pomonella* (Linnaeus).

---

*Most of the common and scientific names in this list are from the 1975 revision approved by the Committee on Common Names of Insects, Entomological Society of America; Special Publication 75-1, D.M. Anderson, Chairman.
**Does not occur in the western hemisphere but are important pests on corn in Asia (Chilo), Africa (Busseola), and Mediterranean area (Sesamia).

THE MOST IMPORTANT CORN INSECTS 577

confused flour beetle, *Tribolium confusum* Jacquelin duVal.
corn blotch leafminer, *Agromyza parvicornis* Loew.
corn earworm, *Heliothis zea* (Boddie).
corn flea beetle, *Chaetocnema pulicaria* Melsheimer.
corn leaf aphid, *Rhopalosiphum maidis* (Fitch).
corn planthopper, *Peregrinus maidis* (Ashmead).
corn root aphid, *Aphis maidiradicis* Forbes.
corn root webworm, *Crambus caliginosellus* Clemens.
corn sap beetle, *Carpophilus dimidiatus* (Fabricius).
corn silk fly, *Euxesta stigmatias* Loew.
lesser corn stalk borer, *Elasmopalpus lignosellus* (Zeller).
differential grasshopper, *Melanoplus differentialis* (Thomas).
dingy cutworm, *Feltia ducens* Walker.
dusky sap beetle, *Carpophilus lugubris* Murray.
English grain aphid, *Macrosiphum avenae* (Fabricius).
European corn borer, *Ostrinia nubilalis* (Hubner).
fall armyworm, *Spodoptera frugiperda* (J.E. Smith).
flat grain beetle, *Cryptolestes pusillus* (Schonherr).
flower thrips, *Frankliniella tritici* (Fitch).
foreign grain beetle, *Ahasverus advena* (Waltl).
garden symphylan, *Scutigerella immaculata* (Newport).
glassy cutworm, *Crymodes devastator* (Brace).
granary weevil, *Sitophilus granarius* (Linnaeus).
grape colaspis, *Colaspis brunnea* (Fabricius).
grass thrips, *Anaphothrips obscurus* (Muller).
hairy fungus beetle, *Typhaea stercorea* (Linnaeus).
housefly, *Musca domestica* Linnaeus.
Indian meal moth, *Plodia interpunctella* (Hubner).
Japanese beetle, *Popillia japonica* Newman.
lantern fly, *Peregrinus maidis* (Ashmead).
lesser cornstalk borer, *Elasmopalpus lignosellus* (Zeller).
maize billbug, *Sphenophorus maidis* Chittenden.
maize noctuid, *Sesamia cretica* Led.**
maize stem borer, *Busseola fusca* Fuller.**
maize weevil, *Sitophilus zeamaize* Motschulsky.
Mediterranean flour moth, *Anagasta kuehniella* (Zeller).
Neotropical corn borer, *Diatraea lineolata* (Walker).
northern corn rootworm, *Diabrotica longicornis* (Say).
pale western cutworm, *Agrotis orthogonia* Morrison.
pink scavenger caterpillar, *Sathrobrota rileyi* (Walsingham).
redlegged grasshopper, *Melanoplus fermurrubrum* (DeGeer).
rice weevil, *Sitophilus orzae* (Linnaeus).
rusty plum aphid, *Hysteroneura setariae* (Thomas).
sand wireworm, *Horistonotus uhlerii* Horn.
sawtoothed grain beetle, *Oryzaephilus surinamensis* (Linnaeus).
seedcorn beetle, *Agonoderus lecontei* and *Clivina impressifrons* Chaudoir.
seedcorn maggot *Hylemya platura* (Meigen).
slender seedcorn beetle, *Clivinia impressifrons* LeConte.
snout beetles, see billbugs
sod webworm, *Crambus trisectus*.
southern corn billbugs, *Sphenophorus callosa*.
southern corn rootworm, *Diabrotica undecimpuncata howardi* Barber.
southern cornstalk borer, *Diatraea crambidoides* (Grote).
southwestern corn borer, *Diatraea grandiosella* (Dyar).

stalk borer, *Papaipema nebris* (Guenee).
striped webworm, *Crambus mutabilis*.
sugarcane beetle, *Euetheola rugiceps* (LeConte).
sugarcane borer, *Diatraea saccharalis* (Fabricius).
toothed flea beetle, *Chaetocnema denticulata* (Illiger).
two spotted spider mite, *Tetranychus urticae* Koch.
two striped grasshopper, *Melanoplus bivittatus* (Say).
variegated cutworm, *Peridroma saucia* (Hubner).
webworms, *Crambus* spp.
western corn rootworm, *Diabrotica virgifera* LeConte
wheat curl mite, *Eriophyes tulipae*.
wheat wireworm, *Agriotes mancus* (Say).
white grubs, *Phyllophaga* spp.

ACKNOWLEDGMENTS

For critical review and helpful suggestions I am indebted to A.J. Ullstrup of Purdue University; W.A. Russell and T.A. Brindley of Iowa State University, and W.D. Guthrie and Leslie C. Lewis, ARS-USDA, cooperating with Iowa State University.

Photo Credits. Figures 3, 14, Ohio agricultural Research and Development Center (14 by Musick); Fig. 4, 19, 20, Univ. of Missouri (4, 19 by Paul S. Szopa and 20 by Lee Jenkins); Fig. 5, 6, 7, 12, 24, ARS-USDA (24 by Wiseman, McMillian, and Widstrom at Southern Grain Research Laboratory Tifton, Ga.); Fig. 9, Saxema and Chada, Oklahoma State Univ.; others from Pioneer Hi-Bred International, Inc., workers and the author.

LITERATURE CITED

Ainslie, G.G. 1919. The larger corn stalk-borer. USDA Farmers Bull. 1025.

----. 1922. Webworms injurious to cereal and forage crops and their control. USDA Farmers Bull. 1258.

----. 1927. The larger sod webworm. USDA Agric. Tech. Bull. 31.

American Phytopathological Society. 1973. A compendium of corn diseases. Advisory Committee to the Maize (Corn) Disease Compendium Project. (M.C. Schurtleff, Chm.) St. Paul, Minn.

Anderson, L.D., and H. Nakakihara. 1968. Toxicity of pesticides to corn earworm on sweet corn in southern California 1962-67. J. Econ. Entomol. 61:1477-1482.

Anonymous. 1937-1938. Ohio Agric. Exp. Stn. Annu. Rep. Bull. 600.

Apple, J.W., and K.K. Patal. 1963. Sequence of attack by northern corn rootworm on the crown roots of corn. Proc. North Central Branch, Entomol. Soc. Am. 18:80-81.

Arant, F.S. 1929. Biology and control of the southern corn rootworm. Alabama Exp. Stn. Bull. 230.

----. 1934. Time of turning legumes and planting corn to avoid injury from the southern corn root worm. Alabama Polytechnic Inst. Agric. Exp. Stn. Cir. 65.

Arbuthnot, R.D. 1953. Present status and potential value of parasites of the European corn borer. Proc. North Central Branch Entomol. Soc. Am. 8:30-32.

Auclair, J.L. 1963. Aphid feeding and nutrition. Annu. Rev. Entomol. 8:439-490.

Babcock, K.W., and A.M. Vance. 1929. The corn borer in Central Europe. USDA Tech. Bull. 135.

Back, E.A., and R.T. Cotton. 1926. Biology of the sawtoothed grain beetle, *Oryzaephilus surinamensis* Linne. J. Agric. Res. 33:435-452.

Baerg, W.J. 1942. Rough-headed corn stalk-beetle. Arkansas Agric. Exp. Stn. Bull. 415.

Baker, W.A., W.G. Bradley, and C.A. Clark. 1949. Biological control of the European corn borer in the United States. USDA Tech. Bull. 983.

Ball, H.J. 1973. Western corn rootworm: A ten-year study of potential resistance to diazinon and phorate in Nebraska. J. Econ. Entomol. 66:1015.

----, and G.T. Weekman. 1963. Differential resistance of corn rootworms to insecticides in Nebraska and adjoining states. J. Econ. Entomol. 56:553-555.

Bancroft, J.B., A.J. Ullstrup, M. Messieha, C.E. Bracker, and T.E. Snazelle. 1966. Some biological and physical properties of a midwestern isolate of maize dwarf mosaic virus. Phytopathology 56:474-478.

Barber, G.W. 1925. A study of the cause of the decrease in the infestation of the European corn borer (*Pyrausta nubalilis,* Hubn.) in the New England area during 1923. Ecology 6:39-47.

----. 1936a. The corn earworm in southeastern Georgia. Georgia Exp. Stn. Bull. 192.

----. 1936b. The cannibalistic habits of the corn ear worm. USDA Tech. Bull. 499.

----. 1941. Hibernation of the corn earworm in southeastern Georgia. USDA Tech. Bull. 791.

----, and F.F. Dicke. 1937. The effectiveness of cultivation as a control for the corn earworm. USDA Bull. 561.

----, and ----. 1939. Effect of temperature and moisture on overwintering pupae of the corn earworm in the northeastern states. J. Agric. Res. 59:711-723.

Batchelder, C.H. 1949. European corn borer location on the corn plant as related to insecticidal control. USDA Bull. 976.

Beck, S.D., J.H. Lilly, and J.F. Stauffer. 1949. Nutrition of the European corn borer, *Pyrausta nubilalis* (Hbn.). I. Development of a satisfactory purified diet for larval growth. Ann. Entomol. Soc. Am. 42:483-496.

----, and E. Smissman. 1960. The European corn borer *Pyrausta nubilalis* and its principal host plant. VIII. Laboratory evaluation of host plant resistance to larval growth and survival. Ann. Entomol. Soc. Am. 53:755-762.

----, and ----. 1961. The European corn borer, *Pyrausta numilalis,* and its principal host plant. IX. Biological activity of chemical analogs of corn resistance Factor A (6-methoxyebnzoxazolinone). Ann. Entomol. Soc. Am. 54:53-61.

----, and J.F. Stauffer. 1950. An aseptic method for rearing European corn borer larvae. J. Econ. Entomol. 43:4-6.

----, and ----. 1957. The European corn borer, *Pyrausta nubilalis* (Hubn.), and its principal host plant. III. Toxic factors influencing larval establishment. Ann. Entomol. Soc. Am. 50:166-170.

Bell, M.E. 1956. Histology of the maize plant in relation to susceptibility of the European corn borer. Iowa State Coll. J. Sci. 31:9-17.

Benton, C., and W.P. Flint. 1938. The comparative attractiveness of various small grains to the chinch bug. USDA Cir. 508.

Bennett, S.E., and L.M. Josephson. 1962. Methods of artificial infestation of corn with the earworm, *Heliothis zea.* J. Econ. Entomol. 60:171-173.

Berry, E.C., J.E. Campbell, C.R. Edwards, J.A. Harding, W.G. Lovely, and G.M. McWhorter. 1972. Further field tests of chemicals for control of the European corn borer. J. Econ. Entomol. 65:1113-1116.

Bigger, J.H. 1928. Hibernation studies of *Colospis (beumea) flavida* (Fab.). J. Econ. Entomol. 21:268-273.

----. 1932. Short rotation fails to prevent attack of *Daibrotica longicornis* Say. J. Econ. Entomol. 25:196-199.

---. 1963. Corn rootworm resistance to chlorinated hydrocarbon insecticides in Illinois. J. Econ. Entomol. 56:118-119.

----, and F.C. Bauer. 1939. Plowing dates as they affected the abundance of corn root aphids at Clayton, Illinois, 1929-1932. J. Am. Soc. Agron. 31:695-697.

----, J.R. Holbert, W.P. Flint, and A.L. Lang. 1938. Resistance of certain corn hybrids to attack of southern corn root worm. J. Econ. Entomol. 31:102-107.

----, and H.B. Petty. 1953. Reduction of corn borer numbers from October to June. Univ. Illinois Bull. 566.

----, R.C. Snelling, and R.A. Blanchard. 1941. Resistance of corn strains to the southern corn rootworm *Diabrotica duodecimpunctata* F. J. Econ. Entomol. 34:605-613.

Blair, B.D., C.A. Triplehorn, and G.W. Ware. 1963. Aldrin resistance in northern corn rootworm adults in Ohio. J. Econ. Entomol. 56:894.

Blanchard, R.A. 1942. Hibernation of the corn earworm in the central and northeastern parts of the United States. USDA Tech. Bull. 838.

----, J.H. Bigger, and R.O. Snelling. 1941. Resistance of corn strains to the corn earworm. J. Am. Soc. Agron. 33:344-350.

----, and W.A. Douglas, 1953. The corn earworm as an enemy of field corn in the eastern states. USDA Farmers' Bull. 1651.

----, A.F. Satterthwait, and R.O. Snelling. 1942. Manual infestation of corn strains as a method of determining differential earworm damage. J. Econ Entomol. 35:508-511.

Blatchley, W.S. 1920. Orthoptera of northeastern America, with especial reference to the faunas of Indiana and Florida. Nature publishing Co., Indianapolis, Ind.

Bottger, G.T. 1940. Preliminary studies of the nutritive requirements of the European corn borer. J. Agric. Res. 60:249-257.

----. 1942. Development of synthetic food media for use in nutrition studies of the European corn borer. J. Agric. Res. 65:493-500.

Bradfute, O.I., T.P. Pirone, P.H. Freytag, and M.C.Y. Lung. 1972. Virus-like particles associated with a leafhopper-transmitted disease of corn in Kentucky. Plant Dis. Rep. 56:652-656.

Brandes, E.W. 1920. Artificial and insect transmission of sugar-cane mosaic. J. Agric. Res. 19:131-138.

Branson, T.F., and P.L. Guss. 1972. Potential for utilizing resistance from relatives of cultivated crops. Proc. North Central Branch Entomol. Soc. Am. 27:91-95.

----, P.L. Guss, and E.E. Ortman. 1969. Toxicity of sorghum roots to larvae of the western corn rootworm. J. Econ. Entomol. 62:1375-1378.

----, and E.E. Ortman. 1970. The host range of larvae of the western corn rootworm: further studies. J. Econ. Entomol. 63:800-803.

Brindley, T.A., and F.F. Dicke. 1963. Significant developments in European corn borer research. Annu. Rev. Entomol. 8:155-176.

----, A.N. Sparks, W.B. Showers, and W.D. Gutherie. 1975. Recent research advances on the European corn borer in North America. Annu. Rev. Entomol. 20:221-239.

Brown, A.W.A. 1968. Insecticide resistance comes of age. Bull. Entomol. Soc. Am. 14:3-9.

Brunson, A.M., and R.H. Painter. 1938. Differential feeding of grasshoppers on corn and sorghums. J. Am. Soc. Agron. 30:334-346.

Burkhardt, C.C. 1952. Feeding and pupating of fall armyworm in corn. J. Econ. Entomol. 45:1035-1037.

Burton, R.L. 1969. Mass rearing the corn earworm in the laboratory. USDA, ARS (Ser.) 33-134.

Caffrey, D.J., and L.H. Worthley. 1927. A progress report on the investigations of the European corn borer. USDA Bull. 1476.

Callahan, P.S. 1957. Oviposition response of the imago of the corn earworm, *Heliothis zea* (Boddie), to various wave lengths of light. Ann. Entomol. Soc. Am. 50:444-452.

Carter, W. 1941. *Peregrinus maidis* (Ashm.) and the transmission of corn mosaic. I. Incubation period and longevity of the virus in the insect. Ann. Antomol. Soc. Am. 34:551-556.

----. 1973. Insects in relation to plant diseases. John Wiley & Sons, New York. 2nd ed. 759 p.

Cartwright, O.L. 1929. The maize billbug in South Carolina. South Carolina Agric. Exp. Stn. Bull. 257.

----. 1930. The rice weevil and associated insects in relation to shuck length and corn varieties. South Carolina Agric. Exp. Stn. Bull. 266.

----. 1934. The southern corn stalk borer, (Diatraea crambidoides (Grote) in South Carolina. South Carolina Agric. Exp. Stn. Bull. 294.

Chamberlain, T.R., and J.A. Callenbach. 1943. Oviposition of June beetles and the survival of their offspring in grasses and legumes. J. Econ. Entomol. 36:681-688.

Chiang, H.C. 1973. Bionomics of northern and western rootworms. Annu. Rev. Entomol. 18:47-72.

----, and F.G. Holdaway. 1960. Relative effectiveness of resistance of field corn to the European corn borer, *Pyrausta nubilalis*, in crop protection and in population control. J. Econ. Entomol. 53:918-924.

Chippendale, G.M., and S.D. Beck. 1964. Nutrition of the European corn borer, *Ostrinia nubilalis* (Hubn.) V. Ascorbic acid as the corn leaf factor. Entomol. Exp. Appl. 7:241-248.

THE MOST IMPORTANT CORN INSECTS 581

Chittenden, F.H. 1916. The pink corn-worm: An insect destructive to corn in the crib. USDA Bull. 363.

Choudhury, M.M., and E. Rosenkranz. 1973. Differential transmission of Mississippi and Ohio corn stunt agents by *Graminella nigrifrons.* Phytopathology 63:127-133.

Christensen, J.J., and C.L. Schneider. 1950. European corn borer (*Pyrausta nubilalis* Hbn.) in relation to shank, stalk, and ear rots of corn. Phytopathology 40:284-291.

Collins, G.N., and J.H. Kempton. 1917. Breeding sweet corn resistant to the corn earworm. J. Agric. Res. 11:549-572.

Coon, B.F. 1951. Japanese beetle damage in field corn. Pennsylvania Agric. Exp. Stn. Prog. Rep. 55.

----, R.C. Miller, and L.W. Aurand. 1948. Correlation between the carotene content of corn and infestation by the corn leaf aphid. Pennsylvania Agric. Exp. Stn. J. Ser. 1436.

Cotton, R.T. 1920. Rice weevil (Calandra)*Sitophilus* oryza. J. Agric. Res. 20.

----. 1938. Control of insects attacking grain in farm storage. USDA Farmers Bull. 1811.

----. 1947. Insect pests of stored grain and grain products. 3rd printing. Burgess Publ. Co., Minneapolis, Minn.

----, H.H. Walkden, G.D. White, and D.A. Wilbur. 1953. Causes of outbreaks of stored-grain insects. North Central Region Publication 35.

Cox, H.C., T.A. Brindley, W.G. Lovely, and J.E. Fahey. 1956b. Granulated insecticides for European corn borer control. J. Econ. Entomol. 49:113-119.

----, W.G. Lovely, and T.A. Brindley. 1956a. Control of the European corn borer with granulated insecticides in 1955. J. Econ. Entomol. 49:834-838.

Crawford, H.G., and G.J. Spencer. 1922. The European corn borer *Pyrausta nubilalis* (Hubn.): Life history in Ontario. Annu. Rep. Entomol. Soc. Ontario 52:22-26.

Crumb, S.E. 1929. Tobacco cutworms. USDA Tech. Bull. 88.

Davis, E.G., J.R. Horton, C.H. Gable, E.V. Walter, and R.A. Blanchard. 1933. The southwestern corn borer. USDA Tech. Bull. 388.

Davis, F.M., C.A. Henderson, and G.E. Scott. 1972. Movements and feeding of larvae of the southwestern corn borer on two stages of corn growth. J.Econ. Entomol. 65:519-521.

----, G.E. Scott, and C.A. Henderson. 1973. Southwestern corn borer: Preliminary screening of corn genotypes for resistance. J. Econ. Entomol. 66:503-506.

Davis, J.J. 1922. Common white grubs. USDA Farmers Bull. 940.

----. 1949. The corn root aphid and methods of controlling it. USDA Farmers Bull. 891.

----, and A.F. Satterthwait. 1916. Life-history studies of *Pseudaletia* (Cirphis) *unipuncta,* the true armyworm. J. Agric. Res. 6:799-812.

Davis, R.E. 1973. Occurrence of a spiroplasma in corn stunt-infected plants in Mexico. Plant Dis. Rep. 57:333-337.

----, R.F. Whitcomb, T.A. Chen, and R.R. Granados. 1972. Current status of the aetiology of corn stunt disease. p. 205-225. *In:* Pathogenic mycoplasmas. CIBA Found. Symp., Jan. 25-27, 1972.

----, and J.F. Worley. 1973. Spiroplasma: Motile, helical microorganism associated with corn stunt disease. Phytopathology 63:403-408.

Dean, G.A. 1913. Mill and stored-grain insects. Kansas Agric. Exp. Stn. Bull. 189.

Decker, G.C. 1941. Protect Iowa's grain crop. Iowa State College 42nd Yearbook (Pt. 4), Ames, Iowa. p. 168-176.

Dicke, F.F. 1932. Studies on the host plants of the European corn borer, *Pyrausta nubilalis* Hubner, in southeastern Michigan. J. Econ. Entomol. 24:868-878.

----. 1939. Seasonal abundance of the corn earworm. J. Agric. Res. 59:237-257.

----, 1950. Response of corn strains to European corn borer infestation. Proc. 5th Annu. Meeting North Central Branch Entomol. Soc. Am. 5:47-49.

----. 1954. Breeding for resistance to European corn borer. p. 44-53 *In:* Ninth Hybrid Corn Industry-Res. Conf. Am. Seed Trade Association.

----. 1969. The corn leaf aphid. Proc. Annu. Corn Sorghum Res. Conf. 24:61-70.

----, and M.T. Jenkins. 1945. Susceptibility of certain strains of field corn in hybrid combinations to damage by corn earworms. USDA Tech. Bull. 898.

Ditman, L.P. 1938. Metabolism in the corn ear worm I. Studies on fat and water. Maryland Agric. Exp. Stn. Bull. 414.

----, and E.N. Cory. 1931. The corn earworm biology and control. Maryland Agric. Exp. Stn. Bull. 328.

Doi, Y., M. Terenaka, K. Yora, and H. Asuyama. 1967. Mycoplasma — or PLT group-like microorganisms found in the phloem elements of plants infected with mulberry dwarf, potato witches' broom, aster yellows, or paulownia witches' broom. Ann. Phytopathol. Soc. Japan. 33:259-266.

Douglas, J.R., J.W. Ingram, K.E. Gibson, and W.E. Peay. 1957. The western bean cutworm as a pest of corn in Idaho. J. Econ. Entomol. 50:543-545.

Douglas, W.A. 1948. The effect of husk extension and tightness on earworm damage to corn. J. Econ. Entomol. 40:661-664.

----, and R.C. Eckhardt. 1957. Dent corn inbreds and hybrids resistant to the corn earworm in the South. USDA Tech. Bull. 1160.

----, C.A. Henderson, and J.A. Langston. 1962. Biology of the pink scavenger caterpillar and its control on corn. J. Econ. Entomol. 55:651-655.

----, and C.E. Smith. 1953. Control of the corn earworm and rice weevil in dent corn with DDT-mineral oil emulsions. J. Econ. Entomol. 46:683-684.

Eden, W.G. 1952. Effect of huskcover of corn on rice weevil damage in Alabama. J. Econ. Entomol. 45:543-544.

----, and F.S. Arant. 1953. Control of corn rootworm in corn. Alabama Agric. Exp. Stn. Leafl. 34.

Edwards, C.R., and E.C. Berry. 1972. Evaluation of five systemic insecticides for control of the European corn borer. J. Econ. Entomol. 62:1129-1132.

Ehler, L.E. 1973. Spider mites associated with grain sorghum and corn in Texas. J. Econ. Entomol. 66:1220-1222.

Eiben, G.J., and D.C. Peters. 1965. Varietal response to rootworm infestation in 1964. Proc. North Central Branch Entomol. Soc. Am. 20:44-46.

Eichmann, R.D. 1940. Corn earworm hibernates in Washington State. J. Econ. Entomol. 33:951-952.

Elliott, C., and F.W. Poos. 1934. Overwintering of *Aplanobacter stewarti*. Science 80:289-290.

----, and F.W. Poos. 1940. Seasonal development, insect vectors, and host range of bacterial wilt of sweet corn. J. Agric. Res. 60:645-686.

Everly, R.T. 1947. Studies on the attractiveness of dent corn to moths of the European corn borer (Abstr.) Ind. Acad. Sci. Proc. for 1946 56:145.

----. 1960a. Loss in corn yield associated with the abundance of the corn leaf aphid, *Rhopalosiphum maidis*, in Indiana. J. Econ. Entomol. 53:924-932.

----. 1960b. Insecticidal control of thrips on corn. Proc. North Central Branch Entomol. Soc. Am. 15:89-91.

----. 1967. Establishment and development of corn leaf aphid populations on inbred and single cross dent corn. Proc. North Central Branch Entomol. Soc. Am. 22:80-84.

----, P. Sandberg, and B. Weaver. 1963. The effect of infestation of the Angoumois grain moth on the germination and vigor of corn. Proc. North Central Branch Entomol. Soc. Am. 18:76-79.

Fahey, J.E., T.A. Brindley, and H.W. Rusk. 1953. Three years' study of DDT residues on corn plants treated for European corn borer control. Iowa State Coll. J. Sci. 28:209-260.

----, R.T. Murphy, and R.D. Jackson. 1965. Residues of chlorinated hydrocarbon insecticides found on corn plants treated for European corn borer control. Iowa State J. Sci. 39:437-447.

----, H.W. Rusk, and H.C. Cox. 1956. Residues on plants treated with DDT granules and emulsions for European corn borer control. J. Econ. Entomol. 49:846-849.

Farrar, M.D. 1953. The granulated type insecticide for soil treatment. J. Econ. Entomol. 46:377-379.

Fitzgerald, P.J., and E.E. Ortman. 1964. Breeding for resistance to western corn rootworm. Proc. 19th Annu. Hybrid Corn Res. Conf. 1-15. American Seed Trade Association.

Fitzgerald, P.J., E.E. Ortman, and T.F. Branson. 1968. Evaluation of mechanical damage to roots of commercial varieties of corn (*Zea mays* L.). Crop Sci. 8:419-421.

Fleming, W.E. 1970. The Japanese beetle in the United States. USDA Handb. 236.

----. 1972. Biology of the Japanese beetle. USDA Tech. Bull. 1449.

Flint, W.P. 1921. Chinch bug resistance shown by certain varieties of corn. J. Econ. Entomol. 14:83-85.

Floyd, E.H., and J.W. Powell. 1958. Some facts influencing the infestation in corn in the field by the rice weevil. J. Econ. Entomol. 51:23-26.

Forbes, S.A. 1893. Noxious and beneficial insects and the state of Illinois. Illinois State Entomologist Report 18.

----. 1904. The more important insect injuries to Indian corn. Illinois Agric. Exp. Stn. Bull. 95.

----. 1915. Recent Illinois work on the corn root-aphis and the control of its injuries. Illinois Agric. Exp. Stn. Bull. 178.

Ford, R.E. 1967. Maize dwarf mosaic virus susceptibility of Iowa native perennial grasses. Phytopathology 57:450-451.

Garman, H., and H.H. Jewett. 1914. The life-history and habits of the corn-ear worm (*Chloridea obsoleta*). Kentucky Agric. Exp. Stn. Bull. 187.

Gillette, C.P. 1912. *Diabrotica virgifera* Lec., a corn rootworm. J. Econ. Entomol. 5:364-366.

Glen, R., K.M. King, and A.P. Arnason. 1943. The identification of wireworms of economic importance in Canada. Can. J. Res. 21:358-387.

Good, N.E. 1936. The flour beetles of the genus *Tribolium*. USDA Tech. Bull. 498.

Granados, R.G. 1970. Sources of corn, *Zea mays* L. resistance to thrips, *Frankliniella occidentalis* (P.) and *F. williami* H., in Mexico. Ph.D. Thesis Kansas State Univ., Manhattan.

----. 1969a. Chemotherapy of the corn stunt disease (abstr.). Phytopathology 59:1556.

----. 1969b. Electron microscopy of plants and insect vectors infected with the corn stunt disease agent. Contrib. Boyce Thompson Inst. 24:173-187.

----, K. Maramorosch, T. Everett, and T.P. Pirone. 1966. Contrib. Boyce Thompson Inst. 23:275-280.

----, ----, and E. Shikata. 1968. Mycoplasma: suspected etiologic agent of corn stunt. Proc. Natl. Acad. Sci. (USA) 60:841-844.

Guthrie, W.D. 1971. Resistance of maize to second-brood European corn borers. 26th Hybrid Corn Industry-Res. Conf. pp. 165-179. Am. Seed Trade Association.

----, and F.F. Dicke. 1972. Resistance of inbred lines of dent corn to leaf feeding by first-brood European corn borers. Iowa State J. Sci. 46:339-357.

----, ----, and C.R. Neiswander. 1960. Leaf and sheath feeding resistance to the European corn borer in eight inbred lines of dent corn. Ohio Agr. Exp. Sta. (Wooster) Res. Bull. 860.

----, J.L. Huggans, and S.M. Chatterji. 1970. Sheath and collar feeding resistance to the second brood European corn borer in six inbred lines of dent corn. Iowa State J. Sci. 44:297-311.

----, E.S. Raun, F.F. Dicke, G.R. Pesho, and S.W. Carter. 1965. Laboratory production of European Corn borer egg masses. Iowa State J. Sci. 40:65-83.

----, and G.H. Stringfield. 1961. Use of test crosses in breeding corn for resistance to the European corn borer. J. Econ. Entomol. 54:784-787.

Hamilton, C.C. 1935. The control of insect pests of lawns and golf courses. New Jersey Exp. Stn. Circ. 347.

Harrison, F.P. 1962. On the control of corn earworm, Heliothis zea, and dusky sap beetle, *Carpophilus lugubris,* in sweet corn. J. Econ. Entomol. 55:671-674.

Hawkins, J.H. 1936. The bionomics and control of wireworms in Maine. Maine Agric Exp. Stn. Bull. 381.

Hawley, I.M. 1922. Insects and other animal pests injurious to field beans in New York. Cornell Univ. Agric. Exp. Stn. Memo. 55.

----, and F.W. Metzger. 1940. Feeding habits of the adult Japanese beetle. USDA Circ. 547.

Headlee, T.J., and J.W. McColloch. 1913. The chinch bug. Kansas Agric. Exp. Stn. Bull. 191.

Henderson, C.A., and F.M. Davis. 1970. Four insecticides tested in the field for control of *Diatraea grandiosella*. J. Econ. Entomol. 63:1495-1497.

Hill, R.E., E. Hixon, and M.H. Muma. 1948. Corn rootworm control tests with benzene hexachloride, DDT, nitrogen fertilizers and crop rotation. J. Econ. Entomol. 41:392-401.

Hinds, W.E. 1914. Reducing insect injury in stored corn. J. Econ. Entomol. 7:203-211.

Holbert, J.R., W.P. Flint, and J.H. Bigger. 1934. Chinch bug resistance in corn — an inherited character. J. Econ. Entomol. 27:121-124.

----, ----, ----, and G.H. Dungan. 1935. Resistance and susceptibility of corn strains to second brood chinch bugs. Iowa State J. Sci. 9:413-425.

Huber, L.L., C.R. Neiswander, and R.M. Salter. 1928. The European corn borer and its environment. Ohio Agric. Exp. Stn. Bull. 429.

----, and G.H. Stringfield. 1942. Aphid infestation of strains of corn as an index of their susceptibility to corn borer attack. J. Agric. Res. 64:283-291.

Hyslop, J.A. 1915. Wireworms attacking cereal and forage crops. USDA Bull. 156.

Ingram, J.W., and E.K. Bynum. 1941. The sugarcane borer. USDA Farmers' Bull. 1884.

----, E.K. Bynum, R. Mathes, W.E. Haley, and L.J. Charpentier. 1951. Pests of sugar cane and their control. USDA Circ. 878.

Isely, D. 1929. The southern corn rootworm. Arkansas Agric. Exp. Stn. Bull. 232.

----. 1935. Relation of hosts to abundance of cotton bollworm. Arkansas Agric. Exp. Stn. Bull. 320.

----, and F.D. Miner. 1944. The lesser cornstalk borer, a pest of fall beans. J. Kansas Entomol. Soc. 17:51-57.

Ishiie, T., Y. Doi, K. Yora, and H. Asuyama. 1967. Suppressive effects of antibiotic of tetracycline group on symptom development of mulberry dwarf disease. Ann. Phytopathology Soc. Japan. 33:267-275.

Jacob, D., and G.M. Chippendale. 1971. Growth and Development of the southwestern corn borer, *Diatraea grandiosella* on meridic diet. Ann. Entomol. Soc. Am. 64:485-488.

James, M.J., and G.L. Greene. 1972. Corn earworm control on sweet corn ears in central and south Florida, 1969-70. J. Econ. Entomol. 65:521-522.

Jennings, C.W., W.A. Russell, and W.D. Guthrie. 1974. Genetics of resistance in maize to first and second brood European corn borer. Crop Sci. 14:394-398.

Jewett, H.H. 1942. Life history of the wireworm *Aeolus mellillus* (Say). Kentucky Agric. Exp. Stn. Bull. 425.

Keaster, A.J., and M.L. Fairchild. 1968. Reduction of corn virus disease incidences and control of southwestern corn borer with systemic insecticides. J. Econ. Entomol. 62:367-369.

Kelly, E.O.G. 1911. Papers on cereal and forage insects. The maize billbug. USDA Bur. Entomol. Bull.95 (Part II)

Kirk, V.M. 1960. Corn insects in South Carolina. South Carolina Agric. Exp. Stn. Bull. 478.

Klun, J.A., and T.A. Brindley. 1966. Role of 6-methoxy-benzoxazolinone in inbred resistance of host plant (maize) to first-brood larvae of European corn borer. J. Econ. Entomol. 59:711-718.

----, W.D. Guthrie, A.R. Hallauer, and W.A. Russell. 1970. Genetic nature of the concentration of 2,4-dihydroxy-7-methoxy 2-H-1-4-benzoxazin-3(4H)-one and resistance to the European corn borer in a diallel set of eleven maize inbreds. Crop Sci. 10:87-90.

----, and J.F. Robinson, 1969. Concentration of two 1,4-benzoxazinones in dent corn at various stages of development of the plant and its relation to resistance of the host plant to the European corn borer. J. Econ. Entomol. 62:214-220.

----, and ----. 1971. European corn borer moth: Sex attractant and sex attractant inhibitors. Ann. Entomol. Soc. Am. 64:1083-1086.

----, C.L. Tipton, and T.A. Brindley. 1967. 2,4-dihydroxy-7-methoxy-1, 4-benzoxazin-3-one (DIMBOA), an active agent in the resistance of maize to the European corn borer. J. Econ. Entomol. 60:1529-1533.

Knapp, J.L., P.A. Hedin, and W.A. Douglas. 1965. Amino acids and reducing sugars in silks of corn resistant or susceptible to corn earworm. Ann. Entomol. Soc. Am. 58:401-402.

Knight, H.H. 1916. The army-worm in New York in 1914 *Leucania unipuncta* Haworth: Order, *Lepidoptera*; Family, *Noctuidae*. Cornell Univ. Agric. Exp. Stn. Bull. 376.

Koehler, B. 1942. Natural mode of entrance of fungi into corn ears and some symptoms that indicate infection. J. Agric. Res. 64:421-442.

Kunkel, L.O. 1921. A possible causative agent for the mosaic disease of corn. Bull. Exp. Stn. Hawaii Sugar Planters Assn. 3:44-58.

----. 1927. The corn mosaic of Hawaii distinct from sugar cane mosaic (abstr.). Phytopathology 17:41.

Kuschel, G. 1961. On problems of synonymy in the *Sitophilus oryzae* complex (3rd contribution Col. Curculionoidea). Centro de Inv. Zool. Univ. Chile. Am. Mag. Natur. Hist. 13:241-244.

Kyle, C.H. 1918. Shuck protection for ear corn. USDA Bull. 708.

Lane, M.C. 1941. Wireworms and their control on irrigated land. U.S.D.A. Farmers' Bull. 186b.

Leiby, R.W. 1920. The larger corn stalk borer in North Carolina. North Carolina Dept. Agric. 41:No. 274.

Leonard, David E. 1966. Biosystematic of the "Leucopterus Complex" of the genus Blissus, Connecticut Agric. Exp. Stn. Bull. 677.

Lillehoj, E.B., W.F. Kwolek, E.E. Vandegraft, M.S. Zuber, O.H. Calbert, N. Widstrom, M.C. Futrell, and A.J. Bockholt, 1975. Aflatoxin production in Aspergillus flavus inoculated ears of corn grown at diverse locations. Crop Sci. 15:267-270.

Lindsay, D.R. 1943. The biology and morphology of *Colaspis flavida* (Say). Iowa State Coll. J. Sci. 18:60-61.

Lingren, P.D., and D.E. Bryan. 1965. Dosage-mortality data on the bollworm, *Heliothis zea,* and the tobacco budworm, *Heliothis virescens,* in Oklahoma. J. Econ. Entomol. 58:14-18.

Loomis, R.S., S.D. Beck, and J.F. Stauffer. 1957. The European corn borer, Pyrausta nubilalis (Hubn.), and its principal host plant. V. A chemical study of host plant resistance. Plant Physiol. 32:379-385.

Lovely, W.G., H.C. Cox, and T.A. Brindley. 1956. Application equipment for granulated insecticides. J. Econ. Entomol. 49:839-846.

Luginbill, P. 1918. The southern corn rootworm and farm practices to control it. USDA Farmers Bull. 950.

----. 1922. Bionomics of the chinch bug. USDA Bull. 1016.

----. 1928. The fall army worm. USDA Tech. Bull. 34.

----. 1950. Habits and control of the fall armyworm. USDA Farmers Bull. 1990.

----, and G.G. Ainslie. 1917. The lesser corn stalk-borer. USDA Bull. 539.

----, and T.R. Chamberlain. 1953. Control of white grubs on cereal and forage crops. USDA Farmers' Bull. 1798.

----, and H.R. Painter. 1953. May beetles of the United States and Canada. USDA Tech. Bull. 1060.

Lynch, R.E., and L.C. Lewis. 1971. Recurrence of the microsporidian *Perezia pyraustae* in the European corn borer *Ostrinia Nubilalis* reared on diet containing Fumidil B. J. Invert. Pathol. 17:243-246.

Maramorosch, K. 1963. The occurrence in Arizona of corn stunt disease and of the leafhopper vector *Dalbulus maidis.* Plant Dis. Rep. 47:858.

----, R.R. Granados, and H. Hirumi. 1970. Mycoplasma disease of plants and insects. Adv. Virus Res. 16:135-193.

----, E. Shikata, and R.R. Granados. 1968. Structures resembling mycoplasma in diseased plants and in insect vectors. Trans. N.Y. Acad. Sci. Ser. 2:841-855.

Marston, A.R. 1931. Breeding European corn borer resistant corn. J. Am. Soc. Agron. 23:950-964.

McCain, F.S., W.G. Eden, B.W. Arthur, and M.C. Carter. 1963. Amino acid content of corn silks in relation to resistance to corn earworm. J. Econ. Entomol. 56:902.

McColloch, J.W. 1921. The corn leaf aphis (*Aphis maidis* Fitch) in Kansas. J. Econ. Entomol. 14:89-94.

McEwen, F.L., and C.Y. Kawanishi. 1967. Insect transmission of corn mosaic: Laboratory studies in Hawaii. J. Econ. Entomol. 60:1413-1417.

McMillian, W.W., and K.J. Starks. 1966. Feeding responses of some noctuid larvae (Lepidoptera) to plant extracts. Ann. Entomol. Soc. Am. 59:516-519.

----, and B.R. Wiseman. 1972. Host plant resistance: A Twentieth Century look at the relationship between *Zea mays* L. and *Heliothis Zea* (Boddie). Florida Agric. Exp. Stn. Monogr. Ser. No. 2.

----, ----, N.W. Widstrom, and E.A. Harrell. 1972. Resistant sweet corn hybrid plus insecticide to reduce losses from corn earworm. J. Econ. Entomol. 65:229-231.

McWhorter, G.M., E.C. Berry, and L.C. Lewis. 1972. Control of the European corn borer with two varieties of *Bacillus thuringiensis*. J. Econ. Entomol. 65:1414-1417.

Mendoza, C.E., and D.C. Peters. 1964. Species differentiation among mature larvae of *Diabrotica undecimpunctata* Howardi, *D. virgifera* and *D. longicornis*. J. Kansas Entomol. 37:123-125.

Messieha, M. 1967. Aphid transmission of maize dwarf mosaic virus. Phytopathology 57:956-959.

Metcalf, Z.P. 1917. Biological investigation of *Spenophorus callosus* Oliv. North Carolina Agric. Exp. Stn. Tech. Bull 13.

Meyers, Marion T., L.L. Huber, C.R. Neiswander, F.D. Richey, and G.H. Stringfield. 1937. Experiments on breeding corn resistant to the European corn borer. USDA Tech. Bull. 583.

Munson, R.E., T.A. Brindley, D.C. Peters, and W.G. Lovely. 1970. Control of both the European corn borer and western corn rootworm with one application of insecticide. J. Econ. Entomol. 63-385-390.

Musick, G.J., and P.J. Tuttle. 1973. Supression of armyworm damage to no-tillage corn with granular carbofuran. J. Econ. Entomol. 66:735-737.

Nault, L.R., and R.H.E. Bradley. 1969. Ann. Entomol. Soc. Am. 62:2, 403-406.

----, M.L. Briones, L.E. Williams, and B.D. Barry. 1967. Relations of the wheat curl mite to kernel red streak of corn. Phytopathology 57:986-989.

----, W.E. Styer, J.K. Knoke, and H.N. Pitre. 1973. Semipersistent transmission of leafhopperborne maize chlorotic dwarf virus. J. Econ. Entomol. 66:1271-1273.

Neiswander, C.R. 1926. The sources of American corn insects. Ph.D. Dissent. Ohio State Univ.

----, and C.A. Triplehorn. 1961. Differential resistance of dent corn strains to the corn leaf aphid, *Rhopalosiphum maidis* (Fitch) in Ohio. Ohio Agr. Exp. Sta. Res. Bull. 898.

Ohio Agricultural Research and Development Center. 1973. A list of references: Maize virus diseases and corn stunt. Maize Virus Information Serv. Library of OARDC, Wooster, Ohio.

Oldfield, G.N. 1970. Mite Transmission of plant viruses. Annu. Rev. Entomol. 15:343-380.

Orlob, G.B., and J.T. Medler. 1961. Biology of Cereal and grass aphids in Wisconsin. Canadian Entomol. 93:703-714.

Ortman, E.E., and E.D. Gerloff. 1970. Rootworm resistance: Problems in measuring and its relationship to performance. p. 161-174. *In*: 25th Corn and Sorghum Res. Conf. Am. Seed Trade Association.

----, D.C. Peters, and P.J. Fitzgerald. 1968. Vertical-pull technique for evaluating tolerance of corn root systems to northern and western rootworms. J. Econ. Entomol. 61:373-375.

Ouye, M.T. 1962. Effect of antimicrobial agents on microorganisms and pink bollworm development. J. Econ. Entomol. 55:854-857.

Painter, R.H. 1928. Notes on the injury to plant cells by chinch bug feeding. Ann. Entomol. Soc. Am. 21:232-242.

----. 1951. Insect resistance in crop plants. The Macmillan Co., New York.

----, and A.M. Brunson. 1940. Differential injury within varieties, inbred lines, and hybrids of field corn caused by the corn earworm *Heliothis armigera* (Hbn.). J. Agric. Res. 61:81-100.

----, R.O. Snelling, and A.M. Brunson. 1935. Hybrid vigor and other factors in relation to chinch bug resistance in corn. J. Econ. Entomol. 28:1025-1030.

Palmer, L.T., and T. Kommedahl. 1969. Root-infecting *Fusarium* species in relation to rootworm infestations in corn. Phytopathology 59:1613-1617.
Parker, J.R., and R.V. Connin. 1964. Grasshoppers their habits and damage. USDA Info. Bull. 287.
Patal, K.K., and J.W. Apple. 1966. Chlorinated hydrocarbon resistant northern corn rootworm in Wisconsin. J. Econ. Entomol. 59:522-525.
Patch, L.H. 1937. Resistance of a single-cross hybrid strain of field corn to European corn borer. J. Econ. Entomol. 30:271-278.
----. 1943. Survival, weight and location of European corn borers feeding on resistant and susceptible field corn. J. Agric. Res. 66:7-19.
----. 1947. Manual infestation of dent corn to study resistance to European corn borer. J. Econ. Entomol. 40:667-671.
----, J.R. Holbert, and R.T. Everly. 1942. Strains of field corn resistant to the survival of the European corn borer. USDA Tech. Bull. 823.
----, and R.T. Everly. 1945. Resistance of dent corn inbred lines to survival of first-generation European corn borer larvae. USDA Tech. Bull. 893.
----, G.W. Still, B.A. App, and C.A. Crooks. 1941. Comparative injury by the European corn borer to open-pollinated and hybrid field corn. J. Agric. Res. 63:355-368.
Penny, L.H., and F.F. Dicke. 1956. Inheritance of resistance in corn to leaf feeding of the European corn borer. Agron. J. 48:200-203.
----, G.E. Scott, and W.D. Guthrie. 1967. Recurrent selection for European corn borer resistance in maize. Crop Sci. 7:407-409.
Pepper, B.B., and C.M. Haenseler. 1944. Control of European corn borer and ear smut on sweet corn with dusts and sprays. New Jersey Agric. Exp. Stn. Circ. 486.
Perkins, W.D., R.L. Jones, A.N. Sparks, B.R. Wiseman, J.W. Snow, and W.W. McMillian. 1973. Artificial diets for mass rearing the corn earworm (*Heliothis zea*). Prod. Res. Rep. 154.
Pesho, G.R., F.F. Dicke, and W.A. Russell. 1965. Resistance of inbred lines of corn (*Zea mays* L.) to the second brood of the European corn borer (*Ostrinia nubilalis* (Hubner). Iowa State J. Sci. 40:85-98.
Peters, L.L., M.L. Fairchild, and M.S. Zuber. 1972a. Effect of corn endosperm containing different levels of amylose on Angoumois grain moth biology. 1. Life cycle, certain physiological responses, and infestation rates. J. Econ. Entomol. 65:576-581.
----, M.S. Zuber, and M.I. Fairchild. 1972b. Effect of corn endosperm containing different levels of amylose on Angoumois grain moth biology. 2. Physical and chemical properties of experimental corn. J. Econ. Entomol. 65:581-584.
Petty, H.B., and J.W. Apple. 1966. Insects: p. 351-417. *In*:W.H. Pierre, S.A. Aldrich, and W.P. Martin (eds.) Advances in corn production: Principles and practices. Iowa State Univ. Press. Ames, Iowa.
Phillips, W.J. 1909. The slender seed-corn ground-beetle. USDA Bur. Entomol. Bull. 85.
----. 1914. Corn-leaf blotch miner. J. Agric. Res. 2:15-41.
----, and G.W. Barber. 1929. A study of the hibernation of the corn earworm in Virginia. Virginia Agric. Exp. Stn. Bull. 40.
----, and ----. 1931. The value of husk protection to corn ears in limiting corn earworm injury. Virginia Agric. Exp. Stn. Tech. Bull. 43.
----, and ----. 1933. Egg-laying habits and fate of eggs of the corn earworm moth and factors affecting them. Virginia Agric. Exp. Stn. Tech. Bull. 47.
----, and ----. 1934. Ear-worm injury in relation to date of planting field corn in central Virginia. Virginia Agric. Exp. Stn. Tech. Bull. 55.
----, and H. Fox. 1924. The rough-headed corn stalk-beetle. USDA Bull. 1267.
----, G.W. Underhill, and F.W. Poos. 1921. The larger corn stalk-borer in Virginia. Virginia Agric. Exp. Stn. Tech. Bull. 22.
Pitre, H.N. 1968. Systemic insecticides for control of the blackfaced leafhopper, *Graminella nigrifrons*, and effect on corn stunt disease. J. Econ. Entomol. 61:765-768.

----, and L.W. Hepner. 1967. Seasonal incidence of indigenous leafhoppers (Homoptera, Cicadellidae) on corn and several winter crops in Mississippi. Ann. Entomol. Soc. Am. 60:1044-1055.

Poos, F.W. 1939. Host plants harboring *Aplanobacter stewarti* without showing external symptoms after inoculation by *Chaetocnema pulicaria*. J. Econ. Entomol. 32:881.

----. 1955. Studies of certain species of *Chaetocnema*. J. Econ. Entomol. 48:555-563.

----, and C. Elliott. 1936. Certain insect vectors of *Aplanobacter stewarti*. J. Agric. Res. 52:585-608.

Quaintance, A.L., and C.T. Brues. 1905. The cotton bollworm. USDA Bull. 50.

Rand, F.V., and L.C. Cash. 1933. Bacterial wilt of corn. USDA Tech. Bull. 362.

Randell, R. 1970. The bionomics of the corn leaf aphid, *Rhapalosiphum maidis* Fitch. Ph.D. thesis. Univ. Illinois.

Raun, E.S., R.E. Hill, and D.L. Keith. 1968. Western bean cutworm. Univ. Nebraska Coll. Agric. Home Econ. Quart. Bull. 14:16-17.

Reddy, G.P.V., and Chippendale, G.M. 1972. Nutritional requirements of the southwestern corn borer, *Zea diatrea grandiosella*. Entomol. Exp. Appl. 15:51-60.

Reid, W.J., Jr. 1940. Biology of the seed-corn maggot in the coastal plain of the south Atlantic States. USDA Tech. Bull. 723.

Rhodes, A.M., and W.H. Luckmann. 1967. Survival and reproduction of the corn leaf aphid on twelve maize genotypes. J. Econ. Entomol. 60:527-530.

Richey, F.D. 1944. Maize hybrids susceptible to earworm. J. Hered. 35:327-328.

Rosenkranz, E. 1969. A new leafhopper-transmissable corn stunt disease agent in Ohio. Phytopathology 59:1344-1346.

Satterthwait, A.F. 1919. How to control billbugs destructive to cereal and forage crops. USDA Farmers Bull. 1003.

Saugstad, E.S., and R.T. Everly. 1967. Overwintering populations of the corn leaf aphid on barley and grasses in Indiana. Proc. North Central Branch Entomol. Soc. Am. 22:69-73.

Schlosberg, M., and W.A. Baker. 1948. Tests of sweet corn lines for resistance to European corn borer larvae. J. Agric. Res. 77:137-156.

Scott, G.E., W.D. Guthrie, and G.R. Pesho. 1967. Effect of second-brood European corn borer infestation on 45 single cross hybrids. Crop Sci. 7:229-230.

Shelford, V.E., and W.P. Flint. 1943. Populations of the chinch bug in the upper Mississippi Valley from 1823-1940. Ecology 24:435-455.

Shepherd, R.J. 1965. Properties of a mosaic virus on corn and Johnson grass and its relation to the sugar cane mosaic virus. Phytopathology 55:1250-1256.

Shotwell, R.L. 1941. Life history and habits of some grasshoppers of economic importance on the Great Plains. USDA Tech. Bull. 774.

----. 1952. Tests with sprays for controlling grasshoppers on farmlands in North Dakota and Texas, 1950-51. USDA Bur. Entomol. Plant Quarantine E-845.

----. 1958. The grasshopper your sharecropper. Missouri Agric. Exp. Stn. Bull. 714.

Sifuentes, J.A., and R.H. Painter. 1964. Inheritance of resistance to western corn rootworm adults in field corn. J. Econ. Entomol. 57:475-477.

Simmons, P., and G.W. Ellington. 1933. Life history of the Angoumois grain moth in Maryland. USDA Tech. Bull. 351.

Slykhuis, J.T. 1953. The relation of *Aceria tulipae* (K.) to streak mosaic and other chlorotic symptoms of wheat. Phytopathology 43:484-485.

Snelling, R.O., R.A. Blanchard, and J.H. Bigger. 1940. Resistance of corn strains to the leaf aphid *Aphis maidis* Fitch. J. Am. Soc. Agron. 32:371-381.

Stahl, C.F. 1927. Corn stripe disease in Cuba not identical with sugar cane mosaic. Trop. Plant Res. Found. Bull. 7.

Stanley, W.W. 1936. Studies of the ecology and control of cutworms in Tennessee. Tennessee Agric Exp. Stn. Bull. 159.

Starks, K.J., and W.W. McMillian. 1967. Resistance in corn to the corn earworm and fall armyworm. Part II. Type of field resistance to the corn earworm. J. Econ. Entomol. 60:920-923.

----, W.W. McMillian, A.A. Sekul, and H.C. Cox. 1965. Corn earworm larval feeding response to corn silk and kernel extracts. Ann. Entomol. Soc. Am. 58:74-76.

Stevens, N.E. 1936. Second experimental forecast of the incidence of bacterial wilt of corn. Plant Dis. Rep. 20:241-244.

Stirrett, G.M. 1938. A field study of the flight, oviposition and establishment periods in the cycle of the European corn borer *Pyrausta nubilalis* Hbn., and the physical factors affecting them. Sci. Agr. 18:355-369, 536-557, 568-585, 656-683.

Stoner, W.N. (Comp.) 1968. Corn (maize) viruses in the Continental United States and Canada. USDA. ARS 33-118.

----, L.E. Williams, and L.J. Alexander. 1964. Transmission by the corn leaf aphid, *Rhopalosiphum maidis* (Fitch) of a virus infecting corn in Ohio. Ohio Agric Exp. Stn., Res. Circ. 136.

Storey, H.H. 1939. Investigations of the mechanism of the transmission of plant viruses by insect vectors. III. The insect's saliva. Proc. Royal Soc. (London) Ser B 127:526-543.

Tate, H.D., and O.S. Bare. 1946. Corn rootworms. Nebraska Agric. Exp. Stn. Bull. 381.

Taylor, L.R., and R.E. Berry. 1968. High altitude migration of aphids in maritime and continental climates. Proc. North Central Branch Entomol. Soc. Am. 23:1, 69.

Tenhet, J.N., and E.W. Howe. 1939. The sand wireworm and its control in South Carolina coastal plain. USDA Tech. Bull. 659.

Thomas, C.A. 1940. The biology and control of wireworms. Pennsylvania Exp. Stn. Bull. 392.

Triplehorn, C.A. 1959. The possible effect of weather on incidence of corn leaf aphid infestation and damage. Proc. North Central Branch Entomol. Soc. Am. 14:28-29.

USDA. 1937-1938. Ohio Agric. Exp. Stn. Annu. Rep. Bull. 600.

----. 1954. Chinch bugs. How to control them. USDA Leafl. 364.

----. 1961. The armyworm and the fall armyworm — how to control them. USDA Leafl. 494.

----. 1962. The European corn borer and how to control it. USDA Farmers Bull. 2084.

----. 1964. Grasshopper control. USDA Farmers Bull. 2193.

----. 1966. Neotropical corn borer (*Zeadiatraea lineolata* (Walker)). USDA Coop. Econ. Inst. Rep. 16:33,823.

VanDerSchaaf, P. 1969. Resistance of corn to laboratory infestation of the larger rice weevil, *Sitophilus zeamais*. J. Econ. Entomol. 62:352-355.

Vanderzant, E.S. 1974. Development significance and application of artificial diets for insects. Annu. Rev. Entomol. 19:139-160.

Vickery, R.A. 1929. Studies on the fall armyworm in the gulf coast district of Texas. USDA Tech. Bull. 138.

Vinal, S.C. 1917. The European corn borer, *Pyrausta nubilalis* Hubner, a recently established pest in Massachusetts. Massachusetts Agric. Exp. Stn. Bull. 178.

----, and D.J. Caffrey. 1919. The European corn borer and its control. Massachusetts Exp. Stn. Bull. 189.

Virtanen, A.I. 1961. Some aspects of factors in the maize plant with toxic effects on insect larvae. Suomen Kemistilehti B34:29-31.

Wahlroos, O., and A.I. Virtanen. 1959. The precursors of 6MBOA in maize and wheat plants: Their isolation and some of their properties. Acta Chem. Scand. 13:1906-1908.

Walkden, H.H. 1950. Cutworms, armyworms and related species attacking cereal and forage crops in the central great plains. USDA Circ. 849.

----, D.A. Wilbur, and H. Gunderson. 1954. Control of stored grain insects in the north central states. Minnesota Agric. Exp. Stn. Bull. 425.

Walter, E.V. 1957. Corn earworm lethal factor in silks of sweet corn. J. Econ. Entomol. 50:105-106.

----, and A.M. Brunson. 1940. Differential susceptibility of corn hybrids to *Aphis maidis*. J. Econ. Entomol. 33:623-628.

Walton, R.R., and G.A. Bieberdorf. 1948. Seasonal history of the southwestern corn borer, *Diatraea grandiosella* Dyar, in Oklahoma; and experiments on methods of control. Oklahoma Agric. Exp. Stn. Tech. Bull. T-32.

----, and C.M. Packard. 1951. The armyworm and its control. USDA Farmer's Bull. 1850.

Webster, F.M. 1913. The western corn rootworm. USDA Bur. Entomol. Bull. 8.

Whitcomb, R.F., and R.E. Davis. 1970. Mycoplasma and phytarboviruses as plant pathogens persistently transmitted by insects. Annu. Rev. Entomol. 15:405-464.

----, J.G. Tully, J.M. Vobe, and P. Saglio. 1973. Spiroplasmas and acholeplasmas: Multiplication in insects. Science 182:1251-1253.

Whitcomb, W.H., J.O. York and P.A. Shockley. 1966. Systemic insecticides for control of southwestern corn borer, *Zea diatraea gandioella*. J. Kansas Entomol. Soc. 39:267-270.

Widstrom, N.W., W.W. McMillian, and B.R. Wiseman. 1970. Resistance in corn to the corn earworm and the fall armyworm. IV. Earworm injury to corn inbreds related to climatic conditions and plant characteristics. J. Econ. Entomol. 63:803-808.

----, L.M. Redlinger, and W.J. Wiser. 1972a. Appraisal of methods for measuring corn kernel resistance to *Sitophilus zeamais*. J. Econ. Entomol. 65:790-792.

----, A.N. Sparks, E.B. Lillehoj, and W.F. Kwolek, 1975. Aflatoxin production and tepidopteran insect injury on corn in Georgia. J. Econ. Entomol. 68:855-856.

----, B.R. Wiseman, and W.W. McMillian. 1972a. Genetic parameters for earworm injury in maize populations with Latin American germ plasm. Crop Sci. 12:358-359.

Wilbur, D.A., H.R. Bryson, and R.H. Painter. 1950. Southwestern corn borer in Kansas. Kansas Agric. Exp. Stn. Bull. 339.

Wildermuth, V.L. 1917. The desert corn flea-beetle. USDA Bull. 436.

----, and E.V. Walter. 1932. Biology and control of the corn leaf aphid with special reference to the southwestern states. USDA Tech. Bull. 306.

Wilson, R.L., and D.C. Peters. 1973. Plant introductions of *Zea mays* as sources of corn rootworm tolerance. J. Econ. Entomol. 66:101-104.

Wiseman, B.R., W.W. McMillian, and N.W. Widstrom. 1970. Husk and kernel resistance among maize hybrids to an insect complex. J. Econ. Entomol. 63:1260-1262.

----, R.H. Painter, and C.E. Wassom. 1967a. Preference of first-instar fall armyworm larvae for corn compared with *Tripsacum dactyloides*. J. Econ. Entomol. 60:1738-1742.

----, C.E. Wassom, and R.H. Painter. 1967b. An unusual feeding habit to measure differences in damage to 81 Latin-American lines of corn by the fall armyworm, *Spodoptera frugiperda* (J.E. Smith). Agron. J. 59:279-281.

Wymore, F.H. 1931. The garden centipede. California Agric. Exp. Stn. Bull. 518.

York, J.O. and W.H. Whitcomb. 1963. Breeding for resistance to the southwestern corn borer. Arkansas Farm Res. 12.

Zuber, M.S., and F.F. Dicke. 1964. Interrelationship of European corn borer plant populations, nitrogen levels and hybrids on stalk quality of corn. Agron. J. 56:401-402.

----, G.J. Musick, and M.L. Fairchild. 1971. A method of evaluating corn strains for tolerance to the western corn rootworm. J. Econ. Entomol. 64:1514-1518.

Chapter 10 Climatic Requirement

ROBERT H. SHAW
Iowa State University
Ames, Iowa

Corn, because of its many divergent types, is grown over a wide range of climatic conditions. Some cultivars grow very short, others up to 6 to 8 m in height; some require 60 to 70 days to mature the grain after emergence, others require 10 to 11 months. Yet, in spite of this range of characteristics, general features of the major production areas can be characterized, and for selected areas, very specific weather relations can be shown.

WORLD-WIDE CORN PRODUCTION AND CLIMATE

A number of workers have attempted to define the limiting climatic conditions for corn production (Finch and Baker, 1917; Jenkins, 1941; Klages, 1942; Wallace and Bressman, 1937; Ward, 1919). Before attempting to define them here, an examination of the areas where corn is produced is necessary.

In Fig. 1 the world-wide corn production for the 1957-61 period is shown (Guidray, n.d.). The bulk of the corn is produced between latitudes 30 and 55°, with relatively little grown at latitudes higher than 47°. Almost half of the world's production is grown in the United States (USDA, Foreign Agric. Serv., 1973). Only two areas produce significant amounts of corn outside the 30 to 55° latitude range, Brazil and Mexico. That grown in Brazil is only a few degrees latitude from 30°, while much of that in Mexico is grown at altitudes several thousand feet above sea level.

In the first edition of this monograph, corn-producing areas were defined in terms of Koppen's climatic classification (Koppen, 1931). Because Koppen's boundary between the humid mesothermal and humid microthermal climates cuts across the middle of the Corn Belt in the United States, a classification that delineates, rather than separates, the Corn Belt seems more appropriate. Trewartha (1968) provides such a classification. His temperate, continental-climate zone, with 4 to 7 months over 10 C (50 F) and with a hot summer (warmest month over 22.2 C/72 F) coincides very closely with the limits of the Corn Belt. The 10 C value marks the threshold of active growth for a number of plants. The so-called Cotton Belt falls in his subtropical classification, with at least 8 months averaging over 10 C. A very high percentage of the corn is produced within these two climatic zones. By Trewartha's classification, typical corn climate is a temperate or subtropical continental, to a transition marine-continental climate, with a relatively long frost-free period. The only significant areas not included would be the tropical areas and some drier climates bordering on the typical climates. Corn is grown in woodland and grassland climates, but, without irrigation, production is limited in the drier areas where the native vegetation was short grass.

Papadakis (1966) developed a crop ecologic classification of climates with 10 major climatic groups. In this classification, there are two subdivisions that would include the Corn Belt: an eastern zone and a western zone, with the dividing line in eastern Iowa. He has also discussed the suitability and limitations of each subdivision for various crops, including corn. (Papadakis, 1970). His study shows very few areas in the world with climates comparable to the Corn Belt. Berbecel and Rogojan (1962) have developed a map for Romania that shows the relative favorability of different areas for corn production.

It is obvious from Fig. 1 that corn has a cold limit. This is a combination cool-temperature, frost-free season limit. Practically no corn is grown where the mean mid-summer temperature is less than 19 C (66 F) or where the average night temperature during the summer months falls much below 13 C (55 F). The greatest production is where the warmest month isotherms range between 21 C (70 F) and 27 C (80 F) and the freeze-free season is 120 to 180 days duration. Within such regions, yields generally are higher with below-normal summer temperatures than with above-normal values. There seems to be no upper temperature limit specific for corn production, but yields usually decrease with higher temperatures. Although corn is generally called a warm weather crop, it is not a hot weather crop.

Corn is grown in areas where the annual precipitation ranges from 25 to over 500 cm. Haise (1958) reported seasonal consumptive use values of 42 to

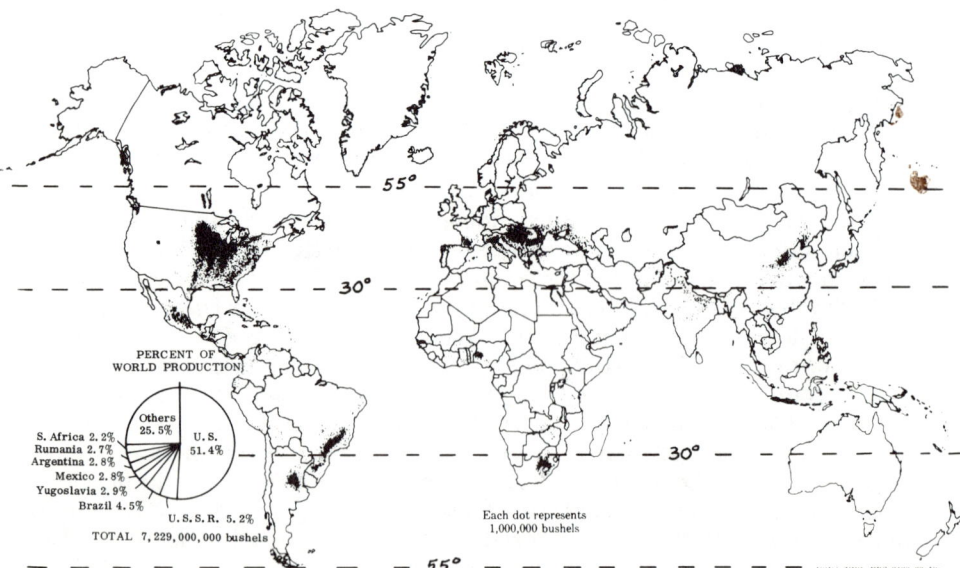

Fig. 1. World-wide corn production. (Guidray, n.d.)

54 cm (16.6 to 21.1 inches) in the Dakotas. Doss et al. (1962) reported that consumptive use averaged near 49 cm (19.1 inches) in Alabama and ranged from 45 to 56 cm (18.0 to 22.1 inches). Vasquez (1961) found a consumptive use value of 47 cm (18.4 inches) for a 125-day period in Puerto Rico. The range of consumptive use by corn is ordinarily 41 to 64 cm (16 to 25 inches) (Hanway, 1966), but amounts as low as 30 cm (12 inches) and as high as 84 cm (33 inches) have been reported (Robins and Rhoades, 1958).

As the temperate and subtropical climates merge into drier steppe climates, moisture demands of corn exceed available moisture, and wheat (*Triticum aestivum* L.) and barley (*Hordeum vulgare* L.) become the important crops. Where corn is grown in a steppe region, yields fluctuate widely with the extreme variations in rainfall. This occurs primarily in areas in the United States, Argentina, South Africa, and the Soviet Union.

A summer rainfall of 15 cm is about the lower limit for corn production without irrigation, but yield responses to irrigation are obtained with much higher summer rainfall, the response depending upon rainfall distribution and soil-moisture reserves. Vasquez (1961) obtained no response from irrigation when 51 cm (20 inches) of rain were well distributed throughout the growing season. There seems to be no upper limit of rainfall under which corn does not grow, but excessive rainfall will decrease yields.

STAGES OF GROWTH AND DEVELOPMENT

Hershey (1934) and Paddick (1944) divided the corn plant development into five different stages, each with its own relation to final yield. Shaw and Loomis (1950) also divided the development of corn into five stages. Hanway (1971) proposed a 10-stage, plant-development system ranging from 0 when the plant tip emerges from the soil to 10 when the plant is physiologically mature. For discussion here, seven different phases will be used. These will be referenced in terms of Hanway's stages (Fig. 2).

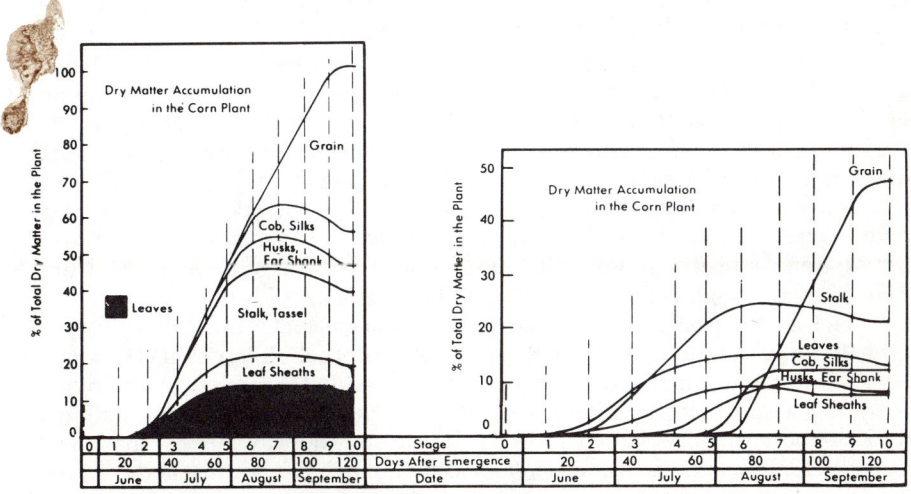

Fig. 2. Dry matter accumulation at various stages of corn development. (Hanway, 1971.)

1) Before planting.
2) Planting to emergence (planting to stage 0).
3) Early vegetative growth from emergence to flower differentiation (stages 0 to 2.5-3).
4) Late vegetative growth from about the beginning of rapid stem elongation (plant height near 50 cm) to tasseling. Tip of tassel emerged (stages 2.5-3 to 4).
5) Tasseling, silking and pollination (stages 4 to 5).
6) Grain production from fertilization to physiological maturity of the grain (stages 6 to 10).
7) Maturation or drying of the grain.

During the very early growth period (about 2 weeks), the growing point of the corn plant is below the soil surface. Under favorable conditions, the entire stem and differentiated tassel are formed under ground in about 2 weeks after seedling emergence (Kiesselbach, 1949). At stage 1.5 the growing point is at the soil surface. The period up to stage 3 includes the seedling stage and early leaf growth up to 5 to 6 weeks after emergence. By the end of the period, the plants have the maximum number of leaves; vascular bundles and ovules on the major embryonic ear shoot are already determined. The potentialities of the plant are established during stage 3 (May and part of June in the Corn Belt), and the period is of considerable theoretical importance in yield prediction. However, weather in the Corn Belt, and probably most areas where corn is grown, is seldom limiting during this period, except as it affects stand, and soil fertility becomes a major factor. There is evidence that even slight water stress can reduce the rate of appearance of floral primordia. Work of Nicholis and May (1963) on barley has shown that, if the stress is mild, and if the period of stress is relatively brief, rate of primordial initiation, upon the relief of stress, is more rapid than in the nonstressed plants and the total number of spikelets formed may be unaffected. Because the corn plant seldom reaches its maximum potentialities under field conditions and has extensive powers of recovery from early season setbacks, yields cannot be predicted accurately from early season observations.

During stage 3 to 4 the leaf area of the plant becomes fully developed and the tip of the tassel emerges at the end of stage 4. The upper internodes of the stalk are elongating rapidly, and the top one or two ears are undergoing rapid enlargement and elongation. Maximum stalk height, stalk diameter, and leaf area may be reached at the end of stage 4.

Stage 5, tasseling, silking, and pollination, is a critical stage in the corn plant. The number of ovules that will be fertilized is being determined. Stress, both moisture and fertility, can reduce yields drastically. In the Corn Belt this stage occurs, on the average, in the latter part of July.

The first 2 weeks of the grain production period are a time of rapid growth of the ear shoot, husks, cobs, and young kernels. The cob has attained nearly full size but little grain weight has been added. From stage 5.5 to 8.5, there is a rapid increase in grain weight. In about a 5-week period, almost 85% of the grain dry weight may be produced (Hanway, 1966; Shaw and Loomis, 1950). By stage 10, physiological maturity has been reached, i.e., the maximum dry weight of grain has been attained (Daynard and Duncan, 1969; Rench and Shaw, 1971). For most Corn Belt cultivars, the silking to physiological maturity time averages 50 to 60 days, but varies some with cultivar, date of planting, location, and nutrition (Peaslee et al., 1971; Rench, 1973). Berbecel and Eftimescu (1972) found that the period from tasseling to maturity ranged from 60 to 62 days for several cultivars grown in Romania.

After physiological maturity is reached, the grain needs to dry down to harvestable values. The length of this period will vary with cultivars, weather conditions, and harvesting methods.

MOISTURE STRESS CONCEPT

In discussing the effects of various climatic factors on corn growth and yield, we need to understand how these various factors affect the moisture supply of the plant. The present-day concept of soil moisture availability recognizes the importance of the amount of moisture in the soil, the texture of the soil (moisture in sand tending to be more available than moisture in clay), and the atmospheric demand for water. A number of different concepts of the availability of soil moisture have been proposed over the years (Halstead, 1954; Pierce, 1958; Thornthwaite and Mather, 1955; Veihmeyer and Hendrickson, 1955). Although the individual relationships proposed seem to differ, they all fit under the theoretical concept of Philip (1957) and Gardner (1960), which includes the previously mentioned factors. The evaluation of the effect of weather factors on moisture availability must use this concept. This can be explained further by use of Fig. 3. More detailed discussions of moisture relations are in Shaw and Burrows (1966) and Shaw and Laing (1966). Figure 3 shows the relationship between soil-moisture content and transpiration for three different types of demand days for corn

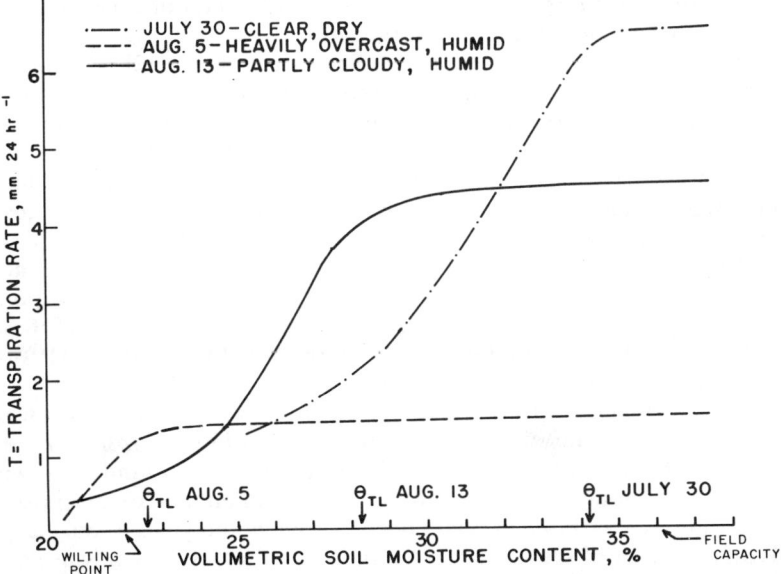

Fig. 3. Daily transpiration for 3 days plotted as a function of soil moisture. (Denmead and Shaw, 1962.)

plants grown in large containers. On a clear, dry day with high atmospheric demand only those plants in containers with a high soil-moisture content were able to meet the demand. On an overcast, humid day the demand was met at soil moisture very near the wilting point. The θ_{TL} points shown are the moisture required for the plant to meet the demand for that particular day. At a soil moisture content lower than this, the plant is under some degree of stress. The stress point varies with the weather, and the relationships shown will vary with different soil types. The use of several factors complicates the explanation of moisture supply, but they are necessary if moisture-stress problems are to be quantitatively evaluated.

The atmospheric demand for water is a function of the energy available (solar radiation), the movement of moisture from the evaporating surface (wind), the dryness of the atmosphere (humidity), and temperature of the air. Temperature alone does not affect evaporation directly, except as it affects the temperature of the evaporating surface, but it does affect the dryness of the atmosphere by varying its capacity to hold water. Radiation is usually considered the major factor in controlling the atmospheric demand.

Moisture stress interrupts photosynthesis and checks growth until turgor is restored by removal of the moisture stress, (de Jager, 1968; Vaadia et al., 1961). Boyer (1970) found that inhibition of photosynthesis in corn plants 4 to 5-weeks-old began at higher leaf potentials in corn than in soybeans [Glycine max (L.) Merr.], i.e., corn was less able to stand severe dessication, but had a higher rate of photosynthesis during dessication. Relative turgidity of the leaves has been used to show the effect of stress on photosynthesis. Barnes and Woolley (1969) stated that the corn plant is under stress when the relative turgidity of the upper leaf is less than 90%. Downey (1971c) showed about the same critical level and found relative net photosynthesis near zero with a relative turgidity near 70%. Data of Shaw and Laing (1966) on soybeans show a similar relationship.

Dale and Shaw (1965), Corsi and Shaw (1971), and Shaw and Felch (1972) have developed a stress index based on the daily balance between soil moisture and atmospheric demand, as shown in Fig. 3. Dale and Shaw (1965) considered any day with a reduction in evapotranspiration from the potential rate as a stress day. The highest correlation with yield was found for the period 6 weeks before to 3 weeks after silking. In the later references the reduction in actual evapotranspiration from the potential evapotranspiration (no moisture stress) is the basis for the index. For example, if actual evapotranspiration for a day was estimated as 0.20, but potential evapotranspiration was 0.30, the stress index for that day would be $1 - 0.20/0.30 = 0.33$. This index, accumulated over a period from 27 June through 31 August has shown a high correlation with corn yields.

The water for the corn crop may come from current, crop-season rainfall, from moisture stored in the soil before planting, or from irrigation. Minor amounts may come from dew. Power et al. (1973) found that, within the range they studied, all sources were effective in enhancing crop production. Rainfall, as a variable to relate to corn growth and yield, is only as good as it is an estimator of the available soil moisture; in fact, it is only as good as it estimates the moisture status of the plant. If 50 to 64 cm (20 to 25 inches) of water are used to produce a high-yielding corn crop, almost half of this may be stored at the beginning of the season in a good soil that has an available moisture capacity of 5 cm/30 cm (2 inches/foot) and if the crop roots to a depth of 152 cm (60 inches). In the Corn Belt, soil moisture reserves at the start of the growing season will vary considerably. The normal situation is to have adequate to excess soil moisture reserves in the eastern part of the Corn Belt and adequate to deficient reserves in the western part. In Iowa, Shaw et al. (1972) found that the average inches of plant-available water in the top 152 cm (60 inches) of soil on 15 April ranged from over 25 cm (10 inches) in eastern and southeastern Iowa to less than 12.5 cm (5 inches) in northwest Iowa. The amount of growing-season rainfall required is closely related to these reserves.

Water use varies with the stage of development of the corn crop. Early in the growing season the loss is primarily evaporation from a bare soil. As the crop cover increases, transpiration becomes an increasingly dominant factor. Denmead and Shaw (1959) found the ratio between evapotranspiration and class A open-pan evaporation shown in Fig. 4. This is similar to the relationship reported by Cackett and Metelerkamp (1964), Downey (1971b), and Mallett (1972), who found maximum ratios ranging from 0.75 to 1.0. Ritchie and Burnett (1971) found a leaf-area index of 2.7 was necessary for cotton and sorghum (*Sorghum bicolor* L.) to reach an evaporation-transpiration rate of 90% of the potential evaporation when soil evaporation was small.

Shaw et al. (1958) found water use by corn averaged 0.25 cm/day for the 15 April to 15 June period and 0.46 cm/day for the 15 June to 1 August period. Over a 3-year period the daily rate of water use for the latter period was positively and highly correlated with final corn yield.

Seasonal water use has varied widely. Harrold and Driebelbis (1951) found evapotranspiration losses from a lysimeter were 44 to 62 cm (17.4 to 24.6 inches) from May through September. Rhoades and Nelson (1955) reported that irrigated corn ordinarily uses 41 to 64 cm (16 to 25 inches) during the growing season. Shaw et al. (1958) found that water use from 15 April to 1 November ranged from less than 44 cm (17.5 inches) to well over 64 cm (25 inches), with yields increasing as the water use increased. Other water-use data were cited earlier.

The amount of water use may vary with stand. At very low stands water use is low. As stands increase, water use increases rapidly up to a point and then changes slowly with increasing stands. There is a point at which increased stands will not increase the utilization of solar energy in evapotranspiration. Olson (1971) found almost the same water use for stands of 35,000, 45,000 and 70,000 plants/ha (14,000, 18,000, and 28,000 plants/acre). This agrees with a field model of evapotranspiration proposed by Viets (1966).

Beer et al. (1967), working in Iowa, found a negative relationship between the amount of water required by irrigation to maintain soil moisture above 60% of the available water-holding capacity and the maximum corn yield obtained with several levels of irrigation (Fig. 5). The less irrigation water required (i.e., the better the natural moisture environment), the higher the yield. Presumably the higher irrigation requirements represented years of higher atmospheric demand situations with more moisture stress occurring, even with high soil moisture levels.

GROWING-DEGREE UNIT CONCEPT

Another factor that will be referred to during discussion of the different stages of development will be growing-degree units or heat units. The actual number of days for corn to reach maturity varies widely with changes in the environment, although cultivars are often designated as a certain number of days to maturity. The growing-degree-unit approach has been proposed as one that provides a more constant maturity index for varying weather conditions as long as the other environmental conditions are not too far from optimum.

The growing-degree-unit (GDU) approach is based on the use of air temperature data, so it is not really a heat unit, but a temperature unit number. It has also been called thermal units (Berbecel et al., 1964). In using it, accumulations of values above a selected base are made. The literature on growing-degree units is extensive. Reviews by Nuttonson (1953), Holmes and Robertson (1959), and Wang (1960, 1963) are available for those interested in more extensive information. Almost all indexes fall under one of the following basic types: 1) exponential (Livingston and Livingston, 1913; Price, 1911), 2) physiological (Brown, 1960; Livingston, 1916), 3) remainder (Gilmore and Rogers, 1958; Holmes and Robertson, 1959, Nuttonson, 1955, 1957), and 4) evapotranspiration (Thornthwaite, 1952).

The exponential index assumes that for a 10 C increase in temperature the growth rate doubles. This method assigns high efficiencies to tem-

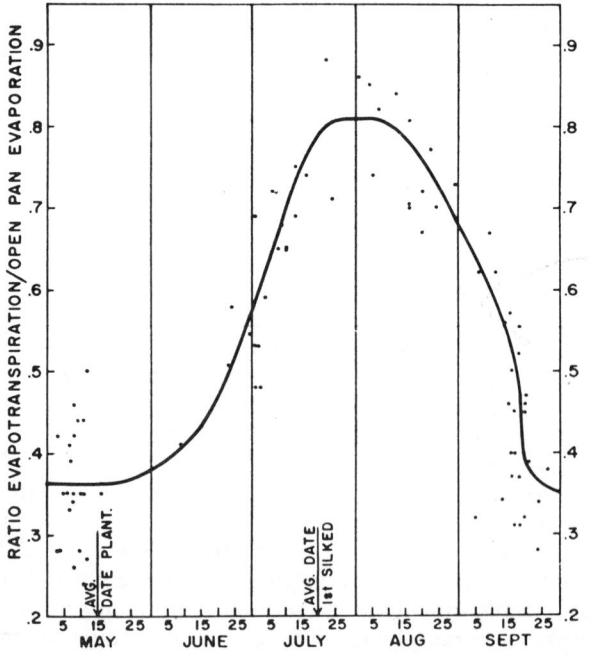

Fig. 4. Ratio of evapotranspiration from corn to open-pan evaporation throughout the growing season. (Denmead and Shaw, 1959.)

CLIMATIC REQUIREMENT

peratures too high for optimum growth. A physiological type of index is one based on the physiological response of plants to temperature and often has been developed from data obtained from controlled conditions. Brown's (1960) equations for soybeans were developed from growth-chamber data. His corn equations (Brown, 1969) were developed from field data and are used to determine maturity ratings for corn in Ontario. The effect of the maximum temperature (F) (T_{max}) on development was obtained from the equation

$$Y_{max} = 1.85\,(T_{max} - 50) - 0.026\,(T_{max} - 50)^2 \quad [1]$$

This assumes a parabolic developmental response curve to temperature. The contribution of nighttime temperatures was calculated by

$$Y_{min} = T_{min} - 40$$

His growing-degree units, H, were then obtained by

$$H = (Y_{max} + Y_{min})/2 \quad [3]$$

The third basic type, the remainder index, accumulates units above a base temperature. In its simplest form, it is calculated by

$$\frac{\text{daily max temp} + \text{daily min temp.}}{2} - 10C = GDU \quad [4]$$

In this example, 10 C was assumed as the base temperature. Several systems are available, but the so-called 30-10 C index (86 to 50 F) will be the primary one discussed here. This is a remainder type index calculated by Eq. 4. Any maximum temperature above 30 C is put in the equation as 30 and any minimum below 10 C is designated as 10. Growing-degree units can be calculated for any stage of development or for the total time from planting or

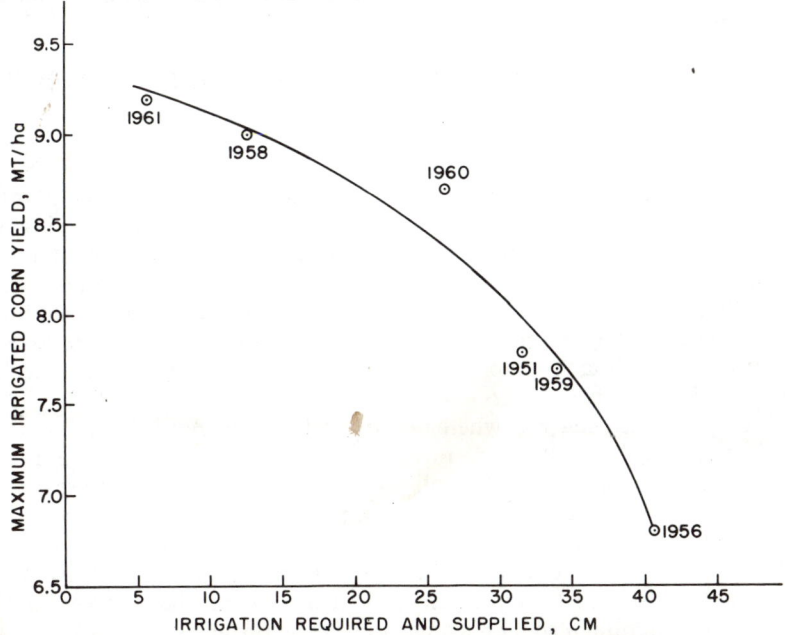

Fig. 5. Relationships between the amount of water required and supplied to maintain soil moisture above 60% of the available water-holding capacity and maximum corn yields obtained on Colo silty clay loam, Ames, Iowa. (Beer et al., 1967.)

emergence to maturity.

Another modification of the remainder system is that of Newman and Blair (1969). They proposed the following modification to the basic daily heat-unit equation:

> When daily mean temperatures average 23.9 C (75 F) or higher and the maximum exceeds 32.2 C (90 F), subtract, from the degree day accumulation that results, the amount the maximum exceeds 32.2 C (90 F) for the day. This largely eliminates the excessive accumulation of degree days in dry, hot climates where corn is usually under water stress during the hot part of the day.

They proposed another modification to be applied in high, cool, mountain climates or under long, cool midsummer days in the northern United States and southern Canada to take into account daily mean temperatures below 15.6 C (60 F):

> When the daily mean is 10 C (50 F) or above, but under 15.6 C (60 F), and the maximum exceeds 18.3 C (65 F), add, to the degree days accumulation the amount the maximum exceeds 18.3 C (65 F).

Cross and Zuber (1972) tested 22 different growing-degree unit methods in Missouri and found that daily measurements gave almost as good results as the use of hourly temperature data. They found the best base temperature for estimation of flowering was 10 C with 30 C optimum. Excess above 30 C was subtracted to account for high temperature stress. Brown (1969), in Ontario, used a threshold of 4.4 C for night temperatures and 10 C for daytime temperatures. Rench (1973), in Iowa, found a base temperature of 7.2 C worked best for the planting to flowering interval and also was the best for the silking to black layer period, although 7.2 C was only slightly better than bases of 4.4 or 10 C.

EFFECT OF WEATHER ON CERTAIN PERIODS OF PLANT GROWTH

Before Planting

The influence of weather on the corn plant starts even before planting. Conditions before planting are especially important in determining soil-moisture reserves. These can reflect a carryover from the previous crop season, or can be from accumulations that may occur during the fall, winter, and early spring. Since evaporation rates in the fall are low, precipitation during this time may be quite efficient for increasing soil-moisture reserves. In many areas, winter precipitation is low, and with a frozen ground, little moisture will enter the soil. In Iowa, Shaw (1965) found that only 25% of the precipitation that occurred when the ground was frozen went into the soil moisture reserve. With the wide range of winter conditions that occur where corn is grown, changes in the soil-moisture reserve will vary greatly, depending on winter precipitation and temperature. Temperature effects are also important in insect and disease problems. Snow cover also affects these problems because of the moderating effect on soil temperature. Early spring precipitation also can be quite effective in increasing the reserves, but the evaporation potential also increases as the spring progresses.

The lower the soil-moisture reserve, the greater is the crop-season rainfall requirement. Thompson (1966, 1969), using regression analyses, found that the optimum preseason rainfall (September through May) for the five highest corn producing states of the Corn Belt was 68 to 71 cm (27 to 28 inches), very near the average for the five states. This amount usually brings the soil moisture to capacity or above capacity but without too much excess.

Freezing weather in the winter improves the tilth of the surface soil (Newlin, 1948) and will help to reduce clods and compaction by machinery. Subsoiling in the Corn Belt has had little effect on yield (Larson and Blake, 1966), but in warmer climates where less soil freezing and thawing occurs, it may help break up the plowsole layer. Although spring rains may help replenish soil-moisture reserves, they may also delay field operations. Spring weather determines the time when field operations can be started. In many areas, a delay in planting reduces the expected yield (Pendleton, 1966); this is probably one of the most important effects of weather before planting.

Planting to Emergence

The period from planting to emergence depends on soil temperature, soil moisture, soil aeration, and seed vigor. Before germination, the seed absorbs water and swells. With warmer temperatures, less water has to be absorbed (Blacklow, 1972), so that germination will start earlier and proceed faster at higher temperatures, assuming water is available. The time from planting to emergence varies widely with environmental conditions and, to a lesser degree, with planting depth (Alessi and Power, 1971). During this stage, development is affected directly by soil temperature and indirectly by air temperatures.

Weather is still a major factor in determining the time of planting. Relatively early planting in the US generally shows higher yields than late planting. For example, Pendleton and Egli (1969) obtained yield decreases when planting was delayed after 30 April in Illinois. Daynard (1972) found that delayed planting decreased the number of days from planting to midsilk and increased the number of days from midsilk to maturity. The total heat units required was reduced only slightly. The optimum planting date will vary with latitude and should consider later critical moisture periods. Early planting may not be the best in all areas of the world.

In the main corn-production areas in the United States, corn planting, until recent years, began when the average air temperature reached near 12 to 14 C (54 to 57 F) (Kincer, 1919). This varied from early February in the south to the middle of May in the north. These dates represented the earliest planted fields and not when the bulk of the corn was planted. In recent years there has been a general move to somewhat earlier planting in the north when air temperatures average near 12 C. The advent of herbicides has aided weed control for the earlier plantings. In past years the bulk of the corn planting was done when the average air temperature was near 16 C (61 F) (Wallace and Bressman, 1937) and averaged 15 May in Iowa. The earlier trend for planting is shown by the Iowa data (USDA Stat. Rep. Serv., 1973), which show an average date of 11 May for the 1968-72 period. At this date the air temperature averages near 14 to 15 C (57 to 59 F).

Germination is affected by soil temperature and soil moisture. Coffman (1923) found that corn germinated best at temperatures above 10 C (50 F) with a sharp decrease in germination below 10 C (50 F). Cummins and Parks

(1961) reported that corn did not germinate at 10 C (50 F).

In tests using a silt loam soil, Wolfe (1927) found that the rapidity of germination increased with increased soil moisture up to 80% of saturation. At 10% saturation, there was no germination because of lack of water, whereas at 100% saturation or above, germination was retarded or prevented because of a lack of oxygen. On a silt-loam soil held at 50 to 60% moisture, a soil temperature of 35 C (95 F) gave slightly more rapid germination than one of 30 C (86 F) and considerably more rapid germination than one of 25 C (77 F).

Blacklow (1972) found that rates of elongation of the radicle and shoot were greatest at about 30 C (86 F) and effectively ceased at constant temperatures of 9 C (48.2 F) and 40 C (104 F). The minimum time for initiation of a radicle and a shoot occurred at 30 C (86 F) with the radicle preceding the shoot. For constant temperatures, the rates of elongation of the radicle and shoot are shown in Fig. 6 (Blacklow, 1972). How well these represent field conditions still needs to be shown. A short duration of high temperatures above 30 C (86 F) in the field may not show as rapid a drop in growth rate as does a constant exposure to this temperature.

In the field at Ames, Iowa, corn emergence has varied from 7 to 15 days after planting in a 6-year period. According to Wallace and Bressman (1937), corn usually emerges in 8 to 10 days at an average temperature of 16 to 18 C (60 to 65 F), but it takes 18 to 20 days at 10 to 13 C (50 to 55.4 F). If the soil is moist and at an average temperature near 21 C (70 F), emergence may occur in 5 to 6 days. Alessi and Power (1971) found emergence was delayed 1 day for each 2.6-cm increase in depth of planting. Since spring soil temperatures decrease with depth (Holmes and Robertson, 1959; Shaw, 1971), delayed emergence is a result of both cooler temperatures and a greater distance involved. Berbecel and Eftimescu (1972) found that the duration from planting to emerging was explained almost entirely by soil temperature. When the available soil moisture for the 20 cm deep surface layer was above a value of 10 mm and the average temperature at the 10-cm depth was accumulated for values > 8 C (46.4 F), the emergence period was almost constant at a sum of 100.9 degrees.

Another factor to consider is that air temperatures are used in many studies, although the soil is the medium in which germination occurs. Air temperatures are often used because of their availability and a lack of soil-temperature data. Newhall (1947) showed that soil temperature closely follows air temperature, i.e., there is little daily heat accumulation in the soil. At shallow depths, the soil may be much warmer than air temperature during periods of intense heating, but on cloudy days, soil temperature at the planting depth closely approximates air temperature. In practical application, this means that bright, warm days in April do not heat up the soil for rapid emergence in early May. To have a short emergence interval, daily heating of the soil must prevail during the period of germination and seedling growth. Shaw (1971) found that maximum soil temperatures in the spring at corn planting depth were slightly cooler than maximum air temperatures, and minimum soil temperatures were slightly warmer than minimum air temperatures, giving average soil and air temperature values very close together.

Other aspects of the effect of weather also need to be considered. Cold,

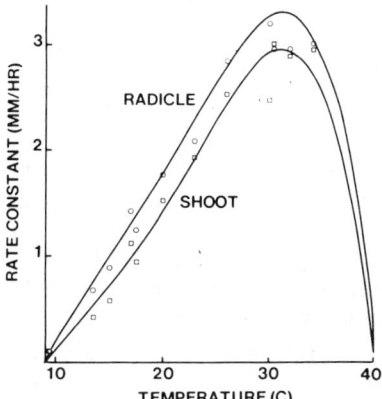

Fig. 6. Rate of elongation of the radicle and shoot of corn to the stage of emergence as a function of soil temperature. (Blacklow, 1972).

wet weather following planting favors the development of pathogens. Seed rots and seedling blight may become prevalent when corn is planted in a cold, wet soil. Germination of corn seed is greatly retarded at 10 C (50 F) or below, but at this temperature certain species of *Pythium* are active (Ullstrup, 1966). However, Ullstrup states seed rot and seedling blights occur relatively infrequently if good seed is planted and proper seed treatment is used. Later planting might be encouraged to reduce seedling diseases, but, as pointed out by Dungan (1944), this method should be used for the reduction of insect and disease damage with caution since delayed planting may decrease the yield and quality of the crop. In the south, Johns and Brown (1941) found that date of planting influenced the amount of damage by southern corn rootworm [*Diabrotica undecimpunctata howardi* (Barber)] and corn borer [*Ostrinia nubilalis* (Hubn). Practices in warmer climates may well vary from those in colder climates where the hybrids used crowd the season available.

Early Vegetative Growth from Emergence to Flower Differentiation

Shortly after emergence, an important change takes place when the plant changes from dependence on stored food to self sufficiency. During the early part of its life, the corn plant requires a limited amount of moisture for the small growth that takes place. This is fortunate because both the initiation and differentiation of vegetative and reproductive primordia in the apical meristem and the enlargement of the cells thus differentiated are very sensitive to water stress (Slatyer, 1969). Stress shortly after emergence decreases the starch and chlorophyll content of seedlings (Maranville and Paulsen, 1970); but if the weather is somewhat dry at this time, the roots will penetrate deeper into the soil, and the plant seems better able to withstand later dry weather, which may more than offset any immediate detrimental effects of stress. Salter and Goode (1967) reported that Russian workers found that stress during the early vegetative stage had little, if any, effect on final yield; deeper, more extensive rooting may be the reason. From now on the plant is subjected to two different environments, the atmosphere and the soil, and the dependency on soil temperature becomes less than during the germination period (Cal and Obendorf, 1972).

Young corn plants are relatively resistant to cold weather, with an air temperature near -1 C (30 F) generally killing exposed above-ground parts (Shaw et al., 1954). Until stage 1.5 when six leaves have fully emerged, the growing point is below the soil surface (Hanway, 1966). Therefore, recovery from a moderate freeze when the growing point is below ground is usually rapid and almost complete. But, occasionally, a late spring freeze may kill early planted corn whose growing point is at or above the soil surface. Hanna (1924) found that air temperatures of -1.7 C (29 F) injured corn and -4.4 C (24 F) killed the corn. Minimum soil temperatures at the 2.5 cm depth will be slightly higher than minimum air temperatures at these temperatures.

In corn plants 2 to 3-weeks-old, Sellschop and Salmon (1928) found that chilling had no marked immediate effect on the plant. From 5 to 10 days after chilling, light yellow bands developed on the leaves, which became filmy and rust colored toward the edges. Bands occurred on those parts of the blades that formed the whorl of the plant at the time of the chilling. In this region the most active growth takes place and the youngest tissues are found. In general, the younger the plant, the greater the injury. They found that 6-week-old plants, chilled at temperatures of 0.5 to 5 C (33 to 41 F) for different lengths of time, recovered and were capable of seed production if less than 25% of the leaf area showed injury soon after chilling. Plants with 25 to 50% injury occasionally recovered, whereas those with more than 50% injury seldom recovered. As the length of chilling increased, the amount of injury also increased. Greater injury occurred in a saturated soil than in a moderately wet soil, and least injury occurred in a moderately dry soil. Purvis and Williamson (1972) found that very young plants were severely injured if flooded, or if in a zero-oxygen atmosphere for more than 1 day. In a flooded soil, the oxygen concentration approached zero in 24 hours.

The responses of very young corn plants to root temperatures have been studied extensively by Grobbelaar (1963). He subjected the roots to a range of temperatures from 5 to 40 C (41 to 104 F), while holding constant the air temperature at 20 C (68 F) and the light intensity. He found that the initiation of crown roots was retarded progressively as root temperature decreased from 20 to 5 C (68 to 41 F). Pronounced effects on shoot growth also were obtained. Optimum shoot growth occurred at a temperature range of 25 to 35 C (77 to 95 F). The shoot apex of the plants, however, also was exposed to the root temperatures concerned up to the age of 20 days. Root-temperature effects on plants older than 20 days, where the shoot apex was also subjected to the air temperature, were nevertheless similar to those encountered with younger plants. An accelerated rate of leaf initiation at 25, 30, and 35 C (77, 87, and 95 F) was shown to be one factor which contributed to the higher growth rate of the shoot. In addition, rate of leaf elongation also seemed to be the most rapid in the range of 25 to 35 C (77 to 95 F). Ultimate size of individual leaves, however, seemed to be favored by temperatures below the optimum range, the longest leaves being obtained at 15 and 20 (59 and 68 F). The total increase in leaf length per plant, however, proceeded most rapidly at 25, 30, and 35 C (77, 87, and 95 F).

Grobbelaar (1963) also found that root temperatures caused pronounced differences in the dry-matter percentage and the water soluble carbohydrate content, presumably sucrose. A progressive increase of water soluble carbohydrate in the shoot was evident at temperatures outside the optimum range. This seemed to be the result of a relatively greater decrease in growth than in photosynthesis. Because of the limited growth, photosynthates could not be fully utilized and, consequently, accumulated in the plant. Root temperatures influenced the proportion of shoots to roots. A relatively greater increase in shoot weight than in root weight occurred as the root temperature was increased from 5 to 40 C (41 to 104F). Since shoot and root growth were practically inhibited at 5 C (41 F), the shoot-root ratio, based on the fresh weight, remained constant. On a dry weight basis, however, a progressive increase in the ratio over time occurred because growth was practically inhibited at this temperature, whereas photosynthesis still continued, but at a slower rate. Root growth at 40 C (104 F) was inhibited while shoot growth proceeded at a retarded rate, which resulted in a progressive increase in shoot-root ratio. Root growth, however, was retarded more than shoot growth at temperatures beyond the range of 20 to 30 C (68 to 86 F) which resulted in an increase in shoot-root ratio.

Root temperature did affect the uptake of nitrate, phosphorus, potassium, calcium, and magnesium. In general, the root temperatures of 5, 10, 15, and 40 C (41, 50, 59, and 104 F) retarded uptake of N, P, and K. Indications of a luxury accumulation by the shoot seemed to exist at temperatures of 20, 25, and 35 C (68, 77, and 95 F). Shoot growth of young and older plants, however, could not be increased at the temperatures of 5, 10, and 15 C (41, 50 and 59 F) by doubling the concentration of macro-elements in the nutrient solution. In fact, doubling this nutrient concentration depressed the growth at the optimum temperature range of young plants.

Grobbelaar (1963) proposed that under his experimental conditions a hampered absorption of water by the roots, which decreased the transpiration rate at temperatures below 20 C (68 F) and at 40 C (104 F), increased the internal diffusion pressure deficit of the plants. This seemed to be responsible for the immediate decrease in growth of the shoot at these temperatures. In addition, the retarded growth at 20, 25, and 35 C (68, 77, and 95 F) may also have been the result of a relatively higher internal diffusion pressure deficit, although no differences in transpiration rate could be determined.

Growth during the early vegetative stage has been related to soil temperatures by several investigators. Willis et al. (1957) found that the most favorable soil temperature at the 10-cm depth for optimum growth rates and yields was around an average daily temperature of 24 C (75 F), slightly lower than Grobbelaar (1963) found at ages of 10 and 20 days. Allmaras and Nelson (1971) found that the optimum soil temperature for growth depended on moisture conditions. When soils were dry, a mulch between the rows aided root and shoot growth even with temperatures below 26 C (79 F), but when soils were moist, treatments that reduced temperatures below 26 C (79 F) consistently reduced growth and yield of dry matter. Van Wijk et al. (1959) found that shoot dry weights were decreased if soil temperatures were

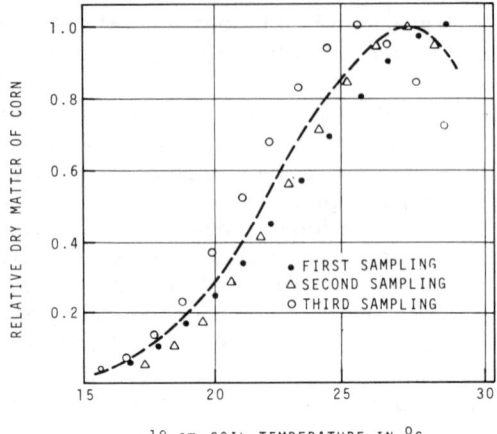

Fig. 7. Relative dry matter production of corn as related to the 10-cm soil temperature in the field. Growth measurements were taken from 13 to 38 days, 22 to 54 days, and 41 to 67 days, respectively, after planting, for the first, second, and third sampling. (Allmaras et al., 1964.)

decreased in Iowa, Minnesota, Ohio, and South Carolina. A mulch increased dry matter in South Carolina, where the early season soil temperature is sometimes above optimum, but decreased dry matter in the other states, where these temperatures are usually below optimum. Adams (1970) increased corn yields 1,000 kg/ha by use of a clear plastic mulch, the effect believed due to faster growth the first 4 to 6 weeks. The relative dry matter production of corn plants, as related to soil temperature, was shown by Allmaras et al. (1964) to be greatest at a daily average temperature of about 27 C (81 Ff (Fig. 7). The final effect on grain yield of early season soil temperature is difficult to evaluate because dry weight and shoot indexes of corn growth are not in phase over a range of temperatures (Allmaras et al., 1964; Arndt, 1945; Walker, 1969). The optimum temperature for shoot production is lower than that for shoot elongation. Because elongation may be a better estimator of leaf area, it may also be a better measure of the effect of soil temperature on grain yield.

Corn growth during the vegetative stage has been found to be related to both air temperature and rainfall. Bair (1942) found that the correlation between growth increments, expressed as dry weight gain of the total plant, and the physiological indexes of Lehenbauer (1914) was high in one year but low in another. Bair explained that the low correlation was due to a poor distribution of rainfall. Hanna (1925) found that the growth was more closely related to air temperature than to any other climatic variable. The best correlation between growth and air temperatures occurred when remainder indices above 10 C (50 F) were used. McCalla et al. (1939) also found that air temperature explained much of the growth rate variation. Loomis (1934) found that the growth rate decreased rapidly as the temperature decreased toward 10 C (50 F). Most rapid growth was made in the late afternoon and early evening and morning, or on cloudy days when the air temperature was high and no water deficits developed. In general, growth rates followed the temperature curve at night and the moisture supply curve during the day.

Over a 3-year period, Kiesselbach (1950) found almost the same amount of growth during the daylight hours as at night. Wallace and Bressman (1937), for short periods of growth, estimated daily growth rates ranging from near 8 cm at a temperature of 18 to 19 C (65 F) to over 17 cm at a temperature of 25 to 26 C (78 F).

Beauchamp and Lathwell (1966) stated that the number of leaf primordia was set between the fourth and sixth leaf on the Canadian cultivars they examined. Adams (1970) reported that the number of leaves was affected by early season soil temperatures. Berbecel and Eftimescu (1972) studied the time involved for leaves to develop and related this to average daily air temperature accumulations above 10 C (50 F). They found that the average accumulation per leaf developed was 32.5 C (89 F) for the period from emergence to tasseling. Ragland et al. (1965) found that the rate of increase in leaf area of corn planted very early was more highly correlated with air temperature than any other element they measured, while that of late planted corn was positively and equally correlated with temperature and relative humidity. Solar radiation, precipitation, black bulb evaporation, and wind were not significantly correlated with leaf area increase. Flooding reduces corn yields; the time and the length of the flooding period affect the yield reduction. In a greenhouse experiment, Mittra and Stickler (1961) found that flooding at the five-leaf stage reduced dry matter 7.5% if flooded 7 days, 34% if flooded 14 days, and 43% if flooded 21 days. Dry matter was harvested 21 days after flooding. Corn was more susceptible than soybeans or grain sorghum. Ritter and Beer (1969) found that flooding when corn was 15 cm in height for 72, 48, and 24 hours reduced corn yields by 32, 22, and 18% respectively, at a low N fertilizer level. At a high nitrogen level, these reductions ranged from 19 to 14% in 1 year to less than 5% the next year.

Correlations between early season weather and yield have generally shown little significance. Wallace (1920) found low correlations between May temperature and yield. He estimated that an average May temperature of 15.6 C (60 F) in central Iowa resulted in average yields, with higher yields occurring at higher temperatures. Rose (1936) also found that correlations between yield and May temperature were relatively low. His results show that May temperatures in the north and northeast sections of the Corn Belt [average May temperature below 15 C (59 F)] were positively correlated with yield, whereas in the southwest section (average May temperature above 16.1 C/(61 F)) they were negatively correlated. For Indiana, Visher (1940) found a positive correlation between yield and May temperature.

In the western part of the Corn Belt, yields have generally increased with increased May rainfall (Rose, 1936; Wallace, 1920). Wallace (1920) estimated, however, that May rainfall above 12.7 cm caused decreased yields. For optimum yields in Indiana, Visher (1940) found that May precipitation should be more than the normal of 10 cm and be accompanied by increased temperature. Too much above normal, however, will cause yield decreases.

Thompson (1963) computed optimum June temperatures for Iowa, Illinois, Indiana, Ohio, and Missouri, assuming normal June rainfall. The optimum temperatures of 21 to 23 C (70 to 74 F) (Fig. 8) were similar to those cited by Wallace (1920). However, correlations between yield and both June temperature and June rainfall were low (Thompson, 1963). A number of

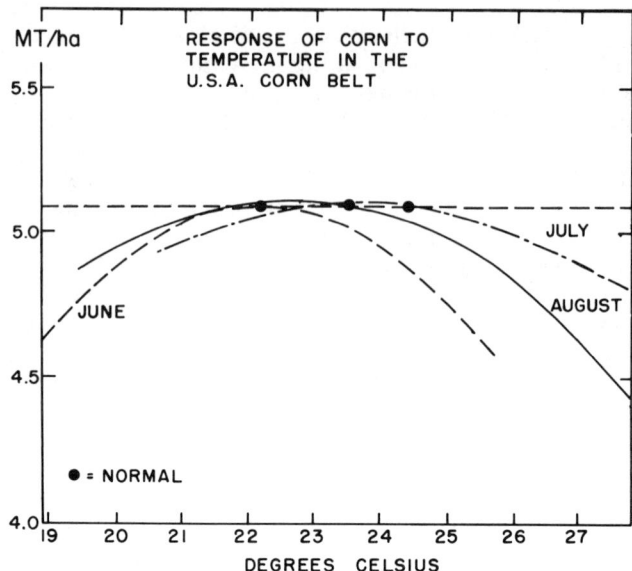

Fig. 8. Average response of corn to summer temperature in five Corn Belt states. (Thompson, 1963).

researchers have examined the relationships between June temperature and yield (Davis and Harrell, 1941; Rose, 1936; Visher, 1940; Wallace, 1920) and generally have found positive correlations where temperatures average below the optimum values and negative correlations where they average above the optimum. Accumulated temperatures above 32 C (90 F) have also shown a negative correlation with yield, except in Ohio and bordering areas where the correlations were positive.

In the southwest part of the Corn Belt, increased rainfall during June gave increased yield (Davis and Harrell, 1941; Rose, 1936; Wallace, 1920), but results have been conflicting in other parts of the Corn Belt. For individual years, the response to June rainfall will be related to the moisture reserve in the soil, June temperature, and subsequent weather. Thompson (1966) determined the response of corn to June rainfall for the five major corn producing states in the Corn Belt, assuming other conditions were normal (Fig. 9). The response curve for June rainfall was relatively flat but did show higher yields with below normal rainfall than with above normal rainfall. For a year with below normal soil moisture reserves, above normal rainfall could be beneficial.

In the previous discussion on early season weather effects on yields, only what would be called macro-effects have been discussed, which averages out the differences among soils. It should be remembered that micro-effects, such as the difference between a poorly drained and a well-drained soil, are important for individual soils, or to individual farmers, and cannot be disregarded. These differences are probably more significant early than later in the season.

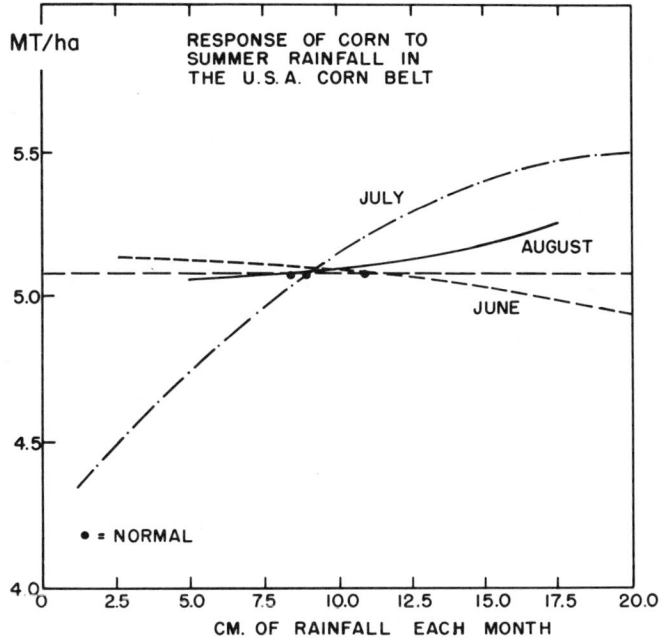

Fig. 9. Average response of corn to summer rainfall in five Corn Belt states. (Thompson, 1966.)

Sopher et al. (1973) stated that, on south Atlantic coastal plain soils, excess moisture and cool temperatures should also be included along with drought measurements early in the growing season. This is probably also true for many of the Corn Belt states.

Late Vegetative Growth From the Beginning of Rapid Stem Elongation to Tasseling

In the late vegetative stage, the relationships between weather and yield have been more marked and significant. In most of the Corn Belt, this growth stage occurs in July before silking in the latter part of the month. Effects of weather at silking will be covered in the next section. Wallace (1920) found that in most states July temperature was negatively correlated with yield. Rose (1936) found that in the northwest, southwest, and southern margins of the Corn Belt the relatively high correlation between July temperature and yield was negative; in the central part of the Corn Belt, he found low, nonsignificant correlations. Thompson (1962, 1963, and 1966) used regression techniques to study the weather-corn yield relationships for several states. For the five major corn states, he found that the optimum average July temperature, assuming July rainfall and the temperature and rainfall for other months were normal, was about 24 C (75 F) (Fig. 8). This is about 1 C below the average normal temperature for the month. Temperatures above normal reduce the yield sharply. The optimum temperature varies with the amount of rainfall (Fig. 10); the optimum average Iowa temperature is near 21 C (70 F) with only 2.5 cm of July rainfall, but is about 28 C (80 F) with 15 cm of

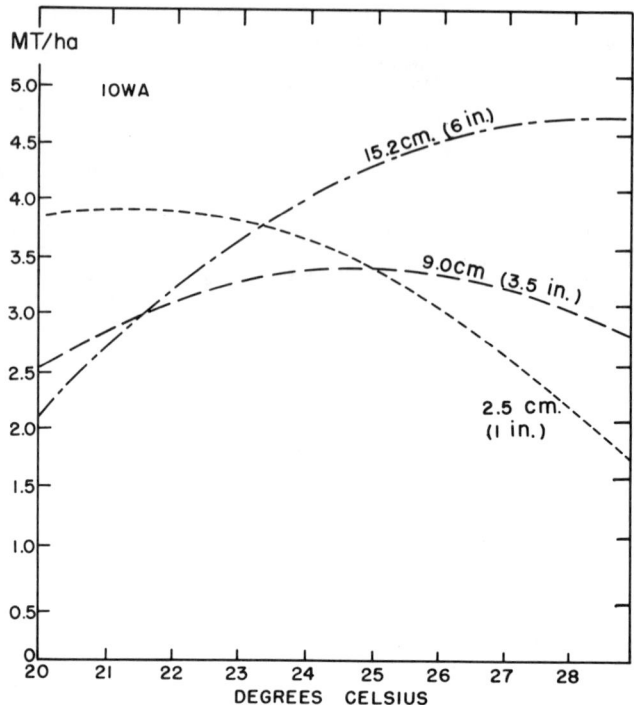

Fig. 10. Relation of corn yield to July temperature with different levels of rainfall in Iowa. (Thompson, 1963.)

rainfall. With more soil moisture available, a higher water demand caused by higher temperatures can be met without stress occurring.

In the late vegetative stage, corn plants grow very rapidly. Water use is greater and, in most areas, rainfall is lower than in June so that the water balance may become very important under low rainfall conditions. The optimum July rainfall (Fig. 9) is much above normal for the Corn Belt states (Hendricks and Scholl, 1943; Thompson, 1966). This is not surprising since atmospheric demand for water is high and the plant requires much more water to meet its needs.

Moisture stress during this period will cause yield reductions. A number of researchers (Claassen and Shaw, 1970a; Denmead and Shaw, 1960; Mallett, 1972; Robins and Domingo, 1953; Wilson, 1968) have studied this problem. In these experiments, corn was grown with a restricted root area and kept well watered, except when stress was imposed. At this stage, water was withheld to get the desired degree of stress. The treatments used subjected corn plants to a rather severe stress for 4 to 6 days. The results, in terms of grain yield reduction per stress day are given in Table 1. Data of Mallett show a linear response of stress up to 8 days duration for all periods.

In the experiment of Denmead and Shaw, the yield reduction was due in part to some fertility stress during the late vegetative period, which decreased the leaf area below the optimum. By preventing this fertility stress in a later experiment, the yield reduction during this period was reduced to 2% per day

Table 1—Percentage of possible grain-yield reduction per stress day
(data to nearest percent)

Growth stage	Researchers				
	Robins and Domingo	Denmead and Shaw	Wilson	Claassen and Shaw	Mallett
Late vegetative	—	3-4	2	2	3
Flowering	6-8	6-8	2-3	3-13	—
Ear filling	3	3	5-6	4-7	4

by Claassen and Shaw. Although not measured, leaf area was much larger than for the experiment of Denmead and Shaw. Downey (1971a) found that stress stunted the growth of the plants to some degree. Hoyt and Bradfield (1962), Eik and Hanway (1966), and Mallett (1972) all have found a linear relationship between corn grain yields and leaf area for leaf-area indexes (LAI) below 3.3. If moisture stress reduces LAI below 3.3, yield reductions will be greater than if a larger leaf area is present. Claassen and Shaw (1970a) found vegetative dry matter was reduced 15 to 17% by a 4-day stress treatment late in the vegetative period.

When corn at a height of 76 cm was flooded, Ritter and Beer (1969) found that 24 hours of flooding, at a low level of nitrogen, reduced yields 14%; this yield reduction increased to 30% with 96 hours of flooding. With a high level of nitrogen in the soil, very little yield reduction occurred even with 96 hours of flooding. When flooded near silking, no reduction in yield occurred at a high nitrogen level, but yield reductions up to 16% occurred with 96 hours of flooding at the low level of nitrogen.

Tasseling, Silking, and Pollination

This is a very critical stage in the corn plant. The number of ovules that will be fertilized is being determined. Both moisture and fertility stress that occur at this stage can have a serious effect on yield, with the exact stage at which it occurs also affecting the yield reduction. Barnes and Woolley (1969) found a 6 to 8% reduction in yield when stress was imposed for a period of a few days at tassel emergence. Their degree of stress cannot be directly compared with that in Table 1 because of a difference in the way stress was imposed.

Claassen and Shaw (1970b) found that stress imposed at 6% silking reduced yield only 3% per day, but at 75% silking the yield reduction was 7% per day with moisture stress. The data in Table 1 indicate that a 6 to 8% reduction per day was most common. A stress imposed at 75% silking and combined with a fertility stress gave a yield reduction of 13% per day with a large reduction in the number of developed kernels. Voladarski and Zinevich (1960) also reported that stress could reduce the number of grains per ear. Berbecel and Eftimescu (1973) found that maximum temperatures above 32 C (90 F) around tasseling and pollination speeded up the differentiation process of the reproductive parts and resulted in higher rates of kernel abortion. If too many kernels are aborted, the total sink size may limit yield, but under normal conditions the number of kernels is not as important as on rice (Yoshida, 1972). The maximum size of the kernel of rice is genetically determined so that a change in number will cause a change in total yield.

Although there is a maximum size of kernels of corn, the size × number factor limit is rarely reached. Prine (1971) also found that a poor light environment at very high plant populations could cause ear abortion.

Barnes and Woolley (1969) subjected a "stress-sensitive" single-eared cultivar and a "stress-resistant" two-eared cultivar to severe moisture stress. At the silking and pollination stage, the two-eared cultivar was more tolerant of stress, with a yield reduction of 14%, compared with a 73% reduction for a single-eared cultivar. The two-eared type also had much greater flexibility in partly avoiding a short period of stress and in taking better advantage of later good weather. Hallauer and Troyer (1972) summarized the data from a number of experiments comparing prolific vs. one-ear hybrids and found that, over a series of environments and planting rates, the prolific hybrids had greater flexibility for adjusting to environmental stress. The prolific hybrids had less genotype × environment interaction and greater stability of performance across environments.

Little information is available regarding the effects of soil temperature at this stage, Adams and Thompson (1973), however, found that decreasing soil temperature from 26 C (78.8 F) to 22 C (71.6 F) during pollination and grain formation had no effect on corn yield, but reducing soil temperature to 23 C (73.4 F) reduced grain sorghum yield about 10% in Texas.

The time at which tasseling and silking occur also are very weather dependent. Wallace and Bressman (1937) cited data showing that a so-called 115-day cultivar took 74 days from planting to tasseling with an average temperature of 20 C (68 F), but only 54 days with an average temperature near 23 C (73 F). Cool nights reduce the rapidity of growth before tasseling. As an average for the 60-day period after planting, they found that for each degree the temperature averaged above 21.1 C (70 F), tasseling was hastened by 2 to 3 days. Data of Shaw and Thom (1951a) and Rench (1973) show similar variations in the length of the period. Rench showed that the length of the period was highly correlated with temperature (growing-degree units), but that soil moisture also needed to be considered. If a cultivar is daylength sensitive, this factor also influences time of tasseling. Berbecel and Eftimescu (1972) estimated the period up to tasseling by the equation,

$$n = \frac{32.5(N-3)}{(t-10)}$$

where:

n is the length in days of the leafing period (from leaf 3 to tasseling),
N is number of leaves specific to hybrid, and
t is mean soil temperature in C for the period at the 10-cm depth.

Allen et al. (1973) found that a hybrid took approximately the same number of langleys (g cal/cm^2) from planting to tasseling, although the number of days varied significantly. Mallett (1972) found that severe stress had little effect on the date of tasseling, although silking was delayed 6 to 8 days. Similar results were found by Shaw and Thom (1951b). Rhoades and Stanley (1973) found that both tasseling and silking occurred earlier as soil-moisture tension was decreased.

Shaw (1949), Du Plessis and Dijkhuis (1967), and Berbecel and Eftimescu (1973) found that stress before and during flowering caused the time between pollen shed and silking to be delayed. With severe stress, silking may be delayed until after all or much of the pollen has been shed, increasing the number of barren stalks and poorly filled ears.

Rench (1973) found that a cultivar maturity rating, as well as temperature and soil moisture, were necessary to predict silking dates. Growing-degree units were a better predictor of silking date than the cultivar maturity rating in number of days, but both still showed considerable variation.

Grain Production From Fertilization to Physiological Maturity of the Grain

During the ear-filling stage, significant reduction in yield can occur from moisture stress. Mallett (1972) subjected corn to stress starting at 10, 20, 30, and 40 days after silking and maintained stress for up to 8 days. Four days of stress caused an average yield reduction of 4.3% per day of stress at each of the times that stress was imposed. In another experiment, Mallett found that the reduction was 4.1% per day of stress. Some yield reductions (Table 1) have been a little higher than this (Claassen and Shaw, 1970b; Wilson, 1968), and others have been a little lower than this (Denmead and Shaw, 1960; Robins and Domingo, 1953). Higher reductions occurred where some degree of fertility stress was confounded with the moisture stress. Data of Claassen and Shaw (1970b) also indicate that less yield reduction occurred as a result of stress late in the season; this reduced damage may have been because of the difficulty of imposing as severe a degree of stress as earlier in the season because of the lower moisture demand situation that occurred.

Barnes and Woolley (1969) obtained a yield reduction of 22% for their two-eared cultivar and 48% for their single-eared cultivar when stress was imposed at the blister-kernel stage. The 22% reduction is comparable to that found by Mallett, but the 48% reduction is much higher. The sensitivity of this particular single-eared cultivar to stress was no doubt a significant factor in its large response to stress.

August weather, which covers the first part of this period, obviously has an effect on yield. Wallace (1920), Rose (1936), Davis and Harrell (1941), Kiesselbach (1950), and Thompson (1963) all have shown that average August temperatures are higher than those associated with optimum corn yields in the Corn Belt. Thompson (1963) (Fig. 8) estimated that optimum temperatures for the five major producing corn states were a little more than 1 C below normal, very close to the value given by Wallace (1920). In Missouri, Bondavalli et al. (1970) found that temperature in the last half of August had a more significant effect on corn yield than the rainfall in the same period, but in the first half of August, rainfall had a more significant effect on yield than temperature. Peters et al. (1971) in an unreplicated experiment, found that a night air temperature of 29.4 C (85 F) for the period from flowering to maturity reduced corn yield almost 40% compared with a cool temperature of 16.6 C (61.8 F). The high temperature produced earlier senescence and maturity and may have induced a water stress on the plants.

In a dry year, with low-soil moisture reserves, increased August rainfall will increase corn yields, but in a wet year, too much August rain may create some problems for harvesting. This may be particularly important on the more poorly drained soils in the wetter areas. Rose (1936) found that August rainfall was positively correlated with yield in most areas in the Corn Belt, but the correlations were low in the central part. Thompson (1962, 1963,

1966) found that increased August rainfall was associated with higher yields (Fig. 9), but correlations with yield were low for the major corn-producing states. August temperature was much more important than August rainfall in his regression techniques to predict corn yields. In a later paper, Thompson (1969) dropped August rainfall from the regression equation and found that August temperature adequately defined the weather for August.

In September the corn crop is progressing toward maturity in the Corn Belt; by the end of the month much of the corn is physiologically mature. In a dry year, stress from either lack of water or high temperatures can reduce yield of long season hybrids; severe stress may cause premature dying, an additional yield loss. In a wet year, soil-moisture reserves for the next season are increased, and the wet weather has little direct effect on yield, but may delay harvesting. An early freeze before physiological maturity may cause serious yield losses, especially if the crop is later maturing. Because of these varied effects, correlations between weather factors in September and yield have been low.

The time of physiological maturity (maximum dry weight) seems well defined in terms of the black-layer development (Daynard and Duncan, 1969; Rench and Shaw, 1971). Kernel moisture at black-layer development or physiological maturity varies with variety. Grain-moisture loss from early ear formation to physiological maturity is poorly correlated with weather factors (Hallauer and Russell, 1961). Degree days showed the best relationship with moisture loss, but results were not consistent. Rench (1973), using six hybrids with 85 to 135 day maturity classifications, found that longer season cultivars took more time to reach black-layer maturity after silking than did the shorter season cultivars. Cooler temperatures increased the length of the period to some extent. Rench found that the best predictor for the length of this period was a cultivar maturity rating; adding a temperature variable increased the correlation slightly. The variation among the six hybrids of the length of the silking to physiological maturity period in days was much less than for the planting to silking interval. Berbecel and Eftimescu (1972) found a similar relation for the two intervals. Since the grain size in corn is loosely restricted, extending the duration of the grain-filling period, or maintaining higher photosynthetic activity during this time, might increase yield (Yoshida, 1972).

Maturation or Drying of the Grain

After physiological maturity, the grain must dry down to a harvestable moisture level. The rate of drying is affected by the weather and the cultivar characteristics. Dodds and Pelton (1967) found that the rate of drying of wheat in the field was influenced by vapor-pressure deficit, hours of sunshine, rate of evaporation, and wind. They found that vapor-pressure deficit was an excellent measure to describe the causes of moisture fluctuation in wheat Hours of sunshine have an effect on temperature and, like wind, can contribute to drying. Schmidt and Hallauer (1966) found that, before physiological maturity, kernel moisture was primarily a physiological process, with the reduction showing some relationship with air temperature. Below 30% moisture, they found that the reduction in kernel moisture was related most closely to the vapor-pressure deficit, with the wet-bulb depression showing almost as good a relationship. Relative humidity gave a poorer relationship, as should be expected when using it as an expression for drying power of the air.

Rain is a major contributor to increases in moisture during the early part of the drying stage, and condensation during periods of high humidity is important during later stages of maturity.

Seasonal Effects

In Fig. 11 the yield decrease due to moisture stress is related to the age of the crop. The shaded area covers the range of most of the results obtained by various experimenters, with extreme values omitted. The line through the shaded area represents average yield reductions obtained. Data before 50 days after planting are not available. This figure is applicable to periods of severe stress of a few days duration. Mallett (1972) found a linear relation between days of stress and total yield reduction for treatments up to 8 days duration. If severe stress occurs for many days, the plants will be killed. In the field, the degree of stress may well vary from day to day depending upon the balance between the soil moisture and the atmospheric demand for water.

The major weather effects at different growth stages on yield have been covered in earlier sections. There are some data, however, which relate yield to total seasonal effects. Total seasonal rainfall has generally not given high correlations with yield. This is partly because of negative correlations for part of the season and positive correlations for other parts of the season. Pengra (1946) found that seasonal rainfall in South Dakota was more highly correlated with corn yields (r = 0.58) than was preseasonal precipitation.

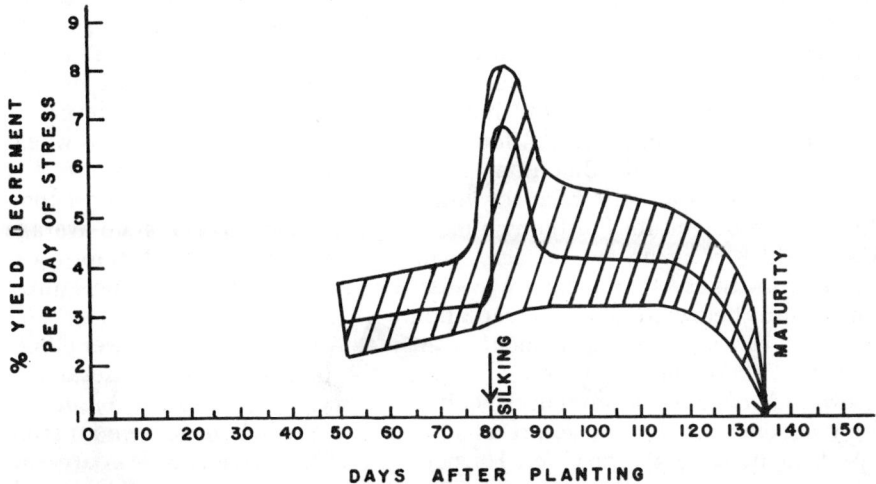

Fig. 11. Schematic diagram of relationship between age of crop and percentage yield decrement due to one day of moisture stress.

In Nebraska, Kiesselbach (1915) found that, for each rise of 0.56 C in mean seasonal temperature during June, July, and August, a 423 kg/ha (6.75 bu/acre) decrease in yield occurred. In more humid areas, this relationship might not be true, at least to such a degree.

Mederski and Jones (1963) found that heated soils gave faster, earlier growth, but reduced mature plant height. Heating increased yields 15% when the soil temperature around the heating cable was maintained near 30 C (86 F). Willis (1956) found that times of emergence, silking, and maturity were hastened by higher soil temperatures. He found that yields increased with increasing soil temperature (10-cm depth) up to about 23 C (73 F), then decreased with a further increase in soil temperature.

L.M. Thompson (personal communication) has found that accumulated degrees above 32.2 C (90 F) are related to yield. By using daily maximum temperatures and accumulating the degrees above 32.2 C (90 F) (i.e., a temperature of 95 would be 5), he found that, for each 5.6 C (10 F) accumulated, yields were reduced 62.7 kg/ha (1 bu/acre) for corn and one-third that amount for soybeans.

Duncan et al. (1973) grew corn with irrigation at Davis and Greenfield, Calif., and Lexington, Ky., all at the same altitude. Yields were highest at Davis, which had the lowest insolation, the highest night temperature, and the highest number of growing-degree days. Other environmental factors also were confounded in these comparisons. Maximum temperature of 37.8 C (100 F) or higher occurred on 25 days of the growing season at Davis. The night temperatures seemed particularly significant. Schwab et al. (1958), however, found that the number of days with maximum temperature above 32.2 C (90 F) was negatively related to irrigated corn yields in Iowa.

Using radiation as a treatment in controlled experiments has nearly always shown positive effects; Pendleton et al. (1967) found that increased light increased corn yields, and Duncan and Hesketh (1968) found that shading reduced yields. Regression analyses of yield and climatic data over several years have frequently shown negative effects of increased radiation. Kiesselbach (1950) found that a seasonal 1% increase in sunshine reduced grain yield by 96 kg/ha (1.5 bu/acre), while an increase of 1 g cal in the seasonal mean daily total radiation reduced the yield by 24 kg/ha (0.4 bu/acre). In areas less dry and hot, one might expect this to be reversed, although McCalla et al. (1939) found a negative correlation between wheat growth and sunlight at Edmonton, Alberta, Canada. In the field, the effect of high radiation, often with relatively high temperatures, can well depend upon the relative magnitudes of the two factors. In Iowa in 1972, a state average corn yield 815 kg/ha (13 bu/acre) higher than ever produced before was produced with excellent moisture, cooler than normal summer temperatures, and slightly below average solar radiation.

In a review on the enrichment of the plant environment, Wittwer (1966) stated, "Carbon dioxide has given the most spectacular yield increases of any growth factor yet discovered in the culture of greenhouse crops." As reported by Waggoner (1969), however, results for corn in the field have been far from spectacular. Recently, however, Harper et al. (1973) have shown a large increase in photosynthetic production in cotton (*Gossypium hirsutum* L.) with high rates of field carbon dioxide enrichment. By using a photosynthesis simulator, Waggoner has estimated that, to increase photosynthesis substantially in full sun, it would be necessary to increase the carbon dioxide concentration above the canopy, as well as at the ground. This may be why addition at ground level only has rarely been successful in the field.

In referring to carbon dioxide concentration in the air, an average value of 300 ppm has often been used. At 41° N lat. during December 1961, the concentration was 315 ppm (Bolin and Keeling, 1963). According to Waggoner (1969), if 315 ppm were present during the growing season, this should increase photosynthesis by 4% in the field. He states that this is the one benefit he has found from air pollution.

Waggoner (1969), with the use of simulators, has examined the effect of wind modification on transpiration and photosynthesis. He found that decreased ventilation would cause a slight decrease in carbon dioxide availability, which should be counteracted by the improved hydration of the plant. Slowing the wind velocity from 1,225 to 22 cm sec^{-1}, decreased the transpiration ratio from 108 to 96. This change in the transpiration ratio summarizes both the advantage of wind management and the modest results that can be anticipated.

ACKNOWLEDGMENT

This is paper No. J-7707 of the Iowa Agriculture and Home Economics Experiment Station, Ames, Iowa, Project 1852.

LITERATURE CITED

Adams, J.E. 1970. Effects of mulches and bed configuration. II. Soil temperature and yield responses of grain sorghum and corn. Agron. J. 62:785-790.

----, and D.O. Thompson. 1973. Soil temperature reduction during pollination and grain formation of corn and grain sorghum. Agron. J. 65:60-63.

Alessi, J., and J.F. Power. 1971. Corn emergence in relation to soil temperature and seeding depth. Agron. J. 63:717-719.

Allen, J.R., G.W. McKee, and J.H. McGahen. 1973. Leaf number and maturity in hybrid corn. Agron. J. 65:233-235.

Allmaras, R.R., W.C. Burrows, and W.E. Larson. 1964. Early growth of corn as affected by soil temperature. Soil Sci. Soc. Am. Proc. 28:271-275.

----, and W.W. Nelson. 1971. Corn (Zea mays L.) root configuration as influenced by some row-interrow variants of tillage and straw mulch management. Soil Sci. Soc. Am. Proc. 35:974-980.

Arndt, C.H. 1945. Temperature-growth relations of roots and hypocotyls of cotton seedlings. Plant Physiol. 20:200-220.

Bair, R.A. 1942. Growth rates of maize. Plant Physiol. 17:619-631.

Barnes, D.L., and D.G. Woolley. 1969. Effect of moisture stress at different stages of growth. I. Comparison of a single-eared and a two-eared corn hybrid. Agron. J. 61:788-790.

Beauchamp, E.G., and D.J. Lathwell. 1966. Effects of root zone temperature on corn leaf morphology. Can. J. Plant Sci. 46:593-601.

Beer, C.E., W.D. Shrader, and R.K. Schwanke. 1967. Interrelationships of plant population, soil moisture and soil fertility in determining corn yields on Colo clay loam at Ames, Iowa. Iowa Agric. Home Econ. Exp. Stn. Res. Bull. 556.

Berbecel, O., and M. Eftimescu. 1972. Effect of agrometeorological conditions on maize growth and development. p. 45-50. Inst. Meteor. Hydrol. Bucharest. (English translation.)

----, and ----. 1973. Effect of agrometeorological conditions on maize growth and development. p. 10-31. Inst. Meteor. Hydrol. Bucharest. (English translation.)

----, ----, E. Gogorici, and I. Rogojan. 1964. The forecast of the vegetative phases of the self sown and cultivated flora. p. 347-358. Culegere de lucrari. Romanian Inst. Meteor. Bucharest. (English summary.)

----, and I. Rogojan. 1962. Evaluarea zonola a conditiilor meteorologice la cultura porumbului. p. 319-330. Culegere de luirari. Romanian Inst. Meteor. Bucharest. (English summary.)

Blacklow, W.M. 1972. Influence of temperature on germination and elongation of the radicle and shoot of corn (Zea mays L.). Crop Sci. 12:647-650.

Bolin, B., and C.D. Keeling. 1963. Large-scale atmospheric mixing as deduced from the seasonal and meridional variations of carbon dioxide. J. Geophys. Res. 68:3899-3920.

Bondavalli, B., D. Colyer, and E.M. Kroth. 1970. Effects of weather, nitrogen and population on corn yield response. Agron. J. 62:669-672.

Boyer, J.S. 1970. Differing sensitivity of photosynthesis to low leaf water potentials in corn and soybeans. Plant Physiol. 46:236-239.

Brown, D.M. 1960. Soybean ecology. I. Development temperature relationships from controlled environment studies. Agron. J. 52:493-496.

----. 1969. Heat units for corn in southern Ontario. Information Leaflet. Ontario Dep. Agric. Food, Canada.

Cackett, K.E., and H.R.R. Metelerkamp. 1964. Evapotranspiration of maize in relation to open-pan evaporation and crop development. Rhod. J. Agric. Res. 2:35-44.

Cal, J.P., and R.L. Obendorf. 1972. Differential growth of corn (Zea mays L.) hybrids seeded at cold root zone temperatures. Crop Sci. 12:572-575.

Claassen, M.M., and R.H. Shaw. 1970a. Water deficit effects on corn. I. Vegetative components. Agron. J. 62:649-652.

----, and ----. 1970b. Water deficit effects on corn. II. Grain components. Agron. J. 62:652-655.

Coffman, F.A. 1923. The minimum temperature for germination of seed. J. Am. Soc. Agron. 15:257-270.

Corsi, W.C., and R.H. Shaw. 1971. Evaluation of stress indices of corn in Iowa. Iowa State J. Sci. 46:79-85.

Cross, H.Z., and M.S. Zuber. 1972. Prediction of flowering dates in maize based on different methods of estimating thermal units. Agron. J. 64:351-355.

Cummins, D.G., and W.L. Parks. 1961. The germination of corn and wheat as affected by various fertilizer salts at different soil temperatures. Soil Sci. Soc. Am. Proc. 25:47-49.

Dale, R.F., and R.H. Shaw. 1965. Effect on corn yields of moisture stress and stand at two fertility levels. Agron. J. 57:475-479.

Davis, F.E., and G.D. Harrell. 1941. Relation of weather and its distribution to corn yields. USDA Tech. Bull. 806.

Daynard, T.B. 1972. Relationships among black-layer formation, grain moisture percentage and heat unit accumulation in corn. Agron. J. 64:716-719.

----, and W.G. Duncan. 1969. The black layer and grain maturity in corn. Crop Sci. 9:473-476.

de Jager, J.M. 1968. Carbon dioxide exchange and photosynthetic activity in forage grasses. Unpublished Ph.D. Thesis. Univ. of Wales, Aberystwyth.

Denmead, O.T., and R.H. Shaw. 1959. Evapotranspiration in relation to the development of the corn crop. Agron. J. 51:725-726.

----, and ----. 1960. The effects of soil moisture stress at different stages of growth on the development and yield of corn. Agron. J. 52:272-274.

----, and ----. 1962. Availability of soil water to plants as affected by soil moisture content and meteorological conditions. Agron. J. 45:385-390.

Dodds, M.E., and W.L. Pelton. 1967. Effect of weather factors on the kernel moisture of a standing crop of wheat. Agron. J. 59:181-184.

Doss, B.D., O.L. Bennett, and D.A. Ashley. 1962. Evapotranspiration by irrigated corn. Agron. J. 54:497-498.

Downey, L.A. 1971a. Effect of gypsum and drought stress on maize (Zea mays L.). I. Growth, light absorption, and yield. Agron. J. 63:569-572.

----. 1971b. Effect of gypsum and drought stress on maize (Zea mays L.). II. Consumptive use of water. Agron. J. 63:597-600.

----. 1971c. Water requirements of maize. J. Aust. Inst. Agric. Sci. 11:32-41.
Duncan, W.G., and J.D. Hesketh. 1968. Net photosynthetic rates, relative leaf growth rates and leaf numbers of 22 races of maize grown at eight temperatures. Crop. Sci. 8:370-374.
----, D.L. Shaver, and W.A. Williams. 1973. Insolation and temperature effects on maize growth and yield. Crop Sci. 13:187-191.
Dungan, G.H. 1944. Yield and bushel weight of corn grain as influenced by time of planting. J. Am. Soc. Agron. 36:166-170.
Du Plessis, D.P., and F.J. Dijkhuis. 1967. The influence of the time lag between pollen shedding and silking on the yield of maize. S. Afr. J. Agric. Sci. 10:667-674.
Eik, K., and J.J. Hanway. 1966. Leaf area in relation to yield of corn grain. Agron. J. 58:16-18.
Finch, V.C., and O.E. Baker. 1917. Geography of the world's agriculture. US Government Printing Office, Washington, D.C.
Gardner, W.R. 1960. Dynamic aspects of water availability to plants. Soil Sci. 89:63-73.
Gilmore, E., and J.S. Rogers. 1958. Heat units as a method of measuring maturity in corn. Agron. J. 50:611-615.
Grobbelaar, W.P. 1963. Responses of young maize plants to root temperatures. Meded. Landbouwhogesch. Wageningen. 63(5) 1-71.
Guidray, Nelson P. n.d. A graphic summary of world agriculture. USDA Misc. Publ. 705. Washington, D.C.
Haise, H.R. 1958. Irrigation. Agronomic trends and problems in the Great Plains. *In* A.G. Norman (ed.) Adv. Agron. 6:47-56. Academic Press, Inc., New York.
Hallauer, A.R., and W.A. Russell. 1961. Effects of selected weather factors on grain moisture reduction from silking to physiologic maturity in corn. Agron. J. 53:225-229.
----, and A.F. Troyer. 1972. Prolific corn hybrids and minimizing risk of stress. 27th Annu. Corn Sorghum Res. Conf. Proc., Am. Seed Trade Assoc., Washington, D.C.
Halstead, M.H. 1954. The fluxes of momentum, heat and water vapor in micrometeorology. The Johns Hopkins Univ., Lab. of Climatology, Publ. in Climatology 7:326-58.
Hanna, W.F. 1924. Growth of corn and sunflowers in relation to climatic conditions. Bot. Gaz. 78:200-214.
----. 1925. The nature of the growth rate in plants. Sci. Agric. 5:133-138.
Hanway, D.G. 1966. Irrigation. p. 155-176. *In* W.H. Pierre, S.A. Aldrich, and W.P. Martin (eds.) Advances in corn production: principles and practices. Iowa State Univ. Press, Ames.
Hanway, J.J. 1971. How a corn plant develops. Iowa Coop Ext. Serv. Spec. Rep. 48 (rev.)
Harper, L.A., D.N. Baker, J.E. Box, Jr., and J.D. Hesketh. 1973. Carbon dioxide and the photosynthesis of field crops: A metered carbon dioxide release in cotton under field conditions. Agron. J. 65:7-11.
Harrold, L.L., and R.F. Dreibelbis. 1951. Agricultural hydrology as evaluated by monolith lysimeters. USDA Tech. Bull. 1050.
Hendricks, W.A., and J.C. Scholl. 1943. The joint effects of precipitation and temperature on corn yields. N.C. Agric. Exp. Stn. Tech. Bull. 74.
Hershey, A.L. 1934. A morphological study of the structure and development of the stem and ears of *Zea mays* L. Unpublished Ph.D. Thesis. Iowa State University, Ames.
Holmes, R.M., and G.W. Robertson. 1959. Heat units and crop growth. Can. Dep. Agric., Ottawa. Publ. 1042.
Hoyt, P., and R. Bradfield. 1962. Effect of varying leaf area on dry matter production in corn. Agron. J. 54:523-525.
Jenkins, M.T. 1941. Influence of climate and weather on growth of corn. USDA Yearb. Agric. p. 308-320.
Johns, D.M., and H.B. Brown. 1941. Effect of date of planting on corn yields, insect infestation and fungous diseases. La. State Bull. 327.
Kiesselbach, T.A. 1915. Transpiration as a factor in crop production. Nebr. Agric. Exp. Stn. Res. Bull. 8.
----. 1949. The structure and reproduction of corn. Nebr. Agric. Exp. Stn. Res. Bull. 161.

----. 1950. Progressive development and seasonal variation of the corn crop. Nebr. Agric. Exp. Stn. Res. Bull. 166.

Kincer, J.B. 1919. Temperature influence on planting and harvest dates. Month. Weather Rev. 47:312-323.

Klages, H.K.W. 1942. Ecological crop geography. The MacMillan Co., New York.

Koppen, W. 1931. Grundries der Klimakunde. Walter de Gruyter, Berlin.

Larson, W.E., and G.R. Blake. 1966. Seedbed and tillage requirements. p. 27-52. *In* W.H. Pierre, S.A. Aldrich, and W.P. Martin (eds.) Advances in corn production: principles and practices. Iowa State Univ. Press, Ames.

Lehenbauer, P.A. 1914. Growth of maize seedlings in relation to temperature. Physiol. Res. 1:247-288.

Livingston, B.E. 1916. Physiological temperature indices for the study of plant growth in relation to climatic conditions. Physiol. Res. 1:399-420.

----, and G.I. Livingston. 1913. Temperature coefficients in plant geography and climatology. Bot. Gaz. 56:349-375.

Loomis, W.E. 1934. Daily growth of maize. Am. J. Bot. 21:1-6.

Mallett, J.B. 1972. The use of climatic data for maize yield predictions. Ph.D. Thesis. Dep. of Crop Sci., Univ. of Natal, Pietermaritzburg, S.A.

Maranville, J.W., and G.M. Paulsen. 1970. Alteration of carbohydrate composition of corn (*Zea mays* L.) seedlings during moisture stress. Agron. J. 62:605-608.

McCalla, A.G., J.R. Weir, and K.W. Neatby. 1939. Effects of temperature and sunlight on the rate of elongation of stems in maize and gladiolus. Can. J. Res. Sect. C. 17:388-409.

Mederski, H.J., and J.B. Jones, Jr. 1963. Effect of soil temperature on corn plant development and yield. I. Studies with a corn hybrid. Soil Sci. Soc. Am. Proc. 27:186-189.

Mittra, M.K., and F.C. Stickler. 1961. Excess water effects on different crops. Trans. Kans. Acad. Sci. 64:275-286.

Newhall, F. 1947. The influence of temperature on the time interval between planting and emergence of corn in Iowa. Unpublished M.S. Thesis. Iowa State Univ., Ames.

Newlin, J.J. 1948. The role of weather in the growth of corn and seed corn in Polk Co. Iowa. Bull. Am. Meteorol. Soc. 29:5-8.

Newman, J.E., and B.O. Blair. 1969. Growing degree days and dent corn maturity. Part II. Mimeo. Agron. Dep. Purdue Univ., Lafayette, Ind.

Nicholis, P.B., and L.H. May. 1963. Studies on the growth of the barley apex. I. Interrelationships between primordium formation apex length, and spikelet development. Aust. J. Biol. Sci. 16:561-571.

Nuttonson, M.Y. 1953. Phenology and thermal environment as a means for a physiological classification of wheat varieties and for predicting maturity dates of wheat. Am. Inst. Crop Ecol., Washington, D.C.

----. 1955. Wheat-climate relationships and the use of phenology in ascertaining the thermal and photo-thermal requirements of wheat. Am. Inst. Crop Ecol., Washington, D.C.

----. 1957. Barley-climate relationships and the use of phenology in ascertaining the thermal and photo-thermal requirements of barley. Am. Inst. Crop Ecol., Washington, D.C.

Olson, T.C. 1971. Yield and water use by different populations of dryland corn, grain sorghum, and forage sorghum in the western Corn Belt. Agron. J. 63:104-106.

Paddick, M.E. 1944. Vegetative development of inbred and hybrid maize. Iowa Agric. Exp. Stn. Res. Bull. 331.

Papadakis, J. 1966. Climates of the world and their agricultural potentialities. Av. Cordoba 4564, Buenos Aires, Argentina.

----. 1970. Agricultural potentialities of world climates. Av. Cordoba 4564, Buenos Aires, Argentina.

Peaslee, D.E., J.L. Ragland, and W.G. Duncan. 1971. Grain filling period of corn as influenced by phosphorus, potassium and the time of planting. Agron. J. 63:561-563.

Pendleton, J.W. 1966. Increasing water use efficiency by crop management. p. 236-258. *In* W.H. Pierre, D. Kirkham, J. Pesek, and R. Shaw (eds.) Plant environment and efficient water use. Am. Soc. Agron. and Soil Sci. Soc. Am., Madison, Wis.

----, and D.B. Egli. 1969. Potential yield fo corn as affected by planting date. Agron. J. 61:70-71.
----, ----, and D.B. Peters. 1967. Response of *Zea mays* L. to a "light rich" field environment. Agron. J. 59:395-397.
Pengra, R.F. 1946. Correlation analysis of precipitation and crop yield data for the sub-humid areas of the northern Great Plains. J. Am. Soc. Agron. 38:848-849.
Peters, D.B., J.W. Pendleton, R.H. Hageman, and C.M. Brown. 1971. Effect of night air temperature on grain yield of corn, wheat and soybeans. Agron. J. 63:809.
Philip, J.R. 1957. The physical principles of soil water movement during the irrigation cycle. 32nd Cong. Int. Comm. Irrig. and Drain. Question 8:8.125-8.154.
Pierce, L.T. 1958. Estimating seasonal and short term fluctuations in evapotranspiration from meadow crops. Bull. Am. Meteorol. Soc. 39:73-78.
Power, J.F., J.J. Bond, W.A. Sellner, and H.M. Olson. 1973. Effect of supplemental water on barley and corn production in a subhumid region. Agron. J. 65:464-467.
Price, H.L. 1911. The application of meteorological data in the study of physiological constants. Va. Agric. Exp. Stn. Ann. Rep. 1909-1910:206-212.
Prine, G.M. 1971. A critical period for ear development in maize. Crop Sci. 11:782-786.
Purvis, A.C., and R.E. Williamson. 1972. Effects of flooding and gaseous composition of the root environment on growth of corn. Agron. J. 64:674-678.
Ragland, J.L., A.L. Hatfield, and G.R. Benoit. 1965. The growth and yield of corn. I. Microclimate effects on the growth rates. Agron. J. 57:217-220.
Rench, W.E. 1973. Climatic influences on and indices of *Zea mays* L. growth and development. Ph.D. Thesis. Iowa State Univ., Ames.
----, and R.H. Shaw. 1971. Black layer development in corn. Agron. J. 63:303-305.
Rhoades, F.M., and R.L. Stanley, Jr. 1973. Response of three corn hybrids to low levels of soil moisture tension in the plow layer. Agron. 65:315-318.
Rhoades, H.F., and L.B. Nelson. 1955. Growing 100-bushel corn with irrigation. USDA Yearb. Agric. p. 394-400.
Ritchie, J.T., and E. Burnett. 1971. Dryland evaporative flux in a subhumid climate. II. Plant influences. Agron. J. 63:56-62.
Ritter, W.F., and C.E. Beer. 1969. Yield reduction by controlled flooding of corn. Trans. ASAE 12:46-50.
Robins, J.S, and C.E. Domingo. 1953. Some effects of severe soil moisture deficits at specific growth stages in corn. Agron. J. 45:618-621.
----, and H.F. Rhoades. 1958. Irrigation of field corn in the west. USDA Leafl. 440.
Rose, J.K. 1936. Corn yield and climate in the Corn Belt. Geogr. Rev. 26:88-102.
Salter, P.J., and J.E. Goode. 1967. Crop responses to water at different stages of growth. Commonw. Agric. Bur. Res. Rev. 2.
Schmidt, J.L., and A.R. Hallauer. 1966. Estimating harvest date of corn in the field. Crop Sci. 6:227-231.
Schwab, G.D., W.D. Shrader, P.R. Nixon, and R.H. Shaw. 1958. Research on irrigation of corn and soybeans at Conesville and Ankeny, Iowa, 1951-1955. Iowa Agric. Home Econ. Exp. Stn. Res. Bull. 458.
Sellschop, J.P.F., and S.C. Salmon. 1928. The influence of chilling above the freezing point on certain crop plants. J. Agric. Res. 37:315-338.
Shaw, R.H. 1949. Studies on corn phenology and maturity in Iowa. Ph.D. Thesis. Iowa State Univ., Ames.
----. 1965. The prediction of soil moisture for the winter period in Iowa. Iowa State J. Sci. 39:337-344.
----. 1971. A comparison of soil and air temperatures in the spring at Ames, Iowa. Iowa State J. Sci. 45:613-620.
----, and W.C. Burrows. 1966. Water supply, use and requirement. p. 121-142. *In* W.H. Pierre, S.A. Aldrich, and W.P. Martin (eds.) Advances in corn production: principles and practices. Iowa State Univ. Press, Ames.
----, and R.E. Felch. 1972. Climatology of a moisture-stress index for Iowa and its relation to corn yields. Iowa State J. Sci. 46:357-368.

----, ----, and E.R. Duncan. 1972. Soil moisture available for plant growth in Iowa. Iowa Agric. Home Econ. Exp. Stn. Spec. Rep. 70.

----, and D.R. Laing. 1966. Moisture stress and plant responses. p. 73-94. *In* W.H. Pierre, Don Kirkham, J. Pesek, and R. Shaw (eds.) Plant environment and efficient water use. Am. Soc. Agron. and Soil Sci. Soc. Am., Madison, Wis.

----, and W.E. Loomis. 1950. Bases for the prediction of corn yields. Plant Physiol. 25:225-244.

----, J.R. Runkles, and G.L. Barger. 1958. Seasonal changes in soil moisture as related to rainfall, soil type and crop growth. Iowa Agric. Home Econ. Exp. Stn. Res. Bull. 457.

----, and H.C.S. Thom. 1951a. On the phenology of field corn, the vegetative period. Agron. J. 43:9-15.

----, and ----. 1951b. On the phenology of field corn, silking to maturity. Agron. J. 43:541-546.

----, ----, and G.L. Barger. 1954. The climate of Iowa. 1. The occurrence of freezing temperatures in the spring and fall. Iowa Agric. Exp. Stn. Spec. Rep. 8.

Slatyer, R.O. 1969. Physiological significance of internal water relations to crop yield. p. 53-83. *In* J.D. Eastin, F.A. Haskins, C.Y. Sullivan, and C.H.M. van Bavel (eds.) Physiological aspects of crop yield. Am. Soc. Agron. and Crop Sci. Soc. Am., Madison, Wis.

Sopher, C.D., R.J. McCracken, and D.D. Mason. 1973. Relationships between drouth and corn yields on selected south Atlantic coastal plain soils. Agron. J. 65:351-354.

Thompson, L.M. 1962. An evaluation of weather factors in the production of corn. Cen. for Agric. Econ. Adjustment Rep. 12T, Iowa State Univ., Ames.

----. 1963. Weather and technology in the production of corn and soybeans. Cen. for Agric. Econ. Development Rep. 17, Iowa State Univ., Ames.

----. 1966. Weather variability and the need for a food reserve. Cen. for Agric. Econ. Development Rep. 26, Iowa State Univ., Ames.

----. 1969. Weather and technology in the production of corn in the U.S. Corn Belt. Agron. J. 61:453-456.

Thornthwaite, C.W. 1952. Climate in relation to planting and irrigation of vegetable crops. The Johns Hopkins Univ. Lab. of Climatology, Seabrook, N.J.

----, and J.R. Mather. 1955. The water budget and its use in irrigation. USDA Yearb. Agric. p. 346-58.

Trewartha, G.T. 1968. An introduction to climate. 4th ed. McGraw Hill, New York.

Ullstrup, A.J. 1966. Diseases of corn and their control. p. 419-446. *In* W.H. Pierre, S.A. Aldrich, and W.P. Martin (eds.) Advances in corn production: principles and practices. Iowa State Univ. Press, Ames.

USDA Foreign Agric. Serv. 1973. World agricultural production and trade. Statistical Rep. World Summaries.

USDA. Statistical Reporting Serv., Iowa Weekly Weather and Crop Bull. 1973. USDC and USDA Stat. Rep. Serv.

Vaadia, R., R.C. Raney, and R.M. Hagen. 1961. Plant water deficits and physiological processes. Annu. Rev. Plant Physiol. 12:265-292.

Van Wijk, W.R., W.E. Larson, and W.C. Burrows. 1959. Soil temperature and the early growth of corn from mulched and unmulched soil. Soil Sci. Soc. Am. Proc. 23:428-434.

Vasquez, R. 1961. Effects of irrigation at different growth stages and nitrogen levels on corn yields in Lajar Valley. P.R. J. Agric. 45:85-105.

Veihmeyer, F.J., and A.H. Hendrickson. 1955. Does transpiration decrease as the soil moisture decreases? Trans. Am. Geophys. Union 36:425-48.

Viets, F.G. 1966. Increasing water use efficiency by soil management. p. 259-274. *In* W.H. Pierre, Don Kirkham, J. Pesek, and R.H. Shaw (eds.) Plant environment and efficient water use. Am. Soc. Agron. and Soil Sci. Soc. Am., Madison, Wis.

Visher, S.S. 1940. Weather influences on crop yields. Econ. Geogr. 16:437-443.

Voladarski, N.I., and L.V. Zinevich. 1960. Drought resistance of corn during ontogeny. Fiziol. Rast. 7:176-179.

Waggoner, P.E. 1969. Environment manipulation for higher yields. p. 343-373. *In* J.D. Eastin, F.A. Haskins, C.Y. Sullivan, and C.H.M. van Bavel (eds.) Physiological aspects of crop yield. Am. Soc. Agron. and Crop Sci. Soc. Am., Madison, Wis.

Wallace, H.A. 1920. Mathematical inquiry into the effect of weather on corn. Monthly Weather Rev. 48:439-446.

----, and E.N. Bressman. 1937. Corn and corn growing. John Wiley & Sons, New York.

Wang, Jen Yu. 1960. A critique of the heat unit approach to plant response studies. Ecology 41:785-790.

----. 1963. Agricultural meteorology. Pacemaker Press, Milwaukee.

Ward, R. DeC. 1919. The larger relations of climate and crops in the United States. Monthly Weather Rev. 47:238-240.

Willis, W.O. 1956. Soil temperature, mulches and corn growth. Ph.D. Thesis. Iowa State Univ., Ames.

----, W.E. Larson, and D. Kirkham. 1957. Corn growth as affected by soil temperature and mulch. Agron. J. 49:323-328.

Wilson, J.H. 1968. Water relations of maize. Pt. 1. Effects of severe soil moisture stress imposed at different stages of growth on grain yields of maize. Rhod. J. Agric. Res. 6:103-105.

Wittwer, S.H. 1966. Carbon dioxide and its role in plant growth. XVII Int. Hort. Congr. Proc. 3:311-322.

Wolfe, T.K. 1927. A study of germination, maturity and yield in corn. Va. State Tech. Bull. 30.

Yoshida, S. 1972. Physiological aspects of grain yield. Annu. Rev. Plant Physiol. 23:437-464.

Chapter **Corn Production**

11 W.E. LARSON
Agricultural Research Service
St. Paul, Minnesota

J.J. HANWAY
Iowa State University
Ames, Iowa

High yields of corn result from selecting the best tillage and fertilization practices; choice of seed and planting procedure; choice of weed, insect, and disease control; and planning of harvesting and marketing programs.

Corn production in the United States per unit area has increased steadily from an average yield of 17.8 quintals/ha (28.4 bu/acre) in 1945 to an average yield of 58.9 quintals/ha (94 bu/acre) in 1973 (Agricultural Statistics, 1973). In 1972 the state of Iowa produced a record average yield of 72.8 quintals/ha (116 bu/acre). While corn yields have increased dramatically, more widespread use of improved technology can result in continued yield increases.

Selection of the optimum management variables must be made every year for each field and must be combined in a practical management scheme. The application of treatments at the right time is of great importance but often overlooked. Exceptional management skills together with a sound technological background are required to grow top yields of corn.

This chapter is designed to give the reader a technical background from which management decisions for corn production can be made.

PLANT GROWTH AND DEVELOPMENT

All normal corn plants follow the same general pattern of plant development (Chandler, 1960; Hanway, 1971; Hanway and Russell, 1969; Jordan et al., 1950; Kiesselbach, 1949; Sayre, 1948). Knowledge of growth sequences during the season is required for the producer to make the best management decisions. Many factors influence the size of the plants, number and size of leaves, length of time between different developmental stages, and the final crop yield (Eisele, 1938; Hanway, 1962a; Jordan et al., 1950; Kiesselbach, 1950; Nelson, 1956).

Corn plant growth depends upon photosynthesis in which carbon dioxide from the air and hydrogen from soil water are transformed by chlorophyll in the leaves into simple sugars. These sugars are then converted into more complex substances (carbohydrates, proteins, etc.), used for growth or stored in the plant. As plants mature, they produce seeds. These seeds, when placed in warm, moist soil, germinate and develop into new seedling plants and the process is repeated. Corn seeds contain adequate stored nutrient and food reserves to supply the seedling plants until they emerge from the soil, initiate a root system, and expose enough leaves to carry on photosynthesis and supply food for the developing plants.

After the seed absorbs water and the young plant begins to grow, the first internode of the stalk elongates until the plant emerges from the soil surface. The growing point of the plant remains at the 2.5 to 5-cm depth until about 3 weeks after plant emergence. During this 3-week period, five or six leaves become fully developed, and all of the leaves, ear shoots, and tassels are initiated at the growing point. Subsequent development and elongation of internodes of the stalk result in a rapid increase in height of the plant. This continues until the tassel is fully emerged from the whorl. As the plant develops and leaves are exposed for photosynthesis, the rate of dry matter accumulation in the plant increases until a rapid, nearly constant daily rate of accumulation is attained which continues until near maturity. As development of the leaves, stalk, and tassel approach completion, the uppermost ear (or ears) develops rapidly. Pollen shedding begins 2 or 3 days after the tassel is fully emerged. Silks normally emerge from the ear and are pollinated within 4 to 10 days after tassel emergence. During the next 12 days, the cob grows rapidly and the fertilized kernels enlarge to the "blister" stage, but increases in grain weight are slight during this period. Then, for 5 or 6 weeks the kernels rapidly increase in weight as starch, protein, and oil are stored in the seeds and the embryonic plant in each kernel develops, By about 50 to 60

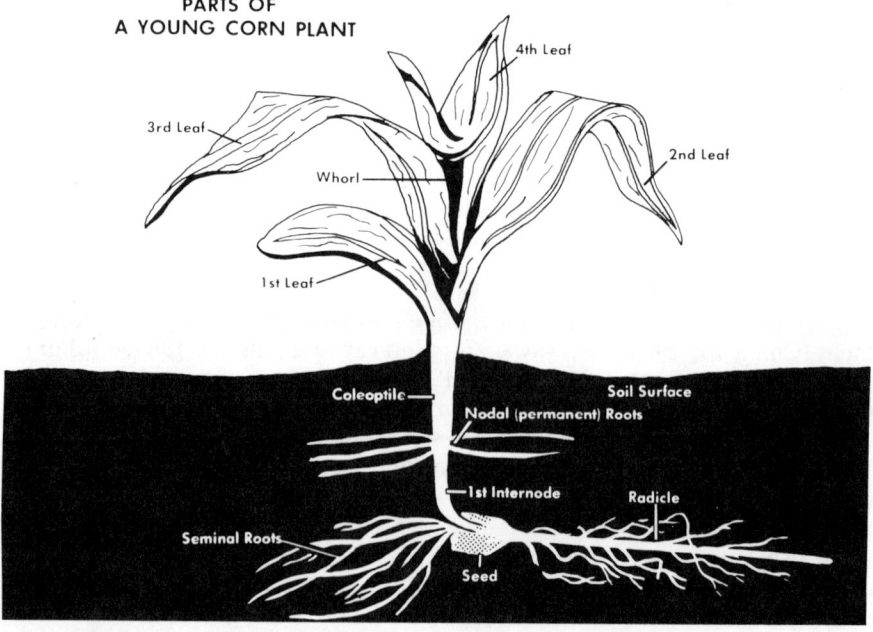

Fig. 1. Parts of a young corn plant (Hanway, 1971).

days after silking the seeds are fully developed and mature. After this, corn begins to lose kernel moisture.

The length of time from planting to maturity may vary from less than 100 to more than 150 days, depending primarily on genetic characteristics of the hybrid. Early maturing hybrids generally are smaller, have fewer leaves, and initiate reproductive stages of development faster than do later maturing hybrids.

The primary root system consisting of the radicle and seminal roots (usually two or five roots) emerges from the basal end of the seed as it germinates (Fig. 1) (Hanway, 1971; Kiesselbach, 1949). These roots serve the plant until about 2 or 3 weeks after plant emergence, after which they cease to develop and usually die. The depth of the primary root system depends upon the depth of planting of the seed. All roots, except the radicle, usually grow at an angle of 25 to 30° from the horizontal. The radicle generally grows in the direction (except up) of seed orientation. The permanent (nodal) root system develops from nodes of the stalk after the seedling emerges from the soil. The number of roots per node increases as each successively higher node emerges. By 2 or 3 weeks after seedling emergence this becomes the major root system of the plant and serves throughout the remainder of the season (Nelson, 1956). Since the first four or five internodes (which are 2.5 to 5 cm below the soil surface at the time of emergence) never elongate and more roots develop at each successive node, a large cluster (umbrella) of roots develops at this group of nodes. Successively higher internodes do elongage to greater lengths so roots that develop from the seventh or eighth node originate above the soil surface and are called brace roots.

The depth of extension of roots in deep soils is a linear function of time until tasseling. From tasseling to the start of grain filling, brace roots develop. During the rapid grain filling stage, total root length and root dry weight do not increase and may decrease before grain matures (Mengel and Barber, 1974).

Figure 2 shows a typical corn root density pattern on a Chalmers silt loam (Typic Argiaquoll) in Indiana 50, 80, and 100 days after shoot emergence (Mengel and Barber, 1974). Root length density increased rapidly between 50 and 80 days and then decreased. Full silking and maximum root length occurred at approximately 80 days.

In addition to the influence of natural profile characteristics, the growth and configuration of root development in soil can be influenced by management practices. Roots can be destroyed by improper cultivation. Fertilizer and lime placement, surface mulch configuration, surface roughness, soil compaction, tillage, and drainage can all influence root development (Allmaras et al., 1973). Root growth and configuration is in response to changes in soil water, temperature, air, nutrient, toxic chemicals, or soil resistance rather than a direct response to the management practice (Allmaras et al., 1973; Pearson, 1966; Taylor, 1971; Willis and Amemiya, 1973). Roots proliferate in zones of high fertility. The pattern of root development varies somewhat with hybrid or inbred lines (Andrew and Solanki, 1966).

Plant growth and development can be divided into five periods each with unique characteristics.

Planting to Emergence

Temperature, water, nutrients, and physical conditions in the surface soil are of utmost importance during this period (Shaw and Thom, 1951a). Under warm, moist conditions seedlings will emerge from the soil within 4 or 5 days after planting. Two weeks or more may be required for seedling emergence under cool conditions. Depth of planting influences the time from planting to emergence not only because seedlings from deeper planting must penetrate a greater depth of soil, but also temperatures at planting time generally decrease as depth in the soil increases. Crusting of the soil surface can physically prevent seedling emergence.

Emergence to Tasseling

During this vegetative period of plant development the "photosynthetic factory" of the plant is established and begins operation at full capacity. As dry matter and nutrient accumulation becomes rapid, demands for all components become large and deficiencies of any factor will limit plant growth and the potential final yield. Since the growing point is below the soil surface for the first 2 or 3 weeks of this period, damage, such as from frost or hail, to the above ground plant may have only a small, insignificant effect on final

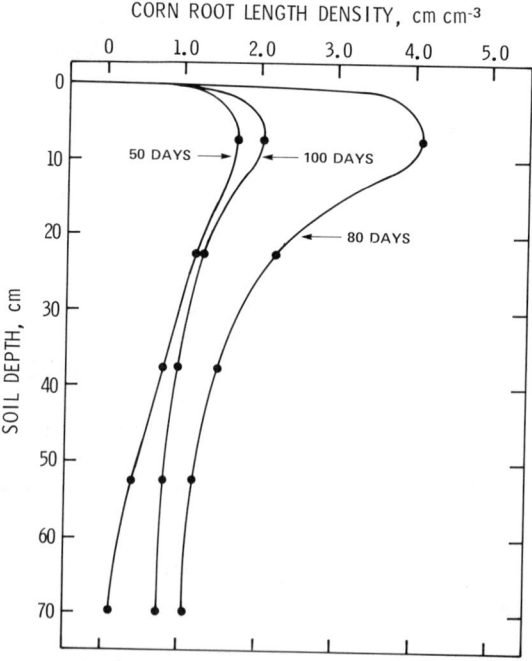

Fig. 2. Corn root length density patterns in a silt loam in Indiana (Mengel and Barber, 1974).

yield. Under cool conditions early season growth responses are commonly obtained from concentrations of nutrients supplied in seed coats or in bands of fertilizer about 5 cm from the seed. Cultural practices before and during this period which reduce or eliminate competition from weeds, improve soil temperature and water conditions are of prime importance. However, cultivation for weed control which destroys part of the root system of the corn plants may be detrimental. Any stress during this period may reduce the size of the "photosynthetic factory" operating later in the season or the number of potential kernels initiated in the ear.

Tasseling to Silking

This period is particularly critical since silks must be pollinated before kernels can develop. Any stress, such as inadequate water, fertility, or light (because of thick planting), may delay silking for 2 or more weeks and may reduce seed set because of inadequate viable pollen at the time of silking. Under severe stress conditions the cobs may be bare. Hot, dry weather resulting in a very high atmospheric demand for water may result in a water stress in the plant even though the soil is not dry throughout the root zone. Severe hail which destroys all or most of the leaves at this stage may result in essentially no grain yield.

Silking to Maturity

This is the period when the corn grain is produced (Shaw and Thom, 1951b). The cob, husks, and shank complete development during the first 2 weeks after silking, after which starch accumulates in the endosperm and dry matter accumulation in the plant is entirely in the kernel. Nutrients are translocated from other plant parts to the grain. A new plant develops in each kernel. In the embryo, the radicle and embryonic leaves are differentiated and the seminal roots are initiated. Dry matter accumulation ceases about 50 to 60 days after silking. Unfavorable environmental or nutritional conditions will result in unfilled kernels and chaffy ears. Water stress, however, will not cause as severe reduction in yield as it will during the tasseling to silking stage.

Dry-down Period

When the kernel is mature, a black layer will develop in the placental region of the kernel and probably cuts off movement of assimilate into the developing floret (Daynard and Duncan, 1969). The kernels will have 30% or more water when physiologically mature. The rate of drying in the field will depend upon climatic conditions and properties of the husk. Drying rates will vary among hybrids. Grain harvested at high moisture levels during the early stages of this period must be artificially dried for safe storage; shelling in mechanical harvesters will be incomplete. Harvesting for silage should be done in the first few days of the period to avoid loss of leaves.

REQUIREMENTS FOR PLANT GROWTH

Soils

The largest concentration of corn in the United States is grown on the deep, medium-textured soils of the Midwest and thus it might be inferred that corn is best suited to these soils (Fig. 3). But corn has wide adaptation to many soils. Significant amounts of corn are grown on light colored, thinner soils of the Lake States; on the Coastal Plain and Southern Piedmont soils of the South Atlantic and Gulf Slope; on the Southwestern Prairie soils; and on the irrigated Alluvial soils of the west. Corn is also grown on widely different soils in South America, Europe, Africa, and Asia (Table 1).

Many studies have related certain soil properties alone, or in a few combinations, to corn growth and grain yields. However, even with modern experimental designs, it is practical to study only a few variables at a given time. The most common properties that have been related to corn yields are available N, P, and K. While corn yields have been related to many soil variables using multiple regression techniques, the soil properties (variables) are usually so intercorrelated that interpretation is difficult. Sopher and McCracken (1973) found that the factors most closely associated with yields on South Atlantic Coastal Plain soils were soil water holding capacities, combinations of clay and sand, extractable P, percent base saturation, properties which control soil acidity, and the amount of charge on the cation exchange complex. Because of previous management, available nutrients were important but not highly limiting.

While it is difficult to identify quantitatively soil factors affecting plant growth, nevertheless, soils in their natural environment vary greatly in

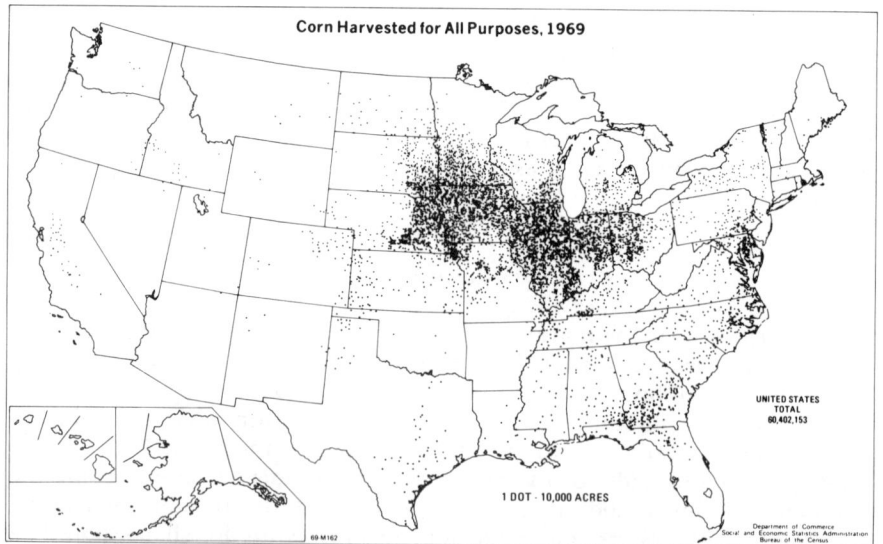

Fig. 3. Corn acreage in United States, 1969 (US Bur. of Census).

Table 1—Leading corn producing countries, 1972†

	Area	Grain yield	Production
	ha	quintals/ha	metric tons
North America			
Canada	533	49.8	2,657
United States	23,205	59.9	139,038
South America			
Argentina	3,600	26.7	9,600
Brazil	10,000	12.9	12,900
Europe			
France	1,920	44.8	8,600
Italy	919	54.8	5,034
Spain	558	40.3	2,250
Bulgaria	700	46.4	3,245
Hungary	1,390	41.2	5,724
Romania	3,264	27.0	8,800
Yugoslavia	2,381	33.2	7,906
Africa			
Egypt	660	38.6	2,550
Nigeria	1,510	8.1	1,219
South Africa	5,200	9.6	5,000
Asia			
China-Mainland	13,000	16.9	22,000
India	5,200	8.7	4,500
Indonesia	2,600	9.5	2,470
Philippines	2,435	8.3	2,015
Thailand	592	22.3	1,320
Turkey	640	16.7	1,070

†Source: Agric. Statistics, 1973.

inherent productivity of corn. For example, in Illinois with a high level of management, corn yields were estimated to range from 9,090 kg/ha on Muscatine silt loam (Aquic Argiudolls) to 2,880 kg/ha on Plainfield sand (Typic Udipsamments) (Odell and Oschwald, 1970). Corn root development in different Illinois soils is illustrated in Fig. 4.

Water

Corn has a water requirement totaling about 40 to 60 cm of evapotranspiration (Downey, 1971). The total water used in evapotranspiration varies considerably with the water available, climatic environment, and the soil and water-management practices. Peters (1960) showed that under Illinois conditions water removal from the soil by evaporation may be as great as from transpiration.

A single stalk in a productive field of corn on a hot, windy day may use as much as 2 liters of water. Total evapotranspiration on days with a high evaporative demand may range from 6 to 9 mm/day. In most areas, however, water use rate at the period of maximum leaf area will probably average 5 to 6 mm/day if sufficient water is available for evaporation and transpiration.

The water needs of corn during the summer in most areas exceed the amount falling as rain (Fig. 5). In the deep soils of the Corn Belt, corn will withdraw water from a 150-cm depth and occasionally from as much as 200 cm. The deep soils of the Corn Belt may store as much as 25 to 30 cm of available water in a 150-cm soil profile (Peters and Bartelli, 1958). Most soils of the Southern Piedmont and Southern Coastal Plain can store less than 20 cm in a 150-cm profile, and corn may withdraw water from only the top 50 to 100 cm (Elkins et al., 1961; Long et al., 1963).

Water deficits at anthesis are most damaging to corn grain yields followed by the early ear and vegetative stages in that order (Barnes and Woolley, 1969; Denmead and Shaw, 1960; Miller and Duley, 1925; Robins and Domingo, 1953). Classen and Shaw (1970) found that a 53% reduction in grain yield resulted from an imposed plant water stress at anthesis while a 30% yield reduction resulted from water stress during the 3-week period after silking. Plant water stress, before or during anthesis, reduced the number of developed kernels while kernel weights were reduced by stress after pollination.

Vincent and Woolley (1972) showed that hybrids with male-sterile cytoplasm could withstand water stress at anthesis better than hybrids with normal cytoplasm.

Corn can be damaged severely by too much water. Schwab et al. (1966) found that corn yields were drastically reduced by two applications of 7.6 cm of water to a poorly drained soil (Typic Haplaquepts). One application was applied when the corn was less than 15 cm tall and the other when it was 61 cm tall. Ritter and Beer (1969) showed that flooding a Cumulic Haplaquoll early in the season was more detrimental to corn grain yields than flooding late in the season. At a high soil nitrogen level yields were decreased in one year by 18% when corn 15 cm tall was flooded for 72 hours and in a second year yields were decreased 6% by flooding for 96 hours. Lal and Taylor (1969) demonstrated that intermittent flooding early in the growing season on a Typic Hapludalfs reduced yields of corn more than did constant water tables of 15 to 30-cm depth. Plants grown under continuously wet conditions often develop greater air space and consequent greater gaseous exchange between leaves and roots (Grable, 1966) which may explain Lal and Taylor's observation.

Corn has been considered reasonably tolerant to low concentrations of O_2 in the soil. Growth damage due to flooding or high water contents on various soils is probably caused by many things including low O_2 or high CO_2 concentrations in the soil air, respiration rate of the plant at the time of flooding, reduced nutrient uptake, and toxic chemicals produced by reducing conditions.

Das and Jat (1972) showed that corn cultivars differ in root porosities when grown under high soil water conditions. Possibly corn cultivars could be bred for adaptation to high soil water conditions. Growth on poorly drained soils may be aggravated by traffic compaction (Larson and Allmaras, 1971).

pH and Salt Tolerance

Corn will grow over a relatively wide pH range as compared with other crops. While many factors are involved, field studies with liming acid soils indicate that maximum corn yields usually occur at pH 6.0 and above. Plant growth in acid soils is influenced by toxic concentrations of aluminum and manganese, and deficient concentrations of calcium. In addition, pH influences reactions affecting virtually all of the major, macro and micronutrients. Microbial reactions within the soil are also affected by pH. Corn is considered a relatively tolerant crop to toxic concentrations of aluminum and manganese (Kamprath and Foy, 1971). Calcium concentrations in acid soils are rarely inadequate for corn growth except, perhaps, on acid sandy soils. Phosphorus availability is usually greater at pH's above about 5.5 to 6. The oxidation of NH_4 to NO_3 in soils is inhibited at very low pH's.

Deficiencies of iron and zinc sometimes occur in corn grown on calcareous soils. Hybrids and inbreds vary in their ability to take up iron and zinc (Kamprath and Foy, 1971).

Corn is considered a crop with a medium salt tolerance. In a grouping of crops according to their relative salt tolerance, Richards (1954) listed corn in a group whose yields would be reduced by 50% in a soil with an EC of 10 in the saturation extract (EC value designates the electrical conductivity values of the saturation extract in mmhos/cm at 25 C.). In a greenhouse study, Hussan et al. (1970) found that production of dry matter by leaves, stems, and tassels of corn decreased as salinity increased from 0 to 16 mmhos/cm but decreases were particularly sharp above an EC value of 8. While soil salinity can have a marked influence on uptake of a number of nutrients,

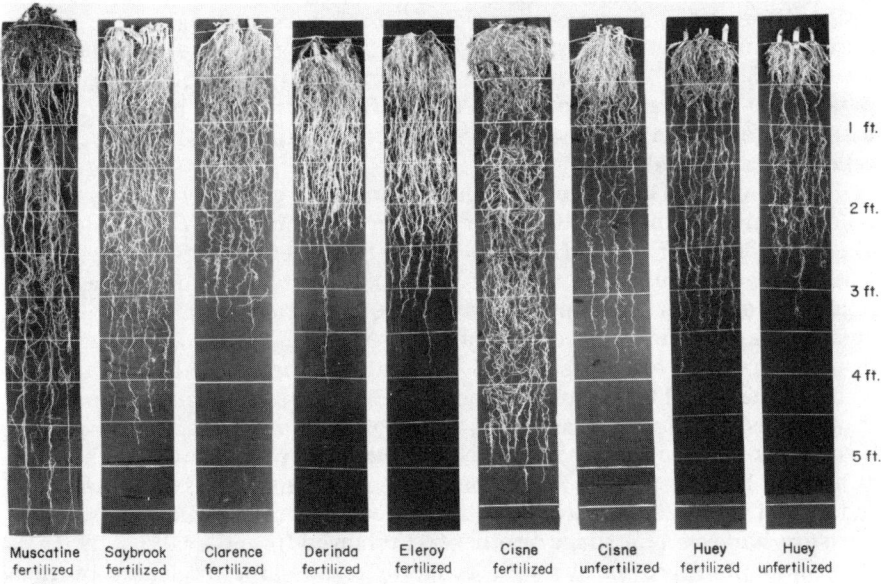

Fig. 4. Soil type and fertilization affects corn root penetration (Fehrenbacher et al., 1967).

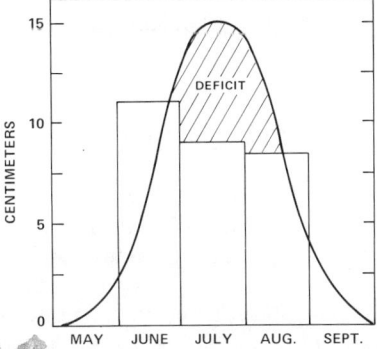

Fig. 5. Water requirement curve for corn in relation to summer rainfall in five Corn Belt states (Thompson, 1962).

decreased dry matter production probably most often results from decreased soil water and increased toxicity of sodium chloride and sulfate in the soil solution.

Mineral Nutrients

In addition to C, H, and O which are derived from water and air, at least 13 nutrient elements (N, P, S, K, Ca, Mg, Fe, Mn, Zn, Cu, B, Cl, and Mo) are essential for normal plant growth and development. Except for impurities of certain of these elements which may be present in the air, rain, or irrigation water, the plant must obtain these 13 elements from the soil or they must be added if their supply in the soil is inadequate. Adequate supplies of each nutrient throughout the growing season are essential for optimum plant growth and development. Availability of these nutrients to plants is influenced by many factors, such as the nutrient and aeration status of the soil, water, temperature conditions, and physical properties of the soil (Nelson, 1968; Olson, 1971; Pierre et al., 1966). Therefore, knowledge of the plants' requirements and methods for supplying the proper conditions becomes essential. Soils are most commonly deficient in nitrogen, phosphorus, and potassium, which are supplied by commercial fertilizers; but fertilization with other elements is necessary in certain areas. Animal manures are an excellent source of many nutrients.

Accumulation of nutrients in plants during the growing season generally follows patterns similar to that of dry matter accumulation (Fig. 6) (Benne et al., 1964; Chandler, 1960; Hanway, 1962b; Jordan et al., 1950; Sayre, 1948). The amounts of nutrients taken up by young plants early in the season are relatively small. As the plants grow, the rate of nutrient uptake per plant increases. Rapid uptake of some elements, such as N and P, continues until near maturity whereas uptake of K is essentially completed by silking time. Much of the N and P and some of the K taken up before grain development is translocated from other plant parts to the grain as it develops. A large part (two-thirds to three-fourths) of the N and P taken up is removed in the grain at harvest. But most of the K remains in the leaves and stalks and is returned to the soil and becomes available to succeeding crops, unless these plant parts are harvested for silage or otherwise removed from the land. Any factor that limits plant growth and yield generally results in a lower total nutrient requirement by the plant. Since many uncontrollable factors which influence plant growth and yield vary from year to year, accurate predictions at the first of the season are not presently possible. Knowledge of the odds of having favorable conditions, however, provides a basis for selecting the most appropriate practices.

Nitrogen (N)

A 100 quintals/ha corn crop will contain at least 200 kg of N in the grain and stover. Only the most fertile soils will supply this much N unless corn follows a legume meadow in the cropping system or N has been applied as manure or N fertilizer.

The degree of N deficiency is reflected in the condition and color of the corn plants. With adequate N, corn leaves are dark green. Early season N deficiency results in spindly, stunted growth with a light yellowish-green color. Later deficiency results in yellowing, especially along the leaf midrib producing a characteristic Y-shaped pattern, followed by browning and death ("firing") of the leaf tissue. Since N is translocated from older to new tissues, the older, lower leaves show these deficiency symptoms first. The loss of functional leaf tissue restricts photosynthesis which results in barren ears or ear tips and/or smaller kernels and in this way limits yield and results in a lower protein content of the grain. Excessive levels of available N do not increase the protein content of the grain appreciably, but may result in accumulation of nitrates in the stalks.

Phosphorus (P)

A 100 quintals/ha corn crop contains about 36 kg of P (84 kg of P_2O_5). At maturity about four-fifths of the total plant P is in the grain. About half of the P in the natural grain has been translocated from other plant parts to the grain as it developed.

Symptoms of P deficiency are not easy to recognize. In very early stages of growth, plants may be stunted and have a dark green color. In some strains of corn a purple color may develop. In later stages P deficiency may delay silking and result in incomplete pollination.

Potassium (K)

A 100 quintals/ha corn crop generally contains at least 190 kg of K (230 kg of K_2O) per ha. Potassium uptake from soils with high levels of available K may be much greater than this. At maturity, usually less than one-fourth of the total plant K is in the grain. Uptake into the plant is essentially complete before rapid grain development begins so K in the grain has been translocated from other plant parts. Potassium in the plant is essential for growth and production of sugars, starches, and proteins, but is not chemically a part of these substances.

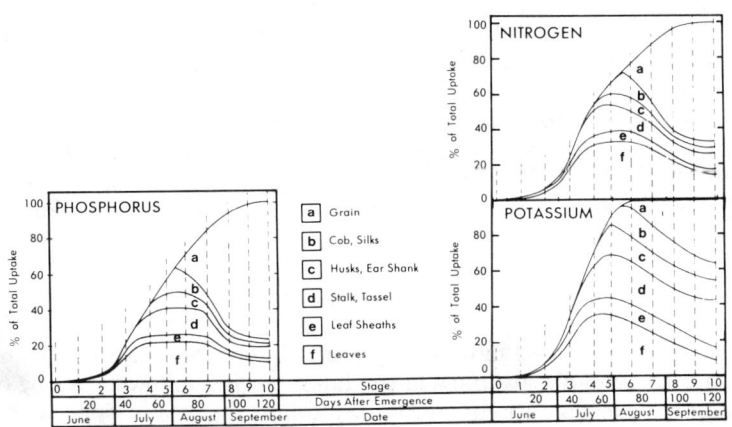

Fig. 6. Nutrient uptake by corn (Hanway, 1971).

Potassium deficiencies develop first and are most extreme in the lower, older leaves since K moves readily in the plant and is always present in the highest concentrations in young, actively growing tissues. In young plants a K deficiency results in yellowish-green leaves. Deficiencies later in the season result in leaf margins and tips gradually turning brown and dying which leads to decreased growth rate and yield potential. Ears from K-deficient plants are small, poorly filled, and chaffy. Stalks of K-deficient plants may be weak and subject to stalk rot which causes stalk breaking and lodging.

CULTURAL PRACTICES

Preplant Tillage Systems

The tillage systems farmers use for corn have been rapidly changing. While the moldboard plow is still the most widely used primary tillage implement, its use on all soils is no longer considered necessary for maximum yields. Researchers have been developing such new tillage systems as the chisel, till, strip, rotary, and no-tillage (Soil Conserv. Soc. Am., 1973). Even when the moldboard plow is used, there has been a general reduction in the amount of secondary tillage prior to planting as well as in the amount of postplanting tillage. There also has been a decided shift toward combining secondary tillage and planting into one trip over the field.

In 1973 more than one-fourth (2.2 million ha) of Iowa's corn and soybeans (*Glycine max* L.) were planted using some form of conservation tillage (Moon, 1973). The chisel system was used on more than 405,000 ha and was the most widely used conservation tillage system.

The search for better soil tillage systems has had the following objectives: (a) reduced power, machinery, and labor cost, (b) less soil compaction, and (c) lower water runoff and erosion.

In addition to modifying the soil environment for enhancing plant growth, a tillage system must provide for the desired control of seed placement, weeds, water infiltration, water evaporation, and erosion. The objective in a tillage system is to manage these interrelated components so as to optimize the control of as many as possible. Since the importance and interrelationships of the components change with the soil, climate, and crop, the optimum tillage system will not be the same for all environments.

The soil attributes influencing growth of corn are nutrient availability, soil air composition, water availability, soil temperature, soil strength (physical resistance to root extension), and toxic chemicals. The effect these attributes have on plant growth and how they can be managed by soil tillage has recently been reviewed in the proceedings of a national conference (Soil Conserv. Soc. Am., 1973).

Moldboard Plow

The traditional and most widely used tillage system for corn consists of plowing with a moldboard plow in the fall or spring followed by one or more secondary tillage operations. Secondary tillage is usually done with a disk harrow, spring-tooth harrow, spike-tooth harrow, field cultivator, or similar implement. There is a strong trend to eliminate much of the secondary tillage as well as to pull two secondary tillage implements in one pass across the field. Sometimes a light tillage implement is pulled immediately ahead of the planter (Bateman and Bowers, 1962). The popularity of the disk harrow and spike-tooth harrow as secondary tillage implements has been declining. A runner or disk-opener planter is generally used for corn.

A modification of the conventional moldboard plow system is to provide only secondary tillage in the row. This is often done by planting in the wheel-tracks of the planter or tractor and this has been called wheel-track planting (Peterson, 1960; Peterson et al., 1958; Rossman and Cook, 1966). In another modification a strip from 15 to 30 cm wide is rotary tilled immediately ahead of the planter.

Griffith et al. (1973) found that in Indiana corn grain yields from spring moldboard plowing methods were equal to or greater than all non-moldboard plowing methods on Runnymede loam (Typic Argiaquolls) over a 4-year period. Yields from strip-rotary and no-tillage were lower than moldboard plowing methods. The moldboard plow usually had a yield advantage over all no-plow systems on Blount silt loam (Aeric Ochraqualfs) and Pewamo silty clay loam (Typic Argiaquolls). On Bedford silt loam (Typic Fragiudults) the 4-year average corn grain yield was about the same for six different tillage systems tested. Under high yield conditions in 1969, yields from no-plow systems were greater than from the two moldboard plow systems. In other years, germination and weed control affected yields and were usually poorer with the no-plow systems. On Tracy sandy loam (Typic Hapludalfs), corn grain yields were about the same for eight tillage methods tested except for till planting which produced superior yields.

In Ohio, corn grain yields from moldboard plow systems on poorly drained soils have been equal to or better than all other tillage systems (Triplett et al., 1970; Triplett et al., 1973). Likewise, in Illinois moldboard plowing with some secondary tillage has produced greater grain yields on poorly drained soils than have other methods tested (Oschwald, 1973).

In Wisconsin, on a number of soils and over several years, corn grain yields from wheel track planting have averaged about the same as from a conventional moldboard plowing system (Peterson, 1960; Peterson et al., 1958). In Iowa on five soil types over several years, corn grain yields from wheel track and conventional moldboard plowing systems were about the same (Larson, 1962).

Fall plowing (moldboard) is commonly practiced in the northern Corn Belt on fine-textured soils, imperfectly to poorly drained, because of improved yields and ease of secondary tillage. Fall plowing has the disadvantage of possible soil damage due to wind erosion in the winter and spring.

Allmaras et al. (1972) found that corn grain yields in southwestern Minnesota and eastern South Dakota averaged about 10% greater from fall plowing as compared with spring plowing in 6 of 8 location years. Soil temperatures were about 2 C greater on fall-plowed than on spring-plowed soil and temperature differences extended as deep as 150 cm. They showed that the net radiation was greater on a rough surface (fall plowed), but only when the sun angle from the horizon was small. Increases in overwinter random roughness resulted in greater soil temperature benefit.

Tillage systems that included fall plowing (moldboard) produced the highest average corn grain yield of the eight systems tested on Runnymede loam (Typic Argiaquolls) in Indiana in 2 years of testing although the differences were not significantly higher than till planting in 1 year. Corn grain yields from fall plowing were not different from spring plowing in 2 years of testing on Tracy sandy loam (Typic Hapludalfs), in 3 years of testing on Blount silt loam (Aeric Ochraqualfs), and in 2 of 3 years on Pewamo silty clay loam (Typic Argiaquolls). Yields from fall plowing were significantly lower in 1 year on Pewamo silty clay loam (Typic Argiaquolls) (Griffith et al., 1973). Fall or winter moldboard plowing is the preferred method of primary tillage on Hoytville silty clay loam (Mollic Ochraqualfs) in Ohio. Corn grain yields averaged 1,199 kg/ha (24 bu/acre) more for fall than spring plowing when the soil was not spring disked and 1,548 kg/ha (31 bu/acre) more when the soils were spring disked (Triplett et al., 1973).

Much of the dissatisfaction with the moldboard plowing system, including several secondary tillage operations, has been because the soil is highly susceptible to water and wind erosion although omission of the secondary tillage after plowing somewhat reduces the soil's erosion potential.

Meyer and Mannering (1961) found that soil water erosion was reduced by 48% and runoff by 37% when tillage practices that omitted all or nearly all of the secondary tillage were used as compared with plowing, two diskings, and spike-tooth harrowing. The experiment was conducted on Russell Silt loam (Typic Hapludalfs) with 5% slope. Simulated rainfall was applied shortly after planting. Wischmeier (1973) found that on a Sidell soil (Typic Argiudolls) with conventional moldboard plowing including secondary tillage, the soil was the most erosive of all systems tested.

Chisel Tillage

Chisel tillage uses the chisel plow with chisel points (straight or twisted) or sweeps (15 to 50 cm wide) attached to shanks on a heavy frame. The points or sweeps usually have an effective spacing of 23 to 38 cm and may be operated from 15 to 30 cm deep in the soil. Chisel tillage may be done in either the spring or fall or both. Often the chisel points are used in the fall, and sweeps with unit planters mounted behind the tillage tool bar are used in the spring. The points are usually operated at a deeper soil depth than the sweeps.

Chisel tillage provides an effective means of reducing wind and water erosion because it leaves some of the crop residues on the surface and often leaves a rougher more porous surface. Depending on machine operations it may be less costly than conventional tillage and is best adapted to soils with good internal drainage.

Four-year averages on five soils in Indiana showed that grain yields from chisel planting were not significantly different from yields obtained from moldboard plowing methods. However, in some years, yields from chisel planting on the soils with good internal drainage (Tracy sandy loam, Typic Hapludalfs; Bedford silt loam, Typic Fragiudults) were higher and on soils with poor drainage (Runnymede loam, Typic Argiaquolls; Blount silt loam, Aeric Ochraqualfs; Pewamo silty clay loam, Typic Argiaquolls) were lower than from moldboard plowing methods. In general, stands were lower on all soils from chisel planting than from moldboard plowing methods and were lowest or nearly lowest of all of the tillage systems tested. Lower stands accounted for some of the lower yields.

Water erosion losses with chisel system using sweeps were only 6% of that from a plow-disk-harrow system when measured during crop stage 1 (planting to 1 month thereafter) on a Bedford silt loam (Typic Fragiudults). Erosion losses seemed larger when chisel plows were equipped with points than when equipped with sweeps (Wischmeier, 1973).

Strip Tillage

In strip tillage the soil is tilled in a band centered near the row with no tillage between rows. In this discussion systems in which the band is tilled with a rotary tiller (strip rotary) and with a sweep (till planting) will be considered. Other machines have also been used to till the strip.

In the strip rotary system a PTO-powered rotary tiller is mounted or pulled in front of the planter. Residues from previous crops are usually chopped. Fertilizer and insecticides can be incorporated in the tilled band.

In the till planting system the tillage tool and planter are mounted on the tractor together. Each unit consists of a wide sweep to displace the top 5 to 7 cm of soil from a preformed ridge. Bars push residues between rows, a packer wheel firms the seed into moist soil, and disks cover the seed with soil. Residues may be chopped prior to planting. Usually ridges are formed during corn growth or after harvest the previous year by cultivating with a disk-hiller (Wittmuss et al., 1971).

In the Indiana experiments of Griffith et al. (1973), corn grain yields from till planting and strip rotary were usually equal to or better than moldboard plowing methods on the medium- and coarse-textured, well-drained soils (Tracy sandy loam, Typic Hapludalfs; Bedford silt loam, Typic Fragiudults) whereas on the poorly drained soils (Pewamo silty clay loam, Typic Argiaquolls; Blount silt loam, Aeric Ochraqualfs) yields were inferior. Yields might be improved on the poorly drained soils by planting on large residue-free ridges that warm rapidly in the spring. On all but the Pewamo soil, stands were about equal to moldboard plowing systems.

A great deal of work in Nebraska has been done developing the till planting system, and it seems most applicable in the western Corn Belt. Lane and Wittmuss (1961) and Wittmuss et al. (1971) reported corn grain yields about equal on till plant and conventional moldboard plow plots in Nebraska. In South Dakota Turnquist et al., (1970) and Olson and Schoelberl (1970) also reported yields from till planting to be about the same as from other methods. Amemiya (1968) found mulch tillage (similar to till planting) produced better corn yields in Iowa in years when the corn showed a moderate to severe water stress. On years with adequate rainfall, yields from mulch tillage and conventional moldboard plowing were about the same on Moody silt loam (Udic Haplustolls) in northwest Iowa. Fisher and Lane (1973) indicated that corn yields from till planting in Nebraska did not experience the extreme lows or highs in bad and good years of conventional systems.

The main advantages of till planting are better erosion control and lower costs (Fisher and Lane, 1973). Weed control may be more of a problem and use of herbicides is necessary (Fisher and Lane, 1973; Griffith et al., 1973; Turnquist et al., 1970).

Wittmus and Swanson (1964) reported comparative annual soil losses from water erosion of 9.0 metric tons/ha with till planting as compared with 31.4 metric tons/ha with a moldboard plow-disk-harrow system.

Turnquist et al. (1970) found that weed control became more difficult with increased years of continuous till-planted corn. Herbicides alone did not provide adequate control. Griffith et al. (1973) found cultivation of till-planted corn difficult because of the packed untilled soil in the row middles.

The strip-rotary system has been used by several investigators with only moderate success (Adams et al., 1970; Griffith et al., 1973). In Griffith et al. (1973), average corn grain yields over a 4-year period from strip rotary were equal to those from conventional moldboard plowing methods on three of five soils tested. Results from strip rotary were favorable on well-drained soils (Typic Hapludalfs, Typic Fragiudults) but were inferior on poorly drained soils (Aeric Ochraqualfs, Typic Argiaquolls).

Strip rotary was used by Adams et al. (1970) to make a seedbed in Coastal bermudagrass (*Cynodon dactylon* L.) and fescue (*Festuca arundinacea* Schreb.) sod on a Cecil sandy loam (Typic Hapludults) with the aim that the grass would not unduly compete with the corn and that the grass would reestablish itself the following year. Corn yields were lower than those from the conventional moldboard plow system. Grasses did establish themselves and produce good yields the following year if irrigation was used during the corn year.

Soil water erosion from till planting and rotary till systems averaged 31 and 16% of that from a moldboard plow-disk-harrow system in an experiment reported by Wischmeier (1973). Tillage practices were applied on the contour and erosion was estimated by using a rainfall simulator at crop stage 1 (planting to 1 month thereafter) on a Bedford silt loam (Typic Fragiudults) on 10% slope.

No-tillage

This system involves only one tillage-planting operation. Usually a nonpowered fluted coulter is placed in front of each planting unit. The coulter tills a strip about 6 cm wide and 8 cm deep. Fertilizer is usually put on or near the soil surface. Vegetation is killed with a herbicide prior to planting. On erosive soils in southern areas a winter cover crop is usually seeded.

Much of the early interest in no-tillage concerned corn production in sods of cool-season perennial grasses. It was thought that with the use of herbicides the sod could be made temporarily dormant and then recover after the corn was harvested to give forage during the fall and winter and the following year. While maintaining a perennial sod during the corn year has not become fully practical, especially without irrigation (Lewis, 1973), the no-tillage system is usually used in combination with a cover crop or following a grass sod. In some areas no-tillage is used to plant corn in residues from previous crops, most commonly corn or soybeans.

In the southeastern United States, on medium to coarse-textured soils with low water-holding capacities, no-tillage has often produced greater corn grain yields while holding erosion to acceptable limits.

Van Doren and Triplett (1973) felt that no-tillage for corn production could be used most widely in geographic areas south of 43° N Lat. on well-drained soils with moderate to severe erosion. Proper use of the no-tillage system allows corn to be grown on slopes where perennial grasses were previously used. It may also be used on poorly drained soils to enable timely planting when plowing is unduly delayed.

In northern areas decreased early corn growth from lowered soil temperatures will limit use of no-tillage (Willis and Amemiya, 1973). Field studies in Ohio, Indiana, and Illinois, all indicate that corn yields from no-tillage have been lower than from moldboard plowing on poorly drained soils (Griffith et al., 1973; Oschwald, 1973; Triplett et al., 1973).

Shear and Moschler (1969) found corn grain yields in Virginia were greater from no-tillage than from conventional tillage in 3 of 6 years with yields equal in other years. The soil was a Lodi loam (porous substrata, Typic Hapludults) on a 2 to 5% slope. In another experiment in Virginia, Moschler et al. (1972) reported that the accumulated corn yield for no-tillage as compared with conventional over a 9-year period on Lodi silt loam was 25.6% greater, over a 6-year period on Davidson clay loam (Rhodic Paleudults) was 13.7% greater, and over a 5-year period on Cecil clay loam (Typic Hapludults) was 39.0% greater.

In West Virginia corn grain and silage yields were greater from no-tillage than from conventional moldboard plowing in 1 of 2 years tested on a Wharton silty clay loam (Aquic Hapludults) on a 15% slope. Severe water erosion occurred on the conventional treatments whereas no erosion occurred from no-tillage in an orchardgrass *(Dactylis glomerata* L.) sod (Bennett et al., 1973).

On an erodible Cecil sandy loam (Typic Hapludults) in Georgia, Carreker et al. (1972) grew 9,700 kg/ha of corn grain using a no-tillage system on fescue (*Festuca arundinacea* Schreb.) sod stunted with herbicide and receiving 448 kg of N/ha and irrigation. Fescue yields the following year were less than 1,000 kg/ha, however. In another experiment on Cecil sandy loam in Georgia, corn grain yield was usually somewhat lower from no-tillage than from rotary strip and mulch tillage, both with and without irrigation. However, coastal bermudagrass sod recovery was good and apparent erosion control was acceptable under lister and no-tillage treatments (Adams et al., 1973).

Average corn grain yields from no-tillage in Indiana over a 4-year period were equal to those from moldboard plowing on well-drained soils (Tracy sandy loam, Typic Hapludalfs; Bedford silt loam, Typic Fragiudults) but ranged from 13 to 30% lower on poorly drained soils (Runnymede loam, Typic Argiaquolls; Blount silt loam, Aeric Ochraqualfs; Pewamo silty clay loam, Typic Argiaquolls). Plant populations were about equal on moldboard plowing systems and no-tillage. Competition from weeds accounted for some of the yield reductions in some years with no-tillage. Soil temperatures were lower and plant heights at 8 weeks after planting were shorter with the moldboard treatments (Griffith et al., 1973).

On first, second, and fourth year corn soil erosion losses from simulated rainfall from no-tillage on Bedford silt loam (Typic Fragiudults) with a 10% slope averaged about 10% of that from conventional moldboard plow-two-disk-harrow systems. The comparisons were made 4 weeks after planting in Indiana (Wischmeier, 1973).

Most of the benefits in corn yield and in erosion from the no-tillage system apparently can be attributed to improved water infiltration and storage. The crop residues on the soil surface protect the soil from sealing over. Thus, one would expect greatest benefit from no-tillage on soils with poor soil structural attributes.

Triplett et al., (1968) compared corn grain yield and water infiltration rate on a moldboard plowed treatment and no cover with no-tillage treatments having 5% cover (stover removed), 48% cover (stover left in place), and 70% cover (double amount of stover) on a Wooster silt loam (Typic Fragiudalfs) in Ohio. Grain yields were significantly higher on the 70% cover treatment than the plowed check. Total water infiltration was greater on treatments having 40 or 80% cover as compared with the plowed-bare treatment. Water content in the soil was greater on the soils with residue cover. The Wooster soils form a surface crust after rainfall and respond to cultivation even if weeds are not present (Triplett et al., 1968).

No-tillage systems in Kentucky on Donerail silt loam (Typic Argiudolls) and Maury silt loam (Typic Paleudults) induced larger amounts of water in the soil than did conventional moldboard plowing methods (Blevins et al., 1971). The increases were apparent to 50-cm depth during most of the season. Similar results have been observed by Jones et al., (1969) in Virginia.

One of the concerns with the no-tillage system has been whether or not surface applications of fertilizer would provide the crop with adequate nutrients during dry seasons. While it has been shown that P and K concentrate near the soil surface in no-tillage systems using surface fertilizer applications (Triplett and Van Doren, 1969; Walker et al., 1970), lower nutrient uptake for corn has not been shown (Singh et al., 1966; Triplett and Van

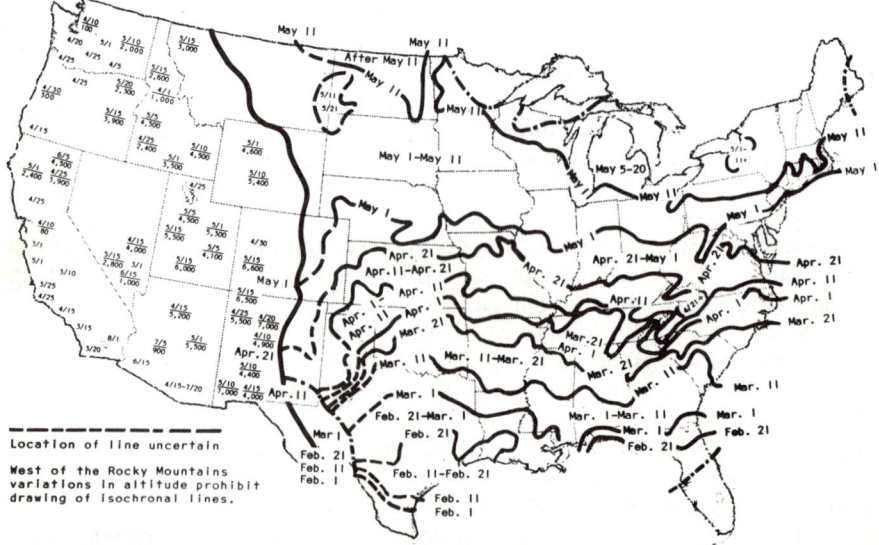

Fig. 7. Average dates when corn should be planted in the United States (Sprague and Larson, 1966).

Doren, 1969). Indeed, in some cases, particularly early in the season, nutrient uptake by corn from no-tillage using surface applications of fertilizer has been greater than from conventional moldboard plowing techniques where fertilizer was incorporated (Singh et al., 1966; Triplett and Van Doren, 1969).

Planting for Grain

Planting Date

A decided trend toward earlier corn planting in the central and northern states has been evident because research has shown corn grain yield advantages. Earlier planting has become more practical because of effective weed control from herbicides, seed treatment, and improved seed quality (Rossman and Cook, 1966). Perhaps the best guide for choosing a corn planting date is to plant when the soil temperature at the 7.6-cm soil depth has reached 15 C for several days. If soil temperature information is not available, a reasonable estimate can be made by averaging the 0700 and noon air temperatures. Even though soil temperatures are adequate, wet soil conditions may prohibit implements from entering fields and delay planting beyond the desirable time. A general guide for optimum corn planting dates in the United States is given in Fig. 7 (Sprague and Larson, 1966).

In Iowa the average percent of corn planted for the years 1970 to 1972, inclusive, was 11, 47, 80, and 97% by 1, 10, 20, and 30 May, respectively (Iowa Crop and Livestock Rep. Serv., 1973).

Hicks et al. (1970) found that corn grain yields from 95 to 105 and 110 to 115-day hybrids were not materially influenced in southern Minnesota by planting from 20 April to about mid May but delaying the planting until after 15 May caused substantially lower yields. Yields of 80 to 85-day hybrids were not decreased by planting as late as 31 May.

In Michigan, Rossman and Cook (1966) found over a 10-year period that 1 to 9 May plantings averaged 9% higher than 12 to 20 May plantings, 16% higher than 22 to 31 May plantings, and 27% higher than 1 to 11 June plantings.

Pendleton and Egli (1969) found that two corn hybrids planted on 19 or 30 April at Urbana, Illinois, outyielded corn planted on 14 or 31 May. Plants started in the greenhouse, transplanted to the field on 19 April, and pollinated 15 June yielded no more grain than the 19 or 30 April planting in the field.

Figure 8 illustrates the large interaction between planting date and population for a short-season and a full-season hybrid in Minnesota.

Planting Depth

Corn planting depth varies widely with soil conditions and climate. Usually the objective is to plant at a soil depth that will optimize soil temperature and soil water and result in rapid and high percentage germination and emergence.

Alessi and Power (1971) found in growth room experiments that from 4 to 24 days were required to achieve 80% emergence, depending upon seed depth and temperature. Temperature had a more pronounced effect than seed depth. In field experiments in North Dakota with adequate soil water, 8 to 13 days were required for near 80% emergence. Increasing the seed depth by 2.5 cm decreased emergence by about 1 day. About 10 days and 68 cumulative degree days (base 10 C) were required for corn emergence when the seed was placed at 7.6-cm depth.

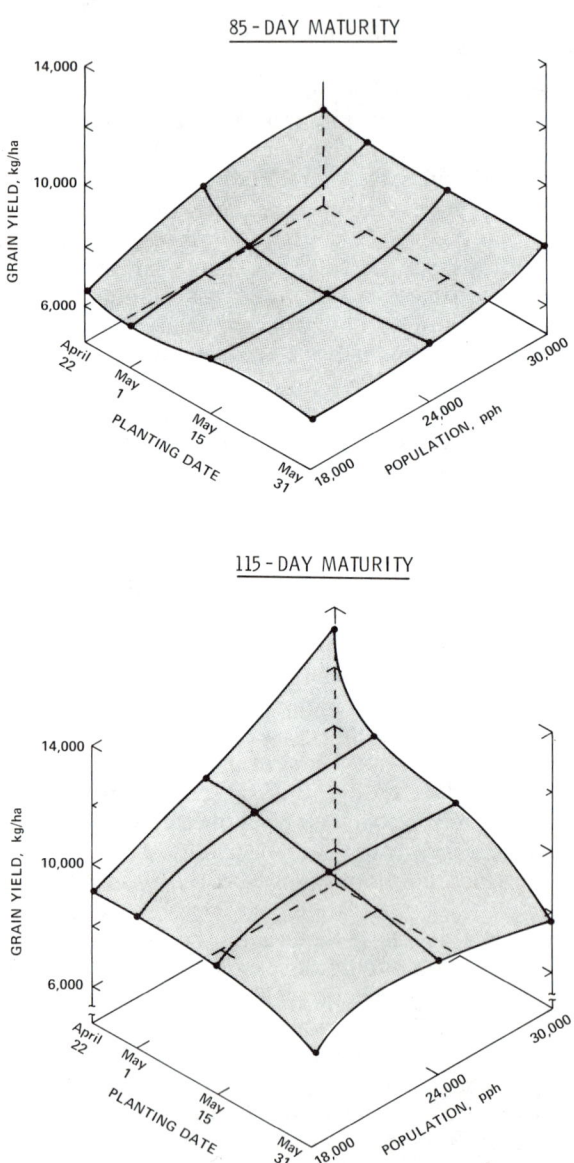

Fig. 8. Corn grain yield as influenced by planting date and population (Hicks et al., 1970).

Table 2—Average distance between seeds at different populations and row spacings

Seeds/ha	cm between rows				
	40	50	60	80	100
			cm between plants in row		
30,000	83.3	66.7	55.6	41.7	33.3
35,000	71.4	57.1	47.6	35.7	28.6
40,000	62.5	50.0	41.7	31.3	25.0
45,000	55.6	44.4	37.0	27.8	22.2
50,000	50.0	40.0	33.3	25.0	20.0
55,000	45.5	36.4	30.3	22.7	18.2
60,000	41.7	33.3	27.8	20.8	16.7
70,000	35.7	28.6	23.8	17.9	14.3
80,000	31.3	25.0	20.8	15.6	12.5
90,000	27.8	22.2	18.5	13.9	11.1
100,000	25.0	20.0	16.7	12.5	10.0

Soil water conditions around the seed are of equal or more concern to the corn grower than is soil temperature. Usually, the seedling depth and soil conditions are manipulated by the grower to enhance water absorption by the seed and to decrease evaporation. While much is known about water conditions required for seed germination, plant emergence, and water loss by evaporation, the situation in the field is so complex that judgement and experience in a given situation is all important. Most corn is probably planted at depths of from 2 to 7 cm. Because soil dries from the surface downward, corn is often planted deeper as the season advances and soil temperatures increase.

Population

Corn populations for maximum economic yield of grain vary with hybrid or genetic materials, row widths, soil fertility, soil water, and climatic affects. A large number of research publications have dealt with determining the optimum plant population for a given hybrid under a certain environment. The results show that optimum populations vary from about 40,000 pph (plants per ha) to over 100,000 pph. The distances between rows and between plants within rows to produce different planting rates are given in Table 2. The most favorable populations tend to be lower in areas south and west of the eastern Corn Belt and higher in the north and east.

Corn plant population in the four Corn Belt States of Indiana, Iowa, Illinois, and Minnesota averaged 47,400 pph (19,200 plants per acre) in 1973 with lower average populations in Nebraska and Missouri (Iowa Crop and Livestock Rep. Serv., 1973).

The leaf area index (LAI) which expresses the ratio of upper leaf surface to ground surface is a useful index of potential plant interception of radiant energy. Because the leaf area per plant varies widely with hybrid and management practice, Brown et al. (1970) suggested that hybrid performance might better be compared with a constant LAI than with a given population. At present, a LAI of about 3.5 seems optimum over a wide variety of conditions. In many areas under a high management level, a population of about 50,000 pph will produce a LAI of about 3.5.

Eik and Hanway (1966) showed that maximum yields of corn grain were obtained with a LAI of 3.3. These conclusions were drawn from a number of experiments with several hybrids grown in 102-cm-spaced rows in Iowa. In North Carolina Nunez and Kamprath (1969) found a LAI of 3.5 was reached at about 50,000 pph as was maximum yields. The LAI increased linearly as plant population increased from 34,500 to 69,000 pph.

The leaf area per plant varies with hybrid, population, and growth conditions. Long season hybrids usually have larger leaf areas per plant than short season hybrids because the long season hybrids have more and larger leaves. Brown et al. (1970) in Georgia found that at a given population the leaf area was 49 dm^2 per plant for D-XL65 and 68 dm^2 per plant for P-309B. At Guelph, Ontario, five hybrids grown at 72,000 pph had a mean leaf area of 40 dm^2 per plant. Hicks and Stucker (1972) found that among 18 hybrids, width of the leaf above the ear varied from 8.6 to 11.2 cm and leaf length varied from 78.1 to 94.1 cm. For two hybrids LAI varied from 2.7 at a population of 39,520 pph to 5.5 at a population of 88,920 pph. Leaf area per plant was 68 dm^2 at the low population and 62 dm^2 at the high population.

Eik and Hanway (1965) found that leaf area per plant varied from 100 to 80 dm^2 per plant when pph varied (9,884 to 59,304). They also present data on the effect of planting geometry and fertilizers on the number and area of leaves per plant.

Nunez and Kamprath (1969) found that leaf area per plant decreased from 76 dm^2 per plant at 34,500 pph to 66 dm^2 per plant at 69,000 pph at one location, and from 68 dm^2 per plant at 34,500 pph to 50 dm^2 per plant at 95,000 pph at a second location. Year and N fertilizer also influenced leaf area per plant and LAI.

Many studies have been reported where yields were related to plant populations, but where LAI was not measured. In Indiana on 11 location-yield trials, grain yields were slightly higher with 54,000 pph than with 69,000 pph (Stivers et al., 1971). Working with two hybrids under irrigation in Missouri for 2 years, Whitaker et al., (1969) found maximum yields at populations of 50,000 to 60,000 pph. In Virginia, Lutz et al., (1971) found maximum yields of 10 hybrids over a 3-year period were reached at about 50,000 pph, although this varied considerably with location, year, and maturity of hybrid. Further south Nunez and Kamprath (1969) found highest yields in North Carolina at 51,750 pph. In Georgia, however, optimum plant populations for two hybrids over a 2-year period varied from 45,000 to 103,000 pph under irrigated conditions and from 45,000 to 71,000 pph under nonirrigated conditions (Brown et al., 1970).

In the western Corn Belt where rainfall often limits growth, populations for maximum yield under nonirrigated conditions are lower than in more humid regions. Under irrigation they are high.

Stickler (1964) found maximum grain yields on nonirrigated land in Kansas at about 40,000 pph while irrigated yields were maximized at 48,000 to 59,000 pph. In western Minnesota and eastern South Dakota, Holt and Timmons (1968) at 32 location-years found that the population-maximum yield relationship shifted to higher populations as rainfall increased on nonirrigated sites. In all cases, however, the slopes of the curves changed gradually, so only small decreases in estimated yields occurred when stands shifted by 5,000 pph on either side of the maximum. For the hybrid used,

39,500 to 44,500 pph produced highest yields in 3 of the 4 years. During the year with the most favorable precipitation, 54,400 pph gave the highest yield.

In northern states and Canada, optimum populations have been higher than further south. In Minnesota at two locations in 2 years when soil water was adequate, yields of corn increased progressively from 44,460 to 74,100 pph, the highest population tested (Hicks et al., 1970). The higher populations were more advantageous with early planting (Fig. 8).

Giesbrecht (1969) found the corn grain response curve of four hybrids had not reached a maximum at 75,000 pph in Manitoba, Canada. Under favorable environmental conditions in Wisconsin, grain yields of four adapted hybrids increased up to 60,000 pph (Andrew and Peek, 1971).

Plant Spacing

Row Spacing

There is general agreement that corn yields increase as row width decreases from approximately 100 to 50 cm, if plant populations are relatively high. The yield increases are not dramatic but in most cases are economical. Yield improvements are greater for narrow row spacings under high yield conditions than at low yield levels.

Row spacings for corn grain production averaged about 92 cm in the Corn Belt in 1972 with more than 20% of the fields having row widths less than 82 cm (Iowa Crop and Livestock Reporting Service, 1973).

Stivers et al. (1971) in Indiana at four widely scattered locations, and 11 location-years, found an average corn grain increase of 7.3% with rows 51-cm wide, and 4.4% with rows 76-cm wide as compared with rows 102-cm wide. Stickler (1964) found corn grain yields in Kansas in 51-cm rows exceeded yields in 102-cm rows by 6% under irrigation and by 5% without irrigation. Corn grain yields were increased an average of 6% when rows were narrowed from 102 to 76 cm in Wisconsin with greater differences for the dry year, late plantings, high populations, and late hybrid (Andrew and Peek, 1971).

Hoff and Mederski (1960) found that the yield superiority of equidistant spacing over 107-cm rows was from 376 to 628 kg/ha. The yield advantage was greater as the population increased to about 50,000 to 60,000 pph.

In two sets of experiments in Canada, small or no increases in yield from narrow rows were found when short season hybrids were grown at high populations (i.e., about 70,000 pph). Hunter et al., (1970) found about a 3% increase in yield in 46-cm rows as compared with 96-cm rows while Giesbrecht (1969) found no corn yield difference among 50, 65, 80, and 95-cm wide rows.

Increased yields with narrow rows particularly at high plant populations apparently can be explained by greater solar energy interception. Denmead et al., (1962) estimated that the energy available for photosynthesis could be 15 to 20% more in an equidistant spacing as compared with the same population in 102-cm rows. Yao and Shaw (1964) found that the ratio of net radiations at the ground to that above the crop decreased as row spacing decreased and that yield and water use efficiency increased as row spacing decreased.

Within-row Spacing

Colville and McGill (1962) found drilling on irrigated corn in Nebraska resulted in corn yields much superior to those from either checking or hill dropping. Yields from drilling ranged from 1,406 to 615 kg/ha greater than checking on individual experiments with yields from hill-dropping intermediate. The population-yield curve for drilled corn was maximum at about 59,000 pph or about 10,000 pph greater than for checked corn in 102-cm rows.

Rossman and Cook (1966) stated that increases in corn grain yield range from 0 to 13% in favor of drilling as compared with check-row hills. The trend is for larger differences at higher populations and at higher yield levels. Using 76-cm rows with 54,000 pph, Erbach et al., (1972) found that intrarow spacing as influenced by two planters did not have a significant influence on corn yield in Iowa.

Planting for Silage

For silage production the objective usually is to maximize total digestible nutrient (TDN) production per unit area. Since corn will continue to photosynthesize, dry matter will continue to accumulate until maturity. After about 2 weeks following silking, all increases in dry matter accumulation are in the grain. Because the grain is the most digestible portion of the plant, maximum TDN is greatest when the corn is mature (black-layer stage). At the black-layer stage, however, corn may be too dry for best ensiling and sometimes the leaves may have dried and dropped. A somewhat longer-season hybrid can be used for silage than for grain production since drying of the grain is not of concern for silage. A hybrid that is physiologically mature at the average frost date should be used. If there is an early frost, grain drying will not be of concern as it would be in grain production. Hybrids that yield well under high populations are preferred for silage.

For maximum silage yield plant populations are usually somewhat, but not greatly, higher than that for corn harvested for grain. The average yield of grain, fodder, and total dry matter for one hybrid as influenced by population is given in Fig. 9. According to the response surface computed by Whitaker et al. (1969) fodder yield was still increasing at 69,000 pph, the highest population tested. Grain yield, however, peaked at about 59,000 pph and total dry matter (grain plus fodder) was still increasing at 69,000 pph.

In six location-year experiments at Blacksburg and Orange, Virginia, corn silage (dry matter) production was usually greater at the medium population level (either 54,400 or 61,800 pph) than at the low level (either 39,500 or 44,500 pph). In one experiment a significantly higher yield was obtained at 74,100 pph than at two lower populations. Yields tended to be higher in 40-cm wide rows than in 80 or 100-cm wide rows. Yields in alternated 25 and 125-cm spaced rows were lower than in uniform spaced rows. Late season hybrids outyielded medium or early season hybrids.

Cummins and Dobson (1973) found that total dry matter production of corn was greater in 51-cm as compared with 102-cm rows and increased from

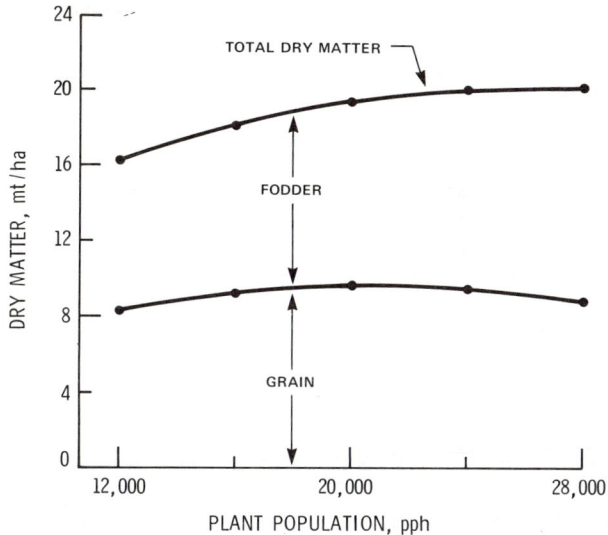

Fig. 9. Effect of corn plant population on yield of grain and stover (Whitaker et al., 1969).

49,000 to 86,000 pph in the Piedmont area of Georgia. However, in vitro dry matter digestibility was highest at populations of 68,000 pph. In the Piedmont there was some reduction in ear content and increase in stalk content as populations increased, but whole plant digestibility was not affected. In the mountain areas as populations increased from 30,000 to 74,000 pph, no change in vitro dry matter digestibility occurred. Row spacing did not change digestibility at either location.

Rutger and Crowder (1967) reported maximum dry matter yields at 88,000 pph in New York but did not find 46-cm wide rows to be superior to 92-cm rows at the 88,000 pph level. Bryant and Blaser (1968) obtained higher yields of corn silage in Virginia at 98,800 than at 39,500 pph, but the 98,800 population did not yield higher than 49,400 or 66,700 pph. Responses to row spacing were inconsistent. Doss et al. (1970) found corn dry matter yields increased about 12% by increasing population from 37,050 to 74,100 pph. Decreasing row spacing from 102 to 51 cm increased dry matter yields of leaves and stalks but decreased ear yields. Rumawas et al. (1971) in Indiana found no difference in dry matter yields due to row spacing (50 and 75 cm).

Mishra et al. (1963) in Illinois found that growing two crops of extremely high population corn (494,000 or 200,000 pph) produced more dry matter and total crude protein than did one crop grown for grain in the conventional manner. However, because silage qualities were poor the authors felt the practice had limited application for silage but might be useful for green chop programs.

Fertilizer

Commercial fertilizers are used extensively to supplement the nutrients present in soils. Use of fertilizers has increased rapidly in recent years and fertilizers now supply more than 100 kg N, 25 kg P, and 50 kg K/ha in many

Table 3—Fertilizer use on corn harvested for grain, selected states, 1973†

	N		P		K	
	Hectares-%	Rate-kg/ha	Hectares-%	Rate-kg/ha	Hectares-%	Rate-kg/ha
Pa.	96	98.0	95	33.9	94	53.7
Ohio	96	112.1	93	38.9	94	75.9
Ind.	98	128.7	98	37.6	98	97.2
Ill.	94	138.2	89	34.4	88	77.0
Mich.	98	97.3	98	37.3	97	70.9
Wis.	98	84.6	96	29.5	97	79.7
Minn.	93	110.0	90	31.4	89	67.4
Iowa	93	122.1	84	29.1	80	50.8
Mo.	95	141.7	86	26.9	85	56.7
S. D.	63	69.2	55	15.1	27	15.0
Neb.	92	156.0	74	20.7	50	19.7
Kans.	96	172.4	65	26.9	42	27.2
Del.	98	113.8	96	36.2	98	93.6
Md.	99	124.0	99	41.6	98	87.6
Va.	100	139.4	98	39.0	98	96.7
N.C.	95	160.9	71	27.6	71	65.8
Ga.	98	131.4	97	25.3	98	69.7
Ky.	98	134.8	92	38.1	91	74.0
Colo.	96	186.8	76	39.5	19	40.3

†From Crop Production, February 1974. Crop Reporting Board, ARS, USDA. Rates were converted from pounds/acre to kg/ha.

states (Table 3) (Hargett, 1973).

The most appropriate time and method of fertilizer application varies with the materials used and the cultural practices and preferences of the individual farmer. Fertilizers containing N, P, and K, applied in bands 5 cm to the side and 3 to 5 cm below the seed at planting, provide early season growth stimulation especially beneficial in cooler regions and for early plantings, even on fertile soils and may be used for maintaining adequate nutrient levels in soils. Nutrient deficiencies of the soils, which limit later season growth, may be corrected by applications that are broadcast and disked in or plowed under prior to planting, injected into the soil as a gas either before or after planting, or sidedressed between the rows after planting. Foliar application shows promise of being an efficient method of nutrient application after leaf development is complete. Alternative methods will be discussed for each nutrient.

Soil tests and plant analyses are used extensively as guides for estimating fertilizer needs for corn (Walsh and Beaton, 1973). Both techniques involve sampling of the soil or plants in the field, preparation of the samples for analysis, analysis of the samples, and interpretation of the analytical results. Before any test can be used effectively, meaningful correlations must be developed between the test results and corn yield response to fertilizer applications. The reliability of the fertilizer recommendations depends upon the adequacy of information available and the techniques used in sampling and testing.

Nitrogen

Important sources of fertilizer N include anhydrous ammonia, urea, ammonium nitrate, N solutions, ammonium sulfate, and ammonium phosphates. Under most conditions, all forms of N fertilizer are equally effective per unit of N. Nitrogen fertilizer can be applied in the fall, winter, or spring before the crop is planted, or it can be applied between rows as long as machinery can pass over the corn. Some N is applied in irrigation waters with either gravity or sprinkler applications. Foliar spraying of urea solutions after leaves are fully developed may be an efficient and effective method of supplying part of the N required.

Because most forms of fertilizer N will quickly be nitrified into the mobile nitrate ion in warm, moist soils, the effectiveness of different forms of fertilizer N when applied to soils will depend upon conditions relating to possible leaching or denitrification of nitrate N. Nitrification and denitrification processes occur in soils at temperatures above 0 C, but are relatively slow at temperatures below 10 C (Brenner and Shaw, 1958). The longer the N is in the soil before it is needed by the crop, the greater the chance for loss. Nitrate forms are frequently inferior under conditions where considerable leaching or waterlogging may occur. Under leaching conditions the soluble nitrate may be removed from the root zone and under waterlogging the nitrate may be denitrified to gaseous forms of N (N_2 or NO). Ammonium sources of N may form unstable ammonium carbonate and be volatilized as ammonia (NH_3) if applied to the surface of alkaline soils. The amounts of available N in soils may be reduced by incorporation into soil organic matter during decomposition of low N crop residues.

The efficiency of fall preplant and sidedress applied N for improving corn yields has been debated for some time. Recent results in humid or irrigated regions all agree that in some years, fall-applied N will be less effective than preplant or sidedress treatments and that sidedressed N may be more effective than preplant N especially on coarse-textured soils.

Spring-applied N produced from 370 to 2,610 kg/ha higher corn yields on clay soils, and 200 to 1,160 kg/ha higher corn grain yields on Brookston (Typic Argiaquolls) and Haldimand soils than fall-applied N during several years in Ontario. Ammonium, urea, and anhydrous ammonia all reacted the same with regards to fall vs. spring applications (Stevenson and Baldwin, 1969). In Illinois, fall-applied N was about 80 to 90% as effective as spring-applied N at application rates of 67 and 134 kg/ha but about equally effective at rates of 201 and 268 kg/ha at three of four locations. Corn grain yields were higher for sidedressing than for spring application in certain years at some locations (Welch et al., 1971). At several locations in southeastern United States, Pearson et al. (1961) found fall-applied N only about one-half as effective as spring-applied N. On a Norfolk soil (Typic Paleudults) in Georgia, Boswell (1971) found sidedressing of N produced greater corn yields than did spring plow down.

On 14 irrigated soils in Nebraska, N sidedressed was clearly superior to either fall or spring application in producing corn grain regardless of chemical N form (Olson et al., 1964). On a sandy, irrigated soil in Wisconsin, Jung et al. (1972) found that N from three sources applied during either the 5th, 6th, 7th, or 8th week after planting was more effective in increasing corn grain and tissue yields than N applied either earlier or later.

Stanford (1965) estimates that the N content of the above-ground portion of the mature corn plant should contain 1.2 to 1.3% N. The total N required for a corn crop can be estimated by multiplying the desired yield (grain plus fodder) by 1.25%. Then the nitrogen fertilizer needed can be calculated from the equation:

$$N_f = \frac{(Y_p \times 0.0125) - N_s}{E}$$

where
N_f = nitrogen fertilizer needed in kg/ha
Y_p = yield potential in kg/ha (grain and fodder)
N_s = available nitrogen from the soil in kg/ha
E = fractional efficiency of nitrogen, or percent recovery of added nitrogen

The fractional efficiency or percent recovery of added mineral N has been found to range between 0.30 and 0.70 (Stanford, 1965) and will probably be near 0.5 when N is applied as a sidedressing on soils that are aerobic and leaching is not excessive.

A major obstacle in using the above equation is the difficulty in obtaining satisfactory values for N_s. At present, soil tests which provide the proposed indexes of N availability are based on the amount of ammonium or nitrate produced on incubation or the amount extracted from a soil by chemical extractants (Bremner, 1965; Jenkinson, 1968; Stanford and Demar, 1970). None of the proposed methods are widely used. A few soil testing laboratories use an incubation or organic matter determination for making N fertilizer recommendations. Many states base their recommendations entirely upon past management of the soil and general knowledge of the soils' responsiveness to N fertilizer (Kurtz and Smith, 1965).

Plant tissue tests can be used as a tool for assessing the N status of corn. The leaf N levels at maximum or 95% of maximum yield as suggested by various researchers, is given in Table 4. While the levels vary for all nutrients, certainly if the nutrient level is at or below the levels given in Table 4, one should suspect possible nutrient deficiencies and perhaps do additional diagnostics.

Voss et al., (1970) point out that the corn grain yield-leaf nutrient content curve rises sharply and then becomes nearly flat near the point of maximum yield. They suggested that the leaf concentration at 95% of maximum yield is a practical level for setting critical leaf concentrations. They emphasize that no exact leaf nutrient concentration level below which corn responds to fertilizer can be defined without defining a number of other variables. Multivariate regression models using %N, %P, %K, and interaction terms as variates did not explain differences in yields from experiments in western Iowa. However, if other production variates, such as past cropping, plant populations, soil yielding potential, and soil water were added, corn yields could be predicted with fair accuracy.

Phosphorus

Phosphorus fertilizer sources for corn are primarily superphosphates and ammonium phosphates. Superphosphates are most widely used as a single source of P. The ordinary superphosphate contains about 8.8% P (20% P_2O_5) and the concentrated superphosphates contain 20 to 22% P (45 to 50% P_2O_5). Ammonium phosphates include a wide variety of materials produced by ammoniation of phosphoric acid. Polyphosphate materials which offer much promise are being developed and made available.

Table 4—Adequate ranges of nutrients by tissue analysis of ear leaf at silking (from Barber and Olson, 1968)

Nutrient	Concentration range
N	2.75-3.25%
P	0.25-0.35%
K	1.75-2.25%
Ca	0.25-0.40%
Mg	0.25-0.40%
S	0.1-0.2 %
Fe	20-250 ppm
Mn	20-150 ppm
Zn	20- 70 ppm
B	4- 20 ppm

Phosphorus applications for corn should either be broadcast applied and plowed under or disked in at the time of tillage in the fall or spring, as a row application at the time of planting, or both. When soluble fertilizer P is added to soil, it reacts rapidly with iron and aluminum in very acid soils and with calcium in other soils reverting to less soluble forms and moves slowly in the soil. Therefore, time of application and placement of P fertilizers are important. If the soil has only a moderate P fixing capacity and adequate amounts of P are applied, P applications can be made safely at some time prior to planting. If the soil has a large P fixing capacity, considerable amounts of P can be applied in a band about 5 cm to the side and 5 cm below the row.

Because a large number of factors enter into the amounts of P used and needed, it is impossible to discuss briefly rates of P needed for corn. Detailed information on P needs can be obtained within each state.

The phosphorus fertilizer needs of corn are usually estimated by use of a soil chemical test and associated corn response together with such other factors as yield desired, solubility of the phosphorus in the fertilizer, and method of application. Most soil testing laboratories use dilute acid-fluoride, dilute hydrochloric or sulfuric acid, water, or sodium bicarbonate as soil P extractants (Olsen and Dean, 1965).

Phosphorus availability in the subsoil may be quite different than in the surface horizon. Accuracy of a phosphate fertilizer recommendation can often be improved by considering a P soil test result from both the surface and subsoil.

Suggested levels of P in corn plants are shown in Table 4. Since the optimum P concentration in corn has been shown to vary with N level, genotype, and environmental factors, these variables should be considered in interpreting plant analysis results (Walsh and Beaton, 1973).

Potassium

Potassium chloride is the main form of K used for corn either as a straight or mixed fertilizer. Potassium sulfate although equally effective is less generally found on the market.

Potassium fertilizer should be applied for corn as a broadcast application and plowed under or disked into the soil before planting, or as a side band placement along the row at the time of corn planting. Potassium may be applied every year or in larger amounts once in several years if the soil is one that does not fix large amounts of K. Barber (1970) found that when 168 kg K/ha were applied, the K effectiveness was about one-half the 2nd year following application and the effectiveness was negligible during the 3rd year. When 660 kg K/ha were applied, the added K was still effective the 4th year after application.

When application rates are relatively low, row-placed K is frequently more efficient than broadcast K. At high rates of application, corn response as influenced by application method is usually slight. Welch et al. (1966) found that surface-applied K in Illinois was 0.33 to 0.47 as efficient, and when plowed down, K was 0.69 to 0.88 as efficient as row application. Surface broadcast applications at rates up to 84 kg K/ha were approximately 73% as effective as row placement in Tennessee. At higher rates, the differences disappeared (Parks and Walker, 1969). In Indiana, Barber (1959b) found no difference between K efficiency when up to 224 kg/ha were plowed down or row applied.

Potassium removal in grain is relatively small; when only grain is harvested much of the potassium taken up by the plants is returned to the soil and made available for following crops. Potassium deficiencies are most likely to occur after forage crops, when K has been removed in the harvested hay or silage.

Availability of soil K is highest in the surface soil (Hanway et al., 1962). Most subsoils are K deficient. Because K is held as an exchangeable ion on soil clay, plant availability is directly proportional to the amount exchangeable. Except in sandy or organic soils, very little of the soil K is in water soluble form so K moves only very short distances in soils unless the sorption capacity of the soil is K saturated. Added fertilizer K is readily absorbed on the clay surfaces.

The need for and amount of K fertilizer for corn is usually estimated by extracting a prepared soil sample with neutral normal ammonium acetate and determining the amount of K therein. Ammonium acetate extracts the K in solution as well as that removed from the soil by exchange with the ammonium ion. Extraction with cold, dilute sulfuric acid has also been used (Pratt, 1965).

The predicted increase in yield of corn grain from different rates of applied K as influenced by the level of exchangeable K in the field-moist 0 to 15-cm layer of soil is given in Fig. 10. This figure was developed from 41 cooperative field experiments in the North Central Region (Hanway et al., 1962).

Optimum levels of K in corn plants, as suggested by different researchers, are shown in Table 4.

Other Nutrients

Zinc deficiencies occur in corn on some soils (Murphy and Walsh, 1972). Deficiencies are most likely to occur on soils when the surface soil has been removed in leveling for irrigation or in terracing and corn is grown on subsoil, on calcareous soils, or on soils high in available P. Zinc may be applied in a mixed fertilizer at planting or be broadcast and plowed under prior to planting. One application may be effective for several years. Recommended

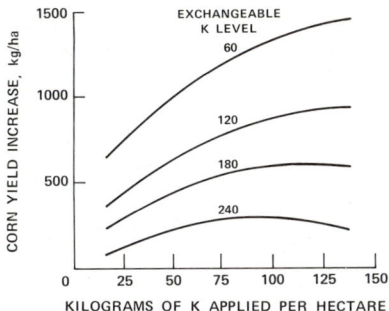

Fig. 10. Predicted increase in yield of corn grain from different rates of applied K as influenced by the level of exchangeable K in the field-moist 0 to 15-cm layer of soil. The values on the curves are exchangeable K in kg/ha (Hanway et al., 1962).

rates of Zn application for corn have ranged from 0.7 to 22 kg/ha depending upon soil conditions and chemical form of Zn fertilizer.

Sulphur deficiencies in crop plants have been reported from many parts of the world. Responses of corn to applied S have occurred on some soils in most of the corn producing areas. These responses generally have occurred on soils that were low in organic matter, sandy in texture, and well-drained (Beaton and Fox, 1971). The need for S fertilizer is closely related to the amounts of N fertilizer applied. Rates of application of 20 to 25 kg S/ha are usually adequate. Sulfate salts of Ca, Mg, NH_4, or K and elemental S are the main sources of S. Sulfur for corn is usually broadcasted or banded, mixed or combined with other fertilizer materials.

Response from application of Fe, B, Mg, and Ca have been observed, but the likelihood of obtaining yield increases of corn from applications of these nutrients is extremely small, except on very unusual soils.

Previous Crop and Erosion Control

Corn yields on some soils may be significantly increased or decreased by the previous crops depending on the soil and weather conditions. Barber (1959a) showed that on a Chalmers silt loam (Typic Argiaquolls) in Indiana, corn yielded more when grown after alfalfa (*Medicago sativa* L.) and smooth bromegrass (*Bromus inermis* Leyss) than when grown after corn. He felt that differences due to water or nutrient levels were small since the soil was always at field capacity or higher in the spring and adequate fertilizer was applied. In a second report, Barber (1972) found that corn yields decreased progressively with years away from a forage crop.

On a Cecil sandy loam (Typic Hapludults) in Georgia, Adams et al. (1970) found that corn yields after sod and annual green manure crops were greater than continuous corn even though as much as 180 kg N/ha had been applied to the soil. The beneficial effects were apparent for 3 to 4 years after the sod. Sod crops used were alfalfa, fescue, and bermudagrass, and green manure crops were rye (*Secale cereale* L.) and vetch *(Vicia)*. Soil physical conditions improved and decreased nematode populations occurred following sod.

A sod crop prior to corn is not always beneficial. Shrader and Pierre (1966) showed that corn yields in Iowa were equal on four soil types when following a grass-legume or when grown continuously to corn if adequate N and other fertilizer were used. The soils were Kenyon silt loam (Typic Hapludolls), Webster silt loam (Typic Haplaquolls), Marshall silt loam (Typic Hapludolls), and Belinda silt loam (Mollic Albaqualfs).

In drier areas corn yields may be decreased after forage crops due to lower soil water or nutrient levels in the soil (Shrader and Pierre, 1966). Voorhees and Holt (1969) showed that the depressions in corn yield after alfalfa could be eliminated by killing the alfalfa in the summer or fall before the corn and thus allowing time for the soil to be recharged with water.

Water erosion of soil cropped to corn has been an extreme hazard to long-time productivity on sloping land. Soil erosion not only carries away the more fertile top soil but also causes stand reduction and loss of water that potentially could be stored in the soil. Extreme erosion may form gullies thereby reducing effective field size. At one time it was not considered desirable to grow corn on sloping soils without meadow crops in the rotation or without strip cropping or terracing. However, new tillage practices have increased the management choices for corn culture without causing excessive erosion.

Mannering et al., (1968) found that water infiltration rates into soils during 1st and 2nd-year corn after a grass-legume meadow were 32 and 26% greater, respectively, than during continuous corn but were not different from continuous corn in subsequent years. Soil losses by erosion from 1st, 2nd, 3rd, and 4th-year corn after meadow were 53, 83, 90, and 97% respectively, of that from continuous corn. Greater soil aggregation accounted for most of the greater infiltration rate and decreased soil loss. The experiment was conducted on a Russell silt loam (Typic Hapludalfs) with a 5% slope.

Mannering and Johnson (1969) found that soil loss and infiltration were not influenced during the first 5 weeks after planting by varying row spacing from 51 to 102 cm at populations of 60,000 pph. However, at 7 to 8 weeks after planting ground cover was increased slightly and soil loss from simulated rainfall was reduced 24% by the 51-cm rows as compared with 102-cm rows. The rows were up and down-slope in their studies. Narrow rows on the contour might be more effective because of greater depression storage of water in tillage tool marks.

An equation for predicting soil erosion loss by rainfall has been developed and is widely used in the United States and foreign countries (Wischmeier and Smith, 1965). The equation is:

$$A = RKLSCP$$

where:

A is the average annual soil loss predicted by the equation,
R is a factor expressing rainfall energy,
K is a factor concerning the inherent erodibility of the soil,
LS is the length and steepness of slope factor,
C is the cropping and management factor, and
P is the supporting conservation practice factor (terracing strip cropping, contouring).

Values for the factors applicable to different regions of the country have been worked out and are available at Soil Conservation and Extension Service offices.

From the equation and suitable factor values, soil loss for most situations can be estimated from the Universal Soil Loss Equation (Wischmeier and Smith, 1965). If water erosion is severe, protective practices can be inserted into the management scheme to bring the soil loss down to safe levels.

Weed Control

The necessity of destroying weeds in cultivated fields has been recognized since cultivation began. Growth of weeds in corn fields reduces corn growth because of competition among the plants for water, light, and nutrients and because of allelopathy (release of phytotoxic substances from weeds). Usually we have thought of weeds as growth of noneconomic plants within planted corn fields. More recently, however, some of our most troublesome weed problems are the growth of a volunteer economic plant within a planted economic crop. Volunteer corn in soybeans (*Glycine max* L.) is an example.

Interference of Weeds with Corn

The interference of weeds with corn and the resultant reduction in yield of corn has been well documented (Behrens and Lee, 1964; Dunham, 1964; Knake, 1962; Wimer and Harland, 1925). The corn grain yield reduction from interference by weeds in a series of 10 experiments in Minnesota ranged from 16 to 93% with an average of 51%. In these trials unweeded treatments were compared with weed-free treatments. However, even when fields are mechanically cultivated the yield reduction due to weeds ranged from 4 to 41% with an average of 17% (Behrens and Lee, 1964).

Some weeds interfere more with corn than others (Knake, 1962) because of differences in water requirement, nutrient uptake, and light competition (Behrens and Lee, 1964). Vengris et al (1955) found a number of weed plants that contain greater concentrations of N, P, K, Ca, and Mg per unit of dry weight than corn. Weeds can exude substances that are phytotoxic to corn (Bell and Koeppe, 1972) but direct evidence of the phytotoxic effect is difficult to obtain. In most reported field experiments allelopathy cannot be ruled out since no way is known to separate competition from allelopathy.

Even when ample amounts of water and nutrients are supplied and light is the same as weeded treatments, weeds and other plants can reduce corn grain yields.

Mechanical Cultivation

Cultivation of corn after planting to destroy or reduce growth of weeds has been accomplished with shovel cultivators, rotary cultivators, disk cultivators, rotary hoes, spike tooth and spring tooth harrows, and other implements. While weed control has been the primary purpose on some soils, other benefits accrue (Meyer and Mannering, 1961; Van Doren and Triplett, 1965).

Weeds are easiest to control mechanically when they are small, if possible even before they can be seen above ground. The most efficient implements for early cultivation after most seedbed preparation methods for corn are the rotary hoe and rotary cultivator.

The rotary hoe and rotary cultivator perform best when the ground is slightly crusted and when most of the weeds have not yet emerged or are not more than 1 cm high. Most annual weeds germinate within 2 cm of the soil's surface, whereas corn is planted deeper. In addition, corn roots penetrate deeper soon after emergence than do most annual weeds. Shallow cultivating tools, such as the rotary hoe, can kill weeds without disturbing the crop. The rotary hoe can pass over corn several times until it is about 15 to 20 cm tall, without severely damaging corn. The rotary cultivator combines tillage wheels near the row with a shovel in the center between the rows. Surface cultivation implements are not effective in controlling perennial or annual weeds once they have established a good root system.

The shovel cultivator is the most common and effective implement used for controlling weeds after the corn gets about 15 cm tall. The shovels should penetrate only deep enough to kill the weeds and not be close to the corn plants. Deep cultivation and cultivation near the row cut the corn roots.

Disk-type cultivators and cultivators with larger than normal sweeps are useful after strip-tillage and no-tillage planting systems or where considerable crop residue or vine weeds occur.

The number of mechanical cultivations with shovel or sweep-type cultivators has generally been decreasing because of greater use of initial cultivating implements (such as the rotary hoe) greater use of herbicides, and improved seedbed tillage methods. Two shovel- or sweep-type cultivations are commonly used; with reduced seedbed preparations, one cultivation is often sufficient. There is usually no advantage in cultivating corn after it is 60 cm tall when weeds have been controlled earlier.

Aiding water intake and reducing runoff are sometimes important functions of corn cultivation. Cultivations where few weeds are present may be beneficial through loosening or breaking the crust on soils that have become compacted or sealed at the surface after rain. On Russell silt loam (Typic Hapludalfs) in Indiana on a 5% slope planted to corn, runoff during a 13.2 cm simulated rain resulted in 4.6 cm of runoff following one cultivation and 7.4 cm of runoff where the soil had not been cultivated. Soil erosion loss was 14.8 and 33.4 metric tons/ha for cultivated and uncultivated, respectively (Meyer and Mannering, 1961).

From their work in Ohio Van Doren and Triplett (1965), classifying soils into crusting and noncrusting types, concluded that on crusting soils without mulch cover, corn grain yields will be improved because of a cultivation even though the field is weed free. The benefit will be greater if the soil the previous year was in a row crop as compared with a sod crop. The benefit is apparently due to greater water intake and penetration. On noncrusting soils a yield benefit is slight if the soil has been tilled with moldboard plow and disked twice and, indeed, yield may be decreased if a reduced tillage system with plowing only was used. The reduced yield is presumably due to greater evaporation.

Table 5—Common and chemical names of herbicides used to control weeds in corn

Common name or designation	Chemical name
Alachlor	2-chloro-2',6'-diethyl-N-(methoxymethyl)acetanilide
Atrazine	2-chloro-4-(ethylamino)-6-(isopropylamino)-s-triazine
Butylate	S-ethyl diisobutylthiocarbamate
Cyanazine	2-[(4-chloro-6-(ethylamino)-s-triazin-2-yl)amino]-2-methylpropionitrile
Dicamba	3,6-dichloro-o-anisic acid
Linuron	3-(3,4-dichlorophenyl)-1-methoxy-1-methylurea
Paraquat	1,1'-dimethyl-4,4'-bipyridinium ion
Propachlor	2-chloro-N-isopropylacetanilide
2,4-D	(2,4-dichlorophenoxy)acetic acid

Pairs of disks, one on each side of and next to the corn row (disk hillers) in combination with sweeps are highly effective when used on the contour for ridging soil around larger corn plants and leaving furrows between the rows. The furrows trap water for later intake into the soil and reduce runoff. In the till-planting system the ridged rows provide a warmer and drier area for planting the next crop (Wittmus et al., 1971).

Chemical Control

The use of selective organic chemicals for weed control was initiated about 30 years ago with the introduction of 2, 4-D (2, 4-dichlorophenoxy acetic acid) as a post-emergence spray for control of broadleaf weeds. Since that time many chemicals have been tested and used, and many more are still in the experimental stages.

Development of selective herbicides has not only drastically changed methods for weed control, but also has made possible development of many other modern practices, such as reduced and no-tillage. The use of herbicides in the present and future will have a large influence on every facet of corn production.

Common and chemical names of herbicides now used to control weeds in corn are given in Table 5. Because new herbicides and use of herbicides are developing so rapidly, we will not attempt to recommend a specific herbicide.

Herbicide applications on corn land are often classified according to the time of application: (a) preplanting treatments are those applied prior (1 or 2 weeks) to planting the crop and are often incorporated into the soil; (b) pre-emergence treatments are those applied just after planting and until corn and weed emergence; (c) post-emergence treatments are applied after the corn emerges and usually until the corn is too tall for use of a field tractor. The time of application depends on the weed problem and herbicide used. Post-emergence treatments made immediately after the last cultivation are commonly called layby applications.

Atrazine and butylate are the most widely used preplanting herbicides. Atrazine effectively controls most broadleaved and grass-weed seedlings. It performs well either incorporated or left on the soil surface. Butylate, which requires incorporation soon after application for best results, is very effective on grass seedlings. The combination of atrazine plus butylate, applied preplanting and incorporated, controls a broader spectrum of weeds than either atrazine or butylate alone. Preplanting incorporated treatments offer the opportunity for applying of herbicides, insecticides, and fertilizers simultaneously.

Some of the most effective and widely used herbicides for pre-emergence application in corn are alachlor, atrazine, propachlor, and cyanazine. Alachlor and propachlor control annual grass weeds and some small-seeded broadleaf weeds. Atrazine controls annual broadleaf weeds and many annual grass weeds. Combinations of alachlor or propachlor with atrazine or cyanazine effectively control essentially all troublesome annual weeds in corn. Linuron, for control of broadleaf weeds, is sometimes used in combination with alachlor or propachlor.

Post-emergence treatments can control (a) weeds that germinate after pre-emergence treatments have lost their effectiveness, (b) susceptible weeds that escape pre-emergence control, and (c) weeds that are not susceptible to pre-emergence control. They can both supplement and complement the use of pre-emergence treatments and other weed-control practices.

Post-emergence applications are only slightly more complicated than pre-emergence operations. Most applications to corn less than 15 to 25 cm tall are made as overall sprays. As the corn grows taller, the chances of injury from overall sprays increase with some herbicides. Directed sprays help minimize the risk.

The most effective and widely used post-emergence treatments in corn are 2,4-D, dicamba, atrazine, and cyanazine. Many kinds of broadleaf weeds are controlled with 2,4-D, perhaps the most widely used post-emergence treatment. Dicamba is similar to 2,4-D in that it controls many kinds of broadleaf weeds. On some species, it is considerably more effective than is 2,4-D.

The effectiveness of soil applied herbicides depends upon a number of soil and climatic conditions.

Most herbicides are absorbed by the soil. Adsorption increases as the amounts of clay increase (Talbert and Fletchall, 1965) and amounts of organic matter increase (Upchurch and Mason, 1962). Type of clay also influences adsorption (Weber et al., 1965). Adsorption decreases herbicide activity and its effectiveness so that larger applications may be necessary. Adsorption on the soil has its benefits, however, in that the leaching by water may be less (Sheets and Danielson, 1959) and longevity of the phytotoxicity of the herbicide in the soil may be increased.

Rainfall after herbicide application disperses the herbicide throughout the soil surface layers where most weed seeds germinate (Splittstoesser and Derscheid, 1962). Thus, rainfall is necessary to make herbicides most effective. Microbial deactivation of herbicides in soils is favored by warm temperatures and moist soil conditions.

The interactions of herbicides with soil and climatic factors are complex and thus require extensive research trials with local conditions.

Corn Harvesting and Storage

In contrast to pre-World War II practices when most of the corn was hand-harvested, nearly all of the corn is mechanically harvested. About three-fourths of the corn in the Corn Belt is mechanically harvested and shelled in the field (Iowa Crop and Livestock Rep. Serv., 1973). In the 1930's the labor requirement for production and harvesting of corn was over 160 man-hours per 10,000 kg. Since 1970 it has taken only 10 man-hours per 10,000 kg on the average and more efficient farmers have a man-hour requirement of about 2 (Agricultural Statistics, 1973). These changing patterns in harvesting have necessitated corresponding changes in handling, drying, and storage procedures.

Harvesting

There are three main types of mechanical harvesting equipment used: the picker, picker-sheller, and corn combine.

The picker snaps the corn from the standing stalks, removes the husks, and delivers the ears into a wagon pulled behind or to one side of the picker. About one-fourth of the corn in the Corn Belt was harvested by the picker in 1972 and the percentage is rapidly decreasing (Agricultural Statistics, 1973). If the corn ears are to be stored in a corn crib, harvesting should be delayed until the grain moisture content is between 20 and 25% depending upon the geographical area and width of the crib used. Field losses increase sharply after the grain has reached physiological maturity because of the increased incidence of dropped ears, stalk lodging, ear rot development, and shelling of the ears on the snapping rolls. Early harvest is particularly advantageous to avoid loss on corn damages from stalk, root, or ear rots and from insects, frosts or storms.

The corn combine is rapidly replacing the picker-sheller and the picker as harvesting machines in the Corn Belt. In 1972, about two-thirds of the acreage was harvested with the corn combine and less than 10% with the picker-sheller (Iowa Crop and Livestock Rep. Serv., 1973). The corn combine and picker-sheller picks, husks, shells, and delivers the corn into hoppers or trailing vehicles all in one operation.

In a survey of 118 corn combines used in Ohio in 1964 to 1966, Byg and Hall (1968) found that the average moisture content of the corn grain at harvest was 25%. Ear loss from pickers averaged 106 kg/ha and from corn combines 212 kg/ha.

Increasing amounts of corn are being harvested at 28 to 32% moisture contents and placed in sealed conventional or bunker silos for feeding to livestock on the farm. This system of management eliminates the cost of drying and may require less handling from field to feed bunk but has the disadvantage that the corn cannot be sold.

About 3.2 million ha (12%) of the total corn acreage is harvested for silage each year in the United States. In the northeastern states nearly one-half of the crop is used in silage (Agricultural Statistics, 1973). By far the greatest amount of silage is made by chopping the entire plant from all rows.

Some new innovations have appeared in making silage from the corn. Some farmers make silage from the ears alone, others from the entire center section of the plant (center-cut), and others which pick corn from one row and chop the entire plant from the adjacent row (pick-chop). Center-cut and pick-chop contains at least 40% grain. The ideal time to harvest silage corn for maximum feeding value is when the kernels are well dented and begin to glaze. By this time some of the lower leaves may be lost but the additional dry matter in the grain will more than offset the loss.

Some interest has developed in a harvester that will remove the aboveground plant, separating the mature shelled grain from the fodder, and delivering each to a separate trailing wagon (Hitzhusen et al., 1970). The fodder can be used with proper supplement as winter feed for animals.

Storage

Four basic methods are used to store corn. They include (a) dry corn, (b) high-moisture corn, (c) cold corn, and (d) chemically treated corn.

Corn that is field-shelled at high moisture contents, often 21 to 28%, must be artificially dried to be stored as dry corn. There are four widely-used different shelled corn drying methods: (a) batch dryers, (b) continuous-flow dryers, (c) bins equipped for in-storage or multiple-layer drying, and (d) bins equipped for batch drying (Iowa Crop and Livestock Rep. Serv., 1973).

About two-thirds of the corn grown in the Corn Belt was artificially dried in 1973. All but a few percent were dried on the farm.

Shelled corn with a moisture content of approximately 17 to 22% can be safely stored in conventional grain bins if the grain temperature is kept below 10 C. A good air distribution system is necessary for adequate storage of cold corn. Obviously, corn cannot be stored this way during warm months.

Shelled corn or ground ear corn can be stored as high-moisture corn (about 20 to 25% for shelled corn and 23 to 28% for ground ear corn) in sealed facilities and have equal or higher feed value than dry corn. For safe storage into warm weather, dry shelled corn must contain not more than 13 to 14% moisture, but it also should be protected against corn storage insects.

Recent research indicates that corn grain, up to 30% moisture, can be stored without mold growth and spoilage by means of chemical treatment. At present propionic acid and mixtures of propionic and acetic acid are the principal materials used but other chemical compounds are being tested. When the chemicals are properly applied, the acid-type preservatives kill the fungi (or molds) and related micro-organisms in grain, and continue almost indefinitely to prevent mold growth. The germ of the grain is also killed, so there is essentially no respiration.

The usual means of applying liquid preservatives is to spray the grain with the preservative as it moves through an auger. All of the grain must be treated with the preservative as an untreated pocket of wet grain can cause extensive spoilage. As this method of storage is new, recent recommendations on the use of chemical preservatives should be obtained from the agricultural extension service.

ACKNOWLEDGMENT

This chapter is a contribution from the Soil and Water Management Research Unit, North Central Region, ARS-USDA, St. Paul, Minn., in cooperation with the Minnesota Agric. Exp. Stn., Paper No. 8904, Sci. Journal Series, and the Iowa Agric. and Home Econ. Exp. Stn., Ames, Project No. 1900.

LITERATURE CITED

1. Adams, W.E., H.D. Morris, and R.N. Dawson. 1970. Effect of cropping systems and nitrogen levels on corn (*Zea mays*) yields in the Southern Piedmont Region. Agron. J. 62(5):655-659.
2. ----, ----, J. Giddens, R.N. Dawson, and G.W. Langdale. 1973. Tillage and fertilization of corn grown on Lespedeza sod. Agron. J. 65(4):653-655.
3. ----, J.E. Pallas, Jr., and R.N. Dawson. 1970. Tillage methods for corn-sod systems in the Southern Piedmont. Agron. J. 62(5):646-649.
4. Agricultural Statistics. 1973. USDA, US Government Printing Office, Washington, D.C.
5. Allessi, J., and J.F. Power. 1971. Corn emergence in relation to soil temperature and seeding depth. Agron. J. 63(5):717-719.
6. Allmaras, R.R., A.L. Black, and R.W. Rickman. 1973. Tillage, soil environment, and root growth. p. 62-86. *In* Conservation tillage: The proceedings of a national conference. Soil Cons. Soc. Am., Ankeny, Iowa.
7. ----, W.W. Nelson, and E.A. Hallauer. 1972. Fall versus spring plowing and related soil heat balance in the western Corn Belt. Minn. Agric. Exp. Stn. Tech. Bull. 283.
8. Amemiya, M. 1968. Tillage-soil water relations of corn as influenced by weather. Agron. J. 60(5):534-537.
9. Andrew, R.H., and J.W. Peek. 1971. Influence of cultural practice and field environment on consistency of corn yields in northern areas. Agron. J. 63(4):628-633.
10. ----, and S.S. Solanki. 1966. Comparative root morphology for inbred lines of corn as related to performance. Agron. J. 58(4):415-418.
11. Barber, S.A. 1959a. The influence of alfalfa, bromegrass, and corn on soil aggregation and crop yield. Soil Sci. Soc. Am. Proc. 23(4):258-259.
12. ----. 1959b. Relation of fertilizer placement to nutrient uptake and crop yield. II. Effect of row potassium, potassium soil levels and precipitation. Agron. J. 51:97-99.
13. ----. 1970. Residual effect of potassium fertilization of continuous corn on Chalmers silt loam. Purdue Univ. Res. Prog. Rep. 377.
14. ----. 1972. Relation of weather to influence of hay crops on subsequent corn yields on a Chalmers silt loam. Agron. J. 64(1):8-10.
15. ----, and R.A. Olson. 1968. Fertilizer use on corn. p. 163-188. *In* L.B. Nelson et al. (eds.) Changing patterns in fertilizer use. Soil Sci. Soc. Am., Inc., Madison, Wis.
16. Barnes, D.L., and D.G. Woolley. 1969. Effect of moisture stress at different stages of growth. I. Comparison of a single-eared and a two-eared corn hybrid. Agron. J. 61(5):788-790.
17. Bateman, H.P., and W. Bowers. 1962. Planning a tillage system for corn. Univ. Ill. Ext. Cir. 846.
18. Beaton, J.D., and R.L. Fox. 1971. Production, marketing, and use of sulfur products. p. 335-379. *In* R.A. Olson et al. (eds.) Fertilizer technology and use. Soil Sci. Soc. Am., Inc., Madison, Wis.
19. Behrens, R., and O.C. Lee. 1964. Weed control. p. 331-352. *In* W.H. Pierre et al. (eds.) Advances in corn production: principles and practices. The Iowa State Univ. Press, Ames.
20. Bell, D.T., and D.E. Koeppe. 1972. Noncompetitive effects of giant foxtail on the growth of corn. Agron. J. 64(3):321-325.
21. Benne, E.J., E. Linden, J.D. Grier, and K. Spike. 1964. Composition of corn plants at different stages of growth and per-acre accumulations of essential nutrients. Mich. Agric. Exp. Stn. Q. Bull. 47:69-85.
22. Bennett, O.L., E.L. Mathias, and P.E. Lundberg. 1973. Crop responses to no-till management practices on hilly terrain. Agron. J. 65(3):488-491.
23. Blevins, R.L., D. Cook, S.H. Phillips, and R.E. Phillips. 1971. Influence of no-tillage on soil moisture. Agron. J. 63(4):593-596.
24. Boswell, F.C. 1971. Comparison of fall, winter, or spring N-P-K fertilizer applications for corn and cotton. Agron. J. 63(6):905-907.

25. Bremner, J.M. 1965. Nitrogen availability indexes. *In* C.A. Black (ed.) Methods of soil analysis. Agronomy 9(2):1324-1345. Am. Soc. Agron., Madison, Wis.
26. ----, and K. Shaw. 1958. Denitrification in soil. II. Factors affecting denitrification. J. Agric. Sci. 51:40-52.
27. Brown, R.H., E.R. Beaty, W.J. Ethredge, and D.D. Hayes. 1970. Influence of row width and plant population on yield of two varieties of corn (*Zea mays* L.). Agron. J. 62(6):767-770.
28. Bryant, H.T., and R.E. Blaser. 1968. Plant constituents of an early and a late corn hybrid as affected by row spacing and plant population. Agron. J. 60(5):557-559.
29. Byg, D.M., and G.E. Hall. 1968. Corn losses and kernel damage in field shelling of corn. Trans. ASAE 11(2):164-166.
30. Carreker, J.R., J.E. Box, Jr., R.N. Dawson, E.R. Beaty, and H.D. Morris. 1972. No-till corn in fescuegrass. Agron. J. 64(4):500-503.
31. Chandler, W.V. 1960. Nutrient uptake by corn in North Carolina. N.C. Agric. Exp. Stn. Tech. Bull. 143.
32. Claassen, M.M., and R.H. Shaw. 1970. Water deficit effects on corn. II. Grain components. Agron. J. 62(5):652-655.
33. Colville, W.L., and D.P. McGill. 1962. Effect of rate and method of planting on several plant characters and yield of irrigated corn. Agron. J. 54(3):235-238.
34. Crop Production. 1974. USDA Statistical Reporting Service. US Government Printing Office, Washington, D.C.
35. Cummins, D.G., and J.W. Dobson, Jr. 1973. Corn for silage as influenced by hybrid maturity, row spacing, plant population, and climate. Agron. J. 65(2):240-243.
36. Das, D.K., and R.L. Jat. 1972. Adaptability of maize to high soil water conditions. Agron. J. 64(6):849-850.
37. Daynard, T.B., and W.G. Duncan. 1969. The black layer and grain maturity in corn. Crop Sci. 9:473-476.
38. Denmead, O.T., L.J. Fritschen, and R.H. Shaw. 1962. Spatial distribution of net radiation in a corn field. Agron. J. 54(6):505-510.
39. ----, and R.H. Shaw. 1960. The effects of soil moisture stress at different stages of growth on the development and yield of corn. Agron. J. 52:272-274.
40. Doss, B.D., C.C. King, and R.M. Patterson. 1970. Yield components and water use by silage corn with irrigation, plastic mulch, nitrogen fertilization, and plant spacing. Agron. J. 62(4):541-543.
41. Downey, L.A. 1971. Effect of gypsum and drought stress on maize (*Zea mays* L.). II. Consumptive use of water. Agron. J. 63(4):597-600.
42. Dunham, R.S. 1964. Losses from weeds. Minn. Agric. Ext. Serv., St. Paul. Spec. Rep. No. 13. p. 1-43.
43. Eik, K., and J.J. Hanway. 1965. Some factors affecting development and longevity of leaves of corn. Agron. J. 57:7-12.
44. ----, and ----. 1966. Leaf area in relation to yield of corn grain. Agron. J. 58(1):16-18.
45. Eisele, H.F. 1938. Influence of environmental factors on the growth of the corn plant under field conditions. Iowa Agric. Exp. Stn. Res. Bull. 229.
46. Elkins, C.B., G.G. Williams, and F.T. Ritchie, Jr. 1961. Soil moisture characteristics of some southern Piedmont soils. USDA, ARS 41-54.
47. Erbach, D.C., D.E. Wilkins, and W.G. Lovely. 1972. Relationships between furrow opener, corn plant spacing, and yield. Agron. J. 64(5): 702-704.
48. Fehrenbacher, J.B., B.W. Ray, and J.D. Alexander. 1967. Root development of corn, soybeans, wheat and meadow in some contrasting Illinois soils. Illinois Research (Spring), Univ. of Ill.
49. Fisher, W.F., and D.E. Lane. 1973. Till-planting. p. 187-194. *In* Conservation tillage: the proceedings of a national conference. Soil Conserv. Soc. of Am., Ankeny, Iowa.
50. Giesbrecht, J. 1969. Effect of population and row spacing on the performance of four corn (*Zea mays* L.) hybrids. Agron. J. 61(3):439-441.
51. Grable, A.R. 1966. Soil aeration and plant growth. Adv. Agron. 18:57-106.

52. Griffith, D.R., J.V. Mannering, H.M. Galloway, S.D. Parsons, and C.B. Richey. 1973. Effect of eight tillage-planting systems on soil temperature, percent stand, plant growth, and yield of corn on five Indiana soils. Agron. J. 65(2):321-326.
53. Hanway, J.J. 1962a. Corn growth and composition in relation to soil fertility: I. Growth of different plant parts and relation between leaf weight and grain yield. Agron. J. 54(2):145-148.
54. ----. 1962b. Corn growth and composition in relation to soil fertility: III. Percentages of N, P, and K in different plant parts in relation to stage of growth. Agron. J. 54(3):222-229.
55. ----. 1971. How a corn plant develops. Iowa State Univ. Spec. Rep. 48 (rev.).
56. ----, S.A. Barber, R.H. Bray, A.C. Caldwell, M. Fried, L.T. Kurtz, K. Lawton, J.T. Pesek, K. Pretty, M. Reed, and F.W. Smith. 1962. North central regional potassium studies. III. Field studies with corn. Iowa State Univ. Res. Bull. 503. North Central Regional Publ. No. 135. p. 407-438.
57. ----, and W.A. Russell. 1969. Dry-matter accumulation in corn (*Zea mays* L.) plants: comparisons among single-cross hybrids. Agron. J. 61(6):947-951.
58. Hargett, N.L. 1973. 1972 fertilizer summary data. Natl. Fertilizer Dev. Cent., TVA, Muscle Shoals, Ala.
59. Hicks, D.R., S.D. Evans, R.D. Frazier, W.E. Lueschen, W.W. Nelson, H.J. Otto, C.J. Overdahl, and R.H. Peterson. 1970. Corn management studies in Minnesota 1967-68-69. Minn. Agric. Exp. Stn. Misc. Rep. 96.
60. ----, and R.E. Stucker. 1972. Plant density effect on grain yield of corn hybrids diverse in leaf orientation. Agron J. 64(4):484-487.
61. Hitzhusen, T.E., S.J. Marley, and W.F. Buchele. 1970. Beefmaker II: Developing a total corn harvester. Agric. Eng. 51:632-634.
62. Hoff, D.J., and H.J. Mederski. 1960. Effect of equidistant corn plant spacing on yield. Agron. J. 52:295-297.
63. Holt, R.F., and D.R. Timmons. 1968. Influence of precipitation, soil water, and plant population interactions on corn grain yields. Agron. J. 60(4):379-381,
64. Hunter, R.B., L.W. Kannenburg, and E.E. Gamble. 1970. Performance of five maize hybrids in varying plant populations and row widths. Agron. J. 62(2):255-256.
65. Hussan, N.A.K., J.V. Drew, D. Knudsen, and R.A. Olsen. 1970. Influence of soil salinity on production of dry matter and uptake and distribution of nutrients in barley and corn. II. Corn (*Zea mays* L.). Agron. J. 62(1):46-48.
66. Iowa Crop and Livestock Reporting Service. 1973. Annual summary. USDA Statistical Reporting Service, Des Moines, Iowa.
67. Jenkinson, D.S. 1968. Chemical tests for potentially available nitrogen in soil. J. Sci. Food Agric. 19:160-168.
68. Jones, J.N., Jr., J.E. Moody, and J.H. Lillard. 1969. Effects of tillage, no tillage, and mulch on soil water and plant growth. Agron. J. 61(5):719-721.
69. Jordan, H.V., K.D. Laird, and D.D. Ferguson. 1950. Growth rates and nutrient uptake by corn in a fertilizer-spacing experiment. Agron. J. 42:261-268.
70. Jung, P.E., Jr., L.A. Peterson, and L.E. Schrader. 1972. Response of irrigated corn to time, rate, and source of applied N on sandy soils. Agron. J. 64(5):668-670.
71. Kamprath, E.J., and C.D. Foy. 1971. Lime-fertilizer-plant interactions in acid soils. p. 105-151. *In* R.A. Olson et al. (ed.) Fertilizer technology and use. Soil Sci. Soc. of Am., Inc., Madison, Wis.
72. Kiesselbach, T.A. 1949. The structure and reproduction of corn. Neb. Agric. Exp. Stn. Res. Bull. 161.
73. ----. 1950. Progressive development and seasonal variations of the corn crop. Nebr. Agric. Exp. Stn. Res. Bull. 166.
74. Knake, E.L. 1962. Losses caused by weeds. Proc. North Central Weed Control Conf. 19:1.
75. Kurtz, L.T., and G.E. Smith. 1965. Nitrogen fertilizer requirements. p. 195-235. *In* W.H. Pierre et al. (eds.) Advances in corn production: principles and practices. The Iowa State Univ. Press, Ames.

76. Lal, R., and G.S. Taylor. 1969. Drainage and nutrient effects in a field lysimeter study: I. Corn yield and soil conditions. Soil Sci. Soc. Am. Proc. 33:937-941.
77. Lane, D.E., and H. Wittmuss. 1961. Nebraska till-plant system. Univ. of Nebraska Ext. Cir. 61:714.
78. Larson, W.E. 1962. Tillage requirements for corn. J. Soil Water Conserv. 17:3-7.
79. ----, and R.R. Allmaras. 1971. Management factors and natural forces as related to compaction. p. 367-427. *In* Compaction of agricultural soils. Am. Soc. Agric. Eng., St. Joseph, Mich.
80. Lewis, W.M. 1973. No-till systems. p. 182-187. *In* Conservation tillage: the proceedings of a national conference. Soil Conserv. Soc. of Am., Ankeny, Iowa.
81. Long, F.L., J.M. Daniels, F.T. Ritchie, Jr., and C.M. Ellerbe. 1963. Soil moisture characteristics of some lower coastal plain soils. USDA, ARS 41-82.
82. Lutz, J.A., Jr., H.M. Camper, and G.D. Jones. 1971. Row spacing and population effects on corn yields. Agron. J. 63(1):12-14.
83. Mannering, J.V., and C.B. Johnson. 1969. Effect of crop row spacing on erosion and infiltration. Agron. J. 61(6):902-905.
84. ----, L.D. Meyer, and C.B. Johnson. 1968. Effect of cropping intensity on erosion and infiltration. Agron. J. 60(2):206-209.
85. Mengel, D.B., and S.A. Barber. 1974. Development and distribution of the corn root system under field conditions. Agron. J. 66:341-344.
86. Meyer, L.D., and J.V. Mannering. 1961. Minimum tillage for corn: Its effect on infiltration and erosion. Agric. Eng. 42:72-75, 86.
87. Miller, M.F., and F.L. Duley. 1925. The effect of varying moisture supply upon the development and composition of the maize plant at different periods of growth. Mo. Agric. Exp. Stn. Res. Bull. 76.
88. Mishra, M.N., J.W. Pendleton, D.L. Mulvaney, and P.E. Johnson. 1963. Investigations of high-population corn for forage and green manure. Agron. J. 55(5):478-480.
89. Moon, W.T. 1973. Conservation tillage jumps in Iowa. Soil Conserv. 39(4):7.
90. Moschler, W.W., G.M. Shear, D.C. Martens, G.D. Jones, and R.R. Wilmouth. 1972. Comparative yield and fertilizer efficiency of no-tillage and conventionally tilled corn. Agron. J. 64(2): 229-231.
91. Murphy, L.S., and L.M. Walsh. 1972. Correction of micronutrient deficiencies with fertilizers. p. 347-387. *In* J.J. Mortvedt et al. (eds.) Micronutrients in agriculture. Soil Sci. Soc. Am., Inc., Madison, Wis.
92. Nelson, L.B. 1956. The mineral nutrition of corn as related to its growth and culture. Adv. Agron. 8:321-375.
93. ----. (ed. comm. chairman.) 1968. Changing patterns in fertilizer use. Soil Sci. Soc. Am., Inc., Madison, Wis.
94. Nunez, R., and E. Kamprath. 1969. Relationships between N response, plant population, and row width on growth and yield of corn. Agron. J. 61(2):279-282.
95. Odell, R.T., and W.R. Oschwald. 1970. Productivity of Illinois soils. Univ. of Illinois Cir. 1016.
96. Olsen, S.R., and L.A. Dean. 1965. Phosphorus. *In* C.A. Black (ed.) Methods of soil analysis. Agronomy 9(2):1035-1049. Am. Soc. Agron., Madison, Wis.
97. Olson, R.A. (ed.) 1971. Fertilizer technology and use. Soil Sci. Soc. Am., Inc., Madison, Wis.
98. ----, A.F. Dreier, C. Thompson, and P.H. Grabouski. 1964. Using fertilizer nitrogen effectively on grain crops. Neb. Agric. Exp. Stn. Res. Bull. SB 479.
99. Olson, T.C., and L.S. Schoeberl. 1970. Corn yields, soil temperature, and water use with four tillage methods in the western Corn Belt. Agron. J. 62(2):229-232.
100. Oschwald, W.R. 1973. Chisel plow and strip tillage systems. p. 194-202. *In* Conservation tillage: the proceedings of a national conference. Soil Conserv. Soc. Am., Ankeny, Iowa.
101. Parks, W.L., and W.M. Walker. 1969. Effect of soil potassium, potassium fertilizer and method of fertilizer placement upon corn yields. Soil Sci. Soc. Am. Proc. 33(3):427-429.
102. Pearson, R.W. 1966. Soil environment and root development. p. 95-126. *In* W.H. Pierre et al. (eds.) Plant environment and efficient water use. Am. Soc. Agron., Madison, Wis.

103. ----, H.V. Jordan, O.L. Bennett, C.E. Scarsbrook, W.E. Adams, and A.W. White. 1961. Residual effects of fall- and spring-applied nitrogen fertilizers on crop yields in the southeastern United States. USDA Tech. Bull. 1254:1-19.
104. Pendleton, J.W., and D.B. Egli. 1969. Potential yield of corn as affected by planting date. Agron. J. 61(1):70-71.
105. Peters, D.B. 1960. Relative magnitude of evaporation and transpiration. Agron. J. 52:536-538.
106. ----, and L.J. Bartelli. 1958. Soil moisture survey of some representative Illinois soil types. USDA, ARS-41.
107. Peterson, A.E. 1960. Advantages of wheel track corn planting. Int. Congr. Soil Sci., Trans. 7th (Madison, Wis.) I:590-597.
108. ----, O.I. Berge, J.T. Murdock, and D.R. Peterson. 1958. Wheel track corn planting. Univ. of Wisconsin Ext. Serv. Cir. 559.
109. Pierre, W.H., S.R. Aldrich, and W.P. Martin (eds.) 1966. Advances in corn production: principles and practices. The Iowa State Univ. Press, Ames.
110. Pratt, P.F. 1965. Potassium. *In* C.A. Black (ed.) Methods of soil analysis. Agronomy 9(2):1022-1030. Amer. Soc. Agron., Madison, Wis.
111. Richards, L.A. (ed.) 1954. Diagnosis and improvement of saline and alkali soils. USDA Agric. Handb. No. 60.
112. Ritter, W.F., and C.E. Beer. 1969. Yield reduction by controlled flooding of corn. Trans. ASAE 12:46-47, 50.
113. Robins, J.S., and C.E. Domingo. 1953. Some effects of severe soil moisture deficits at specific growth stages in corn. Agron. J. 45:618-621.
114. Rossman, E.C., and R.L. Cook. 1966. Soil preparation and date, rate, and pattern of planting. p. 53-101. *In* W.H. Pierre et al. (eds.) Advances in corn production: principles and practices. The Iowa State Univ. Press, Ames.
115. Rumawas, F., B.O. Blair, and R.J. Bula. 1971. Microenvironment and plant characteristics of corn (*Zea mays* L.) planted at two row spacings. Crop Sci. 11:320-323.
116. Rutger, J.N., and L.V. Crowder. 1967. Effect of row width on corn silage yields. Agron. J. 59:475-476.
117. Sayre, J.D. 1948. Mineral accumulation in corn. Plant Physiol. 23:267-281.
118. Schwab, G.O., G.S. Taylor, J.L. Fouss, and E. Stibbe. 1966. Crop response from tile and surface drainage. Soil Sci. Soc. Am. Proc. 30:634-637.
119. Shaw, R.H., and H.C.S. Thom. 1951a. On the phenology of field corn, the vegetative period. Agron. J. 43:9-15.
120. ----, and ----. 1951b. On the phenology of field corn, silking to maturity. Agron. J. 43:541-546.
121. Shear, G.M., and W.W. Moschler. 1969. Continuous corn by the no-tillage and conventional methods: a six-year comparison. Agron. J. 61(4):524-526.
122. Sheets, T.J., and L.L. Danielson. 1959. Herbicides in soils. USDA, ARS 20-9:170-181.
123. Shrader, W.D., and J.J. Pierre. 1966. Soil suitability and cropping systems. p. 1-25. *In* W.H. Pierre et al. (eds.) Advances in corn production: principles and practices. The Iowa State Univ. Press, Ames.
124. Singh, T.A., G.W. Thomas, W.W. Moschler, and D.C. Martens. 1966. Phosphorus uptake by corn (*Zea mays* L.) under no-tillage and conventional practices. Agron. J. 58:147-150.
125. Soil Conservation Society of America. 1973. Conservation tillage: the proceedings of a national conference. Soil Conserv. Soc. Am., Ankeny, Iowa. 241 p.
126. Sopher, C.D., and R.J. McCracken. 1973. Relationships between soil properties, management practices, and corn yields on South Atlantic Coastal Plain soils. Agron. J. 65(4):595-599.
127. Splittstoesser, W.E., and L.A. Derscheid. 1962. Effects of environment upon herbicides applied preemergence. Weeds 10:304-307.
128. Sprague, G.F., and W.E. Larson. 1966. Corn production. USDA Agric. Handb. No. 322.

129. Stanford, G. 1965. Nitrogen requirements of crops for maximum yield. p. 237-257. *In* M.H. McVickar et al. (ed.) Agricultural anhydrous ammonia technology and use. Agric. Ammonia Inst. Memphis, Tenn.; Am. Soc. Agron., Madison, Wis.; and Soil Sci. Soc. Am., Madison, Wis.
130. ----, and W.H. Demar. 1970. Extraction of soil organic nitrogen by autoclaving in water: 3. Diffusable ammonia, an index of soil nitrogen availability. Soil Sci. 109:190-196.
131. Stevenson, C.K., and C.S. Baldwin. 1969. Effect of time and method of nitrogen application and source of nitrogen on the yield and nitrogen content of corn (*Zea mays* L.). Agron. J. 61(3):381-384.
132. Stickler, F.C. 1964. Row width and plant population studies with corn. Agron. J. 56(4):438-441.
133. Stivers, R.K., D.R. Griffith, and E.P. Christmas. 1971. Corn performance in relation to row spacings, populations, and hybrids on five soils in Indiana. Agron. J. 63(4):580-582.
134. Talbert, R.E., and O.H. Fletchall. 1965. The adsorption of some s-triazines in soils. Weeds 13(1):46-52.
135. Taylor, H.M. 1971. Soil conditions as they affect plant establishment, root development, and yield (F) Effects of soil strength on seedling emergence, root growth, and crop yield. p. 292-312. *In* Compaction of agricultural soils. Am. Soc. Agric. Eng., St. Joseph, Mich.
136. Thompson, L.M. 1962. An evaluation of weather factors in the production of corn. CAEA Report 12T. Center for Agric. and Econ. Adjustment. Iowa State Univ. of Science and Technology, Ames. 45 p.
137. Triplett, G.B., Jr., and D.M. Van Doren, Jr. 1969. Nitrogen, phosphorus, and potassium fertilization of non-tilled maize. Agron. J. 61(4):637-639.
138. ----, ----, and S.W. Bone. 1973. An evaluation of Ohio soils in relation to no-tillage corn production. Ohio Agric. Res. and Dev. Cent. Res. Bull. 1068.
139. ----, ----, and W.H. Johnson. 1970. Response of tillage systems as influenced by soil type. Trans. ASAE 13:765-767.
140. ----, ----, and B.L. Schmidt. 1968. Effect of corn (*Zea mays* L.) stover mulch on no-tillage corn yield and water infiltration. Agron. J. 60:236-239.
141. Turnquist, P.K., H. Waelti, and L.A. Mathison. 1970. Till planting corn. S.D. State Univ. Bull. 567.
142. Upchurch, R.D., and D.D. Mason. 1962. The influence of soil organic matter on the phytotoxicity of herbicides. Weeds 10:9-14.
143. Van Doren, D.M., Jr., and G.B. Triplett, Jr. 1965. Corn cultivation . . . Is it needed? Ohio Rep. p. 46-47.
144. ----, and ----. 1973. The future of no-tillage in the U.S.A. Sixth Int. Conf. on Soil Tillage, Wageningen, The Netherlands.
145. Vengris, J.W., W.G. Colby, and M. Drake. 1955. Plant nutrient competition between weeds and corn. Agron. J. 47:213-216.
146. Vincent, G.B., and D.G. Woolley. 1972. Effect of moisture stress at different stages of growth: II. Cytoplasmic male-sterile corn. Agron. J. 64(5):599-602.
147. Voorhees, W.B., and R.F. Holt. 1969. Management of alfalfa to conserve soil moisture. Univ. of Minn. Exp. Stn. Bull. 494.
148. Voss, R.E., J.J. Hanway, and L.C. Dumenil. 1970. Relationship between grain yield and leaf N, P, and K concentrations for corn (*Zea mays* L.) and the factors that influence this relationship. Agron. J. 62(6):726-728.
149. Walker, W.M., J.C. Siemens, and T.R. Peck. 1970. Effect of tillage treatments upon soil tests for soil acidity, soil phosphorus and soil potassium at three soil depths. Soil Sci. Plant Anal. 1:367-375.
150. Walsh, L.M., and J.D. Beaton (eds.) 1973. Soil testing and plant analysis. Soil Sci. Soc. Am., Madison, Wis.
151. Weber, J.B., P.W. Perry, and R.P. Upchurch. 1965. The influence of temperature and time on the adsorption of paraquat, diquat, 2, 4-D, and prometone by clays, charcoal, and an anion exchange resin. Soil Sci. Soc. Am. Proc. 29:678-688.

152. Welch, L.F., P.E. Johnson, G.E. McKibben, L.V. Boone, and J.W. Pendleton. 1966. Relative efficiency of broadcast versus banded potassium for corn. Agron. J. 58:618-621.
153. ----, D.L. Mulvaney, M.G. Oldham, L.V. Boone, and J.W. Pendleton. 1971. Corn yields with fall, spring, and sidedress nitrogen. Agron. J. 63:(1):119-123.
154. Whitaker, F.D., H.G. Heinemann, and W.E. Larson. 1969. Plant population and row spacing influence corn yield. Mo. Agric. Exp. Stn. Res. Bull. 961.
155. Willis, W.O., and M. Amemiya. 1973. Tillage management principles: soil temperature effects. p. 22-42. *In* Conservation tillage: the proceedings of a national conference. Soil Conserv. Soc. Am., Ankeny, Iowa.
156. Wimer, D.C., and M.B. Harland. 1925. The cultivation of corn — weed control versus moisture conservation. Ill. Agric. Exp. Stn. Bull. 259:1-21.
157. Wischmeier, W.H. 1973. Conservation tillage to control water erosion. p. 133-141. *In* Conservation tillage: the proceedings of a national conference. Soil Conserv. Soc. of Am., Ankeny, Iowa.
158. ----, and D.D. Smith. 1965. Predicting rainfall-erosion losses from cropland. USDA Agric. Handb. No. 282.
159. Wittmuss, H.D., D.E. Lane, and B.R. Somerhalder. 1971. Strip till planting of row crops through surface residues. Trans. ASAE 14:60-63, 68.
160. ----, and N.P. Swanson. 1964. Till-planted corn reduces soil losses. Agric. Eng. 45:256-257.
161. Yao, A.Y.M., and R.H. Shaw. 1964. Effect of plant population and planting pattern of corn on the distribution of net radiation. Agron. J. 56(2):165-169.

Chapter 12 Production of Hybrid Corn Seed

WILLIAM F. CRAIG
Funk Seeds International
Bloomington, Illinois

A tremendous amount of research, breeding, and experimental work with corn, starting before the turn of the century, made the development of the hybrid corn seed industry possible. This work was conducted by many outstanding scientists, educational institutions, USDA, and privately owned seed firms in various areas of the USA, although it was centered in what was considered to be at that time, the Corn Belt (Crabb, 1947).

The first commercial hybrids were produced and sold in the early 1920's, and from that modest beginning our present-day sophisticated hybrid corn seed industry has developed.

The first hybrids to be developed were adapted primarily to the central Corn Belt, and by 1933 approximately 1% of the Corn Belt corn acreage was planted with hybrid seed (Airy, 1955). Then came the severe droughts of 1934 and 1936, and because of the superior performance of hybrids in those years, farmers very rapidly began accepting and demanding hybrid seed.

The rapid acceptance by U.S. farmers of hybrid corn varieties in the 1930's and 1940's provided the basis for a large number of firms and individuals to establish themselves in a new and fast growing industry, whereas prior to that time only a very few firms were engaged in the business. This fact, in later years, led to the development of hybrid varieties adapted to virtually every corn growing area of the USA and Canada. Indeed, the development of new hybrid varieties helped to establish profitable corn production on hundreds of thousands of hectares, outside the Corn Belt, where profitable corn production on a commercial scale was theretofore impossible.

Size of the Industry

The area in the USA planted to corn has varied since 1900 from a high of 47 million ha (116 million acres) in 1917 to a low of 26 million ha (64 million acres) in 1969. The area planted to corn in 1974 was 30.5 million ha (77.7 million acres).

Planting rates vary considerably according to plant food applications and available moisture supply, as well as the adaptation of specific hybrids for high plant populations. It is estimated, however, that a minimum of 5 million hl (15 million bu) of hybrid seed were required to plant the 1974 acreage of 30.5 million ha (77.7 million acres). At an estimated average retail price of $30.00/bu, the hybrid corn seed industry has grown to a gross sales volume of $450 million.

Types of Hybrids

In the beginning hybrids produced and sold commercially were almost exclusively double crosses. As production methods became more sophisticated, however, and with the advent of herbicides, insecticides, and the ready availability of economically priced commercial plant foods, the production of single cross hybrid seed became economically feasible. A significant transition from double crosses to single crosses began, within the U.S. Corn Belt, in the late 1950's and continued through the 1960's. Modified single crosses using related line single crosses for female and male seed parents were and are used extensively in hybrid corn seed production. Three-way crosses (single-cross seed parents and inbred pollinators) are also a factor in the industry. While double-cross hybrids are still a significant factor in the market, their importance seems to be decreasing with each succeeding year. It is estimated that they now comprise no more than 20% of the total U.S. market.

Seed Corn Companies

Since the advent of hybrid corn, many different firms and individuals have become involved in the production and sale of hybrid corn seed. Small, privately owned companies may produce and distribute only a few thousand bushels of seed. This size operation is usually dependent upon research and hybrid development carried on by public institutions or upon that conducted by private firms which make a business of producing and selling parent seed (foundation seed) stocks. Smaller companies usually purchase their foundation seed, grow the commercial seed, and sell it directly to farmers in their area. Larger operations usually carry on their own research and development programs, produce their own foundation seed stocks, produce the commercial seed, and distribute it through their own sales organizations. The majority of hybrid corn seed is sold by the various companies to farmer dealers, who, in turn, sell it to farmer customers. It is customary in the industry to deliver to dealers on a consignment basis, and to accept as "returns" seed which remain unsold by the dealer at the end of the planting season. In some geographical areas, sales are made by the producers to jobbers or distributors who seek their own retail dealer outlets. This is a more common practice in areas of relatively low sales potential.

Hybrid corn seed was at first sold in 25 kg (1 bu) packages. In the early 1960's there began a trend to package and distribute in 23 kg (50 lb) packages. Later on in that decade the practice of packaging by kernel count became popular. At the present time, all three methods of packaging are in use throughout the industry.

As the industry has grown towards maturity there has been a marked attrition of companies and a great variation in the growth of individual companies. While there are still many relatively small operations within the industry, there have also developed some large corporations, some of whom have operations in Europe, South America, Canada, and Mexico. The industry's three largest corporations have now captured an estimated share of the U.S. hybrid seed corn market of more than 55%.

PARENT SEED STOCKS

Large quantities of parent seed stocks (foundation seed) are required annually to plant the several hundred thousand hectares of commercial hybrid seed corn produced by seedsmen. In recent years, seed corn producers have devoted increasing attention to developing more effective techniques and procedures to assure adequate supplies of high quality, genetically pure stocks. Some larger companies have parent stock or foundation seed departments responsible only for the production and inventory of parent seed strains needed for commercial seed production.

Seed corn companies must make forecasts of future commercial seed production plans to insure availability of adequate parent stock supplies. Since many seed companies produce and sell a large number of different hybrid varieties, a potential myriad of different parent stock strains must be increased and maintained in sufficient supply to meet commercial seed production requirements.

Seed Stock Increase

Seed stock increase involves the maintenance and increase of inbred lines and the parent seed stocks used to produce commercial hybrids.

Inbred Maintenance

Inbreds are the basic foundation, hence foundation seed in hybrid seed corn production. Inbreds must be maintained and increased under rigid control to insure satisfactory final product performance. Procedures employed may vary among organizations; however, at least two important steps are usually taken. They are 1) establishing and maintaining a base population or breeder seed stock and 2) increasing inbred maintenance seed. It is vital to have a base population, usually bulked self-pollinated seed, which adequately represents the genetic constitution of the inbred. All inbred maintenance increases are made using this base population seed. Increases are usually produced in well isolated blocks by natural random sib mating. In turn, increase maintenance seed stocks are used to plant all parent stock increase isolations.

Use of parents which are variable in phenotype or genotype (Loeffel, 1971) requires special attention in an effective maintenance program. Variability among individuals within the population must be acknowledged and maintenance increases be made in ways to insure accurate perpetuation of this variability from generation to generation. Each parental inbred may present a special case as far as base population and inbred maintenance is concerned.

Parent Stock Increase

Parent stocks are used to plant and produce commercial F_1 hybrid varieties and may consist of any one of three or more basic pedigrees such as 1) inbred, 2) related line cross, and 3) single cross.

In the case of inbred parents, increases are made using inbred maintenance stocks and are produced in isolated increase blocks using random sib mating. Related line cross parents and single cross parents are produced using inbred maintenance stocks in isolated crossing blocks in which the female parent is detasseled or is male sterile.

Procedures and Techniques

Generally equipment and procedures used in planting, detasseling, harvesting, drying, and processing of inbred maintenance and parent stock seed increases are similar to those used in commercial hybrid seed production. Certain important additional steps must be employed to insure maximum genetic purity.

Careful plant and ear selection by trained technicians must be practiced throughout the growing season and at harvest to eliminate individual plants which exhibit phenotypes varying from the established accepted phenotype of the inbred. As much plant selection or rogueing as possible should occur prior to pollination in order to eliminate outcrossing resulting from pollen supplied by undesirable plants. Harvest of parent stocks should occur either by hand in small blocks or by using ear harvest equipment. Harvesting on the ear permits further selection to be practiced by technicians examining ears prior to shelling.

Quality Control

Rigid requirements must be used to maintain genetic purity at maximum levels. "Growouts" of each seed lot are often planted during the winter season to estimate genetic purity prior to use. In many cases, more extensive growouts are conducted, sampling each seed size in each lot, in the summer growing season to obtain additional, more precise estimates of purity.

Observation plantings of parent seed stock lots provide useful information only to the extent that the samples are representative and adequate in size. The growout planting must be of sufficient size to provide an adequate measure of the variation within the lot being sampled.

Genetic purity of parent stocks not only helps insure pure commercial hybrid seed but also reduces cost in rogueing commercial seed fields and ear sorting at harvest.

Parent seed stocks are usually sized just as commercial seed corn. When genetic impurities occur, particularly those caused by outcrosses, they are often concentrated in certain specific seed sizes. As a result, certain seed sizes within a specific lot may have unacceptably high levels of impurities, while other sizes in the same lot are acceptable. By careful selection of specific seed sizes, it is often possible to improve purity of seed stocks used to produce commercial corn.

As a part of quality control of parent seed stocks, germination and measures of mechanical mixture or mislabeling must be identified. Various methods of establishing seed germination are employed and are generally performed on all usable seed sizes of each lot. Growouts will normally identify accidental mixtures or mislabeling of foundation seed which may occur at any point in production, processing, and inventory.

Inventory Control

Large inventories of parent stocks are required. Accurate records of inventory supplies, genetic purity, and germination must be maintained. Many organizations utilize 1,000 viable kernels (MVK) as an inventory unit for parent seed stocks. Whenever possible seed production should be planned on a scale to provide an inventory adequate for several years use. High seed viability can be enhanced by employing controlled environmental storage to be discussed in more detail later in this chapter.

PRODUCING THE SEED CROP

Agronomic practices in seed production fields are basically the same as growing a commercial corn crop. However, there are some additional requirements unique to seed production. Acreages are determined on the basis of projected sales, utilizing yield levels based on past experience. The advent of commercial single cross seed production on a large scale required the need for increased scientific and technical knowledge, specific care, planning, and production technology if the new technique were to be economically successful. Cultural practices within the production fields are planned to reduce risks to a minimum while maximizing yield and seed quality.

Selection and Management of Production Acres

As commercial hybrid seed corn production became a reality in the early 1940's, it brought with it a large number of innovative farmer-seedsmen (Airy et al., 1961). The cooperation of these family farms and companies with their foresight and energies have made the seed corn industry a sound business and helped to make it grow.

Good "corn ground" with adequate fertility and climate, resulted in the localization of most seed production areas to specific geographic areas of the Corn Belt. Expansion of the Corn Belt and increased technical knowledge has created opportunities for seed companies, as they grow in volume, to expand in search of specific seed production areas that provide such necessary factors as degree days, day length, lack of extreme temperatures, and specific farming practices which reduce the seed production risk potential. By matching specific priority needs and area characteristics, risk can be minimized and seed yields and recovery maximized per seed parent hectare planted and harvested.

Most seed producers employ managers for specialized production areas. A good seed production manager selects the best growers with the best land in the area, and uses sound planning and follow through by trained supervisory help; especially during planting, detasseling, and harvesting periods. An area of 1,600 to 2,000 ha (4,000 to 5,000 acres) of seed production responsibility is generally felt to be maximum for an area production manager. With this acreage, regional supervisors, under the direction of the area manager, would be charged with 400 to 800 ha (1,000 to 2,000 acres) responsibilities. During detasseling, additional supervisory help would need to be employed as crew foremen, field foremen, and subregional supervisors.

The correct timing of detasseling is a must to produce hybrids with a high degree of genetic purity. This has to be accomplished while minimizing plant damage so that optimum seed parent yield results. The subject of detasseling is discussed in further detail under the topic of pollen control later in this chapter. The production of seed during the summer months offers opportunities to utilize as supervisors agriculture and science teachers, principals, and other professionally trained personnel during their school vacation periods.

For area production managers to be effective in the management of their respective areas and for coordination of activities, communication is essential. Particularly during detasseling, it is critical that close communication be available so that people and equipment can be effectively used and moved to high priority areas as needed. Many seed companies use a combination of mobile telephones and Citizen's Band radios.

Selecting Contract Growers

Selecting growers within a specific production area is a most important factor. Generally, the selected contract growers will be the most progressive and innovative corn growers in the area, as well as respected farmers and community leaders.

Along with specific character traits of growers, seed producers select suitable soil types with high productivity indices, and which have been maintained in a state of high fertility. Fertility cannot be allowed to be a limiting factor of optimum yields. Tillage and cultural practices must be in line with approved hybrid corn production practices. When approximately 60 to 75% of the commercial corn grown is of single cross hybrids, 80 to 90% of the seed acreage will be devoted to inbred or related line-cross parents. Growers and producers have been able to effectively minimize insect damage, weed competition, and fertility deficiencies. The avoidance of poor soil structure and tilth with their adverse effects on inbred stands remains a real challenge.

Contract growers must be willing to cooperate with seed companies to alter their cultural practices and/or timing, rate, and kind of herbicides or fungicides. Equipment modifications are often necessary. In some areas, it is required that the producer furnish equipment on a lease or rental basis to the growers. In some instances growers cooperate in the purchase and sharing of various specialized pieces of equipment such as unit planters, detasseling, and harvesting equipment.

Contracts

With the advent of single cross hybrid seed production, contract growers were not content to assume the risk of poor stands, high temperature effects on silks and pollen, poor nicks between seed parent silks and male parent pollen due to weather or split date plantings, and poor seed quality. As a result, base guarantees are made which involve payment for complete failure of the seed crop up to a predesignated yield level. Incentive payments are often based on published futures or cash market prices at a specified time and place. In some instances, contracts are based on government based yield calculations.

Usually, the type of contract used, base guarantees, and multiplier factors are based upon anticipated yield levels and degree of difficulty encountered in the production of each individual hybrid. In the case of double-cross or three-way hybrid seed production, fewer guarantees are made with more emphasis placed on yield or incentive payments beyond a certain yield level. Factors such as fertility, herbicide or insecticide costs, seedbed preparation for split date plantings, volunteer removal, and harvesting are items for consideration within the contract. The variations in contracts between companies are considerable.

Isolation of Seed Fields

Isolation is intended to provide assistance in making a hybrid cross that possesses a high degree of genetic purity. Seed fields are planned with isolation of 201 m (40 rods) as the base distance from other corn. Studies confirm (Jones and Brooks, 1950, 1952) 1) that greatest contamination occurs in the 50 to 75 m (10 to 15 rods) nearest contaminating corn, 2) that dilution of contamination by pollen from border rows occurs, 3) that natural barriers may reduce contamination, 4) that an abundant supply of male corn pollen at the right time reduces contamination, 5) that direction from contaminating pollen influences amount of contamination, and 6) the "depth of field" in the direction of contamination source is important.

It has often been said by seedsmen that the best isolation is a perfect nick, that is, when a pollen parent starts shedding just before the seed parent silks start to emerge beyond the husk or tip of the ear shoot. Specific isolation requirements are detailed under the topic of quality control to be covered later in this chapter.

Fertility in Seed Fields

In the past it has been the tendency of contract growers to over fertilize to protect against possible fertility deficiencies, while at the same time striving for a balanced fertility program. With the advent of fertilizer shortages and substantial increases in the cost of basic nutrients, it has become essential to look at fertilizer recommendations carefully. Contract growers use the soil test regularly and apply nutrients necessary to maintain high fertility levels. In many cases, starter or row fertilizer is applied in addition to plow down application for efficiency reasons. Seed production warrants optimum fertility levels in order to maximize rainfall or irrigation, sunlight, and heat units. Another important aspect is that fertility levels need to be kept in balance since inbreds and parent stocks may be more vulnerable to deficiencies than are resulting hybrid crosses due to rooting abilities and genetic differences.

Numerous research studies attest to the importance of optimum fertility levels and balance of nutrients in corn production (Arnold et al., 1974; Bates, 1971; Fink and Wesley, 1974; Fowler, 1967; Gallaher, 1972; Hanway, 1962; Leffler, 1957; Peaslee et al., 1971; Pendleton, 1965; Powell and Webb, 1972; Voss et al., 1970). Continuing soil testing and plant and leaf analysis programs in addition to visual inspections to detect deficiencies are tools that growers and production men use to plan the fertility program (Aldrich and Leng, 1965).

Herbicides, Insecticides, and Fungicides

With today's sophisticated seed production techniques, control of weeds, insects, and diseases within the seed field has become an integral and necessary part of production. Inbreds and related-line parents cannot compete with broadleaf weeds or grasses. For this reason, seed growers rely heavily on specific herbicides and/or combinations for effective weed control with cultivation serving only in a supportive role. Production managers formulate plans with the grower taking into account 1) specific weed problems, 2) grower rotation, 3) soil type and organic matter, 4) equipment, and 5) specific hybrid to be produced.

Insecticides for the control of above and below ground insects are generally a must. The company working with the grower will want to formulate a program that protects against insect damage to stands, the growing plant, and the seed ear. Selection of the material to use will depend upon the specific insect to be controlled as well as the stage of development of the seed crop.

Fungicides have actively become a part of the production program in the 1970's both for control and protection of the more susceptible parent lines to Northern Corn Leaf Blight. Spray programs have been effective in reducing damage on the more susceptible lines. First application should be done when lesions first appear, followed by second or third applications as necessary. Normally, material will remain active on the leaf surface for 7 or more days, depending upon rainfall and humidity conditions. Genetic resistance to disease by multiple or single gene resistance is preferred (Ullstrup, 1970).

Planting the Seed Field

The minimum soil temperature for growth of corn is generally regarded as 10 C (50 F). Most agronomists would also agree that the optimum time for planting corn is as soon as the soil temperature at the 5 cm depth reaches that temperature for a relatively sustained period of time. The literature is replete with studies indicating the advantages of early planting upon yield (Blacklow, 1972; Grogan, 1970; Jones and Mederski, 1963; Mock and Eberhart, 1972; Pendleton and Egli, 1969; Troyer and Brown, 1972).

PRODUCTION OF HYBRID SEED

Fig. 1. A commercial hybrid seed production field planted on a 1:4 pattern of one pollen row (male) to four detasseled rows (female).

Plant population within the seed field is planned to produce maximum yields within limits imposed by the particular germplasm being used and the average rainfall pattern in the production area. Many investigators have studied population effects upon yield (Alessi and Power, 1974; Duncan, 1954; Genter and Camper, 1973; Hicks and Stucker, 1972; Johnson and Tanner, 1972; Stickler, 1964; Williams et al., 1965, 1968) Plant populations in seed fields of today's modern hybrids often approach or attain 20,000 plants for seed parents and often exceed that level for pollen parents, especially with inbred parents where pollen supply may be limited.

Planting Patterns

The most common planting pattern in seed fields today is one row of pollen parent to four rows of seed parent (Fig. 1). In addition to the 1:4 pattern, the 1:2:1:4 pattern is relatively common with single cross seed production. In each case, the seed parent is never more than two rows from the pollen parent and one-half of the seed parent rows are adjacent to a pollen parent in the 1:4 pattern and two-thirds of the seed parents in a 1:2:1:4 pattern as compared to the conventional 2:6 pattern used for double cross production.

Often pollen parents are double planted at a spaced interval to extend the pollen shedding period. This is timed so that peak pollen shed is at the same time the seed parent silk exposure is at its maximum.

Occasionally solid planting of seed parent in 96.5 to 101.6 cm (38 to 40 in) rows is utilized with every fifth between row space, or every fourth seed

Fig. 2. A commercial hybrid seed production field nearing maturity with the pollinator rows (male) removed.

parent row, being interplanted with pollen parent. This accomplishes two purposes: 1) full utilization of land area for seed parent production, and 2) placing the pollen parent closer to the seed parent rows. Solid plantings should be limited to seed parents not so aggressive as to overshadow the pollen parent and thereby delay pollen shed, and to pollen parents of sufficient stalk and root strength so that elbowing and lodging do not result, making it difficult to detassel or remove pollen parents as soon as pollination is complete.

It is now common practice to destroy the pollen parent after pollination is complete by cutting or running it down if it is brittle enough to break (Fig. 2). Destroying the pollen parent at this stage prevents grain formation and possible seed contamination at harvest. Competition with the developing seed parent for nutrients or available moisture is also minimized thus increasing kernel size and seed yield.

Split Date Planting

Split date planting of parents, that is, the planting of the seed parent and pollen parents on different dates, is used so that the two parents reach the flowering stage concurrently (Fig. 3). This has been and continues to be the most popular method of making gross alterations, so that parents of differing maturities are brought together for a timely nick. Split date plantings are made on the basis of, or combination of, days, growth stages, and/or heat units accumulated from the time the first parent was planted. Most success

Fig. 3. Split date seed parent planting. The pollinator rows have emerged while the female rows are just being planted.

has been realized by a combination of heat units and growth stage coupled with good judgement.

Other methods of obtaining smaller adjustments to pollen shed are 1) variable fertilizer rates, 2) variable planting depths between parents, 3) treating pollen parents (Bauman, 1967) to delay germination, 4) growth regulators (Pauli, 1967), and 5) clipping (Cloninger et al., 1974) or flaming (Fowler, 1967) to retard development, either of which can provide 3 to 6 days delay in flowering or extend flowering by as many as 3 to 5 days. Clipping has been used effectively to save a crop when weather conditions have prevented planting the second parent of an intended split after planting one parent. This has been particularly important when it is too risky to replant because too few degree days are left in the season or when a seed shortage exists for one or both parents.

POLLEN CONTROL

Pollen control in the hybrid corn seed production field is an extremely critical factor. This is an essential process to insure hybridization by enforced cross pollination between the female and male parents of the intended cross. Various methods of pollen control in seed fields have been utilized or investigated in recent years, aimed primarily at reducing the cost or easing the difficulty in this critical period while still maintaining the desired genetic purity. Some of the more promising methods include 1) detasseling, 2) cytoplasmic sterility, 3) genic male sterility, and 4) chemical pollen control.

Detasseling the Seed Fields

The detasseling period for the seed producer is probably the most critical and difficult to manage of any of the steps involved in hybrid corn seed production. To assure that each seed field meets the necessary genetic purity standards (see Quality Control section) each tassel from the seed parent rows must be removed prior to shedding and/or before silks emerge on the seed parent plants.

Currently the most widely used method of pollen control involves detasseling or the physical removal of the tassel from the plant, either as a manual operation or in combination with mechanical devices. This is a very costly operation for the seed producer. Newlin (1971) estimated that the seed corn industry spent between $20 million and $25 million for tassel removal in 1971. Increasing wage rates and inflationary costs have no doubt increased this expense to the industry since then.

Manual or Hand Detasseling

During the period of tassel emergence in seed fields each year thousands of workers, usually teenage youth, are employed by seed producers to perform the hand detasseling operation. This activity may be as short as 1 week or up to 5 weeks or more in length, depending upon the volume and spread in seed parent maturities planted within a seed production area.

To have a better understanding of the detasseling operation it is important to consider various factors that influence the job to be done in the field.

1. Tassels must be removed from all plants in a timely manner prior to shedding and silk emergence.
2. When weather conditions favor rapid growth, fields must be covered daily, meaning 7-day workweeks, rain or shine.
3. Some seed parent plant types are easier to detassel than others.
4. Seed parents that begin shedding pollen before fully emerging from the upper leaves, or which silk at about the same time as pollen shed occurs, create difficult detasseling supervision and management problems.
5. Weather conditions can greatly aid or complicate the detasseling season. A windstorm or heavy rain can lodge and tangle the seed parent just as tassels emerge, or severe heat can affect both efficiency of detasselers and emergence of silks and tassels.

Detasselers are usually organized into crews ranging in size from 6 up to 40 or 50 workers. The crew supervisor is responsible for recruiting, transporting, training, managing, and controlling the detasselers in his crew. With larger crews the supervisor will have one or more assistants, sometimes called checkers, to assist in training and managing the job to be done in the field. There customarily will be one supervisor or checker for each 6 to 10 crew members. It is important that each crew member be trained in proper detasseling technique to minimize leaf damage and to ensure that an effective detasseling job is done by each crew member in each seed parent row. The crew supervisor is also responsible for the safety and comfort of the workers while in the field.

Fig. 4. High clearance detasseling cart or personnel carrier used to transport detasselers through the seed fields.

For more effective and efficient labor utilization, detasseling carts or personnel carriers (Fig. 4) are frequently used, especially on the taller growing single cross seed parents. These are motorized, high clearance machines equipped with platforms upon which the workers stand to remove the tassels. The machines move slowly through the field while the detasseler can look down into the plants and more effectively and easily remove the tassels than he can on foot. Usually six or more detasselers will work from each machine so it is important that all detasselers on each machine are equally skilled for maximum effectiveness. During or immediately after heavy rain or windstorms, these machines usually cannot be used and the workers must proceed on foot.

Some seed producers utilize contract detasselers for at least a part of their seed production. With this method the contractor agrees to detassel a specified field area for an established fee paid him by the seed producer. The contractor may work his own hours as long as he meets the producer's established standards for timely removal of the tassels. If he fails to do so the producer will bring in a crew to remove the problem-causing tassels and deduct this expense from the contractor's payment. The contractor must also provide transportation for himself and any other detasselers he might need to help him, as well as the necessary supervision. This method often permits workers employed in other jobs to earn extra income during their free time.

Contract rates paid usually provide the efficient worker more income per hour worked than crew detasselers receive. At the same time contract detasseling is frequently less expensive per seed parent hectare to the seed producer because less supervision and transportation is required for his total production area and fewer detasseling crews need to be employed. However, it is necessary always to have some detasseling crews in a production area to help out in contract fields when and if needed. The producer must also have contract supervisors to train and check on the quality of the work being done by the contractor.

Mechanical Detasseling

The discovery of the susceptibility of T cytoplasm to Race T of Southern Leaf Blight in 1970 (Tatum, 1971; Ullstrup, 1972) caused an almost complete return to hand detasseling as the pollen control method in seed production in 1971. This tremendous increase in demand for detasseling labor coincided with rapidly increasing labor costs and a declining labor supply in the early 1970's. To better control rising production costs, seed producers began to look more closely at mechanical means of tassel removal.

An early study by Airy (1950) reported that at least one producer had looked at mechanical cutting of the tassel as a means of pollen control but had abandoned the idea because of excessive leaf damage resulting in reduced seed yields. However, by the early 1970's detasseling costs and other detasseling management risks were increasing at such a rate that producers were ready to take a new look at mechanical detasseling even though these machines might reduce seed yields. Since 1971 mechanical detasselers have come into widespread use in the industry.

Mechanical detasselers (Fig. 5), mounted on high clearance machines capable of operating even in very muddy fields are of two types:
1. "Cutters" — a rotating cutter blade or knife operating at various planes from horizontal to vertical, adjustable in height, to cut or shred the top of the corn plant including the tassel.
2. "Pullers" — usually two counter rotating small wheels or rollers, adjustable in height, that grasp the tassel and upper leaves pulling them upward in a manner approximating a hand detasseling operation.

The efficiency of mechanical detasselers are affected by many variables in the seed field, most of which are of more importance than the type of mechanical detasseler selected. Some of these variables and their effects, particularly on yield, are discussed under the topic of effect of detasseling on seed yield later in this chapter.

A skillful operator of a mechanical detasseler, working in a uniform seed field in which the tassel is well extruded ahead of pollen shedding, can mechanically remove 70% or more of the tassels in one or two passes over the field with minimum leaf damage (two and one-half leaves or less). However, as conditions become less favorable, percent of tassels removed will decrease and leaf damage will increase. The usual procedure is to delay mechanically detasseling a field as long as possible prior to silk emergence to permit maximum extrusion of tassels so that removal can be effected with a minimum leaf damage. In all cases some hand detasseling is required to remove tassels remaining on missed, late, or short plants, or suckers in the field.

With most seed parents a combination of mechanical and hand detasseling can effect a cost savings per hectare when compared with hand detasseling. An industry survey (Huey, 1971) indicated detasseling costs of

Fig. 5. Series of mechanical detasseling machines working in a seed field. Note tassel emergence on pollinator rows.

$75 to $85/ha, ($30 to $35 per total acre) with mechanical and hand detasseling compared to $124 to $136/ha ($50 to 55 per total acre) for all hand detasseling. A study in 1973 (Funk Seeds International, unpublished data) indicated that mechanical removal of 60 to 70% of seed parent tassels in two field passes plus hand removal of the remainder cost $37 per seed parent ha ($15/acre) less than 100% hand detasseled fields.

However cost savings attained through mechanical detasseling may be offset by seed parent yield reductions if the operation is not carefully managed to minimize leaf damage. The same unpublished Funk study indicated that seed could be produced at a lower cost/kg with a combination of mechanical and hand detasseling as long as yield reduction due to leaf damage did not exceed 10%. If the field conditions were such that a reduction of more than 10% was anticipated, it would result in lower cost to remove all tassels by hand, assuming an adequate labor supply was available. The variable cost factors involved in this study must be continually reviewed, especially in view of the rapidly changing wage and labor supply picture; along with other production factors, such as the importance of producing a high quality genetically pure seed crop for the customer.

Effect of Detasseling on Seed Yield

Prior to the widespread use of T-sterile cytoplasm there was considerable work done with regard to the effect of detasseling and leaf removal on seed yields of double cross seed parents. Removal of the tassel at or near anthesis was associated with an increase in grain yield (Chinwuba, 1961; Dungan and Woodworth, 1939; Grogan, 1956; Hunter et al., 1969), but decreases were also reported (Borgeson, 1943; Kiesselbach, 1945). When tassel removal was accompanied by leaf removal there was a grain yield reduction relative to the number of leaves removed (Airy, 1955; Borgeson, 1943; Dungan and Woodworth, 1939; Hunter et al., 1969). Kiesselbach (1945) summarized literature of hybrids and pointed out that when one, two, and three leaves were removed with the tassel, there were reductions in grain yield of 4.3, 9.3, and 16.4% respectively.

With the demise of T-sterile cytoplasm as a source of sterility in seed production fields in 1970 there was again renewed interest in the effect of detasseling treatments on yield. However, during the years in which T-sterile cytoplasm was being used, production procedures and practices changed somewhat. Production of seed changed from predominantly double cross production to special cross and single cross production. The effects of detasseling on seed yields become more critical where inbred seed parents were involved. The development of mechanical detasseling machines of various types also added a new dimension. When many of these mechanical detasseling machines were introduced to alleviate the problem of hand detasseling in 1971, very little was known about their effect on yield.

Recent work on leaf removal with inbred lines indicate that in general the yield response is similar to single crosses. Hunter et al. (1973) worked with 10 inbred seed parents and found an average increase in yield of 6.9% when tassel alone was removed. When one, two, and three leaves were removed with the tassel, yield reductions averaged 1.5, 4.9, 13.5% relative to the yield where the tassel alone was removed. These figures are not much different from those reported by Kiesselbach (1945) for single crosses. Hunter et al. (1973) did note, however, that there was an interaction among the lines with regard to leaf removal. A removal of three leaves on some lines did not have a detrimental effect on yield while on other lines it reduced yields as much as 28%. K.R. Storer, (1972. Funk Seeds International. Unpublished data.) working with four seed parents (two single crosses, one related line, and one inbred) found yields reduced by 0.8, 3, 11, 18, and 33% for leaf removals of one, two, three, four, and five leaves respectively. With five leaves removed yield reduction ranged from 25 to 40% depending upon the parent.

Recent studies with mechanical detasseling machines have shown varying results. J. Newell (1973. Funk Seeds International. Unpublished data) worked with eight lines and found yield for mechanically detasseled plots to run from 6 to 27.5% below hand pulled plots. Fields were cut or topped twice but additional trips across the field were needed to hand pull or glean remaining tassels. E. Cox (1973. Clyde Black and Sons, Unpublished data) found mechanical detasseling of 14 inbreds reduced yields from 2.5 to 24.4% below hand detasseled treatments. K.R. Storer observed reductions of 19 to 40% for mechanically detasseled treatments when trying to eliminate hand detasseling on four seed parent lines. In some cases these plots were mechanically cut three times with an equivalent removal of four or five whole leaves.

Measurements have been made on the two yield components, weight per kernel and kernel number, to try to determine which are primarily responsible for reduced yield. Results have varied; however, kernel number has generally been the major contributing factor. Hunter et al. (1973) found as much as 20 to 25% of the reduction in yield of inbred lines caused by leaf removal could be attributed to kernel weight. J. Newell found kernel weight accounted for 10 to 50% of the reduction in yield caused by mechanical detasseling. K.R. Storer found that only 6 to 8% of the yield reduction caused by mechanical detasseling could be attributed to kernel weight.

PRODUCTION OF HYBRID SEED

Fig. 6. Helicopter flying low over seed field for the purpose of distributing pollen to receptive silks.

From the studies available, no basis can be found from which to make precise statements regarding the effect of detasseling treatments on seed yields. This is not surprising, however, when the number of variables are considered. Important variables would include: 1) type of detasseling machine, 2) the number of times cut or pulled, 3) the skill and attention of the machine operator, 4) the time of cutting in relation to plant development, 5) the climatic conditions prior, during, and after detasseling operations, and 6) morphological variation among genotypes. These variables, as well as others, plus the interaction among them could alter results greatly.

Seed yields may be increased by "flying" seed fields (Fig. 6) for increased pollen distribution where pollen production may be limited. This practice is commonplace with single cross seed production.

Sterility

Male sterility as a method of pollen control in the production field has progressed significantly since the acceptance of hybrid corn. There are presently three types of sterility either in use or being extensively investigated. The three methods are 1) cytoplasmic, 2) genic, and 3) chemical male sterility. Details on the genetics of cytoplasmic and genic sterility are given in Chapter 4.

Cytoplasmic Male Sterility

The previously mentioned epidemic of southern leaf blight which swept the USA in 1970 brought into disrepute the most satisfactory, practical method of effecting field scale crossing of corn ever found. For about two decades the need to detassel an ever expanding hybrid seed corn hectarage largely was put aside in favor of the conversion of inbred parents to Texas cytoplasmic male sterility. Though other male sterile cytoplasms were available, the T source (Rogers and Edwardson, 1952) proved to be the most satisfactory because more inbreds were completely sterile in T-cytoplasm and fertility restoration was more easily managed genetically in this cytoplasm.

The realization that genetic vulnerability of the corn crop was increased by the nearly complete conversion to T-cytoplasm (National Academy of Sciences, 1972) resulted in a retreat from the extensive use of cytoplasmic male sterility (*cms*) as a substitute for detasseling beginning in 1971. Many other male sterile cytoplasms are known (Beckett, 1971; Duvick, 1965), most of which do not differ from normal cytoplasm in their reaction to leaf diseases (Smith et al., 1971). Of these *cms*-C and *cms*-s are the best known (Duvick, 1972). Since the use of cytoplasmic male sterility is still the most satisfactory technique available to produce hybrids, especially with rising labor costs, it seems likely to become important again in hybrid seed production. However, it remains to be established what, if any, agronomic side affects the various cytoplasms may have (Noble, S.W. 1973. Performance of S-*cms*, C-*cms*, and normal cytoplasm hybrids of corn (*Zea mays* L.). Am. Soc. Agron. Abstr. p. 11.). Although rather universally adopted as a production practice, there was considerable evidence that Texas cytoplasm reduced yields in farmer's fields by a small amount (Duvick, 1965), indicating that each cytoplasm should be tested before introduction.

There are two major ways in which cytoplasmic male sterility could be used to facilitate the field crossing of two stocks. In one case, detasseling is eliminated through the use of completely male sterile inbreds and in the other, cytoplasmic male sterility could be used in certain genotypes to delay pollen shed until the tassel has extended above the leaves followed by cutting with minimal leaf removal. In the latter case, cytoplasms may be used which produce only partial male sterility. In this situation, anther exertion is delayed 1 to 10 days (Duvick, 1965) and usually commences after the tassel is fully extended above the leaves.

The steps in the production of various categories of hybrid seed without detasseling using cytoplasmic male sterility as worked out for *cms*-T will be briefly described. Duvick (1959a, 1965) has reviewed in detail the publications relative to this application.

Most hybrid seed corn produced today is single cross, A × B. If inbred A is nonrestorer genotype (*rf/rf*), or in other words a maintainer genotype, and if it has been put into a male sterile cytoplasm by backcrossing, one can plant blocks of male sterile inbred A-*cms* alternating with blocks of inbred B and produce completely cross-pollinated seed on inbred A-*cms* without detasseling. If inbred B is also a nonrestorer (maintainer) genotype the hybrid planted by the farmer to raise a crop would also be pollen sterile. But if inbred B carries dominant restorer genes (*Rf/Rf*), the hybrid will shed pollen. Providing that a reliable genic restorer system is available for the particular male sterile cytoplasm used, the restorer gene system is likely to be the simplest for the seed producer, but since restored hybrids do not always shed adequate pollen (Duvick, 1959b), it is sometimes risky for the farmer. Consequently, much single cross production is likely to be of a nonrestored genotype in which case 25 to 50% of fertile hybrid seed produced by detasseling is blended with 50 to 75% of the identical hybrid produced by the cytoplasmic male sterile method to produce 25 to 50% plants which will shed normally in the farmer's field. One method of blending which assures a complete mixing is to plant one block of cytoplasmic male sterile inbred A, then the pollinator B and then a block of normal inbred A. Harvesting entails making one trip across the field in A-cms and returning through normal A.

Three-way ((A × B) × C) and sister-line crosses ((A' × A") × C) may be made in a similar way. First nonrestorer genotype A-cms and B, or A'-cms and A", are crossed and then either nonrestorer genotype inbred C for blended production, or restorer genotype inbred C(Rf/Rf) is used to pollinate the cytoplasmic male sterile single cross or sister-line cross seed parent.

Double cross production (A × B) (C × D) can be arranged so that the entire production is without detasseling, but providing normal pollen in the farmer's field. Various modifications are possible in which some detasseling is required and blending may also be required. If inbred A and B are both nonrestorer genotype and A is cytoplasmic male sterile, then the F_1 or single cross (A-cms × B) will be male sterile and it can be planted in alternating blocks with the single cross C × D to produce double cross seed without detasseling. If inbred C and D are nonrestorer genotype, the double cross will also be male sterile and blending with fertile seed from a detasseled block of the same hybrid will be required.

If inbred D has the restorer genotype Rf/Rf, the male single cross C × D would be heterozygous for this gene. The double cross (A × B = rf/rf)(C × D = Rf/rf) will segregate and produce two types of plants in the farmer's field; one half rf/rf and half Rf/rf. All plants would be in male sterile cytoplasm but half the plants would shed pollen. If inbreds C and D were both restorer genotypes, all plants in the double cross would be heterozygous and all would shed pollen.

Finally, if inbred C is cytoplasmic male sterile and nonrestoring genotype (C-cms, rf/rf) and inbred D has the restorer genotype (D Rf/Rf) then this single cross could be made without detasseling, yet as a pollinator in the production of the double cross it would shed pollen and gene segregation would result in 50% of the plants shedding pollen in the farmer's field. To recount, if double cross production without detasseling is desired, and 50% fertility restoration in the crop is acceptable, the pedigree would be (A-cms, rf/rf × B, rf/rf) (C-cms, rf/rf × D, Rf/Rf).

The foregoing discussion assumes that restorer genes operate as in the Texas cytoplasm system (Duvick, 1965) and not as in the USDA or S cytoplasm system described by Buckert (1961). A double cross produced as above using an S-type cytoplasm, might not shed enough fertile pollen in a farmer's field since the pollen shed by Rf_3/rf_3 plants, the restorer gene for cms-S, is only 50% fertile and not 100% as with Rf_1/rf_1, the restorer of cms-T.

It should be noted that the use of the restorer line to restore fertility to cytoplasmic male sterile hybrid corn was covered by a patent (Jones, 1956) assigned to Research Corporation, New York, N.Y., which expired in 1973.

Grogan (1971) has suggested a "multiplasm" approach to the production of hybrid corn to "safeguard against a monocytoplasm, particularly one as a male sterile" to lessen the likelihood of a recurrence of a pest problem such as that encountered with Texas male sterile cytoplasm. Grogan's proposal involves converting the component female lines to two or more cytoplasms with common restorers. According to Grogan,

> The stocks could be maintained separately and mixed equally in hybrid production, or by using a mixture of equal portions of each cytoplasm and reproduced with the normal maintainer counterpart. Periodic checks for proportions of each cytoplasm in the mixture would be possible through the use of differentiating restorer testers. The latter approach would have the advantage of eliminating separate maintenance of stock material, and from another point of view, permit a shift of cytoplasms in the population according to environmental stimuli such as diseases.

The proposal, while simplistic in design, may be difficult to execute in large scale seed production.

Genic Male Sterility

A large number of recessive genes conditioning male sterility are known in corn (Neuffer et al., 1968). A U.S. patent (Patterson, 1973) has been granted Dr. Earl Patterson covering

> Procedures for use of genic male sterility in the production of commercial hybrid maize, including producing and maintaining seed stocks substantially of a homozygous male sterile genotype and stocks substantially of a heterozygous, male sterile allele and male fertile allele, genotype, which include a differentially transmitted variation of chromosomal constitution.

The patent has been assigned to University of Illinois Foundation, Urbana, and a license from University Patents, Inc., is required to produce seed of male sterile corn and use such seed in the conversion of commercial corn lines to male sterile corn seed but not to sell such male sterile corn seed or other corn seed which is covered or whose production is covered by the licensed patents. A royalty will also be applied to the sale of commercial seed reproduced without detasseling employing the "Patterson method."

Ordinarily the male sterile phenotype cannot be propagated in pure form because of male sterility. Homozygous recessive ms/ms plants are generated anew each generation by segregation in an F_2 or backcross as one-fourth or one-half of the progeny. Briefly, the Patterson system allows the generation of a homogeneous ms/ms progeny. If this system has been introduced into inbred A, then a commercial single cross is made by alternating blocks of pure male sterile A (A, ms/ms) with regular inbred B which has the fertility allele Ms/Ms. The single cross will be heterozygous Ms/ms and shed pollen normally in farmer's fields.

In this system, all inbreds are automatically genetic restorers, that is, they carry the dominant Ms/Ms allele for all possible recessive male sterile genes. It is necessary to introduce the ms-allele into an inbred by backcrossing just as with cms, but in addition, it is necessary to breed a maintainer version of each inbred as well, since as already mentioned, all inbreds are restorer genotype. This is the crux of the Patterson patent. He utilized the knowledge that certain modified chromosomes, called duplicate-deficient (Dp-Df), are transmitted through the female, but not through the male, to set up a heterozygote (the maintainer stock) with the dominant male fertile allele on the Dp-Df chromosome and the male sterile allele on a normal chromosome (*Dp-Df Ms/ms*) so that only the *ms* allele is transmitted by the pollen. Thus in the cross *ms/ms* × *Dp-Df Ms/ms* only *ms/ms* plants are produced. After the male sterile and maintainer versions of an inbred have been developed the male sterile version is increased by alternating blocks of the *ms/ms* version with the maintainer version (*Dp-Df Ms/ms*) and harvesting seed from only the *ms/ms* rows. The maintainer version is propagated by selection within the pollinator rows.

Single crosses, sister-related line-parent crosses, three-way crosses, and double crosses may also be produced by the Patterson method. The seed parent single cross (A × B) is produced without detasseling by alternating blocks of A-*ms/ms* with maintainer B-*Dp-Df Ms/ms*. The hybrid so produced will all be male sterile *ms/ms* and may be pollinated with a regular version of inbred C (Ms/Ms) to give a fertile three-way cross. For a double cross, the pollinator single cross (C × D) may be produced without detasseling by crossing C-*ms/ms* × regular inbred D-*Ms/Ms* in which case the hybrid will be fertile and shed pollen, or regular C × D produced by detasseling may be used since only a small amount of seed is required.

Note that the Patterson system requires three versions of each inbred — regular (*Ms/Ms*), male sterile (*ms/ms*), and maintainer (*Dp-Df Ms/ms*) somewhat in the same way that three versions of an inbred are possible with cytoplasmic male sterility — regular, *cms*, and restorer genotype. Note also that just as with cytoplasmic male sterility, restored commercial single crosses may not be substituted for seed parent single crosses in the production of three-way or double crosses.

Successful seed production requires the right version of each inbred be used depending on whether a commercial single cross or a seed parent single cross is desired. Similarly, to multiply the inbred stocks extra care is required. Thus even after the conversion of inbreds which is complicated, time consuming, and expensive; additional expenses in the form of foundation seed production inventory and quality control are required. It remains to be established whether the Patterson system is a panacea to seedsmen.

Chemical Male Sterility

Even prior to practical use of cytoplasmic male sterility, many articles were written concerning the possibilities of development of chemical pollen control agents for corn and other crop plants (Brasfield, 1950; Murneek, 1949; Naylor, 1950). Since these publications, chemical induction of male sterility has been demonstrated in cotton (Eaton, 1957), in cucurbits (Wittwer and Hillyer, 1954), in sunflower (Kiermayer, 1959), in wheat (Rowell and Miller, 1971), in corn (Moore, 1950; Nelson and Rossman, 1958; Rossman, 1958), and in several other crop plants (Heslop-Harrison, 1957; Chopra et al., 1960). Subsequent reviews (Anderson, 1971; Barrons, 1971) have specifically emphasized practical applications and possibilities for development of chemical pollen control agents for corn and sorghum seed production.

A system eliminating pollen-shedding, with no adverse effects on the leaves or grain development of corn or rice, has been reported (Long, J.D., M.B. Weed, D.J. Fitzgerald, and G.E. Barrier. 1973. Chemical control of pollen. Am. Soc. Agron. Abstr. p. 54). Further studies (Laible, 1974) indicate the system may be commercially feasible both economically and with adequate reliability, although further research is necessary.

A selective male gametocide would be of tremendous value as a tool to eliminate tassel removal from seed production fields. No longer would conversion of inbred lines to cytoplasmic or genic male sterility be a necessary investment. Foundation seed operations would not be involved in sterile, maintainer, and restorer seed increases. Even with the elimination of all these problems, new ones of dosage, time, and method of application would take their place. The timeliness of treatment application would probably be critical since the action of most chemicals studied to date (Ericksen and Ross, 1963; Wittwer and Tolbert, 1960) inhibit the early stages of meiosis similar to the case of cytoplasmic male sterility (Artschwager, 1947; Chauhan and Singh, 1966; Dubey and Singh, 1965; Singh and Hadley, 1961). The problem of "shedders" as in cytoplasmic male sterile production would still be present, due to some plants not being treated with enough chemical, or the plant not being at the proper growth stage when treated.

An alternative proposal (Jones and Mangelsdorf, 1951; Jones et al., 1957) would solve these problems. If it were possible to treat genic male sterile plants so they would shed viable pollen then only the foundation seed increase need be treated. No maintainer line would be required and no complicated cytogenetic system would be necessary to assure male-sterile seed parents for the production field. Weather hazards would be eliminated as no treatments or field operations would be necessary in the seed field during pollination time.

This approach would have a built-in safety factor. If a treatment to prevent pollen shed fails, one must either pull or remove tassels mechanically, if time permits, or abandon the field. If a system for restoring pollen shedding ability to male sterile plants were only 50% effective, a satisfactory seed increase could still be obtained. Chemical treatments for fertility restoration is an intriguing possibility but suitable chemicals are currently unknown.

HARVESTING THE SEED CROP

Harvesting of the commercial hybrid corn seed crop is of necessity very closely correlated with operations of processing facilities. The discussion herein follows the pattern generally in use at a typical processing location. There are variations among producers as well as within locations of a single producer. Figure 7 gives a schematic outline of the basic operations of such a facility from harvest through processing and distribution (adapted after Airy, 1955).

Harvesting, husking, and sorting generally are field operations but husking and sorting may be done at the plant. Harvesting, sorting, drying, and shelling may be accomplished at one location with the sizing, cleaning, storage, and distribution at another centralized location.

Maturity and Seed Quality

Harvesting of the seed crop may begin as soon as the corn is physiologically mature. Generally the crop would be harvested when the moisture level of the grain is within the range of 30 to 38% and somewhat depends upon variety, environment, and geographic location. Physiological maturity is regarded as that point when the grain has reached its maximum dry matter accumulation. Much research has been conducted to index maturity by heat unit accumulation or growing degree days (Andrew et al. 1956; Baker, 1970; Gamble, 1971; Gilmore and Rogers, 1958; Gunn and Christensen, 1965; Hallauer and Russell, 1961, 1962; Mederski, H.J., and C.R. Weaver, 1970. Relation of corn development and maturity to growing degree days. Am. Soc. Agron. Abstr. p. 54; Neal, 1968). In general, a high correlation was found between degree days and moisture loss. "Black layer" (Daynard, 1969, 1972; Daynard and Duncan, 1969) formation indicates physiological maturity and serves as a good, reliable, and easily determined indicator thereof. Numerous additional studies have been conducted to study varietal effects upon drying rates (Andrew et al., 1956; Carter and Poneleit, 1973; Crane et al., 1959; Hanway and Russell 1969; Sutton and Stucker, 1974).

Harvest of the seed crop as soon as it is mature 1) reduces risk of freeze injury, 2) avoids excessive field losses from mechanical pickers, 3) reduces risk of delays due to adverse weather conditions, 4) reduces losses from insect damage, and 5) reduces losses from ear and stalk rots and other diseases. All of these factors contribute to the quality of the seed crop not only in appearance but in physical damage.

The effect of freeze damage upon seed germination (Airy, 1955) is a constant major hazard to seed corn producers. Studies reported by Neal (1961) indicated that injury to germination from freeze damage is directly related to kernel moisture as well as intensity and duration of exposure. The higher the

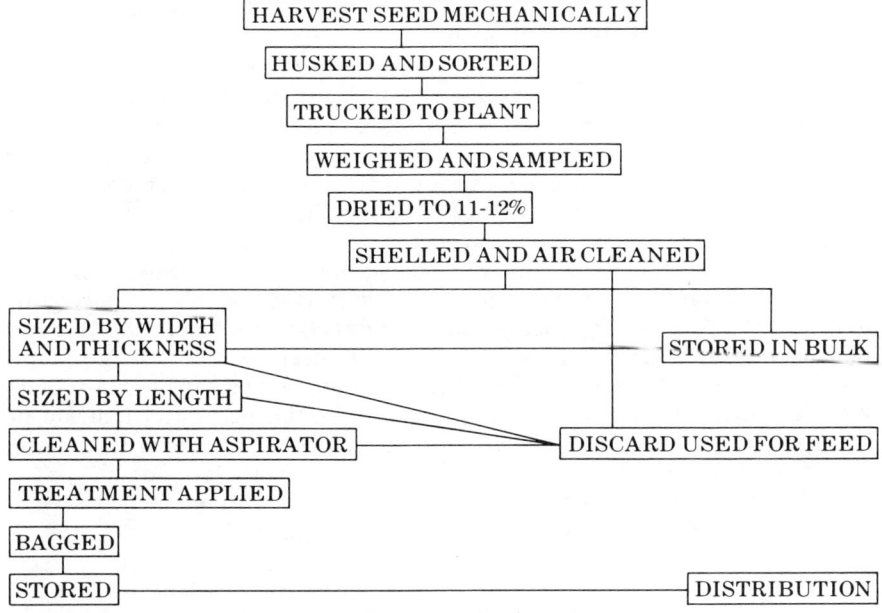

Fig. 7. Typical harvest and processing operations.

moisture, the greater the effect on germination at all levels of freeze treatments. Rossman (1949) summarized his work by concluding that the amount of damage by freezing was dependent on 1) temperature, 2) duration of exposure, 3) moisture of seed, 4) variety, 5) husk protection, 6) stage of maturity, and 7) rate of drying after freezing.

In a study (Jenkinson, R.C. 1973. The effect of simulated early fall frost on the yield and drying rate of corn. M.Sc. thesis, Univ. of Guelph, Ontario), it was found that when paraquat was used as a desiccant at 49% grain moisture the yield reduction was 28%, but had no significant effect on grain moisture at harvest. The early season freezing temperatures of 1974 following late planting as a result of early season wet weather and severe widespread mid-summer drouth resulted in reduced seed yields and some poorer quality seed. The final effects of this unusual growing season are as yet unknown.

Field Operations

The contract grower is responsible for harvesting the seed crop and delivering it to the driers. Harvest rate is determined by drier capacity. A well-coordinated schedule under the direction of the producer's fieldmen is necessary to keep field operations moving smoothly and drier operations at full capacity.

Mechanical harvesters used today are two-row mounted type pickers, two or three row pull-type pickers, and uniharvesters with two, three, or four row heads (Fig. 8). A great deal of emphasis is placed upon reducing mechanical damage from field harvesters by 1) removing pegs from snapping rolls, 2) encouraging the use of new rolls or machines with stripper bars, 3) proper adjustment of rolls, and 4) driving at a slower rate of speed.

The grower delivers the seed to field sorters which are equipped with a husking bed and sorting table (Fig. 9). The husking bed removes the majority of the husks prior to the corn passing over the sorting table. A return conveyor is provided to take unhusked ears back over the husking bed. Usually three or four workers sort the seed as it passes over the table. These workers are instructed and trained to remove any diseased, off-type, or off-color ears from the seed. After sorting, the seed is elevated into a truck and delivered to driers. Some seed corn producers husk and sort the seed after transporting to the plant and before moving to the driers.

In recent years, especially with the development of cage type field shellers, seed has been harvested as shelled corn. It is necessary to have grain moisture at 20% or less to keep mechanical damage at a minimum. It is further necessary to have seed fields free of volunteer corn or off-type plants before this method can be used as there is no opportunity for sorting. Comparative studies (Funk Seeds International, unpublished data) indicate no more, and often less, mechanical damage and greatly reduced field losses from field shelling of most seed parents at grain moisture levels of 20% or under. A portion of this data is presented in Table 1. The shelled seed is delivered to driers where it enters the normal processing flow.

Fig. 8. Mechanically harvesting the seed rows using a four-row uniharvestor.

Fig. 9. The field husking and sorting operation in progress. Workers remove undesirable ears before further processing.

Table 1—Comparison of seed quality factors for two hybrids, each harvested as ear corn and picker shelled, with population and yield values from that seed

Factor	Hybrid A		Hybrid B	
	Picker sheller	Ear harvest	Picker sheller	Ear harvest
% Moisture	15.9	25.0	16.7	26.3
% Damage	4.86	7.55	4.43	7.34
Warm test germ	98.8	99.3	98.3	98.2
Cold test germ.	93.3	93.3	95.3	97.3
Plants/acre	26,300	26,100	26,500	26,600
Yield	148.3	149.6	155.7	155.9

Plant Operations

As the seed is delivered to the plant, it is first weighed, and then a sample is collected as it is being dumped in preparation for movement into the driers. The ear corn sample is taken so that the final sample weight of all the production of a single grower is about 4.5 kg (50 lb) and is representative of the entire hectarage. A sample is secured from each hybrid being grown. If maturity or moisture content within a grower's hectarage varies, more than one sample is secured. The sample is shelled, moisture accurately determined, shelling percentage calculated, and converted to No. 2 grade basis. These data are the basis of payment to the contract grower and represent the net accepted seed.

The seed is then conveyed from the dump immediately to the driers. Seed corn driers vary considerably through the industry with many different systems being used. The majority of the systems today utilize a squirrel-cage or axial-vane fan system, pulling or pushing fresh air through a burner and forcing heated air down or up through bins filled with seed. Most burners today are fired with natural, butane, or propane gas, with a few of the older systems yet using fuel oil.

Drying temperatures vary from 40 to 46 C (105 to 115 F) with temperatures being measured at the point of contact with the seed. The entire drying procedure is very closely monitored since higher temperatures are detrimental to seed quality. Temperatures up to 49 C (120 F) have been reported as noninjurious to seed viability (Harrison and Wright, 1929; Robinson, 1934; Kiesselbach, 1939; Wileman and Ulstrupp, 1945; Dimmock, 1947; McRostie, 1949), however, seed with high initial moisture content is more susceptible to germination damage and is best dried at temperatures of 38 to 43 C (100 to 110 F) (McRostie, 1949; Wileman and Ulstrupp, 1945). It was found (Struve, W.M. 1958. Drying and germinability of maize. Ph.D. thesis, Iowa State College, Ames) that grain harvested with more than about 25% moisture and rapidly dried showed reduced viability and vigor. Injury increased with increase in initial moisture content. Air temperatures of 46 C (Airy, 1955) may be used when moisture content of the seed is 20% or less, when the seed was mature prior to harvest, and when the hybrid is not susceptible to injury by high temperature drying. The seed corn is dried to 11 to 12% moisture at which time it is conveyed to the sheller.

The level to which seed corn is dried and held in storage is critical.

PRODUCTION OF HYBRID SEED

Research (Cal and Obendorf, 1972) has shown that imbibition of low moisture corn kernels (6%) at temperatures of 5 C (41 F) resulted in malformed and delayed seedling growth. Prolonged exposure to this low temperature resulted in reduced survival for certain genetic material. Sensitivity to imbibitional chilling was reversed when initial kernel moisture was 13 or 16%.

PROCESSING THE SEED CROP

Processing of seed corn is a series of activities beginning with shelling the dry seed and ending when the seed is bagged and ready for storage or distribution. The primary steps involved are 1) shelling, 2) cleaning, 3) sizing, 4) treating, 5) bagging, and 6) planter plate selection. These activities are accomplished in plants of all sizes and descriptions ranging from modified and adapted farm buildings to new, modern masonry or metal structures specifically designed and built for the purpose of efficiently handling large volumes of seed (Fig. 10). Individual installations may range in processing volume from a few thousand to several hundred thousand kilograms per year.

Certain objectives must be constantly kept in mind during the processing procedures. First, the seed must be handled in such a manner that a minimum of damage occurs. Any breaks which might occur in the seed coat have a direct and detrimental effect on germinability and vigor of the seedling plant. The degree of the effect depends on the severity and location of the injury (Wortman and Rinke, 1951; Wright, 1948). The effects can be minimized to a degree by the application of seed treatment.

Fig. 10. A modern, fully integrated hybrid corn seed processing plant, storage facility with year around controlled atmosphere conditions, and distribution center located at Bloomington, Illinois.

A second objective is the necessity to achieve maximum plantability within reasonable limitations imposed by equipment available to accomplish the job. The entire objective of sizing seed is to achieve uniformity of kernel size and shape so that seed will drop accurately with plate-type planters. When farmers were planting for stands of from 25 to 35 thousand plants/ha (10 to 14 thousand plants/acre) this factor was of lesser importance. One or two thousand extra plants per acre was not important at those levels. However, the modern corn farmer who is maintaining high soil fertility and planting for maximum production cannot tolerate significant reductions or increases in stand. When planting for stands up to 59 thousand plants/ha (24 thousand plants/acre), 2 thousand more might put a field under serious stress, especially if moisture might be less than average.

Finally, the seed processor must carefully maintain schedules to ensure the orderly movement of seed lots through distribution channels to meet planting schedules. The demands made on the processor have changed and increased as customer requirements have increased and become more specialized. Seed processing in the early days consisted of only drying enough for safe overwinter storage. With the invention and improvement of corn planters the need for sizing became important.

With the advent of hybrid corn, farmers became more critical and demanding of uniform sizing and manufacturers began to make a greater selection of planter plates to meet the farmer's need. Late in the 1950's and early 1960's when hybrid seed, including single crosses, high fertility rates, herbicides, and insecticides became common practice, seed processors were challenged for the best possible in seed sizing and processing.

Shelling the Seed

Shellers in use for seed corn basically are of two types, cage and cone type. They are designed so that when operated at low speeds the seed is more or less rubbed from the cob. As speed increases the action becomes more of a pounding with a drastic increase in kernel damage. It is desirable that the contact edges of cages or cones be well smoothed to reduce damage. Experience has shown that with adequate horsepower shellers can be operated at lower speeds, and reduce kernel damage by keeping the sheller full at all times.

Airy (1955) reported data showing that germination went down in an inverse ratio to the revolutions per minute (rpm) of the sheller cylinder. As cylinder speed increased from 240 rpm to 720 the cold test germinations of treated seed dropped from 79 to 64%, with nearly all of the decrease in the last 120 rpm increment.

The need for improved shellers that would do less kernel damage has long been recognized. Several experimental shellers have been developed and some have been successful in holding down damage, but capacity was so low that they were impractical. One of the most promising is a roller sheller (Brass and Marley, 1973) with about 50% less kernel damage in seed below 20% moisture. The capacity of the test machine was about 37 quintal (150 bu) per hour. Larger models should give correspondingly higher capacities.

Conveying the Seed

Since reduced kernel damage is one of the objectives in processing seed, it follows that each step should be evaluated to determine the amount of damage chargeable to that procedure.

Some of the conveying equipment formerly used was directly responsible for much of the mechanical damage. Grain augers crack and scuff seed. Chain conveyors crush and crack kernels. Single tube elevator legs allow seed to be caught under belts and behind cups. Perhaps the greatest source of damage is dropping seed from great heights into steel bottom bins or onto concrete floors.

Augers have given way to belted conveyors. Elevator legs are of the double tube type and are equipped with either all plastic or plastic lipped buckets. Chain conveyors are seldom used anymore. Breakage from impact on bin bottoms has mostly been eliminated by using grain spirals or ladders to lower seed gently with minimum impact. Any time fall distances are greater than 4.5 m (15 ft) such equipment should be considered. Lateral conveying of seed is done either on rubber belting or with vibrating trough conveyors.

Cleaning the Seed

Seed corn from the sheller contains varying amounts of foreign material, consisting of bits of cob, husk and silk, and pieces of kernels. Storage properties of seed are greatly enhanced by removal of such debris which encourages storage insect problems and the development of hot spots when stored for long periods. Aeration and cooling are improved by removing such material, and storage space required is less.

Machinery used in seed cleaning are combinations of wind chambers and sieving screens and are referred to in the industry as "scalpers." The term scalp in this case was derived from the milling industry and literally means to remove or separate a portion from the whole. Shelled seed is fed into the machine and passes over sloping shaker screens which remove large kernels (over 24/64) and small kernels (through 16.5/64). (Sizing equipment used in the hybrid corn seed industry in USA and abroad is designated in English rather than metric units.) The sieving action removes pieces of cob and fines at the same time. Seed then passes through a blast of air which removes fines, cob, and dust which has escaped the screening action. Debris removed with the air stream is deposited in dust collectors. Grain removed is disposed of as feed or market corn.

In some operations, all seed is cleaned twice. The first time, following shelling, the scalper is set for high volume to remove foreign material preparatory to storage. Before seed enters the sizing line, it is again scalped at a slower rate to remove overs and unders and any remaining foreign matter. A thorough cleaning prior to sizing is essential in that 1) it greatly reduces the amount of dust which may increase housekeeping chores, 2) clean seed flows more freely in the sizing line, and 3) the more undesirable material, cracks, and fines that can be eliminated before sizing, the better the job that can be accomplished by the aspirators at the end of the sizing line.

Sizing the Seed

The term "sizing" as used here should replace the erroneously used term "grading." Sizing as used in the seed industry means separating kernels into uniform lots of sizes and shapes based on width, thickness and length, while grading carries a connotation of quality measurement which does not apply.

Tolerances have been proposed (Bateman, 1972; McKee, 1963) of 3/64 in for width and thickness and 4/64 in for length. These are reasonable tolerances and if seed is sized within these limits will plant with acceptable accuracy with planters in good operating condition. Bateman (1972) suggests kernel drop tolerances of 3% under drop to 7% over drop. This range of accuracy will generally be achieved if seed is sized within the tolerances.

It was early recognized that the large number of seed parents being used in the industry would dictate a large number of size categories in order to get uniformity within a size (Wright, 1941). The advent of single cross hybrids has increased the number of sizes utilized. Because of seed production cost considerations it is the seedsmens' goal to increase recoveries from raw seed to highest possible levels. This means saving some kernel sizes which formerly would have been considered unsalable.

Tables 2 and 3 (Airy, 1955) showing basic screen arrangement being used in sizing are essentially unchanged, therefore are reprinted. A noticeable difference in practice today is that some sizes marked for discard are now utilized for planters not requiring plates. Salvaged seed sizes now often go as low as 16.5/64 to 16/64 in in width. In length size processing short kernels of high quality formerly discarded are now saved.

The types of equipment used today in sizing are essentially the same in operating principle as those previously used, although somewhat more sophisticated. Cylinder sizers used singly or in batteries provide greater capacity than former equipment. Gravity mills have largely been replaced by aspirators. An aspirator using an air separation principle has greater capacity and does not require the degree of finesse in operation. Aspirators do not perform well on mixed size seed lots but are very satisfactory on individual size seed lots. Mixtures or blends of sizes to be used in planters not requiring plates should be aspirated before blending.

Sizing of seed corn accomplishes a purpose important to sales organization. It lends eye appeal. However, with some kernel shapes, extreme uniformity may actually hinder plantability by under dropping. Large flat kernels may fall into this category. In general, sizing of seed today is being done very conscientiously (Bateman, 1972). There is room for progress in reducing kernel damage. Improved machinery design could accomplish this.

Treating the Seed

It is common practice to apply a combination fungicide-insecticide to seed before bagging. Questions have been raised and investigated (Grogan, 1958) as to whether chemicals applied may have a retarding effect on germination. The purpose of treating seed is to protect against seedling diseases and to give short-term protection against storage insects. The recommended

Table 2—Outline of screen arrangement for sizing corn seed, using slot screens first and then dividing each portion with different round hole screens. Screen sizes given in 64ths of an inch†

Slotted screen	Round-hole screen	Length sized	Aspirated‡
Over 14	LR - Large Round 22-25	No	Yes
	LR - Large Round 20-22	Yes	Yes
	MR - Medium Round 18-20	Yes	Yes
	SR - Small Round 17-18	Yes	Yes
Through 14 Over 13	TF - Large Thick Flat 22-25	Yes	Yes
	TF - Medium Thick Flat 20-22	Yes	Yes
	Thick Flat Discard Through 20		
Through 13 Over 12	LF6 - Large Flat 22-24	Yes†	Yes
	MF6 - Medium Flat 20-22	Yes†	Yes
	MF5 - Medium Flat 18-20	Yes	Yes
	Small Flat Discard 17-18		
Through 12	MF4 - Medium Flat 18-20	Yes	Yes
	SF - Small Flat 17-18 or 17½-18½	Yes	Yes
	Tips - Discard 16-17		

†Short kernels reworked and used as a special size only if needed. ‡Discard from Aspirator used for feed only (after Airy, 1955).

Table 3—Outline of screen arrangement for sizing corn seed, using round hole screens first and then dividing each portion with slotted hole screens. Screen sizes given in 64ths of an inch†

Round-hole screen	Slotted-hole screen	Length sized	Aspirated‡
Extra Large Over 24 or 25	Discard for feed		
Large Over 21, 21.5, or 22 Through 24 or 25	LR - Large Round Over 13.5 or 14		
	LF - Large Flat Through 13.5 or 14	Yes†	Yes
Medium Over 18, 18.5 or 19 Through 21, 21.5 or 22	MR - Medium Round Over 13 or 13.5	Yes†	Yes
	MF - Medium Flat Through 13 or 13.5	Yes	Yes
Medium - 1 Over 17.5 or 18 Through 18, 18.5 or 19	MR - Medium Round-1 Over 12.5 or 13	Yes	Yes
	MF - Medium Flat-1 Through 12.5 or 13	Yes	As needed
Small Over 17 or 17.5 Through 17.5 or 18	SR-Small Round Over 12 slot	Yes	Yes
	SF - Small Flat Through 12	Yes	Yes
Extra small or tips Through 17 or 17.5	Discarded for feed		

†Short kernels reworked and used as a special size only if needed. ‡Discard from Aspirator used for feeding purposes only. (After Airy, 1955).

dosage is 600 ppm of actual captan or about 21 g/25 kg (0.75 oz./bu) of seed. A smooth uniform application is desired. Properly treating seed may help offset vulnerability to disease caused by mechanical damage (Wortman and Rinke, 1951; Wright, 1948). A study to evaluate seed treatment effectiveness in controlling seedling diseases, and the effect on plantability (Wright, 1948) indicated that seed treatments are not a panacea for all stand problems.

The most commonly used fungicide is captan. To the basic captan carrier 2 or 3% of an insecticide as methoxychlor or malathion is added. The material is available as wettable powder or as heavy water suspension. It is premixed into a slurry and applied to the seed by machines which meter the material and apply it by a tumbling or spraying action.

Planter Plate Selection

Practically all seed corn producers offer planter plate suggestions for customer convenience. The process of making planter plate checks not only establishes the plates which can best be suggested to the customer, but also serves as a check on the accuracy of sizing. If the check planter will not drop the seed accurately, adjustments can be made on the sizing equipment to improve accuracy. Most plate selection is done on actual planter test stands provided by planter manufacturers. Care must be exercised to check and replace worn parts in test stands just as farmers should do with their equipment.

A new electronically monitored check planter has recently been made available. It is much faster and eliminates the human error factor in counting drop accuracy. This innovation should enable seedsmen to do a better job of planter plate suggestions.

Bagging the Seed

Most seed corn today is packaged by automatic or semiautomatic equipment (Fig. 11). A specified weight of seed is dropped into each bag. The seed is weighed, the bag is hung, filled, and coded with only a service man to keep the operation running smoothly. Today nearly all seed corn in the USA is packaged in multi-ply paper bags, most of which contain a moisture barrier of free polyethylene or a polyethylene coated sheet for protection against exterior moisture. A rough coated or ribbed outer ply with a non-slip coating improves handling and stacking ability.

Traditionally in the U.S. hybrid seed corn has been sold by the bushel but more recently producers have departed from that unit of measure. Some have utilized a unit with a specified number of kernels, usually 80 thousand, regardless of weight. This method has led to some problems as 80 thousand kernels of each size do not weigh the same or require the same space. Bag weights may vary from 18 to 32 kg (40 to 70 lb) and a variety of bag sizes are necessary in order that bags of all sizes will be full. Packaging by kernel count further causes some warehousing problems by creating different size stacks for the same number of bags. Other companies have chosen to package their product in 23 kg (50 lb) units. The approximate kernel count is stamped on each bag.

An important step in processing is collecting and saving a truly representative sample of each seed lot. This is best accomplished by an automatic sampler at the bagging scale which draws off a few kernels as each bag is filled. The sample thus obtained is used by quality control to run germination tests, determine kernel counts, and make damage and inert material checks. State and Federal laws require certain information to be printed on each bag or on a tag affixed to each bag. Requirements vary from state to state and attempts have been made to standardize these requirements.

QUALITY CONTROL

In considering hybrid corn seed quality, many changes have taken place during the past 20 years, and in general for the better. Farmers today, because of higher production goals and increased production costs, are demanding and getting seed of the highest quality. It is the goal of seedsmen to supply, in adequate quantities, a product which will make efficient use of the environmental conditions under which it is grown (Bunch, 1958). From such goals increased sales and profits are insured (Kaerwer, 1971). To help realize such goals each company needs a quality control program to monitor all phases of seed production.

The quality control program should have well defined procedures and standards understood by all levels of management. Data should be collected according to rules and procedures outlined in Rules of Testing Seeds (Anonymous, 1970), the Certification Handbook (Anonymous, 1971), and the

Fig. 11. A typical bagging line with seed being weighed, dropped into bag, and coding information printed, preprinted tag applied and bag sewn closed while traveling on a belt conveyor.

International Rules for Seed Testing (Anonymous, 1966). Often more strict and precise rules and procedures are adopted by company management.

Genetic Purity

Dedicated personnel in research, production, and processing have developed improvements permitting corn yield to improve substantially. Hybrids have been developed with specific inheritance traits resulting in predictable performance under specific conditions. A host of events influence the actual performance in each customer's fields. Those events affecting the quality of the seed are of particular interest to quality control. Genetic purity is the beginning of such a program. Maintenance of genetic purity of each hybrid combination is a challenge to foundation, production, and processing personnel. Quality control programs should satisfy these responsibilities.

Foundation Seed

Maintaining genetic purity starts with pure foundation seed stocks (Gregg, 1972). Having available adequate seed parent and pollen parent foundation seed stocks is not an easy task. Trained and skilled personnel are required to maintain and improve parent stocks on a quality level of near zero tolerance for genetic purity. Procedures and standards followed by certification agencies (Anonymous, 1971; Cowan, 1972) indicate the importance of foundation seed free from weed seed, mechanical damage, and errors that can be perpetuated in commercial seed production. Commercial seed companies generally do not certify all of their foundation seed; however, guidelines developed over the years for the most part exceed or at least equal those of certifying agencies.

Commercial Production

Commercial production of adequate high quality hybrid corn seed to plant the projected acreage of grain and silage in the U.S. and Canada requires both a large acreage and a large number of people. Selection of this land requires careful attention to 1) cropping history, 2) drainage, 3) uniformity, and 4) cultural practices. Special efforts are used to avoid known weed problems and corn planted following corn on the same land.

Minimum standards for isolation of seed corn production fields have been established (Anonymous, 1971) for USA and Canada. Inconsistencies exist among states, but for the most part, 201 m (40 rods) minimum is required between the seed parent of the hybrid being produced and any other corn. This distance can be modified by 1) barrier rows (Fig. 12) and size of field, 2) adequate natural barriers, and 3) differential flowering dates.

The function of barrier rows is to furnish a mass of pollen to dilute pollen from a contaminating source (Jones and Brooks, 1950). Modifying isolation distance by adequate natural barriers (Jones and Brooks, 1952) has been studied. Natural tree barriers were effective in reducing the amount of outcrossing 50% immediately behind the barrier but were less effective at greater distances. Natural barriers are not as effective as border rows of corn.

PRODUCTION OF HYBRID SEED

Differential flowering times is effective in isolation. Full knowledge of parental stock is essential. It is absolutely necessary that silks of the seed parent and pollen in the pollen parent of adjacent fields not be flowering at the same time.

To maintain genetic purity, it is necessary that isolation standards, whatever they might be, should be firmly enforced (Airy, 1950). But even more important, special care should be taken to insure a proper nick of seed parent silks and pollen from the pollinator.

Timely removal of all tassels in seed parent blocks is a major and costly task. Genetic purity of the intended cross is dependent upon the timely removal of tassels in the seed-parent blocks. This is a major and costly task; the compliance standards used closely approximate those established by certifying agencies (Anonymous, 1971). When 5% or more of the seed parent plants have receptive silks, tassels or portions of tassels which have 5 cm of the central spike, branches, or a combination of both shedding pollen, are limited to 1% at any one inspection or a total of 2% for three inspections on different dates.

One method of observing, reporting, and offering assistance in interpreting the standards and rules is the use of independent seed field inspectors. These inspectors are assigned a given acreage, usually no more than 2,000, to be carefully checked to see if genetic purity standards are met. Before detasseling they establish compliance with isolation requirements, and check for volunteer corn or off-type plants in both seed parent and pollinator rows. When detasseling begins these inspectors check all fields at least every other day, many times more frequently. Reports are filed for each check. The object is to keep those responsible informed in order to prevent any violation of standards.

An alternate to, or in some cases an addition to, the seed field inspector is the growout system. Perhaps the most accurate method of checking genetic purity is by the growout, providing an adequate sample of the entire lot being produced is secured. A successful growout is dependent upon 1) a clean, uniform field to minimize variation, and 2) adequate checks so that accurate comparisons can be made. Growouts are most generally made in an area where the crop can be grown in the fall and winter months. Florida and

Fig. 12. A foundation seed field showing additional pollen parent border rows to insure adequate isolation.

Hawaii are the areas most generally used by U.S. seedsmen. In this way, results can be made available before the seed is processed and distributed.

Specialty Corns

Production and quality control procedures for maintaining genetic purity of corn seed other than yellow dent (white, waxy, high lysine, high amylose) requires procedures that are similar but somewhat different than those outlined for dent corn. These specialty corns differ from normal yellow dent in being homozygous recessive at one or more critical loci. Thus contamination arising from any foreign pollen would mask expression of the desired trait.

Production standards are more demanding for specialty corns. Any expression of xenia on the seed kernels must be removed in order to meet genetic purity standards. Isolation standards to minimize contamination have been set at 201 m (660 ft) plus four border rows when the field is 4 ha (10 acres) or less. The distance may be decreased as field size increases and further decreased by increased number of border rows. Bear (1975) discussed purity requirements for waxy and high amylose corn. Most seed companies producing specialty corns will use these basic guidelines for isolation standards.

When contamination does occur, it becomes necessary to establish procedures to remove the off-type kernels in order to maintain genetic purity standards. The primary way of removing contaminated kernels is by hand sorting. In the case of white corn, electronic devices sensitive to minute color changes can be used. The equipment is expensive and its capacity is limited so therefore most of the contaminates are still removed by hand picking.

With waxy corn, a special technique to identify dent contamination can be used (Jugenheimer, 1958). Applying a special iodine solution to the exposed endosperm starch of a nicked kernel causes dent kernels to stain blue while the desired waxy kernel stains a reddish brown.

Physical Quality

Physical quality is measured by the amount and kind of damage present and by the viability and vigor of the seed. These traits are closely correlated with field emergence and stand, uniformity of growth, and ultimate yield. The role of quality control is to monitor, using acceptable tests and procedures, each phase of seed processing from field to the customer.

Seed is said to be in its best physical condition when it is physiologically mature, on the ear on the seed parent plants in the field. Many activities affecting physical quality are performed from the time it leaves the plant in the field until it is planted by the customer. Isolating and identifying the source, nature, and extent of poor physical quality are quality control responsibilities.

Damaged seed, the kind of damage, and its influence on stand and yield has been studied and reported by many (Alberts, 1927; Brown, 1920; Delouche, 1971; Kaerwer, 1971; Tatum and Zuber, 1943; Wortman and Rinke, 1951; Wright, 1948). Their collective conclusions were that seeds are the product of their environment, natural or man made, and that sound seed can be expected to give satisfactory stands resulting in good yields.

Seed left standing in the field after physiological maturity are subjected to conditions tending to lower quality (Matthes and Rushing, 1972) resulting in loss of vigor and viability due to high rates of respiration and increased disease and insect damage. Seed being harvested and dried are subjected to mechanical damage. During shelling, cleaning, sizing, treating, and bagging mechanical damage often occurs. Samples are collected from all areas of concern for analysis.

Sampling

Seed quality is measured by testing a representative sample of a lot or prescribed quantity of seed. Procedures for taking representative samples, and the necessary equipment needed, have been established (Anonymous, 1940, 1970; Justice, 1972). Samples may be taken from standing seed rows during harvesting or during any phase of processing, including bulk or bagged seed (Carter, 1961). Samples should be large enough to enable running the desired test and a retest, if necessary. Storage of samples should be under conditions preserving the original quality as nearly as possible. Many times it will be necessary to reduce the size of the original sample to a working sample of standard size (Justice, 1972).

Purity Analysis

Purity analyses are used to determine the percentage by weight of pure seed in the lot. Concurrently, the percentage of inert matter, noxious weed seeds, and other crops are also determined (Musel, 1961). Equipment and procedures used, as well as, classification of what is pure seed vs. inert matter are described (Anonymous, 1970; Justice, 1972). Once pure seed percentage is determined, the sample may be examined for mechanical damage. Fast green or other stains may be used to make the microscopic cracks more visible. This provides an excellent measure of how carefully seed is being handled as it moves through processing procedures.

Germination

The germination test is most frequently used to measure quality of the product. Germination is defined (Anonymous, 1970) as the ". . . emergence and development from the seed embryo of those essential structures, which for the kind of seed in question, are indicative of the ability to produce a normal plant under favorable conditions." Definite methods and guidelines for the germination of seeds (Anonymous, 1970) and the necessary equipment and procedures (Justice, 1972) have been established. Most of the larger companies have established their own seed testing laboratories, but commercial laboratories are also utilized.

Once a system has been established, correctness and consistency in evaluating the seedlings in the test is of greatest importance. Data collected translates into germination percentage which is placed on the analysis tag, as a requirement of both state and federal seed laws. Standardized rules and procedures help assure that uniform results can be expected when checked by law enforcement laboratories (Colbry et al.,1961).

There is concern that the standard germination test is inadequate to completely describe quality (Fleming, 1966). Standard germination tests are conducted under ideal conditions. Field conditions seldom approximate the ideal. More often some form of stress condition prevails (Delouche, 1973). Further, standard germination tests do not give any indication of the nature of seed deterioration that affects performance before the capacity to germinate is lost.

Additional tests have been developed to measure seed deterioration or vigor. Vigor has been identified as the difference between two lots of seed genetically alike. This difference is not obvious from percent germination (Delouche, 1973; Pollock and Roos, 1972). Seed vigor takes into account the influence of planted seed on emergence, growth, development, and productivity of the plants exclusive of genetic constitution.

a. *Cold Test.* Most seed companies use the cold test as a routine test for vigor. Results indicate the ability of seed to emerge when soil conditions are cold and wet (Everson, 1972; Vaughan, 1971). The amount of mechanical damage and how well the seed is treated is also reflected. Storage conditions for carryover seed and the extent of frost damage can be evaluated by cold test. Procedures for uniformity and value of the cold test as a measurement of quality have been reported (Clark, 1954; Hoppe, 1949, 1956; Hoppe and Middleton, 1950; Moore, 1963; Obendorf, 1972; Potts and Baskin, 1971; Rinke, 1953). Reasons for difficulty in standardizing cold test procedures (Vaughn, 1971) can be different soil types and microorganisms present, and variations in soil moisture.

b. *Tetrozolium Test.* This biochemical test provides quick information on seed viability. The chemical is 2,3,4-triphenyltetrozolium chloride. The colorless tetrazolium, by enzymatic action, turns living tissue red while dead tissue remains colorless. The primary usefulness of the tetrazolium test is that it is a quick test for viability. It is further used to evaluate internal seed injury, insect injury, frost damage, and viability of dormant seed (Bennett and Loomis, 1949; Goodsell, 1948; Moore, 1958; Porter et al., 1947).

c. *Other tests.* Other special tests also have value in determining the quality of the seed crop and should be part of an active quality control program (Vaughn, 1971).

1. Growth rate tests can be incorporated into the regular germination tests.

2. If the germination trays are examined at regular intervals, lots exhibiting the highest "first count" will possess the better seed quality.

3. If counting continues throughout the germination period an index may be calculated by dividing the number of seedlings removed daily by the number of days after planting. The higher the index, the better the quality.

4. Root and shoot growth is a growth rate test and involves the actual measurement of root or shoot length, or both, at a given time after planting. The greater length is indicative of quality. A variation of this test would be weight determinations.

5. A glutamic acid decarboxylase activity (GADA) test measures enzyme activity resulting from the addition of glutamic acid to a ground seed sample. It can be used to measure storability, root growth, and yield potential.

6. Changes in permeability can be measured by soaking seed in distilled water and determining electrical resistance of the water. Low resistance levels indicate seed deterioration.

7. The accelerated aging test is a germination test whereby seeds are exposed to 100% humidity at 42 C (108.6 F) for 84 hours. The percentage survival is an index for longevity of storage and a good measure of seed vigor.

8. Respiration rate is determined by measuring the amount of carbon dioxide given off. More CO_2 released indicates better quality seed.

STORAGE AND DISTRIBUTION

Storage

Seed is stored for only two reasons. First, after harvest and before sale, seed has to be somewhere! The more fundamental reason for storage of seed is to maintain their physiological quality throughout the storage period by minimizing deterioration — Delouche (1968).

Seed is at its highest quality level at physiological maturity and can only deteriorate from that point onward. The best storage conditions can only maintain quality.

The basic requirements for seed storage space are that it be dry, free of rodents, and grain storage insects (Airy, 1955). Storage for both bulk and packaged seed is necessary (Fig. 13, 14). Some facilities have been designed to take advantage of the same space for both. As the bulk seed is moved out for

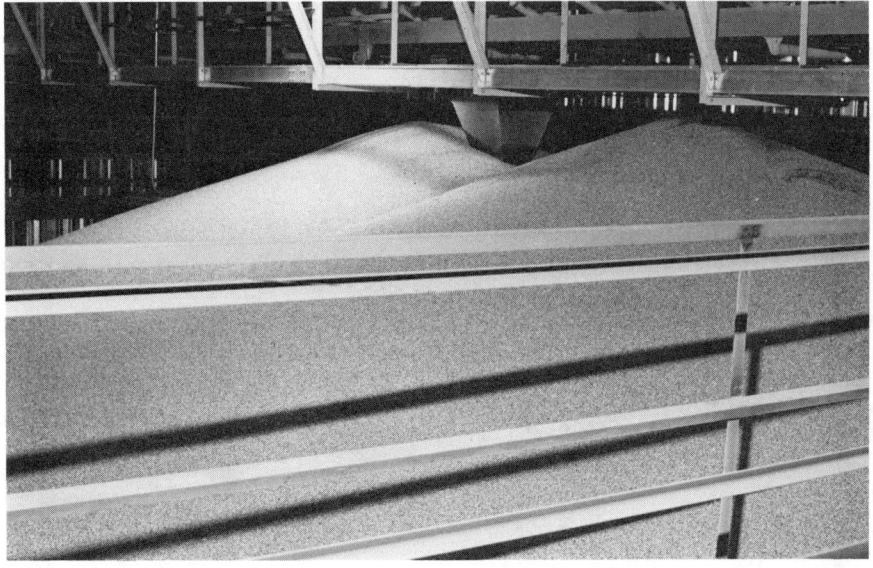

Fig. 13. Bulk seed in storage awaiting processing.

processing, the bagged seed can be moved into the vacated storage areas. Stored bagged seed is normally palletized and moved with fork lift trucks (Fig. 15).

Hybrid corn seed producers must grow additional supplies of commercial seed over and above that which forecasts indicate will be sold in a current sales year. Carryover seed, from year to year, is necessary. Producers normally anticipate growing approximately 20% or more seed than estimated sales requirements in order to fill supply channels, and hedge against possible reduced yields from production acreage. The 1974 production season provides an excellent example. Yields were often severely reduced by drought and some acreage suffered from early freeze damage. Seed supplies for the 1975 planting season would have been critically short had it not been for some quantities of carryover seed.

Storage of seed corn, either bulk or bagged, for prolonged periods at temperatures above 10 C (50 F) leads to deterioration of quality of the seed. For this reason, bulk seed after drying should be cooled, usually by aeration, to as low as 10 C if feasible. Also, storage conditions of surplus bagged seed

Fig. 14. Bagged seed in storage on pallets in preparation for distribution to sales outlets.

during the summer should be at temperatures of 10 C and below, and at a relative humidity in the range of 45 to 55% in order to maintain the desired moisture content.

Storage of foundation seed for many years is desirable, pointing out the need for storage which will maintain seed quality over an extended period of time. Early work (Airy, 1955; Sayre, 1948) indicated the influence of reduced temperatures and moisture level on the maintenance of quality and longevity of stored seed.

Yield performance can be affected by the handling and storage of seed. Yield decreases were reported (Grabe, 1966) for seeds stored under ambient conditions. Storage conditions which delay the rate of deterioration (Delouche, 1963, 1968) have been related to time and condition of storage. Figure 16 illustrates the relation of loss of germinability and the progress of deterioration over time. Considerable deterioration can occur before the germination is affected. The loss of germination is the final consequence of seed deterioration. Several methods of measuring vigor and deterioration have been summarized by N.S. Gill. (1969. Deterioration of corn seed during storage. Ph.D. thesis. Mississippi State College, State College).

Several studies (Delouche, 1968; Grabe, 1963, 1965, 1966; Grabe, D.F. 1965. Agronomic significance of seed deterioration. Am. Soc. Agron. Abstr. p. 40.) have indicated that quality may be maintained at higher levels for longer periods by improved storage conditions. Specifications for equipment and buildings for controlled atmosphere storage and its effects on stored seed are reported (Beck, 1969; Dahlberg, 1967; N.S. Gill, Standfield, 1971, 1972). Studies compared the relative drop with time in ambient storage of cold test, GADA, and yield while the germination was high. All values were high when the seed was stored in controlled atmosphere storage.

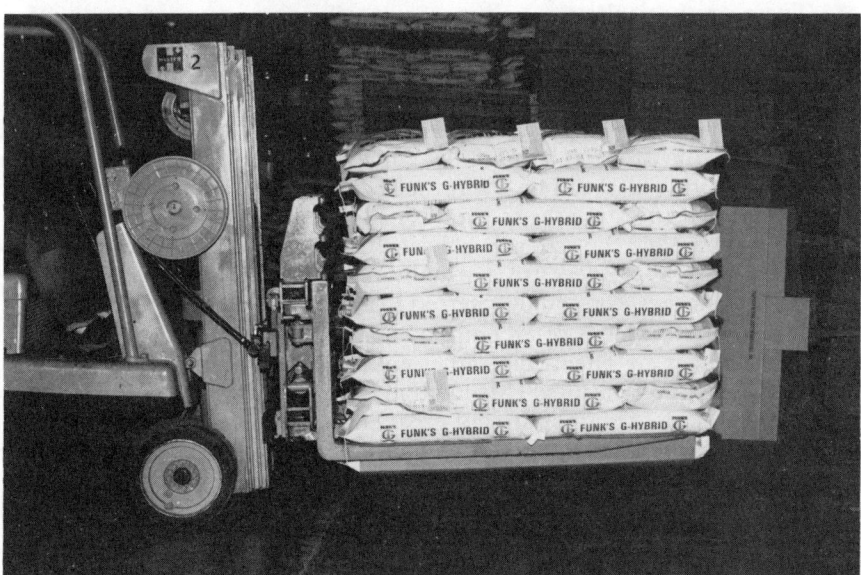

Fig. 15. **Electrically operated fork lift truck handling pallet of seed corn within the storage area.**

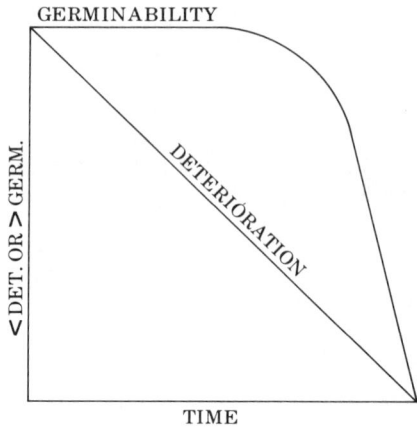

Fig. 16. Relation of loss of germinability and progress of deterioration over time.

The results of research on storage of seeds have been the basis for the design and justification of controlled atmosphere storage warehouses for foundation and commercial seed (Stanfield, 1971). In the case of foundation seed, economic justification lies in the better performance in production fields on commercial seed. With commercial seed, the economic justification is based on the value of inventory maintained in good quality condition as a marketable product.

Sales and Distribution

There are basically two common methods of sales and distribution of seed corn. The first and most widely used method throughout the Corn Belt is the farmer dealer. The second and most commonly used method in lesser concentrated acreage is the jobber.

The farmer dealer, as the name implies, is usually a farmer, and who in addition, is a part-time salesman. This method of distribution depends on the local salesman to 1) call on the farmer, 2) solicit his business, 3) write the order, 4) receive and store the seed on consignment, 5) arrange for pick up or delivery, 6) complete the sale and collect the account, and 7) provide service to the customer as needed. He is usually under the supervision of full-time company employees, commonly referred to as district sales managers.

The jobber-dealer is often employed in related agricultural endeavors, such as fertilizer or chemicals, other seeds, livestock feeds, farm supply stores, or elevator operators. This method of operation is especially adapted to areas where sales volumes are not great enough to justify the farm-to-farm calls of the local salesman, but are compatible with the existing salesman or store operators' business.

ACKNOWLEDGMENTS

The author gratefully acknowledges coworkers in Funk Seeds International, Mr. James Beardsley, Mr. J. Gordon Bidner, Dr. Robert Brawn, Mr. Lorin Jump, Dr. Charles Laible, Mr. Donald McGillivray, Mr. Ralph Primm, Mr. Zenas Stanfield, Dr. Ken Storer, Mr. Kenneth Truelsen, and Mr. Wesley Wilcox, who contributed significantly in the preparation of this manuscript.

LITERATURE CITED

Airy, J.M. 1950. Current problems of detasseling. Proc. 5th Annu. Hybrid Corn Ind-Res. Conf. 5:7-17. Am. Seed Trade Assoc., Washington, D.C.

----. 1955. Production of hybrid corn seed. p. 379-422. In: G.F. Sprague (ed.) Corn and corn improvement. Academic Press, New York.

----, L.A. Tatum, and J.W. Soreson, Jr. 1961. producing seed of hybrid corn and sorghum. p. 145-153. In: Seeds. USDA Yearbook Agric. U.S. Government Printing Office, Washington.

Alberts, H.W. 1927. Effect of pericarp injury on moisture absorption, fungus attack, and viability of corn. J. Am. Soc. Agron. 19:1021-1030.

Aldrich, S.R., and E.R. Leng 1965. Modern corn production. F and W Publishing Corp. Urbana, Illinois.

Alessi, J., and J.F. Power, 1974. Effects of plant population, row spacing, and relative maturity on dry-land corn in the Northern Plains. I. Corn forage and grain yield. Agron. J. 66:316-319.

Anderson, I.C. 1971. Possible practical applications of chemical pollen control in corn and sorghum seed production. Proc. 26th Annu. Corn Sorghum Res. Conf. 26:22-26. Am. Seed Trade Assoc., Washington, D.C.

Andrew, R.H., F.P. Ferwerda, and A.M. Strommen, 1956. Maturation and yield of corn as influenced by climate and production techniques. Agron. J. 48:231-236.

Anonymous. 1940. Federal Seed Act of August 9, 1939. (53 Stat. 1275) with amendments. U.S. Government Printing Office, Washington.

----. 1966. International rules for seed testing. Proc. Int. Seed Test. Assoc. 31.

----. 1970. Rules for testing seeds. Proc. Assoc. Off. Seed Anal. 60 (2).

----. 1971. Certification handbook. Assoc. Off. Seed Cert. Agencies. Publ. 23. Revised 1974.

Arnold, J.M., L.M. Josephson, W.L. Parks, and H.C. Kincer. 1974. Influence of nitrogen, phosphorous, and potassium applications on stalk quality characteristics and yield of corn. Agron. J. 66:605-608.

Artschwager, E. 1947. Pollen degeneration in the male sterile sugarbeet. Special reference to tapetal periplasmodium. J. Agric. Res. 75:191-197.

Baker, R.F. 1970. Relative maturity rating of corn by growing degree days. Proc. 25th Annu. Corn Sorghum Res. Conf. 25:154-160. Am. Seed Trade Assoc., Washington, D.C.

Barrons, K.C. 1971. Possibilities for development of pollen control agents for corn and sorghum. Proc. 26th Annu. Corn Sorghum Res. Conf. 26:27-32. Am. Seed Trade Assoc., Washington, D.C.

Bateman, H.P. 1972. Plant metering, soil and plant factors affecting corn ear populations. Trans. ASAE 15:1013-1020.

Bates, T.E. 1971. Response of corn to small amounts of fertilizer placed with the seed. II. Summary of 22 field trials. Agron. J. 63:369-371.

Bauman, L.F. 1967. Seed coating of delay emergence. Proc. 22nd Annu. Hybrid Corn Ind.-Res. Conf. 22:49-52. Am. Seed Trade Assoc., Washington, D.C.

Bear, R.P. 1975. A marketing program for waxy maize. p. 101-104. In: L.D. Hill (ed.) Corn quality in world markets. Interstate Printers and Publishers, Danville, Illinois.

Beck, J.M. 1969. Systems for controlling relative humidity and temperature. Proc. Miss. Short Course for Seedsmen 12:115-129.

Beckett, J.B. 1971. Classification of male-sterile cytoplasms in Maize (Zea mays L.) Crop Sci. 11:724-727.

Bennett, N., and W.E. Loomis. 1949. Tetrazolium chloride as a test reagent for freezing injury of seed corn. Plant Physiol. 24:162-174.

Blacklow, W.M. 1972. Influence of temperature on germination and elongation of the radicle and shoot of corn (Zea mays L.). Crop Sci. 12:647-650.

Borgeson, C. 1943. Methods of detasseling and yield of hybrid seed corn. J. Am. Soc. Agron. 35:919-922.

Brasfield, T.W. 1950. Commercial possibilities of chemical control. Proc. 5th Annu. Hybrid Corn Ind.-Res. Conf. 5:22-25. Am. Seed Trade Assoc., Washington, D.C.

Brass, R.W., and S.J. Marley. 1973. Roller sheller: low damage corn shelling cylinder. Trans. ASAE 16:64-66.

Brown, E.B. 1920. Relative yields from broken and entire kernels of seed corn J. Am. Soc. Agron. 12:196-197.

Buckert, J.G. 1961. The stage of the genome-plasmon interaction in the restoration of fertility to cytoplasmically pollen-sterile maize. Proc. Natl. Acad. Sci. USA 47:1436-1440.

Bunch, H.D. 1958. Seed research and training at Mississippi State University. Proc. 13th Annu. Hybrid Corn Ind.-Res. Conf. 13:21-34. Am. Seed Trade Assoc., Washington, D.C.

Cal, J.P., and R.L. Obendorf. 1972. Imbibitional chilling injury in Zea mays L. altered by initial kernel moisture and maternal parent. Crop Sci. 12:369-373.

Carter, A.S. 1961. In testing, the sample is all important. p. 414-417. In: Seeds. USDA Yearbook Agric. U.S. Government Printing Office, Washington.

Carter, M.W., and C.G. Poneleit. 1973. Blacklayer maturity and filling period variation among inbred lines of corn (Zea mays L.). Crop Sci. 13:436-439.

Chauhan, S.V.S., and S.P. Singh. 1966. Pollen abortion in male sterile hexaploid wheat ('Norin'), having Aegilops ovata L. cytoplasm. Crop Sci. 6:523-535.

Chinwuba, P.W., C.O. Grogan, and M.S. Zuber. 1961. Interaction of detasseling, sterility, and spacing on yields of maize hybrids. Crop Sci. 1:279-280.

Chopra, V.L., S.K. Jain, and M.S. Swanimanthan. 1960. Studies on the chemical induction of pollen sterility in some crop plants. Ind. J. Genet. and Plant Breed. 20:188-199.

Clark, B.E. 1954. Factors affecting the germination of sweet corn in low temperature laboratory tests. New York Agric. Exp. Stn. Bull. 769.

Cloninger, F.D., M.S. Zuber, and R.D. Horrocks. 1974. Synchronization of flowering in corn (Zea mays L.) by clipping young plants. Agron. J. 66:270-272.

Colbry, V.L., T.F. Swofford, and R.P. Moore. 1961. Tests for germination in the laboratory. p. 433-443. In: Seeds. USDA Yearbook Agric. U.S. Government Printing Office, Washington.

Cowan, R.J. 1972. Seed certification. p. 371-397. In T.T. Kozlowski (ed.) Seed Biology. Vol. 3. Academic Press, New York.

Crabb, A.R. 1947. The hybrid corn makers. Rutgers University Press, New Brunswick. 331 p.

Crane, P.L., S.R. Miles, and J.E. Newman. 1959. Factors associated with varietal differences in rate of field drying in corn. Agron. J. 51:318-320.

Dahlberg, R.W. 1967. Seed warehousing. Proc. Iowa State Univ. Seed Processors' Short Course. September 1967. Iowa State Univ., Ames (mimeo.).

Daynard, T.B. 1969. The 'black layer' — its relationship to grain filling and yield. Proc. 24th Annu. Corn Sorghum Res. Conf. 24:49-54. Am. Seed Trade Assoc., Washington, D.C.

----. 1972. Relationship among black layer formation, grain moisture percentage, and heat unit accumulation in corn. Agron. J. 64:716-719.

----, and W.G. Duncan. 1969. The black layer and grain maturity in corn. Crop Sci. 9:473-476.

Delouche, J.C. 1963. Seed deterioration. Seed World 92(4):14-15.

----. 1968. Physiology of seed storage. Proc. 23rd Annu. Corn Sorghum Res. Conf. 23:89-90. Am. Seed Trade Assoc., Washington, D.C.

----. 1971. Determinants of seed quality. Proc. Mississippi Short Course for Seedsmen. 14:53-67. Mississippi State Univ., State College (mimeo).

----. 1973. The problem of vigor. I. A look at the germination test. Seedsmen's Digest 24:8-24.

Dimmock, F. 1947. The effects of immaturity and artificial drying upon the quality of seed corn. Conn. Dep. Agric. Tech. Bull. 58.

Dubey, D.K., and S.P. Singh, 1965. Mechanism of pollen abortion in three male sterile lines of flax. Crop Sci. 5:121-124.

Duncan, E.R. 1954. Influences of varying plant population, soil fertility, and hybrid on corn yields. Soil Sci. Soc. Am. Proc. 18:437-440.

Dungan, G.H., and C.M. Woodworth. 1939. Loss resulting from pulling leaves with tassels in detasseling corn. J. Am. Soc. Agron. 31:872-875.

Duvick, D.N. 1959a. The use of cytoplasmic male-sterility in hybrid seed production. Econ. Bot. 13:167-195.

----. 1959b. Genetic and environmental interactions with cytoplasmic pollen sterility in corn. Proc. 14th Annu. Hybrid Corn Ind.-Res. Conf. 14:42-52. Am. Seed Trade Assoc., Washington, D.C.

----. 1965. Cytoplasmic pollen sterility in corn. Adv. Genet. 13:1-56.

----. 1972. Potential usefulness of new cytoplasmic male steriles and sterility systems. Proc. 27th Annu. Corn Sorghum Conf. 27:197-201. Am. Seed Trade Assoc., Washington, D.C.

Eaton, F.M. 1957. Selective gametocide opens way to hybrid cotton. Science 126:1174-1175.

Ericksen, A.W., and J.G. Ross. 1963. Irregularities at microsporgenesis in colchicine induced male sterile mutant in *Sorghum vulgare* Pers. Crop Sci. 3:481-483.

Everson, L.E. 1972. The use of test information in quality control and sales. Proc. Mississippi. Short Course for Seedsmen 15:87-92. Mississippi State Univ., State College (mimeo).

Fink, R.J., and D. Wesley, 1974. Corn yield as affected by fertilization and tillage system. Agron. J. 66:70-71.

Fleming, A.A. 1966. Effects of seed age, producer, and storage on corn (*Zea mays* L.) production. Agron. J. 58:227-228.

Fowler, Wayne, 1967. Cultural practices for today's seed fields. Proc. 22nd Annu. Hybrid Corn Ind.-Res. Conf. 22:53-58. Am. Seed Trade Assoc., Washington, D.C.

Gallaher, R.N., W.L. Parks, and L.M. Josephson. 1972. Effects of levels of soil potassium, fertilizer potassium, and season on yield and ear leaf potassium content of corn inbreds. Agron. J. 64:645-647.

Gamble, E.E. 1971. Maturity indexing of corn hybrids. Proc. 26th Annu. Corn Sorghum Res. Conf. 26:180-185. Am. Seed Trade Assoc., Washington, D.C.

Genter, C.F., and H.M. Camper. 1973. Component plant part development in maize as affected by hybrids and population density, Agron. J. 65:669-671.

Gilmore, E.C., Jr., and J.S. Rogers. 1958. Heat units as a method of measuring maturity in corn. Agron. J. 50:611-615.

Goodsell, S.F. 1948. Triphenyltetrozolium chloride for viability determination of frozen seed corn. J. Am. Soc. Agron. 40:432-442.

Grabe D.F. 1963. Seed corn — storage and vigor. Seed World 92 (10):12-14.

----. 1965. Prediction of relative storability of corn seed lots. Proc. Assoc. Off. Seed Anal. 55:92-96.

----. 1966. Significance of seedling vigor in corn. Proc. 26th Annu. Hybrid Corn Ind.-Res. Conf. 26:39-44. Am. Seed Trade Assoc., Washington, D.C.

Gregg, B.R. 1972. Quality control in processing foundation seed. Proc. Mississippi Short Course for Seedsmen. 15:99-111. Mississippi State Univ., State College (mimeo).

Grogan, C.O. 1956. Detasseling responses in corn. Agron. J. 48:247-249.

----. 1958. Interaction of insecticides and fungicides on the germination of stored seed corn. Proc. 13th Annu. Hybrid Corn Ind. Res. Conf. 13:7-12. Am. Seed Trade Assoc., Washington, D.C.

----. 1970. Genetic variability in maize for germination and seedling growth at low temperatures. Proc. 25th Annu. Corn Sorghum Res. Conf. 25:90-98. Am. Seed Trade Assoc., Washington, D.C.

----. 1971. Multiplasm, a proposed method for the utilization of cytoplasms in pest control. Plant Dis. Rep: 55:400-401.

Gunn, R.B., and R. Christensen. 1965. Maturity relationships among early to late hybrids of corn *(Zea mays* L.). Crop Sci. 5:299-302.

Hallauer, A.R., and W.A. Russell. 1961. Effects of selected factors on grain moisture reduction from silking to physiological maturity in corn. Agron. J. 53:225-229.

----, and ----. 1962. Estimates of maturity and its inheritance in maize. Crop Sci. 2:289-294.

Hanway, J.J. 1962. Corn growth and composition in relation to soil fertility. I. Growth of different plant parts and relation between leaf weight and grain yield. Agron. J. 54:145-148.

----, and W.A. Russell. 1969. Dry matter accumulation in corn *(Zea mays* L.) plants; comparisons among singlecross hybrids. Agron. J. 61:947-951.

Harrison, C.M., and A.H. Wright. 1929. Seed corn drying experiments, J. Am. Soc. Agron. 21:994-1000.

Heslop-Harrison, J. 1957. The experimental modification of sex expression in flowering plants. Biol. Rev. Cambridge Phil. Soc. 32:38-90.

Hicks, D.R., and R.E. Stucker. 1972. Plant density effect on grain yield of corn hybrids diverse in leaf orientation. Agron. J. 64:484-487.

Hoppe, P.E. 1949. Differences in *Pythium* injury to corn seedlings at high and low soil temperatures. Phytopathology 39:77-84.

----. 1956. Cold testing. Proc. 11th Annu. Hybrid Corn Ind.-Res. Conf. 11:68-74. Am. Seed Trade Assoc., Washington, D.C.

----, and J.T. Middleton. 1950. Pathogenicity and occurrence in Wisconsin of Pythium species which cause seedling disease in corn. Phytopathology 40:13. (Abstr.).

Huey, J.R. 1971. Experiences and results of mechanical topping versus hand detasseling in 1971. Proc. 26th Annu. Corn Sorghum Res. Conf. 26:144-147. Am. Seed Trade Assoc., Washington, D.C.

Hunter, R.B., T.B. Daynard, D.J. Hume, J.W. Tanner, J.D. Curtis, and L.W. Kannenberg. 1969. Effect of tassel removal on grain yield of corn *Zea mays* L. Crop Sci. 9:405-406.

----, C.G. Mortimore, and L.W. Kannenberg. 1973. Inbred maize performance following tassel and leaf removal. Agron. J. 65:471-472.

Johnson, D.R., and J.W. Tanner. 1972. Comparisons of corn (*Zea mays* L.) inbreds and hybrids grown at equal leaf area index, light penetration, and population. Crop. Sci. 12:482-485.

Jones, D.F. 1956. Production of Hybrid Seed Corn. U.S. Patent 2,753,663.

----, and P.C. Mangelsdorf. 1951. The production of hybrid corn seed without detasseling. Conn. Agric. Exp. Stn. Bull. 550.

----, H.T. Stinson, and U. Khoo. 1957. Pollen restoring genes. Conn. Agric. Exp. Stn. Bull. 610.

Jones, J.B., and H.J. Mederski. 1963. Effect of soil temperature on corn plant development and yield. II. Studies with six inbred lines. Soil Sci. Soc. Am. Proc. 27:186-189.

Jones, M.D., and J.S. Brooks. 1950. Effectiveness of distance and border rows in preventing outcrossing in corn. Okla. Agric. Exp. Stn. Tech. Bull. T-38.

----, and ----. 1952. Effect of tree barriers on outcrossing in corn. Okla. Agric. Expt. Stn. Tech. Bull. T-45.

Jugenheimer, R.W. 1958. Hybrid maize breeding and seed production. FAO Agric. Dev. Paper 62. p. 157. Food Agric. Organization, Rome, Italy.

Justice, O.L. 1972. Essentials of seed testing. p. 301-370. *In*: T.T. Kozlowski (ed.) Seed biology. Vol. 3. Academic Press, New York.

Kaerwer, H.E., Jr. 1971. Quality control and customer service. Proc. Mississippi Short Course for Seedsmen. 14:93-98. Mississippi State Univ., State College (mimeo).

Kiermayer, O. 1959. Induktion mannlick-steriler bluten bei *Helianthus annus* durch 2, 3, 5-tryodobenguesaure (TIBA). Naturrwissenschaften 46(14):457.

Kiesselbach, T.H. 1939. Effect of artificial drying upon the germination of seed corn. J. Am. Soc. Agron. 31:489-496.

----. 1945. The detasseling hazard of hybrid seed corn production. J. Am. Soc. Agron. 37:806-811.

PRODUCTION OF HYBRID SEED

Liable, C.A. 1974. Chemical methods of pollen control. Proc. 29th Annu. Corn Sorghum Res. Conf. 29:174-184. Am. Seed Trade Assoc., Washington, D.C.

Leffler, Allan. 1957. Economic use of fertilizer in connection with hybrid seed corn production. Proc. 12th Annu. Hybrid Corn Ind.-Res. Conf. 12:63-70. Am. Seed Trade Assoc., Washington, D.C.

Loeffel, F.A. 1971. Development and utilization of parental lines. Proc. 26th Annu. Corn Sorghum Res. Conf. 26:209-217. Am. Seed Trade Assoc., Washington, D.C.

Matthes, K., and Rushing, K.W. 1972. Seed drying and conditioning. Proc. Mississippi Short Course for seedsmen. 15:23-37. Mississippi State Univ., State College (mimeo).

McKee, G.W. 1963. Accuracy of seed corn grading and planter plate recommendations. Penn. Agric. Exp. Stn. Prog. Rep. 240.

McRostie, G.P. 1949. Some factors influencing the artificial drying of mature grain corn. Agron. J. 41:425-429.

Mock, J.J., and S.A. Eberhart, 1972. Cold tolerance in adapted maize populations. Crop Sci. 12:466-469.

Moore, R.H. 1950. Several effects of maleic hydrazide on plants. Science 112:52-53.

Moore, R.P. 1958. Tetrozolium tests for determination of injury and viability of seed corn. Proc. 13th Annu. Hybrid Corn Ind.-Res. Conf. 13:13-20. Am. Seed Trade Assoc., Washington, D.C.

----. 1963. Seed vigor or soundness and corn improvement. Proc. 18th Annu. Hybrid Corn Res.-Ind. Conf. 18:72-78. Am. Seed Trade Assoc., Washington, D.C.

Murneek, A.E. 1949. The role of plant growth substances (hormones) in reproduction. Proc. 4th Annu. Hybrid Corn Ind.-Res. Conf. 4:79-85. Am. Seed Trade Assoc., Washington, D.C.

Musel, A.F. 1961. Testing seeds for purity and origin. p. 417-432. *In* Seeds. USDA Yearbook Agric. U.S. Govt. Printing Office, Washington, D.C.

National Academy of Sciences. 1972. Genetic vulnerability of major crops. U.S. Natl. Acad. Sci., Washington. 307 p.

Naylor, A.W. 1950. Observations on the effects of maleic hydrazide on the flowering of tobacco, maize, and cocklebur. Proc. Natl. Acad. Sci. USA 36:230-232.

Neal, N.P. 1961. The influence of freezing temperature of varying intensities and duration of the germination of seed corn at different stages of maturity. Proc. 16th Annu. Hybrid Corn Ind.-Res. Conf. 16:67-73. Am. Seed Trade Assoc., Washington, D.C.

----. 1968. Maturity rating systems for corn hybrids. Proc. 23rd Annu. Corn Sorghum Res. Conf. 23:45-53. Am. Seed Trade Assoc., Washington, D.C.

Nelson, P.M., and E.C. Rossman. 1958. Chemical induction of male sterility in inbred maize by use of gibberellins. Science 127:1500-1501.

Newlin, O.J. 1971. Opportunity and need for better pollen control techniques in corn and sorghum production. Proc. 26th Annu. Corn and Sorghum Res. Conf. 26:7-13. Am. Seed Trade Assoc., Washington, D.C.

Neuffer, M.G., L. Jones, and M.S. Zuber, 1968. The mutants of maize. Crop Sci. Soc. Am., Madison, Wis. 74 p.

Obendorf, R.L. 1972. Factors associated with early germination in corn under cool conditions. Proc. 27th Annu. Corn Sorghum Res. Conf. 27:132-139. Am. Seed Trade Assoc., Washington, D.C.

Patterson, Earl. 1973. Procedures for use of genic male sterility in production of commercial hybrid maize. U.S. Patent 3,710,511.

Pauli, A.W. 1967. Inbred stands-seed vigor and use of hormones. Proc. 22nd Annu. Hybrid Corn Ind.-Res. Conf. 22:45-48. Am. Seed Trade Assoc., Washington, D.C.

Peaslee, D.E., J.L. Ragland, and W.K. Duncan. 1971. Grain filling period as influenced by phosphorus, potassium, and time of planting. Agron. J. 63:561-563.

Pendleton, J.W. 1965. Cultural practices-spacing, etc. Proc. 20th Annu. Hybrid Corn Ind.-Res. Conf. 20:51-58. Am. Seed Trade Assoc., Washington, D.C.

----, and D.B. Egli. 1969. Potential yield of corn as affected by planting date. Agron. J. 61:70-71.

Pollock, B.M., and E.E. Roos. 1972. Seed and seedling vigor. p. 313-387. *In*: T.T. Kozlowski (ed.) Seed biology. Vol. 1. Academic Press, New York.

Porter, R.H., M. Durrell, and H.J. Romm. 1947. The use of 2, 3, 5 - triphenyltetrazolium chloride as a measure of seed germinability. Plant Physiol. 22:149-159.

Potts, H.C., and C.C. Baskin. 1971. General procedures for cold testing. Proc. Mississippi Short Course for Seedsmen 14:75-76. Mississippi State Univ., State College (mimeo).

Powell, R.D., and J.R. Webb. 1972. Effect of high rates of N, P, K fertilizer on corn (*Zea mays* L.) grain yields. Agron. J. 64:653-656.

Rinke, E.H. 1953. Cold test germinations. Proc. 8th Annu. Hybrid Corn Ind.-Res. Conf. 8:54-58. Am. Seed Trade Assoc., Washington, D.C.

Robinson, J.L. 1934. Physiologic factors affecting the germination and growth of seed corn. Iowa Agric. Exp. Stn. Res. Bull. 176.

Rogers, J.S., and J.R. Edwardson. 1952. The utilization of cytoplasmic male sterile inbreds in the production of corn hybrids. Agron. J. 44:8-12.

Rossman, E.C. 1958. Chemical induction of male sterility in inbred corn by use of gibberillins. Proc. 13th Annu. Hybrid Corn Ind.-Res. Conf. 13:40-46. Am. Seed Trade Assoc., Washington, D.C.

----. 1949. Freezing injury of inbred and hybrid maize seed. Agron. J. 41:574-583.

Rowell, P.L., and D.G. Miller. 1971. Induction of male sterility in wheat with 2-chloroethylphosphonic acid (Ethrel). Crop Sci. 11:629-631.

Sayre, J.D. 1948. Storage tests with seed corn. Proc. 3rd Annu. Hybrid Corn Ind.-Res. Conf. 3:57-64. Am. Seed Trade Assoc., Washington, D.C.

Shaw, R.H., and H.C.S. Thom. 1951. On the phenology of field corn, silking to maturity. Agron. J. 43:541-546.

Singh, S.P., and H.H. Hadley. 1962. Pollen abortion in cytoplasmic male sterile sorghum. Crop Sci. 1:430-432.

Smith, D.R., A.L. Hooker, S.M. Lim, and J.B. Beckett. 1971. Disease reaction of thirty sources of cytoplasmic male-sterile corn to *Helminthosporium maydis* race T. Crop Sci. 11:772-773.

Stanfield, Z.A. 1971. Use of seed vigor research data in design and justification of equipment and facilities. Search. 1972.

----. 1972. Conditioned seed storage — justification, design and operation. Proc. Mississippi Short Course for Seedsmen 15:71-75. Mississippi State Univ., State College, (mimeo).

Stickler, F.C. 1964. Row width and plant population studies with corn. Agron. J. 56:438-441.

Sutton, L.M., and R.E. Stucker. 1974. Growing degree days to black layer compared to Minnesota relative maturity rating of corn hybrids. Crop Sci. 14:408-412.

Tatum, L.A. 1971. The southern corn leaf blight epidemic. Science 171:1113-1116.

----, and M.S. Zuber. 1943. Germination of maize under adverse conditions. J. Am. Soc. Agron. 35:48-59.

Troyer, A.F., and W.L. Brown. 1972. Selection for early flowering in corn. Crop Sci. 12:301-304.

Ullstrup, A.J. 1970. A comparison of monogenic and polygenic resistance to *Helminthosporium turcicum* in corn. Proc. 25th Annu. Corn Sorghum Res. Conf. 25:147-153. Am. Seed Trade Assoc., Washington, D.C.

----. 1972. The impacts of the southern corn leaf blight epidemic of 1970-71. Ann. Rev. Phytopathol. 10:37-50.

Vaughn, C. 1971. Practical seed tests and their uses. Proc. Mississippi Short Course for seedsmen 14:69-74. Mississippi State Univ., State College (mimeo).

Voss, R.E., J.J. Hanway, and L.C. Dumenil. 1970. Relationship between grain yield and leaf N, P, and K concentration for corn (*Zea mays* L.) and the factors that influence their relationship. Agron. J. 62:726-728.

Wileman, R.H., and A.J. Ullstrupp. 1945. A study of factors determining safe drying temperatures for seed corn. Ind. Agric. Exp. Stn. Bull. 509.

Williams, W.A., R.S. Loomis, and C.R. Lepley. 1965. Vegetative growth of corn as affected by population density. I. Productivity in relation to interception of solar radiation. Crop Sci. 5:211-215.

----, R.S. Loomis, W.G. Duncan, A.Dovrat, and F. Numez. 1968. Canopy architecture at various population densities and the growth and grain yield of corn. Crop Sci. 8:303-308.

Wittwer, S.H., and I.B. Hillyer. 1954. Chemical induction of male sterility in cucurbits. Science 120:893-894.

----, and N.E. Tolbert. 1960. 2-Chloroethyl trimethylammonium chloride and related compounds as plant growth substances. V. Growth, flowering, and fruiting responses as related to those induced by auxin and gibberellin. Plant Physiol. 35:871-877.

Wortman, L.S., and E.H. Rinke. 1951. Seed corn injury at various stages of processing and its effect upon cold test performance. Agron. J. 43:299-305.

Wright, A.H. 1941. Seed corn grading in relation to planting. Agric. Eng. 22:18-24.

Wright, Harold. 1948. Processing seed corn to avoid injury. Proc. 3rd Ann. Hybrid Corn Ind.-Res. Conf. 3:32-37. Am. Seed Trade Assoc., Washington, D.C.

Chapter 13 Industrial Utilization of Corn

STANLEY A. WATSON
CPC International, Inc.
Argo, Illinois

Corn has achieved a significant level of industrial utilization because it has a combination of desirable virtues: (a) it is produced in large volume; (b) until recently, year-to-year production and prices have been relatively stable; (c) it is comparatively easy to grow and harvest; (d) it is readily storable from season to season; (e) it is broadly useful as a feed and food grain, and has specific industrial value; (f) it is easy and efficiently processed; and (g) it is subject to continual improvement by breeding. Corn has achieved a higher level of industrial utilization than any other cereal grain.

Throughout aboriginal America and in the Colonial Period, when subsistence agriculture was predominant, corn was the basic food crop. And today in areas of the world, such as Central and South America, Central and South Africa, and parts of China, corn is the basic food for subsistence farmers. In the USA, the 18th Century saw a gradual shift toward a more urban population with increased demand for refined wheat flour breads and quantities of animal-derived foods. As a result, here and in other developed countries, corn has become the primary energy-supplying grain for animal feeds. Table 1 shows that 90% of corn utilized in the USA is for animal feeding (farm use plus commercial feeds) (USDA, 1975a, 1975b, 1975c). However, the corn equivalent of the byproduct feeds from corn processing raises the total to 93.5%. This country produces about 57 to 58% of the total world corn crop, about 80% of which is consumed here. In recent years about 25% of USA corn has been exported, the bulk of which is used for animal feeds and industry. In addition to corn raised for grain, over 100 million metric tons of corn silage and forage are produced annually in the USA for animal feeding on 12 to 14% of the 70 to 80 million total harvested acres.

In spite of the dominance of animal feeding in its utilization, corn is still an important food material. It is consumed in its original, unfractionated condition as whole meal for bread and porridge, as popped corn, as tortillas after lime treatment, and as snacks after extruder cooking. Special varieties of corn bred for maximum popping volume is produced in an amount of 200 to 250 thousand metric tons a year in the USA for the popcorn market. The highest per capita consumption of corn foods is 142 kg in Mexico and Central America where corn is indigenous; European countries consume 11 kg/person and the USA 25 kg (USDA, 1976b). In the developed countries nearly all of the corn foods consumed are in the form of specialized products made by fractionating corn by dry or wet milling. The uses for corn that will

Table 1—Volume, types of use and price for corn in the USA. Thousand of metric tons. Yearly figures ending October 1, 1976†

Year	Farm feeding‡	Dry process products	Wet process products§	Sold from Farm — Alcohol distilled spirits	Seed	Export¶	Commercial feeds#	Total disappearance	Avg. farm price, $/metric ton
1975-76†‡‡	53,241	3,550	10,161	635	460	43,132	34,977	145,524	100
1974-75	42,852	3,505	8,255	584	457	29,186	32,285	118,146	119
1973-74	64,800	3,505	7,620	635	432	31,193	43,970	143,347	100
1972-73	50,800	3,505	7,112	737	406	31,574	30,280	141,055	62
1971-72	65,865	3,480	6,655	635	381	19,940	50,117	143,291	42
1970-71	51,920	3,581	6,274	610	432	12,750	33,860	105,463	52
1969-70	55,527	3,505	6,020	787	330	15,190	39,118	119,058	45
1968-69	47,729	3,454	5,715	838	305	13,210	36,298	113,023	42
1967-68	56,467	3,505	5,461	864	330	15,570	31,752	123,460	40
1966-67	53,622	3,505	5,207	838	355	11,837	30,634	105,862	48

†Figures from USDA, 1975a, 1975b, 1976a, except as indicated. ‡Reported as "Retained on farm where grown", but corrected for changes in farm inventory. §Extrapolation of data from USDC, 1974c, and Scallet, 1976. ¶Does not include export of products (1-2%). #A residual term; the difference of total farm sales and first 5 columns under "Sold from Farm", corrected for changes in elevator carryover. Assumed used in commercial livestock and pet feeds. ‡‡Preliminary

Table 2—Typical yellow dent corn composition and hand-dissected kernel parts†

Fraction	Whole kernel†	Starch	Protein	Fat	Sugar	Ash
			%			
Whole kernel (1974)‡	100	71.3	9.9	4.45	—	—
Whole kernel (1946)§	100	73.5	9.0	4.3	1.9	1.5
Endosperm	82.6	87.6	7.9	0.83	0.62	0.33
Germ	11.1	8.0	18.3	33.5	10.5	10.6
Pericarp + tip cap	6.2	7.0	4.3	1.4	—	0.9

†All values on moisture-free (dry) basis. ‡Average composition of corn purchased on the open market for wet milling, 1970-74. §Six yellow dent samples of four commercial hybrids dissected by Earle et al. (1946).

Table 3—Corn ingredients used in manufactured pet foods and farm animal feeds

Ingredient	1,000 metric tons			
	1974	1973	1972	1967
Corn	—	—	10,930	9,373
Cornmeal	—	—	672	376
Hominy feed and meal	—	—	498	583
Corn gluten feed and meal	1,134	1,188	1,120	955
Distillers dried grains and dried solubles‡	430	414	388	406
Total corn ingredients			13,608	12,328

†USDC (1974d).   ‡Van Lanen (1974).

be described in this chapter are worldwide in nature, but utilization figures are largely for the USA because of availability of data.

Milling separates the corn kernel into the three basic parts shown in Table 2. In dry milling, the separated endosperm is the prime product but the oil is separated as an important coproduct. In wet milling, the specific components of each kernel part (starch, protein, oil, fiber, and solubles) are separated into more purified fractions that have a wider scope of specific uses. Aside from food uses, products from corn enter into many industrial products, including paper products, construction materials, textiles, metal castings, pharmaceuticals, ceramics, paints, explosives, and many others. The products of corn are seemingly endless and enter into the everyday life of every person in hundreds of unappreciated ways. A recent survey by the Corn Refiners Association revealed that the average supermarket carries over 1,000 food items in which corn wet milling products or derivatives are ingredients.

FORMULATED ANIMAL FEEDS

In recent years about 40 to 45% of the corn produced has been retained on the farms where it is grown. The proportion retained is affected by the price of corn relative to the price of meat. Since farmers no longer grow their own seed (except those engaged in commercial seed production) it can be assumed that all retained corn is used for livestock feeding. Of the corn sold off the farm (Table 1), 50 to 68 million metric tons is sold annually to custom feeders or feed manufacturers (USDA, 1975b). The 1972 usage of 68.4 million metric tons for all commercial animal feeds less the 11.6 million metric tons of corn and corn meal reported as used in formulated feeds (Table 3) (USDC, 1974d) leaves a remainder of 56.8 million metric tons used for feeding not accounted for in formulated feed sales. Presumably most of this tonnage is used by animal producers who prepare their own feeds and must purchase corn to supplement their own grain production. These farmer-feeders supplement the grain with soybean (*Glycine max* L.) meal and vitamin-mineral premixes. Feeders may use sorghum, barley, and oats (*Sorghum bicolor* L., *Hordeum vulgare* L., *Avena sativa* L.) in place of corn depending on the relative price/value relationships. In the USA, the corn disposition among animal types is approximately as follows: hogs, 37%; beef cattle, 25%; poultry, 17%; dairy cattle, 12%; others, 9% (USDA, 1970).

Although many small hog and dairy farmers still feed whole corn, either shelled or on the ear, most of the corn (and grain sorghum) fed to animals, whether premixed or fed directly, is processed in some manner (Beeson, 1972; Perry, 1972). Fine grinding of corn, aside from the obvious ease of blending with other ingredients, improves the feed conversion efficiency by swine (Beeson, 1972; NRC-42, 1969). However, pelleting of the mixed feed produces about 10% additional improvement in feed efficiency for swine (Jensen and Becker, 1965).

About 75% of the rations fed to chicken broilers and to turkeys, and 50% of the layer rations, are pelleted (Calet, 1965). Aside from the advantages of ease of handling and prevention of segregation of ingredients, pelleted feeds improve feed consumption and growth rate in poultry and give improved feed efficiency. Of all of the cereals, corn shows the most improvement from pelleting (Allred et al., 1957). However, reasons for improvement of corn nutritive value from pelleting are still unclear.

The corn exported from the USA goes mainly to Europe and Japan where it supplements corn from other areas, mainly Latin America, South Africa, and Thailand. Corn usage in all developed countries is similar to uses found in the USA except as modified by the economic situation and competitive cereal or starch crops. Senti and Schaefer (1972) state that the USA, Israel, Jordan, and Lebanon feed 93% of corn to animals, Europe 81%, Brazil 68%, and Syria and Turkey, 60%.

The use of corn and other starchy grains for beef cattle feeding has expanded rapidly in recent years because of the increased efficiency of meat production that results from the feeding of high energy rations. In 1970, 57.5 million metric tons of grains was fed to ruminants of which 31.2 million metric tons was corn and 12.8 million metric tons was grain sorghum. The corn fed to beef cattle and dairy cattle was about equal at 12.6 and 12.5 million metric tons, respectively (Waldo, 1973). Most of the grain fed to beef cattle is preprocessed by a variety of methods including dry grinding, popping, extruding, parching, cooking, and steam-roll flaking (Beeson, 1972; White et al., 1973). The most effective and widely used method is steam-flaking. In this process the corn is cooked 15 min. at 200 F, then flaked to a thickness of about 0.8 mm (1/32 in). The digestibility of starch (energy) in steam-flaked corn is increased by 14% and feed efficiency is increased 12% over unprocessed corn (Beeson, 1972; Hale, 1973; Johnson et al., 1968). Apparently digestion of the starch in the rumen is improved by gelatinization. Optimum gains are achieved only with an adequate source of supplementary dietary nitrogen, such as that in soybean or cottonseed meal and urea. Recent high corn prices and low meat prices have forced a partial retreat from intensive grain feeding of beef cattle back to increased forage feeding. The dairy industry has not widely adopted these special corn processing techniques because milk and butterfat production are not greatly benefited (Williamson, 1972).

Specialization in agricultural production has given rise to a formulated feed industry which has shown remarkable growth in volume and sophistication over the past 50 years. Sales in 1972 of farm animal feeds amounted to $4.67 billion distributed among the animal groups as follows: egg chickens, 12%; broiler chickens, 9.8%; turkey, 2.7%; dairy cattle, 12%; swine, 6.3%; beef cattle, 4.7%; horse and mule, 1.8%; and others (mink, rabbit, laboratory animals, wild bird, etc.), 6.3% (USDC, 1974d). Corn is the

INDUSTRIAL UTILIZATION OF CORN

major ingredient of most of these feeds.

Pet foods, an additional rapidly growing segment of the formulated feed business, achieved sales of $1.4 billion in 1972. A wide variety of nutritious dog and cat foods developed during the last 25 to 30 years have produced convenience and good nutrition (Rhodes, 1975; USDC, 1974d). Corn flakes and cooked, extruded-expanded corn are important components of pet foods.

The major contribution of corn to any feeding system is energy production because about three-fourths of the kernel bulk is starch. However, the protein in whole corn and the corn mill feeds, gluten feed, gluten meal, hominy feed, and distillers grains, is highly utilizable. The protein content of these ingredients together made up about 20% of total protein and 39% of total tonnage of feed mixed commercially in 1972 (USDC, 1974d). But corn protein is deficient in the essential amino acids, tryptophan, and lysine, which are needed for optimum animal growth. Therefore, supplementary sources of protein rich in these two amino acids must be included in the feed formula on a least-cost basis. Oilseed, fish and component meat meals, for this reason, constitute 28% of the entire feed mixture. Soybean meal is, of course, largest because of its ready availability. The achievement of a highly efficient animal agriculture owes its existence to a great expansion of nutritional knowledge in the last 50 years and equally to improvement in corn growing efficiency which released acreage for broad expansion of soybean growing.

WET MILLING

The wet milling process, from its early beginning about 1840, has developed into a highly sophisticated means of separating the components of the corn kernel by chemical and physical methods into a multitude of useful products. Since that date, the wet milling industry has experienced continued improvement in operating efficiency and product quality, continued expansion in variety and utility of products, and growing demand for its products. Rate of growth of this industry in the next quarter century is expected to exceed previous records. Corn wet milling plants may be found in most developed countries as well as in many developing countries. The world wet milling capacity is estimated at over 18 million metric tons (700 million bu.; R.F. Huehner, CPC, private communication). There are approximately 53 wet milling plants in 40 countries on all continents. In the USA there are 19 plants operated by 12 companies. Most USA plants operated at nearly full capacity in 1974 for a record grind of about 7.6 million metric tons (300 million bu.) to meet the strong demand for starch, syrups, and dextrose. From 1974 to the end of 1976, wet milling capacity increased remarkably from a little over 7.6 thousand metric tons (300 thousand bu.) to 11.6 metric tons a year (455 million bu.; B.L. Scallett, Changes in the U.S. corn wet milling industry, Annual Meeting, Am. Soc. Agron., 1976). Prior to this time growth of wet corn processing was about 4% a year. This sudden increase in production was caused by development of a new sweeter syrup, High Fructose Corn Syrup (HFCS; see discussion under Sweetener Products) which can be used in place of sucrose and sells at a lower price. The wet milling grind is expected to continue to grow in response to increasing demand for HFCS and other products to a capacity of 18 to 25 million tons a year (800 to 1,100 million bu.) in the USA by 2010 A.D. (Cantor and Shaffer, 1974). Worldwide wet milling capacity could reach 30 to 40 thousand metric tons by the end of the century.

The Process

The wet milling industry's raw material is shelled corn received at the plants by rail, truck, or barge. It is inspected for freedom from mold, insect or rodent infestation; unfit shipments are rejected. Accepted lots are thoroughly dry cleaned to remove foreign material, dust, and broken corn that will pass through a 1.2-cm (0.48-in) hole. The clean corn is then steeped for 40 to 45 hours at 47 to 53 C to soften it for the initial milling step. The corn is steeped in large wooden, stainless steel or tile tanks arranged in batteries of 6 to 50 tanks (usually 10 to 12) connected in series by pipes. Water is moved by pumps from one steep to another in a countercurrent relationship with respect to age of corn. Because process water must be conserved, the entire process (Fig. 1) is operated in a countercurrent manner. Fresh water is introduced in the final starch-washing step and works its way stepwise, countercurrently to the flow of milled corn, picking up increasing levels of soluble materials. The resulting process water containing 1 to 2% solubles is then used to steep the corn. It is dosed with 0.10 to 0.20% sulfur dioxide and is placed on the corn that has been in the battery the longest time, i.e., the corn that is nearly ready for milling. The water progresses from steep-to-steep over successively newer corn. The oldest water is used to cover the corn just entering the system and then is withdrawn at a solubles concentration of 5 to 7%, moisture-free or dry basis (d.b.). It is evaporated to the thick, brown, corn-steep liquor, 50 to 55% dry substance. It is mainly composed of proteins, peptides, amino acids, lactic acid, and minerals. This fraction represents about 7.5% of the original corn dry substance (Table 4). About half of this material was originally water soluble; the other half was solubilized during steeping by action of native grain enzymes and sulfur dioxide on protein. The solubilizing action of SO_2 on endosperm protein is essential for softening the kernel for optimum starch recovery (Watson and Sanders, 1961). About half of the steepwater dry substance derives from the germ and half from the endosperm. The lactic acid that constitutes nearly one-fourth of the dry substance is produced by growth of lactic acid bacteria indigenous to the steeping system (Watson et al., 1955). The steep temperature is maintained between 47 to 53 C to favor growth of these bacteria. This temperature and the pH of 3.9 to 4.2 produced by the lactic acid inhibits growth of undesirable organisms.

The milling and separation steps are, of course, somewhat different for individual milling plants but, in general, all follow the steps shown in Fig. 1. In the initial milling step, fully steeped corn at about 45% moisture (wet basis) is passed through an attrition (cracking) mill, along with some water, for the purpose of liberating germ with minimum germ breakage. Cracking mills are usually composed of vertical rotating plates bearing rows of pyramidal knobs. Simultaneous to germ release about one-half of the kernel starch (prime starch) is released in this step (Watson, 1967). The germ, which now contains 50 to 60% oil (d.b.), can be separated from the denser components by flotation. A few plants still use the method of skimming the germ from the surface of carefully agitated tanks, but the preferred equipment is continuously operated liquid cyclones (Stavenger and Wuth, 1959; Bier et al., 1974). These cyclones are usually 100 cm high and 15 cm in diameter at the top tapering to a 3-cm underflow. The mixture is forced in at the top with high pressure to provide a centrifugal force to the particles. The lighter germ pieces are removed at the vortex into an overflow pipe while the endosperm pieces and most starch are discharged out the bottom. The germ is washed and dried in preparation for oil recovery.

The underflow from the liquid cyclones is further milled in vertical plate attrition-impact mills (Dill and Ginhaven, 1964) or in impact pin mills (Dowie and Martin, 1962; Watson, 1967). However, one company recovers only the prime starch for further purification, while the remaining bound starch is saccharified and fermented to ethyl alcohol (Smith et al., 1966).

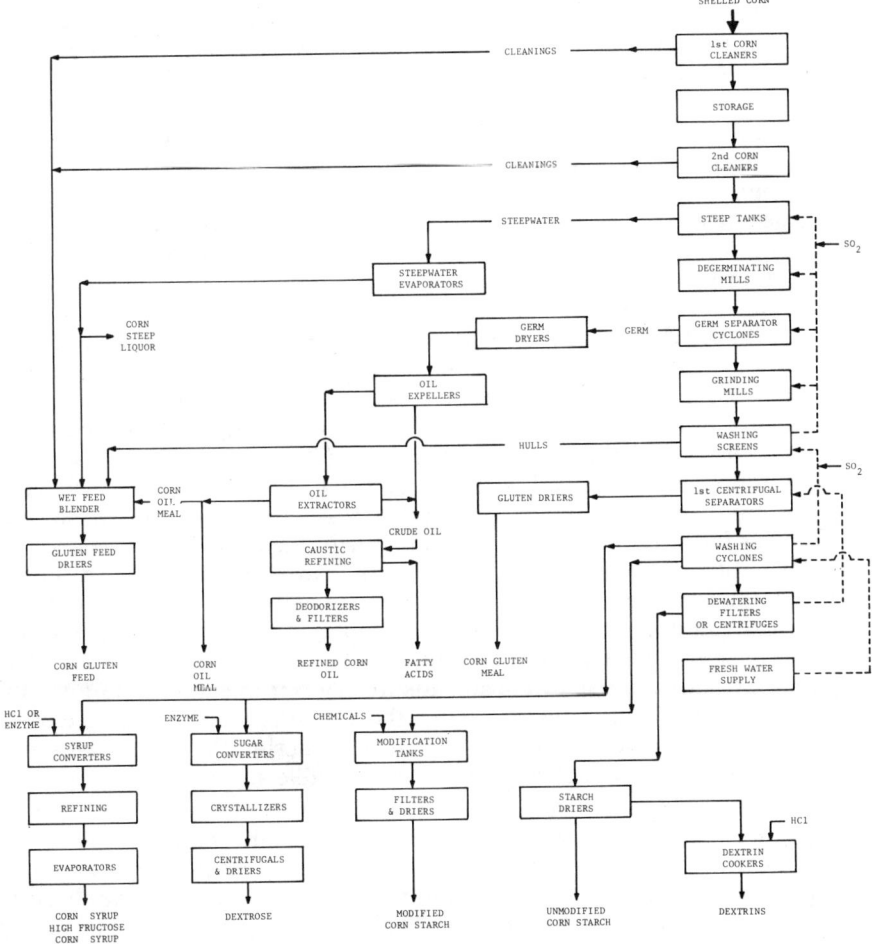

Fig. 1. Wet milling process for corn.

After the endosperm has been finely milled, starch and gluten particles must be separated from pericarp and other larger kernel fragments. This was traditionally accomplished by passing the slurry over a series of shakers equipped with nylon bolting cloth having openings ranging from 55 to 75 microns. Many machines have been devised and used to accomplish this separation (Besso, 1974). The most popular now are the screen pump (Von Titeleboom, 1961) and the screen bend (Fontein, 1954; Chwalek, 1973). In these two devices the starch-fiber slurry is forced over stationary curved metal surfaces perforated by precisely milled slots 70 to 80 microns in width. The fiber discharge is drier than from shakers and the starch filtrate is essentially free of fine fiber pieces. The screen washer (Besso, 1974) claims an additional virtue of releasing some of the bound starch as it is moved across the wedge wire screen by rotors. All methods require at least one reslurry step to produce optimum starch yield. The fiber is then usually dewatered in a continuous screw press and dried alone or in mixture with steepwater to produce corn gluten feed.

The starch and gluten particles in the slurry remaining after fiber removal are separated in any one of several types of continuous centrifugal machines. Usually the disc stack nozzle-type is preferred (Peltzer, 1961; Bier et al., 1974). The lighter gluten particles are discharged with most of the water into the overflow while the heavier starch granules are discharged into the underflow. The gluten is concentrated about 10-fold to 8 to 10% slurry in a second centrifuge equipped for separating water and solids. The gluten is further dewatered by filtration using rotary vacuum filters. The filter cake, containing about 60% moisture, is dried in flash dryers or rotary hot air dryers to produce corn gluten meal. The recovered water is used for fiber washing or steeping (Fig. 1).

The starch slurry must be further purified to a final protein content of about 0.3%. This may be accomplished by diluting the starch slurry with fresh water and passing it through a second separating centrifuge, or by the preferred use of liquid cyclones for the second stage purification (Vegter, 1957). The design of the starch cyclone is similar to the cyclone described for germ separation except it is smaller, i.e., 1 cm in diameter and 15 cm long. A large number of machined nylon cyclones are assembled into a shell which provides for starch slurry introduction, purified starch underflow, and gluten and solubles overflow in parallel for all cyclones in the unit. The advantage of using the cyclone is that they accomplish gluten separation and starch washing simultaneously. Five to seven units are assembled in a series. Between each unit the starch slurry emerging from the discharge ports is diluted with overflow from the unit nearer the fresh water inlet and is pumped to the next unit under high pressure. Thus the starch is washed countercurrently, finally being contacted with the fresh water entering the milling system (Fig. 1). Approximate product yields are given in Table 4.

Starch may be dried by several different methods, but the most economical is the use of flash dryers. To prepare starch slurry for flash drying, it is dewatered in a basket centrifuge, preferably automatically operated for filling with slurry, spinning to dewater and for discharging of the cake into a stream of heated air. The starch is dried in only a few seconds;

Table 4—Yield and composition of products of corn wet milling

Product	Dry product yield	Avg. composition
		% dry basis
Whole Corn	100	
Starch		71.3
Protein		9.87
Fat		4.55
Starch	67.5	
Protein		0.30
Fat		0.02
Germ	7.5	
Starch		7.6
Protein		10.7
Fat		52.0
Oil, Crude†	3.86	
Germ Cake†	3.64	
Starch		15.7
Protein		22.0
Fat		1.0
Gluten Meal	5.8	
Protein		70
Starch		18.5
Fat		6.2
Solubles	7.6	
Protein		46
Fiber	9.5	
Starch		11.5
Protein		11.3
Fat		2.8

†Derived from germ.

flash dryers operate at high volume and low cost. For the manufacture of hydrolyzed products or the manufacture of chemically modified starch products, finished starch slurry from the final cyclone stage is used directly. Although many of the specialized starches are sold in paper bags, most of the large tonnages of commercial starch is transported in hopper cars loaded and unloaded by air fluidization.

Continuing research and engineering developments have transformed the wet milling operation into a highly automated, continuous process. One configuration was recently described by Bier et al. (1974). All phases of the unit operations are monitored by instruments which feed data into central control panels. The process may be operated from these control rooms by activating remote control valves, tank level controls and motor switches, and by automatic measurement of density and flow rates. The combination of continuous flow, precise control, closed stainless steel tanks, and careful sanitation has transformed "corn grinding" into a scientific food processing operation. Continuous progress has been made over many years by the wet milling industry to minimize water and air pollution. Approximately 99% of the initial corn dry substance is recovered as useful products. The unrecoverable remainder is rendered harmless to the environment by modern sewage treatment and smoke abatement procedures (Bensing et al., 1972).

Corn Quality

Corn for wet milling is purchased on the basis of official USA grain standards with U.S. grade No. 2 preferred by most wet millers. Grading criteria include moisture content, weight per bushel (test weight), damage, broken corn, and foreign material. Significant deviations from U.S. grade No. 2 are penalized by a discount schedule designed to compensate the buyer for expected processing troubles, lower starch and/or lower oil yields. Discounting is particularly common in years when an adverse season is experienced over the entire Corn Belt. The corn blight year of 1970 (Anderson et al., 1972) and the ear rot year of 1972 (Tuite et al., 1974) are examples.

Many properties that are not part of the U.S. grade also have important effects on wet milling results. Freeman (1973) has reviewed in depth the effects of both classes of factors. Grade factors and their effects, in order of importance, are: (a) damage causes oil yield reduction; (b) broken corn and foreign material (BCFM) reduces starch and gluten yield because broken corn is put directly into feed; (c) low test weight lowers production rate and causes some starch yield loss. Damage, as defined by U.S. grade standard, is primarily deterioration of germ caused by mold growth. Wet milling of damaged corn kernels results in loss of oil because of severe fragmentation during milling whether or not the mold is from field infection such as corn blight (Anderson et al., 1972) or storage molds (Freeman, et al., 1970). Moisture content is unimportant except as its presence reduces test weight and stimulates microbial infection during storage and shipping. Among nongrade factors, the following are important: (a) low oil and low protein content reduce yield of corn oil and gluten meal, both of which are more valuable than starch; (b) low xanthophyll pigment content lowers pigment content of gluten meal thereby lowering its competitive position (Watson, 1960; Foster, 1965; Beaux et al., 1974); (d) mycotoxin and pesticide residues contamination are unacceptable because of health hazard.

Although mold-damaged corn has always been undesirable for wet milling, the discovery of the aflatoxins and other mycotoxins has emphasized the importance of more careful screening of corn for mold damage. Apparently, corn grown in the northern U.S. Corn Belt is relatively free of aflatoxin as compared to that grown in the southern U.S. Corn Belt (Shotwell et al., 1973a, b). Examination of aliquots from over 1.6 million metric tons of corn arriving at three wet milling plants over a 15-month period showed only 4 carlot samples with positive aflatoxin tests above the FDA tolerance level of 20 ppb (Watson and Yahl, 1971). Furthermore, we subjected two lots of aflatoxin-infected corn to laboratory wet milling and found that the two food products, starch and oil, were essentially free of aflatoxin (Yahl et al., 1971). Although byproducts contained a higher level of aflatoxin in this laboratory test than that of the original corn, careful monitoring of commercially produced byproduct feeds has never shown evidence of aflatoxin. In spite of demonstrated low incidence of aflatoxin in U.S. Midwest corn, the wet milling companies continue to carefully monitor incoming corn for mold and aflatoxin contamination. More recently, corn naturally contaminated with zearalenone was subjected to wet milling with similar results, i.e., no zearalenone in starch or oil but higher levels in feed fractions (Bennett et al., 1974). Zearalenone contamination of commercial corn samples is also negligible.

The recent commercialization of the known property of propionic acid, or mixtures of propionic and acetic acids, to prevent mold growth on wet corn holds the promise of reducing the mold hazard (Sauer and Burroughs, 1974;

Herting and Drury, 1974). Since the feeding value is not affected (Hammel, 1975) and the cost is said to be about the same as drying the corn to 15% moisture, widespread farm use is expected on farms where corn is raised for direct feeding to animals. Wet milling experiments with propionic acid-treated corn indicate unacceptability because odor persists in starch. Also, propionic acid in steepwater would make it toxic for use in antibiotic fermentation media (Freeman, 1973).

Although yellow dent corn is the predominant raw material for wet milling because of its availability, other types of corn have been and are now processed. Flint corn types are processed in Argentina and European countries. After proper steeping, milling results are similar to those obtained with dent corn. White corn is preferred, because of whiter starch, in countries where there is no interest in poultry pigmenting value of gluten meal. Waxy corn and high amylose corn processing are discussed in the section on starch.

A sample of the first opaque-2 (high lysine) hybrid corn developed by Mertz et al. (1964), ($B37o_2$ × $540o_2$) $W64Ao_2$ was wet milled in our laboratory. This primitive opaque-2 hybrid produced starch of good quality but at 10% lower yield than regular corn. Solubles produced during steeping and milling were 80% higher, while gluten yield was 28% lower than for regular corn. Only the gluten was higher in lysine and tryptophan content (Watson and Yahl, 1967). Thus opaque-2 corn does not appear to yield products of significantly increased economic value to compensate for loss in starch and gluten meal yield.

Corn varieties higher in oil content than in regular corn have long been a possibility through recurrent selection breeding (Sprague and Brimhall, 1950), and much progress has been made up to the present in breeding and analytical techniques (Watson and Freeman, 1976). Laboratory wet milling data (Watson, 1967) has shown that every increment of oil above that of normal corn can be recovered as crude oil, but yields of starch and gluten are reduced in proportion to the larger germ size. These results have been confirmed on commercial scale. However, seed corn companies, with one exception, have not found enough incentive to justify breeding for higher oil content. Several possible reasons have been: (a) the high cost of oil analyses necessary to provide the selection for breeding, (b) lack of interest by feeders because of the apparent small gain in feed value, and (c) lack of economic incentive for wet milling companies to pay the large premiums required for contract growing to compensate for lower grain yield of available hybrids and special handling necessary to preserve identity.

The analytical problem appears to be resolvable by using nondestructive wide-line nuclear magnetic resonance (NMR) equipment to determine oil content of single corn kernels. Bauman et al. (1965) demonstrated that the single kernel NMR technique could be effective in selecting for oil content in a segregating F_2 population. Silvela-Sangro (1967) was able to show that continued selection for oil content within segregating populations was effective in increasing oil content. D.E. Alexander (Univ. of Illinois, private communication) has used this technique extensively in the creation of gene pools with high oil content and in selection of agronomically useful inbreds with oil content in the range of 7 to 8%. Meanwhile, Conway et al. (1974) have developed an automated NMR unit capable of analyzing 400 to 500 individual corn kernels each hour with excellent precision. This equipment is expected to become generally available to public and private corn breeders in the near future and should provide sufficient selection pressure to give lines having higher oil content as well as all desired agronomic characters, including normal grain yield (Watson and Freeman, 1976). However, since fatty acid composition of the oil is not measured by the wide-line NMR method, the widely recognized negative correlation between total oil and linoleic acid content (Sniegowski and Baldwin, 1954; Lofland et al., 1954) will depress total degree of unsaturation of the oil unless selection pressure to retain the present level of linoleic acid is applied. That the ratio of linoleic to oleic acid is under genetic control and can be modified by breeding has been adequately demonstrated (Poneleit and Bauman, 1970; de La Roche et al., 1971).

Corn Starch

The major volume product from wet milling is starch (Table 5). It is recovered in purified form in a yield of 67 to 69% with a recovery efficiency of 93 to 96% of the contained starch. In 1974 about half of the corn starch manufactured was sold as starch products and the other half as the hydrolyzed products, corn syrup, and dextrose. Although starch sales (Table 5) are increasing, sweetener sales are increasing more rapidly and are expected to continue accelerating due to the impact of development of sweeter syrups (Cantor and Shaffer, 1974). Of the total USA starch sales of 1.78 million tons, starches other than corn starch, mainly potato and tapioca, amounted to only 84,100 metric tons (185.5 million pounds) in 1972. In Canada, total starch utilized was 197,800 tons of which about 80% was corn starch and 90% was produced in Canada (McNicol et al., 1972).

General Properties and Applications

Industrial corn starch is a fine white powder, 99% pure, containing only about 0.25% protein, less than 0.1% minerals, and 0.65% fat. The fatty components of starch are tightly complexed with the starch molecule and are composed primarily of lysolecithin together with a small amount of phospholipids and fatty acids (Acker and Becker, 1971).

Corn starch is an important manufactured product because of its usefulness in many industrial applications. Although a few uses depend on properties in the dry state, most applications relate to its properties as a cooked, hydrated paste. The industrial and food applications of corn starch are many and space in this chapter allows for only a cursory description. For more detailed information the reader is referred to one of the comprehensive treatises available (Russell, 1973; Schoch, 1970; Whistler and Paschall, 1967).

Corn starch occurs naturally in nearly spherical granules of 5 to 30 microns in diameter (average diameter of 14 microns). The granules are composites of crystalline and amorphous aggregates of two distinct molecular types, amylose and amylopectin, both of which are polymers of d-glucose. Amylose comprises about 27% of the corn starch granule. It is a linear

Table 5—Quantity of wet milling products sold, 1967 and 1972. (USDC, 1974c)

Product description	1,000 metric tons	
	1972	1967
Corn Syrup†		
Type I (20-38 D.E.)‡	174.72	95.48
Type II (38-58 D.E.)	722.39	724.84
Type III (58-73 D.E.)	739.54	395.26
Type IV (73 D.E. and above)	235.82	27.49
Total	1,872.47	1,243.07
Corn Sugar (Dextrose)	612.12	556.97
Total Sweetener Products¶	2,484.59	1,800.04
Corn Starch (Including Milo)§	1,627.58	1,414.75
Maltodextrins	37.59	56.74
Dextrin (Roasted)	38.19	76.52
Total Starch and Dextrins	1,703.36	1,548.01
Coproducts		
Corn Oil#	178.14	145.93
Feed Products		
Steepwater Conc. (50% d.s.)	45.27	35.38
Corn Gluten Feed	1,383.82	1,082.23
Corn Gluten Meal	355.16	396.98
Other By-Products	488.07	154.86

†Dry substance (d.s.) content average 80%. Other products contain about 90% d.s.
‡Dextrose equivalent.   §Milo starch made at one location prior to 1970.   ¶Adjusted to 90% d.s.   #Estimated.

polysaccharide composed of about 500 to 2,000 glucose residues connected by *alpha*-1,4 glucosidic linkages. The more abundant amylopectin, 73% of the granule, is a very large bush-like molecule with short *alpha*-1,4 linked chains connected by *alpha*-1,6 glucosyl branching linkages (Schoch, 1961; French, 1973).

Raw starch granules exhibit a maltese cross birefringence pattern when observed microscopically with plane polarized light. Their most striking characteristic is the ability to absorb water and swell within a narrow specific temperature range known as the gelatinization temperature range (GTR). At the beginning of the range a few granules lose birefringence and at the end-point temperature 98% of the granules have lost birefringence (birefringence end-point temperature BEPT). The GTR is characteristic for each starch species but it is influenced by crop production environment (Freeman et al., 1968). Regular corn starch extracted from corn grown in the USA Midwest typically has a GTR of 62 to 72 C. Chemicals added to a starch slurry and/or chemical alteration of the starch granule can dramatically alter the GTR and the BEPT (Leach, 1965). Raw corn starch absorbs about one part water for two parts starch, but at the BEPT water absorption increases to 20-fold. As a result, the granule swells proportionally and continues to increase in volume as temperature is raised above the BEPT. In this condition, the granule is easily disrupted mechanically or is hydrolytically degraded with acid or enzymes. A direct result of granule swelling is a parallel increase in starch solubility, paste viscosity, and clarity (Leach, 1965).

Adhesion of the swollen granules and binding of water cause a large increase in viscosity. The application of shear to the swollen granules with continued cooking causes marked reduction in viscosity and increased clarity and solubilization. When this paste is cooled, viscosity again quickly increases to an even greater degree accompanied by increased opacity. The transformations are known as gel formation and retrogradation. These properties along with adhesiveness, film-forming ability, and digestibility combined with low cost, are what make corn starch such a useful material in the manufacture of food and industrial products.

Much of the research and development activities of major starch producers has been to modify starch in such a way as to optimize desirable and minimize undesirable properties for a particular commercial application. Most commercial starch products will fall into one of the following categories: (a) Unmodified starch, (b) acid-modified starch, (c) dextrins, (d) oxidized starch, (e) cross-linked starch, (f) chemical derivatives of starch, and (g) pregelatinized starch. Various combinations of these categories are also produced.

Unmodified Starch

Unmodified corn starch, also called mill starch or thick boiling starch, is the largest volume and lowest cost industrial starch product. Russell (1973) estimated that 60% of the 1.59 thousand metric tons (3.5 billion pounds) of all corn starches produced in 1972 was unmodified starch. It is also the raw material from which the modified starches are made. It is used where its properties of high paste viscosity, strong gels, and retrogradation are useful or can be tolerated. Charcoal briquette molding, beneficiation of bauxite ores, and dusting powder are examples. Paper and paper product manufacture comprise the largest industrial uses of raw starch, consuming about 40% of the starch produced. In Canada 60% of the starch is used in paper manufacture (McNicol et al., 1972). In 1972 paperboard mills used 116.2 thousand metric tons (256.2 million pounds) of starch (USDC, 1974h) while all other papers consumed 463.9 thousand metric tons (1,022.8 million pounds (USDC, 1974g). The census of Manufactures (USDC, 1974i) did not separately list starch usage in paper coating and sizing but indicated usage of 95,011 metric tons of glues and adhesives, much of which was starch products. The total usage of 580.1 thousand metric tons (1,279 million pounds) of starch in all paper applications agrees fairly well with Russell's (1973) projected sales of 614.6 thousand metric tons but does not include some tonnage used for specially coated papers. However, it does include some tonnage of other types of starch especially wheat, potato, and tapioca. Nevertheless, the figures indicate that about 40% of all corn starch is used in paper manufacture in the USA. Over 22 million metric tons of paper and 25 million metric tons of paperboard were produced in 1972 and production is increasing at about 4.5% a year.

For most uses in paper, unmodified starch must be thinned by jet cooking or by enzyme thinning as the paste is made at the mill. This improves adhesiveness and increases production rate (Nisson, 1967). Numerous chemical modifications of starch are manufactured to reduce past viscosity, increase adhesiveness, increase retentiveness on fibers, and reduce setback to prolong working time. In food applications chemical modifications are made to improve texture in foods, to increase clarity, to reduce gelatinization temperature, to improve emulsification properties, to increase solubility, etc. A starch can be made that will fit almost any use. New varieties are introduced every year which improve food quality and production efficiency.

Acid Modified Starch

Acid-modified, thin-boiling starches are prepared by heating granular starch slurry with acid. The random hydrolytic action is stopped by neutralizing with sodium hydroxide or sodium carbonate at a point that will give the desired paste thinness. The modified starch is washed and dried. Cooked pastes of acid-modified starches have lower viscosities than mill starch but have a stronger gelling tendency. The lowered viscosity permits use of pastes at higher dry substance. In many uses, such as adhesives for paper lamination and clay coating, the starch must be incorporated at very high solids content. In other uses, such as textile warp sizing applications, where smooth, strong film coatings are needed to protect the fibers during weaving, acid-modified starches are preferred (Sheldneck and Smith, 1967; Compton and Martin, 1967). Of the 124.7 thousand metric tons (275 million pounds) of corn starch used in 1972 by the textile industry, 77,111 metric tons (170 million pounds) was acid modified (Russell, 1973). The greater gelling tendency of acid modified corn starches give desired set and texture in gum confections (Wurzburg, 1970).

Maltodextrins and Pyrodextrins

"Dextrins" are degradation products of starch covering a wide range of properties and made by a variety of methods (Satterthwaite and Iwinski, 1973). Pyrodextrins are produced by the action of dry heat with or without the presence of (hydrochloric) acid; maltodextrins are prepared by the hydrolytic action of *alpha*-amylases on starch pastes to the extent of producing 5 to 20 dextrose equivalent units (D.E.) products of complete solubility but little or no sweetness (Armbruster, 1974; Armbruster and Kooi, 1974) cyclodextrins are ring polymers of 6 to 8 dextrose units produced by the amylase of *Bacillus macerans* (Armbruster and Jacaway, 1972). The latter have the interesting property of complexing with many organic molecules but have not yet reached commercial production.

Maltodextrins are instantly soluble in water, giving clear to hazy solutions of low viscosity. They are completely digestible and completely bland. They contribute body in frozen desserts, soups, etc.' they are excellent carriers for spray drying hygroscopic solids such as instant tea and coffee; they have excellent protective colloid action for fats and are major constituents of coffee whiteners; they are bland diluents for artificial sweeteners, etc. The drop in sales between 1967 and 1972 (Table 5) probably reflects the withdrawal of cyclamates from the market by FDA action. In dry mixes the presence of low D.E. maltodextrins keep package contents free flowing (Ueberbacher, 1970).

Many pyrodextrin products having a wide range of properties and uses are on the market (Satterthwaite and Iwinski, 1973). Combined sales of 37,784 metric tons (83.3 million pounds) in 1972 (Table 4) is about the same volume as for maltodextrins but uses are predominantly in the non-food industries. Pyrodextrins are made by roasting dry starch with a small amount of (hydrochloric) acid catalyst. The initial reaction is primarily hydrolysis of the natural 1-4 and 1-6 linkages, but as time and temperature are increased recombination of fragments occurs through 1-2 and 1-3 as well as the 1-4 and 1-6 linkages producing a highly branched structure with increased solubility. The low conversion products, known as white dextrins, are made in a broad range of solubilities. Clarities of solutions increase and viscosities decrease with increasing solubility. The high conversion yellow dextrins are completely soluble and produce clear, nonhazing solutions of low viscosity.

Major applications for pyrodextrins are in adhesives for fabricating paper products of all kinds and for remoistenable gums such as on postage stamps and packaging tape, sizing and finishing textiles, thickeners for water-soluble, fabric-printing inks, and filler for rug backings (Dux, 1967). Use in foods is limited but is quite acceptable from a regulatory standpoint (Graefe, 1974) even though digestibility of the high conversion types is lower than starch. Uses are found in the pan coating of hard candy and to thicken emulsions.

Oxidized Starches

Another type of thin boiling starch, oxidized corn starch, is prepared by treating starch slurry with sodium hypochlorite solution. The hypochlorite treatment not only bleaches and causes limited hydrolysis but it also forms some carboxyl groups by oxidation of free hydroxyl groups amounting to about 1 per 25 to 50 anhydro glucose units (Graefe, 1974). This results in a lowered gelatinization temperature, improved ease of dispersion, improved paste clarity, and reduced setback or gel formation (Scallett and Sowell, 1967). Paste of oxidized starch, when spread in a thin layer, dries to a clear, adherent, continuous film providing the kind of properties desired for paper sizing, paper clay-coating adhesive, textile warp size for cotton and rayon, and laundry finishing, including aerosol spray starch. Lightly bleached starches find use in tableting certain pharmaceuticals such as aspirin. Of the nearly 74,116 metric tons (166 million pounds) of oxidized starch sold in the USA most is used as adhesive in clay coatings for paper. About 1,814 metric tons (4 million pounds) is used in construction of wallboard, insulation, acoustical tile and the like. However, total starch usage in these three applications amounted to 70,307 metric tons (155 million pounds) in 1972 (Russell, 1973). Dialdehyde starch, a starch severly oxidized with periodate (Mehltretter, 1967), is the only starch product used for imparting wet strength to tissue papers (Russell, 1973). Being toxic, dialdehyde starch cannot be used in foods (Graefe, 1974).

Pregelatinized Starches

The trend toward prepared home and bakery mixes, precooked foods, water-soluble laundry sizes, and many other convenience items, require starch preparations that will form pastes in cold or warm water. These may be prepared by drying pastes on heated rolls, by spray drying, or by simultaneous pasting and drying on the rolls. Any type of modified starch may be prepared in this manner, but waxy starches are preferred because of easier reconstitution in water.

Chemical Derivatives

Bifunctional reagents like glyoxal, epichlorohydrin, phosphorus oxychloride, and metaphosphates can react with two moles of hydroxyl to form linkages binding adjacent starch molecules. The degree of cross-bonding regulates the degree of swelling of the starch granule and the tolerance to viscosity loss in acidic conditions (Leach, 1965). Waxy starch must be cross-linked for use in foods to reduce peak viscosity and limit viscosity loss from shear and acidity while retaining the clarity and lack of setback. Cross-linked waxy starches are now used to thicken fruit pie fillings and prepared foods of all kinds. Heavily cross-bonded corn starch is used for dusting surgical gloves which must be autoclaved; the starch will not gelatinize but is slowly digested in the body (Wurzburg, 1970).

Monofunctional chemical reagents which react to form ester or ether linkages with hydroxyl groups in starch are a means of adding substituent groups which may radically alter the physical and chemical properties of starch. A great number and type of chemical starch modifications are possible. For more details, the reader should consult appropriate chapters in Whistler and Paschall (1965, 1967) and Whistler and BeMiller (1973).

Hydroxyethyl starch, made by reacting ethylene oxide with starch, is probably the largest volume derivative. The hydroxyethyl group lowers gelatinization temperature and makes for stable pastes which produce clear, flexible films on drying. It is popular as a textile warp size and in liquid laundry starches. Hydroxyethyl acid modified corn starch is used in paper sizing and clay coating, and for thickening of high gloss printing inks. A very highly substituted hydroxyethyl starch has been used as a blood plasma extender (Hjermstad, 1973). Hydroxypropyl starch is FDA approved for use in thickening foods such as salad dressings.

Corn starches with cationic groups such as amino alkyl and quaternary ammonium starches have found broad use as internal binders in paper manufacture and have attained a sales volume of 44,648 metric tons (98.4 million pounds) (Russell, 1973). They disperse readily in hot water to give smooth, stable pastes. They are strongly absorbed by the negatively charged paper and, although higher in price, can be used at dosages of 0.5% on a dry pulp basis compared with 2 to 3% for unmodified starch. The cationic starches are also useful as flocculating agents in purification of industrial water and ore beneficiation (Paschall, 1967).

Waxy Starch

Waxy endosperm mutant (wx) was found in China in 1909. In 1936 scientists at the Iowa Agricultural Experiment Station recognized that waxy corn starch properties were somewhat similar to those of tapioca starch and began to develop waxy hybrids. Wet milling possibilities were recognized by Hixon and Sprague (1944), and commercial production began soon thereafter (Shopmeyer et al., 1943). The wet milling properties of waxy corn are very similar to regular corn except that solubles yield is a little higher, while starch and gluten slurries filter and dry appreciably slower. The filtration and drying problems are caused by a small amount of a partially soluble, highly branched polysaccharide known as phytoglycogen. A small amount of amylolytic enzyme added to process streams dissolves phytoglycogen and improves filtration and drying characteristics (Freeman et al., 1975).

In the three decades that have elapsed since waxy corn starch was first isolated and characterized, its excellent properties for specialty food and industrial uses have been fully recognized. Today most wet milling companies sell waxy starch products but prefer to call them "amylopectin starches" because the term "waxy" is misleading (Powell, 1973).

Waxy corn production has climbed steadily as a premium crop grown under contract. It has reached an estimated 635 to 760 thousand metric tons (25 to 30 million bu) a year, a respectable 8 to 10% of the total annual wet milling grind (author's estimate). A number of excellent hybrids are available from at least six seed corn companies. Commercial interest in waxy corn has been recently stimulated by experiments in feeding waxy corn to ruminant animals (Braman et al., 1972). Steers fed waxy corn made faster gains than those fed normal corn. Henderson (1974) has summarized the data from numerous feeding trials and observed a small positive but not statistically significant gain from feeding waxy corn. Nevertheless, this work has stimulated a great deal of activity among commercial seed companies promoting waxy corn.

Waxy starch granules are similar in size and appearance to regular corn starch but are composed entirely of the branched starch fraction, amylopectin. They can be readily identified by observing a reddish purple coloration when stained with dilute I_2-KI solution instead of the intense blue color of regular starch or higher amylose starches. The unique properties of waxy starch pastes are attributable to absence of amylose. The GTR of waxy corn starch is a little higher than for regular starch; the swelling power when cooked at 95 C is 2.66 times greater but with the same degree of solubilization. For this reason unsheared, waxy starch pastes exhibit very high initial viscosity when freshly pasted but on shearing drop below the viscosity of similar regular corn starch pastes, especially under acidic conditions (Leach, 1965).

Waxy starch pastes are "long" and cohesive, whereas regular corn starch pastes are "short" and heavy bodied. Waxy starch pastes have a much greater degree of clarity than normal corn starch and tend to remain clear on cooling because of the absence of retrograding amylose. For the same reason waxy starches do not gel on cooling but remain as fluid, redispersible sols. Waxy starch pastes, dried in thin layers, form translucent films that are readily redissolved. These properties make waxy corn starch useful in many food and industrial uses. However, in nearly all of these uses the properties must be modified by chemical or physical means to increase utility.

Powell (1973) has provided a good review and a comprehensive list of applications for isolated amylopectin and the amylopectin (waxy) starches. Powell summarizes the status of these starches succinctly as follows:

> The unusual optical and rheological properties of these pastes and the greatly reduced tendency to gel in dispersions or crystallize in dried films make possible products and formulations previously unattainable or achieved only with more costly gums."

Although applications for unmodified waxy starches are listed, the actual sales volumes for these uses is probably small. The greater volume of sales is in the more useful and more profitable modified types. The clarity and stability of amylopectin starch gels make them especially suitable for thickening prepared, canned, and frozen foods (Wurzburg, 1970). Chemical cross-linking is necessary to restrict granule swelling and prevent viscosity breakdown on cooking; derivatization with phosphate, acetate, succinate, or hydroxypropyl groups increase paste clarity after repeated cycles of freezing and thawing (Whistler and Paschall, 1967). Starch phosphates and starch acetates of partially crosslinked waxy corn starches are sold in significant quantities for many uses, especially in foods (Whistler and Paschall, 1967). Other chemical modifications of waxy corn starch have been effective for many specialized uses.

Thin boiling, oxidized, and hydroxyalkyl modifications of waxy starch produce improved film clarity for textile sizing and certain types of paper coating. Waxy corn starch is the preferred starting material for maltodextrins because of improved water solubility after drying and greater solution stability and clarity. Waxy pyrodextrins have superior remoistening characteristics. There is no doubt that the volume and variety of waxy starches will continue to grow in response to industry demands.

Higher Amylose Starch

The recessive form of the endosperm mutant *ae* produces starch containing 55 to 60% amylose (Vineyard and Bear, 1952). It was found after a concerted search for corn kernels with higher amylose content. This discovery generated a great deal of breeding, biochemical, and starch application work with *ae* and other recessive endosperm mutants (Zuber, 1965; Senti, 1967). Subsequent breeding has produced modified *ae* types having up to 85% amylose. High amylose hybrid corn varieties have been developed by the Bear Seed Co., Decatur, Ill. The commercial utilization was pioneered by several of the wet milling companies in the early sixties. Wet milling of high amylose corn gives lower starch yields than regular corn (Anderson et al., 1961).

High amylose starch granules are small and often odd shaped. They have a high, broad GTR. Indeed some of the granules don't lose all birefringence even after prolonged boiling; swelling power is only one-fourth that of regular corn starch at 95 C. However, pastes prepared by jet steam cooking alone or in the presence of alkalis, salts, or formaldehyde can be handled commercially if not allowed to cool and can be used where tough, opaque films are valued. High amylose starch has been found to be specifically valuable for sizing glass fibers prior to weaving, for preparation of an edible clear film, and in coating foods for fat holdout and moisture retention (BeMiller, 1973). Although the volume is presently small, continued research may eventually lead to expanded production through lower costs and improvement in the utility of this unique starch.

The amylose and amylopectin can be separated from dispersed pastes of regular corn starch by forming insoluble complexes of amylose with a variety of organic substances, such as *N*-butanol, nitroethane, etc., or by retrogradation or salting out the amylose with sodium sulfate. The amylopectin may be recovered by roll drying (Langlois and Waggoner, 1967).

Several attempts to commercialize one of these processes failed owing to high cost and small markets and were displaced by high amylose starch.

Sweetener Products

The corn wet milling industry began in the mid-19th century as a producer of corn syrups, known for many years as "glucose" or "glucose syrup." Corn syrup has remained an important sweetener ingredient in foods, along with sugar. Corn syrups and crystalline glucose (dextrose) are less sweet than sucrose and therefore have normally brought lower prices. The synonymous names, glucose and dextrose, are applied to the specific six-carbon monosaccharide that is the basic monomer comprising the starch molecule, and is the major source of energy in human and animal nutrition. The name dextrose generally refers to the crystalline commercial product. Recently, the major wet milling companies have developed a means of producing a syrup that has comparable sweetness to sucrose using an enzyme that isomerizes glucose to fructose (Kooi and Smith, 1972; Geyer, 1975). Coming at a time of scarcity and unmitigated high prices of sucrose, the new high fructose corn syrup triggered a new round of expansion of wet milling facilities and entry of new producers into the business. Cantor and Shaffer (1974) have predicted that the advent of these sweeter corn syrups will provide increasing penetration of the total sweetener market from the present 15% to about 30% by the year 2010. If this prediction is realized the market for corn sweeteners will be increased to about 4 million metric tons a year (9 billion pounds).

Corn Syrup

In the time-honored method of manufacturing corn syrup, a 35 to 40% starch slurry is acidified with hydrochloric acid to 0.15 N and heated under pressure to 140 C for various lengths of time depending on the dextrose equivalent (D.E.) desired. To stop the reaction the hydrolyzate is discharged under pressure into a tank where it is neutralized with sodium carbonate. Residual fats and proteins are removed by filtration or centrifugation, color is removed by treatment with activated vegetable carbons, and in some cases residual ash is removed by passage over columns of ion exchange resins. Finally, the water-white hydrolyzate is evaporated to syrups with densities of from 40 to 46° Baume which correspond to 75 to 86% dry substance. The U.S. Department of Commerce (1974c) lists four major types of syrups produced commercially depending on D.E. (Table 5). Shipments to specific industries are listed in Table 6. Canada uses about 67,000 metric tons of syrups with about 40% going into confectionery applications (McNicol et al., 1972).

The syrups consumed in largest volume at the present time are the standard low-priced 42 D.E. acid converted syrup, and the 58 to 73 D.E. high fermentables non-crystallizing syrups produced for the brewing trade (Table 5). The latter syrups are made by converting a medium D.E. (40 to 50) acid syrup with a fungal diastase-type enzyme that will hydrolyze some higher saccharides (Newton, 1970). Syrups with very high maltose are made from acid-thinned paste (15 D.E.) converted with malt. While acid-converted syrup is

Table 6—Corn refiners shipments of corn syrup and dextrose, 1972. (ASCS, 1973)

Type of use	1,000 metric tons†	
	Corn syrup	Dextrose
Bakery, cereal, etc.	260.36	149.08
Confectionary, etc.	399.26	54.87
Ice cream and dairy products	201.31	6.63
Beverages	82.92	8.29
Canned fruits, vegetables, jams, jellies and preserves	169.76	28.87
Multiple and all other food uses	286.79	166.84
Non-food uses	55.87	82.15
Total	1,456.27	496.73

†Dry basis. Assumes a solids content of 80.3% for corn syrup and 92% for dextrose.

composed of a whole range of starch fragments (saccharides), enzyme technology is now available which permits manufacture of syrups with many different proportions of the simple sugars, dextrose, frutose, maltose, maltotriose, and maltotetraose (Pazur, 1965). Combinations of these products with or without sucrose provide an infinite number of food applications (Godziki, 1976).

Corn syrups are water-white, very thick fluids that thin upon heating. They are shipped in drums or tank cars and can be easily pumped on warming. Viscosity increases with increase in total solids concentration and decreases with increasing D.E. Sweetness and fermentability also increase with increase in dextrose content or D.E.

The largest volume uses of corn syrup are in confections, followed by bakery products and dairy products (Table 6) (Agric. Stabilization and Conservation Service, 1973). Syrups provide a body to hard candies giving them chewiness and desirable mouthfeel without excessive sweetness; hygroscopicity of hard candy can be reduced by use of high maltose syrup (Frey, 1967). Syrups produce desirable viscosity in thick fruit syrups such as in canned peaches (Newton, 1970). Corn syrups also act as foam stabilizers in marshmallows and provide desired humectancy to maintain plasticity. Corn syrups are used in baking and in brewing as a source of fermentable carbohydrates and the glucose content produces browning in bread crusts. The more highly converted syrups produce freezing point depression for frozen desserts and osmotic concentration for food preservation. The lower converted products help control the number and size of sugar crystals in fondants, ice cream, icings, and jams and jellies, and sheen in fruit preserves and pie fillings.

About 5% of the corn syrup sales are in non-food uses as bodying agents in inks, shoe polish, textile finishes, adhesive formulations, and pharmaceuticals; in leather tanning; as humectant in tobacco. Low (20 to 30) D.E. syrups can be dried (corn syrup solids) for use in foods where water input must be limited.

High Fructose Corn Syrup

High fructose corn syrup (HFCS) is made from a fully converted dextrose syrup. The enzyme, xylose isomerase, produced by several species of bacteria and actinomyces, was found capable of isomerizing glucose (an aldose) to fructose (a ketose also called levulose), giving a mixture of about 43 to 45% to 57 to 55% glucose (Marshall and Kooi, 1957; Takasaki and Tanabe, 1971). Search for a practical process spawned a feverish research effort from which evolved highly sophisticated processes involving continuous conversion through columns of enzyme bound to solid carriers (Schnyder, 1974; Messing and Filbert, 1975; L. Zittan and S.H. Hemmingsen, Enzymatic isomerization of glucose from laboratory to industrial production. Am. Chem. Soc. meeting, New York, April 1976). To determine usefulness of HFCS in foods, much research has been conducted on food applications.

Carefully controlled taste panel sweetness comparisons have shown that at sugar concentrations used in carbonated beverages (9 to 14%), a 50/50 blend of HFCS and sucrose is sweeter than the same concentration of sucrose (Brooks et al., 1974). Other studies have shown that sweetness comparisons in complicated food systems is often different from comparisons made from water solutions of sugars and is greatly affected by concentration. Hence, HFCS shows equal sweetness to sucrose in some foods; in others it is less sweet. Thus, HFCS has found application in a wide variety of food systems such as confections, baked goods, table syrup, fountain syrups, sweet beverages, catsup and other condiments, pickles, etc. (Wardrip, 1971; Fruin and Scallett, 1975). From its beginning in 1972 of 200 thousand lb., HFCS production in 1976 was estimated to be 2.5 billion pounds (Andres, 1976) and is expected to continue to grow but at a slower rate because of the current low price of sugar. Projections indicate a 5 to 6 billion lb. market in the 1980's (Andres, 1976; Scallett, *op. cit.,* 1976).

Dextrose

Fully converted starch hydrolyzates are now universally prepared from acid or enzyme (*alpha*-amylase) thinned starch pastes by incubation with amylglucosidase, an enzyme of fungal origin. The hydrolyzate is refined and concentrated as for corn syrup and is introduced hot into slowly stirred crystallizers containing a large amount of seed crystals from a previous batch. The mixture is slowly cooled over a 3 to 4-day period. The crystals are recovered and washed in a basket centrifuge and dried (Kooi and Armbruster, 1967). The anhydrous *alpha* form is produced by recrystallizing a solution of dextrose hydrate in a "strike" evaporator pan at high temperature. It is the purest form and finds use in pharmaceuticals. In food systems, such as dark, sweet, or milk chocolate products, only anhydrous dextrose can be used because the presence of water cannot be tolerated (Snyder, 1970).

Confectionery manufacture uses account for about 10% of the dextrose produced (Table 6) mainly as a major component of tableted candies and chewing gum. Dextrose has a negative heat of solution and gives mouth-cooling flavor release as well as sweetness. Other confections including dextrose are gum confections, fondants, and hard candy formulations (Walters, 1975). Another large use for dextrose is in canning and frozen food packs, catsup, frozen desserts, prepared dry mixes, prepared icings, jams, jellies and preserves, pickles, meats, soft drinks, wines, and malt liquors. The largest single use is in baked goods where the dextrose serves as a yeast nutrient, provides some sweetness, and produces crust browning (Snyder, 1970).

In medicine, dextrose (glucose) use in intravenous feeding is well known. It is also used as a tabletting aid and diluent. Over 45,400 metric tons (100 million pounds) of dextrose is converted yearly to sorbitol by hydrogenation and volume is growing steadily. Sorbitol is used in the production of synthetic vitamin C and as an intermediate in other chemical syntheses.

Although dextrose is not as sweet as sucrose, its many functional properties provide unique functional values in foods. Some bottlers use it in blends with sucrose in soft drinks with no noticeable loss of sweetness.

Feed Products

Kernel components, mainly protein and fiber, remaining after recovery of the corn starch and corn oil in the wet milling process amount to about one-third of the original corn dry substance (Table 3). These materials are sold almost exclusively as ingredients in feeds for farm animals and pet foods (Shroder and Heiman, 1970). The gluten fraction is sold as a dry meal standardized at 60% protein. This product contains about 400 mg/kg in xanthophyll pigments at the beginning of a crop year, but xanthophyll content drops gradually through the year reaching about 200 mg/kg by the end of the season. This change reflects gradual destruction of pigments in corn during storage through one season (Watson, 1962). Corn gluten meal is used primarily as a protein concentrate and pigment source in broiler chicken feeds (Kuznitzky, 1968) in areas where yellow skin is a desirable marketing attribute. It is important as a protein source high in essential sulfur amino acids, methionine and cystine (Reiners et al., 1973; Sasse and Baker, 1973), for use in providing adequate amino acid balance with soybean meal. Corn protein purified in such a way as to have bland flavor (Reiners et al., 1972) is a potential source of vegetable protein for food in combination with soy and other proteins.

The wet milling fraction, containing the corn pericarp or "fiber," is composed mostly of cellulose and hemicellulose plus residual starch and protein. It has a digestibility of 80% (total digestible nutrients, TDN) for ruminants and is sold by one wet milling company at a 10% protein level for ruminant and dairy rations. However, most wet millers blend the fiber with steepwater and spent corn germ flakes or expellor cake to produce corn gluten feed. This product has a standardized protein guarantee of 21% and finds use primarily as a major ingredient in dairy and beef cattle rations. Heiman (1961) and Yen et al. (1971) have demonstrated its value at up to 10 to 15% in laying chicken and swine rations, respectively. Versions of corn gluten feed having higher steepwater contents at up to 31.5% protein are available from several companies and may be more suitable for non-ruminant animals because of a lower proportion of fiber (Waldroup, 1969). Corn germ meal containing 22.5% protein (Reiners et al., 1973) is available from wet millers who solvent extract corn germ. It is useful in feeds as an absorbent for liquid ingredients such as molasses, fish solubles, choline, etc. (Shroder and Heiman, 1970).

Corn steep liquor or heavy steepwater is known officially as a feed ingredient by the name "condensed fermented corn extractives." It is sold to a limited extent as a component of liquid feeds for beef cattle (Woods et al., 1969), as a component in beef range blocks (Arner, 1965), and as a fermentation nutrient (discussed under Fermentation Industries). Corn steepwater is rich in phytin (hexaphosphoinositol) from which the vitamin, i-inositol, has been manufactured.

Corn Oil

Corn oil is commercially produced only from corn germ isolated by wet milling or dry milling. Because of this supply restriction combined with strong demand, corn oil usually commands a slightly higher price than the other two major vegetable oils, soybean and cottonseed. The volume of corn oil produced has steadily increased over the last half century as wet and dry milling capacity has expanded. In 1972 the total volume reached 226,800 metric tons (500 million pounds) of which the author estimates that about 180,000 metric tons was derived from wet milling. World corn oil production in 1972 was 300,000 metric tons which is only 1.39% of the total for all vegetable oils. In the USA corn oil consumption is about 12% of all vegetable oils (Jiles, 1973).

Oil is recovered from corn germ by expelling, solvent extraction, or a combination of expelling and extracting as generally described by Norris (1964a). Wet milled germ is preferably expelled from an oil content of 50 to 60% down to 20 to 25% and finally extracted with hexane to a residual oil content of 1 to 2% in the spent corn germ flakes. Most dry corn mills, because of their smaller size, recover oil only be expelling, but one very large dry milling company recovers oil only by extraction (Anonymous, 1975c).

Crude corn oil is composed of 95% triglycerides. Minor components include free fatty acids, waxes, phospholipids, pigments, and odorous compounds (Reiners and Gooding, 1970), which must be removed by refining to achieve an acceptable food product. The process follows well known vegetable oil refining procedures (Norris, 1964b).

In outline, this process involves several steps: (a) forming the sodium soaps of the free fatty acids, (b) removing the emulsion containing the soaps and the phosphatides by centrifugation, (c) removing the waxes by chilling, (d) removing pigments by contact with bleaching clays, (e) removing odors by high vacuum distillation at 225 to 260 C. The fatty acid fraction is recovered by heating the emulsion in the presence of sulfuric acid and is sold as an ingredient for use in beef and poultry rations.

Refined corn oil is 98% triglycerides in which the saturated constituent fatty acids are palmitic (16:0) 11.1%, stearic (18:0) 2.0%, and arachidic (20:0) 0.2%. The unsaturated fatty acids are linoleic (18:2) 61.9%, oleic (18:1) 24.1%, and linolenic (18:3) 0.7%. These fatty acid values were obtained from gas-liquid chromatography analysis of a large number of commercial corn oil samples from a large Illinois refinery. It is well known that variation in the ratio of the two principal fatty acids, oleic and linoleic, is due to genetic background (Jellum and Marion, 1966; Poneleit and Bauman, 1970) and environment (Beadle et al., 1965; Thompson et al., 1973). The linoleic and oleic acid contents of commercial corn oil have gradually changed from 56 and 30% to the present ratio of 62 and 24% over a period of 20 years (Ruark, 1970; Reiners and Gooding, 1970) during which period total oil in corn processed at Illinois wet milling plants dropped from about 4.9 to 4.4%, apparently due to gradual changes in corn hybrids planted. Iodine value of currently produced corn oil averages 127.

The recognized acceptance of corn oil as a food oil is, in part, due to its flavor stability during storage and cooking without added synthetic antioxidants. This stability is the result of an adequate level of natural antioxidants, α- and γ-tocopherols, and a low level of linoelic acid. About 52% of the corn oil in the USA is consumed as salad or cooking oil. Its bland flavor and high smoke point are mainly responsible for its popularity. Another reason for the popularity of corn oil is its high content of unsaturated fatty acids recognized by medical authorities (American Heart Association, 1968; Rathmann et al., 1970) as a dietary method of reducing blood cholesterol levels. The other large use, 45%, as in margarine. Corn oil margarine has grown from its introduction in 1957 to a major food item. Other food uses are minor. The only part used industrially is the 7 to 8% removed from crude oil by refining (USDA, 1975a; Jiles, 1975).

DRY MILLING

The dry corn milling industry began as numerous small mills scattered across the country to supply whole ground corn meal to local clients for household uses. As the population became more urbanized, the demand for food products with longer shelf life resulted in the development of milling systems that produced low fat cornmeal by removal of germ. As the process became more complex, mills became larger and fewer. At latest count (Anonymous, 1969) there were 120 dry corn mills but since that date several mills have been closed including one large mill at Kankakee, Illinois. Many of these mills listed are small whole cornmeal mills, primarily in the rural south, but some are devoted to making whole meal for corn chips and other snack items, tortillas, etc. The degerminated corn mills range in capacity from 1,500 to 2,500 metric tons a day and process at least 80% of the total volume. The total amount of corn utilized for dry milling and breakfast cereals has remained fairly constant at 3,400 to 3,600 thousand metric tons (110 to 120 million bu) a year (Table 1) (USDA, 1974). The most dramatic changes during those years have been in the distribution among the different kinds of products produced.

The Process

As described in an earlier chapter, the normal dent corn kernel is comprised of the following proportional parts: pericarp, tip cap, germ, horny endosperm, and floury endosperm. The degerminated corn dry milling process is designed to make as complete separation of these parts as is economically possible and to retain the horny endosperm portion as discrete pieces. Separate steps in the separation process include cleaning, tempering, degerming, drying, cooling, grading, aspirating, grinding, sifting, and packaging. The flow sheet (Fig. 2) shows how these steps are normally coordinated. However, this diagram is very much simplified because separations are not perfect and fractions have to be remilled, reclassefied, and resifted several times to try to achieve optimum yield and quality. Furthermore, mills differ markedly in the types of equipment used and preference for materials routing (Stivers, 1955; Brekke, 1970; Alexander, 1974).

The initial cleaning steps for the whole corn are of utmost importance because, although high quality corn is purchased for food production, residual contamination must not be allowed to enter the process. Although not shown in Fig. 2, most mills use magnetic separators to remove tramp metal; scourers and aspirators to remove adhering dust, glume, and other light material; screens to remove large and small foreign material; water washing with flotation to remove cob pieces; and settlers to remove stones and nonferous metal. Some mills have used electrostatic separators to remove rat pellets the same size as corn kernels. All mills carefully examine incoming corn for presence of mold and aflatoxin and reject any lots that are suspect (Nesheim, 1973).

The cleaned corn must be tempered with water or steam to condition the corn for milling. The corn is moistened with water in an amount needed to bring the corn up to 20 to 22% moisture and held for a period of time (1 to 3 hours) to allow the water to equilibrate through the kernel. Degermination may be accomplished in a variety of ways but the most popular, especially if flaking grits are desired, is with the Beall degerminator. It is composed of a conical knobbed rotor and a stator shell that is knobbed on the lower convex

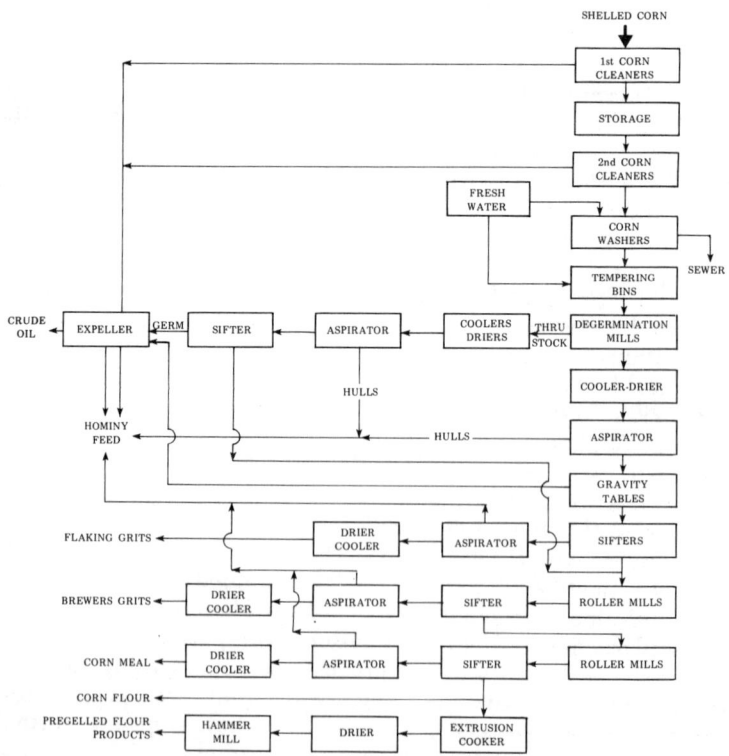

Fig. 2. Dry milling process for corn.

INDUSTRIAL UTILIZATION OF CORN

surface and slotted on the upper surface. Corn enters the small end and broken kernels pass through the slots. Some companies use impact mills or granulators for degermination for different results with and without tempering (Brekke, 1970).

Discharge from the degerminator is dried to 15 to 18% moisture and cooled. It is sized over stacked coarse sifters or reels with the largest material +3 1/2 to +10 mesh passing through aspirators to remove free pericarps and thence to specific gravity tables where germs are separated from endosperm by difference in density. Endosperm fractions and impure germ-endosperm mixtures are passed through corrugated counterrotating roller mills to remove adhering pericarp and/or germ pieces and flatten germ pieces for better screening. Each roller mill set is preceded by an aspirator to remove light material and followed by a sizing device, usually stacked screens of appropriate sizes for the stock being handled. The roller mills are set with differential speeds and a gap appropriate for the material it receives to minimize grit breakage. Endosperm fractions are finished in purifiers designed to remove fine pieces of fiber and are packaged according to grade at 12 to 14% moisture (Easter, 1969).

The Products

Finished germ, containing 18 to 27% oil depending on the mill configuration, is dried to 3% moisture and the corn oil expelled in a standard oil expelling equipment. The germ residue is blended with pericarp fractions, unseparable mixtures of endosperm, pericarp and germ, and corn cleanings to give a commodity feed item known as hominy feed. This product is guaranteed to have a minimum level of 10% protein and 5% oil but generally averages 10.5% protein and 6.5% oil. Hominy feed is used as an ingredient of swine, ruminant, and poultry feeds as a source of energy and good quality protein (Shroder and Heiman, 1970). One large milling company solvent-extracts the germ and has recently announced production of a germ flour containing 25% protein (Anonymous, 1975c) using a process developed by a USDA laboratory (Blessin et al., 1972). This material is expected to become a useful, nutritious food ingredient.

A typical yield of products from dry milling are: hominy feed, 35%; corn oil, 1.0%; grits, meal, and flour, 60.0%; shrinkage, 4.0%. There are no definite specifications which closely define all products which can be manufactured but the classifications generally accepted in the industry (Stivers, 1955; Katz, 1966) are shown in Table 7. The ratio among endosperm fractions varies with the quality of the corn, the milling conditions, and the orders the miller has to fill; most mills are quite flexible in the latter. Fat content is the most generally used criterion of purity. Fat values in Table 7 are minimum values and are affected by milling conditions and corn quality.

The grit fractions are all produced from the horny endosperm, have a bright yellow color or are clear white, and are free of dust and bran pieces. The miller usually tries to optimize yield of flaking grits because of premium price. Each hominy grit will become one corn flake. A common alternate distribution of products, depending on sales needs, is to produce brewers common grits by combining larger than 24-mesh material and milling so that

Table 7—Typical yields and products from dry corn milling. (Katz, 1966)

Product	Yield	U.S. sieve size	Fat
	%		% dry basis
Hominy or cereal flaking grits	12	-3.5 + 6	0.7
Coarse grits	15	-8 + 14	0.7
Regular or medium grits	13	-12 + 18	0.7
Fine grits	10	-18 + 26	0.8
Coarse or granulated meal	3	-28 + 50	1.0-1.2
Cones or dusted meal	3	-50 + 80	0.8-1.5
Corn flour	4	-80 + Pan	1.5-2.5
Corn oil	1	—	100
Hominy feed	37	—	6.5
Shrinkage	2	—	—

95% will pass 12-mesh. All material passing 24-mesh may be sold as cornmeal, but screen size up to 32-mesh is used in some mills.

Quality of grits are judged on color, odor, fat content, and freedom from dust, pericarp, and insect fragments. Flour has long been considered a byproduct which at times was put in feed. However, in the past few years new uses have been found that have channeled all output into more profitable markets (Roberts, 1967). Most mills make specialty products utilizing raw grits, meal, or flour. These raw materials are processed by cooking with steam to the desired degree of starch gelatinization, usually on flaking rolls, dried and ground as required by the user (Alexander, 1974; Anderson et al., 1969; Katz, 1966). The degree of dispersibility of these products is conditioned by the particle size of the starting material, method, time and pH of cooking, and degree of shear.

Uses for Corn Endosperm Products

Utilization of the 2,490 to 2,720 thousand metric tons (5.5 to 6.0 billion pounds) of dry corn products manufactured are distributed among human foods, beverages, pet foods, and industrial products in approximately that order of magnitude (USDC, 1974a, b). Surprisingly about 10% of that volume is utilized as whole meal for food. Probably over 2.27 thousand metric tons are used in pet feeds primarily in the form of flakes either raw or toasted.

The most commonly recognized use, particularly in southern states, is boiled grits served as a side dish with nearly all meals. Some people prefer the bolder flavor of yellow corn products especially in the northern USA, however, most people, especially in southern states, prefer white grits. Over 80% of the table grits sold are white. Corn bread or corn muffins made from cornmeal have long been a favorite bread in all parts of the USA, and more recently these popular dishes have been offered as dry, self-rising mixes for easier preparation. Another favorite food is corn flakes. The flaking grits are cooked to a rubbery consistency with syrup, malt, salt, and flavoring added. After tempering, the cooked grits are flattened between large steel rolls followed by toasting in traveling ovens to a golden brown color (Brockington, 1970).

The rapid growth of dry mix formulations and snack products of various kinds has increased demand for corn flour to the extent that mills sometimes must increase flour yield by milling larger sized fractions. Such flour, of course, has a different texture than normal mill-run flour. Corn flour has been found to be particularly valuable as an ingredient of pancake mixes, baby foods, cookies, biscuits, ice cream cones, ready-to-eat cereals, batter breading mixes, and binders for loaf type sandwich meats (Roberts, 1967).

The high starch content of cornmeals and flour is important in giving a high puff in preparation of extruded snack products in which a delicate corn flavor is desired. On the other hand isolated corn starch is most useful for snacks in which artificial flavors are to be added. The peculiar rubbery consistency of tortillas are probably produced by interaction of starch, protein, and hemicelluloses modified by lime processing. Instantized tortilla flour, which became popular in Mexican-American households, may be made from whole corn or a mixture of meal and flour. Corn chips are generally made from whole corn (Brockington, 1970).

Industrial uses amount to only 5 to 6% of the dry milled corn or about 175 thousand metric tons. Most uses involve pregelatinized corn flour where the more highly purified starches have no particular advantage even though they may be more effective on a weight basis. The major uses are foundry sand core binders (34,000 metric tons); briquette binder (22,700 metric tons); adhesive for gypsum board and other construction materials (22,700 metric tons); wet drilling mud thickeners (11,300 metric tons); refining and pelletizing ores (28,500 metric tons); fermentation (20,400 metric tons); and paperboard (4,500 metric tons) (Alexander, 1974). On the other hand, purified corn starches have the advantage that they can be chemically modified more specifically to fit a particular application and generally have greater adhesive power.

A food item of major importance to the dry miller in recent years has been corn-soy-milk (CSM) (Anderson et al., 1971). This product is a mixture of 68% gelatinized corn meal, 25% defatted soy flour, 5% nonfat dry milk, and 2% minerals and vitamins. A lower cost, but equally nutritious product called Corn-Soy Blend (CSB) (the same as CSM except for omission of milk) has also been produced. The product has been made by companies under contract to U.S. Agency for International Development. Since 1966, when the program began, to 30 Sept. 1974, over 1.17 million metric tons of CSM and 67.3 thousand metric tons of CSB have been distributed to over 100 countries by the USDA. The high cost of dried milk has caused a shift toward CSB. In 1974, 18.5 thousand metric tons of CSM and 86.9 thousand metric tons of CSB were shipped to impoverished people in all parts of the world (USDA, 1974).

Corn endosperm products are also an excellent source of carbohydrate in fermentation as will be discussed in a later section. Brewers grits have long been used as a major brewing adjunct. Some breweries prefer to use a pregelatinized flake if cooking facilities are limited.

Corn Quality

The desire for specific quality characteristics in corn for dry milling has been voiced by numerous industry representatives (Wichser, 1961). The reasons for a great awareness of quality differences are: (a) effective processing depends upon certain physical properties of kernels, large deviations from which may cause erratic mill separations; (b) primary product uses are in foods where purity is highly important; (c) the dry milling industry has less ability to purify products than does the wet milling process.

The oldest quality difference of interest has been yellow vs. white corn. However, since yellow corn is preferred for animal feeding, white corn production on a voluntary basis by farmers has declined below industry needs. As a result some of the larger dry milling companies, particularly in the Corn Belt, have been forced to grow white corn on contract. The amount of the horny part of the endosperm is an important factor in grit yield. The USA dent corn endosperm varies from 52 to 62% horny part, while flint corns (grown in Argentina and Southeast Europe) vary from 62 to 76% horny). Thus the yield of brewers grits containing 1.0% oil as the only grit product can vary from 46 to 65% of the corn milled (Vorwerck and Miecke, 1974). The ratio of horny to floury endosperm is determined by hand dissection, but immersion of kernels in mixtures of solvents (e.g., carbon tetrachloride and kerosene of specific gravity 1.275) provides a quick determination of kernels high in horny endosperm (sinkers) to those low in horny endosperm (floaters) (Wichser, 1961).

The replacement of ear corn harvesting and crib drying by combining artificial drying and bin storage, has been a mixed blessing. On the one hand, elimination of open cribs has reduced incidence of rodent and bird contamination but has produced more fragile kernels. Stress cracks, which are formed in corn endosperm when free kernels are dried from harvest moisture levels of 22 to 28% to safe storage moistures of 15 to 17%, cause fragmentation in the degerminator mills and reduce yield of prime flaking grits. Excessively high drying temperatures and prolonged storage tend to cause migration of fat from germ to endosperm, resulting in high fat in grit fractions (Wyss, 1965). Likewise the trend toward thicker planting with reduced ear sizes produces a larger percentage of small round and large round tip and butt kernels which reduce uniformity of corn to be milled, thereby making processing more difficult (Wichser, 1961).

Mold development in corn stored at moisture levels above 15% has always been a hazard for the corn processing industries because higher levels of mold damage has been tolerated in animal feeds than in foods. However, during the past 10 years the awareness of the highly toxic nature of aflatoxin and other mycotoxins has brought a new awareness of the need to prevent mold damage in corn and other raw agricultural crops. The milling industries, in cooperation with FDA and USDA, have developed a strong program to eliminate aflatoxin-infected corn from commercial channels. The dry milling industry has had a study program on the source and fractionation of aflatoxin-contaminated corn similar to the program described for wet milling. Wichser (private communication, 1970) and Brekke et al. (1975) have reported that dry milling of aflatoxin-contaminated corn produces contaminated endosperm products. Although flaking grits will have lowest aflatoxin level, fractions higher in fat (germ contamination) will be higher in aflatoxin. Thus corn used for dry milling must be essentially free of aflatoxin. It has recently been discovered that *A. flavus* infection can occur in the corn field before harvest due to activities of corn earworms and other insects that may feed on field corn (Anderson et al., 1975; Fennel et al., 1975). In the south the temperature and humidity is much more conducive to mold development. Fortunately corn containing this particular mold can be detec-

INDUSTRIAL UTILIZATION OF CORN

ted by examination with unltraviolet light and every mill is now examining all corn received by this method and by the visual inspection (Fennell et al., 1973; Wichser, 1974, private communication). As mentioned under Wet Milling, treatment of corn with the organic acids, propionic and acetic, prevents growth of mold. However, corn treated in this way cannot be used for dry milling even when the organic acids are removed by drying and the corn blended with normally dried corn; cornmeal made from it exhibits objectionable odors (Wichser, 1974, private communication).

Opaque-2 corn has been of interest to the dry milling industry since its announcement by Mertz et al. (1964), because a more nutritive primary product would be achieved. Wichser (1966) reports that the floury nature of the opaque-2 corn kernels makes them difficult to process in the standard way. Breeding work is under way to develop high lysine types that have a more horny endosperm than the original opaque-2 mutants (Paez and Zuber, 1973). Furthermore, consumer interest in the USA for more nutritious cornmeal is weak, but in Latin America opaque-2 corn is being processed on an increasing commercial scale.

FERMENTATION INDUSTRIES

Developments in the technology of growing microorganisms in pure cultures to produce specifically useful products has become a large and important industry in the last 25 to 30 years in industrialized nations, especially the USA, Japan, and Western Europe. Increasing volumes of corn and corn products are being used to produce a wide variety of chemicals, food ingredients, and beverages (International Union of Pure and Applied Chemistry, 1971; Perlman, 1973). Cultures of yeasts, molds, bacteria, and actinomycetes are grown in deep fermentation tanks and their desired metabolic products recovered by a multitude of sophisticated methods (Casida, 1964). All microorganisms require carbohydrate as an energy source for growth. Corn, cornmeal, corn starch, corn syrup, and dextrose may be used depending on the requirements of the particular organism and the method of recovery or desired purity of the end product.

Fermentation is a rapidly growing industry in all industrialized countries, especially Japan, USA, and most European countries, and in the future it is expected to utilize increasing quantities of corn and corn products. Feed, beverage, and chemical products made by fermentation are growing in number and volume every year. All processors are important users of corn.

Beverages

Beer and distilled liquors are the leading products with respect to volume production and utilization of corn in the USA. In other countries, the use of corn for these beverages is usually a minor ingredient. Maiden (1975) states that ingredients for average British beer are approximately 75.5% from malt, 5.5% from unmalted cereal adjuncts (barley, corn, and wheat), and 19% from brewing sugars and syrups. The British law does not specify any particular ingredients, while in Germany beer must be made from malted barley, hops, yeasts, and water. In 1973 the beer and liquor industries in the

Table 8—Corn utilization in manufacture of alcoholic beverages in 1972

	Beer and malt beverages		Distilled liquors	
	1972‡	1967	1972	1967
Liter × 10⁹§	16.4	12.6	0.715†	0.955†
Corn and corn products metric tons × 10³	685	608.26	635	765
Sugar and syrups metric tons × 10³	125.8	5.1	—	—

†50% alcohol. ‡The 1973 figures are: beer, 4291.9; corn 684.9; sugar and syrup, 125.8.

USA used 1,446 thousand metric tons of corn and corn products (Table 8) (USDC, 1974e and 1974f; Anonymous, 1968, 1974a).

This amounts to about 1.0 to 1.1% of the total corn usage in the USA. The figures for corn and corn products used in beer include common grits purchased from dry millers plus a small volume of corn starch grits supplied by wet millers. Starch grits are made from purified corn starch slurry by heating slightly before filtration to give a filter cake that dries in non-dusting lumps or by compacting powdered starch on steel rolls.

Beer and Malt Beverages

Beer manufacture is basically a process of treating malt to convert and extract the barley starch to fermentable sugars using the amylolytic enzymes present in malt (carefully sprouted and dried barley) followed by yeast fermentation. However, demand for blander, less satiating beers, especially in the USA, has permitted use of more refined carbohydrate sources of two types: (a) dry adjuncts, primarily dry milled corn grits, broken rice, refined corn starch grits, and more recently dextrose; (b) liquid adjuncts, namely corn syrups.

The initial step in beer production is called "mashing" (Fig. 3). A malt-water slurry is slowly heated to extract the enzymes and initiate the digestion of starch and protein in the malt. Corn grits or rice are cooked separately to gelatinize the starch which is then infused into the mash. Corn flakes, having been pregelatinized during roll drying, can be added directly to the mash. When a starch-iodine test shows that all starch has been digested, the mash is filtered (lautered) and washed (sparged) to recover solubles. The extract (wort) is boiled to concentrate the solids, to sterilize and to precipitate excess proteins; hops is added during the boiling to provide the desired bitter flavor and to help precipitate protein. Clarified wort is a solution of 62 to 75% fermentable sugars, primarily glucose and maltose, plus dissolved proteins, minerals, vitamins, etc., extracted from the grain. The wort is cooled to 9 C (48 F) for lager beer (14 C 58 F for ale) and inoculated with the appropriate brewers yeast. After about 8 to 10 days, the yeast has converted the sugars to about equal parts ethyl alcohol and carbon dioxide. The carbon dioxide is drawn off and compressed for later use in carbonation. Alcohol content averages 3.0 to 3.8% for most lager beers but may run up to 6.5% for some ales and stouts. In the final step the beer is aged for 1 to 3 months at 0 to 1 C (32 to 34 F), clarified, carbonated, and packaged.

INDUSTRIAL UTILIZATION OF CORN

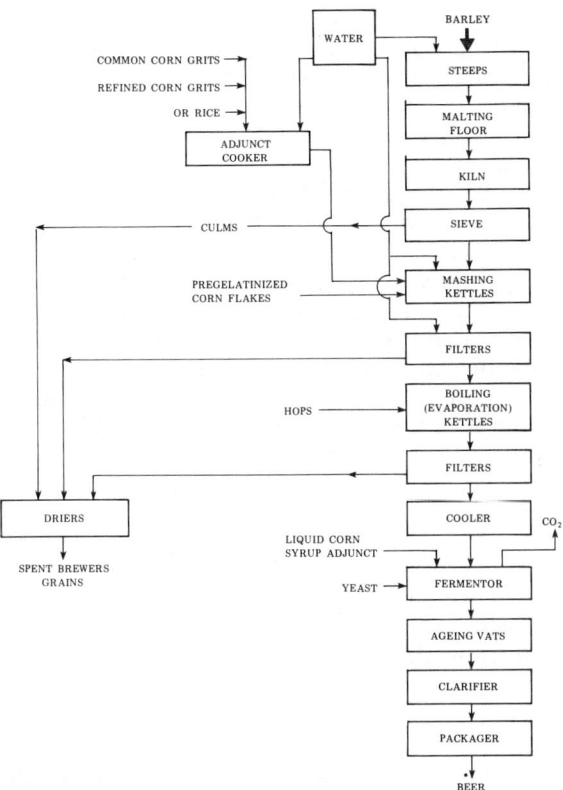

Fig. 3. The brewing process.

As early as 1957 (Mathes, 1957), the wet milling industry demonstrated that corn syrups containing fermentables levels of 65 to 80% (equivalent to wort fermentables) could be used to replace all other brewing adjuncts, 30 to 48% of total extract, with no effect on flavor, aroma, or appearance. The syrup is added to the wort during boiling thereby eliminating the need for cooking and saccharifying the adjunct. Thus lower capital cost for brewery construction and expansion and lower labor costs were made possible by use of corn syrup. Significant use of corn syrups (liquid adjuncts) began about 1965-67 and has since increased about 55% annually (Glienke and Nelson, 1972). In Great Britain, the single-temperature mashing process of 65 C makes the use of corn grits uneconomical, but high fermentable corn syrups are quite useful (Maiden, 1975). The use of corn syrup fills a need of the brewing industry to expand as it meets the growing total and per capita consumption of beer.

Distilled Liquors

In the manufacture of distilled liquors, corn is the major carbohydrate source for grain neutral spirits, corn whiskey, and bourbon. The pulverized grain is cooked (mashed) to gelatinize the starch. It is then cooled to 50 to 60

F and mixed with about 10% malt slurry to convert the starch to sugar. The converted mash is inoculated with yeast and after fermentation is complete the spent grains are removed and washed and the wort distilled to recover the alcohol. The colorless distillate is aged for several years in white oak barrels (charred for bourbon; new for corn whiskey) to remove undesirable flavors. Bourbon is made with a mixture of grains, usually rye and corn, but corn must constitute 51% of the mash bill. Corn whiskey must be made from any corn material; corn, corn grits, corn starch, corn syrup, and corn sugar are the major raw materials. One company recovers a primary starch fraction by wet milling and utilizes the more difficultly recoverable starch in the grain residue for production of grain neutral spirits (Smith et al., 1966). The still bottoms are evaporated to recover solubles then dried with the spent grains to produce a valuable feed component called "distillers dried grains with solubles." Some manufacturers dry and sell the two products separately but major production is in the combined form. This feed contains 27% protein, 8% fat, 8.5% crude fiber and minerals and vitamins (Carpenter, 1970). It is useful as a protein concentrate in a wide variety of animal feeds (Shroder and Heiman, 1970).

The annual production of corn-based whiskies was 700 proof liters (185 million proof gallons) in 1972, down from 252 million in 1967. This reduction was the result of a decline in per capita consumption of whiskey offset by increases in beer, wine, and imported liquor consumption (USDC, 1974f). The use of corn for distilling is down to 558,000 metric tons per year.

Wines

Although most of the sugar in wine fermentation comes from grapes, dextrose may be added prior to fermentation in Eastern USA to compensate for lower sugar levels of Eastern-grown grapes. Such supplementation is forbidden by law in California but is not needed because of the higher sugar content of California grapes. However, specially sweetened wines are made by adding sugar and dextrose. Recently introduced sweet "pop"-wines have achieved significant market penetration. The wine industry uses dextrose at an annual rate of 34 to 39 thousand metric tons (75 to 85 million pounds) per year. Wine consumption in the USA is growing at an annual rate of 4 to 5% per year, and 90% of the consumed product is made in this country.

Antibiotics

Commercial antibiotic production by fermentation was established during World War II and has since grown to a large and important industry. The preferred carbohydrate sources are corn syrup, dextrose, corn starch, lactose, and sucrose. Corn, corn grits, or molasses are less preferred because of purification problems in recovery of the end products. Cornsteep liquor (heavy steepwater) was early found to provide a ready source of soluble nitrogenous nutrients plus unknown growth factors that stimulate antibiotic production (Bowden and Peterson, 1946; Liggett and Koffler, 1948). Were it not for this stimulative effect of cornsteep liquor, the production of penicillin during World War II would have been seriously curtailed. The active ingredient in steepwater has never been identified, and steepwater continues to remain an important source of growth factors and readily available organic nitrogen in nutrient media for penicillin and other fermentations (El-Marsafy et al., 1975). About 15 kg steep liquor (50% dry substance) and 20 kg of carbohydrate are needed for every kilogram of penicillin produced (Casida, 1964).

Total antibiotic production in the USA in 1972 was 7,548 metric tons with a total value of over $410 million (Anonymous, 1974b). Tetracycline is produced in largest volume followed by penicillin, neomycin, bacitracin, and streptomycin. Perlman (1973) lists over 85 different antibiotics. A very large volume, predominantly penicillin, bacitracin, and neomycin, has been used as growth stimulant in animal feeds.

Two vitamins are produced in major tonnages by fermentation: riboflavin (Vitamin B_2), 454 metric tons in 1972 (Anonymous, 1974b); and cobalmide (Vitamin B_{12}) for which no production figures are available. Both fermentations utilize cornsteep liquor and dextrose.

Many other chemicals may be produced by fermentation. Perlman (1973) lists over 18 compounds, but citric acid, glutamic acid, lactic acid, and lysine are the most important (International Union of Pure and Applied Chemistry, 1971). Acetone, n-butanol, and ethanol, formerly large volume chemicals made from corn, are now made from petroleum feedstocks, except for beverage alcohol, while some lactic acid and riboflavin are made by chemical synthesis (Perlman, 1973).

Enzymes

Enzymes, the specific active agents produced by microorganisms in all fermentations, are being produced in purified form for important commercial uses. Perlman (1973) has listed 14 enzyme types being produced commercially while Rubin (1971) has listed over 60 commercial products. In many cases, whole corn or cornmeal supplies the carbohydrate for the production medium, and cornsteep liquor is an important ingredient. Enzymes such as fungal or bacterial *alpha*-amylase, gluco-amylase, and glucose isomerase are important in the production of sweeteners from corn starch. Other enzymes have important uses in food preparation, detergents, meat tenderizers, cheese production, etc.

CORNCOB UTILIZATION

Corncobs are composed chiefly of 80 to 85% cellulose, 19 to 22% hemicellulose, and a little lignin (Donnelly et al., 1973). They can be processed mechanically or chemically to produce distinctly different but useful types of materials. Corncobs are an excellent source of the important chemical, furfural. This chemical is produced on large scale by destructive distillation. In this process the corncobs are soaked with sulfuric acid and heated in a large retort. The pentosan (hemicellulose) is first liberated and hydrolyzed to constituent pentose sugars, xylose (70 to 80%) and arabinose (4 to 8%). Further heating with steam dehydrates the pentose molecule to give furfural, which is distilled and collected for direct use or conversion to other furan chemicals (Dunlop, 1973). The main chemical products for 1972 were as follows: furfural, furfuryl alcohol, tetrahydrofurfuryl alcohol, furan, and tetrahydrofuran. These products are made by two companies in the USA in seven plants having a rated capacity of over 102 thousand metric tons (225 million pounds) of furfural a year using approximately 100 thousand metric tons of corncobs, oat hulls, bagasse, and rice hulls (Anonymous, 1974b). The acidified, charred residues are useful only for fuel value.

Furfural finds uses as a selective solvent in extraction of crude petroleum, purification of butadiene, and in the manufacture of grinding wheels. Furfuryl alcohol polymerizes to produce resins that are resistant to the action of acids, alkalies, and solvents and are relatively non-burning. These resins are used in many fabrication applications and may be copolymerized with urea and formaldehyde. Tetrahydrofurfuryl alcohol is a water miscible solvent useful as a solvent for dyes, resins, and lacquers and is therefore useful in the paint and varnish industry. Tetrahydrofuran is a versatile non-aqueous solvent for many organic compounds and is an intermediate for other chemical products (Dunlop, 1973).

Corncobs are structurally composed of: (a) inner and outer glumes (membranous structures subtending the kernels where they are attached to the cob) (b) the woody ring of lignified conducting tissues, and (c) the inner pith (Clark and Lathrup, 1953). The famous corncob pipes are made by grinding off the inner and outer soft layers followed by filling voids with gypsum and polishing the surfaces of the woody ring. However, pipes are made from special hybrids developed by the University of Missouri and are grown on 2,025 to 4,045 ha (5,000 to 10,000 acres) in one Missouri county. The entire annual world supply of 25 to 30 million pipes is made by three companies (M.S. Zuber, personal communication).

The corncobs of commerce are the cobs that can be collected from the few farmers who still harvest corn with pickers. Cobs dried to 10% moisture are ground, screened, and aspirated to separate the chaffy glumes, the woody ring pieces, and the soft pith (Vander Hooven, 1973). Cobs are extremely tough and difficult to grind. Woody portion granules are at the same time very hard and very absorbent. They are especially useful for blast polishing and cleaning of oil and other contaminants from metal stampings, and castings. Electrical parts can be cleaned without danger of removing insulation. Many finished metal parts are cleaned and deburred by tumbling with cob granules which produce almost no dust even under the most rigorous handling methods. Their absorbancy makes them useful as carriers for pesticides, fertilizers, vitamins, etc. More finely divided fractions are used in hand soaps, cosmetics, and animal litters.

The chaff and pith portions are useful as absorbents for molasses in beef and dairy rations. The pith has been used for removing oil spills from water surfaces with no added pollution effects (Vander Hooven, 1973).

LITERATURE CITED

Acker, L., and G. Beker. 1971. Neuere Utersuchungun uber die Lipide der Getreidestarken. I and II. Die Starke 23, 339-343, 419-424.

Agricultural Stabilization and Conservation Service (ASCS). 1973. Sugar reports. March, No. 250, p. 24. USDA, Washington, D.C.

Alexander, R.J. 1974. Industrial uses of dry milled corn products. *In*: Y. Pomeranz (ed.) Industrial uses of cereals. Am. Assoc. of Cereal Chemists, St. Paul, Minn. 483 p.

Allred, J.B., R.E. Frey, L.S. Jensen, and J. McGinnis. 1957. Studies with chicks on improvement in nutritive value of feed ingredients. Poultry Sci. 36:517-523.

American Heart Association. 1968. Diet and heart disease. Am. Heart Assoc. Publ. EM379. New York.

Anderson, H.W., E.W. Nehring and W.R. Wichser. 1975. Aflatoxin contamination of corn in the field. J. Agric. Food Chem. 23:775-782.

Anderson, R.A., H.F. Conway, V.F. Pfeifer, and E.L. Griffin, Jr. 1969. Gelatinization of corn grits by roll- and extrusion-cooking. Cereal Sci. Today 14:4-7, 11, 12.

----, J.J. Ellis, and E.L. Griffin, Jr. 1972. Effects of Southern corn leaf blight on products of wet milling. Cereal Sci. Today 17:41-44, 53.

----, V.F. Pfeifer, G.N. Bookwalter, and E.L. Griffin, Jr. 1971. Instant CSM food blends for world-wide feeding. Cereal Sci. Today 16:5-11.

----, C. Vojonovich, and E.L. Griffin, Jr. 1961. Wet milling high-amylose corn containing 66-68% amylose starch. Cereal Chem. 38:84-93.

Andres, C. 1976. Sweeteners outlook. Food Process. 37:46-48.

Anonymous. 1968. 106, 974, 397: The total for 1967. Brew. Dig. 44:18.

----. 1969. List of corn mills. Northwest. Miller 276 (9):81-88.

----. 1974a. 138, 449, 266: The total for 1974. Brew. Dig. 49:12.

----. 1974b. Manual of current indicators. Chemical Economics Handb. Stanford Research Institute 565.4011; 565.9830B-C.

----. 1975a. 145, 464, 143: The total for 1975. Brew. Dig. 50:10.

----. 1975c. New 18-25% cereal protein to become commercially available. Food processing, January, p. 19-21. Putnam Publ. Co., Chicago, Ill.

Armbruster, F.C. 1974. Process for producing non-waxy starch hydrolyzates. U.S. Patent 3,853,706.

----, and W.A. Jacaway. 1972. Procedure for production of *alpha*-cyclodextrin. U.S. Patent 3,640,847.

----, and E.R. Kooi. 1974. Low D.E. starch conversion products. U.S. Patent 3,849,194.

Arner, G.L. 1965. Testing protein blocks. Feedstuffs 37 (16):54-57. (17 April).

Bauman, L.F., T.F. Conway, and S.A. Watson. 1965. Inheritance of variations in oil content of individual corn (*Zea mays* L.) kernels. Crop Sci. 5:137-138.

Beadle, J.B., D.E. Just, R.E. Morgan, and R.A. Reiners. 1965. Composition of corn oil. J. Am. Oil Chemists Soc. 43:90-95.

Beaux, Y., J.C. Lasseran, W. Kempf, and A. Guilbot. 1974. European maize source material for wet-milling industries. Die Starke 11:365-400.

Beeson, W.M. 1972. Effect of steam flaking, roasting, popping, and extrusion of grains on their nutritional value for beef cattle. p. 326-337. *In*: Effect of processing the nutritional value of feeds. National Academy of Sciences, Washington, D.C.

BeMiller, J.N. 1973. Starch amylose. p. 545-566. *In*: R.L. Whistler and J.N. BeMiller (eds.) Industrial gums. Academic Press, New York.

Bennett, G.A., E.F. Vandergraft, B.J. Bocan, and S.A. Watson. 1974. Zearalenone in wet milled fractions of contaminated corn. Assoc. of Official Agric. Chemists, Washington, D.C.

Bensing, H.O, D.R. Brown, and S.A. Watson. 1972. Waste utilization and pollution control in wet milling. Cereal Sci. Today 17:304-307.

Besso, R. 1974. Fibre washing. Die Starke 26:342-345.

Bier, T.H., J.C. Elsken, and R.W. Honeychurch. 1974. Integrated starch wet milling process. Die Starke, 26:23-28.

Blessin, C.W., G.E. Inglett, W.J. Garcia, and W.L. Deatherage. 1972. An edible defatted germ flour from a commercial dry milled corn fraction. Cereal Sci. Today 19:224,255.

Bowden, J.P. and W.H. Peterson. 1946. The role of corn steep liquor in production of penicillin. Arch. Biochem. 9:387-399.

Braman, W.L., E.E. Hatfield, F.N. Owens, and J.D. Rincker. 1972. Influence of waxy corn and nitrogen sources for feeding lambs and steers fed all-concentrate rations. J. Animal Sci. 37:1010-1017.

Brekke, O.L. 1970. Corn dry milling industry. G.E. Inglett (ed.) *In*: Corn: culture, processing, products. The AVI Publishing Co., Westport, Conn.

----, A.J. Peplinski, G.E.N. Nelson, and E.L. Griffin, Jr. 1975. Pilot-plant dry milling of corn containing aflatoxin. Cereal Chem. 52:205-211.

Brockington, S.F. 1970. Corn dry milled products. G.E. Inglett (ed.) *In*: Corn: culture, processing, products. The AVI Publishing Co., Westport, Conn.

Brooks, G.A., M.O. Warnecke, and J.E. Long. 1974. Sweetness and sensory properties of dextrose-levulose syrup. p. 97-110. *In*: G.E. Inglett (ed.) Symposium: sweeteners. The AVI Publishing Co., Westport, Conn.

Calet, C. 1965. The relative value of pellets versus mash and grain in poultry nutrition. Worlds Poultry Sci. 21:23-52.

Cantor, S.M., and G.E. Shaffer, Jr. 1974. Food consumption patterns: Pressures toward a three-commodity sweetener system. Cereal Sci. Today 19:266-272, 290-291.

Carpenter, L.E. 1970. Nutrient composition of distillers feeds. Proc. Distillers Feed Res. Council 25:54-61.

Casida, L.E. 1964. Industrial microbiology. John Wiley & Sons, New York. 460 p.

Chwalek, V.P. 1973. Dual screening process for separating starch particles and fibers. U.S. Patent 3,813,298.

Clark, T.F. and E.C. Lathrop, 1953. Corn cobs — their composition, availability, agricultural and industrial uses. USDA Northern Regional Research Laboratory, Peoria, Illinois, Publ. No. AIC-177 (mimeo., revised).

Compton, J. and W.H. Martin. 1967. Starch in the textile industry. p. 147-162. *In*: R.L. Whistler and E.F. Paschall (eds.) Starch: chemistry and technology. Vol. II. Industrial aspects.

Conway, T.F., C.J. Hickman, F.A. Kurtz, and G.A. Winkler. 1974. Computer controlled wide-line nuclear magnetic resonance system for automatically determining total oil content of individual maize seeds. 18th Ampre Congress. Univ. of Nottingham, England.

----, ----, ----, and ----. 1975. Computer controlled wide-line nuclear magnetic resonance system for automatically determining oil content of individual maize seeds, p. 183-184. *In*: P.S. Allen, E.R. Andrew and C.A. Bates (eds.) Magnetic resonance and related phenomena. Vol. 1. North-Holland/American Elsevier, Amsterdam.

de La Roche, I.A., D.E. Alexander, and E.J. Weber. 1971. Inheritance of oleic and linoleic acids in Zea mays L. Crop Sci. 11:856-859.

Dill, R.R., and M.E. Ginhaven. 1964. Impact ring unit for disc mill. U.S. Patent 3,118,624.

Donnelly, B.J., J.L. Helms, and A.H. Lee. 1973. The carbohydrate composition of corn cob hemicelluloses. Cereal Chem. 50:548-552.

Dowie, D.W., and D. Martin. 1962. Wet starch impact milling process. U.S. Patent 3,029,169.

Dunlop, A.P. 1973. The furfural industry. p. 229-236. *In* Y. Pomeranz (ed.) Industrial uses of cereals. Am. Assoc. Cereal Chemists, St. Paul, Minn.

Dux, E.F.W. 1967. Production and use of starch adhesives. p. 538-552. *In*: R.L. Whistler and E.F. Paschall (eds.) Starch: Chemistry and Technology. Academic Press, New York.

Earle, F.R., J.J. Curtis, and J.E. Hubbard. 1946. Composition of the component parts of the corn kernel. Cereal Chem. 44:601-606.

Easter, W.E. 1969. The dry corn milling industry. Bull. Assoc. Operative Millers, July, p. 3112-15.

El-Marsafy, M., M. Abdel-Akher, and H. El-Saied. 1975. Evaluation of various brands of corn steep liquor for penicillin production. Die Starke 27:91-93.

Fennell, D.I., R.J. Bothast, E.B. Lillehoj, and R.E. Peterson. 1973. Bright greenish-yellow fluorescence and associated fungi in white corn naturally contaminated with aflatoxin. Cereal Chem. 50:404-414.

----, E.B. Lillehoj, and W.F. Kwolek. 1975. *Aspergillus flavus* and other fungi associated with insect damaged field corn. Cereal Chem. 52:314-321.

Fontein, Feerk J. 1954. Process and apparatus for separating particles according to size. U.S. Patent 2,916,142.
Foster, G.H. 1965. Drying market corn. Proc. 20th Annu. Hybrid Corn Industry Res. Conf., Chicago, Ill. Am. Seed Trade Assoc., Washington, D.C. p. 75-85.
Freeman, J.E. 1973. Quality factors affecting value of corn for wet milling. Trans. Am. Soc. Agric. Eng. 16:671-678, 682.
----, M. Abdullah, and B.J. Bocan. 1975. An improved process for wet milling corn. U.S. Patent No. 3,928,631.
----, H.J. Heatherwick, and S.A. Watson. 1970. An evaluation of some grain quality tests for corn and effects of germ damage on wet milling results. Proc. of Conference on Research of Corn Quality, Univ. of Illinois, Urbana, Ill., Apr. 28-29, AE-4251.
----, N.W. Kramer, and S.A. Watson. 1968. Gelatinization of starches from corn (Zea mays L.) and sorghum [Sorghum bicolor (L.) Moench]: Effect of genetic and environmental factors. Crop Sci. 8:409-413.
French, D. 1973. Chemical and physical properties of starch. J. Animal Sci. 37:1048-1061.
Frey, R.R. 1967. Hard candy from high maltose corn syrup. U.S. Patent 3,332,783.
Fruin, J.C., and B.L. Scallet. 1975. Isomerized corn syrups in food products. Food Technol. 29:40, 42, 44-45.
Gerry, R.W., 1971. The value of corn gluten feed in rations for chickens. Life Sciences, Spring Issue, University of Maine, Orono, p. 8-10.
Geyer, H.V. 1974. Glucoseisomerase, Neue Moglichkeiten fur die Starke-und Lebensmittelindustrie. Die Starke 26:225-232.
Gleinke, J.F., and B.B. Nelson. 1972. Liquid adjuncts 1972 — a progress report. Modern Brewery Age. p. 10.
Godzicki, M.M. 1975. Engineering 'sugar.' Food Eng. 47(10):EF 12-17.
Graefe, G. 1974. Modifizierte Starken als bistandteil von Lebensmittln. Die Starke 26:145-153.
Hammel, D.L. 1975. Effect of feeding acid preserved grains on performance of pigs. J. Animal Sci. 40:179.
Harden, J.D. 1972. On-line process control for new corn sweetener. Food Eng. 44:59-62.
Heiman, V., 1961. Corn gluten feed in layer and breeder rations. Feed Profile No. 71968, CPC International, Inc., Englewood Cliffs, N.J.
Henderson, H.E. 1974. Survey of recent trials with waxy, high lysine and normal corn. Proc. 29th Corn and Sorghum Res. Conf. Am. Seed Trade Assoc., Washington, D.C. p. 67-80.
Herting, D.C., and E.E. Drury. 1974. Antifungal activity of volatile fatty acids on grains. Cereal Chem. 51:74-83.
Hixon, R.M., and G.F. Sprague. 1944. Waxy starch of maize and other cereals. Ind. Eng. Chem. 34:959-962.
Hjermstad, E.T. 1973. Starch hydroxyethyl ethers and other starch ethers. p. 601-615. In: R.L. Whistler and J.N. BeMiller (eds.) Industrial Gums. Academic Press, New York.
International Union of Pure and Applied Chemistry. 1971. Worldwide survey of fermentation industries, 1967. Technical Reports — Number 3, 17 p.
Jellum, M.D., and J.E. Marion. 1966. Factor affecting oil content and oil composition of corn (Zea mays L.) grain. Crop Sci. 6:41-42.
Jensen, A.H., and D.E. Becker. 1965. Effect of pelleting diets and dietary components in the performance of young pigs. J. Animal Sci. 24:392-397.
Jiles, H. (ed.) 1975. 1974 Commodity yearbook. Commodity Research Bureau, Inc., New York. 385 p.
Johnson, D.E., J.K. Matsushima, and K.L. Knox. 1968. Utilization of flaked vs. cracked corn by steers with observations of starch modification. J. Animal Sci. 27:1431.
Katz, W.W. 1966. Manufacture of corn dry mill products. Symposium on Cereal Flours, St. Louis Section, Am. Assoc. Cereal Chemists.
Kooi, E.R., and F.C. Armbruster. 1967. Production and use of dextrose. In: R.L. Whistler and E.F. Paschall (eds.) Starch: chemistry and technology. Vol. II. Industrial aspects. Academic Press, New York.

----, and R.J. Smith. 1972. Newest natural sweetener: dextrose-levulose syrup from dextrose. Food Technol. 26:57-59.

Kuznitzky, D.D. 1968. Pigmentation potency of xanthophyll sources. Poultry Sci. 47:389-397.

Langlois, D.P., and J.A. Waggoner. 1967. Production and use of amylose. p. 451-497. *In*: Starch: Chemistry and technology. Vol. II. Industrial aspects. Academic Press, New York.

Leach. H.W. 1965. Gelatinization of starch. p. 289-308. *In*: R.L. Whistler and E.F. Paschall (eds.) Starch: Chemistry and Technology. Vol. I. Fundamental Aspects. Academic Press, New York.

Liggett, R.W., and H. Koffler. 1948. Cornsteep liquor in microbiology. Bact. Rev. 124:297-311.

Lofland, H.B., F.W. Quackenbush, and A.M. Brunson. 1954. Distribution of fatty acids in corn oil. J. Am. Oil Chem. Soc. 31:412-414.

McNicol, D.G., K.T. Knechtel, T. Marshall, D.R. Metcalf, A.S. Mills, and P. Wood. 1972. Grain starch utilization study. Canadian Wheat Board, House of Commons, Ottawa, Canada. 155 p.

Maiden, A.M. 1975. Carbohydrates in the brewing industry. Die Starke 27:82-90.

Marshall, R.O., and E.R. Kooi. 1957. Enzymatic inversion of d-glucose to d-fructose. Science 125:648-649.

Mathes, F.E. 1957. Normal brewing practices with use of liquid adjuncts. Technical Proc., 70th Annual Convention of the Master Brewers of America. Miami Beach, Fla. p.1-11.

Mehltretter, C.L. 1967. Production and use of dialdehyde starch. p. 433-444. *In:* R.L. Whistler and E.F. Paschall (eds.) Starch: chemistry and technology. Vol. II. Industrial Aspects. Academic Press, New York.

Mertz, E.T., L.S. Bates, and O.L. Nelson. 1964. Mutant gene that changes protein composition and increases lysine content of maize endosperm. Science 150:1469-1470.

Messing, R.A., and A.M. Filbert, 1975. An immobilized glucose isomerase for the continuous conversion of glucose to fructose. J. Agric. Food Chem. 23:920-923.

Nesheim, R.O. 1973. Aflatoxin as a factor in utilization of corn for human food. Proc. 28th Annu. Corn and Sorghum Res. Conf., Am. Seed Trade Assoc., Washington, D.C. p. 104-109.

Newton, J.M. 1970. Corn syrup. *In*: Symp. proc. products of the wet milling industry in food. Corn Refiners Assoc., Inc., Washington, D.C.

Nissen, E.K. 1967. Starch in the paper industry. p. 121-145. *In*: R.L. Whistler and E.F. Paschall (eds.) Starch: chemistry and technology. Vol. II. Industrial Aspects. Academic Press, New York.

Norris, F.A. 1964a. Extraction of fats and oils. *In*: D. Swern (ed.) Bailey's industrial oil and fat products. Third Edition. Interscience Publ., New York.

----. 1964b. Refining and bleaching. *In*: D. Swern (ed.) Bailey's industrial oil and fat products. Third Edition. Interscience Publ., New York.

NRC-42 Committee on Swine Nutrition. 1969. Cooperative regional studies with growing swine; effects of source of ingredients, form of diet and location on rate and efficiency of gain of growing swine. J. Animal Sci., 29:922-933.

Paez, A.V., and M.S. Zuber. 1973. Inheritance of test weight components in normal, opaque-2 and floury-2 corn (*Zea mays* L.). Crop Sci. 13:417-419.

Paschall, E.F. 1967. Production and use of cationic starches. p. 403-422. *In*: R.L. Whistler and E.F. Paschall (eds.) Starch: chemistry and technology. Vol. II. Industrial aspects.

Pazur, J.H. 1965. Enzymes in synthesis and hydrolysis of starch. *In*: R.L. Whistler and L.E. Paschall (eds.) Chemistry and technology. Vol. I. Fundamental aspects. Academic Press, New York.

Peltzer, A. Sr. 1961. Centrifugal apparatus. U.S. Patent 2,973,896.

Perlman, D. 1973. The fermentation industries, 1973. Am. Soc. Microbiol. News 39:648-654.

Perry, T.W. 1972. Grain processing: Its effect on nutritive value and on the development of ulcers in swine. p. 356-372. *In:* The effect of processing on the nutritional value of feeds. Natl. Academy of Sciences, Washington, D.C.

Poneleit, C.G., and L.F. Bauman. 1970. Diallel analysis of fatty acids in corn (*Zea mays* L.) oil. Crop Sci. 10:338-341.

Powell, E.L. 1973. Starch amylopectin (waxy corn and waxy sorghum). p. 567-576. *In* R.L. Whistler (ed.) Industrial gums. Academic Press, N.Y.

Rathmann, D.M., J.R. Stockton, and D. Melnick. 1970. Dynamic utilization of recent nutritional findings; diet and cardiovascular disease. CRC Critical Rev. Food Technol. 1:331-378.

Reiners, R.A., and C.M. Gooding. 1970. Corn oil. *In*: G.E. Inglett (ed.) Corn: culture, processing, products. The AVI Publishing Co., Westport, Conn.

----, J.B. Hummel, J.C. Pressick, and R.E. Morgan. 1973. Composition of feed products from the wet milling of corn. Cereal Sci. Today 18:372-377.

----, J.C. Pressick, R.L. Urquidi, Leo Morris, E.R. Jensen, and M.O. Warnecke. 1973. Corn proteins for food use. 33rd Annu. meeting, Chicago, Ill. Institute of Food Technologists.

Rhodes, H.E. 1975. Overview of United States pet food industry. Cereal Foods World (Cereal Sci. Today) 20:5-7.

Roberts, H.J. 1967. Corn flour from surplus commodity to premium product. Cereal Sci. Today 12:505-06.

Ruark, R.G. 1970. Corn oil. *In*: Symp. proc. products of the wet milling industry in food. Corn Refiners Assoc., Inc., Washington, D.C.

Rubin, D.H. 1971. Enzymes (industrial). The Mitre Corporation, 200 p.

Russell, C.R. 1973. Industrial uses of corn starch. p. 262-281. *In*: Y. Pomeranz (ed.) Industrial Uses of Cereals. Am. Assoc. of Cereal Chemists, St. Paul, Minn.

Sasse, C.E., and D.H. Baker. 1973. Availability of sulfur amino acids in corn and corn gluten meal for growing chicks. J. Animal Sci. 37:1351-1355.

Satterthwaite, R.W., and D.J. Iwinski. 1973. Starch dextrins. p. 577-599. *In*: R.L. Whistler and J.D. BeMiller (eds.) Industrial Gums. Academic Press, New York.

Sauer, D.B., and R. Burroughs. 1974. Efficacy of various chemicals as grain mold inhibitors. Trans. Proc. Am. Soc. Agric. Eng. 17:557-559.

Scallet, B.L., and E.A. Sowell. 1967. Production and use of hypochlorite oxidized starches. p. 237-253. *In*: R.L. Whistler and E.F. Paschall (eds.) Starch: chemistry and technology. Vol. II. Industrial aspects. Academic Press, New York.

Schnyder, B.J., 1974. Continuous isomerization of glucose to fructose on a commercial basis. Die Starke 26:409-412.

Schoch, T.J. 1961. Starches and amylases. Proc. 1961 annu. meeting of the Am. Soc. Brewing Chemists. p. 83-92. St. Paul, Minn.

----. 1970. Food applications of corn starches. p. 195-219. *In*: G.E. Inglett (ed.) Corn: culture, processing, products. The AVI Publishing Co., Westport, Conn.

Senti, F.R. 1967. High amylose corn starch: its production, properties, and uses. p. 499-522. *In*: R.L. Whistler and E.F. Paschall (eds.) Starch: chemistry and technology. Vol. II. Industrial aspects. Academic Press, New York.

----, and W.C. Schaefer. 1972. Corn. Its importance in food, feed and industrial uses. Cereal Sci. Today 17:352-56.

Shildneck, P., and C.E. Smith. 1967. Production and uses of acid-modified starch. p. 217-235. *In*: R.L. Whistler and E.F. Paschall (eds.) Starch: chemistry and technology. Vol. 11. Industrial aspects. Academic Press, New York.

Shopmeyer, H.H., G.E. Felton, and C.L. Ford. 1943. Waxy cornstarch as a replacement for tapioca. Ind. Eng. Chem. 35:1168-1172.

Shotwell, O.L., M.L. Goulden, and C.W. Hesseltine. 1973a. Aflatoxin: distribution in contaminated corn. Cereal Chem. 51:492-499.

Shotwell, O.L., C.W. Hesseltine, and M.L. Goulden. 1973b. Incidence of aflatoxin associated with Southern corn, 1969-1970. Cereal Sci. Today 18:192-195.

Shroder, J.D., and V. Heiman. 1970. Feed products from corn processing. *In*: G.E. Inglett (ed.) Corn: culture, processing, products. The AVI Publishing Co., Westport, Conn.

Silvela-Sangro, L. 1967. Improvement in recurrent selection methods for oil in maize through nuclear magnetic resonance. Ph.D. thesis. Univ. Illinois (Libr. Congr. Card No. Mic. 67-11921.

Smith, N.B., H.S. McFate, and E.M. Eubanks. 1966. Process for producing starch and alcohol. U.S. Patent 3,236,740.

Sniegowski, M.S., and A.R. Baldwin. 1954. Fatty acid compositions of corn oils in relation to oil contents of the kernels. J. Am. Oil Chemists Soc. 31:414-416.

Snyder, E.C. 1970. Dextrose applications. *In*: Symp. proc. products of the wet milling industry in food. Corn Refiners Assoc., Washington, D.C.

Sprague, G.F., and B. Brimhall. 1950. Relative effectiveness of two systems of selection for oil content of the corn kernel. Agron. J. 42:83-88.

Stavenger, P.L., and D.E. Wuth. 1959. Hydrocyclone control. U.S. Patent 2,913,112.

Stivers, T.E. 1955. American corn milling systems for degermed products. Bull. Assoc. Operative Millers, June, p. 2168-79.

Takasaki, Y., and O. Tanabe. 1971. Enzyme method for converting glucose in glucose syrups to fructose. U.S. Patent 3,616,221.

Thompson, D.L., M.D. Jellum, and C.T. Young. 1973. Effect of controlled temperature environment on oil content and on fatty acid composition of corn oil. J. Am. Oil Chemists Soc., 50:540-542.

Tuite, J., G. Shanor, G. Rambo, J. Foster, and R.W. Caldwell. 1974. The giberella ear rot epidemics of corn in Indiana in 1965 and 1972. Cereal Sci. Today 19:238-241.

Ueberbacher, R.L. 1970. Maltodextrins. *In*: Symp. proc. products of the wet milling industry in food. Corn Refinery Assoc., Washington, D.C.

USDA. 1970. National and state livestock-feed relationships. Statistical Bulletin No. 446 (Suppl.) Economic Research Service, U.S. Govt. Printing Office, Washington, D.C. 117 p.

----. 1974. Report of Title II PL 480 operations famine emergency relief and economic development assistance of foreign countries. World Food Operations Report No. F1-301-4 (103).

----. 1975a. Agricultural statistics. Corn. U.S. Govt. Printing Office, Washington, D.C. p. 28-36.

----. 1975b. Field crops, production, farm use, sales, value, 1973-74. Crop Reporting Board, May 12, 1975. p. 2,3.

----. 1976a. Feed situation. November. Economic Research Service, Washington, D.C. p. 1, 7, 8.

----. 1976b. Reference tables on wheat, corn and coarse grain supply-distribution for individual countries. Foreign Agricultural Service Publ. FG-9-76, May, 1976.

USDC. 1974a. Cereal breakfast foods. 1972 Prelim. Report. Census of Manuf., Bureau of the Census, Washington, D.C. MC72(P)-20D-2.

----. 1974b. Flour and other grain mill products, blended and prepared flour. 1972. Prelim. Report. Census of Manuf. Bureau of the Census, MC72(P)-20D-1.

----. 1974c. Wet corn milling. 1972 Prelim. Report. Census of Manuf. Bureau of the Census, Washington, D.C., MC72(P)-20D-4.

----. 1974d. Dog, cat and other pet food: Prepared feeds, N.E.C., 1972 Prelim. Report. Census of Manuf. Bureau of the Census, Washington, D.C., MC72(P)-20D-5.

----. 1974e. Malt beverages. 1972 Prelim. Report. Census of Manuf. Bureau of the Census, Washington, D.C., MC72(P)-20H-1.

----. 1974f. Distilled liquor, except brandy. 1972 Prelim. Report. Census of Manuf. Bureau of the Census, Washington, D.C. MC72(P)-20H-4.

----. 1974g. Papermills, except building paper. 1972 Prelim. Report. Census of Manuf. Bureau of the Census, Washington, D.C., MC72(P)-26A-2.

----. 1974h. Paperboard mills. 1972 Prelim. Report. Census of Manuf. Bureau of the Census, Washington, D.C., MC72(P)-26A-3.

----. 1974i. Paper coating and glazing. 1972 Prelim. Report. Census of Manuf. Bureau of the Census, Washington, D.C., MC72(P)-26B-1.

VanderHooven, D.I.B. 1973. Industrial and agri-chemical utilization of corn cobs. p. 259-261. *In*: Y. Pomeranz (ed.) Industrial uses of cereals. Am. Assoc. Cereal Chemists, St. Paul, Minn.

Van Lanen, J.M. 1974. Presidents message to conference. Proc. Distillers Feeds Res. Council 29:3-4. Cincinnati, Ohio.

Vegter, H.J. 1957. Corn starch process involving vortical classification. U.S. Patent 2,778,752.

Vineyard, M.L., and R.P. Bear. 1952. Amylose content. Maize Genet. Coop Newsletter 26:5.

Von Titeleboom, M.L.E. 1961. Centrifugal screening type separator and method. U.S. Patent 2,995,246.

Vorwerck, K., and N. Miecke. 1974. Comparison between test results on raw corn and the processing results in practice. Buhler-Miag Nachrichten, Braunschweig, Germany, No. 203E, February 1974. p. 21-26.

Waldo, D.R. 1973. Extent of partition of cereal grain starch digestion in ruminants. J. Animal Sci. 37:1062-1074.

Waldroup, P.W. 1969. Dried corn steep liquor concentrate. Feed Management 20:25-26.

Walters, R.D. 1975. Panning — the specialists specialty. Candy Snack Industry 139 (13):43-51; 140 (1):44-51.

Wardrip, E.K. 1971. High fructose corn syrup. Food Technol. 25:501-504.

Watson, S.A. 1960. Storing and drying corn for the milling industries. Proc. 15th Annu. Hybrid Corn Industry Res. Conf., Chicago, Ill. p. 85-92. Am. Seed Trade Assoc., Washington, D.C.

----. 1962. The yellow carotenoid pigments of corn. Proc. 17th Annu. Hybrid Corn Industry Res. Conf. p. 92-100. Am. Seed Trade Assoc., Washington, D.C.

----. 1967. Manufacture of corn and milo starches. p. 1-52. *In:* R.L. Whistler and E.F. Paschall (eds.) Starch: chemistry and technology. Vol. II. Industrial aspects, Academic Press, New York.

----, and J.E. Freeman. 1975. Breeding corn for higher oil content. Proc. 30th Annu. Hybrid Corn and Sorghum Res. Conf., p. 251-275.

----, Y. Hirata, and C.B. Williams. 1955. A study of the lactic acid fermentation in commercial corn steeping. Cereal Chem. 32:382-394.

----, and E.H. Sanders. 1961. Steeping studies with corn endosperm sections. Cereal Chem. 38:22-33.

----, and K.R. Yahl. 1967. Comparison of the wet milling properties of opaque-2 high lysine corn and normal corn. Cereal Chem. 44:488-498.

----, and ----. 1971. Survey of aflatoxins in commercial supplies of corn and grain sorghum used in wet milling. Cereal Sci. Today 16:153-163.

Whistler, R.L., and J.N. BeMiller (eds.) 1973. Industrial gums. 2nd Ed. Academic Press, New York. 807 p.

----, and E.F. Paschall. 1965. Starch: chemistry and technology. Vol. I. Fundamental aspects. Academic Press, New York. 579 p.

----, and ----. 1967. Starch: chemistry and technology. Vol. II. Industrial aspects. Academic Press, New York. 732 p.

White, T.W., T.W. Perry, B.R. Tonroy, and V.L. Lechtenberg. 1973. Influence of processing on in vitro and in vivo digestibility of corn. J. Animal Sci. 37:1414-1418.

Wichser, W.R. 1961. The world of corn processing. Am. Miller Process. 89 (3):23-24; 89 (4):29-31.

Wichser, W.R. 1966. Comparison of the dry milling properties of opaque-2 and normal dent corn. p. 104-116. *In:* E.T. Mertz and O.E. Nelson (eds.) Proceedings of the high lysine corn conference. Corn Refiners Assoc., Inc., Washington, D.C.

Williamson, J.L. 1972. Effect of grain processing for dairy cattle and other animals. p. 349-357. *In:* The effects of processing on nutritional value of feeds. Natl. Acad. Sci. Washington, D.C.

Woods, W.R., T.J. Klopfenstein, and J.C. Cranfill. 1969. Liquid supplement for beef cattle. J. Animal Sci. 29:177.

Wurzburg, O.B. 1970. Starch and modified starch. *In:* Symp. proc. products of the wet milling industry in food. Corn Refiners Assoc., Washington, D.C.

Wyss, E. 1965. Corn — Its origin and uses. Buehler Diagram, Buehler Brothers Engineering Works, Uzwil, Switzerland. No. 40, Aug./Sept., p. 3-10.

Yahl, K.R., S.A. Watson, R.J. Smith, and R. Barabolok. 1971. Laboratory wet milling of corn containing high levels of aflatoxin and a survey of commercial wet milling products. Cereal Chem. 48:385-391.

Yen, J.T., D.H. Baker, B.G. Harmon, and A.H. Jensen. 1971. Corn gluten feed in swine diets and effect of pelleting on tryptophan availability to pigs and rats. J. Animal Sci. 33:987-991.

Zuber, M.S. 1965. Genetic control of starch development. p. 43-63. *In:* R.L. Whistler, and E.F. Pashall, (eds.) Starch: chemistry and technology. Vol. I. Fundamental aspects. Academic Press, New York.

INDEX

A

Aberrant ratio, 187
Abnormal 10, 21, 171, 187, 189, 227, 274, 283-289
Aceria tulipae, 457, 552
Acetic acid, 751
   corn storage, 662
Acetone, 755
Acyrthosiphon pisum, 550
Additive variance, 325, 331, 332, 342, 344, 346
Aeolothrips bicolor, 543
   fasciatus, 543
Aeolus mellillus, 510
Aflatoxin, 730, 746, 750
Agave, 8
Agonoderus lecontei, 517
Agriotes mancus, 510
Agromyza parvicornis, 544
Agrostis spp., 443
   gladiaria, 509
   ipsilon, 509
Ahasversus advena, 560
Alachlor, 659, 660
Allelic recombination, 258, 259, 260
Alpha amylase, 735, 742, 755
Alternaria ssp., 548
Ambrosia, 8
Amylopectin, 374, 375, 382, 733, 738, 739
Amylopectin starch, 738
Amylose, 374-376, 382, 559, 732, 739
Anaphothrips obscurus, 543
Androgenesis, 187
Andropogoneae, 1, 24
Animal feeds, 723-725
Anthracnose, 408
Antibiotic production
   bacitracin, 755
   neomycin, 755
   penicillin, 755
   streptomycin, 755
Aphis craccuvora, 550
   gossypii, 550
   maidiradicus, 514, 550
Arabinose, 755
Aspergillus sp., 421, 548
   flavus, 395, 418, 750
   glaucus, 418

   niger, 418
   tamarii, 395
Atrazine, 170, 659, 660
Avena sativa, 443, 723

B

B chromosome, 187, 227, 228, 254, 283, 286, 289-293
Bacillus macerans, 735
Bacterial stalk rot, 407-409
Baldulus tripsici, 461
Barley, 18
Beauveria bassiana, 539
Beverages, 751-754
   beer and malt, 752, 753
   distilled liquors, 753, 754
   wines, 754
Black bundle disease, 409-410
Blissus leucopterus, 538
   hirtus, 538
Breakage-fusion-bridge cycle, 275, 276, 277
Brevicoryne brassicae, 550
Bromus inermis, 655
n-Butanol, 755
Butylate, 659, 660

C

Cacahuacintle, 26, 38
Capsicum, 18
Carbohydrates, endosperm, 373-385
   mutants, 375-378
Cephalosporium spp., 548
   acremonium, 409, 410, 420
Ceratomegilla fuscilabrus, 529
Cercospora sorghi, 447
   zeae-maydis, 447
Chaetocnema denticulata, 429, 541
   ectypa, 541
   pulicaria, 429, 541
Chapalote, 7, 38, 39
Charcoal rot, 404, 405
Chemical composition, 334, 363-384
Chemical derivatives, starch, 737
Chemical treatment, stored grain, 662
Chromosomal rearrangements, 260-282
   deficiencies, 280-281
   duplications, 277-280

inversions, 271-277
  rings and rods, 281-282, 288
  translocations, 260-271
Chromosome knobs, 17, 19, 21, 24, 27, 51, 55, 56, 59, 62, 63, 66, 72, 74, 76, 78
Chromosome morphology, 225, 228
Chromosome, number
  euploid series, 232-236
  aneuploidy, 236-239
Chrysopa spp., 529
Cicadulina mbila, 550
  storeyi, 550
  zeae, 550
Citric acid, 755
Cladosporium herberum, 419, 421
Clarkia, 18
Climate and corn production, 591-593
Clivina impressifrons, 517
Cobalmide, 755
Cochliobolus carbonum, 440
  heterostophus, 427
Coelorhachis, 21
Coix lachryma-jobi, 1, 2, 3, 7, 10, 24-25, 42
Colapsis brunnea, 516
Cold testing, 398
Colletotrichum graminicola, 408, 435
Consumption by livestock classes, 723, 724
Corn, antiquity — pollen evidence, 10
  archaeological records, 5-8
  breeding, 305-354
    ear-to-row, 310-312
    early testing, 321-323
    inbreeding and hybridization, 313
    mass selection, 305-307
    recurrent selection, 330-334
    varietal hybridization, 307-310, 320, 321
  chromosome uniformity, 21-27
  evolution, 4-17
  oil, 744, 745
  production data, 631
  quality, dry milling, 749-751
  races, groups of
    Amazon basin, 65-66
    Andean, 68-72
    Central American, 56-58
    Corn Belt dents, 78
    derived southern dents, 77-78
    Great Plains flints and flours, 74
    lowland northern South America, 62-65
    lowland southern South America, 66-68
    Mexican, 54-58
    northern flints, 73, 74
    Pima-Papago, 75
    southeastern flints, 77
    southern dents, 76, 77
    southwestern 12 row, 76
    southwestern semidents, 75, 76
    West Indian, 58-62
  races, lineage, 51-52
  relatives,
    Chionachne, 1
    Coix, 1, 2, 3, 7, 10, 24, 25, 42
    Polytoca, 1
    Schlerachne, 1
    Trilobachne, 1
    Tripsacum, 1, 3, 4, 10, 12, 16, 21, 25, 27, 30, 31, 32, 33
    Zea, 1, 4, 11, 12, 13, 17, 18
  relatives and corn improvement, 35-39
  resistance to insects, 562-574
  starch, 732-740
    acid modified, 735
    general properties, 732, 734
    high amylose, 739
    maltodextrins and pyrodextrins, 735-736
    oxidized, 736
    pregelatinized, 736
    unmodified, 734
    waxy, 737-739
  time of domestication, 10-12, 15-18
Corncob utilization, 755, 756
  furfural, 755, 756
  abrasives, 756
  pipes, 756
Corn steep liquor, 743, 754
Corn-teosinte relations, 19-24
Correlation among inbred traits, 319, 320
Corynebacterium nebraskense, 437
Crambus caliginosellus, 512
  mutabilis, 512
  trisectus, 512
Crepis, 230
Crossing over, 250-260
  chiasmata, 252
  chromatid, 251
  cytological demonstration, 250, 256, 257
  interference, 253, 254
  male and female, 258

INDEX

sister strand exchange, 255
Crymodes devastator, 509
Cryptolester ferrugineus, 557
   pusillus, 556
Cultural practices, 636-655
   fertilization, 649-655
   planting, 643-649
   preplant tillage systems, 636-643
Curvularia eragrostidis, 447
   maculans, 447
Cyanazine, 659, 660
Cyclic hydroxamic acids, 425
Cynaeus augustus, 560
Cynodon spp., 516
   dactylon, 640
Cyperus spp., 538
Cytological map, 126
Cytoplasmic sterility,
   C cms, 433, 688
   S cms, 433, 688
   T cms, 426, 427, 428, 433, 684-686, 688
Cytoplasmic sterility, restorers, 688-690

D

Dactylis glomerata, 641
Dactynotus sp., 550
Dalbulus elimatus, 461, 552
   maidis, 461, 550
DDT, 575
Degree days, 680, 681, 693
Deltocephalus sonorus, 461
Detasseling costs, 684-685
Detasseling, effect on yield, 685-687
Dextrins, 735-736
Dextrose, 742
Diabrotica longicornis, 504, 564
   undecimpunctata, 409, 507, 564, 603
   vergifera, 504, 564
Dialdehyde starch, 736
Diatraea spp., 508
   crambidoides, 530, 533
   grandiosellsa, 530, 534
   lineolata, 530
   saccharalis, 530
Dicamba, 659, 660
DIMBOA, 170, 400, 425, 570,
2,4-D, 659, 660
Diplodia ear rot, 411-413
Diplodia frumenti, 419
   macrospora, 419, 445
   maydis, 135, 395, 397, 400, 401, 411, 413, 548

   zeae, 350
Diplodia stalk rot, 401-403
   control, 403
   infection, 401
Disease control, 394, 395
Disease development, 392-393
   disease-inciting agent, 393
   environment, 392, 393
   host resistance, 393
Diseases,
   downy mildews, 448-453
      crazy top, 450-451
      green ear, 450
      sorghum downy mildew, 451
   ear rots, 411-420
      Diplodia, 411-413
      Fusarium, 415, 416
      Gibberella, 413, 414
      gray, 416-418
      Nigrospora, 414, 415
      other, 418-420
   leaf, 421-448
      anthracnose, 434-435
      bacterial leaf blight, 444-445
      bacterial stripe, 444
      bacterial wilt, 429-432
      brown spot, 437-439
      chocolate leaf spot, 448
      Curvularia leaf spot, 447
      Diplodia leaf spot, 445
      eyespot, 434
      Goss's bacterial wilt, 435-437
      gray leaf spot, 447
      Helminthosporium leaf spot, 440, 441
      Holcus bacterial spot, 443, 444
      northern corn leaf blight, 421-426
      northern leaf spot, 439, 440
      purple sheath spot, 445
      southern leaf blight, 426-429
      southern leaf spot, 441
      zonate leaf spot, 441-443
   non-infectious, 476-483
      genetic abnormalities, 477-479
      induced injuries, 479-481
      nutritional deficiencies, 481-483
   stalk rots,
      anthracnose, 408
      bacterial, 407-409
      black bundle, 409, 410
      charcoal, 404, 405
      Diplodia, 401-403
      Fusarium, 404
      Gibberella, 403, 404

Pythium, 405-407
virus and mycoplasma, 453-462
  bromegrass mosaic, 454-455
  corn mosaic, 455-457
  corn stunt, 460-462
  cucumber mosaic, 455
  maize chlorotic dwarf, 459
  maize dwarf mosaic, 458, 459
  maize leaf fleck, 454
  sugarcane mosaic, 453, 454
  wheat streak mosaic, 457, 458
Disporium sessile, 252
DNA, 232, 256, 257, 275, 278, 279, 292, 293
Dominance variance, 332
Double cross prediction, 323, 324
Drosophila, 24, 226, 232, 249, 250, 251, 254, 256, 257, 259, 266, 271, 273, 277, 274
  ananassae, 244
Dry milling, 745-749
  process, 745-747
  products, 747-748
  uses, 748-749
Drying temperatures, seed corn, 696
Duplicate factors, 129

E

Ear rots, 411-420
Ear-to-row selection, 310-312
  chemical composition, 310
  plant and ear height, 310, 311
  yield, 311, 312
Eirborus tetebrons, 524
Elasmopalpus lignosellus, 534, 536
Elyonurus, 10, 25
Embryology, 89-97
  embryo, 89-97
  endosperm and pericarp, 98-99
  integuments and nucellus, 97-98
  proembryo, 89
  scutellum, 96, 97
Endosperm sugars, 374, 376, 377
Epistatic variance, 333
Erianthus alopecuroides, 471
Eriophyes tulipae, 457, 552
Erosion control, 655-657
Erwinia chrysanthemi, 407
  dissolvens, 407
  stewarti, 409, 430, 548
Ethanol, 755
Ethyl alcohol, 727, 752

Euchlaena mexicana, 424, 437, 470
  (See also *Zea mexicana*)
  perennis, 470
Euetheola rugiceps, 516
Eumicrosoma benefica, 539
European corn borer, 400
  (See also *Ostrinia*)
Exotic germplasm, 341, 342
Extra chromosomal inheritance, 183-185
  chlorophyll variants, 183-184
  cytoplasmic sterility, 184-185

F

Fatty acids, corn oil, 744
Feed products, wet milling, 743, 744
Feltia subgothica, 509
Fermentation industries, 751, 755
  antibiotics, 754-755
  beverages, 752-754
  enzymes, 755
Fertilizer, use of, 649-655
Festuca arundinacea, 640, 641
Formulated animal feeds, 723-725
Frankliniella fusca, 543
  tenuicornis, 543
  occidentalis, 543
  williamsi, 543
Freeze damage, seed corn, 693-694
Furfural, 755
Fusarium kernel rot, 415, 416
Fusarium graminearum, 403, 414
  moniliforme, 397, 404, 411, 477, 547
  roseum, 395
Fusarium stalk rot, 404

G

Gametophyte factors, 188
Gene action, types of, 324-327
  additive, 324, 326, 327, 329
  dominance, 325, 326, 329
  epistasis, 325, 326, 329
  overdominance, 325, 326
Genetic developmental studies, 193-196
Genetic diversity, 320, 321
Genetic drift, 337, 347, 348
Genetic factors,
  anthocyanins and related pigments, 137-145
  chlorophyll and carotinoids, 148-154
  endosperm texture, 130-135
  enzyme electrophoresis, 165-168

gametophyte formation, 170-171
paramutation, 181-182
plant form, 156-164
reaction to insects and diseases, 168-170
Genetic factors, location of, 189-192
deficiencies, 192
hemizygosity, 191
monosomics, 191
test crosses, 189, 190
translocations, 190, 191
trisomics, 192
Genetic fine structure, 204
Genetic gain, 332, 333, 344
Genetic loci, 115-123
Genetic nomenclature, 113-114
Genetic purity, seed corn, 704-706
commercial production, 704-705
foundation seed, 704
specialty corns, 706
Genetic regulatory systems, 173-183
Genetic vulnerability, 52, 79-83
Genotype × environment interactions, 327-329
Germination tests, seed corn, 707-709
cold test, 708
standard, 708
tetrazolium, 708
other, 708-709
Germplasm, tropical, potentially useful, 79-83
Gibberella ear rot, 413, 414
Gibberella moniliforme, 404, 415, 416
saubinetti, 395
var. *subglutinans,* 404, 415
zeae, 395, 400, 403, 410, 411, 414, 425, 548
Gibberella stalk rot, 403-404
Gloeocercospora sorghi, 442
Gluco-amylose, 755
Glucose isomerase, 755
Glutamic acid, 755
Glycine max, 636, 723
Gonatobotrys zeae, 419
Gossypium hirsutum, 616
Graminella nigrifrons, 459, 461, 462, 551, 552, 576
Grasshoppers, 283
Gray ear rot, 416-418
Green manure crops, 655, 656
Growing-degree units, 598-600
Growth and development stages, 593-595
Gynogenesis, 187, 190

H

Harvesting, 660-662
Harvesting, seed crop, 692-697
Haynaldia, 32
Heat units, 598-600
Heliothis zeae, 349, 518-522, 548
Heliothrips femoralis, 543
Helminthosporium sp., 81, 82, 439
carbonum, 169, 420, 439, 440
maydis, 28, 170, 184, 185, 391, 420, 427, 428, 439
race O, 428, 429, 439
race T, 427-429, 433, 439, 440. 445, 684
rostratum, 441
setariae, 440
turcicum, 38, 169, 170, 350, 400, 420, 422, 424, 425, 439
Herbicides, weed control in corn, 659, 660
Heritability, 331
Heterochromatin, 226
High amylose corn, breeding, 384
High fructose corn syrup, 725, 741, 742
High oil, 363
effects of selection, 365, 366
milling, 731
quality, 368-370
Hippodamia convergens, 529
Histology, vegetative plant body, 99-109
leaf, 107-109
root, 104-107
stem, 99-103
Hordeum vulgare, 371, 593, 723
Horistonotus uhlerii, 510, 511
Hybrid seed industry, 671-672
Hybrid seed production, 673-681
contract growers, 676-677
Herbicides, insecticides, and fungicides, 677
isolation, 677
parent seed stocks, 673-675
planting patterns, 679-681
seed production, 675-681
Hylemya platura, 515

I

Inbred line development, 313-317
convergent improvement and backcrossing, 316
gamete selection, 317

homozygous diploids, 315, 316
  pedigree selection, 315
  single-hill, 314, 315
  standard, 313-314
Inbreeding depression, 314, 340, 348
Industrial utilization, 721-723
Insect rearing techniques, 562-564
Insects,
  chemical control, 574-576
  leaf, stalk and ear, 518-543
    armyworm, 534, 535
    chinchbug, 538-540, 573
    corn borers, 530-534
    corn flea beetle, 541-542
    corn leaf aphid, 526-529, 572
    earworm, 518-522, 564
    European corn borer, 522-526, 568, 569
    fall armyworm, 529-530, 572
    grasshoppers, 540, 541, 573
    Japanese beetles, 542, 543.
    lesser cornstalk borers, 536-538
    miscellaneous, 543-546
  relation to diseases, 546-552
  research trends, 502-503
  soil, 503-517
    billbugs, 511, 512
    corn root aphid, 514, 515
    cutworms, 508, 509
    grape colapsis, 516, 517
    rootworms, 503-508, 564, 576
    seed corn beetles, 517
    seed corn maggot, 515
    sugarcane beetle, 516
    webworms, 512, 513
    white grubs, 513, 514
    wireworms, 509-511
  stored grain, 552-561
    angoumois grain moth, 558, 559
    cadelle, 557, 558
    flat grain beetle, 556, 557
    flour beetles, 556
    granary weevil, 554-556
    Indian meal moth, 559, 560
    rice and maize weevil, 553, 554
    saw-toothed grain beetle, 559
    stored grain, control, 561
Inversion heterozygotes, 247
Inversions, 21-23, 226, 227, 271-274
  paracentric, 23, 271-275
  pericentric, 271, 274
Isochromosomes, 238
Isolation requirements, 677
Isotherms, 592

K

Kabatiella zeae, 434
Knob positions, 225, 226, 250, 284

L

Lactic acid, 755
  steepwater, 726
Laspeyeresia pomonella, 574
Leaf area index, 645, 646
Linkage detection, 266, 267
  A-B translocations, 268-271, 289-293
  deficient-duplicates, 268
  hemizygotes, 267
  translocations, 266, 267
  trisomics, 266
Linkage map, 124-125
Linuron, 659, 660
Lipids, kernel, 164-165
Locusta migratoria, 247
Loxagrostis albicosta, 543, 544
Lydella thompsoni, 524
Lysine, 755
Lysiphlebus testaceipes, 529

M

Macrocentrus grandii, 524
Macrophoma zeae, 417, 419
Macrophominia phaseolina, 397, 405, 410
Maize dwarf mosaic, 36, 458, 459
Manisuris, 1, 3, 10, 12, 16, 24, 34, 35
Marmar graminis, 454
  zeae, 455
Mass selection, 305-307, 331
Maydeae, 1, 25, 27, 42, 470
Mechanical detasselers, 684
Medicago sativa, 507, 655
Meiosis, 239-243
  mutants of, 241-243, 251
Melanoplus americana, 540
  bivittatus, 540
  differentialis, 540
  femur-rubrum, 540
  sanguinipes, 540
Melanotus cribulosus, 510
  communis, 510, 511
Minimum tillage, 394, 395, 465, 526, 535, 637-643
Minor elements, requirements, 653-655
Moisture stress, 595-611

INDEX

Monoploids, 233, 234, 235, 248, 314
Monosomes, 30, 188, 191, 238, 239, 249, 250
Mucor spp., 421, 548
Mutation rates, quantitative traits, 314
Mutation studies, 196-204
   chemicals, 197, 199
   gamma ray, 197
   natural, 196, 198
   ultraviolet, 197, 198, 199, 202
   x-rays, 196, 197, 198, 199, 200, 202
Mycosphaerella zeae-maydis, 433
Mycotoxins, 420, 730, 750
Myzus persicae, 454, 550

N

Nal-Tel, 17, 38, 51
Nematode diseases, 472-474
Nematodes, 472-474, 655
Noecentromeres, 285, 286, 287
Neurospora, 251, 254
Nigrospora cob rot, 414, 415
Nigrospora spp., 548
   oryzae, 397, 400, 410, 415
Nitrogen fertilization, 650-652
Nitrogen requirements, 634-635
Nondisjunction, 238
Nonhomologous pairing, 247, 248, 249
Nuclear magnetic resonance, 364, 731, 732
Nucleolus organizer, 225, 229, 230, 231, 264

O

Oligonychus pratensis, 545
   stickneyi, 545
Opaque-2, 371-373
   feeding value, 373
   milling, 731, 751
   performance, 371, 372
Ophiobolus maydis, 427
Opuntia, 8
Orius insidiosus, 522, 524
Orzyaephilus surinamensis, 557
Ostrinia nubilalis, 349, 425, 522-526, 548, 603
Oxalis spp., 468

P

Panicum spp., 447, 458
Paraquat, 659, 660
Parent seed stocks, 673-675
Paspalum spp., 516
Pectobacterium chrysanthemi, 407
Penicillin, 754, 755
Penicillium spp., 548
   oxalicum, 395, 397, 418, 420
Pennisetum glaucum, 448
Pentosans, 755
Peregrinus maidis, 457, 543
Persea, 8
Phaeocytostroma ambigua, 410, 411
Phaseolus, 9, 18
Phenotypic variance, 331, 332, 334, 335
Pheromones, 574
Phosphorus fertilization, 652-653
Phosphorus requirements, 635
Phyllophaga fusca, 513
   hirticula, 513
   rugosa, 513
   tristis, 513
Phyllosticta maydis, 433
Physalospora zeae, 418
   zeicola, 419
Physiologic maturity, 398, 400, 401, 648, 661, 692, 693
Physoderma maydis, 438, 439
Physopella zeae, 472
Phytoalexins, 425
Phytoglycogen, 374, 379, 738
Phytotoxins, 657
Plant growth and development, 625-629
   dry down period, 629
   emergence to tasseling, 628-629
   planting to emergence, 628
   silking to maturity, 629
   tasseling and silking, 629
Plant growth requirements, 630-636
   mineral nutrients, 634-636
   pH and salt tolerance, 632-634
   soils, 630-631
   water, 631-632
Planting, corn, 643-648
   date, 642, 643
   depth, 643-645
   population, 645-647
   row spacing, 647-648
Planting dates, average, 642
Plant tissue tests, 652, 653
Pleiotropism, 128
Plodia interpunctella, 559
Pollen, 10, 17
Pollen control, 681-692
   chemicals, 691-692
   cytoplasmic sterility, 687-690

detasseling, 682-687
genetic sterility, 690-691
Pollo, 17, 38, 39
Polyploidy, 233, 235, 236, 247
Popcorn breeding, 385
Popillia japonica, 542
Potassium fertilization, 653, 654
Potassium requirements, 635-636
Preferential fertilization, 171, 172, 291, 292
Processing seed corn, 697-703, 746
  bagging, 702, 703
  cleaning, 699
  planter plate selection, 702
  shelling, 698
  sizing, 700
  treating, 700-702
Prolificacy, 305, 334, 343
Propachlor, 659, 660
Propionic acid, 730, 731, 751
  corn storage, 662
Protein, grain, 370-373
  quality, 371-373, 382
  quantity, 370-371
Pseudaletia unipuncta, 534
Pseudomonas andropogonis, 444
  alboprecipitans, 409, 445
  lapsa, 407
  syringae, 443, 448
Puccinia polysora, 169, 471
  sorghi, 36, 169, 467-470
Pyrenochaeta terrestris, 410, 411
Pythium root rots, 410, 411
Pythium ssp., 395-397, 603
  aphanidermatum, 405
  arrhenomanes, 410
  debaryanum, 395, 410
  graminocolum, 410
  irregulare, 395
  paerocandrum, 395
  rostratum, 395
  ultimum, 395
  vexans, 395
Pythium stalk rot, 405-407

Q

Quality control, seed corn, 703-709
  genetic purity, 704-706
  physical, 706-709
Quality, milling standards, 730

R

Recurrent selection, 330-339

interpopulation improvement, 337-339
  full sib reciprocal, 338, 345
  reciprocal, 326, 337, 339, 345
intrapopulation improvement, 331
  full sib selection, 331, 335, 336
  half-sib selection, 331, 333, 335, 336
  mass selection, 331, 333, 335
  modified ear-row, 331, 333, 335
  S_1 or S_2 selection, 331, 336, 351, 353
Rhizoctonia zeae, 397, 419
Rhisopus ear rot, 419
Rhizopus ssp., 421, 548
Rhopalomyzus poae, 550
Rhopalosiphum maidis, 38, 526, 550
  padi, 550
  prunifolae, 454
Riboflavin, 755
Ribosomes, 155, 232
Ring and rod chromosomes, 288
RNA, 230, 231, 259, 278
Root rots, 410-411, 504
Rottboellia, 25
Rusts, corn, 467-472
  common rust, 467-470
  southern, 470-471
  tropical, 471
Rye, 253, 283

S

Saccharum officinarum, 437
  spontaneum, 448
Sathrobrota rileyi, 545
Schistocera americana, 540
Sclerophthora graminicola, 448
  macrospora, 451
  rayssiae, 448
Sclerospora maydis, 448
  philippinesis, 448
  sacchari, 448
  sorghi, 448, 451
  spontanea, 449
Sclerotium bataticola, 397
Scuttigerella immaculata, 546
Secale cereale, 655
Seed corn, sales and distribution, 712
Seed quality, 707-709
  gemination, 708-709
  purity analyses, 707
Seed rots and seedling blights, 395-398
  control, 397, 398

INDEX

Selection differential, 331, 346, 347
Selection during inbreeding, 318, 319
Selection index, 351
Setaria spp., 8, 9, 11, 433, 443, 458
    lutescens, 445
    virdis, 450
Simpiesis viridulus, 524
Sitophilus granarius, 554
    oryzae, 553
    zeamaize, 553
Sitotroga cerealella, 558
Smuts, corn, 462-467
    common smut, 462-465
    head smut, 465-466
    false smut, 466-467
Soil tests, 678
    fertilizer requirements, 650-655
Somatic pairing, 244-245
Sorghum spp., 435, 444, 447, 452
    bicolor, 371, 424, 441, 443, 723
    halipensis, 443
    sudanensis, 424
Sphacelotheca reiliana, 466
Sphenophorus, callosa, 511, 512
    maidis, 511, 512
    parvula, 511
    zeae, 511
Spiroplasma, 460, 552
    citri, 461
Spodoptera frugiperda, 529
Stalk and root rots, 398-411
Starch, digestability, 382
Starch paste viscosity, 733
Starch solubility, 733
Steep water, 726
Steeping corn, 726
Storage, 662
Storage rots, 420-421
Storage, seed corn, 709-712
Striga lutea, 475, 476
Sugarcane, 443
Sweet corn breeding, 384
Sweetener products, 740-743
    corn syrup, 740, 741
    dextrose, 742, 743
    high fructose syrup, 741, 742
Synapsis, 244-250

T

Telocentric chromosomes, 238, 290
Tenebroides mauritanicus, 557
Teosinte introgression, 35-37

Teosinte, races of, 1, 4, 5, 10, 12, 16, 17, 18, 19
    Balsas, 36
    Chalco, 22, 25, 26, 36
    Durango, 22
    Florida (Guatemala), 22, 23, 27
    Jutiapa, 26, 36
    Nobogame, 22
    Nojoya, 22, 23
    Xochimilco, 22
Tetranychus urticae, 545
Tetraploids, 235
Theriophis maculata, 550
Tillage systems, preplant, 636-643
    chisel, 638-639
    minimum, 640-643
    moldboard plow, 636-638
    strip, 639-640
Tocopherols, 745
Toxins, 428, 441, 457
Transformation, 189
Translocations, chromosomal, 21, 227, 230, 231, 238, 246, 248-250, 251, 254, 257, 259, 260-271
Tribolium ssp., 556
Trichoderma spp., 397
Trichogramma minutem, 522
Trigonotylus brevipes, 543
Triploids, 235
Tripsacum introgression, 37-39
Tripsacum, species
    australe, 37
    dactyloides, 1, 3, 28, 29, 31, 38, 437, 471, 565
    floredanum, 28, 30, 38
    lanceolatum, 471
    laxum, 471
    pilosum, 471
Trisomics,
    primary, 236, 237, 251, 266, 279
    secondary, 237, 238
    tertiary, 236, 238
Triticum aestivum, 32, 244, 443, 511, 593
Typhaea stercorea, 560

U

Ustilagenoidea virens, 467
Ustilago maydis, 451, 464, 547

V

Varietal hybridization, 307-310, 312
Vicia sp., 655

W

Waxy corn, 376, 383-384
Waxy starch, 737-739
Weather effects on plant growth, 600-617
  before planting, 600-601
  emergence to flower differentiation, 603-609
  fetilization to physiologic maturity, 613-614
  maturation-drying, 614-615
  planting to emergence, 601-603
  seasonal effects, 615-617
  stem elongation to tasseling, 609-611
  tasseling, silking and pollination, 611-613
Weed control, 657-660
  chemical, 659-660
  cultivation, 657-659

Wet milling, capacity, 725
Wet milling, process, 725-745
  component separation, 727-729
  product yields, 729
  steeping, 726
Witchweed, 391, 474-476

X

Xenopus, 232
Xylose, 755

Z

Zea mays, 1, 4, 5
  mexicana, 1, 4
  perennis, 4
Zearalenone, 730
Zea-Tripsacum hybrids, 27-35
Zea-Tripsacum interchange, 266